Strategies for Success

W9-BUC-191

Top Ten Ways to Succeed in Classes that Use Active Learning

By Marilla Svinicki, Ph.D., former Director of the University of Texas Center for Teaching Effectiveness

1. Make the switch from an authority-based conception of learning to a self-regulated conception of learning. Recognize and accept your own responsibility for learning.
2. Be willing to take risks and go beyond what is presented in class or the text.
3. Be able to tolerate ambiguity and frustration in the interest of understanding.
4. See errors as opportunities to learn rather than failures. Be willing to make mistakes in class or in study groups so that you can learn from them.
5. Engage in active listening to what's happening in class.
6. Trust the instructor's experience in designing class activities and participate willingly if not enthusiastically.
7. Be willing to express an opinion or hazard a guess.
8. Accept feedback in the spirit of learning rather than as a reflection of you as a person.
9. Prepare for class physically, mentally, and materially (do the reading, work the problems, etc.).
10. Provide support for your classmate's attempts to learn. The best way to learn something well is to teach it to someone who doesn't understand.

Dr. Dee's Eleventh Rule:
DON'T PANIC! Pushing yourself beyond the comfort zone is scary, but you have to do it in order to improve.

Word Roots for Physiology

Simplify physiology and medicine by learning Latin and Greek word roots. The list below has some of the most common ones.

Using the list, can you figure out what *hyperkalemia* means?*

a- or **an-** without, absence
anti- against
-ase signifies an enzyme
auto self
bi- two
brady- slow
cardio- heart
cephalo- head
cerebro- brain
contra- against
-crine a secretion
crypt- hidden
cutan- skin
-cyte or **cyto-** cell
de- without, lacking
di- two
dys- difficult, faulty
-elle small
-emia in the blood
endo- inside or within
epi- over
erythro- red
exo- outside
extra- outside
gastro- stomach
-gen, -genie produce
gluco-, glyco- sugar or sweet
hemi- half
hemo- blood
hepato- liver
homo- same
hydro- water
hyper- above or excess
hypo- beneath or deficient
inter- between
intra- within
-itis inflammation of
kali- potassium
leuko- white
lipo- fat
lumen inside of a hollow tube
-lysis split apart or rupture
macro- large
micro- small
mono- one
multi- many
myo- muscle
oligo- little, few
para- near, close
patho-, -pathy related to disease
peri- around
poly- many
post- after
pre- before
pro- before
pseudo- false
re- again
retro- backward or behind
semi- half
sub- below
super- above, beyond
supra- above, on top of
tachy- rapid
trans- across, through

* Hyper = excess, kali = potassium, -emia = in the blood, or elevated blood potassium

Owner's Manual

Welcome to Human Physiology! As you begin your study of the human body, one of your main tasks will be to construct for yourself a global view of the body, its systems, and the many processes that keep the systems working. This "big picture" is what physiologists call the integration of systems, and it is a key theme in this book. To integrate information, however, you must do more than simply memorize it. You need to truly understand it and be able to use it to solve problems that you have never encountered before. If you are headed for a career in the health professions, you will do this in the clinics. If you plan a career in biology, you will solve problems in the laboratory, field, or classroom. Analyzing, synthesizing, and evaluating information are skills you need to develop while you are in school, and I hope that the features of this book will help you with this goal.

One of my aims is to provide you not only with information about how the human body functions but also with tips for studying and problem solving. Many of these study aids have been developed with the input of my students, so I think you may find them particularly helpful.

On the following pages, I have put together a brief tour of the special features of the book, especially those that you may not have encountered previously in textbooks. Please take a few minutes to read about them so that you can make optimum use of the book as you study.

Each chapter begins with a list of Learning Outcomes to guide you as you read the chapter. Within the chapters look for the **Running Problem**, **Phys in Action**, and **Try It!** activities. **Phys in Action** are online video clips that I created with the assistance of some of my students. Look for the references to Mastering A&P in the figures with associated Phys in Action clips, and watch Kevin and Michael as they demonstrate physiology in action.

Play Phys in Action
@Mastering Anatomy & Physiology

Pattern recognition is important for all healthcare professionals, so you can begin to develop this skill by learning the key concepts of physiology that repeat over and over as you study different organ systems. Chapter 1 includes two special Focus On features: one on concept mapping, a study strategy that is also used for decision-making in the clinics, and one on constructing and interpreting graphs. The Running Problem in Chapter 1 introduces you to effective ways to find information on the Internet.

Be sure to look for the Essentials and Review figures throughout the book. These figures distill the basics about a topic onto one or two pages, much as the Anatomy Summaries do. My students tell me they find them particularly useful for review when there isn't time to go back and read all the text.

We have also retained the four approaches to learning physiology that proved so popular since this book was first published in 1998.

1. Cellular and Molecular Physiology

Most physiological research today is being done at the cellular and molecular level, and there have been many exciting developments in molecular medicine and physiology in the 10 years since the first edition. For example, now scientists are paying more attention to primary cilia, the single cilium that occurs on most cells of the body. Primary cilia are thought to play a role in some kidney and other diseases. Look for similar links between molecular and cellular biology, physiology, and medicine throughout the book.

FIG. 3.5 Cilia and flagella

(a) Cilia on surface of respiratory epithelium

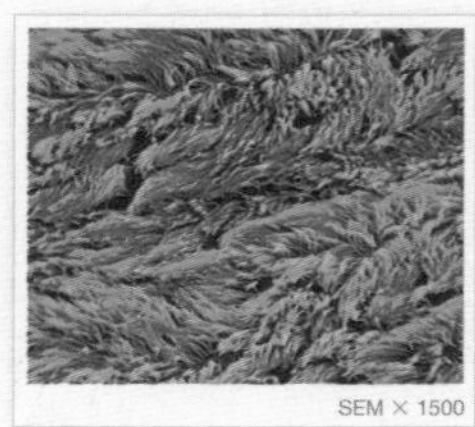
SEM × 1500

2. Physiology as a Dynamic Field

Physiology is a dynamic discipline, with numerous unanswered questions that merit further investigation and research. Many of the "facts" presented in this text are really only our current theories, so you should be prepared to change your mental models as new information emerges from scientific research.

EMERGING CONCEPTS

How tUse this Book

3. An Emphasis tegtion

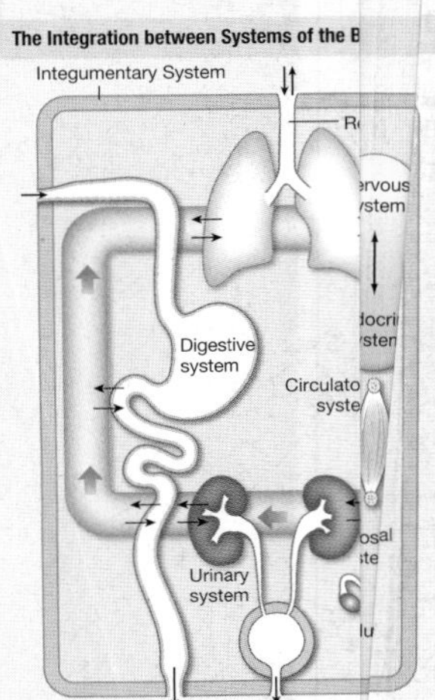
The Integration between Systems of the B
Integumentary System
Digestive system
Urinary system

The organ systems of the body do not work in isolation, although we study them one at a time. To emphasize the integrative nature of physiology, three chapters (Chapters 13, 20, and 25) focus on how the physiological processes of multiple organ systems coordinate with each other, especiahomeostasis is challenged.

4. A Focus on PrSolving

One of the most væ skills students should acquire is the abilit critically and use information to solve problou study physiology, you should be preparecce these skills. You will find a number of featuıbook, such as the Concept Check questions arand Graph Questions. These "test yourself" que designed to challenge your critical thinkiralysis skills. In each chapter, read the Running ßs you work through the text and see if you what you're reading to the clinical scenarid in the problem.

Also, be sure to look at the back of the text, where we have combined the index and glossary to save time when you are looking up unfamiliar words. The appendices have the answers to the Concept Check questions, Figure and Graph Questions, and end-of-chapter questions, as well as reviews of physics, logarithms, and basic genetics. The back end papers include a periodic table of the elements, diagrams of anatomical positions of the body, and tables with conversions and normal values of blood components. Take a few minutes to look at all these features so that you can make optimum use of them.

Level Four Quantitative Problems

30. The following graph represents the disappearance of a drug from the blood as the drug is metabolized and excreted. Based on the graph, what is the half-life of the drug?

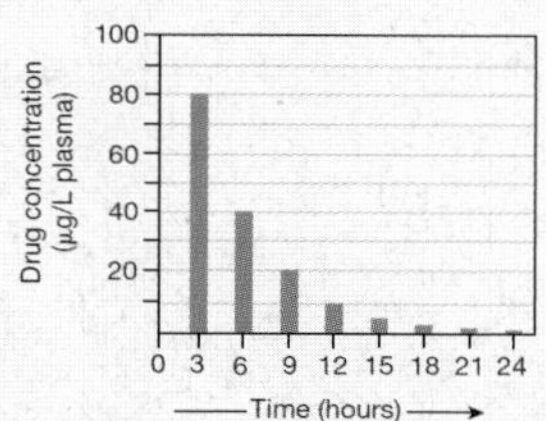
Drug concentration (μg/L plasma)
100
80
60
40
20
0 3 6 9 12 15 18 21 24
Time (hours)

It is my hope that by reading this book, you will develop an integrated view of physiology that allows you to enter your chosen profession with respect for the complexity of the human body and a clear vision of the potential of physiological and biomedical research. May you find physiology as fun and exciting I do. Good luck with your studies!

Warmest regards,
Dr. Dee (as my students call me)
silverthorn@utexas.edu

Phys in ActioTopics:

pp. 130–131 Fig. ⁵ity & Tonicity
pp. 154–155 Fig. ⁵ane Potential
pp. 458–459 Fig.1cardiogram
p. 494 Fig. 15.14 ‹lar Control
p. 545 Fig. 17.7 Tler
p. 549 Fig. 17.10 IPressure
p. 557 Fig. 17.13 ses
p. 573 Fig. 18.7 HOxygen Transport
p. 610 Fig. 19.13 ance
p. 793 Fig. 25.8 Bre & Exercise

Try It Activities:

p. 21 Graphing
p. 135 Membrane Models (Lipid bylayer)
p. 251 Action Potentials
p. 325 Salty-Sweet Taste Experiment
p. 468 Frank-Starling Law of the Heart
p. 605 Insulin
p. 682 Oral Rehydration Therapy

Move Beyond Memorization: Prepare for Tomorrow's Challenges

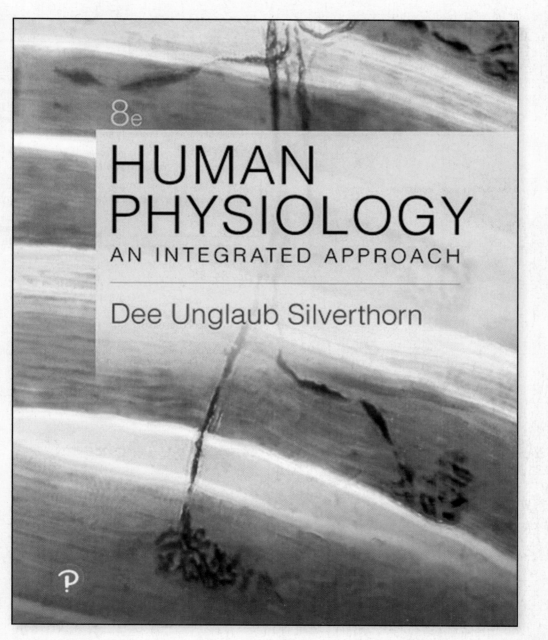

The goals for the **Eighth Edition** of ***Human Physiology: An Integrated Approach*** are to provide an integrated and up-to-date introduction to core concepts in physiology and to equip you with skills for solving real-world problems.

8e

HUMAN PHYSIOLOGY

AN INTEGRATED APPROACH

Dee Unglaub Silverthorn

Pearson

Challenge Yourself: Apply What You Learn

Learning physiology requires that you use information rather than simply memorizing what you think will be on the test. The Eighth Edition text and Mastering™ A&P program provide multiple opportunities for you to practice answering the more challenging types of questions that you are likely to see on a test or exam.

Running Problems explore a real-world disease or disorder that unfolds in short segments throughout the chapter. You can check your understanding by comparing your answers with those in Problem Conclusion at the end of each chapter. Related Coaching Activities can be assigned in Mastering A&P.

RUNNING PROBLEM

The day after Jan's visit, the computerized cytology analysis system rapidly scans the cells on the slide of Jan's cervical tissue, looking for abnormal cell size or shape. The computer is programmed to find multiple views for the cytologist to evaluate. The results of Jan's two Pap tests are shown in **FIGURE 3.14**.

Q6: *Has Jan's dysplasia improved or worsened? What evidence do you have to support your answer?*

Q7: *Use your answer to question 6 to predict whether Jan's HPV infection has persisted or been cleared by her immune system.*

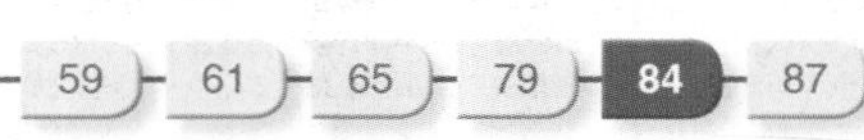

59
61
65
79
84
87

see pp. 84–85

(b) Jan's second Pap test. Are these cells normal or abnormal?

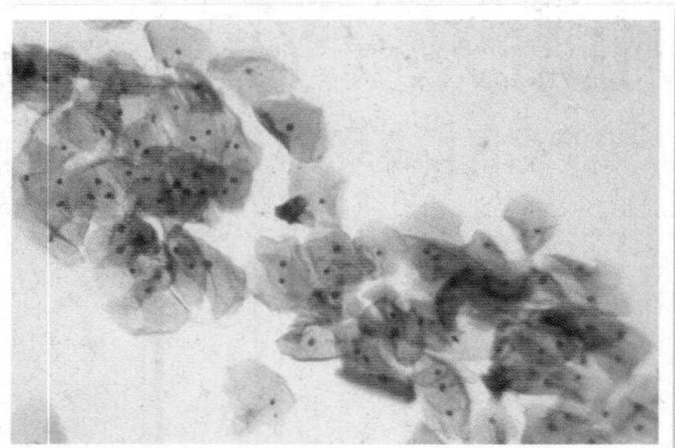

Additional Practice Questions include **Concept Check Questions,** which are placed at intervals throughout the chapter, and Review Questions, which are provided at the end of the chapter and organized into four levels of difficulty. An answer key is in Appendix A.

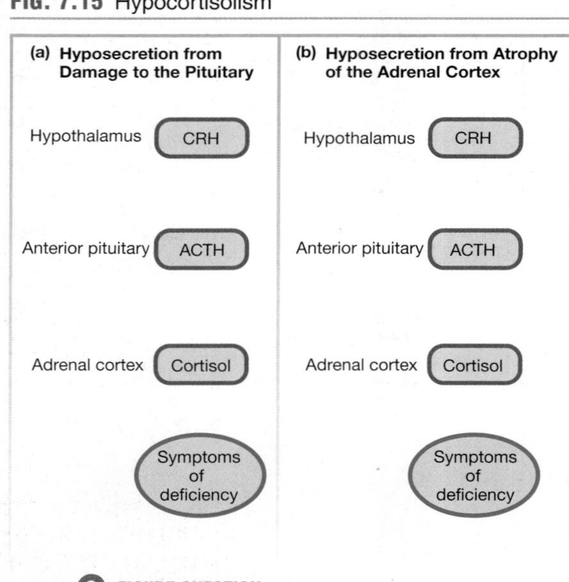

FIG. 7.15 Hypocortisolism
(a) Hyposecretion from Damage to the Pituitary
Hypothalamus
CRH
Anterior pituitary
ACTH
Adrenal cortex
Cortisol
Symptoms of deficiency
(b) Hyposecretion from Atrophy of the Adrenal Cortex
Hypothalamus
CRH
Anterior pituitary
ACTH
Adrenal cortex
Cortisol
Symptoms of deficiency

FIGURE QUESTION

For each condition, use arrows to indicate whether levels of the three hormones in the pathway will be increased, decreased, or unchanged. Draw in negative feedback loops where functional.

Figure Questions challenge you to apply visual literacy skills as you read an illustration or photo. Answers to these questions appear at the end of the text, in Appendix A.

See p. 217

Practice Solving Real-World Problems

NEW! "Try it" boxes present a real-world research problem or classic experiment and guide you through the process of analyzing the data and thinking like a scientist.

NEW! Additional questions for each "Try it" activity are available in **Mastering A&P**. Topics include Graphing (Chapter 1), Cell Membranes (Chapter 5), Action Potentials (Chapter 8), Salty-Sweet Taste Experiment (Chapter 10), Frank-Starling Law of the Heart (Chapter 14), Insulin (Chapter 19) and Oral Rehydration Therapy (Chapter 21).

Instructors: A version of this Try it! Activity can be assigned in Mastering Anatomy & Physiology

TRY IT! Action Potential

What do carnivorous plants and your neurons have in common? Most students learn that action potentials (APs) transmit information rapidly along neurons in an animal's nervous system. While this is true, APs were actually first described in algae! Another plant that uses APs is the Venus flytrap (*Dionaea muscipula*). Because these plants grow in nutrient-poor soil, they are carnivorous. The tips of their two leaves have evolved into *capture organs*, which snap shut when prey, such as a fly, moves over them. Charles Darwin himself, captivated by this phenomenon, encouraged other scientists to describe its mechanism.

In 1873, the English physiologist Sir John Scott Burdon-Sanderson was able to show that electric current flows through the Venus flytrap when a fly touches *trigger hairs* on the inner surface of the capture organs. The hairs act as mechanoreceptors that generate an action potential when bent. The AP closes the leaf tips, trapping the fly inside so the plant can digest it. In a series of experiments, researchers recorded APs in flytrap cells while varying the extracellular concentration of Ca^{2+}.

(a) The capture organ of a Venus flytrap with trigger hairs.

Trigger hairs

(b) Data from Hodick and Sievers, 1986.[1] Arrows indicate when trigger hairs were bent.

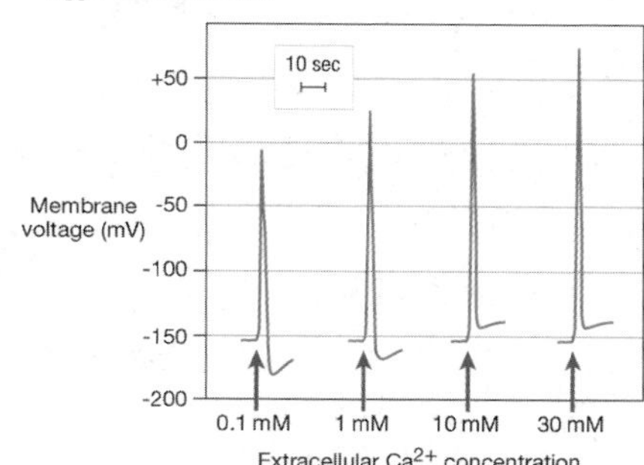
10 sec
+50
0
-50
-100
-150
-200
Membrane voltage (mV)
0.1 mM
1 mM
10 mM
30 mM
Extracellular Ca2+ concentration

GRAPH QUESTIONS

1. Using the results shown in the graph, explain what increasing the concentration of Ca^{2+} does to the flytrap APs.
2. These results suggest that the rising phase of a flytrap AP is primarily due to which ion? Is this ion entering or leaving the cell? How does this compare to APs in your neurons?
3. What experiments could you design to determine which ion is responsible for the repolarization phase of the flytrap's AP?

[1] Hodick, D. & Sievers, A. (1986). The influence of Ca^{2+} on the action potential in mesophyll cells of *Dionaea muscipula* Ellis. *Protoplasma* 133, 83–84.

See p. 251

Graph Questions encourage you to interpret real data presented in graphs. Answers to these questions appear at the end of the text, in Appendix A.

Study More Efficiently Using the Figures

Eye-tracking research has shown that learning and comprehension levels are higher for students who study both the figures and the text together than for students who only read the text. This book offers dozens of illustrations designed to help you learn physiology more efficiently, and make the best use of your study time.

Essentials Figures distill the basics of a topic into one or two pages, helping you to see the big picture of human physiology. Instructors can assign **related Mastering A&P coaching activities** that explore these topics in greater depth.

Selected figures from the text are explored in accompanying **Phys in Action video tutors** and in **coaching activities** in **Mastering A&P.**

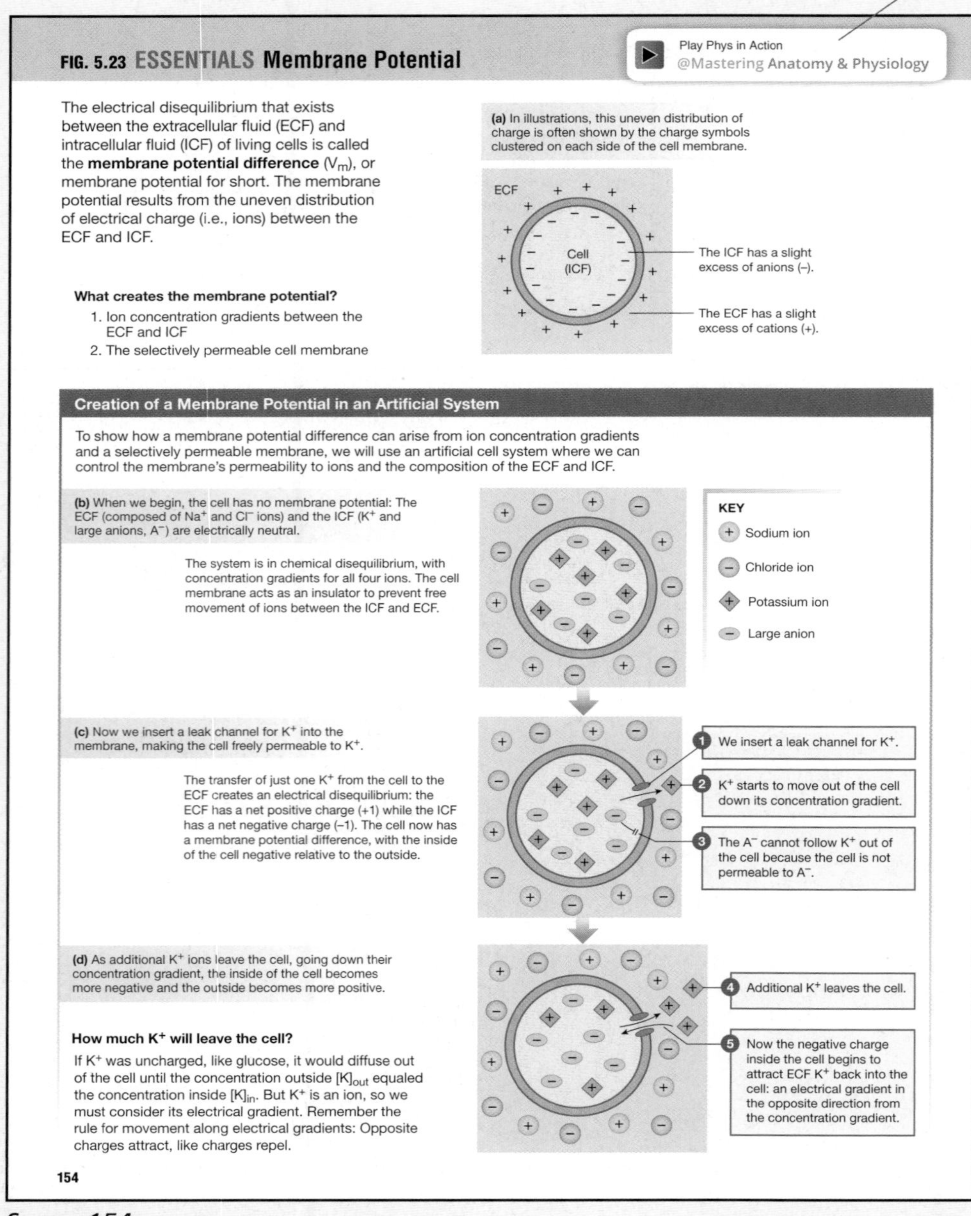

FIG. 5.23 ESSENTIALS Membrane Potential

Play Phys in Action @Mastering Anatomy & Physiology

The electrical disequilibrium that exists between the extracellular fluid (ECF) and intracellular fluid (ICF) of living cells is called the **membrane potential difference** (V_m), or membrane potential for short. The membrane potential results from the uneven distribution of electrical charge (i.e., ions) between the ECF and ICF.

What creates the membrane potential?

1. Ion concentration gradients between the ECF and ICF
2. The selectively permeable cell membrane

(a) In illustrations, this uneven distribution of charge is often shown by the charge symbols clustered on each side of the cell membrane.

ECF
Cell (ICF)
The ICF has a slight excess of anions (–).
The ECF has a slight excess of cations (+).

Creation of a Membrane Potential in an Artificial System

To show how a membrane potential difference can arise from ion concentration gradients and a selectively permeable membrane, we will use an artificial cell system where we can control the membrane's permeability to ions and the composition of the ECF and ICF.

(b) When we begin, the cell has no membrane potential: The ECF (composed of Na^+ and Cl^- ions) and the ICF (K^+ and large anions, A^-) are electrically neutral.

The system is in chemical disequilibrium, with concentration gradients for all four ions. The cell membrane acts as an insulator to prevent free movement of ions between the ICF and ECF.

KEY
Sodium ion
Chloride ion
Potassium ion
Large anion

(c) Now we insert a leak channel for K^+ into the membrane, making the cell freely permeable to K^+.

The transfer of just one K^+ from the cell to the ECF creates an electrical disequilibrium: the ECF has a net positive charge (+1) while the ICF has a net negative charge (–1). The cell now has a membrane potential difference, with the inside of the cell negative relative to the outside.

1 We insert a leak channel for K^+.
2 K^+ starts to move out of the cell down its concentration gradient.
3 The A^- cannot follow K^+ out of the cell because the cell is not permeable to A^-.

(d) As additional K^+ ions leave the cell, going down their concentration gradient, the inside of the cell becomes more negative and the outside becomes more positive.

4 Additional K^+ leaves the cell.
5 Now the negative charge inside the cell begins to attract ECF K^+ back into the cell: an electrical gradient in the opposite direction from the concentration gradient.

How much K^+ will leave the cell?

If K^+ was uncharged, like glucose, it would diffuse out of the cell until the concentration outside $[K]_{out}$ equaled the concentration inside $[K]_{in}$. But K^+ is an ion, so we must consider its electrical gradient. Remember the rule for movement along electrical gradients: Opposite charges attract, like charges repel.

154

See p. 154

Anatomy Summary Figures provide succinct visual overviews of a physiological system from a macro to micro perspective. Whether you are learning the anatomy for the first time or refreshing your memory, these summaries show you the essential features of each system in a single figure.

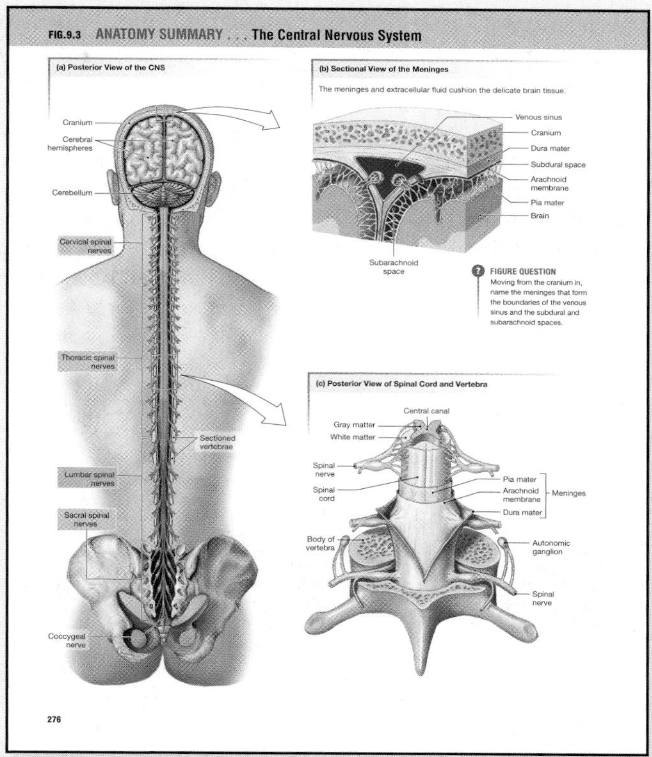

FIG.9.3 ANATOMY SUMMARY . . . The Central Nervous System
(a) Posterior View of the CNS
Cranium
Cerebral hemispheres
Cerebellum
Cervical spinal nerves
Thoracic spinal nerves
Sectioned vertebrae
Lumbar spinal nerves
Sacral spinal nerves
Coccygeal nerve
(b) Sectional View of the Meninges
The meninges and extracellular fluid cushion the delicate brain tissue.
Venous sinus
Cranium
Dura mater
Subdural space
Arachnoid membrane
Pia mater
Brain
Subarachnoid space
FIGURE QUESTION
Moving from the cranium in, name the meninges that form the boundaries of the venous sinus and the subdural and subarachnoid spaces.
(c) Posterior View of Spinal Cord and Vertebra
Central canal
Gray matter
White matter
Spinal nerve
Spinal cord
Pia mater
Arachnoid membrane
Dura mater
Meninges
Body of vertebra
Autonomic ganglion
Spinal nerve
276

See p. 276

Review Figures visually present foundational concepts that you may already be familiar with. You may find it helpful to check out these figures before learning new physiology concepts.

Selected figures from the text can be assigned as **Art-Labeling Activities in Mastering A&P.**

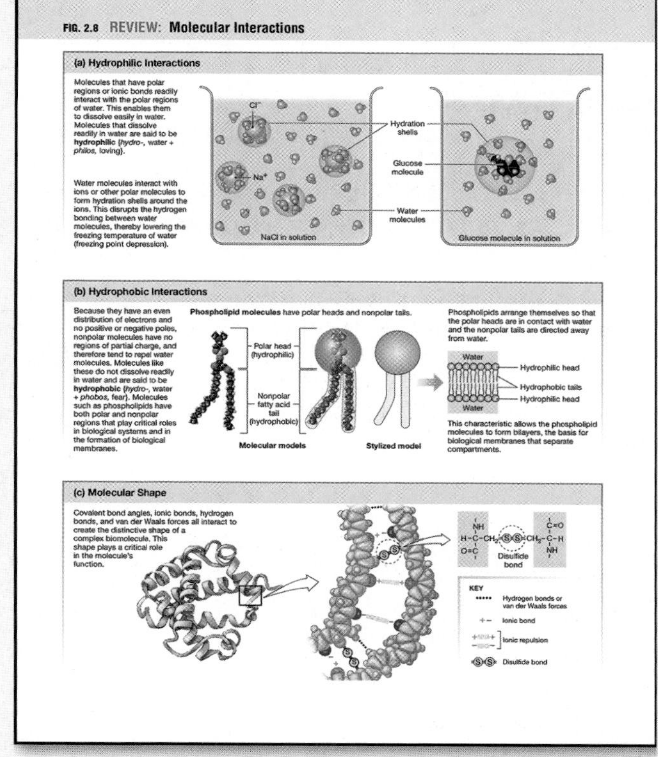

FIG. 2.8 REVIEW: Molecular Interactions
(a) Hydrophilic Interactions
Molecules that have polar regions or ionic bonds readily interact with the polar regions of water. This enables them to dissolve easily in water. Molecules that dissolve readily in water are said to be hydrophilic (hydro-, water + philos, loving).
Water molecules interact with ions or other polar molecules to form hydration shells around the ions. This disrupts the hydrogen bonding between water molecules, thereby lowering the freezing temperature of water (freezing point depression).
Cl⁻
Na⁺
Hydration shells
Glucose molecule
Water molecules
NaCl in solution
Glucose molecule in solution
(b) Hydrophobic Interactions
Because they have an even distribution of electrons and no positive or negative poles, nonpolar molecules have no regions of partial charge, and therefore tend to repel water molecules. Molecules like these do not dissolve readily in water and are said to be hydrophobic (hydro-, water + phobos, fear). Molecules such as phospholipids have both polar and nonpolar regions that play critical roles in biological systems and in the formation of biological membranes.
Phospholipid molecules have polar heads and nonpolar tails.
Polar head (hydrophilic)
Nonpolar fatty acid tail (hydrophobic)
Molecular models
Stylized model
Phospholipids arrange themselves so that the polar heads are in contact with water and the nonpolar tails are directed away from water.
Water
Hydrophilic head
Hydrophobic tails
Hydrophilic head
Water
This characteristic allows the phospholipid molecules to form bilayers, the basis for biological membranes that separate compartments.
(c) Molecular Shape
Covalent bond angles, ionic bonds, hydrogen bonds, and van der Waals forces all interact to create the distinctive shape of a complex biomolecule. This shape plays a critical role in the molecule's function.
Disulfide bond
KEY
Hydrogen bonds or van der Waals forces
Ionic bond
Ionic repulsion
Disulfide bond

See p. 44

Get Online Coaching Through Mastering A&P

Mastering A&P provides tutorials and review questions that you can access before, during, and after class.

Phys in Action! Video Tutors and Coaching Activities help you visualize and master challenging physiological concepts by demonstrating laboratory procedures and real-world applications. Demonstrations include pulmonary function test, tilt table, exercise testing, and more.

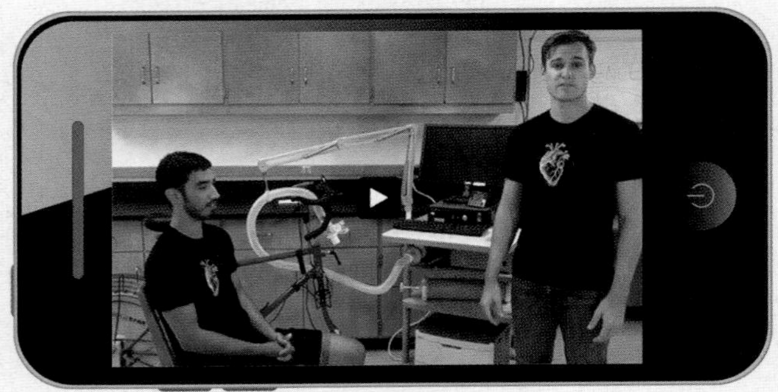

EXPANDED! Interactive Physiology 2.0 Coaching Activities teach complex physiological processes using exceptionally clear animations, interactive tutorials, games, and quizzes. IP2 features new graphics, quicker navigation, and a mobile-friendly design. New topics include Generation of an Action Potential and Cardiac Cycle. IP2 and IP animations can be assigned from the Mastering A&P Item Library or accessed through the Mastering A&P Study Area.

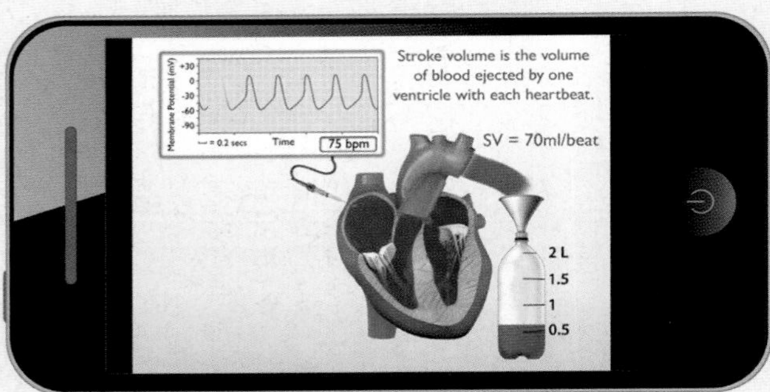

Stroke volume is the volume of blood ejected by one ventricle with each heartbeat.
SV = 70ml/beat
Time
75 bpm
2 L
1.5
1
0.5

Mastering A&P offers thousands of tutorials, activities, and questions that can be used to test yourself, or assigned for homework and practice. Additional highlights include:

- **Nurses Need Physiology Case Studies** guide you through the steps of diagnosing and treating patients in real-world clinical scenarios.
- **A&P Flix Animations** use 3-D, movie-quality graphics to help you visualize complex physiology processes.
- **Dynamic Study Modules** are manageable, mobile-friendly sets of questions with extensive feedback for you to test, learn, and retest yourself on basic concepts.
- **PhysioEx Laboratory Simulations** offer a supplement or substitute for wet labs due to cost, time, or safety concerns.

Access the Complete Textbook Online or Offline with Pearson eText

You can read your textbook without having to add weight to your bookbag.

The Pearson eText mobile app offers offline access and can be downloaded for most iOS and Android phones and tablets from the Apple App or Google Play stores.

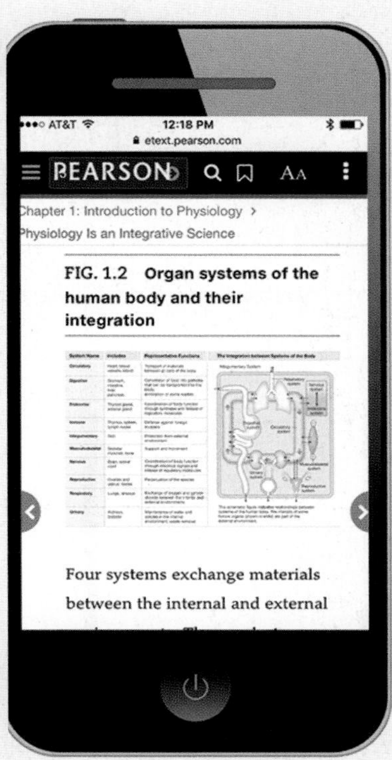

AT&T
12:18 PM
etext.pearson.com
PEARSON
Chapter 1: Introduction to Physiology >
Physiology Is an Integrative Science
FIG. 1.2 Organ systems of the human body and their integration
Four systems exchange materials
between the internal and external

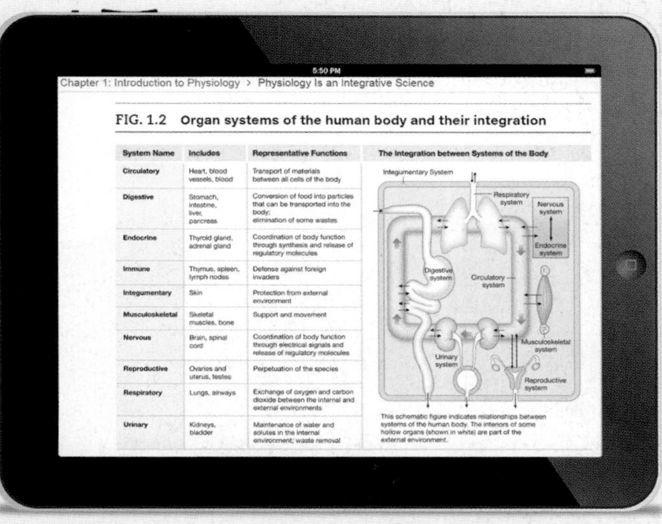

5:50 PM
Chapter 1: Introduction to Physiology > Physiology Is an Integrative Science
FIG. 1.2 Organ systems of the human body and their integration

Powerful interactive and customization functions in the eText platform include instructor and student note-taking, highlighting, bookmarking, search, and links to glossary terms.

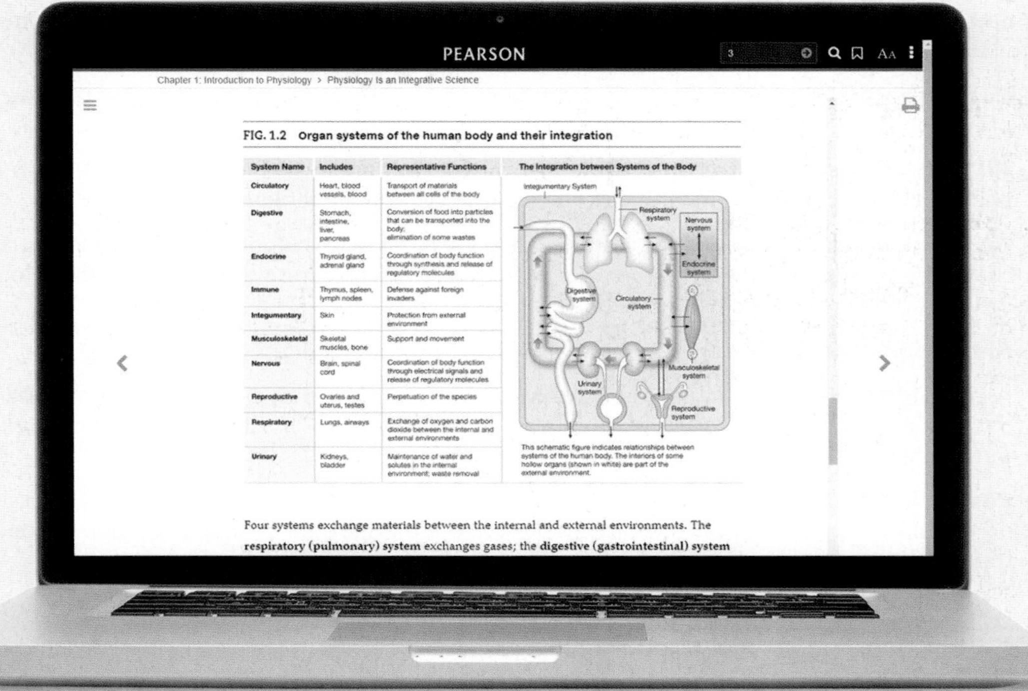

PEARSON
Chapter 1: Introduction to Physiology > Physiology Is an Integrative Science
FIG. 1.2 Organ systems of the human body and their integration
Four systems exchange materials between the internal and external environments. The
respiratory (pulmonary) system exchanges gases; the digestive (gastrointestinal) system

Additional Support for Students & Instructors

NEW! Ready-to-Go Teaching Modules help instructors efficiently make use of the best teaching tools before, during, and after class. Accessed through the Instructor Resources area of Mastering A&P, and curated by author Dee Silverthorn, modules include skill development applications for Human Physiology including Concept Mapping and Graphing.

Learning Catalytics allows students to use their smartphone, tablet, or laptop to respond to questions in class. Visit learningcatalytics.com to learn more.

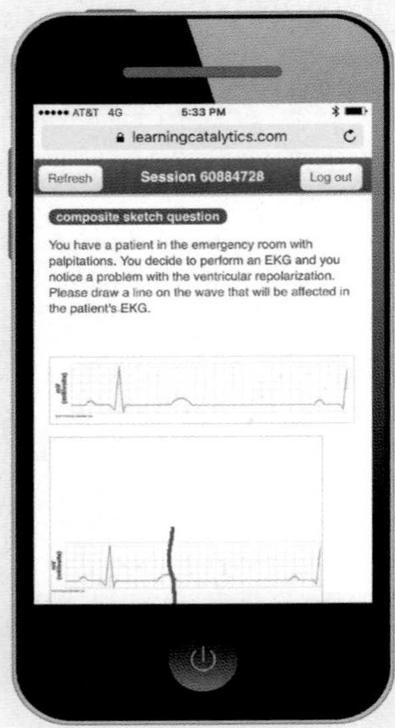
AT&T 4G
5:33 PM
learningcatalytics.com
Refresh
Session 60884728
Log out
composite sketch question
You have a patient in the emergency room with palpitations. You decide to perform an EKG and you notice a problem with the ventricular repolarization. Please draw a line on the wave that will be affected in the patient's EKG.

The Mastering A&P Instructor Resources Area includes the following downloadable tools for instructors who adopt the Eighth Edition for their classes:

- **Customizable PowerPoint® lecture outlines** include customizable images and provide a springboard for lecture prep.
- **All of the figures, photos, and tables from the text** are available in JPEG and PowerPoint® formats, in labelled and unlabelled versions, and with customizable labels and leader lines.
- **Test bank** provides thousands of customizable questions across Bloom's taxonomy levels. Each question is tagged to chapter learning outcomes that can also be tracked within Mastering Anatomy & Physiology assessments. Available in Microsoft® Word and TestGen® formats.
- **Animations and videos** bring human physiology concepts to life.
- **A comprehensive Instructor Resource Manual**, co-authored by Dee Silverthorn and Damian Hill, includes a detailed teaching outline for each chapter, along with a wealth of activities, examples, and analogies that have been thoroughly class-tested with thousands of students.
- **Customizable Study Questions**, co-authored by Dee Silverthorn and Damian Hill, help students focus their reading on the most important points in each chapters and are organized by chapter section headers for easy editing to reflect the material covered in class.

EIGHTH EDITION

HUMAN PHYSIOLOGY

AN INTEGRATED APPROACH

Dee Unglaub Silverthorn, Ph.D.
UNIVERSITY OF TEXAS, AUSTIN

with contributions by

Bruce R. Johnson, Ph.D.
CORNELL UNIVERSITY

and

William C. Ober, M.D.
ILLUSTRATION COORDINATOR

Claire E. Ober, R.N.
ILLUSTRATOR

Anita Impaglizzo,
ILLUSTRATOR

Andrew C. Silverthorn, M.D.
CLINICAL CONSULTANT

Pearson

Courseware Portfolio Manager: *Lauren Harp*
Content Producer: *Deepti Agarwal*
Managing Producer: *Nancy Tabor*
Courseware Director, Content Development: *Barbara Yien*
Courseware Editorial Assistant: *Dapinder Dosanjh*
Rich Media Content Producer: *Nicole Constantine*
Mastering Content Developer, Science: *Lorna Perkins*
Full-Service Vendor: *SPi Global*
Copyeditor: *Alyson Platt*
Art Project Manager: *Stephanie Marquez, Imagineering Art LLC*
Illustrators: *William C. Ober, Anita Impagliazzo, and Cliare E. Ober*
Design Manager: *Maria Guglielmo Walsh*
Interior Designer: *Gary Hespenheide*
Cover Designer: *Gary Hespenheide*
Rights & Permissions Project Manager: *Katrina Mohn, Cenveo Publisher Services*
Rights & Permissions Management: *Ben Ferrini*
Manufacturing Buyer: *Stacey Weinberger, LSC Communications*
Product Marketing Manager: *Wendy Mears*

Cover Photo: *Motor Neuron in Muscle*
Credit: *Kent Wood/Science Source*

Library of Congress Cataloging-in-Publication Data
Catalogue in Publication Data is on file with the Library of Congress

ISBN 10: 0-13-460519-5; ISBN 13: 978-0-13-460519-7 (Student edition)
1 18
ISBN 10: 0-13-470434-7; ISBN 13: 978-0-13-470443-0 (Instructor's Review Copy)

Pearson

www.pearson.com

ABOUT THE AUTHOR

DEE UNGLAUB SILVERTHORN studied biology as an undergraduate at Newcomb College of Tulane University, where she did research on cockroaches. For graduate school, she switched to studying crabs and received a Ph.D. in marine science from the Belle W. Baruch Institute for Marine and Coastal Sciences at the University of South Carolina. Her research interest is epithelial transport, and most recently work in her laboratory has focused on transport properties of the chick allantoic membrane. Her teaching career started in the Physiology Department at the Medical University of South Carolina but over the years she has taught a wide range of students, from medical and college students to those still preparing for higher education. At the University of Texas–Austin, she teaches physiology in both lecture and laboratory settings, and instructs graduate students on developing teaching skills in the life sciences. In 2015 she joined the faculty of the new UT-Austin Dell Medical School. She has received numerous teaching awards and honors, including a 2011 UT System Regents' Outstanding Teaching Award, the 2009 Outstanding Undergraduate Science Teacher Award from the Society for College Science Teachers, the American Physiological Society's Claude Bernard Distinguished Lecturer and Arthur C. Guyton Physiology Educator of the Year, and multiple awards from UT–Austin, including the Burnt Orange Apple Award. The first edition of her textbook won the 1998 Robert W. Hamilton Author Award for best textbook published in 1997–1998 by a University of Texas faculty member. Dee was the president of the Human Anatomy and Physiology Society in 2012–2013, has served as editor-in-chief of Advances in Physiology Education, and is currently chair of the American Physiological Society Book Committee. She works with members of the International Union of Physiological Sciences to improve physiology education in developing countries, and this book has been translated into seven languages. Her free time is spent creating multimedia fiber art and enjoying the Texas hill country with her husband, Andy, and their dogs.

Michael Chirillo, Dee Silverthorn, and Kevin Christmas

About the Illustrators

William C. Ober, M.D. (*art coordinator and illustrator*) received his undergraduate degree from Washington and Lee University and his M.D. from the University of Virginia. He also studied in the Department of Art as Applied to Medicine at Johns Hopkins University. After graduation, Dr. Ober completed a residency in Family Practice and later was on the faculty at the University of Virginia in the Department of Family Medicine and in the Department of Sports Medicine. He also served as Chief of Medicine of Martha Jefferson Hospital in Charlottesville, VA. He is currently a visiting Professor of Biology at Washington & Lee University, where he has taught several courses and led student trips to the Galapagos Islands. He was part of the Core Faculty at Shoals Marine Laboratory, where he taught Biological Illustration for 22 years. The textbooks illustrated by Medical & Scientific Illustration have won numerous design and illustration awards.

Claire E. Ober, R.N. (*illustrator*) practiced pediatric and obstetric nursing before turning to medical illustration as a full-time career. She returned to school at Mary Baldwin College where she received her degree with distinction in studio art. Following a five-year apprenticeship, she has worked as Dr. Ober's partner in Medical and Scientific Illustration since 1986. She was also on the Core Faculty at Shoals Marine Laboratory and co-taught Biological Illustration at both Shoals Marine Lab and at Washington and Lee University.

About the Clinical Consultant

Andrew C. Silverthorn, M.D. is a graduate of the United States Military Academy (West Point). He served in the infantry in Vietnam, and upon his return entered medical school at the Medical University of South Carolina in Charleston. He was chief resident in family medicine at the University of Texas Medical Branch, Galveston, and is currently a family physician in solo practice in Austin, Texas. When Andrew is not busy seeing patients, he may be found on the golf course or playing with his two rescue dogs, Molly and Callie.

About the Contributor

Bruce Johnson, Ph.D. is a Senior Research Associate in the Department of Neurobiology and Behavior at Cornell University. He earned biology degrees at Florida State University (B.A.), Florida Atlantic University (M.S.), and at the Marine Biological Laboratory in Woods Hole (Ph.D.) through the Boston University Marine Program. For three decades, he has led Cornell's highly-praised Principles of Neurophysiology course, in which students receive hands-on instruction in principles and methods in neurophysiology. He is a coauthor of *Crawdad: a CD-ROM Lab Manual for Neurophysiology* and the *Laboratory Manual for Physiology*. Bruce has directed and taught in neuroscience faculty workshops sponsored by NSF (Crawdad), ADInstruments (Crawdad and CrawFly), the Grass Foundation and the Faculty for Undergraduate Neuroscience (FUN). He has also lead workshops and neuroscience courses at the Universities of Copenhagen (Denmark), Cologne (Germany), Ibadan (Nigeria), and the Marine Biological Laboratory. Bruce has been named a Most Influential Faculty Member by the graduating senior class at Cornell and awarded the John M. and Emily B. Clark Award for Distinguished Teaching at Cornell. His other teaching awards include the FUN Educator of the Year Award, FUN Career Service Award, and co-recipient of the 2016 Award for Education in Neuroscience, sponsored by the Society for Neuroscience. He is currently the Editor-in-Chief of the *Journal of Undergraduate Neuroscience Education*. Bruce's research addresses the cellular and synaptic mechanisms of motor network plasticity.

DEDICATION

The 8th edition is dedicated to my colleagues who read every word of the first edition manuscript and provided valuable feedback that helped shape the book.

Park City, Utah, June 1995
(Standing, L to R): Judy Sullivan, Patricia Munn, Dee Silverthorn, Mary Ann Rokitka, Richard Walker, Pat Berger, Norman Scott
(Seated) Shana Ederer, Prentice Hall development editor

NEW TO THIS EDITION

The Eighth Edition of *Human Physiology: An Integrated Approach* builds upon the thorough coverage of integrative and molecular physiology topics that have always been the foundation of this book. The biggest change is a completely revised Chapter 24 on immunology. This field has expanded dramatically since the First Edition published in 1997, and it was time to step back and re-think the presentation of this complicated and complex subject. Neurophysiology is also changing rapidly, requiring multiple updates in Chapters 8 through 11. In nearly every chapter the latest developments in research and medicine meant changes to the presentation of information.

Continuing the revision of the art introduced in the Seventh Edition, we created additional Review and Essentials figures that students can use for quick review as well as new Anatomy Summaries and concept maps. Figures from previous editions that were significantly modified or eliminated are still available to instructors on the Instructor's DVD and in the Instructor Resources area of Mastering A&P.

In addition to the online Phys in Action videos that are referenced in related figures, we have new Try It! activities throughout the book. These activities present data, usually from classic experiments, and ask the students to interpret the results. Topics include Benjamin Franklin's little-known experiment that helped development of the phospholipid bilayer model of the membrane, and the experiments that resulted in oral rehydration therapy for treating cholera.

HIGHLIGHTS OF CONTENT UPDATES

Chapter 1 Introduction to Physiology

- New Focus on Graphing with a new Try It! activity
- Added information on the connectome and microbiome
- Updated information on literature searches and citations

Chapter 2 Molecular Interactions

- Four new element names in the periodic table, inside the back cover of the text
- Added ribbon diagram/Richardson diagram of proteins

Chapter 3 Compartmentation: Cells and Tissues

- Explanations of light and electron microscopy
- New Emerging Concepts box on induced pluripotent stem cells (iPSs)

Chapter 5 Membrane Dynamics

- New Try It! activity on lipid bilayers
- Three Phys in Action video references in Figures 5.4, 5.6, and 5.23

Chapter 6 Communication, Integration, and Homeostasis

- Juxtacrine signaling
- Updated information on NIH Common Fund's *Building Blocks, Biological Pathways, and Networks* Program
- Updated the discussion on cytokine families
- Re-classified receptor-enzymes as catalytic receptors
- GPCR for eicosanoids

Chapter 7 Introduction to the Endocrine System

- Updated information on calcitonin gene-related peptide
- Updated information on melatonin and melatonin-related drugs

Chapter 8 Neurons: Cellular and Network Properties

- Update on mechanisms of axonal transport and associated diseases: dynein, kinesin, fragile X, Alzheimer's, microcephaly
- Try It! activity on action potentials
- New link to online calculator for Nernst and GHK equations
- Added discussion of resistance of extracellular fluid to discussion of resistance to current flow
- Added space constant discussion

Chapter 9 The Central Nervous System

- Added lateral sulcus, insula, cerebral aqueduct
- Re-classification of stages of sleep
- Pericytes in blood-brain barrier formation
- Dopaminergic pathways and addiction

Chapter 10 Sensory Physiology

- New Try It! activity on sweet and salty taste
- Additional information on non-neural sensors and Merkel cells

Chapter 11 Efferent Division: Autonomic and Somatic Motor Control

- Expanded table on properties of autonomic neurotransmitter receptors
- Added N_N and N_M nicotinic subtypes
- Added discussion of sarin nerve gas
- Updated anti-nicotine vaccine
- Etiology of diabetic neuropathy

Chapter 12 Muscles

- Expanded discussion of myosin light chains in striated muscle
- New table with autonomic effects on smooth muscles

Chapter 13 Integrative Physiology I: Control of Body Movement

- Addition information on reflexes and muscle tone
- Updated Parkinson's treatments
- Expanded tetanus Running Problem

Chapter 14 Cardiovascular Physiology

- New Running problem on atypical presentation of myocardial infarction in a woman
- New section and new figure on coronary circulation
- New Try It! activity on Starling's law of the heart
- Added discussion of echocardiography
- Expanded ejection fraction discussion
- New discussion of ion channel subtypes

Chapter 15 Blood Flow and the Control of Blood Pressure

- Updated information on pericytes and their functions
- New discussion of blood-retinal barrier
- Updated discussion of angiogenesis including angiopoietin and angiopoietin/Tie signaling pathway.
- New Review quantitative question on Bernoulli's principle of fluid flow
- New sections on coronary blood flow and cerebral blood flow
- Updated statistics on CV diseases
- Added neurogenic shock

Chapter 16 Blood

- Revised art, includes Figures 16.2, 16.4, 16.6, and 16.7
- Updated information on treatment for sickle cell disease

Chapter 17 Mechanics of Breathing

- Forced vital capacity test
- FEV_1/FVC ratio
- New figure and Figure Question for forced vital capacity test
- Antenatal corticosteroids to prevent NRDS

Chapter 18 Gas Exchange and Transport

- Updated information on action of carbonic anhydrase
- Updated information on hemoglobin-based blood substitutes
- Carotid body plasticity in disease states

Chapter 19 The Kidneys

- New map for factors influencing GFR
- Updated model of organic anion transport, including OAT family transporters
- New figure and table on renal handling of some common substances
- New Try It! activity on glucosuria and the discovery of insulin
- PAH clearance and calculation of renal plasma flow discussion
- New term: renal handling
- New Figure Question
- Updated glomerular filtration barrier to include glomerular capillary glycocalyx, slit diaphragm

Chapter 20 Integrative Physiology II: Fluid and Electrolyte Balance

- New section on role of kidney in hypertension
- New Concept Check question
- Expanded discussion of K^+ handling
- Added zona gomerulosa, paraventricular and supraoptic nuclei
- New section on endocrine pathologies in fluid balance
- New Level 3 Review question on Liddle's syndrome

Chapter 21 The Digestive System

- New Try It! activity on role of the SGLT in treating diarrhea
- New information on cholera vaccine
- Updated discussion on microfold cells
- Added guanylate cyclase-C (GC-C), uroguanylin and guanylin, plecanatide

Chapter 22 Metabolism and Energy Balance

- Updated model for appetite
- Updated pharmacological trials for anorexia
- Latent autoimmune diabetes; also called type 1.5; gestational diabetes (GDM); MODY, maturity-onset diabetes of the young.
- Added mechanism of action of metformin
- Added cardiovascular risk calculator link

Chapter 23 Endocrine Control of Growth and Metabolism

- Expanded discussion of melanocortins and their receptors in the control of food intake.
- Agouti-related protein (AGRP), MC4R receptors
- Added explanation of the role of ghrelin in growth hormone release
- New figure for feedback control of growth hormone release
- Updated discussion on off-label use of growth hormone in adults
- Primary cilia in chrondrocytes and osteocytes act as mechnotransducers
- Role of calcium-sensing receptor and NALCN channel in neuronal excitability
- New figure and discussion of intestinal and renal Ca^{2+} transport
- Skeletal deformaties in ciliopathies
- New figure and discussion of bone remodeling, including RANK, RANKL, osteoprotegerin, osteoid
- New Review question on osteopetrosis

Chapter 24 The Immune System

- 6 NEW figures. Most art significantly revised.
- Added concepts include long-lived plasma cells, mucosa-associated lymphoid tissue (MALT), self-antigens, negative selection, hygiene hypothesis, Zika virus, DAMPS – danger-associated molecular patterns, B cell receptors, regulatory T cells (Tregs)
- Updated information on IgD, contact-dependent signaling

Chapter 26 Reproduction and Development

- Kisspeptin control of GNRH and role in puberty
- Origin of the acrosome
- *Flibanserin* for low libido in women

ACKNOWLEDGMENTS

Writing, editing, and publishing a textbook is a group project that requires the talent and expertise of many people. No one scientist has the detailed background needed in all areas to write a book of this scope, and I am indebted to all my colleagues who so generously share their expertise in each edition. I particularly want to acknowledge Bruce Johnson, Cornell University, Department of Neurobiology and Behavior, a superb neurobiologist and educator, who once again ensured that the chapters on neurobiology are accurate and reflect the latest developments in that rapidly changing field. I would also like to thank Michael Chirillo, a former graduate teaching assistant of mine, for his work developing the Try It! features in between interviewing for and starting a medical residency program. Peter English, a colleague and former student, has also joined the team helping with this revision.

A huge thank you goes to immunologists Natalie Steinel, from UT-Austin Dell Medical School, and Tynan A. Becker, from University of Alaska, for their assistance and critical review of the Chapter 24 revision. Brian Sumner, a 3rd year medical student at the George Washington University School of Medicine, graciously volunteered time out of his busy clinical rotations to read the revised chapter and ensure that it was student-friendly.

The art team of Bill Ober, M.D. and Claire Ober, R.N. has worked with me since the first edition, and I am always grateful for their scientifically astute suggestions and revisions. They were joined in the last edition by Anita Impagliazzo, who brought a fresh eye and new figure ideas.

Instructors and students often contact me directly about the book, and for this edition I would particularly like to thank Allison Brekke, James Mayer, and Dean A. Wiseman for comments and suggestions. Thanks also to my students who keep me informed of the typos that creep in no matter how many people look at the manuscript and pages.

Many other people devoted their time and energy to making this book a reality, and I would like to thank them all, collectively and individually. I apologize in advance to anyone whose name I have omitted.

Reviewers

I am particularly grateful to the instructors who reviewed one or more chapters of the last edition. There were many suggestions in their thoughtful reviews that I was unable to include in the text, but I appreciate the time and thought that went into their comments. The reviewers for this edition include:

Jake Brashears, San Diego City College
Trevor Cardinal, California Polytechnic State University
Michael S. Finkler, Indiana University Kokomo
Victor Fomin, University of Delaware
Jill Gifford, Youngstown State University
David Kurjiaka, Grand Valley State University
Mary Jane Niles, University of San Francisco
Rudy M. Ortiz, University of California, Merced
Jennifer Rogers, University of Iowa
Jia Sun, Imperial Valley College
Alan Sved, University of Pittsburgh

Many other instructors and students took time to write or e-mail queries or suggestions for clarification, for which I thank them. I am always delighted to have input, and I apologize that I do not have room to acknowledge them all individually.

Specialty Reviews

No one can be an expert in every area of physiology, and I am deeply thankful for my friends and colleagues who reviewed entire chapters or answered specific questions. Even with their help, there may be errors, for which I take full responsibility. The specialty reviewers for this edition were:

Natalie Steinel, UT-Austin Dell Medical School
Tynan A. Becker, University of Alaska

Photographs

I would like to thank Kristen Harris, University of Texas who generously provided micrographs from her research.

Supplements

Damian Hill once again worked with me to revise and improve the Instructor Resource Manual that accompanies the book. I believe that supplements should reflect the style and approach of the text, so I am grateful that Damian has continued to be my alter-ego for so many editions. Peter English is helping with Mastering activities this revision.

I would also like to thank my colleagues who helped with the test bank and media supplements for this edition:

Heidi Bustamante, University of Colorado, Boulder
Chad M. Wayne, University of Houston
Margaret Flemming, Austin Community College
Cheryl Neudauer, Minneapolis Community & Technical College

The Development and Production Team

Writing a manuscript is only a first step in the long and complicated process that results in a bound book with all its ancillaries. The team that works with me on book development deserves a lot of credit for the finished product. Gary Hespenheide designed a bright and cheerful cover that continues our tradition of images that show science as art. Anne A. Reid, my long-time developmental editor, is always wonderful to work with, and provides thoughtful suggestions that improve what I wrote.

The team at Pearson Education worked tirelessly to see this edition move from manuscript to bound book. My acquisitions editor, Kelsey Volker Churchman, was joined by Lauren Harp, Senior Acquisitions Editor for the second part of this revision. Ashley Williams and Kate Abderholden, assistant editors, kept track of everyone and everything for us. Chriscelle Palaganas, Program Manager, provided excellent guidance and support throughout the whole production process.

The task of coordinating production fell to Pearson Content Producer Deepti Agarwal. Nathaniel Jones handled composition and project management, and Project Manager Stephanie Marquez at the art house, Imagineering, managed the team that prepared the art for production. Katrina Mohn was the photo researcher who found the wonderful new photos that appear in this edition. Nicole Constantine was the assistant media producer who kept my supplements authors on task and on schedule. Wendy Mears is the product marketing manager who works with the excellent sales teams at Pearson Education and Pearson International, and Derek Perrigo is the Field Marketing Manager for the anatomy and physiology list.

Special Thanks

As always, I would like to thank my students and colleagues who looked for errors and areas that needed improvement. I've learned that awarding one point of extra credit for being the first student to report a typo works really well. My graduate teaching assistants over the years have all played a huge role in my teaching, and their input has helped shape how I teach. Many of them are now faculty members themselves. They include:

Ari Berman, Ph.D.
Lawrence Brewer, Ph.D.
Kevin Christmas, Ph.D.
Michael Chirillo, M.D., Ph.D.
Lynn Cialdella Kam, M.S., M.B.A., Ph.D.
Sarah Davies Kanke, Ph.D.
Peter English, Ph.D.
Carol C. Linder, Ph.D.
Karina Loyo-Garcia, Ph.D.
Jan M. Machart, Ph.D.
Tonya Thompson, M.D.
Patti Thorn, Ph.D.
Justin Trombold, Ph.D.
Kurt Venator, Ph.D.
Kira Wenstrom, Ph.D.

Finally, special thanks to my colleagues in the American Physiological Society, the Human Anatomy & Physiology Society, and the International Union of Physiological Sciences whose experiences in the classroom have enriched my own understanding of how to teach physiology. I would also like to recognize a special group of friends for their continuing support: Penelope Hansen (Memorial University, St. John's), Mary Anne Rokitka (SUNY Buffalo), Rob Carroll (East Carolina University School of Medicine), Cindy Gill (Hampshire College), and Joel Michael (Rush Medical College), as well as Ruth Buskirk, Jeanne Lagowski, Jan M. Machart and Marilla Svinicki (University of Texas).

As always, I thank my family and friends for their patience, understanding, and support during the chaos that seems inevitable with book revisions. The biggest thank you goes to my husband Andy, whose love, support, and willingness to forgo home-cooked meals on occasion help me meet my deadlines.

A Work in Progress

One of the most rewarding aspects of writing a textbook is the opportunity it has given me to meet or communicate with other instructors and students. In the 20 years since the first edition was published, I have heard from people around the world and have had the pleasure of hearing how the book has been incorporated into their teaching and learning.

Because science textbooks are revised every 3 or 4 years, they are always works in progress. I invite you to contact me or my publisher with any suggestions, corrections, or comments about this edition. I am most reachable through e-mail at silverthorn@utexas.edu. You can reach my editor at the following address:

Applied Sciences
Pearson Education
1301 Sansome Street
San Francisco, CA 94111

Dee U. Silverthorn
silverthorn@utexas.edu
University of Texas
Austin, Texas

CONTENTS

CHAPTER 6
Communication, Integration, and Homeostasis 164

UNIT 2 Homeostasis and Control

CHAPTER 9
The Central Nervous System 271

CHAPTER 10
Sensory Physiology 307

UNIT 4 Metabolism, Growth, and Aging

CHAPTER 21
The Digestive System 654

CHAPTER 22
Metabolism and Energy Balance 692

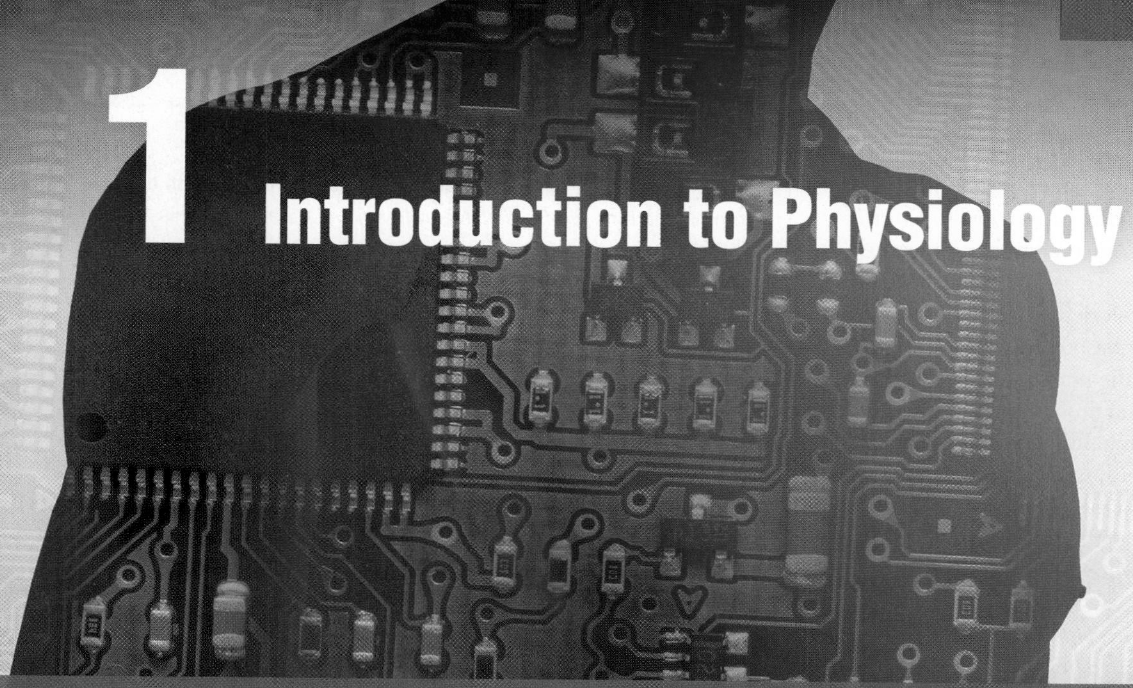

1 Introduction to Physiology

The current tendency of physiological thought is clearly toward an increasing emphasis upon the unity of operation of the Human Body.

Ernest G. Martin, preface to The Human Body *10th edition, 1917*

Welcome to the fascinating study of the human body! For most of recorded history, humans have been interested in how their bodies work. Early Egyptian, Indian, and Chinese writings describe attempts by physicians to treat various diseases and to restore health. Although some ancient remedies, such as camel dung and powdered sheep horn, may seem bizarre, we are still using others, such as blood-sucking leeches and chemicals derived from medicinal plants. The way we use these treatments has changed through the centuries as we have learned more about the human body.

There has never been a more exciting time in human physiology. **Physiology** is the study of the normal functioning of a living organism and its component parts, including all its chemical and physical processes. The term *physiology* literally means "knowledge of nature." Aristotle (384–322 BCE) used the word in this broad sense to describe the functioning of all living organisms, not just of the human body. However, Hippocrates (ca. 460–377 BCE), considered the father of medicine, used the word *physiology* to mean "the healing power of nature," and thereafter the field became closely associated with medicine. By the sixteenth century in Europe, physiology had been formalized as the study of the vital functions of the human body. Currently the term is again used to refer to the study of animals and plants.

Today, we benefit from centuries of work by physiologists who constructed a foundation of knowledge about how the human body functions. Since the 1970s, rapid advances in the fields of cellular and molecular biology have supplemented this work. A few decades ago, we thought that we would find the key to the secret of life by sequencing the human *genome*, which is the collective term for all the genetic information contained in the DNA of a species. However, this deconstructionist view of biology has proved to have its limitations, because living organisms are much more than the simple sum of their parts.

1.1 Physiology Is an Integrative Science

Many complex systems—including those of the human body—possess **emergent properties,** which are properties that cannot be predicted to exist based only on knowledge of the system's individual components. An emergent property is not a property of any single component of the system, and it is greater than the simple sum of the system's individual parts. Emergent properties result from complex, nonlinear interactions of the different components.

For example, suppose someone broke down a car into its nuts and bolts and pieces and laid them out on a floor. Could you predict that, properly assembled, these bits of metal and plastic would become a vehicle capable of converting the energy in gasoline into movement? Who could predict that the right combination of elements into molecules and assemblages of molecules would result in a living organism? Among the most complex emergent properties in humans are emotion, intelligence, and other aspects of brain function. None of these properties can be predicted from knowing the individual properties of nerve cells.

RUNNING PROBLEM **What to Believe?**

Jimmy had just left his first physiology class when he got the text from his mother: *Please call. Need to ask you something.* His mother seldom texted, so Jimmy figured it must be important. "Hi, Mom! What's going on?"

"Oh, Jimmy, I don't know what to do. I saw the doctor this morning and he's telling me that I need to take insulin. But I don't want to! My type of diabetes doesn't need insulin. I think he's just trying to make me see him more by putting me on insulin. Don't you think I'm right?"

Jimmy paused for a moment. "I'm not sure, Mom. He's probably just trying to do what's best for you. Didn't you talk to him about it?"

"Well, I tried but he didn't have time to talk. You're studying these things. Can't you look it up and see if I really need insulin?"

"I guess so. Let me see what I can find out." Jimmy hung up and thought. "Now what?"

2
5
9
12
16
19
24

When the Human Genome Project (*www.genome.gov*) began in 1990, scientists thought that by identifying and sequencing all the genes in human DNA, they would understand how the body worked. However, as research advanced, scientists had to revise their original idea that a given segment of DNA contained one gene that coded for one protein. It became clear that one gene may code for many proteins. The Human Genome Project ended in 2003, but before then researchers had moved beyond genomics to *proteomics*, the study of proteins in living organisms.

Now scientists have realized that knowing that a protein is made by a particular cell does not always tell us the significance of that protein to the cell, the tissue, or the functioning organism. The exciting new areas in biological research are called functional genomics, systems biology, and integrative biology, but fundamentally these are all fields of physiology. The **integration of function** across many **levels of organization** is a special focus of physiology. (To *integrate* means to bring varied elements together to create a unified whole.)

FIGURE 1.1 illustrates levels of organization ranging from the molecular level all the way up to populations of different species living together in *ecosystems* and in the *biosphere*. The levels of organization are shown along with the various subdisciplines of chemistry and biology related to the study of each organizational level. There is considerable overlap between the different fields of study, and these artificial divisions vary according to who is defining them. Notice, however, that physiology includes multiple levels, from molecular and cellular biology to the ecological physiology of populations.

At all levels, physiology is closely tied to anatomy. The structure of a cell, tissue, or organ must provide an efficient physical base for its function. For this reason, it is nearly impossible to study the physiology of the body without understanding the underlying anatomy. Because of the interrelationship of anatomy and physiology, you will find Anatomy Summaries throughout the book.

EMERGING CONCEPTS

The Changing World of Omics

If you read the scientific literature, it appears that contemporary research has exploded into an era of "omes" and "omics." What is an "ome"? The term apparently derives from the Latin word for a mass or tumor, and it is now used to refer to a collection of items that make up a whole, such as a genome. One of the earliest uses of the "ome" suffix in biology is the term *biome*, meaning all organisms living in a major ecological region, such as the marine biome or the desert biome. A genome, for example, is a collection of all the genetic material of an organism. Its physiome describes the organism's coordinated molecular, cellular, and physiological functioning.

The related adjective "omics" describes the research related to studying an "ome." Adding "omics" to a root word has become the cutting-edge way to describe a research field. For example, *pharmacogenomics* (the influence of genetics on the body's response to drugs) is now as important as *genomics*, the sequencing of DNA (the genome). There is even a journal named *OMICS!*

New "omes" emerge every year. The human connectome project (www.neuroscienceblueprint.nih.gov/connectome/) sponsored by the American National Institutes of Health is a collaborative effort by multiple institutions to map all the neural connections of the human brain. NIH also sponsors the human microbiome project (https://commonfund.nih.gov/hmp/overview), whose goal is to study the effects of microbes that normally live on or in the human body. Ignored as unimportant for many years, these microbes are now being shown to have an influence on both health and disease.

These special review features illustrate the anatomy of the physiological systems at different levels of organization.

At the most basic level of organization shown in Figure 1.1, atoms of elements link together to form molecules. Collections of molecules in living organisms form **cells,** the smallest unit of structure capable of carrying out all life processes. A lipid and protein barrier called the **cell membrane** (also called the *plasma membrane*) separates cells from their external environment. Simple organisms are composed of only one cell, but complex organisms have many cells with different structural and functional specializations.

Collections of cells that carry out related functions are called **tissues** {*texere*, to weave}. Tissues form structural and functional units known as **organs** {*organon*, tool}, and groups of organs integrate their functions to create **organ systems.** Chapter 3 reviews the anatomy of cells, tissues, and organs.

The 10 physiological organ systems in the human body are illustrated in **FIGURE 1.2**. Several of the systems have alternate names, given in parentheses, that are based on the organs of the system rather than the function of the system. The **integumentary system** {*integumentum*, covering}, composed of the skin, forms a protective boundary that separates the body's internal environment from the external environment (the outside world). The **musculoskeletal system** provides support and body movement.

Four systems exchange materials between the internal and external environments. The **respiratory (pulmonary) system** exchanges gases; the **digestive (gastrointestinal) system** takes up nutrients and water and eliminates wastes; the **urinary (renal) system** removes excess water and waste material; and the **reproductive system** produces eggs or sperm.

The remaining four systems extend throughout the body. The **circulatory (cardiovascular) system** distributes materials by pumping blood through vessels. The **nervous** and **endocrine systems** coordinate body functions. Note that the figure shows them as a continuum rather than as two distinct systems. Why? Because the lines between these two systems have blurred as we have learned more about the integrative nature of physiological function.

The one system not illustrated in Figure 1.2 is the diffuse **immune system,** which includes but is not limited to the anatomical structures known as the *lymphatic system*. The specialized cells of the immune system are scattered throughout the body. They protect the internal environment from foreign substances by intercepting material that enters through the intestines and lungs

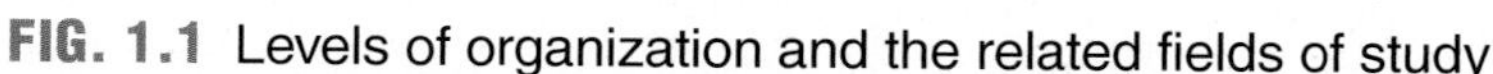

FIG. 1.1 Levels of organization and the related fields of study

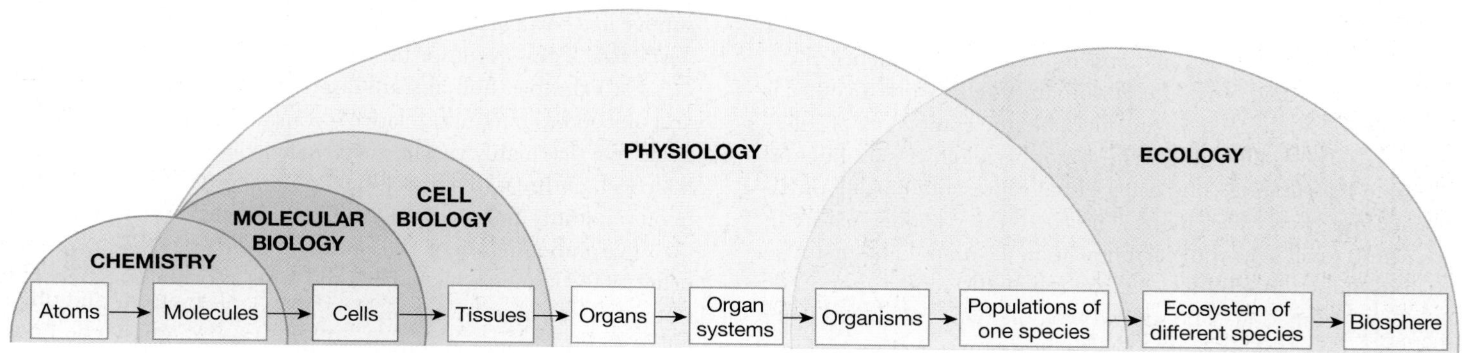
PHYSIOLOGY
ECOLOGY
CELL BIOLOGY
MOLECULAR BIOLOGY
CHEMISTRY
Atoms
Molecules
Cells
Tissues
Organs
Organ systems
Organisms
Populations of one species
Ecosystem of different species
Biosphere

FIG. 1.2 Organ systems of the human body and their integration

FIG. 1.2 Organ Systems of the Human Body and their Integration

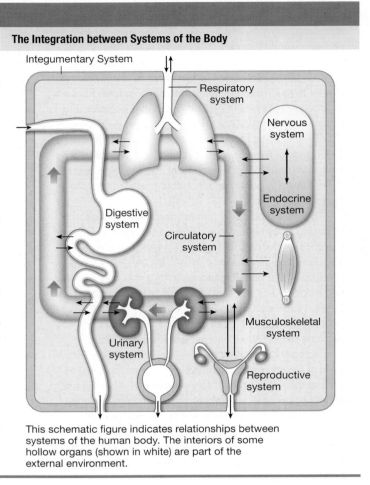

System Name	Includes	Representative Functions
Circulatory	Heart, blood vessels, blood	Transport of materials between all cells of the body
Digestive	Stomach, intestine, liver, pancreas	Conversion of food into particles that can be transported into the body; elimination of some wastes
Endocrine	Thyroid gland, adrenal gland	Coordination of body function through synthesis and release of regulatory molecules
Immune	Thymus, spleen, lymph nodes	Defense against foreign invaders
Integumentary	Skin	Protection from external environment
Musculoskeletal	Skeletal muscles, bone	Support and movement
Nervous	Brain, spinal cord	Coordination of body function through electrical signals and release of regulatory molecules
Reproductive	Ovaries and uterus, testes	Perpetuation of the species
Respiratory	Lungs, airways	Exchange of oxygen and carbon dioxide between the internal and external environments
Urinary	Kidneys, bladder	Maintenance of water and solutes in the internal environment; waste removal

The Integration between Systems of the Body

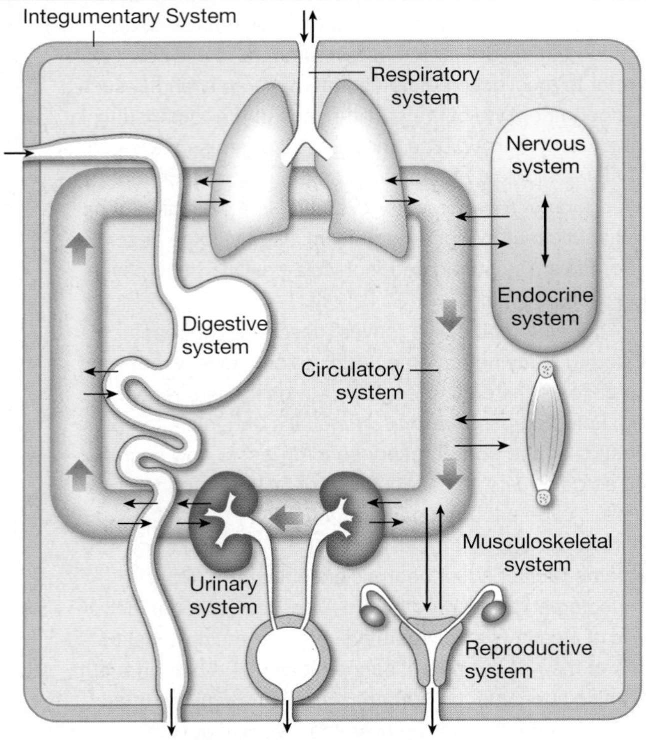

Integumentary System
Respiratory system
Nervous system
Endocrine system
Digestive system
Circulatory system
Musculoskeletal system
Urinary system
Reproductive system

This schematic figure indicates relationships between systems of the human body. The interiors of some hollow organs (shown in white) are part of the external environment.

or through a break in the skin. In addition, immune tissues are closely associated with the circulatory system.

Traditionally, physiology courses and books are organized by organ system. Students study cardiovascular physiology and regulation of blood pressure in one chapter, and then study the kidneys and control of body fluid volume in a different chapter. In the functioning human, however, the cardiovascular and renal systems communicate with each other, so that a change in one is likely to cause a reaction in the other. For example, body fluid volume influences blood pressure, while changes in blood pressure alter kidney function because the kidneys regulate fluid volume. In this book, you will find several integrative physiology chapters that highlight the coordination of function across multiple organ systems.

Understanding how different organ systems work together is just as important as memorizing facts, but the complexity of interactions can be challenging. One way physiologists simplify and integrate information is by using visual representations of physiological processes called maps. The Focus on Mapping feature in this chapter will help you learn how to make maps. The first type of map, shown in **FIGURE 1.3a**, is a schematic representation of structure or function. The second type of map, shown in Figure 1.3b, diagrams a physiological process as it proceeds through time. These process maps are also called *flow charts, and they are frequently used in health care.* You will be able to practice mapping with special end-of-chapter questions throughout the book.

1.2 Function and Mechanism

We define physiology as the normal functioning of the body, but physiologists are careful to distinguish between *function* and *mechanism*. The **function** of a physiological system or event is the "why" of the system or event: Why does a certain response help an animal survive in a particular situation? In other words, what is the *adaptive significance* of this event for this animal?

For example, humans are large, mobile, terrestrial animals, and our bodies maintain relatively constant water content despite living in a dry, highly variable external environment. Dehydration is a constant threat to our well-being. What processes have evolved in our anatomy and physiology that allow us to survive in this hostile environment? One is the production of highly concentrated urine by the kidney, which allows the body to conserve water. This statement tells us *why* we produce concentrated urine but does not tell us *how* the kidney accomplishes that task.

Thinking about a physiological event in terms of its adaptive significance is the **teleological approach** to science. For example, the teleological answer to the question of why red blood cells transport oxygen is "because cells need oxygen and red blood cells bring it to them." This answer explains *why* red blood cells transport oxygen—their function—but says nothing about *how* the cells transport oxygen.

In contrast, most physiologists study physiological processes, or **mechanisms**—the "how" of a system. The **mechanistic approach** to physiology examines process. The mechanistic answer to the question "How do red blood cells transport oxygen?" is "Oxygen binds to hemoglobin molecules in the red blood cells." This very concrete answer explains exactly how oxygen transport occurs but says nothing about the significance of oxygen transport to the animal.

Students often confuse these two approaches to thinking about physiology. Studies have shown that even medical students tend to answer questions with teleological explanations when the more appropriate response would be a mechanistic explanation.[1] Often they do so because instructors ask why a physiological event occurs when they really want to know how it occurs. Staying aware of the two approaches will help prevent confusion.

Although function and mechanism seem to be two sides of the same coin, it is possible to study mechanisms, particularly at the cellular and subcellular level, without understanding their function in the life of the organism. As biological knowledge becomes more complex, scientists sometimes become so involved in studying complex processes that they fail to step back and look at the significance of those processes to cells, organ systems, or the animal. Conversely, it is possible to use teleological thinking incorrectly by saying, "Oh, in this situation the body needs to do this." *This* may be a good solution, but if a mechanism for doing *this* doesn't exist, the situation cannot be corrected.

Applying the concept of integrated functions and mechanisms is the underlying principle in **translational research,** an approach sometimes described as "bench to bedside." Translational research uses the insights and results gained from basic biomedical research on mechanisms to develop treatments and strategies for preventing human diseases. For example, researchers working on rats found that a chemical from the pancreas named *amylin* reduced the rats' food intake. These findings led directly to a translational research study in which human volunteers injected a synthetic form of amylin and recorded their subsequent food intake, but without intentionally modifying their lifestyle.[2] The drug suppressed food intake in humans, and was later approved by the Food and Drug Administration for treatment of diabetes mellitus.

[1] D. R. Richardson. A survey of students' notions of body function as teleologic or mechanistic. *Advan Physiol Educ* 258: 8–10, Jun 1990. Access free at *http://advan.physiology.org*.

[2] S. R. Smith *et al.* Pramlintide treatment reduces 24-h caloric intake and meal sizes and improves control of eating in obese subjects: a 6-wk translational research study. *Am J Physiol Endocrinol Metab* 293: E620–E627, 2007.

RUNNING PROBLEM

When Jimmy got back to his room, he sat down at his computer and went to the Internet. He typed *diabetes* in his search box—and came up with 267 million results. "That's not going to work. What about *insulin*?" Nearly 48 million results. "How in the world am I going to get any answers?" He clicked on the first sponsored ad that advertised "Information for type 2 diabetes." That might be good. His mother had type 2 diabetes. But it was for a pharmaceutical company trying to sell him a drug. "Maybe my physiology prof can help me with this search. I'll ask tomorrow."

Q1: *What search terms could Jimmy have used to get fewer results?*

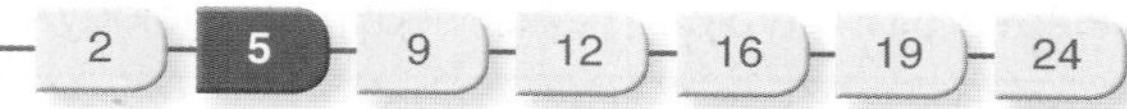

2
5
9
12
16
19
24

At the systems level, we know about most of the mechanics of body function from centuries of research. The unanswered questions today mostly involve integration and control of these mechanisms, particularly at the cellular and molecular levels. Nevertheless, explaining what happens in test tubes or isolated cells can only partially answer questions about function. For this reason, animal and human trials are essential steps in the process of applying basic research to treating or curing diseases.

1.3 Themes in Physiology

"Physiology is not a science or a profession but a point of view."[3] Physiologists pride themselves on relating the mechanisms they study to the functioning of the organism as a whole. For students, being able to think about how multiple body systems integrate their function is one of the more difficult aspects of learning physiology. To develop expertise in physiology, you must do more than simply memorize facts and learn new terminology. Researchers have found that the ability to solve problems requires a conceptual framework, or "big picture," of the field.

This book will help you build a conceptual framework for physiology by explicitly emphasizing the basic biological concepts, or themes, that are common to all living organisms. These concepts form patterns that repeat over and over, and you will begin to recognize them when you encounter them in specific contexts. Pattern recognition is an important skill in healthcare professions, and it will also simplify learning physiology.

In the past few years, three different organizations issued reports to encourage the teaching of biology using these fundamental concepts. Although the descriptions vary in the three reports, five major themes emerge:

1. structure and function across all levels of organization
2. energy transfer, storage, and use
3. information flow, storage, and use within single organisms and within a species of organism

[3] R. W. Gerard. *Mirror to Physiology: A Self-Survey of Physiological Science.* Washington, DC: American Physiology Society, 1958.

FIG. 1.3 Focus on . . . Mapping

Why use maps to study physiology? The answer is simple: maps will help you organize information you are learning in a way that makes sense to you and they will make that information easier to recall on a test. Creating a map requires higher-level thinking about the relationships among items on the map.

Mapping is not just a study technique. Scientists map out the steps in their experiments. Healthcare professionals create maps to guide them while diagnosing and treating patients. You can use mapping for almost every subject you study.

What is a map? Mapping is a nonlinear way of organizing material. A map can take a variety of forms but usually consists of terms (words or short phrases) linked by arrows to indicate associations. You can label the connecting arrows to describe the type of linkage between the terms (structure/function, cause/effect) or with explanatory phrases.

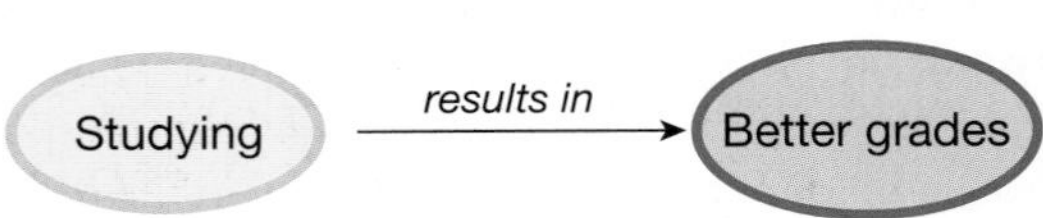
Studying
results in
Better grades

Here are two typical maps used in physiology.

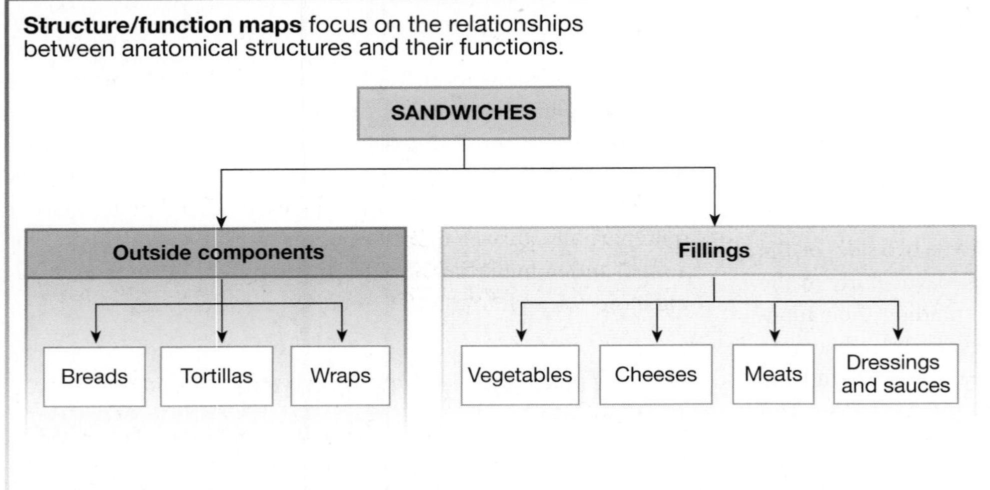
Structure/function maps focus on the relationships between anatomical structures and their functions.
SANDWICHES
Outside components
Fillings
Breads
Tortillas
Wraps
Vegetables
Cheeses
Meats
Dressings and sauces

Process maps or **flow charts** follow normal homeostatic control pathways or the body's responses to abnormal (pathophysiological) events as they unfold over time.

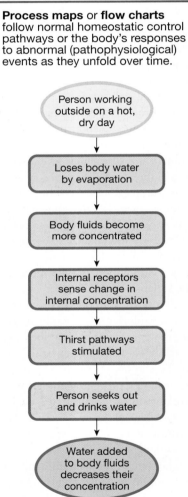

Person working outside on a hot, dry day
↓
Loses body water by evaporation
↓
Body fluids become more concentrated
↓
Internal receptors sense change in internal concentration
↓
Thirst pathways stimulated
↓
Person seeks out and drinks water
↓
Water added to body fluids decreases their concentration

Practice making maps. Many maps appear in this textbook, and they can serve as the starting point for your own maps. However, the real benefit of mapping comes from preparing maps yourself rather than memorizing someone else's maps. Your instructor can help you get started.

The next page walks you through the process of creating a structure-function map.

HINTS

- To help you get started, the end-of-chapter questions in this book include at least one list of terms to map for each chapter.
- Write your terms on individual slips of paper or small sticky notes so that you can rearrange the map more easily.
- Some terms may seem to belong to more than one group. Do not duplicate the item but make a note of it, as this term will probably have several arrows pointing to it or leading away from it.
- If arrows crisscross, try rearranging the terms on the map.
- Use color to indicate similar items.
- Add pictures and graphs that are associated with specific terms in your map.

Electronic mapping. Some people do not like the messiness of hand-drawn maps. There are several electronic ways of making maps, including PowerPoint or free and commercial software programs. Free concept mapping software is available from IHMC CmapTools at http://cmap.ihmc.us. Or search for the term free concept map to find other resources on the Web. A popular commercial program for mapping is Inspiration (www.inspiration.com).

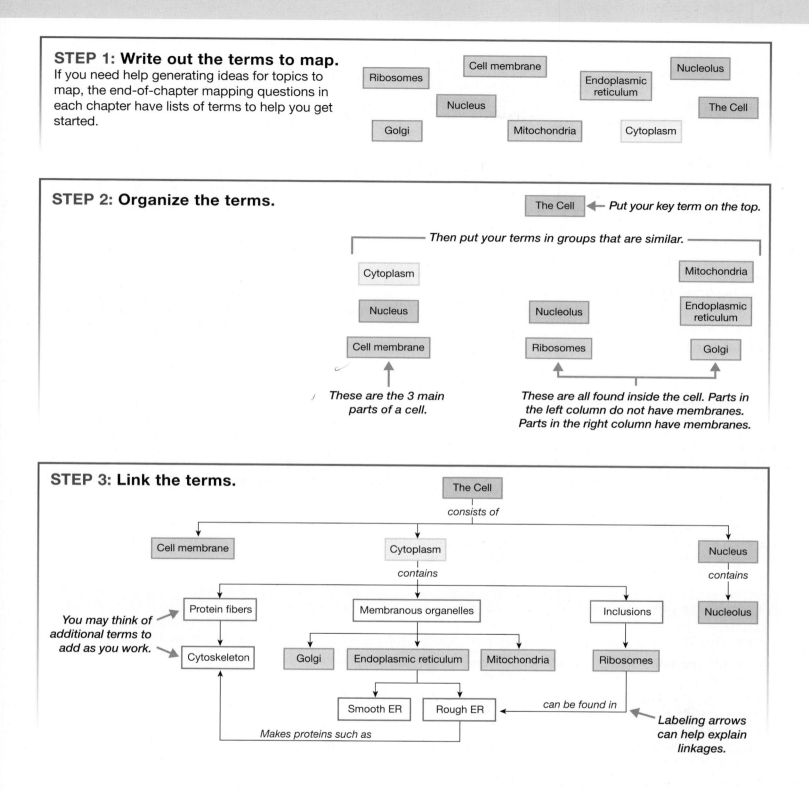

STEP 1: Write out the terms to map.

If you need help generating ideas for topics to map, the end-of-chapter mapping questions in each chapter have lists of terms to help you get started.

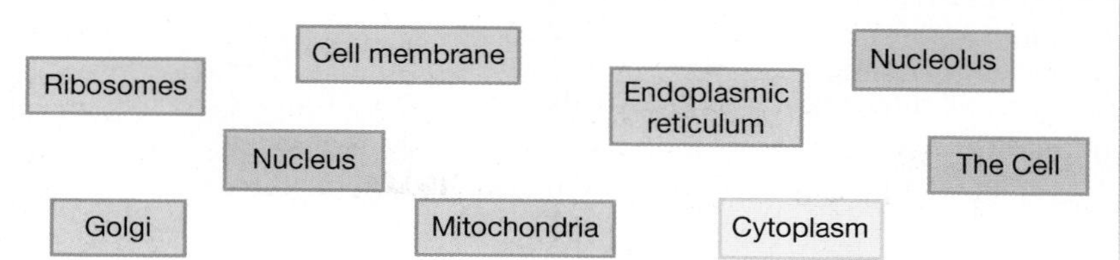

Cell membrane
Nucleolus
Ribosomes
Endoplasmic reticulum
Nucleus
The Cell
Golgi
Mitochondria
Cytoplasm

STEP 2: Organize the terms.

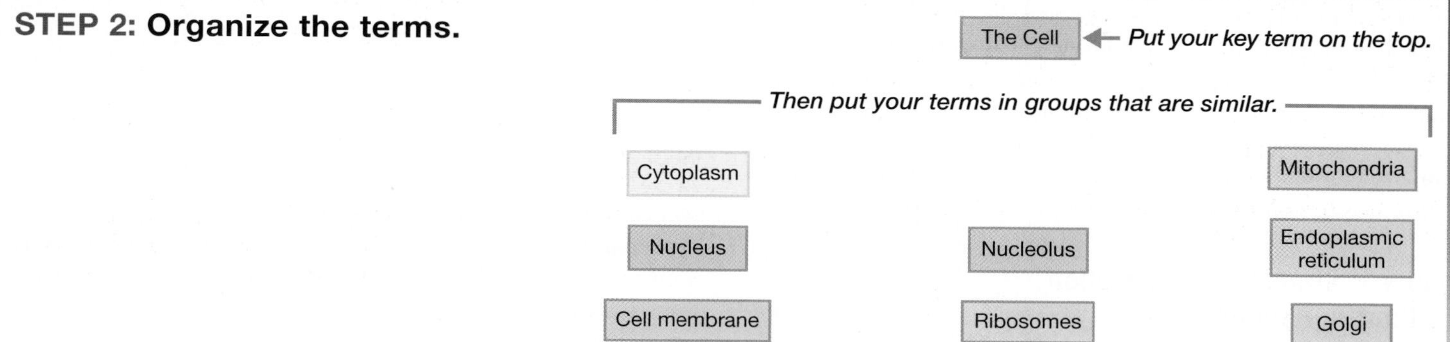

The Cell
Put your key term on the top.
Then put your terms in groups that are similar.
Cytoplasm
Mitochondria
Nucleus
Nucleolus
Endoplasmic reticulum
Cell membrane
Ribosomes
Golgi

These are the 3 main parts of a cell.

These are all found inside the cell. Parts in the left column do not have membranes. Parts in the right column have membranes.

STEP 3: Link the terms.

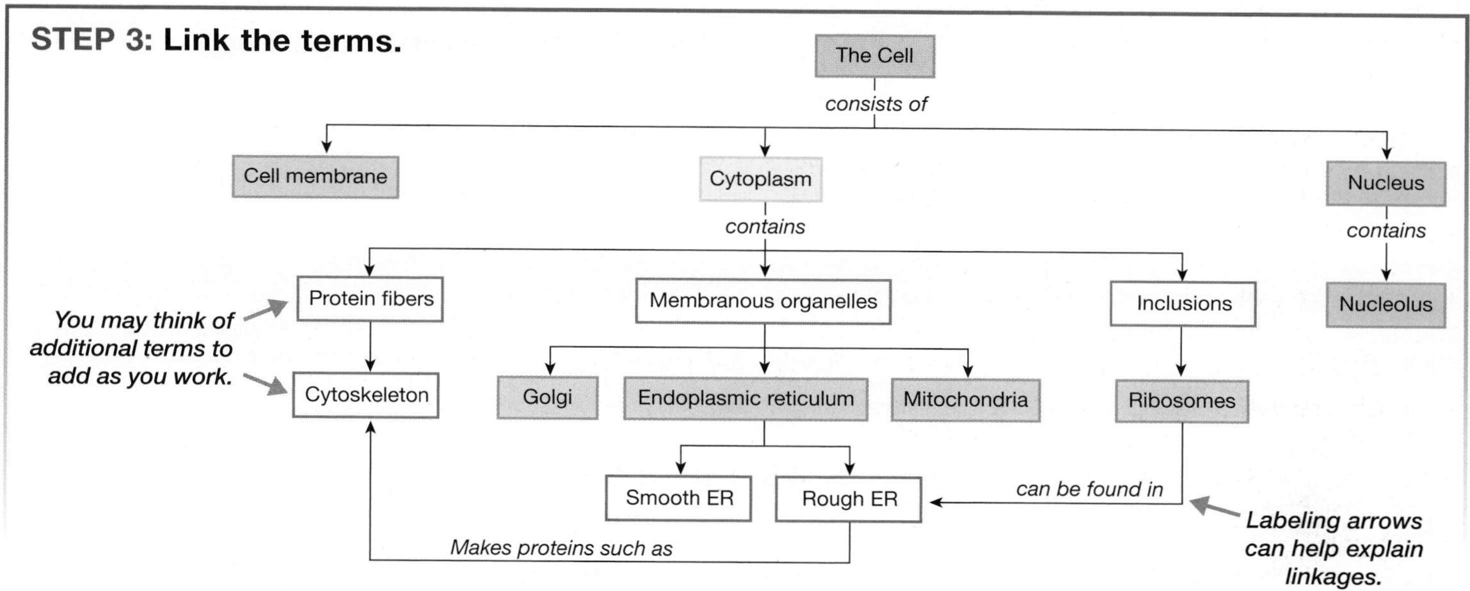

The Cell
consists of
Cell membrane
Cytoplasm
Nucleus
contains
contains
Protein fibers
Membranous organelles
Inclusions
Nucleolus
You may think of additional terms to add as you work.
Cytoskeleton
Golgi
Endoplasmic reticulum
Mitochondria
Ribosomes
Smooth ER
Rough ER
can be found in
Labeling arrows can help explain linkages.
Makes proteins such as

Once you have created your map, sit back and think about it. Are all the items in the right place? You may want to move them around once you see the big picture. Add new concepts or correct wrong links. Review by recalling the main concept and then moving to the more specific details. Ask yourself questions like, What is the cause and what is the effect? What parts are involved? What are the main characteristics?

Science is a collaborative field. A useful way to study with a map is to **trade maps with a classmate** and try to understand each other's maps. Your maps will almost certainly not look the same! It's OK if they are different. Remember that your map reflects the way you think about the subject, which may be different from the way someone else thinks about it. Did one of you put in something the other forgot? Did one of you have an incorrect link between two items?

4. homeostasis and the control systems that maintain it
5. evolution

In addition, all three reports emphasize the importance of understanding how science is done and of the quantitative nature of biology. **TABLE 1.1** lists the core concepts in biology from the three reports.

In this book, we focus on the four themes most related to physiology: structure-function relationships, biological energy use, information flow within an organism, and homeostasis and the control systems that maintain it. The first six chapters introduce the fundamentals of these themes, which you may already be familiar with from earlier biology or chemistry classes. The themes and their associated concepts, with variations, then re-appear over and over in subsequent chapters of this book. Look for them in the summary material at the end of the chapters and in the end-of-chapter questions as well.

Theme 1: Structure and Function Are Closely Related

The integration of structure and function extends across all levels of organization, from the molecular level to the intact body. This theme subdivides into two major ideas: molecular interactions and compartmentation.

Molecular Interactions The ability of individual molecules to bind to or react with other molecules is essential for biological function. A molecule's function depends on its structure and shape, and even a small change to the structure or shape may have significant effects on the function. The classic example of this phenomenon is the change in one amino acid of the hemoglobin protein. (Hemoglobin is the oxygen-carrying pigment of the blood.) This one small change in the protein converts normal hemoglobin to the form associated with sickle cell disease.

Many physiologically significant molecular interactions that you will learn about in this book involve the class of biological molecules called *proteins*. Functional groups of proteins include *enzymes* that speed up chemical reactions, *signal molecules* and the *receptor proteins* that bind signal molecules, and specialized proteins that function as biological pumps, filters, motors, or transporters. Chapter 2 describes molecular interactions involving proteins in more detail.

Interactions between proteins, water, and other molecules influence cell structure and the mechanical properties of cells and tissues. Mechanical properties you will encounter in your study of physiology include *compliance* (ability to stretch), *elastance* (stiffness or the ability to return to the unstretched state), strength, flexibility, and fluidity (*viscosity*).

Compartmentation **Compartmentation** is the division of space into separate compartments. Compartments allow a cell, a tissue, or an organ to specialize and isolate functions. Each level of organization is associated with different types of compartments. At the macroscopic level, the tissues and organs of the body form discrete functional compartments, such as body cavities or the insides of hollow organs. At the microscopic level, cell membranes separate cells from the fluid surrounding them and also create tiny compartments within the cell called organelles. Compartmentation is the theme of Chapter 3.

Theme 2: Living Organisms Need Energy

Growth, reproduction, movement, homeostasis—these and all other processes that take place in an organism require the continuous input of energy. Where does this energy come from, and how is it stored? We will answer those questions and describe some of

TABLE 1.1 Biology Concepts

Scientific Foundations for Future Physicians (HHMI and AAMC)[1]	Vision and Change (NSF and AAAS)[2]	The 2010 Advanced Placement Biology Curriculum (College Board)[3]
Structure/function from molecules to organisms	Structure and function (anatomy and physiology)	Relationship of structure to function
Physical principles applied to living systems Chemical principles applied to living systems	Pathways and transformations of energy and matter	Energy transfer
Biomolecules and their functions	Information flow, exchange, and storage	Continuity and change
Organisms sense and control their internal environment and respond to external change	Systems	Regulation ("a state of dynamic balance")
Evolution as an organizing principle	Evolution	Evolution

[1]*Scientific Foundations for Future Physicians.* Howard Hughes Medical Institute (HHMI) and the Association of American Medical Colleges (AAMC), 2009. *www.aamc.org/scientificfoundations*
[2]*Vision and Change: A Call to Action.* National Science Foundation (NSF) and American Association for the Advancement of Science (AAAS). 2011. *http://visionandchange.org/finalreport.* The report mentioned the integration of science and society as well.
[3]*College Board AP Biology Course Description*, The College Board, 2010. *http://apcentral.collegeboard.com/apc/public/repository/ap-biology-course-description.pdf.* The AP report also included "Interdependence in Nature" and "Science, Technology and Society" as two of their eight themes.

the ways that energy in the body is used for building and breaking down molecules in Chapter 4. In subsequent chapters, you will learn how energy is used to transport molecules across cell membranes and to create movement.

Theme 3: Information Flow Coordinates Body Functions

Information flow in living systems ranges from the transfer of information stored in DNA from generation to generation (genetics) to the flow of information within the body of a single organism. At the organismal level, information flow includes translation of DNA's genetic code into proteins responsible for cell structure and function.

In the human body, information flow between cells *coordinates function*. *Cell-to-cell communication* uses chemical signals, electrical signals, or a combination of both. Information may go from one cell to its neighbors (local communication) or from one part of the body to another (long-distance communication). Chapter 6 discusses chemical communication in the body.

When chemical signals reach their target cells, they must get their information into the cell. Some molecules are able to pass through the barrier of the cell membrane, but signal molecules that cannot enter the cell must pass their message across the cell membrane. How molecules cross biological membranes is the topic of Chapter 5.

Theme 4: Homeostasis Maintains Internal Stability

Organisms that survive in challenging habitats cope with external variability by keeping their internal environment relatively stable, an ability known as **homeostasis** {*homeo-*, similar + *-stasis*, condition}. Homeostasis and regulation of the internal environment are key principles of physiology and underlying themes in each chapter of this book. The next section looks in detail at the key elements of this important theme.

Play BioFlix Animation
@Mastering Anatomy & Physiology

1.4 Homeostasis

The concept of a relatively stable internal environment is attributed to the French physician Claude Bernard in the mid-1800s. During his studies of experimental medicine, Bernard noted the stability of various physiological functions, such as body temperature, heart rate, and blood pressure. As the chair of physiology at the University of Paris, he wrote "La fixité du milieu intérieur est la condition de la vie libre, indépendante." (The constancy of the internal environment is the condition for a free and independent life.)[4] This idea was applied to many of the experimental observations of his day, and it became the subject of discussion among physiologists and physicians.

RUNNING PROBLEM

After his second physiology class, Jimmy introduced himself to his professor and explained his problem. The professor's first suggestion was simple: try to narrow the search. "One of the best ways to search is to combine terms using the connector AND. If you remember set theory from your math class, the connector AND will give you the intersection of the sets. In other words, you'll get only the results that occur in both sets."

Seemed simple enough. Jimmy went back to the Internet and tried *diabetes and insulin.* That search still had 46 million results but on the first page was a link to the American Diabetes Association, *diabetes.org.* Now he was getting somewhere.

Q2: *What kinds of websites should Jimmy be looking for in his results list, and how can he recognize them?*

2
5
9
12
16
19
24

In 1929, an American physiologist named Walter B. Cannon wrote a review for the American Physiological Society.[5] Using observations made by numerous physiologists and physicians during the nineteenth and early twentieth centuries, Cannon proposed a list of variables that are under homeostatic control. We now know that his list was both accurate and complete. Cannon divided his variables into what he described as environmental factors that affect cells (osmolarity, temperature, and pH) and "materials for cell needs" (nutrients, water, sodium, calcium, other inorganic ions, oxygen, as well as "internal secretions having general and continuous effects"). Cannon's "internal secretions" are the hormones and other chemicals that our cells use to communicate with one another.

In his essay, Cannon created the word *homeostasis* to describe the regulation of the body's internal environment. He explained that he selected the prefix *homeo-* (meaning *like* or *similar*) rather than the prefix *homo-* (meaning *same*) because the internal environment is maintained within a range of values rather than at an exact fixed value. He also pointed out that the suffix *-stasis* in this instance means a *condition*, not a state that is static and unchanging. Cannon's homeostasis, therefore, is a state of maintaining "a similar condition," similar to Claude Bernard's relatively constant internal environment.

Some physiologists contend that a literal interpretation of *stasis* {a state of standing} in the word *homeostasis* implies a static, unchanging state. They argue that we should use the word *homeodynamics* instead, to reflect the small changes constantly taking place in our internal environment {*dynamikos*, force or power}. Whether the process is called homeostasis or homeodynamics, the important concept to remember is that the body monitors its internal state and takes action to correct disruptions that threaten its normal function.

[4] C. Bernard. *Leçons sur les phénomènes de la vie communs aux animaux et aux végétaux* (Vol. 1, p. 113), Paris: J.-B. Baillière, 1885. (*http://obvil.paris-sorbonne.fr/corpus/critique/bernard_lecons-phenomenes-vie-I/body-2*)

[5] W. B. Cannon. Organization for physiological homeostasis. *Physiol Rev* 9: 399–443, 1929.

If the body fails to maintain homeostasis of the critical variables listed by Walter Cannon, then normal function is disrupted and a disease state, or **pathological** condition {*pathos*, suffering}, may result. Diseases fall into two general groups according to their origin: those in which the problem arises from internal failure of some normal physiological process, and those that originate from some outside source. Internal causes of disease include the abnormal growth of cells, which may cause cancer or benign tumors; the production of antibodies by the body against its own tissues (autoimmune diseases); and the premature death of cells or the failure of cell processes. Inherited disorders are also considered to have internal causes. External causes of disease include toxic chemicals, physical trauma, and foreign invaders such as viruses and bacteria.

In both internally and externally caused diseases, when homeostasis is disturbed, the body attempts to compensate (FIG. 1.4). If the compensation is successful, homeostasis is restored. If compensation fails, illness or disease may result. The study of body functions in a disease state is known as **pathophysiology.** You will encounter many examples of pathophysiology as we study the various systems of the body.

One very common pathological condition in the United States is **diabetes mellitus,** a metabolic disorder characterized by abnormally high blood glucose concentrations. Although we speak of diabetes as if it were a single disease, it is actually a whole family of diseases with various causes and manifestations. You will learn more about diabetes in the focus boxes scattered throughout

FIG. 1.4 Homeostasis

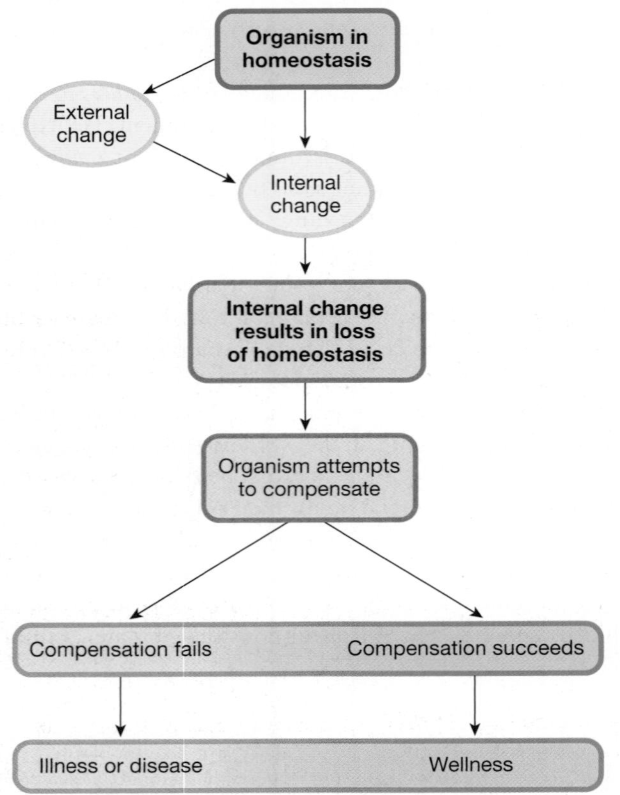
Organism in homeostasis
External change
Internal change
Internal change results in loss of homeostasis
Organism attempts to compensate
Compensation fails
Compensation succeeds
Illness or disease
Wellness

the chapters of this book. The influence of this one disorder on many systems of the body makes it an excellent example of the integrative nature of physiology.

What Is the Body's Internal Environment?

Claude Bernard wrote of the "constancy of the internal environment," but why is constancy so essential? As it turns out, most cells in our bodies are not very tolerant of changes in their surroundings. In this way they are similar to early organisms that lived in tropical seas, a stable environment where salinity, oxygen content, and pH vary little and where light and temperature cycle in predictable ways. The internal composition of these ancient creatures was almost identical to that of seawater. If environmental conditions changed, conditions inside the primitive organisms changed as well. Even today, marine invertebrates cannot tolerate significant changes in salinity and pH, as you know if you have ever maintained a saltwater aquarium.

In both ancient and modern times, many marine organisms relied on the constancy of their external environment to keep their internal environment in balance. In contrast, as organisms evolved and migrated from the ancient seas into estuaries, then into freshwater environments and onto the land, they encountered highly variable external environments. Rains dilute the salty water of estuaries, and organisms that live there must cope with the influx of water into their body fluids. Terrestrial organisms, including humans, face the challenge of dehydration—constantly losing internal water to the dry air around them. Keeping the internal environment stable means balancing water loss with appropriate water intake.

But what exactly is the internal environment of the body? For multicellular animals, it is the watery internal environment that surrounds the cells, a "sea within" the body called the **extracellular fluid (ECF)** {*extra*-, outside of} (FIG. 1.5). Extracellular fluid serves as the transition between an organism's external environment and the **intracellular fluid (ICF)** inside cells {*intra*-, within}. Because extracellular fluid is a buffer zone between cells and the outside world, elaborate physiological processes have evolved to keep its composition relatively stable.

When the extracellular fluid composition varies outside its normal range of values, compensatory mechanisms activate and try to return the fluid to the normal state. For example, when you drink a large volume of water, the dilution of your extracellular fluid triggers a mechanism that causes your kidneys to remove excess water and protect your cells from swelling. Most cells of multicellular animals do not tolerate much change. They depend on the constancy of extracellular fluid to maintain normal function.

Homeostasis Depends on Mass Balance

In the 1960s, a group of conspiracy theorists obtained a lock of Napoleon Bonaparte's hair and sent it for chemical analysis in an attempt to show that he died from arsenic poisoning. Today, a group of students sharing a pizza joke about the garlic odor on their breath. At first glance these two scenarios appear to have little

FIG. 1.5 The body's internal and external environments

(a) Extracellular fluid is a buffer between cells and the outside world.

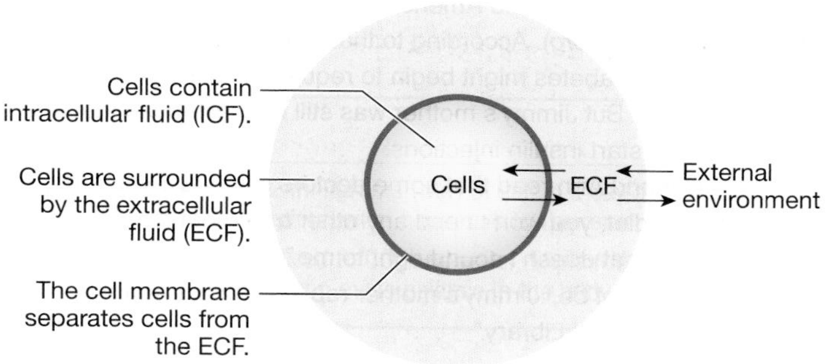
Cells contain intracellular fluid (ICF).
Cells are surrounded by the extracellular fluid (ECF).
The cell membrane separates cells from the ECF.
Cells
ECF
External environment

(b) A box diagram represents the ECF, ICF, and external environment as three separate compartments.

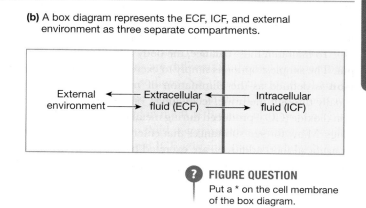

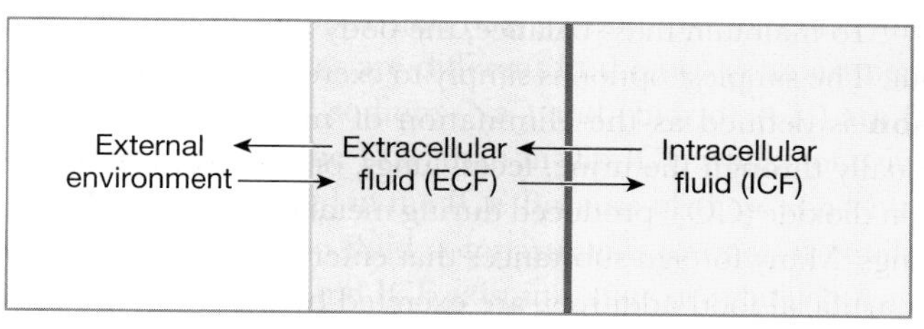
External environment
Extracellular fluid (ECF)
Intracellular fluid (ICF)

FIGURE QUESTION

Put a * on the cell membrane of the box diagram.

in common, but in fact Napoleon's hair and "garlic breath" both demonstrate how the human body works to maintain the balance that we call *homeostasis*.

The human body is an open system that exchanges heat and materials with the outside environment. To maintain homeostasis, the body must maintain mass balance. The **law of mass balance** says that if the amount of a substance in the body is to remain constant, any gain must be offset by an equal loss (**FIG. 1.6a**). The amount of a substance in the body is also called the body's **load,** as in "sodium load."

For example, water loss to the external environment (output) in sweat and urine must be balanced by water intake from the external environment plus metabolic water production (input). The concentrations of other substances, such as oxygen and carbon dioxide, salts, and hydrogen ions (pH), are also maintained through mass balance. The following equation summarizes the law of mass balance:

Total amount of substance x in the body = intake + production − excretion − metabolism

Most substances enter the body from the outside environment, but some (such as carbon dioxide) are produced internally through metabolism (Fig. 1.6b). In general, water and nutrients enter the

FIG. 1.6 Mass balance

(a) Mass balance in an open system requires input equal to output.

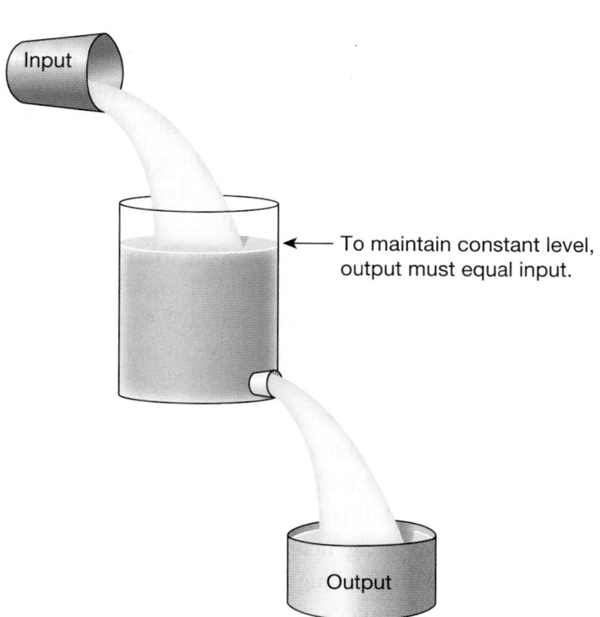
Input
To maintain constant level, output must equal input.
Output

(b) Mass balance in the body

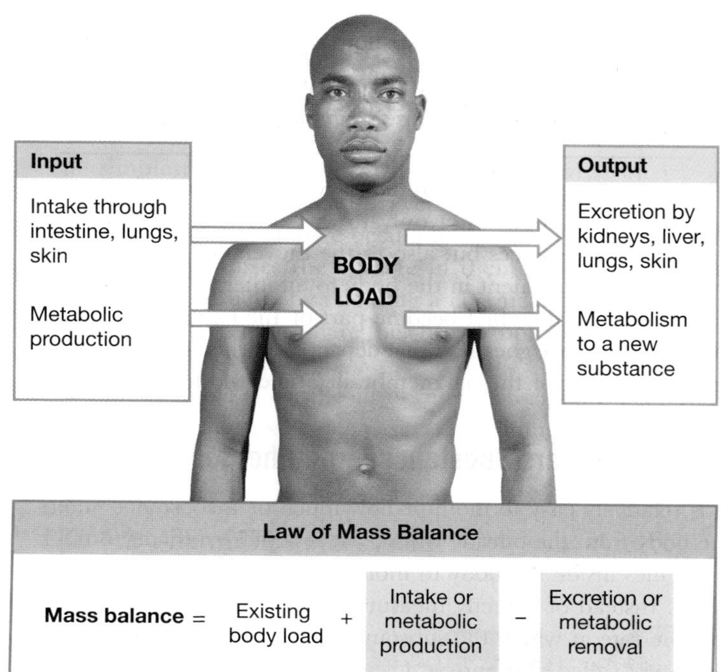
Input
Intake through intestine, lungs, skin
Metabolic production
BODY LOAD
Output
Excretion by kidneys, liver, lungs, skin
Metabolism to a new substance
Law of Mass Balance
Mass balance = Existing body load + Intake or metabolic production − Excretion or metabolic removal

FIG. 1.9 A comparison of local control and reflex control

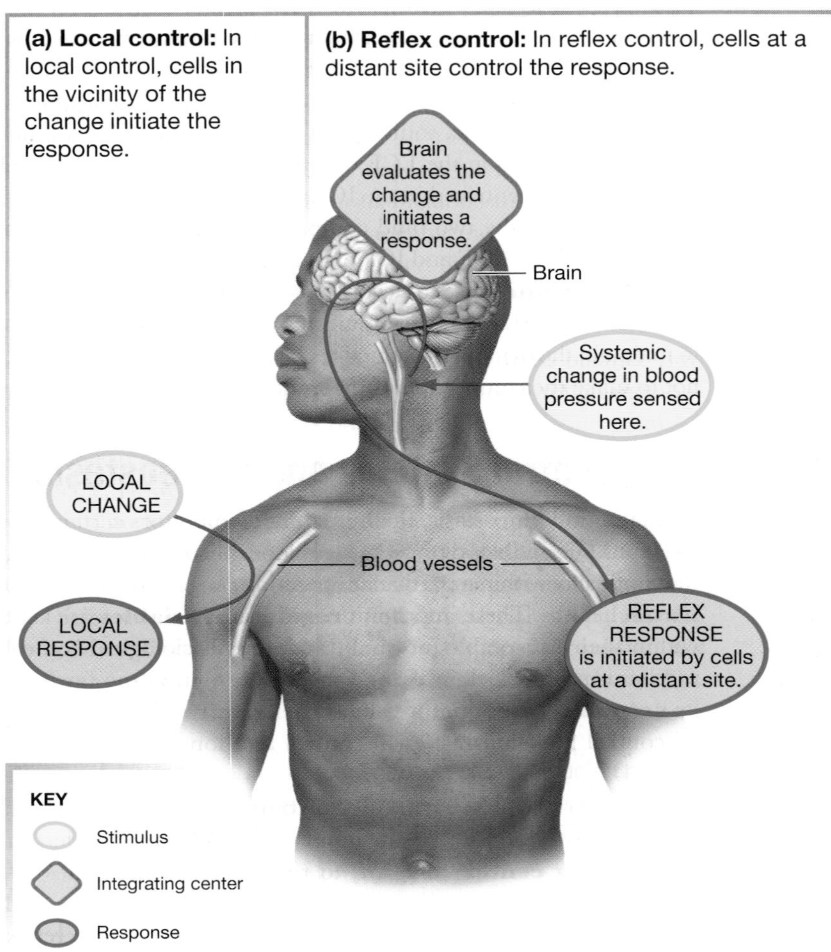
(a) Local control: In local control, cells in the vicinity of the change initiate the response.
(b) Reflex control: In reflex control, cells at a distant site control the response.
Brain evaluates the change and initiates a response.
Brain
Systemic change in blood pressure sensed here.
LOCAL CHANGE
Blood vessels
LOCAL RESPONSE
REFLEX RESPONSE is initiated by cells at a distant site.
KEY
Stimulus
Integrating center
Response

vessels that bring blood to the area sense the lower oxygen concentration and respond by secreting a chemical signal. The signal molecule diffuses to nearby muscles in the blood vessel wall, bringing them a message to relax. Relaxation of the muscles widens (*dilates*) the blood vessel, which increases blood flow into the tissue and brings more oxygen to the area.

Reflex Control Uses Long-Distance Signaling

Changes that are widespread throughout the body, or *systemic* in nature, require more complex control systems to maintain homeostasis. For example, maintaining blood pressure to drive blood flow throughout the body is a systemic issue rather than a local one. Because blood pressure is body-wide, maintaining it requires long-distance communication and coordination. We will use the term *reflex control* to mean any long-distance pathway that uses the nervous system, endocrine system, or both.

A physiological reflex can be broken down into two parts: a response loop and a feedback loop (**FIG. 1.10**). As with the simple control system just described, a **response loop** has three primary components: an *input signal*, an *integrating center* to integrate the signal, and an *output signal*. These three components can be expanded into the following sequence of seven steps to form a pattern that is found with slight variations in all reflex pathways:

Stimulus → sensor → input signal →
integrating center →
output signal → target → response

The input side of the response loop starts with a *stimulus*—the change that occurs when the regulated variable moves out of its desirable range. A specialized **sensor** monitors the variable. If the sensor is activated by the stimulus, it sends an input signal to the integrating center. The integrating center evaluates the information coming from the sensor and initiates an output signal. The output signal directs a target to carry out a response. If successful, the response brings the regulated variable back into the desired range.

In mammals, integrating centers are usually part of the nervous system or endocrine system. Output signals may be chemical signals, electrical signals, or a combination of both. The targets activated by output signals can be any cell of the body.

Response Loops Begin with a Stimulus

To illustrate response loops, let's apply the concept to a simple nonbiological example. Think about an aquarium whose heater is programmed to maintain the water temperature (the regulated variable) at 30 °C (Fig. 1.10). The room temperature is 25 °C. The desired water temperature (30 °C) is the *setpoint* for the regulated variable.

Assume that initially the aquarium water is at room temperature, 25 °C. When you turn the control box on, you set the response loop in motion. The thermometer (sensor) registers a temperature of 25 °C. It sends this information through a wire (input signal) to the control box (integrating center). The control box is programmed to evaluate the incoming temperature signal, compare it with the setpoint for the system (30 °C), and "decide" whether a response is needed to bring the water temperature up to the setpoint. The control box sends a signal through another wire (output signal) to the heater (the target), which turns on and starts heating the water (response). This sequence—from stimulus to response—is the response loop.

This aquarium example involves a variable (temperature) controlled by a single control system (the heater). We can also describe a system that is under dual control. For example, think of a house that has both heating and air conditioning. The owner would like the house to remain at 70 °F (about 21 °C). On chilly autumn mornings, when the temperature in the house falls, the heater turns on to warm the house. Then, as the day warms up, the heater is no longer needed and turns off. When the sun heats the house above the setpoint, the air conditioner turns on to cool the house back to 70 °F. The heater and air conditioner have *antagonistic control* over house temperature because they work in opposition to each other.

FIG. 1.10 The steps in a reflex pathway

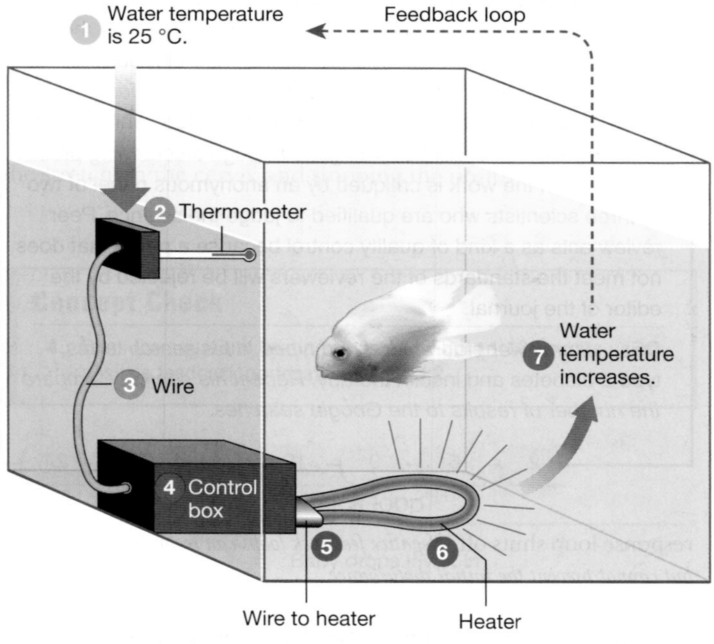
In the aquarium example shown, the control box is set to maintain a water temperature of 30 ± 1 °C.
1 Water temperature is 25 °C.
Feedback loop
2 Thermometer
3 Wire
4 Control box
5
6
7 Water temperature increases.
Wire to heater
Heater

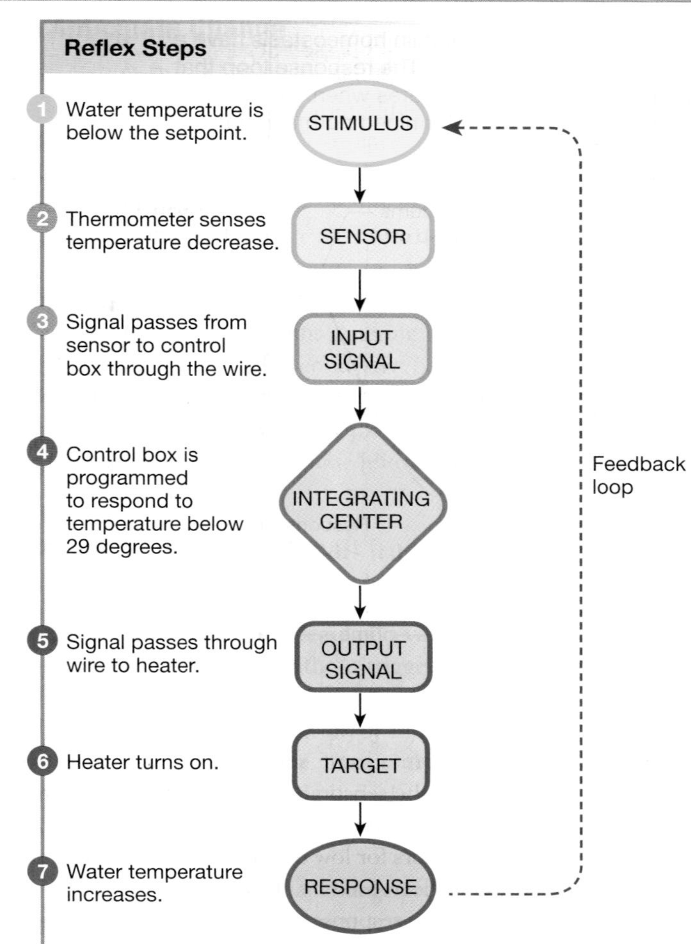
Reflex Steps
1 Water temperature is below the setpoint.
STIMULUS
2 Thermometer senses temperature decrease.
SENSOR
3 Signal passes from sensor to control box through the wire.
INPUT SIGNAL
4 Control box is programmed to respond to temperature below 29 degrees.
INTEGRATING CENTER
5 Signal passes through wire to heater.
OUTPUT SIGNAL
6 Heater turns on.
TARGET
7 Water temperature increases.
RESPONSE
Feedback loop

Similar situations occur in the human body when two branches of the nervous system or two different hormones have opposing effects on a single target.

Concept Check

3. What is the drawback of having only a single control system (a heater) for maintaining aquarium water temperature in some desired range?

Feedback Loops Modulate the Response Loop

The response loop is only the first part of a reflex. For example, in the aquarium just described, the sensor sends temperature information to the control box, which recognizes that the water is too cold. The control box responds by turning on the heater to warm the water. Once the response starts, what keeps the heater from sending the temperature up to, say, 50 °C?

The answer is a **feedback loop,** where the response "feeds back" to influence the input portion of the pathway. In the aquarium example, turning on the heater increases the temperature of the water. The sensor continuously monitors the temperature and sends that information to the control box. When the temperature warms up to the maximum acceptable value, the control box shuts off the heater, thus ending the reflex response.

Negative Feedback Loops Are Homeostatic

For most reflexes, feedback loops are homeostatic—that is, designed to keep the system at or near a setpoint so that the regulated variable is relatively stable. How well an integrating center succeeds in maintaining stability depends on the *sensitivity* of the system. In the case of our aquarium, the control box is programmed to have a sensitivity of ± 1 °C. If the water temperature drops from 30 °C to 29.5 °C, it is still within the acceptable range, and no response occurs. If the water temperature drops below 29 °C (30 − 1), the control box turns the heater on (**FIG. 1.11**). As the water heats up, the control box constantly receives information about the water temperature from the sensor. When the water reaches 31 °C (30 ± 1), the upper limit for the acceptable range, the feedback loop causes the control box to turn the heater off. The water then gradually cools off until the cycle starts all over again. The end result is a regulated variable that *oscillates* {*oscillare*, to swing} around the setpoint.

FIG. 2.1 REVIEW Biochemistry of Lipids

Lipids are biomolecules made mostly of carbon and hydrogen. Most lipids have a backbone of **glycerol** and 1–3 **fatty acids.** An important characteristic of lipids is that they are nonpolar and therefore not very soluble in water. Lipids can be divided into two broad categories.

- **Fats** are solid at room temperature. Most fats are derived from animal sources.
- **Oils** are liquid at room temperature. Most plant lipids are oils.

Fatty Acids

Fatty acids are long chains of carbon atoms bound to hydrogens, with a carboxyl (–COOH) or "acid" group at one end of the chain.

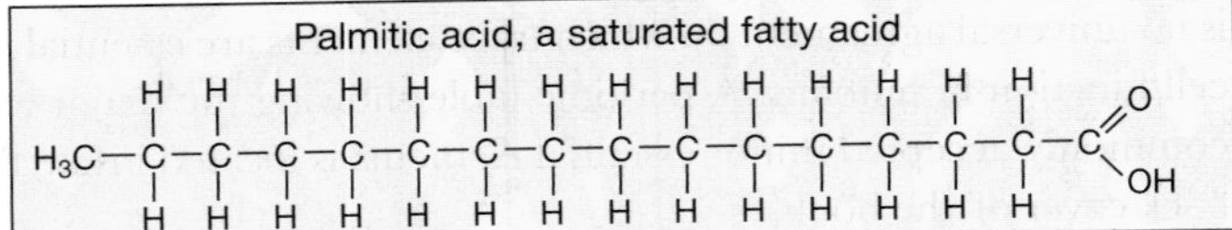

Palmitic acid, a saturated fatty acid
H_3C
OH

Saturated fatty acids have no double bonds between carbons, so they are "saturated" with hydrogens. The more saturated a fatty acid is, the more likely it is to be solid at room temperature.

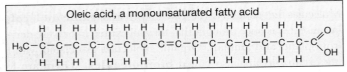

Oleic acid, a monounsaturated fatty acid

Monounsaturated fatty acids have one double bond between two of the carbons in the chain. For each double bond, the molecule has two fewer hydrogen atoms attached to the carbon chain.

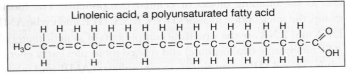

Linolenic acid, a polyunsaturated fatty acid

Polyunsaturated fatty acids have two or more double bonds between carbons in the chain.

Formation of Lipids

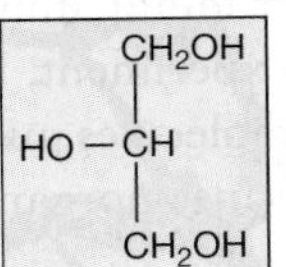

CH_2OH
HO – CH
CH_2OH

Glycerol is a simple 3-carbon molecule that makes up the backbone of most lipids.

Glycerol plus one fatty acid produces a **monoglyceride.**

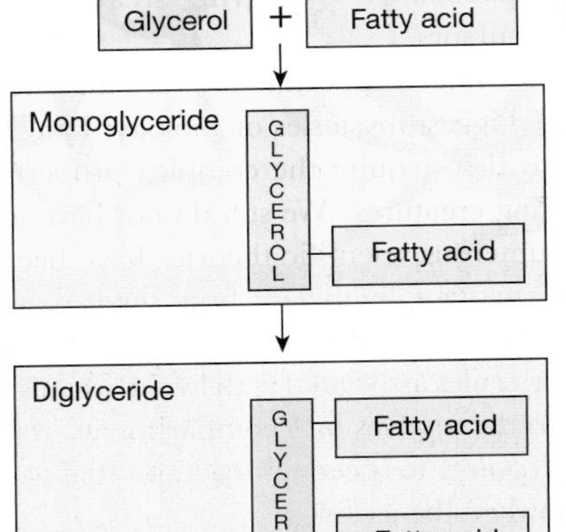

Glycerol
+
Fatty acid
Monoglyceride
GLYCEROL
Fatty acid
Diglyceride
GLYCEROL
Fatty acid
Fatty acid

Glycerol plus two fatty acids produces a **diglyceride.**

Glycerol plus three fatty acids produces a **triglyceride** (triacylglycerol). More than 90% of lipids are in the form of triglycerides.

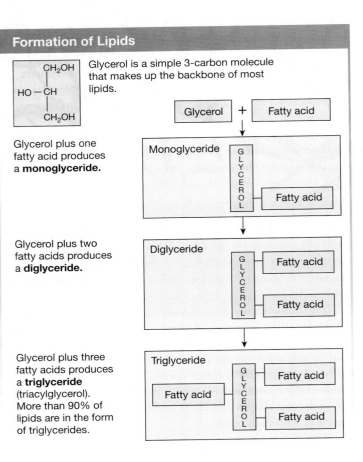

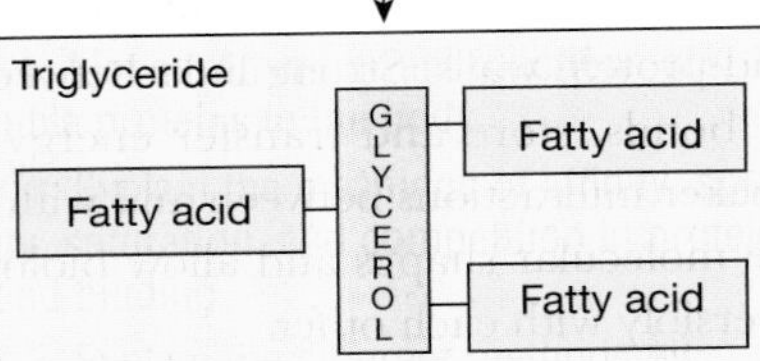

Triglyceride
Fatty acid
GLYCEROL
Fatty acid
Fatty acid

Lipid-Related Molecules

In addition to true lipids, this category includes three types of lipid-related molecules.

Eicosanoids

Eicosanoids {*eikosi*, twenty} are modified 20-carbon fatty acids with a complete or partial carbon ring at one end and two long carbon chain "tails."

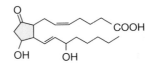

Prostaglandin E_2 (PGE_2)

Eicosanoids, such as thromboxanes, leukotrienes, and prostaglandins, act as regulators of physiological functions.

Steroids

Steroids are lipid-related molecules whose structure includes four linked carbon rings.

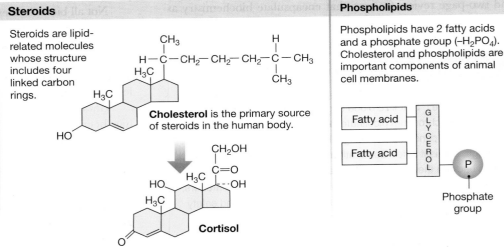

Cholesterol is the primary source of steroids in the human body.

Cortisol

Phospholipids

Phospholipids have 2 fatty acids and a phosphate group ($-H_2PO_4$). Cholesterol and phospholipids are important components of animal cell membranes.

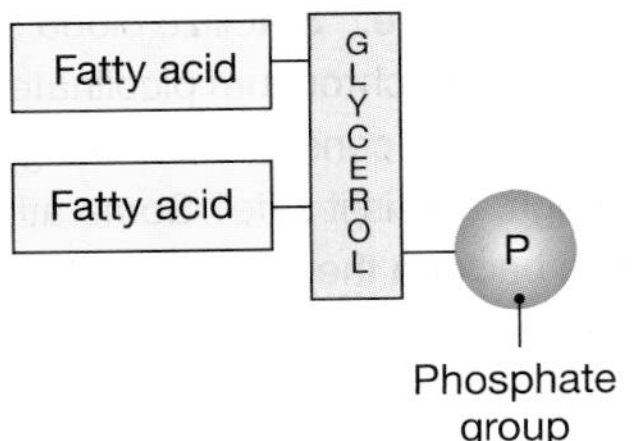

Fatty acid
Fatty acid
GLYCEROL
P
Phosphate group

FIG. 2.2 REVIEW Biochemistry of Carbohydrates

Carbohydrates are the most abundant biomolecule. They get their name from their structure, literally carbon {*carbo-*} with water {*hydro-*}. The general formula for a carbohydrate is $(CH_2O)_n$ or $C_nH_{2n}O_n$, showing that for each carbon there are two hydrogens and one oxygen. Carbohydrates can be divided into three categories: **monosaccharides, disaccharides,** and complex glucose polymers called **polysaccharides.**

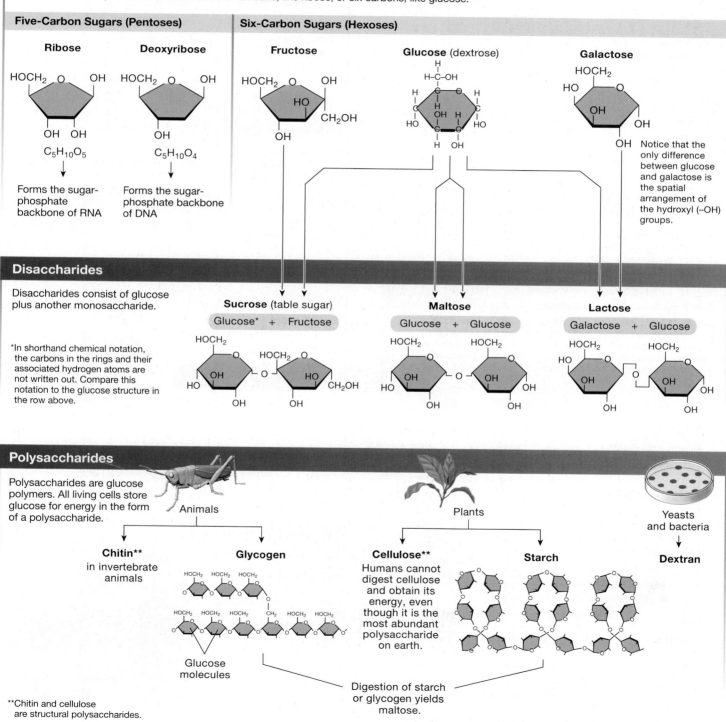

Monosaccharides

Monosaccharides are simple sugars. The most common monosaccharides are the building blocks of complex carbohydrates and have either five carbons, like ribose, or six carbons, like glucose.

Five-Carbon Sugars (Pentoses)

Ribose

HOCH2 O OH
OH OH

$C_5H_{10}O_5$

Forms the sugar-phosphate backbone of RNA

Deoxyribose

HOCH2 O OH
OH

$C_5H_{10}O_4$

Forms the sugar-phosphate backbone of DNA

Six-Carbon Sugars (Hexoses)

Fructose

HOCH2 O OH
HO
CH2OH
OH

Glucose (dextrose)

H
H–C–OH
H C O H
C H C
OH H
HO C C HO
H OH

Galactose

HOCH2
HO O
OH
OH
OH

Notice that the only difference between glucose and galactose is the spatial arrangement of the hydroxyl (–OH) groups.

Disaccharides

Disaccharides consist of glucose plus another monosaccharide.

*In shorthand chemical notation, the carbons in the rings and their associated hydrogen atoms are not written out. Compare this notation to the glucose structure in the row above.

Sucrose (table sugar)

Glucose* + Fructose

HOCH2 O HOCH2 O
OH O HO
HO CH2OH
OH OH

Maltose

Glucose + Glucose

HOCH2 HOCH2
O O
OH O OH
HO OH
OH OH

Lactose

Galactose + Glucose

HOCH2 HOCH2
HO O O
OH O OH
OH
OH OH

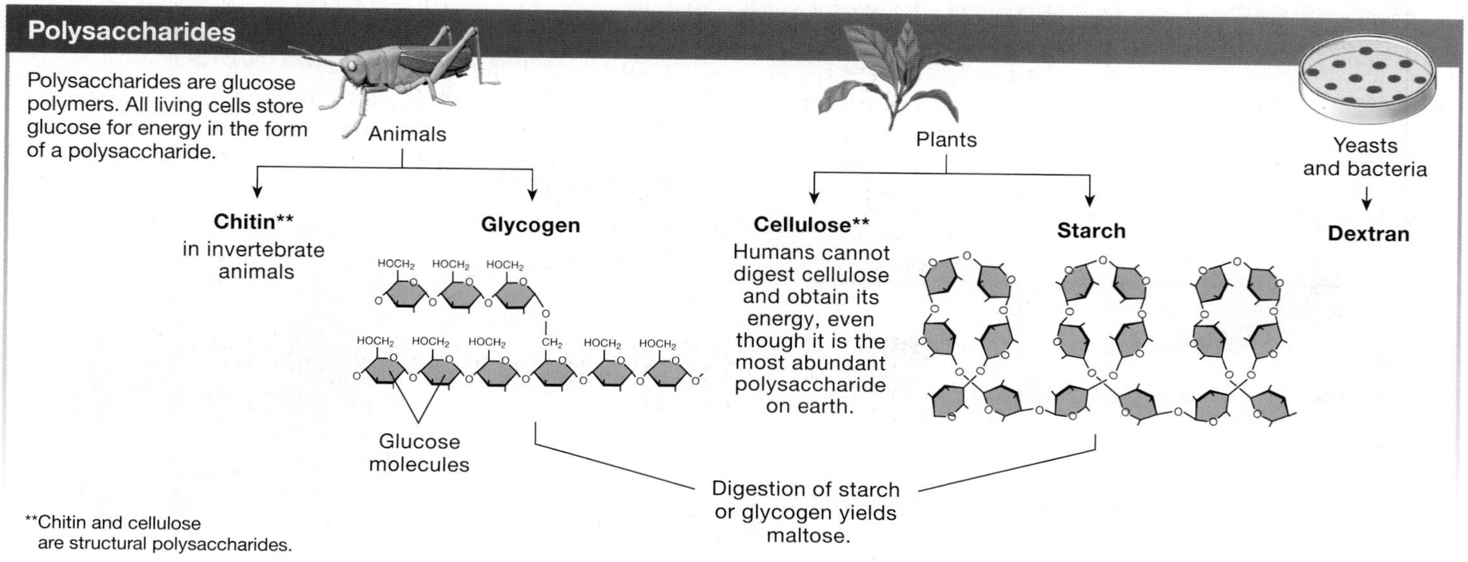

Polysaccharides
Polysaccharides are glucose polymers. All living cells store glucose for energy in the form of a polysaccharide.
Animals
Plants
Yeasts and bacteria
Chitin**
in invertebrate animals
Glycogen
HOCH2 HOCH2 HOCH2
HOCH2 HOCH2 HOCH2 CH2 HOCH2 HOCH2
Glucose molecules
Cellulose**
Humans cannot digest cellulose and obtain its energy, even though it is the most abundant polysaccharide on earth.
Starch
Dextran
Digestion of starch or glycogen yields maltose.
**Chitin and cellulose are structural polysaccharides.

FIG. 2.3 REVIEW Biochemistry of Proteins

Proteins are polymers of smaller building-block molecules called **amino acids.**

Amino Acids

All amino acids have a carboxyl group (–COOH), an amino group (–NH_2), and a hydrogen attached to the same carbon. The fourth bond of the carbon attaches to a variable "R" group.

The nitrogen (N) in the amino group makes proteins our major dietary source of nitrogen.

The R groups differ in their size, shape, and ability to form hydrogen bonds or ions. Because of the different R groups, each amino acid reacts with other molecules in a unique way.

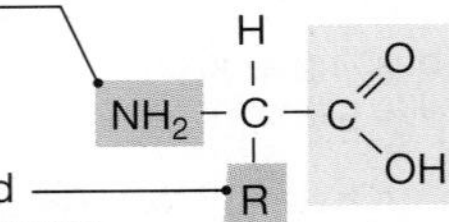

H
O
NH_2 – C – C
OH
R

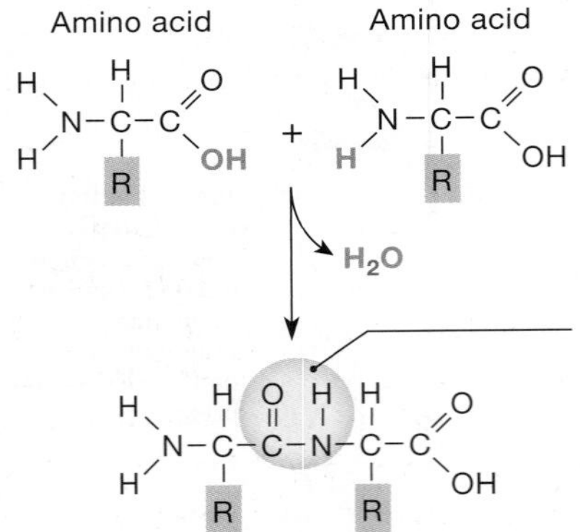

Amino acid
Amino acid
H_2O

In a **peptide bond,** the amino group of one amino acid joins the carboxyl group of the other, with the loss of water.

Amino Acids in Natural Proteins

Twenty different amino acids commonly occur in natural proteins. The human body can synthesize most of them, but at different stages of life some amino acids must be obtained from diet and are therefore considered *essential amino acids*. Some physiologically important amino acids are listed below.

Amino Acid	Three-Letter Abbreviation	One-Letter Symbol
Arginine	Arg	R
Aspartic acid (aspartate)*	Asp	D
Cysteine	Cys	C
Glutamic acid (glutamate)*	Glu	E
Glutamine	Gln	Q
Glycine	Gly	G
Tryptophan	Trp	W
Tyrosine	Tyr	Y

*The suffix -ate indicates the anion form of the acid.

Note:

A few amino acids do not occur in proteins but have important physiological functions.

- *Homocysteine*: a sulfur-containing amino acid that in excess is associated with heart disease
- *γ-amino butyric acid* (gamma-amino butyric acid) or *GABA*: a chemical made by nerve cells
- *Creatine*: a molecule that stores energy when it binds to a phosphate group

Structure of Peptides and Proteins

Primary Structure

The 20 protein-forming amino acids assemble into polymers called peptides. The sequence of amino acids in a peptide chain is called the **primary structure.** Just as the 26 letters of our alphabet combine to create different words, the 20 amino acids can create an almost infinite number of combinations.

Peptides range in length from two to two million amino acids:

- **Oligopeptide** {*oligo-*, few}: 2–9 amino acids
- **Polypeptide:** 10–100 amino acids
- **Proteins:** >100 amino acids

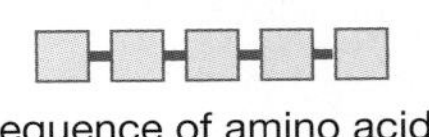

Sequence of amino acids

Secondary Structure

Secondary structure is created primarily by hydrogen bonds between adjacent chains or loops.

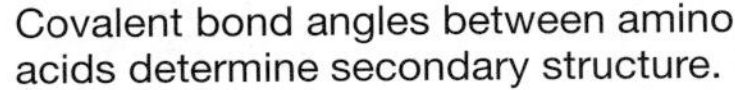

Covalent bond angles between amino acids determine secondary structure.

α-helix

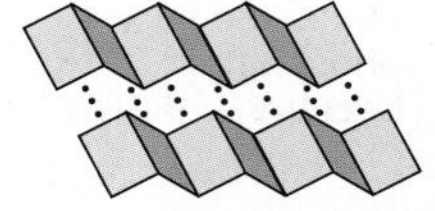

β-strands form sheets

Tertiary Structure

Tertiary structure is the protein's three-dimensional shape.

Fibrous proteins
Collagen

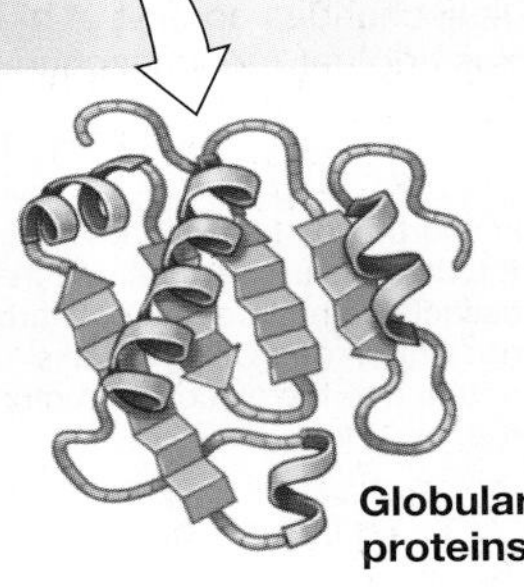

Globular proteins

Tertiary structures can be a mix of secondary structures. Beta-sheets are shown as flat ribbon arrows and alpha helices are shown as ribbon coils.

Quaternary Structure

Multiple subunits combine with noncovalent bonds. Hemoglobin molecules are made from four globular protein subunits.

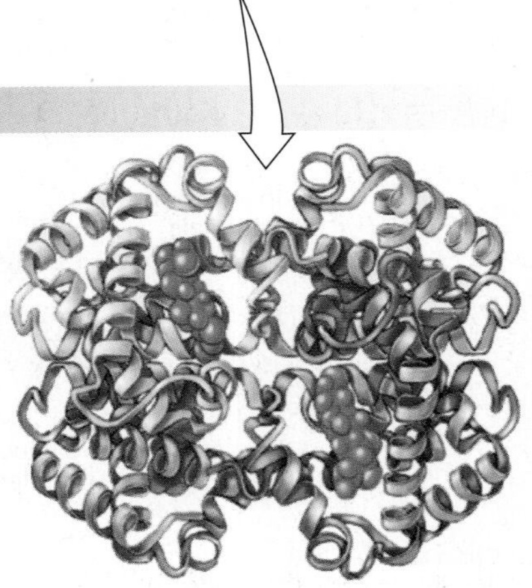

Hemoglobin

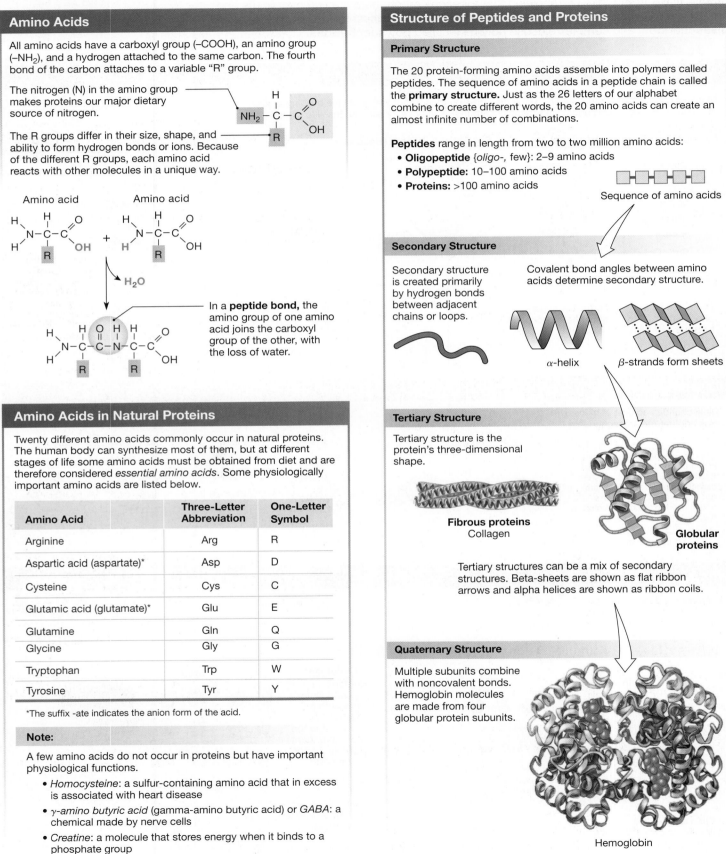

TABLE 2.1 Common Functional Groups

Notice that oxygen, with two electrons to share, sometimes forms a double bond with another atom.

	Shorthand	Bond Structure
Amino	$—NH_2$	$—N(H)H$
Carboxyl (acid)	—COOH	$—C(=O)OH$
Hydroxyl	—OH	—O—H
Phosphate	$—H_2PO_4$	$—O—P(OH)_2=O$

Some combinations of elements, known as **functional groups,** occur repeatedly in biological molecules. The atoms in a functional group tend to move from molecule to molecule as a single unit. For example, *hydroxyl groups,* —OH, common in many biological molecules, are added and removed as a group rather than as single hydrogen or oxygen atoms. Amino groups, $-NH_2$, are the signature of amino acids. The phosphate group, $-H_2PO_4$, plays a role in many important cell processes, such as energy transfer and protein regulation. Addition of a phosphate group is called *phosphorylation*; removal of a phosphate group is *dephosphorylation*.

The most common functional groups are listed in **TABLE 2.1**.

Concept Check

1. List three major essential elements found in the human body.
2. What is the general formula of a carbohydrate?
3. What is the chemical formula of an amino group? Of a carboxyl group?

Electrons Have Four Important Biological Roles

An atom of any element has a unique combination of protons and electrons that determines the element's properties (**FIG. 2.5**). We are particularly interested in the electrons because they play four important roles in physiology:

1. **Covalent bonds.** The arrangement of electrons in the outer energy level (*shell*) of an atom determines an element's ability to bind with other elements. Electrons shared between atoms form strong covalent bonds that bind atoms together to form molecules.
2. **Ions.** If an atom or molecule gains or loses one or more electrons, it acquires an electrical charge and becomes an **ion.** Ions are the basis for electrical signaling in the body. Ions may be single atoms, like the sodium ion Na^+ and chloride ion Cl^-. Other ions are combinations of atoms, such as the bicarbonate ion HCO_3^-. Important ions of the body are listed in **TABLE 2.2**.
3. **High-energy electrons.** The electrons in certain atoms can capture energy from their environment and transfer it to other atoms. This allows the energy to be used for synthesis, movement, and other life processes. The released energy may also be emitted as radiation. For example, bioluminescence in fireflies is visible light emitted by high-energy electrons returning to their normal low-energy state.
4. **Free radicals.** Free radicals are unstable molecules with an unpaired electron. They are thought to contribute to aging and to the development of certain diseases, such as some cancers. Free radicals and high-energy electrons are discussed in Chapter 22.

The role of electrons in molecular bond formation is discussed in the next section. There are four common bond types, two strong and two weak. Covalent and ionic bonds are strong bonds because they require significant amounts of energy to make or break. Hydrogen bonds and van der Waals forces are weaker bonds that require much less energy to break. Interactions between molecules with different bond types are responsible for energy use and transfer in metabolic reactions as well as a variety of other reversible interactions.

Covalent Bonds between Atoms Create Molecules

Molecules form when atoms share pairs of electrons, one electron from each atom, to create **covalent bonds.** These strong bonds require the input of energy to break them apart. It is possible to predict how many covalent bonds an atom can form by knowing how many unpaired electrons are in its outer shell, because an atom is most stable when all of its electrons are paired (**FIG. 2.6**).

For example, a hydrogen atom has one unpaired electron and one empty electron place in its outer shell. Because hydrogen has only one electron to share, it always forms one covalent bond, represented by a single line (—) between atoms. Oxygen has six electrons in an outer shell that can hold eight. That means oxygen can form two covalent bonds and fill its outer shell with

TABLE 2.2 Important Ions of the Body

Cations		Anions	
Na^+	Sodium	Cl^-	Chloride
K^+	Potassium	HCO_3^-	Bicarbonate
Ca^{2+}	Calcium	HPO_4^{2-}	Phosphate
H^+	Hydrogen	SO_4^{2-}	Sulfate
Mg^{2+}	Magnesium		

CHAPTER 2

FIG. 2.4 REVIEW Nucleotides and Nucleic Acids

Nucleotides are biomolecules that play an important role in energy and information transfer. Single nucleotides include the energy-transferring compounds **ATP** (adenosine triphosphate) and **ADP** (adenosine diphosphate), as well as **cyclic AMP,** a molecule important in the transfer of signals between cells. **Nucleic acids** (or nucleotide polymers) such as **RNA** and **DNA** store and transmit genetic information.

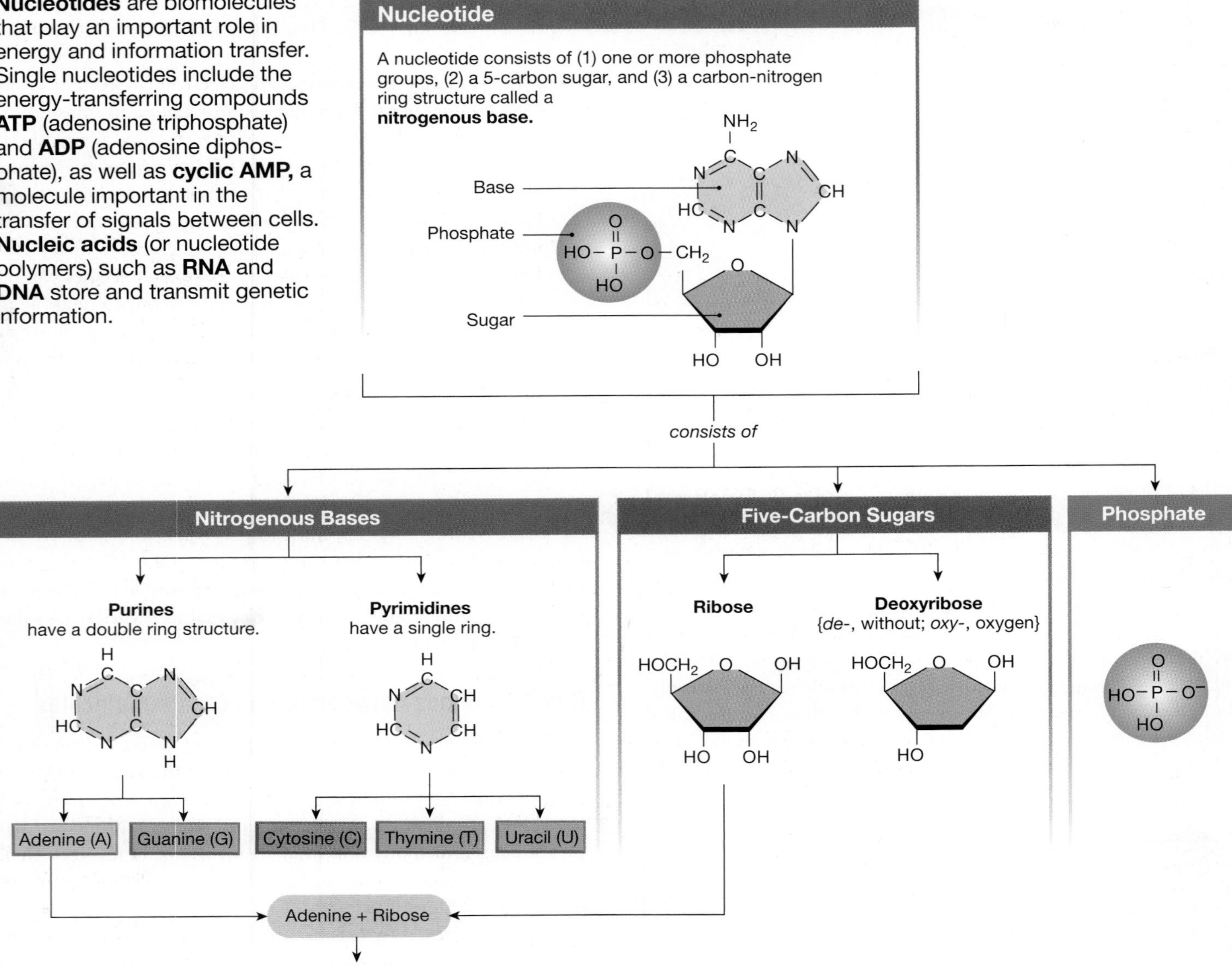

Nucleotide
A nucleotide consists of (1) one or more phosphate groups, (2) a 5-carbon sugar, and (3) a carbon-nitrogen ring structure called a **nitrogenous base.**
Base
Phosphate
Sugar
consists of
Nitrogenous Bases
Purines have a double ring structure.
Pyrimidines have a single ring.
Adenine (A)
Guanine (G)
Cytosine (C)
Thymine (T)
Uracil (U)
Five-Carbon Sugars
Ribose
Deoxyribose {de-, without; oxy-, oxygen}
Phosphate
Adenine + Ribose
Adenosine

Single Nucleotide Molecules

Single nucleotide molecules have two critical functions in the human body: (1) Capture and transfer energy in high-energy electrons or phosphate bonds, and (2) aid in cell-to-cell communication.

Nucleotide	*consists of*	Base	+	Sugar	+	Phosphate Groups	+	Other Component	Function
ATP	=	Adenine	+	Ribose	+	3 phosphate groups			Energy capture and transfer
ADP	=	Adenine	+	Ribose	+	2 phosphate groups			
NAD	=	Adenine	+	2 Ribose	+	2 phosphate groups	+	Nicotinamide	
FAD	=	Adenine	+	Ribose	+	2 phosphate groups	+	Riboflavin	
cAMP	=	Adenine	+	Ribose	+	1 phosphate group			Cell-to-cell communication

Nucleic acids (nucleotide polymers) function in information storage and transmission. The sugar of one nucleotide links to the phosphate of the next, creating a chain of alternating sugar–phosphate groups. The sugar–phosphate chains, or backbone, are the same for every nucleic acid molecule. Nucleotide chains form strands of DNA and RNA.

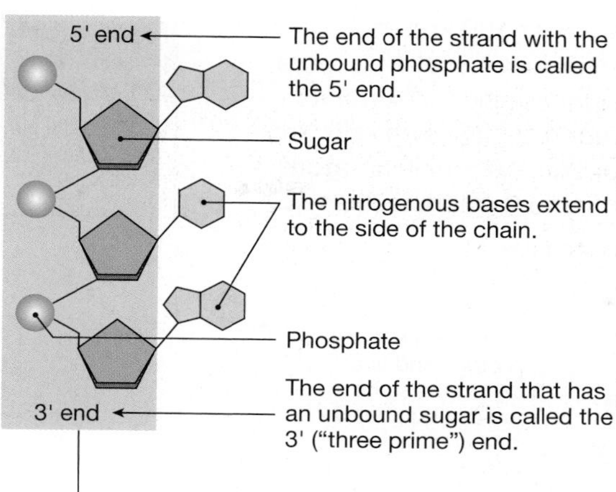

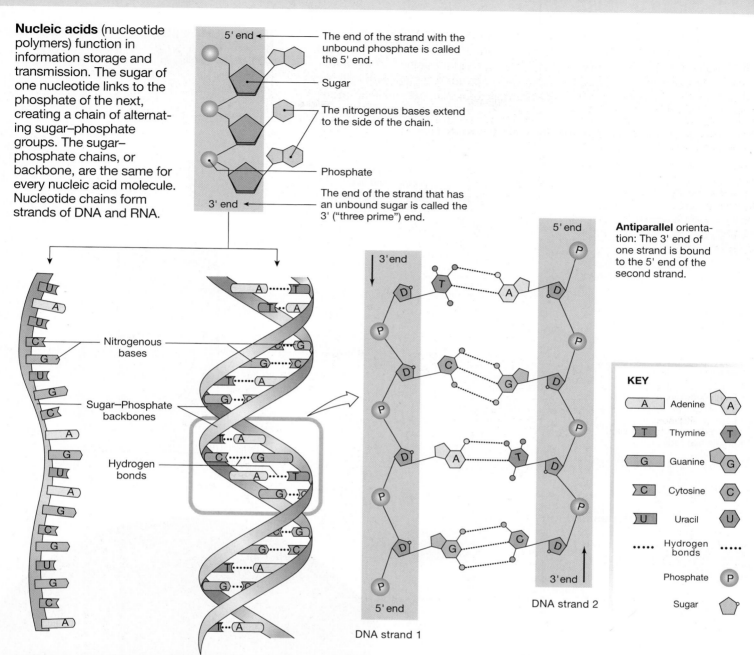

5' end
The end of the strand with the unbound phosphate is called the 5' end.
Sugar
The nitrogenous bases extend to the side of the chain.
Phosphate
The end of the strand that has an unbound sugar is called the 3' ("three prime") end.
3' end

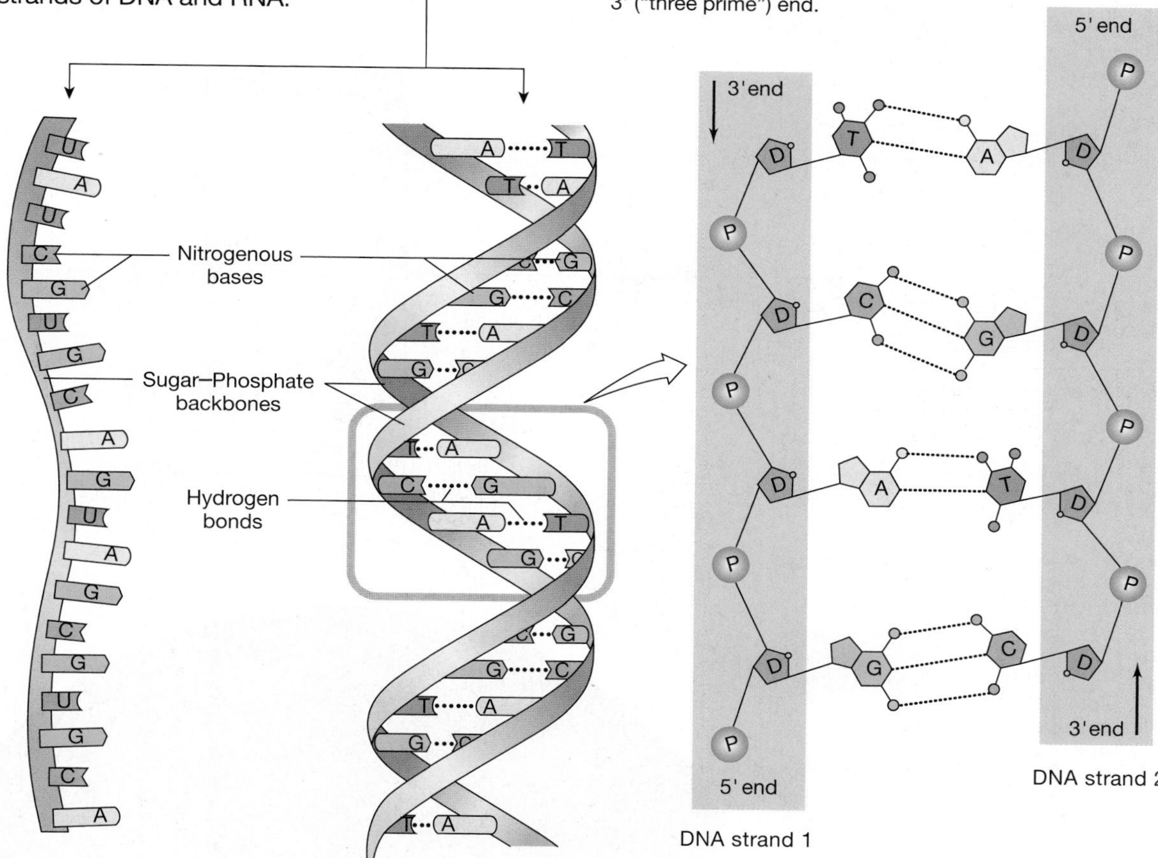

Nitrogenous bases
Sugar–Phosphate backbones
Hydrogen bonds
5' end
3' end
Antiparallel orientation: The 3' end of one strand is bound to the 5' end of the second strand.
DNA strand 1
DNA strand 2

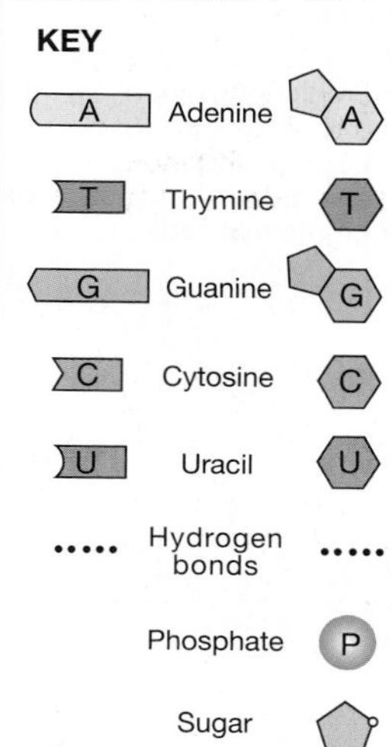

KEY
Adenine
Thymine
Guanine
Cytosine
Uracil
Hydrogen bonds
Phosphate
Sugar

RNA (ribonucleic acid) is a single–strand nucleic acid with ribose as the sugar in the backbone, and four bases—adenine, guanine, cytosine, and uracil.

DNA (deoxyribonucleic acid) is a double helix, a three-dimensional structure that forms when two DNA strands link through hydrogen bonds between complementary base pairs. Deoxyribose is the sugar in the backbone, and the four bases are adenine, guanine, cytosine, and thymine.

Base-Pairing

Bases on one strand form hydrogen bonds with bases on the adjoining strand. This bonding follows very specific rules:

- Because purines are larger than pyrimidines, space limitations always pair a purine with a pyrimidine.
- Guanine (G) forms three hydrogen bonds with cytosine (C).
- Adenine (A) forms two hydrogen bonds with thymine (T) or uracil (U).

Guanine-Cytosine Base Pair

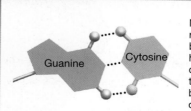

Guanine
Cytosine

More energy is required to break the triple hydrogen bonds of G:::C than the double bonds of A::::T or A::::U.

Adenine-Thymine Base Pair

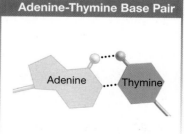

Adenine
Thymine

FIG. 2.5 REVIEW Atoms and Molecules

Elements are the simplest type of matter. There are over 100 known elements,* but only three—oxygen, carbon, and hydrogen—make up more than 90% of the body's mass. These three plus eight additional elements are *major essential elements*. An additional 19 *minor essential elements* are required in trace amounts. The smallest particle of any element is an **atom** {*atomos*, indivisible}. Atoms link by sharing electrons to form molecules.

Major Essential Elements	Minor Essential Elements
H, C, O, N, Na, Mg, K, Ca, P, S, Cl	Li, F, Cr, Mn, Fe, Co, Ni, Cu, Zn, Se, Y, I, Zr, Nb, Mo, Tc, Ru, Rh, La

* A periodic table of the elements can be found inside the back cover of the book.

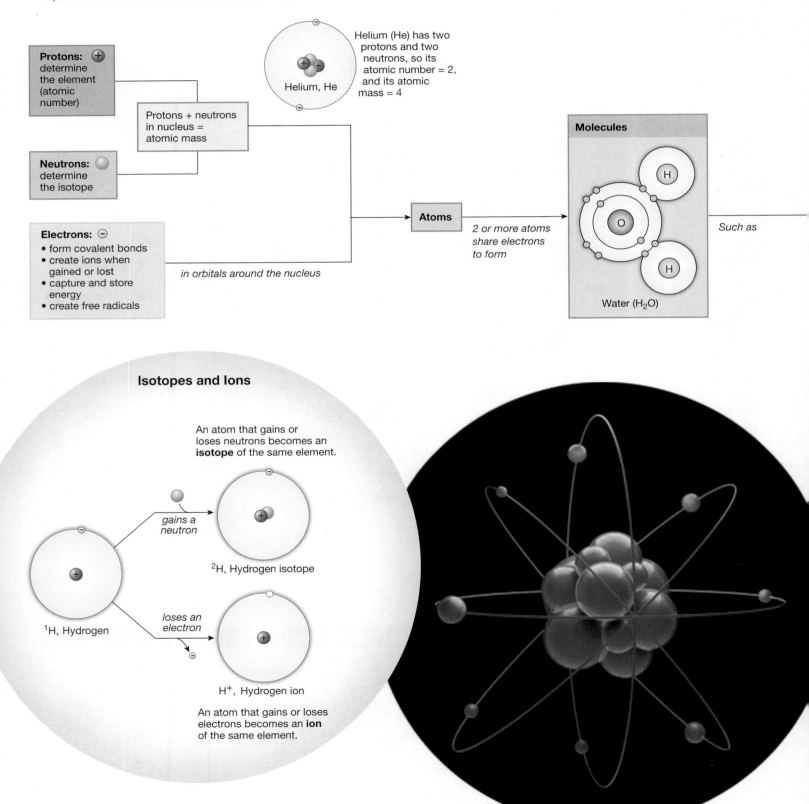

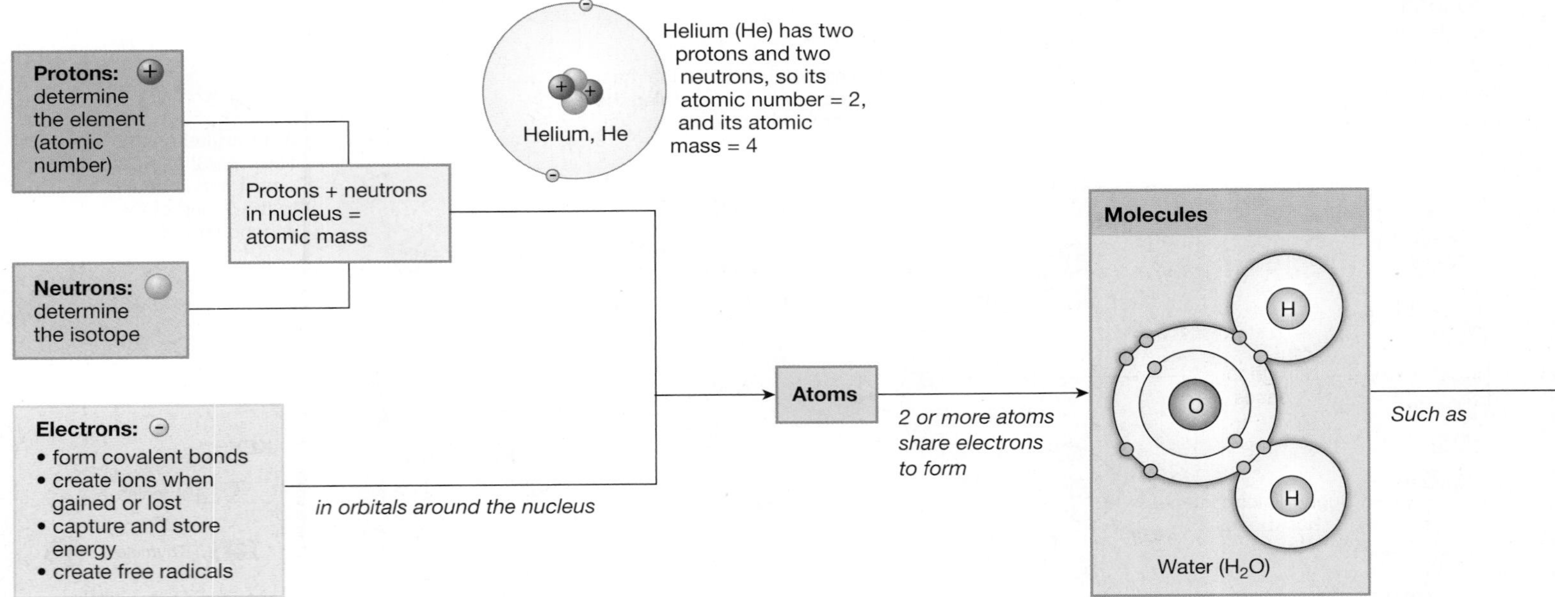
Protons: determine the element (atomic number)
Protons + neutrons in nucleus = atomic mass
Neutrons: determine the isotope
Helium, He
Helium (He) has two protons and two neutrons, so its atomic number = 2, and its atomic mass = 4
Electrons:
• form covalent bonds
• create ions when gained or lost
• capture and store energy
• create free radicals
in orbitals around the nucleus
Atoms
2 or more atoms share electrons to form
Molecules
H
O
H
Water (H_2O)
Such as

Isotopes and Ions

An atom that gains or loses neutrons becomes an **isotope** of the same element.

gains a neutron

^{2}H, Hydrogen isotope

^{1}H, Hydrogen

loses an electron

H^+, Hydrogen ion

An atom that gains or loses electrons becomes an **ion** of the same element.

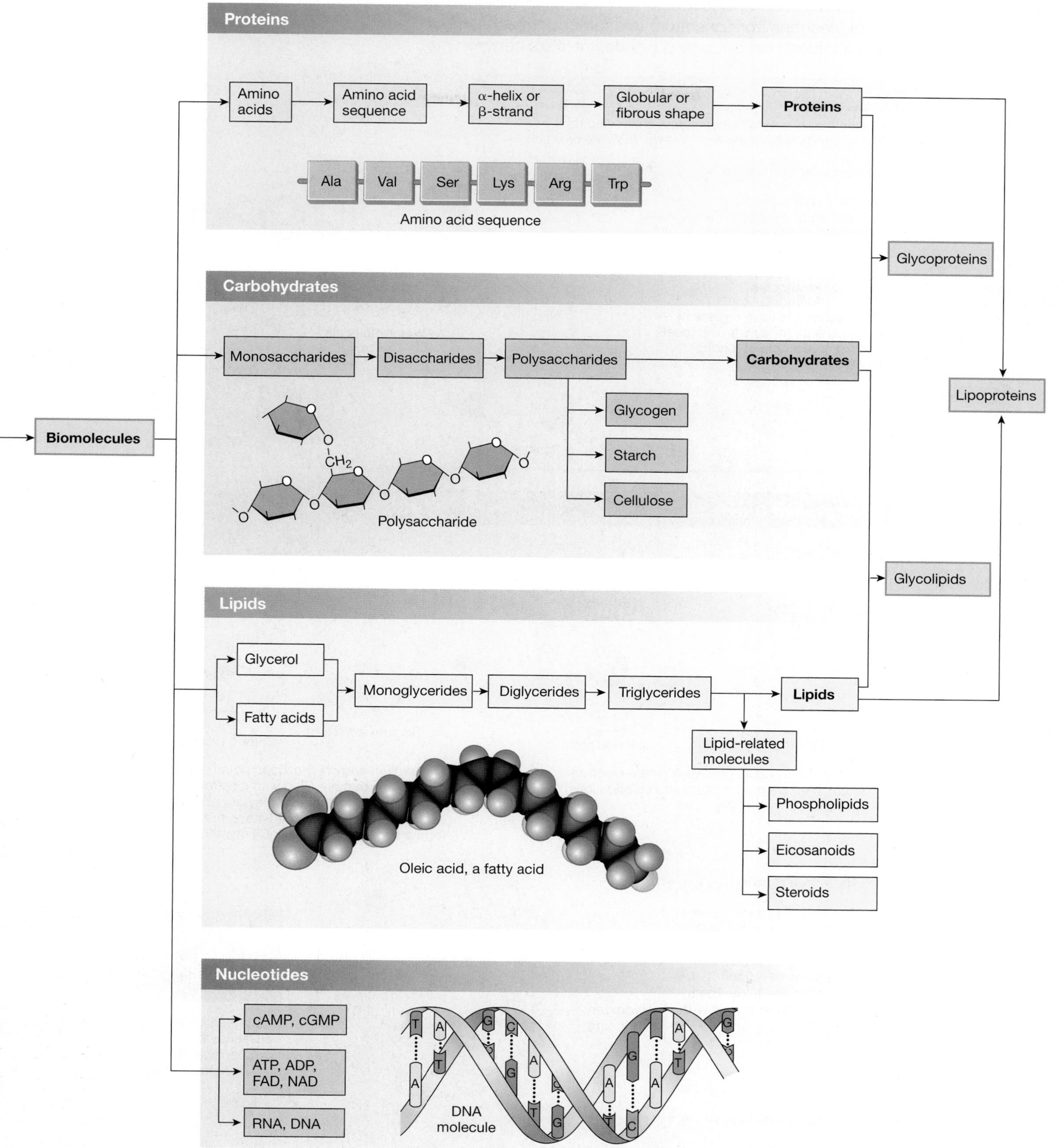

Proteins
Amino acids
Amino acid sequence
α-helix or β-strand
Globular or fibrous shape
Proteins
Ala
Val
Ser
Lys
Arg
Trp
Amino acid sequence
Glycoproteins
Carbohydrates
Monosaccharides
Disaccharides
Polysaccharides
Carbohydrates
Glycogen
Starch
Cellulose
CH2
Polysaccharide
Biomolecules
Lipoproteins
Glycolipids
Lipids
Glycerol
Fatty acids
Monoglycerides
Diglycerides
Triglycerides
Lipids
Lipid-related molecules
Phospholipids
Eicosanoids
Steroids
Oleic acid, a fatty acid
Nucleotides
cAMP, cGMP
ATP, ADP, FAD, NAD
RNA, DNA
DNA molecule

FIG. 2.6 REVIEW Molecular Bonds

When two or more atoms link by sharing electrons, they make units known as **molecules.** The transfer of electrons from one atom to another or the sharing of electrons by two atoms is a critical part of forming **bonds,** the links between atoms.

Bonds

Covalent Bonds

Covalent bonds result when atoms share electrons. These bonds require the most energy to make or break.

(a) Nonpolar Molecules

Nonpolar molecules have an even distribution of electrons. For example, molecules composed mostly of carbon and hydrogen tend to be nonpolar.

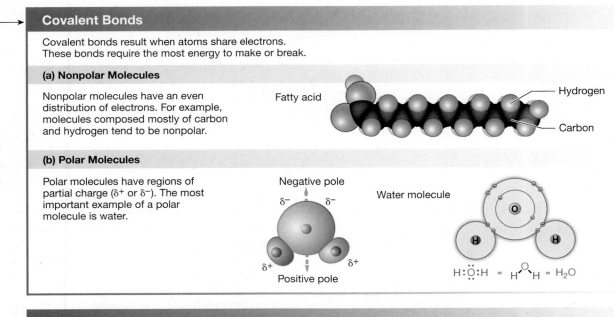

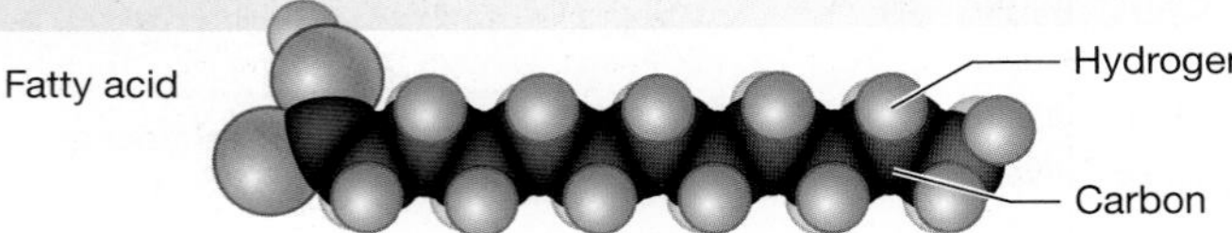
Fatty acid
Hydrogen
Carbon

(b) Polar Molecules

Polar molecules have regions of partial charge (δ^+ or δ^-). The most important example of a polar molecule is water.

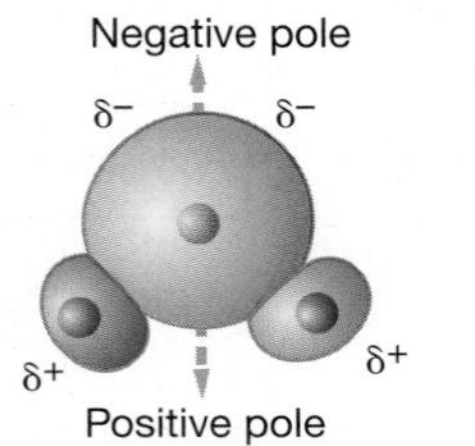
Negative pole
δ−
δ−
δ+
δ+
Positive pole

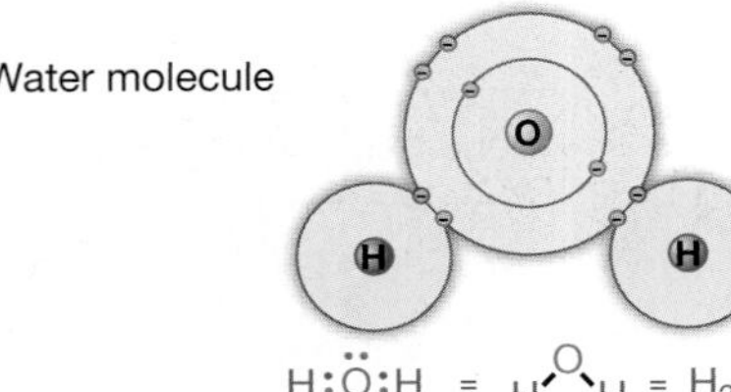
Water molecule
O
H
H
H:Ö:H = H–O–H = H_2O

Noncovalent Bonds

(c) Ionic Bonds

Ionic bonds are electrostatic attractions between ions. A common example is sodium chloride.

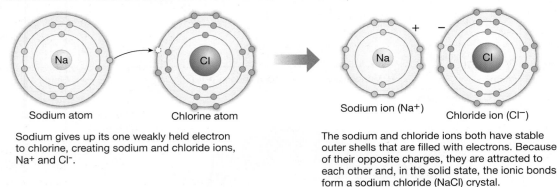

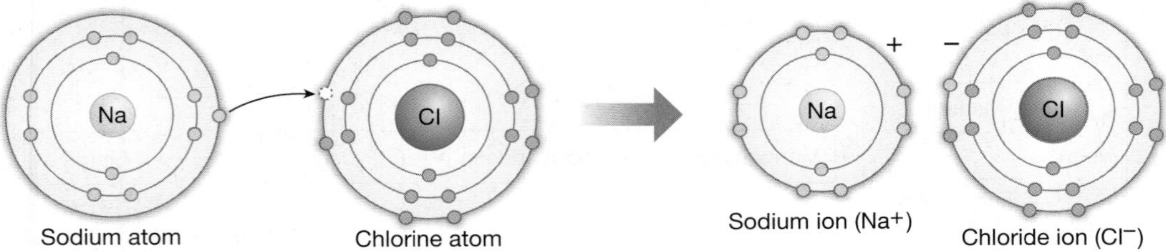
Na
Cl
+
−
Na
Cl
Sodium atom
Chlorine atom
Sodium ion (Na^+)
Chloride ion (Cl^-)

Sodium gives up its one weakly held electron to chlorine, creating sodium and chloride ions, Na^+ and Cl^-.

The sodium and chloride ions both have stable outer shells that are filled with electrons. Because of their opposite charges, they are attracted to each other and, in the solid state, the ionic bonds form a sodium chloride (NaCl) crystal.

(d) Hydrogen Bonds

Hydrogen bonds form between a hydrogen atom and a nearby oxygen, nitrogen, or fluorine atom. So, for example, the polar regions of adjacent water molecules allow them to form hydrogen bonds with one another.

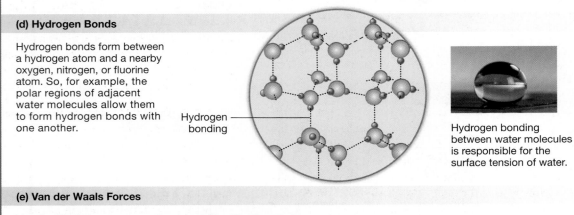

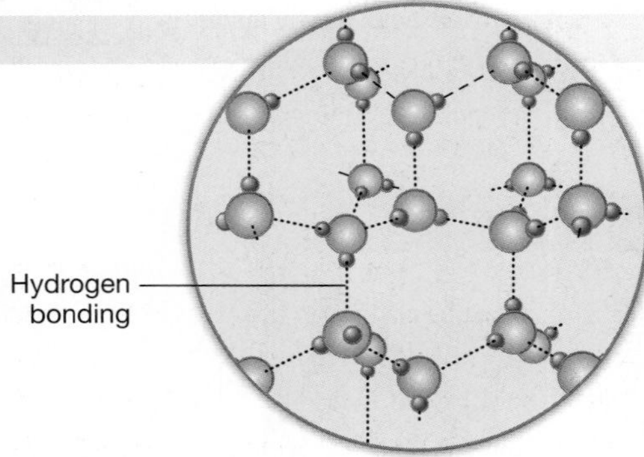
Hydrogen bonding

Hydrogen bonding between water molecules is responsible for the surface tension of water.

(e) Van der Waals Forces

Van der Waals forces are weak, nonspecific attractions between atoms.

RUNNING PROBLEM

What is chromium picolinate? Chromium (Cr) is an essential element that has been linked to normal glucose metabolism. In the diet, chromium is found in brewer's yeast, broccoli, mushrooms, and apples. Because chromium in food and in chromium chloride supplements is poorly absorbed from the digestive tract, a scientist developed and patented the compound chromium picolinate. Picolinate, derived from amino acids, enhances chromium uptake at the intestine. The recommended adequate intake (AI) of chromium for men ages 19–50 is 35 μg/day. (For women, it is 25 μg/day.) As we've seen, Stan takes more than 10 times this amount.

Q1: *Locate chromium on the periodic table of the elements. What is chromium's atomic number? Atomic mass? How many electrons does one atom of chromium have? Which elements close to chromium are also essential elements?*

29
39
40
41
46
48
53

electrons. If adjacent atoms share two pairs of electrons rather than just one pair, a **double bond,** represented by a double line (=), results. If two atoms share three pairs of electrons, they form a triple bond.

Polar and Nonpolar Molecules Some molecules develop regions of partial positive and negative charge when the electron pairs in their covalent bonds are not evenly shared between the linked atoms. When electrons are shared unevenly, the atom(s) with the stronger attraction for electrons develops a slight negative charge (indicated by δ^+), and the atom(s) with the weaker attraction for electrons develops a slight positive charge (δ^-). These molecules are called **polar molecules** because they can be said to have positive and negative ends, or poles. Certain elements, particularly nitrogen and oxygen, have a strong attraction for electrons and are often found in polar molecules.

A good example of a polar molecule is water (H_2O). The larger and stronger oxygen atom pulls the hydrogen electrons toward itself (Fig. 2.6b). This pull leaves the two hydrogen atoms of the molecule with a partial positive charge, and the single oxygen atom with a partial negative charge from the unevenly shared electrons. Note that the net charge for the entire water molecule is zero. The polarity of water makes it a good solvent, and all life as we know it is based on watery, or *aqueous*, solutions.

A **nonpolar molecule** is one whose shared electrons are distributed so evenly that there are no regions of partial positive or negative charge. For example, molecules composed mostly of carbon and hydrogen, such as the fatty acid shown in Figure 2.6a, tend to be nonpolar. This is because carbon does not attract electrons as strongly as oxygen does. As a result, the carbons and hydrogens share electrons evenly, and the molecule has no regions of partial charge.

Noncovalent Bonds Facilitate Reversible Interactions

Ionic bonds, hydrogen bonds, and van der Waals forces are noncovalent bonds. They play important roles in many physiological processes, including pH, molecular shape, and the reversible binding of molecules to each other.

Ionic Bonds Ions form when one atom has such a strong attraction for electrons that it pulls one or more electrons completely away from another atom. For example, a chlorine atom needs only one electron to fill the last of eight places in its outer shell, so it pulls an electron from a sodium atom, which has only one weakly held electron in its outer shell (Fig. 2.6c). The atom that gains electrons acquires one negative charge (−1) for each electron added, so the chlorine atom becomes a chloride ion Cl^-. Negatively charged ions are called **anions**.

An atom that gives up electrons has one positive charge (+1) for each electron lost. For example, the sodium atom becomes a sodium ion Na^+. Positively charged ions are called **cations**.

Ionic bonds, also known as *electrostatic attractions*, result from the attraction between ions with opposite charges. (Remember the basic principle of electricity that says that opposite charges attract and like charges repel.) In a crystal of table salt, the solid form of ionized NaCl, ionic bonds between alternating Na^+ and Cl^- ions hold the ions in a neatly ordered structure.

Hydrogen Bonds A **hydrogen bond** is a weak attractive force between a hydrogen atom and a nearby oxygen, nitrogen, or fluorine atom. No electrons are gained, lost, or shared in a hydrogen bond. Instead, the oppositely charged regions in polar molecules are attracted to each other. Hydrogen bonds may occur between atoms in neighboring molecules or between atoms in different parts of the same molecule. For example, one water molecule may hydrogen-bond with as many as four other water molecules. As a result, the molecules line up with their neighbors in a somewhat ordered fashion (Fig. 2.6d).

Hydrogen bonding between molecules is responsible for the **surface tension** of water. Surface tension is the attractive force between water molecules that causes water to form spherical droplets when falling or to bead up when spilled onto a nonabsorbent surface (Fig. 2.6d). The high cohesiveness {*cohaesus*, to cling together} of water is due to hydrogen bonding and makes it difficult to stretch or deform, as you may have noticed in trying to pick up a wet glass that is "stuck" to a slick table top by a thin film of water. The surface tension of water influences lung function (described in Chapter 17).

Van der Waals Forces **Van der Waals forces** are weak, nonspecific attractions between the nucleus of any atom and the electrons of nearby atoms. Two atoms that are weakly attracted to each other by van der Waals forces move closer together until they are so close that their electrons begin to repel one another. Consequently, van der Waals forces allow atoms to pack closely together and occupy a minimum amount of space. A single van der Waals attraction between atoms is very weak.

CHAPTER 2

RUNNING PROBLEM

One advertising claim for chromium is that it improves the transfer of glucose—the simple sugar that cells use to fuel all their activities—from the bloodstream into cells. In people with diabetes mellitus, cells are unable to take up glucose from the blood efficiently. It seemed logical, therefore, to test whether the addition of chromium to the diet would enhance glucose uptake in people with diabetes. In one Chinese study, diabetic patients receiving 500 micrograms (μg) of chromium picolinate twice a day showed significant improvement in their glucose uptake, but patients receiving 100 micrograms or a placebo did not.

Q2: *If people have a chromium deficiency, would you predict that their blood glucose level would be lower or higher than normal? From the results of the Chinese study, can you conclude that all people with diabetes suffer from a chromium deficiency?*

29
39
40
41
46
48
53

Concept Check

4. Are electrons in an atom or molecule most stable when they are paired or unpaired?
5. When an atom of an element gains or loses one or more electrons, it is called a(n) _____ of that element.
6. Match each type of bond with its description:

(a) covalent bond	1. weak attractive force between hydrogen and oxygen or nitrogen
(b) ionic bond	2. formed when two atoms share one or more pairs of electrons
(c) hydrogen bond	3. weak attractive force between atoms
(d) van der Waals force	4. formed when one atom loses one or more electrons to a second atom

2.2 Noncovalent Interactions

Many different kinds of noncovalent interactions can take place between and within molecules as a result of the four different types of bonds. For example, the charged, uncharged, or partially charged nature of a molecule determines whether that molecule can dissolve in water. Covalent and noncovalent bonds determine molecular shape and function. Finally, noncovalent interactions allow proteins to associate reversibly with other molecules, creating functional pairings such as enzymes and substrates, or signal receptors and molecules.

Hydrophilic Interactions Create Biological Solutions

Life as we know it is established on water-based, or *aqueous*, solutions that resemble dilute seawater in their ionic composition. The adult human body is about 60% water. Na^+, K^+, and Cl^- are the main ions in body fluids, with other ions making up a lesser proportion. All molecules and cell components are either dissolved or suspended in these solutions. For these reasons, it is useful to understand the properties of solutions, which are reviewed in **FIGURE 2.7**.

The degree to which a molecule is able to dissolve in a solvent is the molecule's **solubility:** the more easily a molecule dissolves, the higher its solubility. Water, the biological solvent, is polar, so molecules that dissolve readily in water are polar or ionic molecules whose positive and negative regions readily interact with water. For example, if NaCl crystals are placed in water, polar regions of the water molecules disrupt the ionic bonds between sodium and chloride, which causes the crystals to dissolve (**FIG. 2.8a**). Molecules that are soluble in water are said to be **hydrophilic** {*hydro-*, water + *-philic*, loving}.

In contrast, molecules such as oils that do not dissolve well in water are said to be **hydrophobic** {*-phobic*, hating}. Hydrophobic substances are usually nonpolar molecules that cannot form hydrogen bonds with water molecules. The lipids (fats and oils) are the most hydrophobic group of biological molecules.

When placed in an aqueous solution, lipids do not dissolve. Instead they separate into distinct layers. One familiar example is salad oil floating on vinegar in a bottle of salad dressing. Before hydrophobic molecules can dissolve in body fluids, they must combine with a hydrophilic molecule that will carry them into solution.

For example, cholesterol, a common animal fat, is a hydrophobic molecule. Fat from a piece of meat dropped into a glass of warm water will float to the top, undissolved. In the blood, cholesterol will not dissolve unless it binds to special water-soluble carrier molecules. You may know the combination of cholesterol with its hydrophilic carriers as HDL-cholesterol and LDL-cholesterol, the "good" and "bad" forms of cholesterol associated with heart disease.

Some molecules, such as the phospholipids, have both polar and nonpolar regions (Fig. 2.8b). This dual nature allows them to associate both with each other (hydrophobic interactions) and with polar water molecules (hydrophilic interactions). Phospholipids are the primary component of biological membranes.

Concept Check

7. Which dissolve more easily in water, polar molecules or nonpolar molecules?
8. A molecule that dissolves easily is said to be hydro_____ic.
9. Why does table salt (NaCl) dissolve in water?

Molecular Shape Is Related to Molecular Function

A molecule's shape is closely related to its function. Molecular bonds—both covalent bonds and weak bonds—play a critical role in determining molecular shape. The three-dimensional shape of a molecule is difficult to show on paper, but many molecules have characteristic shapes due to the angles of covalent bonds between the atoms. For example, the two hydrogen atoms of the water molecule shown in Figure 2.6b are attached to the oxygen with a bond angle of 104.5°. Double bonds in long carbon chain fatty acids cause the chains to kink or bend, as shown by the three-dimensional model of oleic acid in Figure 2.5.

Weak noncovalent bonds also contribute to molecular shape. The complex double helix of a DNA molecule, shown in Figure 2.4, results both from covalent bonds between adjacent bases in each strand and the hydrogen bonds connecting the two strands of the helix.

Proteins have the most complex and varied shapes of all the biomolecules. Their shapes are determined both by the sequence of amino acids in the protein chain (the **primary structure** of the protein) plus varied noncovalent interactions as long polypeptide chains loop and fold back on themselves. The stable **secondary structures** of proteins are formed by covalent bond angles between amino acids in the polypeptide chain.

The two common protein secondary structures are the **A-helix** (alpha-helix) spiral and the zigzag shape of **B-sheets** (Fig. 2.3). Adjacent β-*strands* in the polypeptide chain associate into sheetlike structures held together by hydrogen bonding, shown as dotted lines (. . .) in Figure 2.3. The sheet configuration is very stable and occurs in many proteins destined for structural uses. Proteins with other functions may have a mix of β-strands and α-helices. Protein secondary structure is illustrated by ribbon diagrams (or Richardson diagrams), with beta-sheets shown as flat arrows and α-helices as ribbon spirals (Fig. 2.3).

The **tertiary structure** of a protein is its three-dimensional shape, created through spontaneous folding as the result of covalent bonds and noncovalent interactions. Proteins are categorized into two large groups based on their shape: globular and fibrous (see Fig. 2.3). **Globular proteins** can be a mix of α-helices, β-sheets, and amino acid chains that fold back on themselves. The result is a complex tertiary structure that may contain pockets, channels, or protruding knobs. The tertiary structure of globular proteins arises partly from the angles of covalent bonds between amino acids and partly from hydrogen bonds, van der Waals forces, and ionic bonds that stabilize the molecule's shape.

In addition to covalent bonds between adjacent amino acids, covalent **disulfide** (S–S) **bonds** play an important role in the shape of many globular proteins (Fig. 2.8c). The amino acid *cysteine* contains sulfur as part of a *sulfhydryl group* (–SH). Two cysteines in different parts of the polypeptide chain can bond to each other with a disulfide bond that pulls the sections of chain together.

Fibrous proteins can be β-strands or long chains of α-helices. Fibrous proteins are usually insoluble in water and form important structural components of cells and tissues. Examples of fibrous proteins include *collagen*, found in many types of connective tissue, such as skin, and *keratin*, found in hair and nails.

Hydrogen Ions in Solution Can Alter Molecular Shape

Hydrogen bonding is an important part of molecular shape. However, free hydrogen ions, H^+, in solution can also participate in hydrogen bonding and van der Waals forces. If free H^+ disrupts a molecule's noncovalent bonds, the molecule's shape, or *conformation*, can change. A change in shape may alter or destroy the molecule's ability to function.

RUNNING PROBLEM

Chromium is found in several ionic forms. The chromium usually found in biological systems and in dietary supplements is the cation Cr^{3+}. This ion is called trivalent because it has a net charge of +3. The hexavalent cation, Cr^{6+}, with a charge of +6, is used in industry, such as in the manufacturing of stainless steel and the chrome plating of metal parts.

Q3: *How many electrons have been lost from the hexavalent ion of chromium? From the trivalent ion?*

29
39
40
41
46
48
53

The concentration of free H^+ in body fluids, or *acidity*, is measured in terms of **pH. FIGURE 2.9** reviews the chemistry of pH and shows a pH scale with the pH values of various substances. The normal pH of blood in the human body is 7.40, slightly alkaline. Regulation of the body's pH within a narrow range is critical because a blood pH more acidic than 7.00 ($pH < 7.00$) or more alkaline than 7.70 ($pH > 7.70$) is incompatible with life.

Where do hydrogen ions in body fluids come from? Some of them come from the separation of water molecules (H_2O) into H^+ and OH^- ions. Others come from **acids,** molecules that release H^+ when they dissolve in water (Fig. 2.9). Many of the molecules made during normal metabolism are acids. For example, carbonic acid is made in the body from CO_2 (carbon dioxide) and water. In solution, carbonic acid separates into a bicarbonate ion and a hydrogen ion:

$$CO_2 + H_2O \rightleftharpoons H_2CO_3 \text{ (carbonic acid)} \rightleftharpoons H^+ + HCO_3^-$$

Note that when the hydrogen is part of the intact carbonic acid molecule, it does not contribute to acidity. *Only free H^+ contributes to the hydrogen ion concentration.*

We are constantly adding acid to the body through metabolism, so how does the body maintain a normal pH? One answer is buffers. A **buffer** is any substance that moderates changes in pH. Many buffers contain anions that have a strong attraction for H^+ molecules. When free H^+ is added to a buffer solution, the buffer's anions bond to the H^+, thereby minimizing any change in pH.

The bicarbonate anion, HCO_3^-, is an important buffer in the human body. The following equation shows how a sodium bicarbonate solution acts as a buffer when hydrochloric acid (HCl) is added. When placed in plain water, hydrochloric acid separates, or dissociates, into H^+ and Cl^- and creates a high H^+ concentration (low pH). When HCl dissociates in a sodium bicarbonate solution, however, some of the bicarbonate ions combine with some of the H^+ to form undissociated carbonic acid. "Tying up" the added H^+ in this way keeps the free H^+ concentration of the solution from changing significantly and minimizes the pH change.

$H^+ + Cl^-$	+	$HCO_3^- + Na^+$	$\rightleftharpoons$	H_2CO_3	+	$Cl^- + Na^+$
Hydrochloric acid	+	Sodium bicarbonate	$\rightleftharpoons$	Carbonic acid	+	Sodium chloride (table salt)

FIG. 2.7 REVIEW Solutions

Life as we know it is established on water-based, or aqueous, solutions that resemble dilute seawater in their ionic composition. The human body is 60% water. Sodium, potassium, and chloride are the main ions in body fluids. All molecules and cell components are either dissolved or suspended in these saline solutions. For these reasons, the properties of solutions play a key role in the functioning of the human body.

Terminology

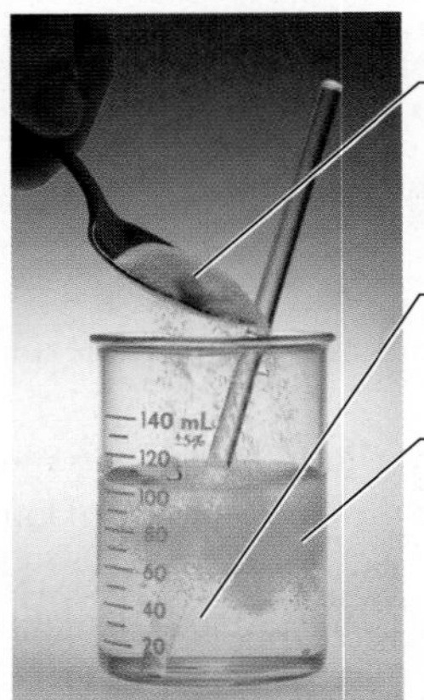

A **solute** is any substance that dissolves in a liquid. The degree to which a molecule is able to dissolve in a solvent is the molecule's solubility. The more easily a solute dissolves, the higher its **solubility**.

A **solvent** is the liquid into which solutes dissolve. In biological solutions, water is the universal solvent.

A **solution** is the combination of solutes dissolved in a solvent. The **concentration** of a solution is the amount of solute per unit volume of solution.

Concentration = solute amount/volume of solution

Expressions of Solute Amount

- **Mass** (weight) of the solute before it dissolves. Usually given in grams (g) or milligrams (mg).
- **Molecular mass** is calculated from the chemical formula of a molecule. This is the mass of one molecule, expressed in atomic mass units (amu) or, more often, in daltons (Da), where 1 amu = 1 Da.

Molecular mass = SUM [atomic mass of each element × the number of atoms of each element]

Example

What is the molecular mass of glucose, $C_6H_{12}O_6$?

Answer

Element	# of Atoms	Atomic Mass of Element
Carbon	6	12.0 amu × 6 = 72
Hydrogen	12	1.0 amu × 12 = 12
Oxygen	6	16.0 amu × 6 = 96

Molecular mass of glucose = 180 amu (or Da)

- **Moles** (mol) are an expression of the number of solute molecules, without regard for their weight. One mole = 6.02×10^{23} atoms, ions, or molecules of a substance. One mole of a substance has the same number of particles as one mole of any other substance, just as a dozen eggs has the same number of items as a dozen roses.
- **Gram molecular weight.** In the laboratory, we use the molecular mass of a substance to measure out moles. For example, one mole of glucose (with 6.02×10^{23} glucose molecules) has a molecular mass of 180 Da and weighs 180 grams. The molecular mass of a substance expressed in grams is called the gram molecular weight.
- **Equivalents** (Eq) are a unit used for ions, where 1 equivalent = molarity of the ion × the number of charges the ion carries. The sodium ion, with its charge of +1, has one equivalent per mole. The hydrogen phosphate ion (HPO_4^{2-}) has two equivalents per mole. Concentrations of ions in the blood are often reported in milliequivalents per liter (mEq/L).

FIGURE QUESTIONS

1. What are the two components of a solution?
2. The concentration of a solution is expressed as:
   (a) amount of solvent/volume of solute
   (b) amount of solute/volume of solvent
   (c) amount of solvent/volume of solution
   (d) amount of solute/volume of solution
3. Calculate the molecular mass of water, H_2O.
4. How much does a mole of KCl weigh?

Expressions of Volume

Volume is usually expressed as liters (L) or milliliters (mL) {*milli-,* 1/1000}. A volume convention common in medicine is the deciliter (dL), which is 1/10 of a liter, or 100 mL.

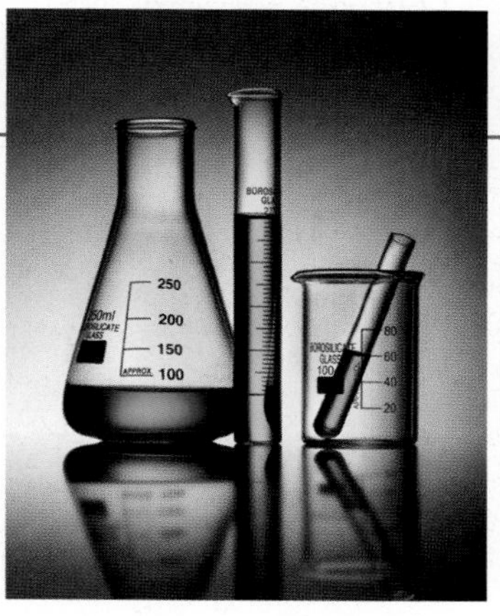

Prefixes

deci- (d)	1/10	1×10^{-1}
milli- (m)	1/1000	1×10^{-3}
micro- (μ)	1/1,000,000	1×10^{-6}
nano- (n)	1/1,000,000,000	1×10^{-9}
pico- (p)	1/1,000,000,000,000	1×10^{-12}

Useful Conversions

- 1 liter of water weighs 1 kilogram (kg) {*kilo-,* 1000}
- 1 kilogram = 2.2 pounds

Expressions of Concentration

- **Percent solutions.** In a laboratory or pharmacy, scientists cannot measure out solutes by the mole. Instead, they use the more conventional measurement of weight. The solute concentration may then be expressed as a percentage of the total solution, or percent solution. A 10% solution means 10 parts of a solute per 100 parts of total solution. Weight/volume solutions, used for solutes that are solids, are usually expressed as g/100 mL solution or mg/dL. An out-of-date way of expressing mg/dL is mg% where % means per 100 parts or 100 mL. A concentration of 20 mg/dL could also be expressed as 20 mg%.

Example	
Solutions used for intravenous (IV) infusions are often expressed as percent solutions. How would you make 500 mL of a 5% dextrose (glucose) solution?	**Answer** 5% solution = 5 g glucose dissolved in water to make a final volume of 100 mL solution. 5 g glucose/100 mL = ? g/500 mL 25 g glucose with water added to give a final volume of 500 mL

- **Molarity** is the number of moles of solute in a liter of solution, and is abbreviated as either mol/L or M. A one molar solution of glucose (1 mol/L, 1 M) contains 6.02×10^{23} molecules of glucose per liter of solution. It is made by dissolving one mole (180 grams) of glucose in enough water to make one liter of solution. Typical biological solutions are so dilute that solute concentrations are usually expressed as **millimoles** per liter (mmol/L or mM).

Example	
What is the molarity of a 5% dextrose solution?	**Answer** 5 g glucose/100 mL = 50 g glucose/1000 mL (or 1 L) 1 mole glucose = 180 g glucose 50 g/L × 1 mole/180 g = 0.278 moles/L or 278 mM

FIGURE QUESTIONS

5. Which solution is more concentrated: a 100 mM solution of glucose or a 0.1 M solution of glucose?
6. When making a 5% solution of glucose, why don't you measure out 5 grams of glucose and add it to 100 mL of water?

FIG. 2.8 REVIEW Molecular Interactions

(a) Hydrophilic Interactions

Molecules that have polar regions or ionic bonds readily interact with the polar regions of water. This enables them to dissolve easily in water. Molecules that dissolve readily in water are said to be **hydrophilic** {*hydro-,* water + *philos,* loving}.

Water molecules interact with ions or other polar molecules to form hydration shells around the ions. This disrupts the hydrogen bonding between water molecules, thereby lowering the freezing temperature of water (freezing point depression).

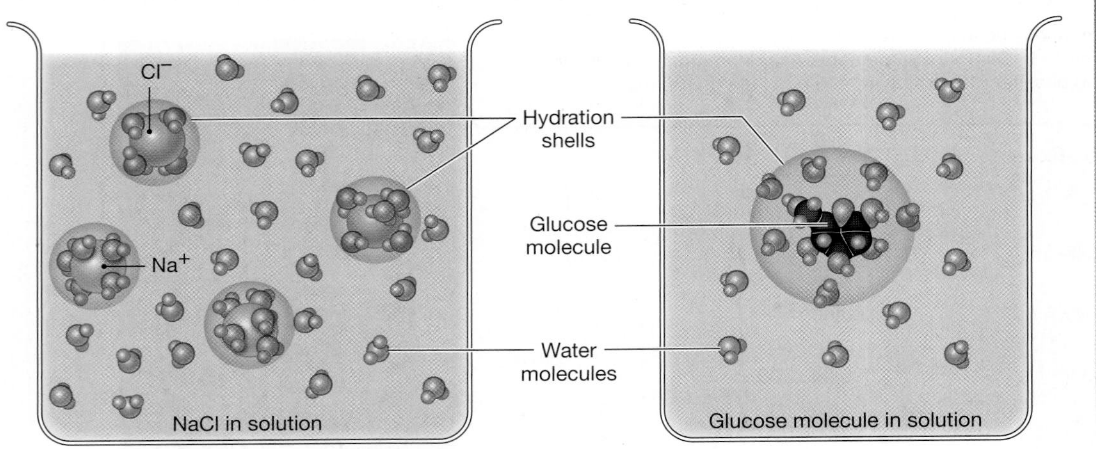
Cl^-
Na^+
Hydration shells
Glucose molecule
Water molecules
NaCl in solution
Glucose molecule in solution

(b) Hydrophobic Interactions

Because they have an even distribution of electrons and no positive or negative poles, nonpolar molecules have no regions of partial charge, and therefore tend to repel water molecules. Molecules like these do not dissolve readily in water and are said to be **hydrophobic** {*hydro-,* water + *phobos,* fear}. Molecules such as phospholipids have both polar and nonpolar regions that play critical roles in biological systems and in the formation of biological membranes.

Phospholipid molecules have polar heads and nonpolar tails.

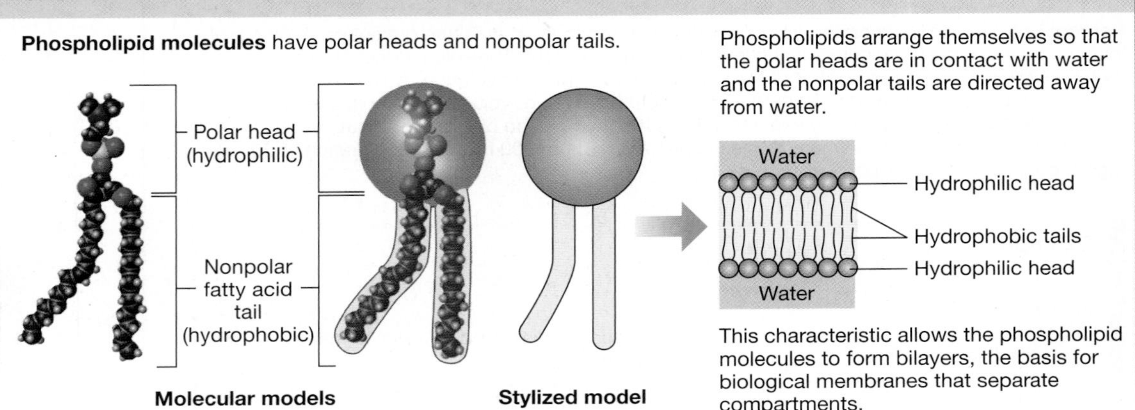
Polar head (hydrophilic)
Nonpolar fatty acid tail (hydrophobic)
Molecular models
Stylized model
Water
Hydrophilic head
Hydrophobic tails
Hydrophilic head
Water

Phospholipids arrange themselves so that the polar heads are in contact with water and the nonpolar tails are directed away from water.

This characteristic allows the phospholipid molecules to form bilayers, the basis for biological membranes that separate compartments.

(c) Molecular Shape

Covalent bond angles, ionic bonds, hydrogen bonds, and van der Waals forces all interact to create the distinctive shape of a complex biomolecule. This shape plays a critical role in the molecule's function.

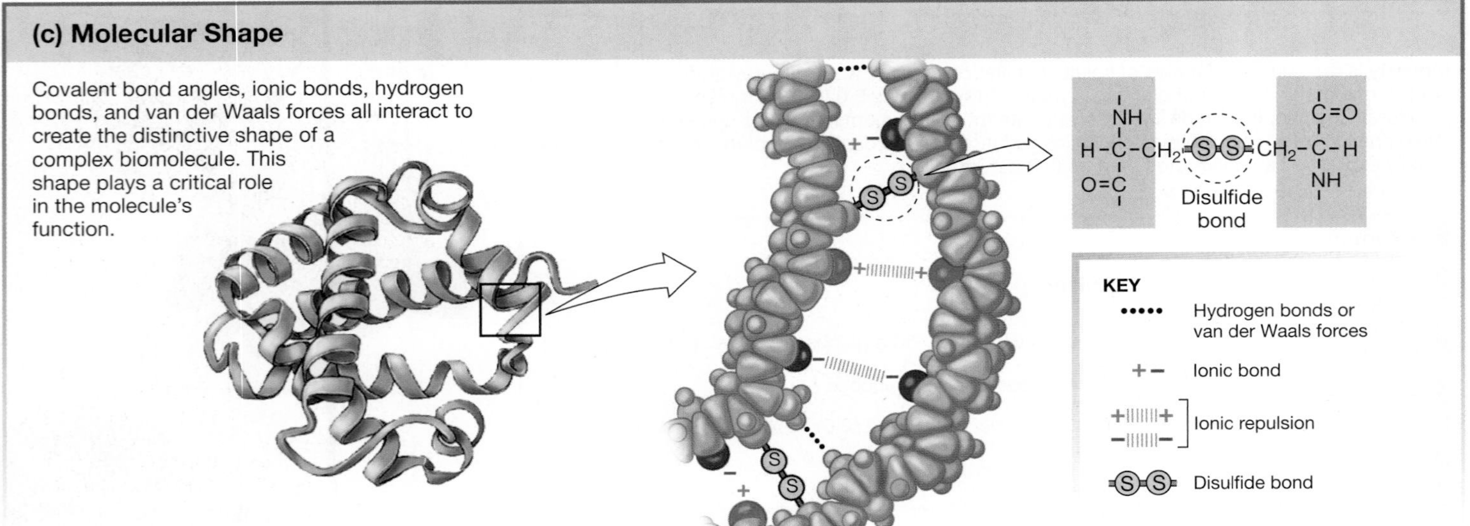
NH
H-C-CH₂-S-S-CH₂-C-H
O=C
C=O
NH
Disulfide bond
KEY
Hydrogen bonds or van der Waals forces
Ionic bond
Ionic repulsion
Disulfide bond

FIG. 2.9 REVIEW pH

Acids and Bases

An **acid** is a molecule that contributes H^+ to a solution.

- The carboxyl group, –COOH, is an acid because in solution it tends to lose its H^+:

$$R\text{–}COOH \rightarrow R\text{–}COO^- + H^+$$

A **base** is a molecule that decreases the H^+ concentration of a solution by combining with free H^+.

- Molecules that produce hydroxide ions, OH^-, in solution are bases because the hydroxide combines with H^+ to form water:

$$R\text{–}OH \rightarrow R^+ + OH^- \rightarrow OH^- + H^+ \rightarrow H_2O$$

- Another molecule that acts as a base is ammonia, NH_3. It reacts with a free H^+ to form an ammonium ion:

$$NH_3 + H^+ \rightarrow NH_4^+$$

pH

The concentration of H^+ in body fluids is measured in terms of **pH.**

- The expression pH stands for "power of hydrogen."

1 $pH = -\log [H^+]$

This equation is read as "pH is equal to the negative log of the hydrogen ion concentration." Square brackets are shorthand notation for "concentration" and by convention, concentration is expressed in mEq/L.

- Using the rule of logarithms that says $-\log x = \log (1/x)$, pH equation (1) can be rewritten as:

2 $pH = \log (1/[H^+])$

This equation shows that pH is inversely related to H^+ concentration. In other words, as the H^+ concentration goes up, the pH goes down.

Example

What is the pH of a solution whose hydrogen ion concentration $[H^+]$ is 10^{-7} meq/L?

Answer

$pH = -\log [H^+]$
$pH = -\log [10^{-7}]$

Using the rule of logs, this can be rewritten as

$pH = \log (1/10^{-7})$

Using the rule of exponents that says $1/10^x = 10^{-x}$

$pH = \log 10^7$

the log of 10^7 is 7, so the solution has a pH of 7.

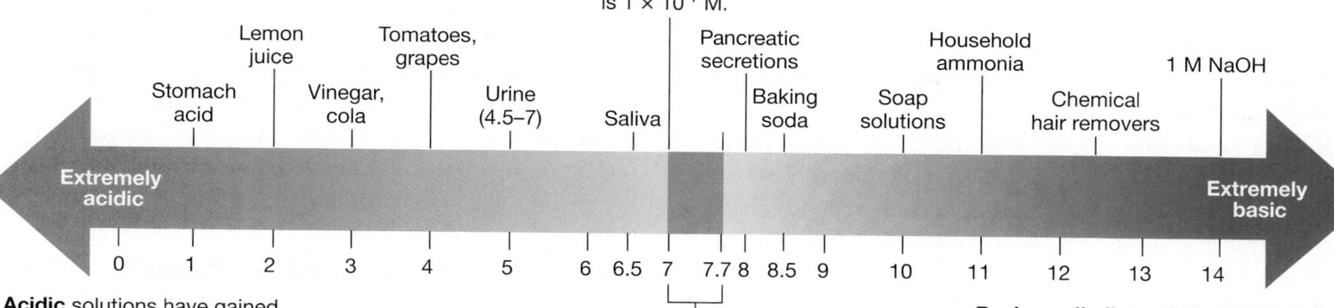

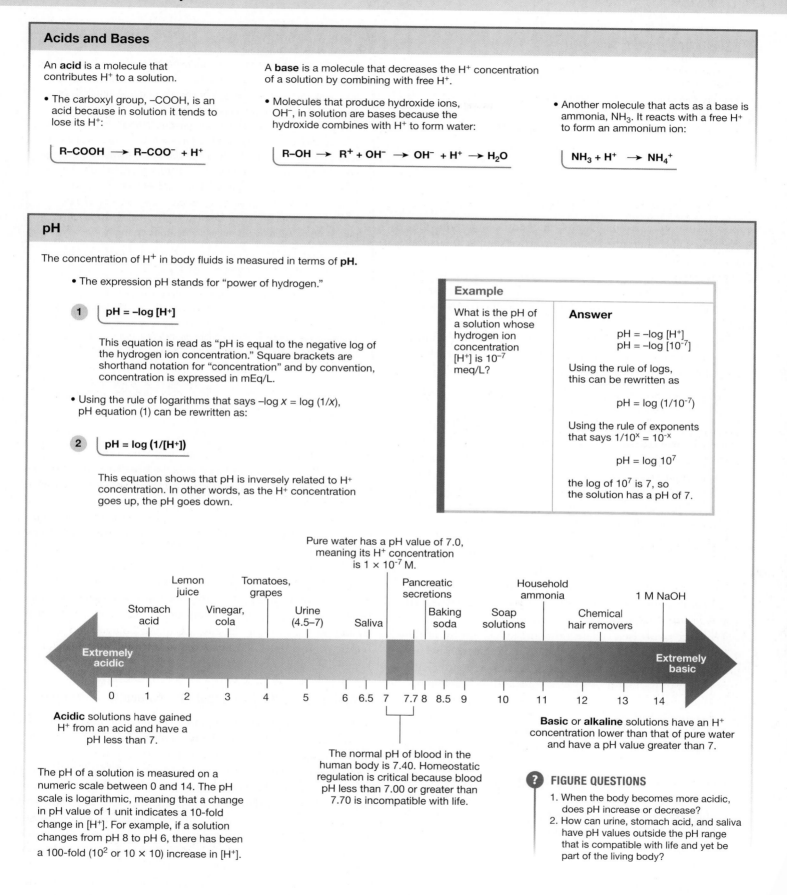

Pure water has a pH value of 7.0, meaning its H^+ concentration is 1×10^{-7} M.
Stomach acid
Lemon juice
Vinegar, cola
Tomatoes, grapes
Urine (4.5–7)
Saliva
Pancreatic secretions
Baking soda
Soap solutions
Household ammonia
Chemical hair removers
1 M NaOH
Extremely acidic
Extremely basic
0
1
2
3
4
5
6
6.5
7
7.7
8
8.5
9
10
11
12
13
14

Acidic solutions have gained H^+ from an acid and have a pH less than 7.

The normal pH of blood in the human body is 7.40. Homeostatic regulation is critical because blood pH less than 7.00 or greater than 7.70 is incompatible with life.

Basic or **alkaline** solutions have an H^+ concentration lower than that of pure water and have a pH value greater than 7.

The pH of a solution is measured on a numeric scale between 0 and 14. The pH scale is logarithmic, meaning that a change in pH value of 1 unit indicates a 10-fold change in $[H^+]$. For example, if a solution changes from pH 8 to pH 6, there has been a 100-fold (10^2 or 10×10) increase in $[H^+]$.

FIGURE QUESTIONS

1. When the body becomes more acidic, does pH increase or decrease?
2. How can urine, stomach acid, and saliva have pH values outside the pH range that is compatible with life and yet be part of the living body?

Concept Check

10. To be classified as an acid, a molecule must do what when dissolved in water?
11. pH is an expression of the concentration of what in a solution?
12. When pH goes up, acidity goes _____.

RUNNING PROBLEM

The hexavalent form of chromium used in industry is known to be toxic to humans. In 1992, officials at California's Hazard Evaluation System and Information Service warned that inhaling chromium dust, mist, or fumes placed chrome and stainless steel workers at increased risk for lung cancer. Officials found no risk to the public from normal contact with chrome surfaces or stainless steel. In 1995 and 2002, a possible link between the biological trivalent form of chromium (Cr^{3+}) and cancer came from *in vitro* studies {*vitrum*, glass—that is, a test tube} in which mammalian cells were kept alive in tissue culture. In these experiments, cells exposed to moderately high levels of chromium picolinate developed potentially cancerous changes.[1]

Q4: *From this information, can you conclude that hexavalent and trivalent chromium are equally toxic?*

[1] D. M. Stearns *et al.* Chromium(III) picolinate produces chromosome damage in Chinese hamster ovary cells. *FASEB J* 9: 1643–1648, 1995.

D. M. Stearns *et al.* Chromium(III) tris(picolinate) is mutagenic at the hypoxanthine (guanine) phosphoribosyltransferase locus in Chinese hamster ovary cells. *Mutat Res Genet Toxicol Environ Mutagen* 513: 135–142, 2002.

29
39
40
41
46
48
53

2.3 Protein Interactions

Noncovalent molecular interactions occur between many different biomolecules and often involve proteins. For example, biological membranes are formed by the noncovalent associations of phospholipids and proteins. Also, glycosylated proteins and glycosylated lipids in cell membranes create a "sugar coat" on cell surfaces, where they assist cell aggregation {*aggregare*, to join together} and adhesion {*adhaerere*, to stick}.

Proteins play important roles in so many cell functions that we can consider them the "workhorses" of the body. Most soluble proteins fall into seven broad categories:

1. **Enzymes.** Some proteins act as **enzymes,** biological catalysts that speed up chemical reactions. Enzymes play an important role in metabolism (discussed in Chapters 4 and 22).
2. **Membrane transporters.** Proteins in cell membranes help move substances back and forth between the intracellular and extracellular compartments. These proteins may form channels in the cell membrane, or they may bind to molecules and carry them through the membrane. Membrane transporters are discussed in detail in Chapter 5.
3. **Signal molecules.** Some proteins and smaller peptides act as hormones and other signal molecules. Different types of signal molecules are described in Chapters 6 and 7.
4. **Receptors.** Proteins that bind signal molecules and initiate cellular responses are called *receptors*. Receptors are discussed along with signal molecules in Chapter 6.
5. **Binding proteins.** These proteins, found mostly in the extracellular fluid, bind and transport molecules throughout the body. Examples you have already encountered include the oxygen-transporting protein *hemoglobin* and the cholesterol-binding proteins, such as LDL, low-density lipoprotein.
6. **Immunoglobulins.** These extracellular immune proteins, also called *antibodies*, help protect the body from foreign invaders and substances. Immune functions are discussed in Chapter 24.
7. **Regulatory proteins.** Regulatory proteins turn cell processes on and off or up and down. For example, the regulatory proteins known as *transcription factors* bind to DNA and alter gene expression and protein synthesis. The details of regulatory proteins can be found in cell biology textbooks.

Although soluble proteins are quite diverse, they do share common features. They all bind to other molecules through noncovalent interactions. The binding, which takes place at a location on the protein molecule called a **binding site,** exhibits important properties that will be discussed shortly: specificity, affinity, competition, and saturation. If binding of a molecule to the protein initiates a process, as occurs with enzymes, membrane transporters, and receptors, we can describe the activity rate of the process and the factors that *modulate*, or alter, the rate.

Any molecule or ion that binds to another molecule is called a **ligand** {*ligare*, to bind or tie}. Ligands that bind to enzymes and membrane transporters are also called **substrates** {*sub*-, below + *stratum*, a layer}. Protein signal molecules and protein transcription factors are ligands. Immunoglobulins bind ligands, but the immunoglobulin-ligand complex itself then becomes a ligand (for details, see Chapter 24).

Proteins Are Selective about the Molecules They Bind

The ability of a protein to bind to a certain ligand or a group of related ligands is called **specificity.** Some proteins are very specific about the ligands they bind, while others bind to whole groups of molecules. For example, the enzymes known as *peptidases* bind polypeptide ligands and break apart peptide bonds, no matter which two amino acids are joined by those bonds. For this reason peptidases are not considered to be very specific in their action. In contrast, *aminopeptidases* also break peptide bonds but are more specific. They will bind only to one end of a protein chain (the end with an unbound amino group) and can act only on the terminal peptide bond.

Ligand binding requires *molecular complementarity*. In other words, the ligand and the protein binding site must be complementary, or compatible. In protein binding, when the ligand and protein come close to each other, noncovalent interactions between the ligand and the protein's binding site allow the two molecules to bind. From studies of enzymes and other binding proteins, scientists have discovered that a protein's binding site and the shape of its ligand do not need to fit one another exactly. When the binding site and the ligand come close to each other, they begin to interact through hydrogen and ionic bonds and van der Waals forces. The protein's binding site then changes shape (*conformation*) to fit more closely to the ligand. This **induced-fit model** of protein-ligand interaction is shown in **FIGURE 2.10**.

Protein-Binding Reactions Are Reversible

The degree to which a protein is attracted to a ligand is called the protein's **affinity** for the ligand. If a protein has a high affinity for a given ligand, the protein is more likely to bind to that ligand than to a ligand for which the protein has a lower affinity.

Protein binding to a ligand can be written using the same notation that we use to represent chemical reactions:

$$P + L \rightleftharpoons PL$$

where P is the protein, L is the ligand, and PL is the bound protein-ligand complex. The double arrow indicates that binding is reversible.

Reversible binding reactions go to a state of **equilibrium,** where the rate of binding ($P + L \rightarrow PL$) is exactly equal to the rate of unbinding, or *dissociation* ($P + L \leftarrow PL$). When a reaction is at equilibrium, the ratio of the product concentration, or protein-ligand complex [PL], to the reactant concentrations [P][L] is always the same. This ratio is called the **equilibrium constant** K_{eq}, and it applies to all reversible chemical reactions:

$$K_{eq} = \frac{[PL]}{[P][L]}$$

The square brackets [ ] around the letters indicate concentrations of the protein, ligand, and protein-ligand complex.

FIG. 2.10 The induced-fit model of protein-ligand (L) binding

In this model of protein binding, the binding site shape is not an exact match to the ligands' (L) shape.

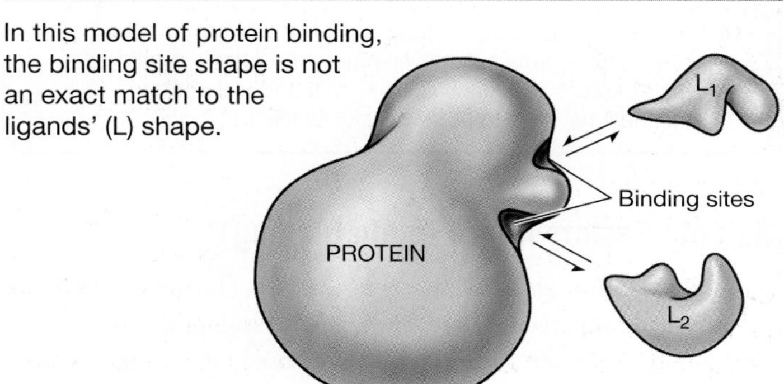
L1
Binding sites
PROTEIN
L2

Binding Reactions Obey the Law of Mass Action

Equilibrium is a dynamic state. In the living body, concentrations of protein or ligand change constantly through synthesis, breakdown, or movement from one compartment to another. What happens to equilibrium when the concentration of P or L changes? The answer to this question is shown in **FIGURE. 2.11**, which begins with a reaction at equilibrium (Fig 2.11a).

In Figure 2.11b, the equilibrium is disturbed when more protein or ligand is added to the system. Now the ratio of [PL] to [P][L] differs from the K_{eq}. In response, the rate of the binding

FIG. 2.11 The law of mass action

The **law of mass action** says that when protein binding is at equilibrium, the ratio of the bound and unbound components remains constant.

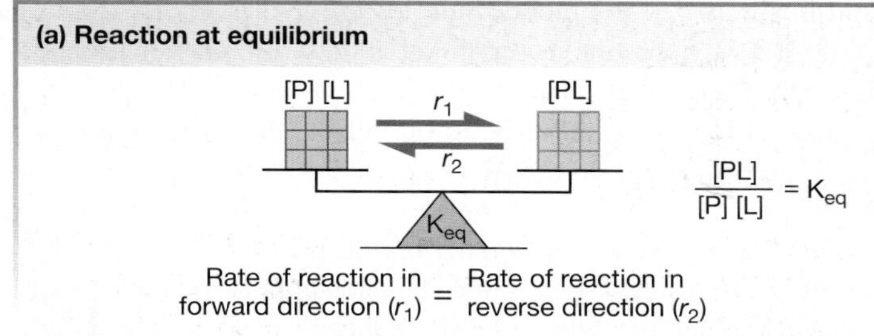
(a) Reaction at equilibrium
[P] [L]
r_1
r_2
[PL]
$\frac{[PL]}{[P][L]} = K_{eq}$
K_{eq}
Rate of reaction in forward direction (r_1) = Rate of reaction in reverse direction (r_2)

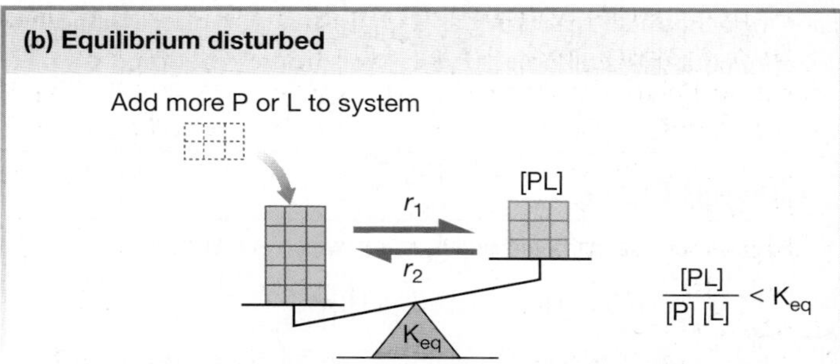
(b) Equilibrium disturbed
Add more P or L to system
r_1
r_2
[PL]
$\frac{[PL]}{[P][L]} < K_{eq}$
K_{eq}

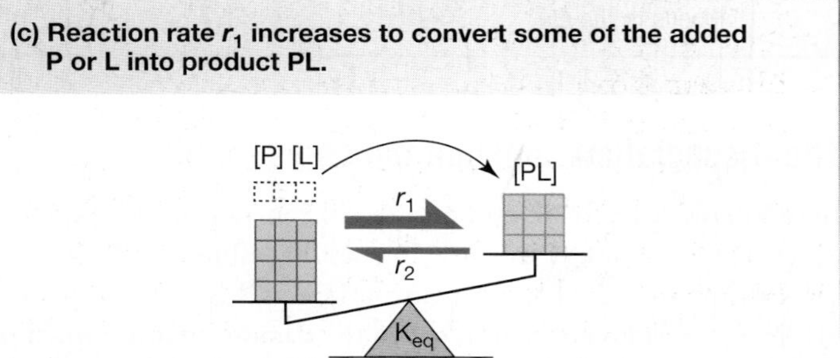
(c) Reaction rate r_1 increases to convert some of the added P or L into product PL.
[P] [L]
[PL]
r_1
r_2
K_{eq}

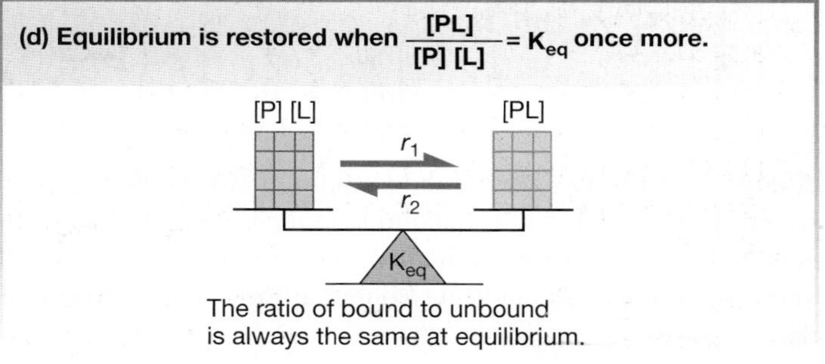
(d) Equilibrium is restored when $\frac{[PL]}{[P][L]} = K_{eq}$ once more.
[P] [L]
[PL]
r_1
r_2
K_{eq}
The ratio of bound to unbound is always the same at equilibrium.

reaction increases to convert some of the added P or L into the bound protein-ligand complex (Fig. 2.11c). As the ratio approaches its equilibrium value again, the rate of the forward reaction slows down until finally the system reaches the equilibrium ratio once more (Fig. 2.11d). [P], [L], and [PL] have all increased over their initial values, but the equilibrium ratio has been restored.

The situation just described is an example of a reversible reaction obeying the **law of mass action,** a simple relationship that holds for chemical reactions whether in a test tube or in a cell. You may have learned this law in chemistry as *Le Châtelier's principle*. In very general terms, the law of mass action says that when a reaction is at equilibrium, the ratio of the products to the substrates is always the same. If the ratio is disturbed by adding or removing one of the participants, the reaction equation will shift direction to restore the equilibrium condition. (Note that the law of mass action is not the same as mass balance [see Chapter 1, p. 10].)

One example of this principle at work is the transport of steroid hormones in the blood. Steroids are hydrophobic, so more than 99% of hormone in the blood is bound to carrier proteins. The equilibrium ratio [PL]/[P][L] is 99% bound:1% unbound hormone. However, only the unbound or "free" hormone can cross the cell membrane and enter cells. As unbound hormone leaves the blood, the equilibrium ratio is disturbed. The binding proteins then release some of the bound hormone until the 99:1 ratio is again restored. The same principle applies to enzymes and metabolic reactions. Changing the concentration of one participant in a chemical reaction has a chain-reaction effect that alters the concentrations of other participants in the reaction.

RUNNING PROBLEM

Stan has been taking chromium picolinate because he heard that it would increase his strength and muscle mass. Then a friend told him that the Food and Drug Administration (FDA) said there was no evidence to show that chromium would help build muscle. In one study,[2] a group of researchers gave high daily doses of chromium picolinate to football players during a two-month training period. By the end of the study, the players who took chromium supplements had not increased muscle mass or strength any more than players who did not take the supplement.

Use Google Scholar (*http://scholar.google.com*) and search for *chromium picolinate and muscle*. Look for articles on body composition or muscle strength in humans before you answer the next question. (Look beyond the first page of results if necessary.)

Q5: *Based on the papers you found, the Hallmark et al. study (which did not support enhanced muscle development from chromium supplements), and the studies that suggest that chromium picolinate might cause cancer, do you think that Stan should continue taking chromium picolinate?*

[2] M. A. Hallmark *et al.* Effects of chromium and resistive training on muscle strength and body composition. *Med Sci Sports Exerc* 28(1): 139–144, 1996.

29
39
40
41
46
48
53

Concept Check

13. Consider the carbonic acid reaction, which is reversible:

$$CO_2 + H_2O \rightleftharpoons H_2CO_3 \rightleftharpoons H^+ + HCO_3^-$$

If the carbon dioxide concentration in the body increases, what happens to the concentration of carbonic acid (H_2CO_3)? What happens to the pH?

The Dissociation Constant Indicates Affinity

In protein-binding reactions, the equilibrium constant is a quantitative representation of the protein's binding affinity for the ligand: high affinity for the ligand means a larger K_{eq}. The reciprocal of the equilibrium constant is called the **dissociation constant** (K_d).

$$K_d = \frac{[P][L]}{[PL]}$$

A large K_d indicates low binding affinity of the protein for the ligand, with more P and L remaining in the unbound state. Conversely, a small K_d means a higher value for [PL] relative to [P] and [L], so a small K_d indicates higher affinity of the protein for the ligand.

If one protein binds to several related ligands, a comparison of their K_d values can tell us which ligand is more likely to bind to the protein. The related ligands compete for the binding sites and are said to be **competitors.** Competition between ligands is a universal property of protein binding.

Competing ligands that mimic each other's actions are called **agonists** {*agonist*, contestant}. Agonists may occur in nature, such as *nicotine*, the chemical found in tobacco, which mimics the activity of the neurotransmitter *acetylcholine* by binding to the same receptor protein. Agonists can also be synthesized using what scientists learn from the study of protein–ligand binding sites. The ability of agonist molecules to mimic the activity of naturally occurring ligands has led to the development of many drugs.

Concept Check

14. A researcher is trying to design a drug to bind to a particular cell receptor protein. Candidate molecule A has a K_d of 4.9 for the receptor. Molecule B has a K_d of 0.3. Which molecule has the most potential to be successful as the drug?

Multiple Factors Alter Protein Binding

A protein's affinity for a ligand is not always constant. Chemical and physical factors can alter, or *modulate*, binding affinity or can even totally eliminate it. Some proteins must be activated before they have a functional binding site. In this section we discuss some of the processes that have evolved to allow activation, modulation, and inactivation of protein binding.

Isoforms Closely related proteins whose function is similar but whose affinity for ligands differs are called **isoforms** of one another. For example, the oxygen-transporting protein *hemoglobin* has multiple isoforms. One hemoglobin molecule has a quaternary structure consisting of four subunits (see Fig. 2.3). In the developing fetus, the hemoglobin isoform has two α (alpha) chains and two γ (gamma) chains that make up the four subunits. Shortly after birth, fetal hemoglobin molecules are broken down and replaced by adult hemoglobin. The adult hemoglobin isoform retains the two α chain isoforms but has two β (beta) chains in place of the γ chains. Both adult and fetal isoforms of hemoglobin bind oxygen, but the fetal isoform has a higher affinity for oxygen. This makes it more efficient at picking up oxygen across the placenta.

Activation Some proteins are inactive when they are synthesized in the cell. Before such a protein can become active, enzymes must chop off one or more portions of the molecule (**FIG. 2.12a**). Protein hormones (a type of signal molecule) and enzymes are two groups that commonly undergo such *proteolytic activation* {*lysis*, to release}. The inactive forms of these proteins are often identified with the prefix *pro-* {before}: prohormone, proenzyme, proinsulin, for example. Some inactive enzymes have the suffix *-ogen* added to the name of the active enzyme instead, as in *trypsinogen*, the inactive form of trypsin.

The activation of some proteins requires the presence of a **cofactor,** which is an ion or small organic functional group. Cofactors must attach to the protein before the binding site will become active and bind to ligand (Fig. 2.12b). Ionic cofactors include Ca^{2+}, Mg^{2+}, and Fe^{2+}. Many enzymes will not function without their cofactors.

Modulation The ability of a protein to bind a ligand and initiate a response can be altered by various factors, including temperature, pH, and molecules that interact with the protein. A factor that influences either protein binding or protein activity is called a **modulator.** There are two basic mechanisms by which modulation takes place. The modulator either (1) changes the protein's ability to bind the ligand or it (2) changes the protein's activity or its ability to create a response. **TABLE 2.3** summarizes the different types of modulation.

Chemical modulators are molecules that bind covalently or noncovalently to proteins and alter their binding ability or their activity. Chemical modulators may activate or enhance ligand binding, decrease binding ability, or completely inactivate the protein so that it is unable to bind any ligand. Inactivation may be either reversible or irreversible.

Antagonists, also called *inhibitors*, are chemical modulators that bind to a protein and decrease its activity. Many are simply molecules that bind to the protein and block the binding site without causing a response. They are like the guy who slips into the front of the movie ticket line to chat with his girlfriend, the cashier. He has no interest in buying a ticket, but he prevents the people in line behind him from getting their tickets for the movie.

TABLE 2.3 Factors That Affect Protein Binding

Essential for Binding Activity	
Cofactors	Required for ligand binding at binding site
Proteolytic activation	Converts inactive to active form by removing part of molecule. Examples: digestive enzymes, protein hormones
Modulators and Factors That Alter Binding or Activity	
Competitive inhibitor	Competes directly with ligand by binding reversibly to active site
Irreversible inhibitor	Binds to binding site and cannot be displaced
Allosteric modulator	Binds to protein away from binding site and changes activity; may be inhibitors or activators
Covalent modulator	Binds covalently to protein and changes its activity. Example: phosphate groups
pH and temperature	Alter three-dimensional shape of protein by disrupting hydrogen or S–S bonds; may be irreversible if protein becomes denatured

Competitive inhibitors are reversible antagonists that compete with the customary ligand for the binding site (Fig. 2.12d). The degree of inhibition depends on the relative concentrations of the competitive inhibitor and the customary ligand, as well as on the protein's affinities for the two. The binding of competitive inhibitors is reversible: increasing the concentration of the customary ligand can displace the competitive inhibitor and decrease the inhibition.

Irreversible antagonists, on the other hand, bind tightly to the protein and cannot be displaced by competition. Antagonist drugs have proven useful for treating many conditions. For example, tamoxifen, an antagonist to the estrogen receptor, is used in the treatment of hormone-dependent cancers of the breast.

Allosteric and covalent modulators may be either antagonists or activators. **Allosteric modulators** {*allos*, other + *stereos*, solid (as a shape)} bind reversibly to a protein at a regulatory site away from the binding site, and by doing so change the shape of the binding site. *Allosteric inhibitors* are antagonists that decrease the affinity of the binding site for the ligand and inhibit protein activity (Fig. 2.12e). *Allosteric activators* increase the probability of protein-ligand binding and enhance protein activity (Fig. 2.12c). For example, the oxygen-binding ability of hemoglobin changes with allosteric modulation by carbon dioxide, H^+, and several other factors (see Chapter 18).

Covalent modulators are atoms or functional groups that bind covalently to proteins and alter the proteins' properties. Like allosteric modulators, covalent modulators may either increase or decrease a protein's binding ability or its activity. One of the most common covalent modulators is the phosphate group. Many proteins in the cell can be activated or inactivated when a phosphate group forms a covalent bond with them, the process known as *phosphorylation*.

FIG. 2.12 ESSENTIALS Protein Activation and Inhibition

Activation

(a) Proteolytic activation: Protein is inactive until peptide fragments are removed.

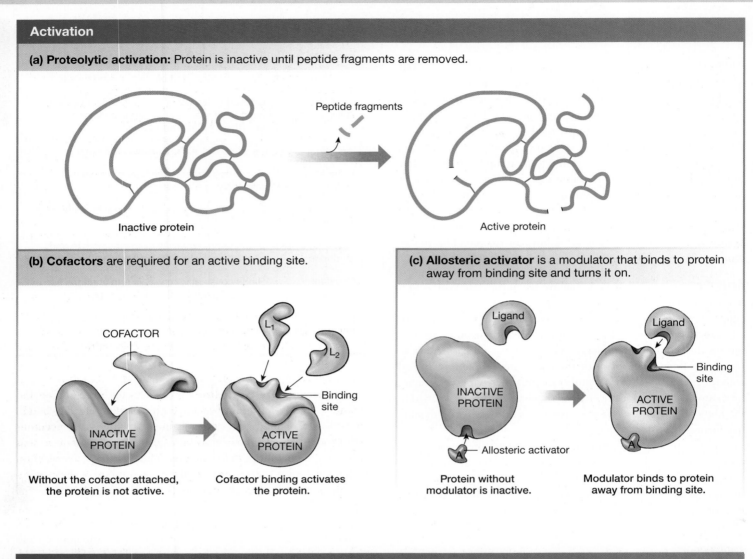

Peptide fragments

Inactive protein

Active protein

(b) Cofactors are required for an active binding site.

COFACTOR

L_1

L_2

Binding site

INACTIVE PROTEIN

ACTIVE PROTEIN

Without the cofactor attached, the protein is not active.

Cofactor binding activates the protein.

(c) Allosteric activator is a modulator that binds to protein away from binding site and turns it on.

Ligand

Ligand

Binding site

INACTIVE PROTEIN

ACTIVE PROTEIN

Allosteric activator

A

Protein without modulator is inactive.

Modulator binds to protein away from binding site.

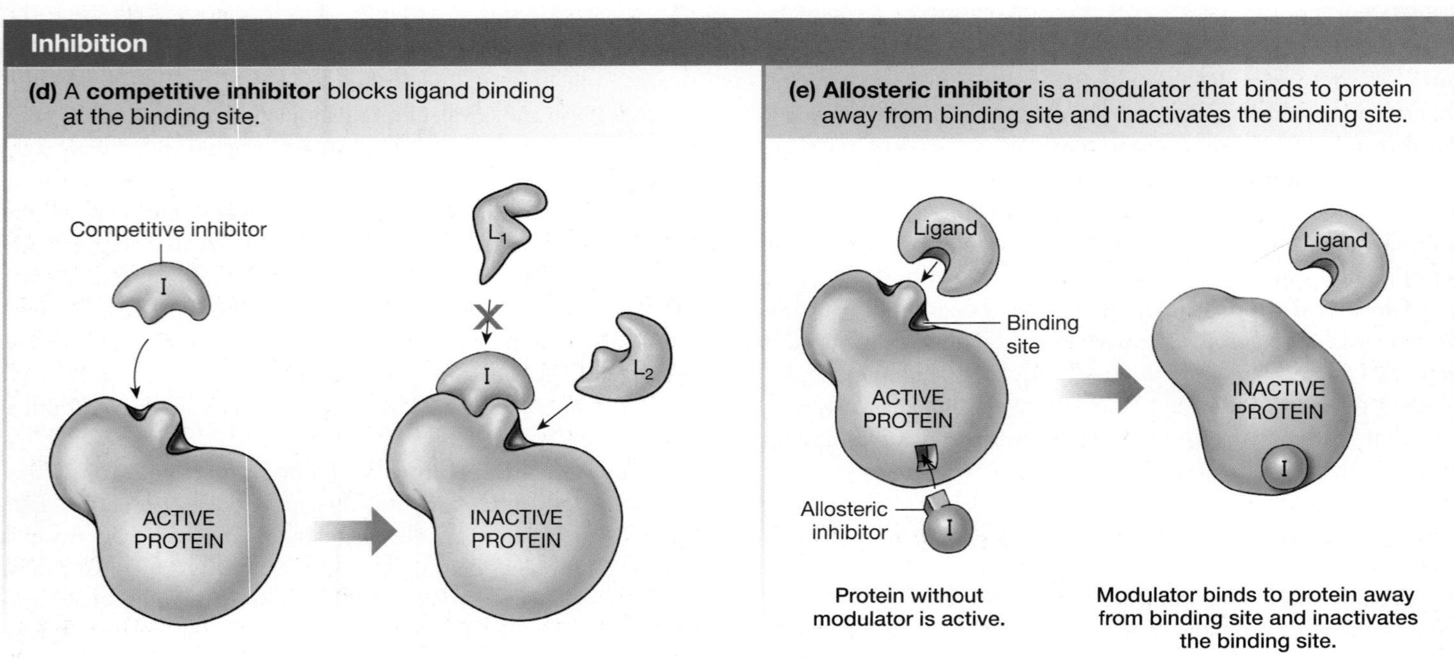
Competitive inhibitor
I
L1
L2
ACTIVE PROTEIN
INACTIVE PROTEIN
Ligand
Binding site
ACTIVE PROTEIN
Allosteric inhibitor
I
Ligand
INACTIVE PROTEIN
I

Inhibition

(d) A **competitive inhibitor** blocks ligand binding at the binding site.

(e) Allosteric inhibitor is a modulator that binds to protein away from binding site and inactivates the binding site.

Protein without modulator is active.

Modulator binds to protein away from binding site and inactivates the binding site.

One of the best known chemical modulators is the antibiotic penicillin. Alexander Fleming discovered this compound in 1928, when he noticed that *Penicillium* mold inhibited bacterial growth in a petri dish. By 1938, researchers had extracted the active ingredient penicillin from the mold and used it to treat infections in humans. Yet it was not until 1965 that researchers figured out exactly how the antibiotic works. Penicillin is an antagonist that binds to a key bacterial protein by mimicking the normal ligand. Because penicillin forms unbreakable bonds with the protein, the protein is irreversibly inhibited. Without the protein, the bacterium is unable to make a rigid cell wall. Without a rigid cell wall, the bacterium swells, ruptures, and dies.

Physical Factors Physical conditions such as temperature and pH (acidity) can have dramatic effects on protein structure and function. Small changes in pH or temperature act as modulators to increase or decrease activity (**FIG. 2.13a**). However, once these factors exceed some critical value, they disrupt the noncovalent bonds holding the protein in its tertiary conformation. The protein loses its shape and, along with that, its activity. When the protein loses its conformation, it is said to be *denatured*.

If you have ever fried an egg, you have watched this transformation happen to the egg white protein *albumin* as it changes from a slithery clear state to a firm white state. Hydrogen ions in high enough concentration to be called acids have a similar effect on protein structure. During preparation of ceviche, the national dish of Ecuador, raw fish is marinated in lime juice. The acidic lime juice contains hydrogen ions that disrupt hydrogen bonds in the muscle proteins of the fish, causing the proteins to become denatured. As a result, the meat becomes firmer and opaque, just as it would if it were cooked with heat.

In a few cases, activity can be restored if the original temperature or pH returns. The protein then resumes its original shape as if nothing had happened. Usually, however, denaturation produces a permanent loss of activity. There is certainly no way to unfry an egg or uncook a piece of fish. The potentially disastrous influence of temperature and pH on proteins is one reason these variables are so closely regulated by the body.

Concept Check

15. Match each chemical to its action(s).

(a) Allosteric modulator	1. Bind away from the binding site
(b) Competitive inhibitor	2. Bind to the binding site
(c) Covalent modulator	3. Inhibit activity only
	4. Inhibit or enhance activity

The Body Regulates the Amount of Protein in Cells

The final characteristic of proteins in the human body is that the amount of a given protein varies over time, often in a regulated fashion. The body has mechanisms that enable it to monitor whether it needs more or less of certain proteins. Complex signaling pathways, many of which themselves involve proteins, direct particular cells to make new proteins or to break down (*degrade*) existing proteins. This programmed production of new proteins (receptors, enzymes, and membrane transporters, in particular) is called **up-regulation.** Conversely, the programmed removal of proteins is called **down-regulation.** In both instances, the cell is directed to make or remove proteins to alter its response.

The amount of protein present in a cell has a direct influence on the magnitude of the cell's response. For example, the graph in Figure 2.13b shows the results of an experiment in which the amount of ligand is held constant while the amount of protein is varied. As the graph shows, an increase in the amount of protein present causes an increase in the response.

As an analogy, think of the checkout lines in a supermarket. Imagine that each cashier is an enzyme, the waiting customers are ligand molecules, and people leaving the store with their purchases are products. One hundred customers can be checked out faster when there are 25 lines open than when there are only 10 lines. Likewise, in an enzymatic reaction, the presence of more protein molecules (enzyme) means that more binding sites are available to interact with the ligand molecules. As a result, the ligands are converted to products more rapidly.

Regulating protein concentration is an important strategy that cells use to control their physiological processes. Cells alter the amount of a protein by influencing both its synthesis and its breakdown. If protein synthesis exceeds breakdown, protein accumulates and the reaction rate increases. If protein breakdown exceeds synthesis, the amount of protein decreases, as does the reaction rate. Even when the amount of protein is constant, there is still a steady turnover of protein molecules.

Reaction Rate Can Reach a Maximum

If the concentration of a protein in a cell is constant, then the concentration of the ligand determines the magnitude of the response. Fewer ligands activate fewer proteins, and the response is low. As ligand concentrations increase, so does the magnitude of the response, up to a maximum where all protein binding sites are occupied.

Figure 2.13c shows the results of a typical experiment in which the protein concentration is constant but the concentration of ligand varies. At low ligand concentrations, the response rate is directly proportional to the ligand concentration. Once the concentration of ligand molecules exceeds a certain level, the protein molecules have no more free binding sites. The proteins are fully occupied, and the rate reaches a maximum value. This condition is known as **saturation.** Saturation applies to enzymes, membrane transporters, receptors, binding proteins, and immunoglobulins.

An analogy to saturation appeared in the early days of television on the *I Love Lucy* show and can be viewed today by searching YouTube. Lucille Ball's character was working at the conveyor belt of a candy factory, loading chocolates into the little paper cups of a candy box. Initially, the belt moved slowly, and she had no difficulty picking up the candy and putting it into the box. Gradually, the belt brought candy to her more rapidly, and she had to increase

FIG. 2.13 ESSENTIALS Factors That Influence Protein Activity

(a) Temperature and pH

Temperature and pH changes may disrupt protein structure and cause loss of function.

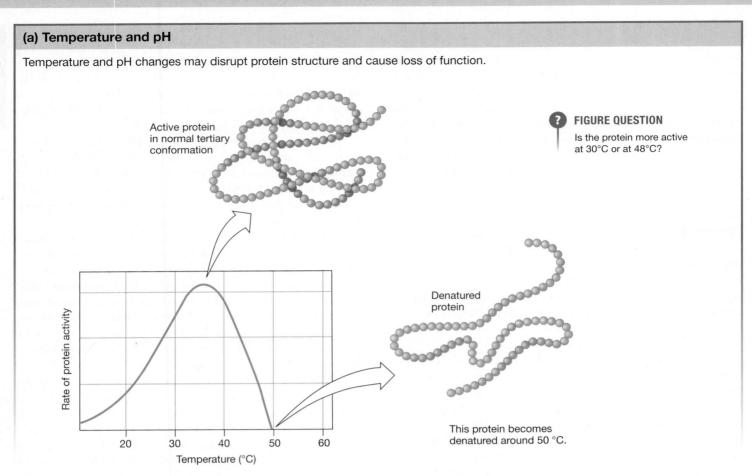

Active protein in normal tertiary conformation

Denatured protein

This protein becomes denatured around 50 °C.

Rate of protein activity

20 30 40 50 60

Temperature (°C)

FIGURE QUESTION

Is the protein more active at 30°C or at 48°C?

(b) Amount of Protein

Reaction rate depends on the amount of protein. The more protein present, the faster the rate.

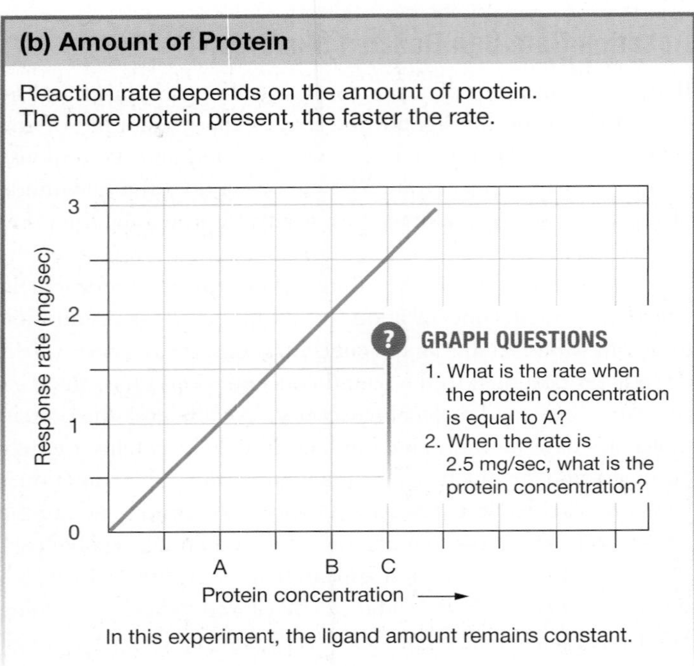
Response rate (mg/sec)
0
1
2
3
A
B
C
Protein concentration
GRAPH QUESTIONS
1. What is the rate when the protein concentration is equal to A?
2. When the rate is 2.5 mg/sec, what is the protein concentration?

In this experiment, the ligand amount remains constant.

(c) Amount of Ligand

If the amount of binding protein is held constant, the reaction rate depends on the amount of ligand, up to the saturation point.

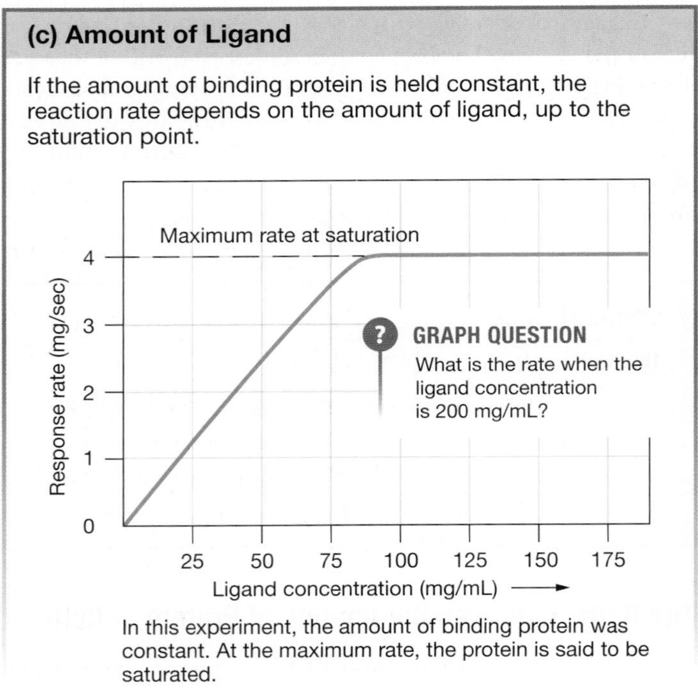
Maximum rate at saturation
Response rate (mg/sec)
0
1
2
3
4
25
50
75
100
125
150
175
Ligand concentration (mg/mL)
GRAPH QUESTION
What is the rate when the ligand concentration is 200 mg/mL?

In this experiment, the amount of binding protein was constant. At the maximum rate, the protein is said to be saturated.

her packing speed to keep up. Finally, the belt brought candy to her so fast that she could not pack it all in the boxes because she was working at her maximum rate. That was Lucy's saturation point. (Her solution was to stuff the candy into her mouth and apron as well as into the box!)

In conclusion, you have now learned about the important and nearly universal properties of soluble proteins: shape-function relationships, ligand binding, saturation, specificity, competition, and activation/inhibition. You will revisit these concepts many times as you work through the organ systems of the body. The body's insoluble proteins, which are key structural components of cells and tissues, are covered in Chapter 3.

Concept Check

16. What happens to the rate of an enzymatic reaction as the amount of enzyme present decreases?

17. What happens to the rate of an enzymatic reaction when the enzyme has reached saturation?

RUNNING PROBLEM **CONCLUSION** Chromium Supplements

In this running problem, you learned that claims of chromium picolinate's ability to enhance muscle mass have not been supported by evidence from controlled scientific experiments. You also learned that studies suggest that some forms of the biological trivalent form of chromium may be toxic. To learn more about current research, go to PubMed (*www.pubmed.gov*) and search for *chromium picolinate*. Compare what you find there with the results of a similar Google search. Should you believe everything you read on the Web? Now compare your answers with those in the summary table.

Question	Facts	Integration and Analysis
Q1: *Locate chromium on the periodic table of elements.*	The periodic table organizes the elements according to atomic number.	N/A
What is chromium's atomic number? Atomic mass?	Reading from the table, chromium (Cr) has an atomic number of 24 and an average atomic mass of 52.	N/A
How many electrons does one atom of chromium have?	Atomic number of an element = number of protons in one atom. One atom has equal numbers of protons and electrons.	The atomic number of chromium is 24; therefore, one atom of chromium has 24 protons and 24 electrons.
Which elements close to chromium are also essential elements?	Molybdenum, manganese, and iron.	N/A
Q2: *If people have chromium deficiency, would you predict that their blood glucose level would be lower or higher than normal?*	Chromium helps move glucose from blood into cells.	If chromium is absent or lacking, less glucose would leave the blood and blood glucose would be higher than normal.
From the result of the Chinese study, can you conclude that all people with diabetes suffer from chromium deficiency?	Higher doses of chromium supplements lowered elevated blood glucose levels, but lower doses have no effect. This is only one study, and no information is given about similar studies elsewhere.	We have insufficient evidence from the information presented to draw a conclusion about the role of chromium deficiency in diabetes.
Q3: *How many electrons have been lost from the hexavalent ion of chromium? From the trivalent ion?*	For each electron lost from an ion, a positively charged proton is left behind in the nucleus of the ion.	The hexavalent ion of chromium, Cr^{6+}, has six unmatched protons and therefore has lost six electrons. The trivalent ion, Cr^{3+}, has lost three electrons.
Q4: *From this information, can you conclude that hexavalent and trivalent chromium are equally toxic?*	The hexavalent form is used in industry and, when inhaled, has been linked to an increased risk of lung cancer. Enough studies have shown an association that California's Hazard Evaluation System and Information Service has issued warnings to chromium workers. Evidence to date for toxicity of trivalent chromium in chromium picolinate comes from studies done on isolated cells in tissue culture.	Although the toxicity of Cr^{6+} is well established, the toxicity of Cr^{3+} has not been conclusively determined. Studies performed on cells in vitro may not be applicable to humans. Additional studies need to be performed in which animals are given reasonable doses of chromium picolinate for an extended period of time.

– Continued next page

RUNNING PROBLEM CONCLUSION *Continued*

Question	Facts	Integration and Analysis
Q5: *Based on the study that did not support enhanced muscle development from chromium supplements and the studies that suggest that chromium picolinate might cause cancer, do you think Stan should continue taking picolinate?*	No research evidence supports a role for chromium picolinate in increasing muscle mass or strength in humans. Other research suggests that chromium picolinate may cause cancerous changes in isolated cells.	The evidence presented suggests that for Stan, there is no benefit from taking chromium picolinate, and there may be risks. Using risk–benefit analysis, the evidence supports stopping the supplements. However, the decision is Stan's personal responsibility. He should keep himself informed of new developments that would change the risk–benefit analysis.

29 – 39 – 40 – 41 – 46 – 48 – 53

CHEMISTRY REVIEW QUIZ

Use this quiz to see what areas of chemistry and basic biochemistry you might need to review. Answers are on p. A-0. The title above each set of questions refers to a review figure on this topic.

Atoms and Molecules (Fig. 2.5)

Match each subatomic particle in the left column with all the phrases in the right column that describe it. A phrase may be used more than once.

1. electron	(a) one has atomic mass of 1 amu
2. neutron	(b) found in the nucleus
3. proton	(c) negatively charged
	(d) changing the number of these in an atom creates a new element
	(e) adding or losing these makes an atom into an ion
	(f) gain or loss of these makes an isotope of the same element
	(g) determine(s) an element's atomic number
	(h) contribute(s) to an element's atomic mass

4. Isotopes of an element have the same number of __________ and __________, but differ in their number of __________. Unstable isotopes emit energy called __________.
5. Name the element associated with each of these symbols: C, O, N, and H.
6. Write the one- or two-letter symbol for each of these elements: phosphorus, potassium, sodium, sulfur, calcium, and chlorine.
7. Use the periodic table of the elements on the inside back cover to answer the following questions:
   (a) Which element has 30 protons?
   (b) How many electrons are in one atom of calcium?
   (c) Find the atomic number and average atomic mass of iodine. What is the letter symbol for iodine?
8. A magnesium ion, Mg^{2+}, has (*gained/lost*) two (*protons/neutrons/electrons*).
9. H^+ is also called a proton. Why is it given that name?
10. Use the periodic table of the elements on the inside back cover to answer the following questions about an atom of sodium.
    (a) How many electrons does the atom have?
    (b) What is the electrical charge of the atom?
    (c) How many neutrons does the average atom have?
    (d) If this atom loses one electron, it would be called a(n) *anion/cation*.
    (e) What would be the electrical charge of the substance formed in (d)?
    (f) Write the chemical symbol for the ion referred to in (d).
    (g) What does the sodium atom become if it loses a proton from its nucleus?
    (h) Write the chemical symbol for the atom referred to in (g).
11. Write the chemical formulas for each molecule depicted. Calculate the molecular mass of each molecule.

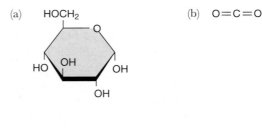

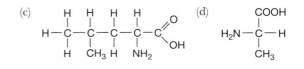

(a) HOCH₂ ring structure: O; HO, OH, OH, OH

(b) O=C=O

(c) H–C(H)(H)–C(H)(CH₃)–C(H)(H)–C(H)(NH₂)–C(=O)OH

(d) H₂N–C(COOH)(CH₃)–H

Lipids (Fig. 2.1)

12. Match each lipid with its best description.

(a) triglyceride	1. most common form of lipid in the body
(b) eicosanoid	2. liquid at room temperature, usually from plants
(c) steroid	3. important component of cell membrane
(d) oil	4. structure composed of carbon rings
(e) phospholipids	5. modified 20-carbon fatty acid

13. Use the chemical formulas given to decide which of the following fatty acids is most unsaturated: (a) $C_{18}H_{36}O_2$ (b) $C_{18}H_{34}O_2$ (c) $C_{18}H_{30}O_2$

Carbohydrates (Fig. 2.2)

14. Match each carbohydrate with its description.

(a) starch	1. monosaccharide
(b) chitin	2. disaccharide, found in milk
(c) glucose	3. storage form of glucose for animals
(d) lactose	4. storage form of glucose for plants
(e) glycogen	5. structural polysaccharide of invertebrates

Proteins (Fig. 2.3)

15. Match these terms pertaining to proteins and amino acids:

(a) the building blocks of proteins	1. essential amino acids
(b) must be included in our diet	2. primary structure
(c) protein catalysts that speed the rate of chemical reactions	3. amino acids
(d) sequence of amino acids in a protein	4. globular proteins
(e) protein chains folded into a ball-shaped structure	5. enzymes
	6. tertiary structure
	7. fibrous proteins

16. What aspect of protein structure allows proteins to have more versatility than lipids or carbohydrates?
17. Peptide bonds form when the _____ group of one amino acid joins the _____ of another amino acid.

Nucleotides (Fig. 2.4)

18. List the three components of a nucleotide.
19. Compare the structure of DNA with that of RNA.
20. Distinguish between purines and pyrimidines.

CHAPTER SUMMARY

This chapter introduces the *molecular interactions* between biomolecules, water, and ions that underlie many of the key themes in physiology. These interactions are an integral part of *information flow, energy storage and transfer*, and the *mechanical properties* of cells and tissues in the body.

2.1 Molecules and Bonds

1. The four major groups of **biomolecules** are carbohydrates, lipids, proteins, and nucleotides. They all contain carbon, hydrogen, and oxygen. (p. 29; Figs. 2.1, 2.2, 2.3, 2.4)
2. Proteins, lipids, and carbohydrates combine to form glycoproteins, glycolipids, or lipoproteins. (p. 29; Fig. 2.5)
3. Electrons are important for covalent and ionic bonds, energy capture and transfer, and formation of free radicals. (p. 33)
4. **Covalent bonds** form when adjacent atoms share one or more pairs of electrons. (p. 33; Fig. 2.6)
5. **Polar molecules** have atoms that share electrons unevenly. When atoms share electrons evenly, the molecule is **nonpolar**. (p. 39; Fig. 2.6)
6. An atom that gains or loses electrons acquires an electrical charge and is called an **ion**. (p. 33; Fig. 2.6)
7. **Ionic bonds** are strong bonds formed when oppositely charged ions are attracted to each other. (p. 39)
8. Weak **hydrogen bonds** form when hydrogen atoms in polar molecules are attracted to oxygen, nitrogen, or fluorine atoms. Hydrogen bonding among water molecules is responsible for the surface tension of water. (p. 39; Fig. 2.6)
9. **Van der Waals forces** are weak bonds that form when atoms are attracted to each other. (p. 39)

2.2 Noncovalent Interactions

10. The universal solvent for biological solutions is water. (p. 40; Figs. 2.7, 2.8a)
11. The ease with which a molecule dissolves in a solvent is called its **solubility** in that solvent. **Hydrophilic** molecules dissolve easily in water, but **hydrophobic** molecules do not. (p. 40)
12. Molecular shape is created by covalent bond angles and weak noncovalent interactions within a molecule. (p. 40; Fig. 2.8)
13. Free H^+ in solution can disrupt a molecule's noncovalent bonds and alter its ability to function. (p. 41)
14. The **pH** of a solution is a measure of its hydrogen ion concentration. The more acidic the solution, the lower its pH. (p. 41; Fig. 2.9)
15. **Buffers** are solutions that moderate pH changes. (p. 41)

2.3 Protein Interactions

16. Most water-soluble proteins serve as enzymes, membrane transporters, signal molecules, receptors, binding proteins, immunoglobulins, or transcription factors. (p. 46)
17. **Ligands** bind to proteins at a binding site. According to the **induced-fit model** of protein binding, the shapes of the ligand and binding site do not have to match exactly. (pp. 46, 47; Fig. 2.10)
18. Proteins are specific about the ligands they will bind. The attraction of a protein to its ligand is called the protein's **affinity** for the ligand. The **equilibrium constant** (K_{eq}) and the **dissociation constant** (K_d) are quantitative measures of a protein's affinity for a given ligand. (pp. 47, 48)

19. Reversible binding reactions go to equilibrium. If equilibrium is disturbed, the reaction follows the **law of mass action** and shifts in the direction that restores the equilibrium ratio. (p. 48; Fig. 2.11)
20. Ligands may compete for a protein's binding site. If competing ligands mimic each other's activity, they are **agonists**. (p. 48)
21. Closely related proteins having similar function but different affinities for ligands are called **isoforms** of one another. (p. 49)
22. Some proteins must be activated, either by **proteolytic activation** or by addition of **cofactors**. (p. 49; Fig. 2.12)
23. **Competitive inhibitors** can be displaced from the binding site, but irreversible **antagonists** cannot. (p. 49; Fig. 2.12)
24. **Allosteric modulators** bind to proteins at a location other than the binding site. **Covalent modulators** bind with covalent bonds. Both types of modulators may activate or inhibit the protein. (p. 49; Fig. 2.12)
25. Extremes of temperature or pH will **denature** proteins. (p. 51; Fig. 2.13)
26. Cells regulate their proteins by **up-regulation** or **down-regulation** of protein synthesis and destruction. The amount of protein directly influences the magnitude of the cell's response. (p. 51; Fig. 2.13)
27. If the amount of protein (such as an enzyme) is constant, the amount of ligand determines the cell's response. If all binding proteins (such as enzymes) become **saturated** with ligand, the response reaches its maximum. (p. 51; Fig. 2.13)

REVIEW QUESTIONS

In addition to working through these questions and checking your answers on p. A-2, review the Learning Outcomes at the beginning of this chapter.

Level One Reviewing Facts and Terms

1. List the four kinds of biomolecules. Give an example of each kind that is relevant to physiology.
2. True or false? All organic molecules are biomolecules.
3. When atoms bind tightly to one another, such as H_2O or O_2, one unit is called a(n) _____.
4. An atom of carbon has four unpaired electrons in an outer shell with space for eight electrons. How many covalent bonds will one carbon atom form with other atoms?
5. Fill in the blanks with the correct bond type.

   In a(n) _____ bond, electrons are shared between atoms. If the electrons are attracted more strongly to one atom than to the other, the molecule is said to be a(n) _____ molecule. If the electrons are evenly shared, the molecule is said to be a(n) _____ molecule.
6. Name two elements whose presence contributes to a molecule becoming a polar molecule.
7. Based on what you know from experience about the tendency of the following substances to dissolve in water, predict whether they are polar or nonpolar molecules: table sugar, vegetable oil.
8. A negatively charged ion is called a(n) _____, and a positively charged ion is called a(n) _____.
9. Define the pH of a solution. If pH is less than 7, the solution is _____; if pH is greater than 7, the solution is _____.
10. A molecule that moderates changes in pH is called a _____.
11. Proteins combined with fats are called _____, and proteins combined with carbohydrates are called _____.
12. A molecule that binds to another molecule is called a(n) _____.
13. Match these definitions with their terms (not all terms are used):

(a) the ability of a protein to bind one molecule but not another	1. irreversible inhibition
(b) the part of a protein molecule that binds the ligand	2. induced fit
(c) the ability of a protein to alter shape as it binds a ligand	3. binding site
	4. specificity
	5. saturation

14. An ion, such as Ca^{2+} or Mg^{2+}, that must be present in order for an enzyme to work is called a(n) _____.
15. A protein whose structure is altered to the point that its activity is destroyed is said to be _____.

Level Two Reviewing Concepts

16. Mapping exercise: Make the list of terms into a map describing solutions.

• concentration	• nonpolar molecule
• equivalent	• polar molecule
• hydrogen bond	• solubility
• hydrophilic	• solute
• hydrophobic	• solvent
• molarity	• water
• mole	

17. A solution in which $[H^+] = 10^{-3}$ M is (acidic/basic), whereas a solution in which $[H^+] = 10^{-10}$ M is (acidic/basic). Give the pH for each of these solutions.
18. Name three nucleotides or nucleic acids, and tell why each one is important.
19. You know that two soluble proteins are isoforms of each other. What can you predict about their structures, functions, and affinities for ligands?
20. You have been asked to design some drugs for the purposes described next. Choose the desirable characteristic(s) for each drug from the numbered list.

(a) Drug A must bind to an enzyme and enhance its activity.	1. antagonist
(b) Drug B should mimic the activity of a normal nervous system signal molecule.	2. competitive inhibitor
(c) Drug C should block the activity of a membrane receptor protein.	3. agonist
	4. allosteric activator
	5. covalent modulator

Level Three Problem Solving

21. You have been summoned to assist with the autopsy of an alien being whose remains have been brought to your lab. The chemical analysis returns with 33% C, 40% O, 4% H, 14% N, and 9% P. From this information, you conclude that the cells contain nucleotides, possibly even DNA or RNA. Your assistant is demanding that you tell him how you knew this. What do you tell him?

22. The harder a cell works, the more CO_2 it produces. CO_2 is carried in the blood according to the following equation:

$$CO_2 + H_2O \rightleftharpoons H^+ + HCO_3^-$$

What effect does hard work by your muscle cells have on the pH of the blood?

Level Four Quantitative Problems

23. Calculate the amount of NaCl you would weigh out to make one liter of 0.9% NaCl. Explain how you would make a liter of this solution.

24. A 1.0 M NaCl solution contains 58.5 g of salt per liter. (a) How many molecules of NaCl are present in 1 L of this solution? (b) How many millimoles of NaCl are present? (c) How many equivalents of Na^+ are present? (d) Express 58.5 g of NaCl per liter as a percent solution.

25. How would you make 200 mL of a 10% glucose solution? Calculate the molarity of this solution. How many millimoles of glucose are present in 500 mL of this solution? (*Hint:* What is the molecular mass of glucose?)

26. The graph shown below represents the binding of oxygen molecules (O_2) to two different proteins, myoglobin and hemoglobin, over a range of oxygen concentrations. Based on the graph, which protein has the higher affinity for oxygen? Explain your reasoning.

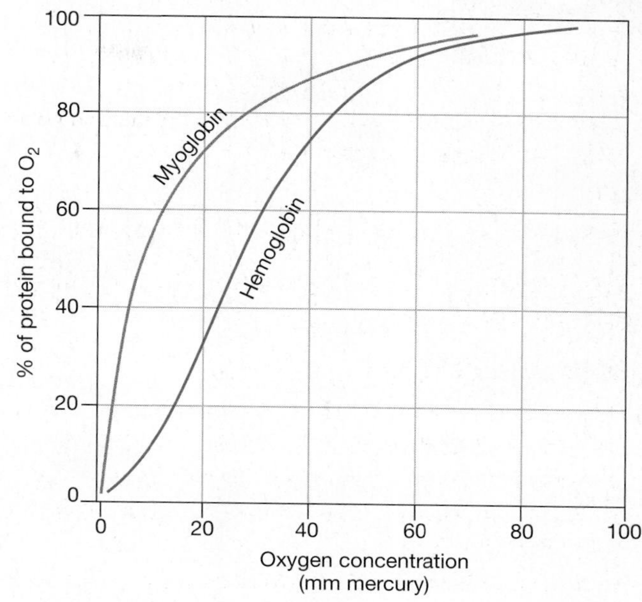

100
80
60
40
20
0
% of protein bound to O_2
Myoglobin
Hemoglobin
0
20
40
60
80
100
Oxygen concentration
(mm mercury)

Answers to Concept Checks, Figure and Graph Questions, and end-of-chapter Review Questions can be found in Appendix A [p. A-1].

3 Compartmentation: Cells and Tissues

Cells are organisms, and entire animals and plants are aggregates of these organisms.

Theodor Schwann, 1839

Cells viewed by microscope

BACKGROUND BASICS

Units of measure: inside back cover

8 Compartmentation

10 Extracellular fluid

40 Hydrophobic molecules

32 Proteins

41 pH

33 Covalent and noncovalent interactions

What makes a compartment? We may think of something totally enclosed, like a room or a box with a lid. But not all compartments are totally enclosed . . . think of the modular cubicles that make up many modern workplaces. And not all functional compartments have walls . . . think of a giant hotel lobby divided into conversational groupings by careful placement of rugs and furniture. Biological compartments come with the same type of anatomic variability, ranging from totally enclosed structures such as cells to functional compartments without visible walls.

The first living compartment was probably a simple cell whose intracellular fluid was separated from the external environment by a wall made of phospholipids and proteins—the cell membrane. **Cells** are the basic functional unit of living organisms, and an individual cell can carry out all the processes of life.

As cells evolved, they acquired intracellular compartments separated from the intracellular fluid by membranes. Over time, groups of single-celled organisms began to cooperate and specialize their functions, eventually giving rise to multicellular organisms. As multicellular organisms evolved to become larger and more complex, their bodies became divided into various functional compartments.

Compartments are both an advantage and a disadvantage for organisms. On the advantage side, compartments separate biochemical processes that might otherwise conflict with one another. For example, protein synthesis takes place in one subcellular compartment while protein degradation is taking place in another. Barriers between compartments, whether inside a cell or inside a body, allow the contents of one compartment to differ from the contents of adjacent compartments. An extreme example is the intracellular compartment called the *lysosome*, with an internal pH of 5 [Fig. 2.9, p. 45]. This pH is so acidic that if the lysosome ruptures, it severely damages or kills the cell that contains it.

The disadvantage to compartments is that barriers between them can make it difficult to move needed materials from one compartment to another. Living organisms overcome this problem with specialized mechanisms that transport selected substances across membranes. Membrane transport is the subject of Chapter 5.

RUNNING PROBLEM **Pap Tests Save Lives**

Dr. George Papanicolaou has saved the lives of millions of women by popularizing the Pap test, a cervical cytology screening method that detects early signs of cancer in the uterine cervix. In the past 50 years, deaths from cervical cancer have dropped dramatically in countries that routinely use the Pap test. In contrast, cervical cancer is a leading cause of death in regions where Pap test screening is not routine, such as Africa and Central America. If detected early, cervical cancer is one of the most treatable forms of cancer. Today, Jan Melton, who had an abnormal Pap test a year ago, returns to Dr. Baird, her family physician, for a repeat test. The results will determine whether she needs to undergo further testing for cervical cancer.

59 – 61 – 65 – 79 – 84 – 87

In this chapter, we explore the theme of compartmentation by first looking at the various compartments that subdivide the human body, from body cavities to the subcellular compartments called organelles. We then examine how groups of cells with similar functions unite to form the tissues and organs of the body. Continuing the theme of molecular interactions, we also look at how different molecules and fibers in cells and tissues give rise to their *mechanical properties:* their shape, strength, flexibility, and the connections that hold tissues together.

3.1 Functional Compartments of the Body

The human body is a complex compartment separated from the outside world by layers of cells. Anatomically, the body is divided into three major body **cavities:** the *cranial cavity* (commonly referred to as the *skull*), the *thoracic cavity* (also called the *thorax*), and the *abdominopelvic cavity* (**FIG. 3.1a**). The cavities are separated from one another by bones and tissues, and they are lined with *tissue membranes.*

The cranial cavity {*cranium*, skull} contains the brain, our primary control center. The thoracic cavity is bounded by the spine and ribs on top and sides, with the muscular *diaphragm* forming the floor. The thorax contains the heart, which is enclosed in a membranous *pericardial sac* {*peri-*, around + *cardium*, heart}, and the two lungs, enclosed in separate *pleural sacs.*

The *abdomen* and *pelvis* form one continuous cavity, the *abdominopelvic cavity.* A tissue lining called the *peritoneum* lines the abdomen and surrounds the organs within it (stomach, intestines, liver, pancreas, gallbladder, and spleen). The kidneys lie outside the abdominal cavity, between the peritoneum and the muscles and bones of the back, just above waist level. The pelvis contains reproductive organs, the urinary bladder, and the terminal portion of the large intestine.

In addition to the body cavities, there are several discrete fluid-filled anatomical compartments. The blood-filled vessels and heart of the circulatory system form one compartment. Our eyes are hollow fluid-filled spheres subdivided into two compartments, the aqueous and vitreous humors. The brain and spinal cord are surrounded by a special fluid compartment known as cerebrospinal fluid (CSF). The membranous sacs that surround the lungs (*pleural sacs*) and the heart (*pericardial sac*) also contain small volumes of fluid (Fig. 3.1a).

The Lumens of Some Organs Are Outside the Body

All hollow organs, such as heart, lungs, blood vessels, and intestines, create another set of compartments within the body. The interior of any hollow organ is called its **lumen** {*lumin*, window}. A lumen may be wholly or partially filled with air or fluid. For example, the lumens of blood vessels are filled with the fluid we call blood.

For some organs, the lumen is essentially an extension of the external environment, and material in the lumen is not truly part of the body's internal environment until it crosses the wall of

FIG. 3.1 ESSENTIALS Body Compartments

BODY COMPARTMENTS

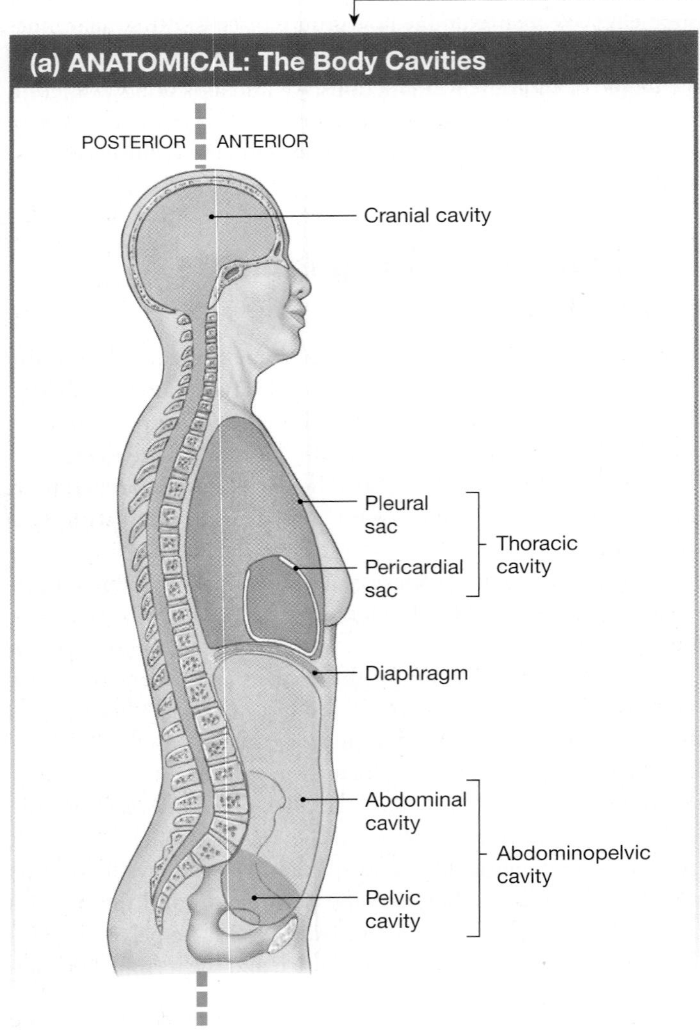

(a) ANATOMICAL: The Body Cavities
POSTERIOR
ANTERIOR
Cranial cavity
Pleural sac
Pericardial sac
Thoracic cavity
Diaphragm
Abdominal cavity
Pelvic cavity
Abdominopelvic cavity

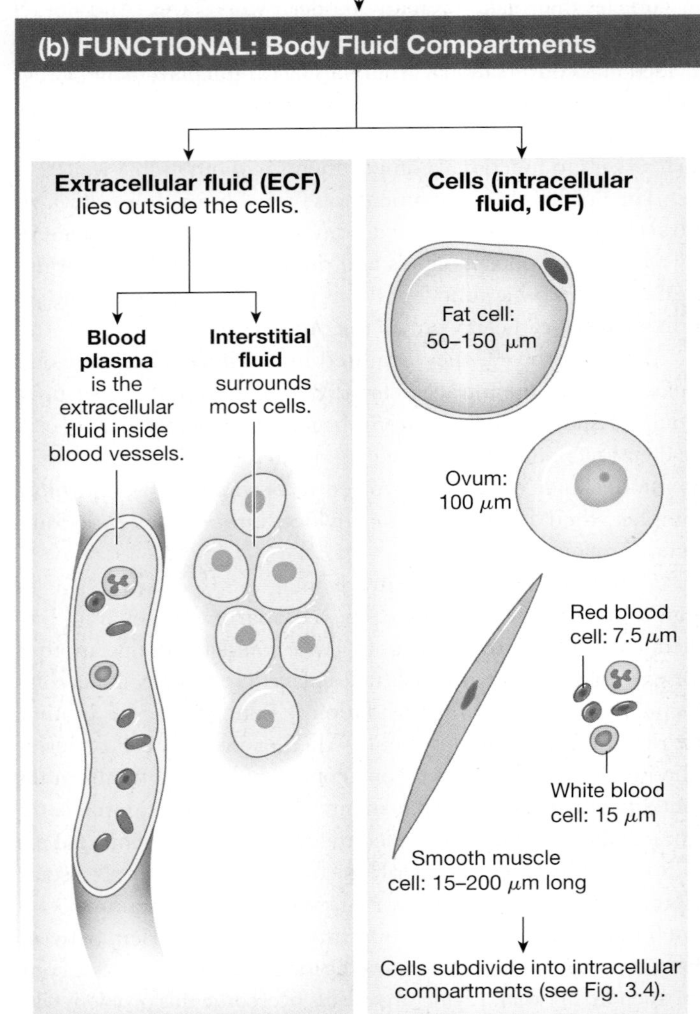

(b) FUNCTIONAL: Body Fluid Compartments
Extracellular fluid (ECF) lies outside the cells.
Blood plasma is the extracellular fluid inside blood vessels.
Interstitial fluid surrounds most cells.
Cells (intracellular fluid, ICF)
Fat cell: 50–150 μm
Ovum: 100 μm
Red blood cell: 7.5 μm
White blood cell: 15 μm
Smooth muscle cell: 15–200 μm long
Cells subdivide into intracellular compartments (see Fig. 3.4).

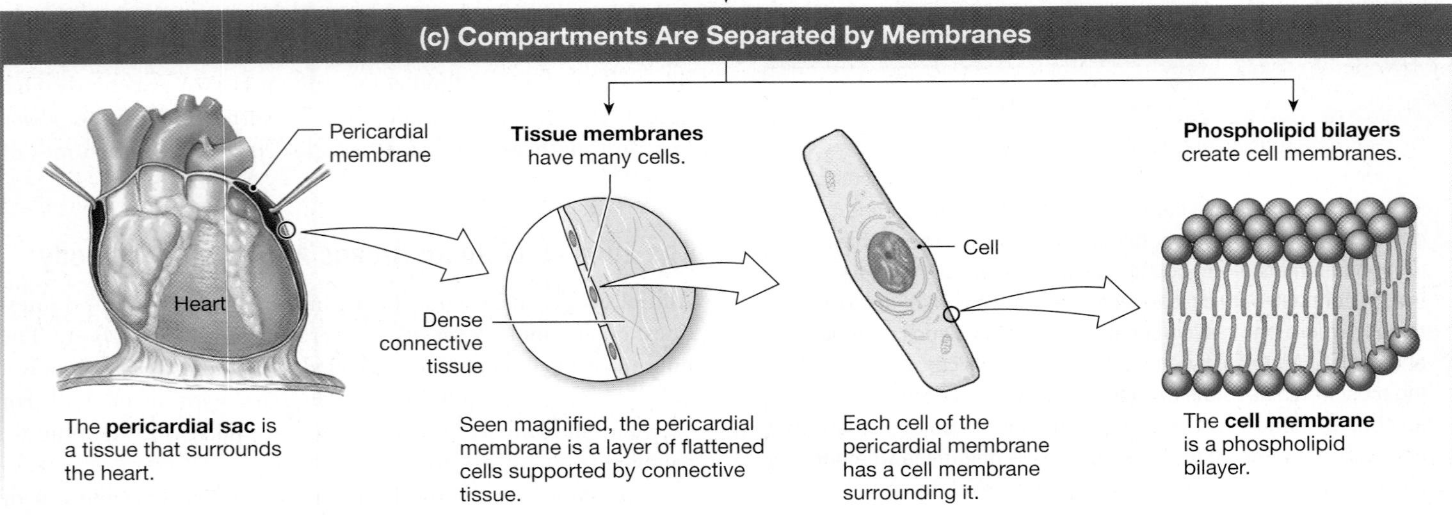

(c) Compartments Are Separated by Membranes
Pericardial membrane
Heart
Tissue membranes have many cells.
Dense connective tissue
Cell
Phospholipid bilayers create cell membranes.

The **pericardial sac** is a tissue that surrounds the heart.

Seen magnified, the pericardial membrane is a layer of flattened cells supported by connective tissue.

Each cell of the pericardial membrane has a cell membrane surrounding it.

The **cell membrane** is a phospholipid bilayer.

RUNNING PROBLEM

Cancer is a condition in which a small group of cells starts to divide uncontrollably and fails to differentiate into specialized cell types. Cancerous cells that originate in one tissue can escape from that tissue and spread to other organs through the circulatory system and the lymph vessels, a process known as *metastasis.*

Q1: *Why does the treatment of cancer focus on killing the cancerous cells?*

59
61
65
79
84
87

the organ. For example, we think of our digestive tract as being "inside" our body, but in reality its lumen is part of the body's external environment [see Fig. 1.2, p. 4]. An analogy would be the hole through a bead. The hole passes through the bead but is not actually inside the bead.

An interesting illustration of this distinction between the internal environment and the external environment in a lumen involves the bacterium *Escherichia coli.* This organism normally lives and reproduces inside the large intestine, an internalized compartment whose lumen is continuous with the external environment. When *E. coli* is residing in this location, it does not harm the host. However, if the intestinal wall is punctured by disease or accident and *E. coli* enters the body's internal environment, a serious infection can result.

Functionally, the Body Has Three Fluid Compartments

In physiology, we are often more interested in functional compartments than in anatomical compartments. Most cells of the body are not in direct contact with the outside world. Instead, their external environment is the extracellular fluid [Fig. 1.5, p. 11]. If we think of all the cells of the body together as one unit, we can then divide the body into two main fluid compartments: (1) the *extracellular fluid* (ECF) outside the cells and (2) the *intracellular fluid* (ICF) within the cells (Fig. 3.1b). The dividing wall between ECF and ICF is the cell membrane. The extracellular fluid subdivides further into **plasma,** the fluid portion of the blood, and **interstitial fluid** {*inter-*, between + *stare*, to stand}, which surrounds most cells of the body.

3.2 Biological Membranes

The word *membrane* {*membrane*, a skin} has two meanings in biology. Before the invention of microscopes in the sixteenth century, a membrane always described a tissue that lined a cavity or separated two compartments. Even today, we speak of *mucous membranes* in the mouth and vagina, the *peritoneal membrane* that lines the inside of the abdomen, the *pleural membrane* that covers the surface of the lungs, and the *pericardial membrane* that surrounds the heart. These visible membranes are tissues: thin, translucent layers of cells.

Once scientists observed cells with a microscope, the nature of the barrier between a cell's intracellular fluid and its external environment became a matter of great interest. By the 1890s, scientists had concluded that the outer surface of cells, the **cell membrane,** was a thin layer of lipids that separated the aqueous fluids of the interior and outside environment. We now know that cell membranes consist of microscopic double layers, or *bilayers,* of phospholipids with protein molecules inserted in them.

In short, the word *membrane* may apply either to a tissue or to a phospholipid-protein boundary layer (Fig. 3.1c). One source of confusion is that tissue membranes are often depicted in book illustrations as a single line, leading students to think of them as if they were similar in structure to the cell membrane. In this section, you will learn more about the phospholipid membranes that create compartments for cells.

The Cell Membrane Separates Cell from Environment

There are two synonyms for the term *cell membrane: plasma membrane* and *plasmalemma.* We will use the term *cell membrane* in this book rather than *plasma membrane* or *plasmalemma* to avoid confusion with the term *blood plasma.* The general functions of the cell membrane include:

1. **Physical isolation.** The cell membrane is a physical barrier that separates intracellular fluid inside the cell from the surrounding extracellular fluid.
2. **Regulation of exchange with the environment.** The cell membrane controls the entry of ions and nutrients into the cell, the elimination of cellular wastes, and the release of products from the cell.
3. **Communication between the cell and its environment.** The cell membrane contains proteins that enable the cell to recognize and respond to molecules or to changes in its external environment. Any alteration in the cell membrane may affect the cell's activities.
4. **Structural support.** Proteins in the cell membrane hold the *cytoskeleton,* the cell's interior structural scaffolding, in place to maintain cell shape. Membrane proteins also create specialized junctions between adjacent cells or between cells and the *extracellular matrix* {*extra-*, outside}, which is extracellular material that is synthesized and secreted by the cells. (**Secretion** is the process by which a cell releases a substance into the extracellular space.) Cell-cell and cell-matrix junctions stabilize the structure of tissues.

How can the cell membrane carry out such diverse functions? Our current model of cell membrane structure provides the answer.

Membranes Are Mostly Lipid and Protein

In the early decades of the twentieth century, researchers trying to decipher membrane structure ground up cells and analyzed their composition. They discovered that all biological membranes consist of a combination of lipids and proteins plus a small amount of carbohydrate. However, a simple and uniform structure did not account for the highly variable properties of membranes found in different types of cells. How could water cross the cell membrane to enter a red blood cell but not be able to enter certain cells of the kidney tubule? The explanation had to lie in the molecular arrangement of the proteins and lipids in the various membranes.

The ratio of protein to lipid varies widely, depending on the source of the membrane (**TBL. 3.1**). Generally, the more metabolically active a membrane is, the more proteins it contains. For example, the inner membrane of a mitochondrion, which contains enzymes for ATP production, is three-quarters protein.

This chemical analysis of membranes was useful, but it did not explain the structural arrangement of lipids and proteins in a membrane. Studies in the 1920s suggested that there was enough lipid in a given area of membrane to create a double layer. The bilayer model was further modified in the 1930s to account for the presence of proteins. With the introduction of electron microscopy, scientists saw the cell membrane for the first time. The 1960s model of the membrane, as seen in electron micrographs, was a "butter sandwich"—a clear layer of lipids sandwiched between two dark layers of protein.

By the early 1970s, freeze-fracture electron micrographs had revealed the actual three-dimensional arrangement of lipids and proteins within cell membranes. Because of what scientists learned from looking at freeze-fractured membranes, S. J. Singer and G. L. Nicolson in 1972 proposed the **fluid mosaic model** of the membrane. **FIGURE 3.2** highlights the major features of this contemporary model of membrane structure.

The lipids of biological membranes are mostly phospholipids arranged in a bilayer so that the phosphate heads are on the membrane surfaces and the lipid tails are hidden in the center of the membrane (Fig. 3.2b). The cell membrane is studded with protein molecules, like raisins in a slice of bread, and the extracellular surface has glycoproteins and glycolipids. All cell membranes are of relatively uniform thickness, about 8 nm.

Membrane Lipids Create a Hydrophobic Barrier

Three main types of lipids make up the cell membrane: phospholipids, sphingolipids, and cholesterol. Phospholipids are made of a glycerol backbone with two fatty acid chains extending to one side and a phosphate group extending to the other [p. 30]. The glycerol-phosphate head of the molecule is polar and thus hydrophilic. The fatty acid "tail" is nonpolar and thus hydrophobic.

When placed in an aqueous solution, phospholipids orient themselves so that the polar heads of the molecules interact with the water molecules while the nonpolar fatty acid tails "hide" by putting the polar heads between themselves and the water. This arrangement can be seen in three structures: the micelle, the liposome, and the phospholipid bilayer of the cell membrane (Fig. 3.2a). **Micelles** are small droplets with a single layer of phospholipids arranged so that the interior of the micelle is filled with hydrophobic fatty acid tails. Micelles are important in the digestion and absorption of fats in the digestive tract.

Liposomes are larger spheres with bilayer phospholipid walls. This arrangement leaves a hollow center with an aqueous core that can be filled with water-soluble molecules. Biologists think that a liposome-like structure was the precursor of the first living cell. Today, liposomes are being used as a medium to deliver drugs and cosmetics.

In medicine, the centers of liposomes are filled with drugs or with fragments of DNA for gene therapy. To make drug delivery more specific, researchers can make *immunoliposomes* that use antibodies to recognize specific types of cancer cells. By targeting drugs to the cells they are treating, researchers hope to increase the effectiveness of the drugs and decrease unwanted side effects.

Phospholipids are the major lipid of membranes, but some membranes also have significant amounts of **sphingolipids**. Sphingolipids also have fatty acid tails, but their heads may be either phospholipids or glycolipids. Sphingolipids are slightly longer than phospholipids.

Cholesterol is also a significant part of many cell membranes. Cholesterol molecules, which are mostly hydrophobic, insert themselves between the hydrophilic heads of phospholipids (Fig. 3.2b). Cholesterol helps make membranes impermeable to small water-soluble molecules and keeps membranes flexible over a wide range of temperatures.

Membrane Proteins May Be Loosely or Tightly Bound to the Membrane

According to some estimates, membrane proteins may be nearly one-third of all proteins coded in our DNA. Each cell has between 10 and 50 different types of proteins inserted into its membranes. Membrane proteins can be described in several different ways. **Integral proteins** are tightly bound to the membrane, and the only way they can be removed is by disrupting the membrane structure with detergents or other harsh methods that destroy the membrane's integrity. Integral proteins include transmembrane proteins and lipid-anchored proteins.

Peripheral proteins {*peripheria*, circumference} attach to other membrane proteins by noncovalent interactions [p. 39] and

TABLE 3.1 Composition of Selected Membranes

Membrane	Protein	Lipid	Carbohydrate
Red blood cell membrane	49%	43%	8%
Myelin membrane around nerve cells	18%	79%	3%
Inner mitochondrial membrane	76%	24%	0%

FIG. 3.2 ESSENTIALS The Cell Membrane

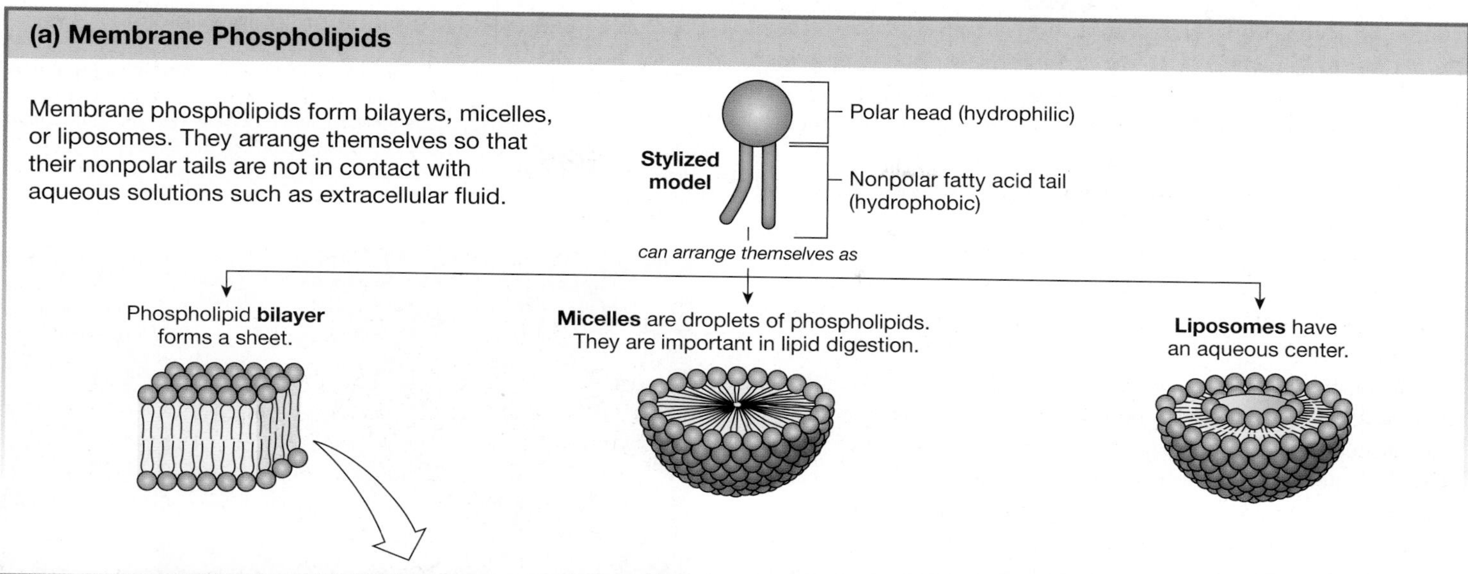
(a) Membrane Phospholipids
Membrane phospholipids form bilayers, micelles, or liposomes. They arrange themselves so that their nonpolar tails are not in contact with aqueous solutions such as extracellular fluid.
Stylized model
Polar head (hydrophilic)
Nonpolar fatty acid tail (hydrophobic)
can arrange themselves as
Phospholipid bilayer forms a sheet.
Micelles are droplets of phospholipids. They are important in lipid digestion.
Liposomes have an aqueous center.

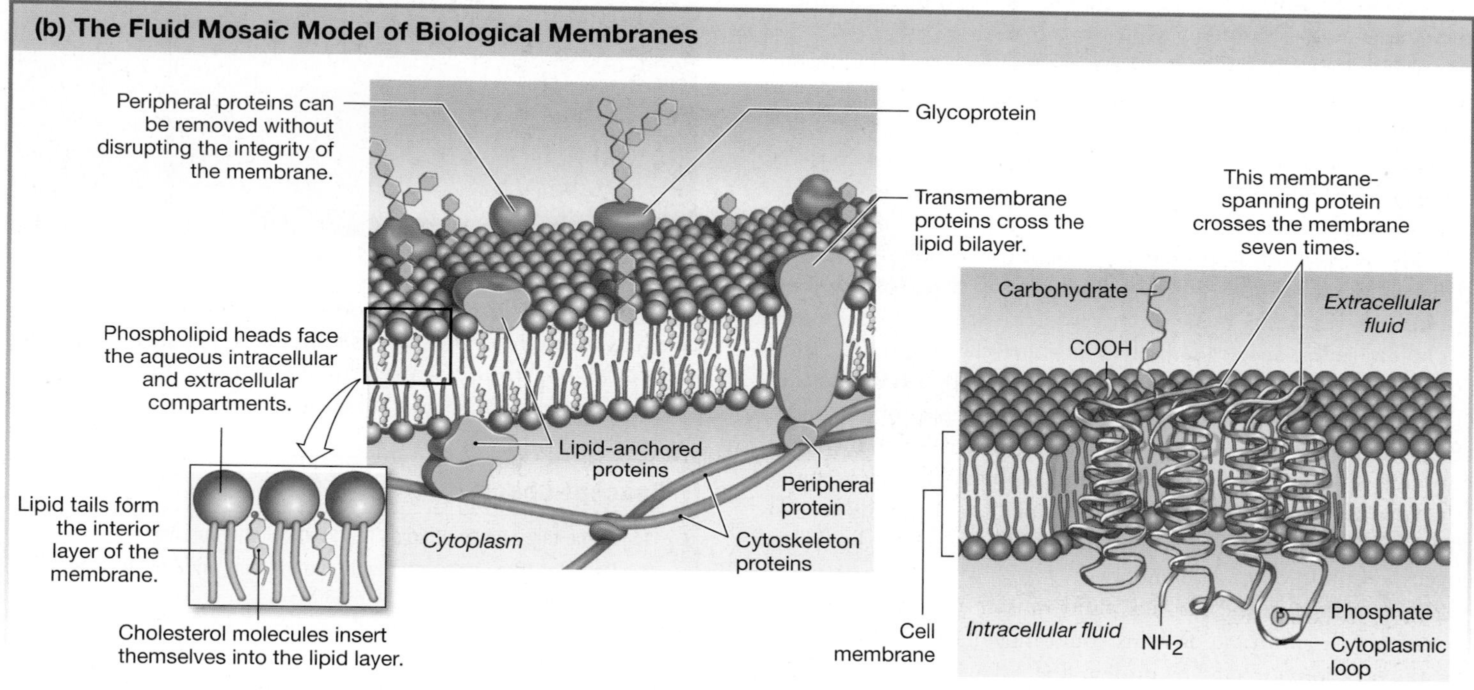
(b) The Fluid Mosaic Model of Biological Membranes
Peripheral proteins can be removed without disrupting the integrity of the membrane.
Glycoprotein
Transmembrane proteins cross the lipid bilayer.
This membrane-spanning protein crosses the membrane seven times.
Phospholipid heads face the aqueous intracellular and extracellular compartments.
Carbohydrate
COOH
Extracellular fluid
Lipid-anchored proteins
Peripheral protein
Cytoplasm
Cytoskeleton proteins
Lipid tails form the interior layer of the membrane.
Cholesterol molecules insert themselves into the lipid layer.
Cell membrane
Intracellular fluid
NH2
Phosphate
Cytoplasmic loop

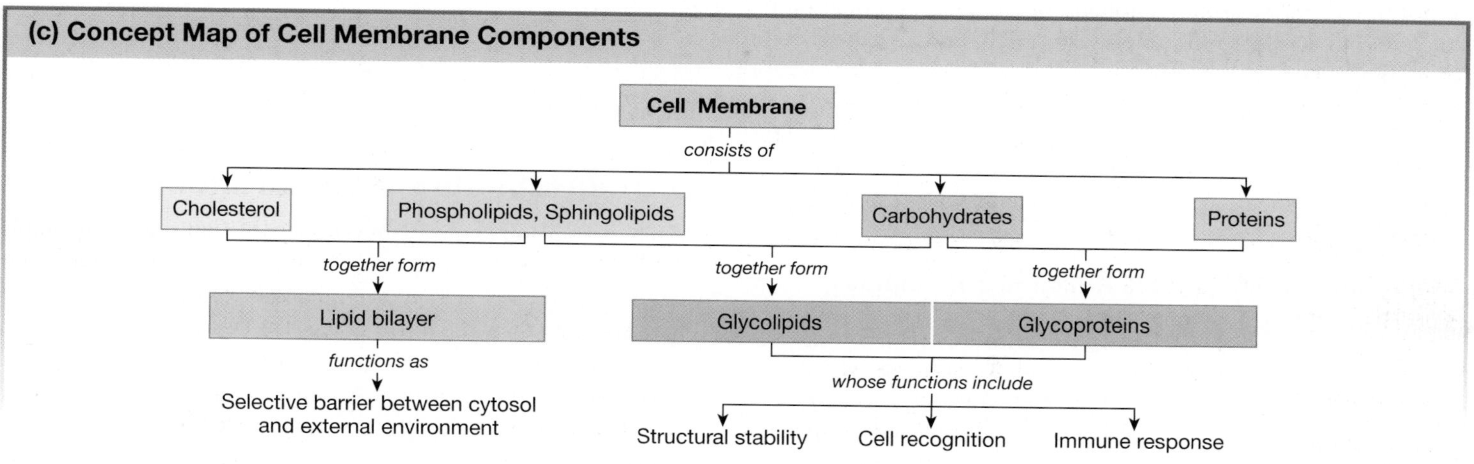
(c) Concept Map of Cell Membrane Components
Cell Membrane
consists of
Cholesterol
Phospholipids, Sphingolipids
Carbohydrates
Proteins
together form
Lipid bilayer
functions as
Selective barrier between cytosol and external environment
together form
together form
Glycolipids
Glycoproteins
whose functions include
Structural stability
Cell recognition
Immune response

can be separated from the membrane by chemical methods that do not disrupt the integrity of the membrane. Peripheral proteins include some enzymes as well as structural binding proteins that anchor the *cytoskeleton* (the cell's internal "skeleton") to the membrane (Fig. 3.2b).

Transmembrane proteins {*trans*-, across} are also called *membrane-spanning* proteins because the protein's chains extend all the way across the cell membrane (Fig. 3.2b). When a protein crosses the membrane more than once, loops of the amino acid chain protrude into the cytoplasm and the extracellular fluid. Carbohydrates may attach to the extracellular loops, and phosphate groups may attach to the intracellular loops. Phosphorylation or dephosphorylation of proteins is one way cells alter protein function [p. 49].

Transmembrane proteins are classified into families according to how many transmembrane segments they have. Many physiologically important membrane proteins have seven transmembrane segments, as shown in Figure 3.2c. Others cross the membrane only once or up to as many as 12 times.

Membrane-spanning proteins are integral proteins, tightly but not covalently bound to the membrane. The 20–25 amino acids in the protein chain segments that pass through the bilayer are nonpolar. This allows those amino acids to create strong noncovalent interactions with the lipid tails of the membrane phospholipids, holding them tightly in place.

Some membrane proteins that were previously thought to be peripheral proteins are now known to be **lipid-anchored proteins** (Fig. 3.2b). Some of these proteins are covalently bound to lipid tails that insert themselves into the bilayer. Others, found only on the external surface of the cell, are held by a **GPI anchor** that consists of a membrane lipid plus a sugar-phosphate chain. (GPI stands for *glycosylphosphatidylinositol*.) Many lipid-anchored proteins are associated with membrane sphingolipids, leading to the formation of specialized patches of membrane called *lipid rafts* (**FIG. 3.3**). The longer tails of the sphingolipids elevate the lipid rafts over their phospholipid neighbors.

According to the original fluid mosaic model of the cell membrane, membrane proteins could move laterally from location to location, directed by protein fibers that run just under the membrane surface. However, researchers have learned that this is not true of all membrane proteins. Some integral proteins are anchored to cytoskeleton proteins (Fig. 3.2b) and are, therefore, immobile. The ability of the cytoskeleton to restrict the movement of integral proteins allows cells to develop *polarity*, in which different faces of the cell have different proteins and therefore different properties. This is particularly important in the cells of the transporting epithelia, as you will see in multiple tissues in the body.

Membrane Carbohydrates Attach to Both Lipids and Proteins

Most membrane carbohydrates are sugars attached either to membrane proteins (glycoproteins) or to membrane lipids (glycolipids). They are found exclusively on the external surface of the cell, where they form a protective layer known as the **glycocalyx** {*glycol*-, sweet + *kalyx*, husk or pod}. Glycoproteins on the cell surface play a key role in the body's immune response. For example, the ABO blood groups are determined by the number and composition of sugars attached to membrane sphingolipids.

FIG. 3.3 Lipid rafts are made of sphingolipids

Sphingolipids (orange) are longer than phospholipids and stick up above the phospholipids of the membrane (black). A lipid-anchored enzyme, placental alkaline phosphatase (yellow), is almost always associated with a lipid raft. Image courtesy of D. E. Saslowsky, J. Lawrence, X. Ren, D. A. Brown, R. M. Henderson, and J. M. Edwardson. Placental alkaline phosphatase is efficiently targeted to rafts in supported lipid bilayers. *J Biol Chem* 277: 26966–26970, 2002.

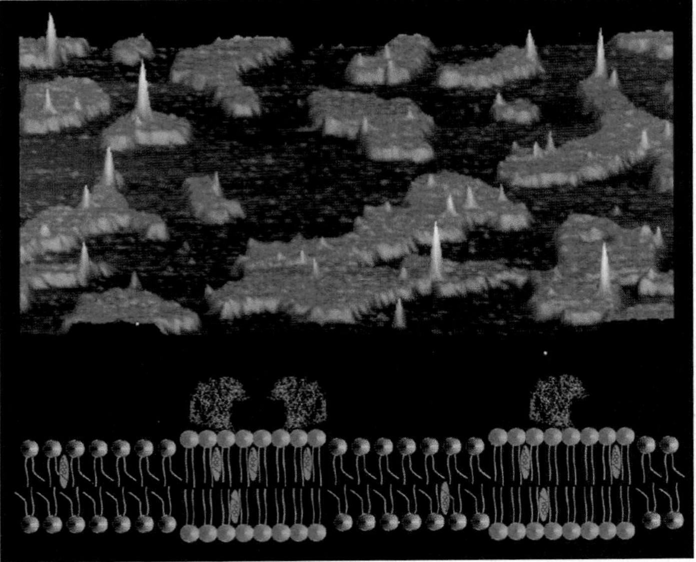

Concept Check

1. Name three types of lipids found in cell membranes.
2. Describe three types of membrane proteins and how they are associated with the cell membrane.
3. Why do phospholipids in cell membranes form a bilayer instead of a single layer?
4. How many phospholipid bilayers will a substance cross passing into a cell?

Figure 3.2c is a summary map organizing the structure of the cell membrane.

3.3 Intracellular Compartments

Much of what we know about cells comes from studies of simple organisms that consist of one cell. But humans are much more complex, with trillions of cells in their bodies. It has been estimated that there are more than 200 different types of cells in the human body, each cell type with its own characteristic structure and function.

During development, cells specialize and take on specific shapes and functions. Each cell in the body inherits identical genetic information in its DNA, but no one cell uses all this information. During

differentiation, only selected genes become active, transforming the cell into a specialized unit. In most cases, the final shape and size of a cell and its contents reflect its function. Figure 3.1b shows five representative cells in the human body. These mature cells look very different from one another, but they all started out alike in the early embryo, and they retain many features in common.

Cells Are Divided into Compartments

We can compare the structural organization of a cell to that of a medieval walled city. The city is separated from the surrounding countryside by a high wall, with entry and exit strictly controlled through gates that can be opened and closed. The city inside the walls is divided into streets and a diverse collection of houses and shops with varied functions. Within the city, a ruler in the castle oversees the everyday comings and goings of the city's inhabitants. Because the city depends on food and raw material from outside the walls, the ruler negotiates with the farmers in the countryside. Foreign invaders are always a threat, so the city ruler communicates and cooperates with the rulers of neighboring cities.

In the cell, the outer boundary is the cell membrane. Like the city wall, it controls the movement of material between the cell interior and the outside by opening and closing "gates" made of protein. The inside of the cell is divided into compartments rather than into shops and houses. Each of these compartments has a specific purpose that contributes to the function of the cell as a whole. In the cell, DNA in the nucleus is the "ruler in the castle," controlling both the internal workings of the cell and its interaction with other cells. Like the city, the cell depends on supplies from its external environment. It must also communicate and cooperate with other cells to keep the body functioning in a coordinated fashion.

FIGURE 3.4a is an overview map of cell structure. The cells of the body are surrounded by the dilute salt solution of the extracellular fluid. The cell membrane separates the inside environment of the cell (the intracellular fluid) from the extracellular fluid.

Play BioFlix Animation
@Mastering Anatomy & Physiology

Internally the cell is divided into the *cytoplasm* and the *nucleus*. The cytoplasm consists of a fluid portion, called *cytosol*; insoluble particles called *inclusions*; insoluble protein fibers; and membrane-bound structures collectively known as *organelles*. Figure 3.4 shows a typical cell from the lining of the small intestine. It has most of the structures found in animal cells.

The Cytoplasm Includes Cytosol, Inclusions, Fibers, and Organelles

The **cytoplasm** includes all material inside the cell membrane except for the nucleus. The cytoplasm has four components:

1. **Cytosol** {*cyto-*, cell + sol(uble)}, or intracellular fluid: The cytosol is a semi-gelatinous fluid separated from the extracellular fluid by the cell membrane. The cytosol contains dissolved nutrients and proteins, ions, and waste products. The other components of the cytoplasm—inclusions, fibers, and organelles—are suspended in the cytosol.
2. **Inclusions** are particles of insoluble materials. Some are stored nutrients. Others are responsible for specific cell functions. These structures are sometimes called the *nonmembranous organelles*.
3. Insoluble **protein fibers** form the cell's internal support system, or **cytoskeleton.**
4. **Organelles**—"little organs"—are membrane-bound compartments that play specific roles in the overall function of the cell. For example, the organelles called mitochondria (singular, *mitochondrion*) generate most of the cell's ATP, and the organelles called lysosomes act as the digestive system of the cell. The organelles work in an integrated manner, each organelle taking on one or more of the cell's functions.

Inclusions Are in Direct Contact with the Cytosol

The inclusions of cells do not have boundary membranes and so are in direct contact with the cytosol. Movement of material between inclusions and the cytosol does not require transport across a membrane. Nutrients are stored as glycogen granules and lipid droplets. Most inclusions with functions other than nutrient storage are made from protein or combinations of RNA and protein.

Ribosomes (Fig. 3.4i) are small, dense granules of RNA and protein that manufacture proteins under the direction of the cell's DNA (see Chapter 4 for details). **Fixed ribosomes** attach to the cytosolic surface of organelles. **Free ribosomes** are suspended free in the cytosol. Some free ribosomes form groups of 10 to 20 known as **polyribosomes.** A ribosome that is fixed one minute may release and become a free ribosome the next. Ribosomes are most numerous in cells that synthesize proteins for export out of the cell.

RUNNING PROBLEM

During a Pap test for cervical cancer, tissue is sampled from the cervix (neck) of the uterus with a collection device that resembles a tiny brush. The cells are rinsed off the brush into preservative fluid that is sent to a laboratory. There the sample is processed onto a glass slide that will be examined first by a computer, then by a trained cytologist. The computer and cytologist look for *dysplasia* {*dys-*, abnormal + *plasia*, growth or cell multiplication}, a change in the size and shape of cells that is suggestive of cancerous changes. Cancer cells can usually be recognized by a large nucleus surrounded by a relatively small amount of cytoplasm. Jan's first Pap test showed all the hallmarks of dysplasia.

Q2: *What is happening in cancer cells that explains the large size of their nucleus and the relatively small amount of cytoplasm?*

59
61
65
79
84
87

FIG. 3.4 REVIEW Cell Structure

(a) This is an overview map of cell structure. The *cell membrane* separates the inside environment of the cell (the intracellular fluid) from the extracellular fluid. Internally the cell is divided into the *cytoplasm* and the *nucleus.* The cytoplasm consists of a fluid portion, called the *cytosol;* membrane-bound structures called *organelles;* insoluble particles called *inclusions;* and protein fibers that create the *cytoskeleton.*

THE CELL

is composed of

Cell membrane

Nucleus

Cytoplasm

Cytosol	Membranous organelles	Inclusions	Protein fibers
Fluid portion of cyto-plasm	• Mitochondria • Endoplasmic reticulum • Golgi apparatus • Lysosomes • Peroxisomes	• Lipid droplets • Glycogen granules • Ribosomes	• Cytoskeleton • Centrioles • Cilia • Flagella

Extracellular fluid

(b) Cytoskeleton

Microvilli increase cell surface area. They are supported by microfilaments.

Microfilaments form a network just inside the cell membrane.

Microtubules are the largest cytoskeleton fiber.

Intermediate filaments include myosin and keratin.

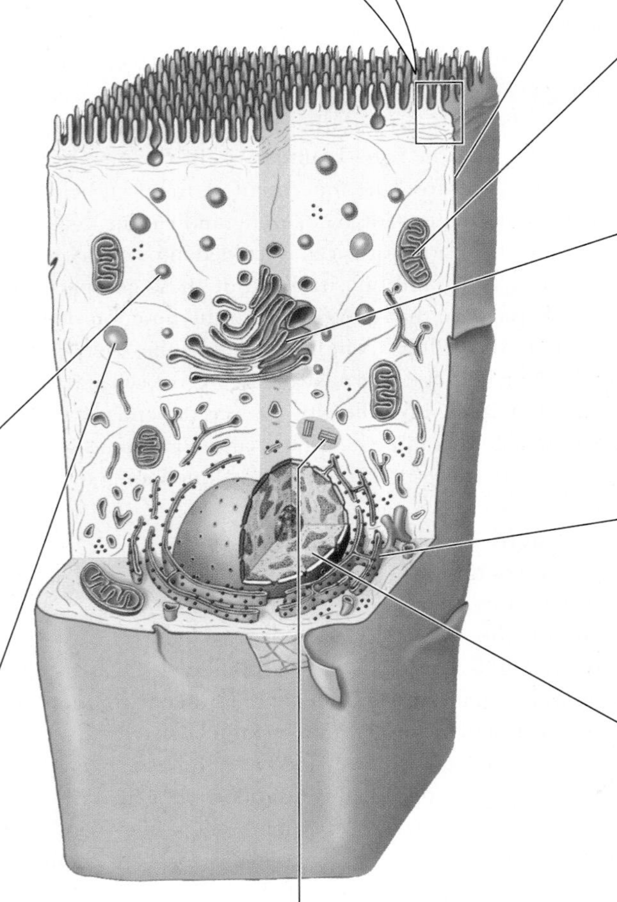

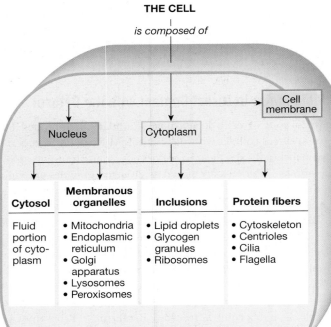

(c) Peroxisomes

Peroxisomes contain enzymes that break down fatty acids and some foreign materials.

(d) Lysosomes

Lysosomes are small, spherical storage vesicles that contain powerful digestive enzymes.

(e) Centrioles

Centrioles are made from microtubules and direct DNA movement during cell division.

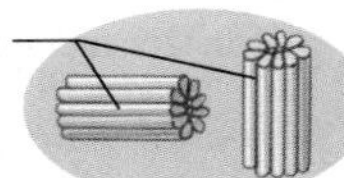
Centrioles

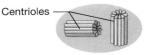

(f) Cell Membrane

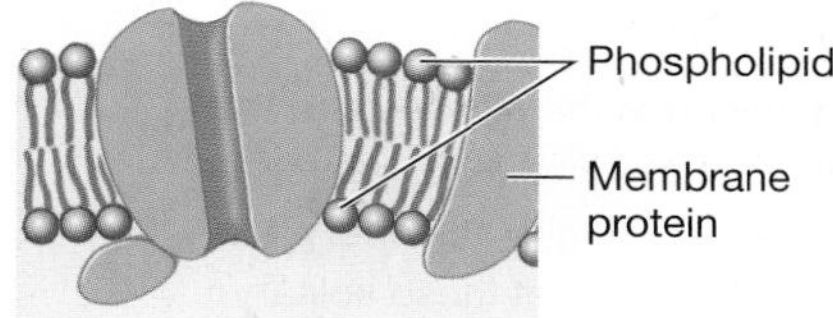

Phospholipid
Membrane protein

The **cell membrane** is a phospholipid bilayer studded with proteins that act as structural anchors, transporters, enzymes, or signal receptors. Glycolipids and glycoproteins occur only on the extracellular surface of the membrane. The cell membrane acts as both a gateway and a barrier between the cytoplasm and the extracellular fluid.

(g) Mitochondria

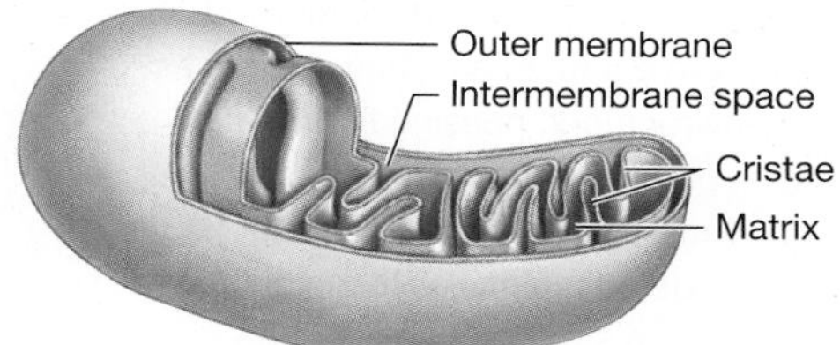

Outer membrane
Intermembrane space
Cristae
Matrix

Mitchondria are spherical to elliptical organelles with a double wall that creates two separate compartments within the organelle. The inner **matrix** is surrounded by a membrane that folds into leaflets called **cristae.** The **intermembrane space,** which lies between the two membranes, plays an important role in ATP production. Mitochondria are the site of most ATP synthesis in the cell.

(h) Golgi Apparatus and Vesicles

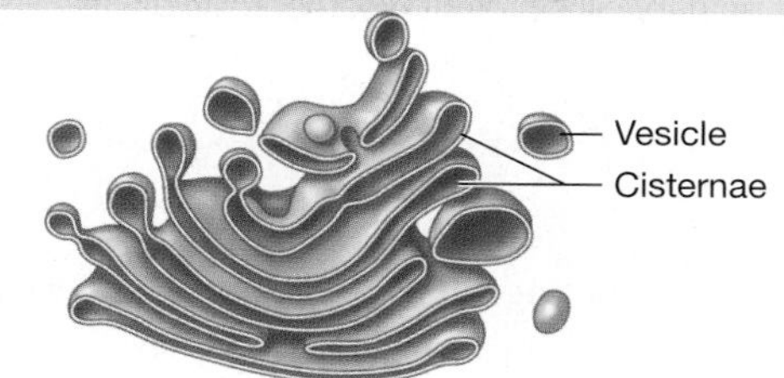

Vesicle
Cisternae

The **Golgi apparatus** consists of a series of hollow curved sacs called **cisternae** stacked on top of one another and surrounded by vesicles. The Golgi apparatus participates in protein modification and packaging.

(i) Endoplasmic Reticulum (ER) and Ribosomes

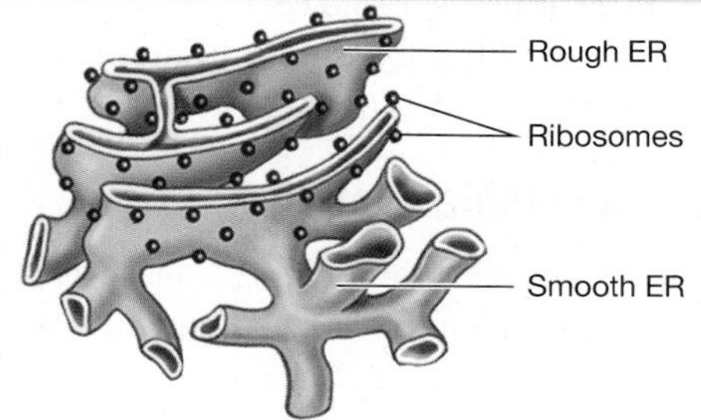

Rough ER
Ribosomes
Smooth ER

The **endoplasmic reticulum (ER)** is a network of interconnected membrane tubes that are a continuation of the outer nuclear membrane. **Rough endoplasmic reticulum** has a granular appearance due to rows of ribosomes dotting its cytoplasmic surface. **Smooth endoplasmic reticulum** lacks ribosomes and appears as smooth membrane tubes. The rough ER is the main site of protein synthesis. The smooth ER synthesizes lipids and, in some cells, concentrates and stores calcium ions.

(j) Nucleus

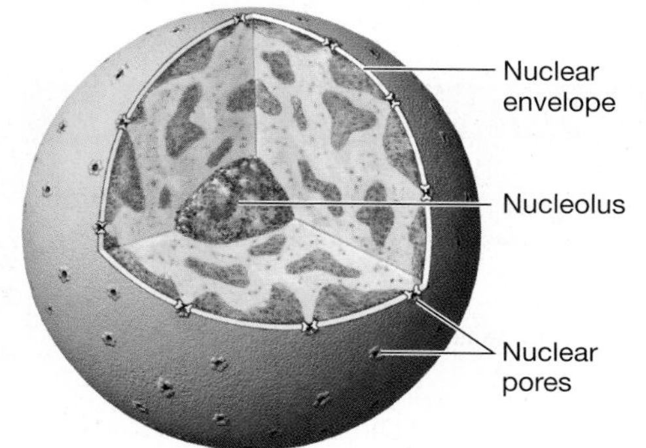

Nuclear envelope
Nucleolus
Nuclear pores

The **nucleus** is surrounded by a double-membrane **nuclear envelope.** Both membranes of the envelope are pierced here and there by **pores** to allow communication with the cytoplasm. The outer membrane of the nuclear envelope connects to the endoplasmic reticulum membrane. In cells that are not dividing, the nucleus appears filled with randomly scattered granular material composed of DNA and proteins. Usually a nucleus also contains from one to four larger dark-staining bodies of DNA, RNA, and protein called **nucleoli.**

Cytoplasmic Protein Fibers Come in Three Sizes

The three families of cytoplasmic protein fibers are classified by diameter and protein composition (**TBL. 3.2**). All fibers are polymers of smaller proteins. The thinnest are **actin fibers,** also called *microfilaments.* Somewhat larger **intermediate filaments** may be made of different types of protein, including *keratin* in hair and skin, and *neurofilament* in nerve cells. The largest protein fibers are the hollow **microtubules,** made of a protein called **tubulin.** A large number of *accessory proteins* are associated with the cell's protein fibers.

The insoluble protein fibers of the cell have two general purposes: structural support and movement. Structural support comes primarily from the cytoskeleton. Movement of the cell or of elements within the cell takes place with the aid of protein fibers and a group of specialized enzymes called *motor proteins.* These functions are discussed in more detail in the sections that follow.

Microtubules Form Centrioles, Cilia, and Flagella

The largest cytoplasmic protein fibers, the microtubules, create the complex structures of centrioles, cilia, and flagella, which are all involved in some form of cell movement. The cell's *microtubule-organizing center,* the **centrosome,** assembles tubulin molecules into microtubules. The centrosome appears as a region of darkly staining material close to the cell nucleus. In most animal cells, the centrosome contains two **centrioles,** shown in the typical cell of Figure 3.4e.

Each centriole is a cylindrical bundle of 27 microtubules, arranged in nine triplets. In cell division, the centrioles direct the movement of DNA strands. Cells that have lost their ability to undergo cell division, such as mature nerve cells, lack centrioles.

Cilia are short, hairlike structures projecting from the cell surface like the bristles of a brush {singular, *cilium,* Latin for eyelash}. Most cells have a single short cilium, but cells lining the upper airways and part of the female reproductive tract are covered with cilia. In these tissues, coordinated ciliary movement creates currents that sweep fluids or secretions across the cell surface.

The surface of a cilium is a continuation of the cell membrane. The core of *motile,* or moving, cilia contains nine pairs of microtubules surrounding a central pair (**FIG. 3.5b**). The microtubules terminate just inside the cell at the *basal body*. These cilia beat rhythmically back and forth when the microtubule pairs in their core slide past each other with the help of the motor protein *dynein.*

Flagella have the same microtubule arrangement as cilia but are considerably longer {singular, *flagellum,* Latin for whip}. Flagella are found on free-floating single cells, and in humans the only flagellated cell is the male sperm cell. A sperm cell has only one flagellum, in contrast to ciliated cells, which may have one surface almost totally covered with cilia (Fig. 3.5a). The wavelike movements of the flagellum push the sperm through fluid, just as undulating contractions of a snake's body push it headfirst through its environment. Flagella bend and move by the same basic mechanism as cilia.

EMERGING CONCEPTS

Single Cilia Are Sensors

Cilia in the body are not limited to the airways and the reproductive tract. Scientists have known for years that most cells of the body contain a single, stationary, or *nonmotile,* cilium, but they thought that these solitary **primary cilia** were mostly evolutionary remnants and of little significance. Primary cilia differ structurally from motile cilia because they lack the central pair of microtubules found in motile cilia (a 9 + 0 arrangement instead of 9 + 2; see Fig. 3.5). Researchers in recent years have learned that primary cilia actually serve a function. They can act as sensors of the external environment, passing information into the cell. For example, primary cilia in photoreceptors of the eye help with light sensing, and primary cilia in the kidney sense fluid flow. Using molecular techniques, scientists have found that these small, insignificant hairs play critical roles during embryonic development as well. Mutations to ciliary proteins cause disorders (*ciliopathies*) ranging from polycystic kidney disease and loss of vision to cancer. The role of primary cilia in other disorders, including obesity, is currently a hot topic in research.

The Cytoskeleton Is a Changeable Scaffold

The cytoskeleton is a flexible, changeable three-dimensional scaffolding of actin microfilaments, intermediate filaments, and microtubules that extends throughout the cytoplasm. Some cytoskeleton protein fibers are permanent, but most are synthesized or disassembled according to the cell's needs. Because of the cytoskeleton's changeable nature, its organizational details are complex and we will not discuss the details.

TABLE 3.2 Diameter of Protein Fibers in the Cytoplasm

	Diameter	Type of Protein	Functions
Microfilaments	7 nm	Actin (globular)	Cytoskeleton; associates with myosin for muscle contraction
Intermediate Filaments	10 nm	Keratin, neurofilament protein (filaments)	Cytoskeleton, hair and nails, protective barrier of skin
Microtubules	25 nm	Tubulin (globular)	Movement of cilia, flagella, and chromosomes; intracellular transport of organelles; cytoskeleton

FIG. 3.5 Cilia and flagella

(a) Cilia on surface of respiratory epithelium

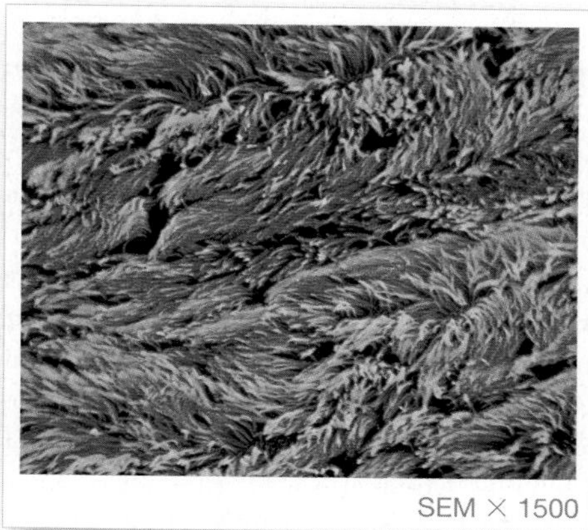
SEM × 1500

This image was taken with a scanning electron microscope (SEM) and then color enhanced. The specimens prepared for scanning electron microscopy are not sectioned. The whole specimen is coated with an electron-dense material, and then bombarded with electron beams. Because some of the electrons are reflected back, a three-dimensional image of the specimen is created.

(b) Cilia and flagella have 9 pairs of microtubules surrounding a central pair.

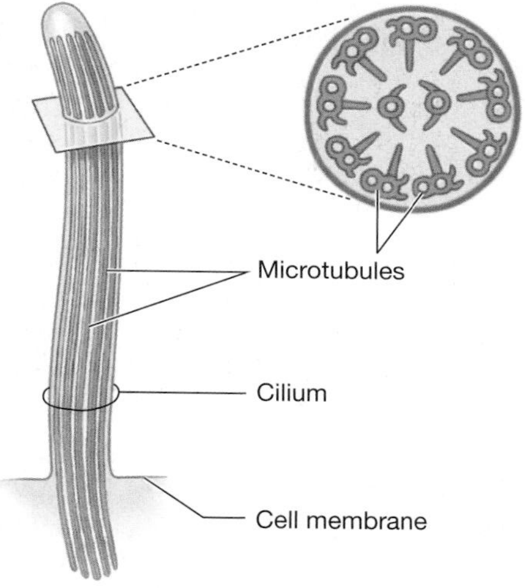
Microtubules
Cilium
Cell membrane

(c) The beating of cilia and flagella creates fluid movement.

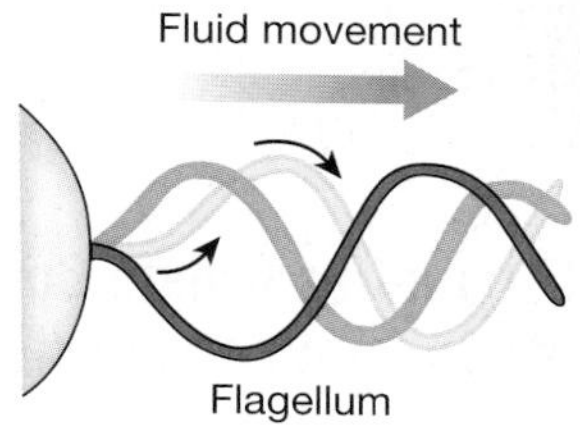
Fluid movement
Flagellum

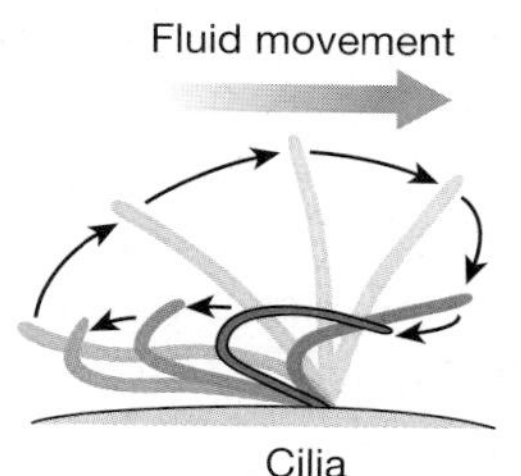
Fluid movement
Cilia

The cytoskeleton has at least five important functions.

1. **Cell shape.** The protein scaffolding of the cytoskeleton provides mechanical strength to the cell and in some cells plays an important role in determining the shape of the cell. Figure 3.4b shows how cytoskeletal fibers help support **microvilli** {*micro*-, small + *villus*, tuft of hair}, fingerlike extensions of the cell membrane that increase the surface area for absorption of materials.
2. **Internal organization.** Cytoskeletal fibers stabilize the positions of organelles. Figure 3.4b illustrates organelles held in place by the cytoskeleton. Note, however, that this figure is only a snapshot of one moment in the cell's life. The interior arrangement and composition of a cell are dynamic, changing from minute to minute in response to the needs of the cell, just as the inside of the walled city is always in motion. One disadvantage of the static illustrations in textbooks is that they are unable to represent movement and the dynamic nature of many physiological processes.
3. **Intracellular transport.** The cytoskeleton helps transport materials into the cell and within the cytoplasm by serving as an intracellular "railroad track" for moving organelles. This function is particularly important in cells of the nervous system, where material must be transported over intracellular distances as long as a meter.
4. **Assembly of cells into tissues.** Protein fibers of the cytoskeleton connect with protein fibers in the extracellular space, linking cells to one another and to supporting material outside the cells. In addition to providing mechanical strength to the tissue, these linkages allow the transfer of information from one cell to another.
5. **Movement.** The cytoskeleton helps cells move. For example, the cytoskeleton helps white blood cells squeeze out of blood vessels and helps growing nerve cells send out long extensions as they elongate. Cilia and flagella on the cell membrane are able to move because of their microtubule cytoskeleton. Special motor proteins facilitate movement and intracellular transport by using energy from ATP to slide or step along cytoskeletal fibers.

Motor Proteins Create Movement

Motor proteins are proteins that convert stored energy into directed movement. Three groups of motor proteins are associated with the cytoskeleton: myosins, kinesins, and dyneins. All three groups use energy stored in ATP to propel themselves along cytoskeleton fibers.

Myosins bind to actin fibers and are best known for their role in muscle contraction (Chapter 12). **Kinesins** and **dyneins** assist the movement of vesicles along microtubules. Dyneins also associate with the microtubule bundles of cilia and flagella to help create their whiplike motion.

Concept Check

5. Name the three sizes of cytoplasmic protein fibers.
6. How would the absence of a flagellum affect a sperm cell?
7. What is the difference between cytoplasm and cytosol?
8. What is the difference between a cilium and a flagellum?
9. What is the function of motor proteins?

Most motor proteins are made of multiple protein chains arranged into three parts: two heads that bind to the cytoskeleton fiber, a neck, and a tail region that is able to bind "cargo," such as an organelle that needs to be transported through the cytoplasm (**FIG. 3.6**). The heads alternately bind to the cytoskeleton fiber, then release and "step" forward using the energy stored in ATP.

Organelles Create Compartments for Specialized Functions

Organelles are subcellular compartments separated from the cytosol by one or more phospholipid membranes similar in structure to the cell membrane. The compartments created by organelles allow the cell to isolate substances and segregate functions. For example, an organelle might contain substances that could be harmful to the cell, such as digestive enzymes. Figures 3.4g, 3.4h, and 3.4i show the four major groups of organelles: mitochondria, the Golgi apparatus, the endoplasmic reticulum, and membrane-bound spheres called **vesicles** {*vesicula*, bladder}.

Mitochondria **Mitochondria** {singular, *mitochondrion*; *mitos*, thread + *chondros*, granule} are unique organelles in several ways. First, they have an unusual double wall that creates two separate compartments within the mitochondrion (Fig. 3.4g). In the center, inside the inner membrane, is a compartment called the **mitochondrial matrix** {*matrix*, female animal for breeding}. The matrix contains enzymes, ribosomes, granules, and surprisingly, its own unique DNA. This **mitochondrial DNA** has a different nucleotide sequence from that found in the nucleus. Because mitochondria have their own DNA, they can manufacture some of their own proteins.

FIG. 3.6 Motor proteins

Motor protein chains form a tail that binds organelles or other cargo, a neck, and two heads that "walk" along the cytoskeleton using energy from ATP.

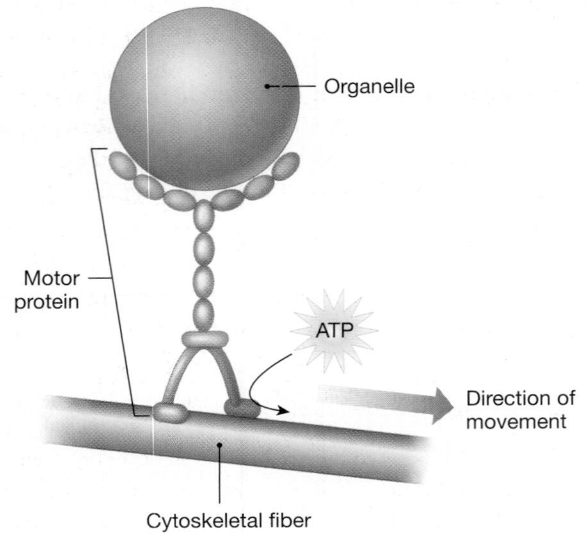
Organelle
Motor protein
ATP
Direction of movement
Cytoskeletal fiber

Why do mitochondria contain DNA when other organelles do not? This question has been the subject of intense scrutiny. According to the *prokaryotic endosymbiont theory*, mitochondria are the descendants of bacteria that invaded cells millions of years ago. The bacteria developed a mutually beneficial relationship with their hosts and soon became an integral part of the host cells. Supporting evidence for this theory is the fact that our mitochondrial DNA, RNA, and enzymes are similar to those in bacteria but unlike those in our own cell nuclei.

The second compartment inside a mitochondrion is the **intermembrane space,** which lies between the outer and inner mitochondrial membranes. This compartment plays an important role in mitochondrial ATP production, so the number of mitochondria in a cell is directly related to the cell's energy needs. For example, skeletal muscle cells, which use a lot of energy, have many more mitochondria than less active cells, such as adipose (fat) cells.

Another unusual characteristic of mitochondria is their ability to replicate themselves even when the cell to which they belong is not undergoing cell division. This process is aided by the mitochondrial DNA, which allows the organelles to direct their own duplication. Mitochondria replicate by budding, during which small daughter mitochondria pinch off from an enlarged parent. For instance, exercising muscle cells that experience increased energy demands over a period of time may meet the demand for more ATP by increasing the number of mitochondria in their cytoplasm.

The Endoplasmic Reticulum The **endoplasmic reticulum (ER)** is a network of interconnected membrane tubes with three major functions: synthesis, storage, and transport of biomolecules (Fig. 3.4i). The name *reticulum* comes from the Latin word for *net* and refers to the netlike arrangement of the tubules. Electron micrographs reveal that there are two forms of endoplasmic reticulum: **rough endoplasmic reticulum (RER)** and **smooth endoplasmic reticulum (SER)**.

The rough endoplasmic reticulum is the main site of protein synthesis. Proteins are assembled on ribosomes attached to the cytoplasmic surface of the rough ER, then inserted into the rough ER lumen, where they undergo chemical modification.

The smooth endoplasmic reticulum lacks attached ribosomes and is the main site for the synthesis of fatty acids, steroids, and lipids [p. 30]. Phospholipids for the cell membrane are produced here, and cholesterol is modified into steroid hormones, such as the sex hormones estrogen and testosterone. The smooth ER of liver and kidney cells detoxifies or inactivates drugs. In skeletal muscle cells, a modified form of smooth ER stores calcium ions (Ca^{2+}) to be used in muscle contraction.

The Golgi Apparatus The **Golgi apparatus** (also known as the Golgi complex) was first described by Camillo Golgi in 1898 (Fig. 3.4h). For years, some investigators thought that this organelle was just a result of the fixation process needed to prepare tissues for viewing under the light microscope. However, we now know from electron microscope studies that the Golgi apparatus is indeed a discrete organelle. It consists of a series of hollow curved sacs, called *cisternae,* stacked on top of one another like a series of hot water bottles and surrounded by vesicles. The Golgi apparatus receives proteins made on the rough ER, modifies them, and packages them into the vesicles.

Cytoplasmic Vesicles Membrane-bound cytoplasmic vesicles are of two kinds: secretory and storage. **Secretory vesicles** contain proteins that will be released from the cell. The contents of most **storage vesicles,** however, never leave the cytoplasm.

Lysosomes {*lysis,* dissolution + *soma,* body} are small storage vesicles that appear as membrane-bound granules in the cytoplasm (Fig. 3.4d). Lysosomes act as the digestive system of the cell. They use powerful enzymes to break down bacteria or old organelles, such as mitochondria, into their component molecules. Those molecules that can be reused are reabsorbed into the cytosol, while the rest are dumped out of the cell. As many as 50 types of enzymes have been identified from lysosomes of different cell types.

Because lysosomal enzymes are so powerful, early workers puzzled over the question of why these enzymes do not normally destroy the cell that contains them. What scientists discovered was that lysosomal enzymes are activated only by very acidic conditions, 100 times more acidic than the normal acidity level in the cytoplasm. When lysosomes first pinch off from the Golgi apparatus, their interior pH is about the same as that of the cytosol, 7.0–7.3. The enzymes are inactive at this pH. Their inactivity serves as a form of insurance. If the lysosome breaks or accidentally releases enzymes, they will not harm the cell.

However, as the lysosome sits in the cytoplasm, it accumulates H^+ in a process that uses energy. Increasing concentrations of H^+ decrease the pH inside the vesicle to 4.8–5.0, and the enzymes are activated. Once activated, lysosomal enzymes can break down biomolecules inside the vesicle. The lysosomal membrane is not affected by the enzymes.

The digestive enzymes of lysosomes are not always kept isolated within the organelle. Occasionally, lysosomes release their enzymes outside the cell to dissolve extracellular support material, such as the hard calcium carbonate portion of bone. In other instances, cells allow their lysosomal enzymes to come in contact with the cytoplasm, leading to self-digestion of all or part of the cell. When muscles *atrophy* (shrink) from lack of use or the uterus diminishes in size after pregnancy, the loss of cell mass is due to the action of lysosomes.

The inappropriate release of lysosomal enzymes has been implicated in certain disease states, such as the inflammation and destruction of joint tissue in *rheumatoid arthritis.* In the inherited conditions known as *lysosomal storage diseases,* lysosomes are ineffective because they lack specific enzymes. One of the best-known lysosomal storage diseases is the fatal inherited condition known as *Tay-Sachs disease.* Infants with Tay-Sachs disease have defective lysosomes that fail to break down glycolipids. Accumulation of glycolipids in nerve cells causes nervous system dysfunction, including blindness and loss of coordination. Most infants afflicted with Tay-Sachs disease die in early childhood. Learn more about Tay-Sachs disease in the Chapter 4 Running Problem.

Peroxisomes are storage vesicles that are even smaller than lysosomes (Fig. 3.4c). For years, they were thought to be a kind of lysosome, but we now know that they contain a different set of enzymes. Their main function appears to be to degrade long-chain fatty acids and potentially toxic foreign molecules.

Peroxisomes get their name from the fact that the reactions that take place inside them generate hydrogen peroxide (H_2O_2), a toxic molecule. The peroxisomes rapidly convert this peroxide to oxygen and water using the enzyme *catalase.* Peroxisomal disorders disrupt the normal processing of lipids and can severely disrupt neural function by altering the structure of nerve cell membranes.

Concept Check

10. What distinguishes organelles from inclusions?
11. What is the anatomical difference between rough endoplasmic reticulum and smooth endoplasmic reticulum? What is the functional difference?
12. How do lysosomes differ from peroxisomes?
13. Apply the physiological theme of compartmentation to organelles in general and to mitochondria in particular.
14. Microscopic examination of a cell reveals many mitochondria. What does this observation imply about the cell's energy requirements?
15. Examining tissue from a previously unknown species of fish, you discover a tissue containing large amounts of smooth endoplasmic reticulum in its cells. What is one possible function of these cells?

The Nucleus Is the Cell's Control Center

The nucleus of the cell contains DNA, the genetic material that ultimately controls all cell processes. Figure 3.4j illustrates the structure of a typical nucleus. Its boundary, or **nuclear envelope,** is a two-membrane structure that separates the nucleus from the cytoplasmic compartment. Both membranes of the envelope are pierced here and there by round holes, or **pores.**

Communication between the nucleus and cytosol occurs through the **nuclear pore complexes,** large protein complexes with a central channel. Ions and small molecules move freely through this channel when it is open, but transport of large molecules such as proteins and RNA is a process that requires energy. Specificity of the transport process allows the cell to restrict DNA to the nucleus and various enzymes to either the cytoplasm or the nucleus.

In electron micrographs of cells that are not dividing, the nucleus appears filled with randomly scattered granular material, or **chromatin,** composed of DNA and associated proteins. Usually a nucleus also contains from one to four larger dark-staining bodies of DNA, RNA, and protein called **nucleoli** {singular, *nucleolus,* little nucleus}. Nucleoli contain the genes and proteins that control the synthesis of RNA for ribosomes.

The process of protein synthesis, modification, and packaging in different parts of the cell is an excellent example of how compartmentation allows separation of function, as shown in **FIGURE 3.7**. RNA for protein synthesis is made from DNA

FIG. 3.7 Protein synthesis demonstrates subcellular compartmentation

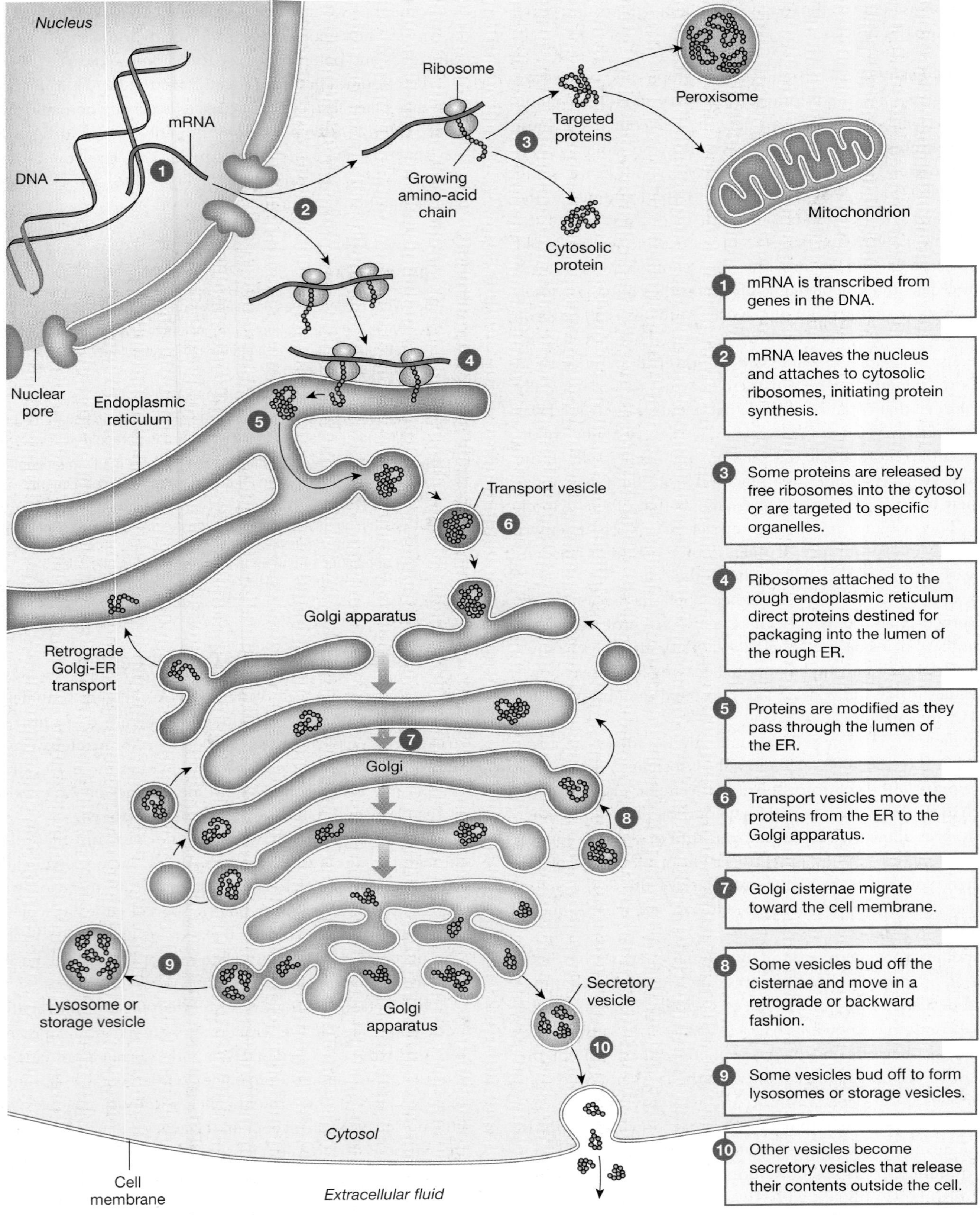
Nucleus
DNA
mRNA
Ribosome
Growing amino-acid chain
Targeted proteins
Peroxisome
Mitochondrion
Cytosolic protein
Nuclear pore
Endoplasmic reticulum
Transport vesicle
Golgi apparatus
Retrograde Golgi-ER transport
Golgi
Lysosome or storage vesicle
Golgi apparatus
Secretory vesicle
Cytosol
Cell membrane
Extracellular fluid
1 mRNA is transcribed from genes in the DNA.
2 mRNA leaves the nucleus and attaches to cytosolic ribosomes, initiating protein synthesis.
3 Some proteins are released by free ribosomes into the cytosol or are targeted to specific organelles.
4 Ribosomes attached to the rough endoplasmic reticulum direct proteins destined for packaging into the lumen of the rough ER.
5 Proteins are modified as they pass through the lumen of the ER.
6 Transport vesicles move the proteins from the ER to the Golgi apparatus.
7 Golgi cisternae migrate toward the cell membrane.
8 Some vesicles bud off the cisternae and move in a retrograde or backward fashion.
9 Some vesicles bud off to form lysosomes or storage vesicles.
10 Other vesicles become secretory vesicles that release their contents outside the cell.

templates in the nucleus 1, then transported to the cytoplasm through the nuclear pores 2. In the cytoplasm, proteins are synthesized on ribosomes that may be free inclusions 3 or attached to the rough endoplasmic reticulum 4. The newly made protein is compartmentalized in the lumen of the rough ER 5 where it is modified before being packaged into a vesicle 6. The vesicles fuse with the Golgi apparatus, allowing additional modification of the protein in the Golgi lumen 7. The modified proteins leave the Golgi packaged in either storage vesicles 8 or secretory vesicles whose contents will be released into the extracellular fluid 10. The molecular details of protein synthesis are discussed elsewhere (see Chapter 4).

3.4 Tissues of the Body

Despite the amazing variety of intracellular structures, no single cell can carry out all the processes of the mature human body. Instead, cells assemble into the larger units we call tissues. The cells in tissues are held together by specialized connections called *cell junctions* and by other support structures. Tissues range in complexity from simple tissues containing only one cell type, such as the lining of blood vessels, to complex tissues containing many cell types and extensive extracellular material, such as connective tissue. The cells of most tissues work together to achieve a common purpose.

The study of tissue structure and function is known as **histology** {*histos*, tissue}. Histologists describe tissues by their physical features: (1) the shape and size of the cells, (2) the arrangement of the cells in the tissue (in layers, scattered, and so on), (3) the way cells are connected to one another, and (4) the amount of extracellular material present in the tissue. There are four primary tissue types in the human body: epithelial, connective, muscle, and *neural*, or nerve. Before we consider each tissue type specifically, let's examine how cells link together to form tissues.

Extracellular Matrix Has Many Functions

Extracellular matrix (usually just called *matrix*) is extracellular material that is synthesized and secreted by the cells of a tissue. For years, scientists believed that matrix was an inert substance whose only function was to hold cells together. However, experimental evidence now shows that the extracellular matrix plays a vital role in many physiological processes, ranging from growth and development to cell death. A number of disease states are associated with overproduction or disruption of extracellular matrix, including chronic heart failure and the spread of cancerous cells throughout the body (*metastasis*).

The composition of extracellular matrix varies from tissue to tissue, and the mechanical properties, such as elasticity and flexibility, of a tissue depend on the amount and consistency of the tissue's matrix. Matrix always has two basic components: proteoglycans and insoluble protein fibers. **Proteoglycans** are glycoproteins, which are proteins covalently bound to polysaccharide chains [p. 29]. Insoluble protein fibers such as *collagen*, *fibronectin*, and *laminin* provide strength and anchor cells to the matrix. Attachments between the extracellular matrix and proteins in the cell membrane or the cytoskeleton are ways cells communicate with their external environment.

The amount of extracellular matrix in a tissue is highly variable. Nerve and muscle tissue have very little matrix, but the connective tissues, such as cartilage, bone, and blood, have extensive matrix that occupies as much volume as their cells. The consistency of extracellular matrix can vary from watery (blood and lymph) to rigid (bone).

Cell Junctions Hold Cells Together to Form Tissues

During growth and development, cells form *cell-cell adhesions* that may be transient or that may develop into more permanent **cell junctions. Cell adhesion molecules (CAMs)** are membrane-spanning proteins responsible both for cell junctions and for transient cell adhesions (**TBL. 3.3**). Cell-cell or cell-matrix adhesions mediated by CAMs are essential for normal growth and development. For example, growing nerve cells creep across the extracellular matrix with the help of *nerve-cell adhesion molecules* (NCAMs). Cell adhesion helps white blood cells escape from the circulation and move into infected tissues, and it allows clumps of platelets to cling to damaged blood vessels. Because cell adhesions are not permanent, the bond between those CAMs and matrix is weak.

Stronger cell junctions can be grouped into three broad categories by function: communicating junctions, occluding junctions {*occludere*, to close up}, and anchoring junctions (**FIG. 3.8**). In animals, the communicating junctions are gap junctions. The occluding junctions of vertebrates are tight junctions that limit movement of materials between cells. The three major types of junctions are described next.

1. **Gap junctions** are the simplest cell-cell junctions (Fig. 3.8b). They allow direct and rapid cell-to-cell communication through cytoplasmic bridges between adjoining cells. Cylindrical proteins called *connexins* interlock to create passageways that look like hollow rivets with narrow channels through their centers. The channels are able to open and close, regulating the movement of small molecules and ions through them.

   Gap junctions allow both chemical and electrical signals to pass rapidly from one cell to the next. They were once thought to occur only in certain muscle and nerve cells, but we now

TABLE 3.3 Major Cell Adhesion Molecules (CAMs)

Name	Examples
Cadherins	Cell-cell junctions such as adherens junctions and desmosomes. Calcium-dependent.
Integrins	Primarily found in cell-matrix junctions. These also function in cell signaling.
Immunoglobulin superfamily CAMs	NCAMs (nerve-cell adhesion molecules). Responsible for nerve cell growth during nervous system development.
Selectins	Temporary cell-cell adhesions.

CHAPTER 3

FIG. 3.8 ESSENTIALS Cell Junctions

Cell junctions connect one cell with another cell (or to surrounding matrix) with membrane-spanning proteins called **cell adhesion molecules,** or **CAMs.**

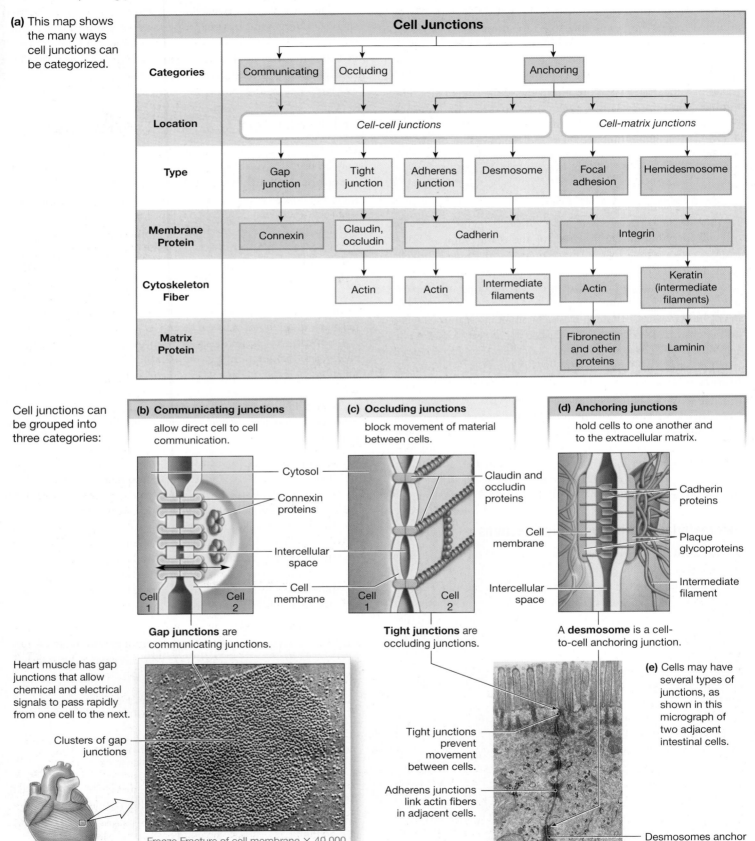

(a) This map shows the many ways cell junctions can be categorized.

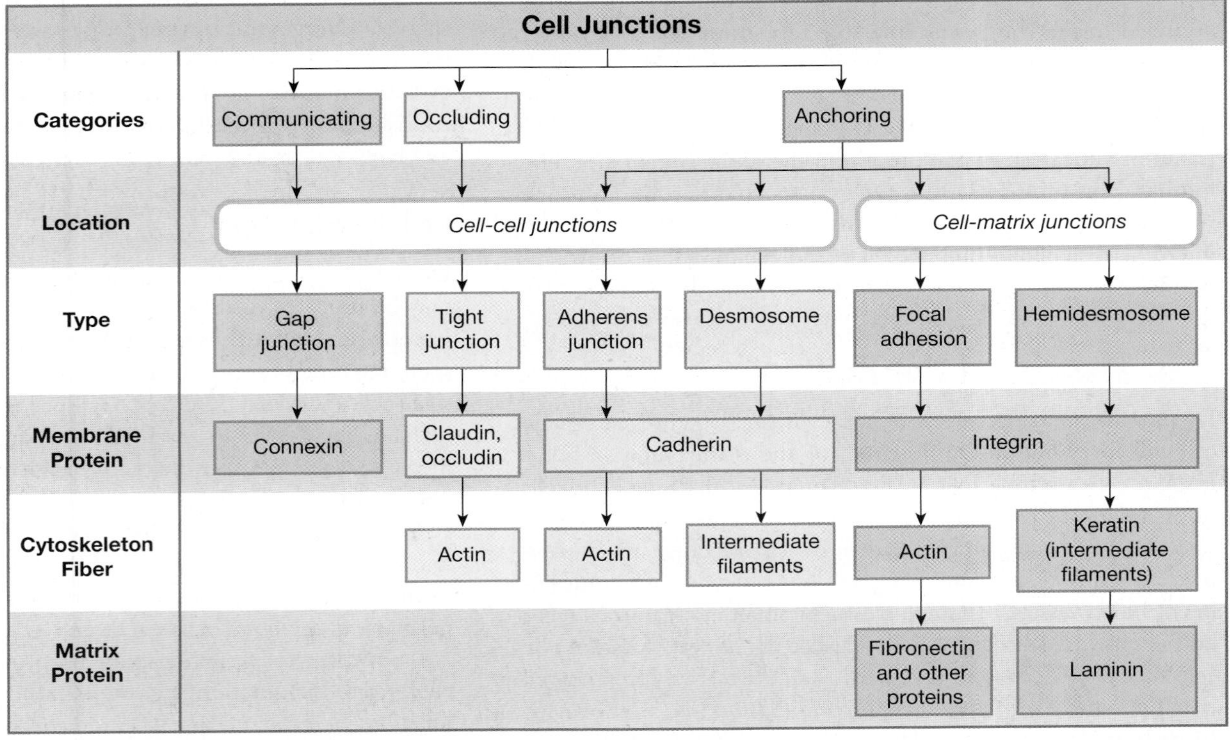

Cell Junctions
Categories
Communicating
Occluding
Anchoring
Location
Cell-cell junctions
Cell-matrix junctions
Type
Gap junction
Tight junction
Adherens junction
Desmosome
Focal adhesion
Hemidesmosome
Membrane Protein
Connexin
Claudin, occludin
Cadherin
Integrin
Cytoskeleton Fiber
Actin
Actin
Intermediate filaments
Actin
Keratin (intermediate filaments)
Matrix Protein
Fibronectin and other proteins
Laminin

Cell junctions can be grouped into three categories:

(b) Communicating junctions allow direct cell to cell communication.

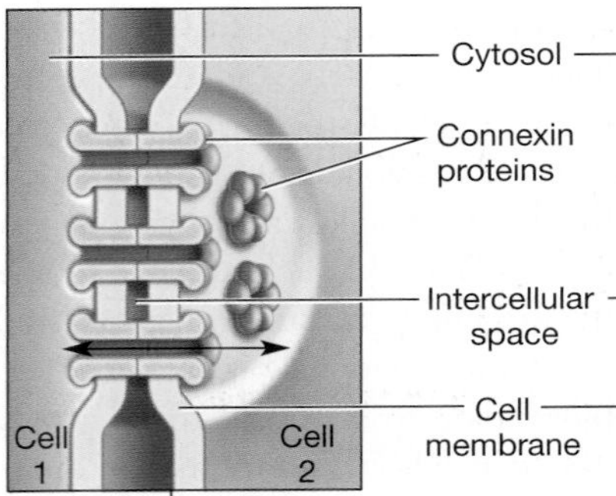

Cytosol

Connexin proteins

Intercellular space

Cell membrane

Cell 1

Cell 2

Gap junctions are communicating junctions.

Heart muscle has gap junctions that allow chemical and electrical signals to pass rapidly from one cell to the next.

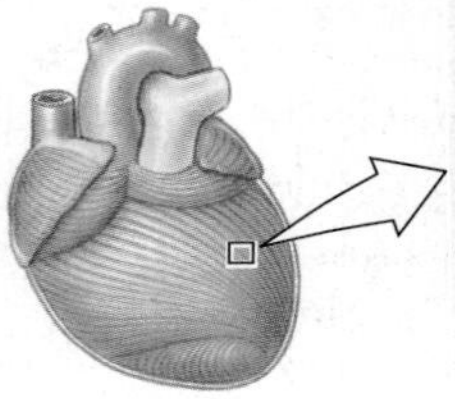

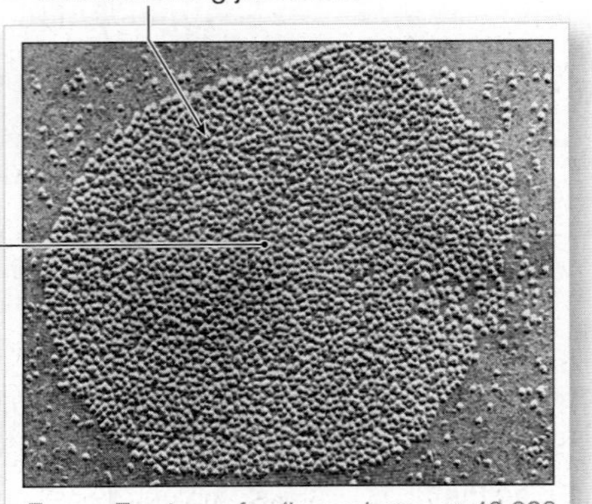

Clusters of gap junctions

Freeze Fracture of cell membrane × 40,000

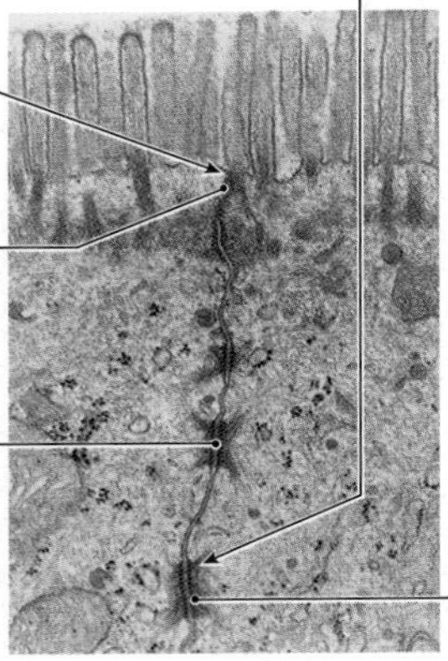

(c) Occluding junctions block movement of material between cells.

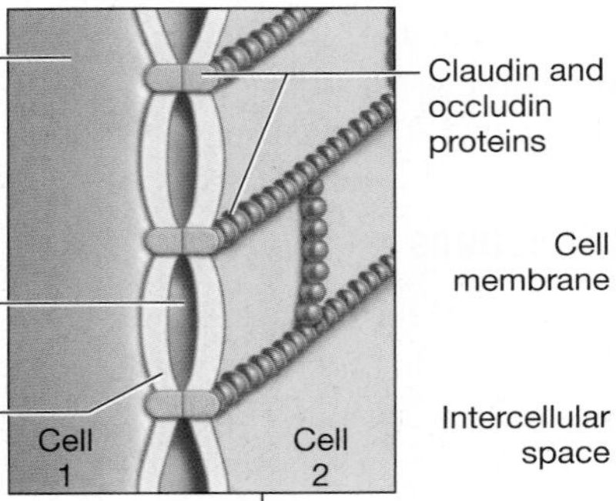

Claudin and occludin proteins

Cell membrane

Intercellular space

Cell 1

Cell 2

Tight junctions are occluding junctions.

(d) Anchoring junctions hold cells to one another and to the extracellular matrix.

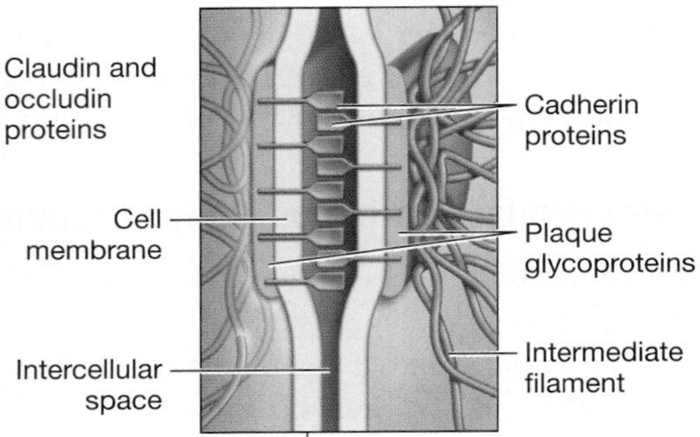

Cadherin proteins

Plaque glycoproteins

Intermediate filament

A **desmosome** is a cell-to-cell anchoring junction.

(e) Cells may have several types of junctions, as shown in this micrograph of two adjacent intestinal cells.

Tight junctions prevent movement between cells.

Adherens junctions link actin fibers in adjacent cells.

Desmosomes anchor cells to each other.

know they are important in cell-to-cell communication in many tissues, including the liver, pancreas, ovary, and thyroid gland.

2. **Tight junctions** are occluding junctions that restrict the movement of material between the cells they link (Fig. 3.8c). In tight junctions, the cell membranes of adjacent cells partly fuse together with the help of proteins called *claudins* and *occludins,* thereby making a barrier. As in many physiological processes, the barrier properties of tight junctions are dynamic and can be altered depending on the body's needs. Tight junctions may have varying degrees of "leakiness."

   Tight junctions in the intestinal tract and kidney prevent most substances from moving freely between the external and internal environments. In this way, they enable cells to regulate what enters and leaves the body. Tight junctions also create the so-called *blood-brain barrier* that prevents many potentially harmful substances in the blood from reaching the extracellular fluid of the brain.

3. **Anchoring junctions** (Fig. 3.8d) attach cells to each other (cell-cell anchoring junctions) or to the extracellular matrix (cell-matrix anchoring junctions). In vertebrates, cell-cell anchoring junctions are created by CAMs called **cadherins,** which connect with one another across the intercellular space. Cell-matrix junctions use CAMs called **integrins.** Integrins are membrane proteins that can also bind to signal molecules in the cell's environment, transferring information carried by the signal across the cell membrane into the cytoplasm.

Anchoring junctions contribute to the mechanical strength of the tissue. They have been compared to buttons or zippers that tie cells together and hold them in position within a tissue. Notice how the interlocking cadherin proteins in Figure 3.8d resemble the teeth of a zipper.

The protein linkage of anchoring cell junctions is very strong, allowing sheets of tissue in skin and lining body cavities to resist damage from stretching and twisting. Even the tough protein fibers of anchoring junctions can be broken, however. If you have shoes that rub against your skin, the stress can shear the proteins connecting the different skin layers. When fluid accumulates in the resulting space and the layers separate, a *blister* results.

Tissues held together with anchoring junctions are like a picket fence, where spaces between the pickets (the cells) allow materials to pass from one side of the fence to the other. Movement of materials between cells is known as the **paracellular** pathway. In contrast, tissues held together with tight junctions are more like a solid brick wall: Very little can pass from one side of the wall to the other between the bricks.

Cell-cell anchoring junctions take the form of either adherens junctions or desmosomes. **Adherens junctions** link actin fibers in adjacent cells together, as shown in the micrograph in Figure 3.8e. **Desmosomes** {*desmos*, band + *soma*, body} attach to intermediate filaments of the cytoskeleton. Desmosomes are the strongest cell-cell junctions. In electron micrographs they can be recognized by the dense glycoprotein bodies, or *plaques,* that lie just inside the cell membranes in the region where the two cells connect (Fig. 3.8d, e). Desmosomes may be small points of contact between two cells (spot desmosomes) or bands that encircle the entire cell (belt desmosomes).

There are also two types of cell-matrix anchoring junctions. **Hemidesmosomes** {*hemi-*, half} are strong junctions that anchor intermediate fibers of the cytoskeleton to fibrous matrix proteins such as laminin. **Focal adhesions** tie intracellular actin fibers to different matrix proteins, such as fibronectin.

The loss of normal cell junctions plays a role in a number of diseases and in metastasis. Diseases in which cell junctions are destroyed or fail to form can have disfiguring and painful symptoms, such as blistering skin. One such disease is *pemphigus,* a condition in which the body attacks some of its own cell junction proteins (*www.pemphigus.org*).

The disappearance of anchoring junctions probably contributes to the metastasis of cancer cells throughout the body. Cancer cells lose their anchoring junctions because they have fewer cadherin molecules and are not bound as tightly to neighboring cells. Once a cancer cell is released from its moorings, it secretes protein-digesting enzymes known as *proteases.* These enzymes, especially those called *matrix metalloproteinases,* (MMPs), dissolve the extracellular matrix so that escaping cancer cells can invade adjacent tissues or enter the bloodstream. Researchers are investigating ways of blocking MMP enzymes to see if they can prevent metastasis.

Now that you understand how cells are held together into tissues, we will look at the four different tissue types in the body: (1) epithelial, (2) connective, (3) muscle, and (4) neural.

Concept Check

16. Name the three functional categories of cell junctions.
17. Which type of cell junction:
    (a) restricts movement of materials between cells?
    (b) allows direct movement of substances from the cytoplasm of one cell to the cytoplasm of an adjacent cell?
    (c) provides the strongest cell-cell junction?
    (d) anchors actin fibers in the cell to the extracellular matrix?

Epithelia Provide Protection and Regulate Exchange

The **epithelial tissues,** or **epithelia** {*epi-*, upon + *thele-*, nipple; singular *epithelium*}, protect the internal environment of the body and regulate the exchange of materials between the internal and external environments (**FIG. 3.9**). These tissues cover exposed surfaces, such as the skin, and line internal passageways, such as the digestive tract. *Any substance that enters or leaves the internal environment of the body must cross an epithelium.*

Some epithelia, such as those of the skin and mucous membranes of the mouth, act as a barrier to keep water in the body and invaders such as bacteria out. Other epithelia, such as those in the kidney and intestinal tract, control the movement of materials between the external environment and the extracellular fluid of the body. Nutrients, gases, and wastes often must cross several different epithelia in their passage between cells and the outside world.

FIG. 3.9 ESSENTIALS Epithelial Tissue

(a) Five Functional Categories of Epithelia

	Exchange	Transporting	Ciliated	Protective	Secretory
Number of Cell Layers	One	One	One	Many	One to many
Cell Shape	Flattened	Columnar or cuboidal	Columnar or cuboidal	Flattened in surface layers; polygonal in deeper layers	Columnar or polygonal
Special Features	Pores between cells permit easy passage of molecules	Tight junctions prevent movement between cells; surface area increased by folding of cell membrane into fingerlike microvilli	One side covered with cilia to move fluid across surface	Cells tightly connected by many desmosomes	Protein-secreting cells filled with membrane-bound secretory granules and extensive RER; steroid-secreting cells contain lipid droplets and extensive SER
Where Found	Lungs, lining of blood vessels	Intestine, kidney, some exocrine glands	Nose, trachea, and upper airways; female reproductive tract	Skin and lining of cavities (such as the mouth) that open to the environment	Exocrine glands, including pancreas, sweat glands, and salivary glands; endocrine glands, such as thyroid and gonads
Key	exchange epithelium	transporting epithelium	ciliated epithelium	protective epithelium	 secretory epithelium

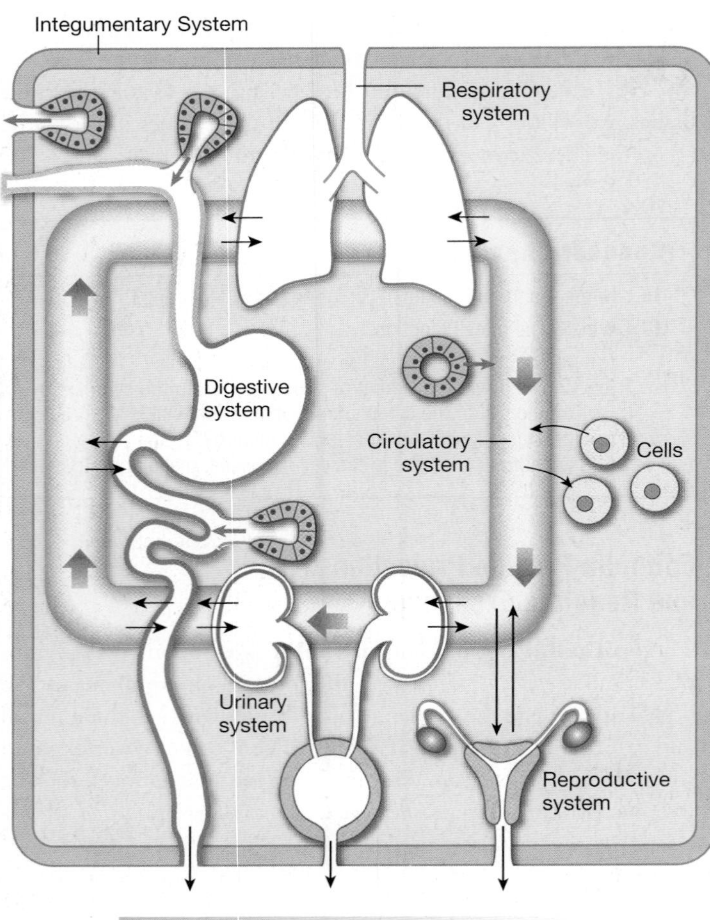

Integumentary System
Respiratory system
Digestive system
Circulatory system
Cells
Urinary system
Reproductive system
KEY
Secretion
Exchange

(b) This diagram shows the distribution of the five kinds of epithelia in the body outlined in the table above.

FIGURE QUESTIONS

1. Where do secretions from endocrine glands go?
2. Where do secretions from exocrine glands go?

(c) Most epithelia attach to an underlying matrix layer called the **basal lamina** or **basement membrane.**

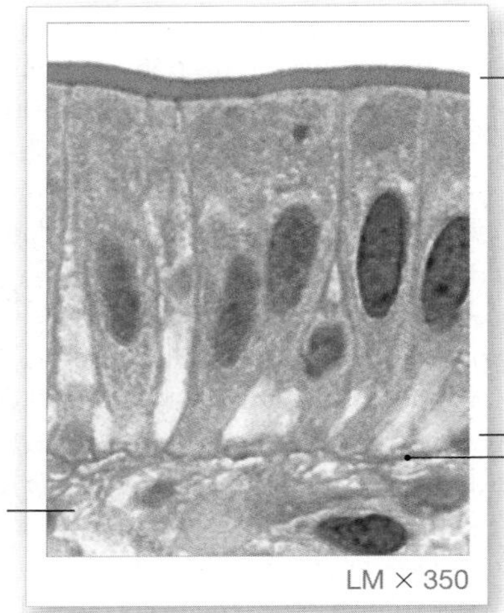

LM × 350
Epithelial cells attach to the basal lamina using cell adhesion molecules.
Basal lamina (basement membrane) is an acellular matrix layer that is secreted by the epithelial cells.
Underlying tissue

A light micrograph (LM) is the photographic image produced when a light microscope directs visible light through a thin section of tissue. The specimen is magnified with both an ocular lens and a revolving nosepiece that holds several objective lenses of progressive magnifying power. Total magnification equals the magnification of the ocular lens times that of the objective lens.

Another type of epithelium is specialized to manufacture and secrete chemicals into the blood or into the external environment. Sweat and saliva are examples of substances secreted by epithelia into the environment. Hormones are secreted into the blood.

Structure of Epithelia Epithelia typically consist of one or more layers of cells connected to one another, with a thin layer of extracellular matrix lying between the epithelial cells and their underlying tissues (Fig. 3.9c). This matrix layer, called the **basal lamina** {*bassus*, low; *lamina*, a thin plate}, or **basement membrane,** is composed of a network of collagen and laminin filaments embedded in proteoglycans. The protein filaments hold the epithelial cells to the underlying cell layers, just as cell junctions hold the individual cells in the epithelium to one another.

The cell junctions in epithelia are variable. Physiologists classify epithelia either as "leaky" or "tight," depending on how easily substances pass from one side of the epithelial layer to the other. In a leaky epithelium, anchoring junctions allow molecules to cross the epithelium by passing through the gap between two adjacent epithelial cells. A typical leaky epithelium is the wall of capillaries (the smallest blood vessels), where all dissolved molecules except for large proteins can pass from the blood to the interstitial fluid by traveling through gaps between adjacent epithelial cells.

In a tight epithelium, such as that in the kidney, adjacent cells are bound to each other by tight junctions that create a barrier, preventing substances from traveling between adjacent cells. To cross a tight epithelium, most substances must enter the epithelial cells and go *through* them. The tightness of an epithelium is directly related to how selective it is about what can move across it. Some epithelia, such as those of the intestine, have the ability to alter the tightness of their junctions according to the body's needs.

Types of Epithelia Structurally, epithelial tissues can be divided into two general types: (1) sheets of tissue that lie on the surface of the body or that line the inside of tubes and hollow organs and (2) secretory epithelia that synthesize and release substances into the extracellular space. Histologists classify sheet epithelia by the number of cell layers in the tissue and by the shape of the cells in the surface layer. This classification scheme recognizes two types of layering—**simple** (one cell thick) and **stratified** (multiple cell layers) {*stratum*, layer + *facere*, to make}—and three cell shapes—**squamous** {*squama*, flattened plate or scale}, **cuboidal,** and **columnar.** However, physiologists are more concerned with the functions of these tissues, so instead of using the histological descriptions, we will divide epithelia into five groups according to their function.

There are five functional types of epithelia: exchange, transporting, ciliated, protective, and secretory (**FIG. 3.10**). *Exchange epithelia* permit rapid exchange of materials, especially gases. *Transporting epithelia* are selective about what can cross them and are found primarily in the intestinal tract and the kidney. *Ciliated epithelia* are located primarily in the airways of the respiratory system and in the female reproductive tract. *Protective epithelia* are found on the surface of the body and just inside the openings of body cavities. *Secretory epithelia* synthesize and release secretory products into the external environment or into the blood.

Figure 3.9b shows the distribution of these epithelia in the systems of the body. Notice that most epithelia face the external environment on one surface and the extracellular fluid on the other. One exception is the endocrine glands and a second is the epithelium lining the circulatory system.

Exchange Epithelia The **exchange epithelia** are composed of very thin, flattened cells that allow gases (CO_2 and O_2) to pass rapidly across the epithelium. This type of epithelium lines the blood vessels and the lungs, the two major sites of gas exchange in the body. In capillaries, gaps or pores in the epithelium also allow molecules smaller than proteins to pass *between* two adjacent epithelial cells, making this a leaky epithelium (Fig. 3.10a). Histologists classify thin exchange tissue as *simple squamous epithelium* because it is a single layer of thin, flattened cells. The simple squamous epithelium lining the heart and blood vessels is also called the **endothelium.**

Transporting Epithelia The **transporting epithelia** actively and selectively regulate the exchange of nongaseous materials, such as ions and nutrients, between the internal and external environments. These epithelia line the hollow tubes of the digestive system and the kidney, where lumens open into the external environment (p. 4). Movement of material from the external environment across the epithelium to the internal environment is called *absorption*. Movement in the opposite direction, from the internal to the external environment, is called *secretion*.

Transporting epithelia can be identified by the following characteristics (Fig. 3.10b):

1. **Cell shape.** Cells of transporting epithelia are much thicker than cells of exchange epithelia, and they act as a barrier as well as an entry point. The cell layer is only one cell thick (a simple epithelium), but cells are cuboidal or columnar.
2. **Membrane modifications.** The **apical membrane,** the surface of the epithelial cell that faces the lumen, has tiny finger-like projections called *microvilli* that increase the surface area available for transport. A cell with microvilli has at least 20 times the surface area of a cell without them. In addition, the **basolateral membrane,** the side of the epithelial cell facing the extracellular fluid, may also have folds that increase the cell's surface area.
3. **Cell junctions.** The cells of transporting epithelia are firmly attached to adjacent cells by moderately tight to very tight junctions. This means that to cross the epithelium, material must move into an epithelial cell on one side of the tissue and out of the cell on the other side.
4. **Cell organelles.** Most cells that transport materials have numerous mitochondria to provide energy for transport processes (discussed further in Chapter 5). The properties of transporting epithelia differ depending on where in the body the epithelia are located. For example, glucose can cross the epithelium of the small intestine and enter the extracellular fluid but cannot cross the epithelium of the large intestine.

The transport properties of an epithelium can be regulated and modified in response to various stimuli. Hormones, for example, affect the transport of ions by kidney epithelium. You

FIG. 3.10 ESSENTIALS Types of Epithelia

(a) Exchange Epithelium

The thin, flat cells of exchange epithelium allow movement through and between the cells.

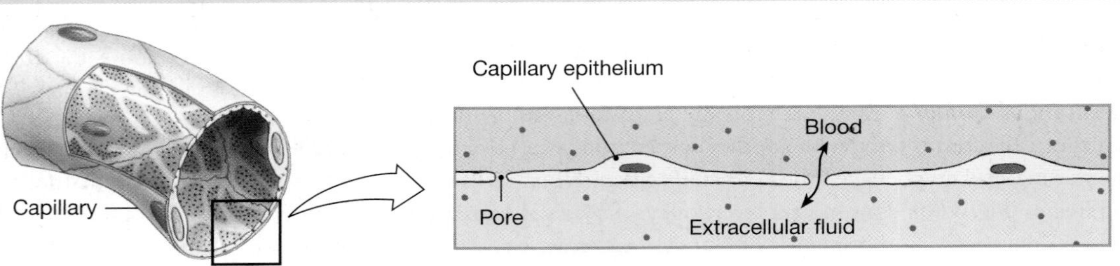

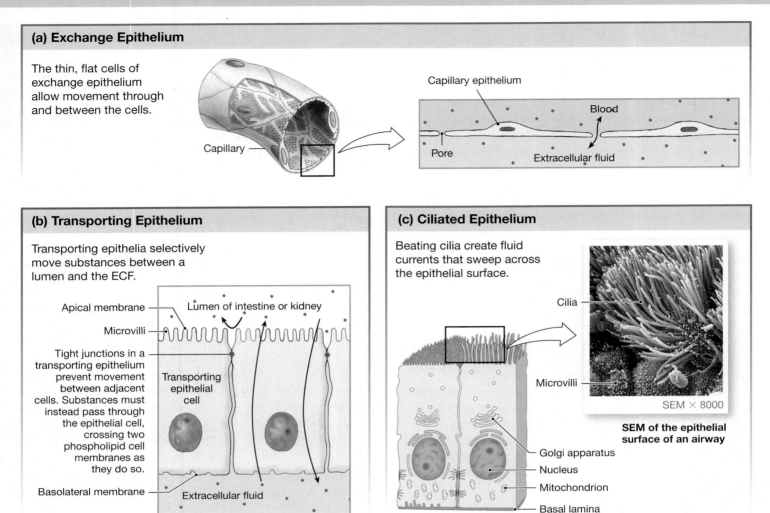

Capillary
Capillary epithelium
Blood
Pore
Extracellular fluid

(b) Transporting Epithelium

Transporting epithelia selectively move substances between a lumen and the ECF.

Apical membrane

Microvilli

Lumen of intestine or kidney

Tight junctions in a transporting epithelium prevent movement between adjacent cells. Substances must instead pass through the epithelial cell, crossing two phospholipid cell membranes as they do so.

Transporting epithelial cell

Basolateral membrane

Extracellular fluid

(c) Ciliated Epithelium

Beating cilia create fluid currents that sweep across the epithelial surface.

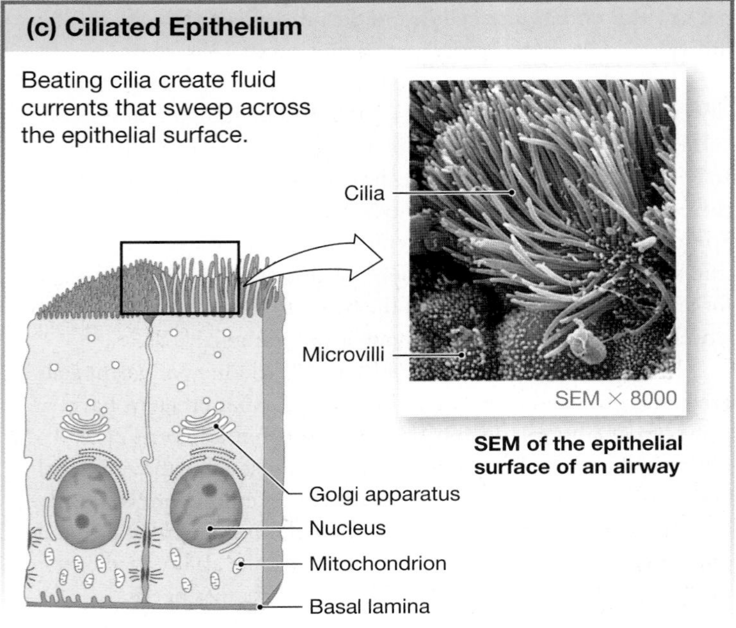

Cilia
Microvilli
SEM × 8000
Golgi apparatus
Nucleus
Mitochondrion
Basal lamina

SEM of the epithelial surface of an airway

(d) Protective Epithelium

Protective epithelia have many stacked layers of cells that are constantly being replaced. This figure shows layers in skin (see Fig. 3.15, *Focus On: The Skin*).

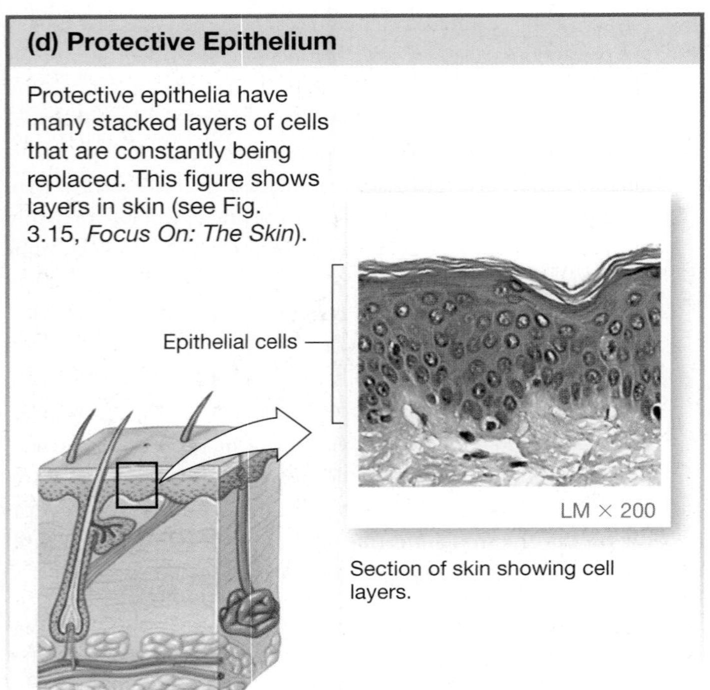

Epithelial cells
LM × 200

Section of skin showing cell layers.

(e) Secretory Epithelium

Secretory epithelial cells make and release a product. Exocrine secretions, such as the mucus shown here, are secreted outside the body. The secretions of endocrine cells (hormones) are released into the blood.

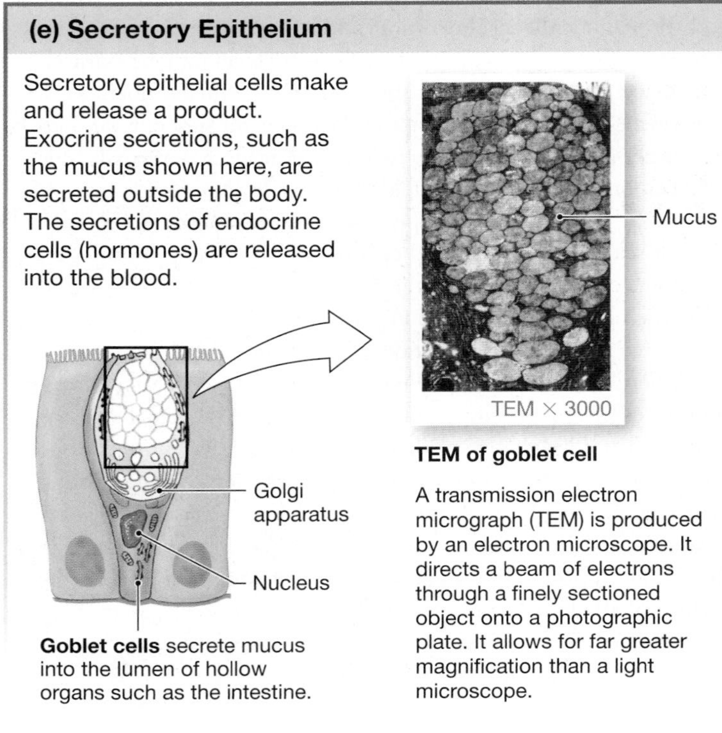

Mucus
TEM × 3000
Golgi apparatus
Nucleus

Goblet cells secrete mucus into the lumen of hollow organs such as the intestine.

TEM of goblet cell

A transmission electron micrograph (TEM) is produced by an electron microscope. It directs a beam of electrons through a finely sectioned object onto a photographic plate. It allows for far greater magnification than a light microscope.

will learn more about transporting epithelia when you study the kidney and digestive systems.

Ciliated Epithelia **Ciliated epithelia** are nontransporting tissues that line the respiratory system and parts of the reproductive tract. The surface of the tissue facing the lumen is covered with cilia that beat in a coordinated, rhythmic fashion, moving fluid and particles across the surface of the tissue (Fig. 3.10c). Injury to the cilia or to their epithelial cells can stop ciliary movement. For example, smoking paralyzes the ciliated epithelium lining the respiratory tract. Loss of ciliary function contributes to the higher incidence of respiratory infection in smokers, when the mucus that traps bacteria can no longer be swept out of the lungs by the cilia.

Protective Epithelia The **protective epithelia** prevent exchange between the internal and external environments and protect areas subject to mechanical or chemical stresses. These epithelia are stratified tissues, composed of many stacked layers of cells (Fig. 3.10d). Protective epithelia may be toughened by the secretion of *keratin* {*keras*, horn}, the same insoluble protein abundant in hair and nails. The *epidermis* {*epi*, upon + *derma*, skin} and linings of the mouth, pharynx, esophagus, urethra, and vagina are all protective epithelia.

Because protective epithelia are subjected to irritating chemicals, bacteria, and other destructive forces, the cells in them have a short life span. In deeper layers, new cells are produced continuously, displacing older cells at the surface. Each time you wash your face, you scrub off dead cells on the surface layer. As skin ages, the rate of cell turnover declines. *Retinoids*, a group of chemicals derived from vitamin A, speed up cell division and surface shedding so treated skin develops a more youthful appearance.

RUNNING PROBLEM

Many kinds of cancer develop in epithelial cells that are subject to damage or trauma. The uterine cervix consists of two types of epithelia. Columnar secretory epithelium with mucus-secreting glands lines the inside of the cervical canal. A protective stratified squamous epithelium covers the outside of the cervix. At the opening of the cervix, these two types of epithelia come together. In many cases, infections caused by the *human papillomavirus* (HPV) cause the cervical cells to develop dysplasia. Dr. Baird ran an HPV test on Jan's first Pap smear, and it was positive for the virus. Today she is repeating the tests to see if Jan's dysplasia and HPV infection have persisted.

Q3: *What other kinds of damage or trauma are cervical epithelial cells normally subjected to?*

Q4: *Which of the two types of cervical epithelia is more likely to be affected by physical trauma?*

Q5: *The results of Jan's first Pap test showed atypical squamous cells of unknown significance (ASC-US). Were these cells more likely to come from the secretory portion of the cervix or from the protective epithelium?*

59 - 61 - 65 - 79 - 84 - 87

Secretory Epithelia **Secretory epithelia** are composed of cells that produce a substance and then secrete it into the extracellular space. Secretory cells may be scattered among other epithelial cells, or they may group together to form a multicellular **gland.** There are two types of secretory glands: exocrine and endocrine.

Exocrine glands release their secretions to the body's external environment {*exo-*, outside + *krinein*, to secrete}. This may be onto the surface of the skin or onto an epithelium lining one of the internal passageways, such as the airways of the lung or the lumen of the intestine (Fig. 3.10e). In effect, an exocrine secretion leaves the body. This explains how some exocrine secretions, like stomach acid, can have a pH that is incompatible with life [Fig. 2.9, p. 45].

Most exocrine glands release their products through open tubes known as **ducts.** Sweat glands, mammary glands in the breast, salivary glands, the liver, and the pancreas are all exocrine glands.

Exocrine gland cells produce two types of secretions. **Serous secretions** are watery solutions, and many of them contain enzymes. Tears, sweat, and digestive enzyme solutions are all serous exocrine secretions. **Mucous secretions** (also called **mucus**) are sticky solutions containing glycoproteins and proteoglycans. Some exocrine glands contain more than one type of secretory cell, and they produce both serous and mucous secretions. For example, the salivary glands release mixed secretions.

Goblet cells, shown in Figure 3.10e, are single exocrine cells that produce mucus. Mucus acts as a lubricant for food to be swallowed, as a trap for foreign particles and microorganisms inhaled or ingested, and as a protective barrier between the epithelium and the environment.

Unlike exocrine glands, **endocrine glands** are ductless and release their secretions, called **hormones,** into the body's extracellular compartment (Fig. 3.9b). Hormones enter the blood for distribution to other parts of the body, where they regulate or coordinate the activities of various tissues, organs, and organ systems. Some of the best-known endocrine glands are the pancreas, the thyroid gland, the gonads, and the pituitary gland. For years, it was thought that all hormones were produced by cells grouped together into endocrine glands. We now know that isolated endocrine cells occur scattered in the epithelial lining of the digestive tract, in the tubules of the kidney, and in the walls of the heart.

FIGURE 3.11 shows the epithelial origin of endocrine and exocrine glands. During embryonic development, epithelial cells grow downward into the supporting connective tissue. Exocrine glands remain connected to the parent epithelium by a duct that transports the secretion to its destination (the external environment). Endocrine glands lose the connecting cells and secrete their hormones into the bloodstream.

FIG. 3.11 Development of endocrine and exocrine glands

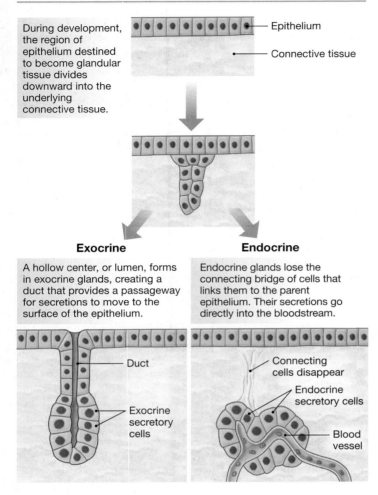

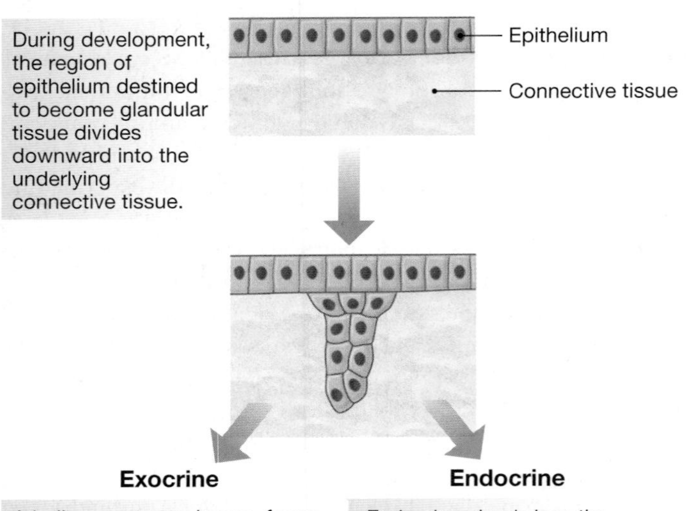
During development, the region of epithelium destined to become glandular tissue divides downward into the underlying connective tissue.
Epithelium
Connective tissue
Exocrine
Endocrine
A hollow center, or lumen, forms in exocrine glands, creating a duct that provides a passageway for secretions to move to the surface of the epithelium.
Endocrine glands lose the connecting bridge of cells that links them to the parent epithelium. Their secretions go directly into the bloodstream.

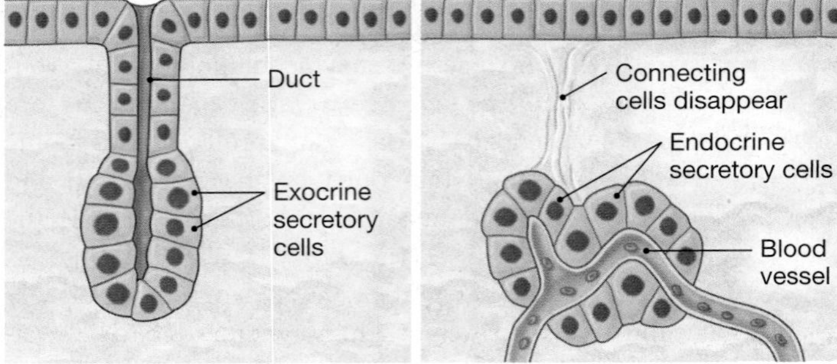
Duct
Exocrine secretory cells
Connecting cells disappear
Endocrine secretory cells
Blood vessel

Concept Check

18. List the five functional types of epithelia.
19. Define secretion.
20. Name two properties that distinguish endocrine glands from exocrine glands.
21. The basal lamina of epithelium contains the protein fiber laminin. Are the overlying cells attached by focal adhesions or hemidesmosomes?
22. You look at a tissue under a microscope and see a simple squamous epithelium. Can it be a sample of the skin surface? Explain.
23. A cell of the intestinal epithelium secretes a substance into the extracellular fluid, where it is picked up by the blood and carried to the pancreas. Is the intestinal epithelium cell an endocrine or an exocrine cell?

Connective Tissues Provide Support and Barriers

Connective tissues, the second major tissue type, provide structural support and sometimes a physical barrier that, along with specialized cells, helps defend the body from foreign invaders such as bacteria. The distinguishing characteristic of connective tissues is the presence of extensive extracellular matrix containing widely scattered cells that secrete and modify the matrix (**FIG. 3.12**). Connective tissues include blood, the support tissues for the skin and internal organs, and cartilage and bone.

Structure of Connective Tissue The extracellular matrix of connective tissue is a **ground substance** of proteoglycans and water in which insoluble protein fibers are arranged, much like pieces of fruit suspended in a gelatin salad. The consistency of ground substance is highly variable, depending on the type of connective tissue (Fig. 3.12a). At one extreme is the watery matrix of blood, and at the other extreme is the hardened matrix of bone. In between are solutions of proteoglycans that vary in consistency from syrupy to gelatinous. The term *ground substance* is sometimes used interchangeably with *matrix*.

Connective tissue cells lie embedded in the extracellular matrix. These cells are described as *fixed* if they remain in one place and as *mobile* if they can move from place to place. **Fixed cells** are responsible for local maintenance, tissue repair, and energy storage. **Mobile cells** are responsible mainly for defense. The distinction between fixed and mobile cells is not absolute, because at least one cell type is found in both fixed and mobile forms.

Extracellular matrix is nonliving, but the connective tissue cells constantly modify it by adding, deleting, or rearranging molecules. The suffix *-blast* {*blastos*, sprout} on a connective tissue cell name often indicates a cell that is either growing or actively secreting extracellular matrix. **Fibroblasts,** for example, are connective tissue cells that secrete collagen-rich matrix. Cells that are actively breaking down matrix are identified by the suffix *-clast* {*klastos*, broken}. Cells that are neither growing, secreting matrix components, nor breaking down matrix may be given the suffix *-cyte*, meaning "cell." Remembering these suffixes should help you remember the functional differences between cells with similar names, such as the osteoblast, osteocyte, and osteoclast, three cell types found in bone.

In addition to secreting proteoglycan ground substance, connective tissue cells produce matrix fibers. Four types of fiber proteins are found in matrix, aggregated into insoluble fibers. **Collagen** {*kolla*, glue + *-genes*, produced} is the most abundant protein in the human body, almost one-third of the body's dry weight. Collagen is also the most diverse of the four protein types, with at least 12 variations. It is found almost everywhere connective tissue is found, from the skin to muscles and bones. Individual collagen molecules pack together to form collagen fibers, flexible but inelastic fibers whose strength per unit weight exceeds that of steel. The amount and arrangement of collagen fibers help determine the mechanical properties of different types of connective tissues.

Three other protein fibers in connective tissue are elastin, fibrillin, and fibronectin. **Elastin** is a coiled, wavy protein that returns to its original length after being stretched. This property is known as *elastance* or *elastic recoil.* Elastin combines with the very thin, straight fibers of **fibrillin** to form filaments and sheets of elastic fibers. These two fibers are important in elastic tissues

FIG. 3.12 ESSENTIALS Connective Tissue

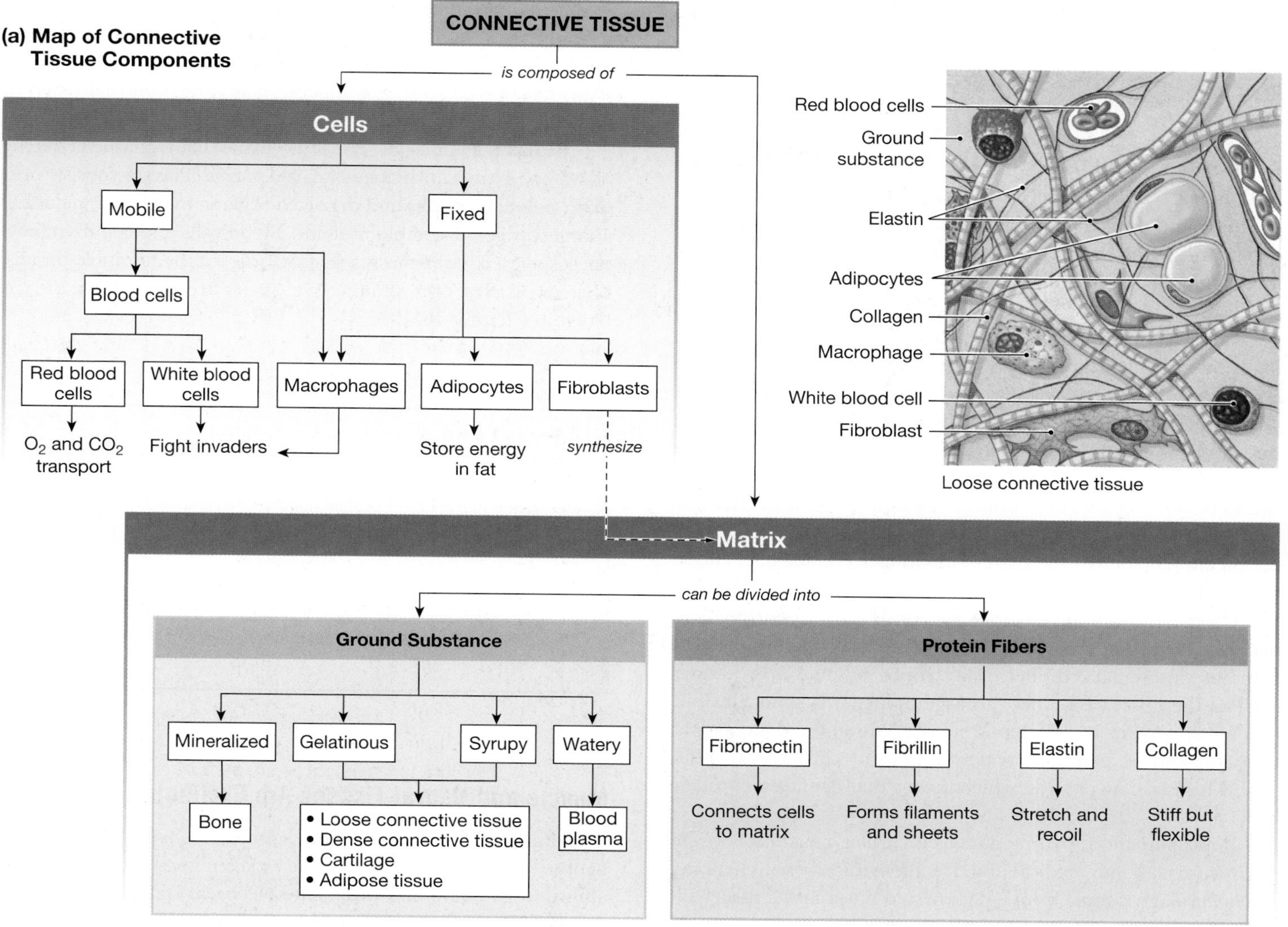
(a) Map of Connective Tissue Components
CONNECTIVE TISSUE
is composed of
Cells
Mobile
Fixed
Blood cells
Red blood cells
White blood cells
Macrophages
Adipocytes
Fibroblasts
O_2 and CO_2 transport
Fight invaders
Store energy in fat
synthesize
Red blood cells
Ground substance
Elastin
Adipocytes
Collagen
Macrophage
White blood cell
Fibroblast
Loose connective tissue
Matrix
can be divided into
Ground Substance
Mineralized
Gelatinous
Syrupy
Watery
Bone
• Loose connective tissue
• Dense connective tissue
• Cartilage
• Adipose tissue
Blood plasma
Protein Fibers
Fibronectin
Fibrillin
Elastin
Collagen
Connects cells to matrix
Forms filaments and sheets
Stretch and recoil
Stiff but flexible

(b) Types of Connective Tissue

Tissue Name	Ground Substance	Fiber Type and Arrangement	Main Cell Types	Where Found
Loose connective tissue	Gel; more ground substance than fibers or cells	Collagen, elastic, reticular; random	Fibroblasts	Skin around blood vessels and organs, under epithelia
Dense, irregular connective tissue	More fibers than ground substance	Mostly collagen; random	Fibroblasts	Muscle and nerve sheaths
Dense, regular connective tissue	More fibers than ground substance	Collagen; parallel	Fibroblasts	Tendons and ligaments
Adipose tissue	Very little ground substance	None	Brown fat and white fat	Depends on age and sex
Blood	Aqueous	None	Blood cells	In blood and lymph vessels
Cartilage	Firm but flexible; hyaluronic acid	Collagen	Chondroblasts	Joint surfaces, spine, ear, nose, larynx
Bone	Rigid due to calcium salts	Collagen	Osteoblasts and osteocytes	Bones

such as the lungs, blood vessels, and skin. As mentioned earlier, **fibronectin** connects cells to extracellular matrix at focal adhesions. Fibronectins also play an important role in wound healing and in blood clotting.

Types of Connective Tissue Figure 3.12b compares the properties of different types of connective tissue. The most common types are loose and dense connective tissue, adipose tissue, blood, cartilage, and bone. By many estimates, connective tissues are the most abundant of the tissue types as they are a component of most organs.

Loose connective tissues (**FIG. 3.13a**) are the elastic tissues that underlie skin and provide support for small glands. **Dense connective tissues** (irregular and regular) provide strength or flexibility. Examples are tendons, ligaments, and the sheaths that surround muscles and nerves. In these dense tissues, collagen fibers are the dominant type. **Tendons** (Fig. 3.13c) attach skeletal muscles to bones. **Ligaments** connect one bone to another. Because ligaments contain elastic fibers in addition to collagen fibers, they have a limited ability to stretch. Tendons lack elastic fibers and so cannot stretch.

Cartilage and bone together are considered supporting connective tissues. These tissues have a dense ground substance that contains closely packed fibers. **Cartilage** is found in structures such as the nose, ears, knee, and windpipe. It is solid, flexible, and notable for its lack of blood supply. Without a blood supply, nutrients and oxygen must reach the cells of cartilage by diffusion. This is a slow process, which means that damaged cartilage heals slowly.

Replacing and repairing damaged cartilage has moved from the research lab into medical practice. Biomedical researchers can take a cartilage sample from a patient and put it into a tissue culture medium to reproduce. Once the culture has grown enough *chondrocytes*—the cells that synthesize the extracellular matrix of cartilage—the cells are seeded into a scaffold. A physician surgically places the cells and scaffold in the patient's knee at the site of cartilage damage so that the chondrocytes can help repair the cartilage. Because the person's own cells are grown and reimplanted, there is no tissue rejection.

The fibrous extracellular matrix of **bone** is said to be *calcified* because it contains mineral deposits, primarily calcium salts, such as calcium phosphate (Fig. 3.13b). These minerals give the bone strength and rigidity. We examine the structure and formation of bone along with calcium metabolism in Chapter 23.

Adipose tissue is made up of **adipocytes,** or fat cells. An adipocyte of **white fat** typically contains a single enormous lipid droplet that occupies most of the volume of the cell (Fig. 3.13e). This is the most common form of adipose tissue in adults.

Brown fat is composed of adipose cells that contain multiple lipid droplets rather than a single large droplet. This type of fat has been known for many years to play an important role in temperature regulation in infants. Until recently it was thought to be almost completely absent in adults. However, modern imaging techniques such as combined CT and PET scans have revealed that adults do have brown fat (discussed in more detail in Chapter 22).

Blood is an unusual connective tissue that is characterized by its watery extracellular matrix called *plasma*. Plasma consists of a dilute solution of ions and dissolved organic molecules, including a large variety of soluble proteins. Blood cells and cell fragments are suspended in the plasma (Fig. 3.13d), but the insoluble protein fibers typical of other connective tissues are absent. We discuss blood in Chapter 16.

Concept Check

24. What is the distinguishing characteristic of connective tissues?
25. Name four types of protein fibers found in connective tissue matrix and give the characteristics of each.
26. Name six types of connective tissues.
27. Blood is a connective tissue with two components: plasma and cells. Which of these is the matrix in this connective tissue?
28. Why does torn cartilage heal more slowly than a cut in the skin?

Muscle and Neural Tissues Are Excitable

The third and fourth of the body's four tissue types—muscle and neural—are collectively called the *excitable tissues* because of their ability to generate and propagate electrical signals called *action potentials*. Both of these tissue types have minimal extracellular matrix, usually limited to a supportive layer called the *external lamina*. Some types of muscle and nerve cells are also notable for their gap junctions, which allow the direct and rapid conduction of electrical signals from cell to cell.

Muscle tissue has the ability to contract and produce force and movement. The body contains three types of muscle tissue: cardiac muscle in the heart; smooth muscle, which makes up most internal organs; and skeletal muscle. Most skeletal muscles attach to bones and are responsible for gross movement of the body. We discuss muscle tissue in more detail in Chapter 12.

Neural tissue has two types of cells. **Neurons,** or nerve cells, carry information in the form of chemical and electrical signals from one part of the body to another. They are concentrated in the brain and spinal cord but also include a network of cells that extends to virtually every part of the body. **Glial cells,** or neuroglia, are the support cells for neurons. We discuss the anatomy of neural tissue in Chapter 8. A summary of the characteristics of the four tissue types can be found in **TABLE 3.4.**

FIG. 3.13 ESSENTIALS Types of Connective Tissue

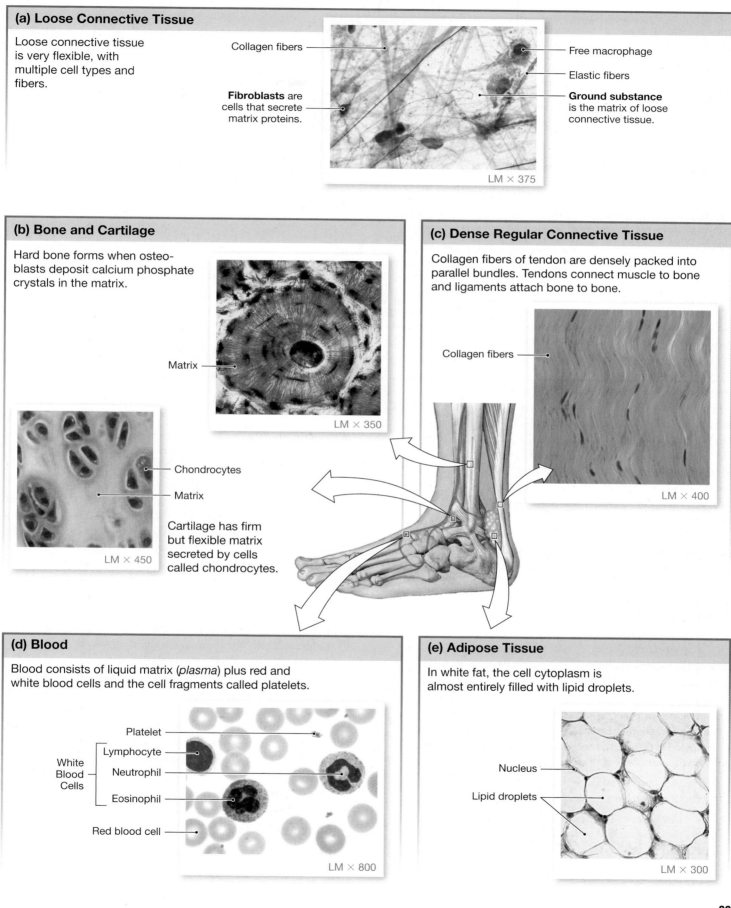

(a) Loose Connective Tissue

Loose connective tissue is very flexible, with multiple cell types and fibers.

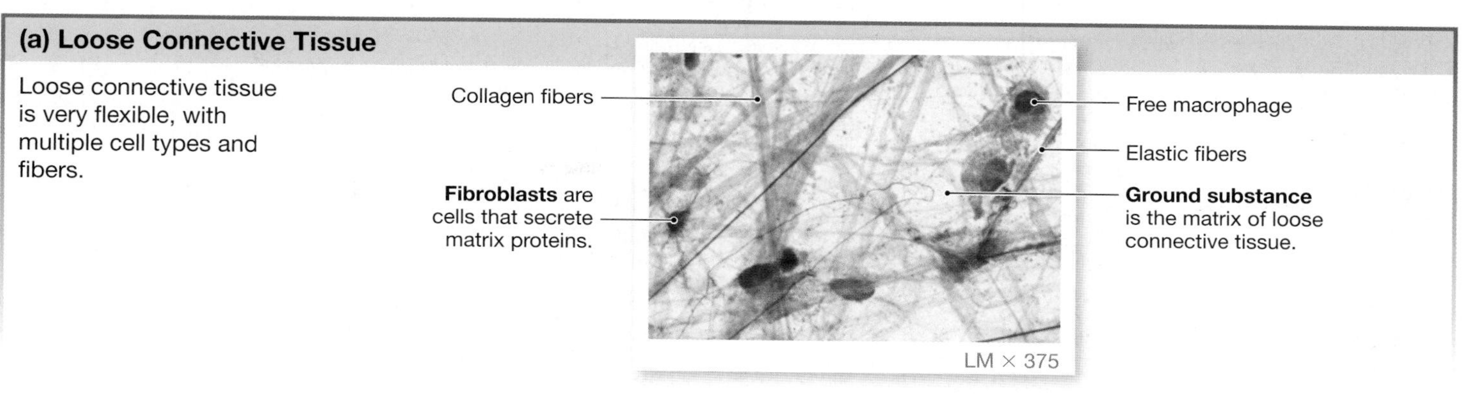

Collagen fibers
Fibroblasts are cells that secrete matrix proteins.
Free macrophage
Elastic fibers
Ground substance is the matrix of loose connective tissue.
LM × 375

(b) Bone and Cartilage

Hard bone forms when osteoblasts deposit calcium phosphate crystals in the matrix.

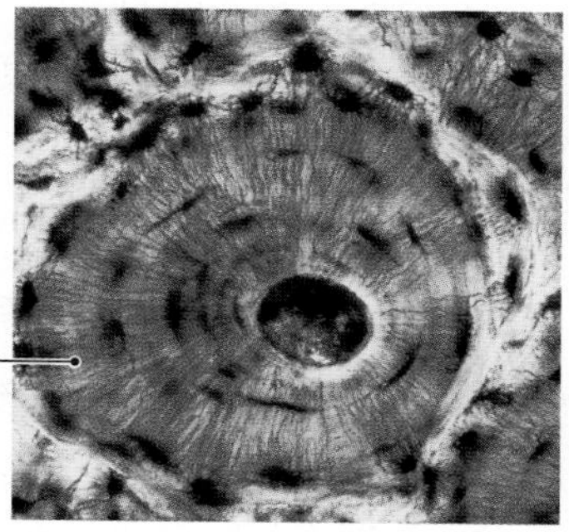

Matrix

LM × 350

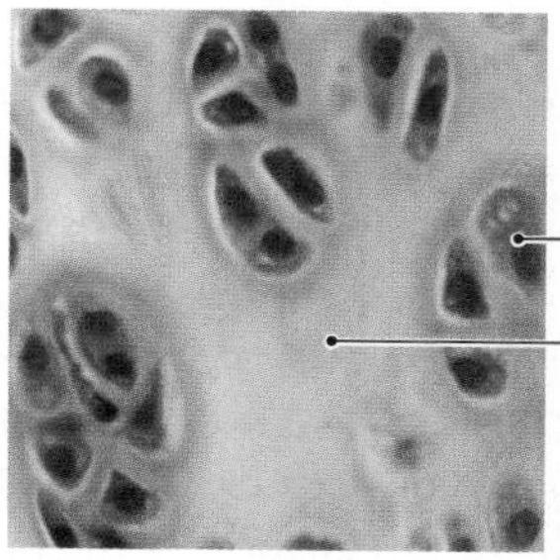

Chondrocytes

Matrix

LM × 450

Cartilage has firm but flexible matrix secreted by cells called chondrocytes.

(c) Dense Regular Connective Tissue

Collagen fibers of tendon are densely packed into parallel bundles. Tendons connect muscle to bone and ligaments attach bone to bone.

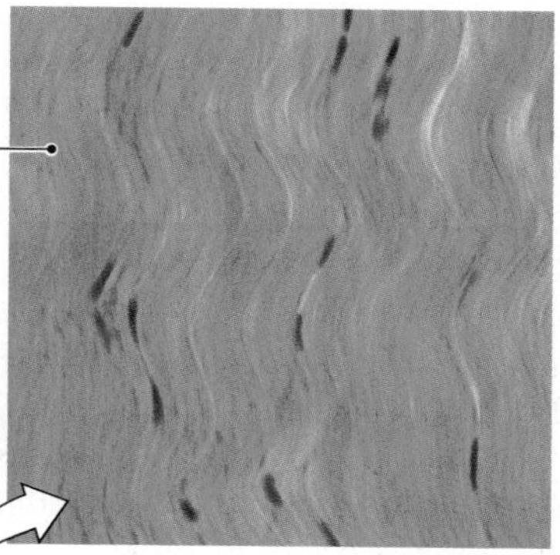

Collagen fibers

LM × 400

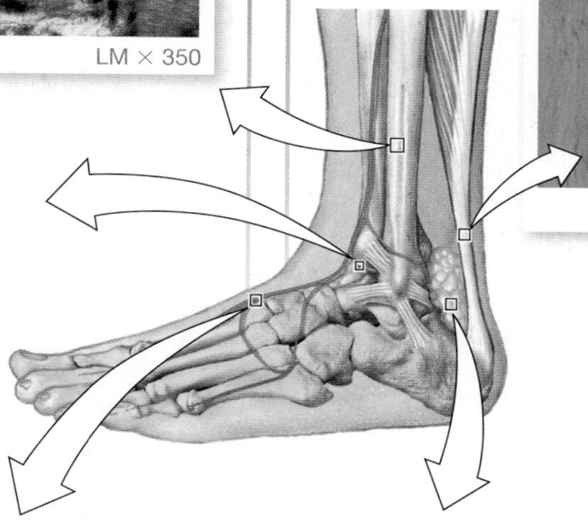

(d) Blood

Blood consists of liquid matrix (*plasma*) plus red and white blood cells and the cell fragments called platelets.

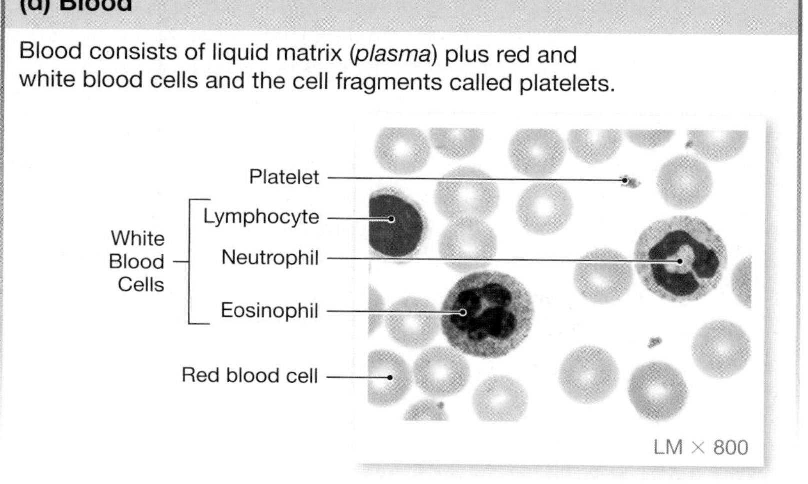

Platelet
White Blood Cells
Lymphocyte
Neutrophil
Eosinophil
Red blood cell
LM × 800

(e) Adipose Tissue

In white fat, the cell cytoplasm is almost entirely filled with lipid droplets.

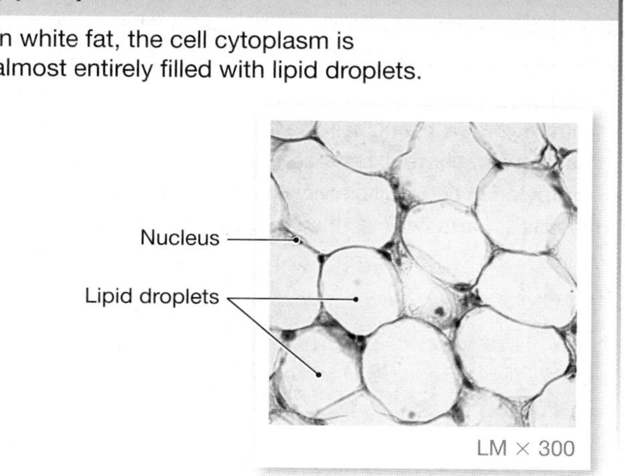

Nucleus
Lipid droplets
LM × 300

TABLE 3.4 Characteristics of the Four Tissue Types

	Epithelial	Connective	Muscle	Nerve
Matrix Amount	Minimal	Extensive	Minimal	Minimal
Matrix Type	Basal lamina	Varied—protein fibers in ground substance that ranges from liquid to gelatinous to firm to calcified	External lamina	External lamina
Unique Features	No direct blood supply	Cartilage has no blood supply	Able to generate electrical signals, force, and movement	Able to generate electrical signals
Surface Features of Cells	Microvilli, cilia	N/A	N/A	N/A
Locations	Covers body surface; lines cavities and hollow organs, and tubes; secretory glands	Supports skin and other organs; cartilage, bone, and blood	Makes up skeletal muscles, hollow organs, and tubes	Throughout body; concentrated in brain and spinal cord
Cell Arrangement and Shapes	Variable number of layers, from one to many; cells flattened, cuboidal, or columnar	Cells not in layers; usually randomly scattered in matrix; cell shape irregular to round	Cells linked in sheets or elongated bundles; cells shaped in elongated, thin cylinders; heart muscle cells may be branched	Cells isolated or networked; cell appendages highly branched and/or elongated

3.5 Tissue Remodeling

Most people associate growth with the period from birth to adulthood. However, cell birth, growth, and death continue throughout a person's life. The tissues of the body are constantly remodeled as cells die and are replaced.

Apoptosis Is a Tidy Form of Cell Death

Cell death occurs two ways, one messy and one tidy. In **necrosis,** cells die from physical trauma, toxins, or lack of oxygen when their blood supply is cut off. Necrotic cells swell, their organelles deteriorate, and finally the cells rupture. The cell contents released this way include digestive enzymes that damage adjacent cells and trigger an inflammatory response. You see necrosis when you have a red area of skin surrounding a scab.

In contrast, cells that undergo *programmed cell death,* or **apoptosis** {ap-oh-TOE-sis or a-pop-TOE-sis; *apo-*, apart, away + *ptosis,* falling}, do not disrupt their neighbors when they die. Apoptosis, also called cell suicide, is a complex process regulated by multiple chemical signals. Some signals keep apoptosis from occurring, while other signals tell the cell to self-destruct. When the suicide signal wins out, chromatin in the nucleus condenses, and the cell pulls away from its neighbors. It shrinks, then breaks up into tidy membrane-bound *blebs* that are gobbled up by neighboring cells or by wandering cells of the immune system.

Apoptosis is a normal event in the life of an organism. During fetal development, apoptosis removes unneeded cells, such as half the cells in the developing brain and the webs of skin between fingers and toes. In adults, cells that are subject to wear and tear from exposure to the outside environment may live only a day or two before undergoing apoptosis. For example, it has been estimated that the intestinal epithelium is completely replaced with new cells every two to five days.

RUNNING PROBLEM

The day after Jan's visit, the computerized cytology analysis system rapidly scans the cells on the slide of Jan's cervical tissue, looking for abnormal cell size or shape. The computer is programmed to find multiple views for the cytologist to evaluate. The results of Jan's two Pap tests are shown in **FIGURE 3.14**.

Q6: *Has Jan's dysplasia improved or worsened? What evidence do you have to support your answer?*

Q7: *Use your answer to question 6 to predict whether Jan's HPV infection has persisted or been cleared by her immune system.*

59 – 61 – 65 – 79 – 84 – 87

Concept Check

29. What are some features of apoptosis that distinguish it from cell death due to injury?

FIG. 3.14 Pap smears of cervical cells

(a) Jan's abnormal Pap test.

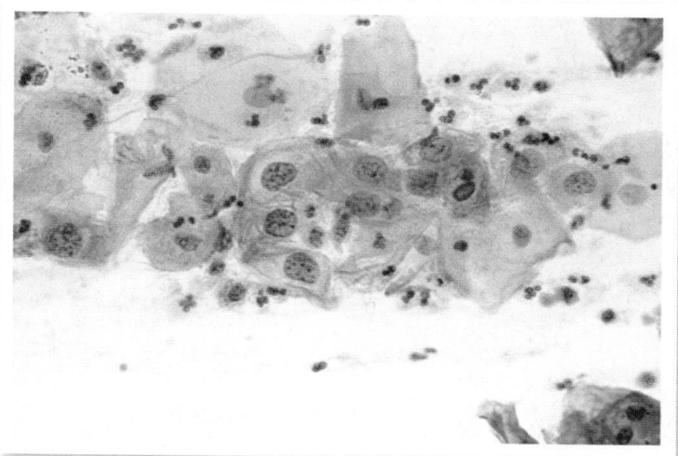

(b) Jan's second Pap test. Are these cells normal or abnormal?

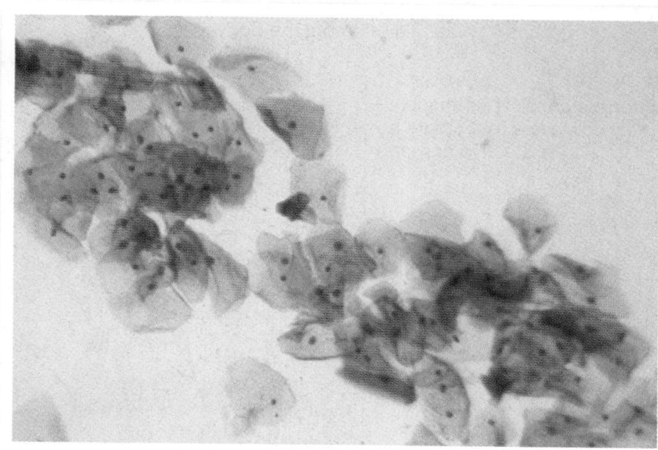

Stem Cells Can Create New Specialized Cells

If cells in the adult body are constantly dying, where do their replacements come from? This question is still being answered and is one of the hottest topics in biological research today. The following paragraphs describe what we currently know.

All cells in the body are derived from the single cell formed at conception. That cell and those that follow reproduce themselves by undergoing the cell division process known as **mitosis** (see Appendix C). The very earliest cells in the life of a human being are said to be **totipotent** {*totus*, entire} because they have the ability to develop into any and all types of specialized cells. Any totipotent cell has the potential to become a functioning organism.

After about day 4 of development, the totipotent cells of the embryo begin to specialize, or *differentiate*. As they do so, they narrow their potential fates and become **pluripotent** {*plures*, many}. Pluripotent cells can develop into many different cell types but not all cell types. An isolated pluripotent cell cannot develop into an organism.

As differentiation continues, pluripotent cells develop into the various tissues of the body. As the cells specialize and mature, many lose the ability to undergo mitosis and reproduce themselves. They can be replaced, however, by new cells created from **stem cells,** less specialized cells that retain the ability to divide.

Undifferentiated stem cells in a tissue that retain the ability to divide and develop into the cell types of that tissue are said to be **multipotent** {*multi*, many}. Some of the most-studied multipotent adult stem cells are found in bone marrow and give rise to blood cells. However, all adult stem cells occur in very small numbers. They are difficult to isolate and do not thrive in the laboratory.

Biologists once believed that nerve and muscle cells, which are highly specialized in their mature forms, could not be replaced when they died. Now research indicates that stem cells for these tissues do exist in the body. However, naturally occurring neural and muscle stem cells are so scarce that they cannot replace large masses of dead or dying tissue that result from diseases such as strokes or heart attacks. Consequently, one goal of stem cell research is to find a source of pluripotent or multipotent stem cells that could be grown in the laboratory. If stem cells could be grown in larger numbers, they could be implanted to treat damaged tissues and *degenerative* diseases, those in which cells degenerate and die. One example of a degenerative disease is Parkinson's disease, in which certain types of nerve cells in the brain die.

EMERGING CONCEPTS

Induced Pluripotent Stems Cells

In 2006 a group of Japanese researchers, led by Shinya Yamanaka, turned mature skin cells from a mouse back into pluripotent stem cells by altering just four genes. Before this work, scientists thought that once a cell had differentiated, it could not go back to a pluripotent state. Yamanaka's **induced pluripotent stem cells (iPS)** cells, changed our model of cell differentiation and provided a way to create stem cells that did not require the use of embryos. In the years since Yamanaka's discovery was announced, we have learned that iPS cells are very helpful disease models for laboratory studies. However, they have proved less successful as a source of stem cells for treating diseases. For his lab's discovery of a way to create iPS cells, Dr. Yamanaka received a Nobel Prize in 2012 (*www.nobelprize.org*).

FIG. 3.15 Focus on . . . The Skin

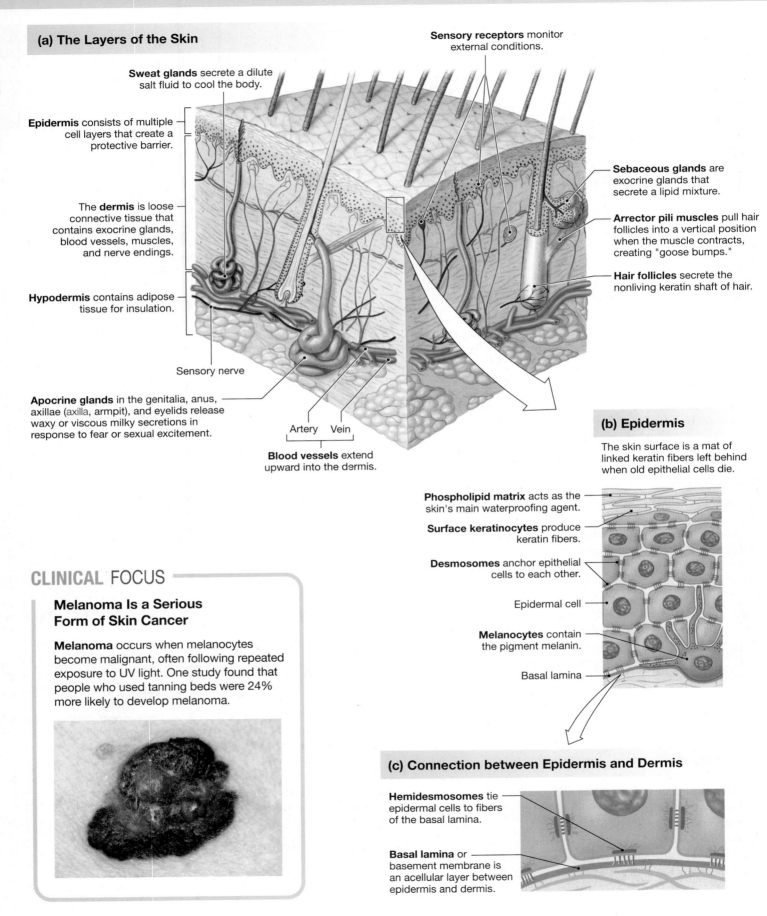

(a) The Layers of the Skin

Sensory receptors monitor external conditions.

Sweat glands secrete a dilute salt fluid to cool the body.

Epidermis consists of multiple cell layers that create a protective barrier.

The **dermis** is loose connective tissue that contains exocrine glands, blood vessels, muscles, and nerve endings.

Hypodermis contains adipose tissue for insulation.

Sensory nerve

Apocrine glands in the genitalia, anus, axillae (axilla, armpit), and eyelids release waxy or viscous milky secretions in response to fear or sexual excitement.

Artery Vein

Blood vessels extend upward into the dermis.

Sebaceous glands are exocrine glands that secrete a lipid mixture.

Arrector pili muscles pull hair follicles into a vertical position when the muscle contracts, creating "goose bumps."

Hair follicles secrete the nonliving keratin shaft of hair.

(b) Epidermis

The skin surface is a mat of linked keratin fibers left behind when old epithelial cells die.

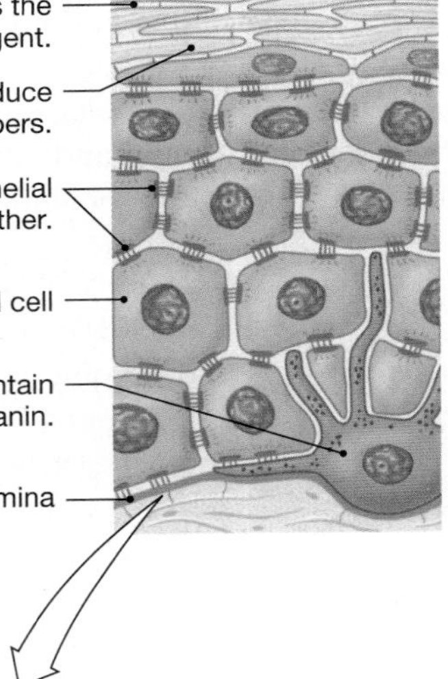

Phospholipid matrix acts as the skin's main waterproofing agent.
Surface keratinocytes produce keratin fibers.
Desmosomes anchor epithelial cells to each other.
Epidermal cell
Melanocytes contain the pigment melanin.
Basal lamina

(c) Connection between Epidermis and Dermis

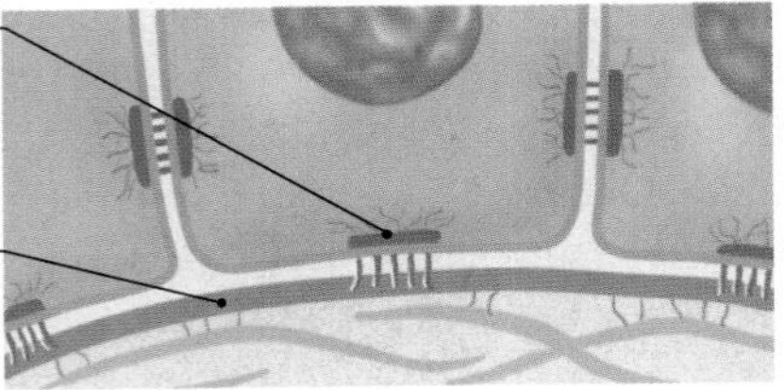

Hemidesmosomes tie epidermal cells to fibers of the basal lamina.
Basal lamina or basement membrane is an acellular layer between epidermis and dermis.

CLINICAL FOCUS

Melanoma Is a Serious Form of Skin Cancer

Melanoma occurs when melanocytes become malignant, often following repeated exposure to UV light. One study found that people who used tanning beds were 24% more likely to develop melanoma.

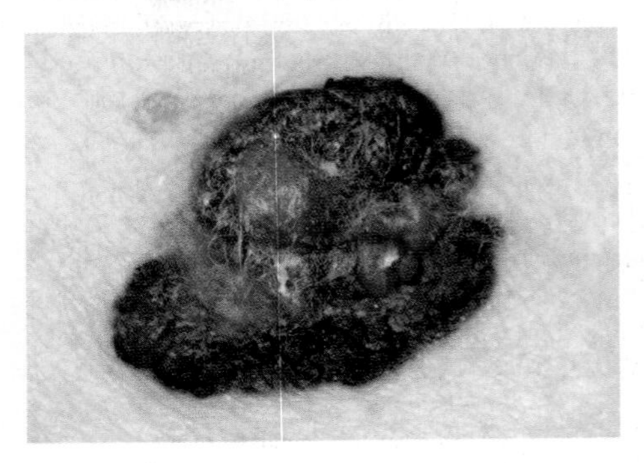

Embryos and fetal tissue are rich sources of stem cells, but the use of embryonic stem cells is controversial and poses many legal and ethical questions. Some researchers hope that adult stem cells will show **plasticity,** the ability to specialize into a cell of a type different from the type for which they were destined.

There are still many challenges facing us before stem cell therapy becomes a standard medical treatment. One is finding a good source of stem cells. A second major challenge is determining the chemical signals that tell stem cells when to differentiate and what type of cell to become. And even once these two challenges are overcome and donor stem cells are implanted, the body may recognize that the new cells are foreign tissue and try to reject them.

Stem cell research is an excellent example of the dynamic and often controversial nature of science. For the latest research findings, as well as pending legislation and laws regulating stem cell research and use, check authoritative websites, such as that sponsored by the U.S. National Institutes of Health (*http://stemcells.nih.gov*).

3.6 Organs

Groups of tissues that carry out related functions may form structures known as **organs.** The organs of the body contain the four types of tissue in various combinations. The skin is an excellent example of an organ that incorporates all four types of tissue into an integrated whole. We think of skin as a thin layer that covers the external surfaces of the body, but in reality it is the heaviest single organ, at about 16% of an adult's total body weight! If it were flattened out, it would cover a surface area of between 1.2 and 2.3 square meters, about the size of a couple of card-table tops. Its size and weight make skin one of the most important organs of the body.

The functions of the skin do not fit neatly into any one chapter of this book, and this is true of some other organs as well. We will highlight several of these organs in special organ *Focus* features throughout the book. These illustrated boxes discuss the structure and functions of these versatile organs so that you can gain an appreciation for the way different tissues combine for a united purpose. The first of these features, *Focus On: The Skin*, appears as **FIGURE 3.15**. *Focus On: The Skin*, provides an illustration of the structure and function of skin.

As we consider the systems of the body in the succeeding chapters, you will see how diverse cells, tissues, and organs carry out the processes of the living body. Although the body's cells have different structures and different functions, they have one need in common: a continuous supply of energy. Without energy, cells cannot survive, let alone carry out all the other processes of daily living. Next, we look at energy in living organisms and how cells capture and use the energy released by chemical reactions.

RUNNING PROBLEM **CONCLUSION** Pap Tests Save Lives

In this running problem, you learned that the Pap test can detect the early cell changes that precede cervical cancer. The diagnosis is not always simple because the change in cell cytology from normal to cancerous occurs along a continuum and can be subject to individual interpretation. In addition, not all cell changes are cancerous. The human papillomavirus (HPV), a common sexually transmitted infection, can also cause cervical dysplasia. In most cases, the woman's immune system overcomes the virus within two years, and the cervical cells revert to normal. A small number of women with persistent HPV infections have a higher risk of developing cervical cancer, however. Studies indicate that 98% of cervical cancers are associated with HPV infection. To learn more about the association between HPV and cervical cancer, go to the National Cancer Institute home page *(www.cancer.gov)* and search for HPV. This site also contains information about cervical cancer. To check your understanding of the running problem, compare your answers with the information in the following summary table.

Question	Facts	Integration and Analysis
Q1: *Why does the treatment of cancer focus on killing the cancerous cells?*	Cancerous cells divide uncontrollably and fail to coordinate with normal cells. Cancerous cells fail to differentiate into specialized cells.	Unless removed, cancerous cells will displace normal cells. This may cause destruction of normal tissues. In addition, because cancerous cells do not become specialized, they cannot carry out the same functions as the specialized cells they displace.
Q2: *What is happening in cancer cells that explains the large size of their nucleus and the relatively small amount of cytoplasm?*	Cancerous cells divide uncontrollably. Dividing cells must duplicate their DNA prior to cell division, and this DNA duplication takes place in the nucleus, leading to the large size of that organelle. (See Appendix C.)	Actively reproducing cells are likely to have more DNA in their nucleus as they prepare to divide, so their nuclei tend to be larger. Each cell division splits the cytoplasm between two daughter cells. If division is occurring rapidly, the daughter cells may not have time to synthesize new cytoplasm, so the amount of cytoplasm is less than in a normal cell.

– Continued next page

RUNNING PROBLEM CONCLUSION *Continued*

Question	Facts	Integration and Analysis
Q3: *What other kinds of damage or trauma are cervical epithelial cells normally subjected to?*	The cervix is the passageway between the uterus and vagina.	The cervix is subject to trauma or damage, such as might occur during sexual intercourse and childbirth.
Q4: *Which of its two types of epithelia is more likely to be affected by trauma?*	The cervix consists of secretory epithelium with mucus-secreting glands lining the inside and protective epithelium covering the outside.	Protective epithelium is composed of multiple layers of cells and is designed to protect areas from mechanical and chemical stress [p. 79]. Therefore, the secretory epithelium with its single-cell layer is more easily damaged.
Q5: *Jan's first Pap test showed atypical squamous cells of unknown significance (ASCUS). Were these cells more likely to come from the secretory portion of the cervix or from the protective epithelium?*	Secretory cells are columnar epithelium. Protective epithelium is composed of multiple cell layers.	Protective epithelium with multiple cell layers has cells that are flat (stratified squamous epithelium). The designation *ASC* refers to these protective epithelial cells.
Q6: *Has Jan's dysplasia improved or worsened? What evidence do you have to support your answer?*	The slide from Jan's first Pap test shows abnormal cells with large nuclei and little cytoplasm. These abnormal cells do not appear in the second test.	The disappearance of the abnormal cells indicates that Jan's dysplasia has resolved. She will return in another year for a repeat Pap test. If it shows no dysplasia, her cervical cells have reverted to normal.
Q7: *Use your answer to question 6 to predict whether Jan's HPV infection has persisted or been cleared by her immune system.*	The cells in the second Pap test appear normal.	Once Jan's body fights off the HPV infection, her cervical cells should revert to normal. Her second HPV test should show no evidence of HPV infection.

59 – 61 – 65 – 79 – 84 – 87

CHAPTER SUMMARY

Cell biology and histology illustrate one of the major themes in physiology: *compartmentation.* In this chapter, you learned how a cell is subdivided into two main compartments—the nucleus and the cytoplasm. You also learned how cells form tissues that create larger compartments within the body. A second theme in this chapter is the *molecular interactions* that create the *mechanical properties* of cells and tissues. Protein fibers of the cytoskeleton and cell junctions, along with the molecules that make up the extracellular matrix, form the "glue" that holds tissues together.

3.1 Functional Compartments of the Body

1. The **cell** is the functional unit of living organisms. (p. 59)
2. The major human body cavities are the cranial cavity (skull), thoracic cavity (thorax), and abdominopelvic cavity. (p. 59; Fig. 3.1a)
3. The **lumens** of some hollow organs are part of the body's external environment. (p. 59)
4. The body fluid compartments are the extracellular fluid (ECF) outside the cells and the intracellular fluid (ICF) inside the cells. The ECF can be subdivided into **interstitial fluid** bathing the cells and **plasma**, the fluid portion of the blood. (p. 61; Fig. 3.1b)

3.2 Biological Membranes

5. The word *membrane* is used both for cell membranes and for tissue membranes that line a cavity or separate two compartments. (p. 61; Fig. 3.1c)
6. The **cell membrane** acts as a barrier between the intracellular and extracellular fluids, provides structural support, and regulates exchange and communication between the cell and its environment. (p. 61)
7. The **fluid mosaic model** of a biological membrane shows it as a **phospholipid bilayer** with proteins inserted into the bilayer. (p. 62; Fig. 3.2b)
8. Membrane lipids include phospholipids, **sphingolipids,** and cholesterol. **Lipid-anchored proteins** attach to membrane lipids. (p. 62)
9. **Transmembrane proteins** are **integral proteins** tightly bound to the phospholipid bilayer. **Peripheral proteins** attach less tightly to either side of the membrane. (p. 63; Fig. 3.2b, c)
10. Carbohydrates attach to the extracellular surface of cell membranes. (p. 62)

3.3 Intracellular Compartments

11. The cytoplasm consists of semi-gelatinous **cytosol** with dissolved nutrients, ions, and waste products. Suspended in the cytosol are the other components of the cytoplasm: insoluble **inclusions** and fibers, which have no enclosing membrane, and **organelles,** which are membrane-enclosed bodies that carry out specific functions. (p. 65; Fig. 3.4a)
12. **Ribosomes** are inclusions that take part in protein synthesis. (p. 65)
13. Insoluble protein fibers come in three sizes: **actin fibers** (also called **microfilaments**), **intermediate filaments,** and **microtubules.** (p. 68; Tbl. 3.2)
14. **Centrioles** that aid the movement of chromosomes during cell division, **cilia** that move fluid or secretions across the cell surface, and **flagella** that propel sperm through body fluids are made of microtubules. (p. 68; Figs. 3.4e, 3.5)
15. The changeable **cytoskeleton** provides strength, support, and internal organization; aids transport of materials within the cell; links cells together; and enables motility in certain cells. (p. 68; Fig. 3.4b)
16. **Motor proteins** such as **myosins, kinesins,** and **dyneins** associate with cytoskeleton fibers to create movement. (p. 69; Fig. 3.6)
17. Membranes around organelles create compartments that separate functions. (p. 70)
18. **Mitochondria** generate most of the cell's ATP. (p. 70; Fig. 3.4g)
19. The **smooth endoplasmic reticulum** is the primary site of lipid synthesis. The **rough endoplasmic reticulum** is the primary site of protein synthesis. (p. 70; Fig. 3.4i)
20. The **Golgi apparatus** packages proteins into vesicles. **Secretory vesicles** release their contents into the extracellular fluid. (p. 70; Fig. 3.4h)
21. **Lysosomes** and **peroxisomes** are small **storage vesicles** that contain digestive enzymes. (p. 71; Figs. 3.4c, d)
22. The **nucleus** contains DNA, the genetic material that ultimately controls all cell processes, in the form of **chromatin.** The double-membrane **nuclear envelope** surrounding the nucleus has **nuclear pore complexes** that allow controlled chemical communication between the nucleus and cytosol. **Nucleoli** are nuclear areas that control the synthesis of RNA for ribosomes. (p. 71; Fig. 3.4j)
23. Protein synthesis is an example of how the cell separates functions by isolating them to separate compartments within the cell (p. 71; Fig. 3.7)

3.4 Tissues of the Body

24. There are four primary tissue types in the human body: epithelial, connective, muscle, and neural. (p. 73)
25. **Extracellular matrix** secreted by cells provides support and a means of cell-cell communication. It is composed of proteoglycans and insoluble protein fibers. (p. 73)
26. Animal cell junctions fall into three categories. **Gap junctions** allow chemical and electrical signals to pass directly from cell to cell. **Tight junctions** restrict the movement of material between cells. **Anchoring junctions** hold cells to each other or to the extracellular matrix. (p. 75; Fig. 3.8)
27. Membrane proteins called **cell adhesion molecules (CAMs)** are essential in cell adhesion and in anchoring junctions. (p. 73; Tbl. 3.3)
28. **Desmosomes** and **adherens junctions** anchor cells to each other. **Focal adhesions** and **hemidesmosomes** anchor cells to matrix. (p. 75; Fig. 3.8)
29. **Epithelial tissues** protect the internal environment, regulate the exchange of material, or manufacture and secrete chemicals. There are five functional types found in the body: exchange, transporting, ciliated, protective, and secretory. (p. 75; Fig. 3.9)
30. **Exchange epithelia** permit rapid exchange of materials, particularly gases. **Transporting epithelia** actively regulate the selective exchange of nongaseous materials between the internal and external environments. **Ciliated epithelia** move fluid and particles across the surface of the tissue. **Protective epithelia** help prevent exchange between the internal and external environments. The **secretory epithelia** release secretory products into the external environment or the blood. (p. 77; Fig. 3.10)
31. **Exocrine glands** release their secretions into the external environment through **ducts. Endocrine glands** are ductless glands that release their secretions, called **hormones,** directly into the extracellular fluid. (p. 79; Fig. 3.9b)
32. **Connective tissues** have extensive extracellular matrix that provides structural support and forms a physical barrier. (p. 80; Fig. 3.12)
33. **Loose connective tissues** are the elastic tissues that underlie skin. **Dense connective tissues,** including **tendons** and **ligaments,** have strength or flexibility because they are made of collagen. **Adipose tissue** stores fat. The connective tissue we call **blood** is characterized by a watery matrix. **Cartilage** is solid and flexible and has no blood supply. The fibrous matrix of **bone** is hardened by deposits of calcium salts. (p. 82; Fig. 3.13)
34. Muscle and neural tissues are called excitable tissues because of their ability to generate and propagate electrical signals called action potentials. **Muscle tissue** has the ability to contract and produce force and movement. There are three types of muscle: cardiac, smooth, and skeletal. (p. 82)
35. **Neural tissue** includes **neurons,** which use electrical and chemical signals to transmit information from one part of the body to another, and support cells known as **glial cells** (neuroglia). (p. 82)

3.5 Tissue Remodeling

36. Cell death occurs by **necrosis,** which adversely affects neighboring cells, and by **apoptosis,** programmed cell death that does not disturb the tissue. (p. 84)
37. **Stem cells** are cells that are able to reproduce themselves and differentiate into specialized cells. Stem cells are most plentiful in embryos but are also found in the adult body. (p. 85)

3.6 Organs

38. **Organs** are formed by groups of tissues that carry out related functions. The organs of the body contain the four types of tissues in various ratios. For example, skin is largely connective tissue. (p. 87)

REVIEW QUESTIONS

In addition to working through these questions and checking your answers on p. A-3, review the Learning Outcomes at the beginning of this chapter.

Level One Reviewing Facts and Terms

1. List the four general functions of the cell membrane.
2. In 1972, Singer and Nicolson proposed the fluid mosaic model of the cell membrane. According to this model, the membrane is composed of a bilayer of _____ and a variety of embedded _____, with _____ on the extracellular surface.
3. What are the two primary types of biomolecules found in the cell membrane?
4. Define and distinguish between inclusions and organelles. Give an example of each.
5. Define cytoskeleton. List five functions of the cytoskeleton.
6. Match each term with the description that fits it best:

(a) cilia (b) centriole (c) flagellum (d) centrosome	1. in human cells, appears as single, long, whip-like tail 2. short, hairlike structures that beat to produce currents in fluids 3. a bundle of microtubules that aid in mitosis 4. the microtubule-organizing center

7. Exocrine glands produce watery secretions (such as tears or sweat) called _____ secretions, or stickier solutions called _____ secretions.
8. Match each organelle with its function:

(a) endoplasmic reticulum (b) Golgi apparatus (c) lysosome (d) mitochondrion (e) peroxisome	1. powerhouse of the cell where most ATP is produced 2. degrades long-chain fatty acids and toxic foreign molecules 3. network of membranous tubules that synthesize biomolecules 4. digestive system of cell, degrading or recycling components 5. modifies and packages proteins into vesicles

9. What process activates the enzymes inside lysosomes?
10. _____ glands release hormones, which enter the blood and regulate the activities of organs or systems.
11. List the four major tissue types. Give an example and location of each.
12. The largest and heaviest organ in the body is the _____.
13. Match each protein to its function. Functions in the list may be used more than once.

(a) cadherin (b) CAM (c) collagen (d) connexin (e) elastin (f) fibrillin (g) fibronectin (h) integrin (i) occludin	1. membrane protein used to form cell junctions 2. matrix glycoprotein used to anchor cells 3. protein found in gap junctions 4. matrix protein found in connective tissue

14. What types of glands can be found within the skin? Name the secretion of each type.
15. The term *matrix* can be used in reference to an organelle or to tissues. Compare the meanings of the term in these two contexts.

Level Two Reviewing Concepts

16. List, compare, and contrast the three types of cell junctions and their subtypes. Give an example of where each type can be found in the body and describe its function in that location.
17. Which would have more rough endoplasmic reticulum: pancreatic cells that manufacture the protein hormone insulin, or adrenal cortex cells that synthesize the steroid hormone cortisol?
18. A number of organelles can be considered vesicles. Define *vesicle* and describe at least three examples.
19. Explain why a stratified epithelium offers more protection than a simple epithelium.
20. Transform this list of terms into a map of cell structure. Add functions where appropriate.

• actin	• microfilament
• cell membrane	• microtubule
• centriole	• mitochondria
• cilia	• nonmembranous organelle
• cytoplasm	• nucleus
• cytoskeleton	• organelle
• cytosol	• peroxisome
• extracellular matrix	• ribosome
• flagella	• rough ER
• Golgi apparatus	• secretory vesicle
• intermediate filament	• smooth ER
• keratin	• storage vesicle
• lysosome	• tubulin

21. Sketch a short series of columnar epithelial cells. Label the apical and basolateral borders of the cells. Briefly explain the different kinds of junctions found on these cells.
22. Arrange the following compartments in the order a glucose molecule entering the body at the intestine would encounter them: interstitial fluid, plasma, intracellular fluid. Which of these fluid compartments is/are considered extracellular fluid(s)?
23. Explain how inserting cholesterol into the phospholipid bilayer of the cell membrane decreases membrane permeability.
24. Compare and contrast the structure, locations, and functions of bone and cartilage.
25. Differentiate between the terms in each set below:
    (a) lumen and wall
    (b) cytoplasm and cytosol
    (c) myosin and keratin
26. When a tadpole turns into a frog, its tail shrinks and is reabsorbed. Is this an example of necrosis or apoptosis? Defend your answer.

27. Match the structures from the chapter to the basic physiological themes in the right column and give an example or explanation for each match. A structure may match with more than one theme.

(a) cell junctions	1. communication
(b) cell membrane	2. molecular interactions
(c) cytoskeleton	3. compartmentation
(d) organelles	4. mechanical properties
(e) cilia	5. biological energy use

28. In some instances, the extracellular matrix can be quite rigid. How might developing and expanding tissues cope with a rigid matrix to make space for themselves?

Level Three Problem Solving

29. One result of cigarette smoking is paralysis of the cilia that line the respiratory passageways. What function do these cilia serve? Based on what you have read in this chapter, why is it harmful when they no longer beat? What health problems would you expect to arise? How does this explain the hacking cough common among smokers?

30. Cancer is abnormal, uncontrolled cell division. What property of epithelial tissues might (and does) make them more prone to developing cancer?

31. What might happen to normal physiological function if matrix metalloproteinases are inhibited by drugs?

Answers to Concept Checks, Figure and Graph Questions, and end-of-chapter Review Questions can be found in Appendix A [p. A-1].

4 Energy and Cellular Metabolism

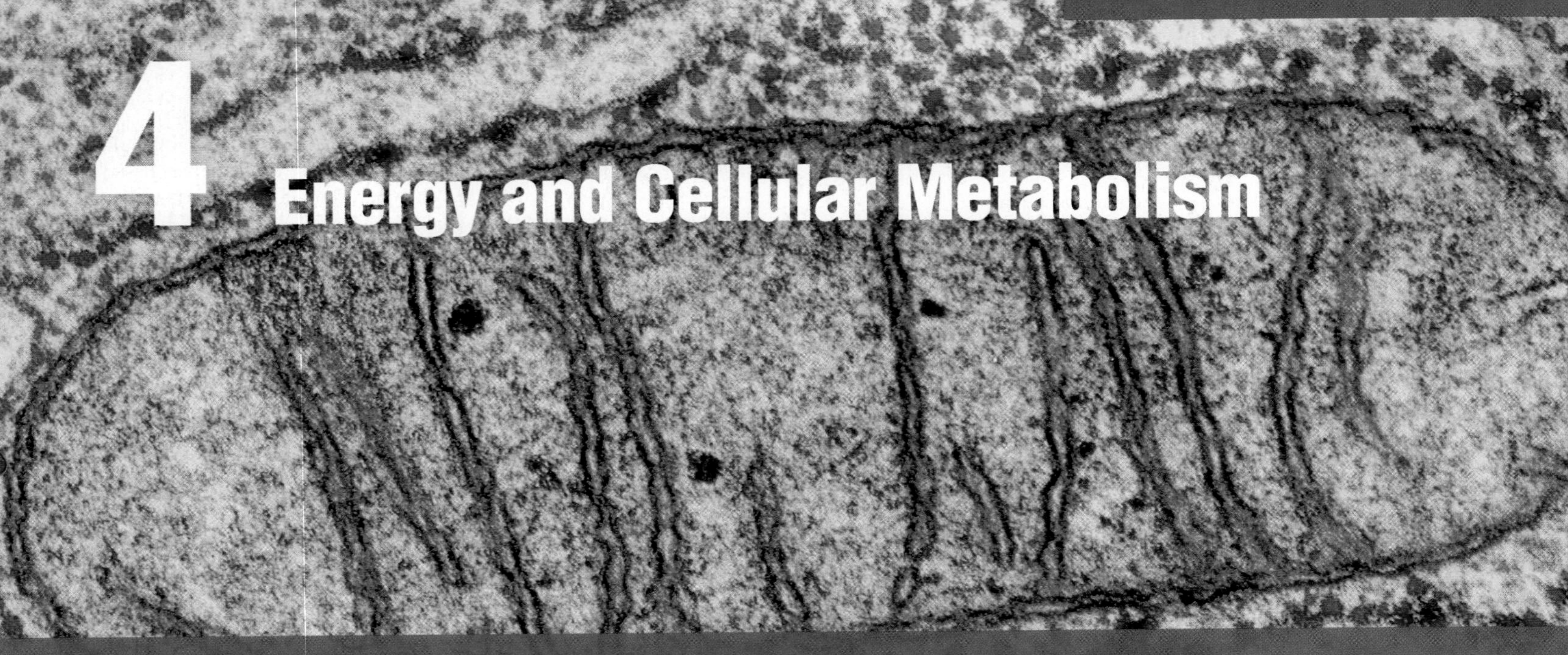

Mitochondrion

There is no good evidence that . . . life evades the second law of thermodynamics, but in the downward course of the energy-flow it interposes a barrier and dams up a reservoir which provides potential for its own remarkable activities.

F. G. Hopkins, 1933. "Some Chemical Aspects of Life," presidential address to the 1933 meeting of British Association for the Advancement of Science

BACKGROUND BASICS

Christine Schmidt, Ph.D., and her graduate students create an engineered matrix and seed them with neurons. They know that if their work is successful and the neurons grow along the scaffold, their work may help people with spinal cord injuries regain function. Just as a child playing with building blocks assembles them into a house, the bioengineer and her students create tissue from cells. In both cases someone familiar with the starting components, building blocks or cells, can predict what the final product will be: blocks make buildings; cells make tissues.

Why then can't biologists, knowing the characteristics of nucleic acids, proteins, lipids, and carbohydrates, explain how combinations of these molecules acquire the remarkable attributes of a living cell? How can living cells carry out processes that far exceed what we would predict from understanding their individual components? The answer is *emergent properties* [p. 2], those distinctive traits that cannot be predicted from the simple sum of the component parts. For example, if you came across a collection of metal pieces and bolts from a disassembled car motor, could you predict (without prior knowledge) that, given an energy source and properly arranged, this collection could create the power to move thousands of pounds?

The emergent properties of biological systems are of tremendous interest to scientists trying to explain how a simple compartment, such as a phospholipid liposome [p. 63], could have evolved into the first living cell. Pause for a moment and see if you can list the properties of life that characterize all living creatures. If you were a scientist looking at pictures and samples sent back from Mars, what would you look for to determine whether life exists there?

Now compare your list with the one in **TABLE 4.1**. Living organisms are highly organized and complex entities. Even a one-celled bacterium, although it appears simple under a microscope, has incredible complexity at the chemical level of organization. It uses intricately interconnected biochemical reactions to acquire, transform, store, and use energy and information. It senses and responds to changes in its internal and external environments and adapts so that it can maintain homeostasis. It reproduces, develops, grows, and dies; and over time, its species evolves.

TABLE 4.1 Properties of Living Organisms

1. Have a complex structure whose basic unit of organization is the cell
2. Acquire, transform, store, and use energy
3. Sense and respond to internal and external environments
4. Maintain homeostasis through internal control systems with feedback
5. Store, use, and transmit information
6. Reproduce, develop, grow, and die
7. Have emergent properties that cannot be predicted from the simple sum of the parts
8. Individuals adapt and species evolve

Energy is essential for the processes we associate with living things. Without energy for growth, repair, and maintenance of the internal environment, a cell is like a ghost town filled with buildings that are slowly crumbling into ruin. Cells need energy to import raw materials, make new molecules, and repair or recycle aging parts. The ability of cells to extract energy from the external environment and use that energy to maintain themselves as organized, functioning units is one of their most outstanding characteristics. In this chapter, we look at the cell processes through which the human body obtains energy and maintains its ordered systems. You will learn how protein interactions [p. 46] apply to enzyme activity and how the subcellular compartments [p. 8] separate various steps of energy metabolism.

RUNNING PROBLEM **Tay-Sachs Disease: A Deadly Inheritance**

In many American ultra-orthodox Jewish communities—in which arranged marriages are the norm—the rabbi is entrusted with an important, life-saving task. He keeps a confidential record of individuals known to carry the gene for Tay-Sachs disease, a fatal, inherited condition that strikes 1 in 3,600 American Jews of Eastern European descent. Babies born with this disease rarely live beyond age 4, and there is no cure. Based on the family trees he constructs, the rabbi can avoid pairing two individuals who carry the deadly gene.

Sarah and David, who met while working on their college newspaper, are not orthodox Jews. Both are aware, however, that their Jewish ancestry might put any children they have at risk for Tay-Sachs disease. Six months before their wedding, they decide to see a genetic counselor to determine whether they are carriers of the gene for Tay-Sachs disease.

93 – 99 – 100 – 104 – 110 – 118

4.1 Energy in Biological Systems

Energy cycling between the environment and living organisms is one of the fundamental concepts of biology. All cells use energy from their environment to grow, make new parts, and reproduce. Plants trap radiant energy from the sun and store it as chemical-bond energy through the process of photosynthesis (**FIG. 4.1**). They extract carbon and oxygen from carbon dioxide, nitrogen from the soil, and hydrogen and oxygen from water to make biomolecules such as glucose and amino acids.

Animals, on the other hand, cannot trap energy from the sun or use carbon and nitrogen from the air and soil to synthesize biomolecules. They must import chemical-bond energy by ingesting the biomolecules of plants or other animals. Ultimately, however, energy trapped by photosynthesis is the energy source for all animals, including humans.

FIG. 4.1 Energy transfer in the environment

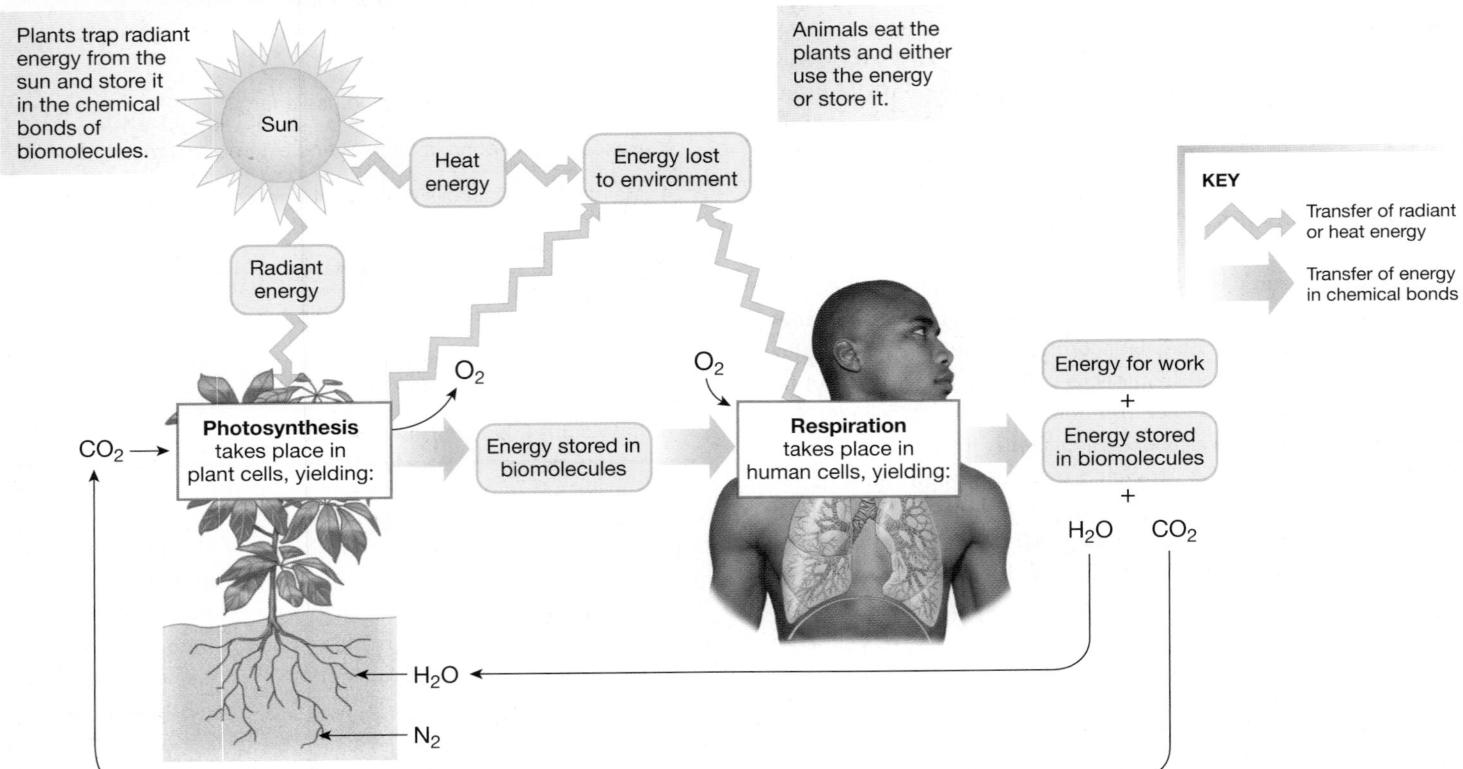
Plants trap radiant energy from the sun and store it in the chemical bonds of biomolecules.
Sun
Heat energy
Energy lost to environment
Animals eat the plants and either use the energy or store it.
KEY
Transfer of radiant or heat energy
Transfer of energy in chemical bonds
Radiant energy
O_2
O_2
Energy for work
CO_2
Photosynthesis takes place in plant cells, yielding:
Energy stored in biomolecules
Respiration takes place in human cells, yielding:
+
Energy stored in biomolecules
+
H_2O
CO_2
H_2O
N_2

Animals extract energy from biomolecules through the process of *respiration*, which consumes oxygen and produces carbon dioxide and water. If animals ingest more energy than they need for immediate use, the excess energy is stored in chemical bonds, just as it is in plants. Glycogen (a glucose polymer) and lipid molecules are the main energy stores in animals [p. 31]. These storage molecules are available for use at times when an animal's energy needs exceed its food intake.

Concept Check

1. Which biomolecules always include nitrogen in their chemical makeup?

Energy Is Used to Perform Work

All living organisms obtain, store, and use energy to fuel their activities. **Energy** can be defined as the capacity to do work, but what is *work*? We use this word in everyday life to mean various things, from hammering a nail to sitting at a desk writing a paper. In biological systems, however, the word means one of three specific things: chemical work, transport work, or mechanical work.

Chemical work is the making and breaking of chemical bonds. It enables cells and organisms to grow, maintain a suitable internal environment, and store information needed for reproduction and other activities. Forming the chemical bonds of a protein is an example of chemical work.

Transport work enables cells to move ions, molecules, and larger particles through the cell membrane and through the membranes of organelles in the cell. Transport work is particularly useful for creating **concentration gradients,** distributions of molecules in which the concentration is higher on one side of a membrane than on the other. For example, certain types of endoplasmic reticulum [p. 70] use energy to import calcium ions from the cytosol. This ion transport creates a high calcium concentration inside the organelle and a low concentration in the cytosol. If calcium is then released back into the cytosol, it creates a "calcium signal" that causes the cell to perform some action, such as muscle contraction.

Mechanical work in animals is used for movement. At the cellular level, movement includes organelles moving around in a cell, cells changing shape, and cilia and flagella beating [p. 68]. At the macroscopic level in animals, movement usually involves muscle contraction. Most mechanical work is mediated by motor proteins that make up certain intracellular fibers and filaments of the cytoskeleton [p. 68].

Energy Comes in Two Forms: Kinetic and Potential

Energy can be classified in various ways. We often think of energy in terms we deal with daily: thermal energy, electrical energy, mechanical energy. We speak of energy stored in chemical bonds. Each type of energy has its own characteristics. However, all types of energy share an ability to appear in two forms: as kinetic energy or as potential energy.

Kinetic energy is the energy of motion {*kinetikos*, motion}. A ball rolling down a hill, perfume molecules spreading through the air, electric charge flowing through power lines, heat warming a frying pan, and molecules moving across biological membranes are all examples of bodies that have kinetic energy.

Potential energy is stored energy. A ball poised at the top of a hill has potential energy because it has the potential to start moving down the hill. A molecule positioned on the high-concentration side of a concentration gradient stores potential energy because it has the potential energy to move down the gradient. In chemical bonds, potential energy is stored in the position of the electrons that form the bond [p. 33]. To learn more about kinetic and potential energy, see Appendix B.

A key feature of all types of energy is the ability of potential energy to become kinetic energy and vice versa.

Energy Can Be Converted from One Form to Another

Recall that a general definition of energy is the capacity to do work. Work always involves movement and therefore is associated with kinetic energy. Potential energy can also be used to perform work, but the potential energy must first be converted to kinetic energy. The conversion from potential energy to kinetic energy is never 100% efficient, and a certain amount of energy is lost to the environment, usually as heat.

The amount of energy lost in the transformation depends on the *efficiency* of the process. Many physiological processes in the human body are not very efficient. For example, 70% of the energy used in physical exercise is lost as heat rather than transformed into the work of muscle contraction.

FIGURE 4.2 summarizes the relationship of kinetic energy and potential energy:

1. Kinetic energy of the moving ball is transformed into potential energy as work is used to push the ball up the ramp (Fig. 4.2a).
2. Potential energy is stored in the stationary ball at the top of the ramp (Fig. 4.2b). No work is being performed, but the capacity to do work is stored in the position of the ball.
3. The potential energy of the ball becomes kinetic energy when the ball rolls down the ramp (Fig. 4.2c). Some kinetic energy is lost to the environment as heat due to friction between the ball and the air and ramp.

In biological systems, potential energy is stored in concentration gradients and chemical bonds. It is transformed into kinetic energy when needed to do chemical, transport, or mechanical work.

Thermodynamics Is the Study of Energy Use

Two basic rules govern the transfer of energy in biological systems and in the universe as a whole. The **first law of thermodynamics,** also known as the *law of conservation of energy*, states that the total amount of energy in the universe is constant. The universe is considered to be a *closed system*—nothing enters and nothing leaves. Energy can be converted from one type to another, but the total amount of energy in a closed system never changes.

The human body is not a closed system, however. As an *open system*, it exchanges materials and energy with its surroundings. Because our bodies cannot create energy, they import it from outside in the form of food. By the same token, our bodies lose energy, especially in the form of heat, to the environment. Energy that stays within the body can be changed from one type to another or can be used to do work.

The **second law of thermodynamics** states that natural spontaneous processes move from a state of order (non-randomness) to a condition of randomness or disorder, also known as **entropy.** Creating and maintaining order in an open system such as the body requires the input of energy. Disorder occurs when open systems lose energy to their surroundings without regaining it. When this happens, we say that the entropy of the open system has increased.

FIG. 4.2 Kinetic and potential energy

(a) Work is used to push a ball up a ramp. Kinetic energy of movement up the ramp is being stored in the potential energy of the ball's position.

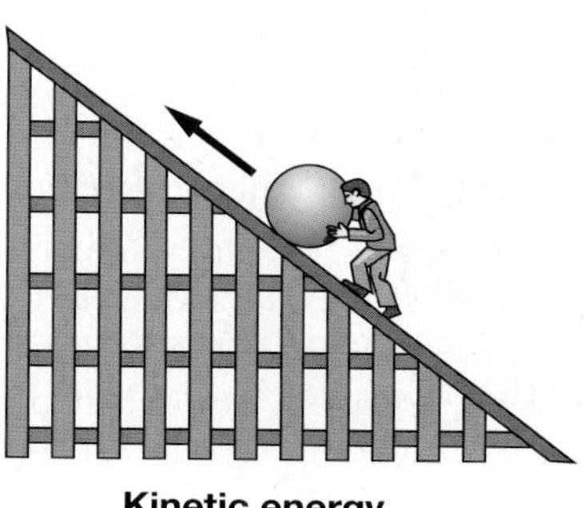

Kinetic energy

(b) The ball sitting at the top of the ramp has potential energy, the potential to do work.

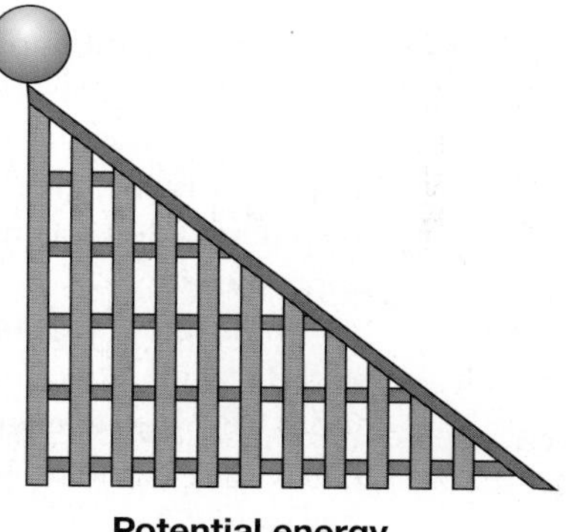

Potential energy

(c) The ball rolling down the ramp is converting the potential energy to kinetic energy. However, the conversion is not totally efficient, and some energy is lost as heat due to friction between the ball, ramp, and air.

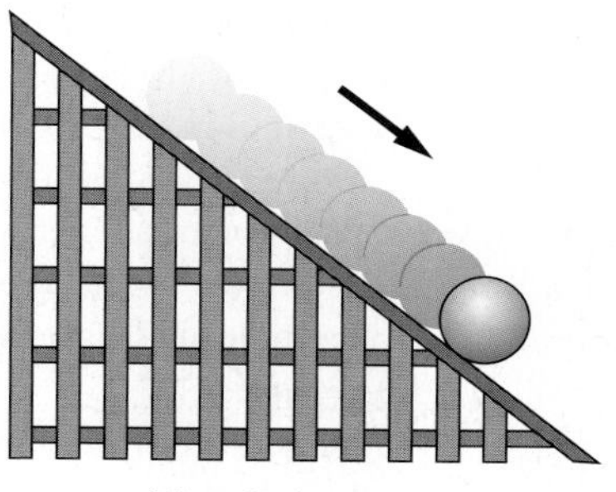

Kinetic energy

The ghost town analogy mentioned earlier illustrates the second law. When people put all their energy into activities away from town, the town slowly falls into disrepair and becomes less organized (its entropy increases). Similarly, without continual input of energy, a cell is unable to maintain its ordered internal environment. As the cell loses organization, its ability to carry out normal functions disappears, and it dies.

In the remainder of this chapter, you will learn how cells obtain energy from and store energy in the chemical bonds of biomolecules. Using chemical reactions, cells transform the potential energy of chemical bonds into kinetic energy for growth, maintenance, reproduction, and movement.

Concept Check

2. Name two ways animals store energy in their bodies.
3. What is the difference between potential energy and kinetic energy?
4. What is entropy?

4.2 Chemical Reactions

Living organisms are characterized by their ability to extract energy from the environment and use it to support life processes. The study of energy flow through biological systems is a field known as **bioenergetics** {*bios*, life + *en*-, in + *ergon*, work}. In a biological system, chemical reactions are a critical means of transferring energy from one part of the system to another.

Energy Is Transferred between Molecules during Reactions

In a **chemical reaction,** a substance becomes a different substance, usually by the breaking and/or making of covalent bonds. A reaction begins with one or more molecules called **reactants** and ends with one or more molecules called **products** (TBL. 4.2). In this discussion, we consider a reaction that begins with two reactants and ends with two products:

$$A + B \rightarrow C + D$$

The speed with which a reaction takes place, the **reaction rate,** is the disappearance rate of the reactants (A and B) or the appearance rate of the products (C and D). Reaction rate is measured as change in concentration during a certain time period and is often expressed as molarity per second (M/sec).

The purpose of chemical reactions in cells is either to transfer energy from one molecule to another or to use energy stored in reactant molecules to do work. The potential energy stored in the chemical bonds of a molecule is known as the **free energy** of the molecule. Generally, complex molecules have more chemical bonds and therefore higher free energies.

For example, a large glycogen molecule has more free energy than a single glucose molecule, which in turn has more free energy than the carbon dioxide and water from which it was synthesized. The high free energy of complex molecules such as glycogen is the reason that these molecules are used to store energy in cells.

TABLE 4.2 Chemical Reactions

Reaction Type	Reactants (Substrates)		Products
Combination	A + B	⟶	C
Decomposition	C	⟶	A + B
Single displacement*	L + MX	⟶	LX + M
Double displacement*	LX + MY	⟶	LY + MX

*X and Y represent atoms, ions, or chemical groups.

To understand how chemical reactions transfer energy between molecules, we should answer two questions. First, how do reactions get started? The energy required to initiate a reaction is known as the *activation energy* for the reaction. Second, what happens to the free energy of the products and reactants during a reaction? The difference in free energy between reactants and products is the *net free energy change of the reaction.*

Activation Energy Gets Reactions Started

Activation energy is the initial input of energy required to bring reactants into a position that allows them to react with one another. This "push" needed to start the reaction is shown in FIGURE 4.3a as the little hill up which the ball must be pushed before it can roll by itself down the slope. A reaction with low activation energy proceeds spontaneously when the reactants are brought together. You can demonstrate a *spontaneous reaction* by pouring a little vinegar onto some baking soda and watching the two react to form carbon dioxide. Reactions with high activation energies either do not proceed spontaneously or else proceed too slowly to be useful. For example, if you pour vinegar over a pat of butter, no observable reaction takes place.

Energy Is Trapped or Released during Reactions

One characteristic property of any chemical reaction is the free energy change that occurs as the reaction proceeds. The products of a reaction have either a lower free energy than the reactants or a higher free energy than the reactants. A change in free energy level means that the reaction has either released or trapped energy.

If the free energy of the products is lower than the free energy of the reactants, as in Figure 4.3b, the reaction releases energy and is called an **exergonic reaction** {*ex*-, out + *ergon*, work}. The energy released by an exergonic, or *energy-producing*, reaction may be used by other molecules to do work or may be given off as

FIG. 4.3 Activation energy in exergonic and endergonic reactions

(a) Activation energy is the "push" needed to start a reaction.

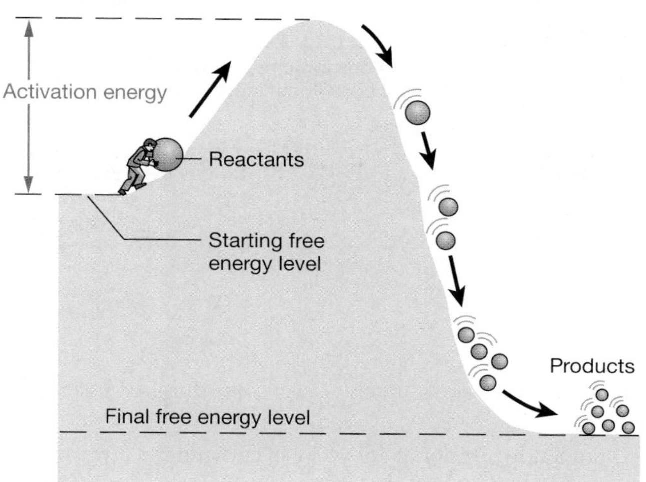
Activation energy
Reactants
Starting free energy level
Products
Final free energy level

(b) Exergonic reactions release energy because the products have less energy than the reactants.

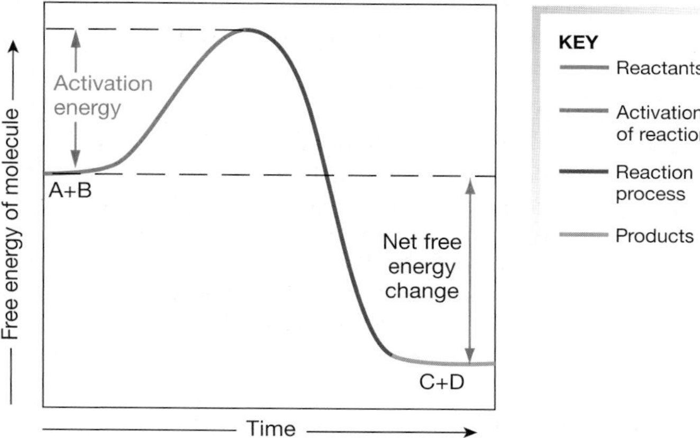
Activation energy
A+B
Net free energy change
C+D
Free energy of molecule
Time
KEY
Reactants
Activation of reaction
Reaction process
Products

(c) Endergonic reactions trap some activation energy in the products, which then have more free energy than the reactants.

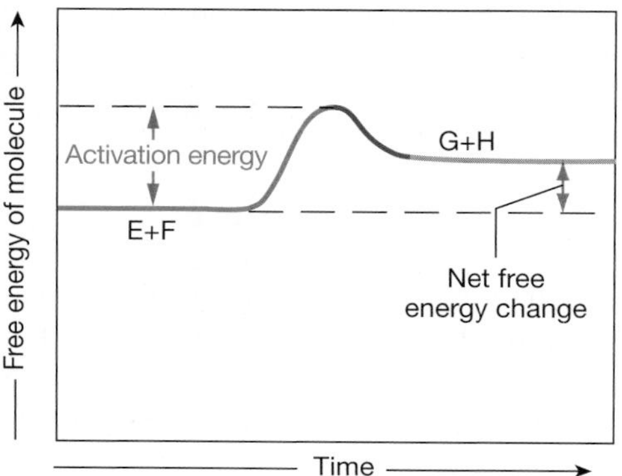
Activation energy
G+H
E+F
Net free energy change
Free energy of molecule
Time

heat. In a few cases, the energy released in an exergonic reaction is stored as potential energy in a concentration gradient.

An important biological example of an exergonic reaction is the combination of ATP and water to form ADP, inorganic phosphate (P_i) and H^+. Energy is released during this reaction when the high-energy phosphate bond of the ATP molecule is broken:

$$ATP + H_2O \rightarrow ADP + P_i + H^+ + \text{energy}$$

Now contrast the exergonic reaction of Figure 4.3b with the reaction represented in Figure 4.3c. In the latter, products retain part of the activation energy that was added, making their free energy greater than that of the reactants. These reactions that require a net input of energy are said to be **endergonic** { *end(o)*, within + *ergon*, work }, or *energy-utilizing*, reactions.

Some of the energy added to an endergonic reaction remains trapped in the chemical bonds of the products. These energy-consuming reactions are often *synthesis* reactions, in which complex molecules are made from smaller molecules. For example, an endergonic reaction links many glucose molecules together to create the glucose polymer glycogen. The complex glycogen molecule has more free energy than the simple glucose molecules used to make it.

If a reaction traps energy as it proceeds in one direction ($A + B \rightarrow C + D$), it releases energy as it proceeds in the reverse direction ($C + D \rightarrow A + B$). (The naming of forward and reverse directions is arbitrary.) For example, the energy trapped in the bonds of glycogen during its synthesis is released when glycogen is broken back down into glucose.

Coupling Endergonic and Exergonic Reactions Where does the activation energy for metabolic reactions come from? The simplest way for a cell to acquire activation energy is to couple an exergonic reaction to an endergonic reaction. Some of the most familiar coupled reactions are those that use the energy released by breaking the high-energy bond of ATP to drive an endergonic reaction:

$$E + F \xrightarrow{ATP \;\rightarrow\; ADP + P_i} G + H$$

In this type of coupled reaction, the two reactions take place simultaneously and in the same location, so that the energy from ATP can be used immediately to drive the endergonic reaction between reactants E and F.

However, it is not always practical for reactions to be directly coupled like this. Consequently, living cells have developed ways to trap the energy released by exergonic reactions and save it for later use. The most common method is to trap the energy in the form of high-energy electrons carried on nucleotides [p. 34]. The nucleotide molecules NADH, $FADH_2$, and NADPH all capture energy in the electrons of their hydrogen atoms (**FIG. 4.4**). NADH and $FADH_2$ usually transfer most of this energy to ATP, which can then be used to drive endergonic reactions.

FIG. 4.4 Energy in biological reactions

Energy released by exergonic reactions can be trapped in the high-energy electrons of NADH, $FADH_2$, or NADPH. Energy that is not trapped is given off as heat.

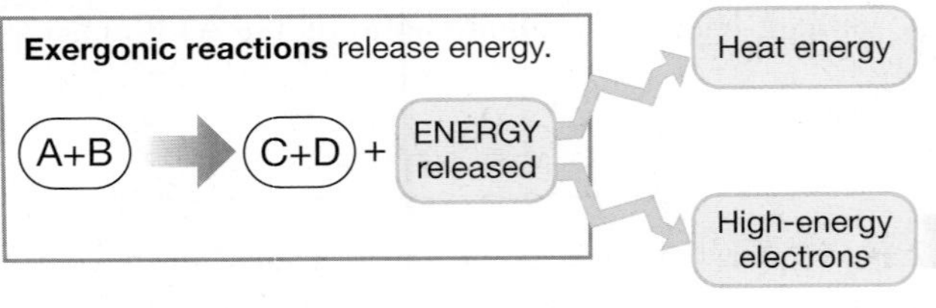

Exergonic reactions release energy.
A+B
C+D +
ENERGY released
Heat energy
High-energy electrons

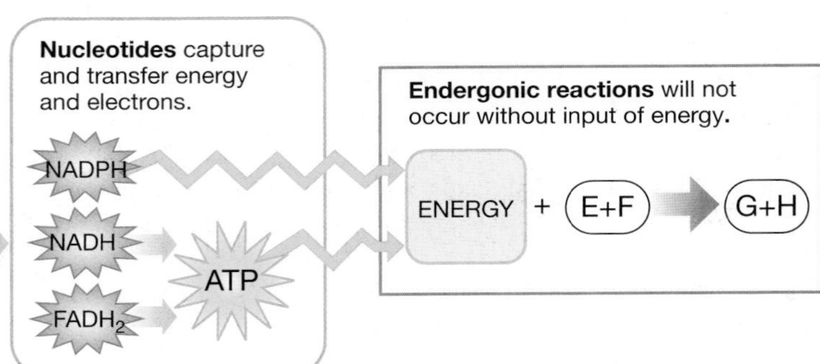

Nucleotides capture and transfer energy and electrons.
NADPH
NADH
$FADH_2$
ATP
Endergonic reactions will not occur without input of energy.
ENERGY + E+F
G+H

Net Free Energy Change Determines Reaction Reversibility

The net free energy change of a reaction plays an important role in determining whether that reaction can be reversed, because the net free energy change of the forward reaction contributes to the activation energy of the reverse reaction. A chemical reaction that can proceed in both directions is called a **reversible reaction.** In a reversible reaction, the forward reaction A + B → C + D and its reverse reaction C + D → A + B are both likely to take place. If a reaction proceeds in one direction but not the other, it is an **irreversible reaction**.

For example, look at the activation energy of the reaction C + D → A + B in **FIGURE 4.5**. This reaction is the reverse of the reaction shown in Figure 4.3b. Because a lot of energy was released in the forward reaction A + B → C + D, the activation energy of the reverse reaction is substantial (Fig. 4.5). As you will recall, the larger the activation energy, the less likely it is that the reaction will proceed spontaneously. Theoretically, all reactions can be reversed with enough energy input, but some reactions release so much energy that they are essentially irreversible.

FIG. 4.5 Some reactions have large activation energies

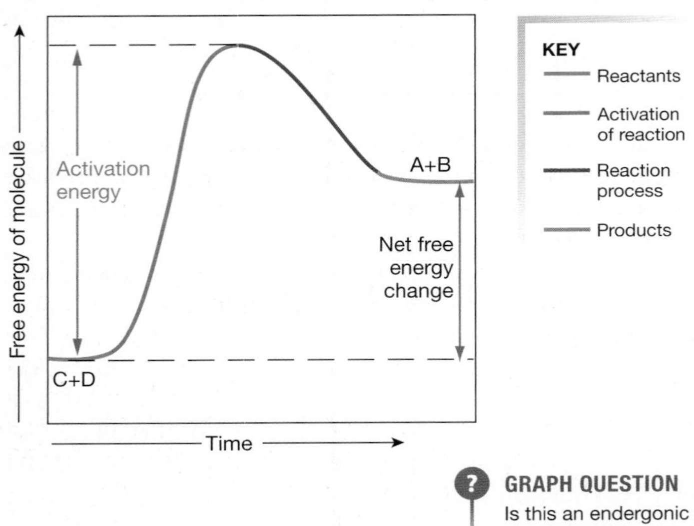

Free energy of molecule
Activation energy
A+B
Net free energy change
C+D
Time
KEY
Reactants
Activation of reaction
Reaction process
Products

GRAPH QUESTION
Is this an endergonic or exergonic reaction?

In your study of physiology, you will encounter a few irreversible reactions. However, most biological reactions are reversible: if the reaction A + B → C + D is possible, then so is the reaction C + D → A + B. Reversible reactions are shown with arrows that point in both directions: A + B $\rightleftharpoons$ C + D. One of the main reasons that many biological reactions are reversible is that they are aided by the specialized proteins known as enzymes.

Concept Check

5. What is the difference between endergonic and exergonic reactions?
6. If you mix baking soda and vinegar together in a bowl, the mixture reacts and foams up, releasing carbon dioxide gas. Name the reactant(s) and product(s) in this reaction.
7. Do you think the reaction of question 6 is endergonic or exergonic? Do you think it is reversible? Defend your answers.

4.3 Enzymes

Enzymes are proteins that speed up the rate of chemical reactions. During these reactions, the enzyme molecules are not changed in any way, meaning they are biological *catalysts*. Without enzymes, most chemical reactions in a cell would go so slowly that the cell would be unable to live. Because an enzyme is not permanently changed or used up in the reaction it catalyzes, we might write it in a reaction equation this way:

$$A + B + \text{enzyme} \rightarrow C + D + \text{enzyme}$$

This way of writing the reaction shows that the enzyme participates with reactants A and B but is unchanged at the end of the reaction. A more common shorthand for enzymatic reactions shows the name of the enzyme above the reaction arrow, like this:

$$A + B \xrightarrow{\text{enzyme}} C + D$$

In enzymatically catalyzed reactions, the reactants A and B are called **substrates**.

CHAPTER 4

RUNNING PROBLEM

Tay-Sachs disease is a devastating condition. Normally, lysosomes in cells contain enzymes that digest old, worn-out parts of the cell. In Tay-Sachs and related lysosomal storage diseases, genetic mutations result in lysosomal enzymes that are ineffective or absent. Tay-Sachs disease patients lack *hexosaminidase A*, an enzyme that digests glycolipids called *gangliosides*. As a result, gangliosides accumulate in nerve cells in the brain, causing them to swell and function abnormally. Infants with Tay-Sachs disease slowly lose muscle control and brain function. There is currently no treatment or cure for Tay-Sachs disease, and affected children usually die before age 4.

Q1: *Hexosaminidase A is also required to remove gangliosides from the light-sensitive cells of the eye. Based on this information, what is another symptom of Tay-Sachs disease besides loss of muscle control and brain function?*

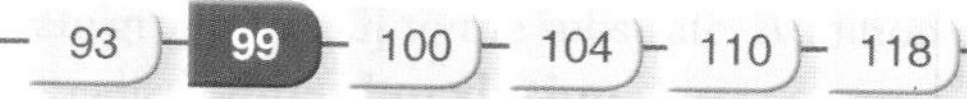

93
99
100
104
110
118

TABLE 4.3 Diagnostically Important Enzymes

Elevated blood levels of these enzymes are suggestive of the pathologies listed.

Enzyme	Related Diseases
Acid phosphatase*	Cancer of the prostate
Alkaline phosphatase	Diseases of bone or liver
Amylase	Pancreatic disease
Creatine kinase (CK)	Myocardial infarction (heart attack), muscle disease
Lactate dehydrogenase (LDH)	Tissue damage to heart, liver, skeletal muscle, red blood cells

*A newer test for a molecule called prostate specific antigen (PSA) has replaced the test for acid phosphatase in the diagnosis of prostate cancer.

Enzymes Are Proteins

Most enzymes are large proteins with complex three-dimensional shapes, although recently researchers discovered that RNA can sometimes act as a catalyst. Like other proteins that bind to substrates, protein enzymes exhibit specificity, competition, and saturation [p. 46].

A few enzymes come in a variety of related forms (isoforms) and are known as **isozymes** {*iso-*, equal} of one another. Isozymes are enzymes that catalyze the same reaction but under different conditions or in different tissues. The structures of related isozymes are slightly different from one another, which causes the variability in their activity. Many isozymes have complex structures with multiple protein chains.

For example, the enzyme *lactate dehydrogenase* (LDH) has two kinds of subunits, named H and M, that are assembled into *tetramers*—groups of four. LDH isozymes include H_4, H_2M_2, and M_4. The different LDH isozymes are tissue specific, including one found primarily in the heart and a second found in skeletal muscle and the liver.

Isozymes have an important role in the diagnosis of certain medical conditions. For example, in the hours following a heart attack, damaged heart muscle cells release enzymes into the blood. One way to determine whether a person's chest pain was indeed due to a heart attack is to look for elevated levels of heart isozymes in the blood. Some diagnostically important enzymes and the diseases of which they are suggestive are listed in **TABLE 4.3**.

Reaction Rates Are Variable

We measure the rate of an enzymatic reaction by monitoring either how fast the products are synthesized or how fast the substrates are consumed. Reaction rate can be altered by a number of factors, including changes in temperature, the amount of enzyme present, and substrate concentrations [p. 51]. In mammals, we consider temperature to be essentially constant. This leaves enzyme amount and substrate concentration as the two main variables that affect reaction rate.

In protein-binding interactions, if the amount of protein (in this case, enzyme) is constant, the reaction rate is proportional to the substrate concentration [see Fig. 2.13b, p. 52]. One strategy cells use to control reaction rates is to regulate the amount of enzyme in the cell. In the absence of appropriate enzyme, many biological reactions go very slowly or not at all. If enzyme is present, the rate of the reaction is proportional to the amount of enzyme and the amount of substrate. If there is so much substrate that all enzyme binding sites are saturated and working at maximum capacity, the reaction rate will reach a maximum [see Fig. 2.13c, p. 52].

This seems simple until you consider a reversible reaction that can go in both directions. In that case, what determines in which direction the reaction goes? The answer is that reversible reactions go to a state of *equilibrium*, where the rate of the reaction in the forward direction ($A + B \rightarrow C + D$) is equal to the rate of the reverse reaction ($C + D \rightarrow A + B$). At equilibrium, there is no net change in the amount of substrate or product, and the ratio $[C][D]/[A][B]$ is equal to the reaction's *equilibrium constant*, K_{eq} [p. 47].

If substrates or products are added or removed by other reactions in a pathway, the reaction rate increases in the forward or reverse direction as needed to restore the ratio $[C][D]/[A][B]$. According to the *law of mass action*, the ratio of [C] and [D] to [A] and [B] is always the same at equilibrium.

Enzymes May Be Activated, Inactivated, or Modulated

Enzyme activity, like the activity of other soluble proteins, can be altered by various factors. Some enzymes are synthesized as inactive molecules (*proenzymes* or *zymogens*) and activated on demand by proteolytic activation [Fig. 2.12a, p. 50]. Others require the binding of inorganic cofactors, such as Ca^{2+} or Mg^{2+} before they become active.

FIG. 4.10 Enzymes control reversibility of metabolic reactions

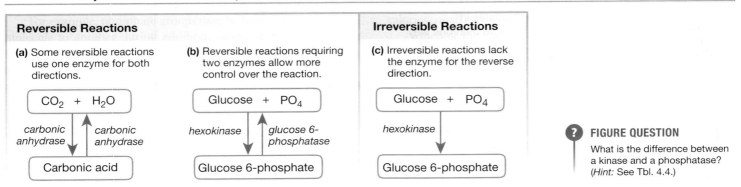

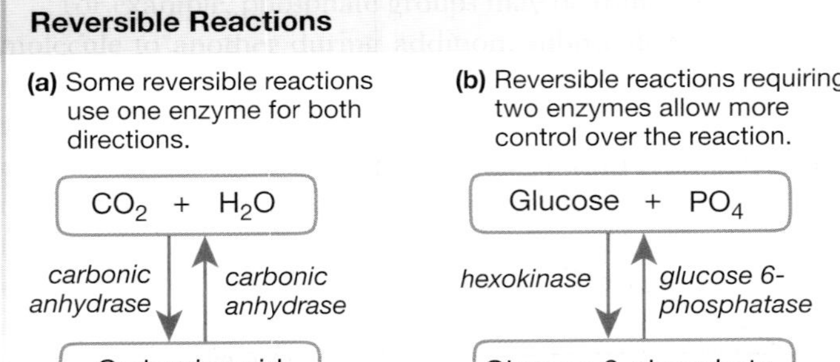
Reversible Reactions
(a) Some reversible reactions use one enzyme for both directions.
CO_2 + H_2O
carbonic anhydrase
carbonic anhydrase
Carbonic acid
(b) Reversible reactions requiring two enzymes allow more control over the reaction.
Glucose + PO_4
hexokinase
glucose 6-phosphatase
Glucose 6-phosphate

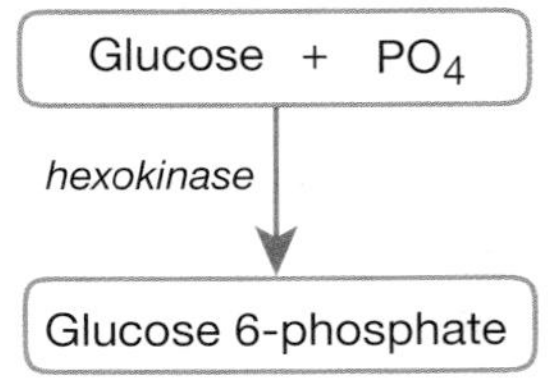
Irreversible Reactions
(c) Irreversible reactions lack the enzyme for the reverse direction.
Glucose + PO_4
hexokinase
Glucose 6-phosphate

FIGURE QUESTION

What is the difference between a kinase and a phosphatase? (*Hint:* See Tbl. 4.4.)

ATP Transfers Energy between Reactions The usefulness of metabolic pathways as suppliers of energy is often measured in terms of the net amount of ATP the pathways can yield. ATP is a nucleotide containing three phosphate groups [p. 34]. One of the three phosphate groups is attached to ADP by a covalent bond in an energy-requiring reaction. Energy is stored in this **high-energy phosphate bond** and then released when the bond is broken during removal of the phosphate group. This relationship is shown by the following reaction:

$$\text{ADP} + \text{P}_i + \text{energy} \rightleftharpoons \text{ADP} \sim \text{P} \ (= \text{ATP})$$

The squiggle ~ indicates a high-energy bond, and P_i is the abbreviation for an inorganic phosphate group. Estimates of the amount of free energy released when a high-energy phosphate bond is broken range from 7 to 12 kcal per mole of ATP.

ATP is more important as a carrier of energy than as an energy-storage molecule. For one thing, the body contains only a limited amount of ATP, estimated at 50 g. But a resting adult human requires 40,000 g (88 pounds!) of ATP to supply the energy for one day's worth of metabolic activity.

RUNNING PROBLEM

In 1989, researchers discovered three genetic mutations responsible for Tay-Sachs disease. This discovery paved the way for a new carrier screening test that detects the presence of one of the three genetic mutations in blood cells rather than testing for lower-than-normal hexosaminidase A levels. David and Sarah will undergo this genetic test.

Q3: *Why might the genetic test for mutations in the Tay-Sachs gene be more accurate than the test that detects decreased amounts of hexosaminidase A?*

Q4: *Can you think of a situation in which the enzyme test might be more accurate than the genetic test?*

93 – 99 – 100 – 104 – 110 – 117

To provide that much ATP, cells constantly recycle ADP from ATP hydrolysis, transferring chemical bond energy from complex biomolecules to the high-energy bonds of ATP, as shown above. In a few cases, the energy is used to make high-energy bonds of a related nucleotide *guanosine triphosphate*, **GTP.** The body stores its energy, therefore, in the chemical bonds of lipids or the glucose polymer glycogen.

The metabolic pathways that yield the most ATP molecules are those that require oxygen—the **aerobic,** or *oxidative*, pathways. **Anaerobic** {*an-*, without + *aer*, air} pathways, which are those that can proceed without oxygen, also produce ATP molecules but in much smaller quantities. The lower ATP yield of anaerobic pathways means that most animals (including humans) are unable to survive for extended periods on anaerobic metabolism alone. In the next section, we consider how biomolecules are metabolized to transfer energy to ATP.

Concept Check

12. Name five ways in which cells regulate the movement of substrates through metabolic pathways.
13. In which part of an ATP molecule is energy trapped and stored? In which part of a NADH molecule is energy stored?
14. What is the difference between aerobic and anaerobic pathways?

Catabolic Pathways Produce ATP

FIGURE 4.11 summarizes the catabolic pathways that extract energy from biomolecules and transfer it to ATP. Aerobic production of ATP from glucose commonly follows two pathways: **glycolysis** {*glycol-*, sweet + *lysis*, dissolve} and the **citric acid cycle** (also known as the tricarboxylic acid cycle). The citric acid cycle was first described by Hans A. Krebs, so it is sometimes called the *Krebs cycle*. Because Dr. Krebs described other metabolic cycles, we will avoid confusion by using the term *citric acid cycle*.

Carbohydrates enter glycolysis in the form of glucose (top of Fig. 4.11). Lipids are broken down into glycerol and fatty acids [p. 30], which enter the pathway at different points: glycerol feeds into glycolysis, and fatty acids are metabolized to acetyl CoA.

FIG. 4.11 ESSENTIALS ATP Production

The catabolic pathways that extract energy from biomolecules and transfer it to ATP are summarized in this overview figure of aerobic respiration of glucose.

Glycolysis and the **citric acid cycle** produce small amounts of ATP directly, but their most important contributions to ATP synthesis are high-energy electrons carried by NADH and $FADH_2$ to the electron transport system in the mitochondria.

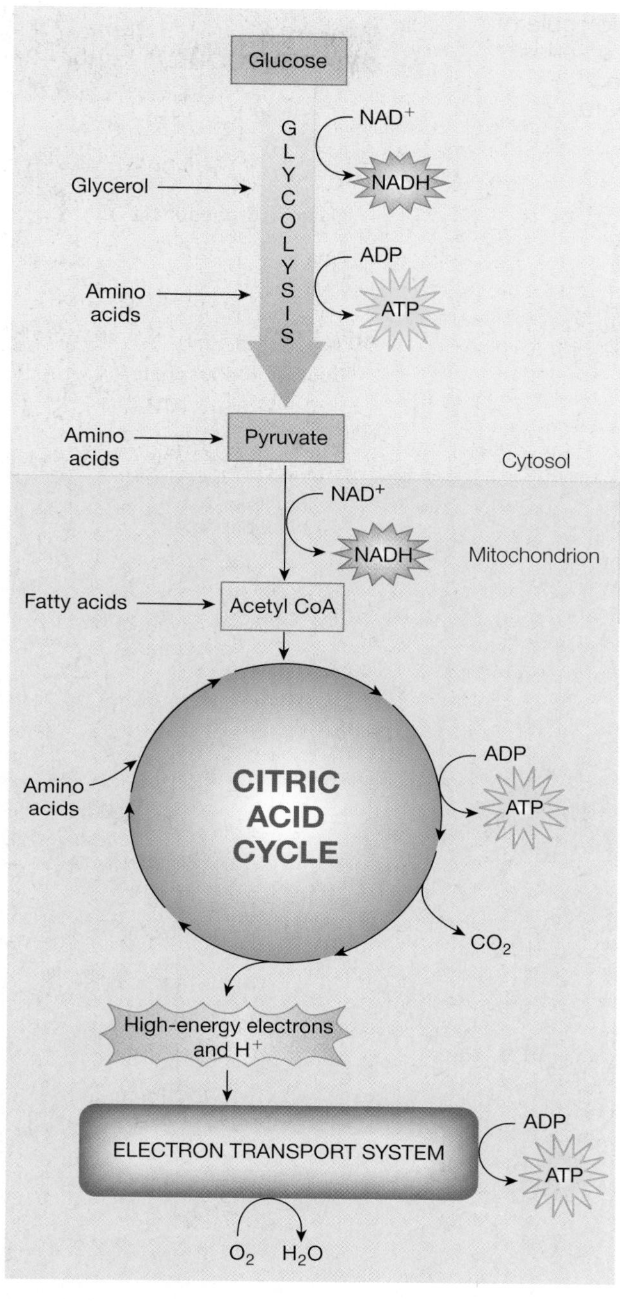

Glucose
GLYCOLYSIS
NAD+
NADH
Glycerol
ADP
ATP
Amino acids
Amino acids
Pyruvate
Cytosol
NAD+
NADH
Mitochondrion
Fatty acids
Acetyl CoA
CITRIC ACID CYCLE
Amino acids
ADP
ATP
CO2
High-energy electrons and H+
ELECTRON TRANSPORT SYSTEM
ADP
ATP
O2
H2O

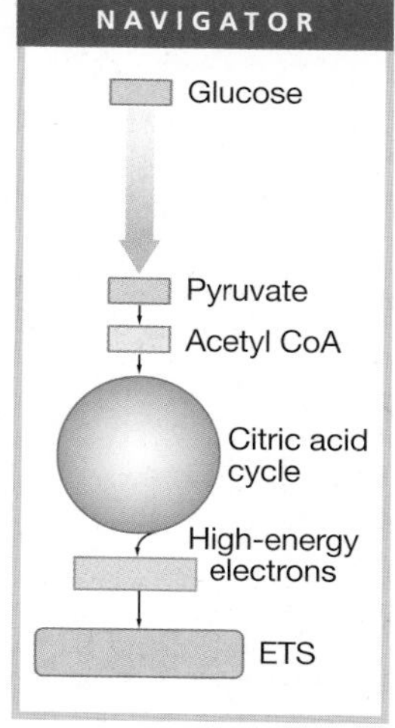

NAVIGATOR
Glucose
Pyruvate
Acetyl CoA
Citric acid cycle
High-energy electrons
ETS

This icon represents the different steps in the metabolic summary figure. Look for it in the figures that follow to help you navigate your way through metabolism.

Aerobic Metabolism of Glucose

The energy production from one glucose molecule can be summarized in the following two equations.

$$\text{Glucose} + O_2 + ADP + P_i \rightarrow CO_2 + H_2O + ATP$$

$$C_6H_{12}O_6 + 6\,O_2 \xrightarrow{30-32\ ADP + P_i \ \rightarrow\ 30-32\ ATP} 6\,CO_2 + 6\,H_2O$$

Proteins are broken down into amino acids, which also enter at various points. Carbons from glycolysis and other nutrients enter the citric acid cycle, which makes a never-ending circle. At each turn, the cycle adds carbons and produces ATP, high-energy electrons, and carbon dioxide.

Both glycolysis and the citric acid cycle produce small amounts of ATP directly, but their most important contribution to ATP synthesis is trapping energy in electrons carried by NADH and $FADH_2$. These compounds transfer the electrons to the **electron transport system** (ETS) in the mitochondria. The electron transport system, in turn, uses energy from those electrons to make the high-energy phosphate bond of ATP. At various points, the process produces carbon dioxide and water. Cells can use the water, but carbon dioxide is a waste product and must be removed from the body.

Because glucose is the only molecule that follows both pathways in their entirety, in this chapter, we look at only glucose catabolism.

- **FIGURE 4.12** summarizes the key steps of glycolysis, the conversion of glucose to pyruvate.
- **FIGURE 4.13** shows how pyruvate is converted to acetyl CoA and how carbons from acetyl CoA go through the citric acid cycle.
- **FIGURE 4.14** illustrates the energy-transferring pathway of the electron transport system.

FIG. 4.12 ESSENTIALS Glycolysis

During glycolysis, one molecule of glucose is converted by a series of enzymatically catalyzed reactions into two pyruvate molecules, producing a net release of energy.

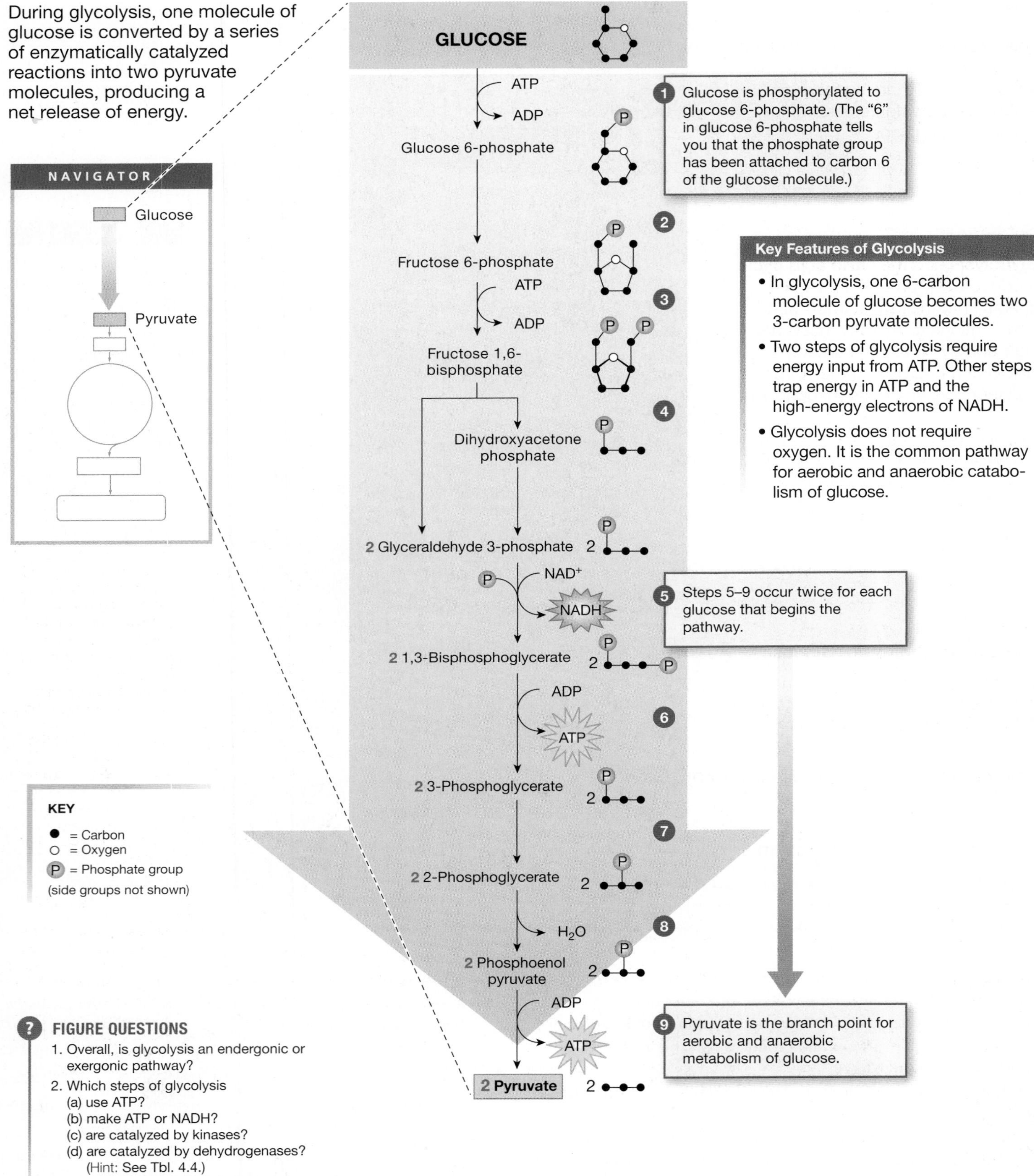

NAVIGATOR
Glucose
Pyruvate
GLUCOSE
ATP
ADP
Glucose 6-phosphate
1 Glucose is phosphorylated to glucose 6-phosphate. (The "6" in glucose 6-phosphate tells you that the phosphate group has been attached to carbon 6 of the glucose molecule.)
2
Fructose 6-phosphate
ATP
ADP
3
Fructose 1,6-bisphosphate
4
Dihydroxyacetone phosphate
2 Glyceraldehyde 3-phosphate 2
P
NAD+
NADH
5 Steps 5–9 occur twice for each glucose that begins the pathway.
2 1,3-Bisphosphoglycerate 2
ADP
ATP
6
2 3-Phosphoglycerate 2
7
2 2-Phosphoglycerate 2
H2O
8
2 Phosphoenol pyruvate 2
ADP
ATP
2 Pyruvate 2
9 Pyruvate is the branch point for aerobic and anaerobic metabolism of glucose.
Key Features of Glycolysis
• In glycolysis, one 6-carbon molecule of glucose becomes two 3-carbon pyruvate molecules.
• Two steps of glycolysis require energy input from ATP. Other steps trap energy in ATP and the high-energy electrons of NADH.
• Glycolysis does not require oxygen. It is the common pathway for aerobic and anaerobic catabolism of glucose.
KEY
● = Carbon
○ = Oxygen
P = Phosphate group
(side groups not shown)

FIGURE QUESTIONS

1. Overall, is glycolysis an endergonic or exergonic pathway?
2. Which steps of glycolysis
   (a) use ATP?
   (b) make ATP or NADH?
   (c) are catalyzed by kinases?
   (d) are catalyzed by dehydrogenases? (Hint: See Tbl. 4.4.)
3. What is the net energy yield (ATP and NADH) for one glucose?

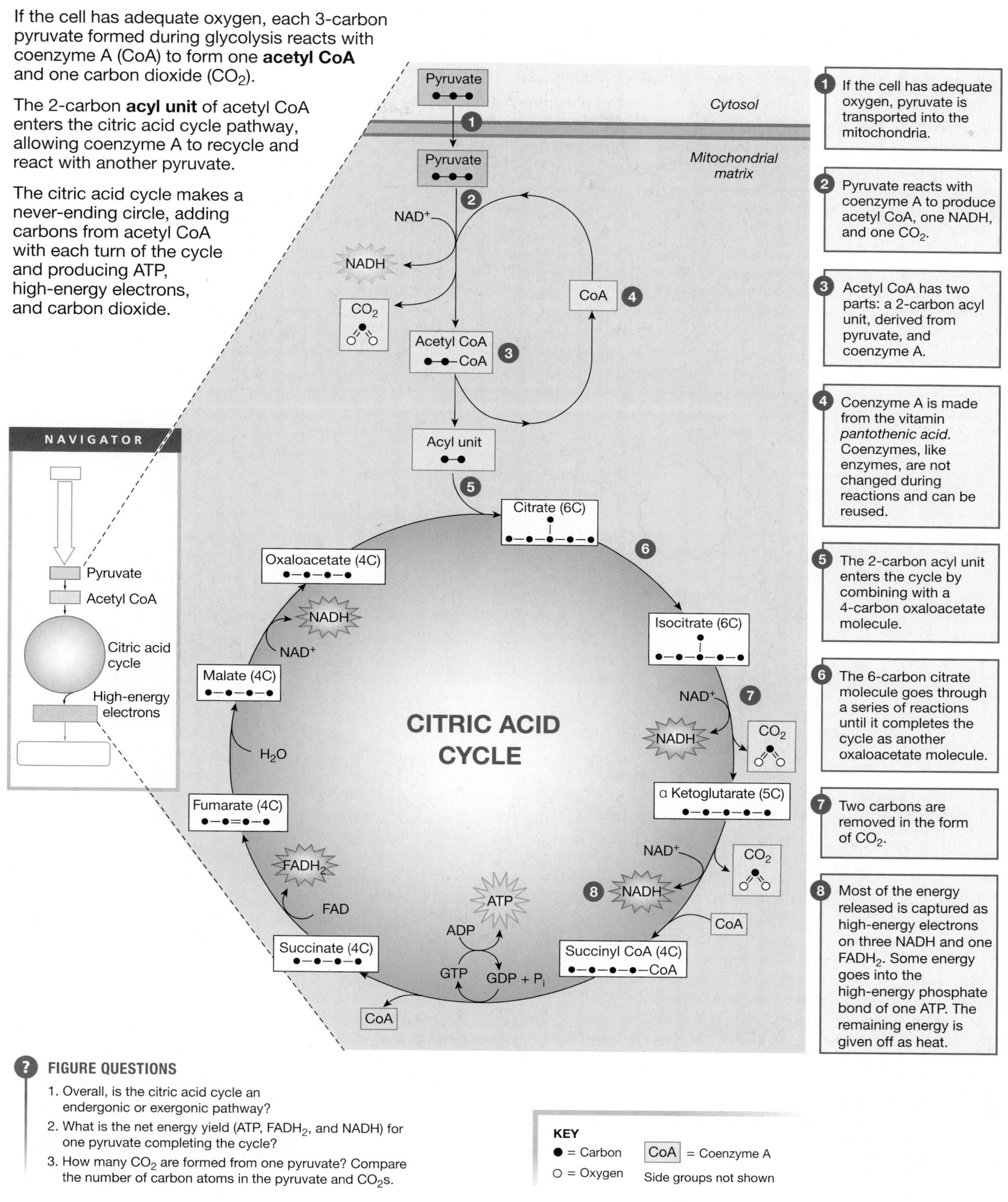

FIG. 4.13 ESSENTIALS Pyruvate, Acetyl CoA, and the Citric Acid Cycle
If the cell has adequate oxygen, each 3-carbon pyruvate formed during glycolysis reacts with coenzyme A (CoA) to form one acetyl CoA and one carbon dioxide (CO_2).
The 2-carbon acyl unit of acetyl CoA enters the citric acid cycle pathway, allowing coenzyme A to recycle and react with another pyruvate.
The citric acid cycle makes a never-ending circle, adding carbons from acetyl CoA with each turn of the cycle and producing ATP, high-energy electrons, and carbon dioxide.
NAVIGATOR
Pyruvate
Acetyl CoA
Citric acid cycle
High-energy electrons
Pyruvate
Cytosol
Pyruvate
Mitochondrial matrix
NAD^+
NADH
CO_2
CoA
Acetyl CoA
CoA
Acyl unit
Citrate (6C)
Oxaloacetate (4C)
NADH
NAD^+
Isocitrate (6C)
Malate (4C)
NAD^+
NADH
CO_2
CITRIC ACID CYCLE
H_2O
α Ketoglutarate (5C)
Fumarate (4C)
NAD^+
CO_2
$FADH_2$
ATP
NADH
FAD
CoA
ADP
Succinate (4C)
Succinyl CoA (4C)
CoA
GTP
GDP + P_i
CoA
1 If the cell has adequate oxygen, pyruvate is transported into the mitochondria.
2 Pyruvate reacts with coenzyme A to produce acetyl CoA, one NADH, and one CO_2.
3 Acetyl CoA has two parts: a 2-carbon acyl unit, derived from pyruvate, and coenzyme A.
4 Coenzyme A is made from the vitamin *pantothenic acid.* Coenzymes, like enzymes, are not changed during reactions and can be reused.
5 The 2-carbon acyl unit enters the cycle by combining with a 4-carbon oxaloacetate molecule.
6 The 6-carbon citrate molecule goes through a series of reactions until it completes the cycle as another oxaloacetate molecule.
7 Two carbons are removed in the form of CO_2.
8 Most of the energy released is captured as high-energy electrons on three NADH and one $FADH_2$. Some energy goes into the high-energy phosphate bond of one ATP. The remaining energy is given off as heat.
FIGURE QUESTIONS
1. Overall, is the citric acid cycle an endergonic or exergonic pathway?
2. What is the net energy yield (ATP, $FADH_2$, and NADH) for one pyruvate completing the cycle?
3. How many CO_2 are formed from one pyruvate? Compare the number of carbon atoms in the pyruvate and CO_2s.
KEY
● = Carbon
○ = Oxygen
CoA = Coenzyme A
Side groups not shown

FIG. 4.14 ESSENTIALS The Electron Transport System

The final step in aerobic ATP production is energy transfer from high-energy electrons of NADH and $FADH_2$ to ATP. This energy transfer requires mitochondrial proteins known as the **electron transport system (ETS),** located in the inner mitochondrial membrane.

- ETS proteins include enzymes and iron-containing **cytochromes**.
- The synthesis of ATP using the ETS is called **oxidative phosphorylation** because the system requires oxygen to act as the final acceptor of electrons and H^+.
- The **chemiosmotic theory** says that potential energy stored by concentrating H^+ in the intermembrane space is used to make the high-energy bond of ATP.

1. NADH and $FADH_2$ release high-energy electrons and H^+ to the ETS. NAD^+ and FAD are coenzymes that recycle.
2. Energy released when pairs of high-energy electrons pass along the transport system is used to concentrate H^+ from the mitochondrial matrix in the intermembrane space. The H^+ concentration gradient is a source of potential energy.
3. By the end of the ETS, the electrons have given up their stored energy.
4. Each pair of electrons released by the ETS combines with two H^+ and an oxygen atom, creating a molecule of water, H_2O.
5. As H^+ move down their concentration gradient through a protein known as **ATP synthase**, the synthase transfers their kinetic energy to the high-energy phosphate bond of ATP.
   - Each 3 H^+ that shuttle through the ATP synthase make a maximum of 1 ATP.
   - A portion of the kinetic energy is released as heat.

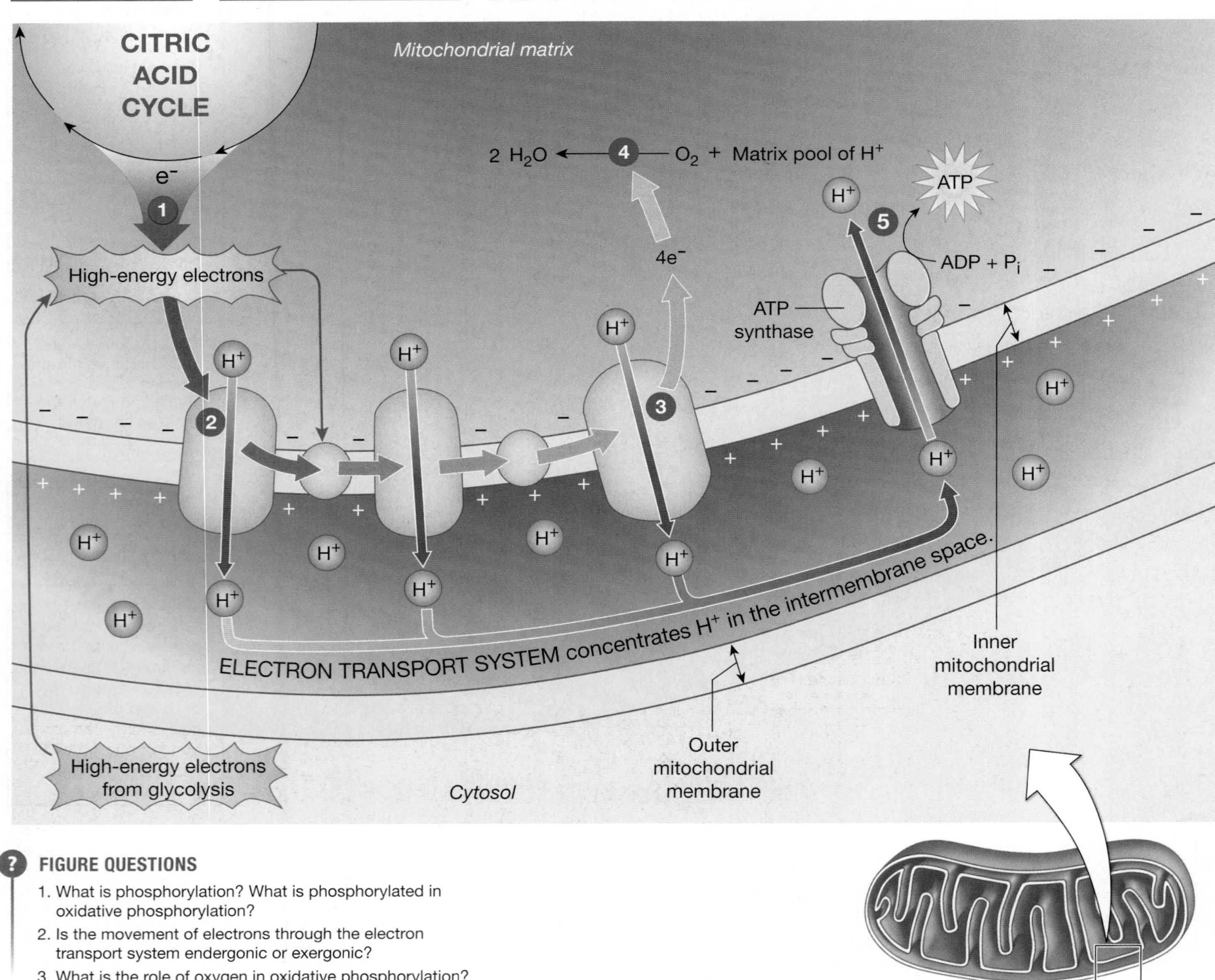

CITRIC ACID CYCLE
Mitochondrial matrix
e⁻
1
High-energy electrons
2 H₂O
4
O₂ + Matrix pool of H⁺
ATP
5
ADP + Pᵢ
4e⁻
ATP synthase
H⁺
2
3
ELECTRON TRANSPORT SYSTEM concentrates H⁺ in the intermembrane space.
Inner mitochondrial membrane
Outer mitochondrial membrane
High-energy electrons from glycolysis
Cytosol

FIGURE QUESTIONS

1. What is phosphorylation? What is phosphorylated in oxidative phosphorylation?
2. Is the movement of electrons through the electron transport system endergonic or exergonic?
3. What is the role of oxygen in oxidative phosphorylation?

We will examine protein and lipid catabolism and synthetic pathways for lipids and glucose when we look at the fate of the nutrients we eat (see Chapter 22).

The aerobic pathways for ATP production are a good example of compartmentation within cells. The enzymes of glycolysis are located in the cytosol, and the enzymes of the citric acid cycle are in the mitochondria. Within mitochondria, concentration of H^+ in the intermembrane compartment stores the energy needed to make the high-energy bond of ATP.

Concept Check

15. Match each component on the left to the molecule(s) it is part of:

(a) amino acids	1. carbohydrates
(b) fatty acids	2. lipids
(c) glycerol	3. polysaccharides
(d) glucose	4. proteins
	5. triglycerides

16. Do endergonic reactions release energy or trap it in the products?

One Glucose Molecule Can Yield 30–32 ATP

Recall from Figure 4.11 that the aerobic metabolism of one glucose molecule produces carbon dioxide, water, and 30–32 ATP. Let's review the role of glycolysis and the citric acid cycle in that ATP production.

In glycolysis (Fig. 4.12), metabolism of one glucose molecule $C_6H_{12}O_6$ has a net yield of two 3-carbon pyruvate molecules, 2 ATP, and high-energy electrons carried on 2 NADH:

$$\text{Glucose} + 2\ \text{NAD}^+ + 2\ \text{ADP} + 2\ \text{P}_i \rightarrow$$
$$2\ \text{Pyruvate} + 2\ \text{ATP} + 2\ \text{NADH} + 2\ \text{H}^+ + 2\ \text{H}_2\text{O}$$

In the next phase, the conversion of pyruvate to acetyl CoA produces one NADH (Fig. 4.13). Carbons from one acetyl CoA going through the citric acid cycle trap energy in 3 NADH molecules, 1 $FADH_2$ and 1 ATP. These steps happen twice for each glucose, giving a total yield of 8 NADH, 2 $FADH_2$, and 2 ATP for the pyruvate-citric acid cycle phase of glucose metabolism.

In the final step, high-energy electrons of NADH and $FADH_2$ passing along the proteins of the electron transport system use their energy to concentrate H^+ in the intermembrane compartment of the mitochondria (Fig. 4.14). When the H^+ move down their concentration gradient through a channel in the ATP synthase, the energy released is transferred to the high-energy phosphate bond of ATP. On average, the NADH and $FADH_2$ from one glucose produce 26–28 ATP.

When we tally the maximum potential energy yield for the catabolism of one glucose molecule through aerobic pathways, the total comes to 30–32 ATP (**FIG. 4.15b**). These numbers are the *potential* maximum because often the mitochondria do not work up to capacity. There are various reasons for this, including the fact that a certain number of H^+ ions leak from the intermembrane space back into the mitochondrial matrix without producing an ATP.

A second source of variability in the number of ATP produced per glucose comes from the two cytosolic NADH molecules produced during glycolysis. These NADH molecules are unable to enter mitochondria and must transfer their electrons through membrane carriers. Inside a mitochondrion, some of these electrons go to $FADH_2$, which has a potential average yield of only 1.5 ATP rather than the 2.5 ATP made by mitochondrial NADH. If cytosolic electrons go to mitochondrial NADH instead, they produce two additional ATP molecules.

Anaerobic Metabolism Makes Two ATP

The metabolism of glucose just described assumes that the cells have adequate oxygen to keep the electron transport system functioning. But what happens to a cell whose oxygen supply cannot keep pace with its ATP demand, such as often happens during strenuous exercise? In that case, the metabolism of glucose shifts from aerobic to anaerobic metabolism, starting at pyruvate (**FIG. 4.16**).

In anaerobic glucose metabolism, pyruvate is converted to lactate instead of being transported into the mitochondria:

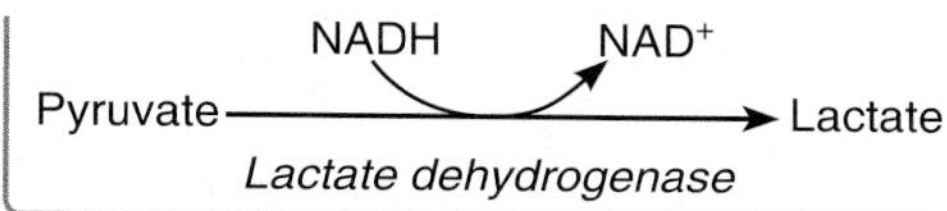
NADH
NAD+
Pyruvate
Lactate
Lactate dehydrogenase

Pyruvate is a branch point for metabolic pathways, like a hub city on a road map. Depending on a cell's needs and oxygen content, pyruvate can be shuttled into the citric acid cycle or diverted into lactate production until oxygen supply improves.

The conversion of pyruvate to lactate changes one NADH back to NAD^+ when a hydrogen atom and an electron are transferred to the lactate molecule. As a result, the net energy yield for the anaerobic metabolism of one glucose molecule is 2 ATP and 0 NADH (Fig. 4.15a), a very puny yield when compared to the 30–32 ATP/glucose that result from aerobic metabolism (Fig. 4.15b). The low efficiency of anaerobic metabolism severely limits its usefulness in most vertebrate cells, whose metabolic energy demand is greater than anaerobic metabolism can provide. Some cells, such as exercising muscle cells, can tolerate anaerobic metabolism for a limited period of time. Eventually, however, they must shift back to aerobic metabolism. Aerobic and anaerobic metabolism in muscle are discussed further in Chapters 12 and 25.

Concept Check

17. How is the separation of mitochondria into two compartments essential to ATP synthesis?

18. Lactate dehydrogenase acts on lactate by (adding or removing?) a(n) __________ and a(n) __________. This process is called (oxidation or reduction?).

19. Describe two differences between aerobic and anaerobic metabolism of glucose.

FIG. 4.15 Energy yields from catabolism of one glucose molecule

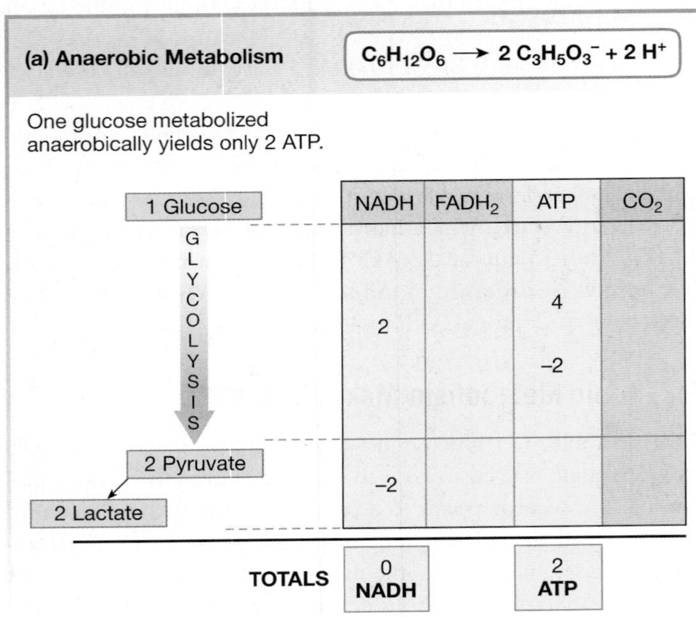
(a) Anaerobic Metabolism
$C_6H_{12}O_6 \longrightarrow 2\ C_3H_5O_3^- + 2\ H^+$
One glucose metabolized anaerobically yields only 2 ATP.
1 Glucose
GLYCOLYSIS
2 Pyruvate
2 Lactate
NADH
FADH2
ATP
CO2
2
4
-2
-2
TOTALS
0 NADH
2 ATP

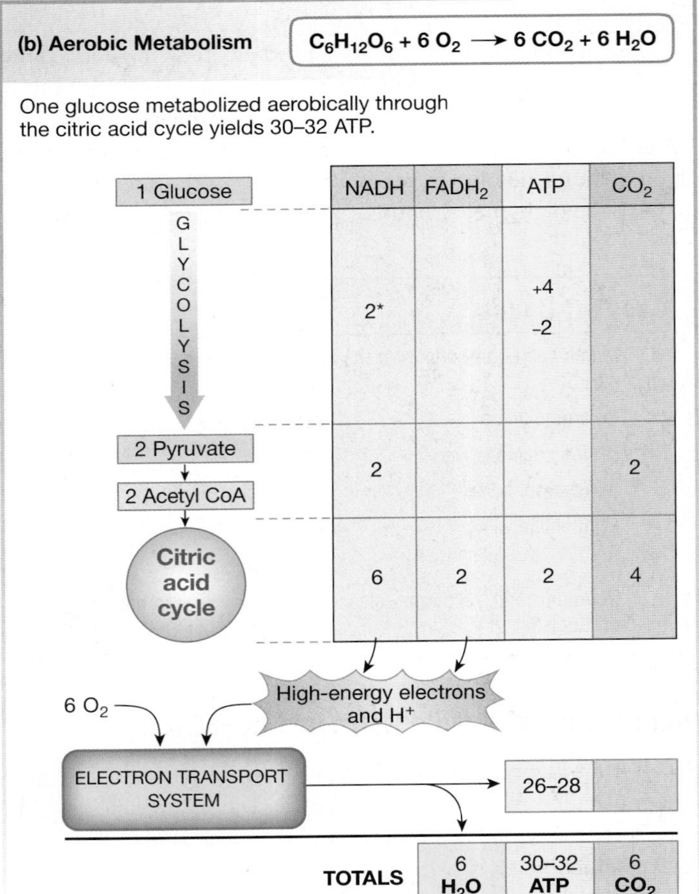
(b) Aerobic Metabolism
$C_6H_{12}O_6 + 6\ O_2 \longrightarrow 6\ CO_2 + 6\ H_2O$
One glucose metabolized aerobically through the citric acid cycle yields 30–32 ATP.
1 Glucose
GLYCOLYSIS
2 Pyruvate
2 Acetyl CoA
Citric acid cycle
NADH
FADH2
ATP
CO2
2*
+4
-2
2
2
6
2
2
4
6 O2
High-energy electrons and H+
ELECTRON TRANSPORT SYSTEM
26–28
TOTALS
6 H2O
30–32 ATP
6 CO2
* Cytoplasmic NADH sometimes yields only 1.5 ATP/NADH instead of 2.5 ATP/NADH.

FIGURE QUESTIONS

1. How many NADH enter the electron transport system when glucose is metabolized to lactate?
2. Some amino acids can be converted to pyruvate. If one amino acid becomes one pyruvate, what is the ATP yield from aerobic metabolism of that amino acid?

Proteins Are the Key to Cell Function

As you have seen, proteins are the molecules that run a cell from day to day. Protein enzymes control the synthesis and breakdown of carbohydrates, lipids, structural proteins, and signal molecules. Protein transporters and pores in the cell membrane and in organelle membranes regulate the movement of molecules into and out of compartments. Other proteins form the structural skeleton of cells and tissues. In these and other ways, protein synthesis is critical to cell function.

The power of proteins arises from their tremendous variability and specificity. Protein synthesis using 20 amino acids can be compared to creating a language with an alphabet of 20 letters. The "words" vary in length from three letters to hundreds of letters, spelling out the structure of thousands of different proteins with different functions. A change in one amino acid during protein synthesis can alter the protein's function, just as changing one letter turns the word "foot" into "food."

The classic example of an amino acid change causing a problem is sickle cell disease. In this inherited condition, when the amino acid valine replaces one glutamic acid in the protein chain, the change alters the shape of hemoglobin. As a result, red blood cells containing the abnormal hemoglobin take on a crescent

RUNNING PROBLEM

David and Sarah had their blood drawn for the genetic test several weeks ago and have been anxiously awaiting the results. Today, they returned to the hospital to hear the news. The tests show that Sarah carries the gene for Tay-Sachs disease but David does not. This means that although some of their children may be carriers of the Tay-Sachs gene like Sarah, none of the children will develop the disease.

Q5: *The Tay-Sachs gene is a recessive gene (t). If Sarah is a carrier of the gene (Tt) but David is not (TT), what is the chance that any child of theirs will be a carrier? (Consult a general biology or genetics text if you need help solving this problem.)*

93 – 99 – 100 – 104 – 110 – 117

FIG. 4.16 Aerobic and anaerobic metabolism

Pyruvate is the branch point between aerobic and anaerobic metabolism of glucose.

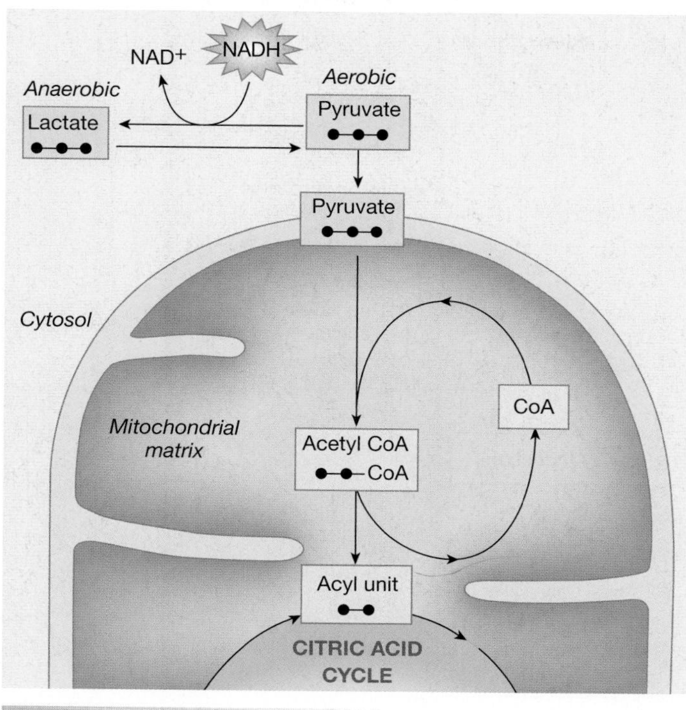
NAD+
NADH
Anaerobic
Aerobic
Lactate
Pyruvate
Pyruvate
Cytosol
Mitochondrial matrix
CoA
Acetyl CoA
CoA
Acyl unit
CITRIC ACID CYCLE

KEY
● = Carbon
○ = Oxygen
CoA = Coenzyme A
H and –OH not shown

FIG. 4.17 The genetic code as it appears in the codons of mRNA

The three-letter abbreviations to the right of the brackets indicate the amino acid each codon represents. The start and stop codons are also marked.

First base of codon	Second base: U		Second base: C		Second base: A		Second base: G		Third base of codon
U	UUU	Phe	UCU	Ser	UAU	Tyr	UGU	Cys	U
U	UUC	Phe	UCC	Ser	UAC	Tyr	UGC	Cys	C
U	UUA	Leu	UCA	Ser	UAA	Stop	UGA	Stop	A
U	UUG	Leu	UCG	Ser	UAG	Stop	UGG	Trp	G
C	CUU	Leu	CCU	Pro	CAU	His	CGU	Arg	U
C	CUC	Leu	CCC	Pro	CAC	His	CGC	Arg	C
C	CUA	Leu	CCA	Pro	CAA	Gln	CGA	Arg	A
C	CUG	Leu	CCG	Pro	CAG	Gln	CGG	Arg	G
A	AUU	Ile	ACU	Thr	AAU	Asn	AGU	Ser	U
A	AUC	Ile	ACC	Thr	AAC	Asn	AGC	Ser	C
A	AUA	Ile	ACA	Thr	AAA	Lys	AGA	Arg	A
A	AUG	Met Start	ACG	Thr	AAG	Lys	AGG	Arg	G
G	GUU	Val	GCU	Ala	GAU	Asp	GGU	Gly	U
G	GUC	Val	GCC	Ala	GAC	Asp	GGC	Gly	C
G	GUA	Val	GCA	Ala	GAA	Glu	GGA	Gly	A
G	GUG	Val	GCG	Ala	GAG	Glu	GGG	Gly	G

(sickle) shape, which causes them to get tangled up and block small blood vessels.

The Protein "Alphabet" One of the mysteries of biology until the 1960s was the question of how only four nitrogenous bases in the DNA molecule—adenine (A), guanine (G), cytosine (C), and thymine (T)—could code for more than 20 different amino acids. If each base controlled the synthesis of one amino acid, a cell could make only four different amino acids. If pairs of bases represented different amino acids, the cell could make 4^2 or 16 different amino acids. Because we have 20 amino acids, this is still not satisfactory. If triplets of bases were the codes for different molecules, however, DNA could create 4^3 or 64 different amino acids. These triplets, called **codons,** are indeed the way information is encoded in DNA and RNA. **FIGURE 4.17** shows the genetic code as it appears in one form of RNA. Remember that RNA substitutes the base uracil (U) for the DNA base thymine [p. 35].

Of the 64 possible triplet combinations, one DNA codon (TAC) acts as the initiator or *start codon* that signifies the beginning of a coding sequence. Three codons serve as terminator or *stop codons* that show where the sequence ends. The remaining 60 triplets all code for amino acids. Methionine and tryptophan have only one codon each, but the other amino acids have between two and six different codons each. Thus, like letters spelling words, the DNA base sequence determines the amino acid sequence of proteins.

Unlocking DNA's Code How does a cell know which of the thousands of bases present in its DNA sequence to use in making a protein? It turns out that the information a cell needs to make a particular protein is contained in a segment of DNA known as a gene. What exactly is a gene? The definition keeps changing, but for this text we will say that a **gene** is a region of DNA that contains the information needed to make a functional piece of RNA, which in turn can make a protein.

FIGURE 4.18 shows the five major steps from gene to RNA to functional protein. First, a section of DNA containing a gene must be activated so that its code can be read **1**. Genes that are continuously being read and converted to RNA messages are said to be *constitutively active*. Usually these genes code for proteins that are essential to ongoing cell functions. Other genes are *regulated*—that is, their activity can be turned on (*induced*) or turned off (*repressed*) by regulatory proteins.

Once a gene is activated, the DNA base sequence of the gene is used to create a piece of RNA in the process known as **transcription** {*trans*, over + *scribe*, to write} (Fig. 4.18 **2**). Human cells have three major forms of RNA: **messenger RNA** (mRNA), **transfer RNA** (tRNA), and **ribosomal RNA** (rRNA). Messenger RNA is processed in the nucleus after it is made **3**. It may either undergo *alternative splicing* (discussed shortly) before leaving the nucleus or be "silenced" and destroyed by enzymes through *RNA interference*. Processed mRNA leaves the nucleus and enters the cytosol. There it works with tRNA and rRNA to direct **translation,** the assembly of amino acids into a protein chain **4**.

FIG. 4.18 ESSENTIALS Overview of Protein Synthesis

The major steps required to convert the genetic code of DNA into a functional protein.

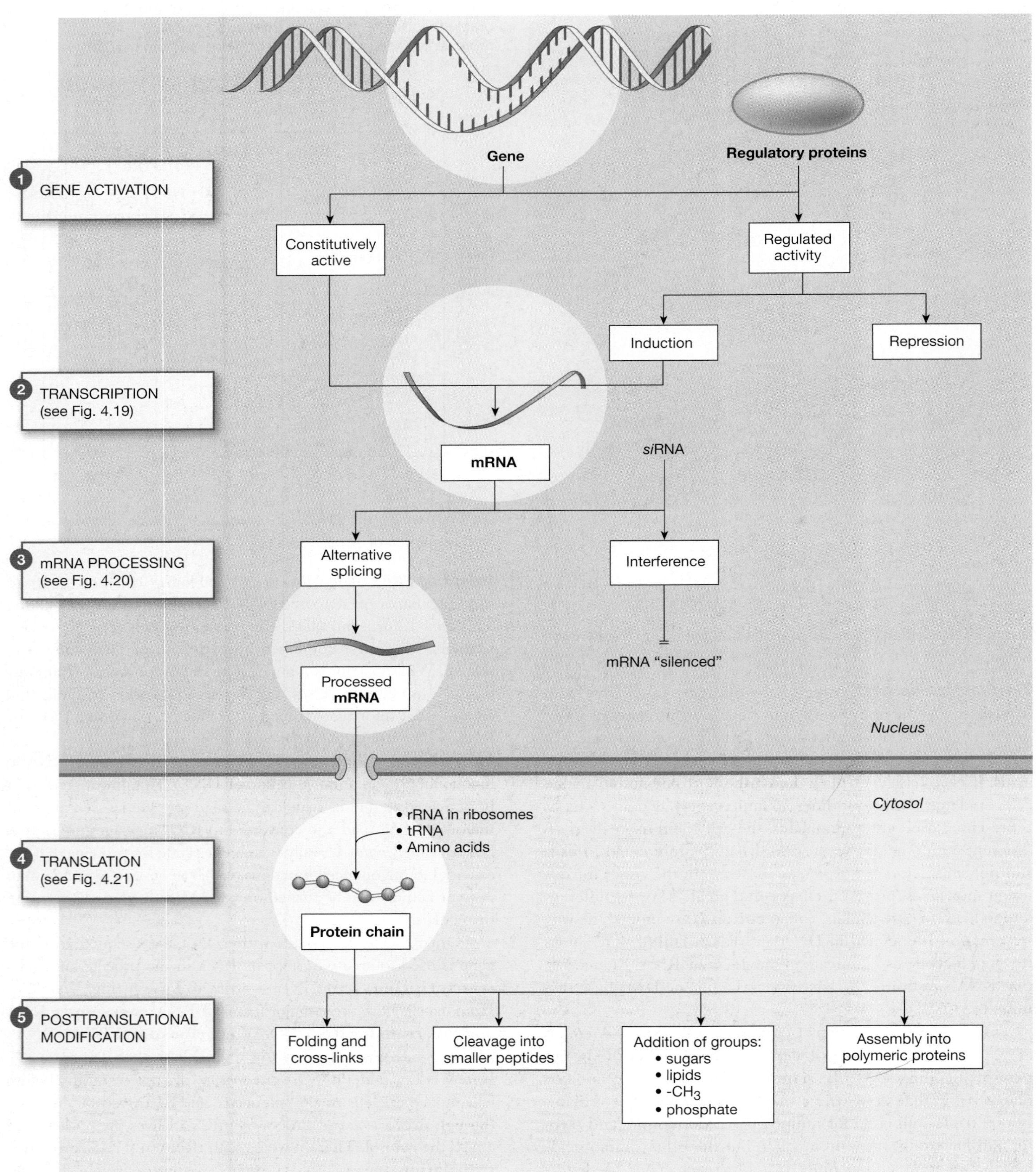
Gene
Regulatory proteins
1 GENE ACTIVATION
Constitutively active
Regulated activity
Induction
Repression
2 TRANSCRIPTION (see Fig. 4.19)
mRNA
siRNA
3 mRNA PROCESSING (see Fig. 4.20)
Alternative splicing
Interference
mRNA "silenced"
Processed mRNA
Nucleus
Cytosol
• rRNA in ribosomes
• tRNA
• Amino acids
4 TRANSLATION (see Fig. 4.21)
Protein chain
5 POSTTRANSLATIONAL MODIFICATION
Folding and cross-links
Cleavage into smaller peptides
Addition of groups:
• sugars
• lipids
• -CH3
• phosphate
Assembly into polymeric proteins

Newly synthesized proteins are then subject to **posttranslational modification** (Fig. 4.18 5). They fold into complex shapes, may be split by enzymes into smaller peptides, or have various chemical groups added to them. The remainder of this chapter looks at transcription, RNA processing, translation, and posttranslational modification in more detail.

Play BioFlix Animation @Mastering Anatomy & Physiology

DNA Guides the Synthesis of RNA

The first steps in protein synthesis are compartmentalized within the nucleus because DNA is a very large molecule that cannot pass through the nuclear envelope. Transcription uses DNA as a template to create a small single strand of RNA that can leave the nucleus (**FIG. 4.19**). The synthesis of RNA from the double-stranded DNA template requires an enzyme known as **RNA polymerase,** plus magnesium or manganese ions and energy in the form of high-energy phosphate bonds:

$$\text{DNA template} + \text{nucleotides A, U, C, G} \xrightarrow{\textit{RNA polymerase, } Mg^{2+} \textit{ or } Mn^{2+}\textit{, and energy}} \text{DNA template} + \text{mRNA}$$

A **promoter** region that precedes the gene must be activated before transcription can begin. Regulatory-protein **transcription factors** bind to DNA and activate the promoter. The active promoter tells the RNA polymerase where to bind to the DNA

FIG. 4.19 Transcription

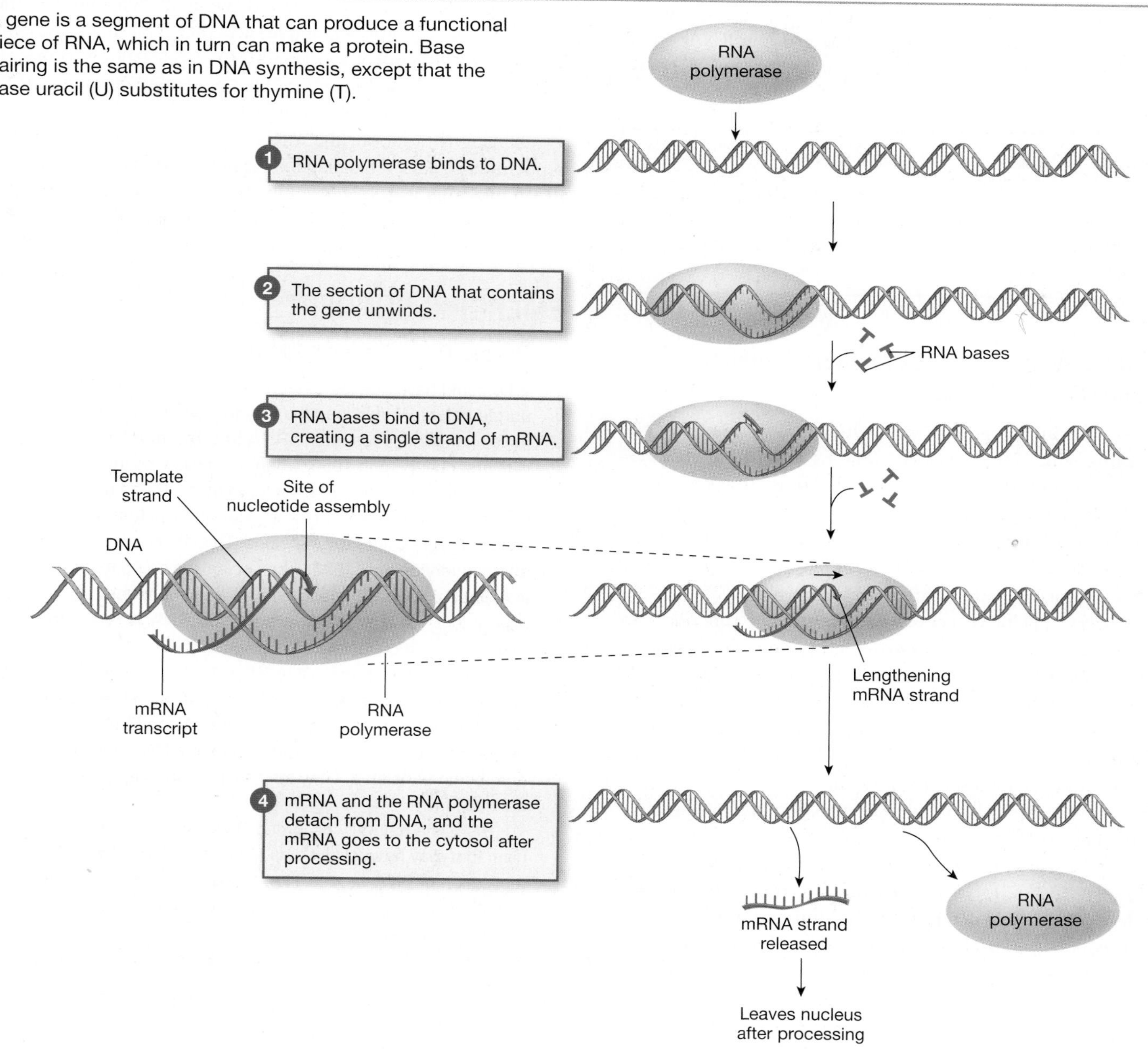
A gene is a segment of DNA that can produce a functional piece of RNA, which in turn can make a protein. Base pairing is the same as in DNA synthesis, except that the base uracil (U) substitutes for thymine (T).
RNA polymerase
1 RNA polymerase binds to DNA.
2 The section of DNA that contains the gene unwinds.
RNA bases
3 RNA bases bind to DNA, creating a single strand of mRNA.
Template strand
Site of nucleotide assembly
DNA
mRNA transcript
RNA polymerase
Lengthening mRNA strand
4 mRNA and the RNA polymerase detach from DNA, and the mRNA goes to the cytosol after processing.
mRNA strand released
RNA polymerase
Leaves nucleus after processing

(Fig. 4.19 1). The polymerase moves along the DNA molecule and "unwinds" the double strand by breaking the hydrogen bonds between paired bases 2. One strand of DNA, called the *template strand*, serves as the guide for RNA synthesis 3. The promoter region is not transcribed into RNA.

During transcription, each base in the DNA template strand pairs with the complementary RNA base (G-C, C-G, T-A, A-U). This pairing of complementary bases is similar to the process by which a double strand of DNA forms (see Appendix C for a review of DNA synthesis). For example, a DNA segment containing the base sequence AGTAC is transcribed into the RNA sequence UCAUG.

As the RNA bases bind to the DNA template strand, they also bond with one another to create a single strand of RNA. During transcription, bases are linked at an average rate of 40 per second. In humans, the largest RNAs may contain as many as 5,000 bases, and their transcription may take more than a minute—a long time for a cellular process. When RNA polymerase reaches the stop codon, it stops adding bases to the growing RNA strand and releases the strand (Fig. 4.19 4).

Concept Check

20. Use the genetic code in Figure 4.17 to write the DNA codons that correspond to the three mRNA stop codons.
21. What does the name RNA polymerase tell you about the function of this enzyme?

Alternative Splicing Creates Multiple Proteins from One DNA Sequence

The next step in the process of protein synthesis is **mRNA processing,** which takes two forms (Fig. 4.18 3). In *RNA interference*, newly synthesized mRNA is inactivated or destroyed before it can be translated into proteins (see the Emerging Concepts box). In **alternative splicing,** enzymes clip segments out of the middle or off the ends of the mRNA strand. Other enzymes then splice the remaining pieces of the strand back together.

Alternative splicing is necessary because a gene contains both segments that encode proteins (**exons**) and noncoding segments called **introns** (**FIG. 4.20**). That means the mRNA initially made from the gene's DNA contains noncoding segments that must be removed before the mRNA leaves the nucleus. The result of alternative splicing is a smaller piece of mRNA that now contains only the coding sequence for a specific protein.

One advantage of alternative splicing is that it allows a single base sequence on DNA to code for more than one protein. The designation of segments as coding or noncoding is not fixed for a given gene. Segments of mRNA that are removed one time can be left in the next time, producing a finished mRNA with a different sequence. The closely related forms of a single enzyme known as *isozymes* are probably made by alternative splicing of a single gene.

After mRNA has been processed, it exits the nucleus through nuclear pores and goes to ribosomes in the cytosol. There mRNA directs the construction of protein.

Concept Check

22. Explain in one or two sentences the relationship of mRNA, nitrogenous bases, introns, exons, mRNA processing, and proteins.

mRNA Translation Links Amino Acids

Protein synthesis requires cooperation and coordination among all three types of RNA: mRNA, rRNA, and tRNA. Upon arrival in the cytosol, processed mRNA binds to ribosomes, which are small particles of protein and several types of rRNA [p. 35]. Each ribosome has two subunits, one large and one small, that come together when protein synthesis begins (**FIG. 4.21** 3). The small ribosomal subunit binds the mRNA, then adds the large subunit so that the mRNA is sandwiched in the middle. Now the ribosome-mRNA complex is ready to begin translation.

During translation, the mRNA codons are matched to the proper amino acid. This matching is done with the assistance of a tRNA molecule (Fig. 4.21 4). One region of each tRNA contains a three-base sequence called an **anticodon** that is complementary

EMERGING CONCEPTS

Purple Petunias and RNAi

Who could have guessed that research to develop a deep purple petunia would lead the way to a whole new area of molecular biology research? **RNA interference (RNAi)** was first observed in 1990, when botanists who introduced purple pigment genes into petunias ended up with plants that were white or striped with white instead of the deeper purple color they expected. This observation did not attract attention until 1998, when scientists doing research in animal biology and medicine had similar problems in experiments on a nematode worm. Now RNAi is one of the newest tools in biotechnology research.

In very simple terms, RNA "silencing" of mRNA is a naturally occurring event accomplished through the production or introduction of *small interfering RNA* (siRNA) molecules. These short RNA strands bind to mRNA and keep it from being translated. They may even target the mRNA for destruction.

RNAi is a naturally occurring RNA-processing mechanism that may have evolved as a means of blocking the replication of RNA viruses. Now researchers are using it to selectively block the production of single proteins within a cell. The scientists' ultimate goal is to create technologies that can be used for the diagnosis and treatment of disease.

FIG. 4.20 mRNA processing

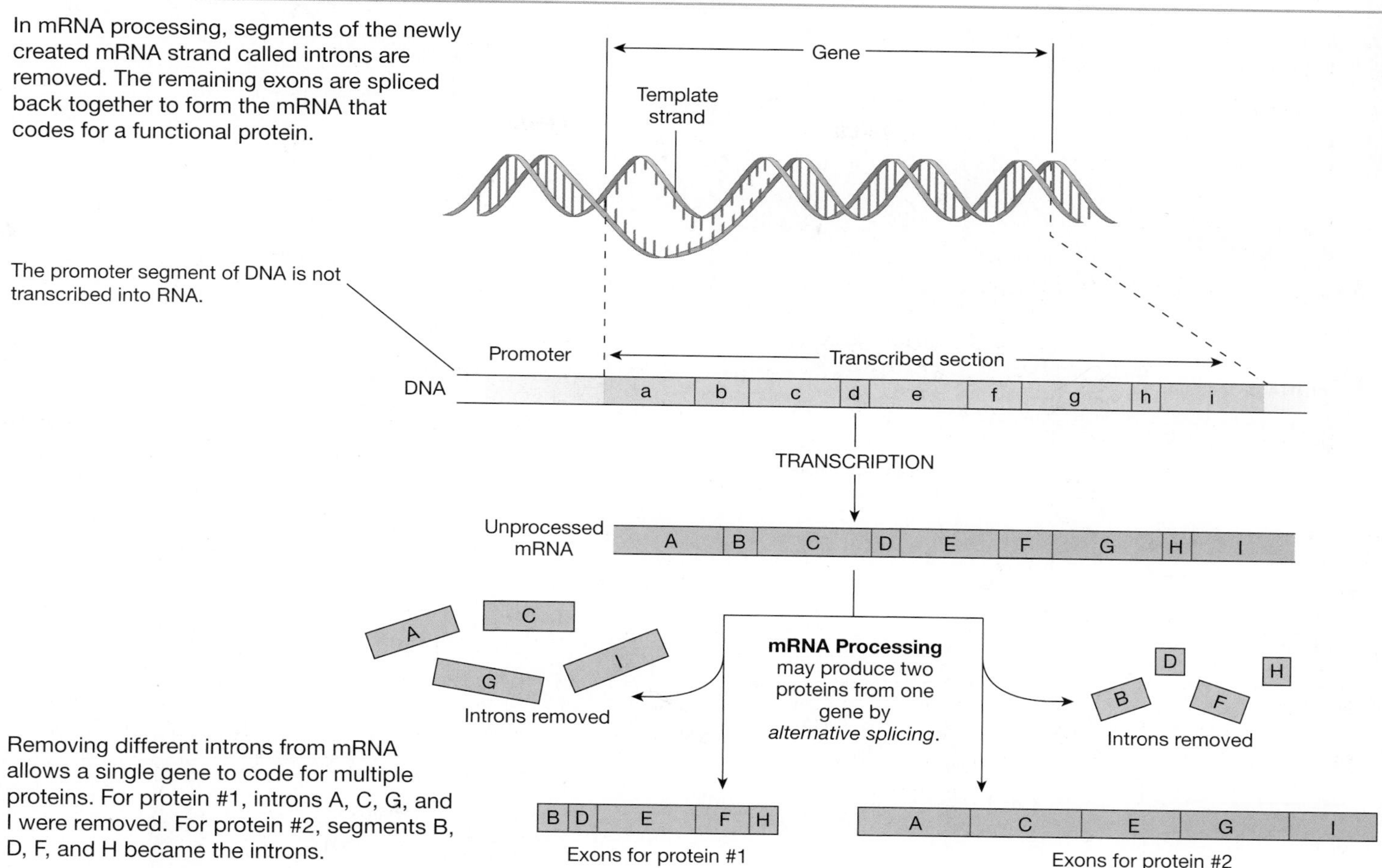
In mRNA processing, segments of the newly created mRNA strand called introns are removed. The remaining exons are spliced back together to form the mRNA that codes for a functional protein.
Gene
Template strand
The promoter segment of DNA is not transcribed into RNA.
Promoter
Transcribed section
DNA
a
b
c
d
e
f
g
h
i
TRANSCRIPTION
Unprocessed mRNA
A
B
C
D
E
F
G
H
I
mRNA Processing may produce two proteins from one gene by alternative splicing.
Introns removed
Introns removed
Removing different introns from mRNA allows a single gene to code for multiple proteins. For protein #1, introns A, C, G, and I were removed. For protein #2, segments B, D, F, and H became the introns.
Exons for protein #1
Exons for protein #2

to an mRNA codon. A different region of the tRNA molecule binds to a specific amino acid.

As translation begins, the anticodons of tRNAs carrying amino acids attach to the complementary codons of ribosomal mRNA. For example, a tRNA with anticodon sequence UUU carries the amino acid lysine. The UUU anticodon pairs with an AAA codon, one of two codons for lysine, on mRNA. The pairing between mRNA and tRNA puts newly arrived amino acids into the correct orientation to link to the growing peptide chain.

Dehydration synthesis links amino acids by creating a *peptide bond* between the amino group ($-NH_2$) of the newly arrived amino acid and the carboxyl end ($-COOH$) of the peptide chain [p. 32]. Once this happens, mRNA releases the "empty" tRNA. The tRNA can then attach to another amino acid molecule with the aid of a cytosolic enzyme and ATP.

When the last amino acid has been joined to the newly synthesized peptide chain, the termination stage has been reached (Fig. 4.21 **5**). The mRNA, the peptide, and the ribosomal subunits separate. The ribosomes are ready for a new round of protein synthesis, but the mRNA is broken down by enzymes known as *ribonucleases*. Some forms of mRNA are broken down quite rapidly, while others may linger in the cytosol and be translated many times.

Protein Sorting Directs Proteins to Their Destination

One of the amazing aspects of protein synthesis is the way specific proteins go from the ribosomes directly to where they are needed in the cell, a process called *protein sorting*. Many newly made proteins carry a *sorting signal*, an address label that tells the cell where the protein should go. Some proteins that are synthesized on cytosolic ribosomes do not have sorting signals. Without a "delivery tag," they remain in the cytosol when they are released from the ribosome [Fig. 3.7, p. 72].

The sorting signal is a special segment of amino acids known as a **signal sequence.** The signal sequence tag directs the protein to the proper organelle, such as the mitochondria or peroxisomes, and allows it to be transported through the organelle membrane. Peptides synthesized on ribosomes attached to the rough endoplasmic reticulum have a signal sequence that directs them through the membrane of the rough ER and into the lumen of this organelle. Once a protein enters the ER lumen, enzymes remove the signal sequence.

Proteins Undergo Posttranslational Modification

The amino acid sequence that comes off a ribosome is the primary structure of a newly synthesized protein [p. 32], but not the final form.

FIG. 4.21 Translation

Translation matches the codons of RNA with amino acids to create a protein.

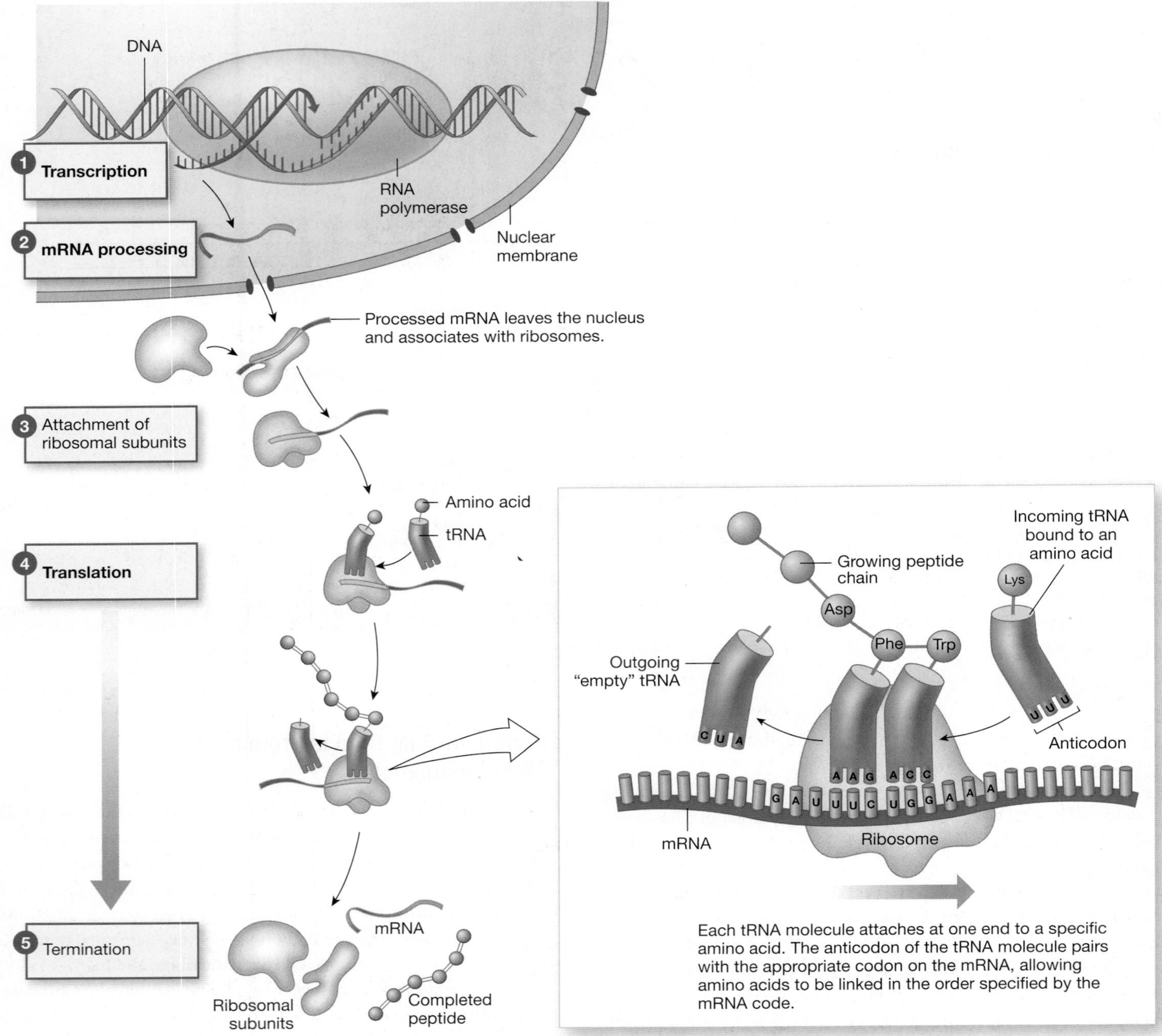
DNA
1 Transcription
RNA polymerase
Nuclear membrane
2 mRNA processing
Processed mRNA leaves the nucleus and associates with ribosomes.
3 Attachment of ribosomal subunits
Amino acid
tRNA
4 Translation
5 Termination
mRNA
Ribosomal subunits
Completed peptide
Incoming tRNA bound to an amino acid
Growing peptide chain
Lys
Asp
Phe
Trp
Outgoing "empty" tRNA
C U A
U U U
Anticodon
A A G
A C C
G A U U U C U G G A A A
mRNA
Ribosome
Each tRNA molecule attaches at one end to a specific amino acid. The anticodon of the tRNA molecule pairs with the appropriate codon on the mRNA, allowing amino acids to be linked in the order specified by the mRNA code.

The newly made protein can now form different types of covalent and noncovalent bonds, a process known as **posttranslational modification.** Cleavage of the amino acid chain, attachment of molecules or groups, and cross-linkages are three general types of posttranslational modification. More than 100 different types of posttranslational modification have been described so far.

In some common forms of posttranslational modification, the amino acid chain can:

1. fold into various three-dimensional shapes. Protein folding creates the tertiary structure of the protein.
2. create cross-links between different regions of its amino acid chain.
3. be cleaved (split) into fragments.
4. add other molecules or groups.
5. assemble with other amino acid chains into a polymeric (many-part) protein. Assembly of proteins into polymers creates the quaternary structure of the protein.

Protein Folding Peptides released from ribosomes are free to take on their final three-dimensional shape. Each peptide first forms its secondary structure, which may be an α-helix or a β-strand [p. 32]. The molecule then folds into its final shape when hydrogen bonds, covalent bonds, and ionic bonds form between amino acids in the chain. Studies show that some protein folding takes place spontaneously, but it is often facilitated by helper proteins called *molecular chaperones*.

The three-dimensional shape of proteins is often essential for proper function. Misfolded proteins, along with other proteins the cell wishes to destroy, are tagged with a protein called *ubiquitin* and sent to *proteasomes*, cylindrical cytoplasmic enzyme complexes that break down proteins.

Cross-Linkage Some protein folding is held in place by relatively weak hydrogen bonds and ionic bonds. However, other proteins form strong covalent bonds between different parts of the amino acid chain. These bonds are often disulfide bonds (S–S) between two cysteine amino acids, which contain sulfur atoms. For example, the three chains of the digestive enzyme chymotrypsin are held together by disulfide bonds.

Cleavage Some biologically active proteins, such as enzymes and hormones, are synthesized initially as inactive molecules that must have segments removed before they become active. The enzyme chymotrypsin must have two small peptide fragments removed before it can catalyze a reaction [Fig. 2.12a, p. 50]. Posttranslational processing also activates some peptide hormones.

Addition of Other Molecules or Groups Proteins can be modified by the addition of sugars (glycosylation) to create glycoproteins, or by combination with lipids to make lipoproteins [p. 29]. The two most common chemical groups added to proteins are phosphate groups, PO_4^{2-} and methyl groups, $-CH_3$. (Addition of a methyl group is called *methylation*.)

Assembly into Polymeric Proteins Many complex proteins have a quaternary structure with multiple subunits, in which protein chains assemble into dimers, trimers, or tetramers. One example is the enzyme lactate dehydrogenase (described on p. 99). Another example is the hemoglobin molecule, with four protein chains [Fig. 2.3, p. 32].

Concept Check

23. What is the removal of a phosphate group called?
24. List three general types of posttranslational modification of proteins.
25. Is hemoglobin a monomer, dimer, trimer, or tetramer?

The many ways that proteins can be modified after synthesis add to the complexity of the human body. We must know not only the sequence of a protein but also how it is processed, where the protein occurs in or outside the cell, and what it does. Scientists working on the Human Genome Project initially predicted that our DNA would code for about 30,000 proteins, but they were not taking into account alternative splicing or posttranslational modifications. Scientists working on the Human Proteome Project (*https://www.hupo.org/human-proteome-project*) are now predicting that we will find more than a million different proteins. The magnitude of this project means that it will continue for many years into the future.

RUNNING PROBLEM **CONCLUSION** Tay-Sachs Disease

In this running problem, you learned that Tay-Sachs disease is an incurable, recessive genetic disorder in which the enzyme that breaks down gangliosides in cells is missing. One in 27 Americans of Eastern European Jewish descent in the United States carries the gene for this disorder. Other high-risk populations include French Canadians, Louisiana "Cajuns," and Irish Americans. By one estimate, about one person in every 250 in the general American population is a carrier of the Tay-Sachs gene.

Blood tests and newer saliva tests can detect the presence of genetic mutations that cause this deadly disease. Check your understanding of this running problem by comparing your answers to those in the summary table. To read more on Tay-Sachs disease, see the NIH reference page (*www.ninds.nih.gov/disorders/taysachs/taysachs.htm*) or the website of the National Tay-Sachs & Allied Diseases Association (*www.ntsad.org*).

Question	Facts	Integration and Analysis
Q1: *What is another symptom of Tay-Sachs disease besides loss of muscle control and brain function?*	Hexosaminidase A breaks down gangliosides. In Tay-Sachs disease, this enzyme is absent, and gangliosides accumulate in cells, including light-sensitive cells of the eye, and cause them to function abnormally.	Damage to light-sensitive cells of the eye could cause vision problems and even blindness.

– Continued next page

Level Two Reviewing Concepts

17. Create maps using the following terms.

Map 1: Metabolism

• acetyl CoA	• glycolysis
• ATP	• high-energy electrons
• citric acid cycle	• lactate
• CO_2	• mitochondria
• cytosol	• NADH
• electron transport system	• oxygen
• $FADH_2$	• pyruvate
• glucose	• water

Map 2: Protein Synthesis

• alternative splicing	• ribosome
• base pairing	• RNA polymerase
• bases (A, C, G, T, U)	• RNA processing
• DNA	• start codon
• exon	• stop codon
• gene	• template strand
• intron	• transcription
• promoter	• transcription factors
• mRNA	• translation
• tRNA	

18. When bonds are broken during a chemical reaction, what are the three possible fates for the potential energy found in those bonds?
19. Match the metabolic processes with the letter of the biological theme that best describes the process:

(a) biological energy use (b) compartmentation (c) molecular interactions	1. Glycolysis takes place in the cytosol; oxidative phosphorylation takes place in mitochondria. 2. The electron transport system traps energy in a hydrogen ion concentration gradient. 3. Proteins are modified in the endoplasmic reticulum. 4. Metabolic reactions are often coupled to the reaction ATP → ADP + P_i. 5. Some proteins have S–S bonds between nonadjacent amino acids. 6. Enzymes catalyze biological reactions.

20. Explain why it is advantageous for a cell to store or secrete an enzyme in an inactive form.
21. Compare the following: (a) the energy yield from the aerobic breakdown of one glucose to CO_2 and H_2O, and (b) the energy yield from one glucose going through anaerobic glycolysis ending with lactate. What are the advantages of each pathway?
22. Briefly describe the processes of transcription and translation. Which organelles are involved in each process?
23. On what molecule does the anticodon appear? Explain the role of this molecule in protein synthesis.
24. Is the energy of ATP's phosphate bond an example of potential or kinetic energy?
25. If ATP releases energy to drive a chemical reaction, would you suspect the activation energy of that reaction to be large or small? Explain.

Level Three Problem Solving

26. Given the following strand of DNA: (1) Find the first start codon in the DNA sequence. *Hint:* The start codon in mRNA is AUG. (2) For the triplets that follow the start codon, list the sequence of corresponding mRNA bases. (3) Give the amino acids that correspond to those mRNA triplets. (See Fig. 4.17.)

    DNA: CGCTACAAGTCACGTACCGTAACGACT

    mRNA:

    Amino acids:

Level Four Quantitative Problems

27. The graph shows the free energy change for the reaction A + B → D. Is this an endergonic or exergonic reaction?

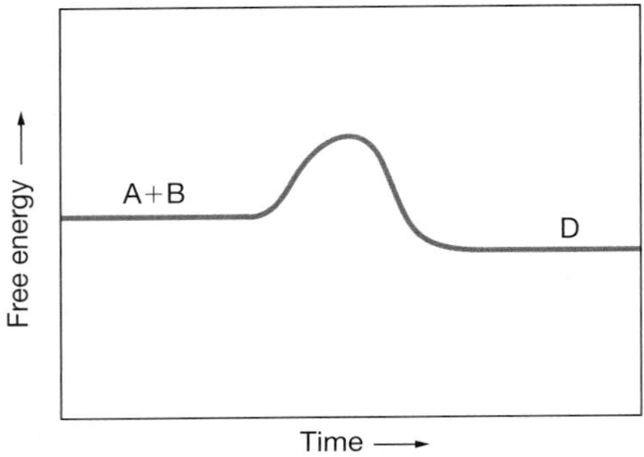
A+B
D
Free energy →
Time →

28. If the protein-coding portion of a piece of processed mRNA is 450 bases long, how many amino acids will be in the corresponding polypeptide? (*Hint:* The start codon is translated into an amino acid, but the stop codon is not.)

Answers to Concept Checks, Figure and Graph Questions, and end-of-chapter Review Questions can be found in Appendix A [p. A-1].

5 Membrane Dynamics

Organisms could not have evolved without relatively impermeable membranes to surround the cell constituents.

E. N. Harvey, in H. Davson and J. F. Danielli's The Permeability of Natural Membranes, *1952*

Transcytosis (purple) across endothelium

BACKGROUND BASICS

In 1992, the medical personnel at isolated Atoifi Hospital in the Solomon Islands of the South Pacific were faced with a dilemma. A patient was vomiting and needed intravenous (IV) fluids, but the hospital's supply had run out, and it would be several days before a plane could bring more. Their solution was to try something they had only heard about—make an IV of coconut water, the sterile solution that forms in the hollow center of developing coconuts. For two days, the patient received a slow drip of fluid into his veins directly from young coconuts suspended next to his bed. He soon recovered and was well enough to go home.*

No one knows who first tried coconut water as an IV solution, although stories have been passed down that both the Japanese and the British used it in the Pacific Theater of Operations during World War II. Choosing the appropriate IV solution is more than a matter of luck, however. It requires a solid understanding of the body's compartments and of the ways different solutes pass between them, topics you will learn about in this chapter.

RUNNING PROBLEM **Cystic Fibrosis**

Over 100 years ago, midwives performed an unusual test on the infants they delivered: The midwife would lick the infant's forehead. A salty taste meant that the child was destined to die of a mysterious disease that withered the flesh and robbed the breath. Today, a similar "sweat test" will be performed in a major hospital—this time, with state-of-the-art techniques—on Daniel Biller, an 18-month-old with a history of weight loss and respiratory problems. The name of the mysterious disease? Cystic fibrosis.

122 – 131 – 138 – 151 – 152 – 159

Homeostasis Does Not Mean Equilibrium

The body has two distinct fluid compartments: the cells and the fluid that surrounds the cells (**FIG. 5.1**). The extracellular fluid (ECF) outside the cells is the buffer between the cells and the environment outside the body. Everything that enters or leaves most cells passes through the ECF.

Water is essentially the only molecule that moves freely between cells and the extracellular fluid. Because of this free movement of water, the extracellular and intracellular compartments reach a state of **osmotic equilibrium** { *osmos*, push or thrust }, in which the fluid concentrations are equal on the two sides of the cell membrane. (Concentration is expressed as amount of solute per volume of solution [Fig. 2.7, p. 42].) Although the overall concentrations of the ECF and intracellular fluid (ICF) are equal, some solutes are more concentrated in one of the two body compartments than in the other (Fig. 5.1d). This means the body is in a state of **chemical disequilibrium**.

Figure 5.1d shows the uneven distribution of major solutes among the body fluid compartments. For example, sodium, chloride, and bicarbonate (HCO_3^-) ions are more concentrated in extracellular fluid than in intracellular fluid. Potassium ions are more concentrated inside the cell. Calcium (not shown in the figure) is more concentrated in the extracellular fluid than in the cytosol, although many cells store Ca^{2+} inside organelles such as the endoplasmic reticulum and mitochondria.

Even the extracellular fluid is not at equilibrium between its two subcompartments, the plasma and the interstitial fluid (IF) [p. 61]. Plasma is the liquid matrix of blood and is found inside the circulatory system. Proteins and other large anions are concentrated in the plasma but cannot cross the leaky exchange epithelium of blood vessels [p. 77], so they are mostly absent from the interstitial fluid (Fig. 5.1d). On the other hand, smaller molecules and ions such as Na^+ and Cl^- are small enough to pass freely between the endothelial cells and therefore have the same concentrations in plasma and interstitial fluid.

The concentration differences of chemical disequilibrium are a hallmark of a living organism, as only the continual input of energy keeps the body in this state. If solutes leak across the cell membranes dividing the intracellular and extracellular compartments, energy is required to return them to the compartment they left. For example, K^+ ions that leak out of the cell and Na^+ ions that leak into the cell are returned to their original compartments by an energy-utilizing enzyme known as the *Na^+-K^+-ATPase*, or the sodium-potassium pump. When cells die and cannot use energy, they obey the second law of thermodynamics [p. 95] and return to a state of randomness that is marked by loss of chemical disequilibrium.

Many body solutes mentioned so far are ions, and for this reason we must also consider the distribution of electrical charge between the intracellular and extracellular compartments. The body as a whole is electrically neutral, but a few extra negative ions are found in the intracellular fluid, while their matching positive ions are located in the extracellular fluid. As a result, the inside of cells is slightly negative relative to the extracellular fluid. This ionic imbalance results in a state of **electrical disequilibrium**. Changes in this disequilibrium create electrical signals. We discuss this topic in more detail later in this chapter.

In summary, note that homeostasis is not the same as equilibrium. The intracellular and extracellular compartments of the body are in osmotic equilibrium, but in chemical and electrical disequilibrium. Furthermore, osmotic equilibrium and the two disequilibria are dynamic *steady states*. The goal of homeostasis is to maintain the dynamic steady states of the body's compartments.

In the remainder of this chapter, we discuss these three steady states, and the role transport mechanisms and the selective permeability of cell membranes play in maintaining these states.

*D. Campbell-Falck *et al.* The intravenous use of coconut water. *Am J Emerg Med 18*: 108–111, 2000.

FIG. 5.1 ESSENTIALS Body Fluid Compartments

(a) The body fluids are in two compartments: the extracellular fluid (ECF) and intracellular fluid (ICF). The ECF and ICF are in osmotic equilibrium but have very different chemical composition.

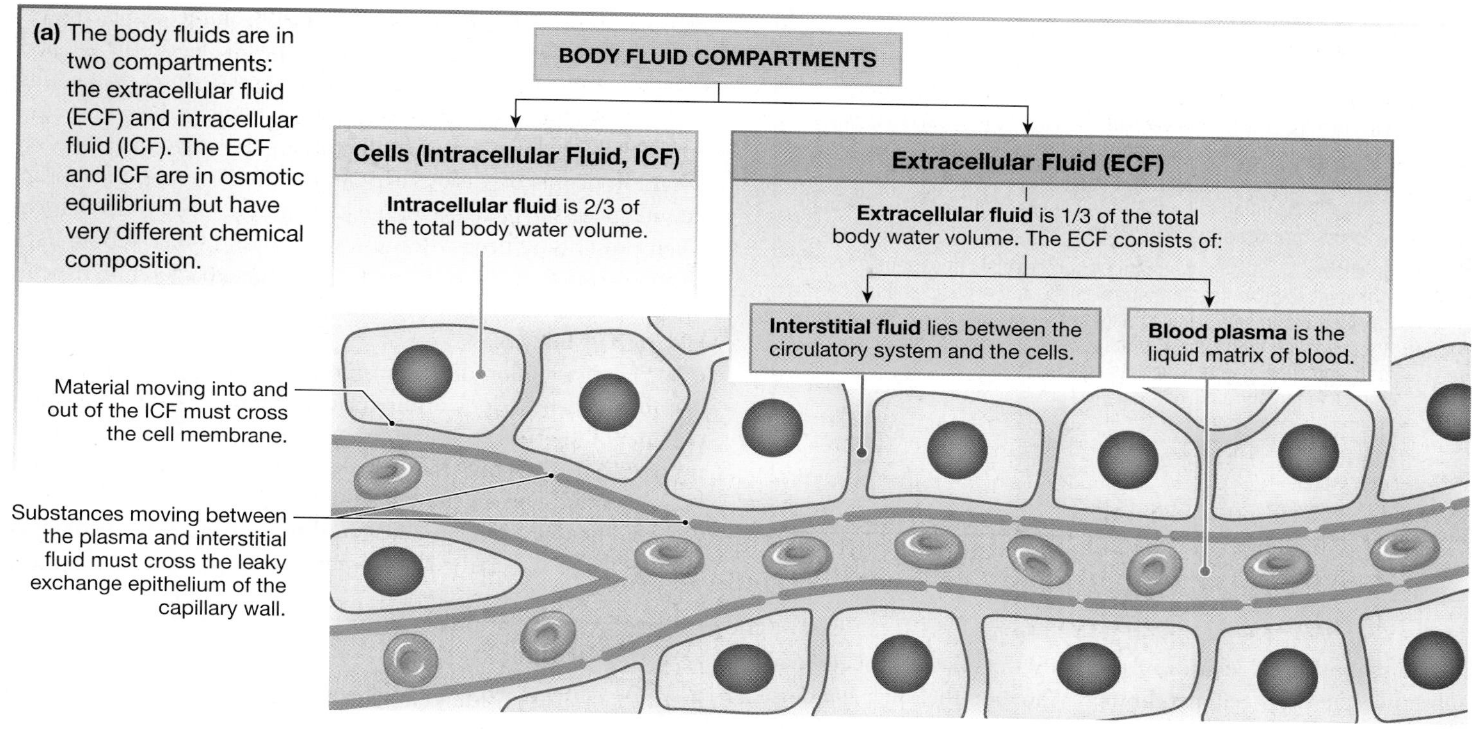

BODY FLUID COMPARTMENTS
Cells (Intracellular Fluid, ICF)
Intracellular fluid is 2/3 of the total body water volume.
Extracellular Fluid (ECF)
Extracellular fluid is 1/3 of the total body water volume. The ECF consists of:
Interstitial fluid lies between the circulatory system and the cells.
Blood plasma is the liquid matrix of blood.
Material moving into and out of the ICF must cross the cell membrane.
Substances moving between the plasma and interstitial fluid must cross the leaky exchange epithelium of the capillary wall.

(b) This figure shows the compartment volumes for the "standard" 70-kg man.

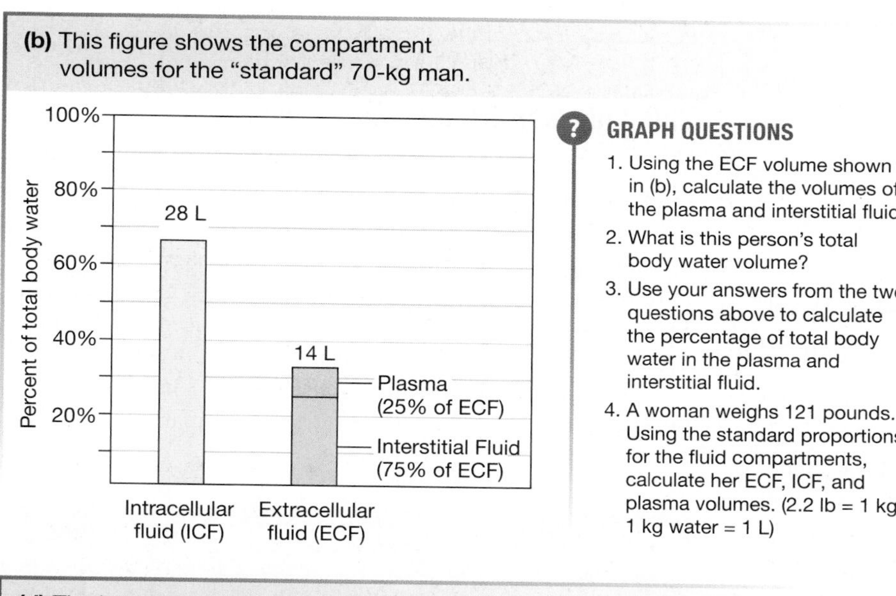

100%
80%
60%
40%
20%
Percent of total body water
28 L
14 L
Plasma (25% of ECF)
Interstitial Fluid (75% of ECF)
Intracellular fluid (ICF)
Extracellular fluid (ECF)

GRAPH QUESTIONS

1. Using the ECF volume shown in (b), calculate the volumes of the plasma and interstitial fluid.
2. What is this person's total body water volume?
3. Use your answers from the two questions above to calculate the percentage of total body water in the plasma and interstitial fluid.
4. A woman weighs 121 pounds. Using the standard proportions for the fluid compartments, calculate her ECF, ICF, and plasma volumes. (2.2 lb = 1 kg. 1 kg water = 1 L)

(c) Fluid compartments are often illustrated with box diagrams like this one.

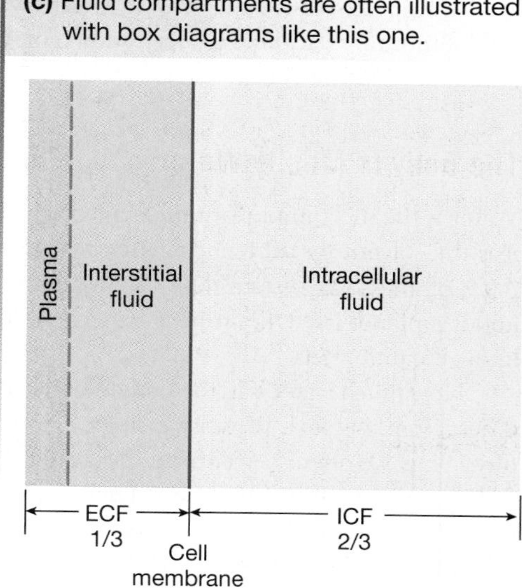

Plasma
Interstitial fluid
Intracellular fluid
ECF 1/3
Cell membrane
ICF 2/3

(d) The body compartments are in a state of chemical disequilibrium. The cell membrane is a selectively permeable barrier between the ECF and ICF.

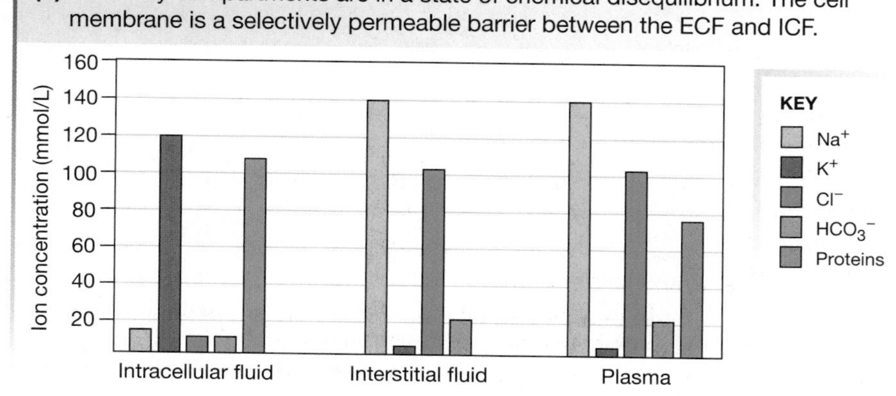

Ion concentration (mmol/L)
160
140
120
100
80
60
40
20
Intracellular fluid
Interstitial fluid
Plasma
KEY
Na^+
K^+
Cl^-
HCO_3^-
Proteins

GRAPH QUESTIONS

5. How does the ion composition of plasma differ from that of the IF?
6. What ions are concentrated in the ECF? In the ICF?

Concept Check

1. Using what you learned about the naming conventions for enzymes [p. 101], explain what the name *Na^+-K^+-ATPase* tells you about this enzyme's actions.
2. The intracellular fluid can be distinguished from the extracellular fluid by the ICF's high concentration of _____ ions and low concentration of _____, _____, and _____ ions.
3. In clinical situations, we monitor homeostasis of various substances such as ions, blood gases, and organic solutes by taking a blood sample and analyzing its plasma. For each of the following substances, predict whether knowing its plasma concentration also tells you its concentration in the ECF and the ICF. Defend your answer.
   (a) Na^+
   (b) K^+
   (c) water
   (d) proteins

5.1 Osmosis and Tonicity

The distribution of solutes in the body depends on whether a substance can cross cell membranes. Water, on the other hand, is able to move freely in and out of nearly every cell in the body by traversing water-filled ion channels and special water channels created by the protein *aquaporin (AQP)*. In this section, we examine the relationship between solute movement and water movement across cell membranes. A sound understanding of this topic provides the foundation for the clinical use of IV fluid therapy.

The Body Is Mostly Water

Water is the most important molecule in the human body because it is the solvent for all living matter. As we look for life in distant parts of the solar system, one of the first questions scientists ask about a planet is, "Does it have water?" Without water, life as we know it cannot exist.

How much water is in the human body? Because one individual differs from the next, there is no single answer. However, in human physiology we often speak of standard values for physiological functions based on "the 70-kg man." These standard values are derived from data published in the mid-twentieth century by the International Commission on Radiological Protection (ICRP). The ICRP was setting guidelines for permissible radiation exposure, and they selected a young (age 20–30) white European male who weighed 70 kilograms (kg) or 154 pounds as their "reference man," or "standard man." In 1984, Reference Man was joined by Reference Woman, a young, 58-kg (127.6 lb) female. The U.S. population is getting larger and heavier, however, and in 1990, the equivalent Reference Man had grown to 77.5 kg and was 8 cm taller.

The 70-kg Reference Man has 60% of his total body weight, or 42 kg (92.4 lb), in the form of water. Each kilogram of water has a volume of 1 liter, so his **total body water** is 42 liters. This is the equivalent of 21 two-liter soft drink bottles!

Adult women have less water per kilogram of body mass than men because women have more adipose tissue. Large fat droplets in adipose tissue occupy most of the cell's volume, displacing the more aqueous cytoplasm [see Fig. 3.13e, p. 83]. Age also influences body water content. Infants have relatively more water than adults, and water content decreases as people grow older than 60.

TABLE 5.1 shows water content as a percentage of total body weight in people of various ages and both sexes. In clinical practice, it is necessary to allow for the variability of body water content when prescribing drugs. Because women and older people have less body water, they will have a higher concentration of a drug in the plasma than will young men if all are given an equal dose per kilogram of body mass.

The distribution of water among body compartments is less variable. When we look at the relative volumes of the body compartments, the intracellular compartment contains about two-thirds (67%) of the body's water (Fig. 5.1b, c). The remaining third (33%) is split between the interstitial fluid (which contains about 75% of the extracellular water) and the plasma (which contains about 25% of the extracellular water).

Concept Check

4. If the 58-kg Reference Woman has total body water equivalent to 50% of her body weight, what are (a) her total body water volume, (b) her ECF and ICF volumes, and (c) her plasma volume?

The Body Is in Osmotic Equilibrium

Water is able to move freely between cells and the extracellular fluid and distributes itself until water concentrations are equal throughout the body—in other words, until the body is in a state of osmotic equilibrium. The movement of water across a membrane in response to a solute concentration gradient is called **osmosis**. In osmosis, water moves to dilute the more concentrated solution. Once concentrations are equal, net movement of water stops.

Look at the example shown in **FIGURE 5.2** in which a selectively permeable membrane separates two compartments of equal

TABLE 5.1 Water Content as Percentage of Total Body Weight by Age and Sex

Age	Male	Female
Infant	65%	65%
1–9	62%	62%
10–16	59%	57%
17–39	61%	51%
40–59	55%	47%
60+	52%	46%

Adapted from I. S. Edelman and J. Leibman, Anatomy of body water and electrolytes, *Am J Med 27*(2): 256–277, 1959.

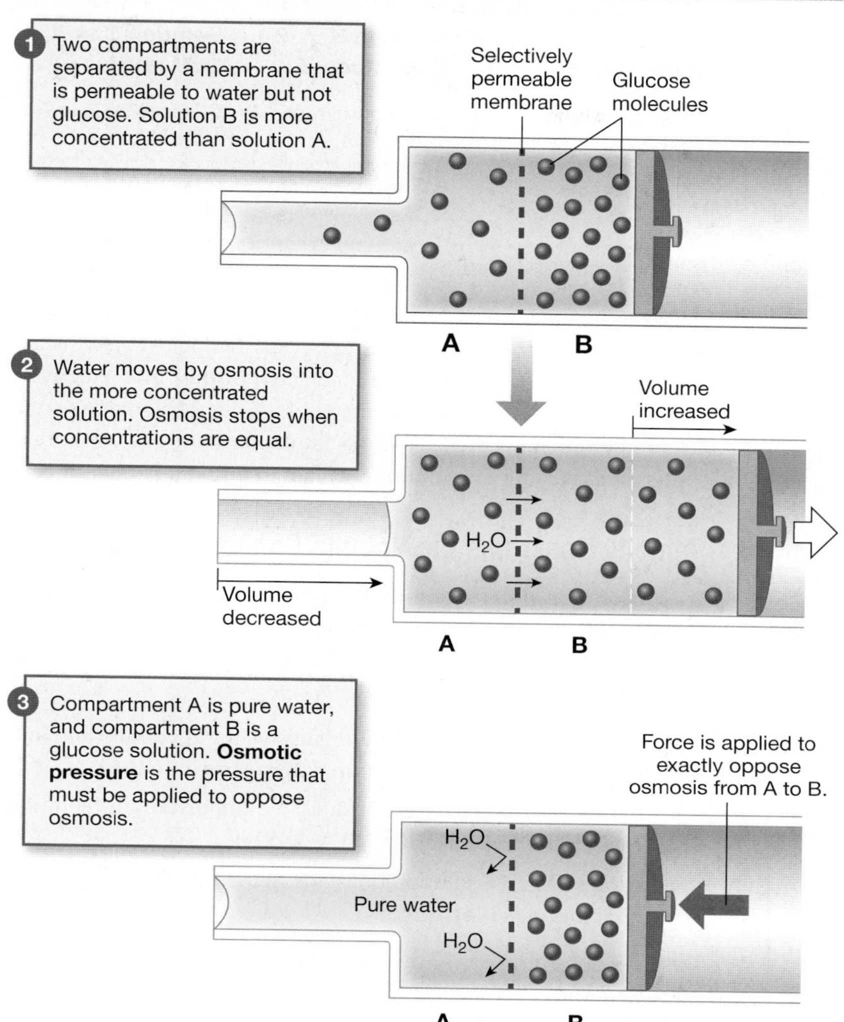

1 Two compartments are separated by a membrane that is permeable to water but not glucose. Solution B is more concentrated than solution A.
Selectively permeable membrane
Glucose molecules
A
B
2 Water moves by osmosis into the more concentrated solution. Osmosis stops when concentrations are equal.
Volume increased
H2O
Volume decreased
A
B
3 Compartment A is pure water, and compartment B is a glucose solution. Osmotic pressure is the pressure that must be applied to oppose osmosis.
Force is applied to exactly oppose osmosis from A to B.
H2O
Pure water
H2O
A
B

FIG. 5.2 Osmosis and osmotic pressure

volume. The membrane is permeable to water but does not allow glucose to cross. In **1**, compartments A and B contain equal volumes of glucose solution. Compartment B has more solute (glucose) per volume of solution and therefore is the more concentrated solution. A concentration gradient across the membrane exists for glucose. However, because the membrane is not permeable to glucose, glucose cannot move to equalize its distribution.

Water, by contrast, can cross the membrane freely. It will move by osmosis from compartment A, which contains the dilute glucose solution, to compartment B, which contains the more concentrated glucose solution. Thus, water moves to dilute the more concentrated solution (Fig. 5.2 **2**).

How can we make quantitative measurements of osmosis? One method is shown in Figure 5.2 **3**. The solution to be measured is placed in compartment B with pure water in compartment A. Because compartment B has a higher solute concentration than compartment A, water will flow from A to B. However, by pushing down on the piston, you can keep water from entering compartment B. The pressure on the piston that exactly opposes the osmotic movement of water into compartment B is known as the **osmotic pressure** of solution B. The units for osmotic pressure, just as with other pressures in physiology, are *atmospheres* (atm) or *millimeters of mercury* (mm Hg). A pressure of 1 mm Hg is equivalent to the pressure exerted on a 1-cm^2 area by a 1-mm-high column of mercury.

Osmolarity Describes the Number of Particles in Solution

Another way to predict the osmotic movement of water quantitatively is to know the concentrations of the solutions with which we are dealing. In chemistry, concentrations are often expressed as *molarity* (*M*), which is defined as number of moles of dissolved solute per liter of solution (mol/L). Recall that one *mole* is 6.02×10^{23} molecules [Fig. 2.7, p. 42].

However, using molarity to describe biological concentrations can be misleading. The important factor for osmosis is the number of osmotically active *particles* in a given volume of solution, not the number of molecules. Because some molecules dissociate into ions when they dissolve in a solution, the number of particles in solution is not always the same as the number of molecules.

For example, one glucose molecule dissolved in water yields one particle, but one NaCl dissolved in water theoretically yields two ions (particles): Na^+ and Cl^-. Water moves by osmosis in response to the total concentration of all *particles* in the solution. The particles may be ions, uncharged molecules, or a mixture of both.

Consequently, for biological solutions we express the concentration as **osmolarity**, the number of osmotically active particles (ions or intact molecules) per liter of solution. Osmolarity is expressed in *osmoles* per liter (osmol/L or OsM) or, for very dilute physiological solutions, milliosmoles/liter (mOsM). To convert between molarity and osmolarity, use the following equation:

$$\text{molarity (mol/L)} \times \text{particles/molecule (osmol/mol)} = \text{osmolarity (osmol/L)}$$

Let us look at two examples, glucose and sodium chloride, and compare their molarities with their osmolarities.

One mole of glucose molecules dissolved in enough water to create 1 liter of solution yields a 1 molar solution (1 M). Because glucose does not dissociate in solution, the solution has only one mole of osmotically active particles:

$$1\text{ M glucose} \times 1\text{ osmole/mole glucose} = 1\text{ OsM glucose}$$

Sodium chloride, however, dissociates when placed in solution. At body temperature, a few NaCl ions fail to separate, so instead of 2 ions per NaCl, the *dissociation factor* is about 1.8. Thus, one mole of NaCl dissociates in solution to yield 1.8 moles of particles (Na^+, Cl^-, and NaCl). The result is a 1.8 OsM solution:

1 mole NaCl/L × 1.8 osmol/mol NaCl = 1.8 osmol/L NaCl

Osmolarity describes only the number of particles in the solution. It says nothing about the composition of the particles. A 1 OsM solution could be composed of pure glucose or pure Na^+ and Cl^- or a mixture of all three solutes.

The normal osmolarity of the human body ranges from 280 to 296 milliosmoles per liter (mOsM). In this book, to simplify calculations, we will round that number up slightly to 300 mOsM.

A term related to osmolarity is osmolality. **Osmolality** is concentration expressed as osmoles of solute per kilogram of water. Because biological solutions are dilute and little of their weight comes from solute, physiologists often use the terms *osmolarity* and *osmolality* interchangeably. Osmolality is usually used in clinical situations because it is easy to estimate people's body water content by weighing them.

Clinicians estimate a person's fluid loss in dehydration by equating weight loss to fluid loss. Because 1 liter of pure water weighs 1 kilogram, a decrease in body weight of 1 kg (or 2.2 lbs.) is considered equivalent to the loss of 1 liter of body fluid. A baby with diarrhea can easily be weighed to estimate its fluid loss. A decrease of 1.1 lbs. (0.5 kg) of body weight is assumed to mean the loss of 500 mL of fluid. This calculation provides a quick estimate of how much fluid needs to be replaced.

Concept Check

5. A mother brings her baby to the emergency room because he has lost fluid through diarrhea and vomiting for two days. The staff weighs the baby and finds that he has lost 2 lbs. If you assume that the reduction in weight is due to water loss, what volume of water has the baby lost (2.2 lbs. = 1 kg)?

Comparing Osmolarities of Two Solutions Osmolarity is a property of every solution. You can compare the osmolarities of different solutions as long as the concentrations are expressed in the same units—for example, as milliosmoles per liter. If two solutions contain the same number of solute particles per unit volume, we say that the solutions are **isosmotic** {*iso*-, equal}. If solution A has a higher osmolarity (contains more particles per unit volume, is more concentrated) than solution B, we say that solution A is **hyperosmotic** to solution B. In the same example, solution B, with fewer osmoles per unit volume, is **hyposmotic** to solution A. TABLE 5.2 shows some examples of comparative osmolarities.

TABLE 5.2 Comparing Osmolarities

Solution A = 1 OsM Glucose	Solution B = 2 OsM Glucose	Solution C = 1 OsM NaCl
A is hyposmotic to B	B is hyperosmotic to A	C is isosmotic to A
A is isosmotic to C	B is hyperosmotic to C	C is hyposmotic to B

Osmolarity is a *colligative* property of solutions, meaning it depends strictly on the *number* of particles per liter of solution. Osmolarity says nothing about what the particles are or how they behave. Before we can predict whether osmosis will take place between any two solutions divided by a membrane, we must know the properties of the membrane and of the solutes on each side of it.

If the membrane is permeable only to water and not to any solutes, water will move by osmosis from a less concentrated (hyposmotic) solution into a more concentrated (hyperosmotic) solution, as illustrated in Figure 5.2. Most biological systems are not this simple, however. Biological membranes are selectively permeable and allow some solutes to cross in addition to water. To predict the movement of water into and out of cells, you must know the *tonicity* of the solution, explained in the next section.

Tonicity Describes the Volume Change of a Cell

Tonicity {*tonikos*, pertaining to stretching} is a physiological term used to describe a solution and how that solution would affect cell volume if the cell were placed in the solution and allowed to come to equilibrium (TBL. 5.3).

- If a cell placed in the solution gains water at equilibrium and swells, we say that the solution is **hypotonic** to the cell.
- If the cell loses water and shrinks at equilibrium, the solution is said to be **hypertonic**.
- If the cell in the solution does not change size at equilibrium, the solution is **isotonic**.

By convention, we always describe the tonicity of the solution relative to the cell. How, then, does tonicity differ from osmolarity?

1. Osmolarity describes the number of solute particles dissolved in a volume of solution. It has units, such as osmoles/liter. The osmolarity of a solution can be measured by a machine called an *osmometer*. Tonicity has no units; it is only a comparative term.
2. Osmolarity can be used to compare any two solutions, and the relationship is reciprocal (solution A is hyperosmotic to solution B; therefore, solution B is hyposmotic to solution A). Tonicity always compares a solution and a cell, and by

TABLE 5.3 Tonicity of Solutions

Solution	Cell Behavior When Placed in the Solution	Description of the Solution Relative to the Cell
A	Cell swells	Solution A is hypotonic
B	Cell doesn't change size	Solution B is isotonic
C	Cell shrinks	Solution C is hypertonic

convention, tonicity is used to describe only the solution—for example, "Solution A is hypotonic to red blood cells."

3. Osmolarity alone does not tell you what happens to a cell placed in a solution. Tonicity by definition tells you what happens to cell volume at equilibrium when the cell is placed in the solution.

This third point is the one that is most confusing to students. Why can't osmolarity be used to predict tonicity? The reason is that the tonicity of a solution depends not only on its concentration (osmolarity) but also on the *nature* of the solutes in the solution.

By nature of the solutes, we mean whether the solute particles can cross the cell membrane. If the solute particles (ions or molecules) can enter the cell, we call them **penetrating solutes**. We call particles that cannot cross the cell membrane **nonpenetrating solutes**. Tonicity depends on the concentration of nonpenetrating solutes only. Let's see why this is true.

First, some preliminary information. The most important nonpenetrating solute in physiology is NaCl. If a cell is placed in a solution of NaCl, the Na^+ and Cl^- ions do not enter the cell. This makes NaCl a nonpenetrating solute. (In reality, a few Na^+ ions may leak across, but they are immediately transported back to the extracellular fluid by the Na^+-K^+-ATPase. For this reason, NaCl is considered a *functionally* nonpenetrating solute.)

By convention, we assume that cells are filled with other types of nonpenetrating solutes. In other words, the solutes inside the cell are unable to leave as long as the cell membrane remains intact. Now we are ready to see why osmolarity alone cannot be used to predict tonicity.

Suppose you know the composition and osmolarity of a solution. How can you figure out the tonicity of the solution without actually putting a cell in it? The key lies in knowing *the relative concentrations of nonpenetrating solutes in the cell and in the solution*. Water will always move until the concentrations of nonpenetrating solutes in the cell and the solution are equal.

Here are the rules for predicting tonicity:

1. *If the cell has a higher concentration of nonpenetrating solutes than the solution,* there will be net movement of water into the cell. The cell swells, and the solution is *hypotonic.*
2. *If the cell has a lower concentration of nonpenetrating solutes than the solution,* there will be net movement of water out of the cell. The cell shrinks, and the solution is *hypertonic.*
3. *If the concentrations of nonpenetrating solutes are the same in the cell and the solution,* there will be no net movement of water at equilibrium. The solution is *isotonic* to the cell.

How does tonicity relate to osmolarity? **FIGURE 5.3** shows the possible combinations of osmolarity and tonicity, and why osmolarity alone cannot predict tonicity. There is one exception to this statement: A hyposmotic solution is always hypotonic, no matter what its composition. The cell will always have a higher concentration of nonpenetrating solutes than the solution, and water will move into the cell (rule 1 above).

As you can see in Figure 5.3, an isosmotic solution may be isotonic or hypotonic. It can never be hypertonic because it can never have a higher concentration of nonpenetrating solutes than the cell. If all solutes in the isosmotic solution are nonpenetrating, then the solution is also isotonic. If there are any penetrating solutes in the isosmotic solution, the solution will be hypotonic.

FIG. 5.3 The relationship between osmolarity and tonicity

The osmolarity of a solution is not an accurate predictor of its tonicity.

TONICITY	OSMOLARITY		
	Hyposmotic	Isosmotic	Hyperosmotic
Hypotonic	✓	✓	✓
Isotonic		✓	✓
Hypertonic			✓

Hyperosmotic solutions may be hypertonic, isotonic, or hypotonic. Their tonicity depends on the relative concentration of nonpenetrating solutes in the solution compared to the cell, as described previously.

Often tonicity is explained using a single cell that is placed into a solution, but here we will use a more physiologically appropriate system: a two-compartment box model that represents the total body divided into ECF and ICF (see Fig. 5.1c). To simplify the calculations, we will use a 3-liter body, with 2 liters in the ICF and 1 liter in the ECF. We assume that the starting osmolarity is 300 mOsM (0.3 OsM) and that solutes in each compartment are nonpenetrating (NP) and cannot move into the other compartment. By defining volumes and concentrations, we can use the equation *solute/volume = concentration* ($S/V = C$) to mathematically determine changes to volumes and osmolarity. *Concentration* is osmolarity.

Always begin by defining the starting conditions. This may be the person's normal state or it may be the altered state that you are trying to return to normal. An example of this would be trying to restore normal volume and osmolarity in a person who has become dehydrated through sweat loss.

FIGURE 5.4 shows the starting conditions for the 3-liter body both as a compartment diagram and in a table. The table format allows you to deal with an example mathematically if you know the volumes and concentration of the body and of the solution added or lost.

The body's volumes and concentration will change as the result of adding or losing solutes, water, or both—the law of mass balance [p. 10]. Additions to the body normally come through the ingestion of food and drink. In medical situations, solutions can be added directly to the ECF through IV infusions. Significant solute and water loss may occur with sweating, vomiting and diarrhea, or blood loss.

Once you have defined the starting conditions, you add or subtract volume and solutes to find the body's new osmolarity. The final step is to determine whether the ECF and ICF volumes change as a result of the water and solute gain or loss. In this last step, you must separate the added solutes into penetrating solutes and nonpenetrating solutes.

In our examples, we use three solutes: NaCl, urea, and glucose. NaCl is considered nonpenetrating. Any NaCl added to the

FIG. 5.4 ESSENTIALS: Osmolarity and Tonicity

For all problems, define your starting conditions. Assume that all initial body solutes are nonpenetrating (NP) and will remain in either the ECF or ICF.

ECF	ICF
300 mosmol NP	600 mosmol NP
1 L	2 L

Use the equation **Solute/volume = concentration (S/V = C)** to solve the problems. You will know two of the three variables and can calculate the third.

Remember that body compartments are in osmotic equilibrium. Once you know the total body's osmolarity (concentration), you also know the ECF and ICF osmolarity because they are the same.

Starting Condition:

We have a 3-liter body that is 300 mOsM. The ECF is 1 liter and the ICF is 2 liters.

Use **S/V = C** to find out how much solute is in each of the two compartments. Rearrange the equation to solve for S: **S = CV.**

1 S_{ICF} = 300 mosmol/L × 2 L = 600 mosmol NP solute in the ICF

2 S_{ECF} = 300 mosmol/L × 1 L = 300 mosmol NP solute in the ECF

We can also do these calculations using the following table format. This table has been filled in with the values for the starting body. Remember that the ECF + ICF must always equal the total body values, and that once you know the total body osmolarity, you know the ECF and ICF osmolarity.

	Total Body	ECF	ICF
Solute (mosmoles)	900 mosmol	300 mosmol	600 mosmol
Volume (L)	3 L	1 L	2 L
Osmolarity (mOsM)	300 mOsM	300 mOsM	300 mOsM

To see the effect of adding a solution or losing fluid, start with this table and add or subtract volume and solute as appropriate. *You cannot add and subtract concentrations. You must use volumes and solute amounts.*

- Work the total body column first, adding or subtracting solutes and volume. Once you calculate the new total body osmolarity, carry that number across the bottom row to the ECF and ICF columns. (The compartments are in osmotic equilibrium.)
- Distribute nonpenetrating solutes to the appropriate compartment. NaCl stays in the ECF. Glucose goes into the cells. Use **V = S/C** to calculate the new compartment volumes.

In the tables below and on the following page, the yellow boxes indicate the unknowns that must be calculated.

Play Phys in Action @Mastering Anatomy & Physiology

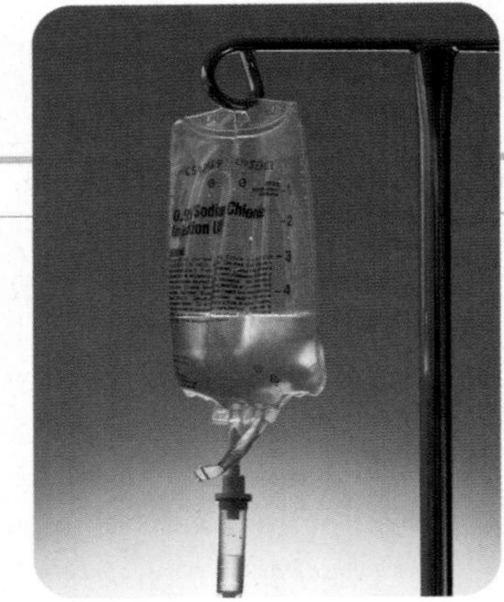

Example 1

Add an IV solution of 1 liter of 300 mOsM NaCl to this body. This solution adds 1 liter of volume and 300 mosmoles of NaCl.

Answer

Work total body first. Add solute and volume, then calculate new osmolarity (yellow box).

	Total Body
Solute (mosmoles)	900 + 300 = 1200 mosmol
Volume (L)	3 + 1 = 4 L
Osmolarity (mOsM)	1200/4 = 300 mOsM

Carry the new osmolarity across to the ECF and ICF boxes (arrows). All of the added NaCl will stay in the ECF, so add that solute amount to the ECF box. ICF solute amount is unchanged. Use V = S/C to calculate the new ECF and ICF volumes (yellow boxes).

	Total Body	ECF	ICF
Solute (mosmoles)	1200 mosmol	300 + 300 = 600	600 mosmol
Volume (L)	4 L	2 L	2 L
Osmolarity (mOsM)	300 mOsM →	300 mOsM →	300 mOsM

The added solution was isosmotic (300 mOsM) and its nonpenetrating concentration was the same as that of the body's (300 mOsM NP). You would predict that the solution was isotonic, and that is confirmed with these calculations, which show no water entering or leaving the cells (no change in ICF volume).

Example 2

Add 2 liters of a 500 mOsM solution. The solution is equal parts NaCl (nonpenetrating) and urea (penetrating), so it has 250 mosmol/L NaCl and 250 mosmol/L urea.

Answer

This solution has both penetrating and nonpenetrating solutes, but only nonpenetrating solutes contribute to tonicity and cause water to shift between compartments.

Before working this problem, answer the following questions:

(a) This solution is __________ osmotic to the 300 mOsM body.

(b) What is the concentration of nonpenetrating solutes [NP] in the solution? _______________

(c) What is the [NP] in the body? ________

(d) Using the rules for tonicity in Table 5.4, will there be water movement into or out of the cells? If so, in what direction?

(e) Based on your answer in (d), this solution is ________ tonic to this body's cells.

Now work the problem using the starting conditions table as your starting point.

What did you add? 2 L of (250 mosmol/L urea and 250 mosmol/L NaCl) = 2 liters of volume + 500 mosmol urea + 500 mosmol NaCl.

Urea does not contribute to tonicity, so we will set the 500 mosmol of urea aside and only add the volume and NaCl in the first step:

Step 1: Add 2 liters and 500 mosmoles NaCl. Do total body column first.

	Total Body
Solute (mosmoles)	900 + 500 = 1400 mosmol
Volume (L)	3 + 2 = 5 L
Osmolarity (mOsM)	1400/5 = 280 mOsM

Step 2: Carry the new osmolarity across to ECF and ICF. All NaCl remains in the ECF so add that solute to the ECF column. Calculate new ECF and ICF volumes.

- Notice that ICF volume + ECF volume = total body volume.

	Total Body	ECF	ICF
Solute (mosmoles)	1400 mosmol	300 + 500 = 800	600
Volume (L)	5 L	2.857 L	2.143 L
Osmolarity (mOsM)	280 mOsM	280 mOsM	280 mOsM

Step 3: Now add the reserved urea solute to the whole body solute to get the final osmolarity. That osmolarity carries over to the ECF and ICF compartments. Urea will distribute itself throughout the body until its concentration everywhere is equal, but it will not cause any water shift between ECF and ICF, so the ECF and ICF volumes remain as they were in Step 2.

	Total Body	ECF	ICF
Solute (mosmoles)	1400 + 500 = 1900		
Volume (L)	5 L	2.857 L	2.143 L
Osmolarity (mOsM)	1900/5 = 380 mOsM	380 mOsM	380 mOsM

Answer the following questions from the values in the table:

(f) What happened to the body osmolarity after adding the solution? ____________ This result means the added solution was ___________osmotic to the body's starting osmolarity.

(g) What happened to the ICF volume? ________________________ This means the added solution was ___________tonic to the cells.

Compare your answers in (f) and (g) to your answers for (a)–(e). Do they match? They should.

If you know the starting conditions of the body and you know the composition of a solution you are adding, you should be able to describe the solution's osmolarity and tonicity relative to the body by asking the questions in (a)–(e). Now test yourself by working Concept Check questions 8 and 9.

body remains in the ECF. Urea is freely penetrating and behaves as if the cell membranes dividing the ECF and ICF do not exist. An added load of urea distributes itself until the urea concentration is the same throughout the body.

Glucose (also called *dextrose*) is an unusual solute. Like all solutes, it first goes into the ECF. Over time, however, 100% of added glucose will enter the cells. When glucose enters the cells, it is phosphorylated to glucose 6-phosphate (G-6-P) and cannot leave the cell again. So although glucose enters cells, it is not freely penetrating because it stays in the cell and adds to the cell's nonpenetrating solutes.

Giving someone a glucose solution is the same as giving them a slow infusion of pure water because glucose 6-phosphate is the first step in the aerobic metabolism of glucose [p. 106]. The end products of aerobic glucose metabolism are CO_2 and water.

The examples shown in Figure 5.4 walk you through the process of adding and subtracting solutions to the body. Ask the following questions when you are evaluating the effects of a solution on the body:

1. What is the osmolarity of this solution relative to the body? (Tbl. 5.2)
2. What is the tonicity of this solution? (Use Fig. 5.3 to help eliminate possibilities.) To determine tonicity, compare the concentration of the nonpenetrating solutes in the solution to the body concentration. (All body solutes are considered to be nonpenetrating.)

For example, consider a solution that is 300 mOsM—isosmotic to a body that is 300 mOsM. The solution's tonicity depends on the concentration of nonpenetrating solutes in the solution. If the solution is 300 mOsM NaCl, the solution's nonpenetrating solute concentration is equal to that of the body. When the solution mixes with the ECF, the ECF nonpenetrating concentration and osmolarity do not change. No water will enter or leave the cells (the ICF compartment), and the solution is isotonic. You can calculate this for yourself by working through Example 1 in Figure 5.4.

Now suppose the 300 mOsM solution has urea as its only solute. Urea is a penetrating solute, so this solution has zero nonpenetrating solutes. When the 300 mOsM urea solution mixes with the ECF, the added volume of the urea solution dilutes the nonpenetrating solutes of the ECF. ($S/V = C$: The same amount of NP solute in a larger volume means a lower NP concentration.)

Now the nonpenetrating concentration of the ECF is less than 300 mOsM. The cells still have a nonpenetrating solute concentration of 300 mOsM, so water moves into the cells to equalize the nonpenetrating concentrations. (Rule: Water moves into the compartment with the higher concentration of NP solutes.) The cells gain water and volume. This means the urea solution is hypotonic, even though it is isosmotic.

Example 2 in Figure 5.4 shows how combining penetrating and nonpenetrating solutes can complicate the situation. This example asks you to describe the solution's osmolarity and tonicity based on its composition before you do the mathematical calculations. This skill is important for clinical situations, when you will not know exact body fluid volumes for the person needing an IV. **TABLE 5.4** lists some rules to help you distinguish between osmolarity and tonicity.

Understanding the difference between osmolarity and tonicity is critical to making good clinical decisions about intravenous fluid therapy. The choice of IV fluid depends on how the clinician wants the solutes and water to distribute between the extracellular and intracellular fluid compartments. If the problem is dehydrated cells, the appropriate IV solution is hypotonic because the cells need fluid. If the situation requires fluid that remains in the extracellular fluid to replace blood loss, an isotonic IV solution is used. In medicine, the tonicity of a solution is usually the most important consideration.

TABLE 5.5 lists some common IV solutions and their approximate osmolarity and tonicity relative to the normal human cell.

TABLE 5.4 Rules for Osmolarity and Tonicity

1. Assume that all intracellular solutes are nonpenetrating.
2. Compare osmolarities before the cell is exposed to the solution. (At equilibrium, the cell and solution are always isosmotic.)
3. Tonicity of a solution describes the volume change of a cell at equilibrium (Tbl. 5.3).
4. Determine tonicity by comparing nonpenetrating solute concentrations in the cell and the solution. Net water movement is into the compartment with the higher concentration of nonpenetrating solutes.
5. Hyposmotic solutions are always hypotonic.

TABLE 5.5 Intravenous Solutions

Solution	Also Known as	Osmolarity	Tonicity
0.9% saline*	Normal saline	Isosmotic	Isotonic
5% dextrose** in 0.9% saline	D5–normal saline	Hyperosmotic	Isotonic
5% dextrose in water	D5W	Isosmotic	Hypotonic
0.45% saline	Half-normal saline	Hyposmotic	Hypotonic
5% dextrose in 0.45% saline	D5–half-normal saline	Hyperosmotic	Hypotonic

* Saline = NaCl
** Dextrose = glucose

What about the coconut water described at the start of the chapter? Chemical analysis shows that it is not an ideal IV solution, although it is useful for emergencies. It is isosmotic to human plasma but is hypotonic, with Na^+ concentrations much lower than normal ECF [Na^+] and high concentrations of glucose and fructose, along with amino acids.

Concept Check

6. Which of the following solutions has/have the most water per unit volume: 1 M glucose, 1 M NaCl, or 1 OsM NaCl?
7. Two compartments are separated by a membrane that is permeable to water and urea but not to NaCl. Which way will water move when the following solutions are placed in the two compartments? (*Hint*: Watch the units!)

Compartment A	Membrane	Compartment B
(a) 1 M NaCl	|	1 OsM NaCl
(b) 1 M urea	|	2 M urea
(c) 1 OsM NaCl	|	1 OsM urea

8. Use the same 3-liter, 300 mOsM body as in Figure 5.4 for this problem. Add 1 liter of 260 mOsM glucose to the body and calculate the new body volumes and osmolarity once all the glucose has entered the cells and been phosphorylated. Before you do the calculations, make the following predictions: This solution is _____osmotic to the body and is _____tonic to the body's cells.
9. Use the same 3-liter, 300 mOsM body as in Figure 5.4 for this problem. A 3-liter person working in the hot sun loses 500 mL of sweat that is equivalent to a 130 mOsM NaCl solution. Assume all NaCl loss comes from the ECF.
   (a) The sweat lost is _____osmotic to the body. This means the osmolarity of the body after the sweat loss will (*increase/decrease/not change*)?
   (b) As a result of this sweat loss, the body's cell volume will (*increase/decrease/not change*)?
   (c) Using the table, calculate what happens to volume and osmolarity as a result of this sweat loss. Do the results of your calculations match your answers in (a) and (b)?
10. You have a patient who lost 1 liter of blood, and you need to restore volume quickly while waiting for a blood transfusion to arrive from the blood bank.
   (a) Which would be better to administer: 5% dextrose in water or 0.9% NaCl in water? (*Hint:* Think about how these solutes distribute in the body.) Defend your choice.
   (b) How much of your solution of choice would you have to administer to return blood volume to normal?

5.2 Transport Processes

Water moves freely between body compartments, but what about other body components? Humans are large complex organisms, and the movement of material within and between body compartments is necessary for communication. This movement requires a variety of transport mechanisms. Some require an outside source of energy, such as that stored in the high-energy bond of ATP [p. 104], while other transport processes use only the kinetic or potential energy already in the system [p. 94]. Movement between compartments usually means a molecule must cross one or more cell membranes. Movement within a compartment is less restricted. For this reason, biological transport is another theme that you will encounter repeatedly as you study the organ systems.

The most general form of biological transport is the **bulk flow** of fluids within a compartment. Although many people equate **fluids** with liquids, in physics both gases and liquids are considered fluids because they flow. The main difference between the two fluids is that gases are compressible because their molecules are so far apart in space. Liquids, especially water, are not compressible. (Think of squeezing on a water balloon.)

In bulk flow, a *pressure gradient* causes fluid to flow from regions of higher pressure to regions of lower pressure. As the fluid flows, it carries with it all of its component parts, including substances dissolved or suspended in it. Blood moving through the circulatory system is an excellent example of bulk flow. The heart acts as a pump that creates a region of high pressure, pushing plasma with its dissolved solutes and the suspended blood cells through the blood vessels. Air flow in the lungs is another example of bulk flow that you will encounter as you study physiology.

Other forms of transport are more specific than bulk flow. When we discuss them, we must name the molecule or molecules that are moving. Transport mechanisms you will learn about in the following sections include diffusion, protein-mediated transport, and vesicular transport.

Play A&P Flix Animation
@Mastering Anatomy & Physiology

RUNNING PROBLEM

Daniel's medical history tells a frightening story of almost constant medical problems since birth: recurring bouts of respiratory infections, digestive ailments, and, for the past six months, a history of weight loss. Then, last week, when Daniel began having trouble breathing, his mother rushed him to the hospital. A culture taken from Daniel's lungs raised a red flag for cystic fibrosis: The mucus from his airways was unusually thick and dehydrated. In cystic fibrosis, this thick mucus causes life-threatening respiratory congestion and provides a perfect breeding ground for infection-causing bacteria.

Q1: *In people with cystic fibrosis, movement of sodium chloride into the lumen of the airways is impaired. Why would failure to move NaCl into the airways cause the secreted mucus to be thick? (*Hint: *Remember that water moves into hyperosmotic regions.)*

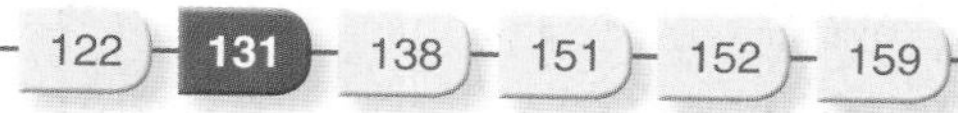
122 – 131 – 138 – 151 – 152 – 159

Cell Membranes Are Selectively Permeable

Many materials move freely within a body compartment, but exchange between the intracellular and extracellular compartments is restricted by the cell membrane. Whether or not a substance enters a cell depends on the properties of the cell membrane and those of the substance. Cell membranes are **selectively permeable**, which means that some molecules can cross them but others cannot.

The lipid and protein composition of a given cell membrane determines which molecules will enter the cell and which will leave [p. 44]. If a membrane allows a substance to pass through it, the membrane is said to be **permeable** to that substance {*permeare*, to pass through}. If a membrane does not allow a substance to pass, the membrane is said to be **impermeable** {*im-*, not} to that substance.

Membrane permeability is variable and can be changed by altering the proteins or lipids of the membrane. Some molecules, such as oxygen, carbon dioxide, and lipids, move easily across most cell membranes. On the other hand, ions, most polar molecules, and very large molecules (such as proteins), enter cells with more difficulty or may not enter at all.

Two properties of a molecule influence its movement across cell membranes: the size of the molecule and its lipid solubility. Very small molecules and those that are lipid soluble can cross directly through the phospholipid bilayer. Larger and less lipid-soluble molecules usually do not enter or leave a cell unless the cell has specific membrane proteins to transport these molecules across the lipid bilayer. Very large lipophobic molecules cannot be transported on proteins and must enter and leave cells in vesicles [p. 70].

There are multiple ways to categorize how molecules move across membranes. One scheme, just described, separates movement according to physical requirements: whether it moves by diffusion directly through the phospholipid bilayer, crosses with the aid of a membrane protein, or enters the cell in a vesicle (**FIG. 5.5**). A second scheme classifies movement according to its energy requirements. **Passive transport** does not require the input of energy other than the potential energy stored in a concentration gradient. **Active transport** requires the input of energy from some outside source, such as the high-energy phosphate bond of ATP.

The following sections look at how cells move material across their membranes. The principles discussed here also apply to movement across intracellular membranes, when substances move between organelles.

5.3 Diffusion

Passive transport across membranes uses the kinetic energy [p. 94] inherent in molecules and the potential energy stored in concentration gradients. Gas molecules and molecules in solution constantly move from one place to another, bouncing off other molecules or off the sides of any container holding them. When molecules start out concentrated in one area of an enclosed space, their motion causes them to spread out gradually until they distribute evenly throughout the available space. This process is known as diffusion.

Diffusion {*diffundere*, to pour out} may be defined as the movement of molecules from an area of higher concentration of the molecules to an area of lower concentration of the molecules.* If you leave a bottle of cologne open and later notice its fragrance across the room, it is because the aromatic molecules in the cologne have diffused from where they are more concentrated (in the bottle) to where they are less concentrated (across the room).

Diffusion has the following seven properties:

1. *Diffusion is a passive process.* By *passive*, we mean that diffusion does not require the input of energy from some outside source. Diffusion uses only the kinetic energy possessed by all molecules.

* Some texts use the term *diffusion* to mean any random movement of molecules, and they call molecular movement along a concentration gradient *net diffusion*. To simplify matters, we will use the term *diffusion* to mean movement down a concentration gradient.

FIG. 5.5 Transport across membranes

Movement of substances across membranes can be classified by the energy requirements of transport (in parentheses) or by the physical pathway (through the membrane layer, through a membrane protein, or in a vesicle).

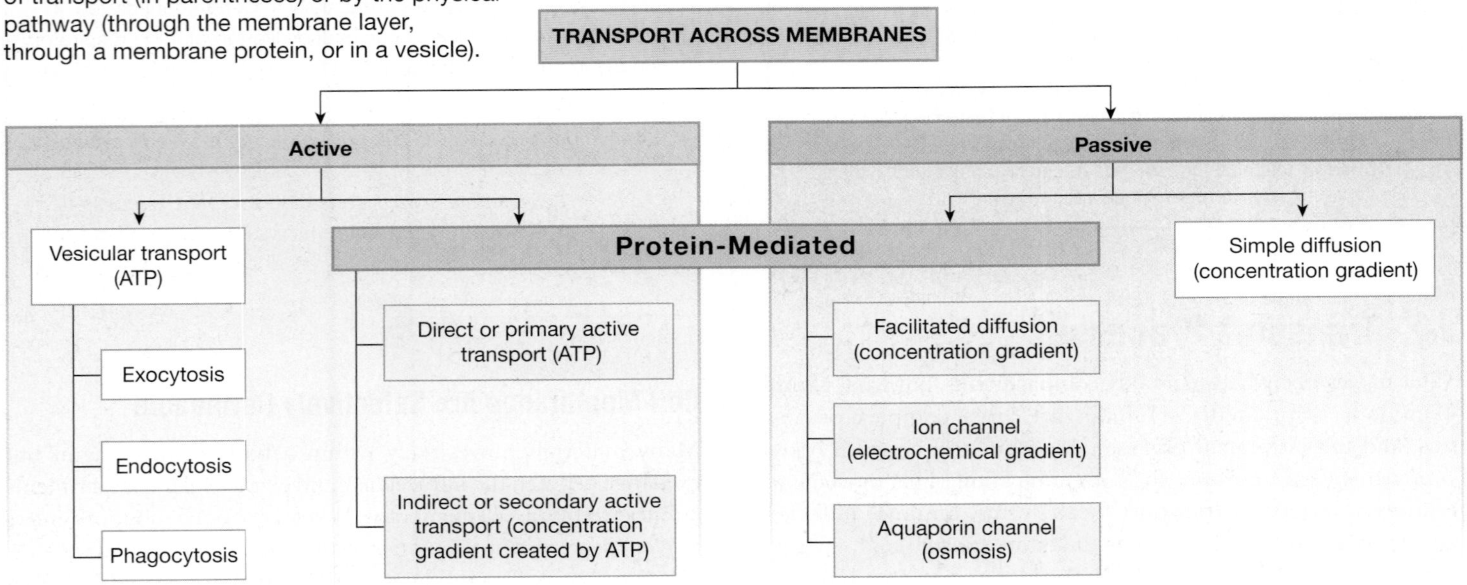

TRANSPORT ACROSS MEMBRANES

Active

Passive

Vesicular transport (ATP)

Exocytosis

Endocytosis

Phagocytosis

Protein-Mediated

Direct or primary active transport (ATP)

Indirect or secondary active transport (concentration gradient created by ATP)

Facilitated diffusion (concentration gradient)

Ion channel (electrochemical gradient)

Aquaporin channel (osmosis)

Simple diffusion (concentration gradient)

2. *Molecules move from an area of higher concentration to an area of lower concentration.* A difference in the concentration of a substance between two places is called a **concentration gradient**, also known as a *chemical gradient.* We say that molecules diffuse *down the gradient,* from higher concentration to lower concentration.

   The rate of diffusion depends on the magnitude of the concentration gradient. The larger the concentration gradient, the faster diffusion takes place. For example, when you open a bottle of cologne, the rate of diffusion is most rapid as the molecules first escape from the bottle into the air. Later, when the cologne has spread evenly throughout the room, the rate of diffusion has dropped to zero because there is no longer a concentration gradient.
3. *Net movement of molecules occurs until the concentration is equal everywhere.* Once molecules of a given substance have distributed themselves evenly, the system reaches equilibrium and diffusion stops. Individual molecules are still moving at equilibrium, but for each molecule that exits an area, another one enters. The *dynamic equilibrium* state in diffusion means that the concentration has equalized throughout the system but molecules continue to move.
4. *Diffusion is rapid over short distances but much slower over long distances.* Albert Einstein studied the diffusion of molecules in solution and found that the time required for a molecule to diffuse from point A to point B is proportional to the square of the distance from A to B. In other words, if the distance doubles from 1 to 2, the time needed for diffusion increases from 1^2 to 2^2 (from 1 to 4).

   What does the slow rate of diffusion over long distances mean for biological systems? In humans, nutrients take 5 seconds to diffuse from the blood to a cell that is 100 μm from the nearest capillary. At that rate, it would take years for nutrients to diffuse from the small intestine to cells in the big toe, and the cells would starve to death.

   To overcome the limitations of diffusion over distance, organisms use various transport mechanisms that speed up the movement of molecules. Most multicellular animals have some form of circulatory system to bring oxygen and nutrients rapidly from the point at which they enter the body to the cells.
5. *Diffusion is directly related to temperature.* At higher temperatures, molecules move faster. Because diffusion results from molecular movement, the rate of diffusion increases as temperature increases. Generally, changes in temperature do not significantly affect diffusion rates in humans because we maintain a relatively constant body temperature.
6. *Diffusion rate is inversely related to molecular weight and size.* Smaller molecules require less energy to move over a distance and therefore diffuse faster. Einstein showed that friction between the surface of a particle and the medium through which it diffuses is a source of resistance to movement. He calculated that diffusion is inversely proportional to the radius of the molecule: the larger the molecule, the slower its diffusion through a given medium. The experiment in **FIGURE 5.6** shows that the smaller and lighter potassium iodide (KI) molecules diffuse more rapidly through the agar gel than the larger and heavier Congo red molecules.
7. *Diffusion can take place in an open system or across a partition that separates two compartments.* Diffusion of cologne within a room is an example of diffusion taking place in an open system. There are no barriers to molecular movement, and the molecules spread out to fill the entire system. Diffusion can also take place between two compartments, such as the intracellular and extracellular compartments, but only if the partition dividing the two compartments allows the diffusing molecules to cross.

For example, if you close the top of an open bottle of cologne, the molecules cannot diffuse out into the room because neither the bottle nor the cap is permeable to the cologne. However, if you replace the metal cap with a plastic bag that has tiny holes in it, you will begin to smell the cologne in the room because the bag is permeable to the molecules. Similarly, if a cell membrane is permeable to a molecule, that molecule can enter or leave the cell by diffusion. If the membrane is not permeable to that particular molecule, the molecule cannot cross.

TABLE 5.6 summarizes these points. An important point to note: ions do not move by diffusion, even though you will read and hear about ions "diffusing across membranes." Diffusion is random molecular motion down a *concentration* gradient. Ion movement is influenced by *electrical* gradients because of the attraction of opposite charges and repulsion of like charges. For this reason, ions move in response to combined electrical and concentration gradients, or *electrochemical gradients.* This electrochemical

FIG. 5.6 Diffusion experiment

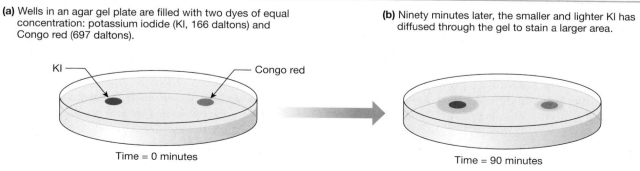

(a) Wells in an agar gel plate are filled with two dyes of equal concentration: potassium iodide (KI, 166 daltons) and Congo red (697 daltons).

(b) Ninety minutes later, the smaller and lighter KI has diffused through the gel to stain a larger area.

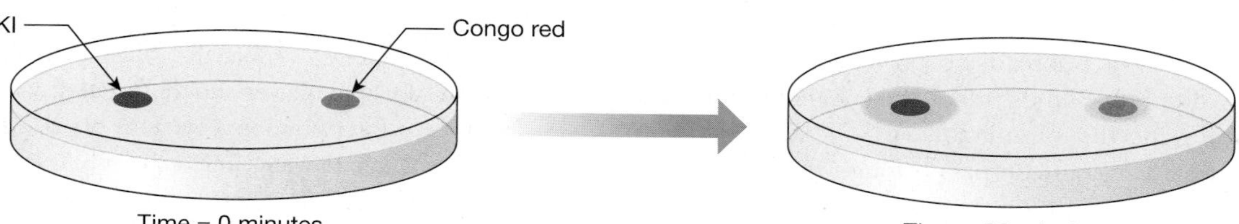
KI
Congo red
Time = 0 minutes
Time = 90 minutes

TABLE 5.6 Rules for Diffusion of Uncharged Molecules

General Properties of Diffusion
1. Diffusion uses the kinetic energy of molecular movement and does not require an outside energy source.
2. Molecules diffuse from an area of higher concentration to an area of lower concentration.
3. Diffusion continues until concentrations come to equilibrium. Molecular movement continues, however, after equilibrium has been reached.
4. Diffusion is faster —along higher concentration gradients. — over shorter distances. — at higher temperatures. — for smaller molecules.
5. Diffusion can take place in an open system or across a partition that separates two systems.
Simple Diffusion across a Membrane
6. The rate of diffusion through a membrane is faster if — the membrane's surface area is larger. — the membrane is thinner. — the concentration gradient is larger. — the membrane is more permeable to the molecule.
7. Membrane permeability to a molecule depends on — the molecule's lipid solubility. — the molecule's size. — the lipid composition of the membrane.

movement is a more complex process than diffusion resulting solely from a concentration gradient, and the two processes should not be confused. We discuss ions and electrochemical gradients in more detail at the end of this chapter.

In summary, diffusion is the passive movement of uncharged molecules down their concentration gradient due to random molecular movement. Diffusion is slower over long distances and slower for large molecules. When the concentration of the diffusing molecules is the same throughout a system, the system has come to chemical equilibrium, although the random movement of molecules continues.

Concept Check

11. If the distance over which a molecule must diffuse triples from 1 to 3, diffusion takes how many times as long?

Lipophilic Molecules Cross Membranes by Simple Diffusion

Diffusion across membranes is a little more complicated than diffusion in an open system. Only lipid-soluble (lipophilic) molecules can diffuse through the phospholipid bilayer. Water and the many vital nutrients, ions, and other molecules that dissolve in water are lipo*phobic* as a rule: they do not readily dissolve in lipids. For these substances, the hydrophobic lipid core of the cell membrane acts as a barrier that prevents them from crossing.

Lipophilic substances that can pass through the lipid center of a membrane move by diffusion. Diffusion directly across the phospholipid bilayer of a membrane is called **simple diffusion** and has the following properties in addition to the properties of diffusion listed earlier.

1. *The rate of diffusion depends on the ability of the diffusing molecule to dissolve in the lipid layer of the membrane.* Another way to say this is that the diffusion rate depends on how permeable the membrane is to the diffusing molecules. Most molecules in solution can mingle with the polar phosphate-glycerol heads of the bilayer [p. 61], but only nonpolar molecules that are lipid-soluble (lipophilic) can dissolve in the central lipid core of the membrane. As a rule, only lipids, steroids, and small lipophilic molecules can move across membranes by simple diffusion.

   One important exception to this statement concerns water. Water, although a polar molecule, may diffuse slowly across some phospholipid membranes. For years, it was thought that the polar nature of the water molecule prevented it from moving through the lipid center of the bilayer, but experiments done with artificial membranes have shown that the small size of the water molecule allows it to slip between the lipid tails in some membranes.

   How readily water passes through the membrane depends on the composition of the phospholipid bilayer. Membranes with high cholesterol content are less permeable to water than those with low cholesterol content, presumably because the lipid-soluble cholesterol molecules fill spaces between the fatty acid tails of the lipid bilayer and thus exclude water. For example, the cell membranes of some sections of the kidney are essentially impermeable to water unless the cells insert special water channel proteins into the phospholipid bilayer. Most water movement across membranes takes place through protein channels.

Instructors: A version of this Try It! Activity can be assigned in @Mastering Anatomy & Physiology

TRY IT! Membrane Models

Most Americans recognize Benjamin Franklin as one of the founding fathers of the United States, but did you know that one of his scientific experiments was the basis for early studies on the composition of the cell membrane? In 1757 Franklin observed how cooking oil released from a ship smoothed the water in its wake, and he began talking with sea captains and fishermen about the calming effect of pouring oil on agitated water. In 1762 he carried out the first of several experiments in which he poured a small amount of oil on a pond's surface and watched as the oil spread out into a very large, thin, almost invisible layer. His calculation of the thickness of the oil layer was remarkably accurate. Franklin's test inspired two Dutch scientists more than 150 years later to do a similar experiment with lipids extracted from the membranes of erythrocytes, or red blood cells (RBCs). Evert Gorter and his student F. Grendel extracted lipids from the RBC cell membranes of humans and several animals. When they dropped the lipid mixture onto water, it spread out to create a thin film that was one molecule thick. Gorter and Grendel then compared the surface area covered by the membrane lipids to the total membrane surface area of the RBCs in the sample. Their data are shown in the graph on the right.

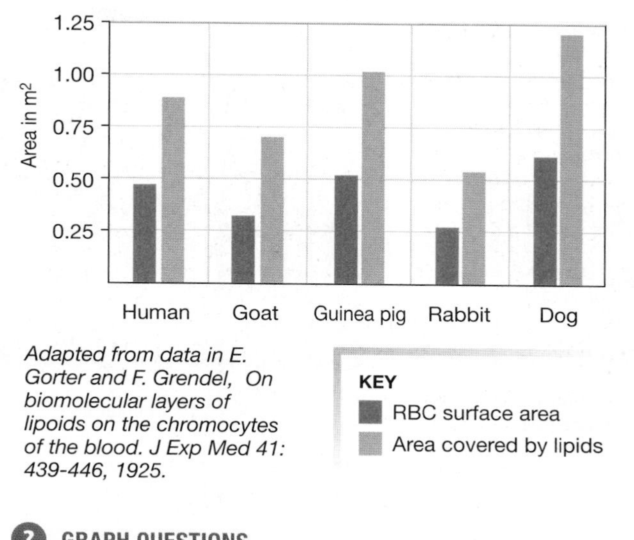
Area in m^2
1.25
1.00
0.75
0.50
0.25
Human
Goat
Guinea pig
Rabbit
Dog

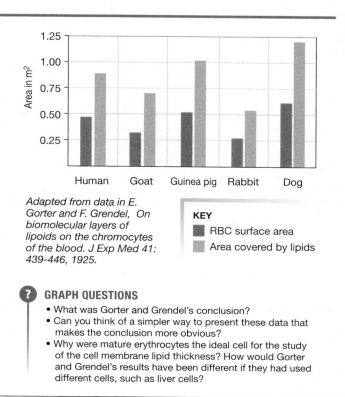

Adapted from data in E. Gorter and F. Grendel, On biomolecular layers of lipoids on the chromocytes of the blood. J Exp Med 41: 439-446, 1925.

KEY
- RBC surface area
- Area covered by lipids

GRAPH QUESTIONS

- What was Gorter and Grendel's conclusion?
- Can you think of a simpler way to present these data that makes the conclusion more obvious?
- Why were mature erythrocytes the ideal cell for the study of the cell membrane lipid thickness? How would Gorter and Grendel's results have been different if they had used different cells, such as liver cells?

2. *The rate of diffusion across a membrane is directly proportional to the surface area of the membrane.* In other words, the larger the membrane's surface area, the more molecules can diffuse across per unit time. This fact may seem obvious, but it has important implications in physiology. One striking example of how a change in surface area affects diffusion is the lung disease emphysema. As lung tissue breaks down and is destroyed, the surface area available for diffusion of oxygen decreases. Consequently, less oxygen can move into the body. In severe cases, the oxygen that reaches the cells is not enough to sustain any muscular activity and the patient is confined to bed.

The rules for simple diffusion across membranes are summarized in Table 5.6. They can be combined mathematically into an equation known as **Fick's law of diffusion**, a relationship that involves the factors just mentioned for diffusion across membranes plus the factor of concentration gradient. In an abbreviated form, Fick's law says that the diffusion rate increases with an increase in surface area, the concentration gradient, or the membrane permeability:

$$\text{rate of diffusion} \propto \text{surface area} \times \text{concentration gradient} \times \text{membrane permeability}$$

FIGURE 5.7 illustrates the principles of Fick's law.

Membrane permeability is the most complex of the terms in Fick's law because several factors influence it:

1. The size (and shape, for large molecules) of the diffusing molecule. As molecular size increases, membrane permeability decreases.
2. The lipid-solubility of the molecule. As lipid solubility of the diffusing molecule increases, membrane permeability to the molecule increases.
3. The composition of the lipid bilayer across which it is diffusing. Alterations in lipid composition of the membrane change how easily diffusing molecules can slip between the individual phospholipids. For example, cholesterol molecules in membranes pack themselves into the spaces between the fatty acids tails and retard passage of molecules through those spaces [Fig. 3.2, p. 63], making the membrane less permeable.

We can rearrange the Fick equation to read:

$$\frac{\text{diffusion rate}}{\text{surface area}} = \text{concentration gradient} \times \text{membrane permeability}$$

This equation now describes the flux of a molecule across the membrane, because **flux** is defined as the diffusion rate per unit surface area of membrane:

$$\text{flux} = \text{concentration gradient} \times \text{membrane permeability}$$

FIG. 5.7 Fick's law of diffusion

Diffusion of an uncharged solute across a membrane is proportional to the concentration gradient of the solute, the membrane surface area, and the membrane permeability to the solute.

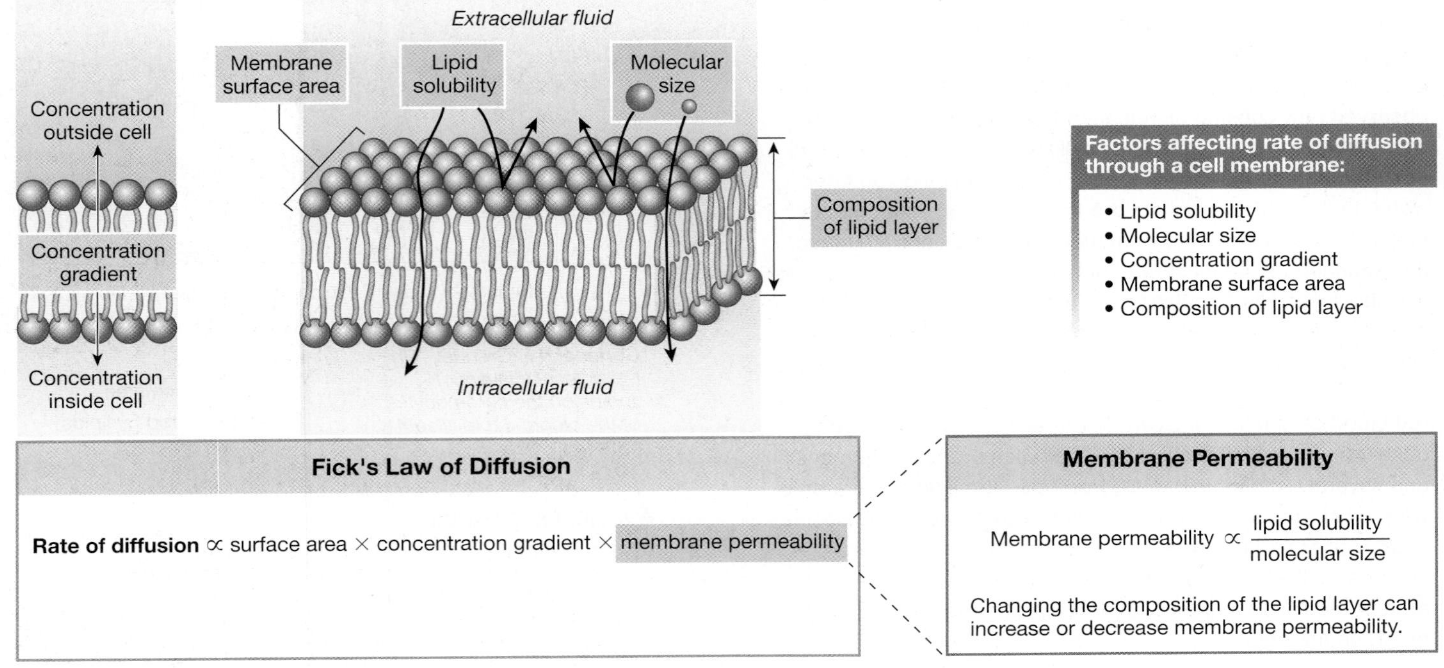

Extracellular fluid
Membrane surface area
Lipid solubility
Molecular size
Concentration outside cell
Concentration gradient
Concentration inside cell
Composition of lipid layer
Intracellular fluid
Factors affecting rate of diffusion through a cell membrane:
• Lipid solubility
• Molecular size
• Concentration gradient
• Membrane surface area
• Composition of lipid layer
Fick's Law of Diffusion
Rate of diffusion ∝ surface area × concentration gradient × membrane permeability
Membrane Permeability
Membrane permeability ∝ lipid solubility / molecular size
Changing the composition of the lipid layer can increase or decrease membrane permeability.

In other words, the flux of a molecule across a membrane depends on the concentration gradient and the membrane's permeability to the molecule.

Remember that the principles of diffusion apply to all biological membranes, not just to the cell membrane. Diffusion of materials in and out of organelles follows the same rules.

Concept Check

12. Where does the energy for diffusion come from?

13. Which is more likely to cross a cell membrane by simple diffusion: a fatty acid molecule or a glucose molecule?

14. What happens to the flux of molecules in each of the following cases?

(a) Molecular size increases.
(b) Concentration gradient increases.
(c) Surface area of membrane decreases.

15. Two compartments are separated by a membrane that is permeable only to water and to yellow dye molecules. Compartment A is filled with an aqueous solution of yellow dye, and compartment B is filled with an aqueous solution of an equal concentration of blue dye. If the system is left undisturbed for a long time, what color will compartment A be: yellow, blue, or green? (Remember, yellow plus blue makes green.) What color will compartment B be?

16. What keeps atmospheric oxygen from diffusing into our bodies across the skin? (*Hint:* What kind of epithelium is skin?)

5.4 Protein-Mediated Transport

In the body, simple diffusion across membranes is limited to lipophilic molecules. The majority of molecules in the body are either lipophobic or electrically charged and therefore cannot cross membranes by simple diffusion. Instead, the vast majority of solutes cross membranes with the help of membrane proteins, a process we call **mediated transport**.

If mediated transport is passive and moves molecules down their concentration gradient, and if net transport stops when concentrations are equal on both sides of the membrane, the process is **facilitated diffusion** (Fig. 5.5). If protein-mediated transport requires energy from ATP or another outside source and moves a substance against its concentration gradient, the process is known as **active transport**.

Membrane Proteins Have Four Major Functions

Protein-mediated transport across a membrane is carried out by membrane-spanning transport proteins. For physiologists, classifying membrane proteins by their function is more useful than classifying them by their structure. Our functional classification scheme recognizes four broad categories of membrane proteins: (1) structural proteins, (2) enzymes, (3) receptors, and (4) transport proteins. **FIGURE 5.8** is a map comparing the structural and functional classifications of membrane proteins. These groupings are not completely distinct, and as you will learn, some membrane proteins have more than one function, such as receptor-channels and receptor-enzymes.

FIG. 5.8 Map of membrane proteins

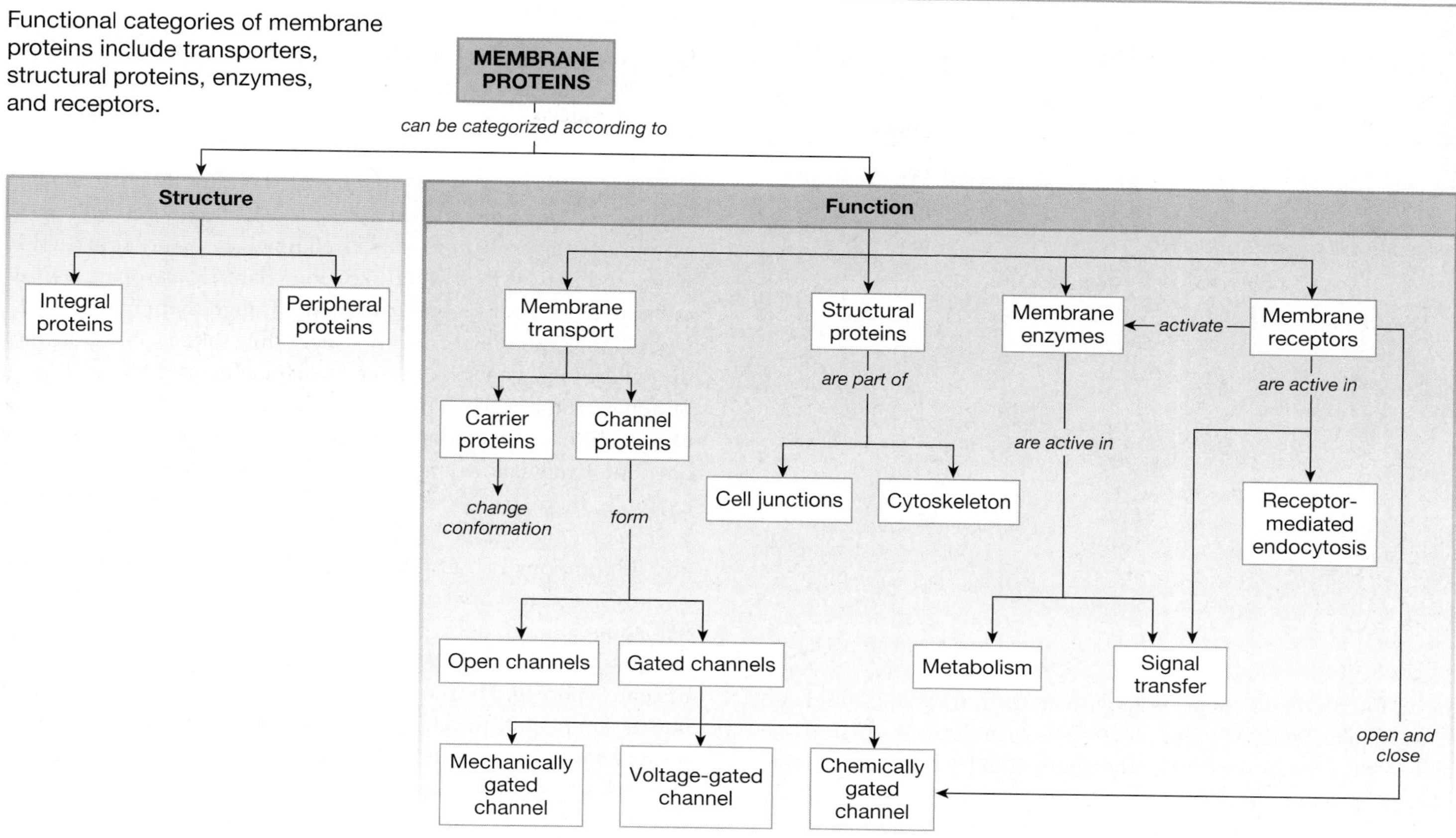

Functional categories of membrane proteins include transporters, structural proteins, enzymes, and receptors.
MEMBRANE PROTEINS
can be categorized according to
Structure
Integral proteins
Peripheral proteins
Function
Membrane transport
Structural proteins
Membrane enzymes
activate
Membrane receptors
are part of
are active in
are active in
Carrier proteins
Channel proteins
change conformation
form
Cell junctions
Cytoskeleton
Receptor-mediated endocytosis
Open channels
Gated channels
Metabolism
Signal transfer
open and close
Mechanically gated channel
Voltage-gated channel
Chemically gated channel

Structural Proteins The **structural proteins** of membranes have three major roles.

1. They help create cell junctions that hold tissues together, such as tight junctions and gap junctions [Fig. 3.8, p. 74].
2. They connect the membrane to the cytoskeleton to maintain the shape of the cell [Fig. 3.2, p. 63]. The microvilli of transporting epithelia are one example of membrane shaping by the cytoskeleton [Fig. 3.4b, p. 66].
3. They attach cells to the extracellular matrix by linking cytoskeleton fibers to extracellular collagen and other protein fibers [p. 68].

Enzymes **Membrane enzymes** catalyze chemical reactions that take place either on the cell's external surface or just inside the cell. For example, enzymes on the external surface of cells lining the small intestine are responsible for digesting peptides and carbohydrates. Enzymes attached to the intracellular surface of many cell membranes play an important role in transferring signals from the extracellular environment to the cytoplasm (see Chapter 6).

Receptors **Membrane receptor proteins** are part of the body's chemical signaling system. The binding of a receptor with its ligand usually triggers another event at the membrane (**FIG. 5.9**). Sometimes the ligand remains on the cell surface, and the receptor-ligand complex triggers an intracellular response. In other instances, the receptor-ligand complex is brought into the cell in a vesicle [p. 70]. Membrane receptors also play an important role in some forms of vesicular transport, as you will learn later in this chapter.

Transport Proteins The fourth group of membrane proteins—**transport proteins**—moves molecules across membranes. There are several different ways to classify transport proteins. Scientists have discovered that the genes for most membrane transport proteins belong to one of two gene "superfamilies": the *ATP-binding cassette (ABC) superfamily* or the *solute carrier (SLC) superfamily*. The ABC family proteins use ATP's energy to transport small molecules or ions across membranes. The 52 families of the SLC superfamily include most facilitated diffusion transporters as well as some active transporters.*

A second way to classify transport recognizes two main types of transport proteins: channels and carriers (**FIG. 5.10**). **Channel proteins** create water-filled passageways that directly link the intracellular and extracellular compartments. **Carrier proteins**, also just called *transporters*, bind to the substrates that they carry but

* The Transporter Classification System, retrieved from *www.tcdb.org*.

FIG. 5.9 Membrane receptors bind extracellular ligands

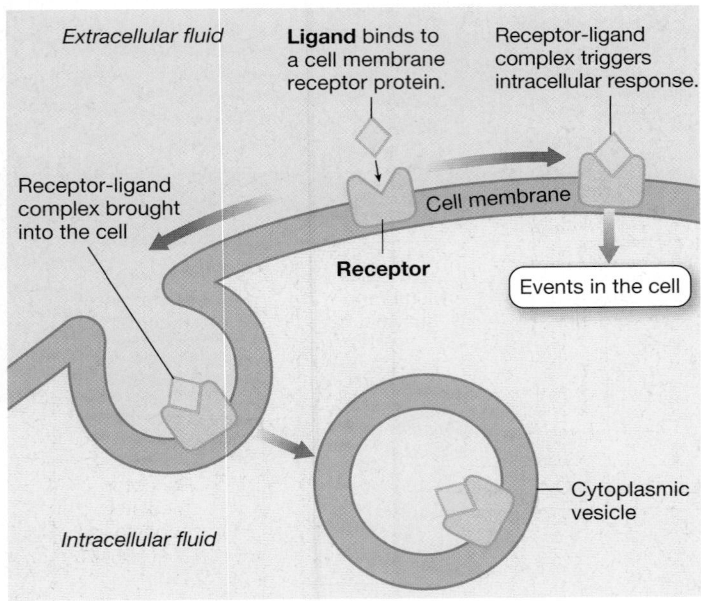
Extracellular fluid
Ligand binds to a cell membrane receptor protein.
Receptor-ligand complex triggers intracellular response.
Receptor-ligand complex brought into the cell
Cell membrane
Receptor
Events in the cell
Cytoplasmic vesicle
Intracellular fluid

never form a direct connection between the intracellular fluid and extracellular fluid. As Figure 5.10 shows, carriers are open to one side of the membrane or the other, but not to both at once the way channel proteins are.

Why do cells need both channels and carriers? The answer lies in the different properties of the two transport proteins. Channel proteins allow more rapid transport across the membrane but generally are limited to moving small ions and water. Carriers, while slower, can move larger molecules than channels can. There is some overlap between the two types, both structurally and functionally. For example, the aquaporin protein AQP has been shown to act both as a water channel and as a carrier for certain small organic molecules.

RUNNING PROBLEM

Cystic fibrosis is a debilitating disease caused by a defect in a membrane channel protein that normally transports chloride ions (Cl^-). The channel—called the **cystic fibrosis transmembrane conductance regulator (CFTR)**—is located in epithelia lining the airways, sweat glands, and pancreas. A gate in the CFTR channel opens when the nucleotide ATP binds to the protein. In the lungs, this open channel transports Cl^- out of the epithelial cells and into the airways. In people with cystic fibrosis, CFTR is nonfunctional or absent. As a result, chloride transport across the epithelium is impaired, and thickened mucus is the result.

Q2: *Is the CFTR a chemically gated, a voltage-gated, or a mechanically gated channel protein?*

122 – 131 – **138** – 151 – 152 – 159

Channel Proteins Form Open, Water-Filled Passageways

Channel proteins are made of membrane-spanning protein subunits that create a cluster of cylinders with a tunnel or *pore* through the center. Nuclear pore complexes [p. 71] and gap junctions (Fig. 3.8b, p. 74) can be considered very large forms of channels. In this text, we restrict use of the term *channel* to smaller channels whose centers are narrow, water-filled pores (**FIG. 5.11**). Movement through these smaller channels is mostly restricted to water and ions. When water-filled ion channels are open, tens of millions of ions per second can whisk through them unimpeded.

Channel proteins are named according to the substances that they allow to pass. Most cells have **water channels** made from a protein called *aquaporin*. In addition, more than 100 types of **ion channels** have been identified. Ion channels may be specific for one ion or may allow ions of similar size and charge to pass. For example, there are Na^+ channels, K^+ channels, and nonspecific *monovalent* ("one-charge") cation channels that transport Na^+, K^+, and lithium ions Li^+. Other ion channels you will encounter frequently in this text are Ca^{2+} channels and Cl^- channels. Ion channels come in many subtypes, or *isoforms*.

The selectivity of a channel is determined by the diameter of its central pore and by the electrical charge of the amino acids that line the channel. If the channel amino acids are positively charged, positive ions are repelled and negative ions can pass through the channel. On the other hand, a cation channel must have a negative charge that attracts cations but prevents the passage of Cl^- or other anions.

Channel proteins are like narrow doorways into the cell. If the door is closed, nothing can go through. If the door is open, there is a continuous passage between the two rooms connected by the doorway. The open or closed state of a channel is determined by regions of the protein molecule that act like swinging "gates."

According to current models, channel "gates" take several forms. Some channel proteins have gates in the middle of the protein's pore. Other gates are part of the cytoplasmic side of the membrane protein. Such a gate can be envisioned as a ball on a chain that swings up and blocks the mouth of the channel (Fig. 5.10a). One type of channel in neurons has two different gates.

Channels can be classified according to whether their gates are usually open or usually closed. **Open channels** spend most of their time with their gate open, allowing ions to move back and forth across the membrane without regulation. These gates may occasionally flicker closed, but for the most part these channels behave as if they have no gates. Open channels are sometimes called either *leak channels* or pores, as in *water pores*.

Gated channels spend most of their time in a closed state, which allows these channels to regulate the movement of ions through them. When a gated channel opens, ions move through the channel just as they move through open channels. When a gated channel is closed, which it may be much of the time, it allows no ion movement between the intracellular and extracellular fluid.

What controls the opening and closing of gated channels? For **chemically gated channels**, the gating is controlled by

FIG. 5.10 ESSENTIALS Membrane Transporters

Membrane transporters are membrane-spanning proteins that help move lipophobic molecules across membranes.

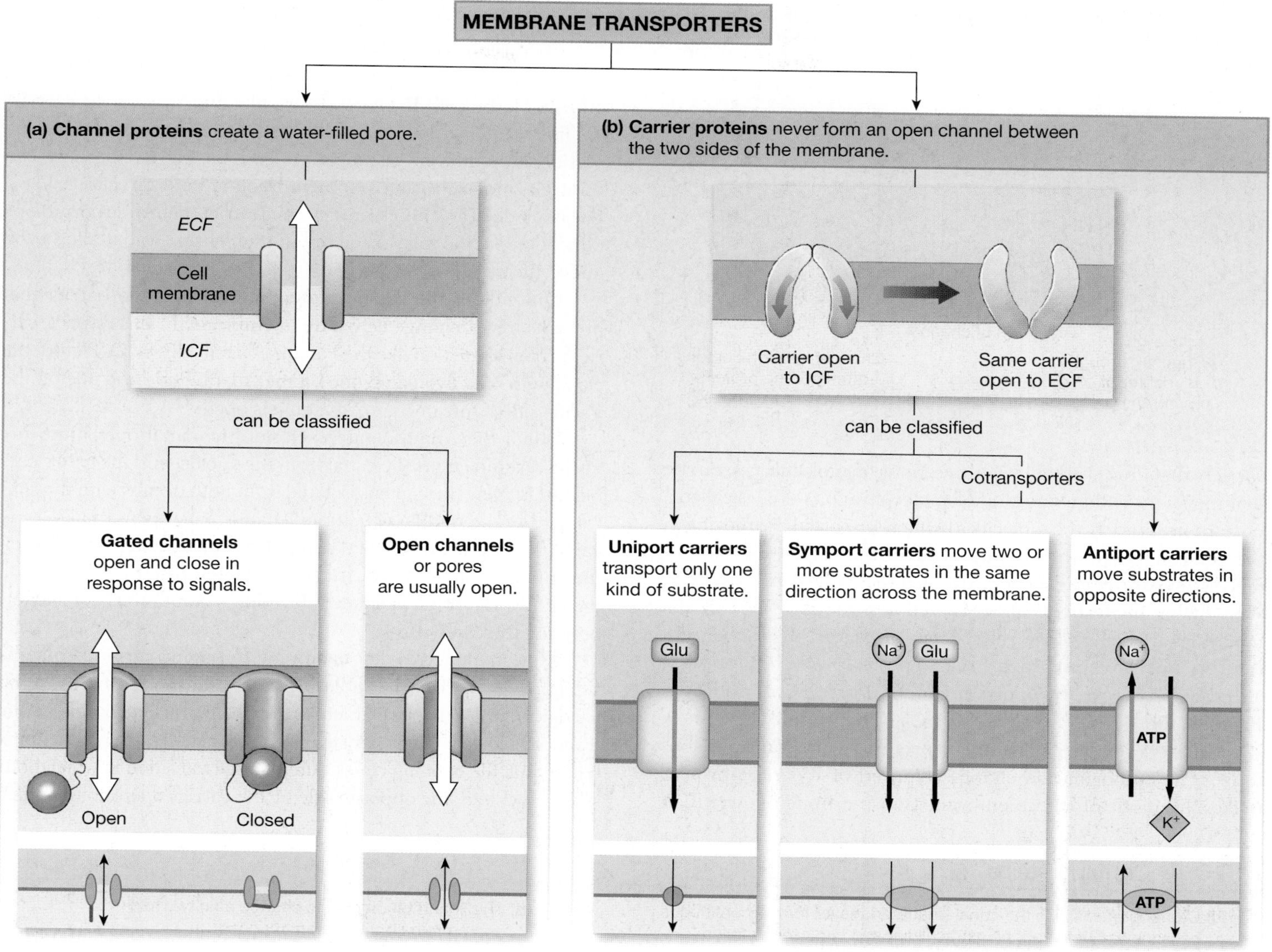

MEMBRANE TRANSPORTERS
(a) Channel proteins create a water-filled pore.
ECF
Cell membrane
ICF
can be classified
Gated channels open and close in response to signals.
Open
Closed
Open channels or pores are usually open.
(b) Carrier proteins never form an open channel between the two sides of the membrane.
Carrier open to ICF
Same carrier open to ECF
can be classified
Cotransporters
Uniport carriers transport only one kind of substrate.
Glu
Symport carriers move two or more substrates in the same direction across the membrane.
Na+
Glu
Antiport carriers move substrates in opposite directions.
Na+
ATP
K+
ATP

Close-up views of transporters are shown in the top two rows and distant views in the bottom row. Primary active transport is indicated by *ATP* on the protein.

intracellular messenger molecules or extracellular ligands that bind to the channel protein. **Voltage-gated channels** open and close when the electrical state of the cell changes. Finally, **mechanically gated channels** respond to physical forces, such as increased temperature or pressure that puts tension on the membrane and pops the channel gate open. You will encounter many variations of these channel types as you study physiology.

> **Concept Check**
>
> **17.** Positively charged ions are called _____, and negatively charged ions are called _____.

Carrier Proteins Change Conformation to Move Molecules

The second type of transport protein is the carrier protein (Fig. 5.10b). Carrier proteins bind with specific substrates and carry them across the membrane by changing conformation. Small organic molecules (such as glucose and amino acids) that are too large to pass through channels cross membranes using carriers. Ions such as Na^+ and K^+ may move by carriers as well as through channels. Carrier proteins move solutes and ions into and out of cells as well as into and out of intracellular organelles, such as the mitochondria.

Some carrier proteins move only one kind of molecule and are known as **uniport carriers**. However, it is common to find

FIG. 5.11 The structure of channel proteins

Many channels are made of multiple protein subunits that assemble in the membrane. Hydrophilic amino acids in the protein line the channel, creating a water-filled passage that allows ions and water to pass through.

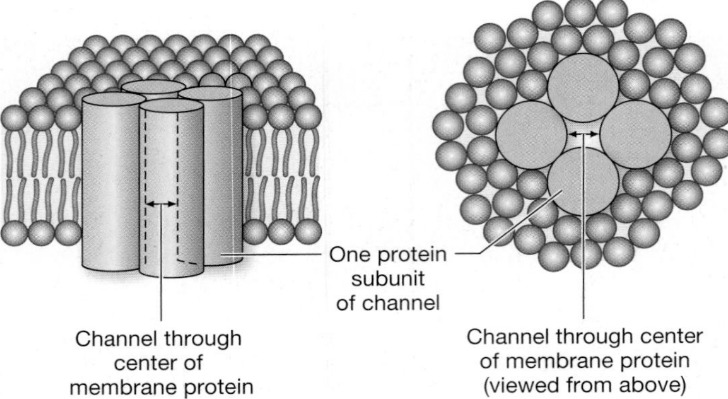
One protein subunit of channel
Channel through center of membrane protein
Channel through center of membrane protein (viewed from above)

carriers that move two or even three kinds of molecules. A carrier that moves more than one kind of molecule at one time is called a **cotransporter**. If the molecules being transported are moving in the same direction, whether into or out of the cell, the carrier proteins are **symport carriers** {*sym-*, together + *portare*, to carry}. (Sometimes, the term *cotransport* is used in place of *symport*.) If the molecules are being carried in opposite directions, the carrier proteins are **antiport carriers** {*anti*, opposite + *portare*, to carry}, also called *exchangers*. Symport and antiport carriers are shown in Figure 5.10b.

Carriers are large, complex proteins with multiple subunits. The conformation change required of a carrier protein makes this mode of transmembrane transport much slower than movement through channel proteins. A carrier protein can move only 1,000 to 1,000,000 molecules per second, in contrast to tens of millions of ions per second that move through a channel protein.

Carrier proteins differ from channel proteins in another way: carriers never create a continuous passage between the inside and outside of the cell. If channels are like doorways, then carriers are like revolving doors that allow movement between inside and outside without ever creating an open hole. Carrier proteins can transport molecules across a membrane in both directions, like a revolving door at a hotel, or they can restrict their transport to one direction, like the turnstile at an amusement park that allows you out of the park but not back in.

One side of the carrier protein always creates a barrier that prevents free exchange across the membrane. In this respect, carrier proteins function like the Panama Canal (**FIG. 5.12**). Picture the canal with only two gates, one on the Atlantic side and one on the Pacific side. Only one gate at a time is open.

When the Atlantic gate is closed, the canal opens into the Pacific. A ship enters the canal from the Pacific, and the gate closes behind it. Now the canal is isolated from both oceans with the ship trapped in the middle. Then the Atlantic gate opens, making the canal continuous with the Atlantic Ocean. The ship sails out of the gate and off into the Atlantic, having crossed the barrier of the land without the canal ever forming a continuous connection between the two oceans.

Movement across the membrane through a carrier protein is similar (Fig. 5.12b). The molecule being transported binds to the carrier on one side of the membrane (the extracellular side in our example). This binding changes the conformation of the carrier protein so that the opening closes. After a brief transition in which both sides are closed, the opposite side of the carrier opens to the other

FIG. 5.12 Carrier proteins

(a) Carrier proteins, like the canal illustrated, never form a continuous passageway between the extracellular and intracellular fluids.

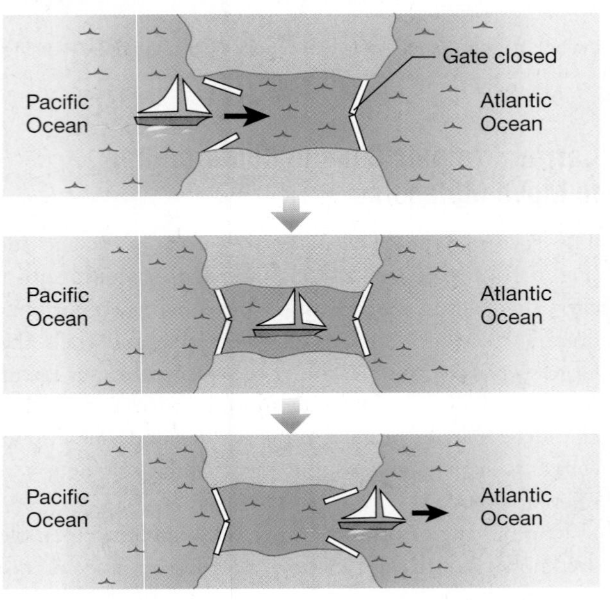
Gate closed
Pacific Ocean
Atlantic Ocean
Pacific Ocean
Atlantic Ocean
Pacific Ocean
Atlantic Ocean

Passage open to one side

Transition state with both gates closed

Passage open to other side

(b) The ligand binding sites change affinity when the protein conformation changes.

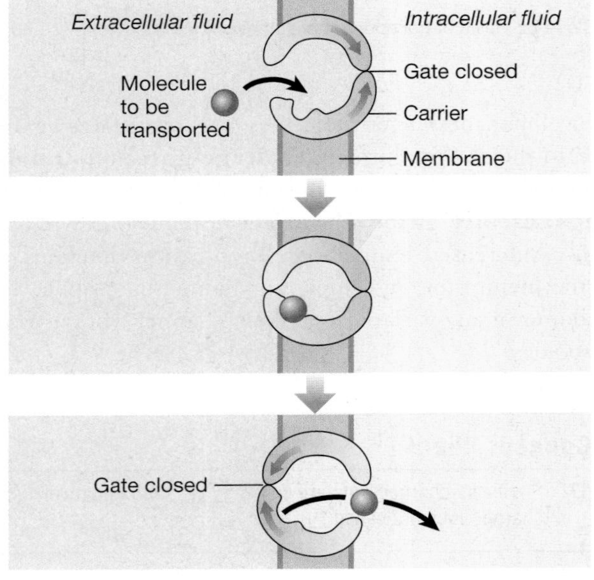
Extracellular fluid
Intracellular fluid
Molecule to be transported
Gate closed
Carrier
Membrane
Gate closed

side of the membrane. The carrier then releases the transported molecule into the opposite compartment, having brought it through the membrane without creating a continuous connection between the extracellular and intracellular compartments.

Carrier proteins can be divided into two categories according to the energy source that powers the transport. As noted earlier, facilitated diffusion is protein-mediated transport in which no outside source of energy except a concentration gradient is needed to move molecules across the cell membrane. Active transport is protein-mediated transport that requires an outside energy source, either ATP or the potential energy stored in a concentration gradient that was created using ATP. We will look first at facilitated diffusion.

> **Concept Check**
>
> **18.** Name four functions of membrane proteins.
>
> **19.** Which kinds of particles pass through open channels?
>
> **20.** Name two ways channels differ from carriers.
>
> **21.** If a channel is lined with amino acids that have a net positive charge, which of the following ions is/are likely to move freely through the channel? Na^+, Cl^-, K^+, Ca^{2+}
>
> **22.** Why can't glucose cross the cell membrane through open channels?

Facilitated Diffusion Uses Carrier Proteins

Some polar molecules appear to move into and out of cells by diffusion, even though we know from their chemical properties that they are unable to pass easily through the lipid core of the cell membrane. The solution to this seeming contradiction is that these polar molecules cross the cell membrane by facilitated diffusion, with the aid of specific carriers. Sugars and amino acids are examples of molecules that enter or leave cells using facilitated diffusion. For example, the family of carrier proteins known as **GLUT transporters** move glucose and related hexose sugars across membranes.

Facilitated diffusion has the same properties as simple diffusion (see Tbl. 5.6). The transported molecules move down their concentration gradient, the process requires no input of outside energy, and net movement stops at equilibrium, when the concentration inside the cell equals the concentration outside the cell (**FIG. 5.13**):

$$[\text{glucose}]_{ECF} = [\text{glucose}]_{ICF}^{*}$$

Facilitated diffusion carriers always transport molecules down their concentration gradient. If the gradient reverses, so does the direction of transport.

Cells in which facilitated diffusion takes place can avoid reaching equilibrium by keeping the concentration of substrate in the cell low. With glucose, for example, this is accomplished by phosphorylation (Fig. 5.13c). As soon as a glucose molecule enters the cell on the GLUT carrier, it is phosphorylated to glucose 6-phosphate, the first step of glycolysis [p. 106]. Addition of the phosphate group prevents buildup of glucose inside the cell and also prevents glucose from leaving the cell.

> **Concept Check**
>
> **23.** Liver cells (hepatocytes) are able to convert glycogen to glucose, thereby making the intracellular glucose concentration higher than the extracellular glucose concentration. In what direction do the hepatic GLUT2 transporters carry glucose when this occurs?

* In this text, the presence of brackets around a solute's name indicates concentration.

FIG. 5.13 Facilitated diffusion of glucose into cells

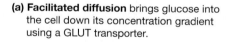

(a) Facilitated diffusion brings glucose into the cell down its concentration gradient using a GLUT transporter.

(b) Diffusion reaches equilibrium when the glucose concentrations inside and outside the cell are equal.

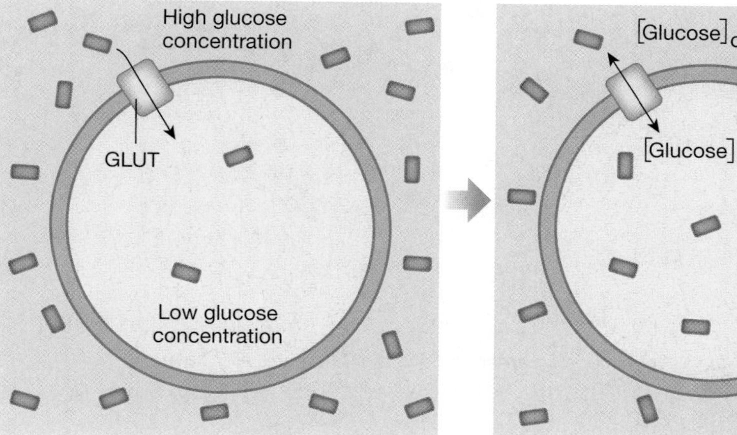
High glucose concentration
GLUT
Low glucose concentration
[Glucose]out =
[Glucose]in

(c) In most cells, conversion of imported glucose into glucose 6-phosphate (G-6-P) keeps intracellular glucose concentrations low so that diffusion never reaches equilibrium.

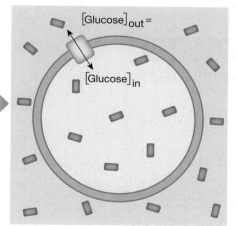

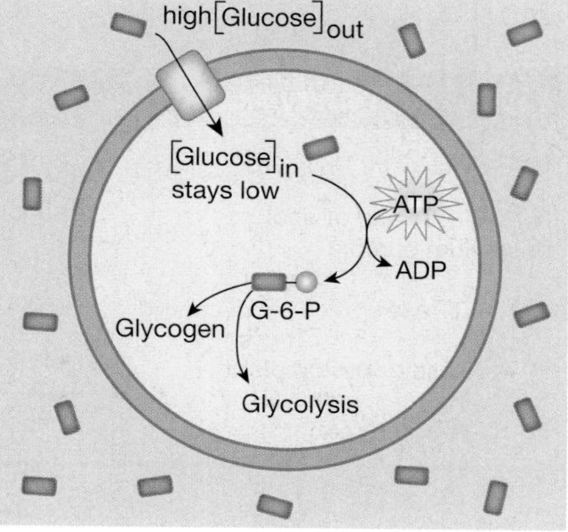
high[Glucose]out
[Glucose]in stays low
ATP
ADP
G-6-P
Glycogen
Glycolysis

Active Transport Moves Substances against Their Concentration Gradients

Active transport is a process that moves molecules *against* their concentration gradient—that is, from areas of lower concentration to areas of higher concentration. Rather than creating an equilibrium state, where the concentration of the molecule is equal throughout the system, active transport creates a state of *dis*equilibrium by making concentration differences more pronounced. Moving molecules against their concentration gradient requires the input of outside energy, just as pushing a ball up a hill requires energy [see Fig. 4.2, p. 95]. The energy for active transport comes either directly or indirectly from the high-energy phosphate bond of ATP.

Active transport can be divided into two types. In **primary (direct) active transport**, the energy to push molecules against their concentration gradient comes directly from the high-energy phosphate bond of ATP. **Secondary (indirect) active transport** uses potential energy [p. 94] stored in the concentration gradient of one molecule to push other molecules against their concentration gradient. All secondary active transport ultimately depends on primary active transport because the concentration gradients that drive secondary transport are created using energy from ATP.

The mechanism for both types of active transport appears to be similar to that for facilitated diffusion. A substrate to be transported binds to a membrane carrier and the carrier then changes conformation, releasing the substrate into the opposite compartment. Active transport differs from facilitated diffusion because the conformation change in the carrier protein requires energy input.

Primary Active Transport Because primary active transport uses ATP as its energy source, many primary active transporters are known as **ATPases**. You may recall that the suffix *-ase* signifies an enzyme, and the stem (ATP) is the substrate upon which the enzyme is acting [p. 101]. These enzymes hydrolyze ATP to ADP and inorganic phosphate (P_i), releasing usable energy in the process. Most of the ATPases you will encounter in your study of physiology are listed in **TABLE 5.7**. ATPases are sometimes called *pumps,* as in the sodium-potassium pump, or Na^+-K^+-ATPase, mentioned earlier in this chapter.

TABLE 5.7 Primary Active Transporters

Names	Type of Transport
Na^+-K^+-ATPase or sodium-potassium pump	Antiport
Ca^{2+}-ATPase	Uniport
H^+-ATPase or proton pump	Uniport
H^+-K^+-ATPase	Antiport

The sodium-potassium pump is probably the single most important transport protein in animal cells because it maintains the concentration gradients of Na^+ and K^+ across the cell membrane (**FIG. 5.14**). The transporter is arranged in the cell membrane so that it pumps 3 Na^+ out of the cell and 2 K^+ into the cell for each ATP consumed. In some cells, the energy needed to move these ions uses 30% of all the ATP produced by the cell. **FIGURE 5.15** illustrates the current model of how the Na^+-K^+-ATPase works.

Secondary Active Transport The sodium concentration gradient, with Na^+ concentration high in the extracellular fluid and low inside the cell, is a source of potential energy that the cell can harness for other functions. For example, nerve cells use the sodium gradient to transmit electrical signals, and epithelial cells use it to drive the uptake of nutrients, ions, and water. Membrane transporters that use potential energy stored in concentration gradients to move molecules are called *secondary active transporters.*

Secondary active transport uses the kinetic energy of one molecule moving down its concentration gradient to push other molecules against their concentration gradient. The cotransported molecules may go in the same direction across the membrane (symport) or in opposite directions (antiport). The most common secondary active transport systems are driven by the sodium concentration gradient.

FIG. 5.14 The sodium-potassium pump, Na^+-K^+-ATPase

The Na^+-K^+-ATPase uses energy from ATP to pump Na^+ out of the cell and K^+ into the cell.

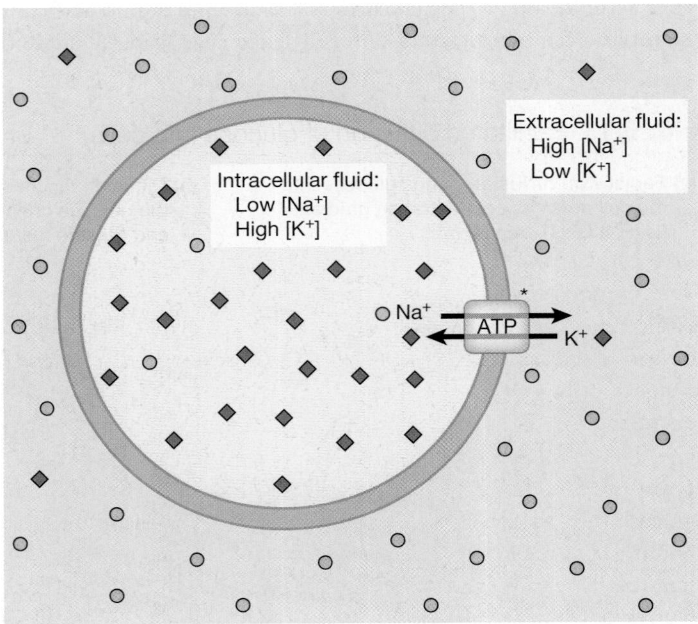
Extracellular fluid:
High [Na+]
Low [K+]
Intracellular fluid:
Low [Na+]
High [K+]
Na+
ATP
*
K+

*In this book, carrier proteins that hydrolyze ATP have the letters ATP written on the membrane protein.

FIG. 5.15 Mechanism of the Na^+-K^+-ATPase

This figure presents one model of how the Na^+-K^+-ATPase uses energy and inorganic phosphate (P_i) from ATP to move ions across a membrane. Phosphorylation and dephosphorylation of the ATPase change its conformation and the binding sites' affinity for ions.

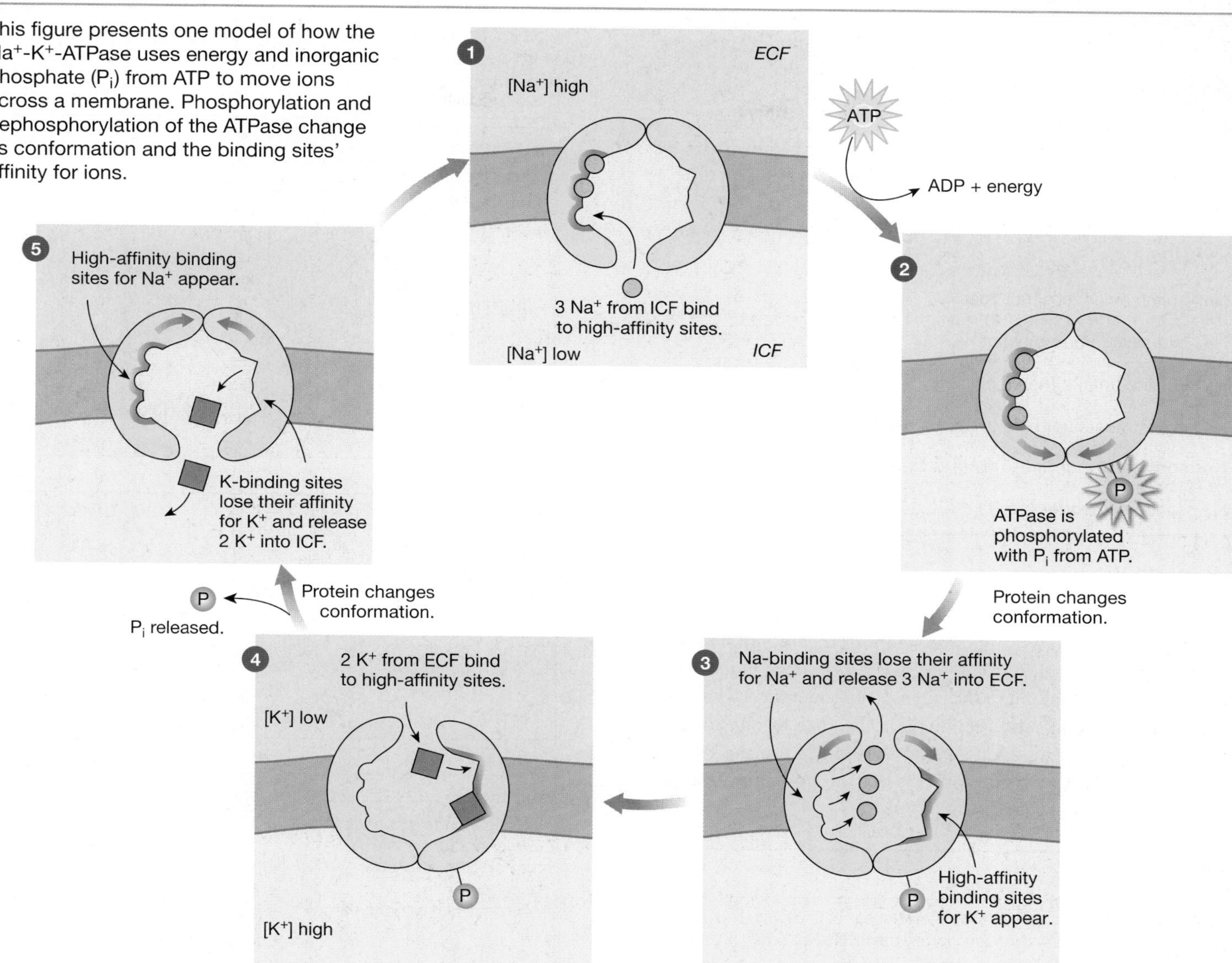

1
ECF
[Na+] high
3 Na+ from ICF bind
to high-affinity sites.
[Na+] low
ICF
ATP
ADP + energy
2
ATPase is
phosphorylated
with Pi from ATP.
Protein changes
conformation.
3
Na-binding sites lose their affinity
for Na+ and release 3 Na+ into ECF.
High-affinity
binding sites
for K+ appear.
P
4
2 K+ from ECF bind
to high-affinity sites.
[K+] low
[K+] high
P
Pi released.
Protein changes
conformation.
5
High-affinity binding
sites for Na+ appear.
K-binding sites
lose their affinity
for K+ and release
2 K+ into ICF.

As one Na^+ moves into the cell, it either brings one or more molecules with it or trades places with molecules exiting the cell. The major Na^+-dependent transporters are listed in **TABLE 5.8**. Notice that the cotransported substances may be either other ions or uncharged molecules, such as glucose. As you study the different systems of the body, you will find these secondary active transporters taking part in many physiological processes.

The mechanism of the **Na^+-glucose secondary active transporter (SGLT)** is illustrated in **FIGURE 5.16**. Both Na^+ and glucose bind to the SGLT protein on the extracellular fluid side. Sodium binds first and causes a conformational change in the protein that creates a high-affinity binding site for glucose (1). When glucose binds to SGLT (2), the protein changes conformation again and opens its channel to the intracellular fluid side (3). Sodium is released to the ICF as it moves down its concentration gradient. The loss of Na^+ from the protein changes the binding site for glucose back to a low-affinity site, so glucose is released and follows Na^+ into the cytoplasm. The net result is the entry of glucose into the cell against its concentration gradient, coupled to the movement of Na^+ into the cell down its concentration gradient. The SGLT transporter can only move glucose into cells because glucose must follow the Na^+ gradient.

In contrast, GLUT transporters are reversible and can move glucose into or out of cells depending on the concentration gradient. For example, when blood glucose levels are high, GLUT transporters on liver cells bring glucose into those cells. During times of fasting, when blood glucose levels fall, liver cells convert their glycogen stores to glucose. When the glucose concentration inside the liver cells builds up and exceeds the glucose concentration in the plasma, glucose leaves the cells on the reversible GLUT transporters. GLUT transporters are found on all cells of the body.

TABLE 5.8 Examples of Secondary Active Transporters

Symport Carriers	Antiport Carriers
Sodium-Dependent Transporters	
Na^+-K^+-$2Cl^-$ (NKCC)	Na^+-H^+ (NHE)
Na^+-glucose (SGLT)	Na^+-Ca^{2+} (NCX)
Na^+-Cl^-	
Na^+-HCO_3^-	
Na^+-amino acids (several types)	
Na^+-bile salts (small intestine)	
Na^+-choline uptake (nerve cells) Na^+-neurotransmitter uptake (nerve cells)	
Nonsodium-Dependent Transporters	
H^+-peptide symporter (pepT)	HCO_3^--Cl^-

If GLUT transporters are everywhere, then why does the body need the SGLT Na^+-glucose symporter? The simple answer is that both SGLT and GLUT are needed to move glucose from one side of an epithelium to the other. Consequently, SGLT transporters are found on certain epithelial cells, such as intestinal and kidney cells, that bring glucose into the body from the external environment. We discuss the process of transepithelial transport of glucose later in this chapter.

Concept Check

24. Name two ways active transport by the Na^+-K^+-ATPase (Fig. 5.15) differs from secondary transport by the SGLT (Fig. 5.16).

Carrier-Mediated Transport Exhibits Specificity, Competition, and Saturation

Both passive and active forms of carrier-mediated transport demonstrate specificity, competition, and saturation—three properties that result from the binding of a substrate to a protein [p. 46].

Specificity Specificity refers to the ability of a transporter to move only one molecule or only a group of closely related molecules [p. 46]. One example of specificity is found in the GLUT family of transporters, which move 6-carbon sugars (*hexoses*), such as glucose, mannose, galactose, and fructose [p. 31], across cell membranes. GLUT transporters have binding sites that recognize and transport hexoses, but they will not transport the disaccharide maltose or any form of glucose that is not found in nature (**FIG. 5.17b**). For this reason we can say that GLUT transporters are specific for naturally occurring 6-carbon monosaccharides.

FIG. 5.16 Sodium-glucose cotransport

The SGLT transporter uses the potential energy stored in the Na^+ concentration gradient to move glucose against its concentration gradient.

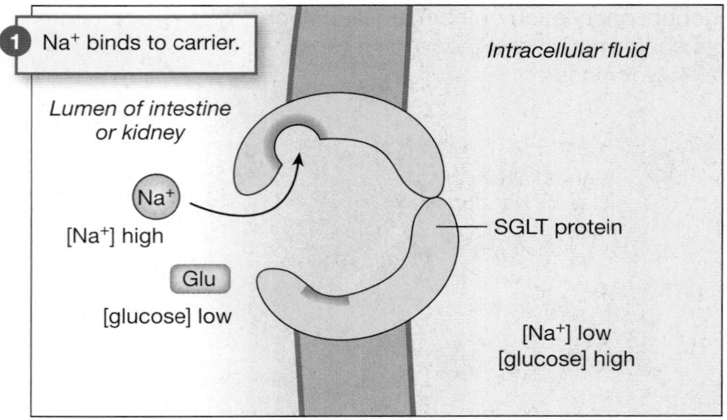

1 Na+ binds to carrier.
Intracellular fluid
Lumen of intestine or kidney
Na+
[Na+] high
SGLT protein
Glu
[glucose] low
[Na+] low
[glucose] high

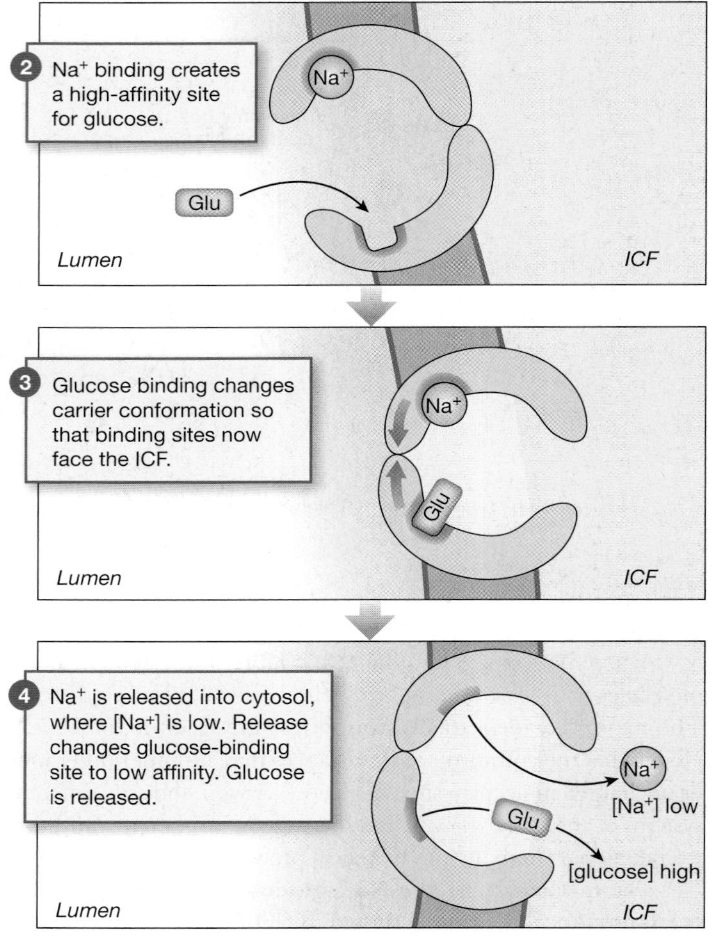

2 Na+ binding creates a high-affinity site for glucose.
Na+
Glu
Lumen
ICF
3 Glucose binding changes carrier conformation so that binding sites now face the ICF.
Na+
Glu
Lumen
ICF
4 Na+ is released into cytosol, where [Na+] is low. Release changes glucose-binding site to low affinity. Glucose is released.
Na+
[Na+] low
Glu
[glucose] high
Lumen
ICF

For many years, scientists assumed that there must be different isoforms of the glucose-facilitated diffusion carrier because they had observed that glucose transport was regulated by hormones in some cells but not in others. However, it was not until the 1980s that the first glucose transporter was isolated. To date,

FIG. 5.17 Transporter saturation and competition

(a) The **GLUT transporter** brings glucose across cell membranes.

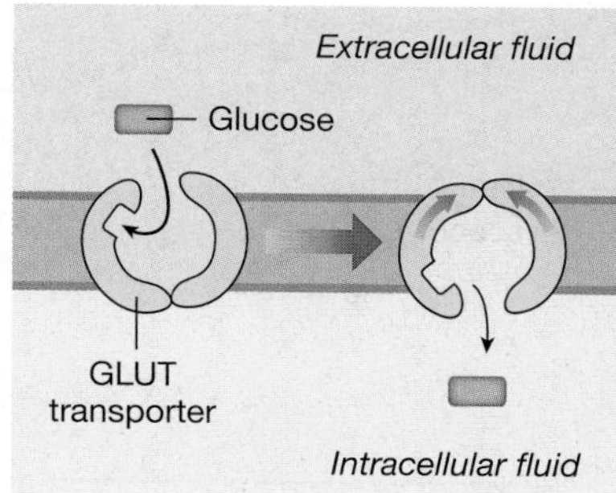
Extracellular fluid
Glucose
GLUT transporter
Intracellular fluid

(b) Maltose is a competitive inhibitor that binds to the GLUT transporter but is not itself carried across the membrane.

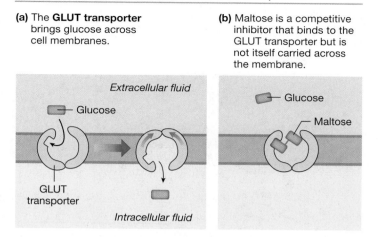

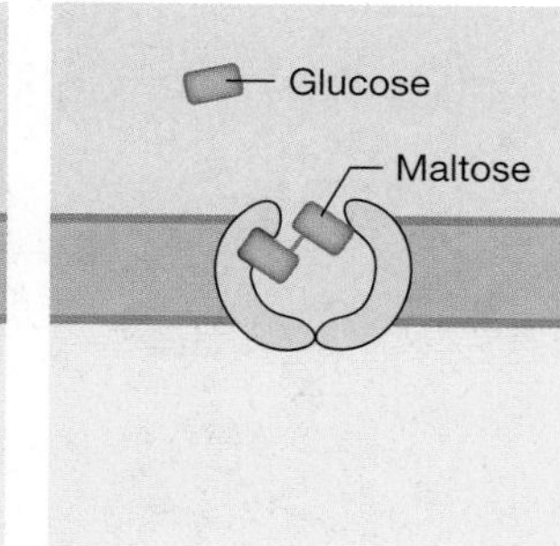
Glucose
Maltose

(c) Saturation. This graph shows that transport can reach a maximum rate when all the carrier binding sites are filled with substrate.

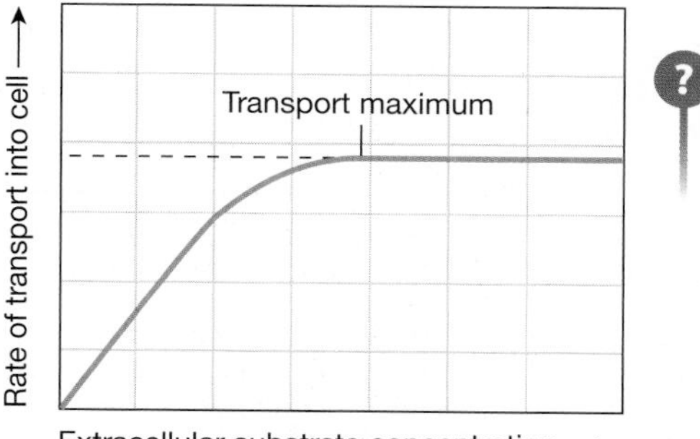
Transport maximum
Rate of transport into cell ⟶
Extracellular substrate concentration ⟶

Transport rate is proportional to substrate concentration until the carriers are saturated.

GRAPH QUESTION
How could the cell increase its transport rate in this example?

(d) Competition. This graph shows glucose transport rate as a function of glucose concentration. In one experiment, only glucose was present. In the second experiment, a constant concentration of galactose was present.

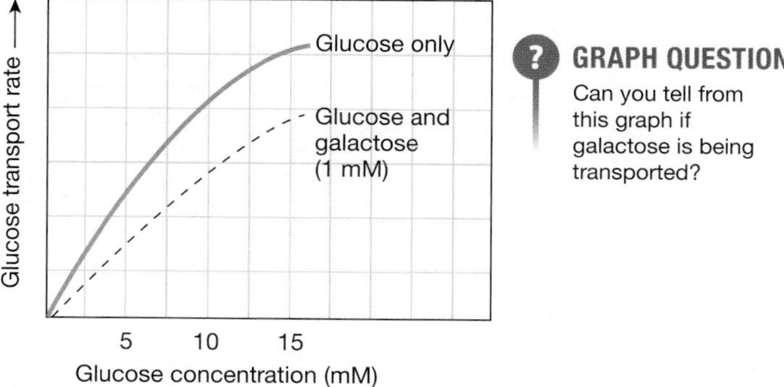
Glucose only
Glucose and galactose (1 mM)
Glucose transport rate ⟶
5
10
15
Glucose concentration (mM)

GRAPH QUESTION
Can you tell from this graph if galactose is being transported?

14 SCL2A (GLUT) genes have been identified. The important GLUT proteins you will encounter in this book include GLUT1, found in most cells of the body; GLUT2, found in liver and in kidney and intestinal epithelia; GLUT3, found in neurons; GLUT4, the insulin-regulated transporter of skeletal muscle; and GLUT5, the intestinal fructose transporter. The restriction of different GLUT transporters to different tissues is an important feature in the metabolism and homeostasis of glucose.

Competition The property of competition is closely related to specificity. A transporter may move several members of a related group of substrates, but those substrates compete with one another for binding sites on the transporter. For example, GLUT transporters move the family of hexose sugars, but each different GLUT transporter has a "preference" for one or more hexoses, based on its binding affinity.

The results of an experiment demonstrating competition are shown in Figure 5.17d. The graph shows glucose transport rate as a function of glucose concentration. The top line (red) shows transport when only glucose is present. The lower line (black) shows that glucose transport decreases if galactose is also present. Galactose competes for binding sites on the GLUT transporters and displaces some glucose molecules. With fewer glucose molecules able to bind to the GLUT protein, the rate of glucose transport into the cell decreases.

Sometimes, the competing molecule is not transported but merely blocks the transport of another substrate. In this case, the competing molecule is a *competitive inhibitor* [p. 49]. In the glucose transport system, the disaccharide maltose is a competitive inhibitor (Fig. 5.17b). It competes with glucose for the binding site, but once bound, it is too large to be moved across the membrane.

Competition between transported substrates has been put to good use in medicine. An example involves gout, a disease caused by elevated levels of uric acid in the plasma. One method of decreasing uric acid in plasma is to enhance its excretion in the urine. Normally, the kidney's *organic anion transporter (OAT)* reclaims urate (the anion form of uric acid) from the urine and returns the acid to the plasma. However, if an organic acid called probenecid is administered to the patient, OAT binds to probenecid instead of to urate, preventing the reabsorption of urate. As a result, more urate leaves the body in the urine, lowering the uric acid concentration in the plasma.

Saturation The rate of substrate transport depends on the substrate concentration and the number of carrier molecules, a property that is shared by enzymes and other binding proteins [p. 51]. For a fixed number of carriers, however, as substrate concentration increases, the transport rate increases up to a maximum, the point at which all carrier binding sites are filled with substrate. At this point, the carriers are said to have reached saturation. At saturation, the carriers are working at their maximum rate, and a further increase in substrate concentration has no effect. Figure 5.17c represents saturation graphically.

As an analogy, think of the carriers as doors into a concert hall. Each door has a maximum number of people that it can allow to enter the hall in a given period of time. Suppose that all the doors together can allow a maximum of 100 people per minute to enter the hall. This is the maximum transport rate, also called the **transport maximum**. When the concert hall is empty, three maintenance people enter the doors every hour. The transport rate is 3 people/60 minutes, or 0.05 people/minute, well under the maximum. For a local dance recital, about 50 people per minute go through the doors, still well under the maximum. When the most popular rock group of the day appears in concert, however, thousands of people gather outside. When the doors open, thousands of people are clamoring to get in, but the doors allow only 100 people/minute into the hall. The doors are working at the maximum rate, so it does not matter whether there are 1,000 or 3,000 people trying to get in. The transport rate saturates at 100 people/minute.

How can cells increase their transport capacity and avoid saturation? One way is to increase the number of carriers in the membrane. This would be like opening more doors into the concert hall. Under some circumstances, cells are able to insert additional carriers into their membranes. Under other circumstances, a cell may withdraw carriers to decrease movement of a molecule into or out of the cell.

All forms of carrier-mediated transport show specificity, competition, and saturation, but as you learned earlier in the chapter, they also differ in one important way: passive mediated transport—better known as facilitated diffusion—requires no input of energy from an outside source. Active transport requires energy input from ATP, either directly or indirectly.

5.5 Vesicular Transport

What happens to the many macromolecules that are too large to enter or leave cells through protein channels or carriers? They move in and out of the cell with the aid of bubble-like *vesicles* [p. 70] created from the cell membrane. Cells use two basic processes to import large molecules and particles: phagocytosis and endocytosis. Some scientists consider phagocytosis to be a type of endocytosis, but mechanistically the two processes are different. Material leaves cells by the process known as exocytosis, a process that is similar to endocytosis run in reverse.

Concept Check

25. What would you call a carrier that moves two substrates in opposite directions across a membrane?
26. In the concert-hall door analogy, we described how the maximum transport rate might be increased by increasing the number of doors leading into the hall. Using the same analogy, can you think of another way a cell might increase its maximum transport rate?

Phagocytosis Creates Vesicles Using the Cytoskeleton

If you studied *Amoeba* in your biology laboratory, you may have watched these one-cell creatures ingest their food by surrounding it and enclosing it within a vesicle that is brought into the cytoplasm. **Phagocytosis** {*phagein*, to eat + *cyte*, cell + *-sis*, process} is the actin-mediated process by which a cell engulfs a bacterium or other particle into a large membrane-bound vesicle called a **phagosome** {*soma*, body} (**FIG. 5.18**). The phagosome pinches off from the cell

FIG. 5.18 Phagocytosis

Phagocytosis uses actin microfilaments and myosin motor proteins to engulf particles in large vesicles.

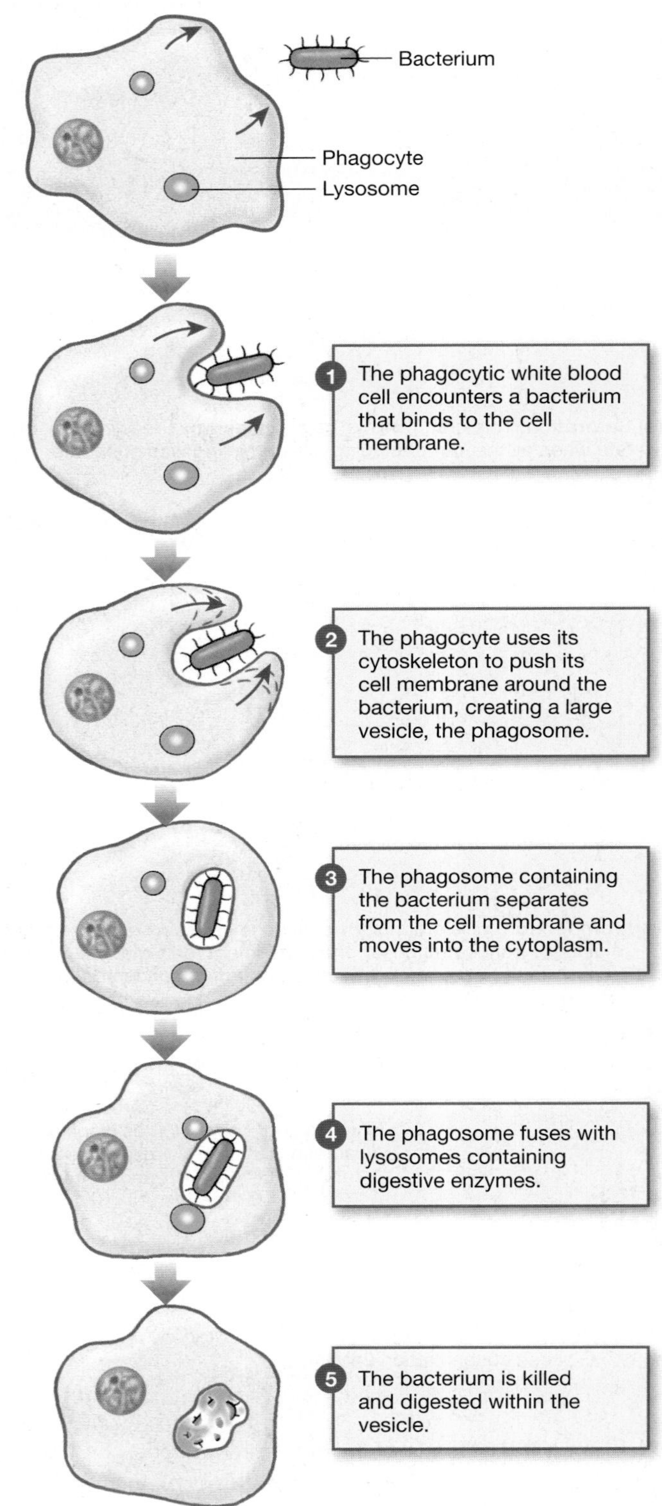
Bacterium
Phagocyte
Lysosome
1 The phagocytic white blood cell encounters a bacterium that binds to the cell membrane.
2 The phagocyte uses its cytoskeleton to push its cell membrane around the bacterium, creating a large vesicle, the phagosome.
3 The phagosome containing the bacterium separates from the cell membrane and moves into the cytoplasm.
4 The phagosome fuses with lysosomes containing digestive enzymes.
5 The bacterium is killed and digested within the vesicle.

membrane and moves to the interior of the cell, where it fuses with a lysosome [p. 71], whose digestive enzymes destroy the bacterium. Phagocytosis requires energy from ATP for the movement of the cytoskeleton and for the intracellular transport of the vesicles. In humans, phagocytosis occurs in certain types of white blood cells called *phagocytes,* which specialize in "eating" bacteria and other foreign particles.

Endocytosis Creates Smaller Vesicles

Endocytosis, the second process by which large molecules or particles move into cells, differs from phagocytosis in two important ways. First, in endocytosis the membrane surface indents rather than pushes out. Second, the vesicles formed from endocytosis are much smaller. In addition, some endocytosis is *constitutive;* that is, it is an essential function that is always taking place. In contrast, phagocytosis must be triggered by the presence of a substance to be ingested.

Endocytosis is an active process that requires energy from ATP. It can be nonselective, allowing extracellular fluid to enter the cell—a process called **pinocytosis** {*pino-*, drink}—or it can be highly selective, allowing only specific molecules to enter the cell. In receptor-mediated endocytosis, a ligand binds to a membrane receptor protein to activate the process.

Receptor-Mediated Endocytosis **Receptor-mediated endocytosis** takes place in regions of the cell membrane known as **coated pits**, indentations where the cytoplasmic side of the membrane has high concentrations of protein. The most common protein found in coated pits is *clathrin,* illustrated in **FIGURE 5.19**. In the first step of the process, extracellular ligands that will be brought into the cell bind to their membrane receptors 1. The receptor-ligand complex migrates along the cell surface until it encounters a coated pit 2. Once the receptor-ligand complex is in the coated pit, the membrane draws inward, or *invaginates* 3, then pinches off from the cell membrane and becomes a cytoplasmic vesicle. The clathrin molecules are released and recycle back to the membrane 4. In the vesicle, the receptor and ligand separate, leaving the ligand inside an *endosome* 5. The endosome moves to a lysosome if the ligand is to be destroyed, or to the Golgi complex if the ligand is to be processed 6.

Meanwhile, the ligand's membrane-bound receptors may be reused in a process known as **membrane recycling**. The vesicle with the receptors moves to the cell membrane 7 and fuses with it 8. The vesicle membrane then is incorporated back into the cell membrane by exocytosis 9. Notice in Figure 5.19 that the cytoplasmic face of the membrane remains the same throughout endocytosis and recycling. The extracellular surface of the cell membrane becomes the inside face of the vesicle membrane.

Receptor-mediated endocytosis transports a variety of substances into the cell, including protein hormones, growth factors, antibodies, and plasma proteins that serve as carriers for iron and cholesterol. Elevated plasma cholesterol levels and cardiovascular disease are associated with abnormalities in receptor-mediated removal of cholesterol from the blood (see Clinical Focus box on LDL: The Lethal Lipoprotein).

CLINICAL FOCUS

LDL: The Lethal Lipoprotein

"Limit the amount of cholesterol in your diet!" has been the recommendation for many years. So why is too much cholesterol bad for you? After all, cholesterol molecules are essential for membrane structure and for making steroid hormones (such as the sex hormones). But elevated cholesterol levels in the blood also lead to heart disease. One reason some people have too much cholesterol in their blood (*hypercholesterolemia*) is not diet but the failure of cells to take up the cholesterol. In the blood, hydrophobic cholesterol is bound to a lipoprotein carrier molecule to make it water soluble. The most common form of carrier is *low-density lipoprotein (LDL).* When the LDL-cholesterol complex (LDL-C) binds to LDL receptors, the complex then enters the cell in a vesicle. When people do not have adequate numbers of LDL receptors on their cell membranes, LDL-C remains in the blood. Hypercholesterolemia due to high levels of LDL-C predisposes these people to develop **atherosclerosis**, also known as hardening of the arteries {*atheroma*, a tumor + *skleros*, hard + *-sis*, condition}. In this condition, the accumulation of cholesterol in blood vessels blocks blood flow and contributes to heart attacks (see Chapter 15).

Caveolae Some endocytosis uses small flask-shaped indentations called **caveolae** ("little caves") rather than clathrin-coated pits to concentrate and bring receptor-bound molecules into the cell. Caveolae are membrane regions with lipid rafts [p. 64], membrane receptor proteins, and specialized membrane proteins named *caveolins* and *cavins.* The receptors in caveolae are lipid-anchored proteins [p. 64]. In many cells, caveolae appear as small indented pockets on the cell membrane, which is how they acquired their name.

Caveolae have several functions: to concentrate and internalize small molecules, to help in the transfer of macromolecules across the capillary endothelium, and to participate in cell signaling. Caveolae appear to be involved in some disease processes, including viral and parasitic infections. Two forms of the disease *muscular dystrophy* are associated with abnormalities in the protein caveolin. Scientists are currently trying to discover more details about the role of caveolae in normal physiology and pathophysiology.

Exocytosis Releases Molecules Too Large for Transport Proteins

Exocytosis is the opposite of endocytosis. In exocytosis, intracellular vesicles move to the cell membrane, fuse with it (Fig. 5.19 8), and then release their contents to the extracellular fluid 9. Cells use exocytosis to export large lipophobic molecules, such as proteins synthesized in the cell, and to get rid of wastes left in lysosomes from intracellular digestion.

FIG. 5.19 ESSENTIALS Endocytosis, Exocytosis, and Membrane Recycling

Membrane removed from the cell surface by endocytosis is recycled back to the cell surface by exocytosis.

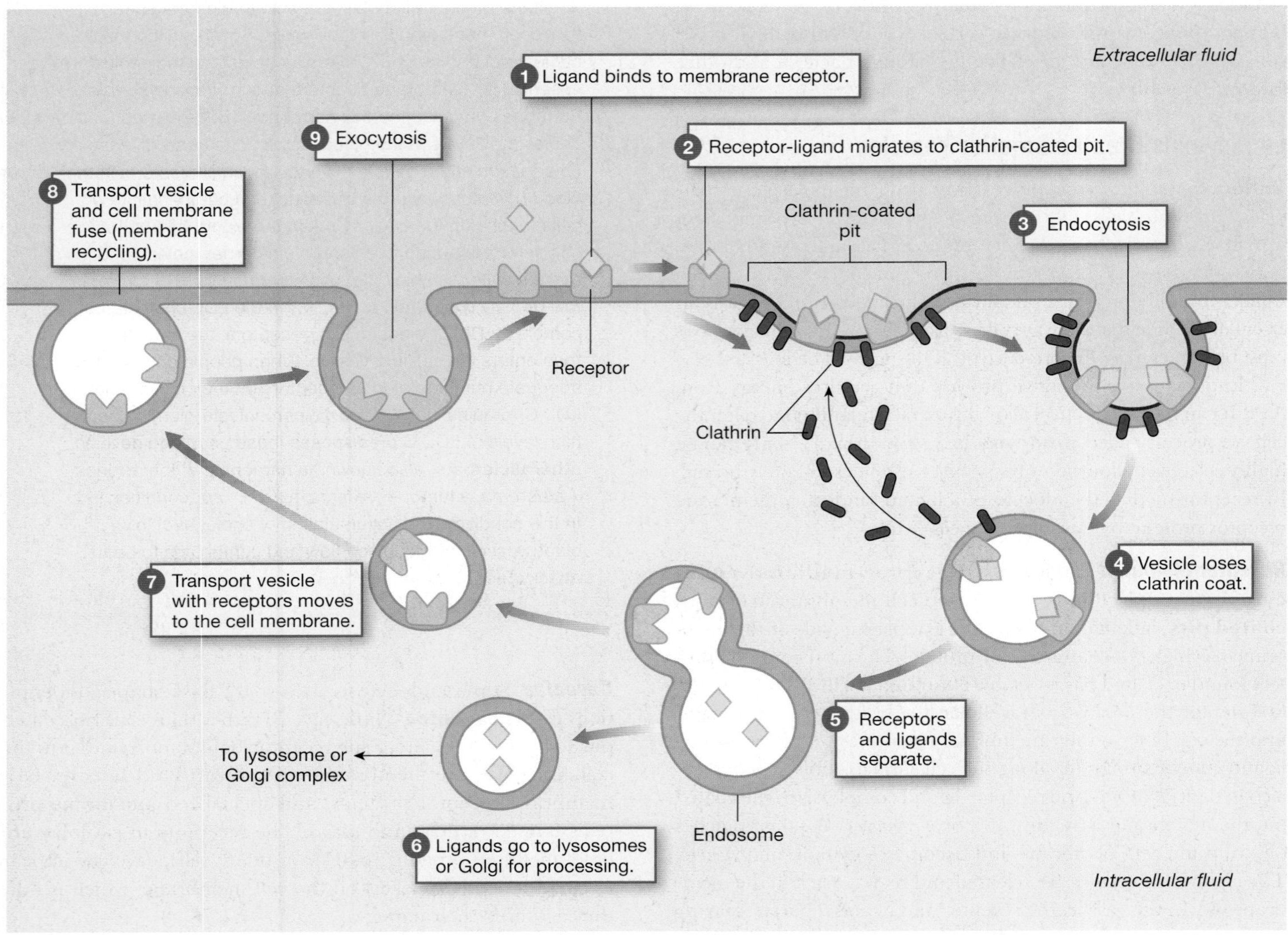

Extracellular fluid
1 Ligand binds to membrane receptor.
2 Receptor-ligand migrates to clathrin-coated pit.
Clathrin-coated pit
3 Endocytosis
Receptor
Clathrin
4 Vesicle loses clathrin coat.
5 Receptors and ligands separate.
Endosome
6 Ligands go to lysosomes or Golgi for processing.
To lysosome or Golgi complex
7 Transport vesicle with receptors moves to the cell membrane.
8 Transport vesicle and cell membrane fuse (membrane recycling).
9 Exocytosis
Intracellular fluid

The process by which the cell and vesicle membranes fuse is similar in a variety of cell types, from neurons to endocrine cells. Exocytosis involves two families of proteins: *Rabs,* which help vesicles dock onto the membrane, and *SNAREs,* which facilitate membrane fusion. In regulated exocytosis, the process usually begins with an increase in intracellular Ca^{2+} concentration that acts as a signal. The Ca^{2+} interacts with a calcium-sensing protein, which in turn initiates secretory vesicle docking and fusion. When the fused area of membrane opens, the vesicle contents diffuse into the extracellular fluid while the vesicle membrane stays behind and becomes part of the cell membrane. Exocytosis, like endocytosis, requires energy in the form of ATP.

Exocytosis takes place continuously in some cells, making it a *constitutive* process. For example, goblet cells [p. 79] in the intestine continuously release mucus by exocytosis, and fibroblasts in connective tissue release collagen [p. 80]. In other cell types, exocytosis is an intermittent process that is initiated by a signal. In many endocrine cells, hormones are stored in secretory vesicles in the cytoplasm and released in response to a signal from outside the cell. Cells also use exocytosis to insert proteins into the cell membrane, as shown in Figure 5.19. You will encounter many examples of exocytosis in your study of physiology.

Concept Check

27. How does phagocytosis differ from endocytosis?
28. Name the two membrane protein families associated with endocytosis.
29. How do cells move large proteins into the cell? Out of the cell?

5.6 Epithelial Transport

All the transport processes described in the previous sections deal with the movement of molecules across a single membrane, that of the cell. However, molecules entering and leaving the body or moving between certain compartments within the body must cross a layer of epithelial cells [p. 75] that are connected to one another by adhesive junctions and tight junctions [p. 75].

The tight junctions of epithelia separate the cell membrane into two regions, or poles. The surface of the epithelial cell that faces the lumen of an organ is called the *apical* { *apex*, the highest point } membrane (**FIG. 5.20**). It is often folded into microvilli that increase its surface area. Below the tight junctions, the three surfaces of the cell that face the extracellular fluid are collectively called the *basolateral* membrane { *basal*, base + *latus*, side }. The apical membrane is also called the *mucosal* membrane. The corresponding term for the basolateral membrane is *serosal* membrane.

Transporting epithelial cells are said to be *polarized* because their apical and basolateral membranes have very different properties. Certain transport proteins, such as the Na^+-K^+-ATPase, are usually found only on the basolateral membrane. Others, like the Na^+-glucose symporter SGLT, are restricted to the apical membrane. This polarized distribution of transporters allows the one-way movement of certain molecules across the epithelium.

Transport of material from the lumen of an organ to the extracellular fluid is called **absorption** (Fig. 5.20). For example, the intestinal epithelium absorbs digested nutrients. When material moves from the ECF to the lumen, the process is called *secretion*. For example, the salivary glands secrete saliva to help moisten the food you eat. Note that the term *secretion* is also used more broadly to mean the release of a substance from a cell.

Epithelial Transport May Be Paracellular or Transcellular

Movement across an epithelium, or **epithelial transport**, may take place either as **paracellular transport** { *para-*, beside } through the junctions between adjacent cells or as **transcellular transport** through the epithelial cells themselves (Fig. 5.20). In "tight" epithelia, the cell-cell junctions act as barriers to minimize the unregulated diffusion of material between the cells, so there is very little paracellular transport. In recent years, however, scientists have learned that some epithelia have the ability to change the "tightness" of their junctions. It appears that some junctional proteins such as *claudins* can form large holes or pores that allow water, ions, and a few small uncharged solutes to move by the paracellular pathway. In certain pathological states, increased movement through the paracellular route is a hallmark of the disease.

In contrast, substances moving by the transcellular route must cross two cell membranes. Molecules cross the first membrane when they move into the epithelial cell from one compartment. They cross the second membrane when they leave the epithelial cell to enter the second compartment. Transcellular transport uses a combination of active and passive transport mechanisms.

Protein-mediated transcellular transport is usually a two-step process, with one "uphill" step that requires energy and one "downhill" step in which the molecule moves passively down its gradient. You will see these steps in the example of glucose transport that follows. Molecules that are too large to be moved by membrane proteins can be transported across the cell in vesicles.

FIG. 5.20 Transporting epithelia are polarized

The apical membrane and the basolateral membrane are the two poles of the cell. Polarized epithelia have different transport proteins on apical and basolateral membranes. This allows selective directional transport across the epithelium. Transport from lumen to ECF is called **absorption.** Transport from ECF to lumen is called **secretion.**

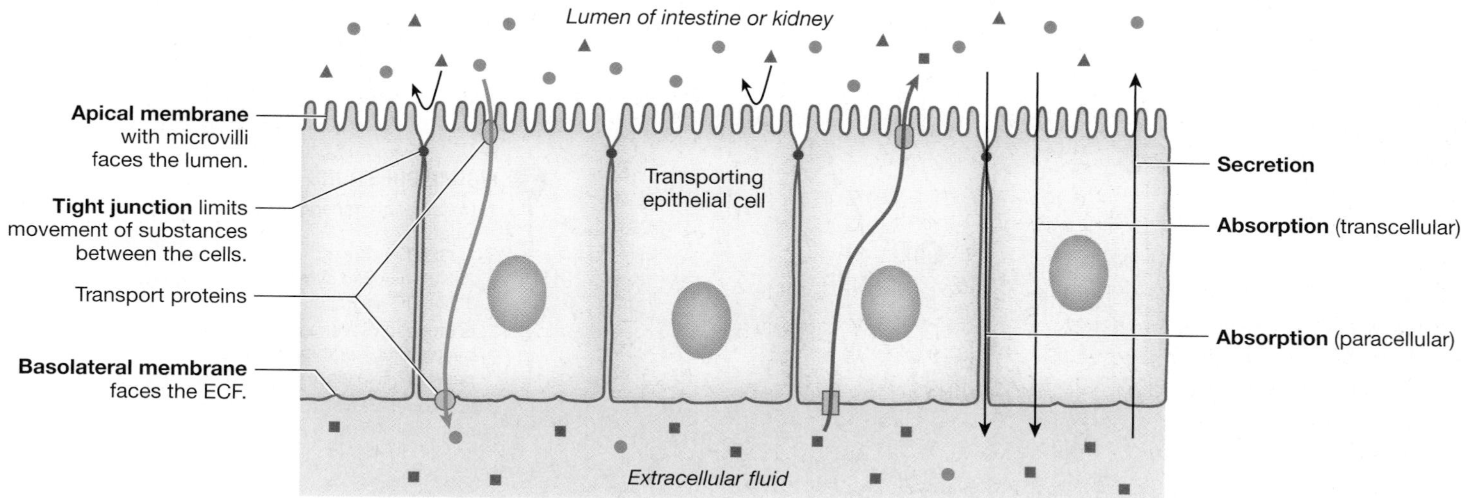
Lumen of intestine or kidney
Apical membrane with microvilli faces the lumen.
Tight junction limits movement of substances between the cells.
Transport proteins
Basolateral membrane faces the ECF.
Transporting epithelial cell
Secretion
Absorption (transcellular)
Absorption (paracellular)
Extracellular fluid

The cells of transporting epithelia can alter their permeability by selectively inserting or withdrawing membrane proteins. Transporters pulled out of the membrane may be destroyed in lysosomes, or they may be stored in vesicles inside the cell, ready to be reinserted into the membrane in response to a signal (another example of membrane recycling). Most epithelial transport you will study in this book involves the transporting epithelia of intestine and kidney, which are specialized to selectively transport molecules into and out of the body.

Transcellular Transport of Glucose Uses Membrane Proteins

The absorption of glucose from the lumen of the kidney tubule or intestine to the extracellular fluid is an important example of directional movement across a transporting epithelium. Transepithelial movement of glucose involves three transport systems: (1) the SGLT-mediated secondary active transport of glucose with Na^+ from the lumen into the epithelial cell at the apical membrane, followed by the movement of Na^+ and glucose out of the cell and into the extracellular fluid on separate transporters; (2) sodium moves out by primary active transport via a Na^+-K^+-ATPase; and (3) glucose leaves the cell by facilitated diffusion on GLUT carriers.

FIGURE 5.21 shows the process in detail. The glucose concentration in the transporting epithelial cell is higher than the glucose concentration in the lumen of the kidney or intestine. For this reason, moving glucose from the lumen into the cell requires the input of energy—in this case, energy stored in the Na^+ concentration gradient. Sodium ions in the lumen bind to the SGLT carrier, as previously described (see Fig. 5.16), and bring glucose with them into the cell. The energy needed to move glucose against its concentration gradient comes from the kinetic energy of Na^+ moving down its concentration gradient (Fig. 5.21 **1**).

Once glucose is in the epithelial cell, it leaves by moving down its concentration gradient on the facilitated diffusion GLUT transporter in the basolateral membrane (Fig. 5.21 **2**). Na^+ is pumped

FIG. 5.21 Transepithelial absorption of glucose

Absorbing glucose from intestinal or kidney tubule lumen involves indirect (secondary) active transport of glucose across the apical membrane and glucose diffusion across the basolateral membrane.

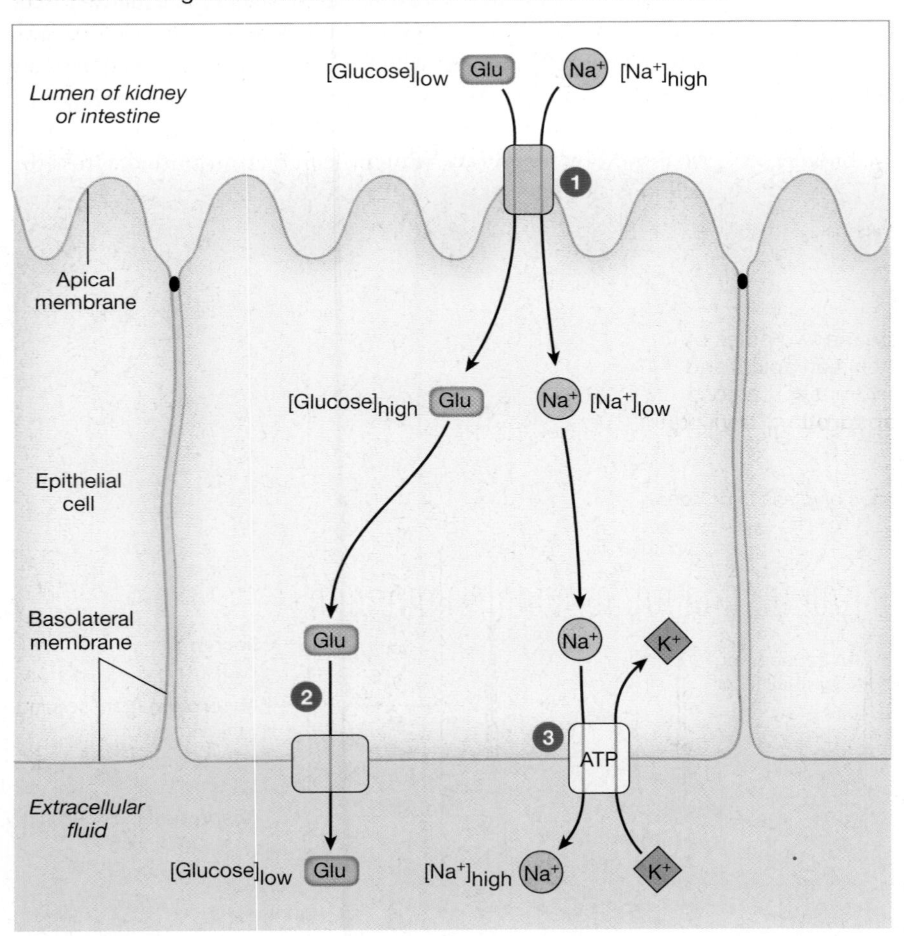
Lumen of kidney or intestine
[Glucose]low
Glu
Na+
[Na+]high
1
Apical membrane
[Glucose]high
Glu
Na+
[Na+]low
Epithelial cell
Basolateral membrane
Glu
Na+
K+
2
3
ATP
Extracellular fluid
[Glucose]low
Glu
[Na+]high
Na+
K+

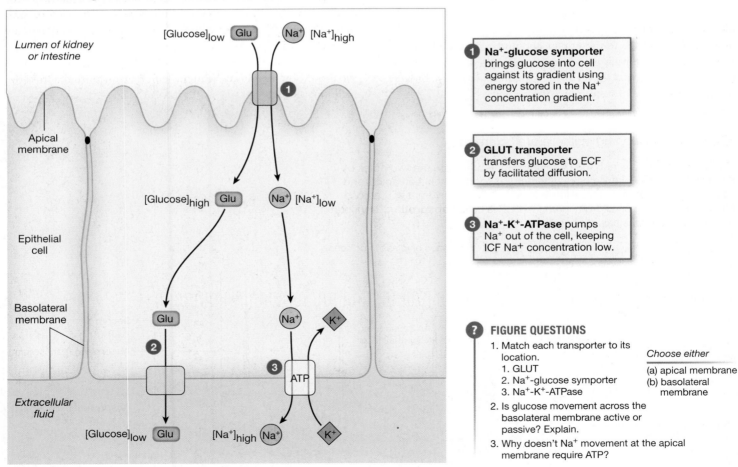

1 **Na^+-glucose symporter** brings glucose into cell against its gradient using energy stored in the Na^+ concentration gradient.

2 **GLUT transporter** transfers glucose to ECF by facilitated diffusion.

3 **Na^+-K^+-ATPase** pumps Na^+ out of the cell, keeping ICF Na^+ concentration low.

FIGURE QUESTIONS

1. Match each transporter to its location.
   1. GLUT
   2. Na^+-glucose symporter
   3. Na^+-K^+-ATPase

   Choose either
   (a) apical membrane
   (b) basolateral membrane
2. Is glucose movement across the basolateral membrane active or passive? Explain.
3. Why doesn't Na^+ movement at the apical membrane require ATP?

RUNNING PROBLEM

The sweat test that Daniel will undergo analyzes levels of NaCl in sweat. Sweat—a mixture of ions and water—is secreted into ducts by the epithelial cells of sweat glands. As sweat moves toward the skin's surface through the ducts, CFTR and Na^+ channels move Cl^- and Na^+ out of the sweat and back into the body. The duct cells are not permeable to water, so that normal reabsorption of NaCl creates sweat with a low salt content. However, without functioning CFTR channels in the epithelium, salt is not reabsorbed. In cystic fibrosis, salt concentrations in the sweat can be four times the normal amount.

Q3: *Based on the information given, is CFTR protein on the apical or basolateral surface of the sweat gland epithelium?*

122 131 138 151 152 159

out of the cell on the basolateral side using Na^+-K^+-ATPase 3. This step requires energy provided by ATP because sodium is more concentrated in the extracellular fluid than in the cell.

The removal of Na^+ from the cell is essential if glucose is to continue to be absorbed from the lumen. The potential energy to run the SGLT symporter comes from the sodium concentration gradient, which depends on low intracellular concentrations of Na^+. If the basolateral Na^+-K^+-ATPase is poisoned with *ouabain* (pronounced wah-bane—a compound related to the heart drug digitalis), Na^+ that enters the cell cannot be pumped out. The Na^+ concentration inside the cell gradually increases until it is equal to that in the lumen. Without a sodium gradient, there is no energy source to run the SGLT symporter, and the absorption of glucose across the epithelium stops.

Transepithelial transport can use ion movement through channels in addition to carrier-mediated transport. For example, the apical membrane of a transporting epithelium may use the Na^+-K^+-$2Cl^-$ (NKCC) symporter to bring K^+ into the cell against its concentration gradient, using energy from the Na^+ gradient. Because the K^+ concentration inside the cell is higher than in the extracellular fluid, K^+ can move out of the cell on the basolateral side through open K^+ leak channels. Na^+ must be pumped out by Na^+-K^+-ATPase. By this simple mechanism, the body can absorb Na^+ and K^+ at the same time from the lumen of the intestine or the kidney.

Concept Check

30. Why does Na^+ movement from the cytoplasm to the extracellular fluid require energy?

31. Ouabain, an inhibitor of the Na^+-K^+-ATPase, cannot pass through cell membranes. What would happen to the transepithelial glucose transport shown in Figure 5.21 if ouabain were applied to the apical side of the epithelium? To the basolateral side of the epithelium?

32. Which GLUT transporter is illustrated in Figure 5.21?

Transcytosis Uses Vesicles to Cross an Epithelium

Some molecules, such as proteins, are too large to cross epithelia on membrane transporters. Instead they are moved across epithelia by **transcytosis**, which is a combination of endocytosis, vesicular transport across the cell, and exocytosis (**FIG. 5.22**). In this process, the molecule is brought into the epithelial cell via receptor-mediated endocytosis. The resulting vesicle attaches to microtubules in the cell's cytoskeleton and is moved across the cell by a process known as **vesicular transport**. At the opposite side of the epithelium, the contents of the vesicle are expelled into the interstitial fluid by exocytosis.

Transcytosis makes it possible for large proteins to move across an epithelium and remain intact. It is the means by which infants absorb maternal antibodies in breast milk. The antibodies are absorbed on the apical surface of the infant's intestinal epithelium and then released into the extracellular fluid.

Concept Check

33. If you apply a poison that disassembles microtubules to a capillary endothelial cell, what happens to transcytosis?

Now that we have considered how solutes move between the body's compartments, we will examine how the transport of ions creates an electrical disequilibrium between the intracellular and extracellular compartments.

FIG. 5.22 Transcytosis across the capillary endothelium

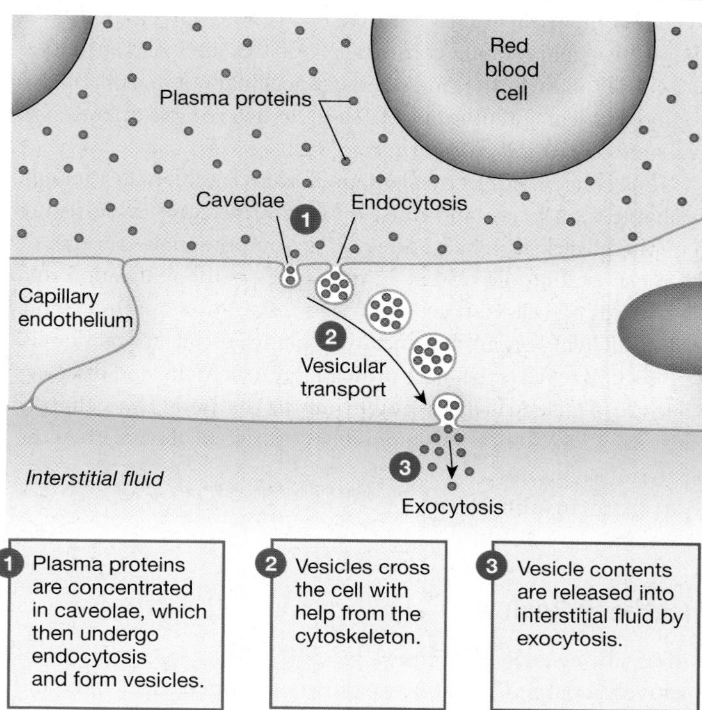

Red blood cell
Plasma proteins
Caveolae
Endocytosis
1
Capillary endothelium
2
Vesicular transport
Interstitial fluid
3
Exocytosis
1 Plasma proteins are concentrated in caveolae, which then undergo endocytosis and form vesicles.
2 Vesicles cross the cell with help from the cytoskeleton.
3 Vesicle contents are released into interstitial fluid by exocytosis.

RUNNING PROBLEM

Three days after Daniel's sweat test, the lab returns the grim results: salt levels in his sweat are more than twice the normal concentration. Daniel is diagnosed with cystic fibrosis. Now, along with antibiotics to prevent lung infections and therapy to loosen the mucus in his airways, Daniel must begin a regimen of pancreatic enzymes to be taken whenever he eats, for the rest of his life. In cystic fibrosis, thick mucus in the pancreatic ducts blocks the secretion of digestive enzymes into the intestine. Without artificial enzymes, he would starve.

Q4: *Why will Daniel starve if he does not take artificial pancreatic enzymes?*

122
131
138
151
152
159

5.7 The Resting Membrane Potential

Many of the body's solutes, including organic compounds such as pyruvate and lactate, are ions and, therefore, carry a net electrical charge. Potassium (K^+) is the major cation within cells, and sodium (Na^+) dominates the extracellular fluid (see Fig. 5.1, p. 123). On the anion side, chloride ions (Cl^-) mostly remain with Na^+ in the extracellular fluid. Phosphate ions and negatively charged proteins are the major anions of the intracellular fluid.

Overall, the body is electrically neutral: for every cation, there is a matching anion. However, ions are not distributed evenly between the ECF and the ICF (**FIG. 5.23a**). The intracellular compartment contains some anions that do not have matching cations, giving the cells a net negative charge. At the same time, the extracellular compartment has the "missing" cations, giving the ECF a net positive charge. One consequence of this uneven distribution of ions is that the intracellular and extracellular compartments are not in electrical equilibrium. Instead, the two compartments exist in a state of *electrical disequilibrium* (p. 122).

The concept of electrical disequilibrium traditionally is taught in chapters on nerve and muscle function because those tissues generate electrical signals known as action potentials. Yet one of the most exciting discoveries in physiology is the realization that other kinds of cells also use electrical signals for communication. In fact, all living organisms, including plants, use electrical signals! This section reviews the basic principles of electricity and discusses what creates electrical disequilibrium in the body. The chapter ends with a look at how the endocrine beta cells of the pancreas use electrical signaling to trigger insulin secretion.

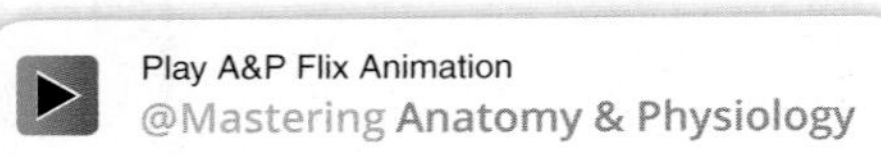
Play A&P Flix Animation
@Mastering Anatomy & Physiology

Electricity Review

Atoms are electrically neutral [p. 36]. They are composed of positively charged protons, negatively charged electrons, and uncharged neutrons, but in balanced proportions, so that an atom is neither positive nor negative. The removal or addition of electrons to an atom creates the charged particles we know as ions. We have discussed several ions that are important in the human body, such as Na^+, K^+, and H^+. For each of these positive ions, somewhere in the body there is a matching electron, usually found as part of a negative ion. For example, when Na^+ in the body enters in the form of NaCl, the "missing" electron from Na^+ can be found on the Cl^-.

Remember the following important principles when you deal with electricity in physiological systems:

1. The **law of conservation of electrical charge** states that the net amount of electrical charge produced in any process is zero. This means that for every positive charge on an ion, there is an electron on another ion. Overall, the human body is electrically neutral.
2. Opposite charges (+ and −) are attracted to each other. The protons and electrons in an atom exhibit this attraction. Like charges (two charges of the same type, such as +/+, or −/−) repel each other.
3. Separating positive charges from negative charges requires energy. For example, energy is required to separate the protons and electrons of an atom.
4. When separated positive and negative charges can move freely toward each other, the material through which they move is called a **conductor**. Water is a good conductor of electrical charge. When separated charges cannot move through the material that separates them, the material is known as an **insulator**. The phospholipid bilayer of the cell membrane is a good insulator, as is the plastic coating on electrical wires.

The word *electricity* comes from the Greek word *elektron*, meaning "amber," the fossilized resin of trees. The Greeks discovered that if they rubbed a rod of amber with cloth, the amber acquired the ability to attract hair and dust. This attraction (called static electricity) arises from the separation of electrical charge that occurs when electrons move from the amber atoms to the cloth. To separate these charged particles, energy (work) must be put into the system. In the case of the amber, work was done by rubbing the rod. In the case of biological systems, the work is usually done by energy stored in ATP and other chemical bonds.

The Cell Membrane Enables Separation of Electrical Charge in the Body

In the body, separation of electrical charge takes place across the cell membrane. This process is shown in Figure 5.23b. The diagram shows an artificial cell system. The cell is filled with positive K^+ and large negative ions. The cell is placed in an aqueous solution of sodium chloride that has dissociated into Na^+ and Cl^-. The phospholipid bilayer of the artificial cell, like the membrane of a real cell, is not permeable to ions, so it acts as an insulator and prevents the ions from moving. Water can freely cross this cell membrane, making the extracellular and intracellular osmotic concentrations equal.

In Figure 5.23b, both the cell and the solution are electrically neutral, and the system is in electrical equilibrium. However, it is not in chemical equilibrium. There are concentration gradients for all four types of ions in the system, and they would all diffuse down their respective concentration gradients if they could cross the cell membrane.

In Figure 5.23c, a leak channel for K^+ is inserted into the membrane. Now the cell is permeable to K^+, but only to K^+. Because of the K^+ concentration gradient, K^+ moves out of the cell. The negative ions in the cell attempt to follow the K^+ because of the attraction of positive and negative charges. But because the membrane is impermeable to negative ions, the anions remain trapped in the cell.

As soon as the first positive K^+ leaves the cell, the electrical equilibrium between the extracellular fluid and intracellular fluid is disrupted: the cell's interior has developed a net charge of -1 while the cell's exterior has a net charge of $+1$. The movement of K^+ out of the cell down its concentration gradient has created an **electrical gradient**—that is, a difference in the net charge between two regions. In this example, the inside of the cell has become negative relative to the outside.

If the only force acting on K^+ were the concentration gradient, K^+ would leak out of the cell until the K^+ concentration inside the cell equaled the K^+ concentration outside. The loss of positive ions from the cell creates an electrical gradient, however. The combination of electrical and concentration gradients is called an **electrochemical gradient**.

Because opposite charges attract each other, the negative proteins inside the cell try to pull K^+ back into the cell (Fig. 5.23d). At some point in this process, the electrical force attracting K^+ into the cell becomes equal in magnitude to the chemical concentration gradient driving K^+ out of the cell. At that point, net movement of K^+ across the membrane stops (Fig. 5.23e). The rate at which K^+ ions move out of the cell down the concentration gradient is exactly equal to the rate at which K^+ ions move into the cell down the electrical gradient. The system has reached *electrochemical equilibrium*.

For any given concentration gradient of a single ion, the membrane potential that exactly opposes the concentration gradient is known as the **equilibrium potential**, or $\mathbf{E}_{\text{ion}}$ (where the subscript *ion* is replaced by the symbol for whichever ion we are looking at). For example, when the concentration gradient is 150 mM K^+ inside and 5 mM K^+ outside the cell, the equilibrium potential for potassium, or E_K is -90 mV.

The equilibrium potential for any ion at 37 °C (human body temperature) can be calculated using the **Nernst equation**:

$$E_{ion} = \frac{61}{z} \log \frac{[\text{ion}]_{out}}{[\text{ion}]_{in}}$$

where 61 is 2.303 RT/F at 37 °C*

z is the electrical charge on the ion ($+1$ for K^+),

$(\text{ion})_{out}$ and $(\text{ion})_{in}$ are the ion concentrations outside and inside the cell, and E_{ion} is measured in mV.

The Nernst equation assumes that the cell in question is freely permeable to only the ion being studied. This is not the usual situation in living cells, however, as you will learn shortly.

All Living Cells Have a Membrane Potential

As the beginning of this chapter pointed out, all living cells are in chemical and electrical disequilibrium with their environment. This electrical disequilibrium, or electrical gradient between the extracellular fluid and the intracellular fluid, is called the **resting membrane potential difference**, or **membrane potential** for short. Although the name sounds intimidating, we can break it apart to see what it means.

1. The *resting* part of the name comes from the fact that an electrical gradient is seen in all living cells, even those that appear to be without electrical activity. In these "resting" cells, the membrane potential has reached a steady state and is not changing.
2. The *potential* part of the name comes from the fact that the electrical gradient created by active transport of ions across the cell membrane is a form of stored, or potential, energy, just as concentration gradients are a form of potential energy. When oppositely charged molecules come back together, they release energy that can be used to do work, in the same way that molecules moving down their concentration gradient can do work (see Appendix B). The work done using electrical energy includes opening voltage-gated membrane channels and sending electrical signals.
3. The *difference* part of the name is to remind you that the membrane potential represents a difference in the amount of electrical charge inside and outside the cell. The word *difference* is usually dropped from the name, as noted earlier, but it is important for remembering what a membrane potential means.

In living systems, we cannot measure absolute electrical charge, so we describe electrical gradients on a relative scale instead. Figure 5.23f compares the two scales. On an absolute scale, the extracellular fluid in our simple example has a net charge of $+1$ from the positive ion it gained, and the intracellular fluid has a net charge of -1 from the negative ion that was left behind. On the number line shown, this is a difference of two units.

In real life, because we cannot measure electrical charge as numbers of electrons gained or lost, we use a device that measures the *difference* in electrical charge between two points. This device artificially sets the net electrical charge of one side of the membrane to 0 and measures the net charge of the second side relative to the first. In our example, resetting the extracellular fluid net charge to 0 on the number line gives the intracellular fluid a net charge of -2. We call the ICF value the resting membrane potential (difference) of the cell.

* R is the ideal gas constant, T is absolute temperature, and F is the Faraday constant. For additional information, see Appendix B.

FIG. 5.23 ESSENTIALS Membrane Potential

The electrical disequilibrium that exists between the extracellular fluid (ECF) and intracellular fluid (ICF) of living cells is called the **membrane potential difference** (V_m), or membrane potential for short. The membrane potential results from the uneven distribution of electrical charge (i.e., ions) between the ECF and ICF.

What creates the membrane potential?

1. Ion concentration gradients between the ECF and ICF
2. The selectively permeable cell membrane

(a) In illustrations, this uneven distribution of charge is often shown by the charge symbols clustered on each side of the cell membrane.

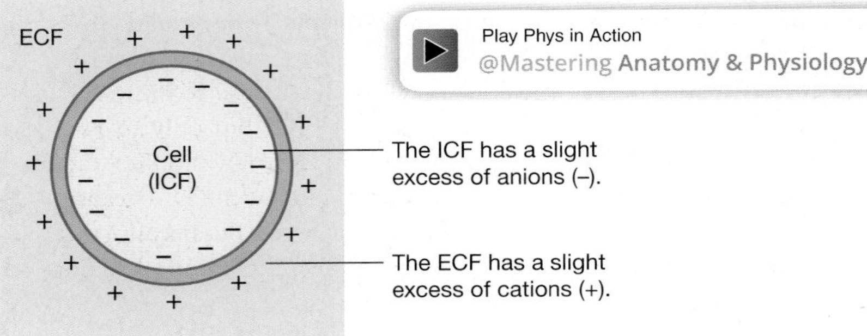
ECF
Cell
(ICF)
Play Phys in Action
@Mastering Anatomy & Physiology
The ICF has a slight excess of anions (–).
The ECF has a slight excess of cations (+).

Creation of a Membrane Potential in an Artificial System

To show how a membrane potential difference can arise from ion concentration gradients and a selectively permeable membrane, we will use an artificial cell system where we can control the membrane's permeability to ions and the composition of the ECF and ICF.

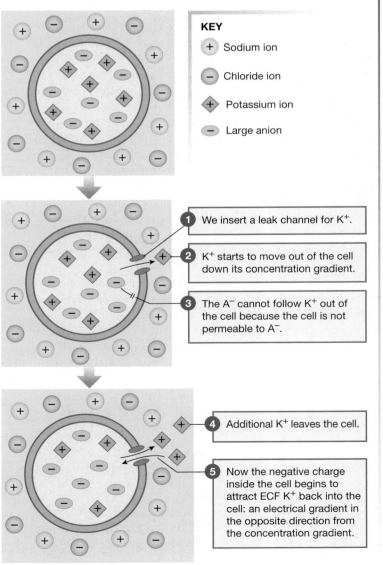

(b) When we begin, the cell has no membrane potential: The ECF (composed of Na^+ and Cl^- ions) and the ICF (K^+ and large anions, A^-) are electrically neutral.

The system is in chemical disequilibrium, with concentration gradients for all four ions. The cell membrane acts as an insulator to prevent free movement of ions between the ICF and ECF.

KEY

- Sodium ion
- Chloride ion
- Potassium ion
- Large anion

(c) Now we insert a leak channel for K^+ into the membrane, making the cell freely permeable to K^+.

The transfer of just one K^+ from the cell to the ECF creates an electrical disequilibrium: the ECF has a net positive charge (+1) while the ICF has a net negative charge (–1). The cell now has a membrane potential difference, with the inside of the cell negative relative to the outside.

1. We insert a leak channel for K^+.
2. K^+ starts to move out of the cell down its concentration gradient.
3. The A^- cannot follow K^+ out of the cell because the cell is not permeable to A^-.

(d) As additional K^+ ions leave the cell, going down their concentration gradient, the inside of the cell becomes more negative and the outside becomes more positive.

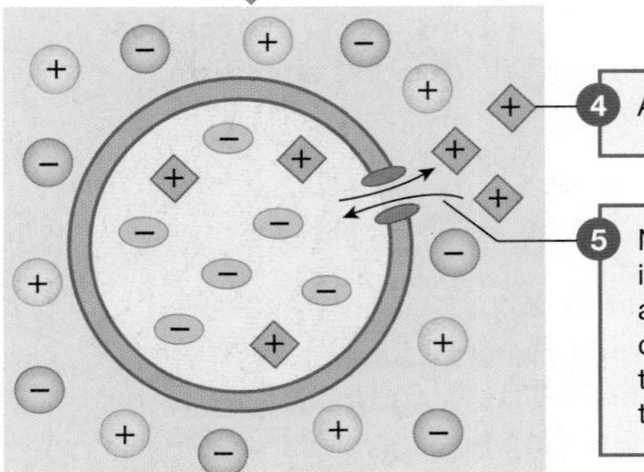

4. Additional K^+ leaves the cell.
5. Now the negative charge inside the cell begins to attract ECF K^+ back into the cell: an electrical gradient in the opposite direction from the concentration gradient.

How much K^+ will leave the cell?

If K^+ was uncharged, like glucose, it would diffuse out of the cell until the concentration outside $[K]_{out}$ equaled the concentration inside $[K]_{in}$. But K^+ is an ion, so we must consider its electrical gradient. Remember the rule for movement along electrical gradients: Opposite charges attract, like charges repel.

Electrochemical Equilibrium

For any given concentration gradient $[\text{Ion}]_{out} - [\text{Ion}]_{in}$ across a cell membrane, there is a membrane potential difference (i.e., electrical gradient) that exactly opposes ion movement down the concentration gradient. At this membrane potential, the cell is at *electrochemical equilibrium*: There is no net movement of ion across the cell membrane.

FIGURE QUESTIONS

1. If the cell in (e) was made freely permeable to only Na^+, which way would the Na^+ move? Would the membrane potential become positive or negative?
2. If it became freely permeable to only Cl^-, which way would Cl^- move? Would the membrane potential become positive or negative?

(e) In this example, the concentration gradient sending K^+ out of the cell is exactly opposed by the electrical gradient pulling K^+ into the cell. This is shown by the arrows that are equal in length but opposite in direction.

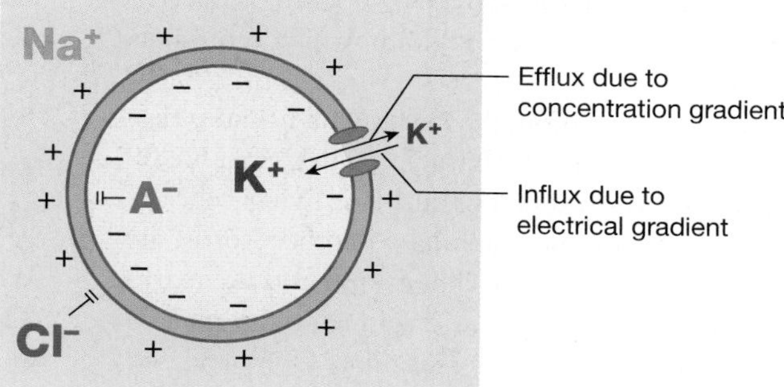

Na⁺
K⁺
A⁻
K⁺
Cl⁻
Efflux due to concentration gradient
Influx due to electrical gradient

Equilibrium Potential

For any ion, the membrane potential that exactly opposes a given concentration gradient is known as the **equilibrium potential** (E_{ion}). To calculate the equilibrium potential for any concentration gradient, we use the Nernst equation:

$$E_{ion} = \frac{61}{z} \log \frac{[\text{ion}]_{out}}{[\text{ion}]_{in}}$$

where z is the charge on the ion. (i.e., $K^+ = +1$)

The Nernst equation is used for a cell that is freely permeable to only one ion at a time. Living cells, however, have limited permeability to several ions. To calculate the actual membrane potential of cells, we use a multi-ion equation called the Goldman-Hodgkin-Katz equation [discussed in Chapter 8].

Approximate Values for Mammalian Cells

	ICF	ECF
K^+	150	5
Na^+	15	145
Cl^-	10	108

Using these values for K^+ and the Nernst equation, the E_K is −90 mV.

FIGURE QUESTIONS (You will need the log function on a calculator.)

3. Calculate the equilibrium potential for Na^+ (E_{Na}).
4. Calculate the E_{Cl}.

Measuring Membrane Potential

(f) In the first example, you saw that the membrane potential results from excess cations in the ECF and excess anions in the ICF. To measure this difference, we can place electrodes in the cell and surrounding fluid (equivalent to the ECF).

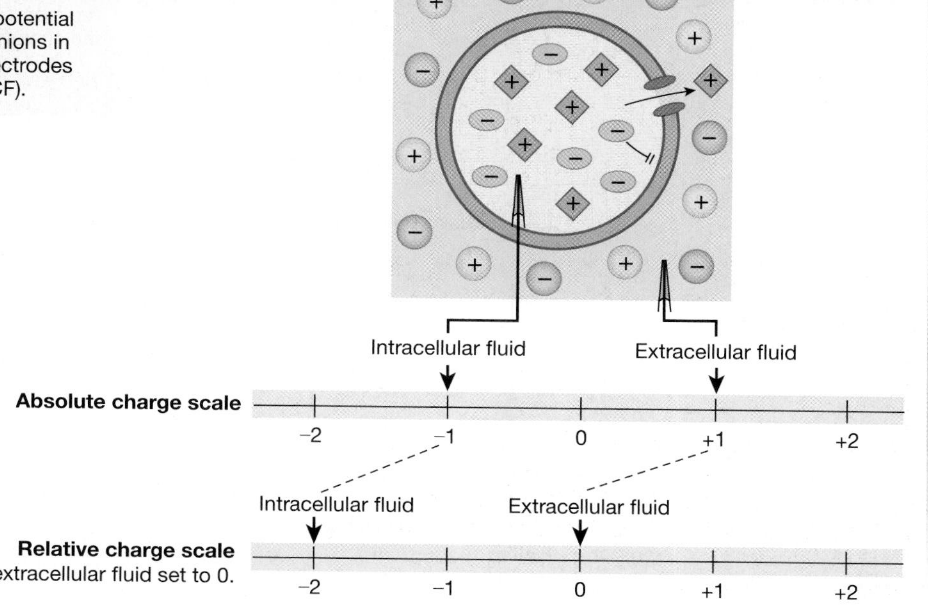

Intracellular fluid
Extracellular fluid
Absolute charge scale
−2
−1
0
+1
+2
Intracellular fluid
Extracellular fluid
Relative charge scale
extracellular fluid set to 0.
−2
−1
0
+1
+2

On a number line, the ECF would be at +1 and the ICF at −1.

In real life, we cannot measure absolute numbers of ions, however. Instead, we measure the difference between the two electrodes. By convention, the ECF is set at 0 mV (the ground). This gives the ICF a relative charge of −2.

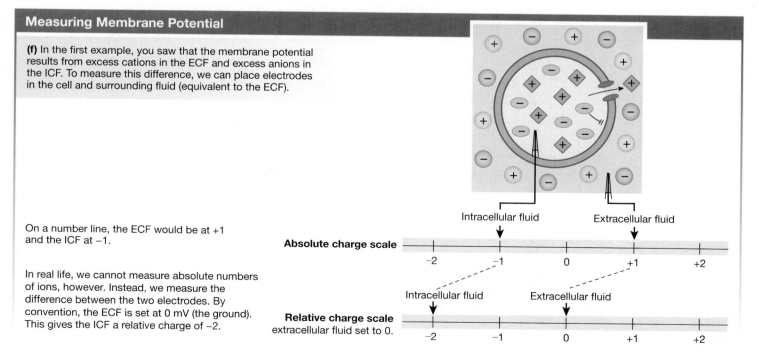

The equipment for measuring a cell's membrane potential is depicted in **FIGURE 5.24**. *Electrodes* are created from hollow glass tubes drawn to very fine points. These *micropipettes* are filled with a liquid that conducts electricity and then connected to a *voltmeter,* which measures the electrical difference between two points in units of either volts (V) or millivolts (mV). A *recording electrode* is inserted through the cell membrane into the cytoplasm of the cell. A *reference electrode* is placed in the external bath, which represents the extracellular fluid.

In living systems, by convention, the extracellular fluid is designated as the *ground* and assigned a charge of 0 mV (Fig. 5.23f). When the recording electrode is placed inside a living cell, the voltmeter measures the membrane potential—in other words, the electrical difference between the intracellular fluid and the extracellular fluid. A recorder connected to the voltmeter can make a recording of the membrane potential versus time.

For resting nerve and muscle cells, the voltmeter usually records a membrane potential between −40 and −90 mV, indicating that the intracellular fluid is negative relative to the extracellular fluid (0 mV). (Throughout this discussion, remember that the extracellular fluid is not really neutral because it has excess positive charges that exactly balance the excess negative charges inside the cell, as shown in Fig. 5.23. The total body remains electrically neutral at all times.)

Play Interactive Physiology 2.0
@Mastering Anatomy & Physiology

The Resting Membrane Potential Is Due Mostly to Potassium

Which ions create the resting membrane potential in animal cells? The artificial cell shown in Figure 5.23c used a potassium channel to allow K^+ to leak across a membrane that was otherwise impermeable to ions. But what processes go on in living cells to create an electrical gradient?

FIG. 5.24 Measuring membrane potential

In the laboratory, a cell's membrane potential is measured by placing one electrode inside the cell and a second in the extracellular bath.

A recording electrode is placed inside the cell.

Cell −70 mV

0 mV

Saline bath

Input

−70 −30 0 +30

The voltmeter measures the difference in electrical charge between the inside of a cell and the surrounding solution. This value is the **membrane potential difference,** or $\mathbf{V_m}$.

Output

The ground (⏚) or reference electrode is placed in the bath and given a value of 0 millivolts (mV).

The membrane potential can change over time.

Membrane potential (mV): +40, +20, 0, −20, −40, −60, −80, −100, −120

Time (msec) →

Membrane potential difference (V_m)

V_m

V_m

Hyperpolarization

Depolarization

If the membrane potential becomes less negative than the resting potential, the cell depolarizes.

Repolarization

If the membrane potential becomes more negative, the cell hyperpolarizes.

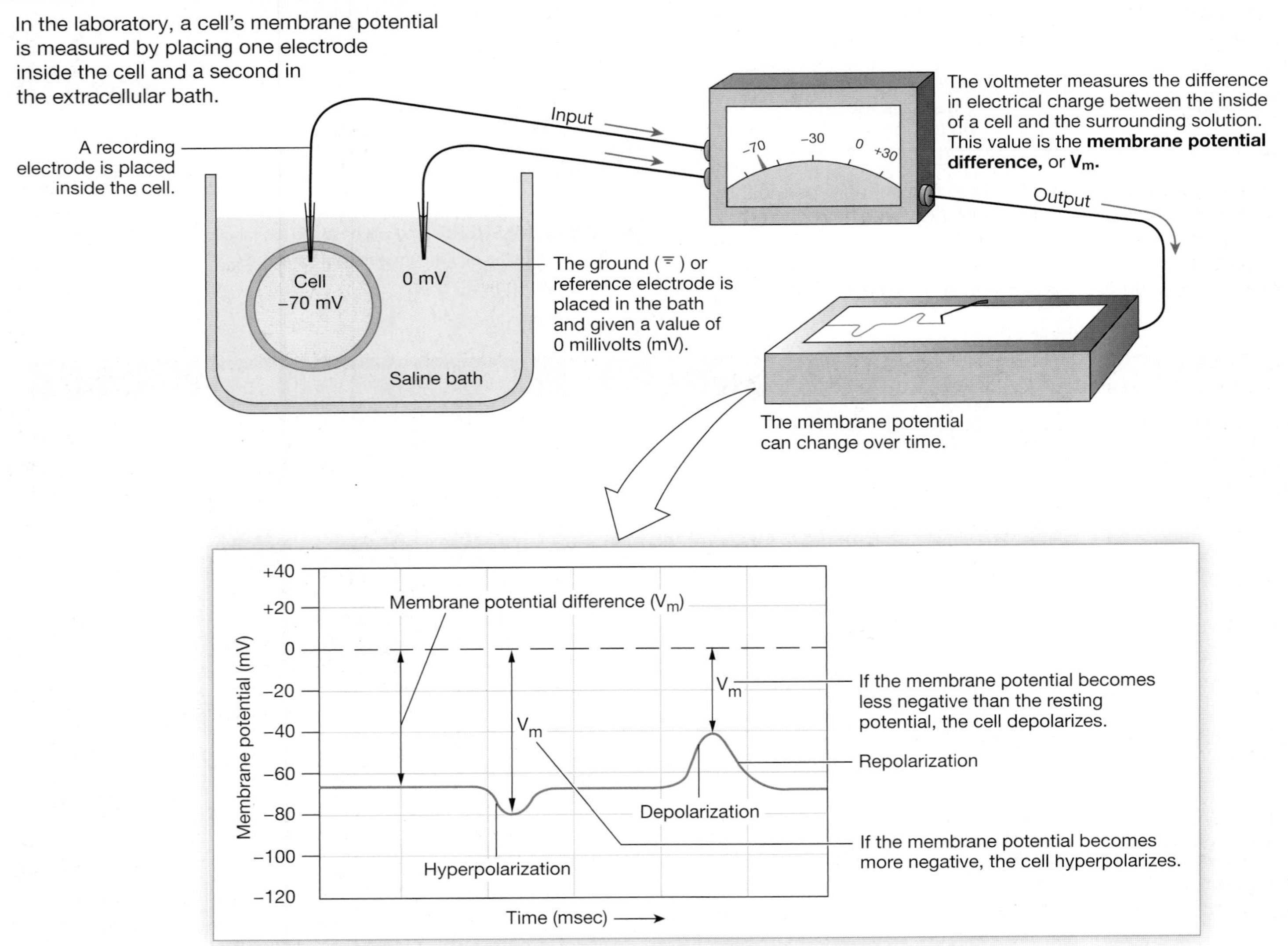

In reality, living cells are not permeable to only one ion. They have open channels and protein transporters that allow ions to move between the cytoplasm and the extracellular fluid. Instead of the Nernst equation, we use a related equation called the *Goldman equation* that considers concentration gradients of the permeable ions and the relative permeability of the cell to each ion. (For more detail on the Goldman equation, see Chapter 8.)

The real cell illustrated in **FIGURE 5.25** has a resting membrane potential of −70 mV. Most cells are about 40 times more permeable to K^+ than to Na^+. As a result, a cell's resting membrane potential is closer to the E_K of −90 mV than to the E_{Na} of +60 mV. A small amount of Na^+ leaks into the cell, making the inside of the cell less negative than it would be if Na^+ were totally excluded. Additional Na^+ that leaks in is promptly pumped out by the Na^+-K^+-ATPase. At the same time, K^+ ions that leak out of the cell are pumped back in. The pump contributes to the membrane potential by pumping 3 Na^+ out for every 2 K^+ pumped in. Because the Na^+-K^+-ATPase helps maintain the electrical gradient, it is called an *electrogenic* pump.

Not all ion transport creates an electrical gradient. Many transporters, like the Na^+-K^+-$2Cl^-$ (NKCC) symporter, are electrically neutral. Some make an even exchange: for each charge that enters the cell, the same charge leaves. An example is the HCO_3^--Cl^- antiporter of red blood cells, which transports these ions in a one-for-one, electrically neutral exchange. Electrically neutral transporters have little effect on the resting membrane potential of the cell.

FIG. 5.25 The resting membrane potential of cells

Most cells in the human body are about 40 times more permeable to K^+ than to Na^+, and the resting membrane potential is about −70 mV. The Na-K-ATPase helps maintain the resting membrane potential by removing Na^+ that leaks into the cell and returning K^+ that has leaked out.

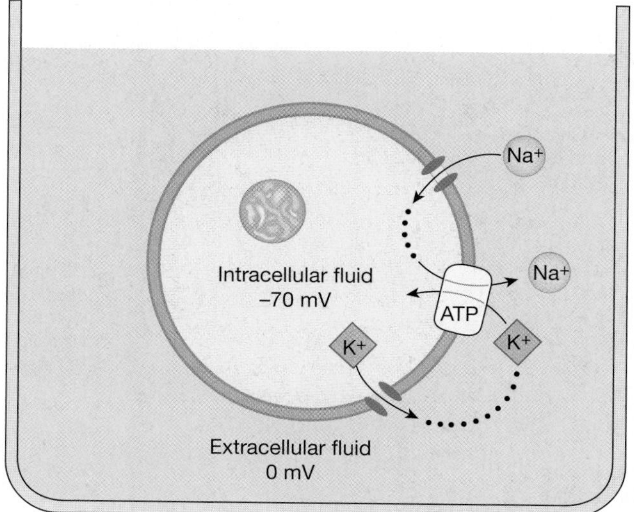
Na+
Na+
Intracellular fluid
−70 mV
ATP
K+
K+
Extracellular fluid
0 mV

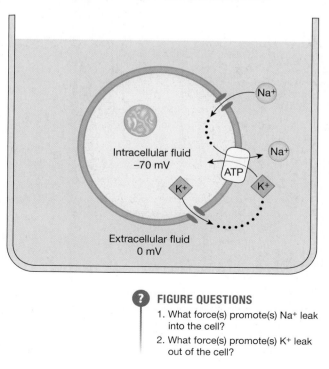

FIGURE QUESTIONS

1. What force(s) promote(s) Na^+ leak into the cell?
2. What force(s) promote(s) K^+ leak out of the cell?

Concept Check

34. What would happen to the resting membrane potential of a cell poisoned with ouabain (an inhibitor of the Na^+-K^+-ATPase)?

Changes in Ion Permeability Change the Membrane Potential

As you have just learned, two factors influence a cell's membrane potential: (1) the concentration gradients of different ions across the membrane and (2) the permeability of the membrane to those ions. If the cell's permeability to an ion changes, the cell's membrane potential changes. We monitor changes in membrane potential using the same recording electrodes that we use to record resting membrane potential.

Figure 5.24 shows a recording of membrane potential plotted against time. The extracellular electrode is set at 0 mV, and the intracellular electrode records the membrane potential difference. The membrane potential (V_m) begins at a steady resting value of −70 mV. When the trace moves upward (becomes less negative), the potential difference between the inside of the cell and the outside (0 mV) is less, and the cell is said to have *depolarized*. A return to the resting membrane potential is termed *repolarization*. If the resting potential becomes more negative, we say the cell has *hyperpolarized*.

A major point of confusion when we talk about changes in membrane potential is the use of the phrases "the membrane potential decreased" or "the membrane potential increased." Normally, we associate "increase" with becoming more positive and "decrease" with becoming more negative—the opposite of what is happening in our cell discussion. The best way to avoid trouble is to speak of the membrane potential becoming more or less negative or the cell depolarizing or hyperpolarizing. Another way to avoid confusion is to add the word *difference* after *membrane potential*. If the membrane potential *difference* is *increasing*, the value of V_m must be moving away from the ground value of 0 and becoming *more negative*. If the membrane potential *difference* is *decreasing*, the value of V_m is moving closer to the ground value of 0 mV and is becoming *less negative*.

What causes changes in membrane potential? In most cases, membrane potential changes in response to movement of one of four ions: Na^+, Ca^{2+}, Cl^-, and K^+. The first three ions are more concentrated in the extracellular fluid than in the cytosol, and the resting cell is minimally permeable to them. If a cell suddenly becomes more permeable to any one of these ions, then those ions will move down their electrochemical gradient into the cell. Entry of Ca^{2+} or Na^+ depolarizes the cell (makes the membrane potential more positive). Entry of Cl^- hyperpolarizes the cell (makes the membrane potential more negative).

Most resting cells are fairly permeable to K^+ but making them more permeable allows even more K^+ to leak out. The cell hyperpolarizes until it reaches the equilibrium potential for K^+. Making the cell *less* permeable to K^+ allows fewer K^+ ions to leak out of the cell. When the cell retains K^+, it becomes more positive and depolarizes. You will encounter instances of all these permeability changes as you study physiology.

It is important to learn that a significant change in membrane potential requires the movement of very few ions. *The concentration gradient does not have to reverse to change the membrane potential.* For example, to change the membrane potential by 100 mV (the size of a typical electrical signal passing down a neuron), only one of every 100,000 K^+ must enter or leave the cell. This is such a tiny fraction of the total number of K^+ ions in the cell that the concentration gradient for K^+ remains essentially unchanged.

5.8 Integrated Membrane Processes: Insulin Secretion

The movement of Na^+ and K^+ across cell membranes has been known to play a role in generating electrical signals in excitable tissues for many years. You will study these processes in detail when you learn about the nervous and muscular systems. Recently, however, we have come to understand that small changes in membrane potential act as signals in nonexcitable tissues, such as endocrine cells. One of the best-studied examples of this process involves the beta cell of the pancreas. Release of the hormone insulin by beta cells demonstrates how membrane processes—such as facilitated diffusion, exocytosis, and the opening and closing of ion channels by ligands and membrane potential—work together to regulate cell function.

The endocrine beta cells of the pancreas synthesize the protein hormone insulin and store it in cytoplasmic secretory vesicles [p. 71]. When blood glucose levels increase, such as after a meal, the beta cells release insulin by exocytosis. Insulin then directs other cells of the body to take up and use glucose, bringing blood concentrations down to pre-meal levels.

A key question about the process that went unanswered until recently was, "How does a beta cell 'know' that glucose levels have gone up and that it needs to release insulin?" The answer, we have now learned, links the beta cell's metabolism to its electrical activity.

FIGURE 5.26a shows a beta cell at rest. Recall from earlier sections in this chapter that gated membrane channels can be opened or closed by chemical or electrical signals. The beta cell has two such channels that help control insulin release. One is a **voltage-gated Ca^{2+} channel**. This channel is closed at the cell's resting membrane potential (5 in Fig. 5.26a). The other is a K^+ leak channel (that is, the channel is usually open) that closes when ATP binds to it. It is called an **ATP-gated K^+ channel (K_{ATP} channel)**. In the resting cell, when glucose concentrations are low, the cell

FIG. 5.26 Insulin secretion and membrane transport

(a) Beta cell at rest. The K_{ATP} channel is open, and the cell is at its resting membrane potential.

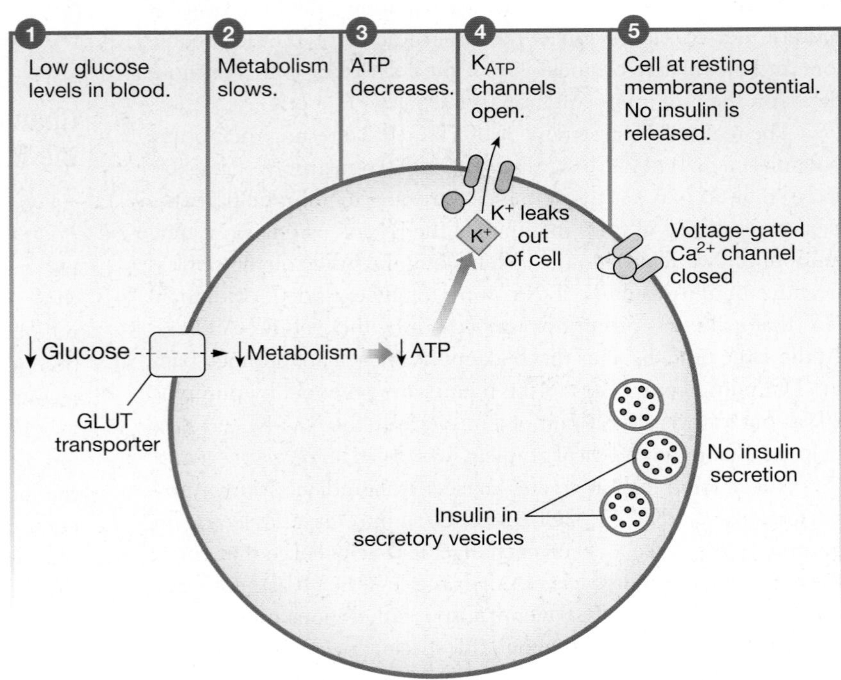
1 Low glucose levels in blood.
2 Metabolism slows.
3 ATP decreases.
4 K_{ATP} channels open.
5 Cell at resting membrane potential. No insulin is released.
K^+ leaks out of cell
Voltage-gated Ca^{2+} channel closed
↓Glucose
↓Metabolism
↓ATP
GLUT transporter
No insulin secretion
Insulin in secretory vesicles

(b) Beta cell secretes insulin. Closure of K_{ATP} channel depolarizes cell, triggering exocytosis of insulin.

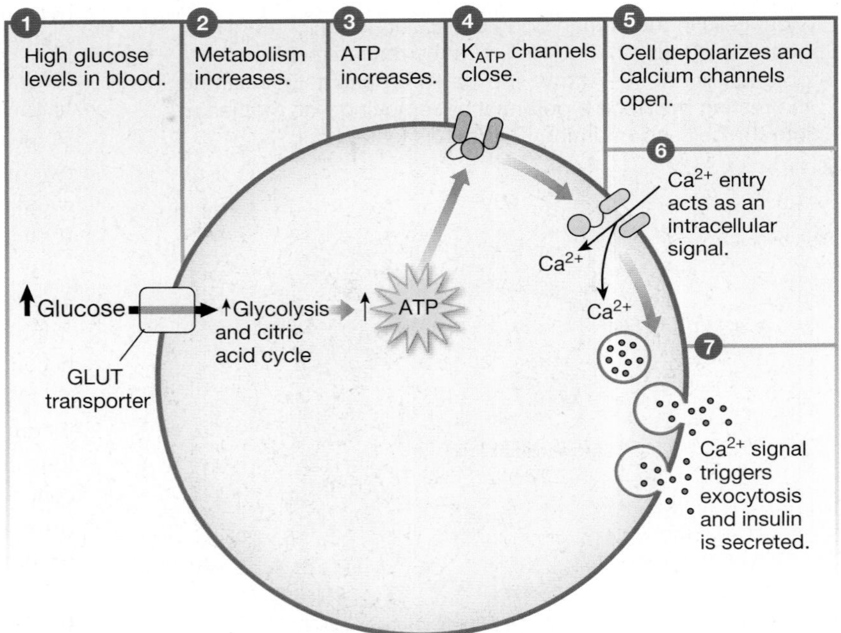
1 High glucose levels in blood.
2 Metabolism increases.
3 ATP increases.
4 K_{ATP} channels close.
5 Cell depolarizes and calcium channels open.
6 Ca^{2+} entry acts as an intracellular signal.
Ca^{2+}
Ca^{2+}
↑Glucose
↑Glycolysis and citric acid cycle
↑ATP
GLUT transporter
7 Ca^{2+} signal triggers exocytosis and insulin is secreted.

FIGURE QUESTIONS

1. Which step shows facilitated diffusion?
2. What kind of gating do the beta cell ion channels have?
3. Does insulin secretion in (b) require energy input from ATP?
4. Why is insulin released by exocytosis and not through a carrier or channel?

makes less ATP (Fig. 5.26a, 1 – 3). There is little ATP to bind to the K_{ATP} channel, and the channel remains open, allowing K^+ to leak out of the cell 4. At the resting membrane potential, the voltage-gated Ca^{2+} channels are closed, and there is no insulin secretion 5.

Figure 5.26b shows how a beta cell secretes insulin in response to an increase in blood glucose. After a meal, plasma glucose levels increase as glucose is absorbed from the intestine 1. Glucose reaching the beta cell diffuses into the cell with the aid of a GLUT transporter. Increased glucose in the cell stimulates the metabolic pathways of glycolysis and the citric acid cycle [p. 105], and ATP production increases 2, 3. When ATP binds to the K_{ATP} channel, the gate to the channel closes, preventing K^+ from leaking out of the cell 4. Retention of K^+ depolarizes the cell 5, which then causes the voltage-sensitive Ca^{2+} channels to open 6. Calcium ions enter the cell from the extracellular fluid, moving down their electrochemical gradient. The Ca^{2+} ions bind to proteins that initiate exocytosis of the insulin-containing vesicles, and insulin is released into the extracellular space 7.

The discovery that cells other than nerve and muscle cells use changes in membrane potential as signals for physiological responses altered our traditional thinking about the role of the resting membrane potential. Next, we will look at other types of signals that the body uses for communication and coordination.

RUNNING PROBLEM **CONCLUSION** Cystic Fibrosis

In this running problem, you learned about cystic fibrosis (CF), one of the most common inherited diseases in the United States. By some estimates, more than 10 million people are symptomless carriers of a mutated CF gene. A person must inherit two copies of the mutated gene, one from each parent, before he or she will develop CF. Although there is no cure for this disease, treatments have become better, and the life span of CF patients continues to improve. Today, the median survival age is nearly 40 years old.

Cystic fibrosis is caused by a defect in the transport of Cl^- into and out of epithelial cells through the CFTR channel protein. In the most common variant of CF, the CFTR channel protein does not fold correctly and cannot be inserted into the cell membrane. Because CFTR channels are found in the epithelial cell membranes of several organs—the sweat glands, lungs, and pancreas—cystic fibrosis may affect many different body processes. Interestingly, the CFTR chloride channel is a member of the ABC transport family and is the only known ion channel in that gene superfamily. Some of the most interesting animal research on cystic fibrosis uses genetically altered mice, called CF mice. These model animals are bred to have CFTR channels with altered functions corresponding to the mutations of the CFTR gene in humans.

To learn more about current research in this disease, go to the Cystic Fibrosis Foundation website (*www.cff.org*) and click the *Our Research* tab. To check your understanding of the running problem, compare your answers with the information in the following table.

Question	Facts	Integration and Analysis
Q1: *Why would failure to transport NaCl into the airways cause the secreted mucus to be thick?*	If NaCl is secreted into the lumen of the airways, the solute concentration of the airway fluid increases. Water moves into compartments with higher osmolarity.	Normally, movement of NaCl creates an osmotic gradient so that water also enters the airway lumen, creating a saline solution that thins the thick mucus. If NaCl cannot be secreted into the airways, there will be no fluid movement to thin the mucus.
Q2: *Is the CFTR a chemically gated, a voltage-gated, or a mechanically gated channel protein?*	Chemically gated channels open when a ligand binds to them. CFTRs open when ATP binds to the channel protein.	ATP is a chemical ligand, which means CFTRs are chemically gated channel proteins.
Q3: *Based on the information given, is the CFTR protein on the apical or basolateral surface of the sweat gland epithelium?*	In normal people, the CFTR channels transport Cl^- from sweat into epithelial cells.	The epithelial surface that faces the lumen of the sweat gland, which contains sweat, is the apical membrane. Therefore, the CFTR proteins are on the apical surface.
Q4: *Why will Daniel starve if he does not take artificial pancreatic enzymes?*	The pancreas secretes mucus and digestive enzymes into ducts that empty into the small intestine. In cystic fibrosis, mucus in the ducts is thick because of lack of Cl^- and fluid secretion. This thick mucus blocks the ducts and prevents digestive enzymes from reaching the small intestine.	Without digestive enzymes, Daniel cannot digest the food he eats. His weight loss over the past six months suggests that this has already become a problem. Taking artificial enzymes will enable him to digest his food.

122 – 131 – 138 – 151 – 152 – 159

CHAPTER SUMMARY

Several key themes come together in this chapter. You learned how the cell membrane creates distinct intracellular and extracellular compartments, illustrating the theme of *compartmentation*. The contents of the intracellular and extracellular compartments differ, but *homeostasis* keeps them in a dynamic steady state. Movement of materials between and within compartments is necessary for *communication* and is accomplished by *bulk flow* and *biological transport*. Flow of solutes and water across cell membranes occurs in response to osmotic, chemical (concentration), or electrical gradients. The cell membrane creates resistance to flow that can be overcome by inserting membrane proteins that act as channels or carriers. Biological transport in the body requires *energy* from concentration gradients or chemical bonds. Finally, the binding of substrates to transporters demonstrates the theme of *protein interactions*.

5.1 Osmosis and Tonicity

1. Most solutes are concentrated in either one compartment or the other, creating a state of **chemical disequilibrium**. (p. 123; Fig. 5.1)
2. Cations and anions are not distributed equally between the body compartments, creating a state of **electrical disequilibrium**. (p. 122)
3. Water moves freely between the cells and extracellular fluid, resulting in a state of **osmotic equilibrium**. (p. 122)
4. The movement of water across a membrane in response to a concentration gradient is called **osmosis**. (p. 124)
5. We express the concentration of biological solutions as **osmolarity**, the number of particles (ions or intact molecules) per liter of solution, in units of milliosmoles per liter (mOsM). (p. 125)
6. **Tonicity** of a solution describes the cell volume change that occurs at equilibrium if the cell is placed in that solution. Cells swell in **hypotonic solutions** and shrink in **hypertonic solutions**. If the cell does not change size at equilibrium, the solution is **isotonic**. (p. 126; Tbl. 5.3)
7. The osmolarity of a solution cannot be used to determine the tonicity of the solution. The relative concentrations of **nonpenetrating solutes** in the cell and in the solution determine tonicity. **Penetrating solutes** contribute to the osmolarity of a solution but not to its tonicity. (p. 127-129; Figs. 5.3, 5.4; Tbl. 5.4)

5.2 Transport Processes

8. In **bulk flow**, a pressure gradient moves a fluid along with its dissolved and suspended materials. (p. 131)
9. The cell membrane is a selectively permeable barrier that restricts free exchange between the cell and the interstitial fluid. The movement of a substance across a membrane depends on the permeability of the membrane to that substance. (p. 131)
10. Movement of molecules across membranes can be classified either by energy requirements or by the physical means the molecule uses to cross the membrane. (p. 132; Fig. 5.5)
11. Lipid-soluble substances can diffuse through the phospholipid bilayer. Less lipid-soluble molecules require the assistance of a membrane protein or vesicle to cross the membrane. (p. 132)
12. **Passive transport** does not require the input of energy. (p. 132)

5.3 Diffusion

13. **Diffusion** is the passive movement of molecules down a chemical (concentration) gradient from an area of higher concentration to an area of lower concentration. Net movement stops when the system reaches **equilibrium**, although molecular movement continues. (p. 133; Tbl. 5.6)
14. Diffusion rate depends on the magnitude of the concentration gradient. Diffusion is slow over long distances, is directly related to temperature, and is inversely related to molecular size. (p. 133)
15. **Simple diffusion** across a membrane is directly proportional to membrane surface area, concentration gradient, and membrane permeability, and inversely proportional to membrane thickness. (p. 134; Fig. 5.7)

5.4 Protein-Mediated Transport

16. Most molecules cross membranes with the aid of membrane proteins. (p. 136)
17. Membrane proteins have four functional roles: **structural proteins** maintain cell shape and form cell junctions; **membrane-associated enzymes** catalyze chemical reactions and help transfer signals across the membrane; **receptor proteins** are part of the body's signaling system; and **transport proteins** move many molecules into or out of the cell. (p. 137; Fig. 5.8)
18. **Channel proteins** form water-filled channels that link the intracellular and extracellular compartments. **Gated channels** regulate movement of substances through them by opening and closing. Gated channels may be regulated by ligands, by the electrical state of the cell, or by physical changes such as pressure. (p. 139; Fig. 5.10)
19. **Carrier proteins** never form a continuous connection between the intracellular and extracellular fluid. They bind to substrates, then change conformation. (p. 137; Fig. 5.12)
20. Protein-mediated diffusion is called **facilitated diffusion**. It has the same properties as simple diffusion. (p. 136; Tbl. 5.6; Fig. 5.13)
21. **Active transport** moves molecules against their concentration gradient and requires an outside source of energy. In **primary (direct) active transport**, the energy comes directly from ATP. **Secondary (indirect) active transport** uses the potential energy stored in a concentration gradient and is indirectly driven by energy from ATP. (p. 136)
22. The most important primary active transporter is the **sodium-potassium-ATPase** (Na^+-K^+-ATPase), which pumps Na^+ out of the cell and K^+ into the cell. (p. 142-143; Figs. 5.14, 5.15)
23. Most secondary active transport systems are driven by the sodium concentration gradient. (p. 142; Tbl. 5.8; Fig. 5.16)
24. All carrier-mediated transport demonstrates **specificity**, **competition**, and **saturation**. Specificity refers to the ability of a transporter to move only one molecule or a group of closely related molecules. Related molecules may compete for a single transporter. Saturation occurs when a group of membrane transporters are working at their maximum rate. (p. 145; Fig. 5.17)

5.5 Vesicular Transport

25. Large macromolecules and particles are brought into cells by phagocytosis or endocytosis. Material leaves cells by exocytosis. When vesicles that come into the cytoplasm by endocytosis are returned to the cell membrane, the process is called **membrane recycling**. (p. 147; Figs 5.18, 5.19)
26. In **receptor-mediated endocytosis**, ligands bind to membrane receptors that concentrate in **coated pits** or **caveolae**. (p. 146-148; Fig. 5.19)

27. In **exocytosis**, the vesicle membrane fuses with the cell membrane before releasing its contents into the extracellular space. Exocytosis requires ATP. (p. 151)

5.6 Epithelial Transport

28. Transporting epithelia have different membrane proteins on their **apical** and **basolateral** surfaces. This polarization allows one-way movement of molecules across the epithelium. (p. 149-150; Figs. 5.20, 5.21)
29. Molecules cross epithelia by moving between the cells by the **paracellular** route or through the cells by the **transcellular** route. (p. 149; Fig. 5.20)
30. Larger molecules cross epithelia by **transcytosis**, which includes **vesicular transport**. (p. 151; Fig. 5.22)

5.7 The Resting Membrane Potential

31. Although the total body is electrically neutral, diffusion and active transport of ions across the cell membrane create an **electrical gradient**, with the inside of cells negative relative to the extracellular fluid. (p. 154-155; Fig. 5.23)
32. The electrical gradient between the extracellular fluid and the intracellular fluid is known as the **resting membrane potential difference**. (p. 156)
33. The movement of an ion across the cell membrane is influenced by the **electrochemical gradient** for that ion. (p. 157)
34. The membrane potential that exactly opposes the concentration gradient of an ion is known as the **equilibrium potential** (E_{ion}). The equilibrium potential for any ion can be calculated using the Nernst equation. (p. 154-155; Fig. 5.23)
35. In most living cells, K^+ is the primary ion that determines the resting membrane potential. (p. 156)
36. Changes in membrane permeability to ions such as K^+, Na^+, Ca^{2+}, or Cl^- alter membrane potential and create electrical signals. (p. 157)

5.8 Integrated Membrane Processes: Insulin Secretion

37. The use of electrical signals to initiate a cellular response is a universal property of living cells. Pancreatic beta cells release insulin in response to a change in membrane potential. (p. 158; Fig. 5.26)

REVIEW QUESTIONS

In addition to working through these questions and checking your answers on p. A-5, review the Learning Outcomes at the beginning of this chapter.

Level One Reviewing Facts and Terms

1. List the four functions of membrane proteins, and give an example of each.
2. Distinguish between active transport and passive transport.
3. Which of the following processes are examples of active transport, and which are examples of passive transport? Simple diffusion, phagocytosis, facilitated diffusion, exocytosis, osmosis, endocytosis.
4. List four factors that increase the rate of diffusion in air.
5. List the three physical methods by which materials enter cells.
6. A cotransporter is a protein that moves more than one molecule at a time. If the molecules are moved in the same direction, the transporters are called __________ carriers; if the molecules are transported in opposite directions, the transporters are called __________ carriers. A transport protein that moves only one substrate is called a(n) __________ carrier.
7. The two types of active transport are __________, which derives energy directly from ATP, and __________, which couples the kinetic energy of one molecule moving down its concentration gradient to the movement of another molecule against its concentration gradient.
8. A molecule that moves freely between the intracellular and extracellular compartments is said to be a(n) __________ solute. A molecule that is not able to enter cells is called a(n) __________ solute.
9. Rank the following individuals in order of how much body water they contain as a percentage of their body weight, from highest to lowest: (a) a 25-year-old, 74-kg male; (b) a 25-year-old, 50-kg female; (c) a 65-year-old, 50-kg female; and (d) a 1-year-old, 11-kg male toddler.
10. What determines the osmolarity of a solution? In what units is body osmolarity usually expressed?
11. What does it mean if we say that a solution is hypotonic to a cell? Hypertonic to the same cell? What determines the tonicity of a solution relative to a cell?
12. Match the membrane channels with the appropriate descriptions. Answers may be used once, more than once, or not at all.

a. chemically gated channel b. open pore c. voltage-gated channel d. mechanically gated channel	1. channel that spends most of its time in the open state 2. channel that spends most of its time in a closed state 3. channel that opens when resting membrane potential changes 4. channel that opens when a ligand binds to it 5. channel that opens in response to membrane stretch 6. channel through which water can pass

13. In your own words, state the four principles of electricity important in physiology.

14. Match each of the following items with its primary role in cellular activity.

a. Na^+-K^+-ATPase	1. ion channel
b. protein	2. extracellular cation
c. unit of measurement for membrane potential	3. source of energy
d. K^+	4. intracellular anion
e. Cl^-	5. intracellular cation
f. ATP	6. millivolts
g. Na^+	7. electrogenic pump
	8. extracellular anion
	9. milliosmoles

15. The membrane potential at which the electrical gradient exactly opposes the concentration gradient for an ion is known as the ion's __________.

16. A material that allows free movement of electrical charges is called a(n) __________, whereas one that prevents this movement is called a(n) __________.

Level Two Reviewing Concepts

17. Create a map of transport across cell membranes using the following terms. You may add additional terms if you wish.

• active transport	• ligand
• carrier	• Na^+-K^+-ATPase
• caveolae	• osmosis
• channel	• passive transport
• clathrin-coated pit	• phospholipid bilayer
• concentration gradient	• receptor-mediated endocytosis
• electrochemical gradient	• secondary active transport
• exocytosis	• simple diffusion
• facilitated diffusion	• small polar molecule
• glucose	• transcytosis
• GLUT transporter	• vesicle
• ion	• vesicular transport
• large polar molecule	• water

18. Draw a large rectangle to represent the total body volume. Using the information in Figure 5.1b, divide the box proportionately into compartments to represent the different body compartments. Use the information in Figure 5.1d and add solutes to the compartments. Use large letters for solutes with higher concentrations and small letters for solutes with low concentrations. Label the cell membranes and the endothelial membrane.

19. What factors influence the rate of diffusion across a membrane? Briefly explain each one.

20. Define the following terms and explain how they differ from one another: specificity, competition, saturation. Apply these terms in a short explanation of facilitated diffusion of glucose.

21. Red blood cells are suspended in a solution of NaCl. The cells have an osmolarity of 300 mOsM, and the solution has an osmolarity of 250 mOsM. (a) The solution is (hypertonic, isotonic, or hypotonic) to the cells. (b) Water would move (into the cells, out of the cells, or not at all).

22. Two compartments are separated by a membrane that is permeable to glucose but not water. Each compartment is filled with 1 M glucose. After six hours, compartment A contains 1.5 M glucose and compartment B contains 0.5 M glucose. What kind of transport occurred? Explain.

23. A 2 M NaCl solution is placed in compartment A and a 2 M glucose solution is placed in compartment B. The compartments are separated by a membrane that is permeable to water but not to NaCl or glucose. Complete the following statements. Defend your answers.

    a. The salt solution is __________osmotic to the glucose solution.
    b. True or false? Water will move from one compartment to another. If water moves, it will move from compartment __________ to compartment __________.

24. Explain the differences between a chemical gradient, an electrical gradient, and an electrochemical gradient.

Level Three Problem Solving

25. Sweat glands secrete into their lumen a fluid that is identical to interstitial fluid. As the fluid moves through the lumen on its way to the surface of the skin, the cells of the sweat gland's epithelium make the fluid hypotonic by removing Na^+ and leaving water behind. Design an epithelial cell that will reabsorb Na^+ but not water. You may place water pores, Na^+ leak channels, K^+ leak channels, and the Na^+-K^+-ATPase in the apical membrane, basolateral membrane, or both.

26. Insulin is a hormone that promotes the movement of glucose into many types of cells, thereby lowering blood glucose concentration. Propose a mechanism that explains how this occurs, using your knowledge of cell membrane transport.

27. The following terms have been applied to membrane carriers: specificity, competition, saturation. Why can these terms also be applied to enzymes? What is the major difference in how enzymes and carriers carry out their work?

28. Integral membrane glycoproteins have sugars added as the proteins pass through the lumen of the endoplasmic reticulum and Golgi complex (p. 115). Based on this information, where would you predict finding the sugar "tails" of the proteins: on the cytoplasmic side of the membrane, the extracellular side, or both? Explain your reasoning.

29. NaCl is a nonpenetrating solute and urea is a penetrating solute for cells. Red blood cells (RBCs) are placed in each of the solutions below. The RBC intracellular concentration of nonpenetrating solute is 300 mOsM. What will happen to the cell volume in each solution? Label the solutions with all the terms that apply: hypertonic, isotonic, hypotonic, hyperosmotic, hyposmotic, isosmotic. Watch units! Assume 1 M NaCl = 2 OsM for simplicity.

    a. 150 mM NaCl plus 150 mM urea
    b. 100 mM NaCl plus 50 mM urea
    c. 100 mM NaCl plus 100 mM urea
    d. 150 mM NaCl plus 100 mM urea
    e. 100 mM NaCl plus 150 mM urea

Level Four Quantitative Problems

30. The addition of dissolved solutes to water lowers the freezing point of water. A 1 OsM solution depresses the freezing point of water by 1.86 °C. If a patient's plasma shows a freezing-point depression of 0.55 °C, what is her plasma osmolarity? (Assume that 1 kg water = 1 L.)

31. The patient in the previous question is found to have total body water volume of 42 L, ECF volume of 12.5 L, and plasma volume of 2.7 L.
    a. What is her intracellular fluid (ICF) volume? Her interstitial fluid volume?
    b. How much solute (osmoles) exists in her whole body? ECF? ICF? plasma? (*Hint:* concentration = solute amount/volume of solution)

32. What is the osmolarity of half-normal saline (= 0.45% NaCl)? [p. 43] Assume that all NaCl molecules dissociate into two ions.

33. If you give 1 L of half-normal saline (see question 32) to the patient in question 31, what happens to each of the following at equilibrium? (*Hint:* NaCl is a nonpenetrating solute.)
    a. Her total body volume
    b. Her total body osmolarity
    c. Her ECF and ICF volumes
    d. Her ECF and ICF osmolarities

34. The following graph shows the results of an experiment in which a cell was placed in a solution of glucose. The cell had no glucose in it at the beginning, and its membrane can transport glucose. Which of the following processes is/are illustrated by this experiment?
    a. diffusion
    b. saturation
    c. competition
    d. active transport

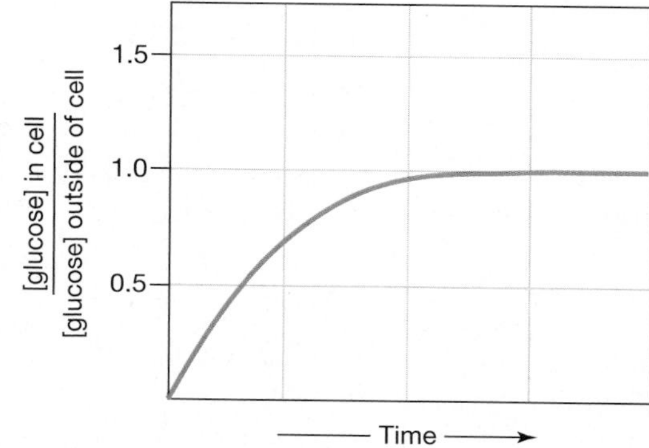

[glucose] in cell / [glucose] outside of cell
1.5
1.0
0.5
Time

Answers to Concept Checks, Figure and Graph Questions, and end-of-chapter Review Questions can be found in Appendix A [p. A-1].

6 Communication, Integration, and Homeostasis

Future progress in medicine will require a quantitative understanding of the many interconnected networks of molecules that comprise our cells and tissues, their interactions, and their regulation.

Overview of the NIH Roadmap, September 30, 2003. NIH Announces Strategy to Accelerate Medical Research Progress.

BACKGROUND BASICS

In 2003, the U.S. National Institutes of Health (NIH) embarked on an ambitious project to promote translation of basic research into new medical treatments and strategies for disease prevention. Contributors to the NIH Common Fund's Building Blocks, Biological Pathways, and Networks program (*http://commonfund.nih.gov/bbpn/index*) from 2004 to 2014 developed tools for research in proteomics and metabolomics, and compiled information on biological pathways to help us understand how cells communicate with one another. In this chapter, we examine the basic patterns of cell-to-cell communication and see how the coordination of function resides in chemical and electrical signals. Each cell in the body can communicate with most other cells. To maintain homeostasis, the body uses a combination of diffusion across small distances; widespread distribution of molecules through the circulatory system; and rapid, specific delivery of messages by the nervous system.

6.1 Cell-to-Cell Communication

In recent years, the amount of information available about cell-to-cell communication has mushroomed as a result of advances in research technology. Signal pathways that once seemed fairly simple and direct are now known to be incredibly complex networks and webs of information transfer, such as the network shown on the opening page of this chapter. In the sections that follow, we distill what is known about cell-to-cell communication into some basic patterns that you can recognize when you encounter them again in your study of physiology. As with many rapidly changing fields, these patterns reflect our current understanding and are subject to modification as scientists learn more about the incredibly complex network of chemical signals that control life processes.

By most estimates, the human body is composed of about 75 *trillion* cells. Those cells face a daunting task—to communicate with one another in a manner that is rapid and yet conveys a tremendous amount of information. Surprisingly, there are only two basic types of physiological signals: electrical and chemical. **Electrical signals** are changes in a cell's membrane potential [p. 153]. **Chemical signals** are molecules secreted by cells into the extracellular fluid. The cells that respond to electrical or chemical signals are called **target cells**, or **targets** for short.

Chemical signals are responsible for most communication within the body. Chemical signals act as *ligands* that bind to proteins to initiate a response. Protein binding of chemical signals obeys the general rules for protein interactions, including *specificity, affinity, competition*, and *saturation* [p. 46].

Our bodies use four basic methods of cell-to-cell communication (**FIG. 6.1**). **Local communication** includes (1) **gap junctions**, which allow direct cytoplasmic transfer of electrical and chemical signals between adjacent cells; (2) **contact-dependent signals**, which occur when surface molecules on one cell membrane bind to surface molecules on another cell's membrane; and (3) chemicals that diffuse through the extracellular fluid to act on cells close by. **Long-distance communication** (4) uses a combination of chemical and electrical signals carried by nerve cells and chemical signals transported in the blood. A given molecule can function as a chemical signal by more than one method. For example, a molecule can act close to the cell that released it (local communication) as well as in distant parts of the body (long-distance communication).

RUNNING PROBLEM **Diabetes Mellitus: A Growing Epidemic**

It is 8:00 A.M. and Marvin Garcia, age 20, is hungry. He came to his family physician's office before breakfast to have a fasting blood glucose test as part of a routine physical examination. In this test, blood is drawn after an overnight fast, and the glucose concentration in the blood is measured. Because he knows he is in good condition, Marvin isn't worried about the results. He is surprised, then, when the nurse practitioner in the doctor's office calls two days later. "Your fasting blood sugar is a bit elevated, Marvin. It is 130 milligrams per deciliter, and normal is 100 or less. Does anyone in your family have diabetes?" "Well, yeah—my dad has it. What exactly is diabetes?"

165 – 168 – 181 – 182 – 186 – 188 – 191

Gap Junctions Create Cytoplasmic Bridges

The simplest form of cell-to-cell communication is the direct transfer of electrical and chemical signals through *gap junctions*, protein channels that create cytoplasmic bridges between adjacent cells (Fig. 6.1a). A gap junction forms from the union of membrane-spanning proteins, called *connexins*, on two adjacent cells [p. 73]. The united connexins create a protein channel (*connexon*) that can open and close. When the channel is open, the connected cells function like a single cell that contains multiple nuclei (a *syncytium*).

When gap junctions are open, ions and small molecules such as amino acids, ATP, and cyclic AMP (cAMP) diffuse directly from the cytoplasm of one cell to the cytoplasm of the next. Larger molecules cannot pass through gap junctions. In addition, gap junctions are the only means by which electrical signals can pass *directly* from cell to cell. Movement of molecules and electrical signals through gap junctions can be modulated or shut off completely.

Gap junctions are not all alike. Scientists have discovered more than 20 different isoforms of connexins that may mix or match to form gap junctions. The variety of connexin isoforms allows gap junction selectivity to vary from tissue to tissue. In mammals, gap junctions are found in almost every cell type, including heart muscle, some types of smooth muscle, lung, liver, and neurons of the brain.

Contact-Dependent Signals Require Cell-to-Cell Contact

Some cell-to-cell communication requires that surface molecules on one cell membrane bind to a membrane protein of another cell (Fig. 6.1b). Such *contact-dependent signaling* occurs in the immune system and during growth and development, such as when nerve cells send out long extensions that must grow from the central axis

FIG. 6.1 ESSENTIALS Communication in the Body

Cell-to-cell communication uses chemical and electrical signaling to coordinate function and maintain homeostasis.

LOCAL COMMUNICATION

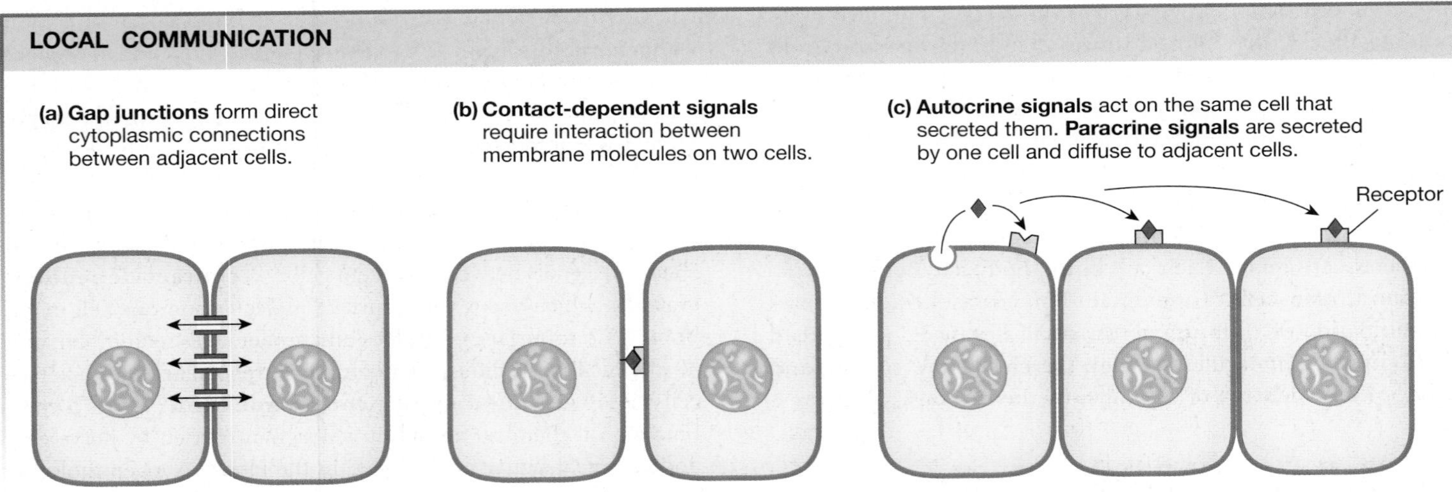

Receptor

(a) Gap junctions form direct cytoplasmic connections between adjacent cells.

(b) Contact-dependent signals require interaction between membrane molecules on two cells.

(c) Autocrine signals act on the same cell that secreted them. **Paracrine signals** are secreted by one cell and diffuse to adjacent cells.

LONG-DISTANCE COMMUNICATION

Long-distance signaling may be electrical signals passing along neurons or chemical signals that travel through the circulatory system.

Endocrine System

(d) Hormones are secreted by endocrine glands or cells into the blood. Only target cells with receptors for the hormone respond to the signal.

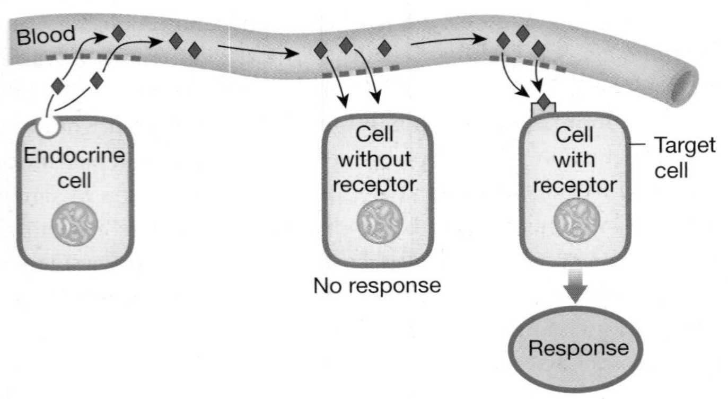

Blood
Endocrine cell
Cell without receptor
No response
Cell with receptor
Target cell
Response

Nervous System

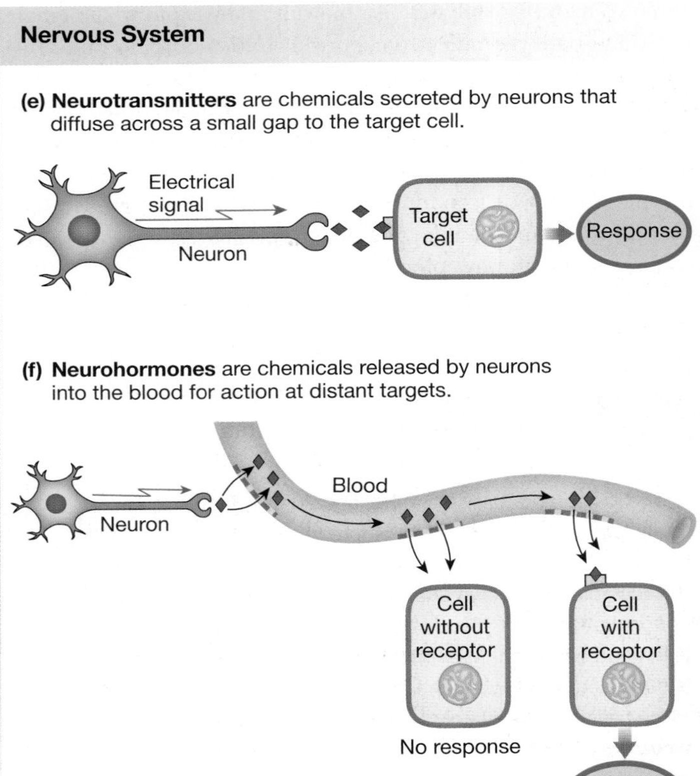

(e) Neurotransmitters are chemicals secreted by neurons that diffuse across a small gap to the target cell.

Electrical signal
Neuron
Target cell
Response

(f) Neurohormones are chemicals released by neurons into the blood for action at distant targets.

Neuron
Blood
Cell without receptor
No response
Cell with receptor
Response

of the body to the *distal* (distant) ends of the developing limbs. **Cell adhesion molecules (CAMs)** first known for their role in cell-to-cell adhesion [p. 73], have now been shown to act as receptors in cell-to-cell signaling. CAMs are linked to the cytoskeleton or to intracellular enzymes. Through these linkages, CAMs transfer signals in both directions across cell membranes. Contact-dependent signaling is also known as *juxtacrine signaling*.

Local Communication Uses Paracrine and Autocrine Signals

Local communication takes place through paracrine and autocrine signaling. A **paracrine signal** {*para-*, beside + *krinen*, to secrete} is a chemical that acts on cells in the immediate vicinity of the cell that secreted the signal. A chemical signal that acts on the cell that secreted it is called an **autocrine signal** {*auto-*, self}. In some cases, a molecule may act as both an autocrine signal and a paracrine signal.

Paracrine and autocrine signal molecules reach their target cells by diffusing through the interstitial fluid (Fig. 6.1c). Because distance is a limiting factor for diffusion, the effective range of paracrine signals is restricted to adjacent cells. A good example of a paracrine molecule is *histamine,* a chemical released from damaged cells. When you scratch yourself with a pin, the red, raised *wheal* that results is due in part to the local release of histamine from the injured tissue. The histamine acts as a paracrine signal, diffusing to capillaries in the immediate area of the injury and making them more permeable to white blood cells and antibodies in the plasma. Fluid also leaves the blood vessels and collects in the interstitial space, causing swelling around the area of injury.

Several important classes of molecules act as local signals. *Cytokines* are regulatory peptides, and *eicosanoids* [p. 30] are lipid-derived paracrine and autocrine signal molecules. We discuss cytokines and eicosanoids in more detail later.

Long-Distance Communication May Be Electrical or Chemical

All cells in the body can release paracrine signals, but most long-distance communication between cells takes place through the nervous and endocrine systems. The endocrine system communicates by using **hormones** {*hormone,* to excite}, chemical signals that are secreted into the blood and distributed all over the body by the circulation. Hormones come in contact with most cells of the body, but only those cells with receptors for the hormone are target cells (Fig. 6.1d).

The nervous system uses a combination of chemical signals and electrical signals to communicate over long distances. An electrical signal travels along a nerve cell (*neuron*) until it reaches the very end of the cell, where it is translated into a chemical signal secreted by the neuron. Chemicals secreted by neurons are called **neurocrine molecules**.

If a neurocrine molecule diffuses from the neuron across a narrow extracellular space to a target cell and has a rapid-onset effect, it is called a **neurotransmitter** (Fig. 6.1e). If a neurocrine acts more slowly as an autocrine or paracrine signal, it is called a **neuromodulator**. If a neurocrine molecule diffuses into the blood for body-wide distribution, it is called a **neurohormone** (Fig. 6.1f). The similarities between neurohormones and classic hormones secreted by the endocrine system bridge the gap between the nervous and endocrine systems, making them a functional continuum rather than two distinct systems.

Cytokines May Act as Both Local and Long-Distance Signals

Cytokines are among the most recently identified communication molecules. Initially the term *cytokine* referred only to peptides that modulate immune responses, but in recent years the definition has been broadened to include a variety of regulatory peptides. Most of these peptides share a similar structure of four or more α-helix bundles [p. 32]. Families of cytokines include *interferons, interleukins, colony-stimulating factors, growth factors, tumor necrosis factors,* and *chemokines.* (Cytokines are discussed in more detail in Chapter 24.)

Cytokines are associated primarily with immune responses, such as inflammation, but they also control cell development and cell differentiation. In development and differentiation, cytokines usually function as autocrine or paracrine signals. In stress and inflammation, some cytokines may act on relatively distant targets and may be transported through the circulation just as hormones are. Many research laboratories are interested in cytokines because of their importance in disease processes.

How do cytokines differ from classic hormones? First, cytokines are not produced by specialized epithelial cells the way hormones are. Instead, any nucleated cell can secrete cytokines at some point in its life span. Second, cytokines are made on demand, in contrast to protein or peptide hormones that are made in advance and stored in the endocrine cell until needed. And finally, the intracellular signal pathways for cytokines are usually different from those for hormones. However, the distinction between cytokines and hormones is sometimes blurry. For example, erythropoietin, the molecule that controls synthesis of red blood cells, is by tradition considered a hormone but functionally fits the definition of a cytokine.

Concept Check

1. Match the communication method on the left with its property on the right.

	Communication is:
(a) autocrine signal	1. electrical
(b) cytokine	2. chemical
(c) gap junction	3. both electrical and chemical
(d) hormone	
(e) neurohormone	
(f) neurotransmitter	
(g) paracrine signal	

2. Which signal molecules listed in the previous question are transported through the circulatory system? Which are released by neurons?
3. A cat sees a mouse and pounces on it. Do you think the internal signal to pounce could have been transmitted by a paracrine signal? Give two reasons to explain why or why not.

CHAPTER 6

6.2 Signal Pathways

Chemical signal molecules are secreted by cells into the extracellular compartment. This is not a very specific way for these signals to find their targets because substances that diffuse through interstitial fluid or that travel through the blood come in contact with many cells. Yet cells do not respond to every signal that reaches them.

Why do some cells respond to a chemical signal while other cells ignore it? The answer lies in the target cell's **receptor proteins** [p. 137]. *A cell can respond to a particular chemical signal only if the cell has the appropriate receptor protein to bind that signal* (Fig. 6.1d).

If a target cell has the receptor for a signal molecule, binding of the signal molecule to the receptor protein initiates a response. All signal pathways share the following features (**FIG. 6.2**):

1. The signal molecule is a *ligand* that binds to a protein receptor. The ligand is also known as a *first messenger* because it brings information to the target cell.
2. Ligand-receptor binding activates the receptor.
3. The receptor in turn activates one or more intracellular signal molecules.
4. The last signal molecule in the pathway creates a response by modifying existing proteins or initiating the synthesis of new proteins.

In the following sections, we describe some basic signal pathways. They may seem complex at first, but they follow patterns that you will encounter over and over as you study the systems of the body. Most physiological processes, from the beating of your heart to learning and memory, use some variation of these pathways. One of the wonders of physiology is the fundamental importance of these signal pathways and the way they have been conserved in animals ranging from worms to humans.

FIG. 6.2 Signal pathways

Most signal pathways consist of the 5 steps shown. Use the shapes and colors of the steps shown here to identify the pattern in later illustrations.

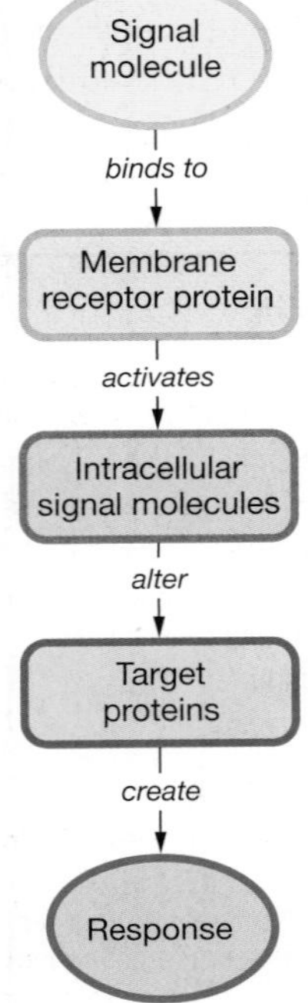
Signal molecule
binds to
Membrane receptor protein
activates
Intracellular signal molecules
alter
Target proteins
create
Response

Receptor Proteins Are Located Inside the Cell or on the Cell Membrane

Protein receptors for signal molecules play an important role in physiology and medicine. About half of all drugs currently in use act on receptor proteins. Target cell receptor proteins may be found in the nucleus, in the cytosol, or on the cell membrane as integral proteins. Where a chemical signal binds to its receptor largely depends on whether that signal molecule is lipophilic or lipophobic (**FIG. 6.3**).

Lipophilic signal molecules enter cells by simple diffusion through the phospholipid bilayer of the cell membrane [p. 134]. Once inside, they bind to *cytosolic receptors* or *nuclear receptors* (Fig. 6.3a). Activation of intracellular receptors often turns on a gene and directs the nucleus to make new mRNA (transcription, [p. 111]). The mRNA then provides a template for synthesis of new proteins (translation, [p. 111]). This process is relatively slow and the cell's response may not be noticeable for an hour or longer. In some instances, the activated receptor can also turn off, or *repress,* gene activity. Many lipophilic signal molecules that follow this pattern are hormones.

Lipophobic signal molecules are unable to enter cells by simple diffusion through the cell membrane. Instead, these signal molecules remain in the extracellular fluid and bind to receptor proteins on the cell membrane (Fig. 6.3b). (Some lipophilic signal molecules also bind to cell membrane receptors in addition to their intracellular receptors.) In general, the response time for pathways linked to membrane receptor proteins is very rapid: responses can be seen within milliseconds to minutes.

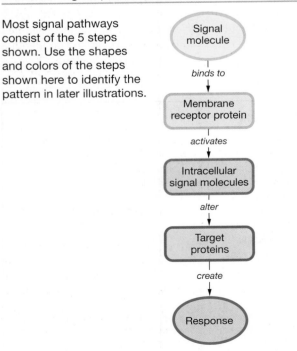

RUNNING PROBLEM

Later that day in the physician's office, the nurse practitioner explains diabetes to Marvin. Diabetes mellitus is a family of metabolic disorders caused by defects in the homeostatic pathways that regulate glucose metabolism. Several forms of diabetes exist, and some can be inherited. One form, called *type 1 diabetes mellitus,* occurs when endocrine cells of the pancreas stop making insulin, a protein hormone involved in blood glucose homeostasis. In another form, type 2 diabetes mellitus, insulin may be present in normal or above-normal levels, but the insulin-sensitive cells of the body do not respond normally to the hormone.

Q1: *In which type of diabetes is the target cell's signal pathway for insulin more likely to be defective?*

Q2: *Insulin is a protein hormone. Would you expect to findits receptor on the cell surface or in the cytoplasm of the target cells?*

165 – **168** – 181 – 182 – 186 – 188 – 191

FIG. 6.3 Target cell receptors may be on the cell surface or inside the cell

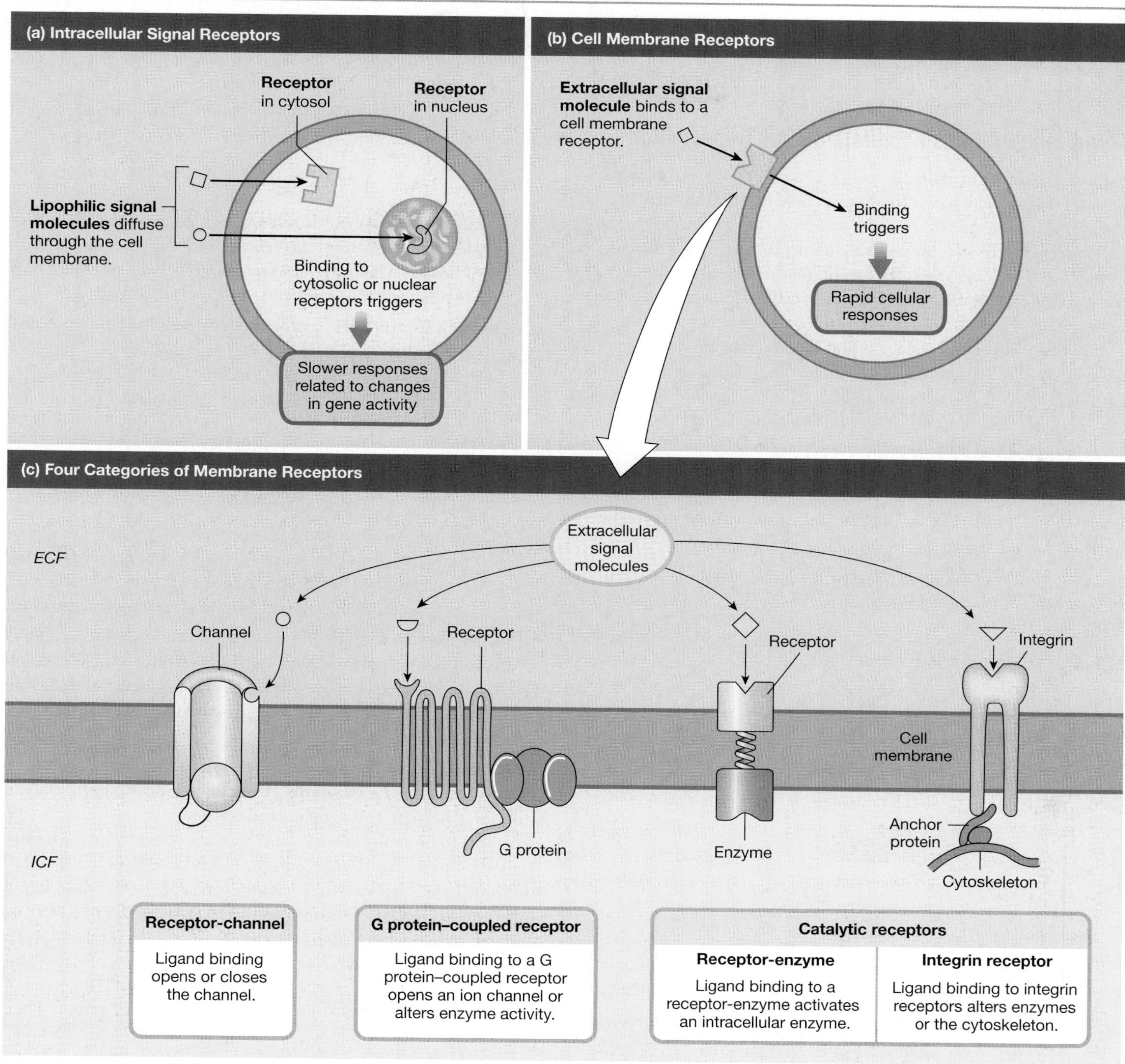
(a) Intracellular Signal Receptors
Receptor in cytosol
Receptor in nucleus
Lipophilic signal molecules diffuse through the cell membrane.
Binding to cytosolic or nuclear receptors triggers
Slower responses related to changes in gene activity
(b) Cell Membrane Receptors
Extracellular signal molecule binds to a cell membrane receptor.
Binding triggers
Rapid cellular responses
(c) Four Categories of Membrane Receptors
ECF
Extracellular signal molecules
Channel
Receptor
Receptor
Integrin
Cell membrane
G protein
Enzyme
Anchor protein
Cytoskeleton
ICF
Receptor-channel
Ligand binding opens or closes the channel.
G protein–coupled receptor
Ligand binding to a G protein–coupled receptor opens an ion channel or alters enzyme activity.
Catalytic receptors
Receptor-enzyme
Ligand binding to a receptor-enzyme activates an intracellular enzyme.
Integrin receptor
Ligand binding to integrin receptors alters enzymes or the cytoskeleton.

We can group membrane receptors into four major categories, illustrated in Figure 6.3c. The simplest receptors are chemically gated (*ligand-gated*) ion channels called *receptor-channels* [p. 138]. Ligand binding opens or closes the channel and alters ion flow across the membrane.

Three other receptor types are shown in Figure 6.3c: *G protein-coupled receptors*, *receptor-enzymes*, and *integrin receptors*. For all three, information from the signal molecule must be passed across the membrane to initiate an intracellular response. This transmission of information from one side of a membrane to the other using membrane proteins is known as *signal transduction*. We will take a closer look at basic signal transduction before returning to the four receptor types that participate in it.

Concept Check

4. List four components of signal pathways.
5. Name three cellular locations of receptors.

Membrane Proteins Facilitate Signal Transduction

Signal transduction is the process by which an extracellular signal molecule activates a membrane receptor that in turn alters intracellular molecules to create a response. The extracellular signal molecule is the first messenger, and the intracellular molecules form a *second messenger system*. The term *signal transduction* comes from the verb *to transduce*, meaning "to lead across" {*trans*, across + *ducere*, to lead}.

A **transducer** is a device that converts a signal from one form into a different form. For example, the transducer in a radio converts radio waves into sound waves (**FIG. 6.4**). In biological systems, membrane proteins act as transducers. They convert the message of extracellular signals into intracellular messenger molecules that trigger a response.

The basic pattern of a biological signal transduction pathway is shown in **FIGURE 6.5a** and can be broken down into the following events.

1. An extracellular signal molecule (the *first messenger*) binds to and activates a membrane receptor.

FIG. 6.4 Signal transduction

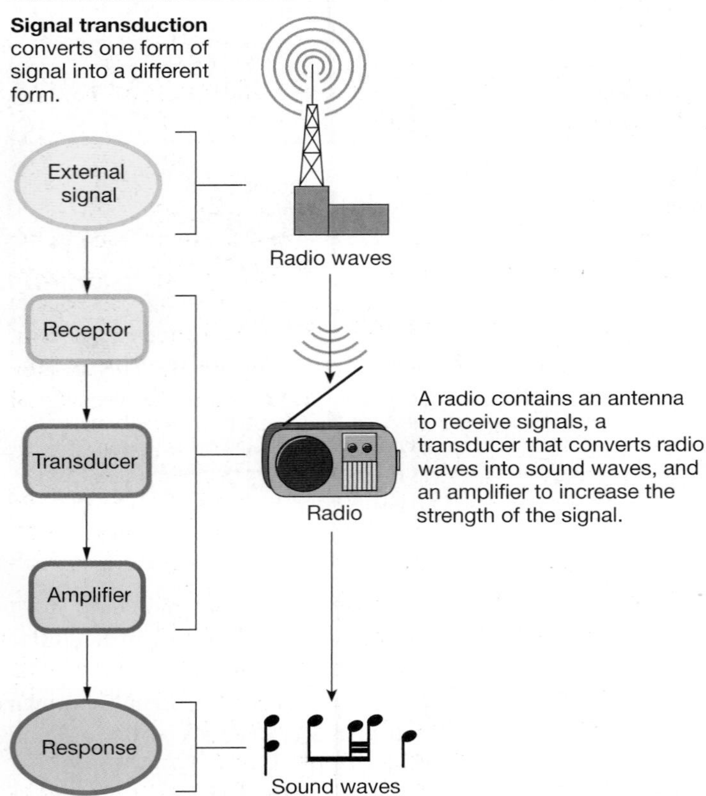
Signal transduction converts one form of signal into a different form.
External signal
Receptor
Transducer
Amplifier
Response
Radio waves
Radio
A radio contains an antenna to receive signals, a transducer that converts radio waves into sound waves, and an amplifier to increase the strength of the signal.
Sound waves

2. The activated membrane receptor turns on its associated proteins and starts an intracellular cascade of **second messengers**.
3. The last second messenger in the cascade acts on intracellular targets to create a response.

Figure 6.5b details the intracellular events in basic signal transduction pathways:

1. Membrane receptors and their associated proteins usually either
   (a) activate **protein kinases**, which are enzymes that transfer a phosphate group from ATP to a protein [p. 102]. Phosphorylation is an important biochemical method of regulating cellular processes.
   (b) activate amplifier enzymes that create intracellular second messengers.
2. Second messenger molecules in turn
   (a) alter the gating of ion channels. Opening or closing ion channels creates electrical signals by altering the cell's membrane potential [p. 157].
   (b) increase intracellular calcium. Calcium binding to proteins changes their function, creating a cellular response.
   (c) change enzyme activity, especially of protein kinases or **protein phosphatases**, enzymes that remove a phosphate group. The phosphorylation or *dephosphorylation* of a protein can change its configuration and create a response.
3. The proteins modified by calcium binding and phosphorylation are responsible for the cell's response to the signal. Examples of responses include increased or decreased enzyme activity and opening or closing of gated ion channels.

Cascades **FIGURE 6.6a** shows how the steps of a signal transduction pathway form a **cascade**. A signaling cascade starts when a stimulus (the signal molecule) converts inactive molecule A (the receptor) to an active form. Active A then converts inactive molecule B into active B, active molecule B in turn converts inactive molecule C into active C, and so on, until at the final step a substrate is converted into a product. Many intracellular signal pathways are cascades. Blood clotting is an important example of an extracellular cascade.

Amplification In signal transduction pathways, the original signal is not only transformed but also amplified {*amplificare*, to make larger}. In a radio, the radio wave signal is also amplified. In cells, **signal amplification** turns one signal molecule into multiple second messenger molecules (Fig. 6.6b).

The process begins when the first messenger ligand combines with its receptor. The receptor-ligand complex turns on an **amplifier enzyme**. The amplifier enzyme activates several molecules, which in turn each activate several more molecules as the cascade proceeds. By the end of the process, the effects of the ligand have been amplified much more than if there were a 1:1 ratio between each step.

FIG. 6.5 Biological signal transduction

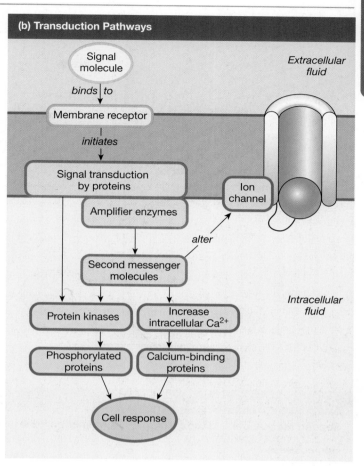

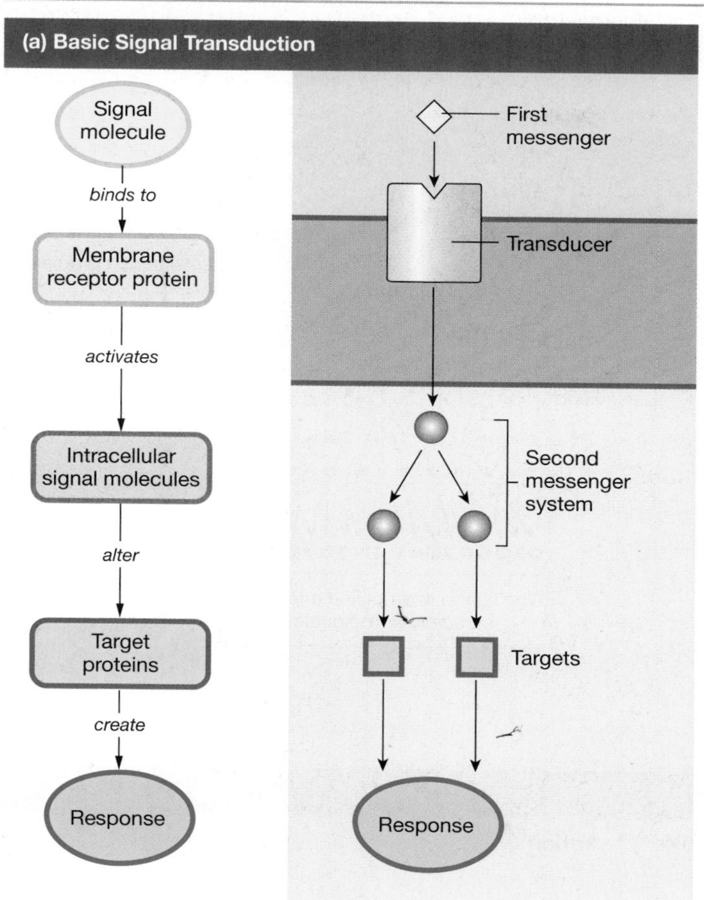
(a) Basic Signal Transduction
Signal molecule
binds to
Membrane receptor protein
activates
Intracellular signal molecules
alter
Target proteins
create
Response
First messenger
Transducer
Second messenger system
Targets
Response
(b) Transduction Pathways
Signal molecule
binds to
Membrane receptor
initiates
Signal transduction by proteins
Amplifier enzymes
Second messenger molecules
alter
Ion channel
Extracellular fluid
Intracellular fluid
Protein kinases
Increase intracellular Ca2+
Phosphorylated proteins
Calcium-binding proteins
Cell response

Amplification gives the body "more bang for the buck" by enabling a small amount of ligand to create a large effect. The most common amplifier enzymes and second messengers are listed in the table in Figure 6.6c.

In the sections that follow, we will examine in more detail the three major types of membrane receptors: receptor-channels, G protein-coupled receptors (GPCR), and catalytic receptors (see Fig. 6.3c). Keep in mind that these receptors may be responding to any of the different kinds of signal molecules—hormones, neurohormones, neurotransmitters, cytokines, or paracrine and autocrine signals.

Concept Check

6. What are the four steps of signal transduction?
7. What happens during amplification? In Figure 6.6b, amplification of one signal molecule binding to the receptor results in how many small dark blue intracellular signal molecules?
8. Why do steroid hormones not require signal transduction and second messengers to exert their action? (*Hint:* Are steroids lipophobic or lipophilic?[p. 44])

The Most Rapid Signal Pathways Change Ion Flow through Channels

The simplest receptors are ligand-gated ion channels. Most of these receptors are neurotransmitter receptors found in nerve and muscle. The activation of **receptor-channels** initiates the most rapid intracellular responses of all receptors. When an extracellular ligand binds to the receptor-channel protein, a channel gate opens or closes, altering the cell's permeability to an ion. Increasing or decreasing ion permeability rapidly changes the cell's membrane potential [p. 157], creating an electrical signal that alters voltage-sensitive proteins (**FIG. 6.7**).

One example of a receptor-channel is the acetylcholine-gated monovalent ("one-charge") cation channel of skeletal muscle. The neurotransmitter *acetylcholine* released from an adjacent neuron binds to the acetylcholine receptor and opens the channel. Both Na^+ and K^+ flow through the open channel, K^+ leaving the cell and Na^+ entering the cell along their electrochemical gradients. The sodium gradient is stronger, however, so net entry of positively charged Na^+ depolarizes the cell. In skeletal muscle, this cascade of intracellular events results in muscle contraction.

Receptor-channels are only one of several ways to trigger ion-mediated cell signaling. Some ion channels are linked

FIG. 6.6 ESSENTIALS Signal Transduction: Cascades and Amplification

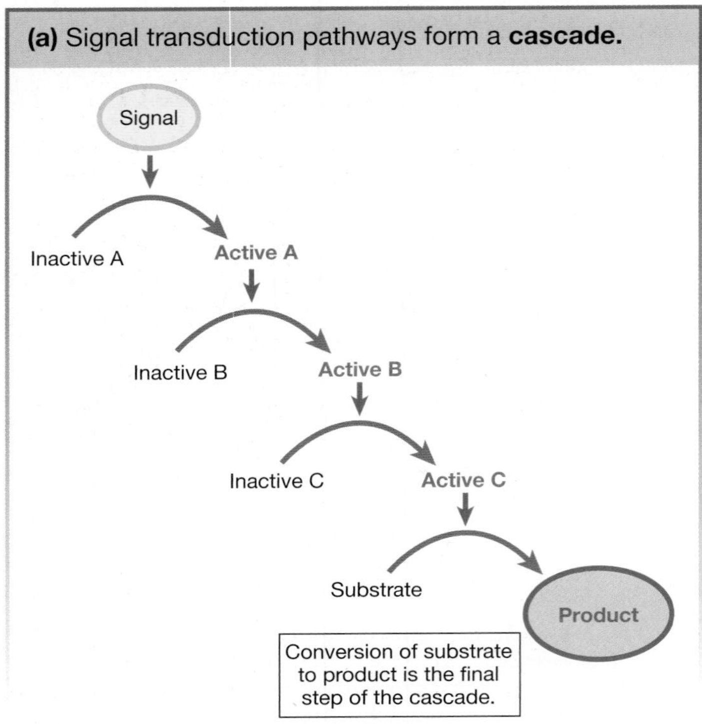
(a) Signal transduction pathways form a **cascade.**
Signal
Inactive A
Active A
Inactive B
Active B
Inactive C
Active C
Substrate
Product
Conversion of substrate to product is the final step of the cascade.

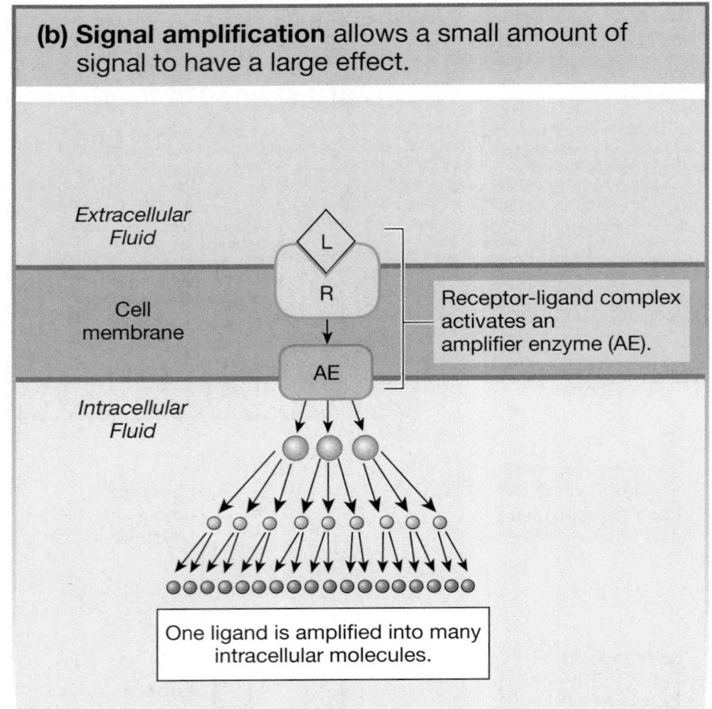
(b) **Signal amplification** allows a small amount of signal to have a large effect.
Extracellular Fluid
L
R
Cell membrane
AE
Intracellular Fluid
Receptor-ligand complex activates an amplifier enzyme (AE).
One ligand is amplified into many intracellular molecules.

(c) Second messenger pathways

Second Messenger	Made from	Amplifier enzyme	Linked to	Action	Effects
Nucleotides					
cAMP	ATP	Adenylyl cyclase (membrane)	GPCR*	Activates protein kinases, especially PKA. Binds to ion channels.	Phosphorylates proteins. Alters channel opening.
cGMP	GTP	Guanylyl cyclase (membrane)	Receptor-enzyme	Activates protein kinases, especially PKG.	Phosphorylates proteins.
		Guanylyl cyclase (cytosol)	Nitric oxide (NO)	Binds to ion channels	Alters channel opening.
Lipid Derived*					
IP_3	Membrane phospholipids	Phospholipase C (membrane)	GPCR	Releases Ca^{2+} from intracellular stores.	See Ca^{2+} effects below.
DAG				Activates protein kinase C.	Phosphorylates proteins.
Ions					
Ca^{2+}				Binds to calmodulin. Binds to other proteins.	Alters enzyme activity. Exocytosis, muscle contraction, cytoskeleton movement, channel opening.

*GPCR = G protein–coupled receptor. IP_3 = Inositol trisphosphate. DAG = diacylglycerol.

FIG. 6.7 Signal transduction using ion channels

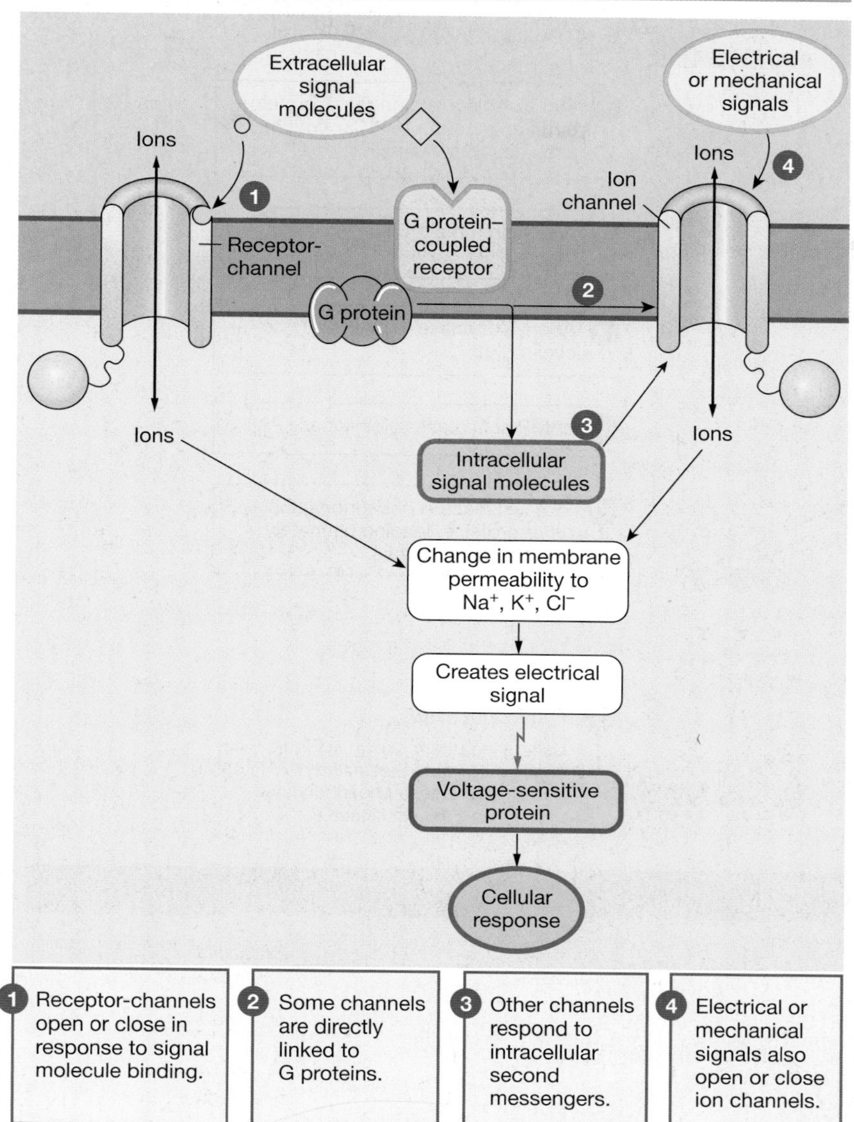

Extracellular signal molecules
Electrical or mechanical signals
Ions
Ions
Ion channel
G protein–coupled receptor
Receptor-channel
G protein
Ions
Ions
Intracellular signal molecules
Change in membrane permeability to Na+, K+, Cl−
Creates electrical signal
Voltage-sensitive protein
Cellular response
1 Receptor-channels open or close in response to signal molecule binding.
2 Some channels are directly linked to G proteins.
3 Other channels respond to intracellular second messengers.
4 Electrical or mechanical signals also open or close ion channels.

to G protein-coupled receptors. When a ligand binds to the G protein receptor, the G protein pathway opens or closes the channel.

Finally, some membrane ion channels are not associated with membrane receptors at all. Voltage-gated channels can be opened directly with a change in membrane potential. Mechanically gated channels open with pressure or stretch on the cell membrane [p. 138]. Intracellular molecules, such as cAMP or ATP, can open or close non-receptor-linked ligand-gated channels. The ATP-gated K^+ channels of the pancreatic beta cell are an example [Fig. 5.26, p. 158].

Most Signal Transduction Uses G Proteins

The **G protein-coupled receptors (GPCRs)** are a large and complex family of membrane-spanning proteins that cross the phospholipid bilayer seven times (see Fig. 6.3c). The cytoplasmic tail of the receptor protein is linked to a three-part membrane transducer molecule known as a **G protein**. Hundreds of G protein-coupled receptors have been identified, and the list continues to grow. The types of ligands that bind to G protein-coupled receptors include hormones, growth factors, olfactory molecules, visual pigments, and neurotransmitters. In 1994, Alfred G. Gilman and Martin Rodbell received a Nobel Prize for the discovery of G proteins and their role in cell signaling (see *http://nobelprize.org/nobel_prizes/medicine/laureates/1994*).

G proteins get their name from the fact that they bind guanosine nucleotides [p. 34]. Inactive G proteins are bound to guanosine diphosphate (GDP). Exchanging the GDP for guanosine triphosphate (GTP) activates the G protein. When G proteins are activated, they either (1) open an ion channel in the membrane or (2) alter enzyme activity on the cytoplasmic side of the membrane.

G proteins linked to amplifier enzymes make up the bulk of all known signal transduction mechanisms. The two most common amplifier enzymes for G protein-coupled receptors are adenylyl cyclase and phospholipase C. The pathways for these amplifier enzymes are described next.

Many Lipophobic Hormones Use GPCR-cAMP Pathways

The **G protein-coupled adenylyl cyclase-cAMP system** was the first identified signal transduction pathway (**FIG. 6.8a**). It was discovered in the 1950s by Earl Sutherland when he was studying the effects of hormones on carbohydrate metabolism. This discovery proved so significant to our understanding of signal transduction that in 1971 Sutherland was awarded a Nobel Prize for his work.

The G protein-coupled adenylyl cyclase-cAMP system is the signal transduction system for many protein hormones. In this system, *adenylyl cyclase* is the amplifier enzyme that converts ATP to the second messenger molecule *cyclic AMP* (cAMP). Cyclic AMP then activates *protein kinase A* (PKA), which in turn phosphorylates other intracellular proteins as part of the signal cascade.

G Protein-Coupled Receptors Also Use Lipid-Derived Second Messengers

Some G protein-coupled receptors are linked to a different amplifier enzyme: phospholipase C (Fig. 6.8b). When a signal molecule activates this G protein-coupled pathway, **phospholipase C (PLC)** converts a membrane phospholipid (*phosphatidylinositol bisphosphate*) into two lipid-derived second messenger molecules: diacylglycerol and inositol trisphosphate.

FIG. 6.8 G protein-coupled signal transduction

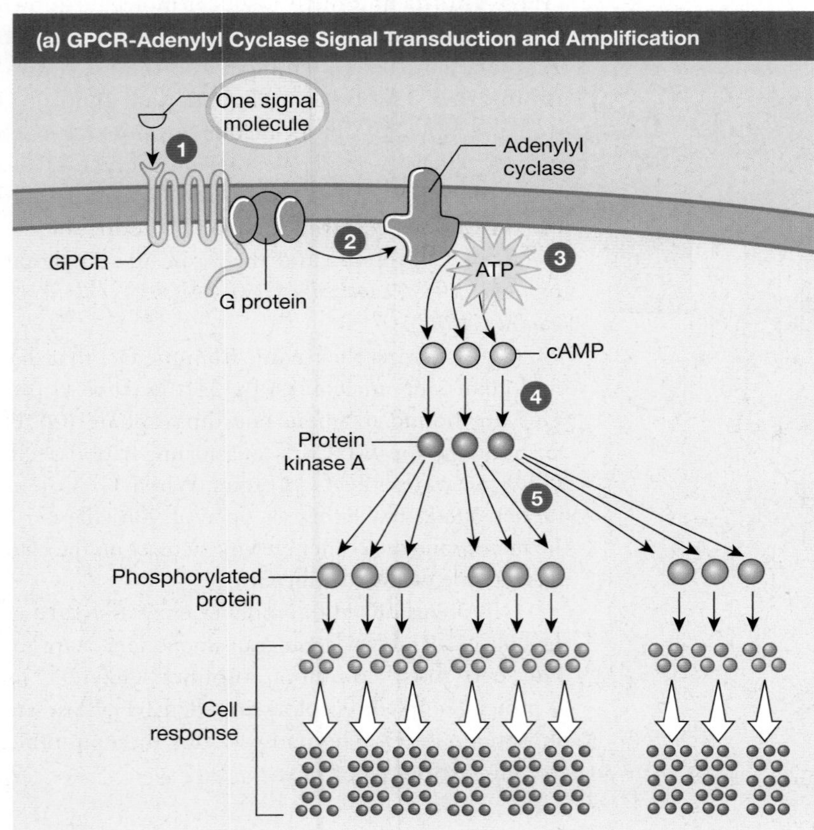
(a) GPCR-Adenylyl Cyclase Signal Transduction and Amplification
One signal molecule
Adenylyl cyclase
GPCR
G protein
ATP
cAMP
Protein kinase A
Phosphorylated protein
Cell response

1. Signal molecule binds to G protein–coupled receptor (GPCR), which activates the G protein.
2. G protein turns on adenylyl cyclase, an amplifier enzyme.
3. Adenylyl cyclase converts ATP to cyclic AMP.
4. cAMP activates protein kinase A.
5. Protein kinase A phosphorylates other proteins, leading ultimately to a cellular response.

FIGURE QUESTION

Using the pattern shown in Figure 6.6a, create a cascade that includes ATP, cAMP, adenylyl cyclase, a phosphorylated protein, and protein kinase A.

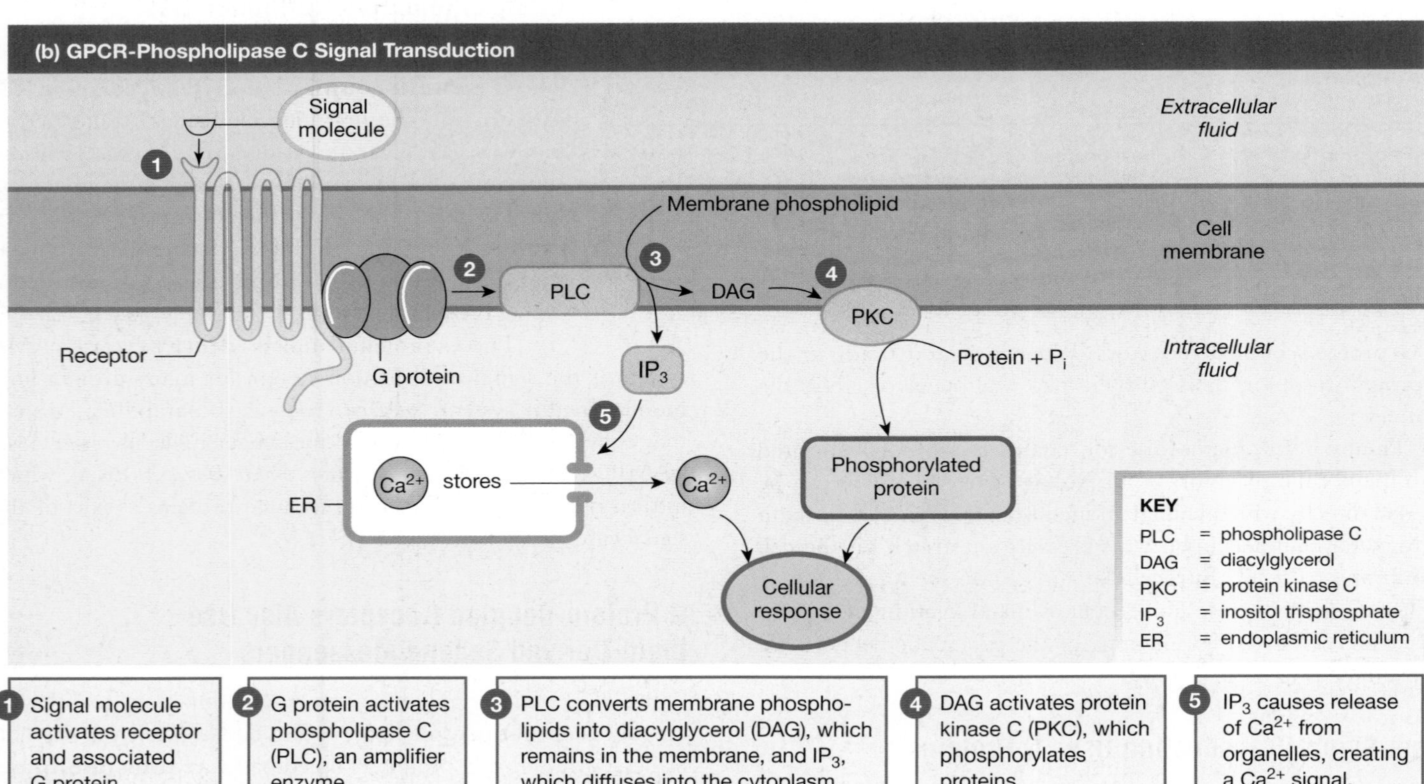
(b) GPCR-Phospholipase C Signal Transduction
Signal molecule
Extracellular fluid
Membrane phospholipid
Cell membrane
PLC
DAG
PKC
Receptor
G protein
IP_3
Protein + P_i
Intracellular fluid
Ca^{2+} stores
Ca^{2+}
ER
Phosphorylated protein
Cellular response
KEY
PLC = phospholipase C
DAG = diacylglycerol
PKC = protein kinase C
IP_3 = inositol trisphosphate
ER = endoplasmic reticulum

1. Signal molecule activates receptor and associated G protein.
2. G protein activates phospholipase C (PLC), an amplifier enzyme.
3. PLC converts membrane phospholipids into diacylglycerol (DAG), which remains in the membrane, and IP_3, which diffuses into the cytoplasm.
4. DAG activates protein kinase C (PKC), which phosphorylates proteins.
5. IP_3 causes release of Ca^{2+} from organelles, creating a Ca^{2+} signal.

Diacylglycerol (DAG) is a nonpolar diglyceride that remains in the lipid portion of the membrane and interacts with **protein kinase C (PKC)**, a Ca^{2+}- activated enzyme associated with the cytoplasmic face of the cell membrane. PKC phosphorylates cytosolic proteins that continue the signal cascade.

Inositol trisphosphate (IP_3) is a water-soluble messenger molecule that leaves the membrane and enters the cytoplasm. There it binds to a calcium channel on the endoplasmic reticulum (ER). IP_3 binding opens the Ca^{2+} channel, allowing Ca^{2+} to diffuse out of the ER and into the cytosol. Calcium is itself an important signal molecule, as discussed later.

Catalytic Receptors Have Enzyme Activity

Catalytic receptors are the newest family of receptors. These **receptor-enzymes** have two regions: a receptor region on the extracellular side of the cell membrane, and an enzyme region on the cytoplasmic side (see Fig. 6.3c). Ligand binding to the receptor activates the intracellular enzyme. Catalytic receptor enzymes include protein kinases, such as *tyrosine kinase* (**FIG. 6.9**), or *guanylyl cyclase,* the amplifier enzyme that converts GTP to **cyclic GMP (cGMP)** [p. 34].

For some catalytic receptors, the extracellular binding region and the intracellular enzyme region are parts of the same protein molecule. An example of this is insulin receptor, which has its own intrinsic tyrosine kinase activity. In other types of catalytic receptors, the enzyme region is a separate protein activated by ligand binding. Cytokine receptors are in this category. There are six major cytokine receptor families, and most are them are associated with a cytosolic enzyme called *Janus family tyrosine kinase,* usually abbreviated as *JAK kinase.*

FIG. 6.9 Receptor-enzymes: The tyrosine kinase receptor

Tyrosine kinase (TK) transfers a phosphate group from ATP to a tyrosine (an amino acid) of a protein.

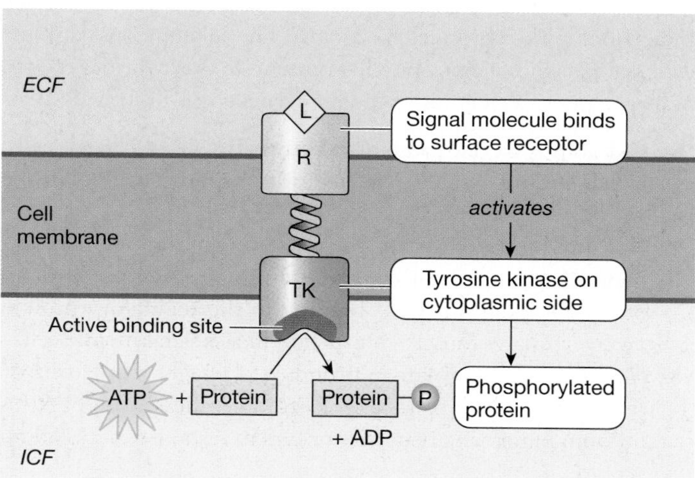
ECF
L
R
Signal molecule binds to surface receptor
Cell membrane
activates
TK
Tyrosine kinase on cytoplasmic side
Active binding site
ATP
+ Protein
Protein P
+ ADP
Phosphorylated protein
ICF

Integrin Receptors Transfer Information from the Extracellular Matrix

The membrane-spanning proteins called *integrins* [p. 75] mediate blood clotting, wound repair, cell adhesion and recognition in the immune response, and cell movement during development. Integrin receptors are currently classified as catalytic receptors but also have properties that are not associated with classic receptors. On the extracellular side of the membrane, integrins bind either to proteins of the extracellular matrix [p. 73] or to ligands such as antibodies and molecules involved in blood clotting. Inside the cell, integrins attach to the cytoskeleton via *anchor proteins* (Fig. 6.3c). Ligand binding to the receptor causes integrins to activate intracellular enzymes or alter the organization of the cytoskeleton.

The importance of integrin receptors is illustrated by inherited conditions in which the receptor is absent. In one condition, platelets—cell fragments that play a key role in blood clotting—lack an integrin receptor. As a result, blood clotting is defective in these individuals.

FIGURE 6.10 is a summary map of basic signal transduction, showing the general relationships among first messengers, membrane receptors, second messengers, and cell responses. The modified proteins that control cell responses can be broadly grouped into four categories:

1. metabolic enzymes
2. motor proteins for muscle contraction and cytoskeletal movement
3. proteins that regulate gene activity and protein synthesis
4. membrane transport and receptor proteins

If you think this list includes almost everything a cell does, you're right!

Concept Check

9. Name the four categories of membrane receptors.
10. What is the difference between a first messenger and a second messenger?
11. Place the following terms in the correct order for a signal transduction pathway:
    (a) cell response, receptor, second messenger, ligand
    (b) amplifier enzyme, cell response, phosphorylated protein, protein kinase, second messenger
12. In each of the following situations, will a cell depolarize or hyperpolarize?
    (a) Cl^- channel opens
    (b) K^+ channel opens
    (c) Na^+ channel opens

6.3 Novel Signal Molecules

The following sections introduce you to some unusual signal molecules that are important in physiology and medicine. They include an ion (Ca^{2+}), three gases, and a family of lipid-derived

FIG. 6.10 ESSENTIALS Summary Map of Signal Transduction

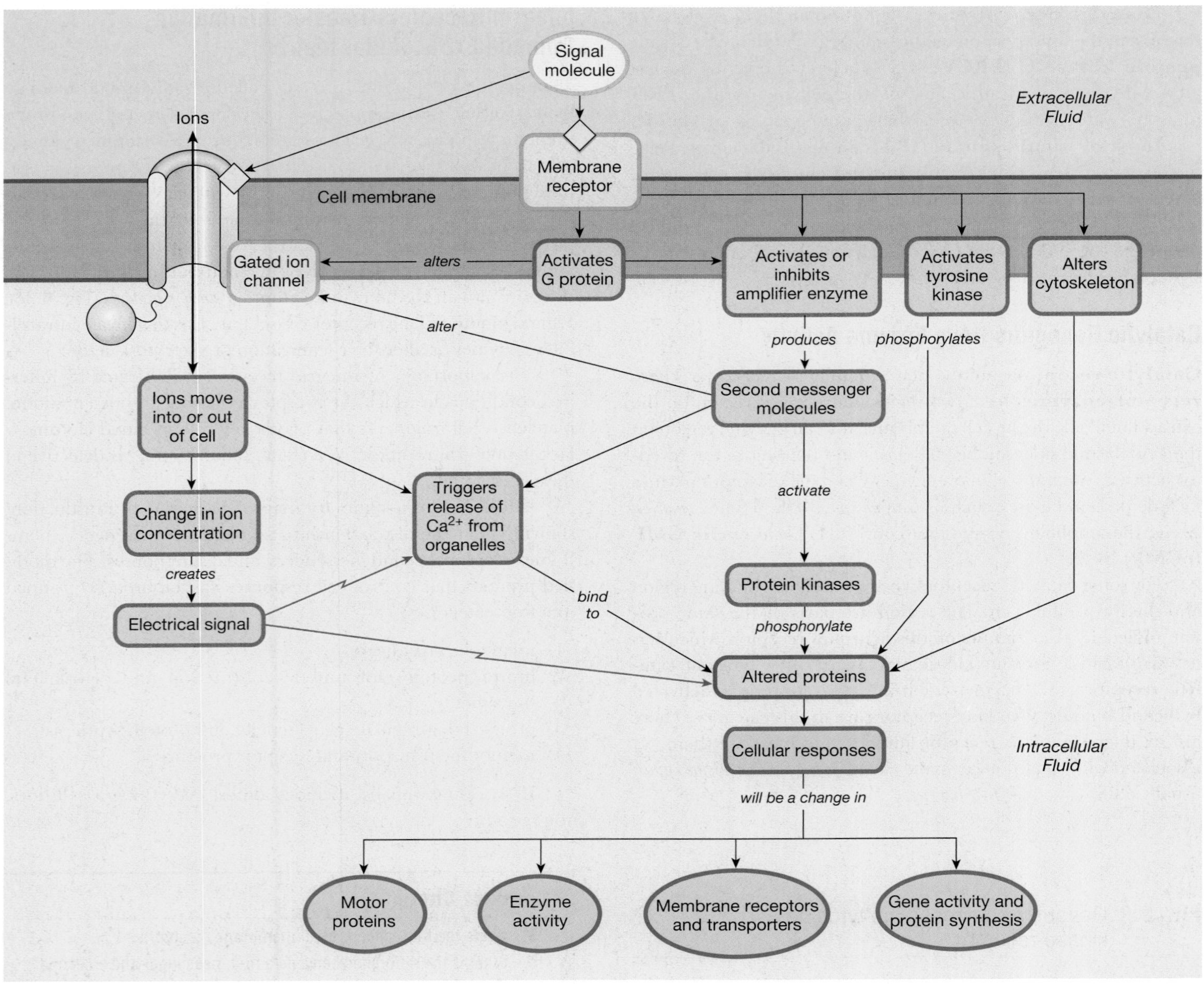

Signal molecule
Extracellular Fluid
Ions
Membrane receptor
Cell membrane
Gated ion channel
alters
Activates G protein
Activates or inhibits amplifier enzyme
Activates tyrosine kinase
Alters cytoskeleton
alter
produces
phosphorylates
Second messenger molecules
Ions move into or out of cell
Triggers release of Ca^{2+} from organelles
activate
Change in ion concentration
creates
Protein kinases
bind to
Electrical signal
phosphorylate
Altered proteins
Cellular responses
Intracellular Fluid
will be a change in
Motor proteins
Enzyme activity
Membrane receptors and transporters
Gene activity and protein synthesis

messengers. The processes controlled by these signal molecules have been known for years, but the control signals themselves were discovered only relatively recently.

Calcium Is an Important Intracellular Signal

Calcium ions are the most versatile ionic messengers (**FIG. 6.11**). Calcium enters the cell through Ca^{2+} channels that may be voltage-gated, ligand-gated, or mechanically gated. Calcium can also be released from intracellular compartments by second messengers, such as IP_3. Most intracellular Ca^{2+} is stored in the endoplasmic reticulum [p. 70], where it is concentrated by active transport.

Release of Ca^{2+} into the cytosol (from any of the sources just mentioned) creates a Ca^{2+} signal, or Ca^{2+} "spark," that can be recorded using special Ca^{2+}-imaging techniques (see the Biotechnology box on calcium signals). The calcium ions combine with cytoplasmic calcium-binding proteins to exert various effects. Several types of calcium-dependent events occur in the cell:

1. Ca^{2+} binds to the protein **calmodulin**, found in all cells. Calcium binding alters enzyme or transporter activity or the gating of ion channels.
2. Calcium binds to other regulatory proteins and alters movement of contractile or cytoskeletal proteins such as microtubules. For example, Ca^{2+} binding to the regulatory protein *troponin* initiates muscle contraction in a skeletal muscle cell.
3. Ca^{2+} binds to regulatory proteins to trigger exocytosis of secretory vesicles [p. 147]. For example, the release of insulin from pancreatic beta cells occurs in response to a calcium signal.

FIG. 6.11 Calcium as an intracellular messenger

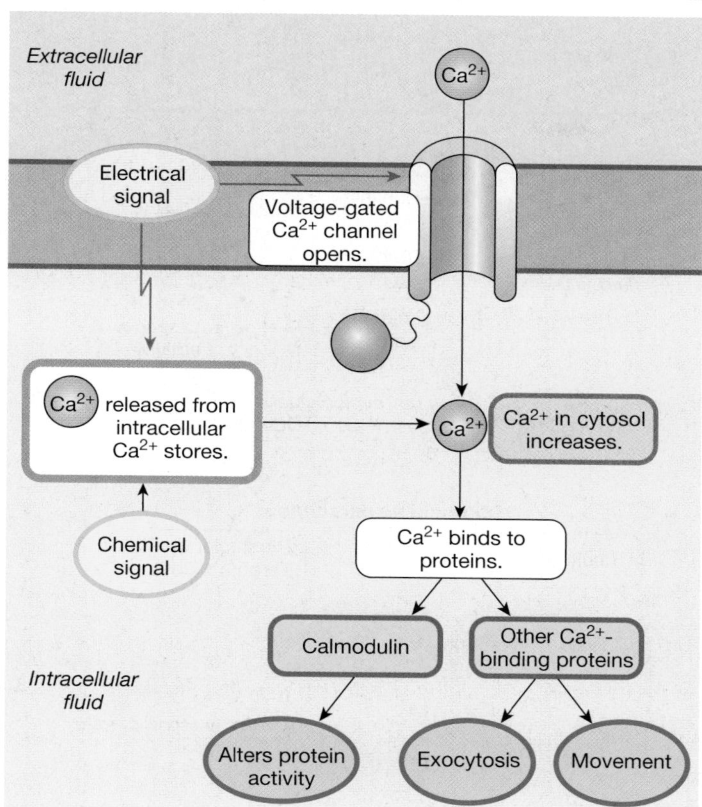
Extracellular fluid
Ca^{2+}
Electrical signal
Voltage-gated Ca^{2+} channel opens.
Ca^{2+} released from intracellular Ca^{2+} stores.
Ca^{2+}
Ca^{2+} in cytosol increases.
Chemical signal
Ca^{2+} binds to proteins.
Calmodulin
Other Ca^{2+}-binding proteins
Intracellular fluid
Alters protein activity
Exocytosis
Movement

4. Ca^{2+} binds directly to ion channels to alter their gating state. An example of this target is a Ca^{2+}-activated K^+ channel found in nerve cells.
5. Ca^{2+} entry into a fertilized egg initiates development of the embryo.

Gases Are Ephemeral Signal Molecules

Soluble gases are short-acting paracrine/autocrine signal molecules that act close to where they are produced. The best-known gaseous signal molecule is **nitric oxide (NO)**, but carbon monoxide and hydrogen sulfide, two gases better known for their noxious effects, can also act as local signals.

For years, researchers knew of a short-lived signal molecule produced by the endothelial cells lining blood vessels. They initially named it *endothelial-derived relaxing factor* (EDRF). This molecule diffuses from the endothelium into adjacent smooth muscle cells, causing the muscle to relax and dilate the blood vessel. Scientists took years to identify EDRF as nitric oxide because it is rapidly broken down, with a half-life of only 2 to 30 seconds. (*Half-life* is the time required for the signal to lose half of its activity.) As a result of this difficult work on NO in the cardiovascular system, Robert Furchgott, Louis Ignarro, and Ferid Murad received the 1998 Nobel Prize for physiology and medicine.

In tissues, NO is synthesized by the action of the enzyme *nitric oxide synthase* (NOS) on the amino acid arginine:

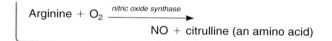

$$\text{Arginine} + O_2 \xrightarrow{\text{nitric oxide synthase}} \text{NO} + \text{citrulline (an amino acid)}$$

The NO produced in this reaction diffuses into target cells, where it binds to intracellular proteins. In many cases, NO binds to the cytosolic form of guanylyl cyclase and causes formation of the second messenger cGMP. In addition to relaxing blood vessels, NO in the brain acts as a neurotransmitter and a neuromodulator.

Concept Check

13. The extracellular fluid Ca^{2+} concentration averages 2.5 mmol/L. Free cytosolic Ca^{2+} concentration is about 0.001 mmol/L. If a cell is going to move calcium ions from its cytosol to the extracellular fluid, will it use passive or active transport? Explain.

BIOTECHNOLOGY

Calcium Signals Glow in the Dark

If you have ever run your hand through a tropical ocean at night and seen the glow of bioluminescent jellyfish, you've seen a calcium signal. Aequorin, a protein complex isolated from jellyfish such as the *Chrysaora fuscescens* shown here, is one of the molecules that scientists use to monitor the presence of calcium ions. When aequorin combines with calcium, it releases light that can be measured by electronic detection systems. Since the first use of aequorin in 1967, researchers have been designing increasingly sophisticated indicators that allow them to follow calcium signals in cells. With the help of molecules called fura, Oregon green, BAPTA, and chameleons, we can now watch calcium ions diffuse through gap junctions and flow out of intracellular organelles.

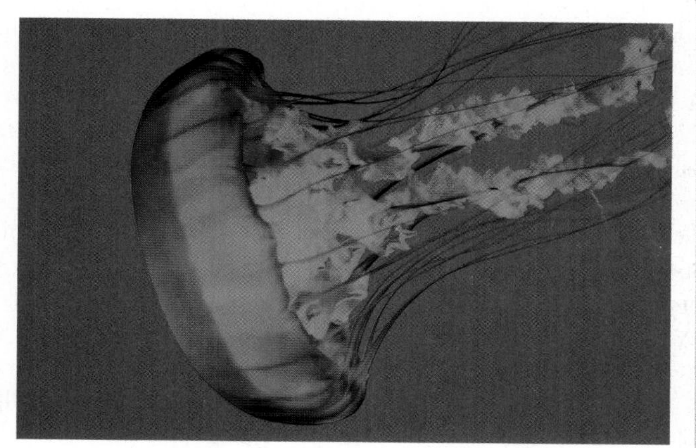

CLINICAL FOCUS

From Dynamite to Medicine

Who would have thought that a component of smog and a derivative of dynamite would turn out to be a biological messenger? Certainly not the peer reviewers who initially rejected Louis Ignarro's attempts to publish his research findings on the elusive gas nitric oxide (NO). The ability of nitrate-containing compounds to relax blood vessels had been known for more than 100 years, ever since workers in Alfred Nobel's dynamite factory complained of headaches caused by nitrate-induced vasodilation. And since the 1860s, physicians have used nitroglycerin to relieve *angina,* heart pain that results from constricted blood vessels. Even today, heart patients carry little nitroglycerin tablets to slide under their tongues when angina strikes. Still, it took years of work to isolate nitric oxide, the short-lived gas that is the biologically active molecule derived from nitroglycerin. Despite our modern technology, direct research on NO is still difficult. Many studies look at its influence indirectly by studying the location and activity of nitric oxide synthase (NOS), the enzyme that produces NO.

FIG. 6.12 The arachidonic acid cascade

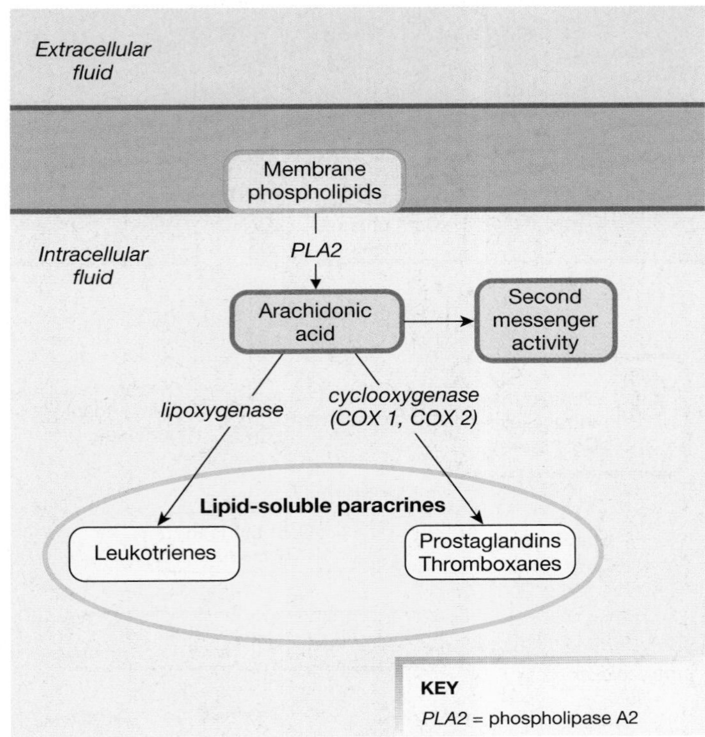
Extracellular fluid
Membrane phospholipids
Intracellular fluid
PLA2
Arachidonic acid
Second messenger activity
lipoxygenase
cyclooxygenase (COX 1, COX 2)
Lipid-soluble paracrines
Leukotrienes
Prostaglandins Thromboxanes
KEY
PLA2 = phospholipase A2

Carbon monoxide (CO), a gas known mostly for its toxic effects, is also a signal molecule produced in minute amounts by certain cells. Like NO, CO activates guanylyl cyclase and cGMP, but it may also work independently to exert its effects. Carbon monoxide targets smooth muscle and neural tissue.

The newest gaseous signal molecule to be described is **hydrogen sulfide (H_2S).** Hydrogen sulfide also acts in the cardiovascular system to relax blood vessels. Garlic is a major dietary source of the sulfur-containing precursors, which may explain studies suggesting that eating garlic has protective effects on the heart.

Some Lipids Are Important Paracrine Signals

One of the interesting developments from sequencing the human genome and using genes to find proteins has been the identification of *orphan receptors,* receptors that have no known ligand. Scientists are trying to work backward through signal pathways to find the ligands that bind to these orphan receptors. It was from this type of research that investigators recognized the importance and universality of *eicosanoids,* lipid-derived paracrine signals that play important roles in many physiological processes.

Eicosanoid signal molecules are derived from arachidonic acid, a 20-carbon fatty acid and act on their target cells using G protein-coupled receptors. The synthesis process for eicosanoids is a network called the *arachidonic acid cascade* (**FIG. 6.12**). For simplicity, we will break the cascade into steps.

Arachidonic acid is produced from membrane phospholipids by the action of an enzyme, **phospholipase A2 (PLA2)**. The activity of PLA2 is controlled by hormones and other signals. Arachidonic acid itself may act directly as a second messenger, altering ion channel activity and intracellular enzymes. It may also be converted into one of several classes of eicosanoid paracrine signals. These lipid-soluble molecules can diffuse out of the cell and combine with G protein-coupled receptors on neighboring cells to exert their action.

There are two major groups of arachidonic acid-derived paracrine molecules to be aware of:

1. **Leukotrienes** are molecules produced by the action of the enzyme *lipoxygenase* on arachidonic acid {*leuko-*, white + *triene*, a molecule with three double bonds between carbon atoms}. Leukotrienes are secreted by certain types of white blood cells. They play a significant role in asthma, a lung condition in which the smooth muscle of the airways constricts, making it difficult to breathe, and in the severe allergic reaction known as *anaphylaxis.* For this reason, pharmaceutical companies have developed drugs to block leukotriene synthesis or action.
2. **Prostanoids** are molecules produced when the enzyme **cyclooxygenase (COX)** acts on arachidonic acid. Prostanoids include **prostaglandins** and **thromboxanes**. These eicosanoids act on many tissues of the body, including smooth muscle in various organs, platelets, kidney, and bone. In addition, prostaglandins are involved in sleep, inflammation, pain, and fever.

The *nonsteroidal anti-inflammatory drugs* (NSAIDs), such as aspirin and ibuprofen, help prevent inflammation by inhibiting COX enzymes and decreasing prostaglandin synthesis. However,

NSAIDs are not specific and may have serious unwanted side effects, such as bleeding in the stomach. The discovery that COX is made as two isozymes, COX1 and COX2, enabled the design of drugs that target a specific COX isozyme. By inhibiting only COX2, the enzyme that produces inflammatory prostaglandins, physicians hoped to treat inflammation with fewer side effects. However, studies have shown that some patients who take COX2 inhibitors and other NSAIDs have increased risk of heart attacks and strokes, so these drugs are not recommended for long-term use.

Arachidonic acid-derived molecules such as eicosanoids are not the only known lipid signal molecules. *Sphingolipids* also act as extracellular signals to help regulate inflammation, cell adhesion and migration, and cell growth and death. Like the eicosanoids, sphingolipids combine with G protein-coupled receptors in the membranes of their target cells.

Concept Check

14. One drug blocks leukotriene action in its target cells. A different drug blocks leukotriene synthesis. Use what you have learned about leukotrienes, signal molecules, and signal transduction to predict what these drugs are acting to have those effects.

6.4 Modulation of Signal Pathways

As you have just learned, signal pathways in the cell can be very complex. Variations among related families of receptors add to the complexity.

Receptors Exhibit Saturation, Specificity, and Competition

Because receptors are proteins, receptor-ligand binding exhibits the general protein-binding characteristics of specificity, competition, and saturation (discussed in [Chapter 2, p. 46]). Similar protein-binding reactions occur in enzymes [Chapter 4, p. 99] and transporters [Chapter 5, p. 144]. Receptors, like enzymes and transporters, also come as families of related *isoforms* [p. 49].

Specificity and Competition: Multiple Ligands for One Receptor Receptors have binding sites for their ligands, just as enzymes and transporters do. As a result, different ligand molecules with similar structures may be able to bind to the same receptor. A classic example of this principle involves two neurocrine molecules responsible for fight-or-flight responses: the neurotransmitter *norepinephrine* and its cousin the neurohormone *epinephrine* (also called *adrenaline*). Both molecules bind to a class of receptors called *adrenergic receptors.* (*Adrenergic* is the adjective relating to adrenaline.) The ability of adrenergic receptors to bind these two signal molecules, but not others, demonstrates the specificity of the receptors.

Epinephrine and norepinephrine also compete with each other for receptor binding sites. Adrenergic receptors come in two major isoforms designated alpha (α) and beta (β). The α isoform has a higher binding affinity for norepinephrine, and the β_2 isoform has a higher affinity for epinephrine.

Agonists and Antagonists When a ligand combines with a receptor, one of two events follows. Either the ligand activates the receptor and elicits a response, or the ligand occupies the binding site and prevents the receptor from responding (FIG. 6.13). A competing ligand that binds and elicits a response is known as an **agonist** of the primary ligand. Competing ligands that bind and block receptor activity are called **antagonists** of the primary ligand.

Pharmacologists use the principle of competing agonists [p. 48] to design drugs that are longer-acting and more resistant to degradation than the *endogenous* ligand produced by the body {*endo-*, within + *-genous*, developing}. One example is the family of modified estrogens (female sex hormones) in birth control pills. These drugs are agonists of naturally occurring estrogens but have chemical groups added to protect them from breakdown and extend their active life.

One Ligand May Have Multiple Receptors

To complicate matters, different cells may respond differently to a single kind of signal molecule. How can one chemical trigger response A in tissue 1 and response B in tissue 2? For most signal molecules, *the target cell response depends on its receptor or its associated intracellular pathways, not on the ligand.*

For many years physiologists were unable to explain the observation that a single signal molecule could have different effects in different tissues. For example, the neurohormone epinephrine dilates blood vessels in skeletal muscle but constricts blood vessels in the intestine. How can that one chemical have opposite effects? The answer became clear when scientists discovered that epinephrine was binding to different adrenergic receptor isoforms in the two tissues.

FIG. 6.13 Receptor agonists and antagonists

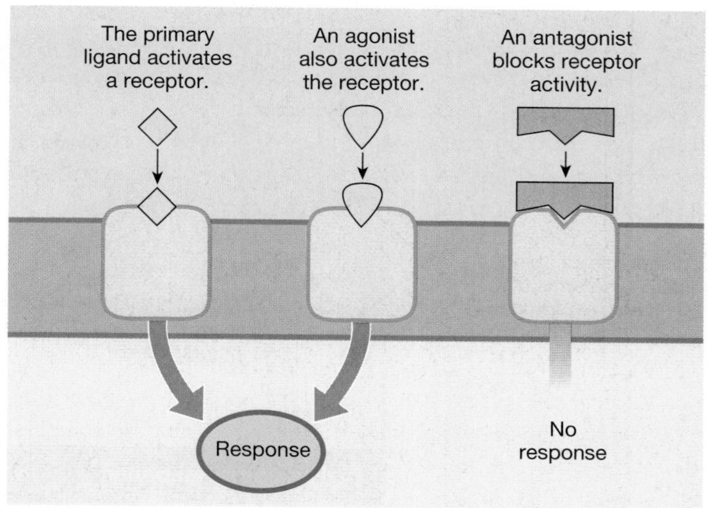
The primary ligand activates a receptor.
An agonist also activates the receptor.
An antagonist blocks receptor activity.
Response
No response

The cellular response that follows activation of a receptor depends on which isoform of the receptor is involved. For example, the α- and β_2-adrenergic receptors for epinephrine are isoforms of each other. When epinephrine binds to α-receptors on intestinal blood vessels, the vessels constrict (**FIG. 6.14**). When epinephrine binds to β_2-receptors on certain skeletal muscle blood vessels, the vessels dilate. The responses of the blood vessels depend on the receptor isoforms and their signal transduction pathways, not on epinephrine. Many drugs now are designed to be specific for only one receptor isoform.

Concept Check

15. What do receptors, enzymes, and transporters have in common that explains why they all exhibit saturation, specificity, and competition?

16. Insulin increases the number of glucose transporters on a skeletal muscle cell but not on the membrane of a liver cell. List two possible mechanisms that could explain how this one hormone can have these two different effects.

Up and Down-Regulation Enable Cells to Modulate Responses

Saturation of proteins refers to the fact that protein activity reaches a maximum rate because cells contain limited numbers of protein molecules [p. 51]. Saturation can be observed with enzymes, transporters, and receptors. A cell's ability to respond to a chemical signal therefore can be limited by the number of receptors for that signal.

A single cell contains between 500 and 100,000 receptors on the surface of its cell membrane, with additional receptors in the cytosol and nucleus. In any given cell, the number of receptors changes over time. Old receptors are withdrawn from the membrane by endocytosis and are broken down in lysosomes. New receptors are inserted into the membrane by exocytosis. Intracellular receptors are also made and broken down. This flexibility permits a cell to vary its responses to chemical signals depending on the extracellular conditions and the internal needs of the cell.

What happens when a signal molecule is present in the body in abnormally high concentrations for a sustained period of time? Initially the increased signal level creates an enhanced response. As this enhanced response continues, the target cells may attempt to bring their response back to normal by either down-regulation or desensitization of the receptors for the signal [p. 51].

Down-regulation is a decrease in receptor number. The cell can physically remove receptors from the membrane through endocytosis [Fig. 5.19, p. 148]. A quicker and more easily reversible way to decrease cell response is *desensitization*, which can be achieved by binding a chemical modulator to the receptor protein. For example, the β-adrenergic receptors described in the previous section can be desensitized by phosphorylation of the receptor.

The result of decreased receptor number or desensitization is a diminished response of the target cell even though the concentration of the signal molecule remains high. Down-regulation and desensitization are one explanation for the development of *drug tolerance*, a condition in which the response to a given dose decreases despite continuous exposure to the drug.

In the opposite situation, when the concentration of a ligand decreases, the target cell may use up-regulation in an attempt to keep its response at a normal level. In **up-regulation**, the target cell inserts more receptors into its membrane. For example, if a neuron is damaged and unable to release normal amounts of neurotransmitter, the target cell may up-regulate its receptors. More receptors make the target cell more responsive to whatever neurotransmitters are present. Up-regulation is also programmed during development as a mechanism that allows cells to vary their responsiveness to growth factors and other signal molecules.

FIG. 6.14 Target response depends on the target receptor

In this example, blood vessels constrict or dilate depending on their receptor type.

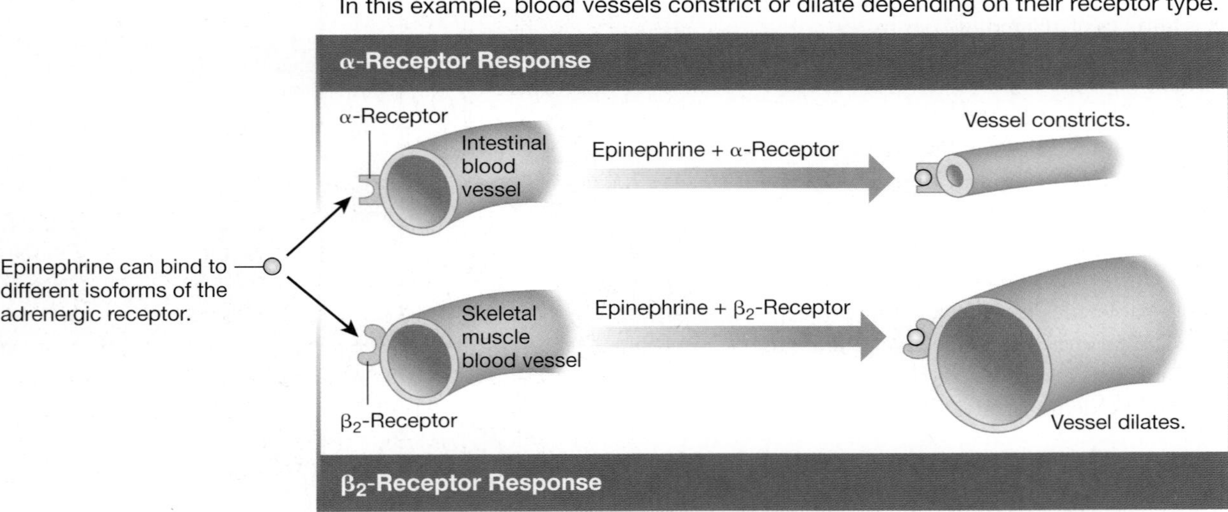
α-Receptor Response
α-Receptor
Intestinal blood vessel
Epinephrine + α-Receptor
Vessel constricts.
Epinephrine can bind to different isoforms of the adrenergic receptor.
Skeletal muscle blood vessel
Epinephrine + β2-Receptor
β2-Receptor
Vessel dilates.
β2-Receptor Response

Concept Check

17. To decrease a receptor's binding affinity, a cell might (select all that apply):
    (a) synthesize a new isoform of the receptor
    (b) withdraw receptors from the membrane
    (c) insert new receptors into the membrane
    (d) use a covalent modulator [*Hint:* p. 49]

Cells Must Be Able to Terminate Signal Pathways

In the body, signals turn on and off, so cells must be able to tell when a signal is over. This requires that signaling processes have built-in termination mechanisms. For example, to stop the response to a calcium signal, a cell removes Ca^{2+} from the cytosol by pumping it either back into the endoplasmic reticulum or out into the extracellular fluid.

Receptor activity can be stopped in a variety of ways. The extracellular ligand can be degraded by enzymes in the extracellular space. An example is the breakdown of the neurotransmitter acetylcholine. Other chemical messengers, particularly neurotransmitters, can be removed from the extracellular fluid through transport into neighboring cells. A widely used class of antidepressant drugs called *selective serotonin reuptake inhibitors* (SSRIs), extends the active life of the neurotransmitter serotonin by slowing its removal from the extracellular fluid.

Once a ligand is bound to its receptor, activity can also be terminated by endocytosis of the receptor-ligand complex [Fig. 5.19, p. 148]. After the vesicle is in the cell, the ligand is removed, and the receptors can be returned to the membrane by exocytosis.

Many Diseases and Drugs Target the Proteins of Signal Transduction

As researchers learn more about cell signaling, they are realizing how many diseases, both inherited and acquired, are linked to problems with signal pathways. Diseases can be caused by alterations in receptors or by problems with G proteins or second messenger pathways (see **TBL. 6.1** for some examples). A single change in the amino acid sequence of a receptor protein can alter the shape of the receptor's binding site, thereby either destroying or modifying its activity.

Pharmacologists are using information about signaling mechanisms to design drugs to treat disease. Some of the alphabet soup of drugs in widespread use are "beta blockers" (β-adrenergic receptor blockers), and calcium-channel blockers for treating high blood pressure; SERMs (selective estrogen receptor modulators) for treating estrogen-dependent cancers; and H2 (histamine type 2) receptor antagonists for decreasing acid secretion in the stomach. You may encounter many of these drugs again if you study the systems in which they are effective.

TABLE 6.1 Some Diseases or Conditions Linked to Abnormal Signaling Mechanisms

Genetically Inherited Abnormal Receptors		
Receptor	**Physiological Alteration**	**Disease or Condition That Results**
Vasopressin receptor (X-linked defect)	Shortens half-life of the receptor	Congenital diabetes insipidus
Calcium sensor in parathyroid gland	Fails to respond to increase in plasma Ca^{2+}	Familial hypercalcemia
Rhodopsin receptor in retina of eye	Improper protein folding	Retinitis pigmentosa
Toxins Affecting Signal Pathways		
Toxin	**Physiological Effect**	**Condition that Results**
Bordetella pertussis toxin	Blocks inhibition of adenylyl cyclase (i.e., keeps it active)	Whooping cough
Cholera toxin	Blocks enzyme activity of G proteins; cell keeps making cAMP	Ions secreted into lumen of intestine, causing massive diarrhea

RUNNING PROBLEM

"My dad takes insulin shots for his diabetes," Marvin says. "What does insulin do?" The nurse practitioner replies that normally insulin helps most cells take up and use glucose. In both types of diabetes, however, fasting blood glucose concentrations are elevated because the cells are not taking up and using glucose normally. If people with type 1 diabetes are given shots of insulin, their blood glucose levels decline. If people with type 2 diabetes are given insulin, blood glucose levels may change very little.

Q3: *In which form of diabetes are the insulin receptors more likely to be up-regulated?*

165 – 168 – 181 – 182 – 186 – 188 – 191

6.5 Homeostatic Reflex Pathways

The cellular signal mechanisms just described are often just one small component of the body's signaling systems that maintain homeostasis. For local control mechanisms, a relatively isolated change occurs in a cell or tissue, and the chemical paracrine or autocrine signals released there are the entire pathway. In more complicated *reflex control pathways* [p. 14], information must be transmitted throughout the body using chemical signals or a combination of electrical and chemical signaling. In the last section of this chapter, we look at some patterns of reflex control pathways you will encounter as you study the different organ systems of the body.

Cannon's Postulates Describe Regulated Variables and Control Systems

Walter Cannon, the father of American physiology, described a number of properties of homeostatic control systems in the 1920s based on his observations of the body in health and disease states.* This was decades before scientists had any idea of how these control systems worked at the cellular and subcellular levels.

Cannon's four postulates are:

1. **The nervous system has a role in preserving the "fitness" of the internal environment.** *Fitness* in this instance means conditions that are compatible with normal function. The nervous system coordinates and integrates blood volume, blood osmolarity, blood pressure, and body temperature, among other regulated variables. (In physiology, a regulated variable is also known as a **parameter** {*para-*, beside + *meter,* measure}).
2. **Some systems of the body are under tonic control {*tonos,* tone}.** To quote Cannon, "An agent may exist which has a moderate activity which can be varied up and down." *Tonic control* is like the volume control on a radio. The radio is always on, but by turning a single knob, you can make the sound level louder or softer. This is one of the more difficult concepts in physiology because we have a tendency to think of responses as being either off or on rather than a response always on that can increase or decrease.

   A physiological example of a tonically controlled system is the minute-to-minute regulation of blood vessel diameter by the nervous system. Increased input from the nervous system decreases vessel diameter, and decreased input from the nervous system increases diameter (**FIG. 6.15a**). In this example, it is the amount of neurotransmitter that determines the vessel's response: more neurotransmitter means a stronger response.
3. **Some systems of the body are under antagonistic control.** Cannon wrote, "When a factor is known which can shift a homeostatic state in one direction, it is reasonable to look for a factor or factors having an opposing effect." Systems that are not under tonic control are often under *antagonistic control,* either by hormones or the nervous system.

   In pathways controlled by the nervous system, neurons from different divisions may have opposing effects. For example, chemical signals from the sympathetic division increase heart rate, but chemical signals from the parasympathetic division decrease it (Fig. 6.15b).

   When chemical signals have opposing effects, they are said to be antagonistic to each other. For example, insulin and glucagon are antagonistic hormones. Insulin decreases the glucose concentration in the blood and glucagon increases it.
4. **One chemical signal can have different effects in different tissues.** Cannon observed correctly that "homeostatic agents antagonistic in one region of the body may be cooperative in another region." However, it was not until scientists learned about cell receptors that the basis for the seemingly contradictory actions of some hormones or nerves became clear. As you learned earlier in this chapter, a single chemical signal can have different effects depending on the receptor and intracellular pathway of the target cell. For example, epinephrine constricts or dilates blood vessels, depending on whether the vessel has α- or β_2-adrenergic receptors (Fig. 6.14).

The remarkable accuracy of Cannon's postulates, now confirmed with cellular and molecular data, is a tribute to the observational skills of scientists in the nineteenth and early twentieth centuries.

* W. B. Cannon. Organization for physiological homeostasis. *Physiological Reviews* 9: 399–443, 1929.

Concept Check

18. What is the difference between tonic control and antagonistic control?
19. How can one chemical signal have opposite effects in two different tissues?

Long-Distance Pathways Maintain Homeostasis

Long-distance reflex pathways are traditionally considered to involve two control systems: the nervous system and the endocrine system. However, cytokines [p. 167] can participate in some long-distance pathways. During stress and systemic inflammatory responses, cytokines work with the nervous and endocrine systems to integrate information from all over the body.

Reflex pathway **response loops** have three major components: *input, integration,* and *output* [p. 14]. These three components can be subdivided into seven more detailed steps, as shown next (**FIG. 6.16**):

Stimulus → sensor or receptor → input signal →
integrating center →
output signal → target → response

RUNNING PROBLEM

"Why is elevated blood glucose bad?" Marvin asks. "The elevated blood glucose itself is not bad after a meal," says the nurse practitioner, "but when it is high after an overnight fast, it suggests that there is something wrong with the way your body is handling its glucose metabolism." When a normal person absorbs a meal containing carbohydrates, blood glucose levels increase and stimulate insulin release. When cells have taken up the glucose from the meal and blood glucose levels fall, secretion of another pancreatic hormone, glucagon, increases. Glucagon increases blood glucose concentrations to keep the level within the homeostatic range.

Q4: *The homeostatic regulation of blood glucose levels by the hormones insulin and glucagon is an example of which of Cannon's postulates?*

165 – 168 – 181 – **182** – 186 – 188 – 191

FIG. 6.15 Tonic and antagonistic control patterns

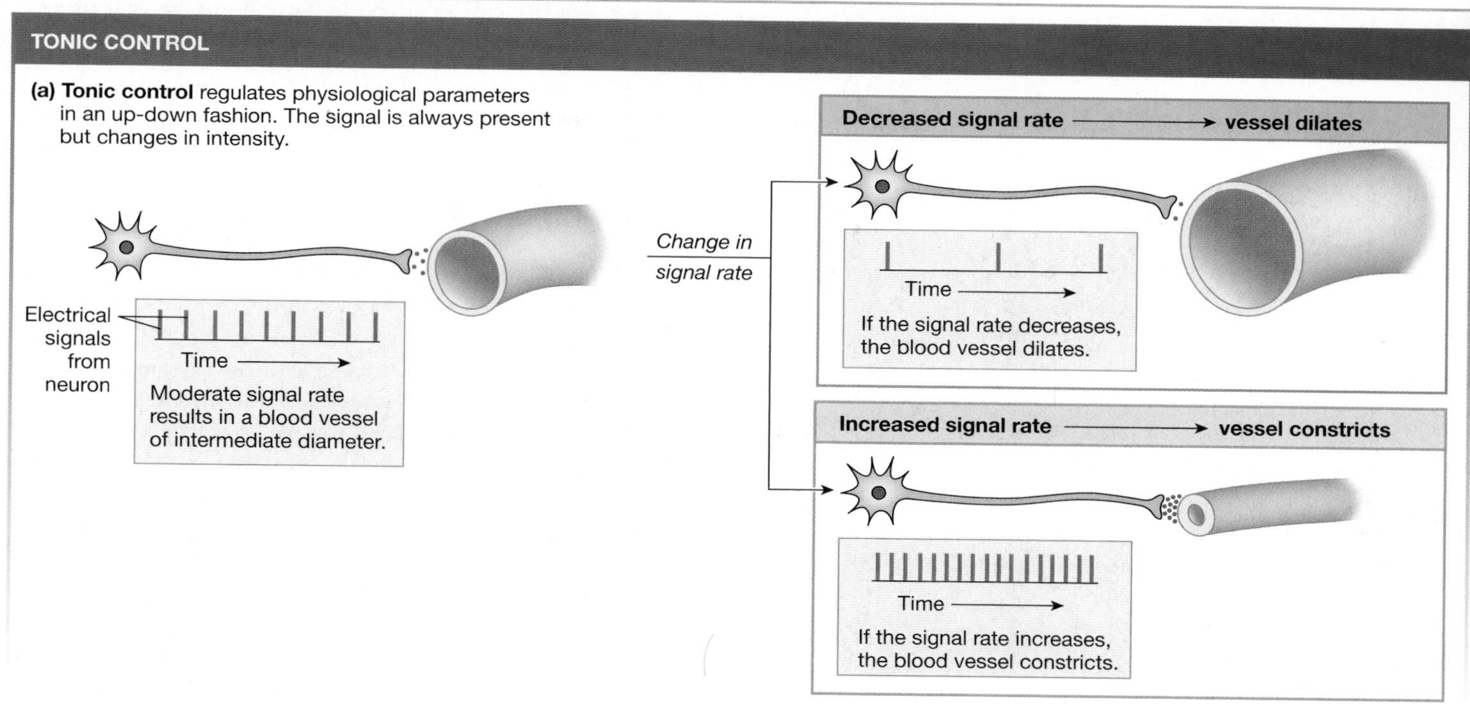
TONIC CONTROL
(a) Tonic control regulates physiological parameters in an up-down fashion. The signal is always present but changes in intensity.
Electrical signals from neuron
Time
Moderate signal rate results in a blood vessel of intermediate diameter.
Change in signal rate
Decreased signal rate → vessel dilates
Time
If the signal rate decreases, the blood vessel dilates.
Increased signal rate → vessel constricts
Time
If the signal rate increases, the blood vessel constricts.

ANTAGONISTIC CONTROL

(b) Antagonistic control uses different signals to send a parameter in opposite directions. In this example, antagonistic neurons control heart rate: some speed it up, while others slow it down.

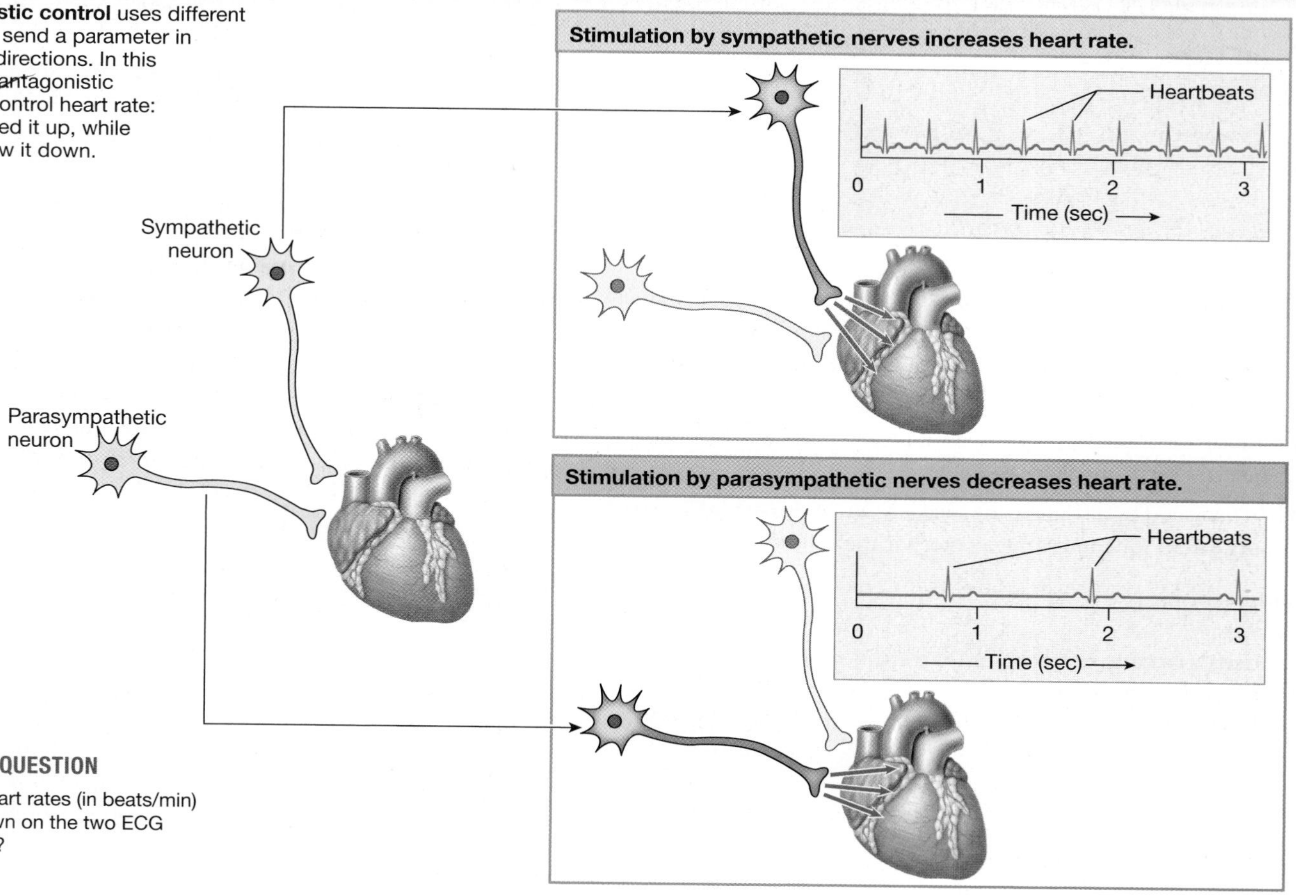

Sympathetic neuron

Parasympathetic neuron

Stimulation by sympathetic nerves increases heart rate.

Heartbeats

0 1 2 3

Time (sec)

Stimulation by parasympathetic nerves decreases heart rate.

Heartbeats

0 1 2 3

Time (sec)

FIGURE QUESTION

What heart rates (in beats/min) are shown on the two ECG tracings?

FIG. 6.16 Steps in a reflex pathway

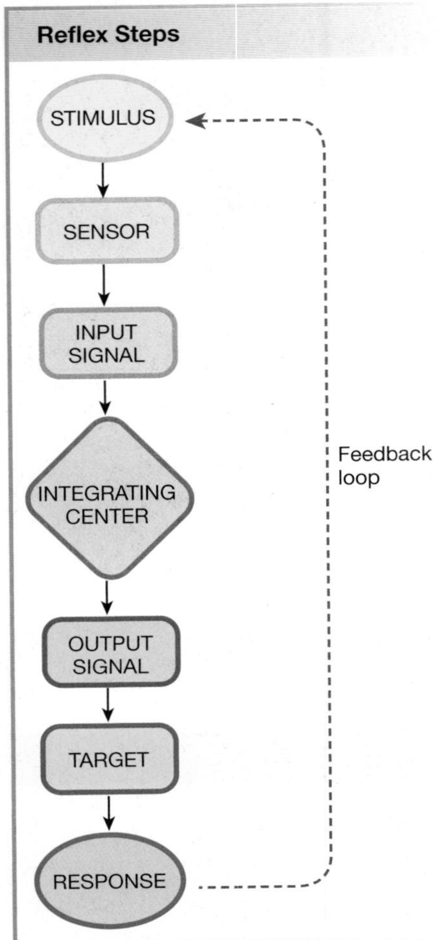
Reflex Steps
STIMULUS
SENSOR
INPUT SIGNAL
INTEGRATING CENTER
OUTPUT SIGNAL
TARGET
RESPONSE
Feedback loop

INPUT:

A **stimulus** is the disturbance or change that sets the pathway in motion. The stimulus may be a change in temperature, oxygen content, blood pressure, or any one of a myriad of other regulated variables.

A **sensor** or sensory receptor continuously monitors its environment for a particular variable.

When activated by a change, the sensor sends an **input (afferent) signal** to the integrating center for the reflex.

INTEGRATION:

The **integrating center** compares the input signal with the **setpoint**, or desired value of the variable. If the variable has moved out of the acceptable range, the integrating center initiates an output signal.

OUTPUT:

The **output (efferent) signal** is an electrical and/or chemical signal that travels to the target.

The **target**, or **effector** {*effectus*, the carrying out of a task} is the cell or tissue that carries out the appropriate **response** to bring the variable back within normal limits.

You will encounter many variations in the number of steps shown. For example, some endocrine reflexes lack a sensor and input signal. Some neural pathways have multiple output signals. Many reflexes have multiple targets and responses.

Now let's look in more detail at each of the reflex steps.

Sensors In the first step in a physiological response loop, a stimulus activates a sensor or receptor. Notice that this is a new and different application of the word *receptor*. Like many other terms in physiology, *receptor* can have different meanings (**FIG. 6.17**). The sensory receptors of a neural reflex are not protein receptors that bind to signal molecules, like those involved in signal transduction. Rather, neural receptors are specialized cells, parts of cells, or complex multicellular receptors (such as the eye) that respond to changes in their environment.

There are many sensory receptors in the body, each located in the best position to monitor the variable it detects. The eyes, ears, and nose are receptors that sense light, sound and motion, and odors, respectively. Your skin is covered with less complex receptors that sense touch, temperature, vibration, and pain. Other sensors are internal: receptors in the joints of the skeleton that send information to the brain about body position, or blood pressure and oxygen receptors in blood vessels that monitor conditions in the circulatory system.

Sensory receptors involved in neural reflexes are divided into central receptors and peripheral receptors. *Central receptors* are located in the brain or are closely linked to the brain. An example is the brain's chemoreceptor for carbon dioxide. *Peripheral receptors* reside elsewhere in the body and include the skin receptors and internal receptors just described.

All sensors have a **threshold**, the minimum stimulus needed to set the reflex response in motion. If a stimulus is below the threshold, no response loop is initiated.

You can demonstrate threshold in a sensory receptor easily by touching the back of your hand with a sharp, pointed object, such as a pin. If you touch the point to your skin lightly enough, you can see the contact between the point and your skin even though you do not feel anything. In this case, the stimulus (pressure from the point of the pin) is below threshold, and the pressure receptors of the skin are not responding. As you press harder, the stimulus reaches threshold, and the receptors respond by sending a signal to the brain, causing you to feel the pin.

Endocrine reflexes that are not associated with the nervous system do not use sensory receptors to initiate their pathways. Instead, endocrine cells act both as sensor and integrating center for the reflex. For example, a pancreatic beta cell sensing and responding directly to changes in blood glucose concentrations is an endocrine cell that is both sensor and integrating center [Fig. 5.26, p. 158].

FIG. 6.17 ESSENTIALS Multiple Meanings of the Word *Receptor*

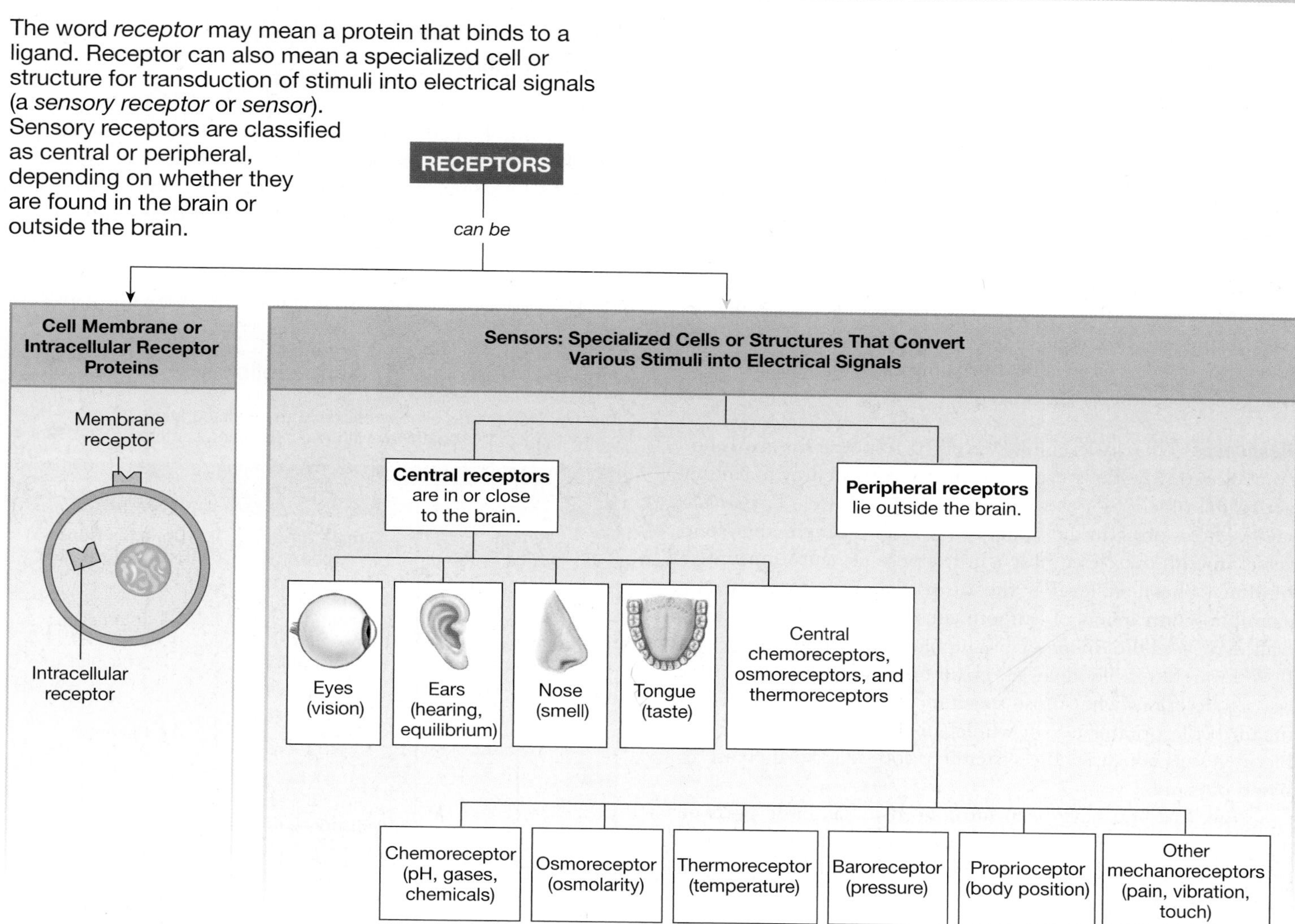

The word *receptor* may mean a protein that binds to a ligand. Receptor can also mean a specialized cell or structure for transduction of stimuli into electrical signals (a *sensory receptor* or *sensor*).
Sensory receptors are classified as central or peripheral, depending on whether they are found in the brain or outside the brain.
RECEPTORS
can be
Cell Membrane or Intracellular Receptor Proteins
Membrane receptor
Intracellular receptor
Sensors: Specialized Cells or Structures That Convert Various Stimuli into Electrical Signals
Central receptors are in or close to the brain.
Peripheral receptors lie outside the brain.
Eyes (vision)
Ears (hearing, equilibrium)
Nose (smell)
Tongue (taste)
Central chemoreceptors, osmoreceptors, and thermoreceptors
Chemoreceptor (pH, gases, chemicals)
Osmoreceptor (osmolarity)
Thermoreceptor (temperature)
Baroreceptor (pressure)
Proprioceptor (body position)
Other mechanoreceptors (pain, vibration, touch)

Input Signal The input signal in a reflex varies depending on the type of reflex. In a neural pathway, such as the pin touch above, the input signal is electrical and chemical information transmitted by a sensory neuron. In an endocrine reflex, there is no input pathway because the stimulus acts directly on the endocrine cell, which serves as both sensor and integrating center.

Integrating Center The integrating center in a reflex pathway is the cell that receives information about the regulated variable and can initiate an appropriate response. In endocrine reflexes, the integrating center is the endocrine cell. In neural reflexes, the integrating center usually lies within the *central nervous system* (CNS), which is composed of the brain and the spinal cord.

If information is coming from a single stimulus, it is a relatively simple task for an integrating center to compare that information with the setpoint and initiate a response if appropriate. Integrating centers really "earn their pay," however, when two or more conflicting signals come in from different sources. The center must evaluate each signal on the basis of its strength and importance and must come up with an appropriate response that integrates information from all contributing receptors. This is similar to the kind of decision-making you must do when on one evening your parents want to take you to dinner, your friends are having a party, there is a television program you want to watch, and you have a major physiology test in three days. It is up to you to rank those items in order of importance and decide what you will do.

Output Signals Output signal pathways are relatively simple. In the nervous system, the output signal is always the electrical and chemical signals transmitted by an efferent neuron. Because all electrical signals traveling through the nervous system are identical, the distinguishing characteristic of the signal is the anatomical pathway of the neuron—the route through which the neuron delivers its signal. For example, the vagus nerve carries neural signals to the heart, and the phrenic nerve carries neural signals to the diaphragm. Output pathways in the nervous system are given the anatomical name of the nerve that carries the signal. For example, we speak of the vagal control of heart rate (*vagal* is the adjective for *vagus*).

In the endocrine system, the anatomical routing of the output signal is always the same—all hormones travel in the blood to their target. Hormonal output pathways are distinguished by the chemical nature of the signal and are therefore named for the hormone that carries the message. For example, the output signal for a reflex integrated through the endocrine pancreas will be either the hormone insulin or the hormone glucagon, depending on the stimulus and the appropriate response.

Targets The targets of reflex control pathways are the cells or tissues that carry out the response. The targets of neural pathways may be any type of muscle, endocrine or exocrine glands, or adipose tissue. Targets of an endocrine pathway are the cells that have the proper receptor for the hormone.

Responses There are multiple levels of response for any reflex control pathway. Let's use the example of a neurotransmitter acting on a blood vessel, as shown in Figure 6.15a. The *cellular response* takes place in the target cell. In this example, the blood vessel smooth muscle contracts in response to neurotransmitter binding. The next level is the *tissue or organ response*. In our example, contraction of smooth muscles in the blood vessel wall decreases the diameter of the blood vessel and decreases flow through this blood vessel. Finally, the more general *systemic response* describes what those specific cellular and tissue events mean to the organism as a whole. In this example, when the blood vessels constrict, the systemic response is an increase in blood pressure.

Now that you have been introduced to the basic parts of a reflex control pathway, we can turn to an analysis of the two primary control systems, the nervous system and the endocrine system.

Concept Check

20. What is the difference between local control and reflex control?

21. Name the seven steps in a reflex control pathway in their correct order.

RUNNING PROBLEM

Marvin is fascinated by the body's ability to keep track of glucose. "How does the pancreas know which hormone to secrete?" he wonders. Special cells in the pancreas called beta cells monitor blood glucose concentrations, and they release insulin when blood glucose increases after a meal. Insulin acts on many tissues of the body so they take up and use glucose.

Q5: *In the insulin reflex pathway, name the stimulus, the sensor, the integrating center, the output signal, the target(s), and the response(s).*

165 - 168 - 181 - 182 - **186** - 188 - 191

Control Systems Vary in Their Speed and Specificity

Physiological reflex control pathways are mediated by the nervous system, the endocrine system, or a combination of the two (**FIG. 6.18**). Reflexes mediated solely by the nervous system or solely by the endocrine system are relatively simple, but some pathways combine neural and endocrine reflexes and can be quite complex. In the most complex pathways, signals pass through three different integrating centers before finally reaching the target tissue. With so much overlap between pathways

FIG. 6.18 Simple and complex reflexes

This figure compares simple reflexes with one integrating center to a complex pathway with two integrating centers.

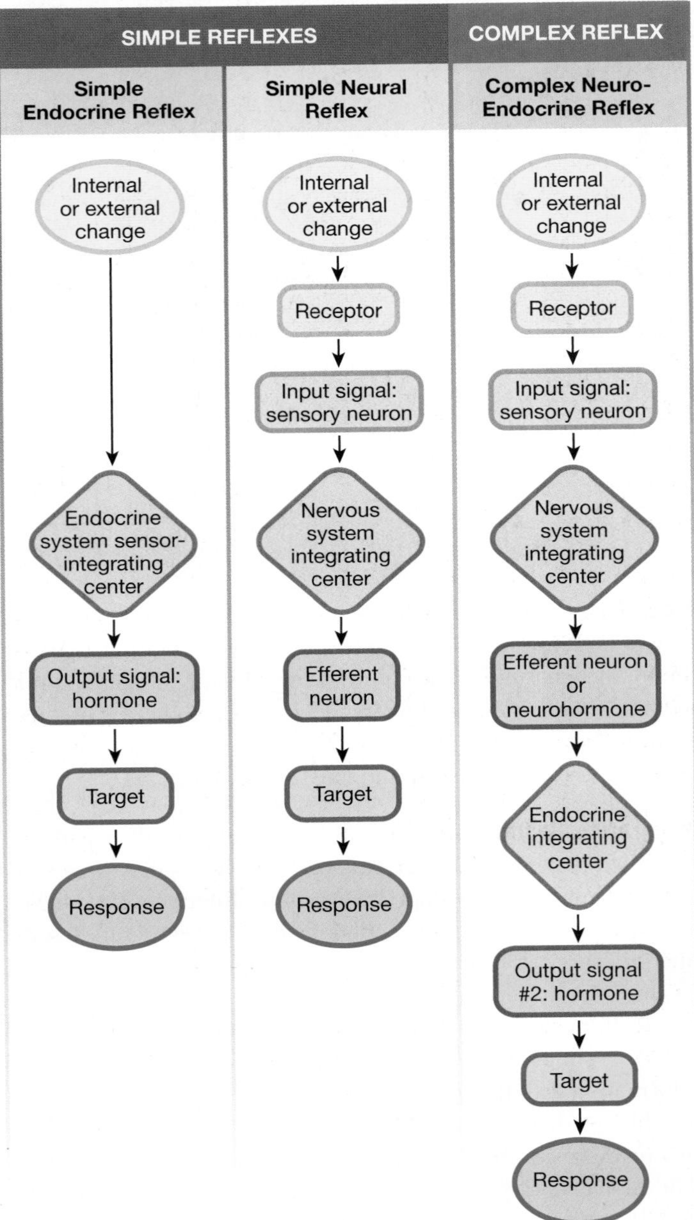
SIMPLE REFLEXES
COMPLEX REFLEX
Simple Endocrine Reflex
Simple Neural Reflex
Complex Neuro-Endocrine Reflex
Internal or external change
Receptor
Input signal: sensory neuron
Endocrine system sensor-integrating center
Nervous system integrating center
Output signal: hormone
Efferent neuron
Efferent neuron or neurohormone
Endocrine integrating center
Output signal #2: hormone
Target
Response

TABLE 6.2 Comparison of Neural and Endocrine Control

Property	Neural Reflex	Endocrine Reflex
Specificity	Each neuron terminates on a single target cell or on a limited number of adjacent target cells.	Most cells of the body are exposed to a hormone. The response depends on which cells have receptors for the hormone.
Nature of the signal	Electrical signal that passes through neuron, then chemical neurotransmitters that carry the signal from cell to cell. In a few cases, signals pass from cell to cell through gap junctions.	Chemical signals secreted in the blood for distribution throughout the body.
Speed	Very rapid.	Distribution of the signal and onset of action are much slower than in neural responses.
Duration of action	Usually very short. Responses of longer duration are mediated by neuromodulators.	Responses usually last longer than neural responses.
Coding for stimulus intensity	Each signal is identical in strength. Stimulus intensity is correlated with increased frequency of signaling.	Stimulus intensity is correlated with amount of hormone secreted.

controlled by the nervous and endocrine systems, it makes sense to consider these systems as parts of a continuum rather than as two discrete systems.

Why does the body need different types of control systems? To answer that question, let us compare endocrine control with neural control. Five major differences are summarized in **TABLE 6.2** and discussed next.

Specificity Neural control is very specific because each neuron has a specific target cell or cells to which it sends its message. Anatomically, we can isolate a neuron and trace it from its origin to where it terminates on its target.

Endocrine control is more general because the chemical messenger is released into the blood and can reach virtually every cell in the body. As you learned in the first half of this chapter, the body's response to a specific hormone depends on which cells have receptors for that hormone and which receptor type they have. Multiple tissues in the body can respond to a hormone simultaneously.

Nature of the Signal The nervous system uses both electrical and chemical signals to send information throughout the body. Electrical signals travel long distances through neurons, releasing chemical signals (neurotransmitters) that diffuse across the small gap between the neuron and its target. In a limited number of instances, electrical signals pass directly from cell to cell through gap junctions.

The endocrine system uses only chemical signals: hormones secreted into the blood by endocrine glands or cells. Neuroendocrine pathways represent a hybrid of the neural and endocrine reflexes. In a neuroendocrine pathway, a neuron creates an electrical signal, but the chemical released by the neuron is a neurohormone that goes into the blood for general distribution.

Concept Check

22. In the simple neural reflex shown in Figure 6.18, which box or boxes represent the brain and spinal cord? (b) Which box or boxes represent the central and peripheral sense organs? (c) In the simple neural reflex, add a dashed line connecting boxes to show how a negative feedback loop would shut off the reflex [p. 15].

Speed Neural reflexes are much faster than endocrine reflexes. The electrical signals of the nervous system cover great distances very rapidly, with speeds of up to 120 m/sec. Neurotransmitters also create very rapid responses, on the order of milliseconds.

Hormones are much slower than neural reflexes. Their distribution through the circulatory system and diffusion from capillary to receptors take considerably longer than signals through neurons. In addition, hormones have a slower onset of action. In target tissues, the response may take minutes to hours before it can be measured.

Why do we need the speedy reflexes of the nervous system? Consider this example. A mouse ventures out of his hole and sees a cat ready to pounce on him and eat him. A signal must go from the mouse's eyes and brain down to his feet, telling him to run back into the hole. If his brain and feet were only 5 micrometers (5 μm = 1/200 millimeter) apart, it would take a chemical signal 20 milliseconds (msec) to diffuse across the space and the mouse could escape. If the brain and feet were 50 μm (1/20 millimeter) apart, diffusion would take 2 seconds and the mouse might get caught. But because the head and feet of a mouse are *centimeters* apart, it would take a chemical signal *three weeks* to diffuse from the mouse's head to his feet. Poor mouse!

Even if the distribution of the chemical signal were accelerated by help from the circulatory system, the chemical message would still take 10 seconds to get to the feet, and the mouse would become cat food. The moral of this tale is that reflexes requiring a speedy response are mediated by the nervous system because they are so much more rapid.

Duration of Action Neural control is of shorter duration than endocrine control. The neurotransmitter released by a neuron combines with a receptor on the target cell and initiates a response. The response is usually very brief, however, because the neurotransmitter is rapidly removed from the vicinity of the receptor by various mechanisms. To get a sustained response, multiple repeating signals must be sent through the neuron.

Endocrine reflexes are slower to start, but they last longer. Most of the ongoing, long-term functions of the body, such as metabolism and reproduction, fall under the control of the endocrine system.

Coding for Stimulus Intensity As a stimulus increases in intensity, control systems must have a mechanism for conveying this information to the integrating center. The signal strength from any one neuron is constant in magnitude and therefore cannot reflect stimulus intensity. Instead, the frequency of signaling through the afferent neuron increases. In the endocrine system, stimulus intensity is reflected by the amount of hormone released: the stronger the stimulus, the more hormone is released.

Complex Reflex Control Pathways Have Several Integrating Centers

FIGURE 6.19 summarizes variations in the neural, neuroendocrine, and endocrine reflex control pathways.

In a simple neural reflex, all the steps of a reflex pathway are present, from sensor to target (Fig. 6.1 **1**). The neural reflex is represented in its simplest form by the knee jerk (or patellar tendon) reflex. A blow to the knee (the stimulus) activates a stretch receptor. A signal travels through an afferent sensory neuron to the spinal cord (the integrating center). If the blow is strong enough (exceeds threshold), a signal travels from the spinal cord through an efferent neuron to the muscles of the thigh (the target or effector). In response, the muscles contract, causing the lower leg to kick outward (the knee jerk).

RUNNING PROBLEM

"OK, just one more question," says Marvin. "You said that people with diabetes have high blood glucose levels. If glucose is so high, why can't it just leak into the cells?"

Q6: *Why can't glucose simply leak into cells when the blood glucose concentration is higher than the intracellular glucose concentration?*

Q7: *What do you think happens to insulin secretion when blood glucose levels fall? What kind of feedback loop is operating?*

165 – 168 – 181 – 182 – 186 – **188** – 191

In a simple endocrine reflex pathway (Fig. 6.19 **6**), some of the steps of the reflex pathway are combined. The endocrine cell acts as both sensor and integrating center so there is no input pathway. The endocrine cell itself monitors the regulated variable and is programmed to initiate a response when the variable goes out of an acceptable range. The output pathway is the hormone, and the target is any cell having the appropriate hormone receptor.

An example of a simple endocrine reflex is secretion of the hormone insulin in response to changes in blood glucose level. The pancreatic beta cells that secrete insulin monitor blood glucose concentrations by using ATP production in the cell as an indicator of glucose availability [Fig. 5.26, p. 158]. When blood glucose increases, intracellular ATP production exceeds the threshold level, and the beta cells respond by secreting insulin into the blood. Any target cell in the body that has insulin receptors responds to the hormone and initiates processes that take glucose out of the blood. The removal of the stimulus acts in a negative feedback manner: the response loop shuts off when blood glucose levels fall below a certain concentration.

Concept Check

23. Match the following terms for parts of the knee jerk reflex to the parts of the simple neural reflex shown in Figure 6.19 **1**: blow to knee, leg muscles, neuron to leg muscles, sensory neuron, brain and spinal cord, stretch receptor, muscle contraction.

The neuroendocrine reflex, shown in Figure 6.19 **2**, is identical to the neural reflex except that the neurohormone released by the neuron travels in the blood to its target, just like a hormone. A simple neuroendocrine reflex is the release of breast milk in response to a baby's suckling. The baby's mouth on the nipple stimulates sensory signals that travel through sensory neurons to the brain (integrating center). An electrical signal in the efferent neuron triggers the release of the neurohormone oxytocin from the brain into the circulation. Oxytocin is carried to the breast, where it causes contraction of smooth muscles in the breast (the target), resulting in the ejection of milk.

In complex pathways, there may be more than one integrating center. Figure 6.19 shows three examples of complex neuroendocrine pathways. The simplest of these, Figure 6.19 **3**, combines a neural reflex with a classic endocrine reflex. An example of this pattern can be found in the control of insulin release.

The pancreatic beta cells monitor blood glucose concentrations directly (Fig. 6.19 **4**), but they are also controlled by the nervous system. During a meal, the presence of food in the stomach stretches the wall of the digestive tract and sends input signals to the brain. The brain in turn sends excitatory output signals to the beta cells, telling them to release insulin. These signals take place even before the food has been absorbed and blood glucose levels have gone up (a *feedforward reflex* [p. 17]). This pathway therefore has two integrating centers (the brain and the beta cells).

There are a number of complex reflex pathways, not all of which are shown in Figure 6.19. One (Fig. 6.19 **4**) uses a

FIG. 6.19 ESSENTIALS Reflex Pathway Patterns

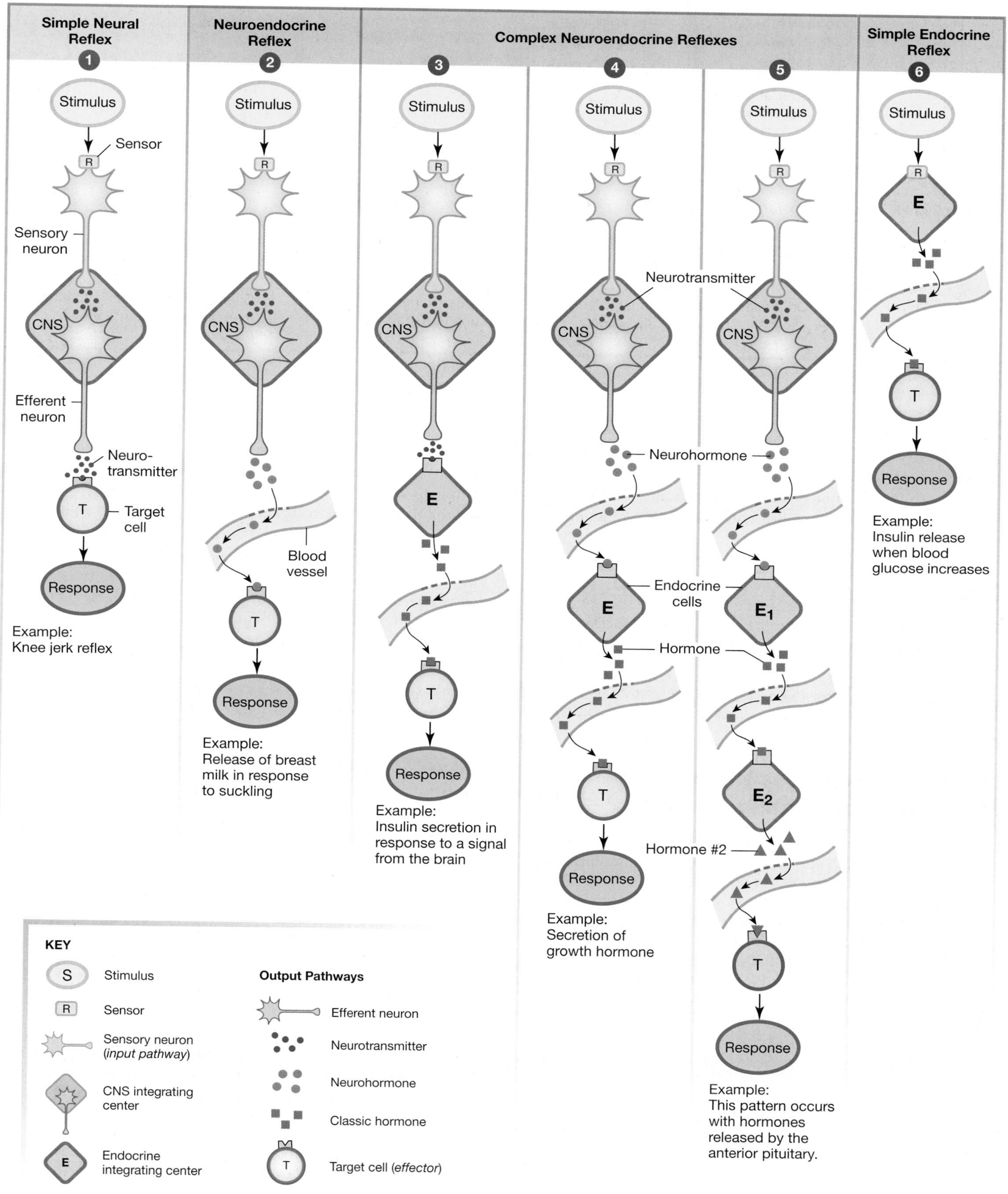
Simple Neural Reflex
1
Stimulus
Sensor
R
Sensory neuron
CNS
Efferent neuron
Neuro-transmitter
T
Target cell
Response
Example:
Knee jerk reflex
Neuroendocrine Reflex
2
Stimulus
R
CNS
Blood vessel
T
Response
Example:
Release of breast milk in response to suckling
Complex Neuroendocrine Reflexes
3
Stimulus
R
CNS
E
T
Response
Example:
Insulin secretion in response to a signal from the brain
4
Stimulus
R
Neurotransmitter
CNS
Neurohormone
Endocrine cells
E
Hormone
T
Response
Example:
Secretion of growth hormone
5
Stimulus
R
CNS
E1
E2
Hormone #2
T
Response
Example:
This pattern occurs with hormones released by the anterior pituitary.
Simple Endocrine Reflex
6
Stimulus
R
E
T
Response
Example:
Insulin release when blood glucose increases
KEY
S
Stimulus
R
Sensor
Sensory neuron (input pathway)
CNS integrating center
E
Endocrine integrating center
Output Pathways
Efferent neuron
Neurotransmitter
Neurohormone
Classic hormone
T
Target cell (effector)

TABLE 6.3 Comparison of Neural, Neuroendocrine, and Endocrine Reflexes

	Neural	Neuroendocrine	Endocrine
Sensor	Special and somatic sensory receptors	Special and somatic sensory receptors	Endocrine cell
Input Signal	Sensory neuron	Sensory neuron	None
Integrating Center	Brain or spinal cord	Brain or spinal cord	Endocrine cell
Output Signal	Efferent neuron (electrical signal and neurotransmitter)	Efferent neuron (electrical signal and neurohormone)	Hormone
Target(s)	Muscles and glands, some adipose tissue	Most cells of the body	Most cells of the body
Response	Contraction and secretion primarily; may have some metabolic effects	Change in enzymatic reactions, membrane transport, or cell proteins	Change in enzymatic reactions, membrane transport, or cell proteins

neurohormone to control the release of a classic hormone. The secretion of growth hormone is an example of this pathway. The most complex neuroendocrine pathways, shown as Figure 6.19 5, include a neurohormone and two classic hormones. This pattern is typical of some hormones released by the anterior pituitary, an endocrine gland located just below the brain (see Chapter 7 for details).

Concept Check

24. Match the following terms with the appropriate parts of the simple neuroendocrine reflex in Fig. 6.19 (terms may be used more than once): food in stomach following a meal, brain and spinal cord, endocrine cells of pancreas, stretch receptors, efferent neuron to pancreas, insulin, adipose cell, blood, sensory neuron.

In describing complex neuroendocrine reflex pathways, we identify only one receptor and input pathway, as indicated in Figure 6.19 5. In the three complex pathways shown, the brain is the first integrating center and the neurohormone is the first output pathway. In Figure 6.19 5, the endocrine target (E_1) of the neurohormone is the second integrating center, and its hormone is the second output pathway. The second endocrine gland in the pathway (E_2) is the third integrating center, and its hormone is the third output pathway. The target of the last signal in the sequence is the effector.

TABLE 6.3 compares the various steps in neural, neuroendocrine, and endocrine reflexes. In the remainder of the text, we use the general patterns shown in Figure 6.19 as a tool for classifying complex reflex pathways. Endocrine and neural pathways play key roles in the maintenance of homeostasis.

RUNNING PROBLEM **CONCLUSION** Diabetes Mellitus

Marvin underwent further tests and was diagnosed with type 2 diabetes. With careful attention to his diet and with a regular exercise program, he has been able to keep his blood glucose levels under control. Diabetes is a growing epidemic in the United States, with more than 29 million diabetics in the United States in 2016 (about 9% of the population). Even scarier is the estimate that another 86 million people are considered "prediabetic"—at significant risk of becoming diabetic. You will learn more about diabetes as you work through the chapters in this book.

To learn more about diabetes now, see the American Diabetes Association website (*www.diabetes.org*) or the Centers for Disease Control and Prevention (*www.cdc.gov/diabetes*).

In this running problem, you learned about glucose homeostasis and how it is maintained by insulin and glucagon. The disease diabetes mellitus is an indication that glucose homeostasis has been disrupted. Check your understanding of this running problem by comparing your answers to the information in the summary table.

Question	Facts	Integration and Analysis
Q1: *In which type of diabetes is the signal pathway for insulin more likely to be defective?*	Insulin is a peptide hormone that uses membrane receptors linked to second messengers to transmit its signal to cells. People with type 1 diabetes lack insulin; people with type 2 diabetes have normal-to-elevated insulin levels.	Normal or high insulin levels suggest that the problem is not with amount of insulin but with the action of the insulin at the cell. The problem in type 2 diabetes could be a defective signal transduction mechanism.

– Continued next page

RUNNING PROBLEM CONCLUSION Continued

Question	Facts	Integration and Analysis
Q2: *Insulin is a protein hormone. Would you expect to find its receptor on the cell surface or in the cytoplasm of the target cells?*	Lipophilic signal molecules have intracellular receptors. Lipophobic molecules have cell membrane receptors.	Proteins are lipophobic so protein hormones like insulin have cell surface receptors.
Q3: *In which form of diabetes are the insulin receptors more likely to be up-regulated?*	Up-regulation of receptors usually occurs if a signal molecule is present in unusually low concentrations [p. 51]. In type 1 diabetes, insulin is not secreted by the pancreas.	In type 1 diabetes, insulin levels are low. Therefore, type 1 is more likely to cause up-regulation of the insulin receptors.
Q4: *The homeostatic regulation of blood glucose levels by the hormones insulin and glucagon is an example of which of Cannon's postulates?*	Cannon's postulates describe the role of the nervous system in maintaining homeostasis, and the concepts of tonic activity, antagonistic control, and different effects of signals in different tissues.	Insulin decreases blood glucose levels, and glucagon increases them. Therefore, the two hormones are an example of an antagonistic control.
Q5: *In the insulin pathway, name the stimulus, the sensor, the integrating center, the output signal, the target(s), and the response(s).*	See the steps of reflex pathways [p. 186].	Stimulus: increase in blood glucose levels; sensor: beta cells of the pancreas that sense the change; integrating center: beta cells; output signal: insulin; targets: any tissues of the body that respond to insulin; responses: cellular uptake and use of glucose.
Q6: *Why can't glucose simply leak into cells when the blood glucose concentration is higher than the intracellular glucose concentration?*	Glucose is lipophobic. Simple diffusion goes across the phospholipid bilayer. Facilitated diffusion uses protein carriers [p. 141].	Because glucose is lipophobic, it cannot cross the membrane by simple diffusion. It must go by facilitated diffusion. If a cell lacks the necessary carriers, facilitated diffusion cannot take place.
Q7: *What do you think happens to the rate of insulin secretion when blood glucose levels fall? What kind of feedback loop is operating?*	The stimulus for insulin release is an increase in blood glucose levels. In negative feedback, the response offsets the stimulus. In positive feedback, the response enhances the stimulus.	An increase in blood glucose concentration stimulates insulin release; therefore, a decrease in blood glucose should decrease insulin release. In this example, the response (lower blood glucose) offsets the stimulus (increased blood glucose), so a negative feedback loop is operating.

165 – 168 – 181 – 182 – 186 – 188 – 191

CHAPTER SUMMARY

Two major themes in physiology stand out in this chapter: *the control of homeostasis* and *communication*. The sensors, integrating centers, and targets of physiological control systems are described in the context of reflex control pathways, which vary from simple to complex. Functional control systems require efficient communication that uses various combinations of chemical and electrical signals. Those signals that cannot enter the cell must use membrane receptor proteins and signal transduction to transfer their information into the cell. The interaction of signal molecules with protein receptors illustrates another fundamental theme of physiology, *molecular interactions*.

6.1 Cell-to-Cell Communication

1. There are two basic types of physiological signals: chemical and electrical. Chemical signals are the basis for most communication within the body. (p. 165)
2. There are four methods of cell-to-cell communication: (1) direct cytoplasmic transfer through gap junctions, (2) contact-dependent signaling, (3) local chemical communication, and (4) long-distance communication. (p. 165; Fig. 6.1)
3. **Gap junctions** are protein channels that connect two adjacent cells. When they are open, chemical and electrical signals pass directly from one cell to the next. (p. 165)
4. **Contact-dependent signals** require direct contact between surface molecules of two cells. (p. 165)
5. Local communication uses **paracrine signals**, chemicals that act on cells close to the cell that secreted the paracrine. A chemical that acts on the cell that secreted it is called an **autocrine signal**. The activity of paracrine and autocrine signal molecules is limited by diffusion distance. (p. 167)
6. Long-distance communication uses **neurocrine molecules** and electrical signals in the nervous system, and **hormones** in the endocrine system. Only cells that possess receptors for a hormone will be **target cells**. (p. 167)
7. **Cytokines** are regulatory peptides that control cell development, differentiation, and the immune response. They serve as both local and long-distance signals. (p. 167)

6.2 Signal Pathways

8. **Chemical signals bind to receptors** and change intracellular signal molecules that direct the response. (p. 168)
9. Lipophilic signal molecules enter the cell and combine with cytoplasmic or nuclear receptors. Lipophobic signal molecules and some lipophilic molecules combine with membrane receptors. (p. 168; Fig. 6.3)
10. **Signal transduction** pathways use membrane receptor proteins and intracellular second messenger molecules to translate signal information into an intracellular response. (p. 170; Fig. 6.5a)
11. Some signal transduction pathways activate **protein kinases**. Others activate **amplifier enzymes** that create **second messenger** molecules. (p. 170; Fig. 6.5b)
12. Signal pathways create intracellular **cascades** that amplify the original signal. (p. 170; Fig. 6.6a)
13. **Ligand-gated ion channels** open or close to create electrical signals. (p. 171; Fig. 6.7)
14. **G proteins** linked to amplifier enzymes are the most prevalent signal transduction system. **G protein-coupled receptors** also alter ion channels. (p. 173; Fig. 6.8)
15. The **G protein-coupled adenylyl cyclase-cAMP-protein kinase A** pathway is the most common pathway for protein and peptide hormones. (p. 173; Fig. 6.8a)
16. In the **G protein-coupled phospholipase C** pathway, the amplifier enzyme **phospholipase C (PLC)** creates two second messengers: **inositol trisphosphate (IP_3)** and **diacylglycerol (DAG)**. IP_3 causes Ca^{2+} release from intracellular stores. Diacylglycerol activates **protein kinase C (PKC)**. (p. 173; Fig. 6.8b)
17. **Receptor-enzymes** activate protein kinases, such as **tyrosine kinase** (Fig. 6.9), or the amplifier enzyme **guanylyl cyclase**, which produces the second messenger **cGMP**. (p. 175)
18. **Integrin** receptors link the extracellular matrix to the cytoskeleton. (p. 175; Fig. 6.3c)

6.3 Novel Signal Molecules

19. Calcium is an important signal molecule that binds to **calmodulin** to alter enzyme activity. It also binds to other cell proteins to alter movement and initiate exocytosis. (p. 176; Fig. 6.11)
20. **Nitric oxide (NO)**, **carbon monoxide (CO)**, and **hydrogen sulfide (H_2S)** are short-lived gaseous signal molecules. NO activates guanylyl cyclase directly. (p. 177)
21. The arachidonic acid cascade creates lipid signal molecules, such as **leukotrienes**, **prostaglandins,** and **thromboxanes**. (p. 178; Fig. 6.12)

6.4 Modulation of Signal Pathways

22. The response of a cell to a signal molecule is determined by the cell's receptor for the signal. (p. 179)
23. Receptors come in related forms called **isoforms**. One ligand may have different effects when binding to different isoforms. (p. 179; Fig. 6.13)
24. A receptor may have multiple ligands. Receptor **agonists** mimic the action of a signal molecule. Receptor **antagonists** block the signal pathway. (p. 179; Fig. 6.14)
25. Receptor proteins exhibit specificity, competition, and saturation. (p. 179)
26. Cells exposed to abnormally high concentrations of a signal for a sustained period of time attempt to bring their response back to normal through down-regulation or by desensitization. In **down-regulation**, the cell decreases the number of receptors. In *desensitization,* the cell decreases the binding affinity of the receptors. **Up-regulation** is the opposite of down-regulation and involves increasing the number of receptors for a signal. (p. 180)
27. Cells have mechanisms for terminating signal pathways, such as removing the signal molecule or breaking down the receptor-ligand complex. (p. 181)
28. Many diseases have been linked to defects in various aspects of signal pathways, such as missing or defective receptors. (p. 181; Tbl. 6.1)

6.5 Homeostatic Reflex Pathways

29. Walter Cannon first stated four basic postulates of homeostasis: (1) The nervous system plays an important role in maintaining homeostasis. (2) Some parameters are under **tonic control**, which allows the parameter to be increased or decreased by a single signal (Fig. 6.15a). (3) Other parameters are under **antagonistic control**, in which one hormone or neuron increases the parameter while another decreases it (Fig. 6.15b). (4) Chemical signals can have different effects in different tissues of the body, depending on the type of receptor present at the target cell. (p. 182; Fig. 6.14)
30. In **reflex control pathways**, an integrating center makes the decision to respond to a change. A chemical or electrical signal to the target cell or tissue then initiates the response. Long-distance reflex pathways involve the nervous and endocrine systems and cytokines. (p. 182)
31. Neural control is faster and more specific than endocrine control but is usually of shorter duration. Endocrine control is less specific and slower to start but is longer lasting and is usually amplified. (p. 184; Tbl. 6.2)
32. Many reflex pathways are complex combinations of neural and endocrine control mechanisms. (p. 186; Figs. 6.18, 6.19)

REVIEW QUESTIONS

In addition to working through these questions and checking your answers on p. A-7, review the Learning Outcomes at the beginning of this chapter.

Level One Reviewing Facts and Terms

1. What are the two routes for long-distance signal delivery in the body?
2. Which two body systems maintain homeostasis by monitoring and responding to changes in the environment?
3. What two types of physiological signals does the body use to send messages? Of these two types, which is available to all cells?
4. In a signal transduction pathway, the signal ligand, also called the first messenger, binds to a(n) ______, which activates and changes intracellular ______.
5. The three main amplifier enzymes are (a) ______, which forms cAMP; (b) ______, which forms cGMP; and (c) ______, which converts a phospholipid from the cell's membrane into two different second messenger molecules.

6. An enzyme known as protein kinase adds the functional group _____ to its substrate, by transferring it from a(n) _____ molecule.
7. Distinguish between central and peripheral receptors.
8. Receptors for signal pathways may be found in the _____, _____, or _____ of the cell.
9. Down-regulation results in a(n) _____ (increased or decreased?) number of receptors in response to a prolonged signal.
10. List two ways a cell may decrease its response to a signal.
11. In a negative feedback loop, the response moves the system in the _____ (same/opposite) direction as the stimulus moves it.

Level Two Reviewing Concepts

12. Explain the relationships of the terms in each of the following sets. Give a physiological example or location if applicable.
    a. gap junctions, connexins, connexon
    b. autocrine signal, paracrine signal
    c. cytokine, neurotransmitter, neurohormone, neuromodulator, hormone
    d. Receptor agonist, receptor antagonist, antagonistic control pathways
    e. transduction, amplification, cascade
13. List and compare the four classes of membrane receptors for signal pathways. Give an example of each.
14. Who was Walter Cannon? Restate his four postulates in your own words.
15. Arrange the following terms in the order of a reflex and give an anatomical example of each step when applicable: input signal, integrating center, output signal, response, sensor, stimulus, target.
16. Compare and contrast the advantages and disadvantages of neural versus endocrine control mechanisms.
17. Would the following reflexes have positive or negative feedback?
    a. glucagon secretion in response to declining blood glucose
    b. increasing milk release and secretion in response to baby's suckling
    c. urgency in emptying one's urinary bladder
    d. sweating in response to rising body temperature
18. Identify the target tissue or organ for each example in question 17.
19. Now identify the integrating center for examples (a), (c), and (d) in question 17.

Level Three Problem Solving

20. In each of the following situations, identify the components of the reflex.
    a. You are sitting quietly at your desk, studying, when you become aware of the bitterly cold winds blowing outside at 30 mph, and you begin to feel a little chilly. You start to turn up the thermostat, remember last month's heating bill, and reach for an afghan to pull around you instead. Pretty soon you are toasty warm again.
    b. While you are strolling through the shopping district, the aroma of cinnamon sticky buns reaches you. You inhale appreciatively but remind yourself that you're not hungry because you had lunch just an hour ago. You go about your business, but 20 minutes later you're back at the bakery, sticky bun in hand, ravenously devouring its sweetness, saliva moistening your mouth.
21. A researcher is studying the smooth muscle of the respiratory system airways. When she exposes the airways to the neurotransmitter acetylcholine, the smooth muscle contracts. When she exposes the airways to the neurohormone epinephrine, the airways relax.
    a. The phenomenon just described is an example of _____ control.
    b. What distinguishes a neurotransmitter from a neurohormone?
    c. Which chemical messenger is secreted in higher concentrations: acetylcholine or epinephrine? Defend your answer.

Level Four Quantitative Problems

22. In a signal cascade for rhodopsin, a photoreceptor molecule, one rhodopsin activates 1,000 molecules of transducin, the next molecule in the signal cascade. Each transducin activates one phosphodiesterase, and each phosphodiesterase converts 4,000 cGMP to GMP.
    a. What is the name of the phenomenon described in this paragraph?
    b. Activation of one rhodopsin will result in the production of how many GMP molecules?

Answers to Concept Checks, Figure and Graph Questions, and end-of-chapter Review Questions can be found in Appendix A [p. A-1].

7 Introduction to the Endocrine System

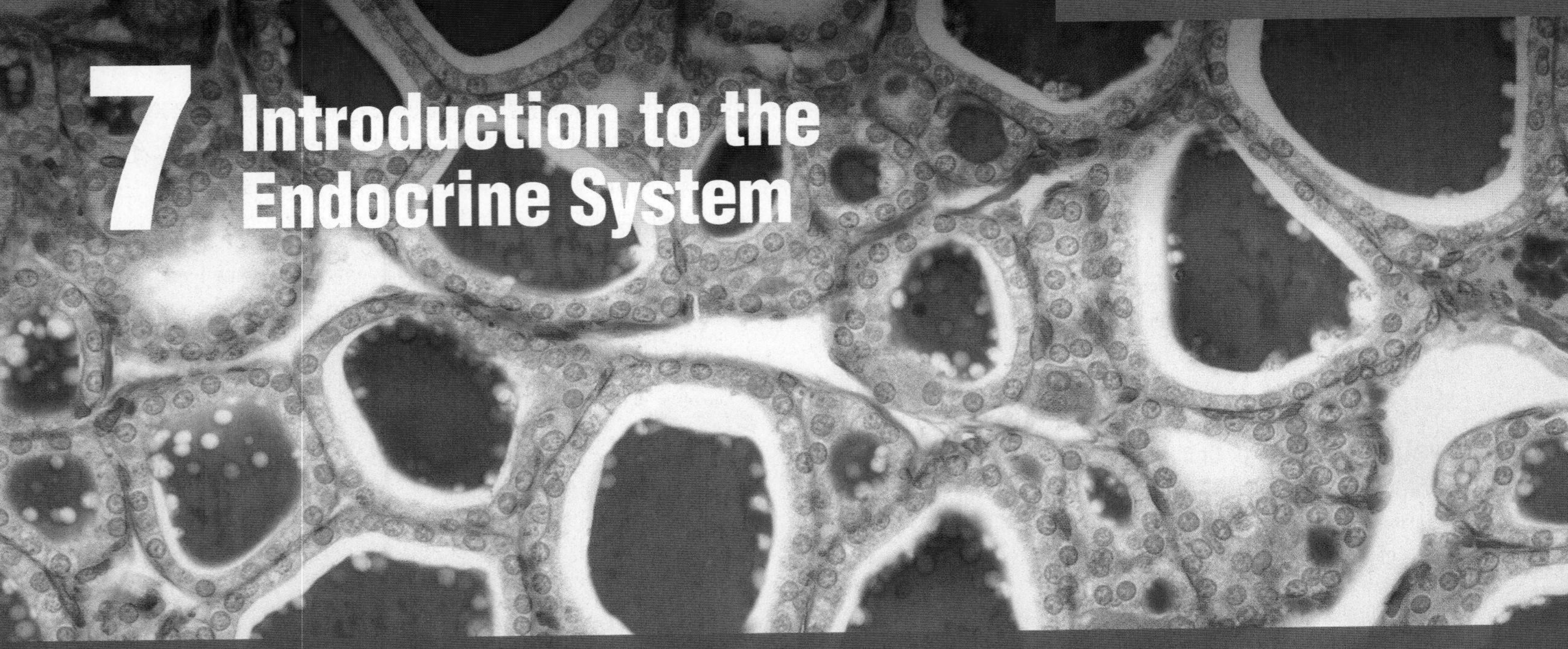

Colloid (red) inside thyroid follicles

The separation of the endocrine system into isolated subsystems must be recognized as an artificial one, convenient from a pedagogical point of view but not accurately reflecting the interrelated nature of all these systems.

Howard Rasmussen, in Williams' Textbook of Endocrinology, *1974*

BACKGROUND BASICS

168 Receptors
32 Peptides and proteins
186 Comparison of endocrine and nervous systems
170 Signal transduction
30 Steroids
46 Specificity

David was seven years old when the symptoms first appeared. His appetite at meals increased, and he always seemed to be hungry. Despite eating more, however, he was losing weight. When he started asking for water instead of soft drinks, David's mother became concerned, and when he wet the bed three nights in a row, she knew something was wrong. The doctor listened to David's symptoms and ordered tests to determine the glucose concentrations of David's blood and urine. The test results confirmed the diagnosis: David had diabetes mellitus. In David's case, the disease was due to lack of insulin, a hormone produced by the pancreas. David was placed on insulin injections, a treatment he would continue for the rest of his life.

One hundred years ago, David would have died not long after the onset of symptoms. The field of **endocrinology**, the study of hormones, was then in its infancy. Most hormones had not been discovered, and the functions of known hormones were not well understood. There was no treatment for diabetes, no birth control pill for contraception. Babies born with inadequate secretion of thyroid hormone did not grow or develop normally.

Today, all that has changed. We have identified a long and growing list of hormones. The endocrine diseases that once killed or maimed can now be controlled by synthetic hormones and sophisticated medical procedures. Although physicians do not hesitate to use these treatments, we are still learning exactly how hormones act on their target cells. This chapter provides an introduction to the basic principles of hormone structure and function. You will learn more about individual hormones as you encounter them in your study of the various systems.

7.1 Hormones

As you have learned, hormones are chemical messengers secreted into the blood by specialized epithelial cells. Hormones are responsible for many functions that we think of as long-term, ongoing functions of the body. Processes that fall mostly under hormonal control include metabolism, regulation of the internal environment (temperature, water balance, ions), and reproduction, growth, and development. Hormones act on their target cells in one of three basic ways: (1) by controlling the rates of enzymatic reactions, (2) by controlling the transport of ions or molecules across cell membranes, or (3) by controlling gene expression and the synthesis of proteins.

RUNNING PROBLEM **Graves' Disease**

The ball slid by the hole and trickled off the green: another bogey. Ben Crenshaw's golf game was falling apart. The 33-year-old professional had won the Masters Tournament only a year ago, but now something was not right. He was tired and weak, had been losing weight, and felt hot all the time. He attributed his symptoms to stress, but his family thought otherwise. At their urging, he finally saw a physician. The diagnosis? Graves' disease, which results in an overactive thyroid gland.

195 – 204 – 212 – 214 – 216 – 217 – 219

Hormones Have Been Known Since Ancient Times

Although the scientific field of endocrinology is relatively young, diseases of the endocrine system have been documented for more than a thousand years. Evidence of endocrine abnormalities can even be seen in ancient art. For example, one pre-Colombian statue of a woman shows a mass on the front of her neck (**FIG. 7.1**). The mass is an enlarged thyroid gland, or *goiter*, a common condition high in the Andes, where the dietary iodine needed to make thyroid hormones was lacking.

The first association of endocrine structure and function was probably the link between the testes and male sexuality. Castration of animals and men was a common practice in both Eastern and Western cultures because it decreased the sex drive and rendered males infertile.

In 1849, A. A. Berthold used this knowledge to perform the first classic experiment in endocrinology. He removed the testes from roosters and observed that the castrated birds had smaller combs, less aggressiveness, and less sex drive than uncastrated roosters. If the testes were surgically placed back into the donor rooster or into another castrated bird, normal male behavior and comb development resumed. Because the reimplanted testes were not connected to nerves, Berthold concluded that the glands must secrete something into the blood that affected the entire body.

FIG. 7.1 An endocrine disorder in ancient art

This pre-Colombian stone carving of a woman shows a mass at her neck. This mass is an enlarged thyroid gland, a condition known as goiter. It was considered a sign of beauty among the people who lived high in the Andes mountains.

Experimental endocrinology did not receive much attention, however, until 1889, when the 72-year-old French physician Charles Brown-Séquard made a dramatic announcement of his sexual rejuvenation after injecting himself with extracts made from bull testes ground up in water. An international uproar followed, and physicians on both sides of the Atlantic began to inject their patients with extracts of many different endocrine organs, a practice known as *organotherapy*.

We now know that the increased virility Brown-Séquard reported was most likely a placebo effect because testosterone is a hydrophobic steroid that cannot be extracted by an aqueous preparation. His research opened the door to hormone therapy, however, and in 1891, organotherapy had its first true success: A woman was treated for low thyroid hormone levels with glycerin extracts of sheep thyroid glands.

As the study of "internal secretions" grew, Berthold's experiments became a template for endocrine research. Once a gland or structure was suspected of secreting hormones, the classic steps for identifying an endocrine gland became:

1. **Remove the suspected gland.** This is equivalent to inducing a state of *hormone deficiency*. If the gland does produce hormones, the animal should start to exhibit anatomical, behavioral, or physiological abnormalities.
2. **Replace the hormone.** This can be done by placing the gland back in the animal or administering an extract of the gland. This *replacement therapy* should eliminate the symptoms of hormone deficiency.
3. **Create a state of hormone excess.** Take a normal animal and implant an extra gland or administer extract from the gland to see if symptoms characteristic of *hormone excess* appear.

Once a gland is identified as a potential source of hormones, scientists purify extracts of the gland to isolate the active substance. They test for hormone activity by injecting animals with the purified extract and monitoring for a response.

CLINICAL FOCUS

Diabetes: The Discovery of Insulin

Type 1 diabetes mellitus, the metabolic condition associated with insulin deficiency, has been known since ancient times. Until the early 1900s physicians had no means of treating the disease, and patients invariably died. It was a series of classic experiments that pinpointed the cause of diabetes. In 1889, Oskar Minkowski at the University of Strasbourg (France) surgically removed the pancreas from dogs and noticed that the dogs developed symptoms of diabetes. He found that implanting pieces of pancreas under the dogs' skin would prevent development of diabetes. Then, in 1921 Frederick G. Banting and Charles H. Best (Toronto, Canada) isolated a substance from pancreatic extracts that reversed the elevated blood glucose levels of diabetes. From there, it was a relatively short process until, in 1922, purified insulin was used in the first clinical trials. Science had found a treatment for a once-fatal disease.

Hormones identified by this technique are sometimes called *classic hormones*. They include hormones of the pancreas, thyroid, adrenal glands, pituitary, and gonads, all discrete endocrine glands that could be easily identified and surgically removed. Not all hormones come from identifiable glands, however, and we have been slower to discover them. For example, many hormones involved in digestion are secreted by endocrine cells scattered throughout the wall of the stomach or intestine, which has made them difficult to identify and isolate. The Anatomy Summary in **FIGURE 7.2** lists the major hormones of the body, the glands or cells that secrete them, and the major effects of each hormone.

What Makes a Chemical a Hormone?

In 1905, the term *hormone* was coined from the Greek verb meaning "to excite or arouse." The traditional definition of a **hormone** is a chemical secreted by a cell or group of cells into the blood for transport to a distant target, where it exerts its effect at very low concentrations. However, as scientists learn more about chemical communication in the body, this definition is continually being challenged.

Hormones Are Secreted by a Cell or Group of Cells Traditionally, the field of endocrinology has focused on chemical messengers secreted by endocrine *glands*, the discrete and readily identifiable tissues derived from epithelial tissue [p. 79]. However, we now know that molecules that act as hormones are secreted not only by classic endocrine glands but also by isolated endocrine cells (hormones of the *diffuse endocrine system*), by neurons (*neurohormones*), and occasionally by cells of the immune system (*cytokines*).

Hormones Are Secreted into the Blood **Secretion** is the movement of a substance from inside a cell to the extracellular fluid or directly into the external environment. According to the traditional definition of a hormone, hormones are secreted into the blood. However, the term *ectohormone* {*ektos*, outside} has been given to signal molecules secreted into the external environment.

Pheromones {*pherein*, to bring} are specialized ectohormones that act on other organisms of the same species to elicit a physiological or behavioral response. For example, sea anemones secrete alarm pheromones when danger threatens, and ants release trail pheromones to attract fellow workers to food sources. Pheromones are also used to attract members of the opposite sex for mating. Sex pheromones are found throughout the animal kingdom, in animals from fruit flies to dogs.

But do humans have pheromones? This question is still a matter of debate. Some studies have shown that human *axillary* (armpit) sweat glands secrete volatile steroids related to sex hormones that may serve as human sex pheromones. In one study, when female students were asked to rate the odors of T-shirts worn by male students, each woman preferred the odor of men who were genetically dissimilar from her. In another study, female axillary secretions rubbed on the upper lip of young women altered the timing of their menstrual cycles. Putative human pheromones are now sold as perfume advertised for attracting the opposite sex, as you will see if you do a web search for *human pheromone*. How humans may sense pheromones is discussed later (see Chapter 10).

Hormones Are Transported to a Distant Target By the traditional definition, a hormone must be transported by the blood to a distant target cell. Experimentally, this property is sometimes difficult to demonstrate. Molecules that are suspected of being hormones but not fully accepted as such are called *candidate hormones*. They are usually identified by the word *factor*. For example, in the early 1970s, the hypothalamic regulating hormones were known as "releasing factors" and "inhibiting factors" rather than releasing and inhibiting hormones.

Currently, **growth factors**, a large group of substances that influence cell growth and division, are being studied to determine if they meet all the criteria for hormones. Although many growth factors act locally as *autocrine* or *paracrine signals* [p. 167], most do not seem to be distributed widely in the circulation. A similar situation exists with the lipid-derived signal molecules called *eicosanoids* [p. 178].

Complicating the classification of signal molecules is the fact that a molecule may act as a hormone when secreted from one location but as a paracrine or autocrine signal when secreted from a different location. For example, in the 1920s, scientists discovered that *cholecystokinin* (CCK) in extracts of intestine caused contraction of the gallbladder. For many years thereafter, CCK was known only as an intestinal hormone. Then in the mid-1970s, CCK was found in neurons of the brain, where it acts as a neurotransmitter or neuromodulator. In recent years, CCK has gained attention because of its possible role in controlling hunger.

Hormones Exert Their Effect at Very Low Concentrations One hallmark of a hormone is its ability to act at concentrations in the nanomolar (10^{-9} M) to picomolar (10^{-12} M) range. Some chemical signals transported in the blood to distant targets are not considered hormones because they must be present in relatively high concentrations before an effect is noticed. For example, histamine released during severe allergic reactions may act on cells throughout the body, but its concentration exceeds the accepted range for a hormone.

As researchers discover new signal molecules and new receptors, the boundary between hormones and nonhormonal signal molecules continues to be challenged, just as the distinction between the nervous and endocrine systems has blurred. Many *cytokines* (p. 167) seem to meet many of the criteria of a hormone. However, experts in cytokine research do not consider cytokines to be hormones because cytokines are synthesized and released on demand, in contrast to classic peptide hormones, which are made in advance and stored in the parent endocrine cell. A few cytokines—for example, *erythropoietin*, the molecule that controls red blood cell production—were classified as hormones before the term *cytokine* was coined, contributing to the overlap between the two groups of signal molecules.

Hormones Act by Binding to Receptors

All hormones bind to target cell receptors and initiate biochemical responses. These responses are the **cellular mechanism of action** of the hormone. As you can see from the table in Figure 7.2, one hormone may act on multiple tissues. To complicate matters, the effects may vary in different tissues or at different stages of development. Or a hormone may have no effect at all in a particular cell. Insulin is an example of a hormone with varied effects. In muscle and adipose tissues, insulin alters glucose transport proteins and enzymes for glucose metabolism. In the liver, it modulates enzyme activity but has no direct effect on glucose transport proteins. In the brain and certain other tissues, glucose metabolism is totally independent of insulin.

Play A&P Flix Animation
@Mastering Anatomy & Physiology

CHAPTER 7

Concept Check

1. Name the membrane transport process by which glucose moves from the extracellular fluid into cells.

Hormone Action Must Be Terminated

Signal activity by hormones and other chemical signals must be of limited duration if the body is to respond to changes in its internal state. For example, insulin is secreted when blood glucose concentrations increase following a meal. As long as insulin is present, glucose leaves the blood and enters cells. However, if insulin activity continues for too long, blood glucose levels can fall so low that the nervous system becomes unable to function properly—a potentially fatal situation. Normally, the body avoids this situation in several ways: by limiting insulin secretion, by removing or inactivating insulin circulating in the blood, and by terminating insulin activity in target cells.

In general, hormones in the bloodstream are *degraded* (broken down) into inactive metabolites by enzymes found primarily in the liver and kidneys. The metabolites are then excreted in either the bile or the urine. The rate of hormone breakdown is indicated by a hormone's **half-life** in the circulation, the amount of time required to reduce the concentration of hormone by one-half. Half-life is one indicator of how long a hormone is active in the body.

Hormones bound to target membrane receptors have their activity terminated in several ways. Enzymes that are always present in the plasma can degrade peptide hormones bound to cell membrane receptors. In some cases, the receptor-hormone complex is brought into the cell by endocytosis, and the hormone is then digested in lysosomes [Fig. 5.19, p. 148]. Intracellular enzymes metabolize hormones that enter cells.

Concept Check

2. What is the suffix in a chemical name that tells you a molecule is an enzyme? [*Hint:* p. 101] Use that suffix to name an enzyme that digests peptides.

FIG. 7.2 ANATOMY SUMMARY . . . Hormones

Location	Hormone	Primary Target(s)
Pineal gland	Melatonin	Brain, other tissues
Hypothalamus	Trophic hormones [P, A] (see Fig. 7.9)	Anterior pituitary
Posterior pituitary (N)	Oxytocin [P]	Breast and uterus
	Vasopressin (ADH) [P]	Kidney
Anterior pituitary	Prolactin [P]	Breast
	Growth hormone (somatotropin) [P]	Liver, many tissues
	Corticotropin (ACTH) [P]	Adrenal cortex
	Thyrotropin (TSH) [P]	Thyroid gland
	Follicle-stimulating hormone [P]	Gonads
	Lutienizing hormone [P]	Gonads
Thyroid gland	Triiodothyronine and thyroxine [A]	Many tissues
	Calcitonin [P]	Bone
Parathyroid gland	Parathyroid hormone [P]	Bone, kidney
Thymus gland	Thymosin, thymopoietin [P]	Lymphocytes
Heart (C)	Atrial natriuretic peptide [P]	Kidneys
Liver (C)	Angiotensinogen [P]	Adrenal cortex, blood vessels
	Insulin-like growth factors [P]	Many tissues
Stomach and small intestine (C)	Gastrin, cholecystokinin, secretin, and others [P]	GI tract and pancreas
Pancreas (G)	Insulin, glucagon, somatostatin, pancreatic polypeptide [P]	Many tissues
Adrenal cortex (G)	Aldosterone [S]	Kidney
	Cortisol [S]	Many tissues
	Androgens [S]	Many tissues
Adrenal medulla (N)	Epinephrine, norepinephrine [A]	Many tissues
Kidney (C)	Erythropoietin [P]	Bone marrow
	1,25 Dihydroxy-vitamin D3 (calciferol) [S]	Intestine
Skin (C)	Vitamin D_3 [S]	Intermediate form of hormone
Testes (male) (G)	Androgens	Many tissues
	Inhibin	Anterior pituitary
Ovaries (female) (G)	Estrogen, progesterone [S]	Many tissues
	Inhibin [P]	Anterior pituitary
	Relaxin (pregnancy) [P]	Uterine muscle
Adipose tissue (C)	Leptin, adiponectin, resistin [P]	Hypothalamus, other tissues
Placenta (pregnant females only) (C)	Estrogen [S]	Many tissues
	Chorionic somatomammotropin [P]	Many tissues
	Chorionic gonadotropin [P]	Corpus luteum

KEY

G = gland
C = endocrine cells
N = neurons

P = peptide
S = steroid
A = amino acid–derived

Main Effects
Circadian rhythms; immune function; antioxidant
Phosphorylates proteins. Alters channel opening
Milk ejection; labor and delivery; behavior
Water reabsorption
Milk production
Growth factor secretion; growth and metabolism
Cortisol release
Thyroid hormone synthesis
Egg or sperm production; sex hormone production
Sex hormone production; egg or sperm production
Metabolism, growth, and development
Plasma calcium levels (minimal effect in humans)
Regulates plasma Ca^{2+} and phosphate levels
Lymphocyte development
Increases Na^+ excretion
Aldosterone secretion; increases blood pressure
Growth
Assist digestion and absorption of nutrients
Metabolism of glucose and other nutrients
Na^+ and K^+ homeostasis
Stress response
Sex drive in females
Fight-or-flight response
Red blood cell production
Increases calcium absorption
Precursor of 1,25-dihydroxycholecalciferol (vitamin D_3)
Sperm production, secondary sex characteristics
Inhibits FSH secretion
Egg production, secondary sex characteristics
Inhibits FSH secretion
Relaxes muscle
Food intake, metabolism, reproduction
Fetal, maternal development
Metabolism
Hormone secretion

7.2 The Classification of Hormones

Hormones can be classified according to different schemes. The scheme used in Figure 7.2 groups them according to their source. A different scheme divides hormones into those whose release is controlled by the brain and those whose release is not controlled by the brain. Another scheme groups hormones according to whether they bind to G protein-coupled receptors, tyrosine kinase-linked receptors, or intracellular receptors, and so on.

A final scheme divides hormones into three main chemical classes: peptide/protein hormones, steroid hormones, and amino acid–derived, or amine, hormones (**TBL. 7.1**). The peptide/protein hormones are composed of linked amino acids. The steroid hormones are all derived from cholesterol [p. 30]. The amino acid–derived hormones, also called *amine hormones,* are modifications of single amino acids, either tryptophan or tyrosine.

Concept Check

3. What is the classic definition of a hormone?
4. Based on what you know about the organelles involved in protein and steroid synthesis [p. 65], what would be the major differences between the organelle composition of a steroid-producing cell and that of a protein-producing cell?

Most Hormones Are Peptides or Proteins

The peptide/protein hormones range from small peptides of only three amino acids to larger proteins and glycoproteins. Despite the size variability among hormones in this group, they are usually called peptide hormones for the sake of simplicity. You can remember which hormones fall into this category by exclusion: If a hormone is not a steroid hormone and not an amino acid derivative, then it must be a peptide or protein.

Peptide Hormone Synthesis, Storage, and Release The synthesis and packaging of peptide hormones into membrane-bound secretory vesicles is similar to that of other proteins. The initial peptide that comes off the ribosome is a large inactive protein known as a preprohormone (**FIG. 7.3** ➊). **Preprohormones** contain one or more copies of a peptide hormone, a *signal sequence* that directs the protein into the lumen of the rough endoplasmic reticulum, and other peptide sequences that may or may not have biological activity.

As the inactive preprohormone moves through the endoplasmic reticulum, the signal sequence is removed, creating a smaller, still-inactive molecule called a **prohormone** (Fig. 7.3 ➍). In the Golgi complex, the prohormone is packaged into secretory vesicles along with *proteolytic* {*proteo-,* protein + *lysis,* rupture} enzymes that chop the prohormone into active hormone and other fragments. This process is called *post-translational modification* [p. 116].

The secretory vesicles containing peptides are stored in the cytoplasm of the endocrine cell until the cell receives a signal for secretion. At that time, the vesicles move to the cell membrane and release their contents by calcium-dependent exocytosis [p. 147].

TABLE 7.1 Comparison of Peptide, Steroid, and Amino Acid–Derived Hormones

			Amine Hormones (Tyrosine Derivatives)	
	Peptide Hormones	**Steroid Hormones**	**Catecholamines**	**Thyroid Hormones**
Synthesis and Storage	Made in advance; stored in secretory vesicles	Synthesized on demand from precursors	Made in advance; stored in secretory vesicles	Made in advance; precursor stored in secretory vesicles
Release from Parent Cell	Exocytosis	Simple diffusion	Exocytosis	Transport protein
Transport in Blood	Dissolved in plasma	Bound to carrier proteins	Dissolved in plasma	Bound to carrier proteins
Half-Life	Short	Long	Short	Long
Location of Receptor	Cell membrane	Cytoplasm or nucleus; some have membrane receptors also	Cell membrane	Nucleus
Response to Receptor-Ligand Binding	Activation of second messenger systems; may activate genes	Activation of genes for transcription and translation; may have nongenomic actions	Activation of second messenger systems	Activation of genes for transcription and translation
General Target Response	Modification of existing proteins and induction of new protein synthesis	Induction of new protein synthesis	Modification of existing proteins	Induction of new protein synthesis
Examples	Insulin, parathyroid hormone	Estrogen, androgens, cortisol	Epinephrine, norepinephrine, dopamine	Thyroxine (T_4)

All of the peptide fragments created from the prohormone are released together into the extracellular fluid, in a process known as *co-secretion* (Fig. 7.3 5).

Post-Translational Modification of Prohormones Studies of prohormone processing have led to some interesting discoveries. Some prohormones, such as that for *thyrotropin-releasing hormone* (TRH), contain multiple copies of the hormone (Fig. 7.3a). Another interesting prohormone is *pro-opiomelanocortin* (Fig. 7.3b). This prohormone splits into three active peptides plus an inactive fragment. In some instances, even the fragments are clinically useful. For example, proinsulin is cleaved into active insulin and an inactive fragment known as *C-peptide* (Fig. 7.3c). Clinicians measure the levels of C-peptide in the blood of diabetics to monitor how much insulin the patient's pancreas is producing.

Transport in the Blood and Half-Life of Peptide Hormones Peptide hormones are water soluble and therefore generally dissolve easily in the extracellular fluid for transport throughout the body. The half-life for peptide hormones is usually quite short, in the range of several minutes. If the response to a peptide hormone must be sustained for an extended period of time, the hormone must be secreted continually.

Cellular Mechanism of Action of Peptide Hormones Because peptide hormones are lipophobic, they are usually unable to enter the target cell. Instead, they bind to surface membrane receptors. The hormone-receptor complex initiates the cellular response by means of a *signal transduction* system (**FIG. 7.4**). Many peptide hormones work through cAMP second messenger systems [p. 173]. A few peptide hormone receptors, such as that of insulin, have tyrosine kinase activity [p. 175] or work through other signal transduction pathways.

The response of cells to peptide hormones is usually rapid because second messenger systems modify existing proteins. The changes triggered by peptide hormones include opening or closing membrane channels and modulating metabolic enzymes or transport proteins. Researchers have recently discovered that some peptide hormones also have longer-lasting effects when their second messenger systems activate genes and direct the synthesis of new proteins.

Steroid Hormones Are Derived from Cholesterol

Steroid hormones have a similar chemical structure because they are all derived from cholesterol (**FIG. 7.5a**). Unlike peptide hormones, which are made in tissues all over the body, steroid hormones are made in only a few organs. The **adrenal cortex**, the outer portion of the adrenal glands {*cortex*, bark}, makes several types of steroid hormones. One **adrenal gland** sits atop each kidney {*ad-*, upon + *renal*, kidney}. The gonads produce sex steroids (estrogens, progesterone, and androgens), and the skin can make vitamin D. In pregnant women, the placenta is also a source of steroid hormones.

FIG. 7.3 ESSENTIALS Peptide Hormone Synthesis and Processing

(a) Peptide Hormone Synthesis

Peptide hormones are made as large, inactive preprohormones that include a signal sequence, one or more copies of the hormone, and additional peptide fragments.

1. Messenger RNA on the ribosomes binds amino acids into a peptide chain called a **preprohormone.** The chain is directed into the ER lumen by a **signal sequence** of amino acids.
2. Enzymes in the ER chop off the signal sequence, creating an inactive **prohormone.**
3. The prohormone passes from the ER through the Golgi complex.
4. Secretory vesicles containing enzymes and prohormone bud off the Golgi. The enzymes chop the prohormone into one or more active peptides plus additional peptide fragments.
5. The secretory vesicle releases its contents by exocytosis into the extracellular space.
6. The hormone moves into the circulation for transport to its target.

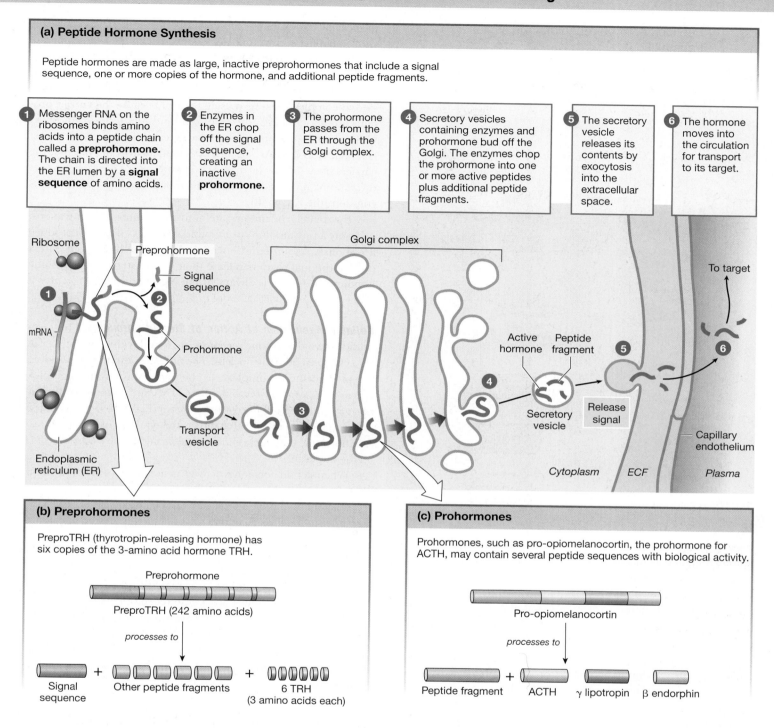

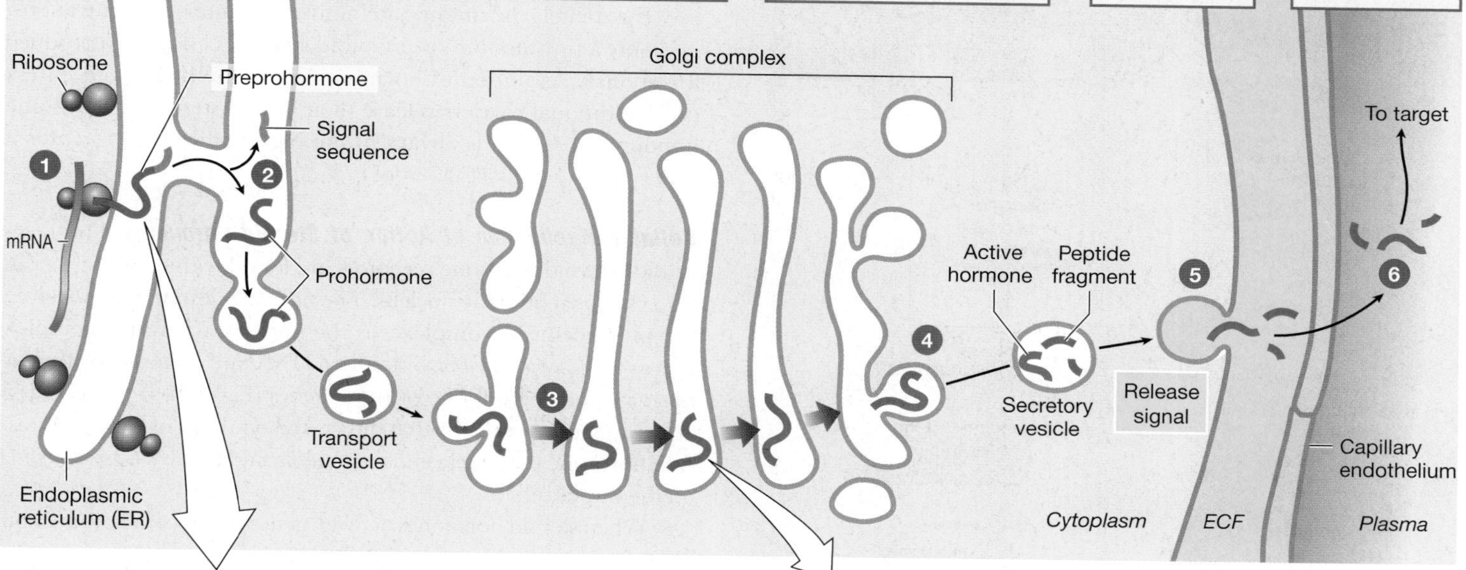

Ribosome
Preprohormone
Signal sequence
mRNA
Prohormone
Endoplasmic reticulum (ER)
Transport vesicle
Golgi complex
Active hormone
Peptide fragment
Secretory vesicle
Release signal
To target
Capillary endothelium
Cytoplasm
ECF
Plasma

(b) Preprohormones

PreproTRH (thyrotropin-releasing hormone) has six copies of the 3-amino acid hormone TRH.

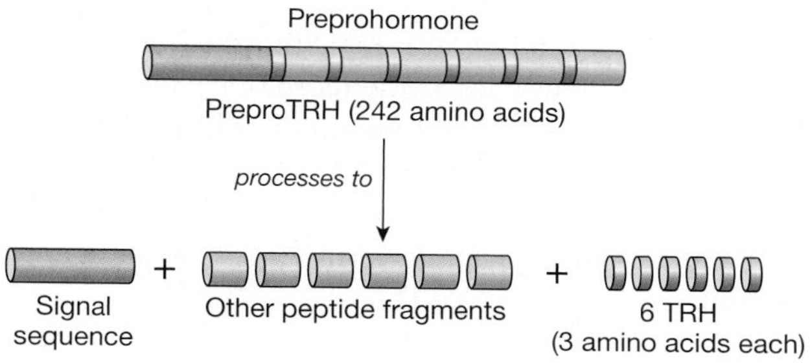

Preprohormone
PreproTRH (242 amino acids)
processes to
Signal sequence
+
Other peptide fragments
+
6 TRH (3 amino acids each)

(c) Prohormones

Prohormones, such as pro-opiomelanocortin, the prohormone for ACTH, may contain several peptide sequences with biological activity.

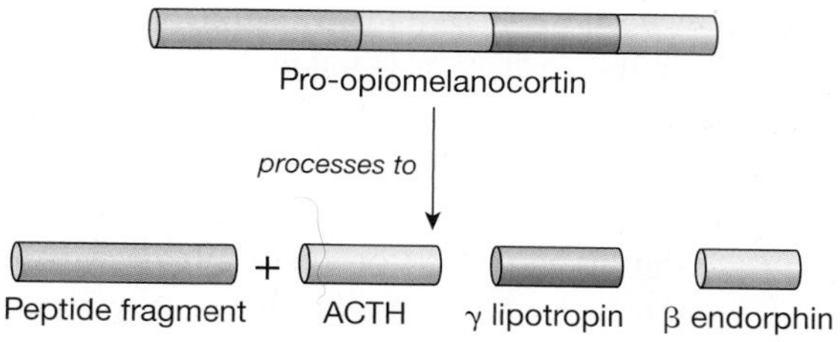

Pro-opiomelanocortin
processes to
Peptide fragment
+
ACTH
γ lipotropin
β endorphin

(d) Prohormones Process to Active Hormone Plus Peptide Fragments

The peptide chain of insulin's prohormone folds back on itself with the help of disulfide (S—S) bonds. The prohormone cleaves to insulin and C-peptide.

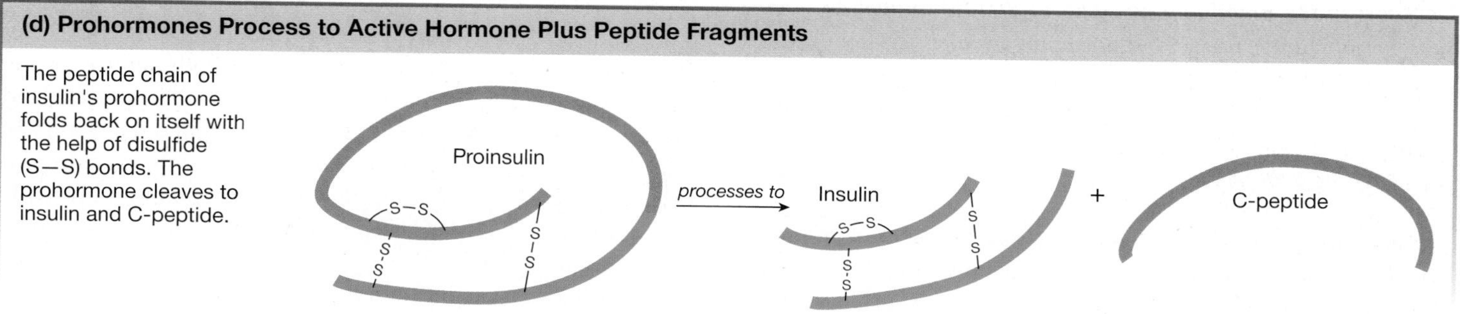

Proinsulin
processes to
Insulin
+
C-peptide

FIG. 7.4 Peptide hormone receptors and signal transduction

Peptide hormones (H) cannot enter their target cells and must combine with membrane receptors (R) that initiate signal transduction processes.

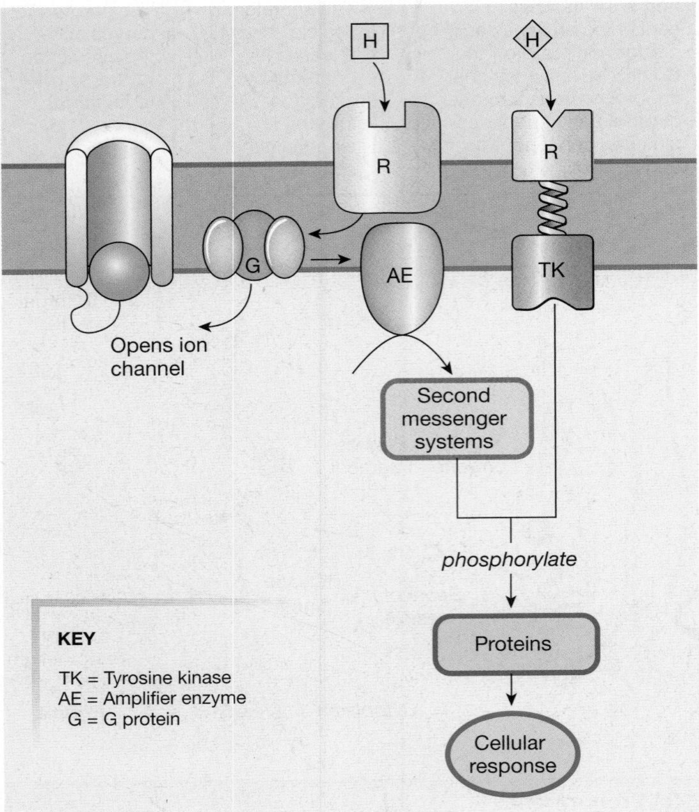
H
H
R
R
G
AE
TK
Opens ion channel
Second messenger systems
phosphorylate
Proteins
Cellular response
KEY
TK = Tyrosine kinase
AE = Amplifier enzyme
G = G protein

Steroid Hormone Synthesis and Release Cells that secrete steroid hormones have unusually large amounts of smooth endoplasmic reticulum, the organelle in which steroids are synthesized. Steroids are lipophilic and diffuse easily across membranes, both out of their parent cell and into their target cell. This property also means that steroid-secreting cells cannot store hormones in secretory vesicles. Instead, they synthesize their hormone as it is needed. When a stimulus activates the endocrine cell, precursors in the cytoplasm are rapidly converted to active hormone. The hormone concentration in the cytoplasm rises, and the hormones move out of the cell by simple diffusion.

Transport in the Blood and Half-Life of Steroid Hormones Like their parent cholesterol, steroid hormones are not very soluble in plasma and other body fluids. For this reason, most of the steroid hormone molecules found in the blood are bound to protein carrier molecules (Fig. 7.5b ❶). Some hormones have specific carriers, such as *corticosteroid-binding globulin.* Others simply bind to general plasma proteins, such as *albumin.*

The binding of a steroid hormone to a carrier protein protects the hormone from enzymatic degradation and results in an extended half-life. For example, **cortisol**, a hormone produced by the adrenal cortex, has a half-life of 60–90 minutes. (Compare this with epinephrine, an amino acid–derived hormone whose half-life is measured in seconds.)

Although binding steroid hormones to protein carriers extends their half-life, it also blocks their entry into target cells. The carrier-steroid complex remains outside the cell because the carrier proteins are lipophobic and cannot diffuse through the membrane. Only an unbound hormone molecule can diffuse into the target cell (Fig. 7.5b ❷). As unbound hormone leaves the plasma, the carriers obey the law of mass action and release hormone so that the ratio of unbound to bound hormone in the plasma remains constant [the K_d, p. 48].

Fortunately, hormones are active in minute concentrations, and only a tiny amount of unbound steroid is enough to produce a response. As unbound hormone leaves the blood and enters cells, additional carriers release their bound steroid so that some unbound hormone is always in the blood and ready to enter a cell.

Cellular Mechanism of Action of Steroid Hormones The best-studied steroid hormone receptors are found within cells, either in the cytoplasm or in the nucleus. The ultimate destination of steroid receptor-hormone complexes is the nucleus, where the complex acts as a *transcription factor,* binding to DNA and either activating or *repressing* (turning off) one or more genes (Fig. 7.5b ❸). Activated genes create new mRNA that directs the synthesis of new proteins. Any hormone that alters gene activity is said to have a *genomic effect* on the target cell.

When steroid hormones activate genes to direct the production of new proteins, there is usually a lag time between hormone-receptor binding and the first measurable biological effects. This lag can be as much as 90 minutes. Consequently, steroid hormones do not mediate reflex pathways that require rapid responses.

In recent years, researchers have discovered that several steroid hormones, including estrogens and aldosterone, have cell membrane receptors linked to signal transduction pathways, just as peptide hormones do. These receptors enable those steroid hormones to initiate rapid **nongenomic responses** in addition to their slower genomic effects. With the discovery of nongenomic effects of steroid hormones, the functional differences between steroid and peptide hormones seem almost to have disappeared.

Some Hormones Are Derived from Single Amino Acids

The amino acid–derived, or amine, hormones are small molecules created from either tryptophan or tyrosine, both notable for the carbon ring structures in their R-groups [p. 32]. The pineal gland hormone **melatonin** is derived from tryptophan (see *Focus On: The Pineal Gland,* Fig. 7.16) but the other amino acid–derived hormones—the catecholamines and thyroid hormones—are made from tyrosine (**FIG. 7.6**). Catecholamines are a modification of a single tyrosine molecule. The thyroid hormones combine two tyrosine molecules with iodine atoms.

FIG. 7.5 ESSENTIALS Steroid Hormones

Most steroid hormones are made in the adrenal cortex or gonads (ovaries and testes). Steroid hormones are not stored in the endocrine cell because of their lipophilic nature. They are made on demand and diffuse out of the endocrine cell.

(a) Cholesterol is the parent compound for all steroid hormones.

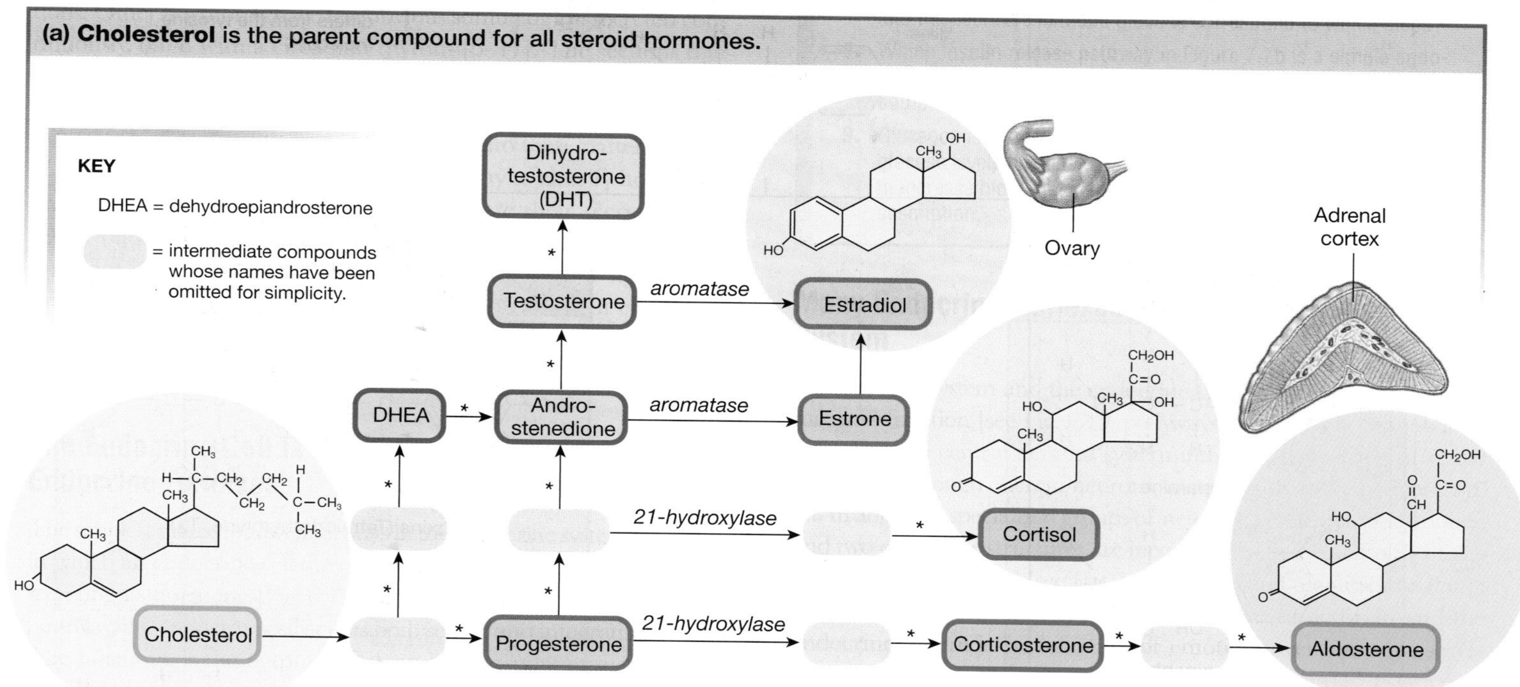

KEY
DHEA = dehydroepiandrosterone
= intermediate compounds whose names have been omitted for simplicity.
Dihydro-testosterone (DHT)
Testosterone
aromatase
Estradiol
Ovary
Adrenal cortex
DHEA
Andro-stenedione
Estrone
21-hydroxylase
Cortisol
Cholesterol
Progesterone
Corticosterone
Aldosterone

*Each step is catalyzed by an enzyme, but only two enzymes are shown in this figure.

(b) Steroid hormones act primarily on intracellular receptors.

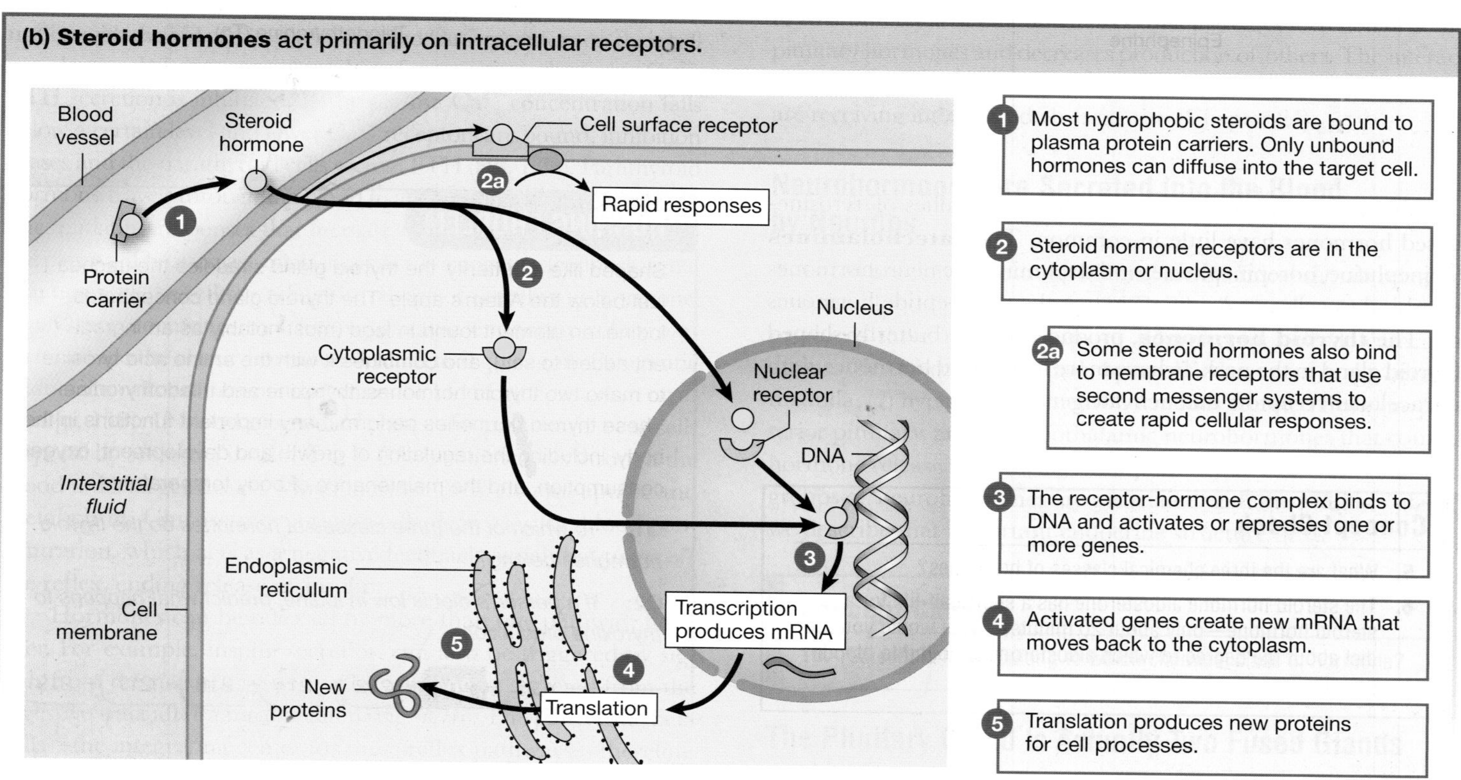

Blood vessel
Steroid hormone
Cell surface receptor
2a
Rapid responses
1
Protein carrier
2
Nucleus
Cytoplasmic receptor
Nuclear receptor
DNA
Interstitial fluid
3
Endoplasmic reticulum
Cell membrane
Transcription produces mRNA
5
4
New proteins
Translation

1 Most hydrophobic steroids are bound to plasma protein carriers. Only unbound hormones can diffuse into the target cell.

2 Steroid hormone receptors are in the cytoplasm or nucleus.

2a Some steroid hormones also bind to membrane receptors that use second messenger systems to create rapid cellular responses.

3 The receptor-hormone complex binds to DNA and activates or represses one or more genes.

4 Activated genes create new mRNA that moves back to the cytoplasm.

5 Translation produces new proteins for cell processes.

RUNNING PROBLEM

Ben Crenshaw was diagnosed with Graves' disease, one form of hyperthyroidism. The goal of treatment is to reduce thyroid hormone activity, and Ben's physician offered him several alternatives. One treatment involved drugs that prevent the thyroid gland from using iodine. Another treatment was a single dose of radioactive iodine that destroys the thyroid tissue. A third treatment was surgical removal of all or part of the thyroid gland. Ben elected initially to use the thyroid-blocking drug. Several months later he was given radioactive iodine.

Q5: *Why is radioactive iodine (rather than some other radioactive element, such as cobalt) used to destroy thyroid tissue?*

195 – 204 – 212 – **214** – 216 – 217 – 219

receptor. Instead, they may act through different metabolic pathways, or one hormone may decrease the number of receptors for the opposing hormone. For example, evidence suggests that growth hormone decreases the number of insulin receptors, providing part of its functional antagonistic effects on blood glucose concentration.

The synergistic, permissive, and antagonistic interactions of hormones make the study of endocrinology both challenging and intriguing. With this brief survey of hormone interactions, you have built a solid foundation for learning more about hormone interactions.

7.5 Endocrine Pathologies

As one endocrinologist said, "There are no good or bad hormones. A balance of hormones is important for a healthy life . . . . Unbalance leads to diseases."* We can learn much about the normal functions of a hormone by studying the diseases caused by hormone imbalances. There are three basic patterns of endocrine pathology: hormone excess, hormone deficiency, and abnormal responsiveness of target tissues to a hormone.

To illustrate endocrine pathologies, we will use a single example, that of cortisol production by the adrenal cortex (see Fig. 7.11b). This is a complex reflex pathway that starts with the secretion of corticotropin-releasing hormone (CRH) from the hypothalamus. CRH stimulates release of adrenocorticotropin (ACTH) from the anterior pituitary. ACTH in turn controls the synthesis and release of cortisol from the adrenal cortex. As in other homeostatic reflex pathways, negative feedback shuts off the pathway. As cortisol increases, it acts as a negative feedback signal, causing the pituitary and hypothalamus to decrease their output of ACTH and CRH, respectively.

* W. König, preface to *Peptide and Protein Hormones,* New York: VCH Publishers, 1993.

Hypersecretion Exaggerates a Hormone's Effects

If a hormone is present in excessive amounts, the normal effects of the hormone are exaggerated. Most instances of hormone excess are due to **hypersecretion**. There are numerous causes of hypersecretion, including benign tumors (*adenomas*) and cancerous tumors of the endocrine glands. Occasionally, nonendocrine tumors secrete hormones.

Any substance coming from outside the body is referred to as *exogenous* {*exo-*, outside}, and sometimes a patient may exhibit signs of hypersecretion as the result of medical treatment with an exogenous hormone or agonist. In this case, the condition is said to be *iatrogenic,* or physician-caused {*iatros,* healer + *-gen,* to be born}. It seems simple enough to correct the hormone imbalance by stopping treatment with the exogenous hormone, but this is not always the case.

In our example, exogenous cortisol administered as a drug acts as a negative feedback signal, just as cortisol produced within the body would, shutting off the production of CRH and ACTH (**FIG. 7.13**). Without the trophic "nourishing" influence of ACTH, the body's own cortisol production shuts down. If the pituitary remains suppressed and the adrenal cortex is deprived of ACTH long enough, the cells of both glands shrink and lose their ability to manufacture ACTH and cortisol. The loss of cell mass is known as **atrophy** {*a-*, without + *trophikós,* nourishment}. If the cells of an endocrine gland atrophy because of exogenous hormone administration, they may be very slow or totally unable to regain normal function when the treatment with exogenous hormone is stopped.

As you may know, steroid hormones such as cortisol can be used to treat poison ivy and severe allergies. However, when treatment is complete, the dosage must be tapered off gradually to allow the pituitary and adrenal gland to work back up to normal hormone production. As a result, packages of steroid pills direct patients ending treatment to take six pills one day, five the day after that, and so on. Low-dose, over-the-counter steroid creams usually do not pose a risk of feedback suppression when used as directed.

FIG. 7.13 Negative feedback from exogenous hormone

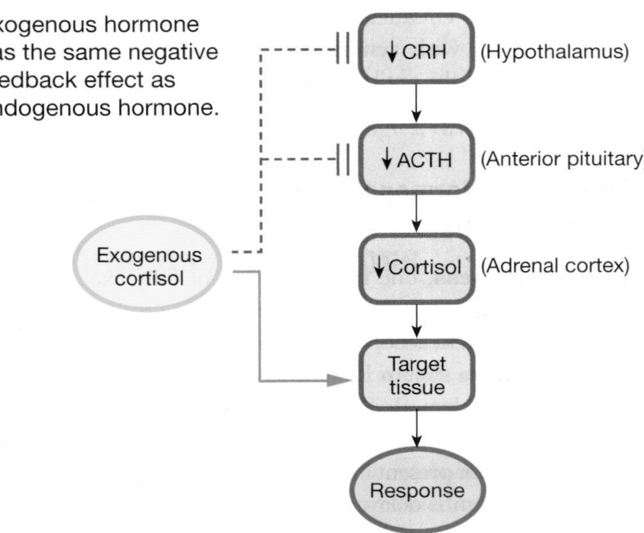
Exogenous hormone has the same negative feedback effect as endogenous hormone.
↓CRH
(Hypothalamus)
↓ACTH
(Anterior pituitary)
Exogenous cortisol
↓Cortisol
(Adrenal cortex)
Target tissue
Response

Hyposecretion Diminishes or Eliminates a Hormone's Effects

Symptoms of hormone deficiency occur when too little hormone is secreted (**hyposecretion**). Hyposecretion may occur anywhere along the endocrine control pathway, in the hypothalamus, pituitary, or other endocrine glands. For example, hyposecretion of thyroid hormone may occur if there is insufficient dietary iodine for the thyroid gland to manufacture the iodinated hormone. The most common cause of hyposecretion pathologies is atrophy of the gland due to some disease process.

Negative feedback pathways are affected in hyposecretion, but in the opposite direction from hypersecretion. The absence of negative feedback causes trophic hormone levels to rise as the trophic hormones attempt to make the defective gland increase its hormone output. For example, if the adrenal cortex atrophies as a result of tuberculosis, cortisol production diminishes. The hypothalamus and anterior pituitary sense that cortisol levels are below normal, so they increase secretion of CRH and ACTH, respectively, in an attempt to stimulate the adrenal gland into making more cortisol.

Receptor or Second Messenger Problems Cause Abnormal Tissue Responsiveness

Endocrine diseases do not always arise from problems with endocrine glands. They may also be triggered by changes in the responsiveness of target tissues to the hormones. In these situations, the target tissues show abnormal responses even though hormone levels may be within the normal range. Changes in the target tissue response are usually caused by abnormal interactions between the hormone and its receptor or by alterations in signal transduction pathways.

Down-Regulation If hormone secretion is abnormally high for an extended period of time, target cells may *down-regulate* (decrease the number of) their receptors in an effort to diminish their responsiveness to excess hormone. **Hyperinsulinemia** {*hyper-*, elevated + insulin + *-emia*, in the blood} is a classic example of down-regulation in the endocrine system. In this disorder, sustained high levels of insulin in the blood cause target cells to remove insulin receptors from the cell membrane. Patients suffering from hyperinsulinemia may show signs of diabetes despite their high blood insulin levels.

Receptor and Signal Transduction Abnormalities Many forms of inherited endocrine pathologies can be traced to problems with hormone action in the target cell. Endocrinologists once believed that these problems were rare, but they are being recognized more frequently as scientists increase their understanding of receptors and signal transduction mechanisms.

Some pathologies are due to problems with the hormone receptor [Tbl. 6.1, p. 181]. If a mutation alters the protein sequence of the receptor, the cellular response to receptor-hormone binding may be altered. In other mutations, the receptors may be absent or completely nonfunctional. For example, in *androgen insensitivity syndrome*, androgen receptors are nonfunctional in the male fetus because of a genetic mutation. As a result, androgens produced by the developing fetus are unable to influence development of the genitalia. The result is a child who appears to be female but lacks a uterus and ovaries.

Genetic alterations in signal transduction pathways can lead to symptoms of hormone excess or deficiency. In the disease called *pseudohypoparathyroidism* {*pseudo-*, false + *hypo-*, decreased + parathyroid + *-ism*, condition or state of being}, patients show signs of low parathyroid hormone even though blood levels of the hormone are normal or elevated. These patients have inherited a defect in the G protein that links the hormone receptor to the cAMP amplifier enzyme, adenylyl cyclase. Because the signal transduction pathway does not function, target cells are unable to respond to parathyroid hormone, and signs of hormone deficiency appear.

CHAPTER 7

Diagnosis of Endocrine Pathologies Depends on the Complexity of the Reflex

Diagnosis of endocrine pathologies may be simple or complicated, depending on the complexity of the reflex. For example, consider a simple endocrine reflex, such as that for parathyroid hormone. If there is too much or too little hormone, the problem can arise in only one location: the parathyroid glands (see Fig. 7.7a). However, with complex hypothalamic-pituitary-endocrine gland reflexes, the diagnosis can be much more difficult.

If a pathology (deficiency or excess) arises in the last endocrine gland in a complex reflex pathway, the problem is considered to be a **primary pathology**. For example, if a tumor in the adrenal cortex begins to produce excessive amounts of cortisol, the resulting condition is called *primary hypersecretion*. If dysfunction occurs in the anterior pituitary, the problem is a **secondary pathology**. For example, if the pituitary is damaged because of head trauma and ACTH secretion diminishes, the resulting cortisol deficiency is considered to be *secondary hyposecretion* of cortisol. Pathologies of hypothalamic trophic hormones occur infrequently; they would be considered *tertiary* hyposecretion and hypersecretion.

The diagnosis of pathologies in complex endocrine pathways depends on understanding negative feedback in the control pathway. **FIGURE 7.14** shows three possible causes of excess cortisol secretion. To determine which is the correct *etiology* (cause) of the disease in a particular patient, the clinician must assess the levels of the three hormones in the control pathway.

If cortisol levels are high but levels of both trophic hormones are low, the problem is a primary disorder (Fig. 7.14a). There are two possible explanations for elevated cortisol: endogenous cortisol hypersecretion or the exogenous administration of cortisol for therapeutic reasons (see Fig. 7.13). In both cases, high levels of cortisol act as a negative feedback signal that shuts off production of CRH and ACTH. The pattern of high cortisol with low trophic hormone levels points to a primary disorder.

When the problem is endogenous—an adrenal tumor that is secreting cortisol in an unregulated fashion—the normal control pathways are totally ineffective. Although negative feedback shuts off production of the trophic hormones, the tumor is not dependent on them for cortisol production, so cortisol secretion continues in their absence. The tumor must be removed or suppressed before cortisol secretion can be controlled.

FIG. 7.16 Focus on . . . The Pineal Gland

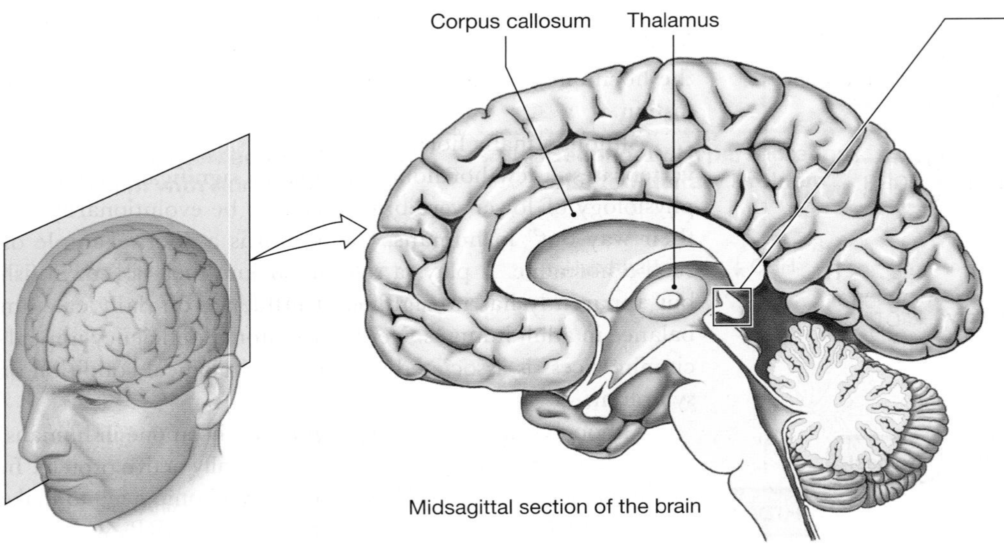

Corpus callosum
Thalamus
Midsagittal section of the brain

The **pineal gland** is a pea-sized structure buried deep in the brain of humans. Nearly 2000 years ago, this "seat of the soul" was thought to act as a valve that regulated the flow of vital spirits and knowledge into the brain. By 1950, however, scientists had decided that it was a vestigial structure with no known function.

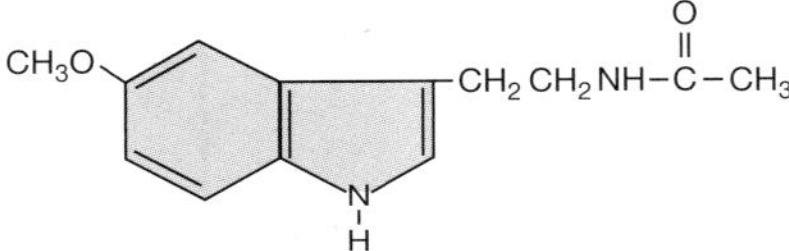

CH3O
CH2 CH2 NH–C–CH3
O
N
H

Melatonin is an amino acid–derived hormone made from tryptophan.

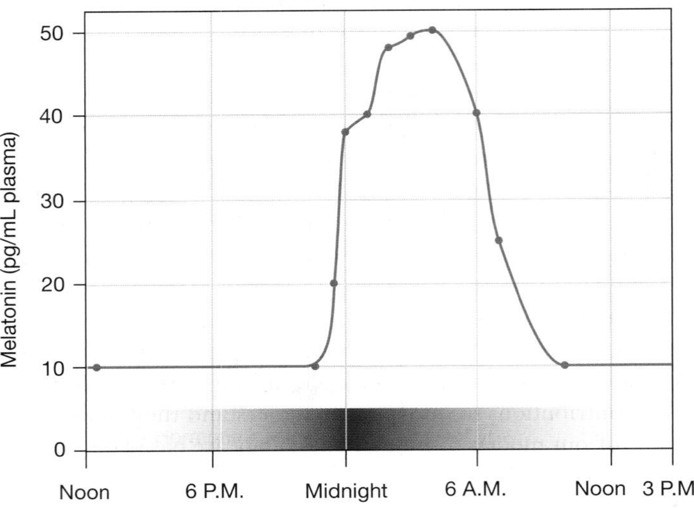

Melatonin (pg/mL plasma)
50
40
30
20
10
0
Noon
6 P.M.
Midnight
6 A.M.
Noon
3 P.M.

Melatonin is the "darkness hormone," secreted at night as we sleep. It is the chemical messenger that transmits information about light-dark cycles to the brain center that governs the body's biological clock.

(Based on J. Arendt, Melatonin, *Clin Endocrinol* 29: 205–229, 1988.)

About 1957, one of the wonderful coincidences of scientific research occurred. An investigator heard about a factor in beef pineal glands that could lighten the skin of amphibians. Using the classical methodology of endocrinology, he obtained pineal glands from a slaughterhouse and started making extracts. His biological assay consisted of dropping pineal extracts into bowls of live tadpoles to see if their skin color blanched. Several years and hundreds of thousands of pineal glands later, he had isolated a small amount of melatonin.

Sixty years later, we are still learning about the functions of melatonin in humans. In addition to its role in sleep-wake cycles and the body's internal clock, scientists have evidence that melatonin has other actions in the nervous system, where it binds to MT_1 and MT_2 G protein-coupled receptors. Some studies using mouse models suggest that melatonin may help slow the progression of Alzheimer's disease. Melatonin has also been linked to sexual function, the onset of puberty, and depression in the darker winter months (seasonal affective disorder, or SAD).

In 2017, there were about 90 active clinical trials in the United States testing the efficacy of melatonin in treating disorders ranging from sleep disturbances and depression to cancer. Most of these tests are Phase II and Phase III clinical trials. Phase II trials are usually placebo-controlled, double-blind studies. Phase III trials include more patients and some uncontrolled studies. Some Phase III studies are "open-label," meaning that the patients and healthcare providers know what drug is being administered. To learn more about clinical trials, see clinicaltrials.gov, a registry of clinical trials around the world.

carried out in fish or frogs or rats, to name a few. For example, the pineal gland hormone *melatonin* (**FIG. 7.16**) was discovered through research using tadpoles. Many small nonhuman vertebrates have short life cycles that facilitate studying aging or reproductive physiology. Genetically altered mice (transgenic or knockout mice) have provided researchers valuable information about proteomics.

Opponents of animal research argue that scientists should not experiment with animals at all and should use only cell cultures and computer models. Cell cultures and models are valuable tools and can be helpful in the initial stages of medical research, but at some point new drugs and procedures must be tested on intact organisms prior to clinical trials in humans. Responsible scientists follow guidelines for appropriate animal use and limit the number of animals killed to the minimum needed to provide valid data.

In this chapter, we have examined how the endocrine system with its hormones helps regulate the slower processes in the body. As you will see, the nervous system takes care of the more rapid responses needed to maintain homeostasis.

RUNNING PROBLEM **CONCLUSION** Graves' Disease

In this running problem, you learned that in Graves' disease, thyroid hormone levels are high because an immune-system protein mimics TSH. You also learned that the thyroid gland concentrates iodine for synthesis of thyroid hormones, and that radioactive iodine can concentrate in the gland and destroy the thyroid cells. Ben Crenshaw's treatment for Graves' disease was successful. He went on to win the Masters Tournament for a second time in 1995 and he still plays golf professionally today.

Graves' disease is the most common form of hyperthyroidism. Other famous people who have suffered from it range from hip-hop star Missy Elliott to former U.S. President George H. W. Bush and First Lady Barbara Bush. To learn more about Graves' disease and other thyroid conditions, visit the Endocrine Society's Hormone Foundation website at *www.hormone.org* or the American Thyroid Association at *www.thyroid.org*. Check your answers to the problem questions by comparing them to the information in the following summary table.

Question	Facts	Integration and Analysis
Q1: *To which of the three classes of hormones do thyroid hormones belong?*	The three classes are peptides, steroids, and amino-acid derivatives.	Thyroid hormones are made from the amino acid tyrosine, making them amino-acid derivatives.
Q2: *If a person's diet is low in iodine, predict what happens to thyroxine production.*	The thyroid gland concentrates iodine and combines it with the amino acid tyrosine to make thyroid hormones.	If iodine is lacking in the diet, a person is unable to make thyroid hormones.
Q3: *In a normal person, when thyroid hormone levels in the blood increase, will negative feedback increase or decrease the secretion of TSH?*	Negative feedback shuts off response loops.	Normally negative feedback decreases TSH secretion.
Q4: *In a person with a hyperactive gland that is producing too much thyroid hormone, would you expect the level of TSH to be higher or lower than in a normal person?*	Thyroid hormone is the negative feedback signal.	If thyroid hormone is high, you would expect strong negative feedback and even lower levels of TSH.
Q5: *Why is radioactive iodine (rather than some other radioactive element, such as cobalt) used to destroy thyroid tissue?*	The thyroid gland concentrates iodine to make thyroid hormones.	Radioactive iodine is concentrated in the thyroid gland and therefore selectively destroys that tissue. Other radioactive elements distribute more widely throughout the body and may harm normal tissues.
Q6: *If levels of TSH are low and thyroxine levels are high, is Graves' disease a primary disorder or a secondary disorder (one that arises as a result of a problem with the anterior pituitary or the hypothalamus)? Explain your answer.*	In secondary hypersecretion disorders, you would expect the levels of the anterior pituitary trophic hormones to be elevated.	In Graves' disease, TSH from the anterior pituitary is very low. Therefore, the oversecretion of thyroid hormones is not the result of elevated TSH. This means that Graves' disease is a primary disorder that is caused by a problem in the thyroid gland itself.
Q7: *Antibodies are proteins that bind to the TSH receptor. From that information, what can you conclude about the cellular location of the TSH receptor?*	Receptors may be membrane receptors or intracellular receptors. Proteins cannot cross the cell membrane.	The TSH receptor is a membrane receptor. It uses the cAMP second messenger pathway for signal transduction.
Q8: *In Graves' disease, why doesn't negative feedback shut off thyroid hormone production before it becomes excessive?*	In normal negative feedback, increasing levels of thyroid hormone shut off TSH secretion. Without TSH stimulation, the thyroid stops producing thyroid hormone.	In Graves' disease, high levels of thyroid hormone have shut off endogenous TSH production. However, the thyroid gland still produces hormone in response to the binding of antibody to the TSH receptor. In this situation, negative feedback fails to correct the problem.

195 - 204 - 212 - 214 - 216 - 217 - 219

List 2

• ACTH	• oxytocin
• anterior pituitary	• peptide/protein
• blood	• portal system
• endocrine cell	• posterior pituitary
• gonadotropins	• prolactin
• growth hormone	• releasing hormone
• hypothalamus	• trophic hormone
• inhibiting hormone	• TSH
• neurohormone	• vasopressin
• neuron	

Level Three Problem Solving

27. The terms *specificity, receptors,* and *down-regulation* can be applied to many physiological situations. Do their meanings change when applied to the endocrine system? What chemical and physical characteristics do hormones, enzymes, transport proteins, and receptors have in common that makes specificity important?

28. Dexamethasone is a drug used to suppress the secretion of adrenocorticotrophic hormone (ACTH) from the anterior pituitary. Two patients with hypersecretion of cortisol are given dexamethasone. Patient A's cortisol secretion falls to normal as a result, but patient B's cortisol secretion remains elevated. Draw maps of the reflex pathways for these two patients (see Fig. 7.11b for a template) and use the maps to determine which patient has primary hypercortisolism. Explain your reasoning.

29. Some early experiments for male birth control pills used drugs that suppressed gonadotropin (FSH and LH) release. However, men given these drugs stopped taking them because the drugs decreased testosterone secretion, which decreased the men's sex drive and caused impotence.

    a. Use the information given in Figure 7.9 to draw the GnRH-FSH/LH-testosterone reflex pathway. Use the pathway to show how suppressing gonadotropins decreases sperm production and testosterone secretion.

    b. Researchers subsequently suggested that a better treatment would be to give men extra testosterone. Draw another copy of the reflex pathway to show how testosterone could suppress sperm production without the side effect of impotence.

Level Four Quantitative Problems

30. The following graph represents the disappearance of a drug from the blood as the drug is metabolized and excreted. Based on the graph, what is the half-life of the drug?

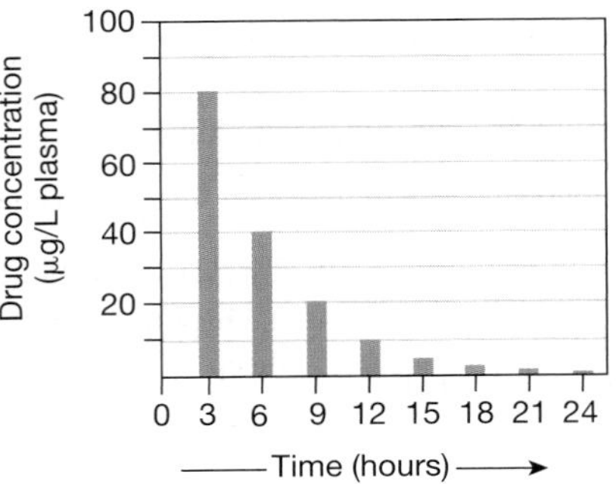
Drug concentration (µg/L plasma)
100
80
60
40
20
0 3 6 9 12 15 18 21 24
Time (hours)

31. The following graph shows plasma TSH concentration in three groups of subjects. Which pattern would be consistent with the following pathologies? Explain your reasoning.

    a. primary hypothyroidism
    b. primary hyperthyroidism
    c. Secondary hyperthyroidism

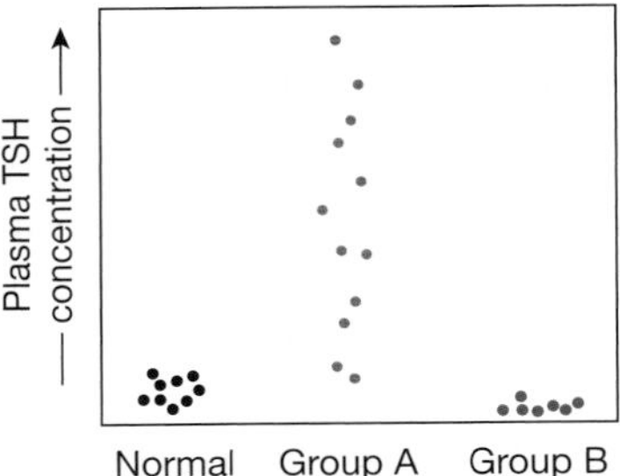
Plasma TSH concentration
Normal
Group A
Group B

32. Based on what you have learned about the pathway for insulin secretion, draw and label a graph showing the effect of plasma glucose concentration on insulin secretion.

Answers to Concept Checks, Figure and Graph Questions, and end-of-chapter Review Questions can be found in Appendix A [p. A-1].

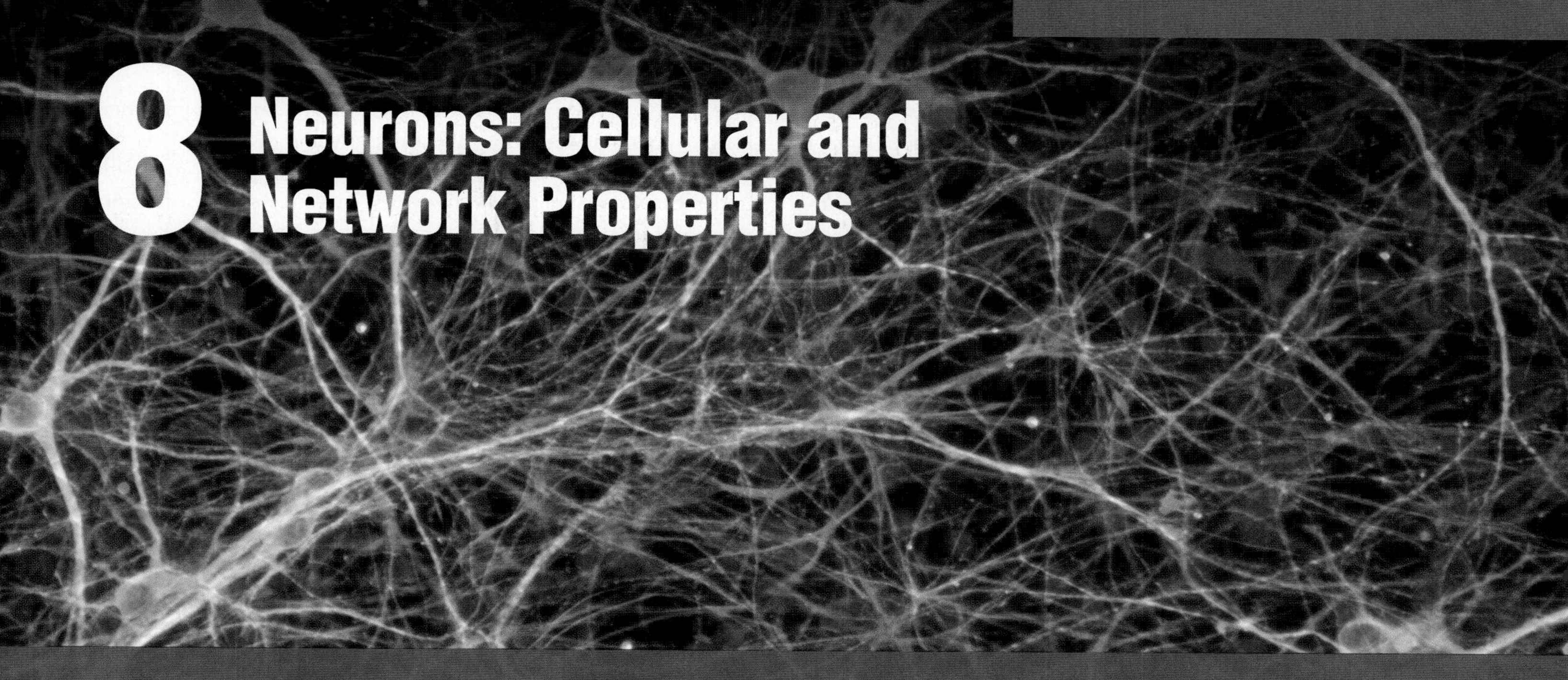

8 Neurons: Cellular and Network Properties

The future of clinical neurology and psychiatry is intimately tied to that of molecular neural science.

Eric R. Kandel, James H. Schwartz, and Thomas M. Jessell, in the preface to their book, Principles of Neural Science*, 2000*

Neurons (blue) and glial cells (red)

BACKGROUND BASICS

In an eerie scene from a science fiction movie, white-coated technicians move quietly through a room filled with bubbling cylindrical fish tanks. As the camera zooms in on one tank, no fish can be seen darting through aquatic plants. The lone occupant of the tank is a gray mass with a convoluted surface like a walnut and a long tail that appears to be edged with beads. Floating off the beads are hundreds of fine fibers, waving softly as the oxygen bubbles weave through them. This is no sea creature. . . . It is a brain and spinal cord, removed from its original owner and awaiting transplantation into another body. Can this be real? Is this scenario possible? Or is it just the creation of an imaginative movie screenwriter?

The brain is regarded as the seat of the soul, the mysterious source of those traits that we think of as setting humans apart from other animals. The brain and spinal cord are also integrating centers for homeostasis, movement, and many other body functions. They are the control center of the **nervous system,** a network of billions of nerve cells linked together in a highly organized manner to form the rapid control system of the body.

Nerve cells, or **neurons,** carry electrical signals rapidly and, in some cases, over long distances. They are uniquely shaped cells, and most have long, thin extensions, or **processes,** that can extend up to a meter in length. In most pathways, neurons release chemical signals, called **neurotransmitters,** into the extracellular fluid to communicate with neighboring cells. In a few pathways, neurons are linked by *gap junctions* [p. 165], allowing electrical signals to pass directly from cell to cell.

Using electrical signals to release chemicals from a cell is not unique to neurons. For example, pancreatic beta cells generate an electrical signal to initiate exocytosis of insulin-containing storage vesicles [p. 159]. Single-celled protozoa and plants also employ electrical signaling mechanisms, in many cases using the same types of ion channels as vertebrates do. Scientists sequencing ion channel proteins have found that many of these channel proteins have been highly conserved during evolution, indicating their fundamental importance.

Although electrical signaling is universal, sophisticated neural networks are unique to animal nervous systems. Reflex pathways in the nervous system do not necessarily follow a straight line from one neuron to the next. One neuron may influence multiple neurons, or many neurons may affect the function of a single neuron. The intricacy of neural networks and their neuronal components underlies the emergent properties of the nervous system. **Emergent properties** are complex processes, such as consciousness, intelligence, and emotion that cannot be predicted from what we know about the properties of individual nerve cells and their specific connections. The search to explain emergent properties makes neuroscience one of the most active research areas in physiology today.

Neuroscience, like many other areas of science, has its own specialized language. In many instances, multiple terms describe a single structure or function, which potentially can lead to confusion. **TABLE 8.1** lists some neuroscience terms used in this book, along with their common synonyms, which you may encounter in other publications.

RUNNING PROBLEM **Mysterious Paralysis**

"Like a polio ward from the 1950s" is how Guy McKhann, M.D., a neurology specialist at the Johns Hopkins School of Medicine, describes a ward of Beijing Hospital that he visited on a trip to China in 1986. Dozens of paralyzed children—some attached to respirators to assist their breathing—filled the ward to overflowing. The Chinese doctors thought the children had Guillain-Barré syndrome (GBS), a rare paralytic condition, but Dr. McKhann wasn't convinced. There were simply too many stricken children for the illness to be the rare Guillain-Barré syndrome. Was it polio—as some of the Beijing staff feared? Or was it another illness, perhaps one that had not yet been discovered?

224 – 226 – 233 – 248 – 251 – 254 – 264 – 265

8.1 Organization of the Nervous System

The nervous system can be divided into two parts (**FIG. 8.1**). The **central nervous system (CNS)** consists of the **brain** and the **spinal cord**. The **peripheral nervous system (PNS)** consists of **sensory (afferent) neurons** and **efferent neurons**. Information flow through the nervous system follows the basic pattern of a reflex [p. 14]:

stimulus → sensor → input signal → integrating center → output signal → target → response.

Sensory receptors throughout the body continuously monitor conditions in the internal and external environments. These

TABLE 8.1 Synonyms in Neuroscience

Term Used in This Book	Synonym(s)
Action potential	AP, spike, nerve impulse, conduction signal
Autonomic nervous system	Visceral nervous system
Axon	Nerve fiber
Axonal transport	Axoplasmic flow
Axon terminal	Synaptic knob, synaptic bouton, presynaptic terminal
Axoplasm	Cytoplasm of an axon
Cell body	Cell soma, nerve cell body
Cell membrane of an axon	Axolemma
Glial cells	Neuroglia, glia
Interneuron	Association neuron
Rough endoplasmic reticulum	Nissl substance, Nissl body
Sensory neuron	Afferent neuron, afferent

FIG. 8.1 ESSENTIALS The Organization of the Nervous System

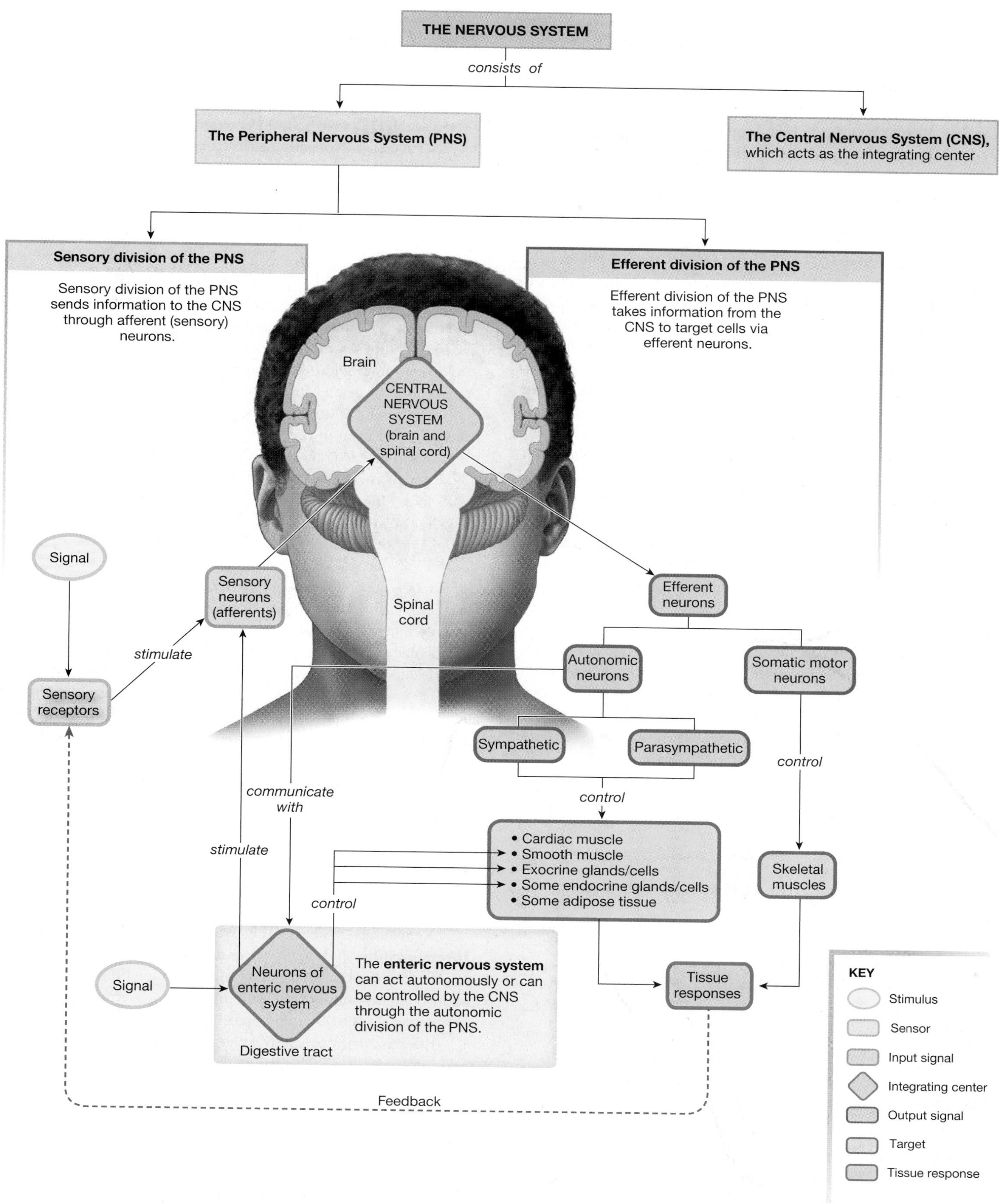

THE NERVOUS SYSTEM
consists of
The Peripheral Nervous System (PNS)
The Central Nervous System (CNS), which acts as the integrating center
Sensory division of the PNS
Sensory division of the PNS sends information to the CNS through afferent (sensory) neurons.
Efferent division of the PNS
Efferent division of the PNS takes information from the CNS to target cells via efferent neurons.
Brain
CENTRAL NERVOUS SYSTEM (brain and spinal cord)
Spinal cord
Signal
Sensory receptors
stimulate
Sensory neurons (afferents)
Efferent neurons
Autonomic neurons
Somatic motor neurons
Sympathetic
Parasympathetic
control
communicate with
stimulate
control
• Cardiac muscle
• Smooth muscle
• Exocrine glands/cells
• Some endocrine glands/cells
• Some adipose tissue
control
Skeletal muscles
Signal
Neurons of enteric nervous system
The enteric nervous system can act autonomously or can be controlled by the CNS through the autonomic division of the PNS.
Digestive tract
Tissue responses
Feedback
KEY
Stimulus
Sensor
Input signal
Integrating center
Output signal
Target
Tissue response

sensors send information along sensory neurons to the CNS, which is the integrating center for neural reflexes. CNS neurons integrate information that arrives from the sensory division of the PNS and determine whether a response is needed.

If a response is needed, the CNS sends output signals that travel through efferent neurons to their targets, which are mostly muscles and glands. Efferent neurons are subdivided into the **somatic motor division**, which controls skeletal muscles, and the **autonomic division**, which controls smooth and cardiac muscles, exocrine glands, some endocrine glands, and some types of adipose tissue. Terminology used to describe efferent neurons can be confusing. The expression *motor neuron* is sometimes used to refer to all efferent neurons. However, clinically, the term *motor neuron* (or *motoneuron*) is often used to describe somatic motor neurons that control skeletal muscles.

The autonomic division of the PNS is also called the *visceral nervous system* because it controls contraction and secretion in the various internal organs {*viscera*, internal organs}. Autonomic neurons are further divided into **sympathetic** and **parasympathetic branches,** which can be distinguished by their anatomical organization and by the chemicals they use to communicate with their target cells. Many internal organs receive innervation from both types of autonomic neurons, and it is common for the two divisions to exert *antagonistic control* over a single target [p. 182].

In recent years, a third division of the nervous system has received considerable attention. The **enteric nervous system** is a network of neurons in the walls of the digestive tract. It is frequently controlled by the autonomic division of the nervous system, but it is also able to function autonomously as its own integrating center. You will learn more about the enteric nervous system when you study the digestive system.

It is important to note that the CNS can initiate activity without sensory input, such as when you decide to text a friend. Also, the CNS need not create any measurable output to the efferent divisions. For example, thinking and dreaming are complex higher-brain functions that can take place totally within the CNS.

Concept Check

1. Organize the following terms describing functional types of neurons into a map or outline: afferent, autonomic, brain, central, efferent, enteric, parasympathetic, peripheral, sensory, somatic motor, spinal, sympathetic.

8.2 Cells of the Nervous System

The nervous system is composed primarily of two cell types: neurons—the basic signaling units of the nervous system—and support cells known as *glial cells* (or glia or neuroglia).

Neurons Carry Electrical Signals

The neuron, or nerve cell, is the functional unit of the nervous system. (A *functional unit* is the smallest structure that can carry out the functions of a system.) Neurons are uniquely shaped cells with long processes that extend outward from the *nerve cell body*. These processes are usually classified as either **dendrites,** which receive incoming signals, or **axons,** which carry outgoing information. The shape, number, and length of axons and dendrites vary from one neuron to the next, but these structures are an essential feature that allows neurons to communicate with one another and with other cells. Neurons may be classified either structurally or functionally (**FIG. 8.2**).

Structurally, neurons are classified by the number of processes that originate from the cell body. The model neuron that is commonly used to teach how a neuron functions is *multipolar*, with many dendrites and branched axons (Fig. 8.2e). Multipolar neurons in the CNS look different from multipolar efferent neurons (Fig. 8.2d). In other structural neuron types, the axons and dendrites may be missing or modified. *Pseudounipolar* neurons have the cell body located off one side of a single long process that is called the axon (Fig. 8.2a). (During development, the dendrites fused and became part of the axon.) *Bipolar* neurons have a single axon and single dendrite coming off the cell body (Fig. 8.2b). *Anaxonic* neurons lack an identifiable axon but have numerous branched dendrites (Fig. 8.2c).

Because physiology is concerned chiefly with function, however, we will classify neurons according to their functions: sensory (afferent) neurons, interneurons, and efferent (somatic motor and autonomic) neurons. Sensory neurons carry information about temperature, pressure, light, and other stimuli from sensory receptors to the CNS. Peripheral sensory neurons are pseudounipolar, with cell bodies located close to the CNS and very long processes that extend out to receptors in the limbs and internal organs. In these sensory neurons, the cell body is out of the direct path of signals passing along the axon (Fig. 8.2a). In contrast, sensory neurons in the nose and eye are much smaller bipolar neurons. Signals that begin at the dendrites travel through the cell body to the axon (Fig. 8.2b).

Neurons that lie entirely within the CNS are known as **interneurons** (short for *interconnecting neurons*). They come in a variety of forms but often have quite complex branching processes that allow them to communicate with many other neurons (Fig. 8.2c, d). Some interneurons are quite small compared to the model neuron.

Efferent neurons, both somatic motor and autonomic, are generally very similar to the neuron in Figure 8.2e. The axons may divide several times into branches called **collaterals** {*col-*, with + *lateral*, something on the side}. Efferent neurons have enlarged endings called **axon terminals**. Many autonomic neurons also have enlarged regions along the axon called **varicosities** [see Fig. 11.7, p. 362]. Both axon terminals and varicosities store and release neurotransmitter.

RUNNING PROBLEM

Guillain-Barré syndrome is a relatively rare paralytic condition that strikes after a viral infection or an immunization. There is no cure, but usually the paralysis slowly disappears, and lost sensation slowly returns as the body repairs itself. In classic Guillain-Barré, patients can neither feel sensations nor move their muscles.

Q1: *Which division(s) of the nervous system may be involved in Guillain-Barré syndrome?*

224
226
233
248
251
254
264
265

FIG. 8.2 ESSENTIALS Neuron anatomy

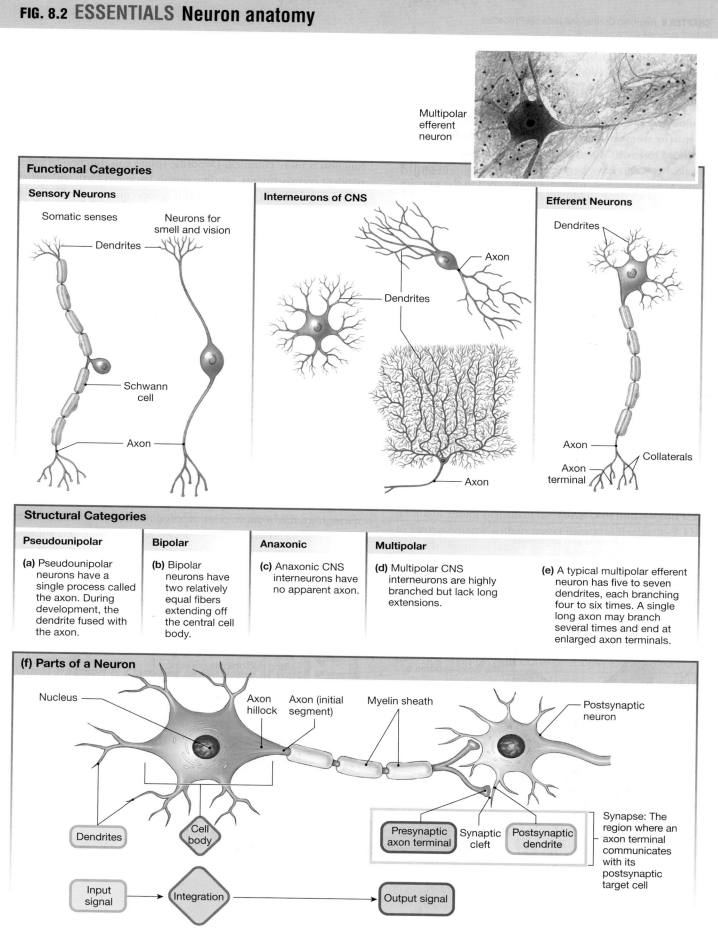

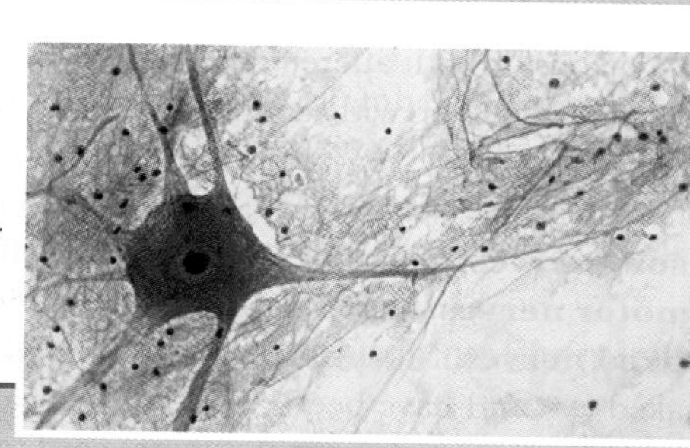

Multipolar efferent neuron

Functional Categories

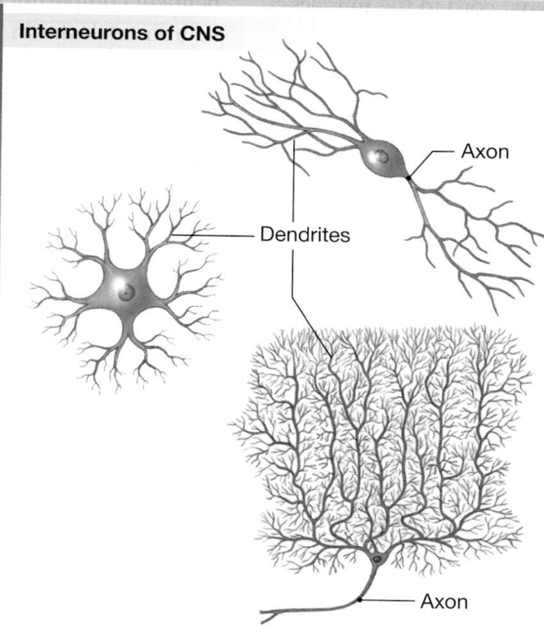

Sensory Neurons
Somatic senses
Neurons for smell and vision
Dendrites
Schwann cell
Axon
Interneurons of CNS
Axon
Dendrites
Axon
Efferent Neurons
Dendrites
Axon
Collaterals
Axon terminal

Structural Categories

Pseudounipolar	Bipolar	Anaxonic	Multipolar	
(a) Pseudounipolar neurons have a single process called the axon. During development, the dendrite fused with the axon.	**(b)** Bipolar neurons have two relatively equal fibers extending off the central cell body.	**(c)** Anaxonic CNS interneurons have no apparent axon.	**(d)** Multipolar CNS interneurons are highly branched but lack long extensions.	**(e)** A typical multipolar efferent neuron has five to seven dendrites, each branching four to six times. A single long axon may branch several times and end at enlarged axon terminals.

(f) Parts of a Neuron

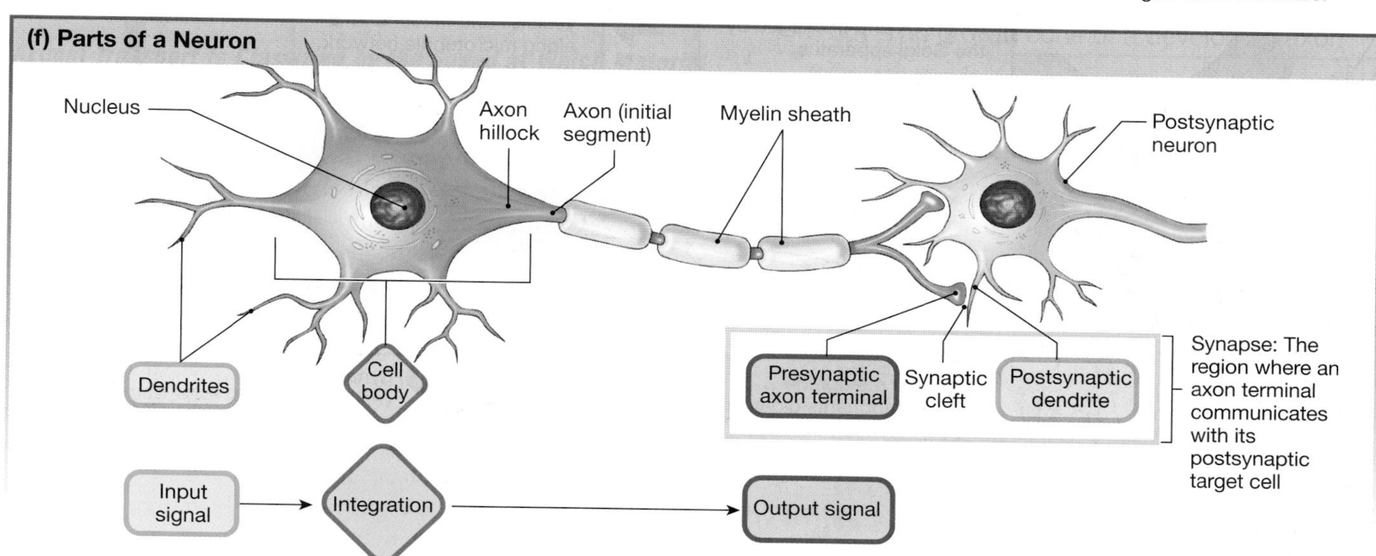

Nucleus
Axon hillock
Axon (initial segment)
Myelin sheath
Postsynaptic neuron
Dendrites
Cell body
Presynaptic axon terminal
Synaptic cleft
Postsynaptic dendrite
Synapse: The region where an axon terminal communicates with its postsynaptic target cell
Input signal
Integration
Output signal

FIG. 8.5 ESSENTIALS Glial cells

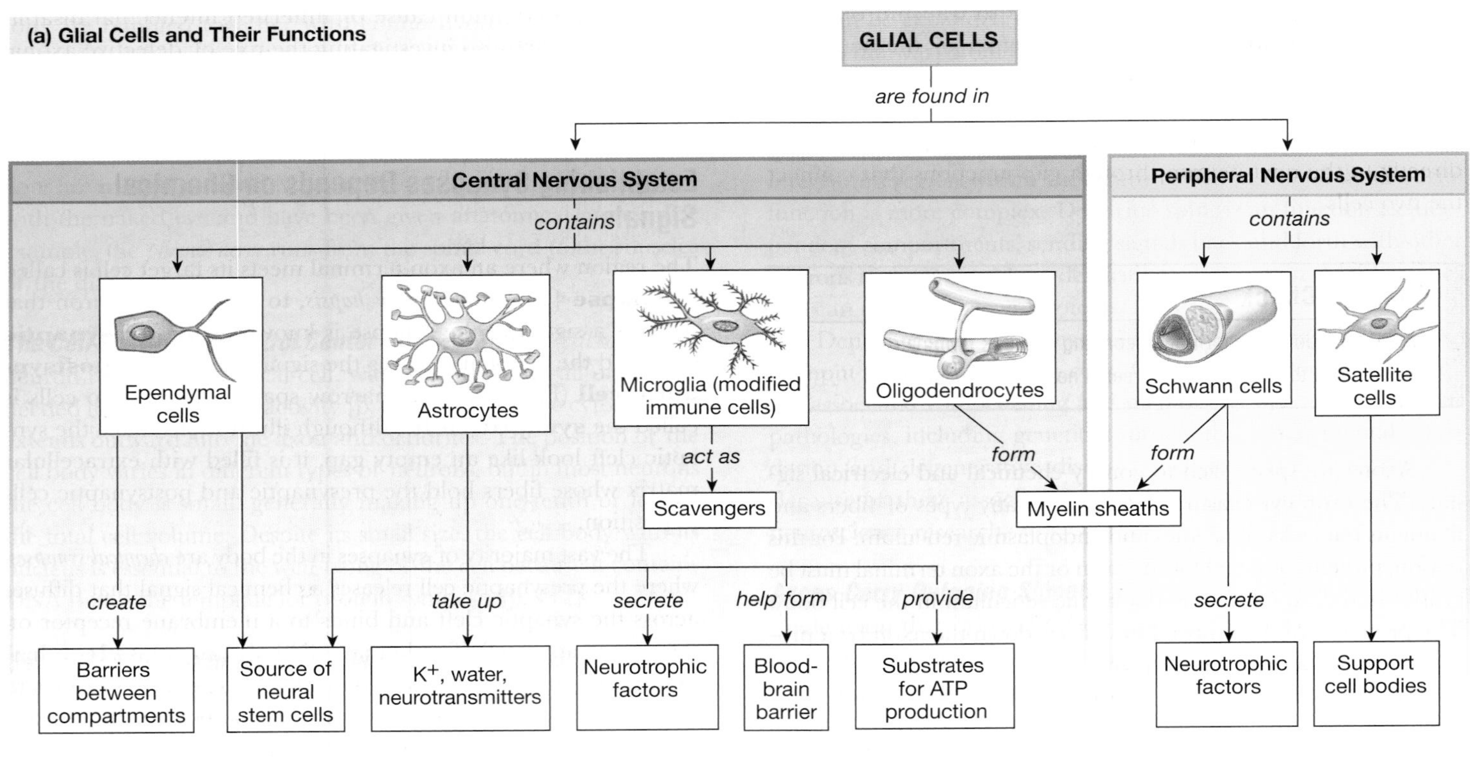
(a) Glial Cells and Their Functions
GLIAL CELLS
are found in
Central Nervous System
contains
Ependymal cells
Astrocytes
Microglia (modified immune cells)
Oligodendrocytes
Peripheral Nervous System
contains
Schwann cells
Satellite cells
act as
Scavengers
form
form
Myelin sheaths
create
take up
secrete
help form
provide
secrete
Barriers between compartments
Source of neural stem cells
K+, water, neurotransmitters
Neurotrophic factors
Blood-brain barrier
Substrates for ATP production
Neurotrophic factors
Support cell bodies

(b) Glial Cells of the Central Nervous System

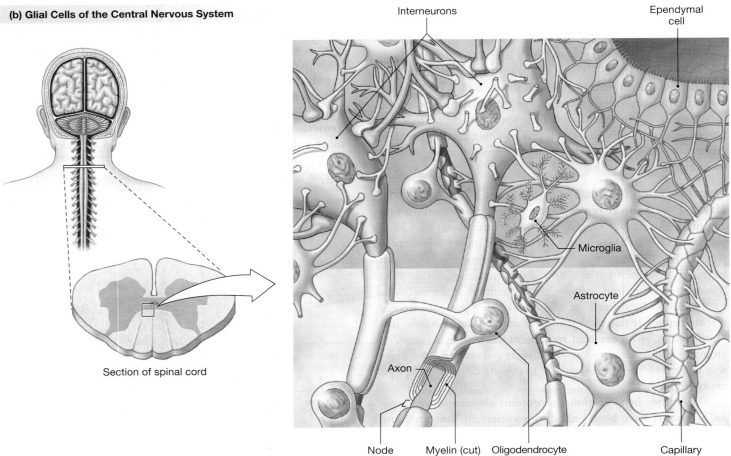

Section of spinal cord

Interneurons

Ependymal cell

Microglia

Astrocyte

Axon

Node

Myelin (cut)

Oligodendrocyte

Capillary

(c) Each Schwann cell forms myelin around a small segment of one axon.

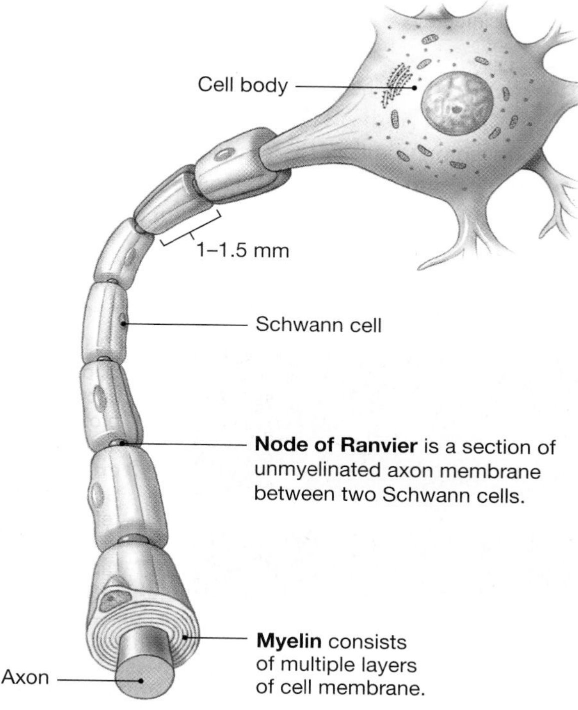
Cell body
1–1.5 mm
Schwann cell
Node of Ranvier is a section of unmyelinated axon membrane between two Schwann cells.
Myelin consists of multiple layers of cell membrane.
Axon

(d) Myelin Formation in the Peripheral Nervous System

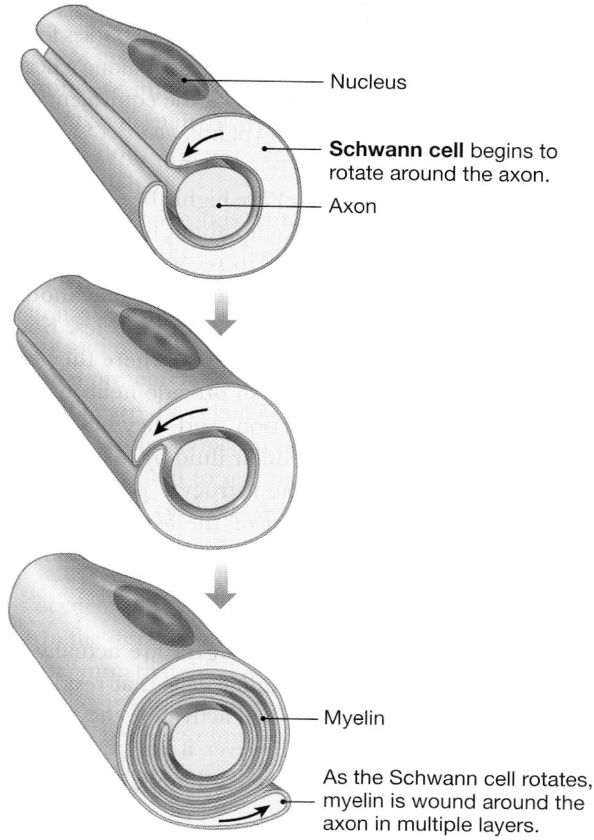
Nucleus
Schwann cell begins to rotate around the axon.
Axon
Myelin
As the Schwann cell rotates, myelin is wound around the axon in multiple layers.

where target cells are moved to an unusual location in the embryo, the axons in many instances are still able to find their targets by "sniffing out" the target's chemical scent. Growth cones depend on many different types of signals to find their way: growth factors, molecules in the extracellular matrix, and membrane proteins on the growth cones and on cells along the path. For example, *integrins* [p. 75] on the growth cone membrane bind to *laminins*, protein fibers in the extracellular matrix. *Nerve-cell adhesion molecules* (NCAMs) [p. 73] interact with membrane proteins of other cells.

Once an axon reaches its target cell, a synapse forms. However, synapse formation must be followed by electrical and chemical activity, or the synapse will disappear. The survival of neuronal pathways depends on **neurotrophic factors** {*trophikos*, nourishment} secreted by neurons and glial cells. There is still much we have to learn about this complicated process, and it is an active area of physiological research.

This "use it or lose it" scenario is most dramatically reflected by the fact that the infant brain is only about one-fourth the size of the adult brain. Further brain growth is due not to an increase in the number of cells but to an increase in size and number of axons, dendrites, and synapses. This development depends on electrical signaling between sensory pathways, interneurons, and efferent neurons.

Babies who are neglected or deprived of sensory input may experience delayed development ("failure to thrive") because of the lack of nervous system stimulation. On the other hand, there is no evidence that extra stimulation in infancy enhances intellectual development, despite a popular movement to expose babies to art, music, and foreign languages before they can even walk.

Once synapses form, they are not fixed for life. Variations in electrical activity can cause rearrangement of the synaptic connections, a process that continues throughout life. Maintaining synapses is one reason that older adults are urged to keep learning new skills and information.

Play BioFlix Animation
@Mastering Anatomy & Physiology

Concept Check

4. Draw a chain of three neurons that synapse on one another in sequence. Label the presynaptic and postsynaptic ends of each neuron, the cell bodies, dendrites, axons, and axon terminals.

Glial Cells Provide Support for Neurons

Glial cells {*glia*, glue} are the unsung heroes of the nervous system, outnumbering neurons by 10–50 to 1. For many years, scientists thought that the primary function of glial cells was physical support, and that glial cells had little influence on information processing. That view has changed. Although glial cells do not participate directly in the transmission of electrical signals over long distances, they do communicate with neurons and provide important biochemical and structural support.

Schwann Cells and Oligodendrocytes Neural tissue secretes very little extracellular matrix [p. 165], so glial cells create structural

8.3 Electrical Signals in Neurons

Nerve and muscle cells are described as *excitable tissues* because of their ability to propagate electrical signals rapidly in response to a stimulus. We now know that many other cell types generate electrical signals to initiate intracellular processes [See insulin secretion, p. 158], but the ability of nerve and muscle cells to send a constant electrical signal over long distance is characteristic of electrical signaling in these tissues.

The Nernst Equation Predicts Membrane Potential for a Single Ion

Recall that all living cells have a resting membrane potential difference (V_m) [p. 153] that represents the separation of electrical charge across the cell membrane. Two factors influence the membrane potential:

1. *The uneven distribution of ions* across the cell membrane. Normally, sodium (Na^+), chloride (Cl^-), and calcium (Ca^{2+}) are more concentrated in the extracellular fluid than in the cytosol. Potassium (K^+) is more concentrated in the cytosol than in the extracellular fluid.
2. *Differing membrane permeability to those ions.* The resting cell membrane is much more permeable to K^+ than to Na^+ or (Ca^{2+}) This makes K^+ the major ion contributing to the resting membrane potential.

The *Nernst equation* describes the membrane potential that would result if the membrane were permeable to only one ion [p. 153]. For any given ion concentration gradient, this membrane potential is called the *equilibrium potential* of the ion (E_{ion}):

$$E_{ion}\text{ (in mV)} = \frac{61}{z}\log\frac{[\text{ion}]_{out}}{[\text{ion}]_{in}}$$

where:

61 is 2.303 RT/F* at 37 °C,

z is the electrical charge on the ion (+1 for K^+), and

$[\text{ion}]_{out}$ and $[\text{ion}]_{in}$ are the ion concentrations outside and inside the cell.

The equilibrium potential for an ion is the membrane potential at which the electrical and chemical forces acting on the ion are equal and opposite.

* R is the ideal gas constant, T is absolute temperature, and F is the Faraday constant. For additional information on these values, see Appendix B.

When we use the estimated intracellular and extracellular concentrations for K^+ (**TBL. 8.2**) in the Nernst equation, the equation predicts a potassium equilibrium potential, or E_K of −90 mV. However, an average value for the resting membrane potential of neurons is −70 mV (inside the cell relative to outside), more positive than predicted by the potassium equilibrium potential. This means that other ions must be contributing to the membrane potential. Neurons at rest are slightly permeable to Na^+, and the leak of positive Na^+ into the cell makes the resting membrane potential slightly more positive than it would be if the cell were permeable only to K^+.

Play Interactive Physiology 2.0
@Mastering Anatomy & Physiology

Concept Check

7. Given the values in Table 8.2, use the Nernst equation to calculate the equilibrium potential for Ca^{2+}. Express the concentrations as powers of 10 and use your knowledge of logarithms [p. A-00] to try the calculations without a calculator.

The GHK Equation Predicts Membrane Potential Using Multiple Ions

In living systems, several different ions contribute to the membrane potential of cells. The **Goldman-Hodgkin-Katz (GHK) equation** calculates the membrane potential that results from the contribution of all ions that can cross the membrane. The GHK equation includes membrane permeability values because the permeability of an ion influences its contribution to the membrane potential. If the membrane is not permeable to a particular ion, that ion does not affect the membrane potential.

For mammalian cells, we assume that Na^+, K^+, and Cl^- are the three ions that influence membrane potential in resting cells. Each ion's contribution to the membrane potential is proportional to its ability to cross the membrane. The GHK equation for cells that are permeable to Na^+, K^+, and Cl^- is

$$V_m = 61\,log\frac{P_K[K^+]_{out} + P_{Na}[Na^+]_{out} + P_{Cl}[Cl^-]_{in}}{P_K[K^+]_{in} + P_{Na}[Na^+]_{in} + P_{Cl}[Cl^-]_{out}}$$

where:

V_m is the resting membrane potential in mV at 37 °C,

61 is 2.303 RT/F at 37 °C,

TABLE 8.2 Ion Concentrations and Equilibrium Potentials

Ion	Extracellular Fluid (mM)	Intracellular Fluid (mM)	E_{ion} at 37 °C
K^+	5 mM (normal: 3.5–5)	150 mM	−90 mV
Na^+	145 mM (normal: 135–145)	15 mM	+60 mV
Cl^-	108 mM (normal: 100–108)	10 mM (normal: 5–15)	−63 mV
Ca^{2+}	1 mM	0.0001 mM	See Concept Check question 7

P is the relative permeability of the membrane to the ion shown in the subscript, and

$[ion]_{out}$ and $[ion]_{in}$ are the ion concentrations outside and inside the cell.

Although this equation looks quite intimidating, it can be simplified into words to say: Resting membrane potential (V_m) is determined by the combined contributions of the (concentration gradient × membrane permeability) for each ion.

Notice that if the membrane is not permeable to an ion, the permeability value for that ion is zero, and the ion drops out of the equation. For example, cells at rest normally are not permeable to Ca^{2+} and, therefore, Ca^{2+} is not part of the GHK equation.

The GHK equation predicts resting membrane potentials based on given ion concentrations and membrane permeabilities. If permeabilities for Na^+ and Cl^- are zero, the equation reverts back to the Nernst equation for K^+. The GHK equation explains how the cell's slight permeability to Na^+ makes the resting membrane potential more positive than the E_K determined with the Nernst equation. The GHK equation can also be used to predict what happens to membrane potential when ion concentrations or membrane permeabilities change. To see this for yourself, try the free Nernst/Goldman equation simulator developed by the University of Arizona and available at *http://www.nernstgoldman.physiology.arizona.edu/*.

Ion Movement Creates Electrical Signals

The resting membrane potential of living cells is determined primarily by the K^+ concentration gradient and the cell's resting permeability to K^+, Na^+, and Cl^-. A change in either the K^+ concentration gradient or ion permeabilities changes the membrane potential. If you know numerical values for ion concentrations and permeabilities, you can use the GHK equation to calculate the new membrane potential.

In medicine, you usually will not have numerical values, however, so it is important to be able to think conceptually about the relationship between ion concentrations, permeabilities, and membrane potential. For example, at rest, the cell membrane of a neuron is only slightly permeable to Na^+. If the membrane suddenly increases its Na^+ permeability, Na^+ enters the cell, moving down its electrochemical gradient [p. 153]. The addition of positive Na^+ to the intracellular fluid *depolarizes* the cell membrane and creates an electrical signal.

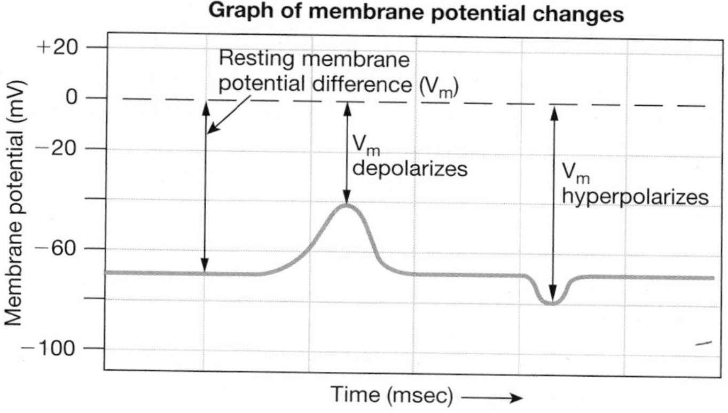
Graph of membrane potential changes
+20
0
−20
−60
−100
Membrane potential (mV)
Resting membrane potential difference (Vm)
Vm depolarizes
Vm hyperpolarizes
Time (msec)

The movement of ions across the membrane can also *hyperpolarize* a cell. If the cell membrane suddenly becomes more permeable to K^+, positive charge is lost from inside the cell, and the cell becomes more negative (hyperpolarizes). A cell may also hyperpolarize if negatively charged ions, such as Cl^-, enter the cell from the extracellular fluid.

Concept Check

8. Would a cell with a resting membrane potential of −70 mV depolarize or hyperpolarize in the following cases? (You must consider both the concentration gradient and the electrical gradient of the ion to determine net ion movement.)
   (a) Cell becomes more permeable to Ca^{2+}.
   (b) Cell becomes less permeable to K^+.
9. Would the cell membrane depolarize or hyperpolarize if a small amount of Na^+ leaked into the cell?

It is important to understand that a change in membrane potential from −70 mV to a positive value, such as +30 mV *does not mean that the ion concentration gradients have reversed!* A significant change in membrane potential occurs with the movement of very few ions. For example, to change the membrane potential by 100 mV, only 1 of every 100,000 K^+ must enter or leave the cell. This is such a tiny fraction of the total number of K^+ in the cell that the intracellular concentration of K^+ remains essentially unchanged even though the membrane potential has changed by 100 mV.

To appreciate how a tiny change can have a large effect, think of getting one grain of beach sand into your eye. There are so many grains of sand on the beach that the loss of one grain is not significant, just as the movement of one K^+ across the cell membrane does not significantly alter the concentration of K^+. However, the electrical signal created by moving a few K^+ across the membrane has a significant effect on the cell's membrane potential, just as getting that one grain of sand in your eye creates significant discomfort.

Gated Channels Control the Ion Permeability of the Neuron

How does a cell change its ion permeability? The simplest way is to open or close existing channels in the membrane. Neurons contain a variety of gated ion channels that alternate between open and closed states, depending on the intracellular and extracellular conditions [p. 138]. A slower method for changing membrane permeability is for the cell to insert new channels into the membrane or remove some existing channels.

Ion channels are usually named according to the primary ion(s) they allow to pass through them. There are four major types of selective ion channels in the neuron: (1) Na^+ channels, (2) K^+ channels, (3) Ca^{2+} channels, and (4) Cl^- channels. Other channels are less selective, such as the *monovalent cation channels* that allow both Na^+ and K^+ to pass.

The ease with which ions flow through a channel is called the channel's **conductance** (G) {*conductus*, escort}. Channel conductance varies with the gating state of the channel and with the channel protein isoform. Some ion channels, such as the K^+ *leak channels* that are the major determinant of resting membrane potential, spend most of their time in an open state. Other channels have gates that open or close in response to particular stimuli. Most gated channels fall into one of three categories [p. 138]:

1. **Mechanically gated ion channels** are found in sensory neurons and open in response to physical forces such as pressure or stretch.
2. **Chemically gated ion channels** in most neurons respond to a variety of ligands, such as extracellular neurotransmitters and neuromodulators or intracellular signal molecules.
3. **Voltage-gated ion channels** respond to changes in the cell's membrane potential. Voltage-gated Na^+ and K^+ channels play an important role in the initiation and conduction of electrical signals along the axon.

Not all voltage-gated channels behave in exactly the same way. The voltage level for channel opening varies from one channel type to another. For example, some channels we think of as leak channels are actually voltage-gated channels that remain open in the voltage range of the resting membrane potential.

The speed with which a gated channel opens and closes also differs among different types of channels. Channel opening to allow ion flow is called channel *activation*. For example, voltage-gated Na^+ channels and voltage-gated K^+ channels of axons are both activated by cell depolarization. The Na^+ channels open very rapidly, but the K^+ channels are slower to open. The result is an initial flow of Na^+ across the membrane, followed later by K^+ flow.

Many channels that open in response to depolarization close only when the cell repolarizes. The gating portion of the channel protein has an electrical charge that moves the gate between open and closed positions as membrane potential changes. This is like a spring-loaded door: It opens when you push on it, then closes when you release it.

Some channels also spontaneously *inactivate*. Even though the activating stimulus that opened them continues, the channel "times out" and closes. This is similar to doors with an automatic timed open-close mechanism. The door opens when you hit the button, then after a certain period of time, it closes itself, whether you are still standing in the doorway or not. An inactivated channel returns to its normal closed state shortly after the membrane repolarizes. The specific mechanisms underlying channel inactivation vary with different channel types.

Each major channel type has several to many subtypes with varying properties, and the list of subtypes gets longer each year. Within each subtype there may be multiple isoforms that express different opening and closing *kinetics* {*kinetikos*, moving} as well as associated proteins that modify channel properties. In addition, channel activity can be modulated by chemical factors that bind to the channel protein, such as phosphate groups.

CLINICAL FOCUS

Mutant Channels

Ion channels are proteins, and like other proteins they may lose or change function if their amino acid sequence is altered. **Channelopathies** {*pathos*, suffering} are inherited diseases caused by mutations in ion channel proteins. The most common channelopathy is *cystic fibrosis*, which results from defects in Cl^- channel. Because ion channels are so closely linked to the electrical activity of cells, many channelopathies manifest themselves as disorders of the excitable tissues (nerve and muscle). By studying defective ion channels, scientists have now shown that some disease states are actually families of related diseases with different causes but similar symptoms. For example, the condition known as *long Q-T syndrome* (LQTS; named for changes in the electrocardiogram test) is a cardiac problem characterized by an irregular heartbeat, fainting, and sometimes sudden death. Scientists have identified eight different gene mutations in K^+, Na^+, or Ca^{2+} channels that result in various subtypes of LQTS. Other well-known channelopathies include some forms of epilepsy and *malignant hyperthermia*, where ion flow through defective channels raises body temperature to potentially lethal levels.

Current Flow Obeys Ohm's Law

When ion channels open, ions may move into or out of the cell. The flow of electrical charge carried by an ion is called the ion's **current**, abbreviated $\mathbf{I_{ion}}$. The direction of ion movement depends on the *electrochemical* (combined electrical and concentration) gradient of the ion [p. 153]. Potassium ions usually move out of the cell. Na^+, Cl^-, and Ca^{2+} usually flow into the cell. The net flow of ions across the membrane depolarizes or hyperpolarizes the cell, creating an electrical signal.

Current flow, whether across a membrane or inside a cell, obeys a rule known as **Ohm's law** [p. A-00]. Ohm's law says that current flow (I) is directly proportional to the electrical potential difference (in volts, V) between two points and inversely proportional to the resistance (R) of the system to current flow: $I = V \times 1/R$ or $I = V/R$. In other words, as resistance R increases, current flow I decreases. (You will encounter a variant of Ohm's law when you study fluid flow in the cardiovascular and respiratory systems.)

Resistance in biological flow is the same as resistance in everyday life: It is a force that opposes flow. Electricity is a form of energy and, like other forms of energy, it dissipates as it encounters resistance. As an analogy, think of rolling a ball along the floor. A ball rolled across a smooth wood floor encounters less resistance than a ball rolled across a carpeted floor. If you throw both balls with the same amount of energy, the ball that encounters less resistance retains energy longer and travels farther along the floor.

In biological electricity, resistance to current flow comes from two main sources: the resistance of the cell membrane (R_m) and the internal resistance of the cytoplasm (R_i). (The extracellular fluid also creates resistance R_o, but it is so small compared to R_m and R_i that it is usually ignored.) The phospholipid bilayer of the cell membrane is normally an excellent insulator, and a membrane with no open ion channels has very high resistance and low conductance. If ion channels open, ions (current) flow across the membrane if there is an electrochemical gradient for them. Opening ion channels therefore decreases the membrane resistance.

The internal resistance of most neurons is determined by the composition of the cytoplasm and the diameter of the cell. Cytoplasmic composition is relatively constant. Internal resistance decreases as cell diameter increases, so larger diameter neurons have lower resistance. The membrane resistance and internal resistance together determine how far current will flow through a cell before the energy is dissipated and the current dies. The combination of resistances (R_m, R_i, and R_o) creates the *length constant* for a given neuron and can be calculated mathematically. The length constant is sometimes called the *space constant*.

Voltage changes across the membrane can be classified into two basic types of electrical signals: graded potentials and action potentials (**TBL. 8.3**). **Graded potentials** are variable-strength signals that travel over short distances and lose strength as they travel through the cell. They are used for short-distance communication. If a depolarizing graded potential is strong enough when it reaches an integrating region within a neuron, the graded potential initiates an action potential. **Action potentials** are very brief, large depolarizations that travel for long distances through a neuron without losing strength. Their function is rapid signaling over long distances, such as from your toe to your brain.

Graded Potentials Reflect Stimulus Strength

Graded potentials in neurons are depolarizations or hyperpolarizations that occur in the dendrites and cell body or, less frequently, near the axon terminals. These changes in membrane potential are called "graded" because their size, or *amplitude* {*amplitudo*, large}, is directly proportional to the strength of the triggering event. A large stimulus causes a strong graded potential, and a small stimulus results in a weak graded potential.

In neurons of the CNS and the efferent division, graded potentials occur when chemical signals from other neurons open chemically gated ion channels, allowing ions to enter or leave the neuron. Mechanical stimuli (such as stretch) or chemical stimuli open ion channels in some sensory neurons. Graded potentials may also occur when an open channel closes, decreasing the movement of ions through the cell membrane. For example, if K^+ leak channels close, fewer K^+ leave the cell. The retention of K^+ depolarizes the cell.

Concept Check

10. Match each ion's movement with the type of graded potential it creates.

(a) Na^+ entry
(b) Cl^- entry
(c) K^+ exit
(d) Ca^{2+} entry

1. depolarizing
2. hyperpolarizing

FIGURE 8.7a shows a graded potential that begins when a stimulus opens monovalent cation channels on the cell body of a neuron. Sodium ions move into the neuron, bringing in electrical energy.

TABLE 8.3 Comparison of Graded Potential and Action Potential in Neurons

	Graded Potential	Action Potential
Type of Signal	Input signal	Regenerating conduction signal
Occurs Where?	Usually dendrites and cell body	Trigger zone through axon
Types of Gated Ion Channels Involved	Mechanically, chemically, or voltage-gated channels	Voltage-gated channels
Ions Involved	Usually Na^+, K^+, Ca^{2+}	Na^+ and K^+
Type of Signal	Depolarizing (e.g., Na^+) or hyperpolarizing (e.g., Cl^-)	Depolarizing
Strength of Signal	Depends on initial stimulus; can be summed	All-or-none phenomenon; cannot be summed
What Initiates the Signal?	Entry of ions through gated channels	Above-threshold graded potential at the trigger zone opens ion channels
Unique Characteristics	No minimum level required to initiate	Threshold stimulus required to initiate
	Two signals coming close together in time will sum	Refractory period: two signals too close together in time cannot sum
	Initial stimulus strength is indicated by frequency of a series of action potentials	

FIG. 8.7 ESSENTIALS Graded potentials

(a) Graded Potentials

Graded potentials decrease in strength as they spread out from the point of origin due to current leak out of the cell.

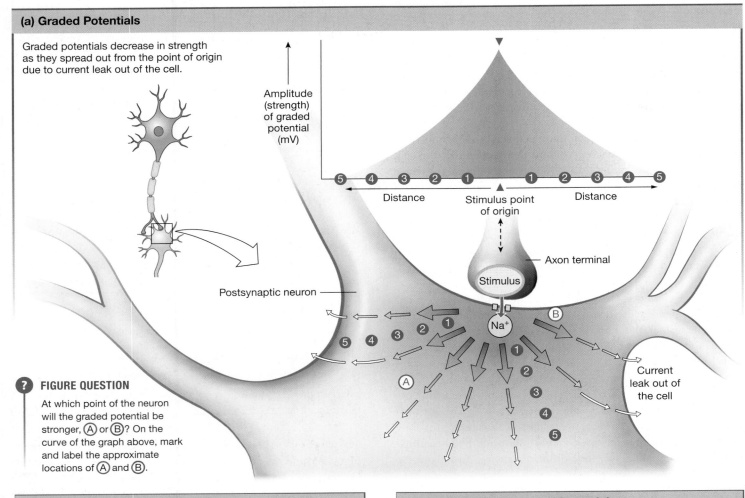

Amplitude (strength) of graded potential (mV)

5 4 3 2 1 1 2 3 4 5

Distance

Stimulus point of origin

Distance

Axon terminal

Stimulus

Postsynaptic neuron

Na^+

A

B

Current leak out of the cell

FIGURE QUESTION

At which point of the neuron will the graded potential be stronger, Ⓐ or Ⓑ? On the curve of the graph above, mark and label the approximate locations of Ⓐ and Ⓑ.

(b) Subthreshold Graded Potential

A graded potential starts above threshold (T) at its initiation point but decreases in strength as it travels through the cell body. At the trigger zone, it is below threshold and, therefore, does not initiate an action potential.

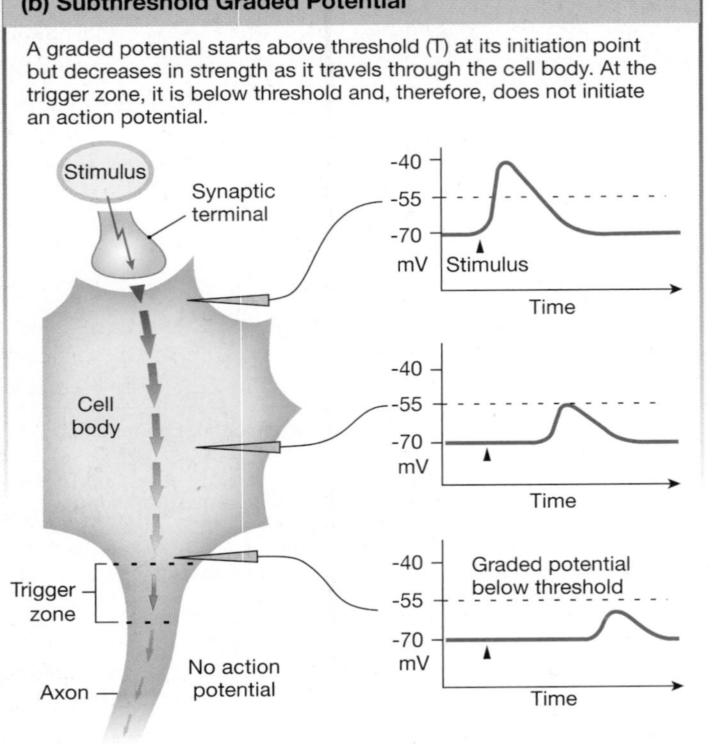

Stimulus
Synaptic terminal
Cell body
Trigger zone
Axon
No action potential
-40
-55
-70
mV
Stimulus
Time
Graded potential below threshold

(c) Suprathreshold Graded Potential

A stronger stimulus at the same point on the cell body creates a graded potential that is still above threshold by the time it reaches the trigger zone, so an action potential results.

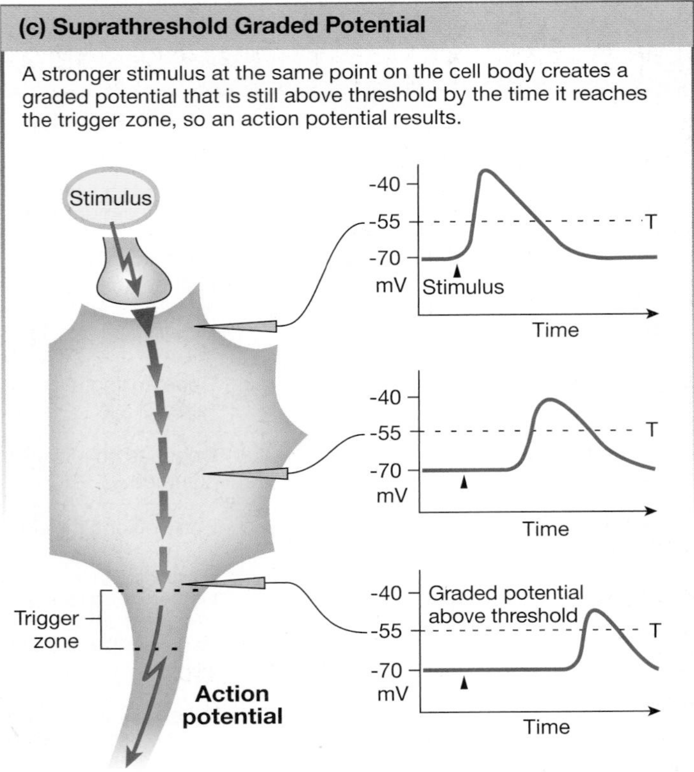

Stimulus
Trigger zone
Action potential
-40
-55
-70
mV
T
Stimulus
Time
Graded potential above threshold

The positive charge carried in by the Na^+ spreads as a wave of depolarization through the cytoplasm, just as a stone thrown into water creates ripples or waves that spread outward from the point of entry. The wave of depolarization that moves through the cell is known as **local current flow**. By convention, current in biological systems is the net movement of *positive* electrical charge.

The strength of the initial depolarization in a graded potential is determined by how much charge enters the cell, just as the size of waves caused by a stone tossed in water is determined by the size of the stone. If more Na^+ channels open, more Na^+ enters, and the graded potential has higher initial amplitude. The stronger the initial amplitude, the farther the graded potential can spread through the neuron before it dies out.

Why do graded potentials lose strength as they move through the cytoplasm? Two factors play a role:

1. *Current leak.* The membrane of the neuron cell body has open leak channels that allow positive charge to leak out into the extracellular fluid. Some positive ions leak out of the cell across the membrane as the depolarization wave moves through the cytoplasm, decreasing the strength of the signal moving down the cell.
2. *Cytoplasmic resistance.* The cytoplasm provides resistance to the flow of electricity, just as water creates resistance that diminishes the waves from the stone. The combination of current leak and cytoplasmic resistance means that the strength of the signal inside the cell decreases over distance.

Graded potentials that are strong enough eventually reach the region of the neuron known as the **trigger zone**. In efferent neurons and interneurons, the trigger zone is the *axon hillock* and the very first part of the axon, a region known as the **initial segment**. In sensory neurons, the trigger zone is immediately adjacent to the receptor, where the dendrites join the axon (see Fig. 8.2).

Concept Check

11. Identify the trigger zones of the neurons illustrated in Figure 8.2, if possible.

The trigger zone is the integrating center of the neuron and contains a high concentration of voltage-gated Na^+ channels in its membrane. If graded potentials reaching the trigger zone depolarize the membrane to the threshold voltage, voltage-gated Na^+ channels open, and an action potential begins. If the depolarization does not reach threshold, the graded potential simply dies out as it moves into the axon.

Because depolarization makes a neuron more likely to fire an action potential, depolarizing graded potentials are considered to be *excitatory*. A hyperpolarizing graded potential moves the membrane potential farther from the threshold value and makes the neuron less likely to fire an action potential. Consequently, hyperpolarizing graded potentials are considered to be *inhibitory*.

Figure 8.7b shows a neuron with three recording electrodes placed at intervals along the cell body and trigger zone. A single stimulus triggers a *subthreshold* graded potential, one that is below threshold by the time it reaches the trigger zone. Although the cell is depolarized to −40 mV at the site where the graded potential begins, the current decreases as it travels through the cell body. As a result, the graded potential is below threshold by the time it reaches the trigger zone. (For the typical mammalian neuron, threshold is about −55 mV.) The stimulus is not strong enough to depolarize the cell to threshold at the trigger zone, and the graded potential dies out without triggering an action potential.

Figure 8.7c shows *suprathreshold* graded potential, one that is strong enough to cause an action potential. A stronger initial stimulus on the cell body initiates a stronger depolarization and current flow. Although this graded potential also diminishes with distance as it travels through the neuron, its higher initial strength ensures that it is above threshold at the trigger zone. In this example, the graded potential triggers an action potential. The ability of a neuron to respond to a stimulus and fire an action potential is called the cell's **excitability**.

Action Potentials Travel Long Distances

Action potentials, also known as *spikes*, are electrical signals of uniform strength that travel from a neuron's trigger zone to the end of its axon. In action potentials, voltage-gated ion channels in the axon membrane open sequentially as electrical current passes down the axon. As a result, additional Na^+ entering the cell reinforce the depolarization, which is why an action potential does not lose strength over distance the way a graded potential does. Instead, the action potential at the end of an axon is identical to the action potential that started at the trigger zone: a depolarization of about 100 mV amplitude. The high-speed movement of an action potential along the axon is called **conduction** of the action potential.

Action potentials are sometimes called **all-or-none** phenomena because they either occur as a maximal depolarization (if the stimulus reaches threshold) or do not occur at all (if the stimulus is below threshold). The strength of the graded potential that initiates an action potential has no influence on the amplitude of the action potential.

When we talk about action potentials, it is important to realize that there is no single action potential that moves through the cell. The action potential that occurs at the trigger zone is like the movement in the first domino of a series of dominos standing on end (**FIG. 8.8a**). As the first domino falls, it strikes the next, passing on its kinetic energy. As the second domino falls, it passes kinetic energy to the third domino, and so on. If you could take a snapshot of the line of falling dominos, you would see that as the first domino is coming to rest in the fallen position, the next one is almost down, the third one most of the way down, and so forth, until you reach the domino that has just been hit and is starting to fall.

In an action potential, a wave of electrical energy moves down the axon. Instead of getting weaker over distance, action potentials are replenished along the way so that they maintain constant amplitude. As the action potential passes from one part of the axon to

TABLE 8.4 Major Neurocrines*

Chemical	Receptor	Type	Receptor Location	Key Agonists, Antagonists, and Potentiators**
Acetylcholine (ACh)	Cholinergic			
	Nicotinic (nAChR)	ICR‡ (Na^+, K^+)	Skeletal muscles, autonomic neurons, CNS	**Agonist:** nicotine **Antagonists:** curare, α-bungarotoxin
	Muscarinic (M)	GPCR	Smooth and cardiac muscle, endocrine and exocrine glands, CNS	**Agonist:** muscarine **Antagonist:** atropine
Amines				
Norepinephrine (NE) Epinephrine (E)	Adrenergic (α, β)	GPCR	Smooth and cardiac muscle, glands, CNS	**Antagonists:** α-receptors: ergotamine, phentolamine. β-receptors: propranolol
Dopamine (DA)	Dopamine (D)	GPCR	CNS	**Agonist:** bromocriptine **Antagonists:** antipsychotic drugs
Serotonin (5-hydroxytryptamine, 5-HT)	Serotonergic (5-HT)	ICR (Na^+, K^+), GPCR	CNS	**Agonist:** sumatriptan **Antagonist:** LSD
Histamine	Histamine (H)	GPCR	CNS	**Antagonists:** ranitidine (Zantac®) and cimetidine (Tagamet®)
Amino Acids				
Glutamate	Glutaminergic ionotropic (iGluR)			
	AMPA	ICR (Na^+, K^+)	CNS	**Agonist:** quisqualate
	NMDA	ICR (Na^+, K^+)	CNS	**Potentiator:** serine
	Glutaminergic metabotropic (mGluR)	GPCR	CNS	**Potentiator:** glycine
GABA (γ-aminobutyric acid)	GABA	ICR (Cl^-), GPCR	CNS	**Antagonist:** picrotoxin **Potentiators:** alcohol, barbiturates
Glycine	Glycine (GlyR)	ICR (Cl^-)	CNS	**Antagonist:** strychnine
Purines				
Adenosine	Purine (P)	GPCR	CNS	
Gases				
Nitric oxide (NO)	None	N/A	N/A	

*This table does not include the numerous peptides that can act as neurocrines.
**This list does not include many chemicals that are used as agonists and antagonists in physiological research.
‡ICR = ion channel-receptor; GPCR = G protein-coupled receptor; AMPA = α-amino-3-hydroxy-5-methylisoxazole-4-proprionic acid; NMDA = N-methyl-D-aspartate; LSD = lysergic acid diethylamide; N/A = not applicable.

two precursors is a simple enzymatic reaction that takes place in the axon terminal. Neurons that secrete ACh and receptors that bind ACh are described as **cholinergic**.

Cholinergic receptors come in two main subtypes: **nicotinic**, named because *nicotine* is an agonist, and **muscarinic**, for which *muscarine*, a compound found in some fungi, is an agonist. Cholinergic nicotinic receptors are receptor-channels found on skeletal muscle, in the autonomic division of the PNS, and in the CNS. Nicotinic receptors are monovalent cation channels through which both Na^+ and K^+ can pass. Sodium entry into cells exceeds K^+ exit because the electrochemical gradient for Na^+ is stronger. As a result, net Na^+ entry depolarizes

CLINICAL FOCUS

Myasthenia Gravis

What would you think was wrong if suddenly your eyelids started drooping, you had difficulty watching moving objects, and it became difficult to chew, swallow, and talk? What disease attacks these skeletal muscles but leaves the larger muscles of the arms and legs alone? The answer is {*myo-*, muscle + *asthenes*, weak + *gravis*, severe}, an autoimmune disease in which the body fails to recognize the acetylcholine (ACh) receptors on skeletal muscle as part of "self." The immune system then produces antibodies to attack the receptors. The antibodies bind to the ACh receptor protein and change it in some way that causes the muscle cell to pull the receptors out of the membrane and destroy them. This destruction leaves the muscle with fewer ACh receptors in the membrane. Even though neurotransmitter release is normal, the muscle target has a diminished response that exhibits as muscle weakness. Currently medical science does not have a cure for myasthenia gravis, although various drugs can help control its symptoms. To learn more about this disease, visit the website for the Myasthenia Gravis Foundation of America at *www.myasthenia.org*.

the postsynaptic cell and makes it more likely to fire an action potential.

Cholinergic muscarinic receptors come in five related subtypes. They are all G protein-coupled receptors linked to second messenger systems. The tissue response to activation of a muscarinic receptor varies with the receptor subtype. These receptors occur in the CNS and on targets of the autonomic parasympathetic division of the PNS.

Amines The amine neurotransmitters are all active in the CNS. Like the amine hormones [p. 202], these neurotransmitters are derived from single amino acids. **Serotonin**, also called *5-hydroxytryptamine* or 5-HT, is made from the amino acid tryptophan. *Histamine*, made from histidine, plays a role in allergic responses in addition to serving as a neurotransmitter.

The amino acid tyrosine is converted to **dopamine**, **norepinephrine**, and **epinephrine**. Norepinephrine is the major neurotransmitter of the PNS autonomic sympathetic division. All three tyrosine-derived molecules can also function as neurohormones.

Neurons that secrete norepinephrine are called **adrenergic neurons**, or, more properly, **noradrenergic neurons**. The adjective *adrenergic* does not have the same obvious link to its neurotransmitter as *cholinergic* does to *acetylcholine*. Instead, the adjective derives from the British name for epinephrine, *adrenaline*. In the early part of the twentieth century, British researchers thought that sympathetic neurons secreted adrenaline (epinephrine), hence the modifier *adrenergic*. Although our understanding has changed, the name persists. Whenever you see reference to "adrenergic control" of a function, you must make the connection to a neuron secreting norepinephrine.

Adrenergic receptors are divided into two classes: α (alpha) and β (beta), with multiple subtypes of each. Like cholinergic muscarinic receptors, adrenergic receptors are linked to G proteins. The two subtypes of adrenergic receptors work through different second messenger pathways. The action of epinephrine on β-receptors in dog liver led E. W. Sutherland to the discovery of cyclic AMP and the concept of second messenger systems as transducers of extracellular messengers [p. 173].

Concept Check

18. When pharmaceutical companies design drugs, they try to make a given drug as specific as possible for the particular receptor subtype they are targeting. For example, a drug might target adrenergic β_1- receptors rather than all adrenergic α and β receptors. What is the advantage of this specificity?

Amino Acids Several amino acids function as neurotransmitters in the CNS. **Glutamate** is the primary excitatory neurotransmitter of the CNS, and **aspartate** is an excitatory neurotransmitter in selected regions of the brain. *Excitatory neurotransmitters* depolarize their target cells, usually by opening ion channels that allow flow of positive ions into the cell.

The main inhibitory neurotransmitter in the brain is **gamma-aminobutyric acid (GABA)**. The *inhibitory neurotransmitters* hyperpolarize their target cells by opening Cl^- channels and allowing Cl^- to enter the cell.

Glutamate also acts as a neuromodulator. The action of glutamate at a particular synapse depends on which of its receptor types occurs on the target cell. Metabotropic glutaminergic receptors act through GPCRs. Two ionotropic glutamate receptors are receptor-channels.

AMPA receptors are ligand-gated monovalent cation channels similar to nicotinic acetylcholine channels. Glutamate binding opens the channel, and the cell depolarizes because of net Na^+ influx. AMPA receptors are named for their agonist *α-amino-3-hydroxy-5-methylisoxazole-4-proprionic acid.*

NMDA receptors are named for the glutamate agonist *N-methyl-D-aspartate*. They are unusual for several reasons. First, they are nonselective cation channels that allow Na^+, K^+, and Ca^{2+} to pass through the channel. Second, channel opening requires both glutamate binding and a change in membrane potential. The NMDA receptor-channel's action is described in the section on long-term potentiation later in this chapter.

RUNNING PROBLEM

Dr. McKhann decided to perform nerve conduction tests on some of the paralyzed children in Beijing Hospital. He found that although the rate of conduction along the children's nerves was normal, the strength of the summed action potentials traveling down the nerve was greatly diminished.

Q4: *Is the paralytic illness that affected the Chinese children a demyelinating condition? Why or why not?*

224 – 226 – 233 – 248 – **251** – 254 – 264 – 265

Glycine and the amino acid D-*serine* potentiate, or enhance, the excitatory effects of glutamate at one type of glutamate receptor. D-serine is made and released by glial cells as well as neurons, which illustrates the role that glial cells can play in altering synaptic communication.

Peptides The nervous system secretes a variety of peptides that act as neurotransmitters and neuromodulators in addition to functioning as neurohormones. These peptides include **substance P**, involved in some pain pathways, and the **opioid peptides** (**enkephalins** and **endorphins**) that mediate pain relief, or *analgesia* {*an-*, without *algos*, pain}. Peptides that function as both neurohormones and neurotransmitters include *cholecystokinin* (CCK), *vasopressin* (AVP), and *atrial natriuretic peptide* (ANP). Many peptide neurotransmitters are co-secreted with other neurotransmitters.

Purines *Adenosine*, *adenosine monophosphate* (AMP), and *adenosine triphosphate* (ATP) can all act as neurotransmitters. These molecules, known collectively as *purines* [p. 34], bind to *purinergic* receptors in the CNS and on other excitable tissues such as the heart. The purinergic receptors are all G protein-coupled receptors.

Gases One of the most interesting neurotransmitters is *nitric oxide* (NO), an unstable gas synthesized from oxygen and the amino acid arginine. Nitric oxide acting as a neurotransmitter diffuses freely into a target cell rather than binding to a membrane receptor [p. 177]. Once inside the target cell, nitric oxide binds to target proteins. With a half-life of only 2–30 seconds, nitric oxide is elusive and difficult to study. It is also released from cells other than neurons and often acts as a paracrine signal.

Recent work suggests that *carbon monoxide* (CO) and hydrogen sulfide (H_2S), both known as toxic gases, are produced by the body in tiny amounts to serve as neurotransmitters.

Lipids Lipid neurocrine molecules include several eicosanoids [p. 30] that are the endogenous ligands for *cannabinoid receptors*. The CB_1 cannabinoid receptor is found in the brain, and the CB_2 receptor is found on immune cells. The receptors were named for one of their exogenous ligands, Δ^9-*tetrahydrocannabinoid* (THC), which comes from the plant *Cannabis sativa*, more commonly known as marijuana. Lipid neurocrine signals all bind to G protein-coupled receptors.

RUNNING PROBLEM

Dr. McKhann then asked to see autopsy reports on some of the children who had died of their paralysis at Beijing Hospital. In the reports, pathologists noted that the patients had normal myelin but damaged axons. In some cases, the axon had been completely destroyed, leaving only a hollow shell of myelin.

Q5: *Do the results of Dr. McKhann's investigation suggest that the Chinese children had classic GBS? Why or why not?*

224 – 226 – 233 – 248 – 251 – **254** – 264 – 265

BIOTECHNOLOGY

Of Snakes, Snails, Spiders, and Sushi

What do snakes, marine snails, and spiders have to do with neurophysiology? They all provide neuroscientists with compounds for studying synaptic transmission, extracted from the neurotoxic venoms these creatures use to kill their prey. The Asian snake *Bungarus multicinctus* provides us with α-bungarotoxin, a long-lasting poison that binds tightly to nicotinic acetylcholine receptors. The fish-hunting cone snail, *Conus geographus*, and the funnel web spider, *Agelenopsis aperta*, use toxins that block different types of voltage-gated Ca^{2+} channels. One of the most potent poisons known, however, comes from the Japanese puffer fish, a highly prized delicacy whose flesh is consumed as sushi. The puffer has tetrodotoxin (TTX) in its gonads. This neurotoxin blocks Na^+ channels on axons and prevents the transmission of action potentials, so ingestion of only a tiny amount can be fatal. The Japanese chefs who prepare the puffer fish, or *fugu*, for consumption are carefully trained to avoid contaminating the fish's flesh as they remove the toxic gonads. There's always some risk involved in eating *fugu*, though—one reason that traditionally the youngest person at the table is the first to sample the dish.

Neurotransmitters Are Released from Vesicles

When we examine the axon terminal of a presynaptic cell with an electron microscope, we find many small **synaptic vesicles** filled with neurotransmitter that is released on demand (**FIG. 8.18**). Some vesicles are "docked" at active zones along the membrane closest to the synaptic cleft, waiting for a signal to release their contents. Other vesicles act as a reserve pool, clustering close to the docking sites. Axon terminals also contain mitochondria to produce ATP for metabolism and transport. In this section, we discuss general patterns of neurotransmitter synthesis, storage, release, and termination of action.

Neurotransmitter Synthesis Neurotransmitter synthesis takes place both in the nerve cell body and in the axon terminal. Polypeptides must be made in the cell body because axon terminals do not have the organelles needed for protein synthesis. Protein synthesis follows the usual pathway [p. 112]. The large *propeptide* that results is packaged into vesicles along with the enzymes needed to modify it. The vesicles then move from the cell body to the axon terminal by fast axonal transport. Inside the vesicle, the propeptide is broken down into smaller active peptides—a pattern similar to the preprohormone-prohormone-active hormone process in endocrine cells [p. 199]. For example, one propeptide contains the amino acid sequences for three active peptides that are co-secreted: ACTH, gamma (γ-)lipotropin, and beta (β-)endorphin.

Smaller neurotransmitters, such as acetylcholine, amines, and purines, are synthesized and packaged into vesicles in the axon terminal. The enzymes needed for their synthesis are made in the cell body and released into the cytosol. The dissolved enzymes are then brought to axon terminals by slow axonal transport.

Neurotransmitter Release Neurotransmitters in the axon terminal are stored in vesicles, so their release into the synaptic cleft takes place by exocytosis [p. 147]. From what we can tell, exocytosis in neurons is similar to exocytosis in other types of cells, but much faster. Neurotoxins that block neurotransmitter release, including tetanus and botulinum toxins, exert their action by inhibiting specific proteins of the cell's exocytotic apparatus.

FIGURE 8.19a shows how neurotransmitters are released by exocytosis. When the depolarization of an action potential reaches the axon terminal, the change in membrane potential sets off a sequence of events **1**. The axon terminal membrane has voltage-gated Ca^{2+} channels that open in response to depolarization **2**. Calcium ions are more concentrated in the extracellular fluid than in the cytosol, and so they move into the cell down their electrochemical gradient.

Ca^{2+} entering the cell binds to regulatory proteins and initiates exocytosis **3**. The membrane of the synaptic vesicle fuses with the cell membrane, aided by multiple membrane proteins. The fused area opens, and neurotransmitter inside the synaptic vesicle moves into the synaptic cleft **4**. The neurotransmitter molecules diffuse across the gap to bind with membrane receptors on the postsynaptic cell. When neurotransmitters bind to their receptors, a response is initiated in the postsynaptic cell **5**. Each synaptic vesicle contains the same amount of neurotransmitter, so measuring the magnitude of the target cell response is an indication of how many vesicles released their content.

In the classic model of exocytosis, the membrane of the vesicle becomes part of the axon terminal membrane (Fig. 5.19, p. 148). To prevent a large increase in membrane surface area, the membrane is recycled by endocytosis of vesicles at regions away from the active sites (Fig. 8.3). The recycled vesicles are then refilled with newly made neurotransmitter.

The transporters that concentrate neurotransmitter into vesicles are H^+-dependent antiporters (p. 140). The vesicles use H^+-ATPases to concentrate H^+ inside the vesicle, then exchange H^+ for the neurotransmitter.

Recently, a second model of secretion has emerged. In this model, called the **kiss-and-run pathway**, synaptic vesicles fuse to the presynaptic membrane at a complex called the **fusion pore**. This fusion opens a small channel that is just large enough for neurotransmitter to pass through. Then, instead of opening the fused area wider and incorporating the vesicle membrane into the cell membrane, the vesicle pulls back from the fusion pore and returns to the pool of vesicles in the cytoplasm.

Concept Check

19. Which organelles are needed to synthesize proteins and package them into vesicles?
20. What is the function of mitochondria in a cell?
21. How do mitochondria get to the axon terminals?

FIG. 8.18 A chemical synapse

The axon terminal contains mitochondria and synaptic vesicles filled with neurotransmitter. The postsynaptic membrane has receptors for neurotransmitter that diffuses across the synaptic cleft.

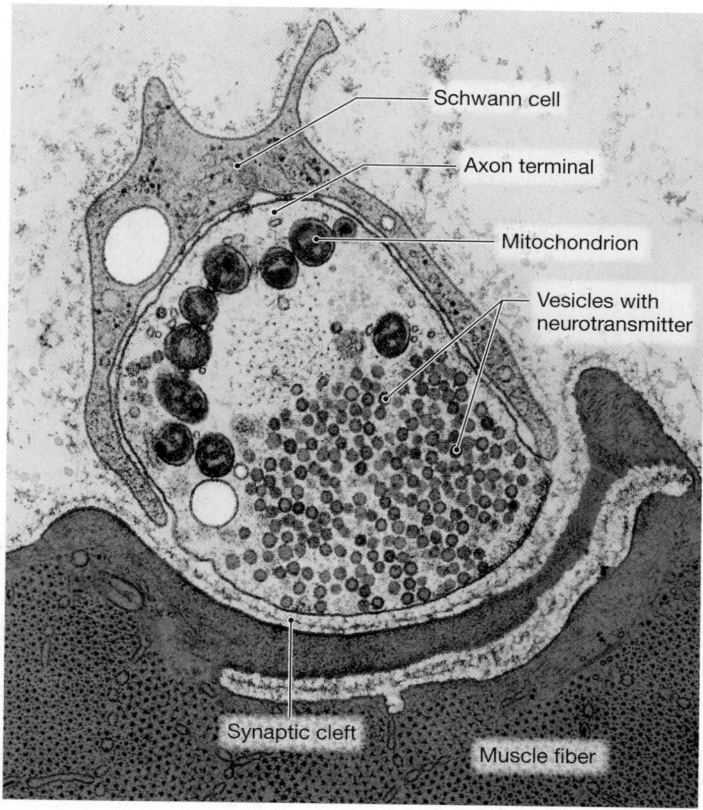
Schwann cell
Axon terminal
Mitochondrion
Vesicles with neurotransmitter
Synaptic cleft
Muscle fiber

Concept Check

22. In an experiment on synaptic transmission, a synapse was bathed in a Ca^{2+}-free medium that was otherwise equivalent to extracellular fluid. An action potential was triggered in the presynaptic neuron. Although the action potential reached the axon terminal at the synapse, the usual response of the postsynaptic cell did not occur. What conclusion did the researchers draw from these results?
23. Classify the H^+-neurotransmitter exchange as facilitated diffusion, primary active transport, or secondary active transport. Explain your reasoning.

Termination of Neurotransmitter Activity A key feature of neural signaling is its short duration, due to the rapid removal or inactivation of neurotransmitter in the synaptic cleft. Recall that ligand binding to a protein is reversible and goes to a state of equilibrium, with a constant ratio of unbound to bound ligand [p. 47]. If unbound neurotransmitter is removed from the synapse, the receptors release bound neurotransmitter, terminating its activity, to keep the ratio of unbound/bound transmitter constant.

Removal of unbound neurotransmitter from the synaptic cleft can be accomplished in various ways (Fig. 8.19b). Some neurotransmitter molecules simply diffuse away from the synapse, becoming

FIG. 8.19 ESSENTIALS Synaptic communication

Cell-to-cell communication uses chemical and electrical signaling to coordinate function and maintain homeostasis.

(a) Neurotransmitter Release

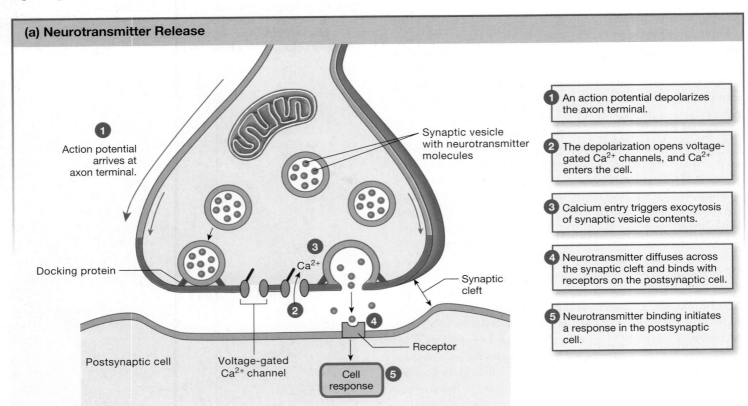

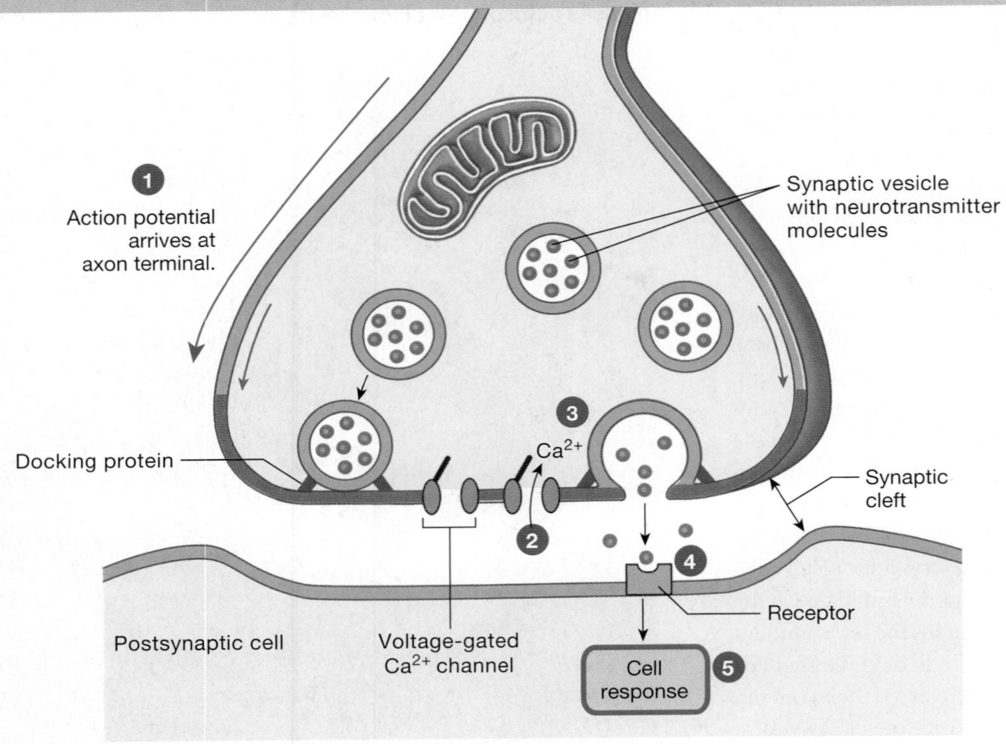

1 Action potential arrives at axon terminal.
Synaptic vesicle with neurotransmitter molecules
Docking protein
3
Ca^{2+}
2
Synaptic cleft
4
Receptor
Postsynaptic cell
Voltage-gated Ca^{2+} channel
Cell response 5

1. An action potential depolarizes the axon terminal.
2. The depolarization opens voltage-gated Ca^{2+} channels, and Ca^{2+} enters the cell.
3. Calcium entry triggers exocytosis of synaptic vesicle contents.
4. Neurotransmitter diffuses across the synaptic cleft and binds with receptors on the postsynaptic cell.
5. Neurotransmitter binding initiates a response in the postsynaptic cell.

(b) Neurotransmitter Termination

Neurotransmitter action terminates when the chemicals are broken down, are taken up into cells, or diffuse away from the synapse.

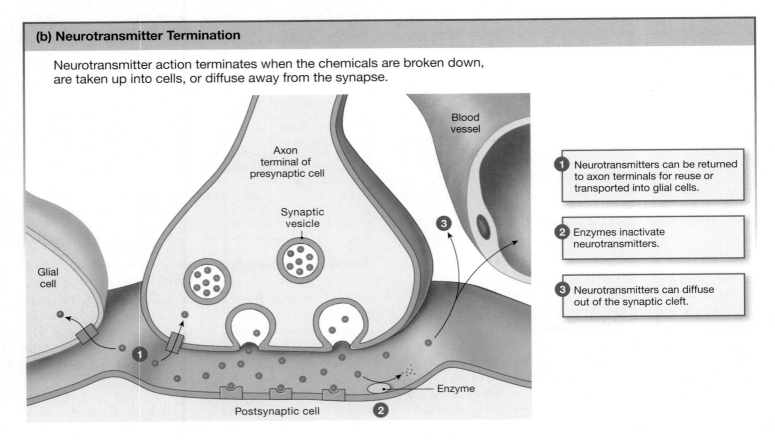

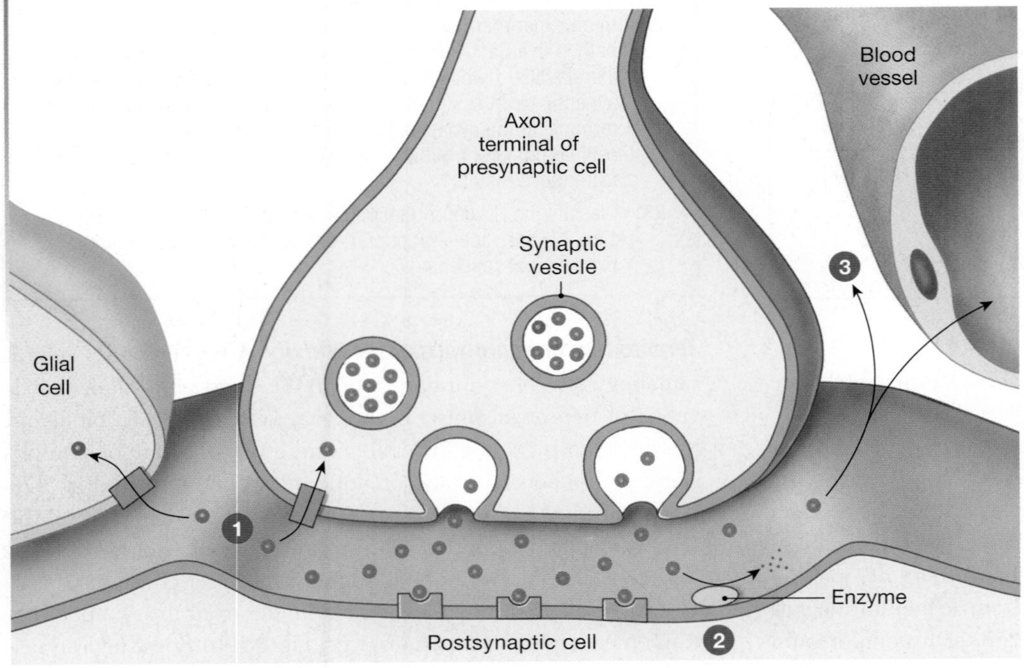

Blood vessel
Axon terminal of presynaptic cell
Synaptic vesicle
3
Glial cell
1
Enzyme
Postsynaptic cell
2

1. Neurotransmitters can be returned to axon terminals for reuse or transported into glial cells.
2. Enzymes inactivate neurotransmitters.
3. Neurotransmitters can diffuse out of the synaptic cleft.

separated from their receptors. Other neurotransmitters are inactivated by enzymes in the synaptic cleft. For example, acetylcholine (ACh) in the extracellular fluid is rapidly broken down into choline and acetyl CoA by the enzyme **acetylcholinesterase (AChE)** in the extracellular matrix and in the membrane of the postsynaptic cell (**FIG. 8.20**). Choline from degraded ACh is transported back into the presynaptic axon terminal on a Na^+-dependent cotransporter. Once back in the axon terminal, it can be used to make new acetylcholine.

Many neurotransmitters are removed from the extracellular fluid by transport back into the presynaptic cell or into adjacent neurons or glia. For example, norepinephrine action is terminated when the intact neurotransmitter is transported back into the presynaptic axon terminal. Norepinephrine uptake uses an Na^+-dependent cotransporter. Once back in the axon terminal, norepinephrine is either transported back into vesicles or broken down by intracellular enzymes such as *monoamine oxidase* (MAO), found in mitochondria. Neurotransmitters and their components can be recycled to refill empty synaptic vesicles.

Concept Check

24. One class of antidepressant drugs is called selective serotonin reuptake inhibitors (SSRIs). What do these drugs do to serotonin activity at the synapse?

25. How does the axon terminal make acetyl CoA for acetylcholine synthesis? (*Hint:* see p. 107.)

26. Is Na^+-dependent neurotransmitter reuptake facilitated diffusion, primary active transport, or secondary active transport? Explain your reasoning.

Stronger Stimuli Release More Neurotransmitter

A single action potential arriving at the axon terminal releases a constant amount of neurotransmitter. Neurons therefore can use the frequency of action potentials to transmit information about the duration and strength of the stimuli that activated them. Duration of a stimulus is coded by the duration of a series of repeated action potentials. A stronger stimulus causes more action potentials per second to arrive at the axon terminal, which in turn may result in more neurotransmitter release.

For example, let's consider how a sensory neuron tells the CNS the intensity of an incoming stimulus. An above-threshold graded potential reaching the trigger zone of the sensory neuron does not trigger just one action potential. Instead, even a small graded potential that is above threshold triggers a burst of action potentials (**FIG. 8.21a**). As graded potentials increase in strength (amplitude), they trigger more frequent action potentials (Fig. 8.21b).

Electrical signaling patterns in the CNS are more variable. Brain neurons show different electrical personalities by firing action potentials in a variety of patterns, sometimes spontaneously, without an external stimulus to bring them to threshold. For example, some neurons are *tonically active* [p. 182], firing regular trains of action potentials (beating pacemakers). Other neurons exhibit *bursting*, bursts of action potentials rhythmically alternating with intervals of quiet (rhythmic pacemakers).

These different firing patterns in CNS neurons are created by ion channel variants that differ in their activation and inactivation voltages, opening and closing speeds, and sensitivity to neuromodulators. This variability makes brain neurons more dynamic and complicated than the simple somatic motor neuron we use as our model.

FIG. 8.20 Synthesis and recycling of acetylcholine

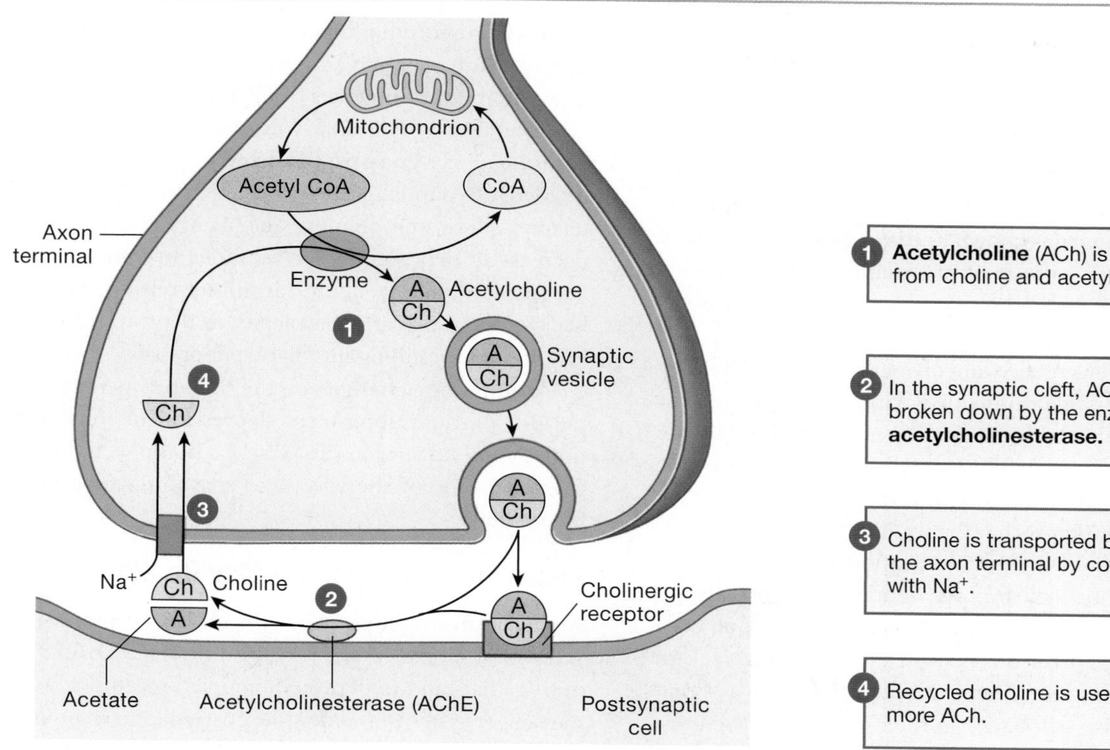
Mitochondrion
Acetyl CoA
CoA
Axon terminal
Enzyme
Acetylcholine
A
Ch
1
Synaptic vesicle
4
Ch
3
Na+
Ch
A
Choline
2
Cholinergic receptor
Acetate
Acetylcholinesterase (AChE)
Postsynaptic cell

1 **Acetylcholine** (ACh) is made from choline and acetyl CoA.

2 In the synaptic cleft, ACh is rapidly broken down by the enzyme **acetylcholinesterase.**

3 Choline is transported back into the axon terminal by cotransport with Na^+.

4 Recycled choline is used to make more ACh.

FIG. 8.21 Coding the strength of a stimulus

The frequency of action potential firing indicates the strength of a stimulus.

(a) Weak stimulus releases little neurotransmitter.

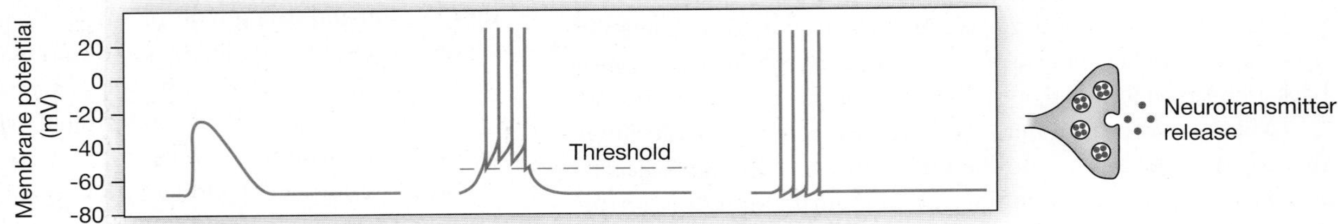
Membrane potential (mV)
20
0
-20
-40
-60
-80
Threshold
Neurotransmitter release

(b) Strong stimulus causes more action potentials and releases more neurotransmitter.

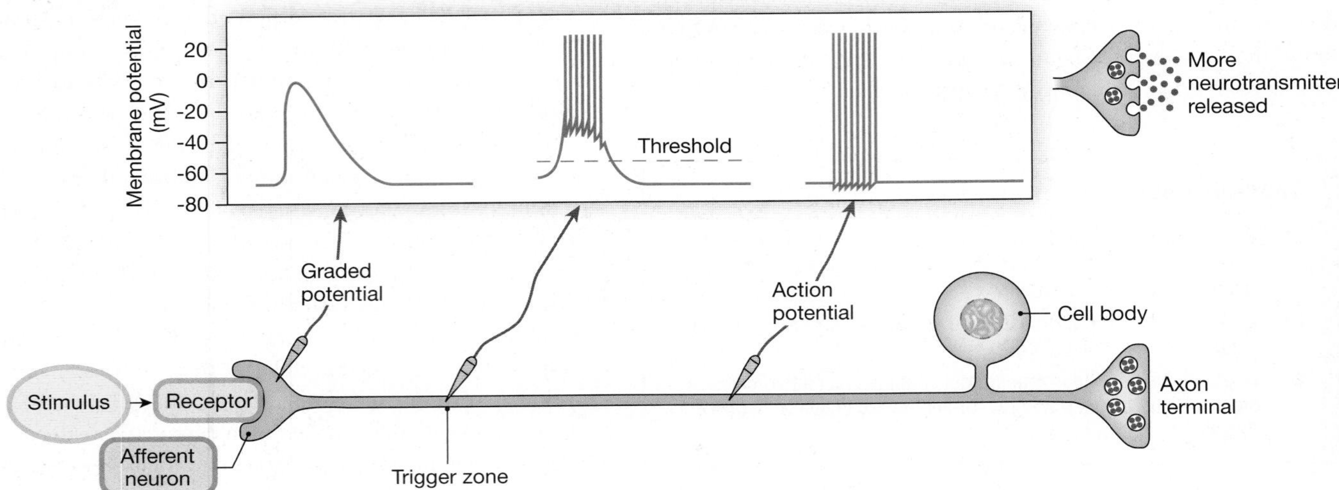
Membrane potential (mV)
20
0
-20
-40
-60
-80
Threshold
More neurotransmitter released
Graded potential
Action potential
Cell body
Stimulus
Receptor
Afferent neuron
Trigger zone
Axon terminal

8.5 Integration of Neural Information Transfer

Communication between neurons is not always a one-to-one event as we have been describing. Frequently, the axon of a presynaptic neuron branches, and its collaterals (branches) synapse on multiple target neurons. This pattern is known as **divergence** (**FIG. 8.22a**). On the other hand, when a group of presynaptic neurons provide input to a smaller number of postsynaptic neurons, the pattern is known as **convergence** (Fig. 8.22b).

Combination of convergence and divergence in the CNS may result in one postsynaptic neuron with synapses from as many as 10,000 presynaptic neurons (Fig. 8.22c). For example, the Purkinje neurons of the CNS have highly branched dendrites so that they can receive information from many neurons (Fig. 8.22d).

In addition, we now know that the traditional view of chemical synapses as sites of one-way communication, with all messages moving from presynaptic cell to postsynaptic cell, is not always correct. In the brain, there are some synapses where cells on both sides of the synaptic cleft release neurotransmitters that act on the opposite cell. Perhaps more importantly, we have learned that many postsynaptic cells "talk back" to their presynaptic neurons by sending neuromodulators that bind to presynaptic receptors. Variations in synaptic activity play a major role in determining how communication takes place in the nervous system.

The ability of the nervous system to change activity at synapses is called **synaptic plasticity** {*plasticus*, that which may be molded}. Synaptic plasticity occurs primarily in the CNS. Short-term plasticity may enhance activity at the synapse (facilitation) or decrease it (depression). For example, in some cases of sustained activity at a synapse, neurotransmitter release decreases over time because the axon cannot replenish its neurotransmitter supply rapidly enough, resulting in synaptic depression.

Sometimes changes at the synapse persist for significant periods of time (long-term depression or long-term potentiation, described later in this section). In the sections that follow, we examine some of the ways that communication at synapses can be modified.

Postsynaptic Responses May Be Slow or Fast

A neurotransmitter combining with its receptor sets in motion a series of responses in the postsynaptic cell (**FIG. 8.23**). Neurotransmitters that bind to G protein-coupled receptors linked to second messenger systems initiate slow postsynaptic responses.

FIG. 8.22 ESSENTIALS Divergence and convergence

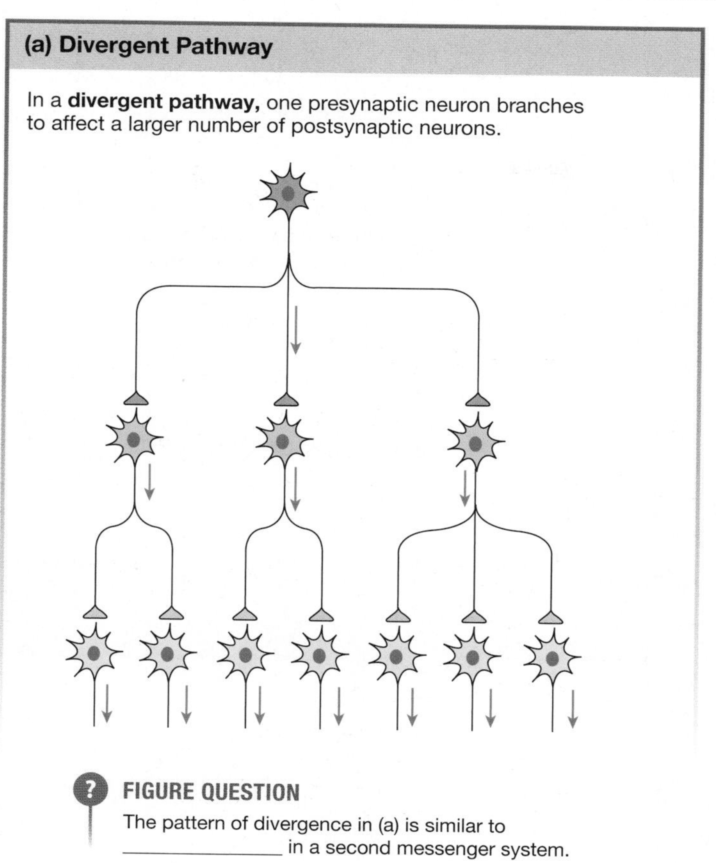
(a) Divergent Pathway
In a **divergent pathway,** one presynaptic neuron branches to affect a larger number of postsynaptic neurons.

FIGURE QUESTION

The pattern of divergence in (a) is similar to ________________ in a second messenger system.

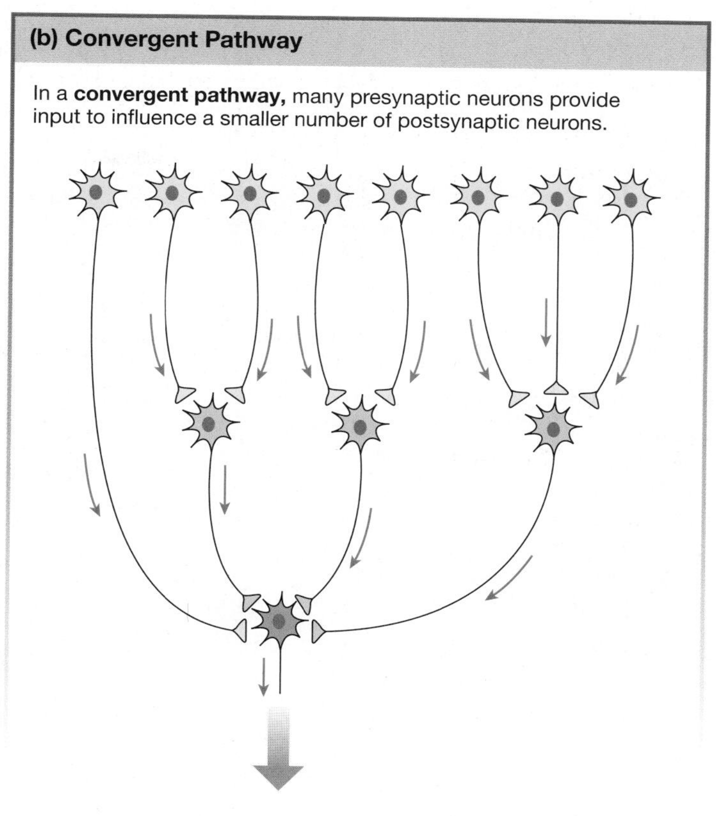
(b) Convergent Pathway
In a **convergent pathway,** many presynaptic neurons provide input to influence a smaller number of postsynaptic neurons.

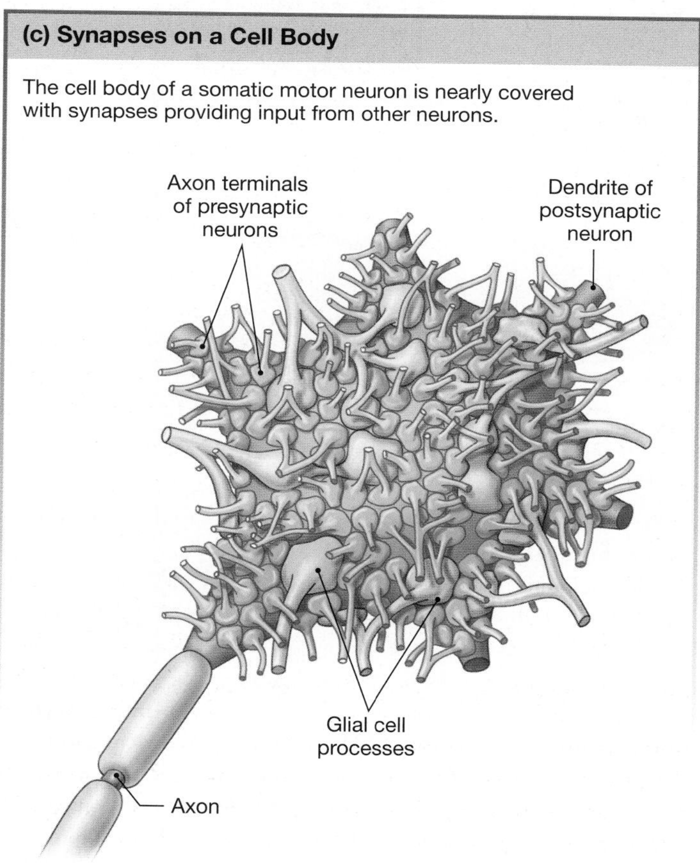
(c) Synapses on a Cell Body
The cell body of a somatic motor neuron is nearly covered with synapses providing input from other neurons.
Axon terminals of presynaptic neurons
Dendrite of postsynaptic neuron
Glial cell processes
Axon

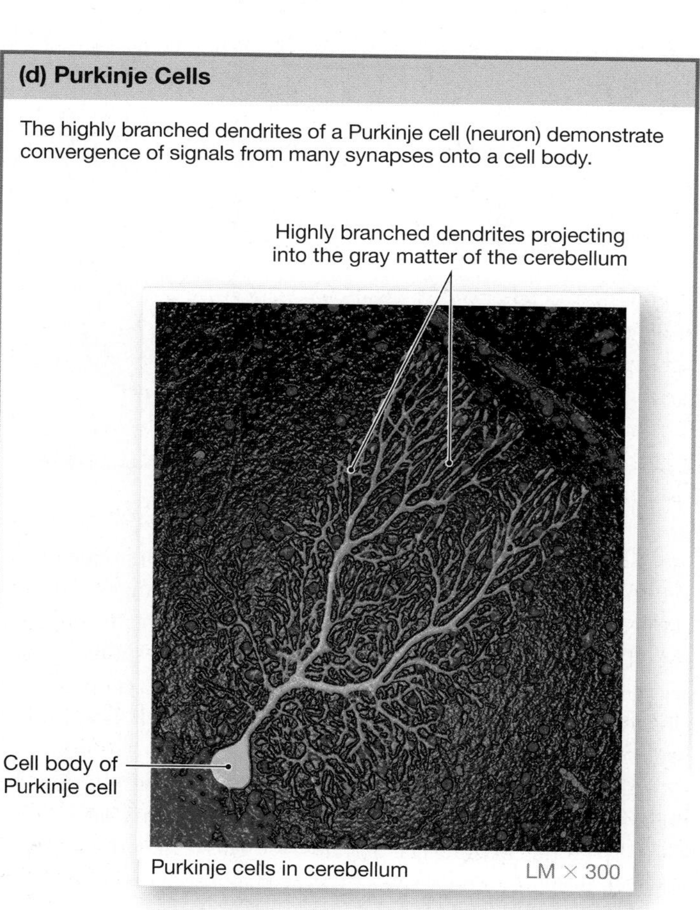
(d) Purkinje Cells
The highly branched dendrites of a Purkinje cell (neuron) demonstrate convergence of signals from many synapses onto a cell body.
Highly branched dendrites projecting into the gray matter of the cerebellum
Cell body of Purkinje cell

Purkinje cells in cerebellum LM × 300

FIG. 8.23 ESSENTIALS Fast and slow postsynaptic responses

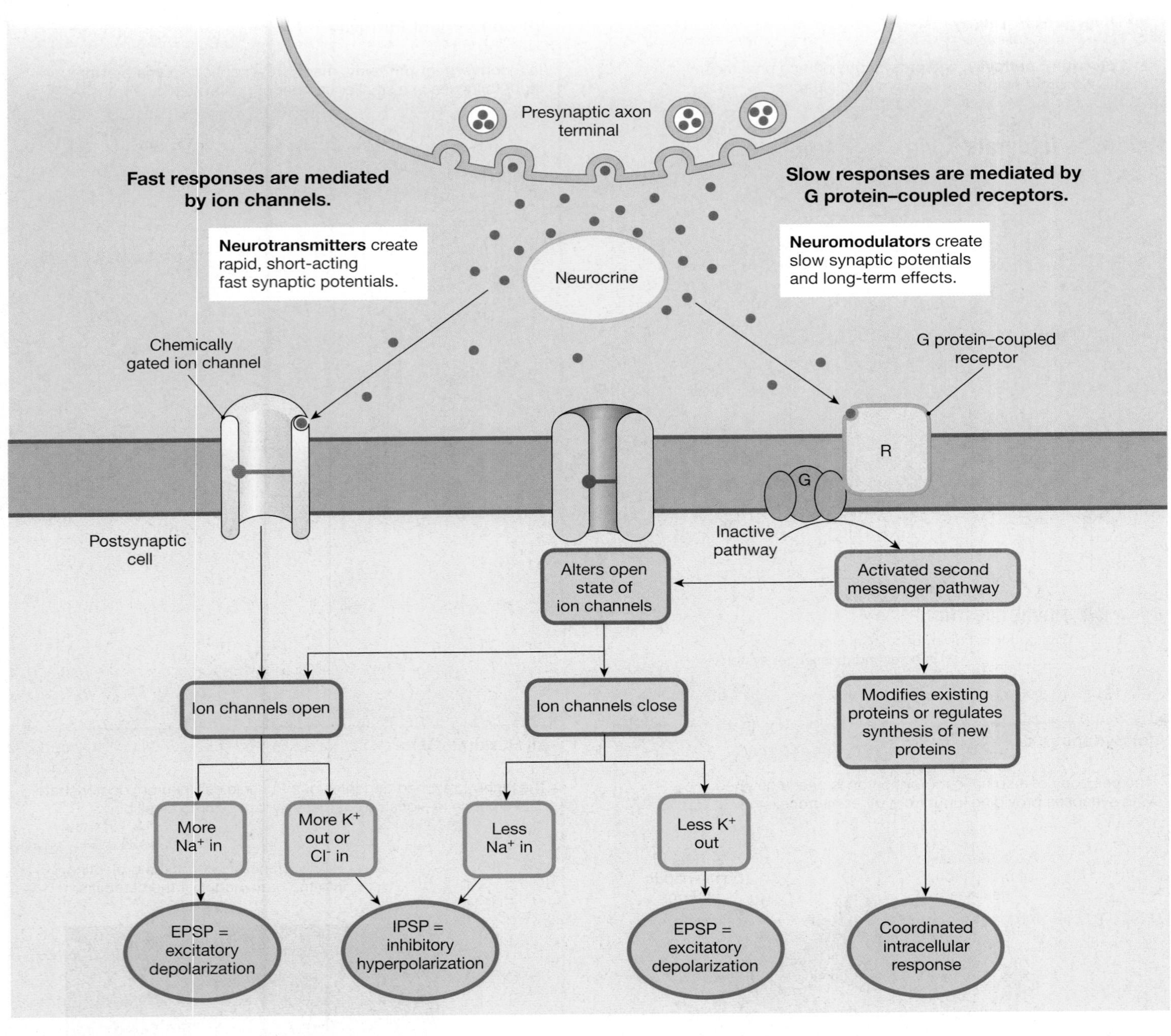

Presynaptic axon terminal
Fast responses are mediated by ion channels.
Slow responses are mediated by G protein–coupled receptors.
Neurotransmitters create rapid, short-acting fast synaptic potentials.
Neuromodulators create slow synaptic potentials and long-term effects.
Neurocrine
Chemically gated ion channel
G protein–coupled receptor
R
G
Postsynaptic cell
Inactive pathway
Alters open state of ion channels
Activated second messenger pathway
Ion channels open
Ion channels close
Modifies existing proteins or regulates synthesis of new proteins
More Na+ in
More K+ out or Cl- in
Less Na+ in
Less K+ out
EPSP = excitatory depolarization
IPSP = inhibitory hyperpolarization
EPSP = excitatory depolarization
Coordinated intracellular response

Some second messengers act from the cytoplasmic side of the cell membrane to open or close ion channels. Changes in membrane potential resulting from these alterations in ion flow are called **slow synaptic potentials** because the response of the second messenger pathway takes longer than the direct opening or closing of a channel. In addition, the response itself lasts longer, usually seconds to minutes.

Slow postsynaptic responses are not limited to altering the open state of ion channels. Neurotransmitters acting on GPCRs may also modify existing cell proteins or regulate the production of new cell proteins. These types of slow response have been linked to the growth and development of neurons and to the mechanisms underlying long-term memory.

Fast synaptic responses are always associated with the opening of ion channels. In the simplest response, the neurotransmitter binds to and opens a receptor-channel on the postsynaptic cell, allowing ions to move between the postsynaptic cell and the extracellular fluid. The resulting change in membrane potential is called a **fast synaptic potential** because it begins quickly and lasts only a few milliseconds.

If the synaptic potential is depolarizing, it is called an **excitatory postsynaptic potential (EPSP)** because it makes the cell more likely to fire an action potential. If the synaptic potential is hyperpolarizing, it is called an **inhibitory postsynaptic potential (IPSP)** because hyperpolarization moves the membrane potential away from threshold and makes the cell less likely to fire an action potential.

Pathways Integrate Information from Multiple Neurons

When two or more presynaptic neurons converge on the dendrites or cell body of a single postsynaptic cell, the response of the postsynaptic cell is determined by the summed input from the presynaptic neurons. **FIGURE 8.24c** shows the three-dimensional reconstruction of dendritic spines of a postsynaptic neuron, with numerous excitatory and inhibitory synapses providing input. The summed input from these synapses determines the activity of the postsynaptic neuron.

The combination of several nearly simultaneous graded potentials is called **spatial summation**. The word *spatial* {*spatium*, space} refers to the fact that the graded potentials originate at different locations (spaces) on the neuron.

Figure 8.24d illustrates spatial summation when three presynaptic neurons releasing excitatory neurotransmitters ("excitatory neurons") converge on one postsynaptic neuron. Each neuron's EPSP is too weak to trigger an action potential by itself, but if the three presynaptic neurons fire simultaneously, the sum of the three EPSPs is above threshold and creates an action potential.

Spatial summation is not always excitatory. If summation prevents an action potential in the postsynaptic cell, the summation is called **postsynaptic inhibition**. This occurs when presynaptic neurons release inhibitory neurotransmitter. For example, Figure 8.24e shows three presynaptic neurons, two excitatory and one inhibitory, converging on a postsynaptic cell. The neurons fire, creating one IPSP and two EPSPs that sum as they reach the trigger zone. The IPSP counteracts the two EPSPs, creating an integrated signal that is below threshold. As a result, no action potential is generated at the trigger zone.

Temporal Summation Summation of graded potentials does not always require input from more than one presynaptic neuron. Two subthreshold graded potentials from the same presynaptic neuron can be summed if they arrive at the trigger zone close enough together in time. Summation that occurs from graded potentials overlapping in time is called **temporal summation** {*tempus*, time}. Let's see how this can happen.

Figure 8.24a shows recordings from an electrode placed in the trigger zone of a neuron. A stimulus (X_1) starts a subthreshold graded potential on the cell body at the time marked on the *x*-axis. The graded potential reaches the trigger zone and depolarizes it, as shown on the graph (A_1), but not enough to trigger an action potential. A second stimulus (X_2) occurs later, and its subthreshold graded potential (A_2) reaches the trigger zone sometime after the first. The interval between the two stimuli is so long that the two graded potentials do not overlap. Neither potential by itself is above threshold, so no action potential is triggered.

In Figure 8.24b, the two stimuli occur closer together in time. As a result, the two subthreshold graded potentials arrive at the trigger zone at almost the same time. The second graded potential adds its depolarization to that of the first, causing the trigger zone to depolarize to threshold.

In many situations, graded potentials in a neuron incorporate both temporal and spatial summation. The summation of graded potentials demonstrates a key property of neurons: *postsynaptic integration*. When multiple signals reach a neuron, postsynaptic integration creates a signal based on the relative strengths and durations of the signals. If the integrated signal is above threshold, the neuron fires an action potential. If the integrated signal is below threshold, the neuron does not fire.

CHAPTER 8

Concept Check

27. In Figure 8.24e, assume the postsynaptic neuron has a resting membrane potential of −70 mV and a threshold of −55 mV. If the inhibitory presynaptic neuron creates an IPSP of −5 mV and the two excitatory presynaptic neurons have EPSPs of 10 and 12 mV, will the postsynaptic neuron fire an action potential?

28. In the graphs of Figure 8.24a, b, why doesn't the membrane potential change at the same time as the stimulus?

Synaptic Activity Can Be Modified

The examples of synaptic integration we just discussed all took place on the postsynaptic side of a synapse, but the activity of presynaptic cells can also be altered, or *modulated*. When a modulatory neuron terminates on a presynaptic cell, the IPSP or EPSP created by the modulatory neuron can alter the action potential reaching the axon terminals of the presynaptic cell and modulate neurotransmitter release. In *presynaptic facilitation*, input from an excitatory neuron increases neurotransmitter release by the presynaptic cell.

If modulation of a neuron decreases its neurotransmitter release, the modulation is called *presynaptic inhibition*. Presynaptic inhibition may be global or selective. In global presynaptic inhibition (Fig. 8.24f), input on the dendrites and cell body of a neuron decreases neurotransmitter release by all collaterals and all target cells of the neuron are affected equally.

In selective modulation, one collateral can be inhibited while others remain unaffected. Selective presynaptic alteration of neurotransmitter release provides a more precise means of control than global modulation. For example, Figure 8.24g shows selective presynaptic modulation of a single collateral's axon terminal so that only its target cell fails to respond.

Synaptic activity can also be altered by changing the target (postsynaptic) cell's responsiveness to neurotransmitter. This may be accomplished by changing the structure, affinity, or number of neurotransmitter receptors. Modulators can alter all of these parameters by influencing the synthesis of enzymes, membrane transporters, and receptors. Most neuromodulators act through second messenger systems that alter existing channels, and their effects last much longer than do those of neurotransmitters. One signal molecule can act as either a neurotransmitter or a neuromodulator, depending on its receptor (Fig. 8.23).

FIG. 8.24 ESSENTIALS Integration of synaptic signaling

Temporal Summation

Temporal summation occurs when two graded potentials from one presynaptic neuron occur close together in time.

(a) No summation. Two subthreshold graded potentials will not initiate an action potential if they are far apart in time.

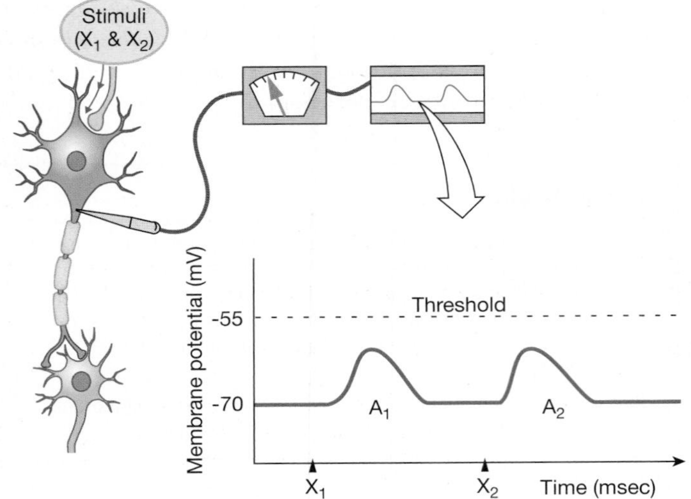

Stimuli (X_1 & X_2)
Membrane potential (mV)
-55
-70
Threshold
A_1
A_2
X_1
X_2
Time (msec)

(b) Summation causing action potential. If two subthreshold potentials arrive at the trigger zone within a short period of time, they may sum and initiate an action potential.

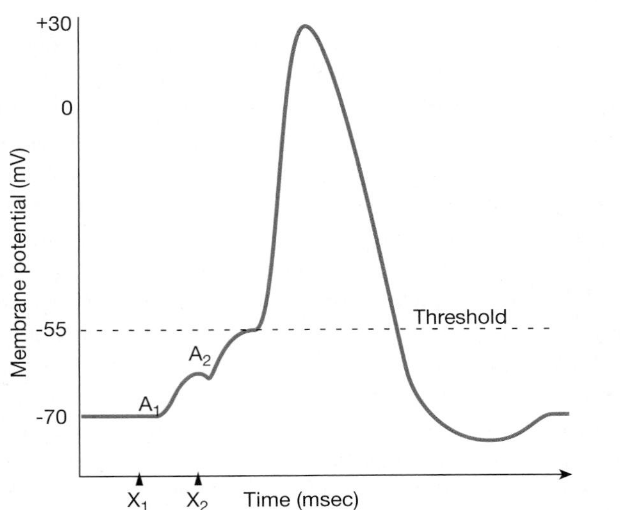

+30
0
-55
-70
Membrane potential (mV)
Threshold
A_1
A_2
X_1
X_2
Time (msec)

Spatial Summation

Spatial summation occurs when the currents from nearly simultaneous graded potentials combine.

(c) Multiple presynaptic neurons provide input on the dendrites and cell body of the postsynaptic neurons.

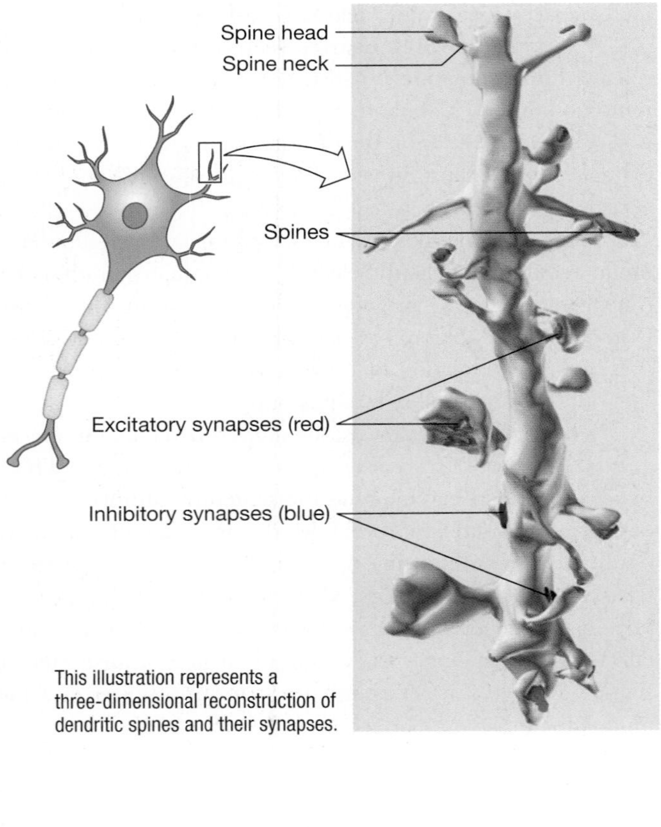

Spine head

Spine neck

Spines

Excitatory synapses (red)

Inhibitory synapses (blue)

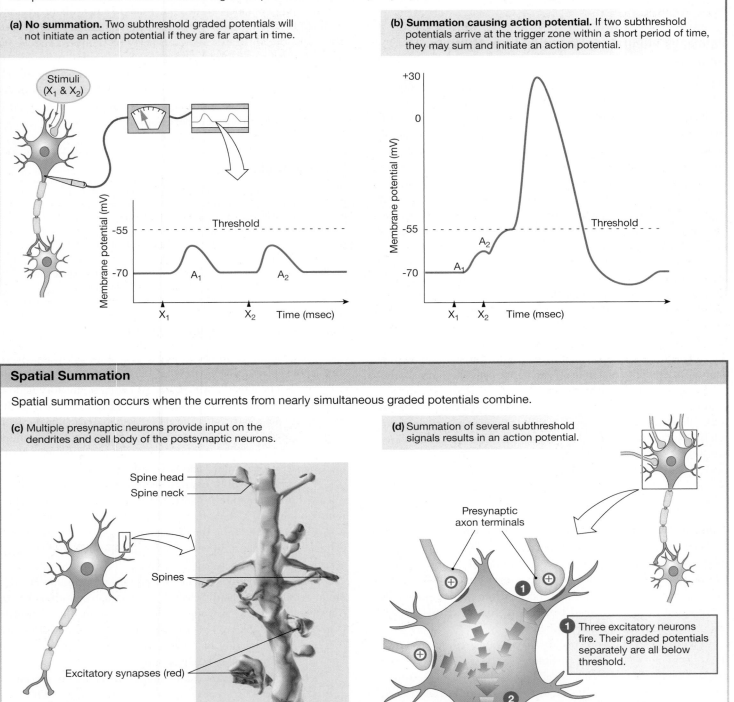

This illustration represents a three-dimensional reconstruction of dendritic spines and their synapses.

(d) Summation of several subthreshold signals results in an action potential.

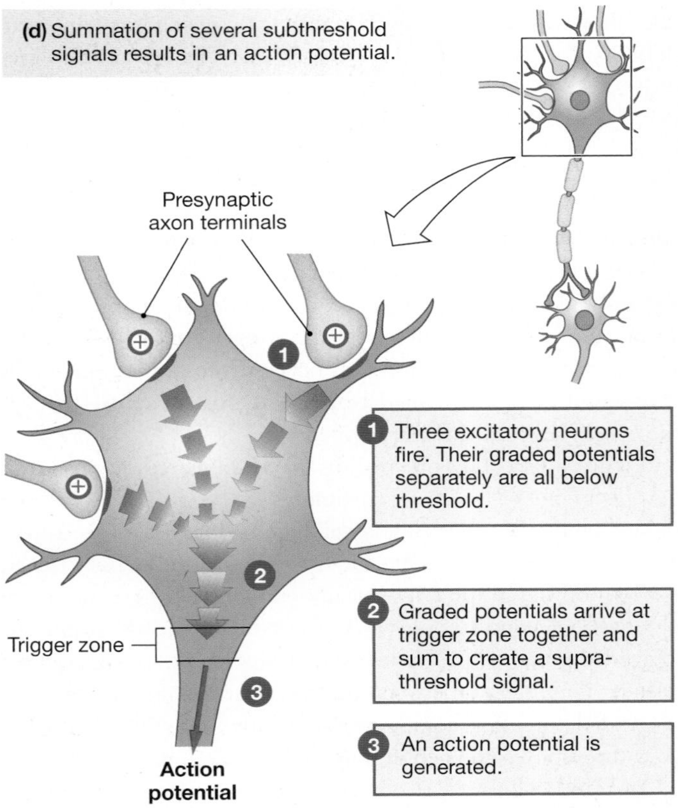

Presynaptic axon terminals

Trigger zone

Action potential

1. Three excitatory neurons fire. Their graded potentials separately are all below threshold.
2. Graded potentials arrive at trigger zone together and sum to create a suprathreshold signal.
3. An action potential is generated.

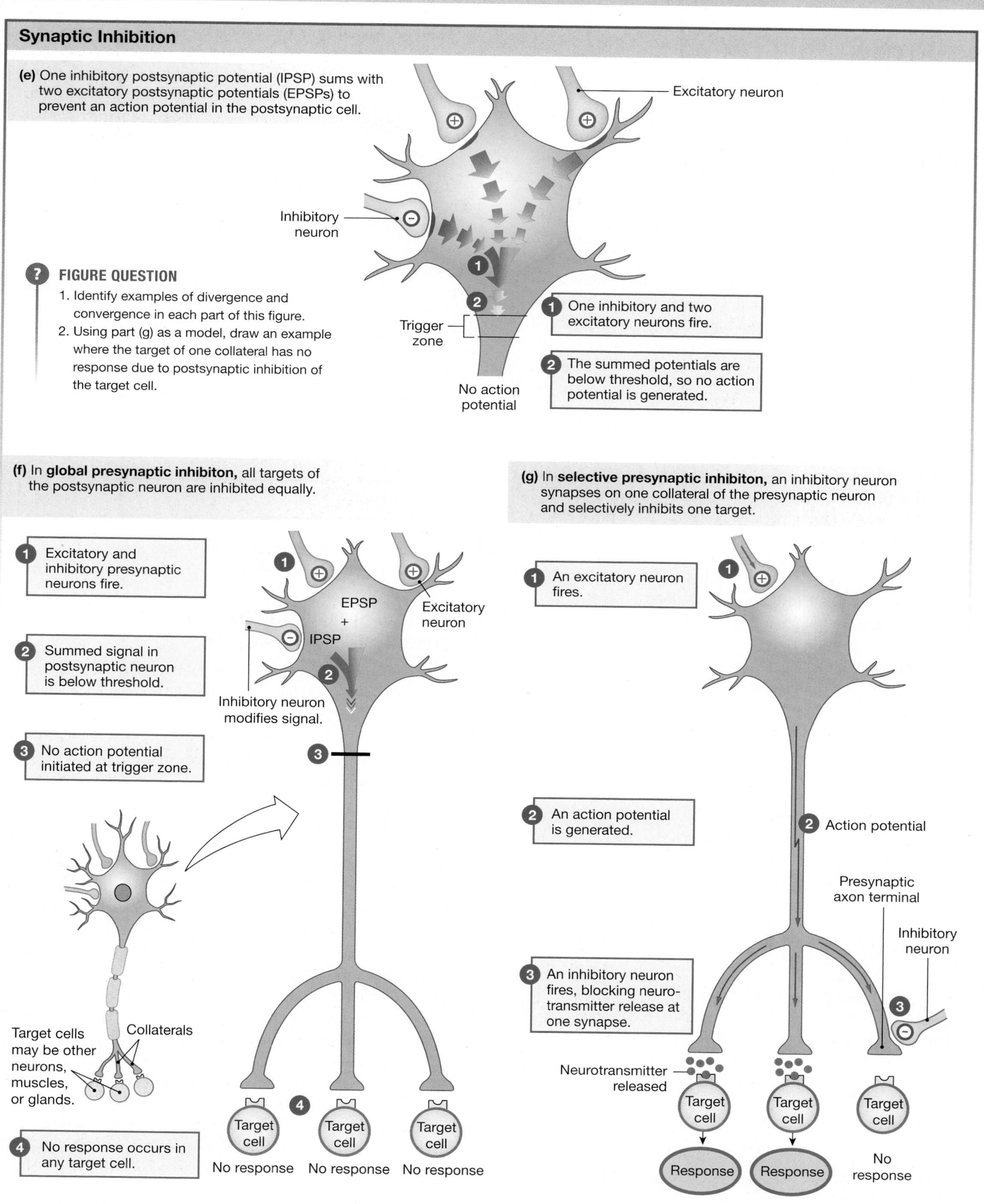

Synaptic Inhibition
(e) One inhibitory postsynaptic potential (IPSP) sums with two excitatory postsynaptic potentials (EPSPs) to prevent an action potential in the postsynaptic cell.
Excitatory neuron
Inhibitory neuron
Trigger zone
No action potential
1 One inhibitory and two excitatory neurons fire.
2 The summed potentials are below threshold, so no action potential is generated.
FIGURE QUESTION
1. Identify examples of divergence and convergence in each part of this figure.
2. Using part (g) as a model, draw an example where the target of one collateral has no response due to postsynaptic inhibition of the target cell.
(f) In global presynaptic inhibiton, all targets of the postsynaptic neuron are inhibited equally.
1 Excitatory and inhibitory presynaptic neurons fire.
2 Summed signal in postsynaptic neuron is below threshold.
3 No action potential initiated at trigger zone.
4 No response occurs in any target cell.
EPSP
+
IPSP
Excitatory neuron
Inhibitory neuron modifies signal.
Target cells may be other neurons, muscles, or glands.
Collaterals
Target cell
No response
No response
No response
(g) In selective presynaptic inhibiton, an inhibitory neuron synapses on one collateral of the presynaptic neuron and selectively inhibits one target.
1 An excitatory neuron fires.
2 An action potential is generated.
3 An inhibitory neuron fires, blocking neuro-transmitter release at one synapse.
Action potential
Presynaptic axon terminal
Inhibitory neuron
Neurotransmitter released
Target cell
Response
Response
No response

Concept Check

29. Why are axon terminals sometimes called "biological transducers"?

Long-Term Potentiation Alters Synapses

Two of the "hot topics" in neurobiology today are **long-term potentiation (LTP)** {*potentia*, power} and *long-term depression* (LTD), where activity at a synapse brings about sustained changes in the quality or quantity of the synaptic connections. Many times changes in synaptic transmission, such as the facilitation and inhibition we just discussed, are of limited duration. However, if synaptic activity persists for longer periods, the neurons may adapt through LTP and LTD. Long-lasting changes in brain synapses are probably responsible for acquired behaviors, such as addiction and spatial memory.

Long-term potentiation and *long-term depression* are the most studied types of synaptic plasticity. Our understanding of LTP and LTD is changing rapidly, however, and the mechanisms are not the same in different brain areas. The descriptions below reflect some of what we currently know about long-term adaptations of synaptic transmission.

A key element in long-term changes in the CNS is the amino acid glutamate, the main excitatory neurotransmitter in the CNS. As you learned previously, glutamate has two types of receptor channels: AMPA receptors and NMDA receptors. The NMDA receptor has an unusual property. First, at resting membrane potentials, the NMDA channel is blocked by both a gate and an Mg^{2+} ion. Glutamate binding opens the ligand-activated gate, but ions cannot flow past the Mg^{2+}. However, if the cell depolarizes, the Mg^{2+} blocking the channel is expelled, and then ions can flow through the channel. Thus, the NMDA channel opens only when the receptor is bound to glutamate *and* the cell is depolarized.

In long-term potentiation, when presynaptic neurons release glutamate, the neurotransmitter binds to both AMPA and NMDA receptors on the postsynaptic cell (**FIG. 8.25** 1). Binding to the AMPA receptor opens a cation channel, and net Na^+ entry depolarizes the cell 2. Simultaneously, glutamate binding to the NMDA receptor opens the channel gate, and depolarization of the cell creates electrical repulsion that knocks the Mg^{2+} out of the NMDA channel 3. Once the NMDA channel is open, Ca^{2+} enters the cytosol 4.

RUNNING PROBLEM

Dr. McKhann suspected that the disease afflicting the Chinese children—which he named *acute motor axonal polyneuropathy* (AMAN)—might be triggered by a bacterial infection. He also thought that the disease initiated its damage of axons at neuromuscular junctions, the synapses between somatic motor neurons and skeletal muscles.

Q6: *Based on information provided in this chapter, name other diseases involving altered synaptic transmission.*

224 – 226 – 233 – 248 – 251 – 254 – **264** – 265

The Ca^{2+} signal initiates second messenger pathways 5. As a result of these intracellular pathways, the postsynaptic cell becomes more sensitive to glutamate, possibly by inserting more glutamate receptors in the postsynaptic membrane (up-regulation, p. 51). In addition, the postsynaptic cell releases a paracrine that acts on the presynaptic cell to enhance glutamate release 6.

Long-term depression seems to have two components: a change in the number of postsynaptic receptors and a change in the isoforms of the receptor proteins. In the face of continued neurotransmitter release from presynaptic neurons, the postsynaptic neurons withdraw AMPA receptors from the cell membrane by endocytosis [p. 147], a process similar to down-regulation of receptors in the endocrine system [p. 180]. In addition, different protein subunits are inserted into the AMPA receptor proteins, changing current flow through the ion channels.

Researchers believe that long-term potentiation and depression are related to the neural processes for learning and memory, and to changes in the brain that occur with clinical depression and other mental illnesses. The clinical link makes LTP and LTD hot topics in neuroscience research.

Concept Check

30. Why would depolarization of the membrane drive Mg^{2+} from the channel into the extracellular fluid?

Disorders of Synaptic Transmission Are Responsible for Many Diseases

Synaptic transmission is the most vulnerable step in the process of signaling through the nervous system. It is the point at which many things go wrong, leading to disruption of normal function. Yet, at the same time, the receptors at synapses are exposed to the extracellular fluid, making them more accessible to drugs than intracellular receptors are. In recent years, scientists have linked a variety of nervous system disorders to problems with synaptic transmission. These disorders include Parkinson's disease, schizophrenia, and depression. The best understood diseases of the synapse are those that involve the *neuromuscular junction* between somatic motor neurons and skeletal muscles. One example of neuromuscular junction pathology is *myasthenia gravis* (see Clinical Focus box on p. 253). Diseases resulting from synaptic transmission problems within the CNS have proved more difficult to study because they are more difficult to isolate anatomically.

Drugs that act on synaptic activity, particularly synapses in the CNS, are the oldest known and most widely used of all pharmacological agents. Caffeine, nicotine, and alcohol are common drugs in many cultures. Some of the drugs we use to treat conditions such as schizophrenia, depression, anxiety, and epilepsy act by influencing events at the synapse. In many disorders arising in the CNS, we do not yet fully understand either the cause of the disorder or the drug's mechanism of action. This subject is one major area of pharmacological research, and new classes of drugs are being formulated and approved every year.

FIG. 8.25 Long-term potentiation

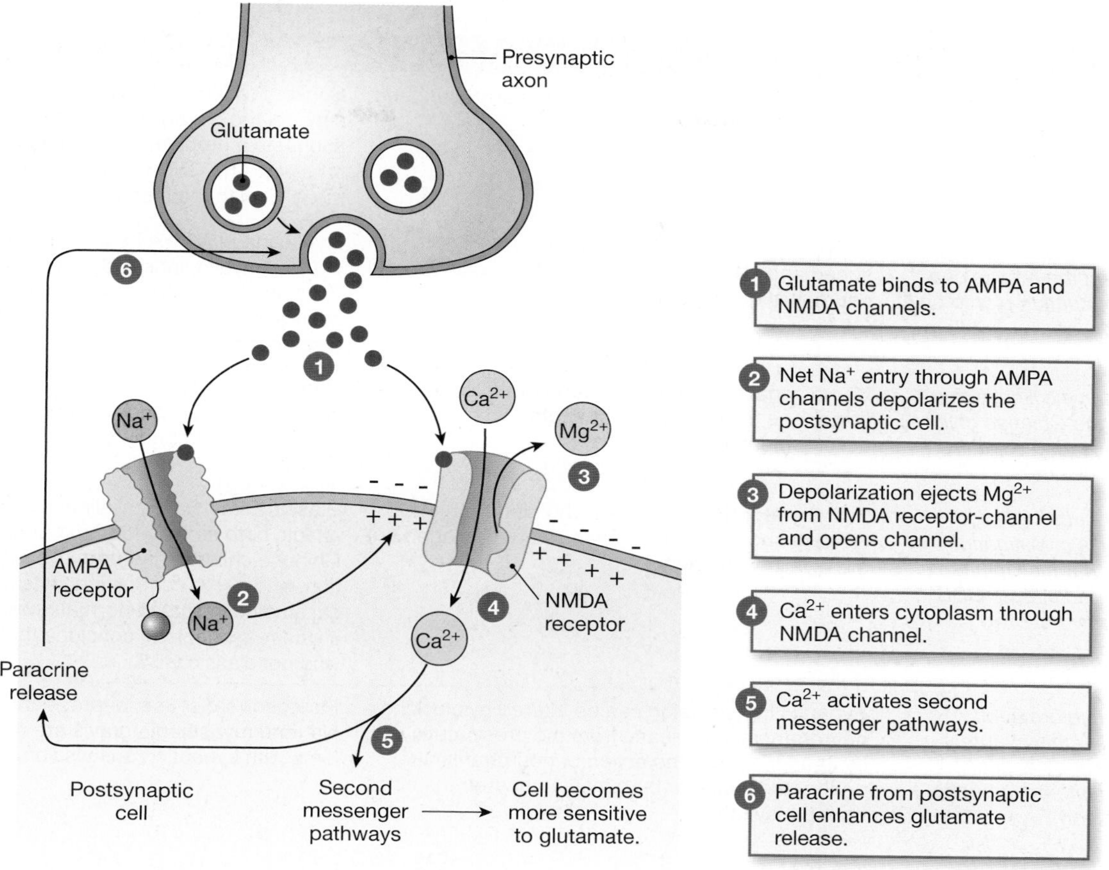

Presynaptic axon
Glutamate
6
1
Na+
Ca2+
Mg2+
3
AMPA receptor
2
Na+
4
Ca2+
NMDA receptor
Paracrine release
5
Postsynaptic cell
Second messenger pathways
Cell becomes more sensitive to glutamate.

1. Glutamate binds to AMPA and NMDA channels.
2. Net Na^+ entry through AMPA channels depolarizes the postsynaptic cell.
3. Depolarization ejects Mg^{2+} from NMDA receptor-channel and opens channel.
4. Ca^{2+} enters cytoplasm through NMDA channel.
5. Ca^{2+} activates second messenger pathways.
6. Paracrine from postsynaptic cell enhances glutamate release.

RUNNING PROBLEM **CONCLUSION** Mysterious Paralysis

In this running problem you learned about *acute motor axonal polyneuropathy* (AMAN), a baffling paralytic illness that physicians thought might be a new disease. Although its symptoms resemble those of classic Guillain-Barré syndrome, AMAN is not a demyelinating disease and it affects only somatic motor neurons. In both classic GBS and AMAN, the body's immune system makes antibodies against nervous system components. This similarity led experts eventually to conclude that AMAN is a subtype of GBS. The classic form of GBS is also known as *acute inflammatory demyelinating polyneuropathy* (AIDP). AIDP is more common in Europe and North America; AMAN is the predominant form of GBS in China, Japan, and South America. Interestingly, the 2016 outbreak of the Zika virus in Latin America and French Polynesia was associated with an increase in neurological complications resembling Guillain-Barré syndrome. On further investigation, scientists noted the same symptoms seen in the Chinese children: normal conduction velocity but decreased strength of action potentials, pointing to the AMAN subtype of GBS. Now check your understanding of this problem by comparing your answers to the information in the following summary table.

Question	Facts	Integration and Analysis
Q1: *Which division(s) of the nervous system may be involved in GBS?*	The nervous system is divided into the central nervous system and the afferent (sensory) and efferent subdivisions of the peripheral nervous system. Efferent neurons are either somatic motor neurons, which control skeletal muscles, or autonomic neurons, which control glands and smooth and cardiac muscle.	Patients with classic GBS can neither feel sensations nor move their muscles. This suggests a problem in both afferent and somatic motor neurons. However, it is also possible that there is a problem in the CNS integrating center. You do not have enough information to determine which division is affected.

Continued

RUNNING PROBLEM CONCLUSION *Continued*

Question	Facts	Integration and Analysis
Q2: *Do you think the paralysis found in the Chinese children affected both sensory (afferent) and somatic motor neurons? Why or why not?*	The Chinese children can feel a pin prick but cannot move their muscles.	Sensory (afferent) function is normal if they can feel the pin prick. Paralysis of the muscles suggests a problem with somatic motor neurons, with the CNS centers controlling movement, or with the muscles themselves.
Q3: *In classic GBS, what would you expect the results of a nerve conduction test to be?*	Nerve conduction tests measure conduction speed and strength. In classic GBS, myelin around neurons is destroyed.	Myelin insulates axons and increases speed. Without myelin, ions leak out of the axon. Thus, in classic GBS you would expect decreased conduction speed or blocked conduction.
Q4: *Is the paralytic illness that affected the Chinese children a demyelinating condition? Why or why not?*	Nerve conduction tests showed normal conduction speed but decreased strength of the summed action potentials.	Myelin loss should decrease conduction speed as well as block conduction. Therefore, this illness is probably not a demyelinating disease.
Q5: *Do the results of Dr. McKhann's investigation suggest that the Chinese children had classic GBS? Why or why not?*	Autopsy reports on children who died from the disease showed that the axons were damaged but the myelin was normal.	Classic GBS is a demyelinating disease that affects both sensory and motor neurons. The Chinese children had normal sensory function, and nerve conduction tests and histological studies indicated normal myelin. Therefore, it was reasonable to conclude that the disease was not classic GBS.
Q6: *Based on information provided in this chapter, name other diseases involving altered synaptic transmission.*	Synaptic transmission can be altered by blocking neurotransmitter release from the presynaptic cell, by interfering with the action of neurotransmitter on the target cell, or by removing neurotransmitter from the synapse.	Parkinson's disease, depression, schizophrenia, and myasthenia gravis are related to problems with synaptic transmission.

224 – 226 – 233 – 248 – 251 – 254 – 264 – 265

CHAPTER SUMMARY

This chapter introduces the nervous system, one of the major control systems responsible for maintaining *homeostasis*. The divisions of the nervous system correlate with the steps in a reflex pathway. Sensory receptors monitor regulated variables and send input signals to the central nervous system through sensory (afferent) neurons. Output signals, both electrical and chemical, travel through the efferent divisions (somatic motor and autonomic) to their targets throughout the body. Information transfer and *communication* depend on electrical signals that pass along neurons, on *molecular interactions* between signal molecules and their receptors, and on signal transduction in the target cells.

1. The **nervous system** is a complex network of neurons that form the rapid control system of the body. (p. 224)
2. **Emergent properties** of the nervous system include consciousness, intelligence, and emotion. (p. 224)

8.1 Organization of the Nervous System

3. The nervous system is divided into the **central nervous system (CNS)**, composed of the **brain** and **spinal cord**, and the **peripheral nervous system (PNS)**. (p. 224; Fig. 8.1)
4. The peripheral nervous system has **sensory (afferent) neurons** that bring information into the CNS, and **efferent neurons** that carry information away from the CNS back to various parts of the body. (p. 224)
5. The efferent neurons include **somatic motor neurons**, which control skeletal muscles, and **autonomic neurons**, which control smooth and cardiac muscles, glands, and some adipose tissue. (p. 226)
6. Autonomic neurons are subdivided into **sympathetic** and **parasympathetic** branches. (p. 226)

8.2 Cells of the Nervous System

7. Neurons have a **cell body** with a nucleus and organelles to direct cellular activity, **dendrites** to receive incoming signals, and an **axon** to transmit electrical signals from the cell body to the **axon terminal**. (p. 226; Fig. 8.2)
8. **Interneurons** are neurons that lie entirely within the CNS. (p. 226; Fig. 8.2c, d)
9. Material is transported between the cell body and axon terminal by **axonal transport**. (p. 229; Fig. 8.3)
10. The region where an axon terminal meets its target cell is called a **synapse**. The target cell is called the **postsynaptic cell**,

and the neuron that releases the chemical signal is known as the **presynaptic cell**. The region between these two cells is the **synaptic cleft**. (p. 229; Fig. 8.2f)

11. Developing neurons find their way to their targets by using chemical signals. (p. 229)
12. **Glial cells** provide physical support and communicate with neurons. **Schwann cells** and **satellite cells** are glial cells associated with the peripheral nervous system. **Oligodendrocytes**, **astrocytes**, **microglia**, and **ependymal cells** are glial cells found in the CNS. Microglia are modified immune cells that act as scavengers. (p. 230; Fig. 8.5)
13. Schwann cells and oligodendrocytes form insulating **myelin sheaths** around neurons. The **nodes of Ranvier** are sections of uninsulated membrane occurring at intervals along the length of an axon. (p. 232; Fig. 8.5c)
14. **Neural stem cells** that can develop into new neurons and glia are found in the ependymal layer as well as in other parts of the nervous system. (p. 233)

8.3 Electrical Signals in Neurons

15. The **Nernst equation** describes the membrane potential of a cell that is permeable to only one ion. (p. 234)
16. Membrane potential is influenced by the concentration gradients of ions across the membrane and by the permeability of the membrane to those ions. (p. 234)
17. The **Goldman-Hodgkin-Katz (GHK) equation** predicts membrane potential based on ion concentration gradients and membrane permeability for multiple ions. (p. 234)
18. The permeability of a cell to ions changes when ion channels in the membrane open and close. Movement of only a few ions significantly changes the membrane potential. (p. 235)
19. Gated ion channels in neurons open or close in response to chemical or mechanical signals or in response to depolarization of the cell membrane. Channels also close through inactivation. (p. 235)
20. Current flow (I) obeys **Ohm's law:** I = voltage/resistance. **Resistance** to current flow comes primarily from the cell membrane, which is a good insulator, and from the cytoplasm. **Conductance** (G) is the reciprocal of resistance: G = 1/R. (p. 236)
21. **Graded potentials** are depolarizations or hyperpolarizations whose strength is directly proportional to the strength of the triggering event. Graded potentials lose strength as they move through the cell. (p. 237; Tbl. 8.3; Fig. 8.7)
22. The wave of depolarization that moves through a cell is known as **local current flow**. (p. 239)
23. **Action potentials** are rapid electrical signals that travel undiminished in amplitude (strength) down the axon from the cell body to the axon terminals. (p. 239)
24. Action potentials begin in the **trigger zone** if a single graded potential or the sum of multiple graded potentials exceeds the **threshold** voltage. (p. 238; Fig. 8.7c)
25. Depolarizing graded potentials make a neuron more likely to fire an action potential. Hyperpolarizing graded potentials make a neuron less likely to fire an action potential. (p. 239)
26. Action potentials are uniform, **all-or-none** depolarizations that can travel undiminished over long distances. (p. 239)
27. The rising phase of the action potential is due to increased Na^+ permeability. The falling phase of the action potential is due to increased K^+ permeability. (p. 240; Fig. 8.9)
28. The voltage-gated Na^+ channels of the axon have a fast **activation gate** and a slower **inactivation gate**. (p. 242; Fig. 8.10)
29. Very few ions cross the membrane during an action potential. The Na^+-K^+-ATPase eventually restores Na^+ and K^+ to their original compartments. (p. 242)
30. Once an action potential has begun, there is a brief period of time known as the **absolute refractory period** during which a second action potential cannot be triggered, no matter how large the stimulus. Because of this, action potentials cannot be summed. (p. 243; Fig. 8.11)
31. During the **relative refractory period**, a higher-than-normal graded potential is required to trigger an action potential. (p. 243)
32. The myelin sheath around an axon speeds up conduction by increasing membrane resistance and decreasing current leakage. Larger-diameter axons conduct action potentials faster than smaller-diameter axons do. (p. 245)
33. The apparent jumping of action potentials from node to node is called **saltatory conduction**. (p. 247; Fig. 8.16)
34. Changes in blood K^+ concentration affect resting membrane potential and the conduction of action potentials. (p. 249; Fig. 8.17)

8.4 Cell-to-Cell Communication in the Nervous System

35. In **electrical synapses**, an electrical signal passes directly from the cytoplasm of one cell to another through gap junctions. **Chemical synapses** use neurotransmitters to carry information from one cell to the next, with the neurotransmitters diffusing across the synaptic cleft to bind with receptors on target cells. (pp. 249–250)
36. Neurotransmitters come in a variety of forms. **Cholinergic** neurons secrete **acetylcholine**. **Adrenergic neurons** secrete **norepinephrine**. **Glutamate**, **GABA**, **serotonin**, **adenosine**, and **nitric oxide** are other major neurotransmitters. (p. 252; Tbl. 8.4)
37. Neurotransmitter receptors are either ligand-gated ion channels (ionotropic receptors) or G protein-coupled receptors (metabotropic receptors). (p. 253)
38. Neurotransmitters are synthesized in the cell body or in the axon terminal. They are stored in **synaptic vesicles** and are released by exocytosis when an action potential reaches the axon terminal. (p. 254; Fig. 8.19a)
39. Neurotransmitter action is rapidly terminated by reuptake into cells, diffusion away from the synapse, or enzymatic breakdown. (p. 256; Fig. 8.19b)
40. Information about the strength and duration of a stimulus is conveyed by the amount of neurotransmitter released. Increased frequency of action potentials releases more neurotransmitter. (p. 257; Fig. 8.21)

8.5 Integration of Neural Information Transfer

41. When a presynaptic neuron synapses on a larger number of postsynaptic neurons, the pattern is known as **divergence**. When several presynaptic neurons provide input to a smaller number of postsynaptic neurons, the pattern is known as **convergence.** (p. 258; Fig. 8.22)
42. Synaptic transmission can be modified in response to activity at the synapse, a process known as **synaptic plasticity**. (p. 258)
43. G protein-coupled receptors either create **slow synaptic potentials** or modify cell metabolism. Ion channels create **fast synaptic potentials**. (p. 260; Fig. 8.23)

44. The summation of simultaneous graded potentials from different neurons is known as **spatial summation**. The summation of graded potentials that closely follow each other sequentially is called **temporal summation**. (p. 261; Fig. 8.25)
45. **Presynaptic modulation** of an axon terminal allows selective modulation of collaterals and their targets. **Postsynaptic modulation** occurs when a modulatory neuron synapses on a postsynaptic cell body or dendrites. (p. 261; Fig. 8.24)
46. **Long-term potentiation** and **long-term depression** are mechanisms by which neurons change the strength of their synaptic connections. (p. 264; Fig. 8.25)

REVIEW QUESTIONS

In addition to working through these questions and checking your answers on p. A-9, review the Learning Outcomes at the beginning of this chapter.

Level One Reviewing Facts and Terms

1. List the three functional classes of neurons, and explain how they differ structurally and functionally.
2. Somatic motor neurons control _____, and _____ neurons control smooth and cardiac muscles, glands, and some adipose tissue.
3. Autonomic neurons are classified as either _____ or _____ neurons.
4. Match each term with its description:

(a) axon (b) dendrite (c) afferent (d) efferent (e) trigger zone	1. process of a neuron that receives incoming signals 2. sensory neuron, transmits information to CNS 3. long process that transmits signals to the target cell 4. region of neuron where action potential begins 5. neuron that transmits information from CNS to the rest of the body

5. Name the two primary cell types found in the nervous system.
6. Draw a typical neuron and label the cell body, axon, dendrites, nucleus, trigger zone, axon hillock, collaterals, and axon terminals. Draw mitochondria, rough endoplasmic reticulum, Golgi complex, and vesicles in the appropriate sections of the neuron.
7. Axonal transport refers to the
   (a) release of neurotransmitters into the synaptic cleft.
   (b) use of microtubules to send secretions from the cell body to the axon terminal.
   (c) movement of organelles and cytoplasm up and down the axon.
   (d) movement of the axon terminal to synapse with a new postsynaptic cell.
   (e) none of these.
8. Match the numbers of the appropriate characteristics with the two types of potentials. Characteristics may apply to one or both types.

(a) action potential (b) graded potential	1. all-or-none 2. can be summed 3. amplitude decreases with distance 4. exhibits a refractory period 5. amplitude depends on strength of stimulus 6. has no threshold

9. Arrange the following events in the proper sequence:
   (a) Efferent neuron reaches threshold and fires an action potential.
   (b) Afferent neuron reaches threshold and fires an action potential.
   (c) Effector organ responds by performing output.
   (d) Integrating center reaches decision about response.
   (e) Sensory organ detects change in the environment.
10. List the four major types of ion channels found in neurons. Are they chemically gated, mechanically gated, or voltage-gated?
11. Match the glial cell(s) on the right to the functions on the left. There may be more than one correct answer for each function.

(a) modified immune cells (b) help form the blood-brain barrier (c) form myelin (d) separate CNS fluid compartments (e) found in peripheral nervous system (f) found in ganglia	1. astrocytes 2. ependymal cells 3. microglia 4. oligodendrocytes 5. satellite cells 6. Schwann cells

12. An action potential is (circle all correct answers)
   (a) a reversal of the Na^+ and K^+ concentrations inside and outside the neuron.
   (b) the same size and shape at the beginning and end of the axon.
   (c) initiated by inhibitory postsynaptic graded potentials.
   (d) transmitted to the distal end of a neuron and causes release of neurotransmitter.
13. Choose from the following ions to fill in the blanks correctly: Na^+, K^+, Ca^{2+}, Cl^-
   (a) The resting cell membrane is more permeable to _____ than to _____. Although _____ contribute little to the resting membrane potential, they play a key role in generating electrical signals in excitable tissues.
   (b) The concentration of _____ is 12 times greater outside the cell than inside.
   (c) The concentration of _____ is 30 times greater inside the cell than outside.
   (d) An action potential occurs when _____ enter the cell.
   (e) The resting membrane potential is due to the high _____ permeability of the cell.
14. What is the myelin sheath?
15. List two factors that enhance conduction speed.
16. List three ways neurotransmitters are removed from the synapse.
17. Draw and label a graph of an action potential. Below the graph, draw the positioning of the K^+ and Na^+ channel gates during each phase.

Level Two Reviewing Concepts

18. What causes the depolarization phase of an action potential? (Circle all that apply.)
    (a) K^+ leaving the cell through voltage-gated channels
    (b) K^+ being pumped into the cell by the Na^+-K^+-ATPase
    (c) Na^+ being pumped into the cell by the Na^+-K^+-ATPase
    (d) Na^+ entering the cell through voltage-gated channels
    (e) opening of the Na^+ channel inactivation gate
19. Name any four neurotransmitters, their receptor(s), and tell whether the receptor is an ion channel or a GPCR.
20. Create a map showing the organization of the nervous system using the following terms, plus any terms you choose to add:

• afferent signals	• neuron
• astrocyte	• neurotransmitter
• autonomic division	• oligodendrocyte
• brain	• parasympathetic division
• CNS	• peripheral division
• efferent neuron	• satellite cell
• ependymal cell	• Schwann cell
• glands	• sensory division
• glial cells	• somatic motor division
• integration	• spinal cord
• interneuron	• stimulus
• microglia	• sympathetic division
• muscles	• target

21. Arrange the following terms to describe the sequence of events after a neurotransmitter binds to a receptor on a postsynaptic neuron. Terms may be used more than once or not at all.
    (a) action potential fires at axon hillock
    (b) trigger zone reaches threshold
    (c) cell depolarizes
    (d) exocytosis
    (e) graded potential occurs
    (f) ligand-gated ion channel opens
    (g) local current flow occurs
    (h) saltatory conduction occurs
    (i) voltage-gated Ca^{2+} channels open
    (j) voltage-gated K^+ channels open
    (k) voltage-gated Na^+ channels open
22. Match the best term (hyperpolarize, depolarize, repolarize) to the following events. The cell in question has a resting membrane potential of -70 mV.
    (a) membrane potential changes from -70 mV to -50 mV
    (b) membrane potential changes from -70 mV to -90 mV
    (c) membrane potential changes from $+20$ mV to -60 mV
    (d) membrane potential changes from -80 mV to -70 mV
23. A neuron has a resting membrane potential of -70 mV. Will the neuron hyperpolarize or depolarize when each of the following events occurs? (More than one answer may apply; list all those that are correct.)
    (a) Na^+ enters the cell
    (b) K^+ leaves the cell
    (c) Cl^- enters the cell
    (d) Ca^{2+} enters the cell
24. If all action potentials within a given neuron are identical, how does the neuron transmit information about the strength and duration of the stimulus?
25. The presence of myelin allows an axon to (choose all correct answers):
    (a) produce more frequent action potentials.
    (b) conduct impulses more rapidly.
    (c) produce action potentials of larger amplitude.
    (d) produce action potentials of longer duration.
26. Define, compare, and contrast the following concepts:
    (a) threshold, subthreshold, suprathreshold, all-or-none, overshoot, undershoot
    (b) graded potential, EPSP, IPSP
    (c) absolute refractory period, relative refractory period
    (d) afferent neuron, efferent neuron, interneuron
    (e) sensory neuron, somatic motor neuron, sympathetic neuron, autonomic neuron, parasympathetic neuron
    (f) fast synaptic potential, slow synaptic potential
    (g) temporal summation, spatial summation
    (h) convergence, divergence

Level Three Problem Solving

27. If human babies' muscles and neurons are fully developed and functional at birth, why can't they focus their eyes, sit up, or learn to crawl within hours of being born? (*Hint:* Muscle strength is not the problem.)
28. The voltage-gated Na^+ channels of a neuron open when the neuron depolarizes. If depolarization opens the channels, what makes them close when the neuron is maximally depolarized?
29. One of the pills that Ji takes for high blood pressure caused his blood K^+ level to decrease from 4.5 mM to 2.5 mM. What happens to the resting membrane potential of his liver cells? (Circle all that are correct.)
    (a) decreases
    (b) increases
    (c) does not change
    (d) becomes more negative
    (e) becomes less negative
    (f) fires an action potential
    (g) depolarizes
    (h) hyperpolarizes
    (i) repolarizes
30. Characterize each of the following stimuli as being mechanical, chemical, or thermal:
    (a) bath water at 106 °F
    (b) acetylcholine
    (c) a hint of perfume
    (d) epinephrine
    (e) lemon juice
    (f) a punch on the arm
31. An unmyelinated axon has a much greater requirement for ATP than a myelinated axon of the same diameter and length. Can you explain why?

Level Four Quantitative Problems

32. The GHK equation is sometimes abbreviated to exclude chloride, which plays a minimal role in membrane potential for most cells. In addition, because it is difficult to determine absolute membrane permeability values for Na^+ and K^+, the equation is revised to use the ratio of the two ion permeabilities, expressed as $\alpha = P_{Na}/P_K$:

$$V_m = 61 \log \frac{[K^+]_{out} + \alpha[Na^+]_{out}}{[K^+]_{in} + \alpha[Na^+]_{in}}$$

Thus, if you know the relative membrane permeabilities of the two ions and their intracellular (ICF) and extracellular (ECF) concentrations, you can predict the membrane potential for a cell.

Using a calculator with log function or the free online Nernst/Goldman equation simulator from the University of Arizona (*www.nernstgoldman.physiology.arizona.edu/*), do the following calculations.

(a) A resting cell has an alpha (α) value of 0.025 and the following ion concentrations:

Na^+: ICF = 5 mM, ECF = 135 mM
K^+: ICF = 150 mM, ECF = 4 mM

What is the cell's membrane potential?

(b) The Na^+ permeability of the cell in (a) suddenly increases so that α = 20. Now what is the cell's membrane potential?

(c) Mrs. Nguyen has high blood pressure, and her physician puts her on a drug whose side effect decreases her plasma (ECF) K^+ from 4 mM to 2.5 mM. Using the other values in (a), calculate the membrane potential with decreased plasma K^+.

(d) The physician prescribes a potassium supplement for Mrs. Nguyen, who decides that if two pills are good, four must be better. Her plasma (ECF) K^+ now goes to 6 mM. What happens to her membrane potential?

33. In each of the following scenarios, will an action potential be produced? The postsynaptic neuron has a resting membrane potential of −70 mV.

(a) Fifteen neurons synapse on one postsynaptic neuron. At the trigger zone, 12 of the neurons produce EPSPs of 2 mV each, and the other three produce IPSPs of 3 mV each. The threshold for the postsynaptic cell is −50 mV.

(b) Fourteen neurons synapse on one postsynaptic neuron. At the trigger zone, 11 of the neurons produce EPSPs of 2 mV each, and the other three produce IPSPs of 3 mV each. The threshold for the postsynaptic cell is −60 mV.

(c) Fifteen neurons synapse on one postsynaptic neuron. At the trigger zone, 14 of the neurons produce EPSPs of 2 mV each, and the other one produces an IPSP of 9 mV. The threshold for the postsynaptic cell is −50 mV.

Answers to Concept Checks, Figure and Graph Questions, and end-of-chapter Review Questions can be found in Appendix A [p. A-1].

9 The Central Nervous System

Neuronal assemblies have important properties that cannot be explained by the additive qualities of individual neurons.

O. Hechter, in Biology and Medicine into the 21st Century, *1991*

Purkinje cells in the cerebellum

BACKGROUND BASICS

Directions of the body: inside back cover

51 Up-regulation
134 Diffusion through membranes
73 Cell-to-cell junctions
229 Synapses
231 Glial cells
77 Transporting epithelium
237 Ionic basis of action potentials
167 Neurotransmitters and neuromodulators
207 Posterior pituitary
260 Fast and slow synaptic potentials
264 Long-term potentiation

In vertebrate brain evolution, the most dramatic change is seen in the *forebrain* region {*fore*, in front}, which includes the **cerebrum** {*cerebrum*, brain; adjective *cerebral*}. In fish, the forebrain is a small bulge dedicated mainly to processing olfactory information about odors in the environment (Fig. 9.1d). In birds and rodents, part of the forebrain has enlarged into a cerebrum with a smooth surface (Fig. 9.1e).

In humans, the cerebrum is the largest and most distinctive part of the brain, with deep grooves and folds (Fig. 9.1f). More than anything else, the cerebrum is what makes us human. All evidence indicates that it is the part of the brain that allows reasoning and cognition.

The other brain structure whose evolution is obvious in the vertebrates is the *cerebellum*, a region of the *hindbrain* devoted to coordinating movement and balance. Birds (Fig. 9.1e) and humans (Fig. 9.1f) both have well-developed cerebellar structures. The cerebellum, like the cerebrum, is readily identifiable in these animals by its grooves and folds.

In this chapter, we begin with an overview of CNS anatomy and functions. We then look at how neural networks create the higher brain functions of thought and emotion.

9.3 Anatomy of the Central Nervous System

The vertebrate CNS consists of the brain and the spinal cord. As you learned in the previous section, brains increase in complexity and degree of specialization as we move up the phylogenetic tree from fish to humans. However, if we look at the vertebrate nervous system during development, a basic anatomical pattern emerges. In all vertebrates, the CNS consists of layers of neural tissue surrounding a fluid-filled central cavity lined with epithelium.

The CNS Develops from a Hollow Tube

In the very early embryo, cells that will become the nervous system lie in a flattened region called the **neural plate**. As development proceeds (at about day 20 of human development), neural plate cells along the edge migrate toward the midline (**FIG. 9.2a**).

By about day 23 of human development, the neural plate cells have fused with each other, creating a **neural tube** (Fig. 9.2b). *Neural crest cells* from the lateral edges of the neural plate now lie dorsal to the neural tube. The lumen of the neural tube will remain hollow and become the central cavity of the CNS.

The cells lining the neural tube will either differentiate into the epithelial *ependyma* [p. 233] or remain as undifferentiated *neural stem cells*. The outer cell layers of the neural tube will become the neurons and glia of the CNS. Neural crest cells will become the sensory and motor neurons of the peripheral nervous system.

By week 4 of human development, the anterior portion of the neural tube has begun to specialize into the regions of the brain (Fig. 9.2c). Three divisions are obvious: a **forebrain**, a **midbrain**, and a **hindbrain**. The tube posterior to the hindbrain will become the spinal cord. At this stage, the portion of the forebrain that will become the cerebrum is not much larger than the other regions of the brain.

As development proceeds, the growth of the cerebrum begins to outpace that of the other regions (Fig. 9.2d). By week 6, the CNS has formed the seven major divisions that are present at birth. Six of these regions are in the brain—(1) the cerebrum, (2) the *diencephalon*, (3) the midbrain, (4) and (5) the cerebellum and *pons*, (6) the *medulla oblongata*—and the seventh is the spinal cord. The cerebrum and diencephalon develop from the forebrain. The cerebellum, pons, and medulla oblongata are divisions of the hindbrain.

By week 6 the central cavity (lumen) of the neural tube has begun to enlarge into the hollow **ventricles** {*ventriculus*, belly} of the brain. There are two *lateral ventricles* (the first and second) and two *descending ventricles* (the third and fourth). The central cavity of the neural tube also becomes the *central canal* of the spinal cord.

By week 11, the cerebrum is noticeably enlarged (Fig. 9.2e), and at birth, the cerebrum is the largest and most obvious structure we see when looking at a human brain (Fig. 9.2f). The fully developed cerebrum surrounds the diencephalon, midbrain, and pons, leaving only the cerebellum and medulla oblongata visible below it. Because of the flexion (bending) of the neural tube early in development (see Fig. 9.2c), some directional terms have different meanings when applied to the brain (Fig. 9.2g).

The CNS Is Divided into Gray Matter and White Matter

The central nervous system, like the peripheral nervous system, is composed of neurons and supportive glial cells. Interneurons are those neurons completely contained within the CNS. Sensory (afferent) and efferent neurons link interneurons to peripheral receptors and effectors.

When viewed on a macroscopic level, the tissues of the CNS are divided into gray matter and white matter (**FIG. 9.3c**). **Gray matter** consists of unmyelinated nerve cell bodies, dendrites, and axons. The cell bodies are assembled in an organized fashion in both the brain and the spinal cord. They form layers in some parts of the brain and in other parts cluster into groups of neurons that have similar functions. Clusters of cell bodies in the brain and spinal cord are known as *nuclei*. Nuclei are usually identified by specific names—for example, the *lateral geniculate nucleus*, where visual information is processed.

White matter is mostly myelinated axons and contains very few neuronal cell bodies. Its pale color comes from the myelin sheaths that surround the axons. Bundles of axons that connect different regions of the CNS are known as **tracts**. Tracts in the central nervous system are equivalent to nerves in the peripheral nervous system.

The consistency of the brain and spinal cord is soft and jellylike. Although individual neurons and glial cells have highly

FIG. 9.2 ESSENTIALS Development of the Human Nervous System

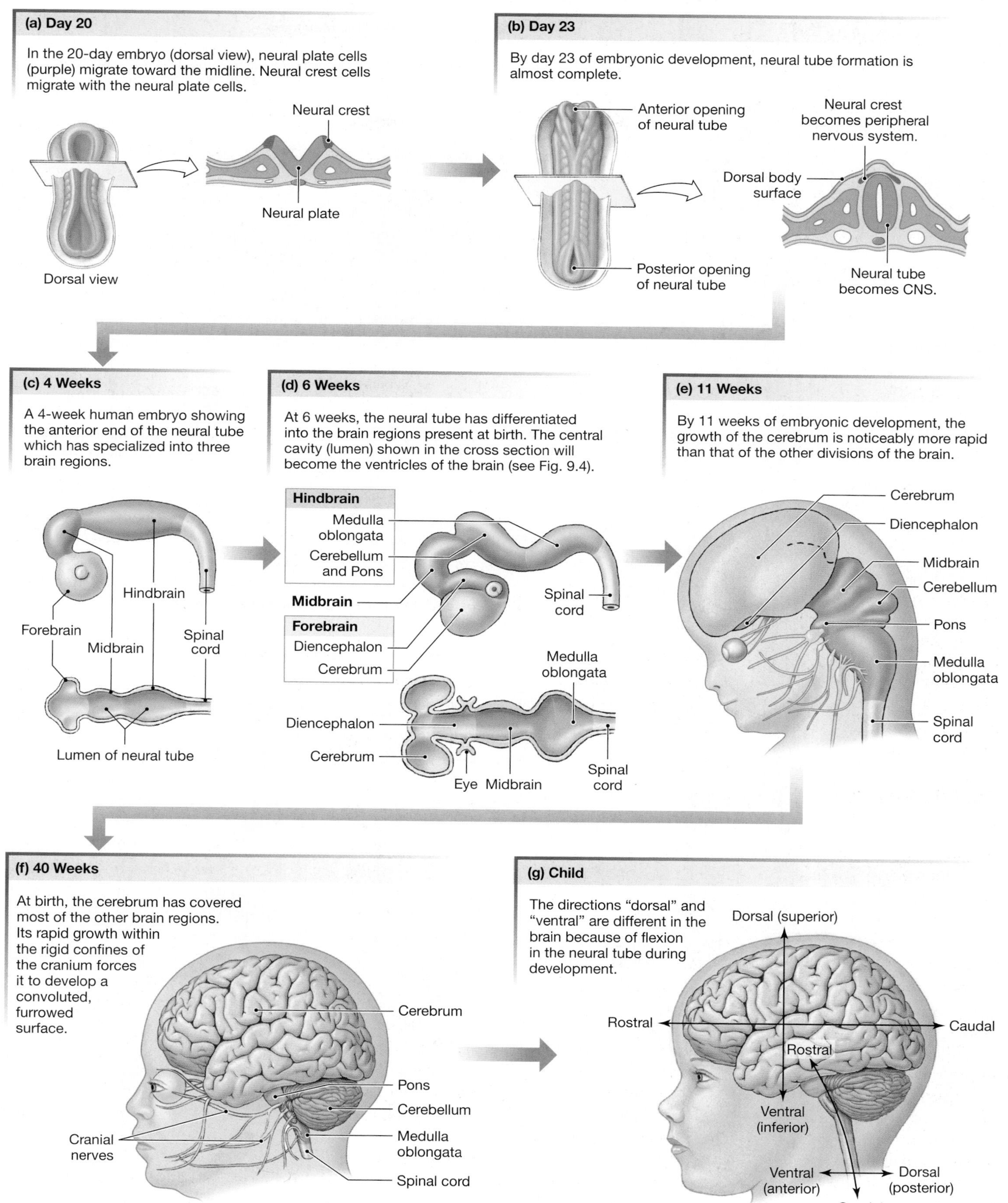
(a) Day 20
In the 20-day embryo (dorsal view), neural plate cells (purple) migrate toward the midline. Neural crest cells migrate with the neural plate cells.
Neural crest
Neural plate
Dorsal view
(b) Day 23
By day 23 of embryonic development, neural tube formation is almost complete.
Anterior opening of neural tube
Posterior opening of neural tube
Neural crest becomes peripheral nervous system.
Dorsal body surface
Neural tube becomes CNS.
(c) 4 Weeks
A 4-week human embryo showing the anterior end of the neural tube which has specialized into three brain regions.
Hindbrain
Forebrain
Midbrain
Spinal cord
Lumen of neural tube
(d) 6 Weeks
At 6 weeks, the neural tube has differentiated into the brain regions present at birth. The central cavity (lumen) shown in the cross section will become the ventricles of the brain (see Fig. 9.4).
Hindbrain
Medulla oblongata
Cerebellum and Pons
Midbrain
Forebrain
Diencephalon
Cerebrum
Spinal cord
Medulla oblongata
Diencephalon
Cerebrum
Eye
Midbrain
Spinal cord
(e) 11 Weeks
By 11 weeks of embryonic development, the growth of the cerebrum is noticeably more rapid than that of the other divisions of the brain.
Cerebrum
Diencephalon
Midbrain
Cerebellum
Pons
Medulla oblongata
Spinal cord
(f) 40 Weeks
At birth, the cerebrum has covered most of the other brain regions. Its rapid growth within the rigid confines of the cranium forces it to develop a convoluted, furrowed surface.
Cerebrum
Pons
Cerebellum
Medulla oblongata
Spinal cord
Cranial nerves
(g) Child
The directions “dorsal” and “ventral” are different in the brain because of flexion in the neural tube during development.
Dorsal (superior)
Rostral
Caudal
Rostral
Ventral (inferior)
Ventral (anterior)
Dorsal (posterior)
Caudal

FIG.9.3 Anatomy Summary . . . The Central Nervous System

(a) Posterior View of the CNS

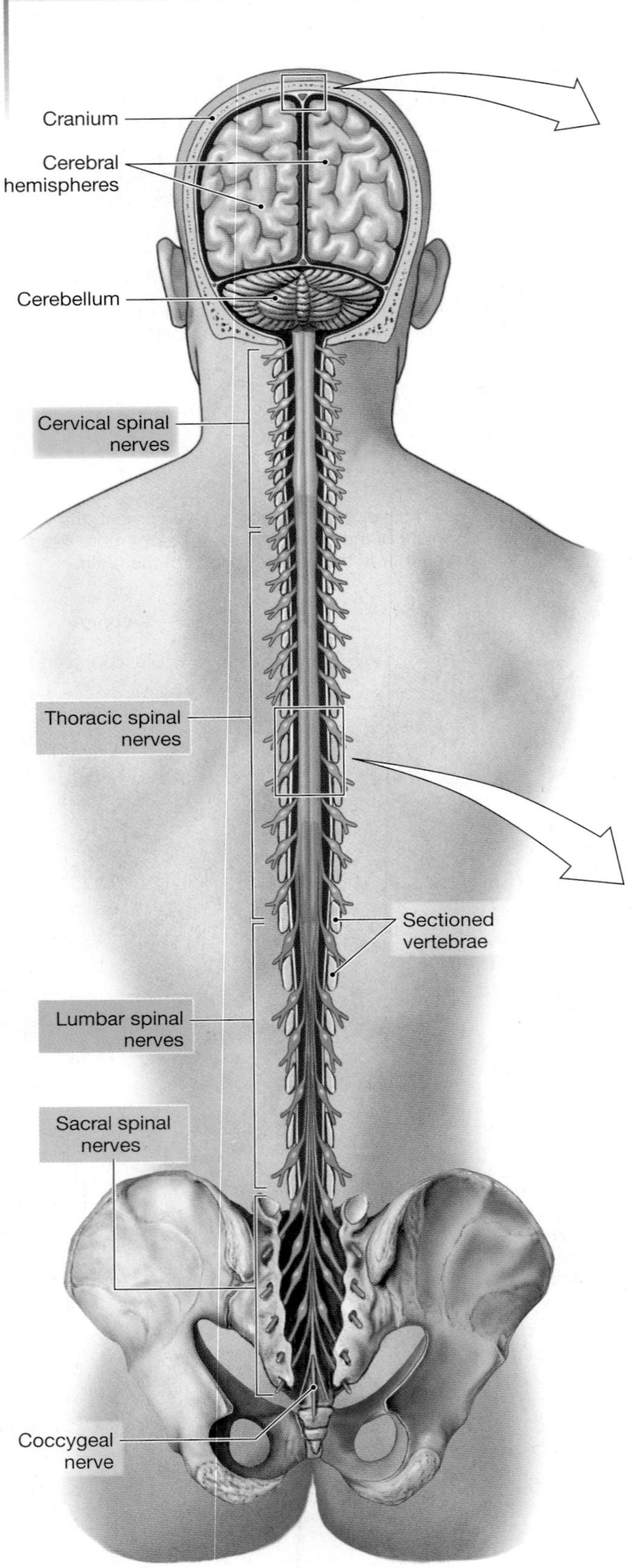

Cranium
Cerebral hemispheres
Cerebellum
Cervical spinal nerves
Thoracic spinal nerves
Sectioned vertebrae
Lumbar spinal nerves
Sacral spinal nerves
Coccygeal nerve

(b) Sectional View of the Meninges

The meninges and extracellular fluid cushion the delicate brain tissue.

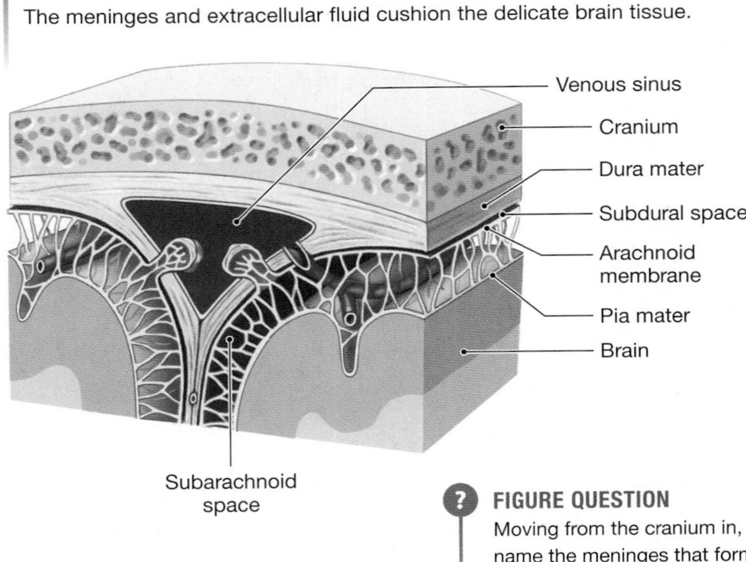

Venous sinus
Cranium
Dura mater
Subdural space
Arachnoid membrane
Pia mater
Brain
Subarachnoid space

FIGURE QUESTION

Moving from the cranium in, name the meninges that form the boundaries of the venous sinus and the subdural and subarachnoid spaces.

(c) Posterior View of Spinal Cord and Vertebra

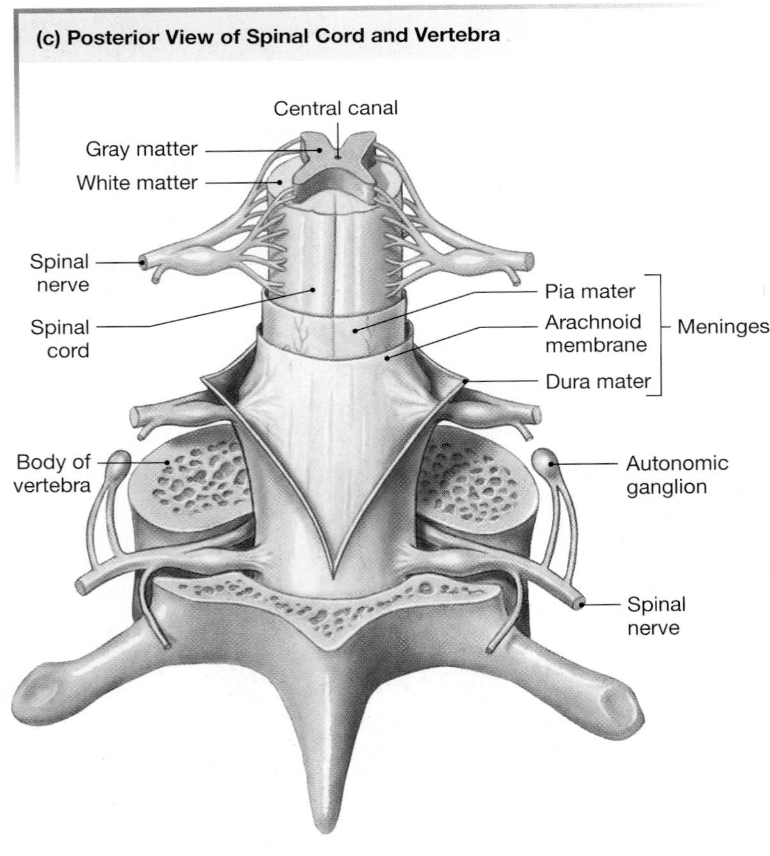

Central canal
Gray matter
White matter
Spinal nerve
Spinal cord
Pia mater
Arachnoid membrane
Dura mater
Meninges
Body of vertebra
Autonomic ganglion
Spinal nerve

organized internal cytoskeletons that maintain cell shape and orientation, neural tissue has minimal extracellular matrix and must rely on external support for protection from trauma. This support comes in the form of an outer casing of bone, three layers of connective tissue membrane, and fluid between the membranes (Fig. 9.3b, c).

Concept Check

2. Name the four kinds of glial cells found in the CNS, and describe the function(s) of each [p. 231].

Bone and Connective Tissue Support the CNS

In vertebrates, the brain is encased in a bony **skull**, or **cranium** (Fig. 9.3a), and the spinal cord runs through a canal in the **vertebral column**. The body segmentation that is characteristic of many invertebrates can still be seen in the bony **vertebrae** (singular *vertebra*), which are stacked on top of one another and separated by disks of connective tissue. Nerves of the peripheral nervous system enter and leave the spinal cord by passing through notches between the stacked vertebrae (Fig. 9.3c).

Three layers of membrane, collectively called the **meninges** {singular *meninx*, membrane}, lie between the bones and tissues of the central nervous system. These membranes help stabilize the neural tissue and protect it from bruising against the bones of the skeleton. Starting from the bones and moving toward the neural tissue, the membranes are (1) the dura mater, (2) the arachnoid membrane, and (3) the pia mater (Fig. 9.3b, c).

The **dura mater** {*durare*, to last + *mater*, mother} is the thickest of the three membranes (think *durable*). It is associated with veins that drain blood from the brain through vessels or cavities called *sinuses*. The middle layer, the **arachnoid** {*arachnoides*, cobweblike} **membrane**, is loosely tied to the inner membrane, leaving a *subarachnoid space* between the two layers. The inner membrane, the **pia mater** {*pius*, pious + *mater*, mother}, is a thin membrane that adheres to the surface of the brain and spinal cord. Arteries that supply blood to the brain are associated with this layer.

The final protective component of the CNS is extracellular fluid, which helps cushion the delicate neural tissue. The cranium has an internal volume of 1.4 L, of which about 1 L is occupied by the cells. The remaining volume is divided into two distinct extracellular compartments: the blood (100–150 mL), and the *cerebrospinal fluid* and interstitial fluid (250–300 mL). The cerebrospinal fluid and interstitial fluid together form the extracellular environment for neurons. Interstitial fluid lies inside the pia mater. Cerebrospinal fluid is found in the ventricles and in the space between the pia mater and the arachnoid membrane. The cerebrospinal and interstitial fluid compartments communicate with each other across the leaky junctions of the pial membrane and the ependymal cell layer lining the ventricles.

Concept Check

3. What is a ganglion? What is the equivalent structure in the CNS?
4. Peripheral nerves are equivalent to what organizational structure in the CNS?

The Brain Floats in Cerebrospinal Fluid

Cerebrospinal fluid (CSF) is a salty solution that is continuously secreted by the **choroid plexus**, a specialized region on the walls of the ventricles (**FIG. 9.4b**). The choroid plexus is remarkably similar to kidney tissue and consists of capillaries and a transporting epithelium [p. 77] derived from the ependyma. The choroid plexus cells selectively pump sodium and other solutes from plasma into the ventricles, creating an osmotic gradient that draws water along with the solutes (Fig. 9.4c).

From the ventricles, cerebrospinal fluid flows into the **subarachnoid space** between the pia mater and the arachnoid membrane, surrounding the entire brain and spinal cord in fluid (Fig. 9.4b). The cerebrospinal fluid flows around the neural tissue and is finally absorbed back into the blood by special **villi** {singular *villus*, shaggy hair} on the arachnoid membrane in the cranium (Fig. 9.4d). The rate of fluid flow through the central nervous system is sufficient to replenish the entire volume of cerebrospinal fluid about three times a day.

Cerebrospinal fluid serves two purposes: physical protection and chemical protection. The brain and spinal cord float in the thin layer of fluid between the membranes. The buoyancy of cerebrospinal fluid reduces the weight of the brain nearly fold. Lighter weight translates into less pressure on blood vessels and nerves attached to the CNS.

The cerebrospinal fluid also provides protective padding. When there is a blow to the head, the CSF must be compressed before the brain can hit the inside of the cranium. However, water is minimally compressible, which helps CSF cushion the brain. For a dramatic demonstration of the protective power of cerebrospinal fluid, shake a block of tofu (representing the brain) in an empty jar. Then shake a second block of tofu in a jar completely filled with water to see how cerebrospinal fluid safeguards the brain.

In addition to physically protecting the delicate tissues of the CNS, cerebrospinal fluid creates a closely regulated extracellular environment for the neurons. The choroid plexus is selective about which substances it transports into the ventricles, and, as a result, the composition of cerebrospinal fluid is different from that of the plasma. The concentration of K^+ is lower in the cerebrospinal fluid, and the concentration of H^+ is higher than in plasma. The concentration of Na^+ in CSF is similar to that in the blood. Cerebrospinal fluid normally contains very little protein and no blood cells.

Cerebrospinal fluid exchanges solutes with the interstitial fluid of the CNS and provides a route by which wastes can be

FIG.9.4 Anatomy Summary . . . Cerebrospinal Fluid

(a) Ventricles of the Brain

The first and second ventricles form the lateral ventricles. They connect to the third ventricle through narrow openings. The *cerebral aqueduct* then leads from the third ventricle in the diencephalon to the fourth ventricle in the brainstem. The fourth ventricle narrows to become the central canal of the spinal cord. Compare the frontal view here to the cross section in Figure 9.10a.

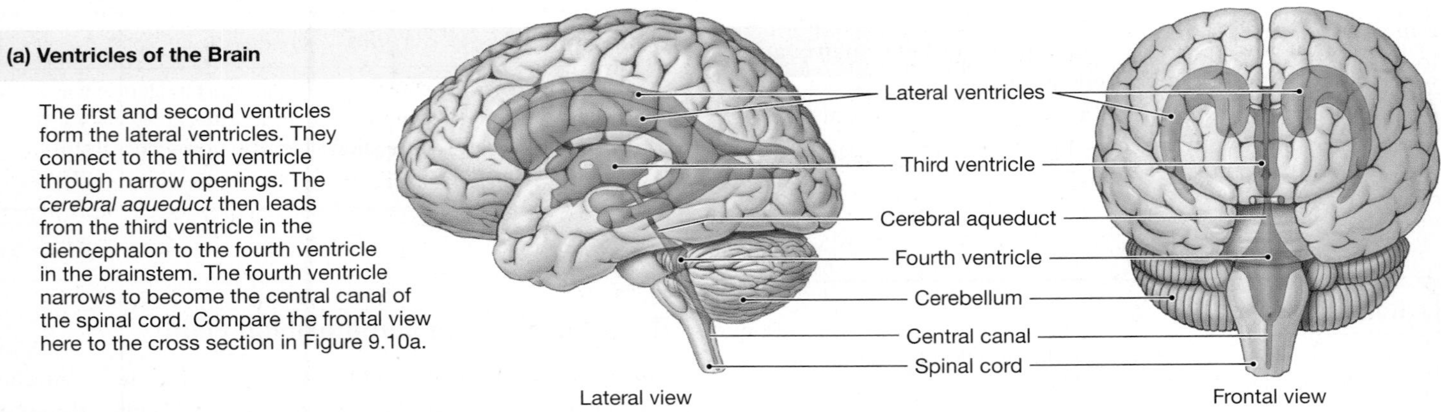

Lateral ventricles
Third ventricle
Cerebral aqueduct
Fourth ventricle
Cerebellum
Central canal
Spinal cord
Lateral view
Frontal view

(b) Cerebrospinal Fluid Secretion

Cerebrospinal fluid is secreted into the ventricles and flows throughout the subarachnoid space, where it cushions the central nervous system.

(c) Choroid Plexus

The choroid plexus transports ions and nutrients from the blood into the cerebrospinal fluid.

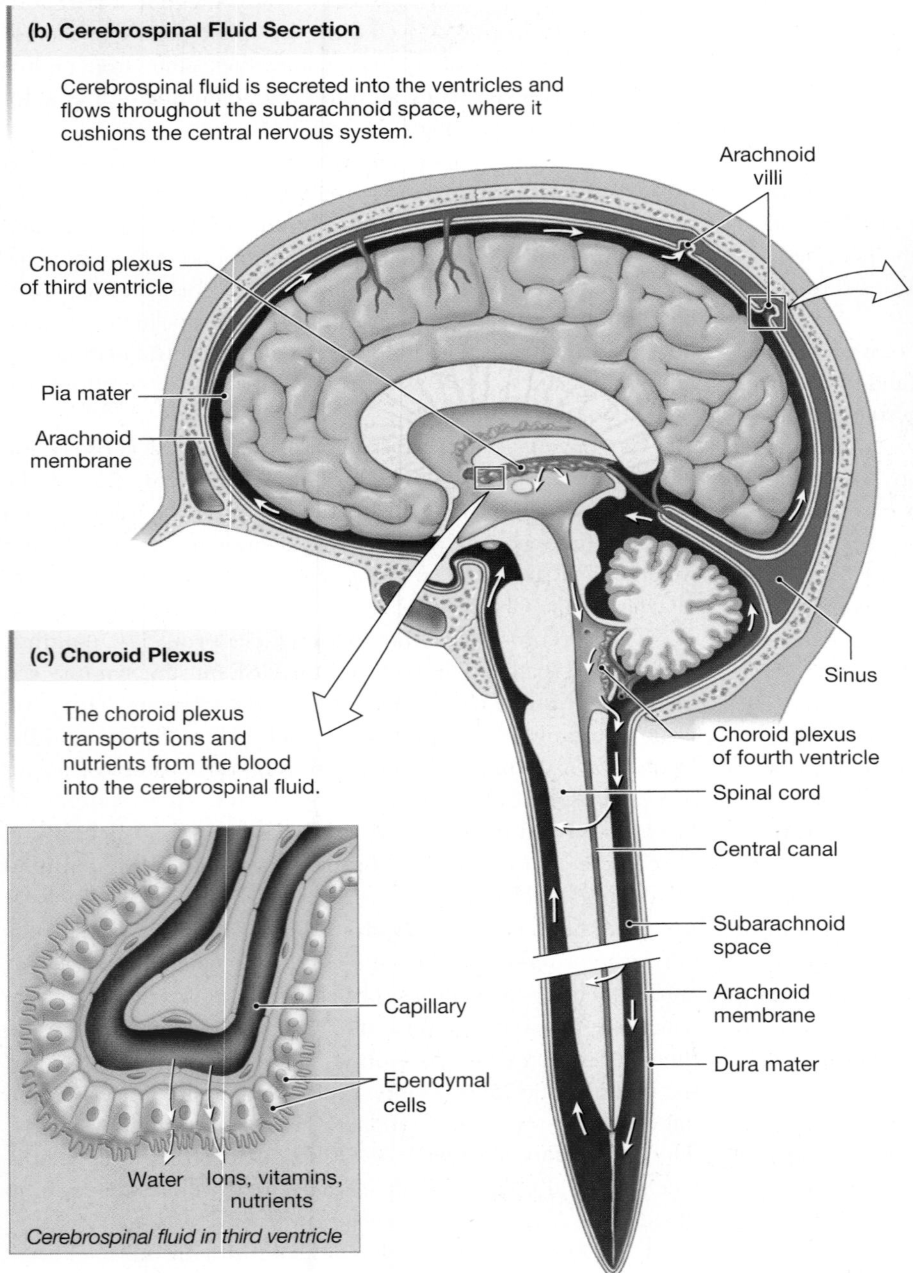

Arachnoid villi
Choroid plexus of third ventricle
Pia mater
Arachnoid membrane
Sinus
Choroid plexus of fourth ventricle
Spinal cord
Central canal
Subarachnoid space
Arachnoid membrane
Dura mater
Capillary
Ependymal cells
Water
Ions, vitamins, nutrients
Cerebrospinal fluid in third ventricle

(d) Cerebrospinal Fluid Reabsorption

Cerebrospinal fluid is reabsorbed into the blood at fingerlike projections of the arachnoid membrane called villi.

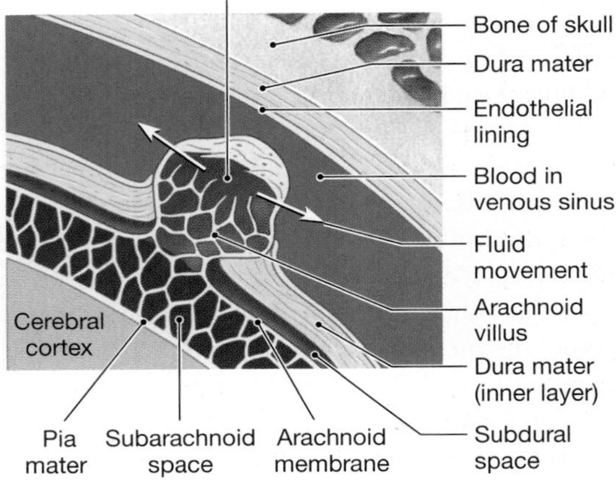

Cerebrospinal fluid
Bone of skull
Dura mater
Endothelial lining
Blood in venous sinus
Fluid movement
Arachnoid villus
Dura mater (inner layer)
Subdural space
Cerebral cortex
Pia mater
Subarachnoid space
Arachnoid membrane

FIGURE QUESTIONS

1. Physicians may extract a sample of cerebrospinal fluid when they suspect an infection in the brain. Where is the least risky and least difficult place for them to insert a needle through the meninges? (See Fig. 9.4b.)
2. The aqueduct of Sylvius is the narrow passageway between the third and fourth ventricles. What happens to CSF flow if the aqueduct becomes blocked by infection or tumor, a condition known as aqueductal stenosis {*stenos*, narrow}? On a three-dimensional imaging study of the brain, how would you distinguish aqueductal stenosis from a blockage of CSF flow in the subarachnoid space near the frontal lobe?

removed. Clinically, a sample of cerebrospinal fluid is presumed to be an indicator of the chemical environment in the brain. This sampling procedure, known as a *spinal tap* or *lumbar puncture*, is generally done by withdrawing fluid from the subarachnoid space between vertebrae at the lower end of the spinal cord. The presence of proteins or blood cells in cerebrospinal fluid suggests an infection.

Concept Check

5. If the concentration of H^+ in cerebrospinal fluid is higher than that in the blood, what can you say about the pH of the CSF?
6. Why is rupturing a blood vessel running between the meninges potentially a surgical emergency?
7. Is cerebrospinal fluid more like plasma or more like interstitial fluid? Defend your answer.

The Blood-Brain Barrier Protects the Brain

The final layer of protection for the brain is a functional barrier between the interstitial fluid and the blood. This barrier is necessary to isolate the body's main control center from potentially harmful substances in the blood and from blood-borne pathogens such as bacteria. To achieve this protection, most of the 400 miles of brain capillaries create a functional **blood-brain barrier** (**FIG. 9.5**). Although not a literal barrier, the highly selective permeability of brain capillaries shelters the brain from toxins and from fluctuations in hormones, ions, and neuroactive substances such as neurotransmitters in the blood.

Why are brain capillaries so much less permeable than other capillaries? In most capillaries, leaky cell-cell junctions and pores allow free exchange of solutes between the plasma and interstitial fluid [p. 73]. In brain capillaries, however, the endothelial cells form tight junctions with one another, junctions that prevent solute movement between the cells. Tight junction formation is induced

FIG. 9.5 The blood-brain barrier

(a) This cerebral angiogram shows the extensive blood supply to the brain, which has about 400 miles of capillaries.

Anterior cerebral artery
Posterior cerebral artery
Middle cerebral artery
Circle of Willis
Internal carotid artery

(b) Neurons are protected from harmful substances in the blood because brain capillaries are not leaky.

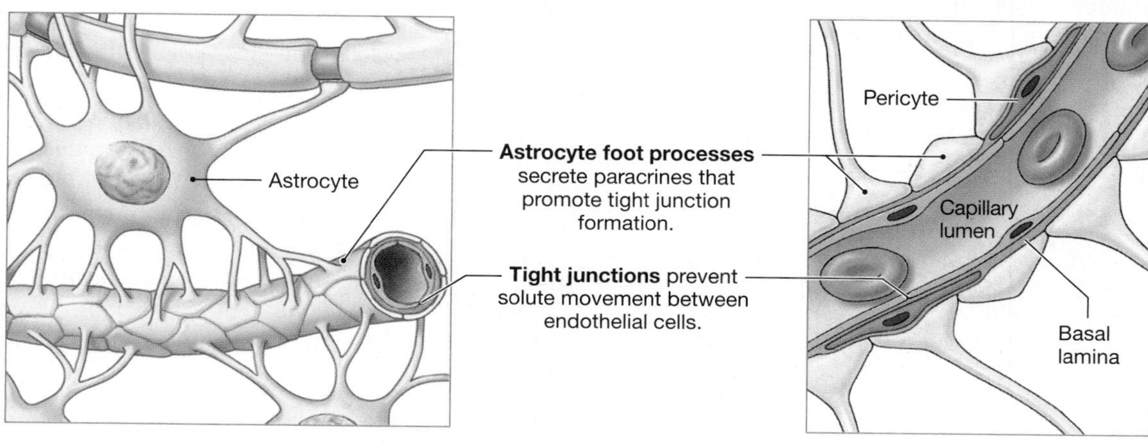
Astrocyte
Astrocyte foot processes secrete paracrines that promote tight junction formation.
Tight junctions prevent solute movement between endothelial cells.
Pericyte
Capillary lumen
Basal lamina

by paracrine signals from adjacent contractile cells called *pericytes* and from astrocytes whose foot processes surround the capillary. As a result, it is the brain tissue itself that creates the blood-brain barrier.

The selective permeability of the blood-brain barrier can be attributed to its transport properties. The capillary endothelium uses selected membrane carriers and channels to move nutrients and other useful materials from the blood into the brain interstitial fluid. Other transporters move wastes from the interstitial fluid into the plasma. Any water-soluble molecule that is not transported on one of these carriers cannot cross the blood-brain barrier.

One interesting illustration of how the blood-brain barrier works is seen in *Parkinson's disease*, a neurological disorder in which brain levels of the neurotransmitter dopamine are too low because dopaminergic neurons are either damaged or dead. Dopamine administered in a pill or injection is ineffective because it is unable to cross the blood-brain barrier. The dopamine precursor *L-dopa*, however, is transported across the cells of the blood-brain barrier on an amino acid transporter [p. 142]. Once neurons have access to L-dopa in the interstitial fluid, they metabolize it to dopamine, thereby allowing the deficiency to be treated.

RUNNING PROBLEM

Ben was diagnosed with infantile spasms, or West syndrome, a form of epilepsy characterized by the onset of head-drop seizures at 4 to 7 months and by arrested or deteriorating mental development. Ben was started on a month-long regimen of adrenocorticotropin (ACTH) [p. 211] shots plus an anti-epileptic drug called vigabatrin to control the seizures. Scientists are unsure why ACTH is so effective in controlling this type of seizure. They have found that, among its effects, it increases myelin formation, increases blood-brain barrier integrity, and enhances binding of the neurotransmitter GABA at synapses. Vigabatrin prolongs synaptic activity of GABA by slowing its breakdown. As expected, Ben's seizures disappeared completely before the month of treatment ended, and his development began to return to a normal level.

Q1: *How might a leaky blood-brain barrier lead to a cascade of action potentials that trigger a seizure?*

Q2: *GABA opens Cl^- channels on the postsynaptic cell. What does this do to the cell's membrane potential? Does GABA make the cell more or less likely to fire action potentials?*

Q3: *Why is it important to limit the duration of ACTH therapy, particularly in very young patients? [p. 214]*

272 – **280** – 296 – 297 – 300 – 301

The blood-brain barrier effectively excludes many water-soluble substances, but smaller lipid-soluble molecules can diffuse through the cell membranes [p. 134]. This is one reason some antihistamines make you sleepy but others do not. Older antihistamines were lipid-soluble amines that readily crossed the blood-brain barrier and acted on brain centers controlling alertness. The newer drugs are much less lipid soluble and as a result do not have the same sedative effect.

A few areas of the brain lack a functional blood-brain barrier, and their capillaries have leaky endothelium like most of the rest of the body. In these areas of the brain, the function of adjacent neurons depends in some way on direct contact with the blood. For instance, the hypothalamus releases neurosecretory hormones that must pass into the capillaries of the *hypothalamic-hypophyseal portal system* for distribution to the anterior pituitary [p. 209].

Another region that lacks the blood-brain barrier is the vomiting center in the medulla oblongata. These neurons monitor the blood for possibly toxic foreign substances, such as drugs. If they sense something harmful, they initiate a vomiting reflex. Vomiting removes the contents of the digestive system and helps eliminate ingested toxins.

Neural Tissue Has Special Metabolic Requirements

A unique property of the central nervous system is its specialized metabolism. Neurons require a constant supply of oxygen and glucose to make ATP for active transport of ions and neurotransmitters. To supply these substrates, about 15% of the blood pumped by the heart goes to the brain and is distributed through the extensive cerebral vascular system (Fig. 9.5a). Disruption of blood flow or low levels of oxygen or glucose in the blood can have devastating effects on brain function.

The brain has such a high demand for oxygen that at any moment it is using about one-fifth of the body's oxygen supply. Oxygen passes freely across the blood-brain barrier to reach neurons and glial cells. If blood flow to the brain is interrupted, a person loses consciousness in seconds, and brain damage occurs after only a few minutes without oxygen.

Under normal circumstances, the only energy source for neurons is glucose, which is one reason that blood glucose homeostasis is critical. Glucose is transported from the plasma across the blood-brain barrier and into the CSF by membrane transporters. It is used directly by neurons for aerobic metabolism. Glucose is also taken up by astrocytes and converted to lactate [p. 109] that neurons can use for ATP production.

By some estimates, the brain is responsible for about half of the body's glucose consumption. Consequently, the body uses several homeostatic pathways to ensure that glucose concentrations in the blood always remain adequate to meet the brain's demand. If glucose homeostasis fails, progressive **hypoglycemia** (low blood glucose levels) leads to confusion, unconsciousness, and eventually death.

CLINICAL FOCUS

Diabetes: Hypoglycemia and the Brain

Neurons are picky about their food. Under most circumstances, the only biomolecule that neurons use for energy is glucose. Surprisingly, this can present a problem for diabetic patients, whose problem is too much glucose in the blood. In the face of sustained high blood glucose, the cells of the blood-brain barrier down-regulate [p. 51] their glucose transporters. Then, if the patient's blood glucose level falls below normal (hypoglycemia) because of excess insulin or failing to eat, the neurons of the brain may not be able to obtain glucose fast enough to sustain their electrical activity. The individual may exhibit confusion, irritability, and slurred speech as brain function begins to fail. Prompt administration of sugar, either by mouth or intravenous infusion is necessary to prevent permanent damage. In extreme cases, hypoglycemia can cause coma or even death.

Now that you have a broad overview of the central nervous system, we will examine the structure and function of the spinal cord and brain in more detail.

Concept Check

8. Oxidative phosphorylation takes place in which organelle?
9. Name the two metabolic pathways for aerobic metabolism of glucose. What happens to NADH produced in these pathways?
10. In the late 1800s, the scientist Paul Ehrlich injected blue dye into the bloodstream of animals. He noticed that all tissues except the brain stained blue. He was not aware of the blood-brain barrier, so what conclusion do you think he drew from his results?
11. In a subsequent experiment, a student of Ehrlich's injected the dye into the cerebrospinal fluid of the same animals. What do you think he observed about staining in the brain and in other body tissues?

9.4 The Spinal Cord

The spinal cord is the major pathway for information flowing back and forth between the brain and the skin, joints, and muscles of the body. In addition, the spinal cord contains neural networks responsible for locomotion. If the spinal cord is severed, there is loss of sensation from the skin and muscles as well as *paralysis*, loss of the ability to voluntarily control muscles.

FIG. 9.6 Organization of the spinal cord

The spinal cord contains nuclei with cell bodies of efferent neurons and tracts of axons going to and from the brain.

(a) One segment of spinal cord, ventral view, showing its pair of nerves

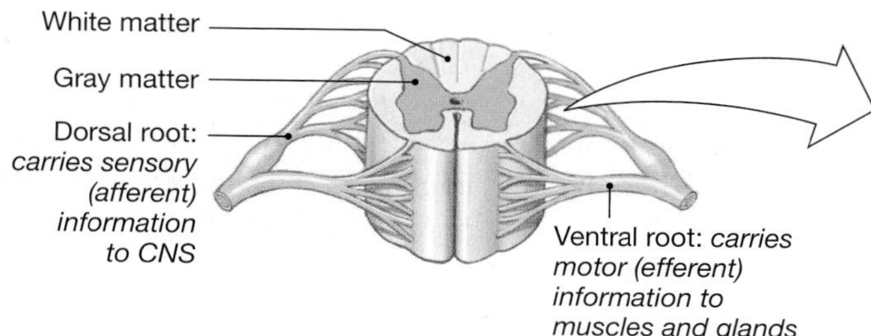
White matter
Gray matter
Dorsal root: *carries sensory (afferent) information to CNS*
Ventral root: *carries motor (efferent) information to muscles and glands*

(b) Gray matter consists of sensory and motor nuclei.

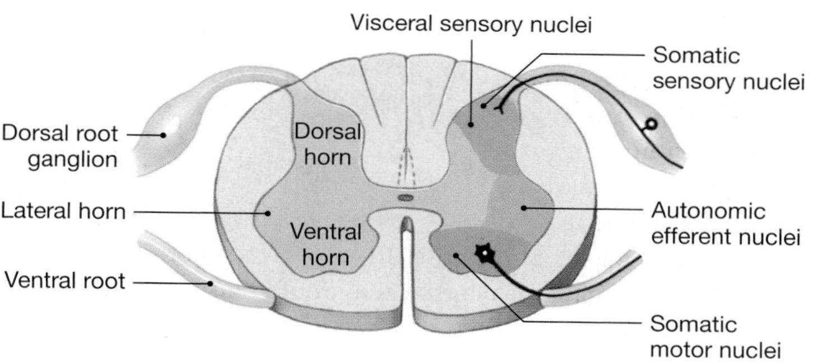
Visceral sensory nuclei
Somatic sensory nuclei
Dorsal root ganglion
Dorsal horn
Lateral horn
Ventral horn
Ventral root
Autonomic efferent nuclei
Somatic motor nuclei

(c) White matter in the spinal cord consists of tracts of axons carrying information to and from the brain.

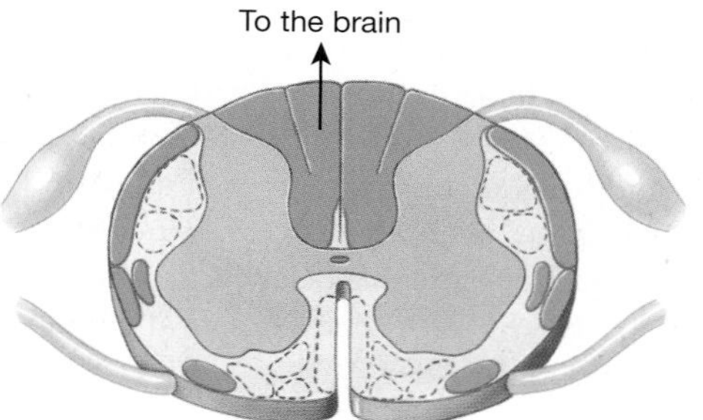
To the brain

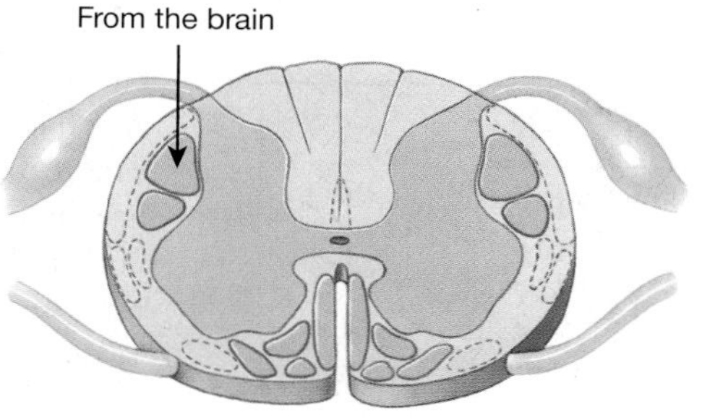
From the brain

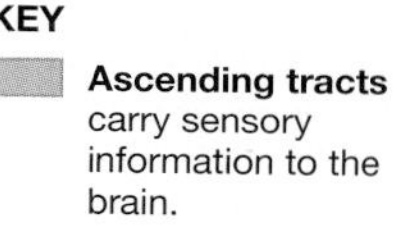
KEY
Ascending tracts carry sensory information to the brain.
Descending tracts carry commands to motor neurons.

CHAPTER 9

The spinal cord is divided into four regions: *cervical, thoracic, lumbar,* and *sacral*, named to correspond to the adjacent vertebrae (see Fig. 9.3a). Each spinal region is subdivided into segments, and each segment gives rise to a bilateral pair of **spinal nerves**. Just before a spinal nerve joins the spinal cord, it divides into two branches called **roots** (**FIG. 9.6a**).

The **dorsal root** of each spinal nerve is specialized to carry incoming sensory information. The **dorsal root ganglia**, swellings found on the dorsal roots just before they enter the cord (Fig. 9.6b), contain cell bodies of sensory neurons. The **ventral root** carries information from the CNS to muscles and glands.

In cross section, the spinal cord has a butterfly- or H-shaped core of gray matter and a surrounding rim of white matter. Sensory fibers from the dorsal roots synapse with interneurons in the **dorsal horns** of the gray matter. The dorsal horn cell bodies are organized into two distinct nuclei, one for somatic information and one for visceral information (Fig. 9.6b).

The **ventral horns** of the gray matter contain cell bodies of motor neurons that carry efferent signals to muscles and glands. The ventral horns are organized into somatic motor and autonomic nuclei. Efferent fibers leave the spinal cord via the ventral root.

The white matter of the spinal cord is the biological equivalent of fiber-optic cables that telephone companies use to carry our communications systems. White matter can be divided into a number of **columns** composed of tracts of axons that transfer information up and down the cord. **Ascending tracts** take sensory information to the brain. They occupy the dorsal and external lateral portions of the spinal cord (Fig. 9.6c). **Descending tracts** carry mostly efferent (motor) signals from the brain to the cord. They occupy the ventral and interior lateral portions of the white matter. **Propriospinal tracts** {*proprius*, one's own} are those that remain within the cord.

The spinal cord can function as a self-contained integrating center for simple *spinal reflexes*, with signals passing from a sensory neuron through the gray matter to an efferent neuron (**FIG. 9.7**). In addition, spinal interneurons may route sensory information to the brain through ascending tracts or bring commands from the brain to motor neurons. In many cases, the interneurons also modify information as it passes through them. Reflexes play a critical role in the coordination of body movement [the subject of Chapter 13].

> **Concept Check**
>
> **12.** What are the differences between horns, roots, tracts, and columns of the spinal cord?
>
> **13.** If a dorsal root of the spinal cord is cut, what function will be disrupted?

FIG. 9.7 Spinal reflexes

In a spinal reflex, sensory information entering the spinal cord is acted on without input from the brain. However, sensory information about the stimulus may be sent to the brain.

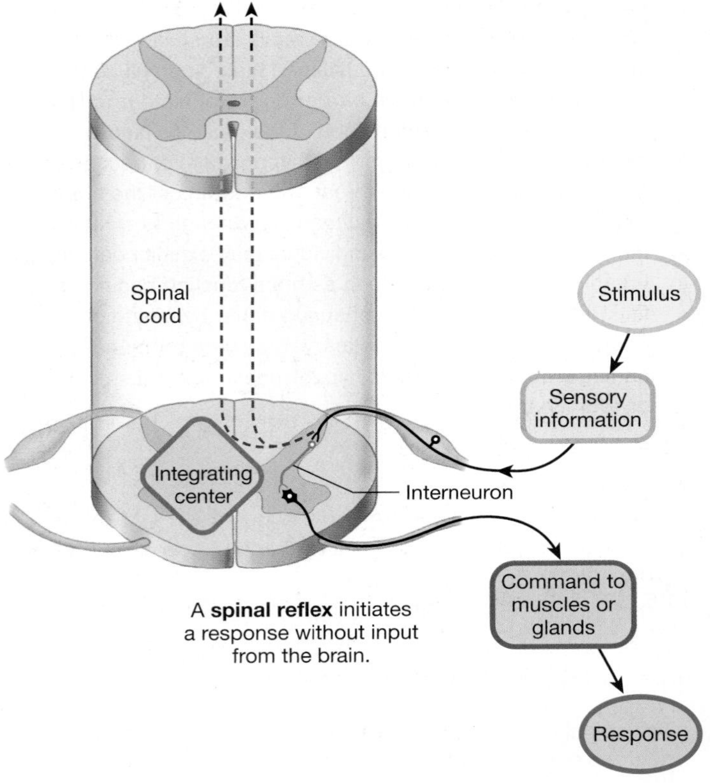
Spinal cord
Stimulus
Sensory information
Integrating center
Interneuron
Command to muscles or glands
A **spinal reflex** initiates a response without input from the brain.
Response

9.5 The Brain

Thousands of years ago, Aristotle declared that the heart was the seat of the soul. However, most people now agree that the brain is the organ that gives the human species its unique attributes. The challenge facing today's scientists is to understand how circuits formed by millions of neurons result in complex behaviors such as speaking, writing a symphony, or creating imaginary worlds for an interactive computer game. Brain function may be the ultimate emergent property [p. 2]. The question remains whether we will ever be able to decipher how emotions such as happiness and love arise from the chemical and electrical signals passing along circuits of neurons.

It is possible to study the brain at many levels of organization. The most reductionist view looks at the individual neurons and at what happens to them in response to chemical or electrical signals. A more integrative study might look at groups of neurons and how they interact with one another in *circuits, pathways*, or networks. The most complicated approach starts with a behavior or physiological response and works backward to dissect the neural circuits that create the behavior or response.

For centuries, studies of brain function were restricted to anatomical descriptions. However, when we study the brain we see no tidy 1:1 relationship between structure and function. An adult human brain has a mass of about 1400 g and contains an estimated 85 billion neurons. When you consider that each one of these billions of neurons may receive as many as 200,000 synapses, the number of possible neuronal connections is mind boggling. To complicate matters even more, those synapses are not fixed and are constantly changing.

A basic principle to remember when studying the brain is that one function, even an apparently simple one such as bending your finger, will involve multiple brain regions (as well as the spinal cord). Conversely, one brain region may be involved in several functions at the same time. In other words, understanding the brain is not simple and straightforward.

FIGURE 9.8 is an Anatomy Summary to follow as we discuss major brain regions, moving from the most primitive to the most complex. Of the six major divisions of the brain present at birth (see Fig. 9.2e), only the medulla, cerebellum, and cerebrum are visible when the intact brain is viewed in profile. The remaining three divisions (diencephalon, midbrain, and pons) are covered by the cerebrum.

The Brain Stem Is the Oldest Part of the Brain

The **brain stem** is the oldest and most primitive region of the brain and consists of structures that derive from the embryonic midbrain and hindbrain. The brain stem can be divided into white matter and gray matter, and in some ways, its anatomy is similar to that of the spinal cord. Some ascending tracts from the spinal cord pass through the brain stem, while other ascending tracts synapse there. Descending tracts from higher brain centers also travel through the brain stem on their way to the spinal cord.

Pairs of peripheral nerves branch off the brain stem, similar to spinal nerves along the spinal cord (Fig. 9.8f). Eleven of the 12 **cranial nerves** (numbers II–XII) originate along the brain stem. (The first cranial nerve, the olfactory nerve, enters the forebrain.) Cranial nerves carry sensory and motor information for the head and neck (**TBL. 9.1**).

The cranial nerves are described according to whether they include sensory fibers, efferent fibers, or both (mixed nerves). For example, cranial nerve X, the **vagus nerve** {*vagus*, wandering}, is a mixed nerve that carries both sensory and motor fibers for many internal organs. An important component of a clinical neurological examination is testing the functions controlled by these nerves.

The brain stem contains numerous discrete groups of nerve cell bodies, or nuclei. Many of these nuclei are associated with the **reticular formation**, a diffuse collection of neurons that extends throughout the brain stem. The name *reticular* means "network" and comes from the crisscrossed axons that branch profusely up into superior sections of the brain and down into the spinal cord. Nuclei in the brain stem are involved in many basic processes, including arousal and sleep, muscle tone and stretch reflexes, coordination of breathing, blood pressure regulation, and modulation of pain.

Concept Check

14. Are the following white matter or gray matter? (a) ascending tracts, (b) reticular formation, (c) descending tracts.

15. Using the information from Table 9.1, describe the types of activities you might ask a patient to perform if you wished to test the function of each cranial nerve.

16. In anatomical directional terminology, the cerebrum, which is located next to the top of the skull, is said to be ____ to the brain stem.

Starting at the spinal cord and moving toward the top of the skull, the brain stem consists of the medulla oblongata, the pons, and the midbrain (Fig. 9.8f). Some authorities include the cerebellum as part of the brain stem. The diamond-shaped fourth ventricle runs through the interior of the brain stem. The *cerebral aqueduct* connects it to the third ventricle in the diencephalon at its superior end. The inferior end of the fourth ventricle tapers to become the central canal of the spinal cord (see Fig. 9.4a).

Medulla The **medulla oblongata**, frequently just called the *medulla* {*medulla*, marrow; adjective *medullary*}, is the transition from the spinal cord into the brain proper (Fig. 9.8f). Its white matter includes ascending **somatosensory tracts** {*soma*, body} that bring sensory information to the brain, and descending **corticospinal tracts** that convey information from the cerebrum to the spinal cord.

About 90% of corticospinal tracts cross the midline to the opposite side of the body in a region of the medulla known as the **pyramids**. As a result of this crossover, each side of the brain controls the opposite side of the body. Gray matter in the medulla includes nuclei that control many involuntary functions, such as blood pressure, breathing, swallowing, and vomiting.

Pons The **pons** {*pons*, bridge; adjective *pontine*} is a bulbous protrusion on the ventral side of the brain stem above the medulla and below the midbrain. Because its primary function is to act as a relay station for information transfer between the cerebellum and cerebrum, the pons is often grouped with the cerebellum. The pons also coordinates the control of breathing along with centers in the medulla.

Midbrain The third region of the brain stem, the **midbrain**, or *mesencephalon* {*mesos*, middle}, is a relatively small area that lies between the lower brain stem and the diencephalon. The primary

FIG. 9.8 Anatomy Summary . . . The Brain

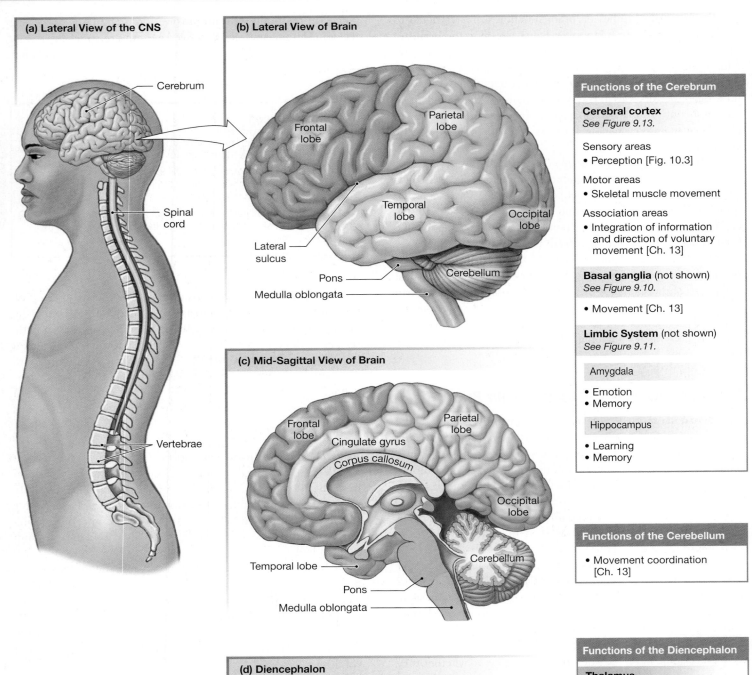

(a) Lateral View of the CNS

Cerebrum
Spinal cord
Vertebrae

(b) Lateral View of Brain

Frontal lobe
Parietal lobe
Temporal lobe
Occipital lobe
Lateral sulcus
Pons
Cerebellum
Medulla oblongata

Functions of the Cerebrum

Cerebral cortex
See Figure 9.13.

Sensory areas
- Perception [Fig. 10.3]

Motor areas
- Skeletal muscle movement

Association areas
- Integration of information and direction of voluntary movement [Ch. 13]

Basal ganglia (not shown)
See Figure 9.10.

- Movement [Ch. 13]

Limbic System (not shown)
See Figure 9.11.

Amygdala
- Emotion
- Memory

Hippocampus
- Learning
- Memory

(c) Mid-Sagittal View of Brain

Frontal lobe
Cingulate gyrus
Parietal lobe
Corpus callosum
Occipital lobe
Cerebellum
Temporal lobe
Pons
Medulla oblongata

Functions of the Cerebellum

- Movement coordination [Ch. 13]

(d) Diencephalon

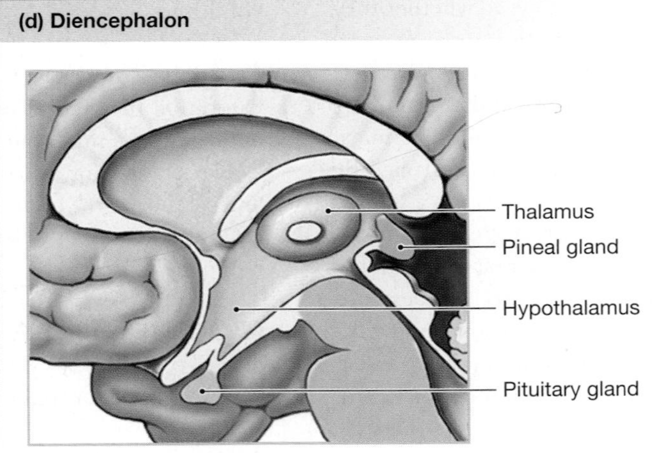
Thalamus
Pineal gland
Hypothalamus
Pituitary gland

Functions of the Diencephalon

Thalamus
- Integrating center and relay station for sensory and motor information

Pineal gland
- Melatonin secretion [Fig. 7.16]

Hypothalamus
See Table 9.2.
- Homeostasis [Ch. 11]
- Behavioral drives

Pituitary gland
- Hormone secretion [Fig. 7.8]

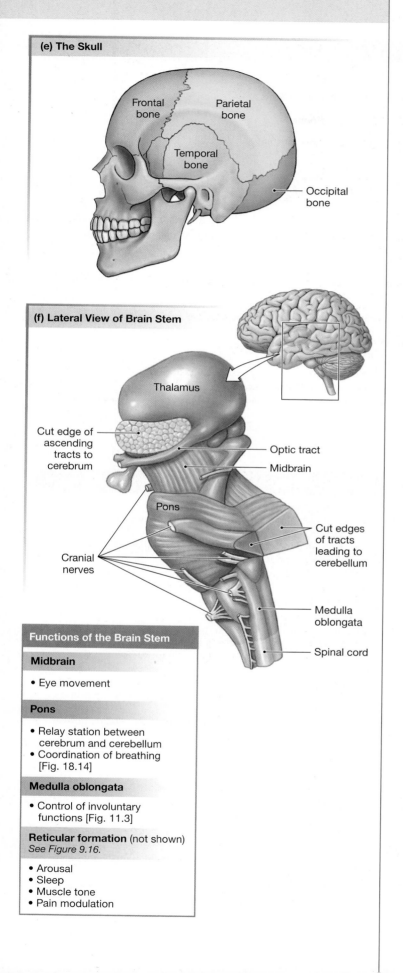

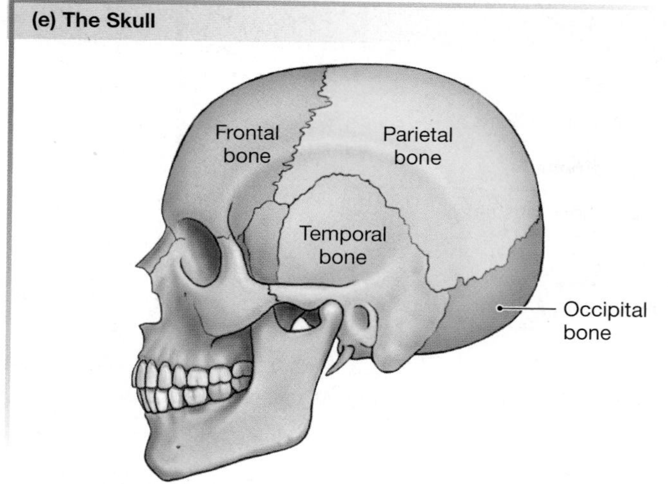
(e) The Skull
Frontal bone
Parietal bone
Temporal bone
Occipital bone

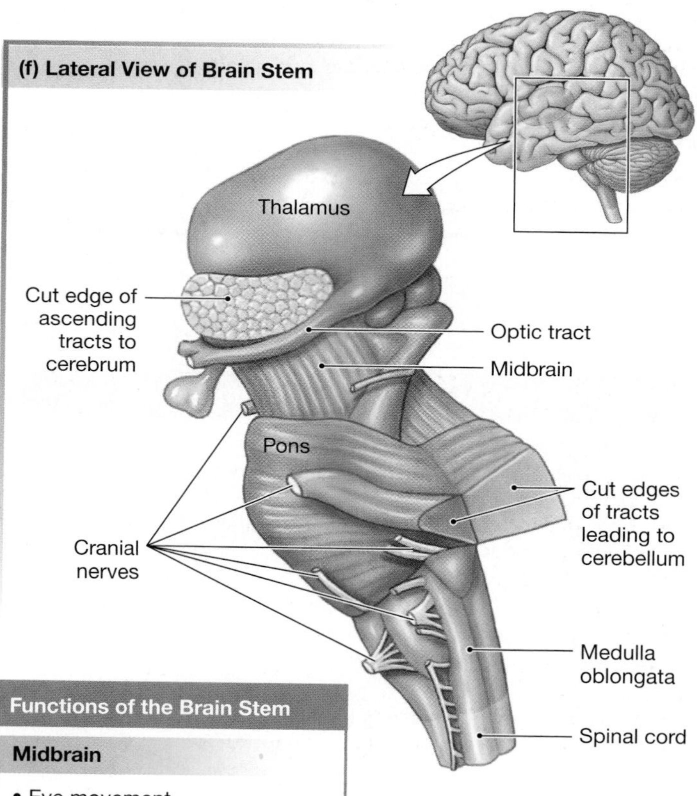
(f) Lateral View of Brain Stem
Thalamus
Cut edge of ascending tracts to cerebrum
Optic tract
Midbrain
Pons
Cut edges of tracts leading to cerebellum
Cranial nerves
Medulla oblongata
Spinal cord

Functions of the Brain Stem

Midbrain

- Eye movement

Pons

- Relay station between cerebrum and cerebellum
- Coordination of breathing [Fig. 18.14]

Medulla oblongata

- Control of involuntary functions [Fig. 11.3]

Reticular formation (not shown)
See Figure 9.16.

- Arousal
- Sleep
- Muscle tone
- Pain modulation

function of the midbrain is control of eye movement, but it also relays signals for auditory and visual reflexes.

The Cerebellum Coordinates Movement

The **cerebellum** is the second largest structure in the brain (Fig. 9.8a–c). It is located inside the base of the skull, just above the nape of the neck. The name *cerebellum* {adjective *cerebellar*} means "little brain," and, indeed, most of the nerve cells in the brain are in the cerebellum. The specialized function of the cerebellum is to process sensory information and coordinate the execution of movement. Sensory input into the cerebellum comes from somatic receptors in the periphery of the body and from receptors for equilibrium and balance located in the inner ear. The cerebellum also receives motor input from neurons in the cerebrum. [See Chapters 10 and 13 for additional information.]

The Diencephalon Contains the Centers for Homeostasis

The **diencephalon**, or "between-brain," lies between the brain stem and the cerebrum. It is composed of two main sections, the thalamus and the hypothalamus, and two endocrine structures, the pituitary and pineal glands (**FIG. 9.9**).

Most of the diencephalon is occupied by many small nuclei that make up the **thalamus** {*thalamus*, bedroom; adjective *thalamic*}. The thalamus receives sensory fibers from the optic tract, ears, and spinal cord as well as motor information from the cerebellum. It projects fibers to the cerebrum, where the information is processed.

The thalamus is often described as a relay station because almost all sensory information from lower parts of the CNS passes through it. Like the spinal cord, the thalamus can modify information passing through it, making it an integrating center as well as a relay station.

The **hypothalamus** lies beneath the thalamus. Although the hypothalamus occupies less than 1% of total brain volume, it is the center for homeostasis and contains centers for various behavioral drives, such as hunger and thirst. Output from the hypothalamus also influences many functions of the autonomic division of the nervous system, as well as a variety of endocrine functions (**TBL. 9.2**).

The hypothalamus receives input from multiple sources, including the cerebrum, the reticular formation, and various sensory receptors. Output from the hypothalamus goes first to the thalamus and eventually to multiple effector pathways.

Two important endocrine structures are located in the diencephalon: the pituitary gland and the pineal gland [p. 218]. The posterior pituitary (*neurohypophysis*) is a down-growth of the hypothalamus and secretes neurohormones that are synthesized in hypothalamic nuclei. The anterior pituitary (*adenohypophysis*) is a true endocrine gland. Its hormones are regulated by hypothalamic neurohormones secreted into the hypothalamic-hypophyseal portal system. Later in this chapter, we discuss the pineal gland, which secretes the hormone melatonin.

TABLE 9.1 The Cranial Nerves

Number	Name	Type	Primary Function
I	Olfactory	Sensory	Olfactory (smell) information from nose
II	Optic	Sensory	Visual information from eyes
III	Oculomotor	Motor	Eye movement, pupil constriction, lens shape
IV	Trochlear	Motor	Eye movement
V	Trigeminal	Mixed	Sensory information from face, mouth; motor signals for chewing
VI	Abducens	Motor	Eye movement
VII	Facial	Mixed	Sensory for taste; efferent signals for tear and salivary glands, facial expression
VIII	Vestibulocochlear	Sensory	Hearing and equilibrium
IX	Glossopharyngeal	Mixed	Sensory from oral cavity, baro- and chemoreceptors in blood vessels; efferent for swallowing, parotid salivary gland secretion
X	Vagus	Mixed	Sensory and efferents to many internal organs, muscles, and glands
XI	Spinal accessory	Motor	Some muscles in neck and shoulder
XII	Hypoglossal	Motor	Tongue muscles

Note: Mnemonic for remembering the cranial nerves in order: **O**h **O**nce **O**ne **T**akes **T**he **A**natomy **F**inal, **V**ery **G**ood **V**acations **S**ound **H**eavenly.

FIG. 9.9 The diencephalon

The diencephalon lies between the brain stem and the cerebrum. It consists of thalamus, hypothalamus, pineal gland, and pituitary gland.

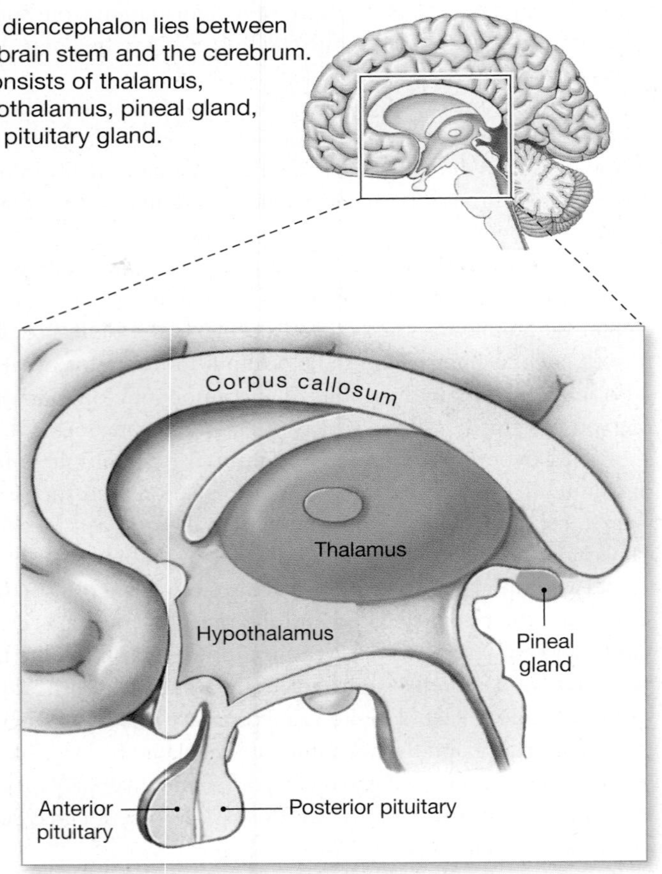
Corpus callosum
Thalamus
Hypothalamus
Pineal gland
Anterior pituitary
Posterior pituitary

TABLE 9.2 Functions of the Hypothalamus

1. Activates sympathetic nervous system
   - Controls catecholamine release from adrenal medulla (as in fight-or-flight reaction)
   - Helps maintain blood glucose concentrations through effects on endocrine pancreas
   - Stimulates shivering and sweating
2. Maintains body temperature
3. Controls body osmolarity
   - Motivates thirst and drinking behavior
   - Stimulates secretion of vasopressin [p. 207]
4. Controls reproductive functions
   - Directs secretion of oxytocin (for uterine contractions and milk release)
   - Directs trophic hormone control of anterior pituitary hormones FSH and LH [p. 211]
5. Controls food intake
   - Stimulates satiety center
   - Stimulates feeding center
6. Interacts with limbic system to influence behavior and emotions
7. Influences cardiovascular control center in medulla oblongata
8. Secretes trophic hormones that control release of hormones from anterior pituitary gland

Concept Check

17. Starting at the spinal cord and moving up, name the subdivisions of the brain stem.
18. What are the four primary structures of the diencephalon?

The Cerebrum Is the Site of Higher Brain Functions

As noted earlier in the chapter, the cerebrum is the largest and most distinctive part of the human brain and fills most of the cranial cavity. It is composed of two hemispheres connected primarily at the **corpus callosum** (Figs. 9.8c and 9.9), a distinct structure formed by axons passing from one side of the brain to the other. This connection ensures that the two hemispheres communicate and cooperate with each other. Each cerebral hemisphere is divided into four lobes, named for the bones of the skull under which they are located: *frontal*, *parietal*, *temporal*, and *occipital* (Fig. 9.8b, c, e).

The surface of the cerebrum in humans and other primates has a furrowed, walnut-like appearance, with grooves called *sulci* {singular *sulcus*, a furrow} dividing convolutions called *gyri* {singular *gyrus*, a ring or circle}. During development, the cerebrum grows faster than the surrounding cranium, causing the tissue to fold back on itself to fit into a smaller volume. The degree of folding is directly related to the level of processing of which the brain is capable. Less-advanced mammals, such as rodents, have brains with a relatively smooth surface. The human brain, on the other hand, is so convoluted that if it were inflated enough to smooth the surfaces, it would be three times as large and would need a head the size of a beach ball.

Gray Matter and White Matter Cerebral gray matter can be divided into three major regions: the cerebral cortex, the basal ganglia, and the limbic system. The **cerebral cortex** {*cortex*, bark or rind; adjective *cortical*, plural *cortices*} is the outer layer of the cerebrum, only a few millimeters thick (**FIG. 9.10a**). Neurons of the cerebral cortex are arranged in anatomically distinct vertical columns and horizontal layers (Fig. 9.10b). It is within these layers that our higher brain functions arise.

The second region of cerebral gray matter consists of the **basal ganglia** (Fig. 9.10a), which are involved in the control of movement. The basal ganglia are also called the *basal nuclei*. Neuroanatomists prefer to reserve the term *ganglia* for clusters of nerve cell bodies outside the CNS, but the term *basal ganglia* is commonly used in clinical settings.

FIG. 9.10 Gray matter of the cerebrum

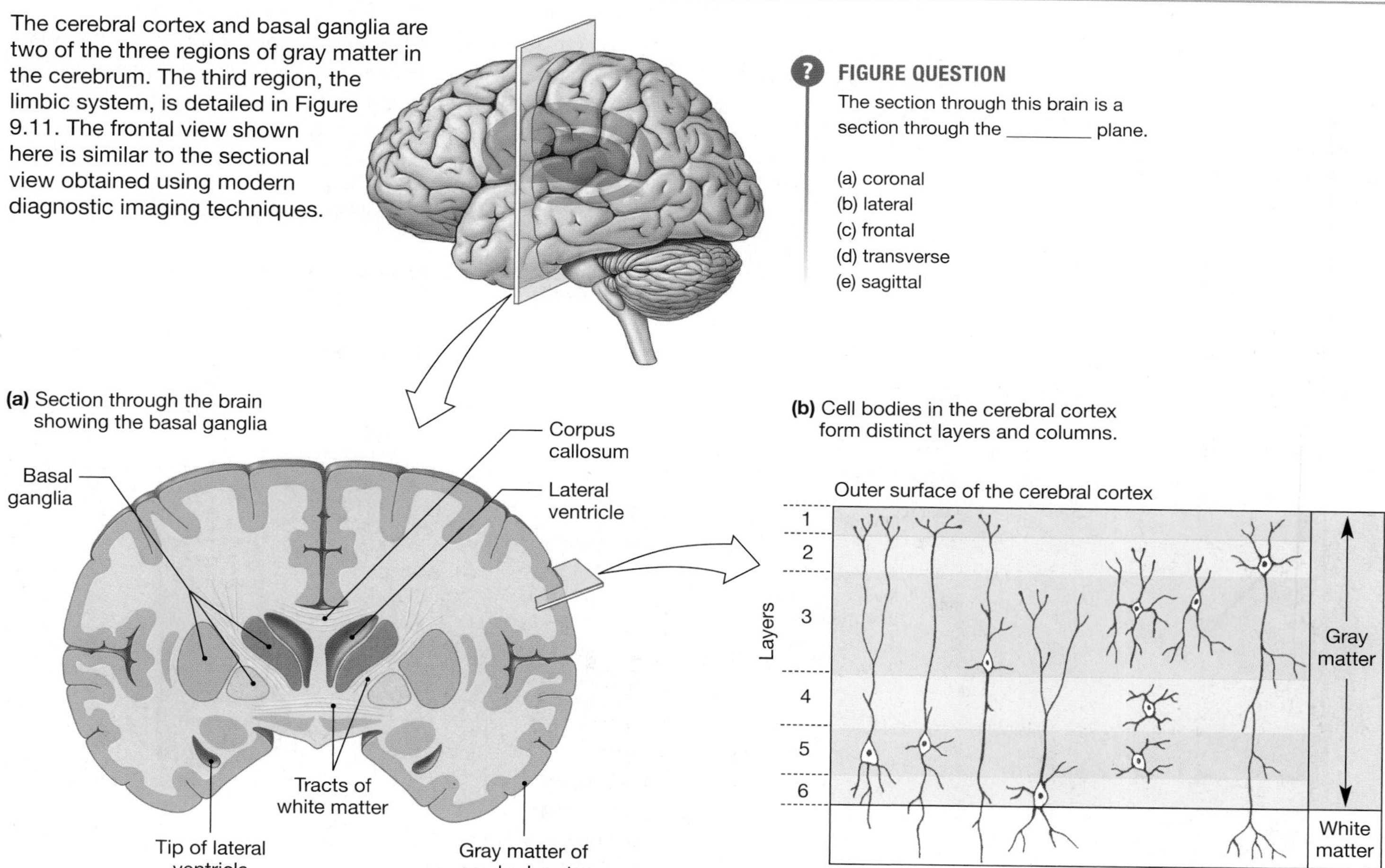
The cerebral cortex and basal ganglia are two of the three regions of gray matter in the cerebrum. The third region, the limbic system, is detailed in Figure 9.11. The frontal view shown here is similar to the sectional view obtained using modern diagnostic imaging techniques.
FIGURE QUESTION
The section through this brain is a section through the ________ plane.
(a) coronal
(b) lateral
(c) frontal
(d) transverse
(e) sagittal
(a) Section through the brain showing the basal ganglia
Basal ganglia
Corpus callosum
Lateral ventricle
Tracts of white matter
Tip of lateral ventricle
Gray matter of cerebral cortex
(b) Cell bodies in the cerebral cortex form distinct layers and columns.
Outer surface of the cerebral cortex
Layers
1
2
3
4
5
6
Gray matter
White matter

The third region of the cerebrum is the **limbic system** {*limbus*, a border}, which surrounds the brain stem (**FIG. 9.11**). The limbic system represents probably the most primitive region of the cerebrum. It acts as the link between higher cognitive functions, such as reasoning, and more primitive emotional responses, such as fear. The major areas of the limbic system are the **amygdala** and **cingulate gyrus**, which are linked to emotion and memory, and the **hippocampus**, which is associated with learning and memory.

White matter in the cerebrum is found mostly in the interior (Fig. 9.10a). Bundles of fibers allow different regions of the cortex to communicate with one another and transfer information from one hemisphere to the other, primarily through the corpus callosum. According to some estimates, the corpus callosum may have as many as 200 million axons passing through it! Information entering and leaving the cerebrum goes along tracts that pass through the thalamus (with the exception of olfactory information, which goes directly from olfactory receptors to the cerebrum).

Concept Check

19. Name the anatomical location in the brain where neurons from one side of the body cross to the opposite side.
20. Name the divisions of the brain in anatomical order, starting from the spinal cord.

FIG. 9.11 The limbic system

The limbic system includes the amygdala, hippocampus, and cingulate gyrus. Anatomically, the limbic system is part of the gray matter of the cerebrum. The thalamus is shown for orientation purposes and is not part of the limbic system.

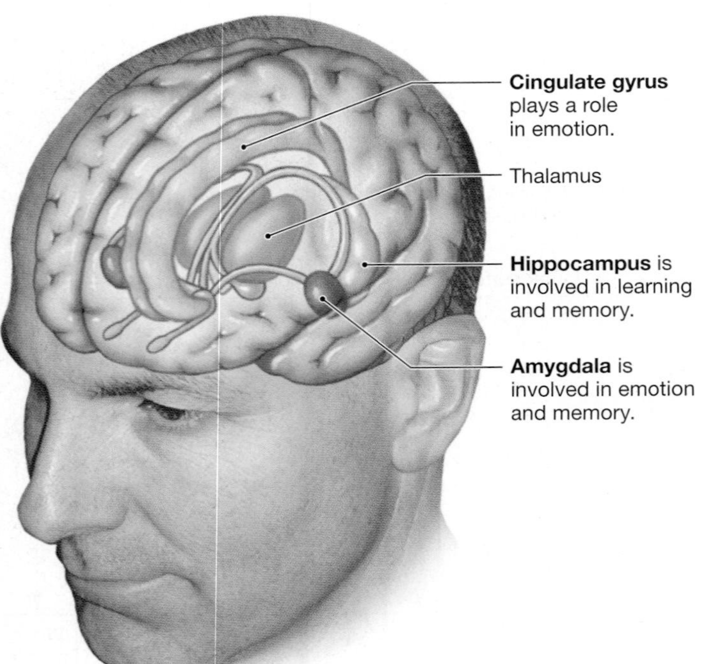
Cingulate gyrus plays a role in emotion.
Thalamus
Hippocampus is involved in learning and memory.
Amygdala is involved in emotion and memory.

9.6 Brain Function

From a simplistic view, the CNS is an information processor much like a computer. For many functions, it follows a basic reflex pathway [p. 14]. The brain receives sensory input from the internal and external environments, integrates and processes the information, and, if appropriate, creates a response (**FIG. 9.12a**). What makes the brain more complicated than this simple reflex pathway, however, is its ability to generate information and output signals *in the absence of external input*. Modeling this intrinsic input requires a more complex diagram.

Larry Swanson of the University of Southern California presents one approach to modeling brain function in his book *Brain Architecture: Understanding the Basic Plan* (2nd edition, Oxford University Press, 2011). He describes three systems that influence output by the motor systems of the body: (1) the **sensory system**, which monitors the internal and external environments and initiates reflex responses; (2) a **cognitive system** that resides in the cerebral cortex and is able to initiate voluntary responses; and (3) a **behavioral state system**, which also resides in the brain and governs sleep-wake cycles and other intrinsic behaviors. Information about the physiological or behavioral responses created by motor output feeds back to the sensory system, which in turn communicates with the cognitive and behavioral state systems (Fig. 9.12b).

In most of the physiological organ systems of the body that you will study, simple reflex pathways initiated through the sensory system and executed by motor output are adequate to explain homeostatic control mechanisms. However, the cognitive and behavioral state systems remain potential sources of influence. At its simplest, this influence may take the form of voluntary

FIG. 9.12 Simple and complex pathways in the brain

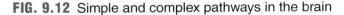

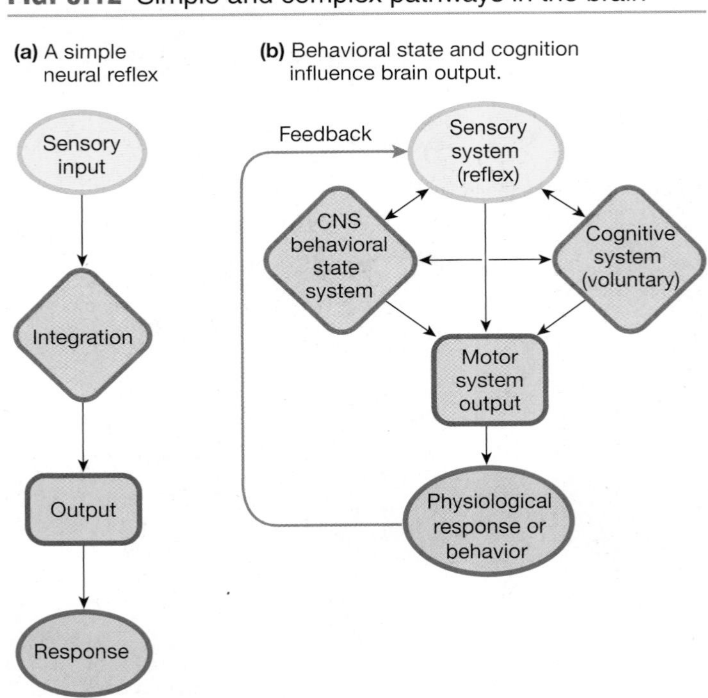
(a) A simple neural reflex
Sensory input
Integration
Output
Response
(b) Behavioral state and cognition influence brain output.
Feedback
Sensory system (reflex)
CNS behavioral state system
Cognitive system (voluntary)
Motor system output
Physiological response or behavior

behaviors, such as breath-holding, that override automatic functions. More subtle and complicated interactions include the effect of emotions on normal physiology, such as stress-induced heart palpitations, and the role of circadian rhythms in jet lag and shift work.

In the sections that follow, we take a brief look at sensory and motor systems in the brain. We conclude this chapter with a discussion of some aspects of the behavioral state system and the cognitive system, such as circadian rhythms, sleep-wake cycles, emotion, learning, and memory.

The Cerebral Cortex Is Organized into Functional Areas

The cerebral cortex serves as an integrating center for sensory information and a decision-making region for many types of motor output. If we examine the cortex from a functional viewpoint, we can divide it into three specializations: (1) **sensory areas** (also called sensory fields), which receive sensory input and translate it into perception (awareness); (2) **motor areas**, which direct skeletal muscle movement; and (3) **association areas** (association cortices), which integrate information from sensory and motor areas and can direct voluntary behaviors (**FIG. 9.13**). Information passing along a pathway is usually processed in more than one of these areas.

The functional areas of the cerebral cortex do not necessarily correspond to the anatomical lobes of the brain. For one thing, functional specialization is not symmetrical across the cerebral cortex: each lobe has special functions not shared by the matching lobe on the opposite side. This **cerebral lateralization** of function is sometimes referred to as *cerebral dominance*, more popularly known as left brain–right brain dominance (**FIG. 9.14**). Language and verbal skills tend to be concentrated on the left side of the brain, with spatial skills concentrated on the right side. The left brain is the dominant hemisphere for right-handed people, and it appears that the right brain is the dominant hemisphere for many left-handed people.

Even these generalizations are subject to change, however. Neural connections in the cerebrum, like those in other parts of the nervous system, exhibit a certain degree of plasticity. For example, if a person loses a finger, the regions of motor and sensory cortex previously devoted to control of the finger do not go dormant. Instead, adjacent regions of the cortex extend their functional fields and take over the parts of the cortex that are no longer used by the absent finger. Similarly, skills normally associated with one side of the cerebral cortex can be developed in the other hemisphere, as when a right-handed person with a broken hand learns to write with the left hand.

Much of what we know about functional areas of the cerebral cortex comes from study of patients who have either inherited

FIG. 9.13 Functional areas of the cerebral cortex

The cerebral cortex contains sensory areas for perception, motor areas that direct movement, and association areas that integrate information.

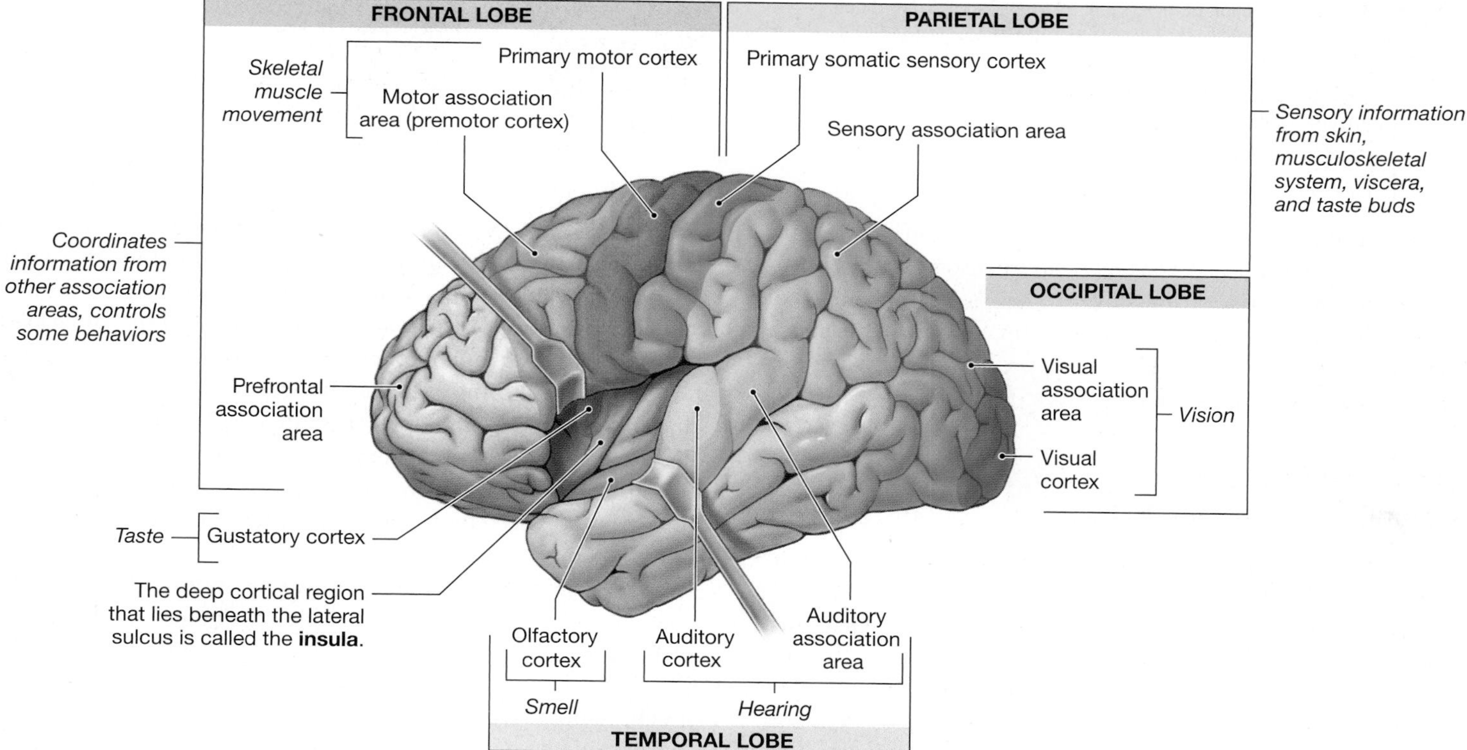
FRONTAL LOBE
Skeletal muscle movement
Primary motor cortex
Motor association area (premotor cortex)
PARIETAL LOBE
Primary somatic sensory cortex
Sensory association area
Sensory information from skin, musculoskeletal system, viscera, and taste buds
Coordinates information from other association areas, controls some behaviors
OCCIPITAL LOBE
Prefrontal association area
Visual association area
Visual cortex
Vision
Taste
Gustatory cortex
The deep cortical region that lies beneath the lateral sulcus is called the **insula**.
Olfactory cortex
Auditory cortex
Auditory association area
Smell
Hearing
TEMPORAL LOBE

FIG. 9.14 Cerebral lateralization

The distribution of functional areas in the two cerebral hemispheres is not symmetrical.

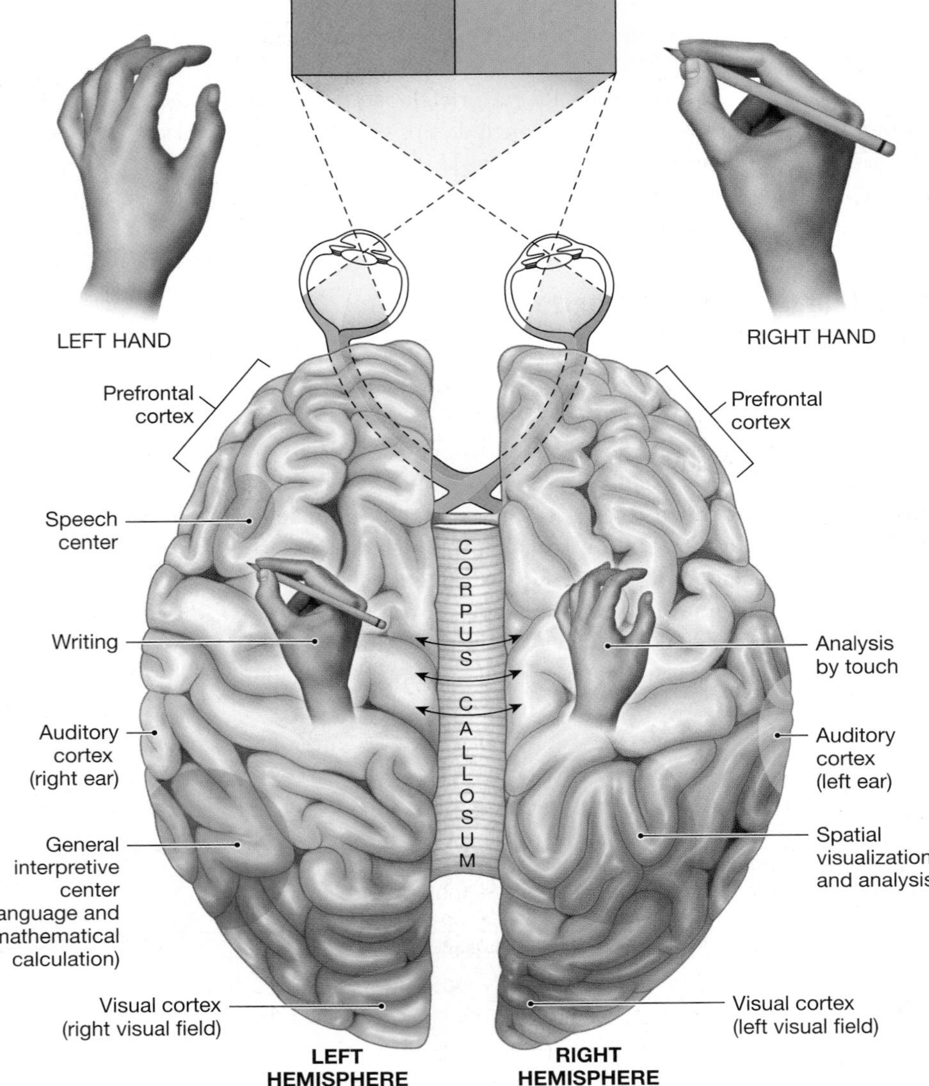
LEFT HAND
RIGHT HAND
Prefrontal cortex
Prefrontal cortex
Speech center
Writing
CORPUS CALLOSUM
Analysis by touch
Auditory cortex (right ear)
Auditory cortex (left ear)
General interpretive center (language and mathematical calculation)
Spatial visualization and analysis
Visual cortex (right visual field)
Visual cortex (left visual field)
LEFT HEMISPHERE
RIGHT HEMISPHERE

FIGURE QUESTIONS

1. What would a person see if a stroke destroyed all function in the right visual cortex?
2. What is the function of the corpus callosum?
3. Many famous artists, including Leonardo da Vinci and Michelangelo, were left-handed. How is this related to cerebral lateralization?

neurological defects or suffered wounds in accidents or war. In some instances, surgical lesions made to treat some medical condition, such as uncontrollable epilepsy, have revealed functional relationships in particular brain regions. Imaging techniques such as *positron emission tomography* (PET) scans provide noninvasive ways for us to watch the human brain at work (**TBL. 9.3**).

The Spinal Cord and Brain Integrate Sensory Information

The sensory system monitors the internal and external environments and sends information to neural integrating centers, which in turn initiate appropriate responses. In its simplest form, this pathway is the classic reflex, illustrated in Figure. 9.12a. The simplest reflexes can be integrated in the spinal cord, without input from higher brain centers (see Fig. 9.7). However, even simple spinal reflexes usually send sensory information to the brain, creating perception of the stimulus. Brain functions dealing with perception are the most difficult to study because they require communication between the subject and the investigator—the subject must be able to tell the investigator what he or she is seeing, hearing, or feeling.

Sensory information from the body travels in ascending pathways to the brain. Information about muscle and joint position and movement goes to the cerebellum as well as to the cerebral cortex, allowing the cerebellum to assist with automatic subconscious coordination of movement. Most sensory information continues on to the cerebral cortex, where five sensory areas process information.

The **primary somatic sensory cortex** (also called the *somatosensory cortex*) in the parietal lobe is the termination point of pathways from the skin, musculoskeletal system, and viscera (Fig. 9.13). The somatosensory pathways carry information about touch, temperature, pain, itch, and body position. Damage to this part of the brain leads to reduced sensitivity of the skin on

TABLE 9.3 Selected Brain Imaging Techniques

In Vitro Techniques	
Horseradish peroxidase (HRP)	HRP enzyme is brought into axon terminals by endocytosis and transported by retrograde axonal transport to the cell body and dendrites. Completion of the enzyme-substrate reaction makes the entire neuron visible under a microscope.
Brainbow mice	Transgenic mice in which fluorescent proteins have been inserted into the neurons. Neurons light up in a rainbow of colors depending on which proteins they are expressing. *www.jax.org/news-and-insights/2013/december/an-expanded-brainbow-tool-kit-for-fluorescently-labelling-cells-in-mice*
CLARITY: Clear, lipid-exchanged, anatomically rigid, imaging/immunostaining-compatible tissue hydrogel	Intact brain samples are made transparent by a technique that removes lipids and embeds the sample in a plastic matrix. Allows easier three-dimensional reconstructions of neural networks. *www.nature.com/news/see-through-brains-clarify-connections-1.12768*
In Vivo Imaging of Living Brain Activity	
Electroencephalography (EEG)	Brain electrical activity from many neurons is measured by electrodes placed on the scalp (see Fig. 9.17a).
Positive emission tomography (PET)	Glucose is tagged with a radioactive substance that emits positively charged particles. Metabolically active cells using glucose light up more (see Fig. 9.20). *www.radiologyinfo.org/en/info.cfm?pg=pet*
Functional magnetic resonance imaging (fMRI)	Active brain tissue has increased blood flow and uses more oxygen. Hydrogen nuclei in water create a magnetic signal that indicates more active regions. *www.nature.com/news/brain-imaging-fmri-2-0-1.10365*

the opposite side of the body because sensory fibers cross to the opposite side of the midline as they ascend through the spine or medulla.

The special senses of vision, hearing, taste, and olfaction (smell) each have different brain regions devoted to processing their sensory input (Fig. 9.13). The **visual cortex**, located in the occipital lobe, receives information from the eyes. The **auditory cortex**, located in the temporal lobe, receives information from the ears. The **olfactory cortex**, a small region in the temporal lobe, receives input from chemoreceptors in the nose. The **gustatory cortex**, deeper in the brain near the edge of the frontal lobe, receives sensory information from the taste buds. [The sensory systems are described in detail in Chapter 10.]

Sensory Information Is Processed into Perception

Once sensory information reaches the appropriate cortical area, information processing has just begun. Neural pathways extend from sensory areas to appropriate association areas, which integrate somatic, visual, auditory, and other stimuli into *perception*, the brain's interpretation of sensory stimuli.

Often the perceived stimulus is very different from the actual stimulus. For instance, photoreceptors in the eye receive light waves of different frequencies, but we perceive the different wave energies as different colors. Similarly, the brain translates pressure waves hitting the ear into sound and interprets chemicals binding to chemoreceptors as taste or smell.

One interesting aspect of perception is the way our brain fills in missing information to create a complete picture, or translates a two-dimensional drawing into a three-dimensional shape (**FIG. 9.15**). Thus, we sometimes perceive what our brains expect to perceive. Our perceptual translation of sensory stimuli allows the information to be acted upon and used in voluntary motor control or in complex cognitive functions such as language.

The Motor System Governs Output from the CNS

The motor output component of the nervous system is associated with the efferent division of the nervous system [Fig. 8.1, p. 225]. Motor output can be divided into three major types: (1) skeletal muscle movement, controlled by the somatic motor division; (2) neuroendocrine signals, which are neurohormones

FIG. 9.15 Perception

The brain has the ability to interpret sensory information to create the perception of (a) shapes or (b) three-dimensional objects.

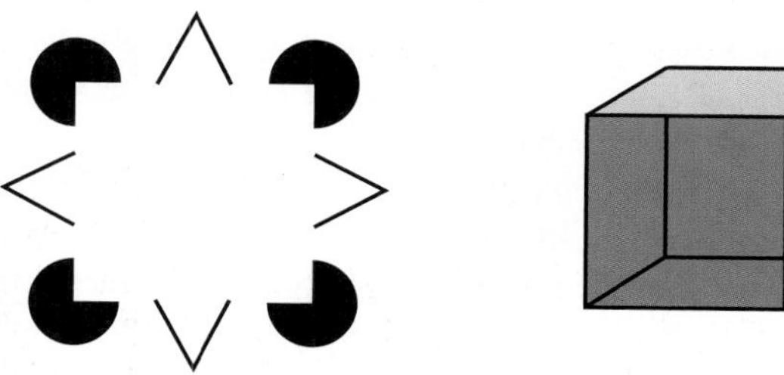

(a) What shape do you see? **(b)** What is this object?

secreted into the blood by neurons located primarily in the hypothalamus and adrenal medulla; and (3) *visceral* responses, the actions of smooth and cardiac muscle or endocrine and exocrine glands. Visceral responses are governed by the autonomic division of the nervous system.

Information about skeletal muscle movement is processed in several regions of the CNS. Simple stimulus-response pathways, such as the knee jerk reflex, are processed either in the spinal cord or in the brain stem. Although these reflexes do not require integration in the cerebral cortex, they can be modified or overridden by input from the cognitive system.

Voluntary movements are initiated by the cognitive system and originate in the **primary motor cortex** and **motor association area** in the frontal lobes of the cerebrum (Fig. 9.13). These regions receive input from sensory areas as well as from the cerebellum and basal ganglia. Long output neurons called *pyramidal cells* project axons from the motor areas through the brain stem to the spinal cord. Other pathways go from the cortex to the basal ganglia and lower brain regions. Descending motor pathways cross to the opposite side of the body, which means that damage to a motor area manifests as paralysis or loss of function on the opposite side of the body. [Chapter 13 discusses motor pathways in more detail.]

Neuroendocrine and visceral responses are coordinated primarily in the hypothalamus and medulla. The brain stem contains the control centers for many of the automatic life functions we take for granted, such as breathing and blood pressure. It receives sensory information from the body and relays motor commands to peripheral muscles and glands.

The hypothalamus contains centers for temperature regulation, eating, and control of body osmolarity, among others. The responses to stimulation of these centers may be neural or hormonal reflexes or a behavioral response. Stress, reproduction, and growth are also mediated by the hypothalamus by way of multiple hormones. You will learn more about these reflexes in later chapters as we discuss the various systems of the body.

Sensory input is not the only factor determining motor output by the brain. The behavioral state system can modulate reflex pathways, and the cognitive system exerts both voluntary and involuntary control over motor functions.

The Behavioral State System Modulates Motor Output

The behavioral state system is an important modulator of sensory and cognitive processing. Many neurons in the behavioral state system are found in regions of the brain outside the cerebral cortex, including parts of the reticular formation in the brain stem, the hypothalamus, and the limbic system.

The neurons collectively known as the **diffuse modulatory systems** originate in the reticular formation in the brain stem and project their axons to large areas of the brain (**FIG. 9.16**). There are four modulatory systems that are generally classified according to the neurotransmitter they secrete: *noradrenergic* (norepinephrine), *serotonergic* (serotonin), *dopaminergic* (dopamine), and *cholinergic* (acetylcholine). The diffuse modulatory systems regulate brain function by influencing attention, motivation, wakefulness, memory, motor control, mood, and metabolic homeostasis.

The dopaminergic system is one of the best-studied because of its role in the movement disorder called Parkinson's disease. As mentioned earlier, dopamine is unable to cross the blood-brain barrier so drugs to supplement dopamine must be given as precursors that can be transported. Dopaminergic pathways also have been implicated in addictive behaviors and the brain's "reward centers."

One function of the behavioral state system is control of levels of consciousness and sleep-wake cycles. **Consciousness** is the body's state of arousal or awareness of self and environment. Experimental evidence shows that the **reticular activating system**, a diffuse collection of neurons in the reticular formation, plays an essential role in keeping the "conscious brain" awake.

If connections between the reticular formation and the cerebral cortex are disrupted surgically, an animal becomes comatose. Other evidence for the importance of the reticular formation in states of arousal comes from studies showing that general anesthetics depress synaptic transmission in that region of the brain. Presumably, blocking ascending pathways between the reticular formation and the cerebral cortex creates a state of unconsciousness.

One way to define arousal states is by the pattern of electrical activity created by the cortical neurons. The measurement of brain activity is recorded by a procedure known as **electroencephalography** (see Table 9.3). Surface electrodes placed on or in the scalp detect depolarizations of the cortical neurons in the region just under the electrode. The complete cessation of brain waves is one of the clinical criteria for determining death.

Why Do We Sleep?

In humans, our major rest period is marked by a behavior known as **sleep**, defined as an easily reversible state of inactivity characterized by lack of interaction with the external environment. Most mammals and birds show the same stages of sleep as humans, telling us that sleep is a very ancient property of vertebrate brains. Depending on how sleep is defined, it appears that even invertebrates such as flies go through rest periods that could be described as sleep.

Why we need to sleep is one of the unsolved mysteries in neurophysiology, and a question that may have more than one answer. Some explanations that have been proposed include to conserve energy, to avoid predators, to allow the body to repair itself, and to process memories. Some of the newest research indicates that sleep is important for clearing wastes out of the cerebrospinal fluid, particularly some of the proteins that build up in degenerative neurological diseases such as Alzheimer's.

FIG. 9.16 Diffuse modulatory systems

The neurons collectively known as the diffuse modulatory systems originate in the reticular formation of the brain stem and project their axons to large areas of the brain. The four systems are named for their neurotransmitters.

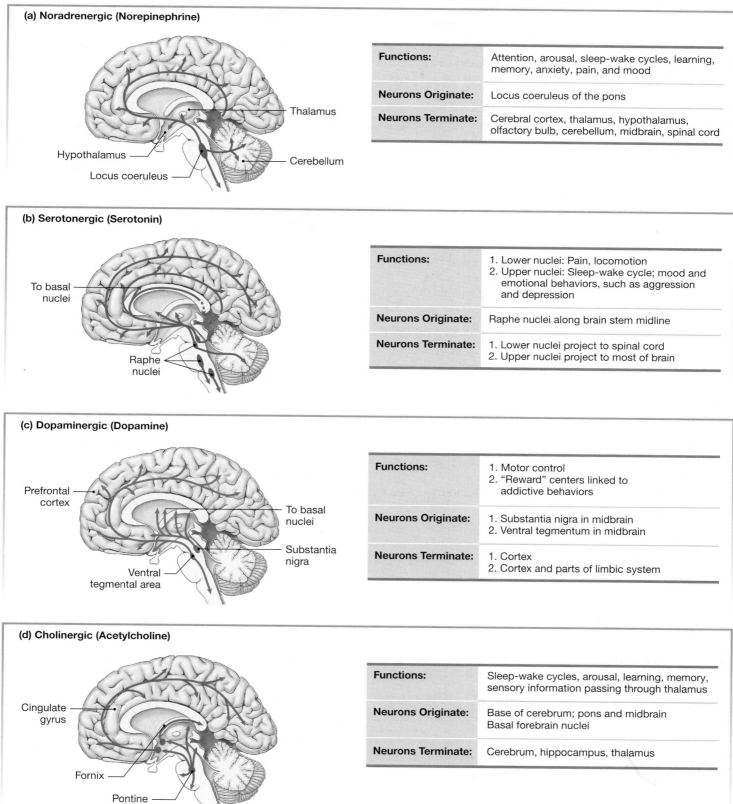

(a) Noradrenergic (Norepinephrine)

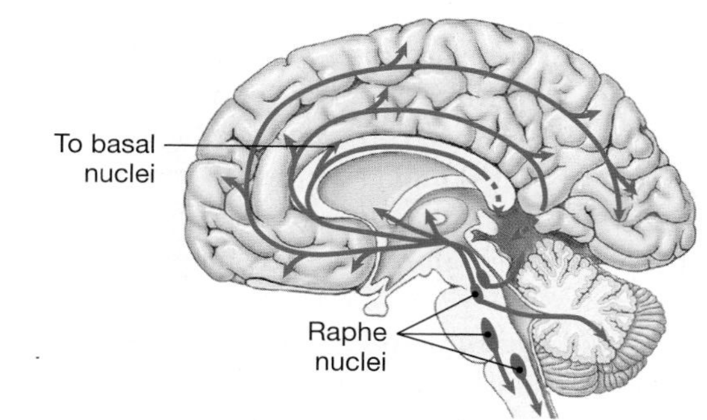
Thalamus
Hypothalamus
Cerebellum
Locus coeruleus

Functions:	Attention, arousal, sleep-wake cycles, learning, memory, anxiety, pain, and mood
Neurons Originate:	Locus coeruleus of the pons
Neurons Terminate:	Cerebral cortex, thalamus, hypothalamus, olfactory bulb, cerebellum, midbrain, spinal cord

(b) Serotonergic (Serotonin)

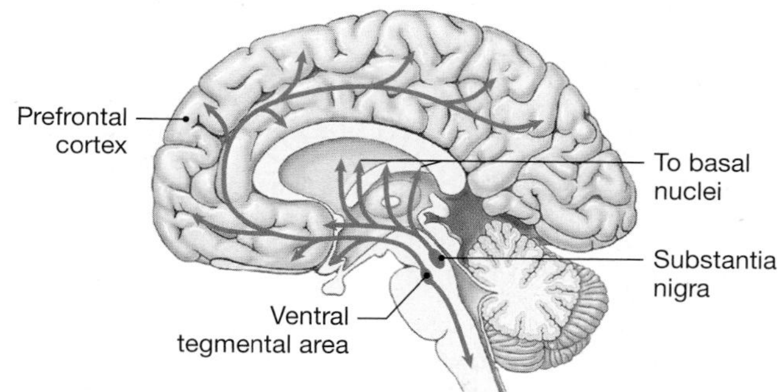
To basal nuclei
Raphe nuclei

Functions:	1. Lower nuclei: Pain, locomotion 2. Upper nuclei: Sleep-wake cycle; mood and emotional behaviors, such as aggression and depression
Neurons Originate:	Raphe nuclei along brain stem midline
Neurons Terminate:	1. Lower nuclei project to spinal cord 2. Upper nuclei project to most of brain

(c) Dopaminergic (Dopamine)

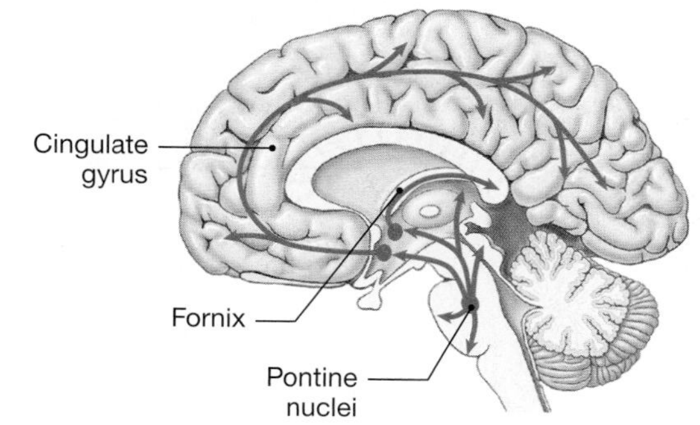
Prefrontal cortex
To basal nuclei
Substantia nigra
Ventral tegmental area

Functions:	1. Motor control 2. "Reward" centers linked to addictive behaviors
Neurons Originate:	1. Substantia nigra in midbrain 2. Ventral tegmentum in midbrain
Neurons Terminate:	1. Cortex 2. Cortex and parts of limbic system

(d) Cholinergic (Acetylcholine)

Cingulate gyrus
Fornix
Pontine nuclei

Functions:	Sleep-wake cycles, arousal, learning, memory, sensory information passing through thalamus
Neurons Originate:	Base of cerebrum; pons and midbrain Basal forebrain nuclei
Neurons Terminate:	Cerebrum, hippocampus, thalamus

EMERGING CONCEPTS

Brain Glymphatics

The traditional view of brain fluid circulation shows cerebrospinal fluid (CSF) secreted in the ventricles and moving by bulk flow through the subarachnoid space until being reabsorbed into venous blood at the arachnoid villi (Fig. 9.4). Removal of waste products from the interstitial fluid surrounding neurons and glial cells was thought to be limited movement from the interstitial fluid into the CSF. Then in 2012 a group of scientists found that radiolabeled solutes injected into the subarachnoid CSF appeared in the brain interstitial fluid, suggesting some previously undiscovered route for CSF flow back into brain tissue. This CSF movement occurs by a *paravascular* route, moving along the *outside* of blood vessels, aided by water movement across astrocytes. The scientists proposed the name *glymphatics* for the system, for glial-dependent lymphatic-like function. Clearance of brain metabolites, including proteins, by the glymphatics seems to occur mostly during sleep. This pathway for removal of brain waste products has been suggested as a reason why we sleep. Glymphatics are now the subject of studies asking whether buildup of brain proteins in certain disease, such as Alzheimer's, may be the result of poor glymphatics function.

There is good evidence supporting the link between sleep and memory. A number of studies have demonstrated that sleep deprivation impairs our performance on tasks and tests, one reason for not pulling "all-nighters." At the same time, 20–30 minute "power naps" have also been shown to improve memory, and they can help make up a sleep deficit.

Physiologically, what distinguishes being awake from various stages of sleep? From studies, we know that the sleeping brain consumes as much oxygen as the awake brain, so sleep is a metabolically active state. Sleep is divided into stages, each marked by identifiable, predictable events associated with characteristic somatic changes and EEG patterns. The stages of sleep were revised in 2016 by the American Academy of Sleep Medicine.

In awake states, many neurons are firing but not in a coordinated fashion (**FIG. 9.17a**). An *electroencephalogram*, or **EEG**, of the waking-alert (eyes open) state shows a rapid, irregular pattern with no dominant waves. In awake-resting (eyes closed) states, sleep, or coma, electrical activity of the neurons begins to synchronize into waves with characteristic patterns. The more synchronous the firing of cortical neurons, the larger the amplitude of the waves. Accordingly, the awake-resting state, called stage W, is characterized by low-amplitude, high-frequency waves.

As the person falls asleep and the state of arousal lessens, the frequency of the waves decreases. The two major sleep phases are rapid eye movement sleep, or REM sleep, and non-REM sleep. Non-REM sleep is subdivided into **stages N1**, **N2**, and **N3**. Stage N3 sleep is also called **slow-wave sleep** or **deep sleep**. It is indicated on the EEG by the presence of *delta waves*, high-amplitude, low-frequency waves of long duration that sweep across the cerebral cortex (Fig. 9.17a). During this phase of the sleep cycle, sleepers adjust body position without conscious commands from the brain to do so.

In contrast, **rapid eye movement (REM) sleep** (stage R) is marked by an EEG pattern closer to that of an awake person, with low-amplitude, high-frequency waves. During REM sleep, brain activity inhibits motor neurons to skeletal muscles, paralyzing them. Exceptions to this pattern are the muscles that move the eyes and those that control breathing. The control of homeostatic

FIG. 9.17 Electroencephalograms (EEGs) and the sleep cycle

(a) Recordings of electrical activity in the brain during awake-resting and sleep periods show characteristic patterns.

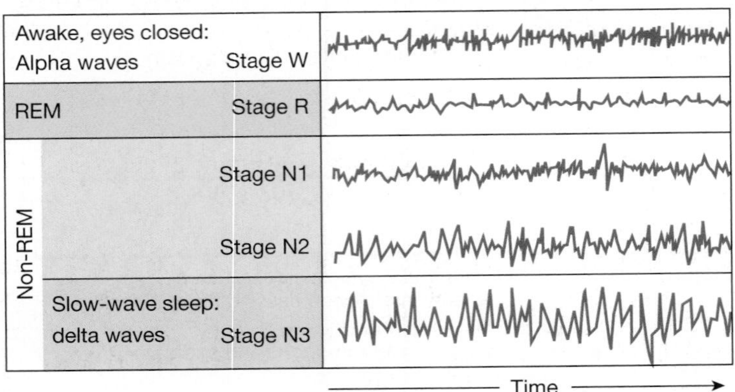

Awake, eyes closed:
Alpha waves
Stage W
REM
Stage R
Stage N1
Stage N2
Non-REM
Slow-wave sleep:
delta waves
Stage N3
Time

KEY

Amplitude
Frequency

(b) The deepest sleep occurs in the first three hours.

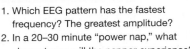

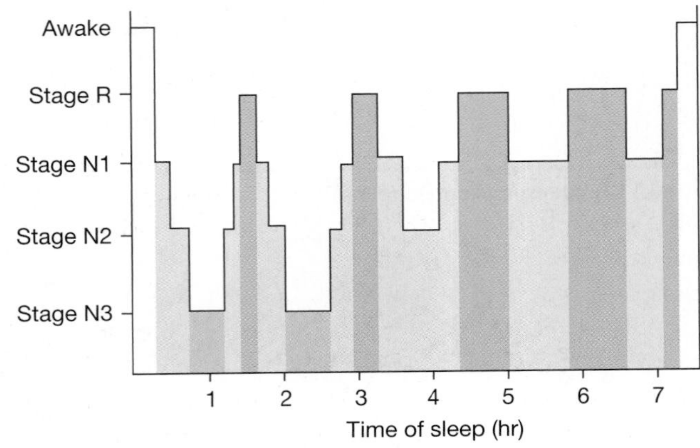

Awake
Stage R
Stage N1
Stage N2
Stage N3
1
2
3
4
5
6
7
Time of sleep (hr)

FIGURE QUESTIONS

1. Which EEG pattern has the fastest frequency? The greatest amplitude?
2. In a 20–30 minute "power nap," what sleep stages will the napper experience?

functions is depressed during REM sleep, and body temperature falls toward ambient temperature.

REM sleep is the period during which most dreaming takes place. The eyes move behind closed lids, as if following the action of the dream. Sleepers are most likely to wake up spontaneously from periods of REM sleep.

A typical eight-hour sleep consists of repeating cycles, as shown in Fig. 9.17b. In the first hour, the person moves from wakefulness through stages N1 and N3 and finally into a deep sleep (stage N3; first blue area in Fig. 9.17b). The sleeper then cycles between deep sleep and REM sleep (stage R), with stages N1 and N2 occurring in between. Near the end of an eight-hour sleep period, a sleeper spends the most time in stage N1 and REM sleep, until finally awakening for the day.

If sleep is a neurologically active process, what is it that makes us sleepy? The possibility of a sleep-inducing factor was first proposed in 1913, when scientists found that cerebrospinal fluid from sleep-deprived dogs could induce sleep in normal animals. Since then, a variety of sleep-inducing factors have been identified. Curiously, many of them are also substances that enhance the immune response, such as interleukin-1, interferon, serotonin, and tumor necrosis factor. As a result of this finding, some investigators have suggested that one answer to the puzzle of the biological reason for sleep is that we need to sleep to enhance our immune response. Whether or not that is a reason for why we sleep, the link between the immune system and sleep induction may help explain why we tend to sleep more when we are sick.

Another clue to what makes us sleepy come from studies on *caffeine* and its methylxanthine cousins *theobromine* and *theophylline* (found in chocolate and tea). These chemicals are probably the most widely consumed psychoactive drugs, known since ancient times for their stimulant effect. Molecular research has revealed that the methylxanthines are receptor antagonists for *adenosine*, a molecule composed of the nitrogenous base adenine plus the sugar ribose [p. 34]. The discovery that the stimulant effect of caffeine comes from its blockade of adenosine receptors has led scientists to investigate adenosine's role in sleep-wake cycles. Evidence suggests that adenosine accumulates in the extracellular fluid during waking hours, increasingly suppressing activity of the neurons that promote wakefulness.

Sleep disorders are relatively common, as you can tell by looking at the variety of sleep-promoting agents available over the counter in drugstores. Among the more common sleep disorders are *insomnia* (the inability to go to sleep or remain asleep long enough to awake refreshed), sleep apnea, and sleepwalking. *Sleep apnea* {*apnoos*, breathless} is a condition in which the sleeper awakes when the airway muscles relax to the point of obstructing normal breathing.

Sleepwalking, or *somnambulism* {*somnus*, sleep + *ambulare*, to walk}, is a sleep behavior disorder that for many years was thought to represent the acting out of dreams. However, most dreaming occurs during REM sleep (stage 1), while sleepwalking takes place during deep sleep (stage 4). During sleepwalking episodes, which may last from 30 seconds to 30 minutes, the subject's eyes are open and registering the surroundings. The subject is able to avoid bumping into objects, can negotiate stairs, and in some cases is reported to perform such tasks as preparing food or folding clothes. The subject usually has little if any conscious recall of the sleepwalking episode upon awakening.

Sleepwalking is most common in children, and the frequency of episodes declines with age. There is also a genetic component, as the tendency to sleepwalk runs in families. To learn more about the different sleep disorders, see the NIH website for the National Center for Sleep Disorder Research (*www.nhlbi.nih.gov/about/org/ncsdr/*).

Concept Check

21. During sleep, relay neurons in the thalamus reduce information reaching the cerebrum by altering their membrane potential. Are these neurons more likely to have depolarized or hyperpolarized? Explain your reasoning.

Physiological Functions Exhibit Circadian Rhythms

All organisms (even plants) have alternating daily patterns of rest and activity. These alternating activity patterns, like many other biological cycles, generally follow a 24-hour light-dark cycle and are known as *circadian rhythms* [p. 17]. When an organism is placed in conditions of constant light or darkness, these activity rhythms persist, apparently cued by an internal clock.

In mammals, the primary "clock" resides in networks of neurons located in the **suprachiasmatic nucleus (SCN)** of the hypothalamus, with secondary clocks influencing the behavior of different tissues. A very simple interpretation of how the biological clock works is that clock cycling results from a complex feedback loop in which specific genes turn on and direct protein synthesis. The proteins accumulate, turn off the genes, and then are themselves degraded. As the proteins disappear, the genes turn back on and the cycle begins again. The SCN clock has intrinsic activity that is synchronized with the external environment by sensory information about light cycles received through the eyes.

Circadian rhythms in humans can be found in most physiological functions and usually correspond to the phases of our sleep-wake cycles. For example, body temperature and cortisol secretion both cycle on a daily basis [Fig. 1.14, p. 18]. Melatonin from the pineal gland [p. 218] also is strongly linked to light-dark cycling: melatonin is sometimes called the "darkness hormone" because its secretion increases in the evening. The suprachiasmatic nucleus has melatonin receptors, supporting the hypothesis that melatonin can modulate clock cycling.

Disruption of circadian rhythms, such as occurs with shift work and jet lag, can have detrimental effects on mental and physical health. Sleep disturbances, depression, seasonal affective depressive disorder, diabetes, and obesity have all been linked to abnormal circadian rhythms. Jet lag, which occurs when people shift their light-dark cycles by travel across time zones, is a common manifestation of the effect of circadian rhythms on daily function. Melatonin treatments and exposure to natural daylight in the new location are the only treatments shown to have any significant effect on jet lag.

RUNNING PROBLEM

About 6 months after the start of treatment, Ben's head-drop seizures returned, and his development began to decline once again. An EEG following Ben's relapse did not demonstrate the erratic wave patterns specific to infantile spasms but did show abnormal activity in the right cortex. A neurologist ordered a positron emission tomography (PET) scan to determine the focus of Ben's seizure activity.

Ben received an injection of radioactively labeled glucose. He was then placed in the center of a PET machine lined with radiation detectors that created a map of his brain showing areas of high and low radioactivity. Those parts of his brain that were more active absorbed more glucose and thus emitted more radiation when the radioactive compound began to decay.

Q4: *What is the rationale for using radioactively labeled glucose (and not some other nutrient) for the PET scan?*

272
280
296
297
300
301

Emotion and Motivation Involve Complex Neural Pathways

Emotion and motivation are two aspects of brain function that probably represent an overlap of the behavioral state system and cognitive system. The pathways involved are complex and form closed circuits that cycle information among various parts of the brain, including the hypothalamus, limbic system, and cerebral cortex. We still do not understand the underlying neural mechanisms, and this is a large and active area of neuroscience research.

Emotions are difficult to define. We know what they are and can name them, but in many ways they defy description. One characteristic of emotions is that they are difficult to voluntarily turn on or off. The most commonly described emotions, which arise in different parts of the brain, are anger, aggression, sexual feelings, fear, pleasure, contentment, and happiness.

The limbic system, particularly the region known as the *amygdala*, is the center of emotion in the human brain. Scientists have learned about the role of this brain region through experiments in humans and animals. When the amygdala is artificially stimulated in humans, as it might be during surgery for epilepsy, patients report experiencing feelings of fear and anxiety. Experimental lesions that destroy the amygdala in animals cause the animals to become tamer and to display hypersexuality. As a result, neurobiologists believe that the amygdala is the center for basic instincts such as fear and aggression.

The pathways for emotions are complex (**FIG. 9.18**). Sensory stimuli feeding into the cerebral cortex are constructed in the brain to create a representation (perception) of the world. After information is integrated by the association areas, it is passed on to the limbic system. Feedback from the limbic system to the cerebral cortex creates awareness of the emotion, while descending pathways to the hypothalamus and brain stem initiate voluntary behaviors and unconscious responses mediated by autonomic, endocrine, immune, and somatic motor systems.

FIG. 9.18 Emotions affect physiology

The association between stress and increased susceptibility to viruses is an example of an emotionally linked immune response.

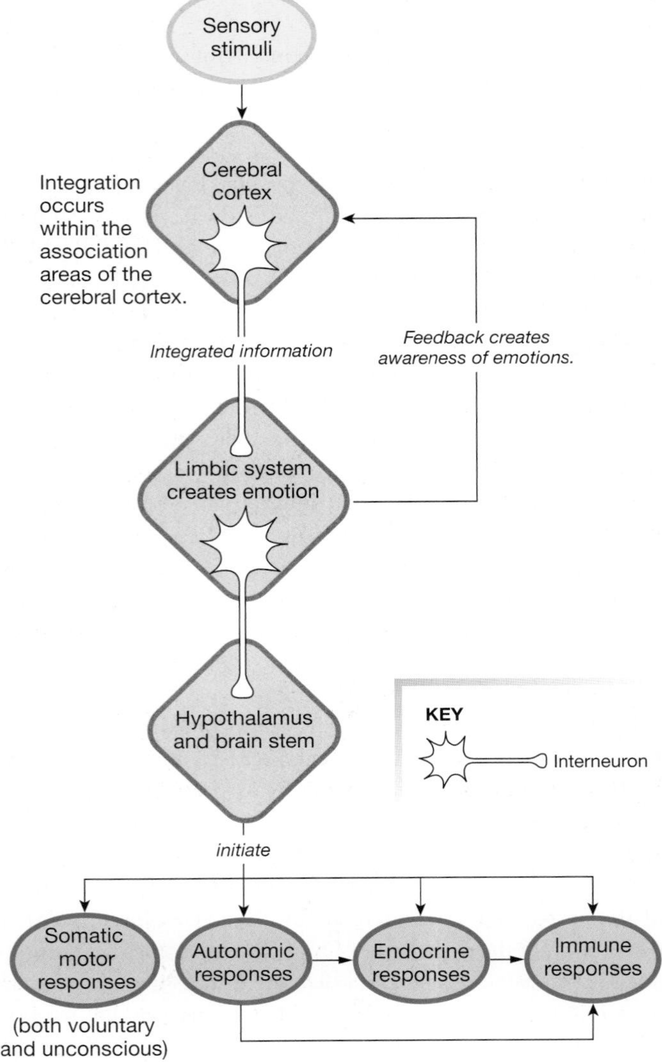
Sensory stimuli
Cerebral cortex
Integration occurs within the association areas of the cerebral cortex.
Integrated information
Feedback creates awareness of emotions.
Limbic system creates emotion
Hypothalamus and brain stem
KEY
Interneuron
initiate
Somatic motor responses
(both voluntary and unconscious)
Autonomic responses
Endocrine responses
Immune responses

The physical result of emotions can be as dramatic as the pounding heart of a fight-or-flight reaction or as insidious as the development of an irregular heartbeat. The links between mind and body are difficult to study and will take many years of research to understand.

Motivation is defined as internal signals that shape voluntary behaviors. Some of these behaviors, such as eating, drinking, and having sex, are related to survival. Others, such as curiosity and having sex (again), are linked to emotions. Some motivational states are known as **drives** and generally have three properties in common: (1) they create an increased state of CNS arousal or alertness, (2) they create goal-oriented behavior, and (3) they are capable of coordinating disparate behaviors to achieve that goal.

Motivated behaviors often work in parallel with autonomic and endocrine responses in the body, as you might expect with behaviors originating in the hypothalamus. For example, if you eat salty popcorn, your body osmolarity increases. This stimulus acts on the thirst center of the hypothalamus, motivating you to seek something to drink. Increased osmolarity also acts on an endocrine center in the hypothalamus, releasing a hormone that increases water retention by the kidneys. In this way, one stimulus triggers both a motivated behavior and a homeostatic endocrine response.

Some motivated behaviors can be activated by internal stimuli that may not be obvious even to the person in whom they are occurring. Eating, curiosity, and sex drive are three examples of behaviors with complex stimuli underlying their onset. We may eat, for example, because we are hungry or because the food looks good or because we do not want to hurt someone's feelings. Many motivated behaviors stop when the person has reached a certain level of satisfaction, or **satiety**, but they may also continue *despite* feeling satiated.

Pleasure is a motivational state that is being intensely studied because of its relationship to *addictive behaviors*, such as drug use. Animal studies have shown that pleasure is a physiological state that is accompanied by increased activity of the neurotransmitter dopamine in certain parts of the brain. Drugs that are addictive, such as cocaine and nicotine, act by enhancing the effectiveness of dopamine, thereby increasing the pleasurable sensations perceived by the brain. As a result, use of these drugs rapidly becomes a learned behavior.

Interestingly, not all behaviors that are addictive are pleasurable. For example, there are a variety of compulsive behaviors that involve self-mutilation, such as pulling out hair by the roots. Fortunately, many behaviors can be modulated, given motivation.

Moods Are Long-Lasting Emotional States

Moods are similar to emotions but are longer-lasting, relatively stable subjective feelings related to one's sense of well-being. Moods are difficult to define at a neurobiological level, but evidence obtained in studying and treating mood disorders suggests that mood disturbances reflect changes in CNS function, such as abnormal neurotransmitter release or reception in different brain regions.

Mood disorders are estimated to be the fourth leading cause of illness in the world today. **Depression** is a mood disturbance that affects nearly 10% of the United States population each year. It is characterized by sleep and appetite disturbances and alterations of mood and libido that may seriously affect the person's ability to function at school or work or in personal relationships. Many people do not realize that depression is not a sign of mental or moral weakness, or that it can be treated successfully with drugs and psychotherapy. (For detailed information about depression, go to *www.nlm.nih.gov/medlineplus/depression.html.*)

The drug therapy for depression has changed in recent years, but all the major categories of antidepressant drugs alter some aspect of synaptic transmission. The older *tricyclic antidepressants*, such as amitriptyline, block reuptake of norepinephrine into the presynaptic neuron, thus extending the active life of the neurotransmitter. The antidepressants known as *selective serotonin reuptake inhibitors* (SSRIs) and *serotonin/norepinephrine reuptake inhibitors* (SNRIs) slow down the removal of serotonin and norepinephrine from the synapse. As a result of uptake inhibition, the neurotransmitter lingers in the synaptic cleft longer than usual, increasing transmitter-dependent activity in the postsynaptic neuron. Other antidepressant drugs alter brain levels of dopamine. The effectiveness of these different classes of antidepressant drugs suggests that norepinephrine, serotonin, and dopamine are all involved in brain pathways for mood and emotion.

Interestingly, patients need to take antidepressant drugs for several weeks before they experience their full effect. This delay suggests that the changes taking place in the brain are long-term modulation of pathways rather than simply enhanced fast synaptic responses. Several studies in humans and animal models provide evidence that antidepressants promote the growth of new neurons, which would also explain the delayed onset of full action.

The causes of major depression are complex and probably involve a combination of genetic factors, the serotonergic and noradrenergic diffuse modulatory systems, trophic factors such as *brain-derived neurotrophic factor* (BDNF), and stress. The search to uncover the biological basis of disturbed brain function is a major focus of neuroscience research today.

Some research into brain function has become quite controversial, particularly that dealing with sexuality and the degree to which behavior in general is genetically determined in humans. We will not delve deeply into any of these subjects because they are complex and would require lengthy explanations to do them justice. Instead, we will look briefly at some of the recent models proposed to explain the mechanisms that are the basis for higher cognitive functions.

Learning and Memory Change Synaptic Connections in the Brain

For many years, motivation, learning, and memory (all of which are aspects of the cognitive state) were considered to be in the realm of psychology rather than biology. Neurobiologists in decades past were more concerned with the network and cellular aspects of neuronal function. In recent years, however, the two fields have overlapped more and more. Scientists have discovered that the underlying basis for cognitive function seems to be explainable in terms of cellular

RUNNING PROBLEM

Ben's halted development is a feature unique to infantile spasms. The abnormal portions of the brain send out continuous action potentials during frequent seizures and ultimately change the interconnections of brain neurons. The damaged portions of the brain harm normal portions to such an extent that medication or surgery should be started as soon as possible. If intervention is not begun early, the brain can be permanently damaged and development will never recover.

Q5: *The brain's ability to change its synaptic connections as a result of neuronal activity is called ____________.*

272 – 280 – 296 – **297** – 300 – 301

events that influence plasticity—events such as long-term potentiation [p. 264]. The ability of neurons to change their responsiveness or alter their connections with experience is fundamental to the two cognitive processes of learning and memory.

Learning Is the Acquisition of Knowledge

How do you know when you have learned something? Learning can be demonstrated by behavioral changes, but behavioral changes are not required for learning to occur. Learning can be internalized and is not always reflected by overt behavior while the learning is taking place. Would someone watching you read your textbook or listen to a professor's lecture be able to tell whether you had learned anything?

Learning can be classified into two broad types: associative and nonassociative. **Associative learning** occurs when two stimuli are associated with each other, such as Pavlov's classic experiment in which he simultaneously presented dogs with food and rang a bell. After a period of time, the dogs came to associate the sound of the bell with food and began to salivate in anticipation of food whenever the bell was rung. Another form of associative learning occurs when an animal associates a stimulus with a given behavior. An example would be a mouse that gets a shock each time it touches a certain part of its cage. It soon associates that part of the cage with an unpleasant experience and avoids the area.

Nonassociative learning is a change in behavior that takes place after repeated exposure to a single stimulus. This type of learning includes habituation and sensitization, two adaptive behaviors that allow us to filter out and ignore background stimuli while responding more sensitively to potentially disruptive stimuli. In **habituation**, an animal shows a decreased response to an irrelevant stimulus that is repeated over and over. For example, a sudden loud noise may startle you, but if the noise is repeated over and over again, your brain begins to ignore it. Habituated responses allow us to filter out stimuli that we have evaluated and found to be insignificant.

Sensitization is the opposite of habituation, and the two behaviors combined help increase an organism's chances for survival. In sensitization learning, exposure to a noxious or intense stimulus causes an enhanced response upon subsequent exposure. For example, people who become ill while eating a certain food may find that they lose their desire to eat that food again. Sensitization is adaptive because it helps us avoid potentially harmful stimuli. At the same time, sensitization may be maladaptive if it leads to the hypervigilant state known as *post-traumatic stress disorder* (PTSD).

Memory Is the Ability to Retain and Recall Information

Memory is the ability to retain and recall information. Memory is a very complex function, but scientists have tried to classify it in different ways. We think of several types of memory: short-term and long-term, reflexive and declarative. Processing for different types of memory appears to take place through different pathways. With noninvasive imaging techniques such as MRI and PET scans, researchers have been able to track brain activity as individuals learned to perform tasks.

Memories are stored throughout the cerebral cortex in pathways known as **memory traces**. Some components of memories are stored in the sensory cortices where they are processed. For example, pictures are stored in the visual cortex, and sounds in the auditory cortex.

Learning a task or recalling a task already learned may involve multiple brain circuits that work in parallel. This *parallel processing* helps provide backup in case one of the circuits is damaged. It is also believed to be the means by which specific memories are generalized, allowing new information to be matched to stored information. For example, a person who has never seen a volleyball will recognize it as a ball because the volleyball has the same general characteristics as all other balls the person has seen.

In humans, the hippocampus seems to be an important structure in both learning and memory. Patients who have part of the hippocampus destroyed to relieve a certain type of epilepsy also have trouble remembering new information. When given a list of words to repeat, they can remember the words as long as their attention stays focused on the task. If they are distracted, however, the memory of the words disappears, and they must learn the list again. Information stored in long-term memory before the operation is not affected. This inability to remember newly acquired information is a defect known as **anterograde amnesia** {*amnesia*, oblivion}.

Memory has multiple levels of storage, and our memory bank is constantly changing (**FIG. 9.19**). When a stimulus comes into the CNS, it first goes into **short-term memory**, a limited storage area that can hold only about 7 to 12 pieces of information at a time. Items in short-term memory disappear unless an effort, such as repetition, is made to put them into a more permanent form.

Working memory is a special form of short-term memory processed in the prefrontal lobes. This region of the cerebral cortex is devoted to keeping track of bits of information long enough to put them to use in a task that takes place after the information

FIG. 9.19 Memory processing

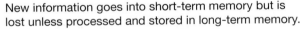

New information goes into short-term memory but is lost unless processed and stored in long-term memory.

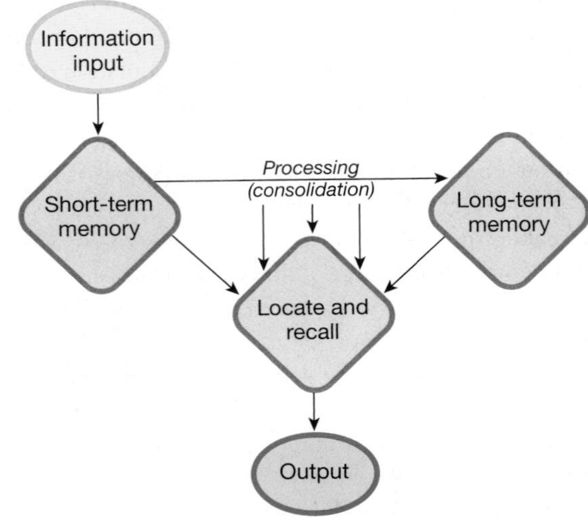
Information input
Short-term memory
Processing (consolidation)
Long-term memory
Locate and recall
Output

has been acquired. Working memory in these regions is linked to long-term memory stores, so that newly acquired information can be integrated with stored information and acted on.

For example, suppose you are trying to cross a busy road. You look to the left and see that there are no cars coming for several blocks. You then look to the right and see that there are no cars coming from that direction either. Working memory has stored the information that the road to the left is clear, and so using this stored knowledge about safety, you are able to conclude that there is no traffic from either direction and it is safe to cross the road.

In people with damage to the prefrontal lobes of the brain, this task becomes more difficult because they are unable to recall whether the road is clear from the left once they have looked away to assess traffic coming from the right. Working memory allows us to collect a series of facts from short- and long-term memory and connect them in a logical order to solve problems or plan actions.

Long-term memory is a storage area capable of holding vast amounts of information. Think of how much information humans needed to remember in centuries past, when books were scarce and most history was passed down by word of mouth. Wandering bards and troubadours kept long epic poems and ballads, such as *The Odyssey* and *Beowulf*, stored in their memory banks, to be retrieved at will.

The processing of information that converts short-term memory into long-term memory is known as **consolidation** (Fig. 9.19). Consolidation can take varying periods of time, from seconds to minutes. Information passes through many intermediate levels of memory during consolidation, and in each of these stages, the information can be located and recalled.

As scientists studied the consolidation of short-term memory into long-term memory, they discovered that the process involves changes in neuronal excitability or synaptic connections in the circuits involved in learning. In some cases, new synapses form; in others, the effectiveness of synaptic transmission is altered through long-term potentiation or through long-term depression. These changes are evidence of plasticity and show us that the brain is not "hard-wired."

Long-term memory has been divided into two types that are consolidated and stored using different neuronal pathways (TBL. 9.4). **Reflexive (implicit) memory**, which is automatic and does not require conscious processes for either creation or recall, involves the amygdala and the cerebellum. Information stored in reflexive memory is acquired slowly through repetition. Motor skills fall into this category, as do procedures and rules.

TABLE 9.4 Types of Long-Term Memory

Reflexive (Implicit) Memory	Declarative (Explicit) Memory
Recall is automatic and does not require conscious attention	Recall requires conscious attention
Acquired slowly through repetition	Depends on higher-level thinking skills such as inference, comparison, and evaluation
Includes motor skills and rules and procedures	Memories can be reported verbally
Procedural memories can be demonstrated	

For example, you do not need to think about putting a period at the end of each sentence or about how to pick up a fork. Reflexive memory has also been called *procedural memory* because it generally concerns how to do things. Reflexive memories can be acquired through either associative or nonassociative learning processes, and these memories are stored.

Declarative (explicit) memory, on the other hand, requires conscious attention for its recall. Its creation generally depends on the use of higher-level cognitive skills such as inference, comparison, and evaluation. The neuronal pathways involved in this type of memory are in the temporal lobes. Declarative memories deal with knowledge about ourselves and the world around us that can be reported or described verbally.

Sometimes, information can be transferred from declarative memory to reflexive memory. The quarterback on a football team is a good example. When he learned to throw the football as a small boy, he had to pay close attention to gripping the ball and coordinating his muscles to throw the ball accurately. At that point of learning to throw the ball, the process was in declarative memory and required conscious effort as the boy analyzed his movements.

With repetition, however, the mechanics of throwing the ball were transferred to reflexive memory: they became a reflex that could be executed without conscious thought. That transfer allowed the quarterback to use his conscious mind to analyze the path and timing of the pass while the mechanics of the pass became automatic. Athletes often refer to this automaticity of learned body movements as *muscle memory*.

Memory is an individual thing. We process information on the basis of our experiences and perception of the world. Because people have widely different experiences throughout their lives, it follows that no two people will process a given piece of information in the same way. If you ask a group of people about what happened during a particular event such as a lecture or an automobile accident, no two descriptions will be identical. Each person processed the event according to her or his own perceptions and experiences. Experiential processing is important to remember when studying in a group situation, because it is unlikely that all group members learn or recall information the same way.

Memory loss and the inability to process and store new memories are devastating medical conditions. In younger people, memory problems are usually associated with trauma to the brain from accidents. In older people, strokes and progressive *dementia* {*demens*, out of one's mind} are the main causes of memory loss.

Alzheimer's disease is a progressive neurodegenerative disease of cognitive impairment that accounts for about half the cases of dementia in the elderly. Alzheimer's is characterized by memory loss that progresses to a point where the patient does not recognize family members. Over time, even the personality changes, and in the final stages, other cognitive functions fail so that patients cannot communicate with caregivers.

Diagnosis of Alzheimer's is usually made through the patient's declining performance on cognitive function examinations. Scientists are studying whether tests for specific proteins in the cerebrospinal fluid or advanced imaging studies can reveal if a person has the disease, but the data are inconclusive at this stage. Currently, the only definitive diagnosis of Alzheimer's comes after death, when brain tissue can be examined for neuronal degeneration, extracellular plaques made of β-*amyloid protein*, and intracellular tangles of *tau*, a protein that is normally associated with microtubules.

The presence of amyloid plaques and tau tangles is diagnostic, but the underlying cause of Alzheimer's is unclear. There is a known genetic component, and other theories include oxidative stress and chronic inflammation. Currently there is no proven prevention or treatment, although drugs that are acetylcholine agonists or acetylcholinesterase inhibitors slow the progression of the disease.

By one estimate, Alzheimer's affects about 5.2 million Americans, with the number expected to rise as baby boomers age. The forecast of 16 million Americans with Alzheimer's by the year 2050 has put this disease in the forefront of neurobiological research.

Although pathological memory loss is a concern, the ability to forget is also important for our mental health. Post-traumatic stress disorder is an example of where forgetting would be beneficial.

Language Is the Most Elaborate Cognitive Behavior

One of the hallmarks of an advanced nervous system is the ability of one member of a species to exchange complex information with other members of the same species. Although found predominantly in birds and mammals, this ability also occurs in certain insects that convey amazingly detailed information by means of sound (crickets), touch and sight (bees), and odor (ants). In humans, the exchange of complex information takes place primarily through spoken and written language. Because language is considered the most elaborate cognitive behavior, it has received considerable attention from neurobiologists.

Language skills require the input of sensory information (primarily from hearing and vision), processing in various centers in the cerebral cortex, and the coordination of motor output for vocalization and writing. In most people, the centers for language ability are found in the left hemisphere of the cerebrum. Even 70% of people who are either left-handed (right-brain dominant) or ambidextrous use their left brain for speech. The ability to communicate through speech has been divided into two processes: the combination of different sounds to form words (vocalization) and the combination of words into grammatically correct and meaningful sentences.

The model presented here is a simplified version of what scientists now know is a very complex function that involves many regions of the cerebral cortex. Traditionally, integration of spoken language in the human brain has been attributed to two regions in the cerebral cortex: **Wernicke's area** at the junction of the parietal, temporal, and occipital lobes and **Broca's area** in the posterior part of the frontal lobe, close to the motor cortex (**FIG. 9.20**). Most of what we know about these areas comes from studies of people with brain lesions (because nonhuman animals are not capable of speech). Even primates that communicate on the level of a small child through sign language and other visual means do not have the physical ability to vocalize the sounds of human language.

Input into the language areas comes from either the visual cortex (reading) or the auditory cortex (listening). Sensory input from either cortex goes first to Wernicke's area, then to Broca's area. After integration and processing, output from Broca's area to the motor cortex initiates a spoken or written action.

If damage occurs to Wernicke's area, a person may have difficulty understanding spoken or visual information. The person's own speech may be nonsense because the person is unable to retrieve words. This condition is known as **receptive aphasia** {a-, not + *phatos*, spoken} because the person is unable to understand sensory input.

Damage to Broca's area causes an **expressive aphasia**, or *Broca aphasia*. People with Broca aphasia understand simple, unambiguous spoken and written language but have difficulty interpreting complicated sentences with several elements linked together. This difficulty appears to be a deficit in short-term memory. These people also have difficulty speaking or writing in normal syntax. Their response to a question may consist of appropriate words strung together in random order.

Mechanical forms of aphasia occur as a result of damage to the motor cortex. Patients with this type of damage find themselves unable to physically shape the sounds that make up words, or unable to coordinate the muscles of their arm and hand to write.

RUNNING PROBLEM

The PET scan revealed two abnormal spots, or loci (plural of locus), on Ben's right hemisphere, one on the parietal lobe and one overlapping a portion of the primary motor cortex. Because the loci triggering Ben's seizures were located on the same hemisphere and were in the cortex, Ben was a candidate for a hemispherectomy, removal of the cortex of the affected hemisphere. Surgeons removed 80% of his right cerebral cortex, sparing areas crucial to vision, hearing, and sensory processing. Normally the motor cortex would be spared as well, but in Ben's case a seizure locus overlapped much of the region.

Q6: *In which lobes are the centers for vision, hearing, and sensory processing located?*

Q7: *Which of Ben's abilities might have suffered if his left hemisphere had been removed instead?*

Q8: *By taking only the cortex of the right hemisphere, what parts of the cerebrum did surgeons leave behind?*

Q9: *Why were the surgeons careful to spare Ben's right lateral ventricle?*

272 – 280 – 296 – 297 – 300 – 301

FIG. 9.20 Language processing

People with damage to Wernicke's area do not understand spoken or written communication. Those with damage to Broca's area understand but are unable to respond appropriately.

(a) Speaking a Written Word

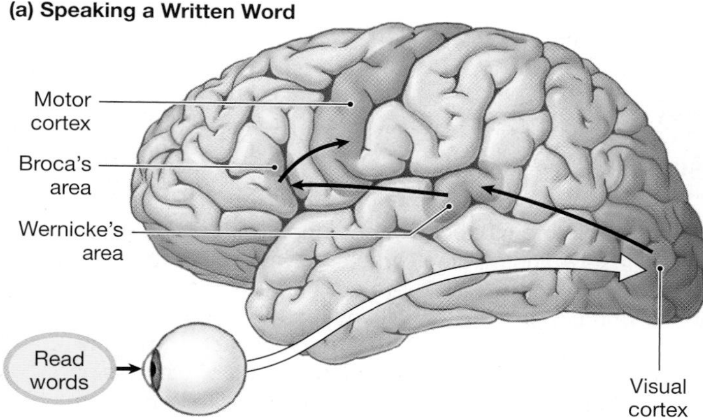
Motor cortex
Broca's area
Wernicke's area
Read words
Visual cortex

(b) Speaking a Heard Word

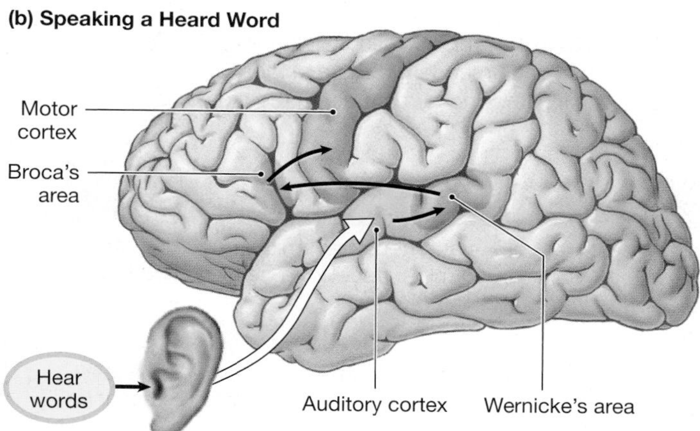
Motor cortex
Broca's area
Hear words
Auditory cortex
Wernicke's area

(c) PET Scan of the Brain at Work

In PET scans, neurons take up radio-labeled glucose. The most active areas show up as red-yellow regions.

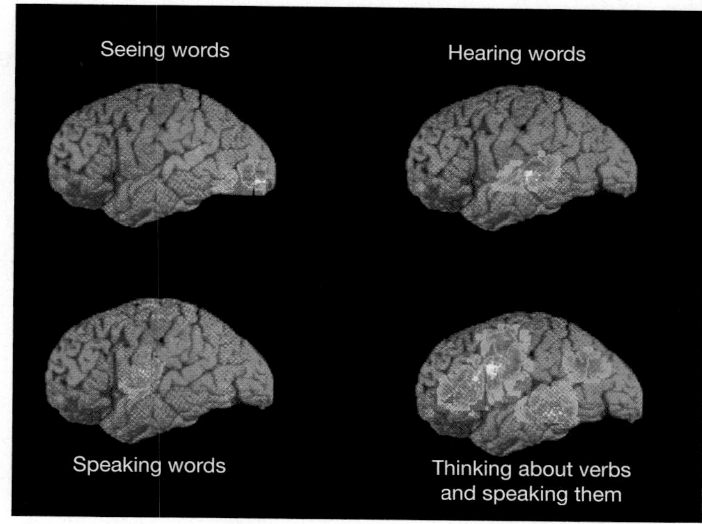
Seeing words
Hearing words
Speaking words
Thinking about verbs and speaking them

Sight, hearing, speaking, thinking. Colored Positron Emission Tomography (PET) scans of areas of the human brain activated by different tasks. The left-side of the brain is seen. At upper left, sight activates the visual area in the occipital cortex at the back of the brain. At upper right, hearing activates the auditory area in the superior temporal cortex of the brain. At lower left, speaking activates the speech centers in the insula and motor cortex. At lower right, thinking about verbs and speaking them generates high activity, including in hearing, speaking, temporal and parietal areas. PET scans detect blood flow.

Personality Is a Combination of Experience and Inheritance

One of the most difficult aspects of brain function to translate from the abstract realm of psychology into the physical circuits of neurobiology is the combination of attributes we call **personality**. What is it that makes us individuals? The parents of more than one child will tell you that their offspring were different from birth, and even in the womb. If we all have the same brain structure, what makes us different?

This question fascinates many people. The answer that is evolving from neurobiology research is that we are a combination of our experiences and the genetic constraints we inherit. One complicating factor is the developmental aspect of "experience," as scientists are showing that exposure of developing embryos to hormones while still in the womb can alter brain pathways.

What we learn or experience and what we store in memory create a unique pattern of neuronal connections in our brains. Sometimes, these circuits malfunction, creating depression, schizophrenia, or any number of other personality disturbances. Psychiatrists for many years attempted to treat these disorders as if they were due solely to events in the person's life, but now we know that there is a genetic component to many of these disorders.

Schizophrenia {*schizein*, to split + *phren*, the mind} is an example of a brain disorder that has both a genetic and an environmental basis. In the American population as a whole, the risk of developing schizophrenia is about 1%. However, if one parent has schizophrenia, the risk increases to 10%, and if an identical twin is schizophrenic, the risk for the other twin is about 50%. The cause of schizophrenia is not currently known. However, as with many other conditions involving altered mental states, schizophrenia can be treated with drugs that influence neurotransmitter release and activity in the brain. To learn more about diagnosis and treatment of schizophrenia, see the NIH website *https://medlineplus.gov/schizophrenia.html.*

We still have much to learn about repairing damage to the CNS. One of the biggest tragedies in life is the intellectual and personality changes that sometimes accompany traumatic brain injury. Physical damage to the delicate circuits of the brain, particularly to the frontal lobe, can create a whole new personality. The person who exists after the injury may not be the same personality who inhabited that body before the injury. Although the change may not be noticeable to the injured person, it can be devastating to the victim's family and friends. Perhaps as we learn more about how neurons link to one another, we will be able to find a means of restoring damaged networks and preventing the lasting effects of head trauma and brain disorders.

RUNNING PROBLEM **CONCLUSION** Infantile Spasms

Ben has remained seizure-free since the surgery and shows normal development in all areas except motor skills. He remains somewhat weaker and less coordinated on his left side, the side opposite (*contralateral*) to the surgery. Over time, the weakness should subside with the aid of physical therapy. Ben's recovery stands as a testament to the incredible plasticity of the brain. Apart from the physical damage caused to the brain, a number of children with epilepsy have developmental delays that stem from the social aspects of their disorder. Young children with frequent seizures often have difficulty socializing with their peers because of overprotective parents, missed school days, and the fear of people who do not understand epilepsy. Their problems can extend into adulthood, when people with epilepsy may have difficulty finding employment or driving if their seizures are not controlled. There are numerous examples of adults who undergo successful epilepsy surgery but are still unable to fully enter society because they lack social and employment skills. Not surprisingly, the rate of depression is much higher among people with epilepsy. To learn more about this disease, start with the Epilepsy Foundation (*www.epilepsyfoundation.org*).

Note: This Running Problem was developed by Susan E. Johnson while she was an undergraduate student at the University of Texas at Austin studying for a career in the biomedical sciences.

Question	Facts	Integration and Analysis
Q1: *How might a leaky blood-brain barrier lead to action potentials that trigger a seizure?*	Neurotransmitters and other chemicals circulating freely in the blood are normally separated from brain tissue by the blood-brain barrier.	Ions and neurotransmitters entering the brain might depolarize neurons and trigger action potentials.
Q2: *What does GABA do to the cell's membrane potential? Does GABA make the cell more or less likely to fire action potentials?*	GABA opens Cl^- channels.	Cl^- entering a neuron hyperpolarizes the cell and makes it less likely to fire action potentials.
Q3: *Why is it important to limit the duration of ACTH therapy?*	Exogenous ACTH acts in a short negative feedback loop, decreasing the output of CRH from the hypothalamus and ACTH production by the anterior pituitary. [See Fig. 7.13, p. 214.]	Long-term suppression of endogenous hormone secretion by ACTH can cause CRH- and ACTH-secreting neurons to atrophy, resulting in a lifelong cortisol deficiency.
Q4: *What is the rationale for using radioactively labeled glucose (and not some other nutrient) for the PET scan?*	Glucose is the primary energy source for the brain.	Glucose usage is more closely correlated to brain activity than any other nutrient in the body. Areas of abnormally high glucose usage are suggestive of overactive cells.
Q5: *The brain's ability to change its synaptic connections as a result of neuronal activity is called _______.*	Changes in synaptic connections as a result of neuronal activity are an example of plasticity.	N/A
Q6: *In which lobes are the centers for vision, hearing, and sensory processing located?*	Vision is processed in the occipital lobe, hearing in the temporal lobe, and sensory information in the parietal lobe.	N/A

Continued

RUNNING PROBLEM CONCLUSION *Continued*

Question	Facts	Integration and Analysis
Q7: *Which of Ben's abilities might have suffered if his left hemisphere had been removed instead?*	In most people, the left hemisphere contains Wernicke's area and Broca's area, two centers vital to speech. The left brain controls right-sided sensory and motor functions.	Patients who have undergone left hemispherectomies have difficulty with speech (abstract words, grammar, and phonetics). They show loss of right-side sensory and motor functions.
Q8: *By taking only the cortex of the right hemisphere, what parts of the cerebrum did surgeons leave behind?*	The cerebrum consists of gray matter in the cortex and interior nuclei, white matter, and the ventricles.	The surgeons left behind the white matter, interior nuclei, and ventricles.
Q9: *Why were the surgeons careful to spare Ben's right lateral ventricle?*	The walls of the ventricles contain the choroid plexus, which secretes cerebrospinal fluid (CSF). CSF plays a vital protective role by cushioning the brain.	CSF protection is particularly important following removal of portions of brain tissue because the potential damage from jarring of the head is much greater.

272 – 280 – 296 – 297 – 300 – 301

CHAPTER SUMMARY

The brain is the primary control center of the body, and (as you will learn in later chapters) homeostatic responses in many organ systems are designed to maintain brain function. The ability of the brain to create complex thoughts and emotions in the absence of external stimuli is one of its *emergent properties*.

9.1 Emergent Properties of Neural Networks

1. Neural networks create **affective** and **cognitive behaviors**. (p. 272)
2. The brain exhibits **plasticity**, the ability to change connections as a result of experience. (p. 272)

9.2 Evolution of Nervous Systems

3. Nervous systems evolved from a simple network of neurons to complex brains. (p. 272; Fig. 9.1)
4. The **cerebrum** is responsible for thought and emotion. (p. 274)

9.3 Anatomy of the Central Nervous System

5. The central nervous system consists of layers of cells around a fluid-filled central cavity and develops from the **neural tube** of the embryo. (p. 274; Fig. 9.2)
6. The **gray matter** of the CNS consists of unmyelinated nerve cell bodies, dendrites, and axon terminals. The cell bodies either form layers in parts of the brain or else cluster into groups known as **nuclei**. (p. 274)
7. Myelinated axons form the **white matter** of the CNS and run in bundles called **tracts**. (p. 274)
8. The brain and spinal cord are encased in the **meninges** and the bones of the **cranium** and vertebrae. The meninges are the **pia mater**, the **arachnoid membrane**, and the **dura mater**. (p. 277; Fig. 9.3)
9. The **choroid plexus** secretes **cerebrospinal fluid (CSF)** into the **ventricles** of the brain. Cerebrospinal fluid cushions the tissue and creates a controlled chemical environment. (p. 277; Fig. 9.4)
10. Tight junctions in brain capillaries create a **blood-brain barrier** that prevents possibly harmful substances in the blood from entering the interstitial fluid. (p. 279; Fig. 9.5)
11. The normal fuel source for neurons is glucose, which is why the body closely regulates blood glucose concentrations. (p. 280)

9.4 The Spinal Cord

12. Each segment of the spinal cord is associated with a pair of **spinal nerves**. (p. 282)
13. The **dorsal root** of each spinal nerve carries incoming sensory information. The **dorsal root ganglia** contain the nerve cell bodies of sensory neurons. (p. 282; Fig. 9.6)
14. The **ventral roots** carry information from the central nervous system to muscles and glands. (p. 282)
15. **Ascending tracts** of white matter carry sensory information to the brain, and **descending tracts** carry efferent signals from the brain. **Propriospinal tracts** remain within the spinal cord. (p. 282)
16. **Spinal reflexes** are integrated in the spinal cord. (p. 282; Fig. 9.7)

9.5 The Brain

17. The brain has six major divisions: cerebrum, diencephalon, midbrain, cerebellum, pons, and medulla oblongata. (p. 283; Fig. 9.8)
18. The **brain stem** is divided into medulla oblongata, pons, and midbrain (mesencephalon). **Cranial nerves** II to XII originate here. (p. 283; Fig. 9.8f; Tbl. 9.1)

19. The **reticular formation** is a diffuse collection of neurons that play a role in many basic processes. (p. 283)
20. The **medulla oblongata** contains **somatosensory** and **corticospinal tracts** that convey information between the cerebrum and spinal cord. Most tracts cross the midline in the **pyramid** region. The medulla contains control centers for many involuntary functions. (p. 283)
21. The **pons** acts as a relay station for information between the cerebellum and cerebrum. (p. 283)
22. The **midbrain** controls eye movement and relays signals for auditory and visual reflexes. (p. 283)
23. The **cerebellum** processes sensory information and coordinates the execution of movement. (p. 285)
24. The **diencephalon** is composed of the thalamus and hypothalamus. The **thalamus** relays and modifies sensory and motor information going to and from the cerebral cortex. (p. 285; Fig. 9.9)
25. The **hypothalamus** contains centers for behavioral drives and plays a key role in homeostasis by its control over endocrine and autonomic function. (p. 285; Tbl. 9.2)
26. The **pituitary gland** and **pineal gland** are endocrine glands located in the diencephalon. (p. 285)
27. The cerebrum is composed of two hemispheres connected at the **corpus callosum**. Each cerebral hemisphere is divided into **frontal, parietal, temporal**, and **occipital lobes**. (p. 287)
28. Cerebral gray matter includes the **cerebral cortex**, basal ganglia, and limbic system. (p. 287; Fig. 9.10)
29. The **basal ganglia** help control movement. (p. 287)
30. The **limbic system** acts as the link between cognitive functions and emotional responses. It includes the **amygdala** and **cingulate gyrus**, linked to emotion and memory, and the **hippocampus**, associated with learning and memory. (p. 288; Fig. 9.11)

9.6 Brain Function

31. Three brain systems influence motor output: a **sensory system**, a **cognitive system**, and a **behavioral state system**. (p. 288; Fig. 9.12)
32. Higher brain functions, such as reasoning, arise in the cerebral cortex. The cerebral cortex contains three functional specializations: **sensory areas, motor areas**, and **association areas**. (p. 289; Fig. 9.13)
33. Each hemisphere of the cerebrum has developed functions not shared by the other hemisphere, a specialization known as **cerebral lateralization**. (p. 289; Fig. 9.14)
34. Sensory areas receive information from sensory receptors. The **primary somatic sensory cortex** processes information about touch, temperature, and other somatic senses. The **visual cortex, auditory cortex, gustatory cortex**, and **olfactory cortex** receive information about vision, sound, taste, and odors, respectively. (p. 290)
35. **Association areas** integrate sensory information into perception. **Perception** is the brain's interpretation of sensory stimuli. (p. 291)
36. Motor output includes skeletal muscle movement, neuroendocrine secretion, and visceral responses. (p. 292)
37. Motor areas direct skeletal muscle movement. Each cerebral hemisphere contains a **primary motor cortex** and **motor association area**. (p. 292)
38. The **behavioral state system** controls states of arousal and modulates the sensory and cognitive systems. (p. 292)
39. The **diffuse modulatory systems** of the reticular formation influence attention, motivation, wakefulness, memory, motor control, mood, and metabolic homeostasis. (p. 292; Fig. 9.16)
40. The **reticular activating system** keeps the brain **conscious**, or aware of self and environment. Electrical activity in the brain varies with levels of arousal and can be recorded by **electroencephalography**. (p. 292; Fig. 9.17)
41. **Sleep** is an easily reversible state of inactivity with characteristic stages. The two major phases of sleep are **REM (rapid eye movement) sleep** and **slow-wave sleep** (non-REM sleep). The physiological reason for sleep is uncertain. (p. 292)
42. **Circadian rhythms** are controlled by an internal clock in the **suprachiasmatic nucleus** of the hypothalamus. (p. 295)
43. The limbic system is the center of **emotion** in the human brain. Emotional events influence physiological functions. (p. 296; Fig. 9.18)
44. **Motivation** arises from internal signals that shape voluntary behaviors related to survival or emotions. Motivational **drives** create goal-oriented behaviors. (p. 296)
45. **Moods** are long-lasting emotional states. Many mood disorders can be treated by altering neurotransmission in the brain. (p. 297)
46. Learning is the acquisition of knowledge about the world around us. **Associative learning** occurs when two stimuli are associated with each other. **Nonassociative learning** is a change in behavior that takes place after repeated exposure to a single stimulus. (p. 298)
47. In **habituation**, an animal shows a decreased response to a stimulus that is repeated over and over. In **sensitization**, exposure to a noxious or intense stimulus creates an enhanced response on subsequent exposure. (p. 298)
48. **Memory** has multiple levels of storage and is constantly changing. Information is first stored in **short-term memory** but disappears unless consolidated into long-term memory. (p. 298; Fig. 9.19)
49. **Long-term memory** includes **reflexive memory**, which does not require conscious processes for its creation or recall, and **declarative memory**, which uses higher-level of cognitive skills for formation and requires conscious attention for its recall. (p. 299; Tbl. 9.4)
50. The **consolidation** of short-term memory into long-term memory appears to involve changes in the synaptic connections of the circuits involved in learning. (p. 299)
51. Language is considered the most elaborate cognitive behavior. The integration of spoken language in the human brain involves information processing in **Wernicke's area** and **Broca's area**. (p. 300; Fig. 9.20)

REVIEW QUESTIONS

In addition to working through these questions and checking your answers on p. A-10, review the Learning Outcomes at the beginning of this chapter.

Level One Reviewing Facts and Terms

1. The ability of human brains to change circuit connections and function in response to sensory input and past experience is known as _____.
2. _____ behaviors are related to feeling and emotion. _____ behaviors are related to thinking.
3. The part of the brain called the _____ is what makes us human, allowing human reasoning and cognition.
4. In vertebrates, the central nervous system is protected by the bones of the _____ and _____.
5. Name the meninges, beginning with the layer next to the bones.
6. List and explain the purposes of cerebrospinal fluid (CSF). Where is CSF made?
7. Compare the CSF concentration of each of the following substances with its concentration in the blood plasma.
   a. H^+
   b. Na^+
   c. K^+
8. The only fuel source for neurons under normal circumstances is _____. Low concentration of this fuel in the blood is termed _____. To synthesize enough ATP to continually transport ions, the neurons consume large quantities of _____. To supply these needs, about _____% of the blood pumped by the heart goes to the brain.
9. What is the blood-brain barrier, and what is its function?
10. How are gray matter and white matter different from each other, both anatomically and functionally?
11. Name the cerebral cortex areas that (a) direct perception, (b) direct movement, and (c) integrate information and direct voluntary behaviors.
12. What does *cerebral lateralization* refer to? What functions tend to be centered in each hemisphere?
13. Match each of the following areas with its function.

a. medulla oblongata	1. coordinates execution of movement
b. pons	2. is composed of the thalamus and hypothalamus
c. midbrain	3. controls arousal and sleep
d. reticular formation	4. fills most of the cranium
e. cerebellum	5. contains control centers for blood pressure and breathing
f. diencephalon	6. relays and modifies information going to and from the cerebrum
g. thalamus	7. transfers information to the cerebellum
h. hypothalamus	8. contains integrating centers for homeostasis
i. cerebrum	9. relays signals and visual reflexes, plus eye movement

14. Name the 12 cranial nerves in numerical order and their major functions.
15. Name and define the two major phases of sleep. How are they different from each other?
16. List several homeostatic reflexes and behaviors influenced by output from the hypothalamus. What is the source of emotional input into this area?
17. The _____ region of the limbic system is believed to be the center for basic instincts (such as fear) and learned emotional states.
18. What are the broad categories of learning? Define habituation and sensitization. What anatomical structure of the cerebrum is important in both learning and memory?
19. What two centers of the cortex are involved in integrating spoken language?

Level Two Reviewing Concepts

20. **Mapping exercise:** Map the following terms describing CNS anatomy. You may draw pictures or add terms if you wish.

• arachnoid membrane	• ascending tracts
• blood-brain barrier	• brain
• capillaries	• cell bodies
• cerebrospinal fluid	• cervical nerves
• choroid plexus	• cranial nerves
• descending tracts	• dorsal root
• dorsal root ganglion	• dura mater
• ependyma	• gray matter
• lumbar nerves	• meninges
• nuclei	• pia mater
• propriospinal tracts	• sacral nerves
• spinal cord	• thoracic nerves
• ventral root	• ventricles
• vertebral column	• white matter

21. Trace the pathway that the cerebrospinal fluid follows through the nervous system.
22. What are the three brain systems that regulate motor output by the CNS?
23. Explain the role of Wernicke's and Broca's areas in language.
24. Compare and contrast the following concepts:
   a. diffuse modulatory systems, reticular formation, limbic system, and reticular activating system
   b. different forms of memory
   c. nuclei and ganglia
   d. tracts, nerves, horns, nerve fibers, and roots

25. What properties do motivational states have in common?
26. What changes occur at synapses as memories are formed?
27. Replace each question mark in the following table with the appropriate word(s):

Cerebral Area	Lobe	Functions
Primary somatic sensory cortex	?	Receives sensory information from peripheral receptors
?	Occipital	Processes information from the eyes
Auditory cortex	Temporal	?
?	Temporal	Receives input from chemoreceptors in the nose
Motor cortices	?	?
Association areas	NA	?

28. Given the wave shown below, draw (a) a wave having a lower frequency, (b) a wave having a larger amplitude, (c) a wave having a higher frequency. (*Hint:* See Fig. 9.17, p. 294.)

Level Three Problem Solving

29. Mr. Andersen, a stroke patient, experiences expressive aphasia. His savvy therapist, Cheryl, teaches him to sing to communicate his needs. What signs did he exhibit before therapy? How do you know he did not have receptive aphasia? Using what you have learned about cerebral lateralization, hypothesize why singing worked for him.
30. A study was done in which 40 adults were taught about the importance of using seat belts in their cars. At the end of the presentation, all participants scored at least 90% on a comprehensive test covering the material taught. The people were also secretly videotaped entering and leaving the parking lot of the class site. Twenty subjects entered wearing their seat belts; 22 left wearing them. Did learning occur? What is the relationship between learning and actually buckling the seat belts?
31. In 1913, Henri Pieron kept a group of dogs awake for several days. Before allowing them to sleep, he withdrew cerebrospinal fluid from the sleep-deprived animals. He then injected this CSF into normal, rested dogs. The recipient dogs promptly went to sleep for periods ranging from two hours to six hours. What conclusion can you draw about the possible source of a sleep-inducing factor? What controls should Pieron have included?
32. A 2002 study presented the results of a prospective study [p. 23] done in Utah.* The study began in 1995 with cognitive assessment of 1889 women whose mean age was 74.5 years. Investigators asked about the women's history of taking calcium, multivitamin supplements, and postmenopausal hormone replacement therapy (estrogen or estrogen/progesterone). Follow-up interviews in 1998 looked for the development of Alzheimer's disease in the study population. Data showed that 58 of 800 women who had not used hormone replacement therapy developed Alzheimer's, compared with 26 of 1066 women who had used hormones.
    a. Can the researchers conclude from the data given that hormone replacement therapy decreases the risk of developing Alzheimer's? Should other information be factored into the data analysis?
    b. How applicable are these findings to American women as a whole? What other information might you want to know about the study subjects before you draw any conclusions?
33. A young woman having a seizure was brought into the emergency room. Her roommate said that the woman had taken the street drug Ecstasy the night before and that she had been drinking a lot of water. A blood test showed that her plasma Na^+ was very low: 120 mM (normal 135–145), and her plasma osmolality was 250 mosmoles/kg (normal 280–296). Why would her low osmolality and low Na^+ concentration disrupt her brain function and cause seizures?

* P. P. Zandi *et al.* Hormone replacement therapy and incidence of Alzheimer disease in older women: The Cache County study. *JAMA* 288: 2123–2129, 2002 Nov. 6.

Answers to Concept Checks, Figure and Graph Questions, and end-of-chapter Review Questions can be found in Appendix A [p. A-1].

10 Sensory Physiology

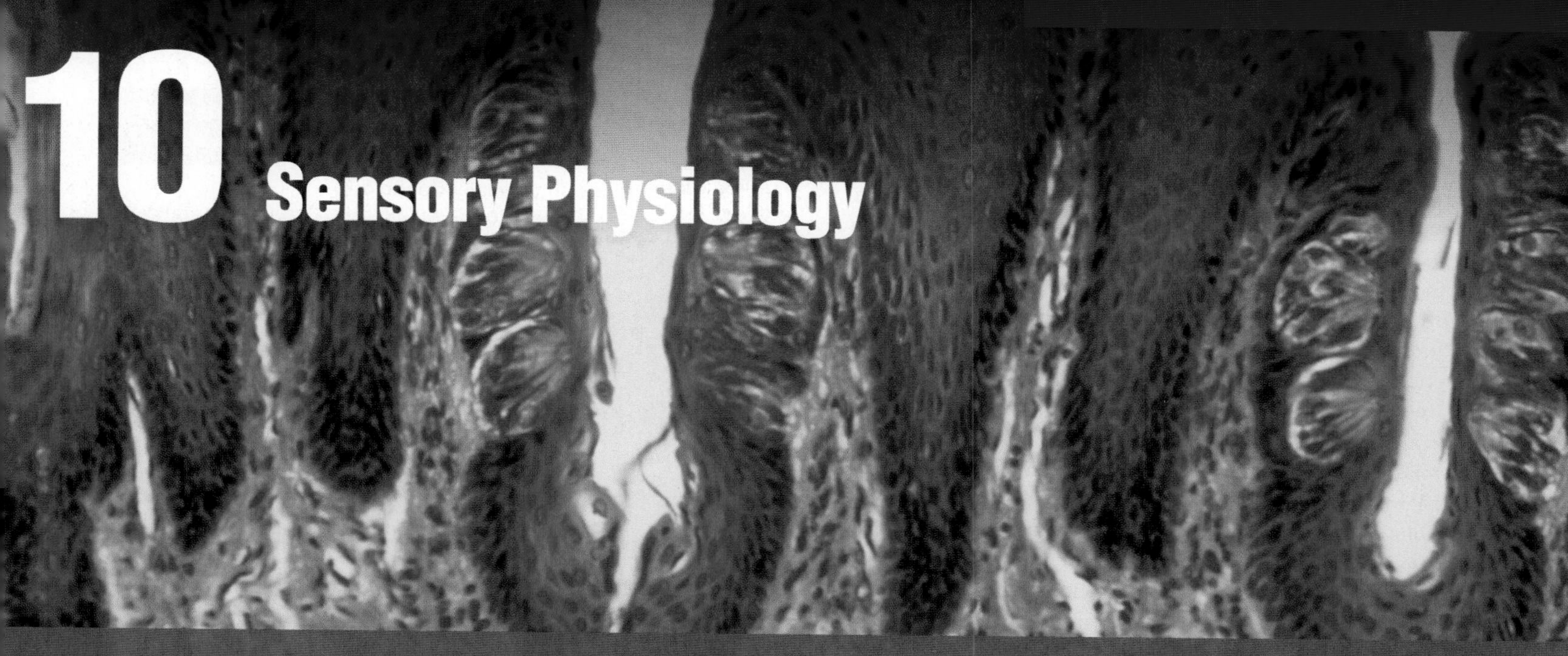

Nature does not communicate with man by sending encoded messages.

Oscar Hechter, in Biology and Medicine into the 21st Century, *1991*

Gustatory papillae with taste buds

BACKGROUND BASICS

Imagine floating in the dark in an indoor tank of buoyant salt water: there is no sound, no light, and no breeze. The air and water are the same temperature as your body. You are in a sensory deprivation chamber, and the only sensations you are aware of come from your own body. Your limbs are weightless, your breath moves in and out effortlessly, and you feel your heart beating. In the absence of external stimuli, you turn your awareness inward to hear what your body has to say.

In decades past, flotation tanks for sensory deprivation were a popular way to counter the stress of a busy world. These facilities are now returning to popularity, and they illustrate the role of the afferent division of the nervous system: to provide us with information about the environment outside and inside our bodies. Sometimes we perceive sensory signals when they reach a level of conscious awareness, but other times signals are processed completely at the subconscious level (**TBL. 10.1**). Stimuli that usually do not reach conscious awareness include changes in muscle stretch and tension as well as a variety of internal parameters that the body monitors to maintain homeostasis, such as blood pressure and pH. The responses to these stimuli constitute many of the subconscious reflexes of the body, and you will encounter them in later chapters as we explore the processes that maintain physiological homeostasis.

In this chapter, we are concerned primarily with sensory stimuli whose processing reaches the conscious level of perception. These stimuli are associated with the **special senses** of vision, hearing, taste, smell, and equilibrium, and the **somatic senses** of touch, temperature, pain, itch, and proprioception. **Proprioception,** which is defined as the awareness of body movement and position in space, is mediated by muscle and joint sensory receptors called **proprioceptors** and may be either unconscious or conscious. If you close your eyes and raise your arm above your head, you are aware of the arm's position because of proprioceptor activation.

We first consider general properties of sensory pathways. We then look at the unique receptors and pathways that distinguish the different sensory systems from one another.

TABLE 10.1 Information Processing by the Sensory Division

Stimulus Processing Usually Conscious	
Special Senses	**Somatic Senses**
Vision	Touch
Hearing	Temperature
Taste	Pain
Smell	Itch
Equilibrium	Proprioception
Stimulus Processing Usually Subconscious	
Somatic Stimuli	**Visceral Stimuli**
Muscle length and tension	Blood pressure
Proprioception	Distension of gastrointestinal tract
	Blood glucose concentration
	Internal body temperature
	Osmolarity of body fluids
	Lung inflation
	pH of cerebrospinal fluid
	pH and oxygen content of blood

RUNNING PROBLEM **Ménière's Disease**

On December 23, 1888, Vincent Van Gogh, the legendary French painter, returned to his room in a boardinghouse in Arles, France, picked up a knife, and cut off his own ear. A local physician, Dr. Felix Ray, examined Van Gogh that night and wrote that the painter had been "assailed by auditory hallucinations" and in an effort to relieve them, "mutilated himself by cutting off his ear." A few months later, Van Gogh committed himself to a lunatic asylum. By 1890, Van Gogh was dead by his own hand. Historians have postulated that Van Gogh suffered from epilepsy, but some American neurologists disagree. They concluded that the painter's strange attacks of dizziness, nausea, and overwhelming tinnitus (ringing or other sounds in the ears), which he described in desperate letters to his relatives, are more consistent with Ménière's disease, a condition that affects the inner ear. Today, Anant, a 20-year-old college student, will be examined by an otolaryngologist (ear-nose-throat specialist) to see if his periodic attacks of severe dizziness and nausea are caused by the same condition that might have driven Van Gogh to suicide.

308 - 312 - 335 - 338 - 341 - 347 - 351

10.1 General Properties of Sensory Systems

All sensory pathways have certain elements in common. They begin with a stimulus, in the form of physical energy that acts on a sensory receptor. The receptor, or sensor, is a *transducer* that converts the stimulus into an intracellular signal, which is usually a change in membrane potential. If the stimulus is above *threshold*, action potentials pass along a sensory neuron to the central nervous system (CNS), where incoming signals are integrated. Some stimuli pass upward to the cerebral cortex, where they reach conscious perception, but others are acted on subconsciously, without our awareness. At each synapse along the pathway, the nervous system can modulate and shape the sensory information.

Sensory systems in the human body vary widely in complexity. The simplest systems are single sensory neurons with branched dendrites that function as sensors, such as pain and itch receptors. The most complex systems include multicellular **sense organs,** such as the ear and the eye. The cochlea of the ear contains about 16,000 sensory receptors and more than a million associated parts, and the human eye has about 126 million sensory receptors.

Receptors Are Sensitive to Particular Forms of Energy

Sensory receptors vary widely in complexity, ranging from the branched endings of a single sensory neuron to complex nonneural cells that act as sensors. The simplest sensory receptors consist of a neuron with naked ("free") nerve endings (**FIG. 10.1a**). In more complex receptors, the nerve endings are encased in connective tissue (Fig. 10.1b). The axons of both simple and complex neural receptors may be myelinated or unmyelinated.

The nonneural sensors include some of the most highly specialized receptors, such as *hair cells* of the ear (Fig. 10.1c). Nonneural sensors are usually highly organized cells that synapse onto sensory neurons. When activated, the nonneural sensor releases a chemical signal that initiates an action potential in the associated sensory neuron. Both neural and nonneural receptors develop from the same embryonic tissue.

Nonneural *accessory structures* are critical to the operation of many sensory systems. For example, the lens and cornea of the eye help focus incoming light onto photoreceptors. The hairs on our arms help **somatosensory receptors** sense movement in the air millimeters above the skin surface. Accessory structures often enhance the information-gathering capability of the sensory system.

Receptors can be divided into four major groups, based on the type of stimulus to which they are most sensitive (**TBL. 10.2**). **Chemoreceptors** respond to chemical ligands that bind to the receptor (taste and smell, for example). **Mechanoreceptors** respond to various forms of mechanical energy, including pressure, vibration, gravity, acceleration, and sound (hearing, for example). **Thermoreceptors** respond to temperature, and **photoreceptors** for vision respond to light.

FIG. 10.1 Simple, complex, and nonneural sensory receptors

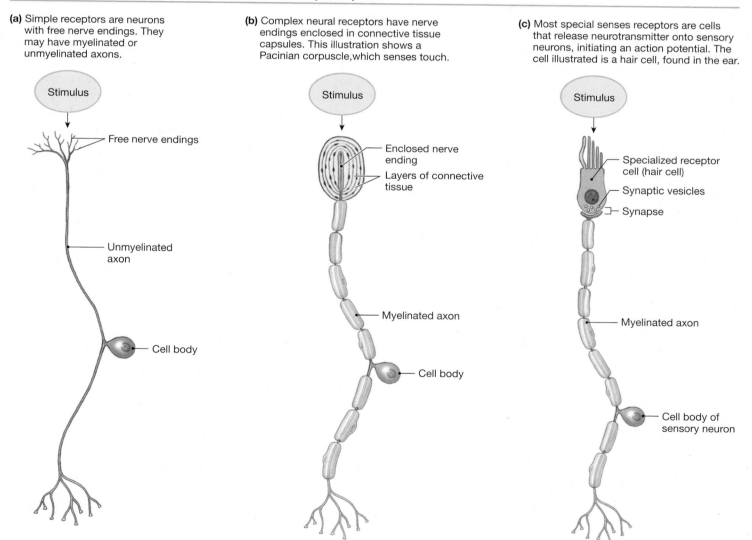

(a) Simple receptors are neurons with free nerve endings. They may have myelinated or unmyelinated axons.

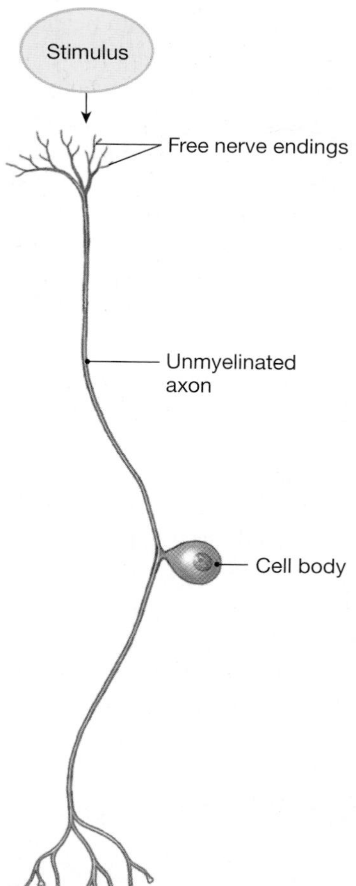
Stimulus
Free nerve endings
Unmyelinated axon
Cell body

(b) Complex neural receptors have nerve endings enclosed in connective tissue capsules. This illustration shows a Pacinian corpuscle,which senses touch.

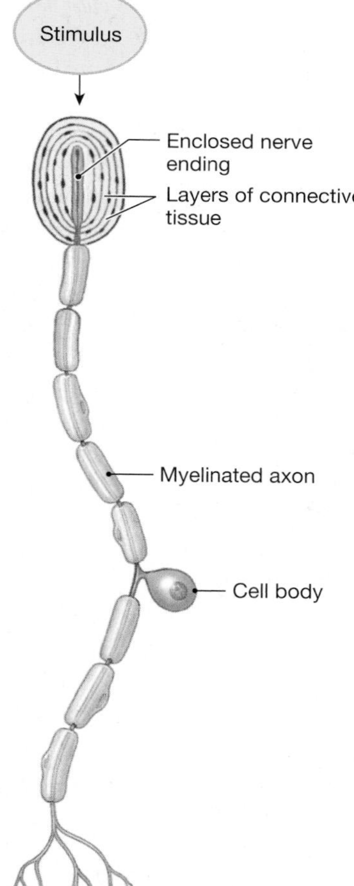
Stimulus
Enclosed nerve ending
Layers of connective tissue
Myelinated axon
Cell body

(c) Most special senses receptors are cells that release neurotransmitter onto sensory neurons, initiating an action potential. The cell illustrated is a hair cell, found in the ear.

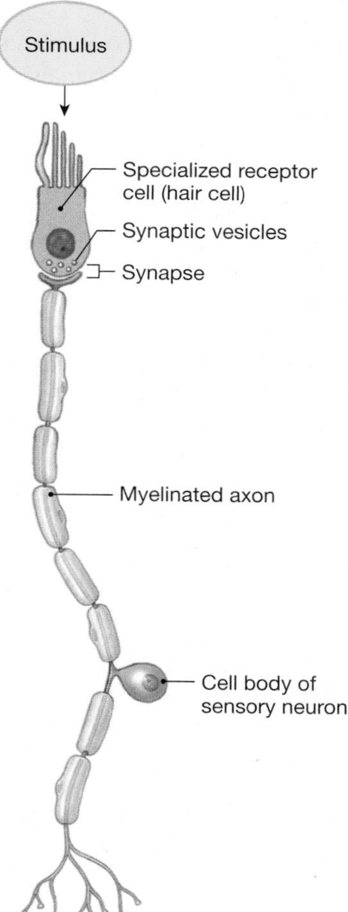
Stimulus
Specialized receptor cell (hair cell)
Synaptic vesicles
Synapse
Myelinated axon
Cell body of sensory neuron

TABLE 10.2 Types of Sensory Receptors

Type of Receptor	Examples of Stimuli
Chemoreceptors	Oxygen, pH, various organic molecules such as glucose
Mechanoreceptors	Pressure (baroreceptors), cell stretch (osmoreceptors), vibration, acceleration, sound
Photoreceptors	Photons of light
Thermoreceptors	Varying degrees of heat

Concept Check

1. What advantage do myelinated axons provide?
2. What accessory role does the outer ear (the pinna) play in the auditory system?
3. For each of the somatic and visceral stimuli listed in Table 10.1, which of the following receptor types is the appropriate transducer: mechano-, chemo-, photo-, or thermoreceptors?

Sensory Transduction Converts Stimuli into Graded Potentials

How do receptors convert diverse physical stimuli, such as light or heat, into electrical signals? The first step is **transduction,** the conversion of stimulus energy into information that can be processed by the nervous system [p. 170]. In many receptors, the opening or closing of ion channels converts mechanical, chemical, thermal, or light energy directly into a change in membrane potential. Some sensory transduction mechanisms include signal transduction and second messenger systems that initiate the change in membrane potential.

Each sensory receptor has an **adequate stimulus,** a particular form of energy to which it is most responsive. For example, thermoreceptors are more sensitive to temperature changes than to pressure, and mechanoreceptors respond preferentially to stimuli that deform the cell membrane. Although sensors are specific for one form of energy, they can respond to most other forms if the intensity is high enough. Photoreceptors of the eye respond most readily to light, for instance, but a blow to the eye may cause us to "see stars," an example of mechanical energy of sufficient force to stimulate the photoreceptors.

Sensory receptors can be incredibly sensitive to their preferred form of stimulus. For example, a single photon of light stimulates certain photoreceptors, and a single *odorant* molecule may activate the chemoreceptors involved in the sense of smell. The minimum stimulus required to activate a receptor is known as the **threshold,** just as the minimum depolarization required to trigger an action potential is called the threshold [p. 184].

How is a physical or chemical stimulus converted into a change in membrane potential? The stimulus opens or closes ion channels in the receptor membrane, either directly or indirectly (through a second messenger). In most cases, channel opening results in net influx of Na^+ or other cations into the receptor, depolarizing the membrane. In a few cases, the response to the stimulus is hyperpolarization when K^+ leaves the cell. In the case of vision, the stimulus (light) closes cation channels to hyperpolarize the receptor.

The change in sensory receptor membrane potential is a graded potential [p. 237] called a **receptor potential.** In some cells, the receptor potential initiates an action potential that travels along the sensory fiber to the CNS. In other cells, receptor potentials influence neurotransmitter secretion by the receptor cell, which in turn alters electrical activity in an associated sensory neuron.

A Sensory Neuron Has a Receptive Field

Somatic sensory and visual neurons are activated by stimuli that fall within a specific physical area known as the neuron's **receptive field.** For example, a touch-sensitive neuron in the skin responds to pressure that falls within its receptive field. In the simplest case, one receptive field is associated with one sensory neuron (the **primary sensory neuron** in the pathway), which in turn synapses on one CNS neuron (the **secondary sensory neuron**). (Primary and secondary sensory neurons are also known as *first-order* and *second-order neurons.*) Receptive fields frequently overlap with neighboring receptive fields.

In addition, sensory neurons of neighboring receptive fields may exhibit *convergence* [p. 258], in which multiple presynaptic neurons provide input to a smaller number of postsynaptic neurons (**FIG. 10.2**). Convergence allows multiple simultaneous subthreshold stimuli to sum at the postsynaptic (secondary) neuron. When multiple primary sensory neurons converge on a single secondary sensory neuron, their individual receptive fields merge into a single, large *secondary receptive field*, as shown in Figure 10.2a.

The size of secondary receptive fields determines how sensitive a given area is to a stimulus. For example, sensitivity to touch is demonstrated by a **two-point discrimination test.** In some regions of skin, such as that on the arms and legs, two pins placed within 20 mm of each other are interpreted by the brain as a single pinprick. In these areas, many primary neurons converge on a single secondary neuron, so the secondary receptive field is very large (Fig. 10.2a).

In contrast, more sensitive areas of skin, such as the fingertips, have smaller receptive fields, with as little as a 1:1 relationship between primary and secondary sensory neurons (Fig. 10.2b). In these regions, two pins separated by as little as 2 mm can be perceived as two separate touches.

The CNS Integrates Sensory Information

Sensory information from much of the body enters the spinal cord and travels through ascending pathways to the brain. Some

FIG. 10.2 Receptive fields of sensory neurons

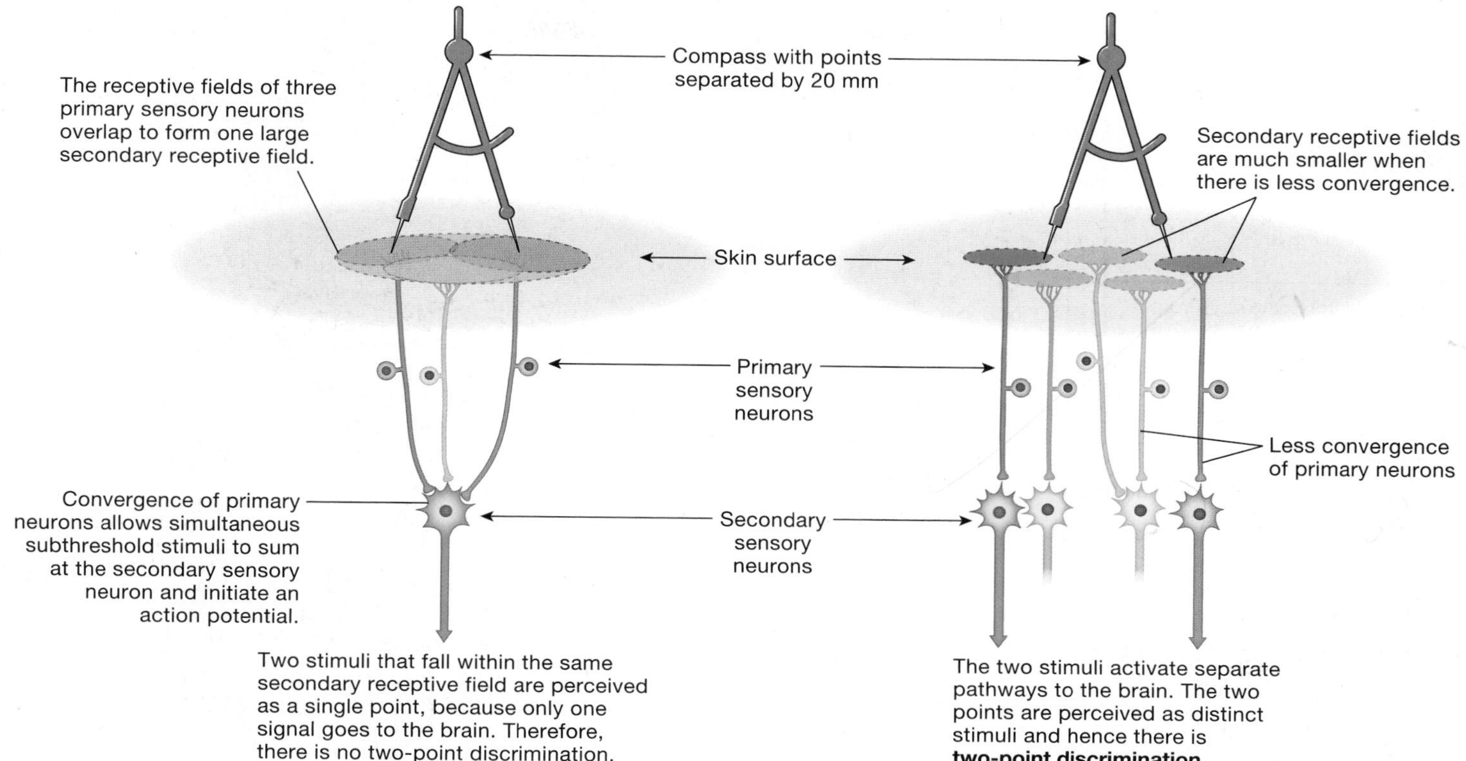

(a) Convergence creates large receptive fields.
(b) Small receptive fields are found in more sensitive areas.
Compass with points separated by 20 mm
The receptive fields of three primary sensory neurons overlap to form one large secondary receptive field.
Secondary receptive fields are much smaller when there is less convergence.
Skin surface
Primary sensory neurons
Less convergence of primary neurons
Convergence of primary neurons allows simultaneous subthreshold stimuli to sum at the secondary sensory neuron and initiate an action potential.
Secondary sensory neurons
Two stimuli that fall within the same secondary receptive field are perceived as a single point, because only one signal goes to the brain. Therefore, there is no two-point discrimination.
The two stimuli activate separate pathways to the brain. The two points are perceived as distinct stimuli and hence there is **two-point discrimination.**

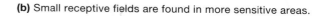

sensory information goes directly into the brain stem via the cranial nerves [p. 283]. Sensory information that initiates visceral reflexes is integrated in the brain stem or spinal cord and usually does not reach conscious perception. An example of an unconscious visceral reflex is the control of blood pressure by centers in the brain stem.

Each major division of the brain processes one or more types of sensory information (**FIG. 10.3**). For example, the midbrain receives visual information, and the medulla oblongata receives input for sound and taste. Information about balance and equilibrium is processed primarily in the cerebellum. These pathways, along with those carrying somatosensory information, project to the thalamus, which acts as a relay and processing station before passing the information on to the cerebrum.

Only *olfactory* {*olfacere*, to sniff} information is not routed through the thalamus. The sense of smell, a type of chemoreception, is considered one of the oldest senses, and even the most primitive vertebrate brains have well-developed regions for processing olfactory information. Information about odors travels from the nose through the first cranial nerve [p. 283] and *olfactory bulb* to the olfactory cortex in the cerebrum. Perhaps it is because of this direct input to the cerebrum that odors are so closely linked to memory and emotion. Most people have experienced encountering a smell that suddenly brings back a flood of memories of places or people from the past.

One interesting aspect of CNS processing of sensory information is the **perceptual threshold,** the level of stimulus intensity necessary for you to be aware of a particular sensation. Stimuli bombard your sensory receptors constantly, but your brain can filter out and ignore some stimuli. You experience a change in perceptual threshold when you "tune out" the radio while studying or when you "zone out" during a lecture. In both cases, the noise is adequate to stimulate sensory neurons in the ear, but neurons higher in the pathway dampen the perceived signal so that it does not reach the conscious brain.

Decreased perception of a stimulus, or *habituation,* is accomplished by *inhibitory modulation* [p. 261]. Inhibitory modulation diminishes a suprathreshold stimulus until it is below the perceptual threshold. It often occurs in the secondary and higher neurons of a sensory pathway. If the modulated stimulus suddenly becomes important, such as when the professor asks you a question, you can consciously focus your attention and overcome the inhibitory modulation. At that point, your conscious brain seeks to retrieve and recall recent sound input from your subconscious so that you can answer the question.

FIG. 10.3 Sensory pathways in the brain

Most pathways pass through the thalamus on their way to the cerebral cortex.

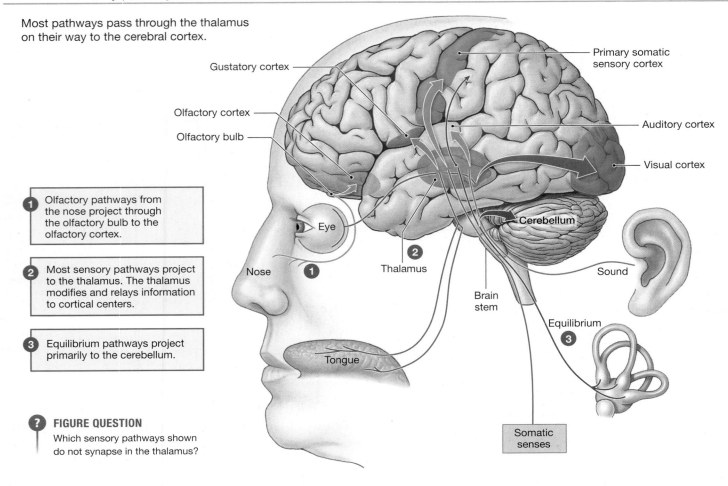

1. Olfactory pathways from the nose project through the olfactory bulb to the olfactory cortex.
2. Most sensory pathways project to the thalamus. The thalamus modifies and relays information to cortical centers.
3. Equilibrium pathways project primarily to the cerebellum.

FIGURE QUESTION

Which sensory pathways shown do not synapse in the thalamus?

RUNNING PROBLEM

Ménière's disease—named for its discoverer, the nineteenth-century French physician Prosper Ménière—is associated with a buildup of fluid in the inner ear and is also known as *endolymphatic hydrops* { *hydro-*, water }. Symptoms of Ménière's disease include episodic attacks of vertigo, nausea, and tinnitus, accompanied by hearing loss and a feeling of fullness in the ears. *Vertigo* is a false sensation of spinning movement that patients often describe as dizziness.

Q1: *In which part of the brain is sensory information about equilibrium processed?*

308
312
335
338
341
347
351

Coding and Processing Distinguish Stimulus Properties

If all stimuli are converted to action potentials in sensory neurons and all action potentials are identical, how can the CNS tell the difference between, say, heat and pressure, or between a pinprick to the toe and one to the hand? The attributes of the stimulus must somehow be preserved once the stimulus enters the nervous system for processing. This means that the CNS must distinguish four properties of a stimulus: (1) its nature, or **modality,** (2) its location, (3) its intensity, and (4) its duration.

Sensory Modality The modality of a stimulus is indicated by which sensory neurons are activated and by where the pathways of the activated neurons terminate in the brain. Each receptor type is most sensitive to a particular modality of stimulus. For example, some neurons respond most strongly to touch; others respond to changes in temperature. Each sensory modality can be subdivided into qualities. For instance, color vision is divided into red, blue, and green according to the wavelengths that most strongly stimulate the different visual receptors.

In addition, the brain associates a signal coming from a specific group of receptors with a specific modality. This 1:1 association of a receptor with a sensation is called **labeled line coding.** Stimulation of a cold receptor is always perceived as cold, whether the actual stimulus was cold or an artificial depolarization of the receptor. The blow to the eye that causes us to "see" a flash of light is another example of labeled line coding.

Location of the Stimulus The location of a stimulus is also coded according to which receptive fields are activated. The sensory regions of the cerebrum are highly organized with respect to incoming signals, and input from adjacent sensory receptors is processed in adjacent regions of the cortex. This arrangement preserves the topographical organization of receptors on the skin, eye, or other regions in the processing centers of the brain.

For example, touch receptors in the hand project to a specific area of the cerebral cortex. Experimental stimulation of that area of the cortex during brain surgery is interpreted as a touch to the hand, even though there is no contact. Similarly, one type of the *phantom limb pain* reported by amputees occurs when secondary sensory neurons in the spinal cord become hyperactive, resulting in the sensation of pain in a limb that is no longer there.

Auditory information is an exception to the localization rule, however. Neurons in the ears are sensitive to different frequencies of sound, but they have no receptive fields and their activation provides no information about the location of the sound. Instead, the brain uses the timing of receptor activation to compute a location, as shown in **FIGURE 10.4**.

A sound originating directly in front of a person reaches both ears simultaneously. A sound originating on one side reaches the closer ear several milliseconds before it reaches the other ear. The brain registers the difference in the time it takes for the sound stimuli to reach the two sides of the auditory cortex and uses that information to compute the sound's source.

Lateral inhibition, which increases the contrast between activated receptive fields and their inactive neighbors, is another way of isolating the location of a stimulus. **FIGURE 10.5** shows this process for a pressure stimulus to the skin. A pin pushing on the skin activates three primary sensory neurons, each of which releases neurotransmitters onto its corresponding secondary neuron.

However, the three secondary neurons do not all respond in the same fashion. The secondary neuron closest to the stimulus (neuron B) suppresses the response of the secondary neurons lateral to it (i.e., on either side), where the stimulus is weaker, and simultaneously allows its own pathway to proceed without interference. The inhibition of neurons farther from the stimulus enhances the contrast between the center and the sides of the receptive field, making the sensation more easily localized. In the visual system, lateral inhibition sharpens our perception of visual edges.

The pathway in Figure 10.5 also is an example of **population coding,** the way multiple receptors function together to send the CNS more information than would be possible from a single receptor. By comparing the input from multiple receptors, the CNS can make complex calculations about the quality and spatial and temporal characteristics of a stimulus.

FIG. 10.4 Localization of sound

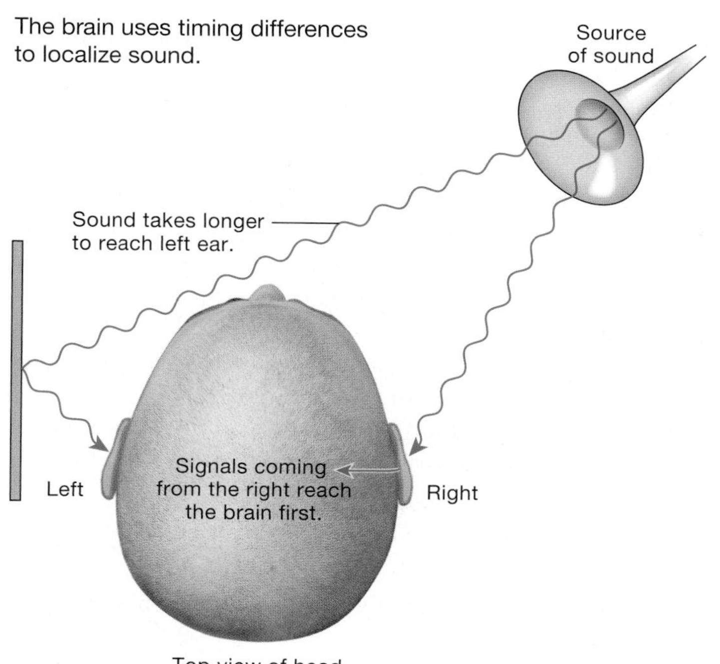
The brain uses timing differences to localize sound.
Source of sound
Sound takes longer to reach left ear.
Signals coming from the right reach the brain first.
Left
Right
Top view of head

CHAPTER 10

Concept Check

4. In Figure 10.5, what kind(s) of ion channel might open in neurons A and C that would depress their responsiveness: Na^+, K^+, Ca^{2+}, or Cl^-?

Intensity of the Stimulus The intensity of a stimulus cannot be directly calculated from a single sensory neuron action potential because a single action potential is "all-or-none." Instead, stimulus intensity is coded in two types of information: the number of receptors activated (another example of population coding) and the frequency of action potentials coming from those receptors, called *frequency coding*.

Population coding for intensity occurs because the threshold for the preferred stimulus is not the same for all receptors. Only the most sensitive receptors (those with the lowest thresholds) respond to a low-intensity stimulus. As a stimulus increases in intensity, additional receptors are activated. The CNS then translates the number of active receptors into a measure of stimulus intensity.

For individual sensory neurons, intensity discrimination begins at the receptor. If a stimulus is below threshold, the primary sensory neuron does not respond. Once stimulus intensity exceeds threshold, the primary sensory neuron begins to fire action potentials. As stimulus intensity increases, the receptor potential amplitude (strength) increases in proportion, and the frequency of action potentials in the primary sensory neuron increases, up to a maximum rate (**FIG. 10.6**).

Duration of the Stimulus The duration of a stimulus is coded by the duration of action potentials in the sensory neuron. In general, a longer stimulus generates a longer series of action potentials in the primary sensory neuron. However, if a stimulus persists, some receptors **adapt,** or cease to respond. Receptors fall into one of two classes, depending on how they adapt to continuous stimulation.

Tonic receptors are slowly adapting receptors that fire rapidly when first activated, then slow and maintain their firing

FIG. 10.5 Lateral inhibition

Lateral inhibition enhances contrast and makes a stimulus easier to perceive. The responses of primary sensory neurons A, B, and C are proportional to the intensity of the stimulus in each receptor field. Secondary sensory neuron B inhibits secondary neurons A and C, creating greater contrast between B and its neighbors.

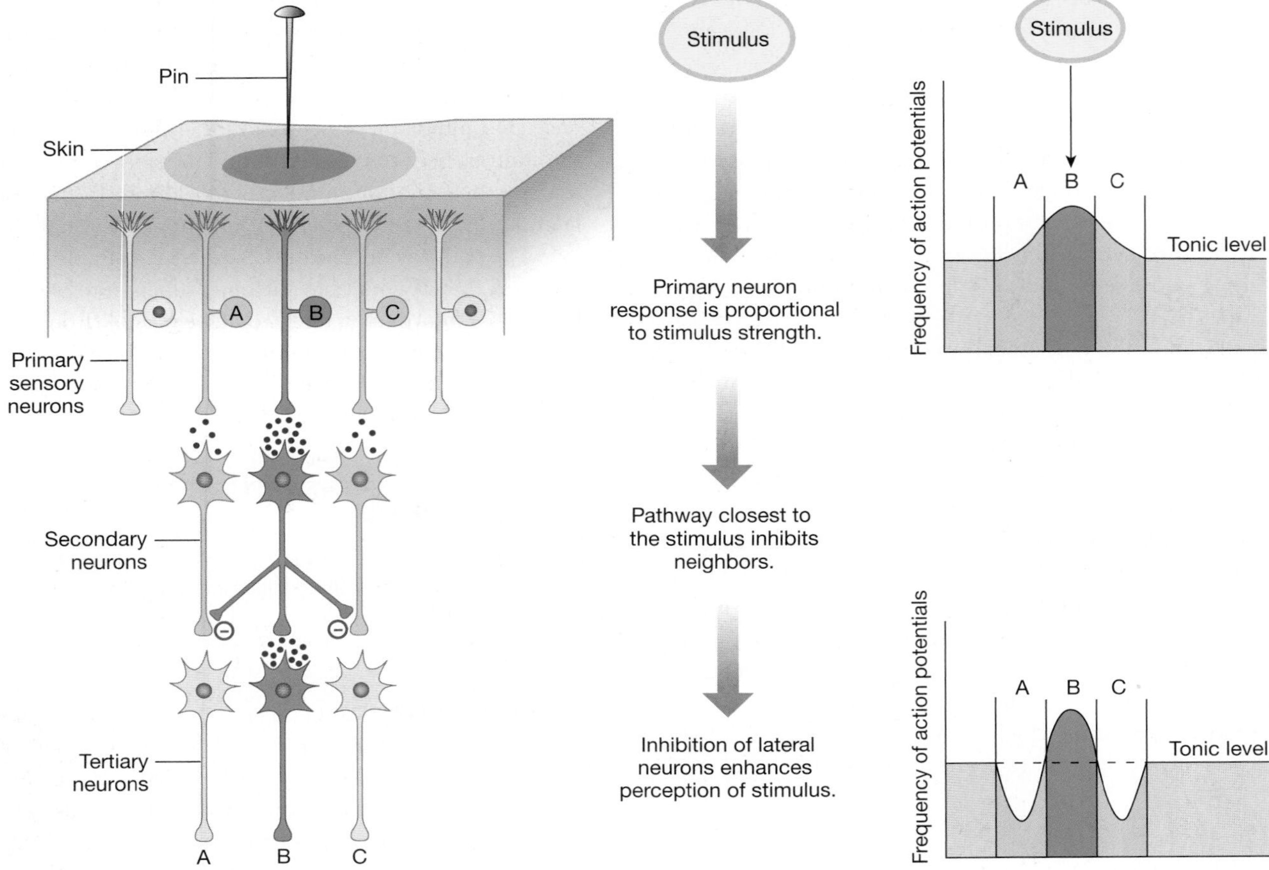
Pin
Skin
Primary sensory neurons
A
B
C
Secondary neurons
Tertiary neurons
A
B
C
Stimulus
Primary neuron response is proportional to stimulus strength.
Pathway closest to the stimulus inhibits neighbors.
Inhibition of lateral neurons enhances perception of stimulus.
Stimulus
Frequency of action potentials
A B C
Tonic level
Frequency of action potentials
A B C
Tonic level

as long as the stimulus is present (**FIG. 10.7a**). Pressure-sensitive baroreceptors, irritant receptors, and some tactile receptors and proprioceptors fall into this category. In general, the stimuli that activate tonic receptors are parameters that must be monitored continuously by the body.

In contrast, **phasic receptors** are rapidly adapting receptors that fire when they first receive a stimulus but cease firing if the strength of the stimulus remains constant (Fig. 10.7b). Phasic receptors are attuned specifically to *changes* in a parameter. Once a stimulus reaches a steady intensity, phasic receptors adapt to the new steady state and turn off. This type of response allows the body to ignore information that has been evaluated and found not to threaten homeostasis or well-being.

Our sense of smell is an example of a sense that uses phasic receptors. For example, you can smell your cologne when you put it on in the morning, but as the day goes on your olfactory receptors adapt and are no longer stimulated by the cologne molecules. You no longer smell the fragrance, yet others may comment on it.

Adaptation of phasic receptors allows us to filter out extraneous sensory information and concentrate on what is new, different, or essential. In general, once adaptation of a phasic receptor has occurred, the only way to create a new signal is to either increase the intensity of the excitatory stimulus or remove the stimulus entirely and allow the receptor to reset.

The molecular mechanism for sensory receptor adaptation depends on the receptor type. In some receptors, K^+ channels in the receptor membrane open, causing the membrane to repolarize and stopping the signal. In other receptors, Na^+ channels quickly inactivate. In yet other receptors, biochemical pathways alter the receptor's responsiveness.

Accessory structures may also decrease the amount of stimulus reaching the receptor. In the ear, for example, tiny muscles contract and dampen the vibration of small bones in response to loud noises, thus decreasing the sound signal before it reaches auditory receptors.

To summarize, the specificity of sensory pathways is established in several ways:

1. Each receptor is most sensitive to a particular type of stimulus.

FIG. 10.6 Coding for stimulus intensity and duration

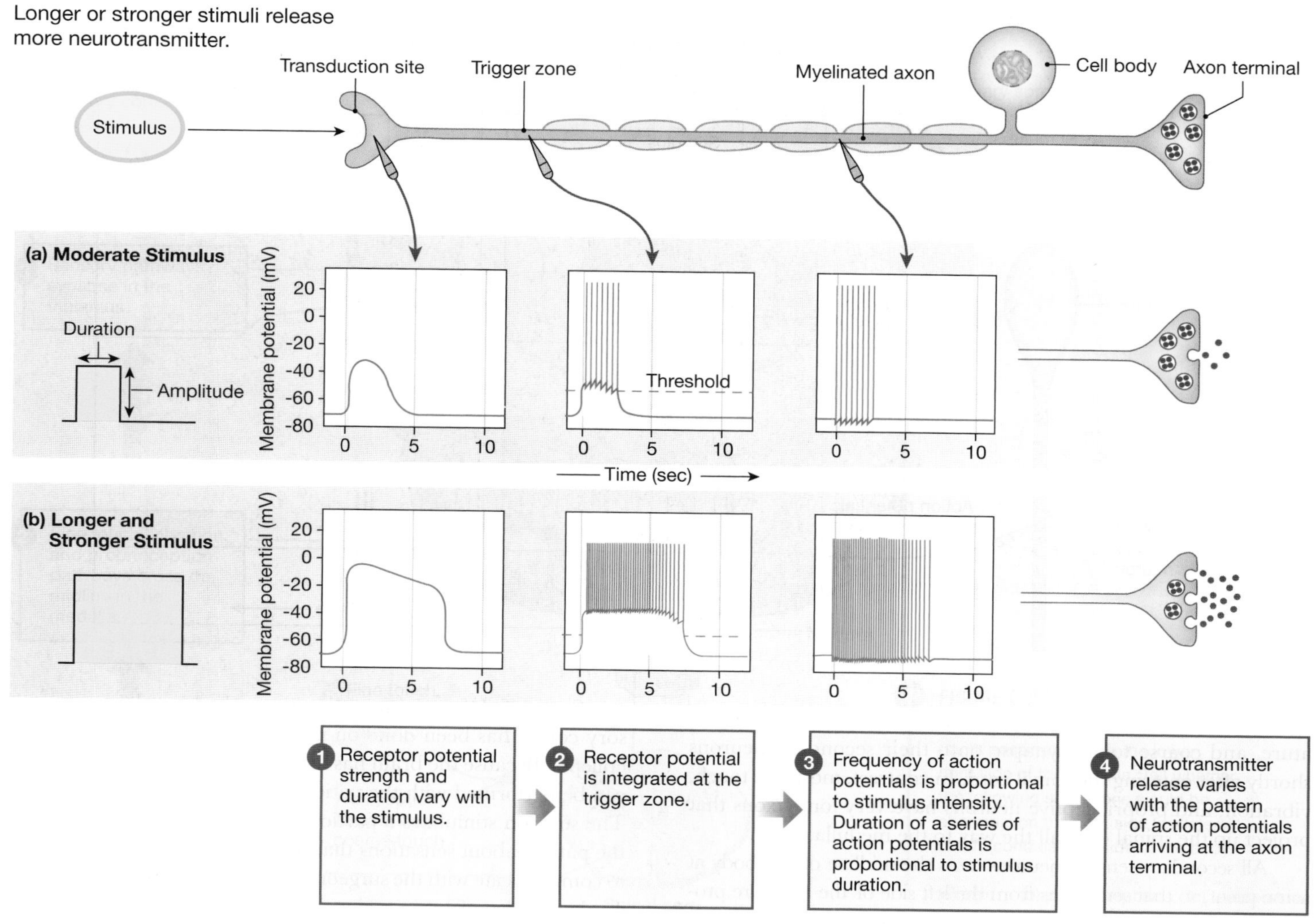
Longer or stronger stimuli release more neurotransmitter.
Transduction site
Trigger zone
Myelinated axon
Cell body
Axon terminal
Stimulus
(a) Moderate Stimulus
Duration
Amplitude
Membrane potential (mV)
Threshold
Time (sec)
(b) Longer and Stronger Stimulus
1 Receptor potential strength and duration vary with the stimulus.
2 Receptor potential is integrated at the trigger zone.
3 Frequency of action potentials is proportional to stimulus intensity. Duration of a series of action potentials is proportional to stimulus duration.
4 Neurotransmitter release varies with the pattern of action potentials arriving at the axon terminal.

2. A stimulus above threshold initiates action potentials in a sensory neuron that projects to the CNS.
3. Stimulus intensity and duration are coded in the pattern of action potentials reaching the CNS.
4. Stimulus location and modality are coded according to which receptors are activated or (in the case of sound) by the timing of receptor activation.
5. Each sensory pathway projects to a specific region of the cerebral cortex dedicated to a particular receptive field. The brain can then tell the origin of each incoming signal.

Concept Check

5. How do sensory receptors communicate the intensity of a stimulus to the CNS?
6. What is the adaptive significance of irritant receptors that are tonic instead of phasic?

10.2 Somatic Senses

There are four somatosensory modalities: touch, proprioception, temperature, and *nociception*, which includes pain and itch. (We discuss details of proprioception in Chapter 13.)

Pathways for Somatic Perception Project to the Cortex and Cerebellum

Receptors for the somatic senses are found both in the skin and in the viscera. Receptor activation triggers action potentials in the associated *primary sensory neuron*. Primary sensory neurons in the peripheral nervous system are *pseudounipolar* neurons [p. 226] whose nerve cell bodies lie in the *dorsal root ganglia* [p. 282] alongside the spinal cord. Their axon terminals synapse in the CNS onto interneurons that serve as the *secondary sensory neurons*. The location of the synapse between a primary neuron and its secondary neuron varies according to the type of receptor (**FIG. 10.8**).

CHAPTER 10

FIG. 10.9 The somatosensory cortex

Each body part is represented next to the area of the sensory cortex that processes stimuli for that body part. This mapping was created by two neurosurgeons, W. Penfield and T. Rasmussen, in 1950 and is called a homunculus (little man).

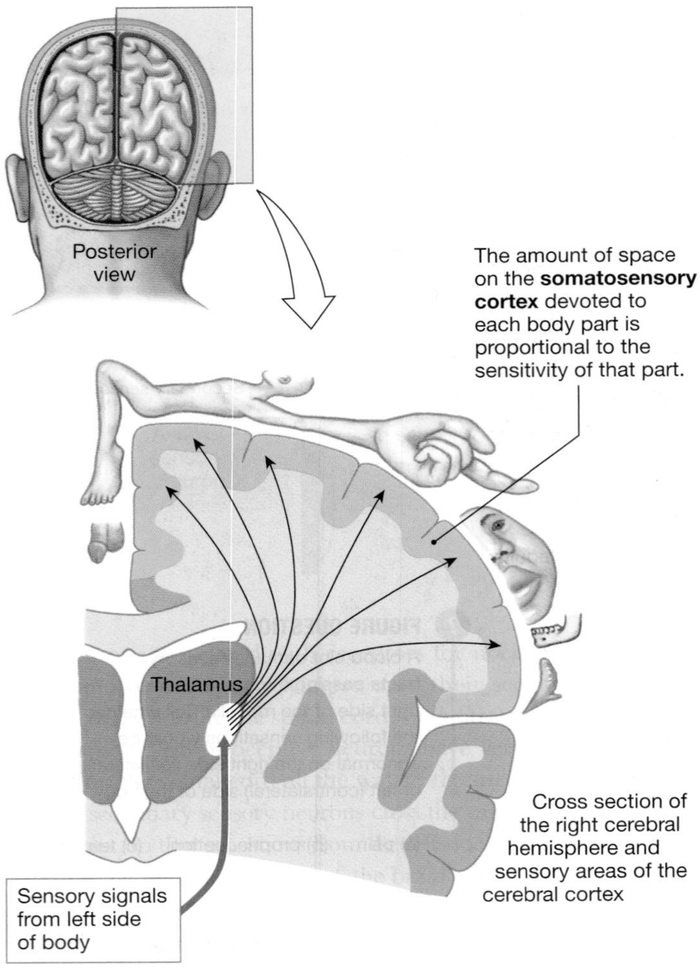
Posterior view
The amount of space on the **somatosensory cortex** devoted to each body part is proportional to the sensitivity of that part.
Thalamus
Cross section of the right cerebral hemisphere and sensory areas of the cerebral cortex
Sensory signals from left side of body

Pacinian corpuscles are composed of nerve endings encapsulated in layers of connective tissue (see Fig. 10.1b). They are found in the subcutaneous layers of skin and in muscles, joints, and internal organs. The concentric layers of connective tissue in the corpuscles create large receptive fields.

Pacinian corpuscles respond best to high-frequency vibrations, whose energy is transferred through the connective tissue capsule to the nerve ending, where the energy opens mechanically gated ion channels [p. 138]. Recent research using knockout mice indicates that another sensory receptor, the Merkel receptor, also uses mechanically gated ion channels to respond to touch.

Pacinian corpuscles are rapidly adapting phasic receptors, and this property allows them to respond to a change in touch but then ignore it. For example, you notice your shirt when you first put it on, but the touch receptors soon adapt. Properties of the remaining touch receptors depicted in Figure 10.10—Meissner's corpuscles, Ruffini corpuscles, and Merkel receptors—are summarized in the table of that figure.

Merkel receptors are an example of nonneural sensors in the somatic senses. The Merkel receptor consists of a *Merkel cell* that synapses onto a primary sensory neuron. Merkel receptors are found in the greatest density in the fingertips and are responsible for the high sensitivity of tactile reception in these areas.

Skin Temperature Receptors Are Free Nerve Endings

Temperature receptors, or *thermoreceptors*, are found throughout the body in skin, muscles, internal organs, and the CNS because of the importance of temperature homeostasis. In the skin, thermoreceptors are free nerve endings that terminate in the subcutaneous layers. **Cold receptors** are sensitive primarily to temperatures lower than body temperature. **Warm receptors** are stimulated by temperatures in the range extending from normal body temperature (37 °C) to about 45 °C. Above that temperature, pain receptors are activated, creating a sensation of painful heat. Thermoreceptors in the brain play an important role in thermoregulation.

The receptive field for a thermoreceptor is about 1 mm in diameter, and the receptors are scattered across the body. There are considerably more cold receptors than warm ones. Temperature receptors slowly adapt between 20 and 40 °C. Their initial response tells us that the temperature is changing, and their sustained response tells us about the ambient temperature. Outside the 20–40 °C range, where the likelihood of tissue damage is greater, the receptors do not adapt, and painful sensations begin to overlap with the thermal sensations.

Thermoreceptors use a family of cation channels called **transient receptor potential (TRP) channels** to initiate an action potential. TRP channels also play a key role in the transduction of painful or irritating stimuli (discussed next) and provide a link between thermal sensation and pain.

Nociceptors Initiate Protective Responses

Nociceptors {*nocere*, to injure} are neurons with free nerve endings (Fig. 10.1a) that respond to a variety of strong noxious stimuli (chemical, mechanical, or thermal) that cause or have the potential to cause tissue damage. Nociceptors are found in the skin, joints, muscles, bones and various internal organs, but not in the central nervous system. Activation of nociceptor pathways initiates adaptive, protective responses. For example, discomfort from overuse of muscles and joints warns us to take it easy and avoid additional damage to these structures. Afferent signals from nociceptors are carried to the CNS in two types of primary sensory fibers: *Aδ (A-delta) fibers* and *C fibers* (**TBL. 10.3**). The most common sensation carried by these pathways is perceived as pain, but when histamine or some other stimulus activates a subtype of C fiber, we perceive the sensation we call itch.

Pain is a subjective perception, the brain's interpretation of sensory information transmitted along pathways that begin at nociceptors. Pain is highly individual and multidimensional, and may vary with a person's emotional state. The discussion here is limited to the sensory processing of nociception.

FIG. 10.10 Sensory receptors in the skin

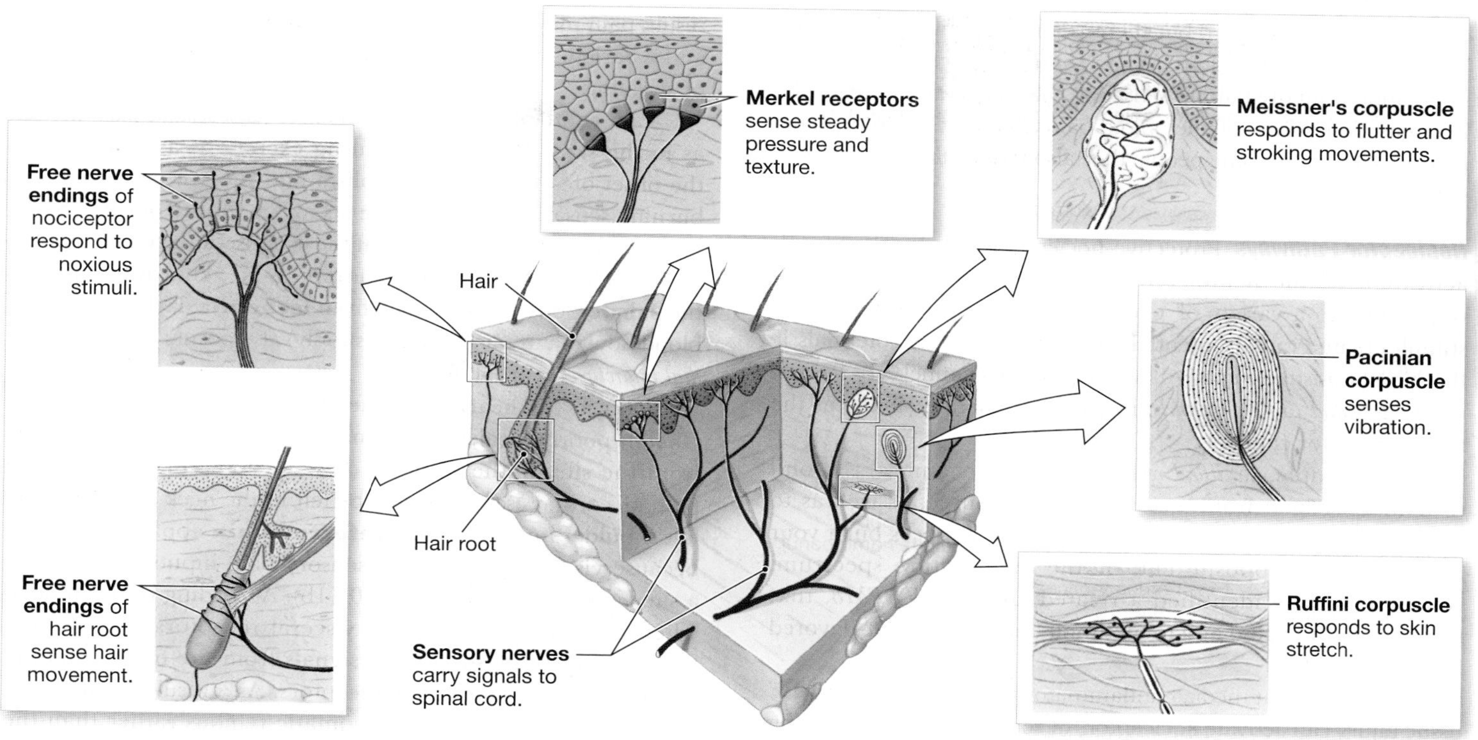
Free nerve endings of nociceptor respond to noxious stimuli.
Hair
Merkel receptors sense steady pressure and texture.
Meissner's corpuscle responds to flutter and stroking movements.
Pacinian corpuscle senses vibration.
Hair root
Free nerve endings of hair root sense hair movement.
Sensory nerves carry signals to spinal cord.
Ruffini corpuscle responds to skin stretch.

Receptor	Primary Sensory	Secondary Sensory	Synapse with...	Adaptation
Free Nerve Endings	Temperature, noxious stimuli, hair movement	Around the hair roots and under surface of skin	Unmyelinated nerve endings	Variable
Meissner's Corpuscles	Flutter, stroking	Superficial layers of skin	Nerve endings encapsulated in connective tissue	Rapid
Pacinian Corpuscles	Vibration	Deep layers of skin	Single nerve ending surrounded by layers of connective tissue	Rapid
Ruffini Corpuscles	Stretch of skin	Deep layers of skin	Nerve endings encapsulated in connective tissue	Slow
Merkel Receptors	Steady pressure, texture	Superficial layers of skin	Epidermal cell synapses with enlarged nerve ending	Slow

Fast pain, described as sharp and localized, is rapidly transmitted to the CNS by the small, myelinated Aδ fibers. **Slow pain,** described as duller and more diffuse, is carried on small, unmyelinated C fibers. The timing distinction between the two is most obvious when the stimulus originates far from the CNS, such as when you stub your toe. You first experience a quick stabbing sensation (fast pain), followed shortly by a dull throbbing (slow pain).

Itch (*pruritus*) comes only from nociceptors in the skin and is characteristic of many rashes and other skin conditions. However, itch can also be a symptom of a number of systemic diseases,

TABLE 10.3 Classes of Somatosensory Nerve Fibers

Fiber Type	Fiber Characteristics	Speed of Conduction	Associated with
Aβ (beta)	Large, myelinated	30–70 m/sec	Mechanical stimuli
Aδ (delta)	Small, myelinated	12–30 m/sec	Cold, fast pain, mechanical stimuli
C	Small, unmyelinated	0.5–2 m/sec	Slow pain, heat, cold, mechanical stimuli

FIG. 10.12 The gate control model

In the gate control model of pain modulation, nonpainful stimuli can diminish the pain signal.

(a) In absence of input from C fibers, a tonically active inhibitory interneuron suppresses pain pathway.

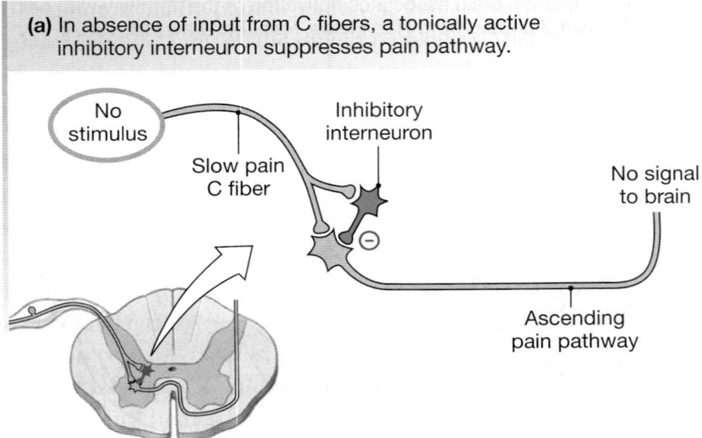

No stimulus
Inhibitory interneuron
Slow pain C fiber
No signal to brain
Ascending pain pathway

(b) With strong pain, C fiber stops inhibition of the pathway, allowing a strong signal to be sent to the brain.

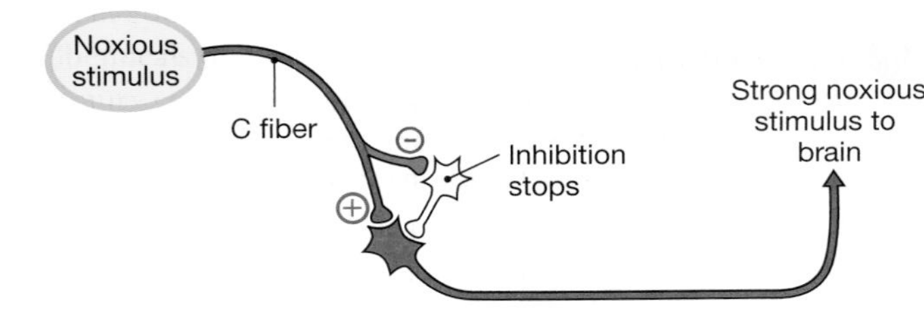

Noxious stimulus
C fiber
Inhibition stops
Strong noxious stimulus to brain

(c) Pain can be modulated by simultaneous somatosensory input.

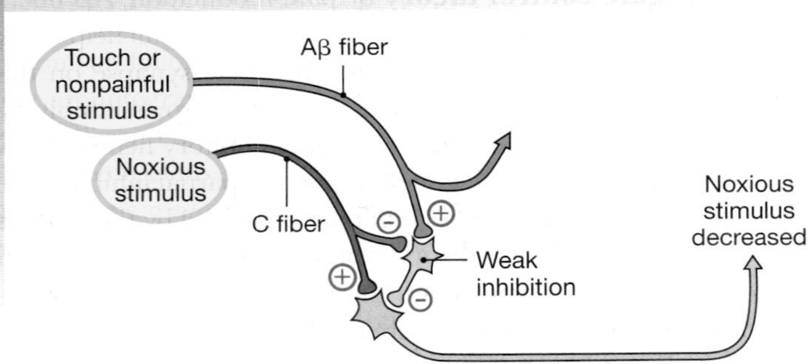

Touch or nonpainful stimulus
Aβ fiber
Noxious stimulus
C fiber
Weak inhibition
Noxious stimulus decreased

believed to be partly responsible for pain suppression by descending pathways from the brain to the spinal cord. The endogenous opioid **β-endorphin** is produced from the same prohormone as ACTH (adrenocorticotropin) in neuroendocrine cells of the hypothalamus [Fig. 7.3b, p. 201]. Although opioid drugs are effective at relieving pain, a person taking them for long periods of time may develop tolerance and need larger and larger doses to obtain the same effect.

As a result, scientists are exploring alternative drugs and strategies for pain relief. Some chronic pain may be caused by sensitization of nociceptive nerve endings near a site of injury when the body releases chemicals in response to the damage. Non-narcotic anti-inflammatory drugs such as aspirin and COX2 inhibitors can often relieve pain, but even over-the-counter doses may have adverse side effects. New research is focused on blocking TRP channels in the sensitized nociceptor nerve endings.

For people with severe chronic pain, possible treatments include electrically stimulating inhibitory pain pathways to the brain, or in extreme cases, surgically severing sensory nerves at the dorsal root. Acupuncture can also be effective, although the physiological reason for its effectiveness is not clear. The leading theory on how acupuncture works proposes that properly placed acupuncture needles trigger the release of endorphins by the brain.

Concept Check

7. What is the adaptive advantage of a spinal reflex?
8. Rank the speed of signal transmission through the following fiber types, from fastest to slowest: (a) small diameter, myelinated fiber; (b) large diameter, myelinated fiber; (c) small diameter, unmyelinated fiber.
9. Your sense of smell uses phasic receptors. What other receptors (senses) adapt to ongoing stimuli?

10.3 Chemoreception: Smell and Taste

The five special senses—smell, taste, hearing, equilibrium, and vision—are concentrated in the head region. Like somatic senses, the special senses rely on receptors to transform information about the environment into patterns of action potentials that can be interpreted by the brain. Smell and taste are both forms of *chemoreception*, one of the oldest senses from an evolutionary perspective. Unicellular bacteria use chemoreception to sense their environment, and primitive animals without formalized nervous systems use chemoreception to locate food and mates. It has been hypothesized that chemoreception evolved into chemical synaptic communication in animals.

Olfaction Is One of the Oldest Senses

Imagine waking up one morning and discovering a whole new world around you, a world filled with odors that you never dreamed existed—scents that told you more about your surroundings than you ever imagined from looking at them. This is exactly what happened to a young patient of Dr. Oliver Sacks (an account is in *The Man Who Mistook His Wife for a Hat and Other Clinical Tales*). Or imagine skating along the sidewalk without a helmet, only to fall and hit your head. When you regain consciousness, the world has lost all odor: no smell of grass or perfume or garbage. Even your food has lost much of its taste, and you now eat only to survive because eating has lost its pleasure.

We do not realize the essential role that our sense of smell plays in our lives until a head cold or injury robs us of the ability to smell. **Olfaction** allows us to discriminate among millions of different odors. Even so, our noses are not nearly as sensitive as those of many other animals whose survival depends on olfactory cues. The **olfactory bulb,** the extension of the forebrain that receives input from the primary olfactory neurons, is much better developed in vertebrates whose survival is more closely linked to chemical monitoring of their environment (**FIG. 10.13a**).

FIG. 10.13 Anatomy Summary . . . The Olfactory System

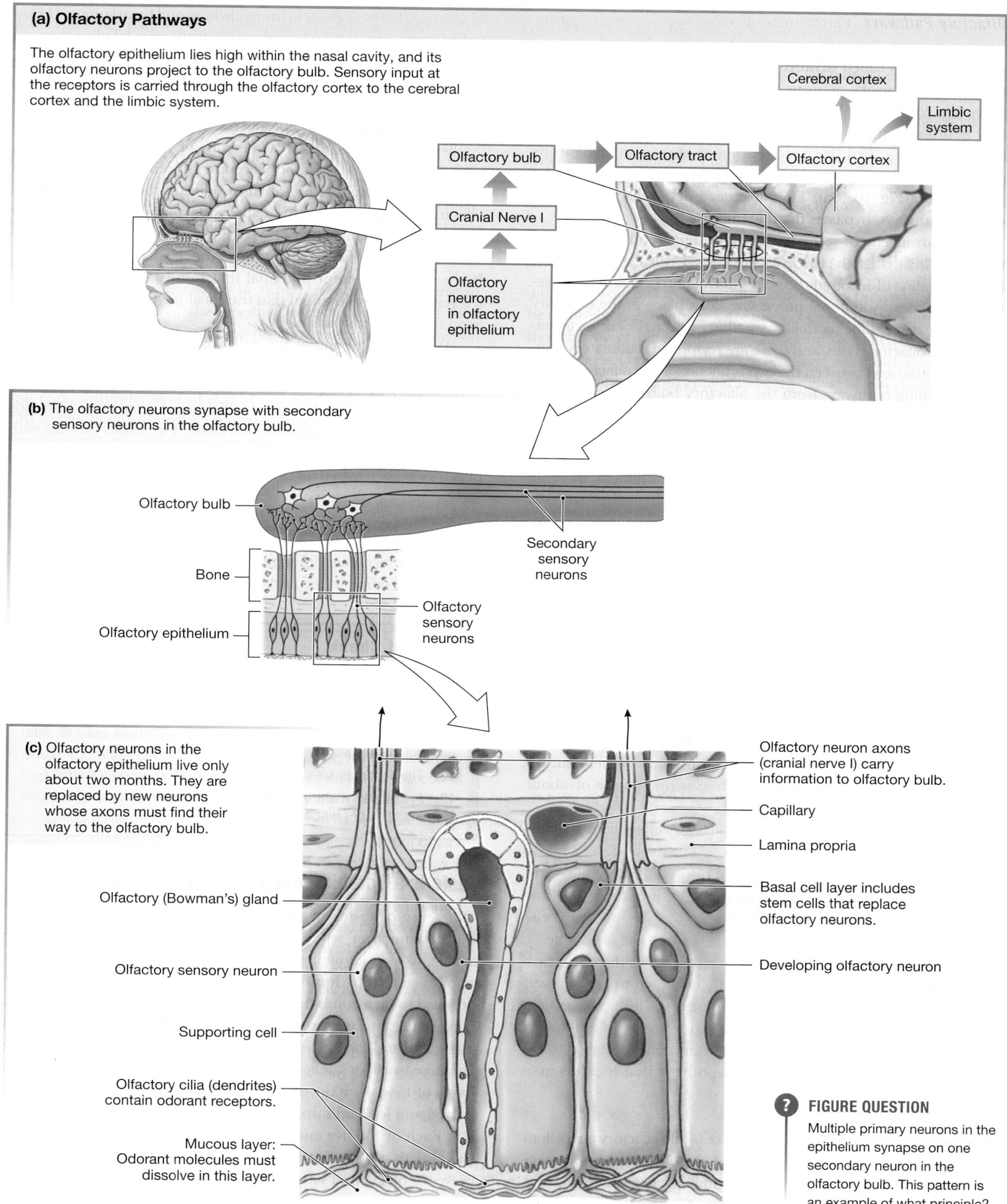

(a) Olfactory Pathways
The olfactory epithelium lies high within the nasal cavity, and its olfactory neurons project to the olfactory bulb. Sensory input at the receptors is carried through the olfactory cortex to the cerebral cortex and the limbic system.
Cerebral cortex
Limbic system
Olfactory bulb
Olfactory tract
Olfactory cortex
Cranial Nerve I
Olfactory neurons in olfactory epithelium
(b) The olfactory neurons synapse with secondary sensory neurons in the olfactory bulb.
Olfactory bulb
Bone
Olfactory epithelium
Secondary sensory neurons
Olfactory sensory neurons
(c) Olfactory neurons in the olfactory epithelium live only about two months. They are replaced by new neurons whose axons must find their way to the olfactory bulb.
Olfactory neuron axons (cranial nerve I) carry information to olfactory bulb.
Capillary
Lamina propria
Basal cell layer includes stem cells that replace olfactory neurons.
Olfactory (Bowman's) gland
Developing olfactory neuron
Olfactory sensory neuron
Supporting cell
Olfactory cilia (dendrites) contain odorant receptors.
Mucous layer: Odorant molecules must dissolve in this layer.

FIGURE QUESTION

Multiple primary neurons in the epithelium synapse on one secondary neuron in the olfactory bulb. This pattern is an example of what principle?

FIG. 10.14 ESSENTIALS Taste

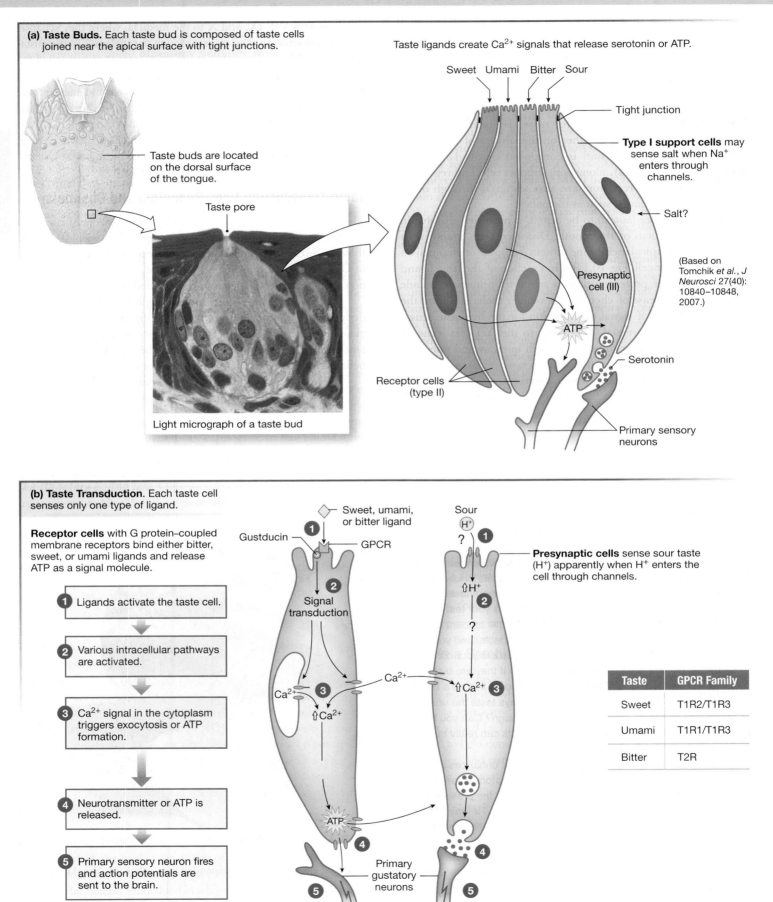

(a) Taste Buds. Each taste bud is composed of taste cells joined near the apical surface with tight junctions.

Taste buds are located on the dorsal surface of the tongue.

Taste pore

Light micrograph of a taste bud

Taste ligands create Ca^{2+} signals that release serotonin or ATP.

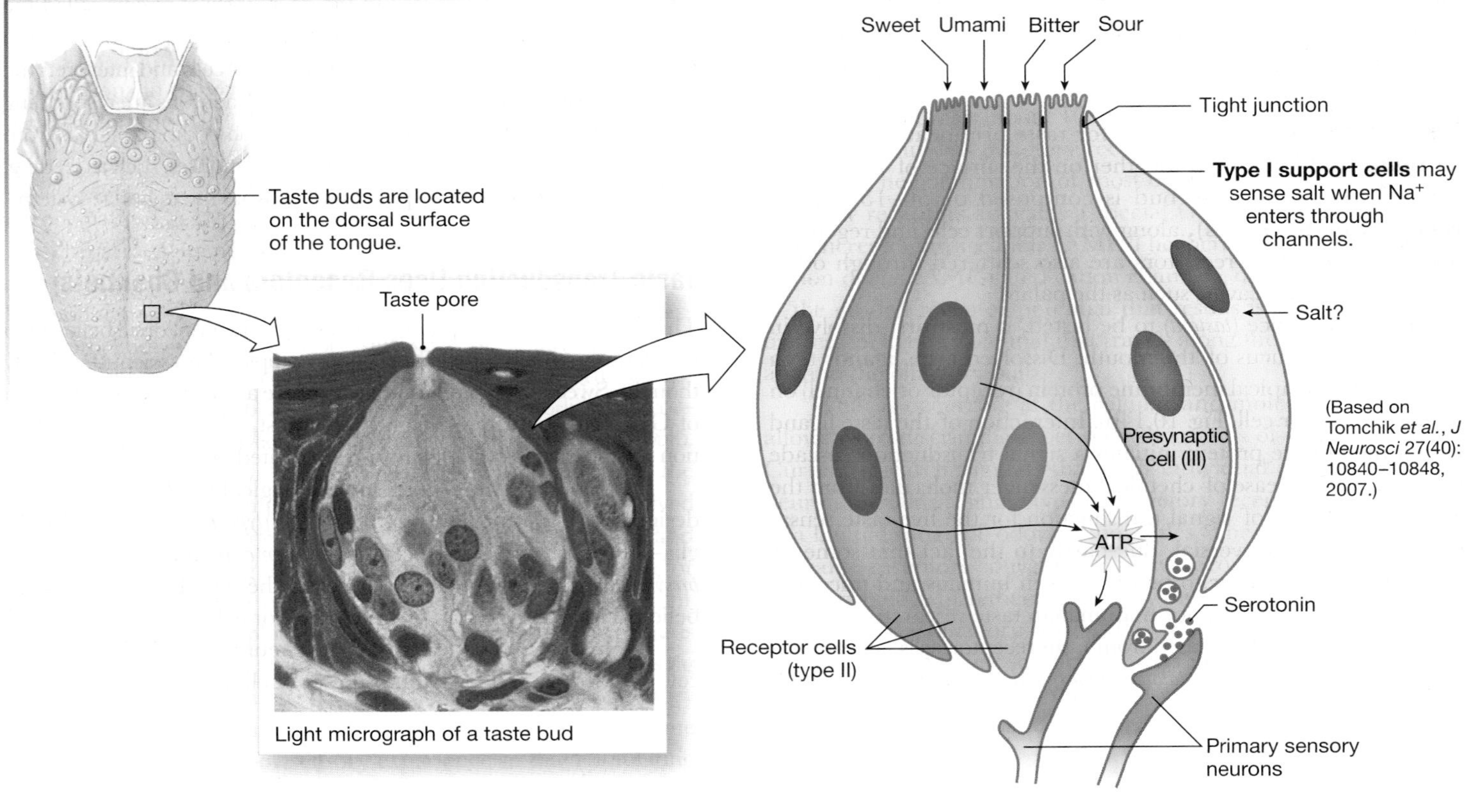

Sweet
Umami
Bitter
Sour

Tight junction

Type I support cells may sense salt when Na^+ enters through channels.

Salt?

Presynaptic cell (III)

ATP

Serotonin

Receptor cells (type II)

Primary sensory neurons

(Based on Tomchik *et al.*, *J Neurosci* 27(40): 10840–10848, 2007.)

(b) Taste Transduction. Each taste cell senses only one type of ligand.

Receptor cells with G protein–coupled membrane receptors bind either bitter, sweet, or umami ligands and release ATP as a signal molecule.

1. Ligands activate the taste cell.
2. Various intracellular pathways are activated.
3. Ca^{2+} signal in the cytoplasm triggers exocytosis or ATP formation.
4. Neurotransmitter or ATP is released.
5. Primary sensory neuron fires and action potentials are sent to the brain.

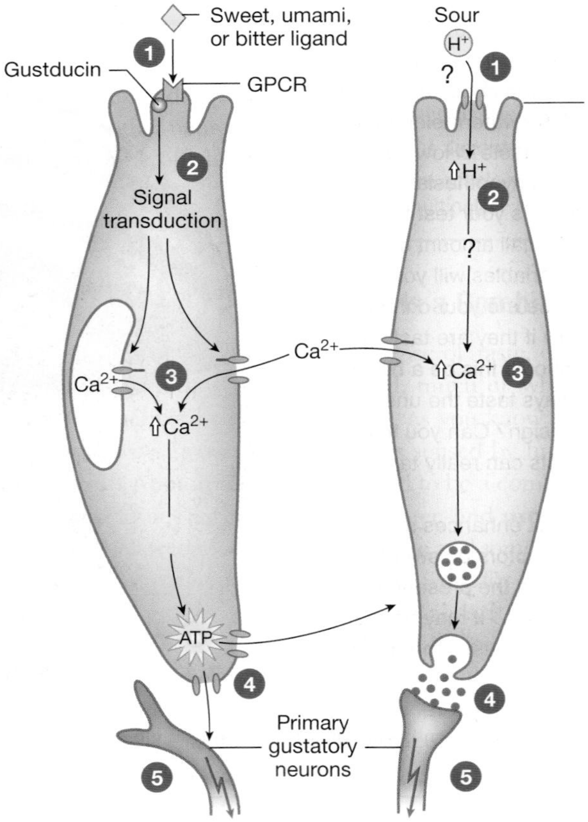

Sweet, umami, or bitter ligand

Gustducin

GPCR

Signal transduction

Ca^{2+}

⇧Ca^{2+}

ATP

Primary gustatory neurons

Sour

H^+

?

⇧H^+

?

⇧Ca^{2+}

Presynaptic cells sense sour taste (H^+) apparently when H^+ enters the cell through channels.

Taste	GPCR Family
Sweet	T1R2/T1R3
Umami	T1R1/T1R3
Bitter	T2R

tip protrudes into the oral cavity through a *taste pore* (Fig. 10.14a). In a given bud, tight junctions link the apical ends of adjacent cells together, limiting movement of molecules between the cells. The apical membrane of a TRC is modified into microvilli to increase the amount of surface area in contact with the environment.

Sweet, Bitter, and Umami Tastes The ***type II* taste receptor cells** respond to sweet, bitter, and umami sensations. These cells express multiple G protein-coupled receptors (GPCR) on their apical surfaces (Fig. 10.14b). Sweet and umami tastes are associated with T1R receptors with different combinations of subunits. Bitter taste uses about 30 variants of T2R receptors.

The type II cell receptors activate a special G protein called **gustducin,** which in turn activates multiple signal transduction pathways. Some of these pathways release Ca^{2+} from intracellular stores, while others open cation channels and allow Ca^{2+} to enter the cell. Calcium signals then initiate ATP release from the type II cells.

ATP in type II cells is not released through secretory vesicles. Instead it leaves the cell through a wide-pore channel protein named *CALHM1*, which stands for *calcium homeostasis modulator 1*. ATP then acts as a paracrine signal on both sensory neurons and neighboring presynaptic cells. This communication between neighboring taste receptor cells creates complex interactions.

Sour Taste The ***type III presynaptic cells*** respond to sour tastes. Models of transduction mechanisms for sour tastes are complicated by the fact that increasing H^+, the sour taste signal, also changes pH. There is evidence that H^+ acts on ion channels of the presynaptic cell from both extracellular and intracellular sides of the membrane. The intracellular pathways remain uncertain. Ultimately, H^+-mediated depolarization of the presynaptic cell results in serotonin release by exocytosis. Serotonin in turn excites the primary sensory neuron.

Salt Taste The cells responsible for salt taste have not been definitively identified, but some evidence suggests that salt reception may reside in the type 1 support cells. Signal transduction for salt taste in humans is equally unclear, complicated by the fact that mice have two different mechanisms, but humans appear to have only one. In the current model for salty taste, Na^+ enters the taste receptor cell through an apical ion channel, such as the *epithelial* Na^+ *channel* (ENaC, pronounced ēē-knack). Sodium entry depolarizes the cell, setting off a series of events that culminate with the primary sensory neuron firing an action potential.

The mechanisms of taste transduction are a good example of how our models of physiological function must periodically be revised as new research data are published. For many years, the widely held view of taste transduction was that an individual taste receptor cell could sense more than one taste, with cells differing in their sensitivities. However, gustation research using molecular biology techniques and knockout mice currently indicates that each taste receptor cell is sensitive to only one taste.

Nontraditional Taste Sensations The sensations we call taste are not all mediated by the traditional taste receptors. For years, physiologists thought fat in the diet was appealing because of its texture, and food experts use the phrase "mouth feel" to describe the sensation of eating something fatty, such as ice cream, that seems to coat the inside of the mouth. But now it appears that the tongue may have taste receptors for fats.

Research in rodents has identified a membrane receptor called *CD36* that lines taste pores and binds fats. Activation of this receptor helps trigger the feedforward digestive reflexes that prepare the digestive system for a meal. Currently evidence is lacking for a similar receptor in humans, but "fatty" may turn out to be a sixth taste sensation. Other candidates for new taste sensations include carbonation (dissolved CO_2) and Ca^{2+}, another essential element obtained through the diet. Evidence suggests that carbonation is sensed by sour receptors, which use a membrane-anchored enzyme called *carbonic anhydrase* to convert dissolved CO_2 to bicarbonate ion and H^+. The H^+ then activates the same receptor as H^+ from sour-tasting sources.

Some additional taste sensations are related to somatosensory pathways rather than taste receptor cells. Nerve endings in the mouth have TRP receptors and carry spicy sensations through the *trigeminal nerve* (CN V). *Capsaicin* from chili peppers, menthol from mint, and molecules in cinnamon, mustard oil, and many Indian spices activate these receptors to add to our appreciation of the food we eat. The newest taste sensation, proposed by Japanese researchers, is *kokumi*, which means "rich taste." Kokumi is created by peptides that have no taste by themselves but which enhance the "mouthfulness" or sensation of thickness of foods.

What would you say to the idea of taste buds in your gut? Scientists have known for years that the stomach and intestines have the ability to sense the composition of a meal and secrete appropriate hormones and enzymes. Gut chemoreception is mediated by the same receptors and signal transduction mechanisms that occur in taste buds on the tongue. Studies have found the T1R receptor proteins for sweet and umami tastes as well as the G protein gustducin in various cells of rodent and human intestines. It appears that chemoreceptors for "tasting" the environment are also found in other places in the body, including the upper airways and on sperm.

An interesting psychological aspect of taste is the phenomenon named **specific hunger.** Humans and other animals that are lacking a particular nutrient may develop a craving for that substance. **Salt appetite,** representing a lack of Na^+ in the body, has been recognized for years. Hunters have used their knowledge of this specific hunger to stake out salt licks because they know that animals will seek them out. Salt appetite is directly related to Na^+ concentration in the body and cannot be assuaged by ingestion of other cations, such as Ca^{2+} or K^+. Other appetites, such as cravings for chocolate, are more difficult to relate to specific nutrient needs and probably reflect complex mixtures of physical, psychological, environmental, and cultural influences.

Concept Check

14. With what essential nutrient is the umami taste sensation associated?

15. Map or diagram the neural pathway from a presynaptic taste receptor cell to the gustatory cortex.

CHAPTER 10

FIG. 10.15 Anatomy Summary . . . The Ear

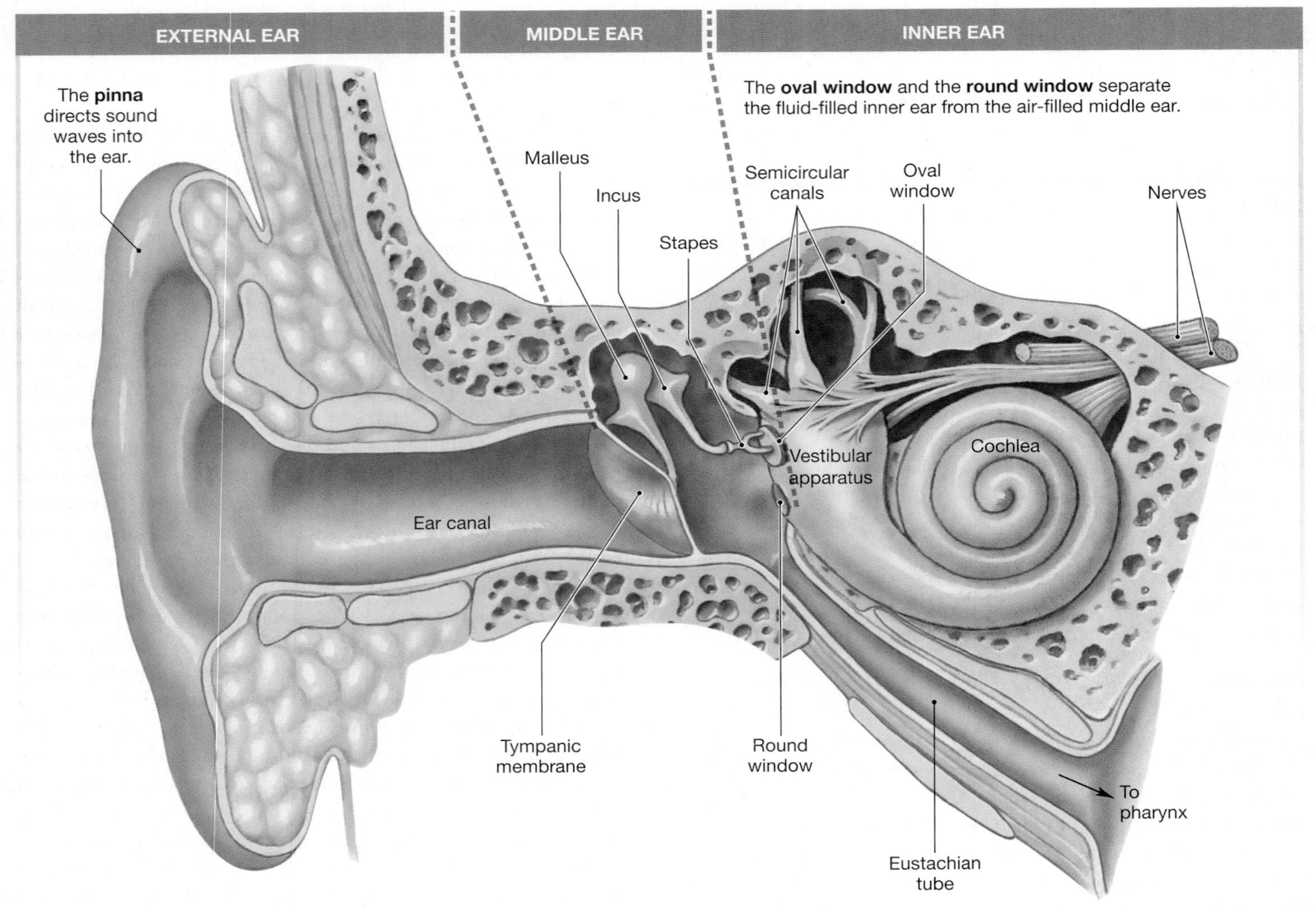
EXTERNAL EAR
MIDDLE EAR
INNER EAR
The pinna directs sound waves into the ear.
The oval window and the round window separate the fluid-filled inner ear from the air-filled middle ear.
Malleus
Incus
Stapes
Semicircular canals
Oval window
Nerves
Vestibular apparatus
Cochlea
Ear canal
Tympanic membrane
Round window
To pharynx
Eustachian tube

10.4 The Ear: Hearing

The ear is a sense organ that is specialized for two distinct functions: hearing and equilibrium. It can be divided into external, middle, and inner sections, with the neurological elements housed in and protected by structures in the inner ear. The vestibular complex of the inner ear is the primary sensor for equilibrium. The remainder of the ear is used for hearing.

The *external ear* consists of the outer ear, or **pinna,** and the **ear canal** (**FIG. 10.15**). The pinna is another example of an important accessory structure to a sensory system, and it varies in shape and location from species to species, depending on the animals' survival needs. The ear canal is sealed at its internal end by a thin membranous sheet of tissue called the **tympanic membrane,** or *eardrum*.

The tympanic membrane separates the external ear from the *middle ear,* an air-filled cavity that connects with the pharynx through the **Eustachian tube.** The Eustachian tube is normally collapsed, sealing off the middle ear, but it opens transiently to allow middle ear pressure to equilibrate with atmospheric pressure during chewing, swallowing, and yawning. Colds or other infections that cause swelling can block the Eustachian tube and result in fluid buildup in the middle ear. If bacteria are trapped in the middle ear fluid, the ear infection known as *otitis media* { *oto*-, ear + -*itis*, inflammation + *media*, middle } results.

Three small bones of the middle ear conduct sound from the external environment to the inner ear: the **malleus** { hammer }, the **incus** { anvil }, and the **stapes** { stirrup }. The three bones are connected to one another with the biological equivalent of hinges. One end of the malleus is attached to the tympanic membrane, and the stirrup end of the stapes is attached to a thin membrane that separates the middle ear from the inner ear.

The inner ear consists of two major sensory structures. The *vestibular apparatus* with its *semicircular canals* is the sensory transducer for our sense of equilibrium, described in the following section. The **cochlea** of the inner ear contains sensory receptors for hearing. On external view the cochlea is a membranous tube that lies coiled like a snail shell within a bony cavity. Two membranous disks, the **oval window** (to which the stapes is attached) and the **round window,** separate the liquid-filled cochlea from the air-filled middle ear. Branches of cranial nerve VIII, the *vestibulocochlear nerve,* lead from the inner ear to the brain.

Hearing Is Our Perception of Sound

Hearing is our perception of the energy carried by *sound waves*, which are pressure waves with alternating peaks of compressed air and valleys in which the air molecules are farther apart (**FIG. 10.16a**). The classic question about hearing is, "If a tree falls in the forest with no one to hear, does it make a noise?" The physiological answer is no, because noise, like pain, is a perception that results from the brain's processing of sensory information. A falling tree emits sound waves, but there is no noise unless someone or something is present to process and perceive the wave energy as sound.

Sound is the brain's interpretation of the frequency, amplitude, and duration of sound waves that reach our ears. Our brains translate **frequency** of sound waves (the number of wave peaks that pass a given point each second) into the **pitch** of a sound. Low-frequency waves are perceived as low-pitched sounds, such as the rumble of distant thunder. High-frequency waves create high-pitched sounds, such as the screech of fingernails on a blackboard.

Sound wave frequency (Fig. 10.16b) is measured in waves per second, or **hertz (Hz)**. The average human ear can hear sounds over the frequency range of 20–20,000 Hz, with the most acute hearing between 1000–3000 Hz. Our hearing is not as acute as that of many other animals, just as our sense of smell is less acute. Bats listen for ultra-high-frequency sound waves (in the kilohertz range) that bounce off objects in the dark. Elephants and some birds can hear sounds in the infrasound (very-low-frequency) range.

Loudness is our interpretation of sound intensity and is influenced by the sensitivity of an individual's ear. The intensity of a sound wave is a function of the wave height, or **amplitude** (Fig. 10.16b). Intensity is measured on a logarithmic scale in units called **decibels (dB)**. Each 10-dB increase represents a fold increase in intensity.

Normal conversation has a typical noise level of about 60 dB. Sounds of 80 dB or more can damage the sensitive hearing receptors of the ear, resulting in hearing loss. A typical heavy metal rock concert has noise levels around 120 dB, an intensity that puts listeners in immediate danger of damage to their hearing. The amount of damage depends on the duration and frequency of the noise as well as its intensity.

FIG. 10.16 Sound waves

(a) Sound waves alternate peaks of compressed air and valleys where the air is less compressed.

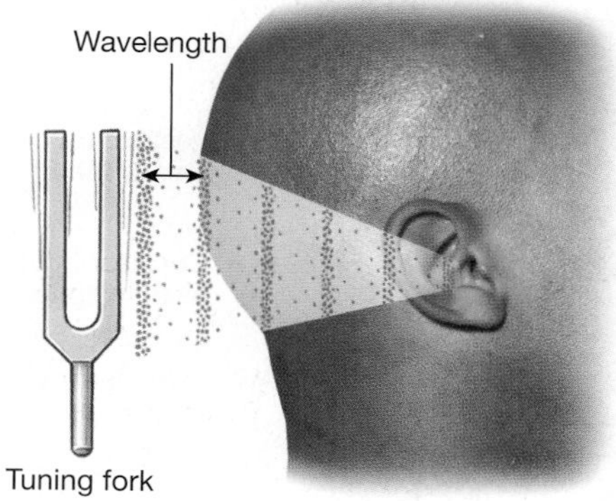
Wavelength
Tuning fork

(b) Sound waves are distinguished by their frequency, measured in hertz (Hz), and amplitude, measured in decibels (dB).

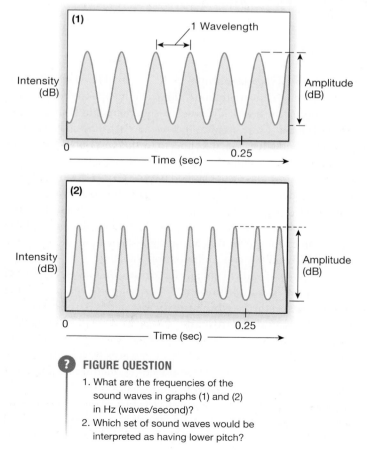

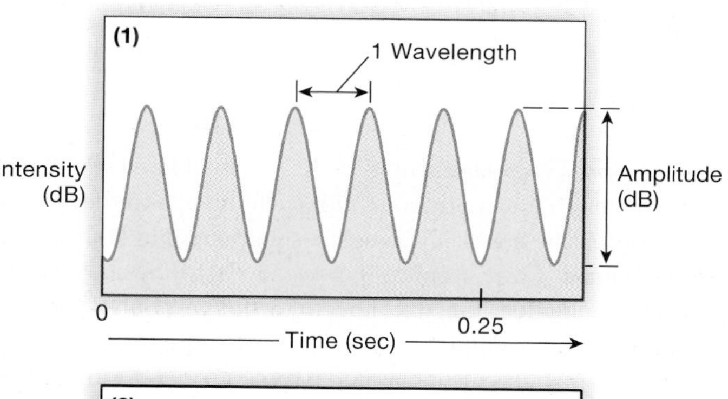
(1)
1 Wavelength
Intensity (dB)
Amplitude (dB)
0
0.25
Time (sec)

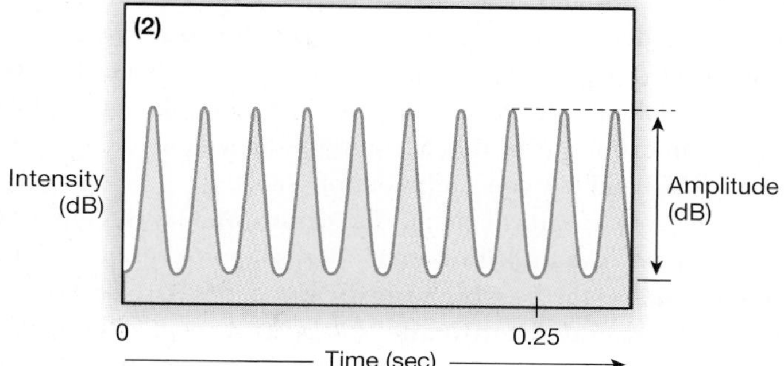
(2)
Intensity (dB)
Amplitude (dB)
0
0.25
Time (sec)

FIGURE QUESTION

1. What are the frequencies of the sound waves in graphs (1) and (2) in Hz (waves/second)?
2. Which set of sound waves would be interpreted as having lower pitch?

Concept Check

16. What is a kilohertz?

Sound Transduction Is a Multistep Process

Hearing is a complex sense that involves multiple transductions. Energy from sound waves in the air becomes mechanical vibrations, then fluid waves in the cochlea. The fluid waves open ion channels in *hair cells*, the sensory receptors for hearing. Ion flow into hair cells creates electrical signals that release neurotransmitter (chemical signal), which in turn triggers action potentials in the primary auditory neurons.

These transduction steps are shown in **FIGURE 10.17**. Sound waves striking the outer ear are directed down the ear canal until they hit the tympanic membrane and cause it to vibrate (first transduction). The tympanic membrane vibrations are transferred to the malleus, the incus, and the stapes, in that order. The arrangement of the three connected middle ear bones creates a "lever" that multiplies the force of the vibration (*amplification*) so that very little sound energy is lost due to friction. If noise levels are so high that there is danger of damage to the inner ear, small muscles in the middle ear can pull on the bones to decrease their movement and thereby dampen sound transmission to some degree.

FIG. 10.17 Sound transmission through the ear

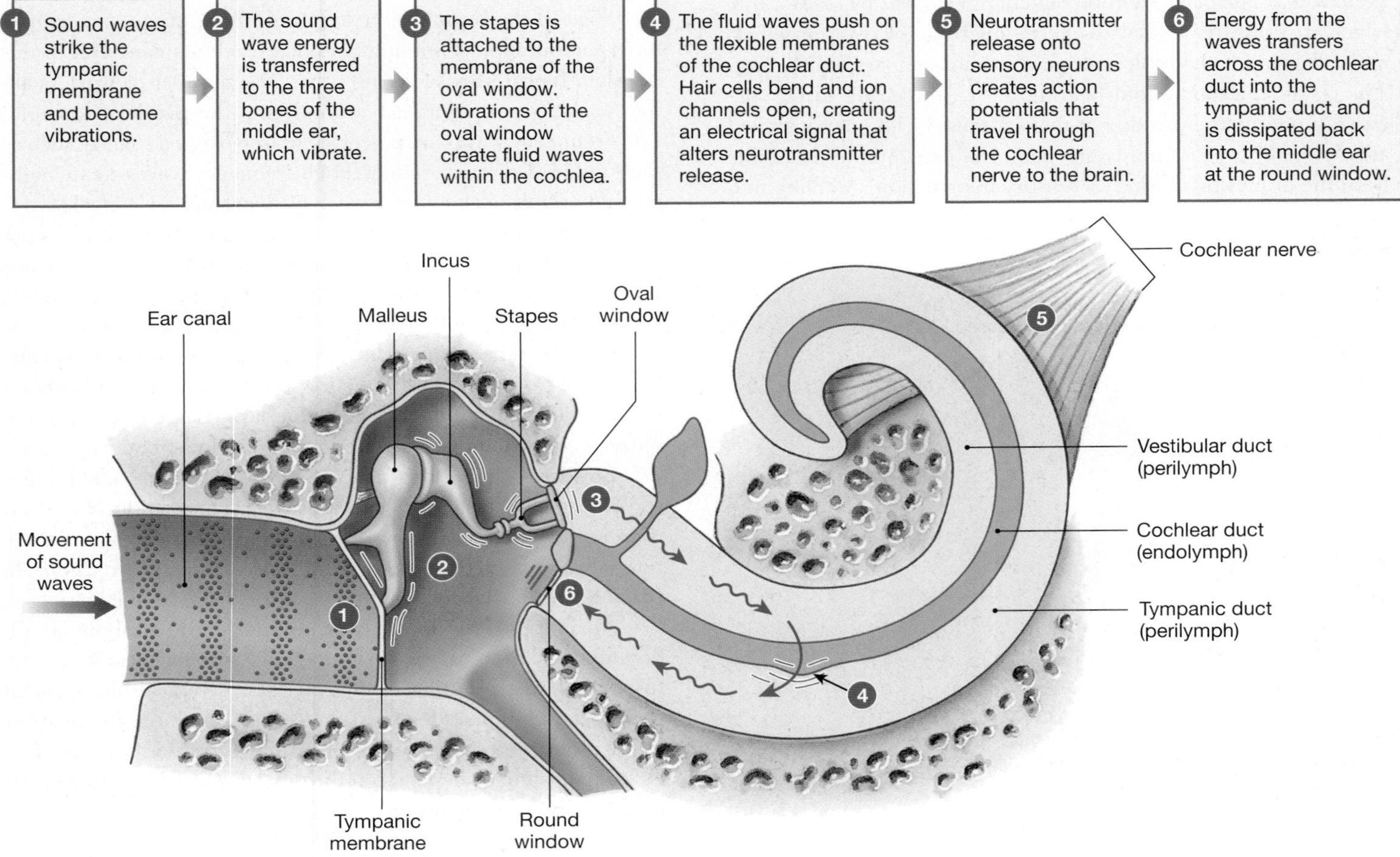

1 Sound waves strike the tympanic membrane and become vibrations.
2 The sound wave energy is transferred to the three bones of the middle ear, which vibrate.
3 The stapes is attached to the membrane of the oval window. Vibrations of the oval window create fluid waves within the cochlea.
4 The fluid waves push on the flexible membranes of the cochlear duct. Hair cells bend and ion channels open, creating an electrical signal that alters neurotransmitter release.
5 Neurotransmitter release onto sensory neurons creates action potentials that travel through the cochlear nerve to the brain.
6 Energy from the waves transfers across the cochlear duct into the tympanic duct and is dissipated back into the middle ear at the round window.
Cochlear nerve
Incus
Oval window
Ear canal
Malleus
Stapes
Vestibular duct (perilymph)
Cochlear duct (endolymph)
Tympanic duct (perilymph)
Movement of sound waves
Tympanic membrane
Round window

As the stapes vibrates, it pulls and pushes on the thin tissue of the oval window, to which it is attached. Vibrations at the oval window create waves in the fluid-filled channels of the cochlea (second transduction). As waves move through the cochlea, they push on the flexible membranes of the *cochlear duct* and bend sensory **hair cells** inside the duct. The wave energy dissipates back into the air of the middle ear at the round window.

Movement of the cochlear duct opens or closes ion channels on hair cell membranes, creating electrical signals (third transduction). These electrical signals alter neurotransmitter release (fourth transduction). Neurotransmitter binding to the primary auditory neurons initiates action potentials (fifth transduction) that send coded information about sound through the *cochlear branch of the vestibulocochlear nerve* (cranial nerve VIII) and the brain.

The Cochlea Is Filled with Fluid

As we have just seen, the transduction of wave energy into action potentials takes place in the cochlea of the inner ear. Uncoiled, the cochlea can be seen to be composed of three parallel, fluid-filled channels: (1) the **vestibular duct,** or *scala vestibuli* {*scala*, stairway; *vestibulum*, entrance}; (2) the central **cochlear duct,** or *scala media* {*media*, middle}; and (3) the **tympanic duct,** or *scala tympani* {*tympanon*, drum} (**FIG. 10.18**). The vestibular and tympanic ducts are continuous with each other, and they connect at the tip of the cochlea through a small opening known as the **helicotrema** {*helix*, a spiral + *trema*, hole}. The cochlear duct is a dead-end tube, but it connects to the vestibular apparatus through a small opening.

The fluid in the vestibular and tympanic ducts is similar in ion composition to plasma and is known as **perilymph.** The cochlear duct is filled with **endolymph** secreted by epithelial cells in the duct. Endolymph is unusual because it is more like intracellular fluid than extracellular fluid in composition, with high concentrations of K^+ and low concentrations of Na^+.

The cochlear duct contains the **organ of Corti,** composed of four rows of hair cell receptors plus support cells. The organ of Corti sits on the **basilar membrane** and is partially covered by the **tectorial membrane** {*tectorium*, a cover}, both flexible tissues that move in response to fluid waves passing through the vestibular duct (Fig. 10.18). As the waves travel through the cochlea, they displace basilar and tectorial membranes, creating up-and-down oscillations that bend the hair cells.

Hair cells, like taste receptor cells, are nonneural receptor cells. The apical surface of each hair cell is modified into 50–100 stiffened cilia known as **stereocilia,** arranged in ascending height

FIG. 10.18 Anatomy Summary . . . The Cochlea

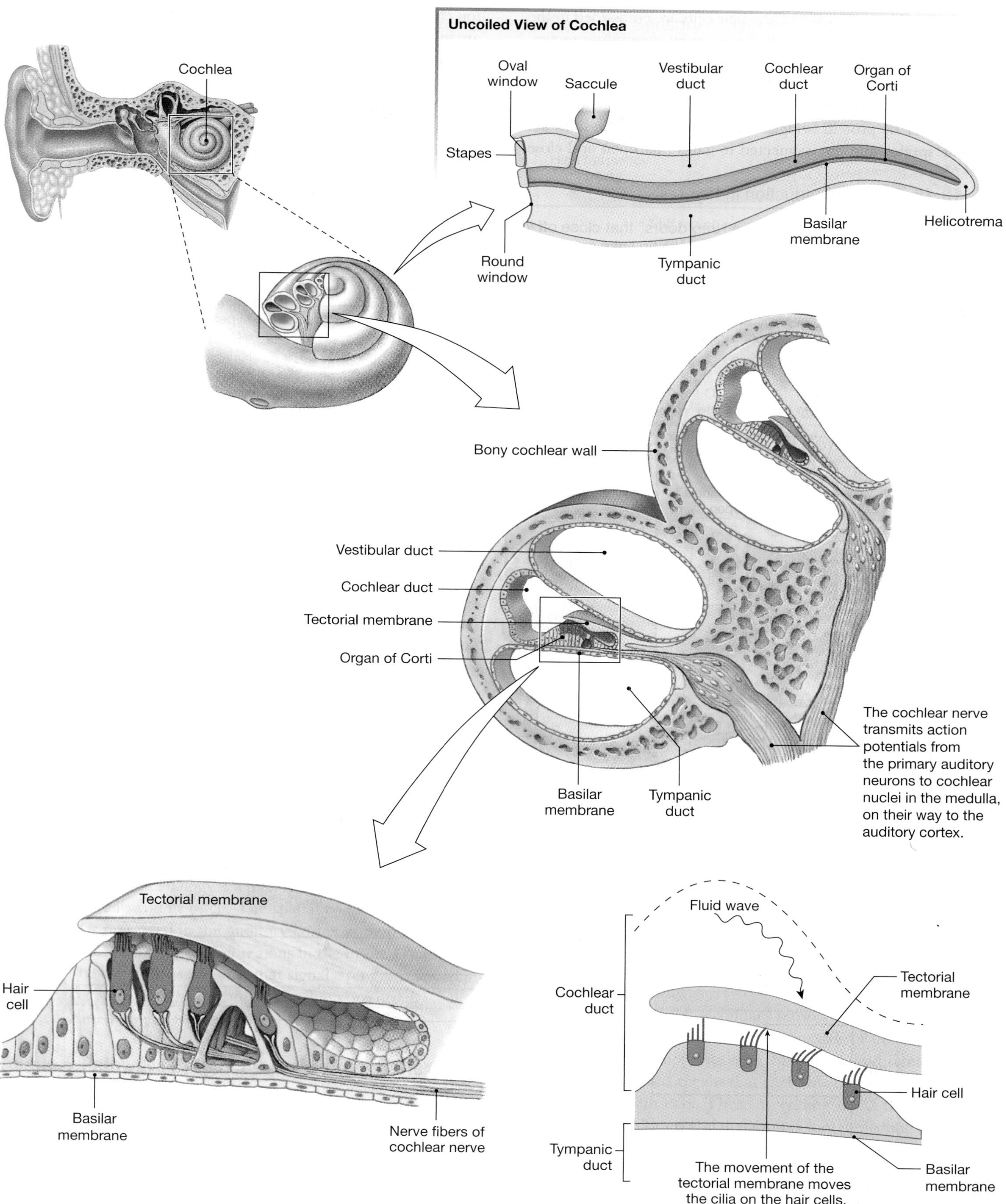

Cochlea
Uncoiled View of Cochlea
Oval window
Saccule
Vestibular duct
Cochlear duct
Organ of Corti
Stapes
Helicotrema
Basilar membrane
Round window
Tympanic duct
Bony cochlear wall
Vestibular duct
Cochlear duct
Tectorial membrane
Organ of Corti
Basilar membrane
Tympanic duct
The cochlear nerve transmits action potentials from the primary auditory neurons to cochlear nuclei in the medulla, on their way to the auditory cortex.
Tectorial membrane
Hair cell
Basilar membrane
Nerve fibers of cochlear nerve
Fluid wave
Cochlear duct
Tectorial membrane
Hair cell
Tympanic duct
The movement of the tectorial membrane moves the cilia on the hair cells.
Basilar membrane

FIG. 10.21 The auditory pathways

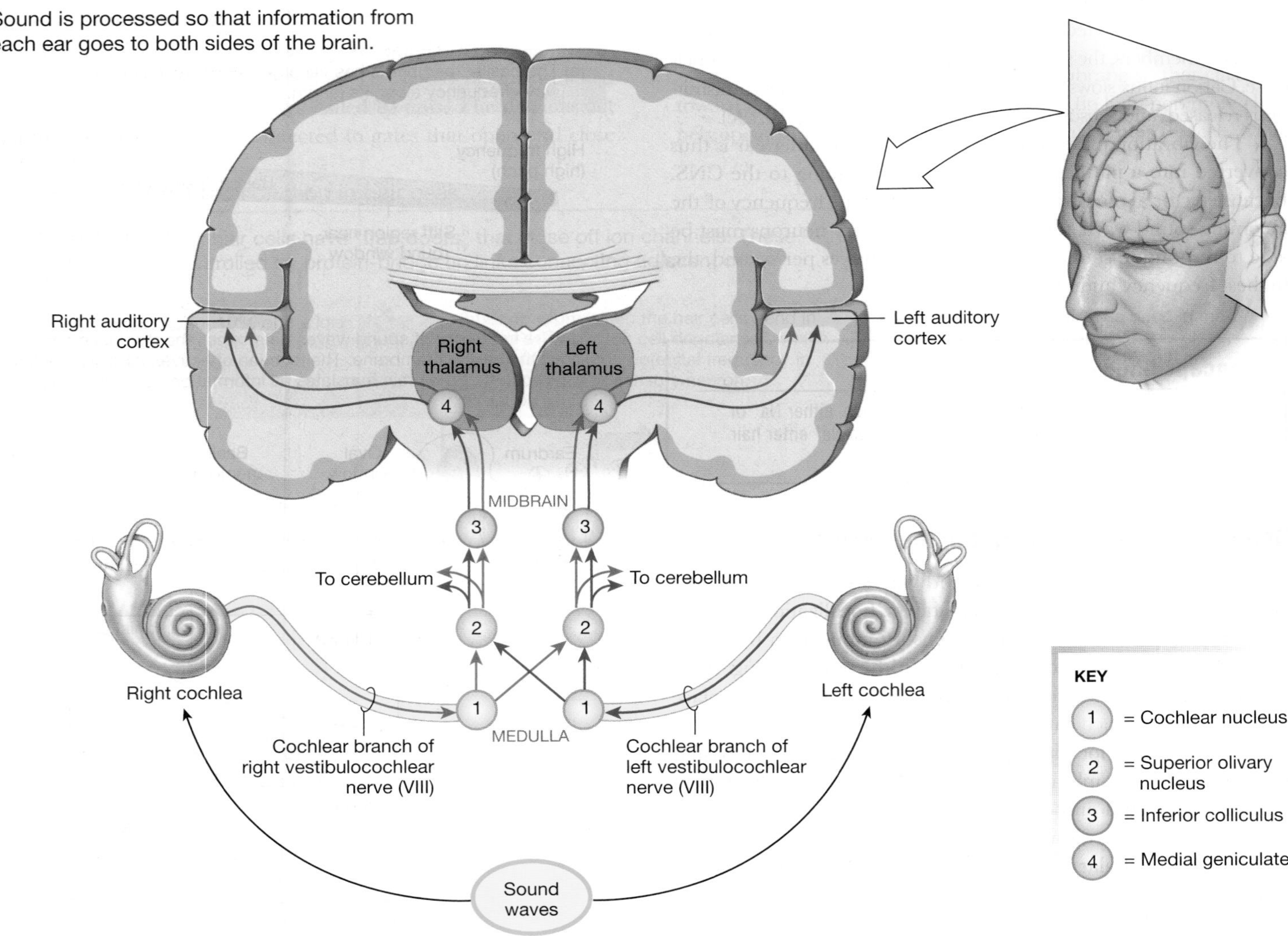
Sound is processed so that information from each ear goes to both sides of the brain.
Right auditory cortex
Left auditory cortex
Right thalamus
Left thalamus
4
4
MIDBRAIN
3
3
To cerebellum
To cerebellum
2
2
1
1
MEDULLA
Right cochlea
Left cochlea
Cochlear branch of right vestibulocochlear nerve (VIII)
Cochlear branch of left vestibulocochlear nerve (VIII)
Sound waves
KEY
1 = Cochlear nucleus
2 = Superior olivary nucleus
3 = Inferior colliculus
4 = Medial geniculate

is coming from directly in front of a person, it will not reach both ears at the same time (see Fig. 10.4). The brain records the time differential for sound arriving at the ears and uses complex computation to create a three-dimensional representation of the sound source.

Hearing Loss May Result from Mechanical or Neural Damage

There are three forms of hearing loss: conductive, central, and sensorineural. In *conductive hearing loss,* sound cannot be transmitted either through the external ear or the middle ear. The causes of conductive hearing loss range from an ear canal plugged with earwax (*cerumen*), to fluid in the middle ear from an infection, to diseases or trauma that impede vibration of the malleus, incus, or stapes. Correction of conductive hearing loss includes microsurgical techniques in which the bones of the middle ear can be reconstructed.

Central hearing loss results either from damage to the neural pathways between the ear and cerebral cortex or from damage to the cortex itself, as might occur from a stroke. This form of hearing loss is relatively uncommon.

Sensorineural hearing loss arises from damage to the structures of the inner ear, including death of hair cells as a result of loud noises. The loss of hair cells in mammals is currently irreversible. Birds and lower vertebrates, however, are able to regenerate hair cells to replace those that die. This discovery has researchers exploring strategies to duplicate the process in mammals. In recent experiments scientists treated organ of Corti support cells with a mixture of drugs and growth factors to create stem cells that could develop into hair cells. This work was a first step in the complicated process of creating new hair cells in the body.

The incidence of hearing loss in younger people is increasing because of prolonged exposure to rock music and environmental noises. Ninety percent of hearing loss in the elderly—called *presbycusis* {*presbys*, old man + *akoustikos*, able to be heard}—is sensorineural. Currently, the primary treatment for sensorineural hearing loss is the use of hearing aids, but amazing results have been obtained with cochlear implants. *Cochlear implants* consist of a microphone, tiny computerized speech processor, and transmitter that fit behind the ear like a conventional hearing aid plus a receiver and 8–24 electrodes surgically placed under the skin. The speech processor converts sound into electrical impulses that are passed along directly to the auditory nerve, bypassing any damaged areas in the inner ear.

Hearing is probably our most important social sense. Suicide rates are higher among deaf people than among those who have lost their sight. More than any other sense, hearing connects us to other people and to the world around us.

Concept Check

18. Map or diagram the pathways followed by a sound wave entering the ear, starting in the air at the outer ear and ending on the auditory cortex.
19. Why is somatosensory information projected to only one hemisphere of the brain but auditory information is projected to both hemispheres? (*Hint:* See Figs. 10.4 and 10.8.)
20. Would a cochlear implant help a person who suffers from nerve deafness? From conductive hearing loss?

10.5 The Ear: Equilibrium

Equilibrium is a state of balance, whether the word is used to describe ion concentrations in body fluids or the position of the body in space. The special sense of equilibrium has two components: a dynamic component that tells us about our movement through space, and a static component that tells us if our head is not in its normal upright position. Sensory information from the inner ear combined with signals from joint and muscle proprioceptors tell our brain where body parts are in relation to one another and to the environment. Visual information also plays an important role in equilibrium, as you know if you have ever gone to one of the 360° movie theaters where the scene tilts suddenly to one side and the audience tilts with it!

Our sense of equilibrium is mediated by hair cells lining the fluid-filled vestibular apparatus of the inner ear. These nonneural receptors respond to changes in rotational, vertical, and horizontal acceleration and positioning. The hair cells function just like those of the cochlea, but gravity and acceleration rather than sound waves provide the force that moves the stereocilia. Vestibular hair cells have a single long cilium called a **kinocilium** { *kinein*, to move } located at one side of the ciliary bundle. The kinocilium creates a reference point for the direction of bending.

When the cilia bend, tip links between them open and close ion channels. Movement in one direction causes the hair cells to depolarize; with movement in the opposite direction, they hyperpolarize. This is similar to what happens in cochlear hair cells (see Fig. 10.19).

The Vestibular Apparatus Provides Information about Movement and Position

The **vestibular apparatus,** also called the *membranous labyrinth*, is an intricate series of interconnected fluid-filled chambers. (In Greek mythology the labyrinth was a maze that housed a monster called the Minotaur.) In humans, the vestibular apparatus consists of two saclike **otolith organs**—the **saccule** and the **utricle**—along with three **semicircular canals** that connect to the utricle at their bases (**FIG. 10.22a**). The otolith organs tell us about *linear acceleration* and head position. The three semicircular canals sense *rotational acceleration* in various directions.

The vestibular apparatus, like the cochlear duct, is filled with high-K^+, low-Na^+ endolymph secreted by epithelial cells. Like cerebrospinal fluid, endolymph is secreted continuously and drains from the inner ear into the venous sinus in the dura mater of the brain.

If endolymph production exceeds the drainage rate, buildup of fluid in the inner ear may increase fluid pressure within the vestibular apparatus. Excessive accumulation of endolymph is believed to contribute to *Ménière's disease,* a condition marked by episodes of dizziness and nausea. If the organ of Corti in the cochlear duct is damaged by fluid pressure within the vestibular apparatus, hearing loss may result.

The Semicircular Canals Sense Rotational Acceleration

The three semicircular canals of the vestibular apparatus monitor rotational acceleration. They are oriented at right angles to one another, like three planes that come together to form the corner of a box (Fig. 10.22a). The horizontal canal monitors rotations that we associate with turning, such as an ice skater's spin or shaking your head left and right to say "no." The posterior canal monitors left-to-right rotation, such as the rotation when you tilt your head toward your shoulders or perform a cartwheel. The superior canal is sensitive to forward and back rotation, such as nodding your head front to back or doing a somersault.

At one end of each canal is an enlarged chamber, the **ampulla** { bottle }, which contains a sensory structure known as

RUNNING PROBLEM

Anant reports to the otolaryngologist that he never knows when his attacks of dizziness will strike and that they last from 10 minutes to an hour. They often cause him to vomit. He also reports that he has a persistent low buzzing sound in one ear and that he does not seem to hear low tones as well as he could before the attacks started. The buzzing sound (tinnitus) often gets worse during his dizzy attacks.

Q2: *Subjective tinnitus occurs when an abnormality somewhere along the anatomical pathway for hearing causes the brain to perceive a sound that does not exist outside the auditory system. Starting from the ear canal, name the auditory structures in which problems may arise.*

308 – 312 – 335 – 338 – 341 – 347 – 351

RUNNING PROBLEM

Although many vestibular disorders can cause the symptoms Anant is experiencing, two of the most common are positional vertigo and Ménière's disease. In *positional vertigo,* calcium crystals normally embedded in the otolith membrane of the maculae become dislodged and float toward the semicircular canals. The primary symptom of positional vertigo is brief episodes of severe dizziness brought on by a change in position, such as moving to the head-down yoga position called "downward-facing dog." People with positional vertigo often say they feel dizzy when they lie down or turn over in bed.

Q3: *When a person with positional vertigo changes position, the displaced crystals float toward the semicircular canals. Why would this cause dizziness?*

Q4: *Compare the symptoms of positional vertigo and Ménière's disease. On the basis of Anant's symptoms, which condition do you think he has?*

308
312
335
338
341
347
351

FIG. 10.24 External anatomy of the eye

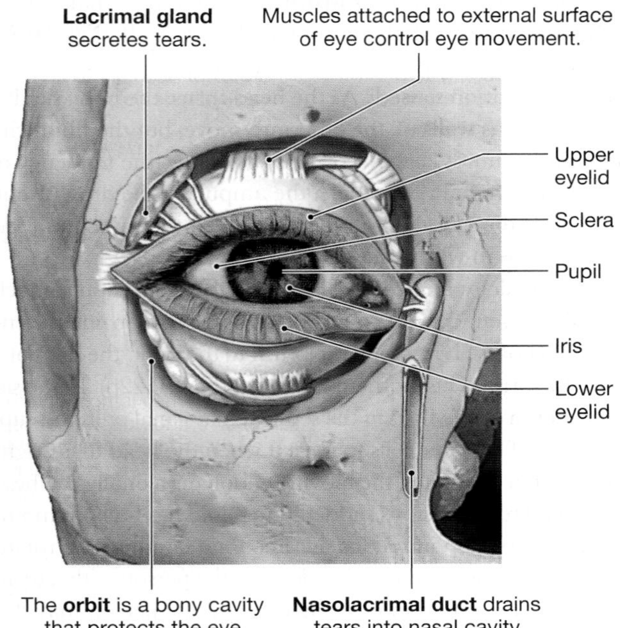
Lacrimal gland secretes tears.
Muscles attached to external surface of eye control eye movement.
Upper eyelid
Sclera
Pupil
Iris
Lower eyelid
The **orbit** is a bony cavity that protects the eye.
Nasolacrimal duct drains tears into nasal cavity.

10.6 The Eye and Vision

The eye is a sensory organ that functions much like a camera. It focuses light on a light-sensitive surface (the retina) using a lens and an aperture or opening (the pupil) whose size can be adjusted to change the amount of entering light. **Vision** is the process through which light reflected from objects in our environment is translated into a mental image. This process can be divided into three steps:

1. Light enters the eye, and the lens focuses the light on the retina.
2. Photoreceptors of the retina transduce light energy into an electrical signal.
3. Neural pathways from retina to brain process electrical signals into visual images.

The Skull Protects the Eye

The external anatomy of the eye is shown in **FIGURE 10.24**. Like sensory elements of the ears, the eyes are protected by a bony cavity, the *orbit,* which is formed by facial bones of the skull. Accessory structures associated with the eye include six *extrinsic eye muscles,* skeletal muscles that attach to the outer surface of the eyeball and control eye movements. Cranial nerves III, IV, and VI innervate these muscles.

The upper and lower *eyelids* close over the anterior surface of the eye, and the *lacrimal apparatus,* a system of glands and ducts, keeps a continuous flow of tears washing across the exposed surface so that it remains moist and free of debris. Tear secretion is stimulated by parasympathetic neurons from cranial nerve VII.

The **pupil** is an opening through which light can pass into the interior of the eye. Pupil size varies with the contraction and relaxation of a ring of smooth *pupillary muscle.* The pupil appears as the black spot inside the colored ring of pigment known as the **iris.** The pigments and other components of the iris determine eye color.

The eye itself is a hollow sphere divided into two compartments (chambers) separated by a lens (**FIG. 10.25**). The **lens,** suspended by ligaments called **zonules,** is a transparent disk that focuses light. The anterior chamber in front of the lens is filled with **aqueous humor** {*humidus,* moist}, a low-protein, plasma-like fluid secreted by the *ciliary epithelium* supporting the lens. Behind the lens is a much larger chamber, the **vitreous chamber,** filled mostly with the **vitreous body** {*vitrum,* glass; also called the *vitreous humor*}, a clear, gelatinous matrix that helps maintain the shape of the eyeball. The outer wall of the eyeball, the **sclera,** is composed of connective tissue.

Normally, aqueous humor secreted by the ciliary epithelium flows through the pupil and drains out through the *canal of Schlemm* in the anterior chamber of the eye (Fig. 10.25). If outflow is blocked, the aqueous humor accumulates, causing *intraocular* (within the eyeball) pressure to build up. Increased intraocular pressure is one risk factor for the eye disease *glaucoma,* characterized by degeneration of the optic nerve. Treatments to decrease intraocular pressure due to excess aqueous humor production either decrease fluid secretion or increase outflow.

Glaucoma is the leading cause of blindness worldwide. Not everyone with elevated pressure develops glaucoma, however, and a significant number of people with glaucoma have normal intraocular pressure. Research suggests that the optic nerve degeneration in glaucoma may be due to nitric oxide or apoptosis-inducing factors, and studies in these areas are underway.

FIG. 10.25 Anatomy Summary . . . The Eye

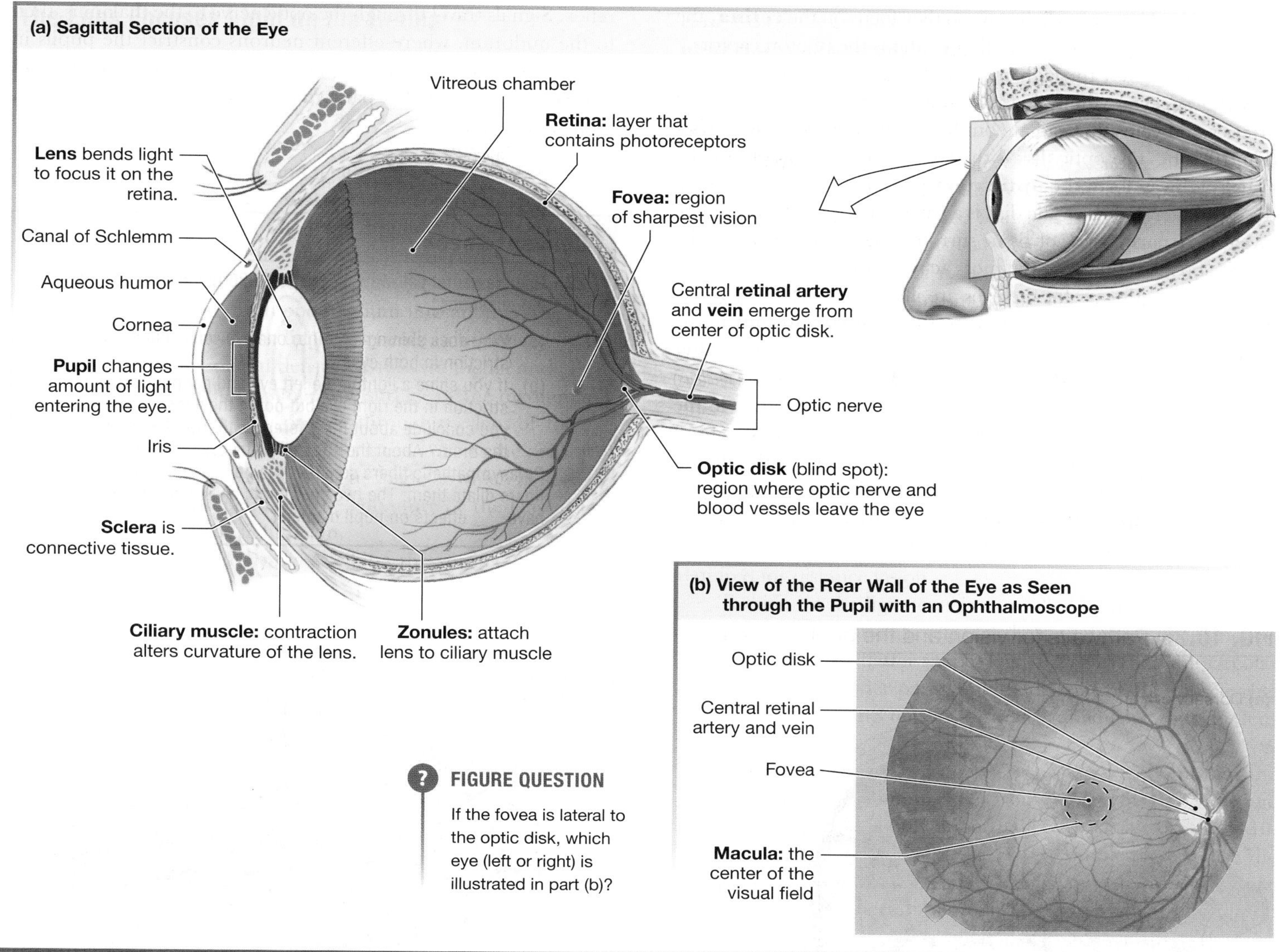

(a) Sagittal Section of the Eye
Vitreous chamber
Retina: layer that contains photoreceptors
Lens bends light to focus it on the retina.
Fovea: region of sharpest vision
Canal of Schlemm
Aqueous humor
Central retinal artery and vein emerge from center of optic disk.
Cornea
Pupil changes amount of light entering the eye.
Optic nerve
Iris
Optic disk (blind spot): region where optic nerve and blood vessels leave the eye
Sclera is connective tissue.
Ciliary muscle: contraction alters curvature of the lens.
Zonules: attach lens to ciliary muscle
(b) View of the Rear Wall of the Eye as Seen through the Pupil with an Ophthalmoscope
Optic disk
Central retinal artery and vein
Fovea
Macula: the center of the visual field
FIGURE QUESTION
If the fovea is lateral to the optic disk, which eye (left or right) is illustrated in part (b)?

Concept Check

24. What functions do the aqueous humor serve?

Light Enters the Eye through the Cornea

In the first step of the visual pathway, light from the environment enters the anterior surface of the eye through the **cornea,** a transparent disk of tissue that is a continuation of the sclera. This light is modified two ways. First, the amount of light that reaches photoreceptors is modulated by changes in the size of the pupil. Second, the light is focused by changes in the shape of the lens.

The human eye functions over a 100,000-fold range of light intensity. Most of this ability comes from the sensitivity of the photoreceptors, but the pupils assist by regulating the amount of light that falls on the retina. In bright sunlight, the pupils narrow to about 1.5 mm in diameter when a parasympathetic pathway constricts the circular *pupillary sphincter muscles*. In the dark, the opening of the pupil dilates to 8 mm, a 28-fold increase in pupil area. Dilation occurs when radial *dilator muscles* lying perpendicular to the circular sphincter muscles contract under the influence of sympathetic neurons.

In addition to regulating the amount of light that hits the retina, the pupils create what is known as **depth of field.** A simple example comes from photography. Imagine a picture of a puppy sitting in the foreground amid a field of wildflowers. If only the puppy and the flowers immediately around her are in focus, the picture is said to have a shallow depth of field. If the puppy and the wildflowers all the way back to the horizon are in focus, the picture has full depth of field. Full depth of field is created by constricting the pupil (or the diaphragm on a camera) so that only a narrow beam of light enters the eye. In this way, a greater depth of the image is focused on the retina.

After passing through the opening of the pupil, light strikes the lens, which has two convex surfaces. The cornea and lens

FIG. 10.27 ESSENTIALS Optics of the Eye

Light passing through a curved surface will bend or refract.

(a) A **concave lens** scatters light rays.

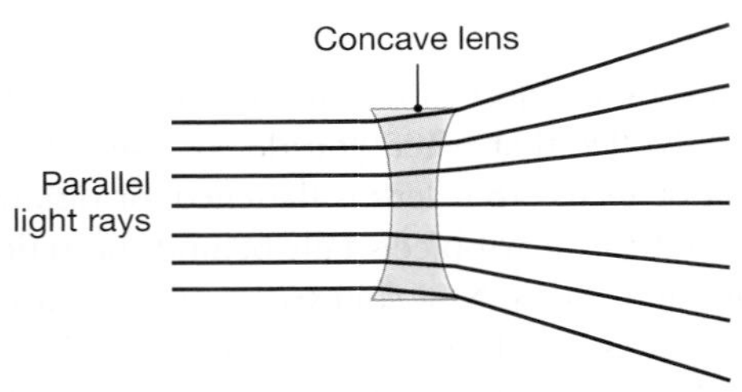
Concave lens
Parallel
light rays

(b) A **convex lens** causes light rays to converge.

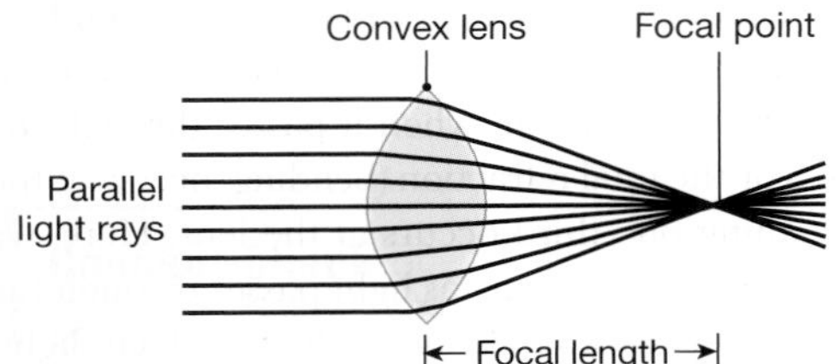
Convex lens
Focal point
Parallel
light rays
Focal length

The **focal length** of the lens is the distance from the center of the lens to the **focal point.**

For clear vision, the focal point must fall on the retina.

(c) Parallel light rays pass through a flattened lens, and the focal point falls on the retina.

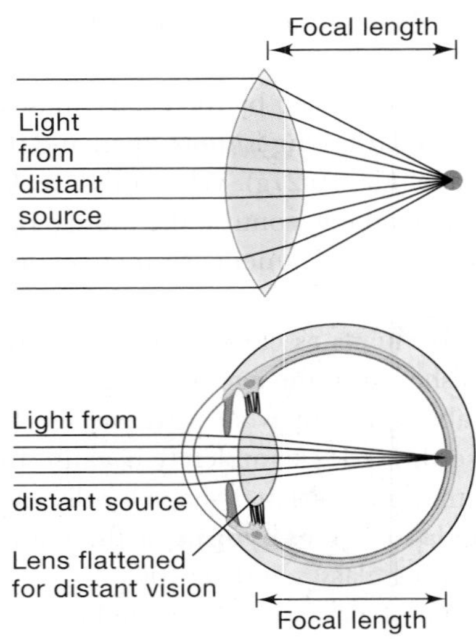
Focal length
Light
from
distant
source
Light from
distant source
Lens flattened
for distant vision
Focal length

(d) For close objects, the light rays are no longer parallel. The lens and its focal length have not changed, but the object is seen out of focus because the light beam is not focused on the retina.

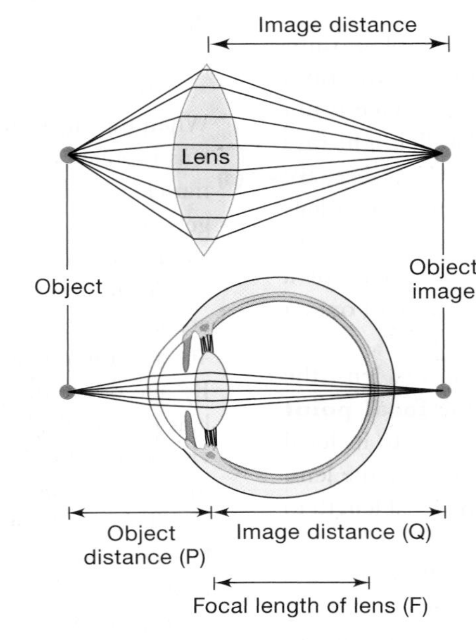
Image distance
Lens
Object
Object
image
Object
distance (P)
Image distance (Q)
Focal length of lens (F)

(e) To keep an object in focus as it moves closer, the lens becomes more rounded.

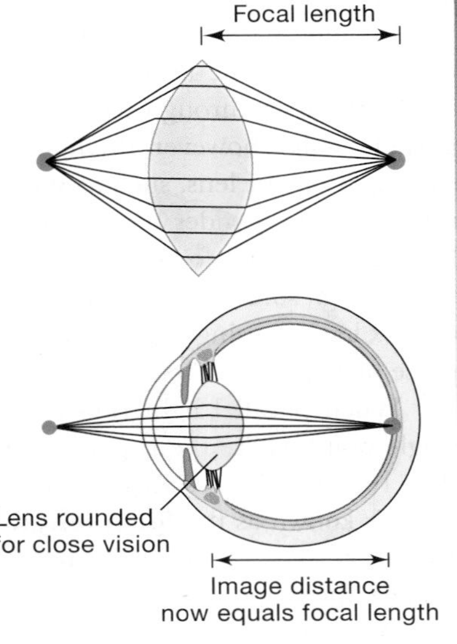
Focal length
Lens rounded
for close vision
Image distance
now equals focal length

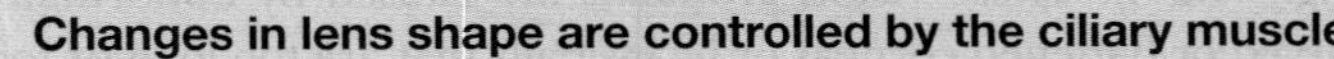

Changes in lens shape are controlled by the ciliary muscle.

(f) The lens is attached to the ciliary muscle by inelastic ligaments (zonules).

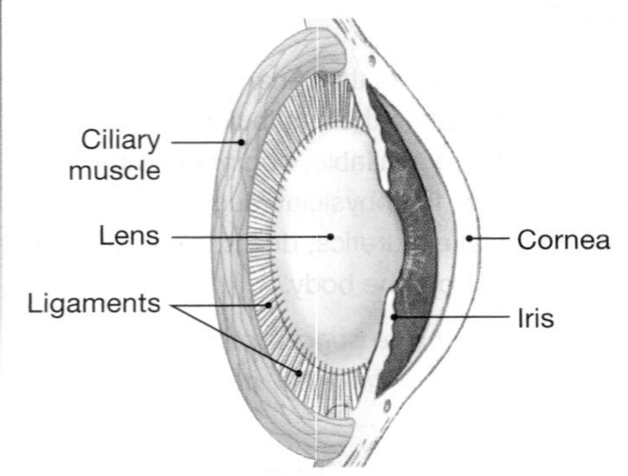
Ciliary
muscle
Lens
Ligaments
Cornea
Iris

(g) When ciliary muscle is relaxed, the ligaments pull on and flatten the lens.

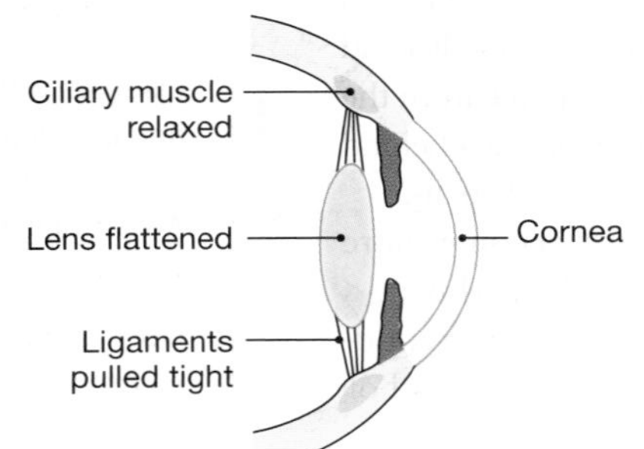
Ciliary muscle
relaxed
Lens flattened
Ligaments
pulled tight
Cornea

(h) When ciliary muscle contracts, it releases tension on the ligaments and the lens becomes more rounded.

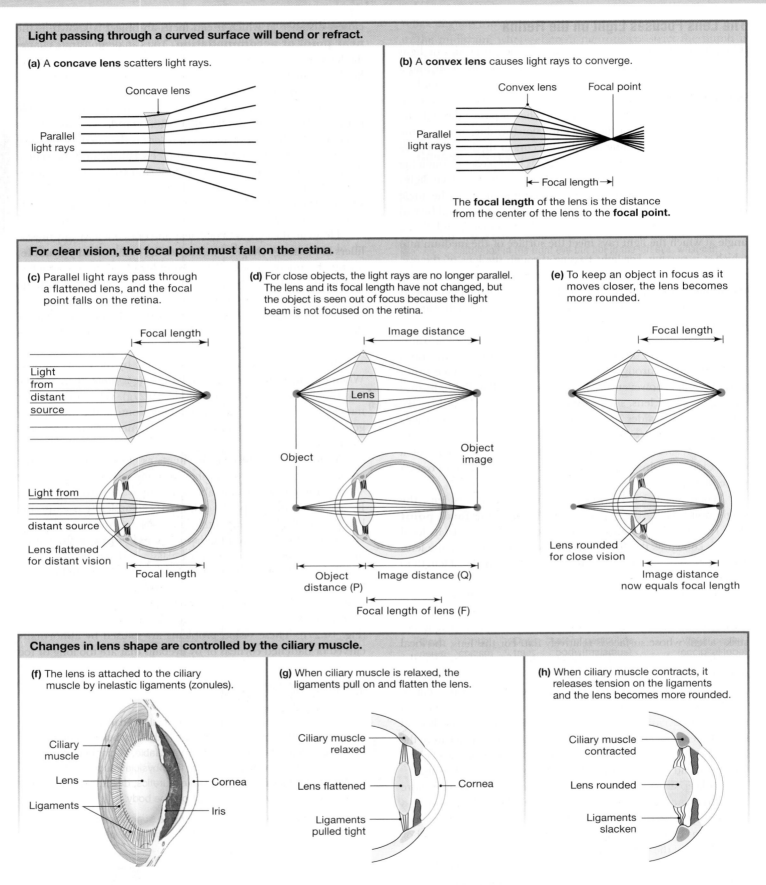

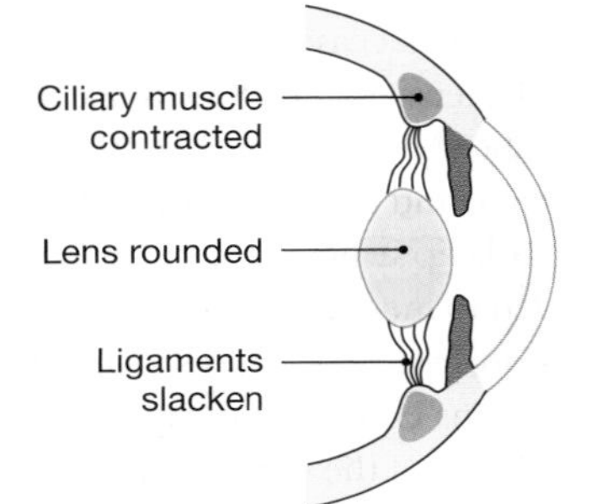
Ciliary muscle
contracted
Lens rounded
Ligaments
slacken

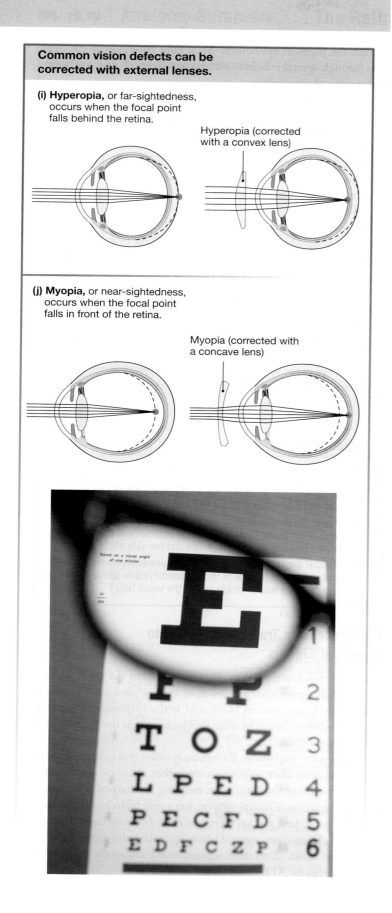

Common vision defects can be corrected with external lenses.

(i) Hyperopia, or far-sightedness, occurs when the focal point falls behind the retina.

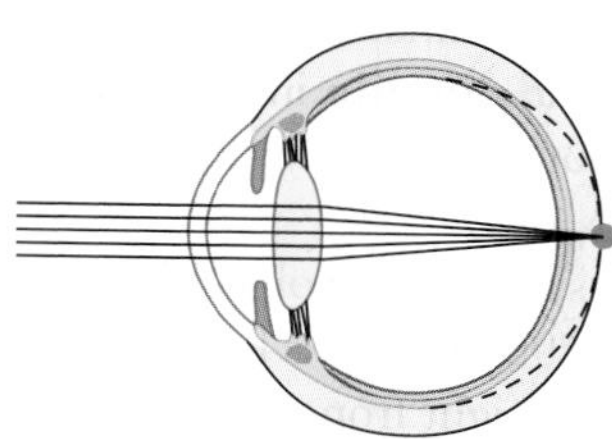

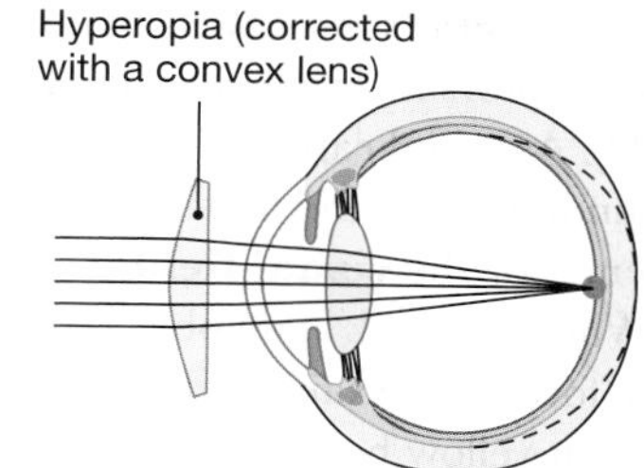
Hyperopia (corrected with a convex lens)

(j) Myopia, or near-sightedness, occurs when the focal point falls in front of the retina.

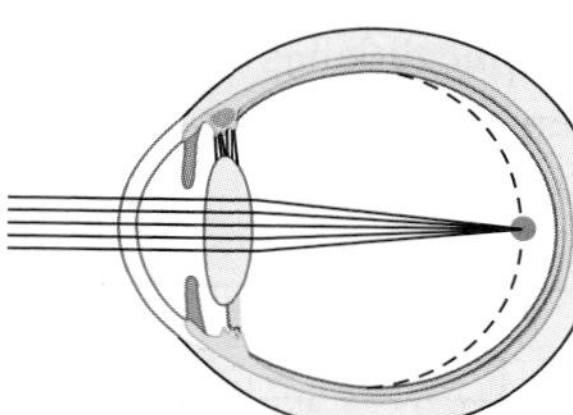

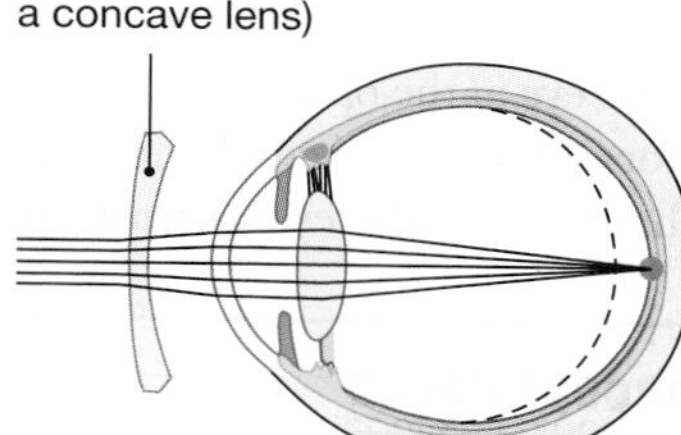
Myopia (corrected with a concave lens)

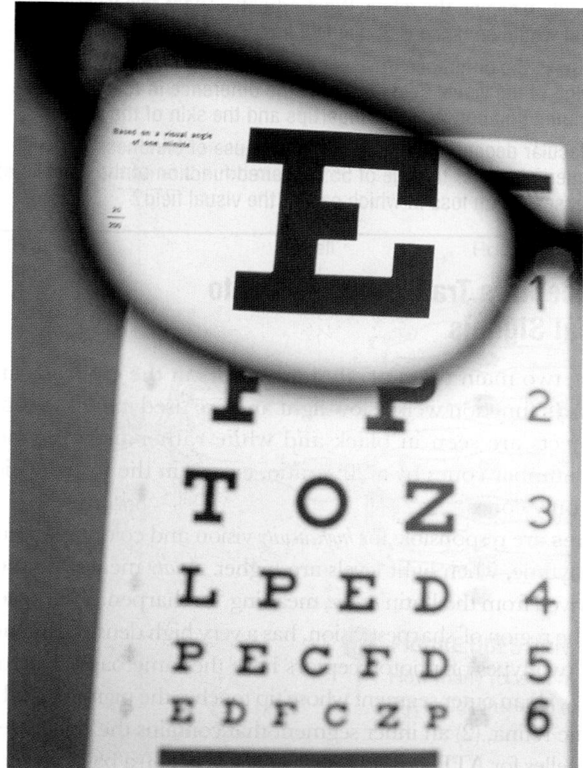
E
1
F P
2
T O Z
3
L P E D
4
P E C F D
5
E D F C Z P
6

short. Placing a lens with the appropriate curvature in front of the eye changes the refraction of light entering the eye and corrects the problem. A third common vision problem, **astigmatism,** is usually caused by a cornea that is not a perfectly shaped dome, resulting in distorted images.

Concept Check

27. If a person's cornea, which helps focus light, is more rounded than normal (has a greater curvature), is this person more likely to be hyperopic or myopic? (*Hint:* See Fig. 10.27.)

28. The relationship between the focal length of a lens (F), the distance between an object and the lens (P), and the distance from the lens to the object's image (Q) is expressed as $1/F = 1/P + 1/Q$.

(a) If the focal length of a lens does not change but an object moves closer to the lens, what happens to the image distance Q?

(b) If an object moves closer to the lens and the image distance Q must stay the same for the image to fall on the retina, what must happen to the focal length F of the lens? For this change in F to occur, should the lens become flatter or more rounded?

29. (a) Explain how convex and concave corrective lenses change the refraction of light.

(a) Which type of corrective lens should be used for myopia, and why? For hyperopia?

Phototransduction Occurs at the Retina

Once light hits the retina, photoreceptors convert the light energy into electrical signals. Light energy is part of the electromagnetic spectrum, which ranges from high-energy, very-short-wavelength waves such as X-rays and gamma rays to low-energy, lower-frequency microwaves and radio waves (**FIG. 10.28**). However, our brains can perceive only a small portion of this broad energy

FIG. 10.28 The electromagnetic spectrum

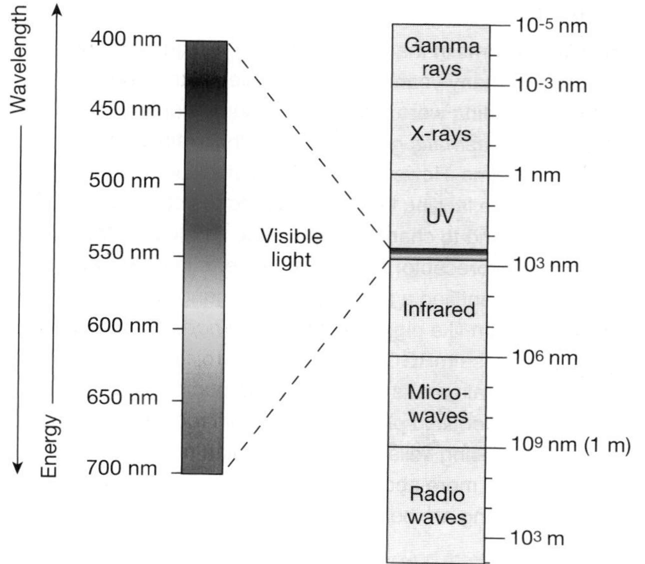
Wavelength
Energy
400 nm
450 nm
500 nm
550 nm
600 nm
650 nm
700 nm
Visible light
Gamma rays
X-rays
UV
Infrared
Micro-waves
Radio waves
10-5 nm
10-3 nm
1 nm
103 nm
106 nm
109 nm (1 m)
103 m

FIG. 10.30 Photoreceptors: Rods and cones

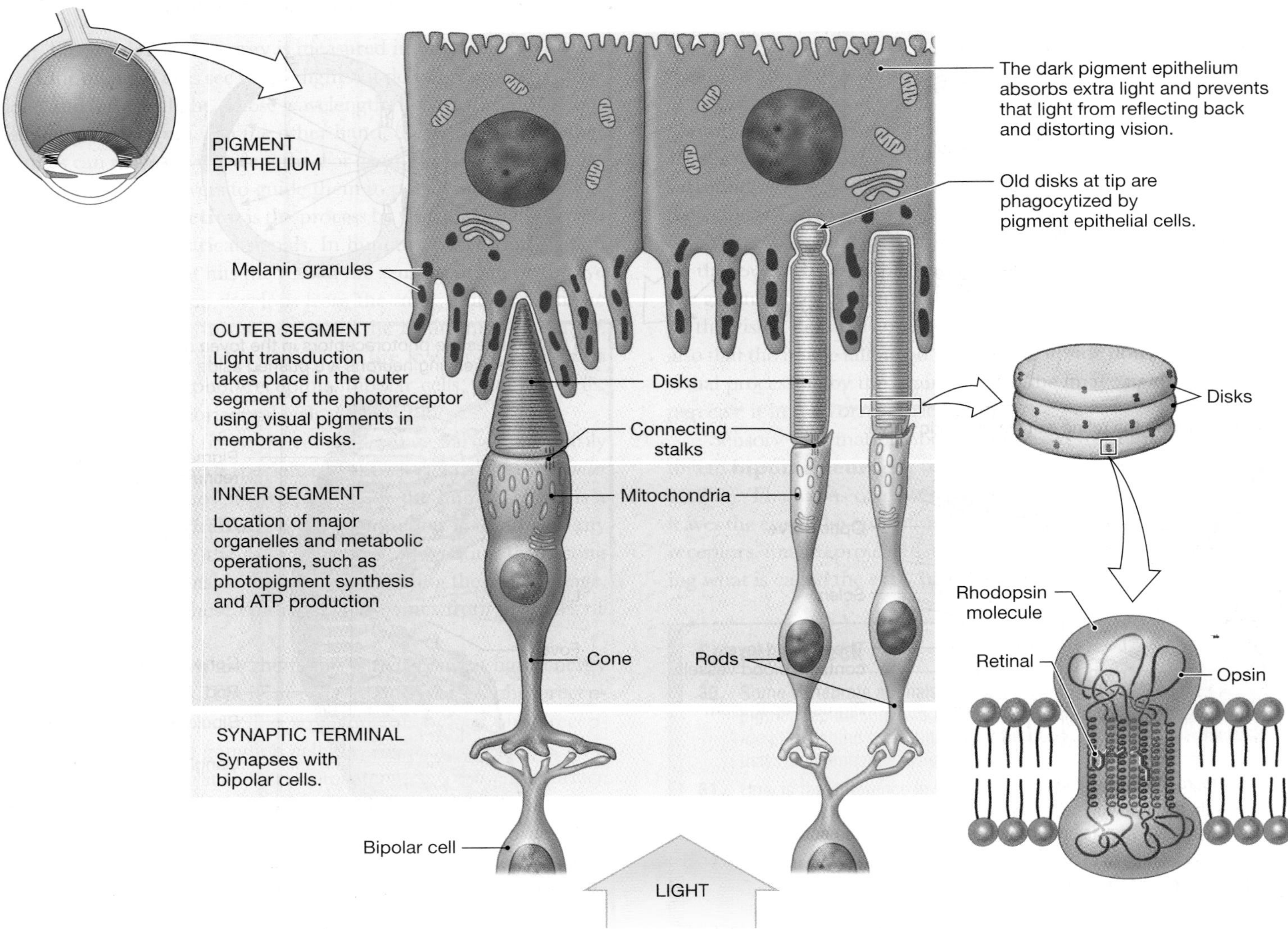
PIGMENT
EPITHELIUM
The dark pigment epithelium absorbs extra light and prevents that light from reflecting back and distorting vision.
Old disks at tip are phagocytized by pigment epithelial cells.
Melanin granules
OUTER SEGMENT
Light transduction takes place in the outer segment of the photoreceptor using visual pigments in membrane disks.
Disks
Connecting stalks
Disks
INNER SEGMENT
Location of major organelles and metabolic operations, such as photopigment synthesis and ATP production
Mitochondria
Rhodopsin molecule
Retinal
Opsin
Cone
Rods
SYNAPTIC TERMINAL
Synapses with bipolar cells.
Bipolar cell
LIGHT

these layers actually separate from the cell membrane and form free-floating membrane disks. In the cones, the disks stay attached.

Light-sensitive **visual pigments** are bound to the disk membranes in outer segments of photoreceptors. These visual pigments are transducers that convert light energy into a change in membrane potential. Rods have one type of visual pigment, **rhodopsin.** Cones have three different pigments that are closely related to rhodopsin.

The visual pigments of cones are excited by different wavelengths of light, allowing us to see in color. White light is a combination of colors, as demonstrated when you separate white light by passing it through a prism. The eye contains cones for red, green, and blue light. Each cone type is stimulated by a range of light wavelengths but is most sensitive to a particular wavelength (FIG. 10.31). Red, green, and blue are the three primary colors that make the colors of visible light, just as red, blue, and yellow are the three primary colors that make different colors of paint.

The color of any object we are looking at depends on the wavelengths of light reflected by the object. Green leaves reflect green light, and bananas reflect yellow light. White objects reflect most wavelengths. Black objects absorb most wavelengths, which is one reason they heat up in sunlight while white objects stay cool.

Our brain recognizes the color of an object by interpreting the combination of signals coming to it from the three different color cones. The details of color vision are still not fully understood, and there is some controversy about how color is processed in the cerebral cortex. **Color-blindness** is a condition in which a person inherits a defect in one or more of the three types of cones and has difficulty distinguishing certain colors. Probably the best-known form of color-blindness is red-green, in which people have trouble telling red and green apart.

Concept Check

33. Why is our vision in the dark in black and white rather than in color?

FIG. 10.31 Light absorption by visual pigments

There are three types of cone pigment, each with a characteristic light absorption spectrum. Rods are for black and white vision in low light.

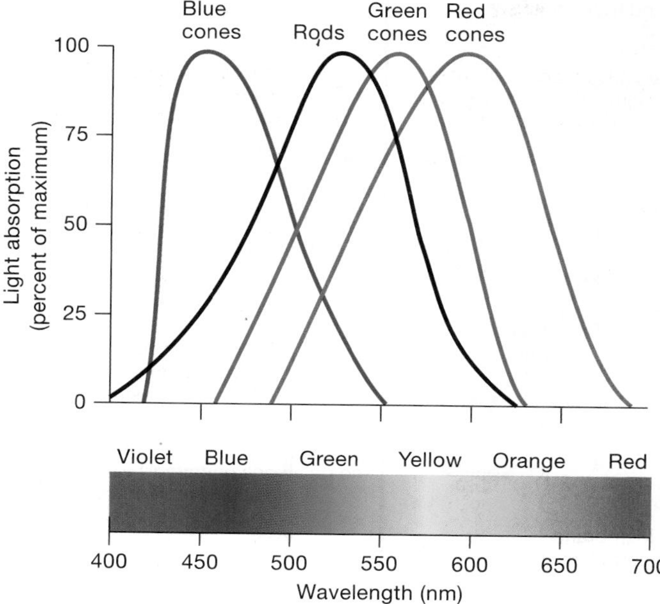
Blue cones
Rods
Green cones
Red cones
100
75
50
25
0
Light absorption (percent of maximum)
Violet
Blue
Green
Yellow
Orange
Red
400
450
500
550
600
650
700
Wavelength (nm)

FIGURE QUESTION

1. Which pigment absorbs light over the broadest spectrum of wavelengths?
2. Over the narrowest?
3. Which cone pigment absorbs the most light at 500 nm?

Phototransduction The process of phototransduction is similar for rhodopsin (in rods) and the three color pigments (in cones). Rhodopsin is composed of two molecules: **opsin,** a protein embedded in the membrane of the rod disks, and **retinal,** a vitamin A derivative that is the light-absorbing portion of the pigment (see Fig. 10.30). In the absence of light, retinal binds snugly into a binding site on the opsin (Fig. 10.32). When activated by as little as one photon of light, retinal changes shape to a new configuration. The activated retinal no longer binds to opsin and is released from the pigment in the process known as **bleaching.**

How does rhodopsin bleaching lead to action potentials traveling through the optical pathway? To understand the pathway, we must look at other properties of the rods. Electrical signals in cells occur as a result of ion movement between the intracellular and extracellular compartments. Rods contain three main types of cation channels: **cyclic nucleotide-gated (CNG) channels** that allow Na^+ and Ca^{2+} to enter the rod, K^+ channels that allow K^+ to leak out of the rod, and voltage-gated Ca^{2+} channels in the synaptic terminal that help regulate exocytosis of neurotransmitter.

When a rod is in darkness and rhodopsin is not active, cyclic GMP (cGMP) levels in the rod are high, and both CNG and K^+ channels are open (**FIG. 10.32** 1). Sodium and Ca^{2+} ion influx is greater than K^+ efflux, so the rod stays depolarized to an average membrane potential of −40 mV (instead of the more usual −70 mV). At this slightly depolarized membrane potential, the voltage-gated Ca^{2+} channels are open and there is tonic (continuous) release of the neurotransmitter glutamate from the synaptic portion of the rod onto the adjacent bipolar cell.

When light activates rhodopsin, a second-messenger cascade is initiated through the G protein **transducin** (Fig. 10.32 2). (Transducin is closely related to gustducin, the G protein found in type II taste receptor cells.) The transducin second-messenger cascade decreases the concentration of cGMP, which closes the CNG channels. As a result, cation influx slows or stops.

With decreased cation influx and continued K^+ efflux, the inside of the rod hyperpolarizes, and glutamate release onto the bipolar neurons decreases. Bright light closes all CNG channels and stops all neurotransmitter release. Dimmer light causes a response that is graded in proportion to the light intensity.

After activation, retinal diffuses out of the rod and is transported into the pigment epithelium. There it reverts to its inactive form before moving back into the rod and being reunited with opsin (Fig. 10.32 3). The recovery of rhodopsin from bleaching can take some time and is a major factor in the slow adaptation of the eyes when moving from bright light into the dark.

Concept Check

34. Draw a map or diagram to explain phototransduction. Start with bleaching and end with release of neurotransmitter.

Signal Processing Begins in the Retina

We now move from the cellular mechanism of light transduction to the processing of light signals by the retina and brain, the third and final step in our vision pathway. Signal processing in the retina is an excellent example of convergence [p. 258], in which multiple neurons synapse onto a single

RUNNING PROBLEM

Anant's condition does not improve with the low-salt diet and diuretics, and he continues to suffer from disabling attacks of vertigo with vomiting. In severe cases of Ménière's disease, surgery is sometimes performed when less invasive treatments have failed. In one surgical procedure for the disease, a drain is inserted to relieve pressure in the endolymph by removing some of the fluid. If that fails to provide relief, as a last resort the vestibular nerve can be severed. This surgery is difficult to perform, as the vestibular nerve lies near many other important nerves, including facial nerves and the auditory nerve. Patients who undergo this procedure are advised that the surgery can result in deafness if the cochlear nerve is inadvertently severed.

Q6: *Why would severing the vestibular nerve alleviate Ménière's disease?*

308 – 312 – 335 – 338 – 341 – 347 – 351

The receptive field of a ganglion cell near the fovea is quite small. Only a few photoreceptors are associated with each ganglion cell, and so visual acuity is greatest in these areas. At the edge of the retina, multiple photoreceptors converging onto a single ganglion cell results in vision that is not as sharp (Fig. 10.33a).

An analogy for this arrangement is to think of pixels on your computer screen. Assume that two screens have the same number of "photoreceptors," as indicated by a maximal screen resolution of 1280 × 1024 pixels. If screen A has one photoreceptor becoming one "ganglion cell" pixel, the actual screen resolution is 1280 × 1024, and the image is very clear. If eight photoreceptors on screen B converge into one ganglion cell pixel, then the actual screen resolution falls to 160 × 128, resulting in a very blurry and perhaps indistinguishable image.

Receptive fields of ganglion cells are roughly circular (unlike the irregular shape of somatic sensory receptive fields) and are divided into sections: a round center and its doughnut-shaped **surround** (Fig. 10.33b). This organization allows each ganglion cell to use contrast between the center and its surround to interpret visual information. Strong contrast between the center and surround elicits a strong excitatory response (a series of action potentials) or a strong inhibitory response (no action potentials) from the ganglion cell. Weak contrast between center and surround gets an intermediate response.

There are two types of ganglion cell receptive fields. In an *on-center/off-surround field,* the associated ganglion cell responds most strongly when light is brightest in the center of the field (Fig. 10.33c). If light is brightest in the off-surround region of the field, the on-center/off-surround field ganglion cell is inhibited and stops firing action potentials. The reverse happens with *off-center/on-surround fields.*

What happens if light is uniform across a receptive field? In that case, the ganglion cell responds weakly. Thus, the retina uses *contrast* rather than absolute light intensity to recognize objects in the environment. One advantage of using contrast is that it allows better detection of weak stimuli.

Scientists have now identified multiple types of ganglion cells in the primate retina. The two predominant types, which account for 80% of retinal ganglion cells, are M cells and P cells. Large *magnocellular* ganglion cells, or **M cells,** are more sensitive to information about movement. Smaller *parvocellular* ganglion cells, or **P cells,** are more sensitive to signals that pertain to form and fine detail, such as the texture of objects in the receptive field. A recently discovered subtype of ganglion cell, the *melanopsin retinal ganglion cell,* apparently also acts as a photoreceptor to relay information about light cycles to the suprachiasmatic nucleus, which controls circadian rhythms [p. 17].

Processing Beyond the Retina Once action potentials leave ganglion cell bodies, they travel along the optic nerves to the CNS for further processing. As noted earlier, the optic nerves enter the brain at the optic chiasm. At this point, some nerve fibers from each eye cross to the other side of the brain for processing. **FIGURE 10.34** shows how information from the right side of each eye's visual field is processed on the left side of the brain, and information from the left side of the field is processed on the right side of the brain.

FIG. 10.34 Binocular vision

The left visual field of each eye is projected to the visual cortex on the right side of the brain, and the right visual field is projected to the left visual cortex. Objects seen by both eyes fall within the binocular zone and are perceived in three dimensions. Objects seen with only one eye fall outside the binocular zone and are perceived in only two dimensions.

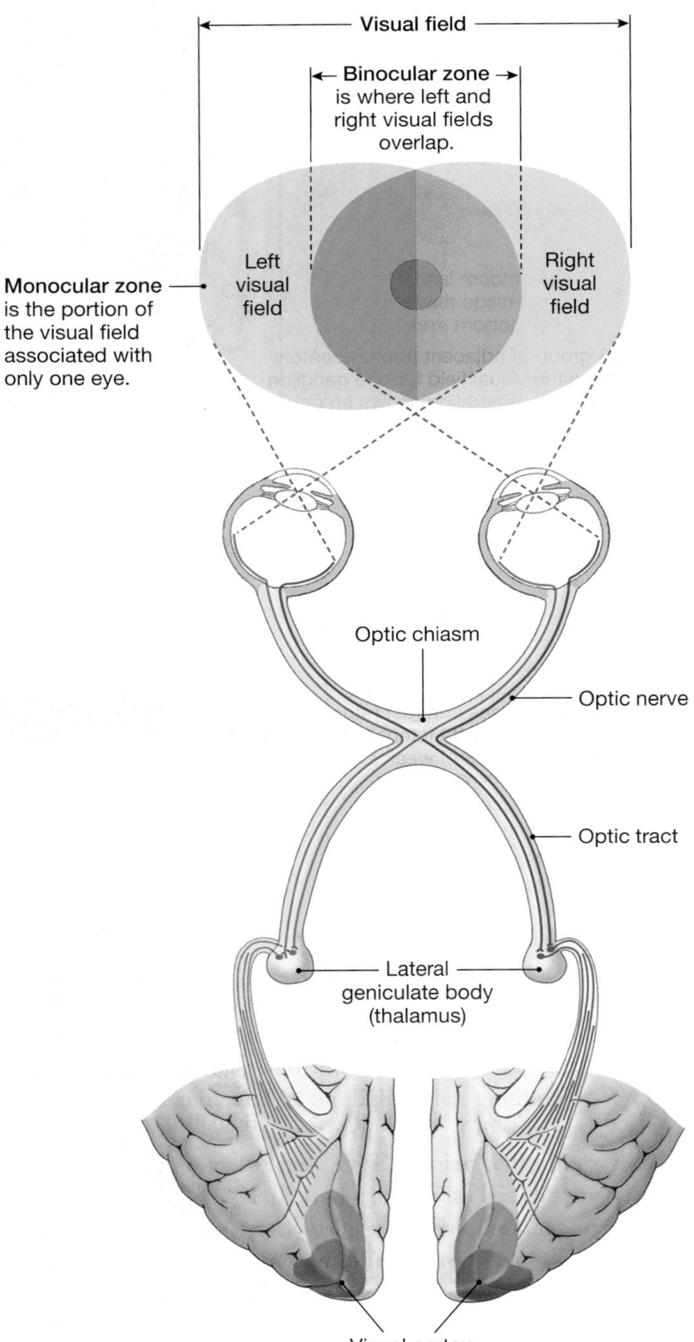
Visual field
Binocular zone is where left and right visual fields overlap.
Monocular zone is the portion of the visual field associated with only one eye.
Left visual field
Right visual field
Optic chiasm
Optic nerve
Optic tract
Lateral geniculate body (thalamus)
Visual cortex

The central portion of the visual field, where left and right sides of each eye's visual field overlap, is the **binocular zone.** The two eyes have slightly different views of objects in this region, and the brain processes and integrates the two views to create three-dimensional representations of the objects. Our sense of depth perception—that is, whether one object is in front of or behind another—depends on binocular vision. Objects that fall within the visual field of only one eye are in the **monocular zone** and are viewed in two dimensions.

Once axons leave the optic chiasm, some fibers project to the midbrain, where they participate in control of eye movement or coordinate with somatosensory and auditory information for balance and movement (see Fig. 10.26). Most axons, however, project to the lateral geniculate body of the thalamus, where the optic fibers synapse onto neurons leading to the visual cortex in the occipital lobe.

The lateral geniculate body is organized in layers that correspond to the different parts of the visual field, which means that information from adjacent objects is processed together. This **topographical organization** is maintained in the visual cortex, with the six layers of neurons grouped into vertical columns. Within each portion of the visual field, information is further sorted by form, color, and movement.

The cortex merges monocular information from the two eyes to give us a binocular view of our surroundings. Information from on/off combinations of ganglion cells is translated into sensitivity to line orientation in the simplest pathways, or into color, movement, and detailed structure in the most complex. Each of these attributes of visual stimuli is processed through a separate pathway, creating a network whose complexity we are just beginning to unravel.

RUNNING PROBLEM **CONCLUSION** Méniére's Disease

Anant was told about the surgical options but elected to continue medical treatment for a little longer. Over the next two months, his Ménière's disease gradually resolved. The cause of Ménière's disease is still unknown, which makes treatment difficult. To learn more about treatments that are available to alleviate Ménière's disease, do an Internet search. Now check your understanding of this running problem by comparing your answers to those in the summary table.

Question	Facts	Integration and Analysis
Q1: *In which part of the brain is sensory information about equilibrium processed?*	The major equilibrium pathways project to the cerebellum. Some information is also processed in the cerebrum.	N/A
Q2: *Subjective tinnitus occurs when an abnormality occurs somewhere along the anatomical pathway for hearing. Starting from the ear canal, name the auditory structures in which problems may arise.*	The middle ear consists of malleus, incus, and stapes, bones that vibrate with sound. The hearing portion of the inner ear consists of hair cells in the fluid-filled cochlea. The cochlear (auditory) nerve leads to the brain.	Subjective tinnitus could arise from a problem with any of the structures named. Abnormal bone growth can affect the middle ear bones. Excessive fluid accumulation in the inner ear will affect the hair cells. Neural defects may cause the cochlear nerve to fire spontaneously, creating the perception of sound.
Q3: *When a person with positional vertigo changes position, the displaced crystals float toward the semicircular canals. Why would this cause dizziness?*	The ends of the semicircular canals contain sensory cristae, each crista consisting of a cupula with embedded hair cells. Displacement of the cupula creates a sensation of rotational movement.	If the floating crystals displace the cupula, the brain will perceive movement that is not matched to sensory information coming from the eyes. The result is vertigo, an illusion of movement.
Q4: *Compare the symptoms of positional vertigo and Ménière's disease. On the basis of Anant's symptoms, which condition do you think he has?*	The primary symptom of positional vertigo is brief dizziness following a change in position. Ménière's disease combines vertigo with tinnitus and hearing loss.	Anant complains of dizzy attacks typically lasting up to an hour that come on without warning, making it more likely that Anant has Ménière's disease.
Q5: *Why is limiting salt (NaCl) intake suggested as a treatment for Ménière's disease?*	Ménière's disease is characterized by too much endolymph in the inner ear. Endolymph is an extracellular fluid.	Reducing salt intake should also reduce the amount of fluid in the extracellular compartment because the body will retain less water. Reduction of ECF volume may decrease fluid accumulation in the inner ear.
Q6: *Why would severing the vestibular nerve alleviate Ménière's disease?*	The vestibular nerve transmits information about balance and rotational movement from the vestibular apparatus to the brain.	Severing the vestibular nerve prevents false information about body rotation from reaching the brain, thus alleviating the vertigo of Ménière's disease.

308 - 312 - 335 - 338 - 341 - 347 - 351

20. List the following structures in the sequence in which a beam of light entering the eye will encounter them: (a) aqueous humor, (b) cornea, (c) lens, (d) pupil, (e) retina.
21. The three primary colors of vision are __________, __________, and __________. White light containing these colors stimulates photoreceptors called __________. Lack of the ability to distinguish some colors is called __________.
22. List six types of cells found in the retina, and briefly describe their functions.

Level Two Reviewing Concepts

23. Compare and contrast the following:
    (a) the special senses with the somatic senses
    (b) different types of touch receptors with respect to structure, size, and location
    (c) transmission of sharp localized pain with transmission of dull and diffuse pain (include the particular fiber types involved as well as the presence or absence of myelin in your discussion)
    (d) the forms of hearing loss
    (e) convergence of retinal neurons with convergence of primary somatic sensory neurons
24. Draw three touch receptors having overlapping receptive fields (see Fig. 10.2) and number the fields 1–3. Draw a primary and secondary sensory neuron for each receptor so they have separate ascending pathways to the cortex. Use the information in your drawing to answer this question: How many different regions of the skin can the brain distinguish using input from these three receptors?
25. Describe the neural pathways that link pain with emotional distress, nausea, and vomiting.
26. Trace the neural pathways involved in olfaction. What is G_{olf}?
27. Compare the current models of signal transduction in taste buds for salty/sour ligands and sweet/bitter/umami ligands.
28. Put the following structures in the order in which a sound wave would encounter them: (a) pinna, (b) cochlear duct, (c) stapes, (d) ion channels, (e) oval window, (f) hair cells/stereocilia, (g) tympanic membrane, (h) incus, (i) vestibular duct, (j) malleus
29. Draw the structures and receptors of the vestibular apparatus for equilibrium. Label the components. Briefly describe how they function to notify the brain of movement.
30. Map the following terms related to vision. Add terms if you wish.

Map 1

• accommodation reflex	• depth of field	• lens
• binocular vision	• field of vision	• macula
• blind spot	• focal point	• optic chiasm
• ciliary muscle	• fovea	• optic disk
• cornea	• iris	• optic nerve
• cranial nerve III	• lateral geniculate	• phototransduction
• pupillary reflex	• visual cortex	• zonules
• retina	• visual field	

Map 2: The Retina

• amacrine cells	• ganglion cells	• pigment epithelium
• bipolar cells	• horizontal cells	• retinal
• bleaching	• melanin	• rhodopsin
• cGMP	• melanopsin	• rods
• cones	• opsin	• transducin

31. Explain how accommodation by the eye occurs. What is the loss of accommodation called?
32. List four common visual problems, and explain how they occur.
33. Explain how the intensity and duration of a stimulus are coded so that the stimulus can be interpreted by the brain. (Remember, action potentials are all-or-none phenomena.)
34. Make a table of the special senses. In the first row, write these stimuli: sound, standing on the deck of a rocking boat, light, a taste, an aroma. In row 2, describe the location of the receptor for each sense. In row 3, describe the structure or properties of each receptor. In a final row, name the cranial nerve(s) that convey(s) each sensation to the brain. (p. 319)

Level Three Problem Solving

35. You are prodding your blindfolded lab partner's arm with two needle probes (with her permission). Sometimes she can tell you are using two probes. But when you probe less sensitive areas, she thinks there is just one probe. Which sense are you testing? Which receptors are being stimulated? Explain why she sometimes feels only one probe.
36. Consuming alcohol depresses the nervous system and vestibular apparatus. In a sobriety check, police officers use this information to determine if an individual is intoxicated. What kinds of tests can you suggest that would show evidence of this inhibition?
37. Often, children are brought to medical attention because of speech difficulties. If you were a clinician, which sense would you test first in such patients, and why?
38. A clinician shines a light into a patient's left eye, and neither pupil constricts. Shining the light into the right eye elicits a normal consensual reflex. What problem in the reflex pathway could explain these observations?
39. An optometrist wishes to examine a patient's retina. Which of the following classes of drugs might dilate the pupil? Explain why you did or did not select each choice.
    (a) a sympathomimetic { *mimicus*, imitate }
    (b) a muscarinic antagonist
    (c) a cholinergic agonist
    (d) an anticholinesterase
    (e) a nicotinic agonist
40. The iris of the eye has two sets of antagonistic muscles, one for dilation and one for constriction. One set of muscles is radial (radiating from the center of the pupil), and the other set is circular. Draw an iris and pupil, and arrange the muscles so that contraction of one set causes pupillary constriction and contraction of the other set causes dilation.
41. As people age, their ability to see at night decreases. What changes in the retina might explain this?

Level Four Quantitative Problems

42. The relationship between focal length (F) of a lens, object distance (P), and image distance or focal point (Q) is $1/F = 1/P + 1/Q$. Assume the distance from lens to retina is 20 mm.
    (a) For a distant object, P = infinity (∞) and $1/\infty = 0$. If Pavi sees a distant object in focus, what is the focal length of her lens in meters?
    (b) If the object moves to 1 foot in front of Pavi's lens and the lens does not change shape, what is the image distance (1 in. = 2.54 cm)? What must happen to Pavi's lens for the closer image to come into focus?

Answers to Concept Checks, Figure and Graph Questions, and end-of-chapter Review Questions can be found in Appendix A [p. A-1].

11 Efferent Division: Autonomic and Somatic Motor Control

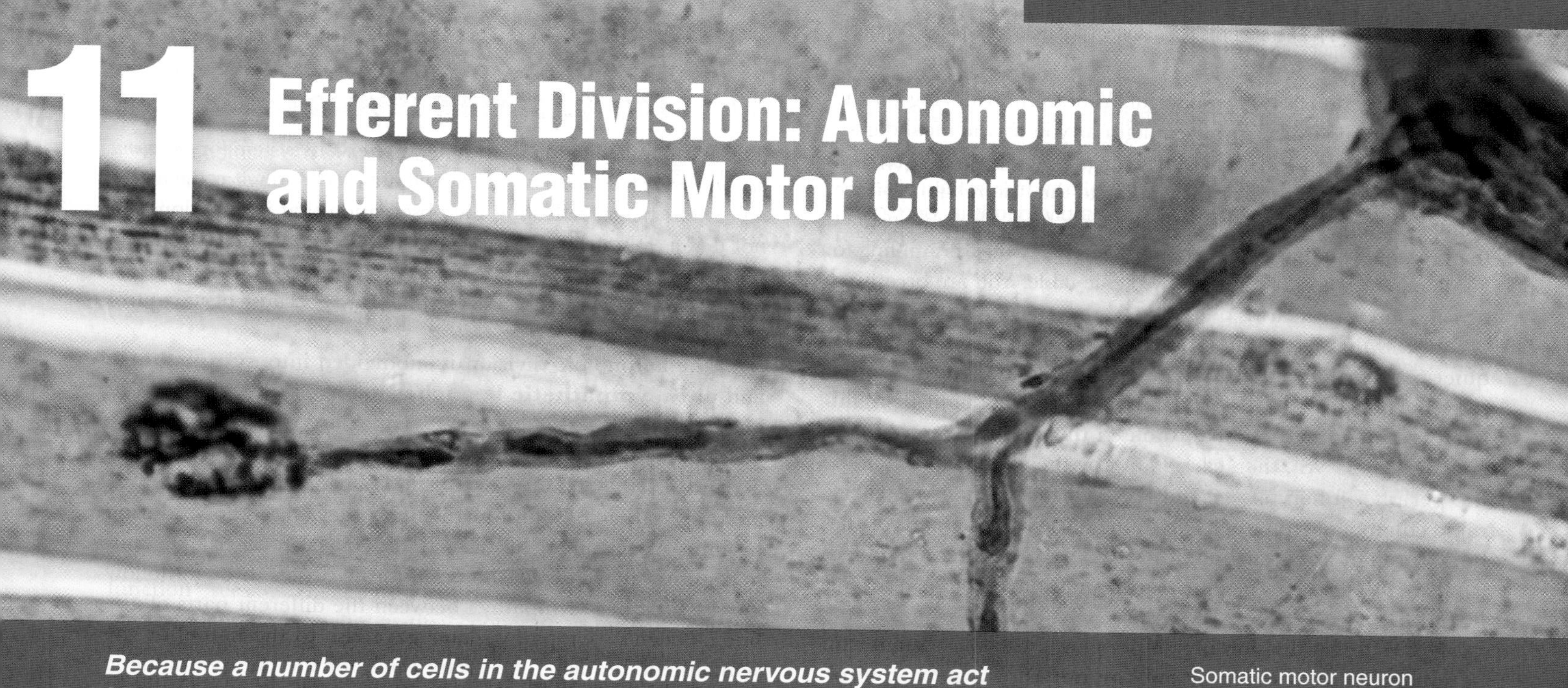

Somatic motor neuron synapsing on muscle fiber

Because a number of cells in the autonomic nervous system act in conjunction, they have relinquished their independence to function as a coherent whole.

Otto Appenzeller and Emilio Oribe, in The Autonomic Nervous System, *1997*

BACKGROUND BASICS

RUNNING PROBLEM

Neuroscientists have learned that addictive behaviors develop because certain chemicals act as positive reinforcers in the brain, creating physical and psychological dependence. Nicotine is an addictive drug that enhances dopamine release in the brain's reward centers and creates pleasurable sensations. Over time, the brain also begins to associate the social aspects of cigarette smoking with pleasure, a conditioned response that makes quitting difficult. If smokers do stop smoking, they may suffer from unpleasant physical withdrawal symptoms, including lethargy, hunger, and irritability.

Q1: *To avoid withdrawal symptoms, people continue to smoke, resulting in chronically elevated nicotine levels in their blood. Nicotine binds to nicotinic acetylcholine receptors (nAChRs). What is the usual response of cells that are chronically exposed to elevated concentrations of a signal molecule?* [Hint: *p. 180*]

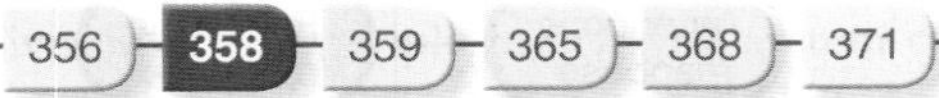
356
358
359
365
368
371

Antagonistic Control Is a Hallmark of the Autonomic Division

The sympathetic and parasympathetic branches of the autonomic nervous system display all four of Walter Cannon's properties of homeostasis: (1) preservation of the fitness of the internal environment, (2) up-down regulation by tonic control, (3) antagonistic control, and (4) chemical signals with different effects in different tissues [p. 182].

Many internal organs are under *antagonistic control,* in which one autonomic branch is excitatory and the other branch is inhibitory (see the table on the right side of Fig. 11.5). For example, sympathetic innervation increases heart rate, while parasympathetic stimulation decreases it. Consequently, heart rate can be regulated by altering the relative proportions of sympathetic and parasympathetic control.

Exceptions to dual antagonistic innervation include the sweat glands and the smooth muscle in most blood vessels. These tissues are innervated only by the sympathetic branch and rely strictly on tonic (up-down) control.

The two autonomic branches are usually antagonistic in their control of a given target tissue, but they sometimes work cooperatively on different tissues to achieve a common goal. For example, blood flow for penile erection is under control of the parasympathetic branch, while muscle contraction for sperm ejaculation is directed by the sympathetic branch.

In some autonomic pathways, the neurotransmitter receptor determines the response of the target tissue. For instance, most blood vessels contain one type of *adrenergic receptor* [p. 253] that causes smooth muscle contraction (vasoconstriction). However, some blood vessels also contain a second type of adrenergic receptor that causes smooth muscle relaxation (vasodilation). Both receptors are activated by the catecholamines norepinephrine and epinephrine [p. 204]. In these blood vessels, the adrenergic receptor, not the chemical signal, determines the response [p. 179].

Concept Check

3. Define homeostasis.

Autonomic Pathways Have Two Efferent Neurons in Series

All autonomic pathways (sympathetic and parasympathetic) consist of two neurons in a series (**FIG. 11.4**). The first neuron, called the **preganglionic neuron,** originates in the central nervous system and projects to an **autonomic ganglion** outside the CNS. There, the preganglionic neuron synapses with the second neuron in the pathway, the **postganglionic neuron**. This neuron has its cell body in the ganglion and projects its axon to the target tissue. (A *ganglion* is a cluster of nerve cell bodies that lie outside the CNS. The equivalent in the CNS is a *nucleus* [p. 274].)

Divergence [p. 258] is an important feature of autonomic pathways. On average, one preganglionic neuron entering a ganglion synapses with 8 or 9 postganglionic neurons. Some synapse on as many as 32 neurons! Each postganglionic neuron may then innervate a different target, meaning that a single signal from the CNS can affect a large number of target cells simultaneously.

In the traditional view of the autonomic division, autonomic ganglia were simply a way station for the transfer of signals from preganglionic neurons to postganglionic neurons. We now know, however, that ganglia are more than a simple collection of axon terminals and nerve cell bodies: they also contain neurons that lie completely within them. These neurons enable the autonomic ganglia to act as mini-integrating centers, receiving sensory input from the periphery of the body and modulating outgoing autonomic signals to target tissues. Presumably, this arrangement means that a reflex could be integrated totally within a ganglion, with no involvement of the CNS. That pattern of control is known to exist in the enteric nervous system [p. 224], which is discussed with the digestive system [Chapter 21].

FIG. 11.4 Autonomic pathways

Autonomic pathways consist of two neurons that synapse in an autonomic ganglion.

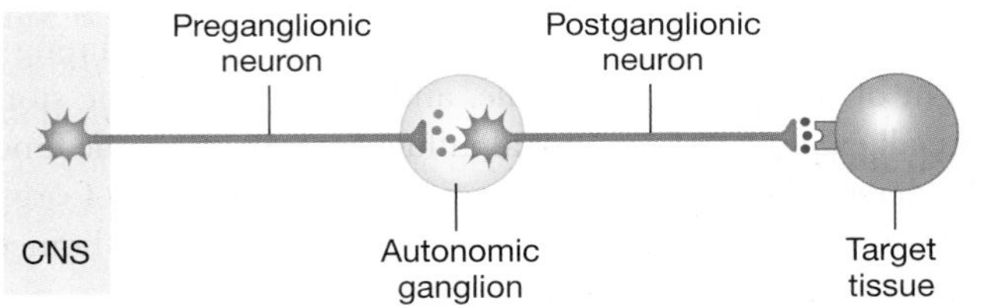
Preganglionic neuron
Postganglionic neuron
CNS
Autonomic ganglion
Target tissue

Sympathetic and Parasympathetic Branches Originate in Different Regions

How, then, do the two autonomic branches differ anatomically? The main anatomical differences are (1) the pathways' point of origin in the CNS and (2) the location of the autonomic ganglia. As **FIGURE 11.5** shows, most sympathetic pathways (red) originate in the thoracic and lumbar regions of the spinal cord. *Sympathetic ganglia* are found primarily in two ganglion chains that run along either side of the bony vertebral column, with additional ganglia along the descending aorta. Long nerves (axons of postganglionic neurons) project from the ganglia to the target tissues. Because most sympathetic ganglia lie close to the spinal cord, sympathetic pathways generally have short preganglionic neurons and long postganglionic neurons.

Many parasympathetic pathways (shown in blue in Fig. 11.5) originate in the brain stem, and their axons leave the brain in several cranial nerves [p. 283]. Other parasympathetic pathways originate in the sacral region (near the lower end of the spinal cord) and control pelvic organs. In general, parasympathetic ganglia are located either on or near their target organs. Consequently, parasympathetic preganglionic neurons have long axons, and parasympathetic postganglionic neurons have short axons.

Parasympathetic innervation goes primarily to the head, neck, and internal organs. The major parasympathetic tract is the **vagus nerve** (cranial nerve X), which contains about 75% of all parasympathetic fibers. This nerve carries both sensory information from internal organs to the brain and parasympathetic output from the brain to organs.

Vagotomy, a procedure in which the vagus nerve is surgically cut, was an experimental technique used in the nineteenth and early twentieth centuries to study the effects of the autonomic nervous system on various organs. For a time, vagotomy was the preferred treatment for stomach ulcers because removal of parasympathetic innervation to the stomach decreased the secretion of stomach acid. However, this procedure had many unwanted side effects and has been abandoned in favor of drug therapies that treat the problem more specifically.

Concept Check

4. A nerve that carries both sensory and motor information is called a(n) ________ nerve.
5. Name the four regions of the spinal cord in order, starting from the brain stem.

The Autonomic Nervous System Uses a Variety of Chemical Signals

Chemically, the sympathetic and parasympathetic branches can be distinguished by their neurotransmitters and receptors, using the following rules and **FIGURE 11.6**:

1. Both sympathetic and parasympathetic preganglionic neurons release acetylcholine (ACh) onto *nicotinic cholinergic receptors* (nAChR) on the postganglionic cell [p. 252].
2. Most postganglionic sympathetic neurons secrete norepinephrine (NE) onto *adrenergic receptors* on the target cell.
3. Most postganglionic parasympathetic neurons secrete acetylcholine onto *muscarinic cholinergic receptors* (mAChR) on the target cell.

However, there are some exceptions to these rules. A few sympathetic postganglionic neurons, such as those that terminate on sweat glands, secrete ACh rather than norepinephrine. These neurons are therefore called *sympathetic cholinergic neurons.*

A small number of autonomic neurons secrete neither norepinephrine nor acetylcholine and are known as *nonadrenergic, noncholinergic neurons.* Some of the chemicals they use as neurotransmitters include substance P, somatostatin, vasoactive intestinal peptide (VIP), adenosine, nitric oxide, and ATP. The nonadrenergic, noncholinergic neurons are assigned to either the sympathetic or parasympathetic branch according to where their preganglionic fibers leave the nerve cord. A few autonomic neurons *co-secrete* more than one neurotransmitter simultaneously.

Autonomic Pathways Control Smooth and Cardiac Muscle and Glands

The targets of autonomic neurons are smooth muscle, cardiac muscle, many exocrine glands, a few endocrine glands, lymphoid tissues, the liver, and some adipose tissue. The synapse between a postganglionic autonomic neuron and its target cell is called the **neuroeffector junction** (recall that targets are also called effectors).

The structure of an autonomic synapse differs from the model synapse [Fig. 8.2f, p. 227]. Autonomic postganglionic axons end

RUNNING PROBLEM

Shanika's doctor congratulates her for trying once more to stop smoking. He explains that quitting is most likely to be successful if the smoker uses a combination of behavioral modification strategies and drug therapy. Currently, there are three types of pharmacological treatments used for nicotine addiction: nicotine replacement, bupropion, and varenicline. Bupropion inhibits reuptake of the monoamines (dopamine, serotonin, and norepinephrine) by neurons, mimicking the effects of nicotine. Varenicline binds to nAChRs. Nicotinic receptors are found throughout the nervous system, and evidence suggests that activation of nAChR by nicotine in certain regions of the brain plays a key role in nicotine addiction.

Q2: *Cholinergic receptors are classified as either nicotinic or muscarinic, on the basis of the agonist molecules that bind to them. What happens to a postsynaptic cell when nicotine rather than ACh binds to a nicotinic cholinergic receptor?*

356 – 358 – **359** – 365 – 368 – 371

FIG. 11.5 ESSENTIALS The Autonomic Nervous System

The autonomic nervous system can be divided into two divisions: the sympathetic division and the parasympathetic division.

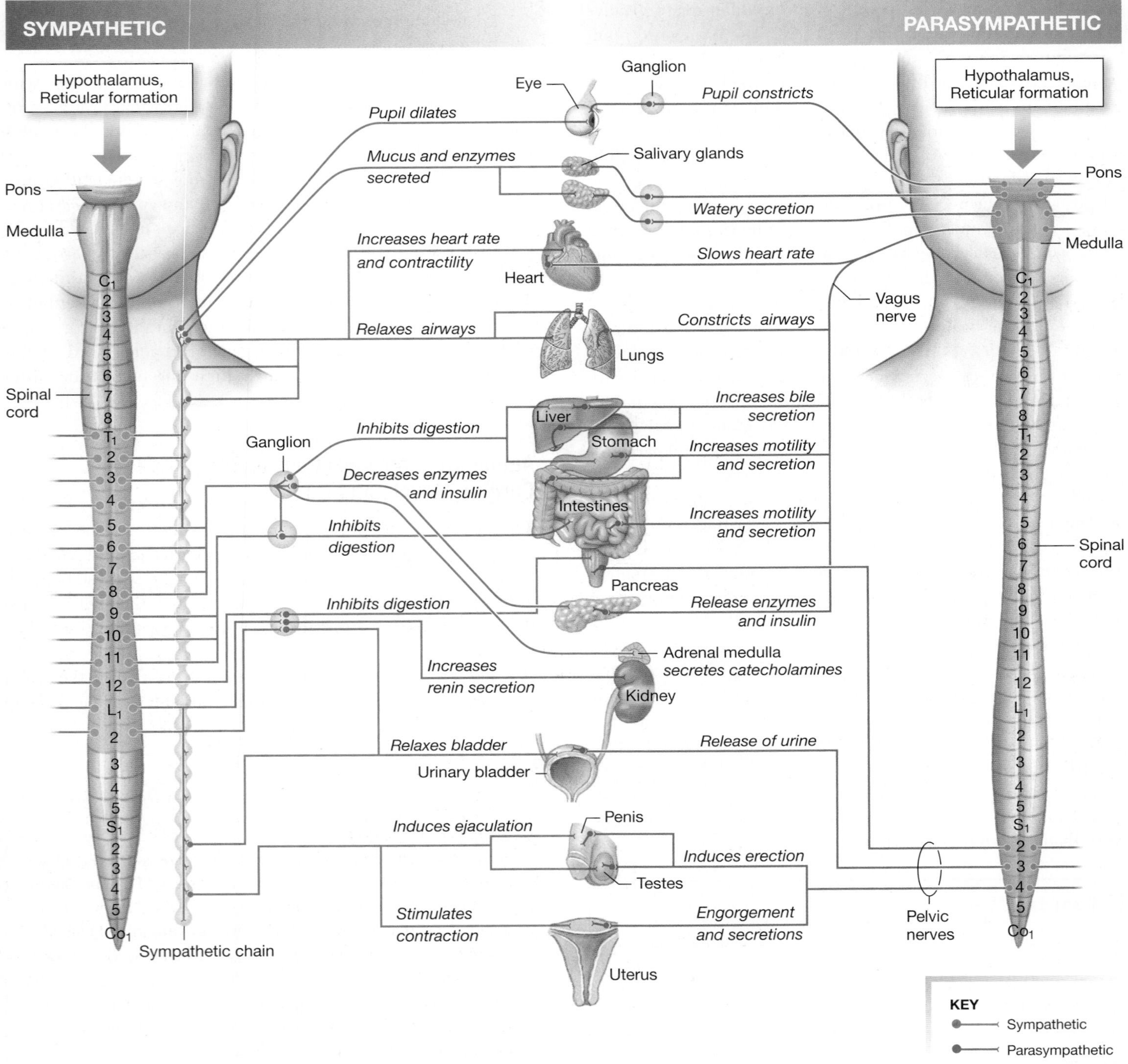

SYMPATHETIC
PARASYMPATHETIC
Hypothalamus, Reticular formation
Pons
Medulla
Spinal cord
Sympathetic chain
Eye
Ganglion
Pupil dilates
Pupil constricts
Mucus and enzymes secreted
Salivary glands
Watery secretion
Increases heart rate and contractility
Heart
Slows heart rate
Vagus nerve
Relaxes airways
Constricts airways
Lungs
Liver
Stomach
Increases bile secretion
Inhibits digestion
Increases motility and secretion
Decreases enzymes and insulin
Intestines
Inhibits digestion
Increases motility and secretion
Pancreas
Inhibits digestion
Release enzymes and insulin
Adrenal medulla secretes catecholamines
Increases renin secretion
Kidney
Relaxes bladder
Urinary bladder
Release of urine
Induces ejaculation
Penis
Induces erection
Testes
Stimulates contraction
Engorgement and secretions
Uterus
Pelvic nerves
KEY
Sympathetic
Parasympathetic

Characteristic	Sympathetic	Parasympathetic
Origin in the CNS	Thoracic and lumbar segments	Brainstem and sacral segments
Ganglion Location	Close to spinal cord	On or close to targets
Pathways	Short preganglionic, long postganglionic neurons	Long preganglionic, short postganglionic neurons

Sympathetic and Parasympathetic Responses

Although the sympathetic and parasympathetic divisions frequently innervate the same organs and tissues, they often have opposing effects.

Effector Organ	Sympathetic Response	Adrenergic Receptor	Parasympathetic Response **
Pupil of eye	Dilates	α	Constricts
Salivary glands	Mucus, enzymes	α and β_2	Watery secretion
Heart	Increases rate and force of contraction	β_1	Slows rate
Arterioles and veins	Constricts Dilates	α β_2	– –
Lungs	Bronchioles dilate	β_2*	Bronchioles constrict
Digestive tract	Decreases motility and secretion	α, β_2	Increases motility and secretion
Exocrine pancreas	Decreases enzyme secretion	α	Increases enzyme secretion
Endocrine pancreas	Inhibits insulin secretion	α	Stimulates insulin secretion
Adrenal medulla	Secretes catecholamines	–	– –
Kidney	Increases renin secretion	β_1	– –
Urinary bladder	Urinary retention	α, β_2	Release of urine
Adipose tissue	Fat breakdown	β_3	– –
Male and female sex organs	Ejaculation (male)	α	Erection
Uterus	Depends on stage of cycle	α, β_2	Depends on stage of cycle
Lymphoid tissue	Generally inhibitory	α, β_2	– –
	*Hormonal epinephrine only		**All parasympathetic responses are mediated by muscarinic receptors.

FIGURE QUESTIONS

1. What is an advantage of having ganglia in the sympathetic chain linked to each other?
2. Which organs have antagonistic control by sympathetic and parasympathetic divisions? Which have cooperative control, with sympathetic and parasympathetic division each contributing to a function?

FIG. 11.6 Sympathetic and parasympathetic neurotransmitters and receptors

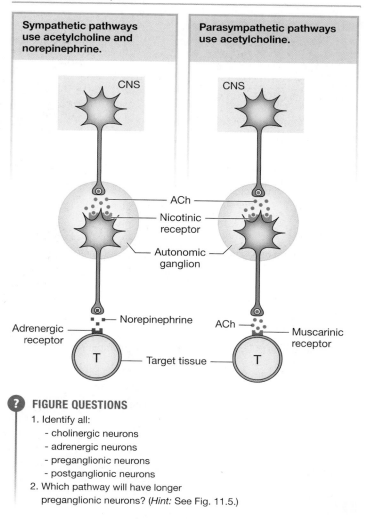

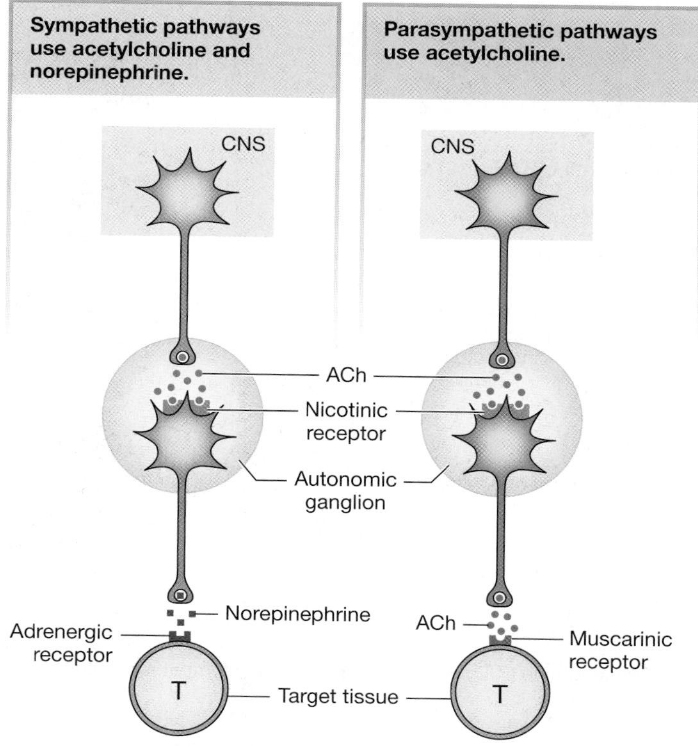
Sympathetic pathways use acetylcholine and norepinephrine.
Parasympathetic pathways use acetylcholine.
CNS
CNS
ACh
Nicotinic receptor
Autonomic ganglion
Norepinephrine
ACh
Adrenergic receptor
Muscarinic receptor
T
Target tissue
T

FIGURE QUESTIONS

1. Identify all:
   - cholinergic neurons
   - adrenergic neurons
   - preganglionic neurons
   - postganglionic neurons
2. Which pathway will have longer preganglionic neurons? (*Hint:* See Fig. 11.5.)

with a series of swollen areas at their distal ends, like beads spaced out along a string (**FIG. 11.7a**). Each of these swellings, known as a **varicosity** {*varicosus*, abnormally enlarged or swollen}, contains vesicles filled with neurotransmitter.

The branched ends of the axon lie across the surface of the target tissue, but the underlying target cell membrane does not possess clusters of neurotransmitter receptors in specific sites. Instead, the neurotransmitter is simply released into the interstitial fluid to diffuse to wherever the receptors are located. The result is a less-directed form of communication than that which occurs between a somatic motor neuron and a skeletal muscle. The diffuse release of autonomic neurotransmitter means that a single postganglionic neuron can affect a large area of target tissue.

The release of autonomic neurotransmitters is subject to modulation from a variety of sources. For example, sympathetic varicosities contain receptors for hormones and for paracrine signals such as histamine. These modulators may either facilitate or inhibit neurotransmitter release. Some preganglionic neurons co-secrete neuropeptides along with acetylcholine. The peptides act as

FIG. 11.7 Autonomic synapses

(a) Autonomic varicosities release neurotransmitter over the surface of target cells.

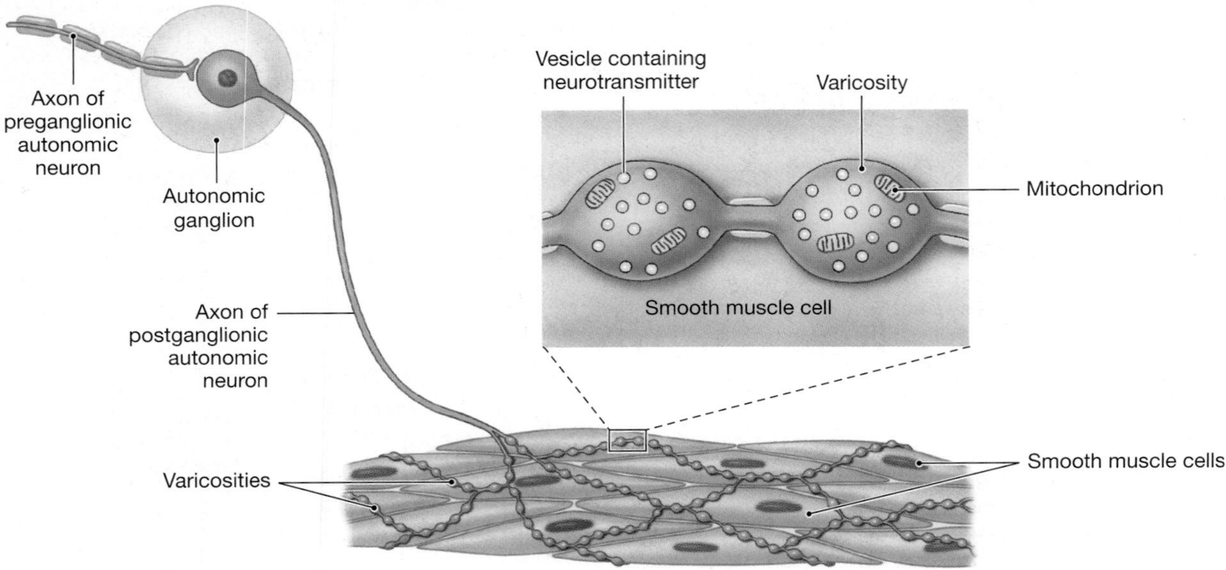

Axon of preganglionic autonomic neuron
Autonomic ganglion
Vesicle containing neurotransmitter
Varicosity
Mitochondrion
Smooth muscle cell
Axon of postganglionic autonomic neuron
Varicosities
Smooth muscle cells

(b) Norepinephrine (NE) release and removal at a sympathetic neuroeffector junction

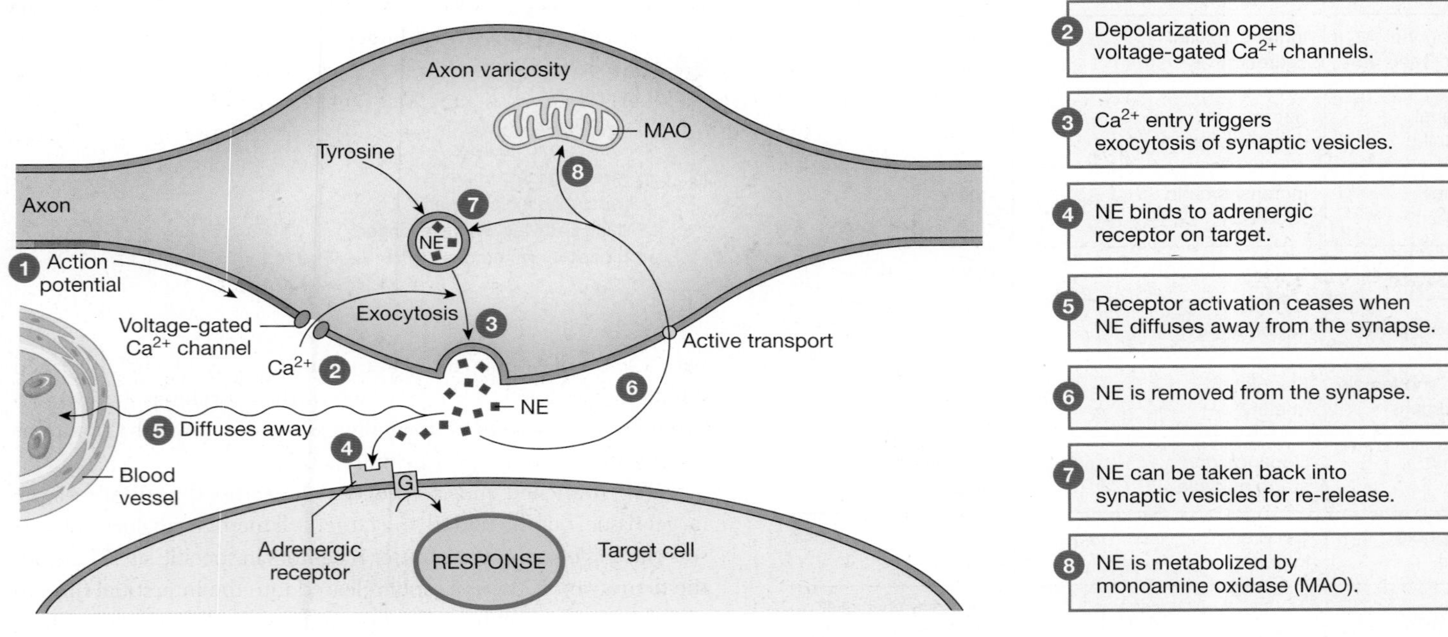

Axon varicosity
MAO
Tyrosine
Axon
NE
1 Action potential
Voltage-gated Ca2+ channel
Exocytosis 3
Ca2+ 2
Active transport
6
7
8
NE
5 Diffuses away
4
G
Blood vessel
Adrenergic receptor
RESPONSE
Target cell
1 Action potential arrives at the varicosity.
2 Depolarization opens voltage-gated Ca2+ channels.
3 Ca2+ entry triggers exocytosis of synaptic vesicles.
4 NE binds to adrenergic receptor on target.
5 Receptor activation ceases when NE diffuses away from the synapse.
6 NE is removed from the synapse.
7 NE can be taken back into synaptic vesicles for re-release.
8 NE is metabolized by monoamine oxidase (MAO).

neuromodulators, producing slow synaptic potentials that modify the activity of postganglionic neurons (p. 260).

Autonomic Neurotransmitters Are Synthesized in the Axon

The primary autonomic neurotransmitters, acetylcholine and norepinephrine, can be synthesized in the axon varicosities (Fig. 11.7b). Both are small molecules easily synthesized by cytoplasmic enzymes. Neurotransmitter made in the varicosities is packaged into synaptic vesicles for storage.

Neurotransmitter release follows the pattern found in other cells: depolarization—calcium signal—exocytosis [p. 147]. When an action potential arrives at the varicosity, voltage-gated Ca^{2+} channels open, Ca^{2+} enters the neuron, and the synaptic vesicle contents are released by exocytosis. Once neurotransmitters are released into the synapse, they either diffuse through the interstitial fluid until they encounter a receptor on the target cell or drift away from the synapse.

TABLE 11.1 Postganglionic Autonomic Neurotransmitters

	Sympathetic Division	Parasympathetic Division
Neurotransmitter	Norepinephrine (NE)	Acetylcholine (ACh)
Receptor Types	α- and β-adrenergic	Nicotinic and muscarinic cholinergic
Synthesized from	Tyrosine	Acetyl CoA + choline
Inactivation Enzyme	Monoamine oxidase (MAO) in mitochondria of varicosity	Acetylcholinesterase (AChE) in synaptic cleft
Varicosity Membrane Transporters for	Norepinephrine	Choline

The concentration of neurotransmitter in the synapse is a major factor in autonomic control of a target: more neurotransmitter means a longer or stronger response. The concentration of neurotransmitter in a synapse is influenced by its rate of breakdown or removal (Fig. 11.7b). Neurotransmitter activation of its receptor terminates when the neurotransmitter (1) diffuses away, (2) is metabolized by enzymes in the extracellular fluid, or (3) is actively transported into cells around the synapse. The uptake of neurotransmitter by varicosities allows neurons to reuse the chemicals.

These steps are shown for norepinephrine in Figure 11.7b. Norepinephrine is synthesized in the varicosity from the amino acid tyrosine. Once released into the synapse, norepinephrine may combine with an adrenergic receptor on the target cell, diffuse away, or be transported back into the varicosity. Inside the neuron, recycled norepinephrine is either repackaged into vesicles or broken down by **monoamine oxidase (MAO)**, the main enzyme responsible for degradation of catecholamines. ([See Fig. 8.20, p. 257] for a similar figure on acetylcholine.)

TABLE 11.1 compares the characteristics of the two primary autonomic neurotransmitters.

Autonomic Receptors Have Multiple Subtypes

The autonomic nervous system uses only a few neurotransmitters, but it diversifies its actions by having multiple receptor subtypes with different second messenger pathways (**TBL. 11.2**). Target tissues of the sympathetic division have two types of adrenergic receptors with multiple subtypes. Most targets of the parasympathetic division have one of two subtypes of muscarinic cholinergic receptors. As previously noted, postganglionic autonomic neurons have nicotinic cholinergic receptors on their postsynaptic membranes (N_N subtype).

TABLE 11.2 Properties of Autonomic Neurotransmitter Receptors

Receptor	Found in	Sensitivity	Effect on Second Messenger
Adrenergic Receptors			
α_1	Most sympathetic target tissues	NE > E*	Increases IP_3 and intracellular Ca^{2+}; increases PKC§
α_2	Gastrointestinal tract and pancreas	NE > E	Decreases cAMP
β_1	Heart muscle, kidney	NE = E	Increases cAMP
β_2	Certain blood vessels and smooth muscle of some organs	E > NE	Increases cAMP
β_3	Adipose tissue	NE > E	Increases cAMP
Cholinergic Receptors			
N_N	Postganglionic autonomic neurons		Opens nonspecific monovalent cation channels
N_M	Skeletal muscle		Opens nonspecific monovalent cation channels
M_1, M_3, M_5	Nervous system and parasympathetic target tissues**		Increases IP_3 and intracellular Ca^{2+}; increases PKC§
M_2, M_4	Nervous system and parasympathetic target tissues**		Decreases cAMP; opens K^+ channels

* NE = norepinephrine, E = epinephrine
** M_2 and M_3 are the primary muscarinic receptors on parasympathetic targets
§ IP_3 = inositol trisphosphate, PKC = protein kinase C

TABLE 11.3 Agonists and Antagonists of Neurotransmitter Receptors

Receptor Type	Neurotransmitter	Agonist	Antagonists	Indirect Agonists/Antagonists
Cholinergic	Acetylcholine			AChE* *inhibitors:* neostigmine
Muscarinic		Muscarine	Atropine, scopolamine	
Nicotinic		Nicotine	α-bungarotoxin (muscle only), TEA (tetraethylammonium; ganglia only), curare	
Adrenergic	Norepinephrine (NE), epinephrine			*Stimulate NE release:* ephedrine, amphetamines; *Prevents NE uptake:* cocaine
Alpha (α)		Phenylephrine	"Alpha-blockers"	
Beta (β)		Isoproterenol, albuterol	"Beta-blockers": propranolol (β_1 and β_2), metoprolol (β_1 only)	

*AChE = acetylcholinesterase

on smooth muscle of the prostate gland and bladder. Relaxing these muscles helps relieve the urinary symptoms of prostatic enlargement.

Primary Disorders of the Autonomic Nervous System Are Relatively Uncommon

Diseases and malfunction of the autonomic nervous system are relatively rare. Direct damage (trauma) to hypothalamic control centers may disrupt the body's ability to regulate water balance or temperature. Generalized sympathetic dysfunction, or *dysautonomia,* may result from systemic diseases such as cancer and diabetes mellitus. There are also some conditions, such as *multiple system atrophy,* in which the CNS control centers for autonomic functions degenerate.

In many cases of sympathetic dysfunction, the symptoms are manifested most strongly in the cardiovascular system, when diminished sympathetic input to blood vessels results in abnormally low blood pressure. Other prominent symptoms of sympathetic pathology include urinary *incontinence* {*in-*, unable + *continere,* to contain}, which is the loss of bladder control, or *impotence,* which is the inability to achieve or sustain a penile erection.

Occasionally, patients suffer from primary autonomic failure when sympathetic neurons degenerate. In the face of continuing diminished sympathetic input, target tissues up-regulate [p. 51], putting more receptors into the cell membrane to maximize the cell's response to available norepinephrine. This increase in receptor abundance leads to *denervation hypersensitivity,* a state in which the administration of exogenous adrenergic agonists causes a greater-than-expected response.

CLINICAL FOCUS

Diabetes: Autonomic Neuropathy

Primary disorders of the autonomic division are rare, but the secondary condition known as **diabetic autonomic neuropathy** is quite common. This complication of diabetes often begins as a sensory neuropathy, with tingling and loss of sensation in the hands and feet. In some patients, pain is the primary symptom. About 30% of diabetic patients go on to develop autonomic neuropathies, manifested by dysfunction of the cardiovascular, gastrointestinal, urinary, and reproductive systems (abnormal heart rate, constipation, incontinence, impotence). The cause of diabetic neuropathies is complicated. Patients who have chronically elevated blood glucose levels are more likely to develop neuropathies, but there is no single metabolic responsible. Contributing factors that lead to damage or loss of myelinated and unmyelinated nerve fibers include oxidative stress, inflammation, and disruptions to glucose metabolism. Currently, there is no prevention for diabetic neuropathies other than controlling blood glucose levels, and no cure. The primary recourse for patients is taking drugs that treat the symptoms.

FIG. 11.9 ESSENTIALS Efferent Divisions of the Nervous System

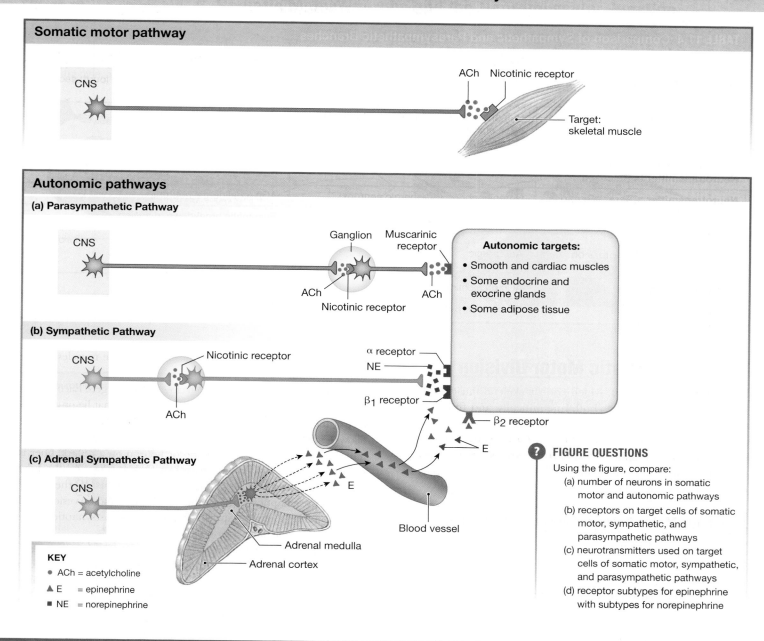

Somatic motor pathway

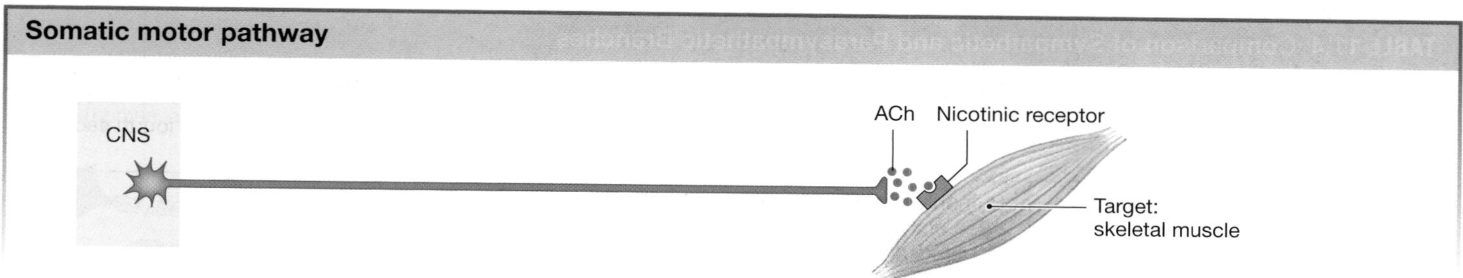

CNS
ACh
Nicotinic receptor
Target:
skeletal muscle

Autonomic pathways

(a) Parasympathetic Pathway

CNS
Ganglion
Muscarinic receptor
ACh
Nicotinic receptor
ACh

Autonomic targets:

- Smooth and cardiac muscles
- Some endocrine and exocrine glands
- Some adipose tissue

(b) Sympathetic Pathway

CNS
Nicotinic receptor
ACh
α receptor
NE
β_1 receptor
β_2 receptor
E

(c) Adrenal Sympathetic Pathway

CNS
E
Blood vessel
Adrenal medulla
Adrenal cortex

KEY

- ● ACh = acetylcholine
- ▲ E = epinephrine
- ■ NE = norepinephrine

FIGURE QUESTIONS

Using the figure, compare:

(a) number of neurons in somatic motor and autonomic pathways

(b) receptors on target cells of somatic motor, sympathetic, and parasympathetic pathways

(c) neurotransmitters used on target cells of somatic motor, sympathetic, and parasympathetic pathways

(d) receptor subtypes for epinephrine with subtypes for norepinephrine

Summary of Sympathetic and Parasympathetic Branches

As you have seen in this discussion, the branches of the autonomic nervous system share some features but are distinguished by others. Many of these features are summarized in **FIGURE 11.9** and compared in **TABLE 11.4**.

1. Both sympathetic and parasympathetic pathways consist of two neurons (preganglionic and postganglionic) in series. One exception to this rule is the adrenal medulla, in which postganglionic sympathetic neurons have been modified into a neuroendocrine organ.
2. All preganglionic autonomic neurons secrete acetylcholine onto nicotinic receptors. Most sympathetic neurons secrete norepinephrine onto adrenergic receptors. Most parasympathetic neurons secrete acetylcholine onto muscarinic receptors.
3. Sympathetic pathways originate in the thoracic and lumbar regions of the spinal cord. Parasympathetic pathways leave the CNS at the brain stem and in the sacral region of the spinal cord.
4. Most sympathetic ganglia are located close to the spinal cord (are *paravertebral*). Parasympathetic ganglia are located close to or in the target tissue.
5. The sympathetic branch controls functions that are useful in stress or emergencies (fight-or-flight). The parasympathetic branch is dominant during rest-and-digest activities.

12 Muscles

Skeletal muscle

A muscle is . . . an engine, capable of converting chemical energy into mechanical energy. It is quite unique in nature, for there has been no artificial engine devised with the great versatility of living muscle.

Ralph W. Stacy and John A. Santolucito, in Modern College Physiology, *1966*

BACKGROUND BASICS

82 Tendons
102 Kinases and phosphatases
99 Isozymes
109 Anaerobic and aerobic metabolism
106 Glycolysis
182 Tonic control
179 Nitric oxide
239 Threshold
261 Summation
226 Autonomic neurons
368 Somatic motor neurons
368 Neuromuscular junction

It was his first time to be the starting pitcher. As he ran from the bullpen onto the field, his heart was pounding and his stomach felt as if it were tied in knots. He stepped onto the mound and gathered his thoughts before throwing his first practice pitch. Gradually, as he went through the familiar routine of throwing and catching the baseball, his heart slowed and his stomach relaxed. It was going to be a good game.

The pitcher's pounding heart, queasy stomach, and movements as he runs and throws all result from muscle contraction. Our muscles have two common functions: to generate motion and to generate force. Our skeletal muscles also generate heat and contribute significantly to the homeostasis of body temperature. When cold conditions threaten homeostasis, the brain may direct our muscles to shiver, creating additional heat.

The human body has three types of muscle tissue: skeletal muscle, cardiac muscle, and smooth muscle. Most **skeletal muscles** are attached to the bones of the skeleton, enabling these muscles to control body movement. **Cardiac muscle** {*kardia*, heart} is found only in the heart and moves blood through the circulatory system. Skeletal and cardiac muscles are classified as **striated muscles** {*stria*, groove} because of their alternating light and dark bands seen under the light microscope (**FIG. 12.1a, b**).

Smooth muscle is the primary muscle of internal organs and tubes, such as the stomach, urinary bladder, and blood vessels. Its primary function is to influence the movement of material into, out of, and within the body. An example is the passage of food through the gastrointestinal tract. Viewed under the

RUNNING PROBLEM **Periodic Paralysis**

This morning, Paul, age 6, gave his mother the fright of her life. One minute he was happily playing in the backyard with his new beagle puppy. The next minute, after sitting down to rest, he could not move his legs. In answer to his screams, his mother came running and found her little boy unable to walk. Panic-stricken, she scooped him up, brought him into the house, and dialed 9-1-1. But as she hung up the phone and prepared to wait for the paramedics, Paul got to his feet and walked over to her. "I'm OK now, Mom," he announced. "I'm going outside."

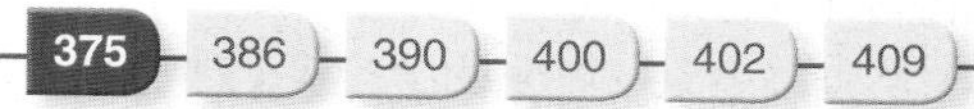
375
386
390
400
402
409

FIG. 12.1 The three types of muscles

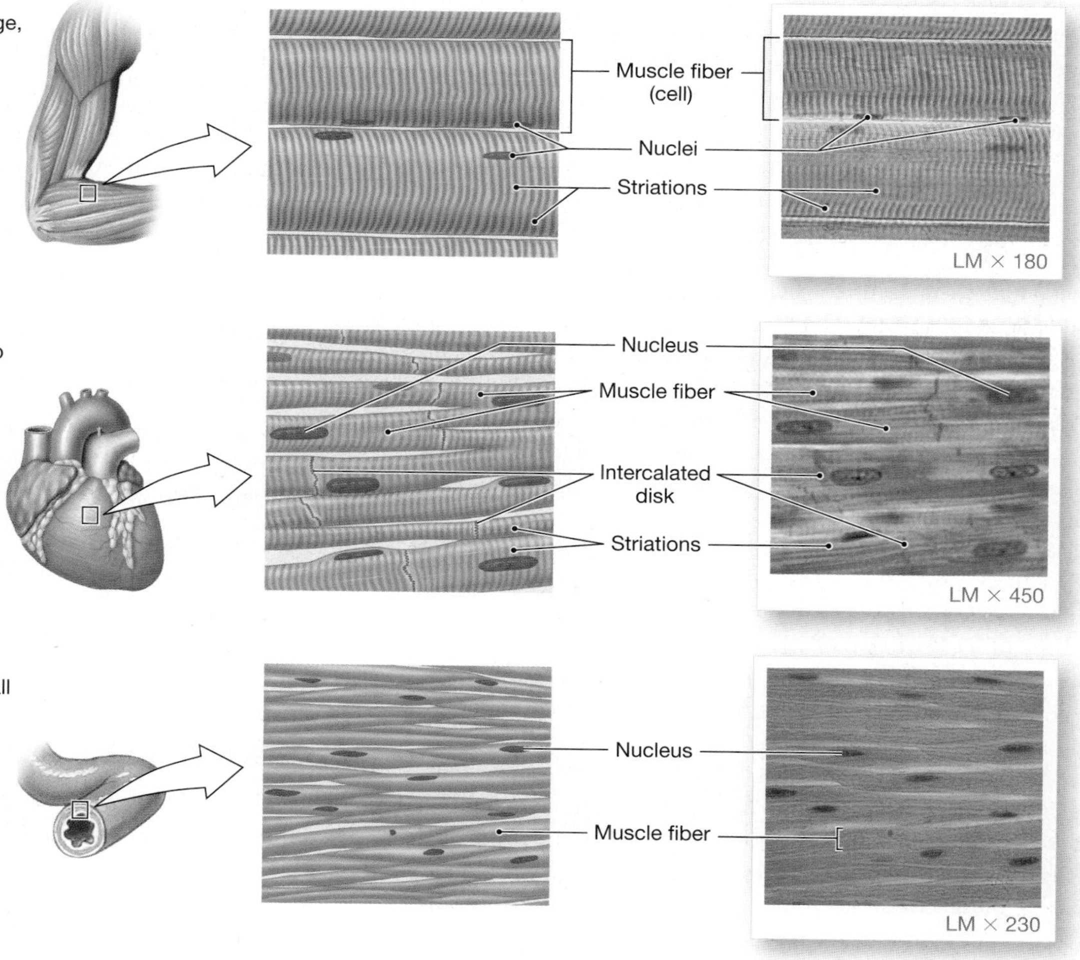

(a) Skeletal muscle fibers are large, multinucleate cells that appear striped or striated under the microscope.

Muscle fiber (cell) — Nuclei — Striations — LM × 180

(b) Cardiac muscle fibers are also striated but they are smaller, branched, and uninucleate. Cells are joined in series by junctions called intercalated disks.

Nucleus — Muscle fiber — Intercalated disk — Striations — LM × 450

(c) Smooth muscle fibers are small and lack striations.

Nucleus — Muscle fiber — LM × 230

microscope, smooth muscle lacks the obvious cross-bands of striated muscles (Fig. 12.1c). Its lack of banding results from the less organized arrangement of contractile fibers within the muscle cells.

Skeletal muscles are often described as voluntary muscles, and smooth and cardiac muscle as involuntary. However, this is not a precise classification. Skeletal muscles can contract without conscious direction, and we can learn a certain degree of conscious control over some smooth and cardiac muscle.

Skeletal muscles are unique in that they contract only in response to a signal from a somatic motor neuron. They cannot initiate their own contraction, and their contraction is not influenced directly by hormones.

In contrast, cardiac and smooth muscle have multiple levels of control. Their primary extrinsic control arises through autonomic innervation, but some types of smooth and cardiac muscle can contract spontaneously, without signals from the central nervous system. In addition, the activity of cardiac and some smooth muscle is subject to modulation by the endocrine system. Despite these differences, smooth and cardiac muscle share many properties with skeletal muscle.

In this chapter, we discuss skeletal and smooth muscle anatomy and contraction, and conclude by comparing the properties of skeletal muscle, smooth muscle, and cardiac muscle. All three muscle types have certain properties in common. The signal to initiate muscle contraction is an intracellular calcium signal, and movement is created when a motor protein called *myosin* uses energy from adenosine triphosphate (ATP) to change its conformation. The details of these processes vary with the different muscle types.

12.1 Skeletal Muscle

Skeletal muscles make up the bulk of muscle in the body and constitute about 40% of total body weight. They position and move the skeleton, as their name suggests. Skeletal muscles are usually attached to bones by **tendons** made of collagen [p. 80]. The **origin** of a muscle is the end of the muscle that is attached closest to the trunk or to the more stationary bone. The **insertion** of the muscle is the more *distal* {*distantia*, distant} or more mobile attachment.

When the bones attached to a muscle are connected by a flexible joint, contraction of the muscle moves the skeleton. The muscle is called a **flexor** if the centers of the connected bones are brought closer together when the muscle contracts, and the movement is called *flexion*. The muscle is called an **extensor** if the bones move away from each other when the muscle contracts, and the movement is called *extension*.

Most joints in the body have both flexor and extensor muscles, because a contracting muscle can pull a bone in one direction but cannot push it back. Flexor-extensor pairs are called **antagonistic muscle groups** because they exert opposite effects. **FIGURE 12.2** shows a pair of antagonistic muscles in the

FIG. 12.2 Antagonistic muscles

Antagonistic muscle groups move bones in opposite directions. Muscle contraction can pull on a bone but cannot push a bone away.

(a) Flexion moves bones closer together. For example, when doing an arm curl, the radius and ulna move towards the humerus.

(b) Extension moves bones away from each other. For example, when doing a push-up, the radius and ulna move away from the humerus.

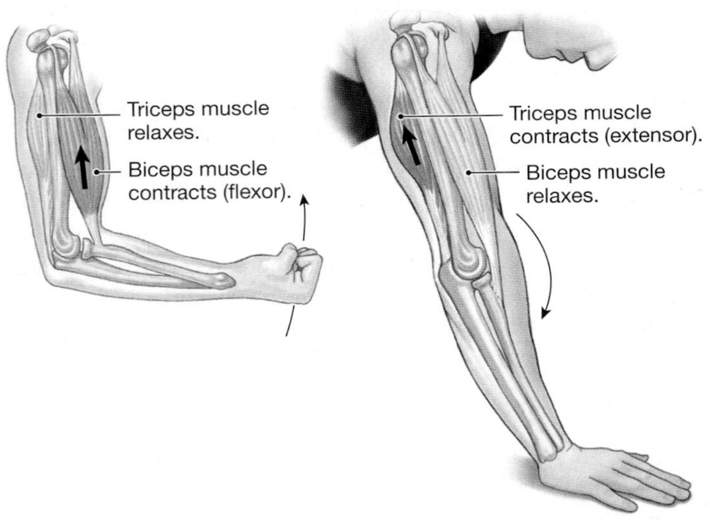
Triceps muscle relaxes.
Biceps muscle contracts (flexor).
Triceps muscle contracts (extensor).
Biceps muscle relaxes.

arm: the *biceps brachii* {*brachion*, arm}, which acts as the flexor, and the *triceps brachii*, which acts as the extensor. When you do a "dumbbell curl" with a weight in your hand, the biceps muscle contracts and the hand and forearm move toward the shoulder. When you do a push-up, lifting your body weight by straightening your arm, the triceps contracts, and the flexed forearm moves away from the shoulder. In each case, when one muscle contracts and shortens, the antagonistic muscle must relax and lengthen.

Concept Check

1. Identify as many pairs of antagonistic muscle groups in the body as you can. If you cannot name them, point out the probable location of the flexor and extensor of each group.

Skeletal Muscles Are Composed of Muscle Fibers

Muscles function together as a unit. A skeletal muscle is a collection of muscle cells, or **muscle fibers**, just as a nerve is a collection of neurons. Each skeletal muscle fiber is a long, cylindrical cell with up to several hundred nuclei near the surface of the fiber (see Anatomy Summary, **FIG. 12.3a**). Skeletal muscle fibers are the largest cells in the body, created by the fusion of many individual embryonic muscle cells. Committed stem cells called **satellite cells** lie just outside the muscle fiber membrane. Satellite cells become active and differentiate into muscle when needed for muscle growth and repair.

The fibers in a given muscle are arranged with their long axes in parallel (Fig. 12.3a). Each skeletal muscle fiber is sheathed in connective tissue, with groups of adjacent muscle fibers bundled together into units called **fascicles**. Collagen, elastic fibers, nerves, and blood vessels are found between the fascicles. The entire muscle is enclosed in a connective tissue sheath that is continuous with the connective tissue around the muscle fibers and fascicles and with the tendons holding the muscle to underlying bones.

Muscle Fiber Anatomy Muscle physiologists, like neurobiologists, use specialized vocabulary (**TBL. 12.1**). The cell membrane of a muscle fiber is called the **sarcolemma** {*sarkos*, flesh + *lemma*, shell}, and the cytoplasm is called the **sarcoplasm**. The main intracellular structures in striated muscles are **myofibrils** {*myo-*, muscle}, highly organized bundles of contractile and elastic proteins that carry out the work of contraction.

Skeletal muscle fibers also contain extensive **sarcoplasmic reticulum (SR)**, a form of modified endoplasmic reticulum that wraps around each myofibril like a piece of lace (Figs. 12.3b, 12.4). The sarcoplasmic reticulum consists of longitudinal tubules with enlarged end regions called the **terminal cisternae** {*cisterna*, a reservoir}. The sarcoplasmic reticulum concentrates and sequesters Ca^{2+} {*sequestrare*, to put in the hands of a trustee} with the help of a Ca^{2+}*-ATPase* in the SR membrane. Calcium release from the SR creates calcium signals that play a key role in contraction in all types of muscle.

The terminal cisternae are adjacent to and closely associated with a branching network of **transverse tubules**, also known as **t-tubules** (**FIG. 12.4**). One t-tubule and its two flanking terminal cisternae are called a *triad*. The membranes of t-tubules are a continuation of the muscle fiber membrane, which makes the lumen of t-tubules continuous with the extracellular fluid.

To understand how this network of t-tubules deep inside the muscle fiber communicates with the outside, take a lump of soft clay and poke your finger into the middle of it. Notice how the outside surface of the clay (analogous to the surface membrane of the muscle fiber) is now continuous with the sides of the hole that you poked in the clay (the membrane of the t-tubule).

T-tubules allow action potentials to move rapidly from the cell surface into the interior of the fiber so that they reach the terminal cisternae nearly simultaneously. Without t-tubules, the action potential would reach the center of the fiber only by conduction of the action potential through the cytosol, a slower and less direct process that would delay the response time of the muscle fiber.

The cytosol between the myofibrils contains many glycogen granules and mitochondria. Glycogen, the storage form of glucose found in animals, is a reserve source of energy. Mitochondria contain the enzymes for oxidative phosphorylation of glucose and other biomolecules, so they produce much of the ATP for muscle contraction [p. 70].

Myofibrils Are Muscle Fiber Contractile Structures

One muscle fiber contains a thousand or more myofibrils that occupy most of the intracellular volume, leaving little space for cytosol and organelles (Fig. 12.3b). Each myofibril is composed of several types of proteins organized into repeating contractile structures called *sarcomeres*. Myofibril proteins include the motor protein *myosin*, which forms thick filaments; the microfilament *actin* [p. 68], which creates *thin filaments;* the regulatory proteins *tropomyosin* and *troponin;* and two giant accessory proteins, *titin* and *nebulin*.

Myosin {*myo-*, muscle} is a motor protein with the ability to create movement [p. 69]. Various isoforms of myosin occur in different types of muscle and help determine the muscle's speed of contraction. One myosin molecule is composed of two identical protein chains, each with one large *heavy chain* plus two smaller *light chains*.

The heavy chains of the myosin molecule are organized into three regions: a pair of tadpole-like heads, stiff rodlike sections that intertwine to form a tail, and an elastic neck region that joins the head to the tail (Fig. 12.3e). The neck creates a hinge that allows the heads to swivel around their point of attachment.

The myosin light chains wrap around the lower neck region of the myosin heads and add rigidity to the hinge. They have a regulatory function as well. Phosphorylation of myosin light chains in striated muscles enhances the force of muscle contraction by mechanisms that are still being investigated.

The heavy chains of the myosin heads form the *motor domain* that uses energy from the high-energy phosphate bond of ATP to create movement. Because myosin acts as an enzyme to hydrolyze ATP, the motor domain is considered a **myosin ATPase**. The heavy chains of the myosin heads also contain the binding sites for actin.

In skeletal muscle, about 250 myosin molecules join to create a **thick filament**. Each thick filament is arranged so that the myosin heads are clustered at each end of the filament, and the central region of the filament is a bundle of myosin tails (see Fig. 12.3e).

Actin {*actum*, to do} is a protein that makes up the **thin filaments** of the muscle fiber. One actin molecule is a globular protein (*G-actin*), represented in Figure 12.3f by a round ball. Usually, multiple G-actin molecules polymerize to form long chains or filaments, called *F-actin*. In skeletal muscle, two F-actin polymers twist together like a double strand of beads, creating the thin filaments of the myofibril.

Most of the time, the parallel thick and thin filaments of the myofibril are connected by myosin **crossbridges** that span the space between the filaments. Each G-actin molecule has a single *myosin-binding site,* and each myosin head has one

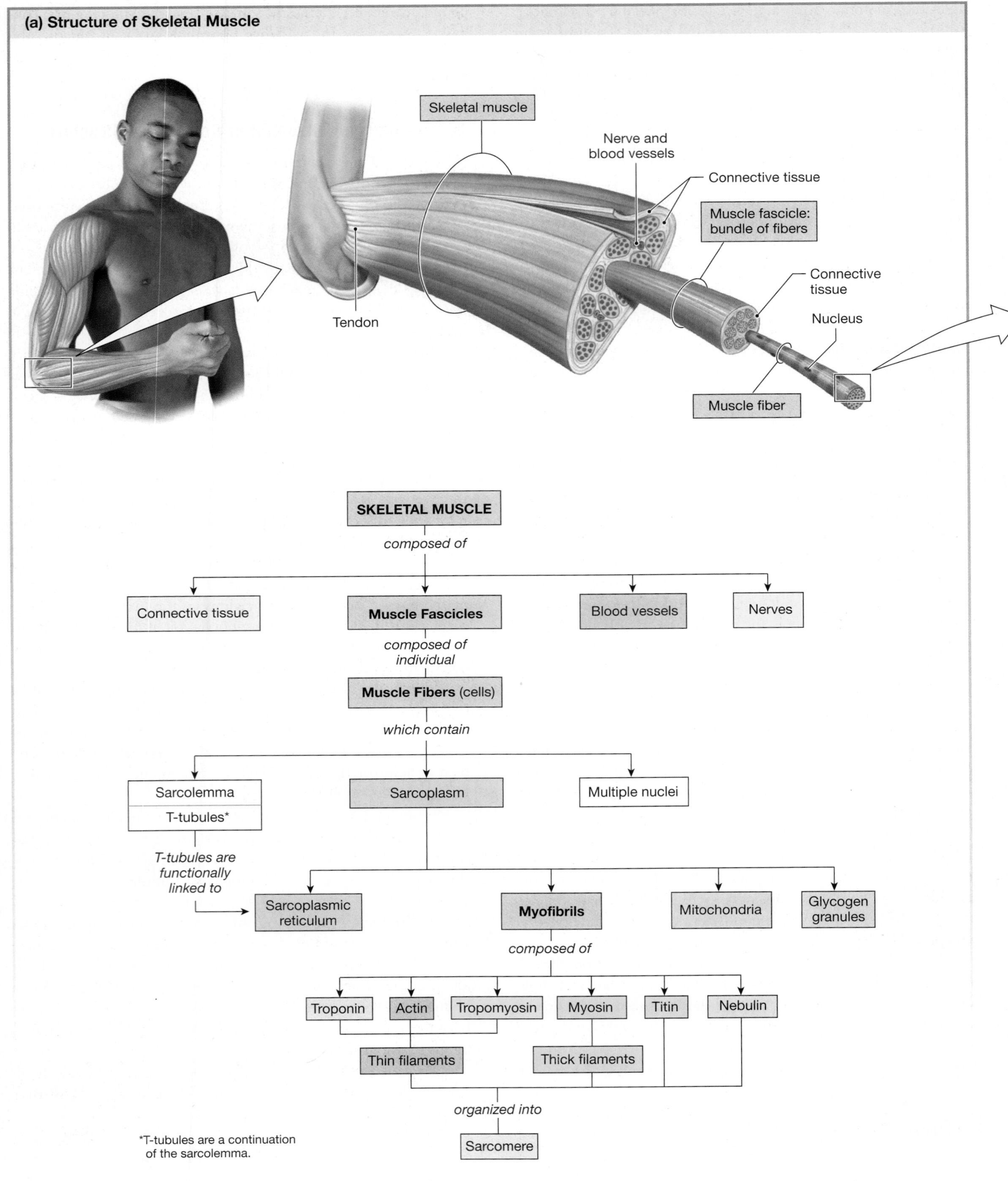

(a) Structure of Skeletal Muscle
Skeletal muscle
Nerve and blood vessels
Connective tissue
Muscle fascicle: bundle of fibers
Connective tissue
Tendon
Nucleus
Muscle fiber
SKELETAL MUSCLE
composed of
Connective tissue
Muscle Fascicles
Blood vessels
Nerves
composed of individual
Muscle Fibers (cells)
which contain
Sarcolemma
T-tubules*
Sarcoplasm
Multiple nuclei
T-tubules are functionally linked to
Sarcoplasmic reticulum
Myofibrils
Mitochondria
Glycogen granules
composed of
Troponin
Actin
Tropomyosin
Myosin
Titin
Nebulin
Thin filaments
Thick filaments
organized into
Sarcomere
*T-tubules are a continuation of the sarcolemma.

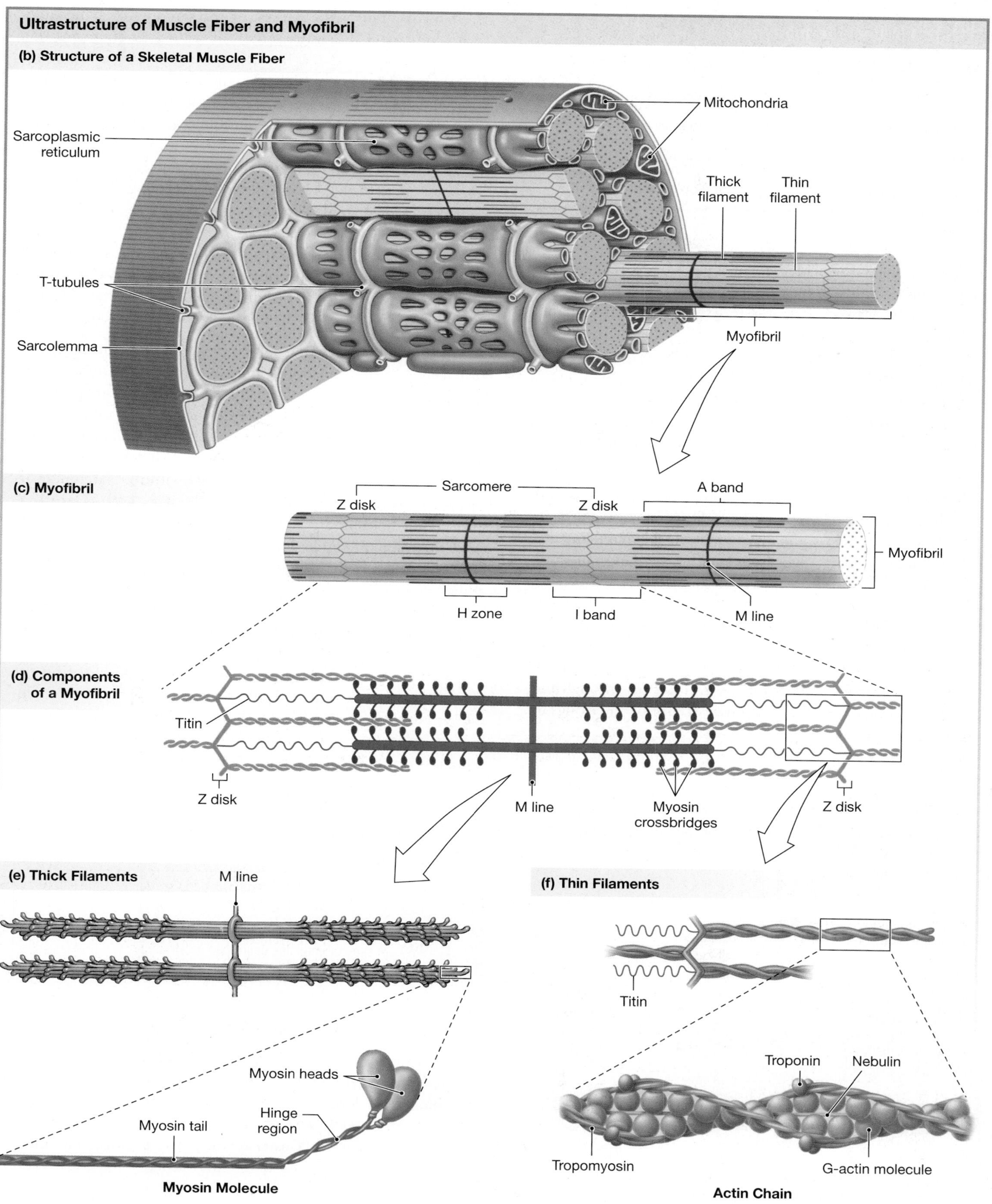

Ultrastructure of Muscle Fiber and Myofibril
(b) Structure of a Skeletal Muscle Fiber
Mitochondria
Sarcoplasmic reticulum
Thick filament
Thin filament
T-tubules
Sarcolemma
Myofibril
(c) Myofibril
Sarcomere
Z disk
Z disk
A band
Myofibril
H zone
I band
M line
(d) Components of a Myofibril
Titin
Z disk
M line
Myosin crossbridges
Z disk
(e) Thick Filaments
M line
Myosin heads
Hinge region
Myosin tail
Myosin Molecule
(f) Thin Filaments
Titin
Troponin
Nebulin
Tropomyosin
G-actin molecule
Actin Chain

TABLE 12.1 Muscle Terminology

General Term	Muscle Equivalent
Muscle cell	Muscle fiber
Cell membrane	Sarcolemma
Cytoplasm	Sarcoplasm
Modified endoplasmic reticulum	Sarcoplasmic reticulum

actin-binding site. Crossbridges form when the myosin heads of thick filaments bind to actin in the thin filaments (Fig. 12.3d). Crossbridges have two states: low-force (relaxed muscles) and high-force (contracting muscles).

Under a light microscope, the arrangement of thick and thin filaments in a myofibril creates a repeating pattern of alternating light and dark bands (Figs. 12.1a, 12.3c). One repeat of the pattern forms a **sarcomere** {*sarkos*, flesh + *-mere*, a unit or segment}, the contractile unit of the myofibril. Each sarcomere has the following elements (**FIG. 12.5**):

1. **Z disks**. One sarcomere is composed of two Z disks and the filaments found between them. Z disks are zigzag protein structures that serve as the attachment site for thin filaments. The abbreviation *Z* comes from *zwischen*, the German word for "between."
2. **I bands**. These are the lightest color bands of the sarcomere and represent a region occupied only by thin filaments. The abbreviation *I* comes from *isotropic*, a description from early microscopists meaning that this region reflects light uniformly under a polarizing microscope. A Z disk runs through the middle of every I band, so each half of an I band belongs to a different sarcomere.

FIG. 12.4 T-tubules

T-tubules are extensions of the cell membrane (sarcolemma) that associate with the ends (terminal cisternae) of the sarcoplasmic reticulum.

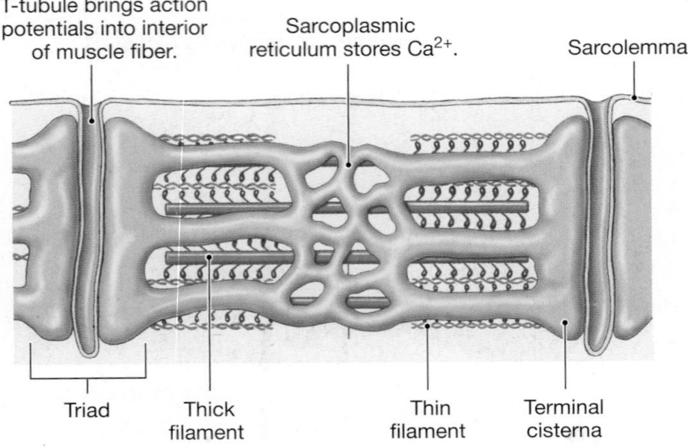
T-tubule brings action potentials into interior of muscle fiber.
Sarcoplasmic reticulum stores Ca^{2+}.
Sarcolemma
Triad
Thick filament
Thin filament
Terminal cisterna

3. **A band**. This is the darkest of the sarcomere's bands and encompasses the entire length of a thick filament. At the outer edges of the A band, the thick and thin filaments overlap. The center of the A band is occupied by thick filaments only. The abbreviation *A* comes from *anisotropic* {*an-*, not}, meaning that the protein fibers in this region scatter light unevenly.
4. **H zone**. This central region of the A band is lighter than the outer edges of the A band because the H zone is occupied by thick filaments only. The *H* comes from *helles*, the German word for "clear."
5. **M line**. This band represents proteins that form the attachment site for thick filaments, equivalent to the Z disk for the thin filaments. Each M line divides an A band in half. *M* is the abbreviation for *mittel*, the German word for "middle."

In three-dimensional array, the actin and myosin molecules form a lattice of parallel, overlapping thin and thick filaments, held in place by their attachments to the Z-disk and M-line proteins, respectively (Fig. 12.5b). When viewed end-on, each thin filament is surrounded by three thick filaments, and six thin filaments encircle each thick filament (Fig. 12.5c, rightmost circle).

The proper alignment of filaments within a sarcomere is ensured by two proteins: titin and nebulin (**FIG. 12.6**). **Titin** is a huge elastic molecule and the largest known protein, composed of more than 25,000 amino acids. A single titin molecule stretches from one Z disk to the neighboring M line. To get an idea of the immense size of titin, imagine that one titin molecule is an 8-foot-long piece of the very thick rope used to tie ships to a wharf. By comparison, a single actin molecule would be about the length and weight of a single eyelash.

Titin has two functions: (1) it stabilizes the position of the contractile filaments and (2) its elasticity returns stretched muscles to their resting length. Titin is helped by **nebulin,** an inelastic giant protein that lies alongside thin filaments and attaches to the Z disk. Nebulin helps align the actin filaments of the sarcomere.

Concept Check

2. Why are the ends of the A band the darkest region of the sarcomere when viewed under the light microscope?
3. What is the function of t-tubules?
4. Why are skeletal muscles described as striated?

Muscle Contraction Creates Force

The contraction of muscle fibers is a remarkable process that enables us to create force to move or to resist a load. In muscle physiology, the force created by contracting muscle is called **muscle tension**. The **load** is a weight or force that opposes contraction of a muscle. **Contraction,** the creation of tension in a muscle, is an active process that requires energy input from ATP. **Relaxation** is the release of tension created by a contraction.

FIG. 12.5 ESSENTIALS The Sarcomere

Organization of a Sarcomere

The Z disk (not shown in part (c)) has accessory proteins that link the thin filaments together, similar to the accessory proteins shown for the M line. Myosin heads are omitted for simplicity.

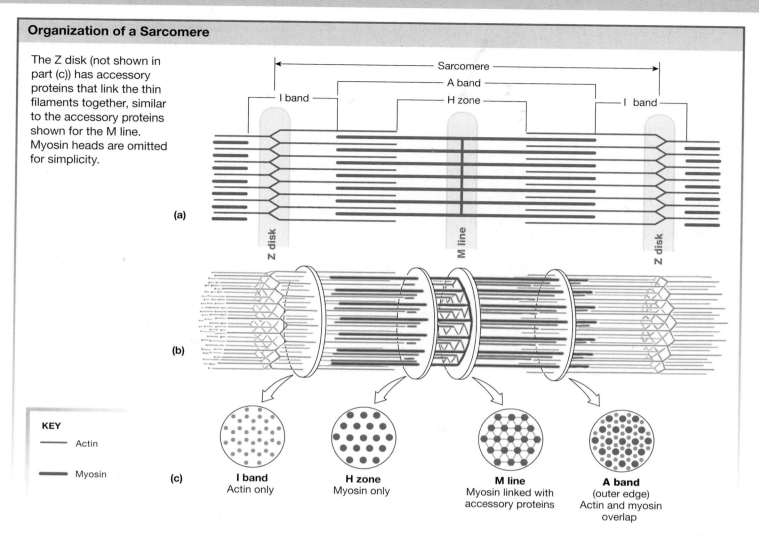

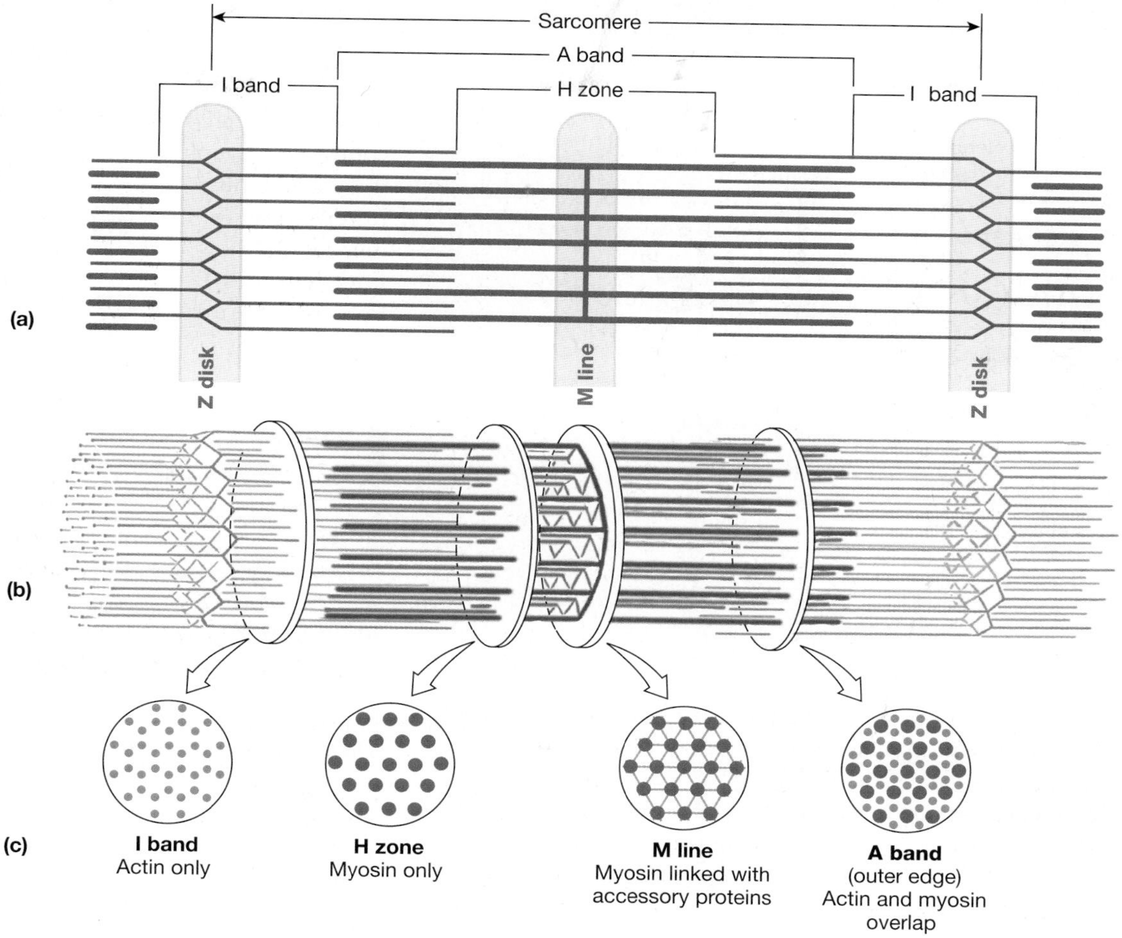

Sarcomere
A band
I band
H zone
I band
Z disk
M line
Z disk

(a)

(b)

KEY
Actin
Myosin

(c)

I band
Actin only

H zone
Myosin only

M line
Myosin linked with accessory proteins

A band
(outer edge)
Actin and myosin overlap

The sarcomere shortens during contraction.

As contraction takes place, actin and myosin do not change length but instead slide past one another.

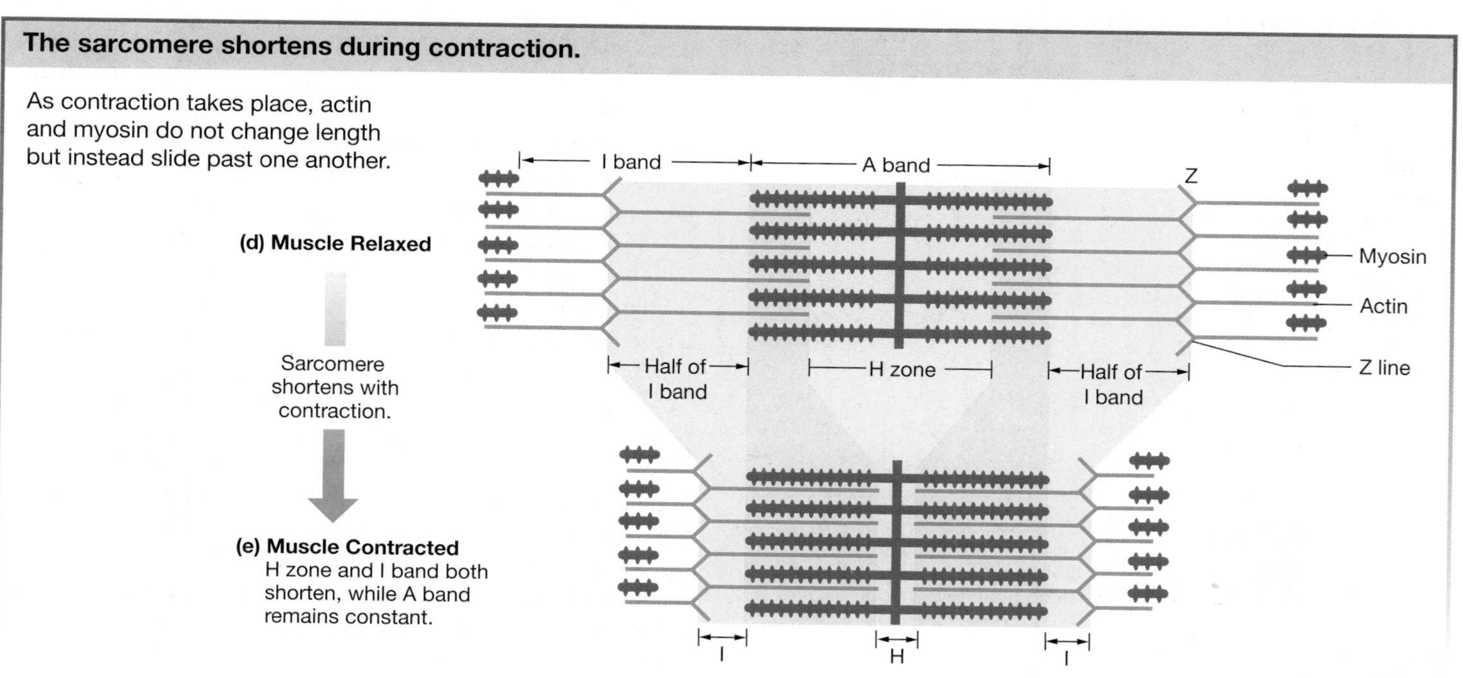
I band
A band
Z
(d) Muscle Relaxed
Myosin
Actin
Z line
Half of I band
H zone
Half of I band
Sarcomere shortens with contraction.
(e) Muscle Contracted
H zone and I band both shorten, while A band remains constant.
I
H
I

FIG. 12.6 Titin and nebulin

Titin and nebulin are giant accessory proteins. Titin spans the distance from one Z disk to the neighboring M line. Nebulin, lying along the thin filaments, attaches to a Z disk but does not extend to the M line.

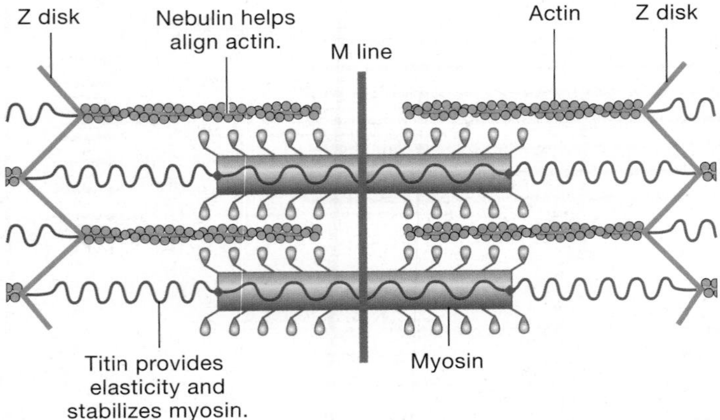
Z disk
Nebulin helps align actin.
M line
Actin
Z disk
Titin provides elasticity and stabilizes myosin.
Myosin

3. At the molecular level, a **contraction-relaxation cycle** can be explained by the *sliding filament theory of contraction*. In intact muscles, one contraction-relaxation cycle is called a muscle *twitch*.

In the sections that follow, we start with the sliding filament theory for muscle contraction. From there, we look at the integrated function of a muscle fiber as it undergoes excitation-contraction coupling. The skeletal muscle section ends with a discussion of the innervation of muscles and how muscles move bones around joints.

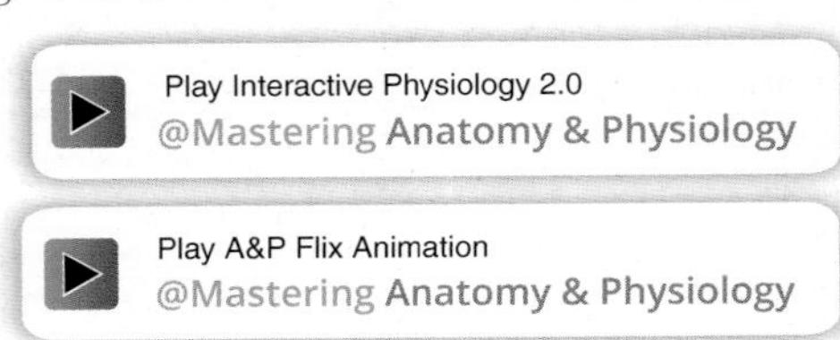
Play Interactive Physiology 2.0
@Mastering Anatomy & Physiology
Play A&P Flix Animation
@Mastering Anatomy & Physiology

Concept Check

5. What are the three anatomical elements of a neuromuscular junction?
6. What is the chemical signal at a neuromuscular junction?

FIGURE 12.7 maps the major steps leading up to skeletal muscle contraction.

1. **Events at the neuromuscular junction** convert an acetylcholine signal from a somatic motor neuron into an electrical signal in the muscle fiber [p. 382].
2. **Excitation-contraction (E-C) coupling** is the process in which muscle action potentials are translated into calcium signals. The calcium signals in turn initiate a contraction-relaxation cycle.

FIG. 12.7 Summary map of muscle contraction

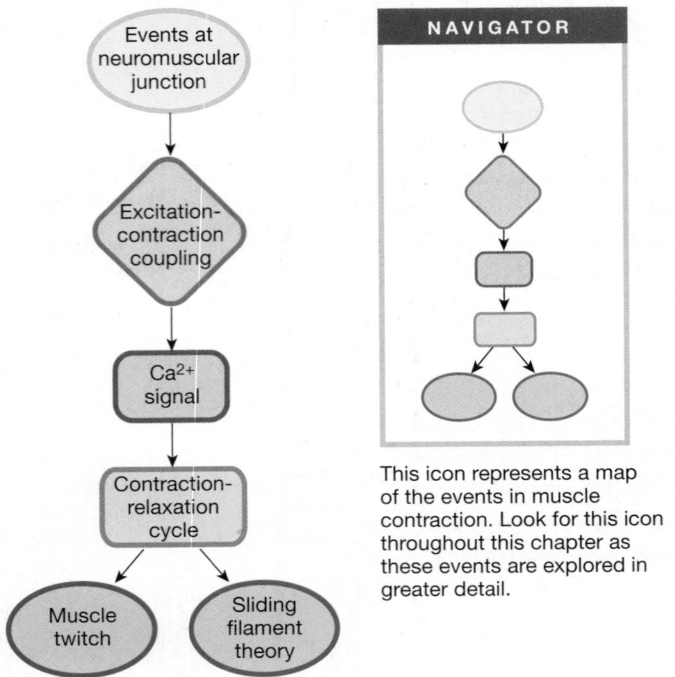
Events at neuromuscular junction
Excitation-contraction coupling
Ca^{2+} signal
Contraction-relaxation cycle
Muscle twitch
Sliding filament theory
NAVIGATOR
This icon represents a map of the events in muscle contraction. Look for this icon throughout this chapter as these events are explored in greater detail.

Actin and Myosin Slide Past Each Other during Contraction

In previous centuries, scientists observed that when muscles move a load, they shorten. This observation led to early theories of contraction, which proposed that muscles were made of molecules that curled up and shortened when active, then relaxed and stretched at rest, like elastic in reverse. The theory received support when myosin was found to be a helical molecule that shortened upon heating (the reason meat shrinks when you cook it).

In 1954, however, scientists Andrew Huxley and Rolf Niedergerke discovered that the length of the A band of a myofibril remains constant during contraction. Because the A band represents the myosin filament, Huxley and Niedergerke realized that shortening of the myosin molecule could not be responsible for contraction. Subsequently, they proposed an alternative model, the **sliding filament theory of contraction**. In this model, overlapping actin and myosin filaments of fixed length slide past one another in an energy-requiring process, resulting in muscle contraction.

If you examine a myofibril at its resting length, you see that within each sarcomere, the ends of the thick and thin filaments overlap slightly (Fig. 12.5d). In the relaxed state, a sarcomere has a large I band (thin filaments only) and an A band whose length is the length of the thick filament.

When the muscle contracts, the thick and thin filaments slide past each other. The Z disks of the sarcomere move closer together as the sarcomere shortens (Fig. 12.5e). The I band and H zone—regions where actin and myosin do not overlap in resting muscle—almost disappear.

Despite shortening of the sarcomere, the length of the A band remains constant. These changes are consistent with the sliding of thin actin filaments along the thick myosin filaments as the actin filaments move toward the M line in the center of the sarcomere. It is from this process that the sliding filament theory of contraction derives its name.

The sliding filament theory explains how a muscle can contract and create force without creating movement. For example, if you push on a wall, you are creating tension in many muscles of your body without moving the wall. According to the sliding filament theory, tension generated in a muscle fiber is directly proportional to the number of high-force crossbridges between the thick and thin filaments.

Myosin Crossbridges Move Actin Filaments

The movement of myosin crossbridges provides force that pushes the actin filament during contraction. The process can be compared to a competitive sailing team, with many people holding the rope that raises a heavy mainsail. When the order to raise the mainsail comes, each person on the team begins pulling on the rope, hand over hand, grabbing, pulling, and releasing repeatedly as the rope moves past.

In muscle, myosin heads bind to actin molecules, which are the "rope." A calcium signal initiates the **power stroke**, when myosin crossbridges swivel and push the actin filaments toward the center of the sarcomere. At the end of a power stroke, each myosin head releases actin, then swivels back and binds to a new actin molecule, ready to start another contractile cycle. During contraction, the heads do not all release at the same time or the fibers would slide back to their starting position, just as the mainsail would fall if the sailors all released the rope at the same time.

The power stroke repeats many times as a muscle fiber contracts. The myosin heads bind, push, and release actin molecules over and over as the thin filaments move toward the center of the sarcomere.

Myosin ATPase Where does energy for the power stroke come from? The answer is ATP. Myosin converts the chemical bond energy of ATP into the mechanical energy of crossbridge motion.

Myosin is an ATPase (*myosin ATPase*) that hydrolyzes ATP to ADP and inorganic phosphate (P_i). The energy released by ATP hydrolysis is trapped by myosin and stored as potential energy in the angle between the myosin head and the long axis of the myosin filament. Myosin heads in this position are said to be "cocked," or ready to rotate. The potential energy of the cocked heads becomes kinetic energy in the power stroke that moves actin.

Calcium Signals Initiate Contraction

How does a calcium signal turn muscle contraction on and off? The answer is found in **troponin (TN)**, a calcium-binding complex of three proteins. Troponin controls the positioning of an elongated protein polymer, **tropomyosin** {*tropos*, to turn}.

In resting skeletal muscle, tropomyosin wraps around actin filaments and partially covers actin's myosin-binding sites (**FIG. 12.8a**). This is tropomyosin's blocking or "off" position. Weak, low-force actin-myosin binding can still take place, but myosin is blocked from completing its power stroke, much as the safety latch on a gun keeps the cocked trigger from being pulled. Before contraction can occur, tropomyosin must be shifted to an "on" position that uncovers the remainder of actin's myosin-binding site.

The off-on positioning of tropomyosin is regulated by troponin. When contraction begins in response to a calcium signal (1 in Fig. 12.8b), one protein of the complex—**troponin C**—binds reversibly to Ca^{2+} 2. The calcium-troponin C complex pulls tropomyosin completely away from actin's myosin-binding sites 3. This "on" position enables the myosin heads to form strong, high-force crossbridges and carry out their power strokes 4, moving the actin filament 5. Contractile cycles repeat as long as the binding sites are uncovered.

For muscle relaxation to occur, Ca^{2+} concentrations in the cytosol must decrease. By the law of mass action [p. 48], when

FIG. 12.8 Troponin and tropomyosin

(a) Relaxed state.
Myosin head cocked. Tropomyosin partially blocks binding site on actin. Myosin is weakly bound to actin.

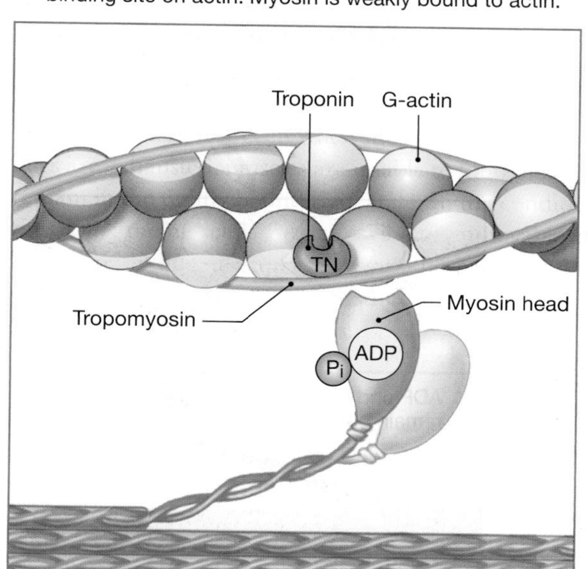
Troponin
G-actin
TN
Tropomyosin
Myosin head
ADP
Pi

(b) Initation of contraction.
A calcium signal initiates contraction.

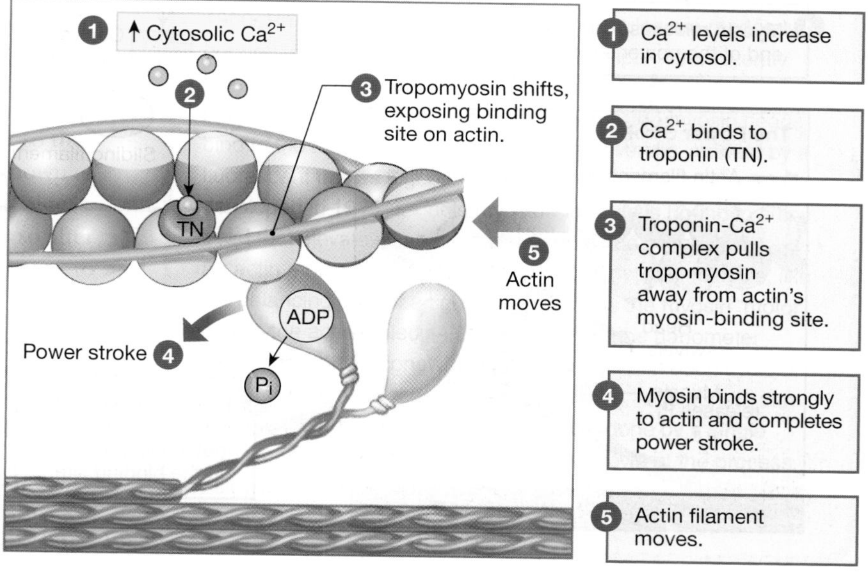
1 ↑ Cytosolic Ca2+
2
3 Tropomyosin shifts, exposing binding site on actin.
TN
5 Actin moves
ADP
Power stroke 4
Pi
1 Ca2+ levels increase in cytosol.
2 Ca2+ binds to troponin (TN).
3 Troponin-Ca2+ complex pulls tropomyosin away from actin's myosin-binding site.
4 Myosin binds strongly to actin and completes power stroke.
5 Actin filament moves.

a muscle fiber is called *excitation-contraction coupling.* E-C coupling has four major events:

1. Acetylcholine (ACh) is released from the somatic motor neuron.
2. ACh initiates an action potential in the muscle fiber.
3. The muscle action potential triggers calcium release from the sarcoplasmic reticulum.
4. Calcium combines with troponin to initiate contraction.

Now let's look at these steps in detail.

Acetylcholine released into the synapse at a neuromuscular junction binds to ACh receptor-channels on the motor end plate of the muscle fiber (**FIG. 12.10a** 1) [p. 368]. When the ACh-gated channels open, they allow both Na^+ and K^+ to cross the membrane. However, Na^+ influx exceeds K^+ efflux because the electrochemical driving force is greater for Na^+ [p. 153]. The addition of net positive charge to the muscle fiber depolarizes the membrane, creating an **end-plate potential (EPP)**. Normally, end-plate potentials always reach threshold and initiate a muscle action potential (Fig. 12.10a 2).

The action potential travels across the surface of the muscle fiber and into the t-tubules by the sequential opening of voltage-gated Na^+ channels. The process is similar to the conduction of action potentials in axons, although action potentials in skeletal muscle are conducted more slowly than action potentials in myelinated axons [p. 247].

The action potential that moves down the t-tubules causes Ca^{2+} release from the sarcoplasmic reticulum (Fig. 12.10b 3, 4). Free cytosolic Ca^{2+} levels in a resting muscle are normally quite low, but after an action potential, they increase about 100 fold. As you've learned, when cytosolic Ca^{2+} levels are high, Ca^{2+} binds to troponin, tropomyosin moves to the "on" position 5, and contraction occurs 6.

At the molecular level, transduction of the electrical signal into a calcium signal requires two key membrane proteins. The t-tubule membrane contains a voltage-sensing **L-type calcium channel** protein ($Ca_v1.1$) called a **dihydropyridine (DHP) receptor** (Fig. 12.10b 3). The DHP receptors, found only in skeletal muscle, are mechanically linked to Ca^{2+} channels in the adjacent sarcoplasmic reticulum. These SR **Ca^{2+} release channels** are also known as **ryanodine receptors (RyR)**.

RUNNING PROBLEM

Paul had experienced mild attacks of muscle weakness in his legs before, usually in the morning. Twice the weakness had come on after exposure to cold. Each attack had disappeared within minutes, and Paul seemed to suffer no lasting effects. On the advice of Paul's family doctor, Mrs. Leong takes her son to see a specialist in muscle disorders, who suspects a condition called periodic paralysis. The periodic paralyses are a family of disorders caused by Na^+ or Ca^{2+} ion channel mutations in the membranes of skeletal muscle fibers. The specialist believes that Paul has a condition in which defective voltage-gated Na^+ channels fail to inactivate after they open.

Q1: *When Na^+ channels on the muscle membrane open, which way does Na^+ move?*

Q2: *What effect would continued movement of Na^+ have on the membrane potential of muscle fibers?*

375 – 386 – 390 – 400 – 402 – 409

When the depolarization of an action potential

Play A&P Flix Animation
@Mastering Anatomy & Physiology

reaches a DHP receptor, the receptor changes conformation. The conformation change opens the RyR Ca^{2+} release channels in the sarcoplasmic reticulum (Fig. 12.10b 4). Stored Ca^{2+} then flows down its electrochemical gradient into the cytosol, where it initiates contraction.

Scientists used to believe that the calcium channel we call the DHP receptor did not form an open channel for calcium entry from the ECF. However, in recent years, it has become apparent that there is a small amount of Ca^{2+} movement through the DHP receptor, described as *excitation-coupled Ca^{2+} entry.* However, skeletal muscle contraction will still take place if there is no ECF Ca^{2+} to come through the channel, so the physiological role of excitation-coupled Ca^{2+} entry is unclear.

Relaxation To end a contraction, calcium must be removed from the cytosol. The sarcoplasmic reticulum pumps Ca^{2+} back into its lumen using a **Ca^{2+}-ATPase** [p. 142]. As the free cytosolic Ca^{2+} concentration decreases, the equilibrium between bound and unbound Ca^{2+} is disturbed and calcium releases from troponin. Removal of Ca^{2+} allows tropomyosin to slide back and block actin's myosin-binding site. As the crossbridges release, the muscle fiber relaxes with the help of elastic fibers in the sarcomere and in the connective tissue of the muscle.

Timing of E-C Coupling The graphs in **FIGURE 12.11** show the timing of electrical and mechanical events during E-C coupling. The somatic motor neuron action potential is followed by the skeletal muscle action potential, which in turn is followed by contraction. A single contraction-relaxation cycle in a skeletal muscle fiber is known as a **twitch**. Notice that there is a short delay—the **latent period**—between the muscle action potential and the beginning of muscle tension development. This delay represents the time required for calcium release and binding to troponin.

Once contraction begins, muscle tension increases steadily to a maximum value as crossbridge interaction increases. Tension then decreases in the relaxation phase of the twitch. During relaxation, elastic elements of the muscle return the sarcomeres to their resting length.

A single action potential in a muscle fiber evokes a single twitch (Fig. 12.11, bottom graph). However, muscle twitches vary from fiber to fiber in the speed with which they develop tension (the rising slope of the twitch curve), the maximum tension they achieve (the height of the twitch curve), and the duration of the twitch (the width of the twitch curve). You will learn about factors that affect these parameters in upcoming sections. First, we discuss how muscles produce ATP to provide energy for contraction and relaxation.

Play Interactive Physiology 2.0
@Mastering Anatomy & Physiology

Concept Check

12. Which part of contraction requires ATP? Does relaxation require ATP?
13. What events are taking place during the latent period before contraction begins?

 Excitation-Contraction Coupling and Relaxation

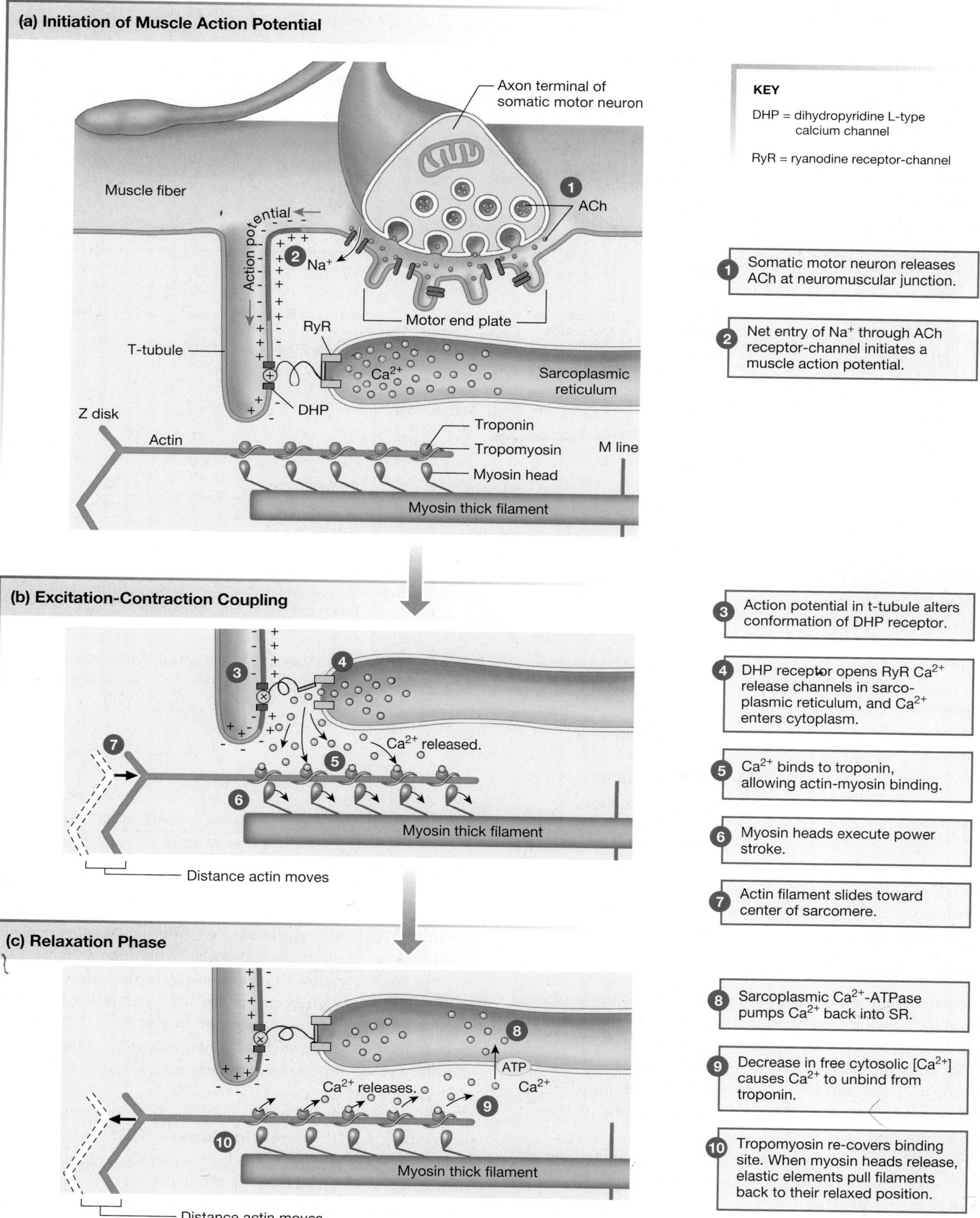
(a) Initiation of Muscle Action Potential
Axon terminal of somatic motor neuron
Muscle fiber
Action potential
Na+
ACh
Motor end plate
RyR
T-tubule
Ca2+
Sarcoplasmic reticulum
DHP
Z disk
Actin
Troponin
Tropomyosin
M line
Myosin head
Myosin thick filament
(b) Excitation-Contraction Coupling
Ca2+ released.
Myosin thick filament
Distance actin moves
(c) Relaxation Phase
ATP
Ca2+
Ca2+ releases.
Myosin thick filament
Distance actin moves

KEY

DHP = dihydropyridine L-type calcium channel

RyR = ryanodine receptor-channel

1. Somatic motor neuron releases ACh at neuromuscular junction.
2. Net entry of Na^+ through ACh receptor-channel initiates a muscle action potential.
3. Action potential in t-tubule alters conformation of DHP receptor.
4. DHP receptor opens RyR Ca^{2+} release channels in sarcoplasmic reticulum, and Ca^{2+} enters cytoplasm.
5. Ca^{2+} binds to troponin, allowing actin-myosin binding.
6. Myosin heads execute power stroke.
7. Actin filament slides toward center of sarcomere.
8. Sarcoplasmic Ca^{2+}-ATPase pumps Ca^{2+} back into SR.
9. Decrease in free cytosolic [Ca^{2+}] causes Ca^{2+} to unbind from troponin.
10. Tropomyosin re-covers binding site. When myosin heads release, elastic elements pull filaments back to their relaxed position.

FIG. 12.11 Timing of E-C coupling

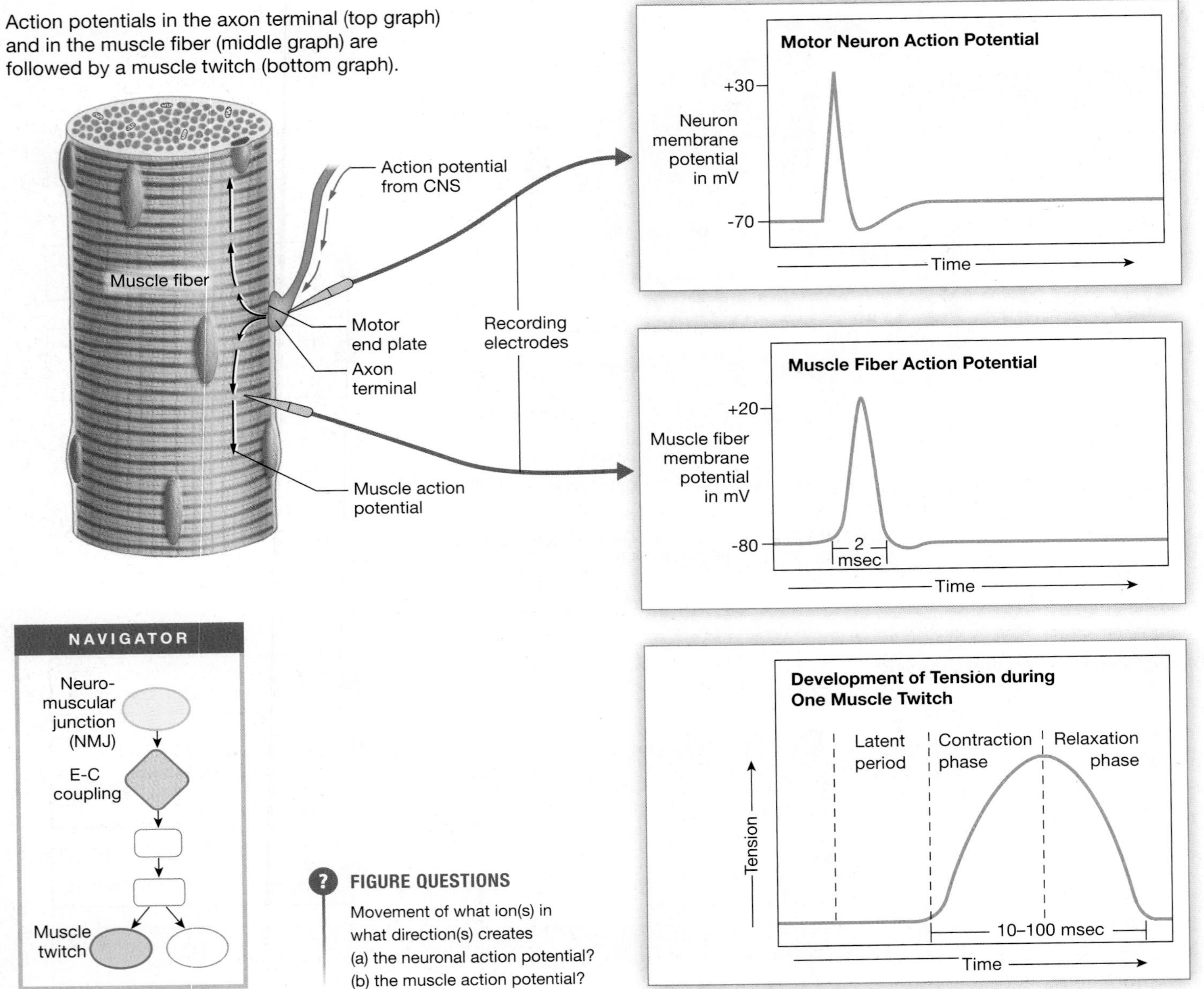

Action potentials in the axon terminal (top graph) and in the muscle fiber (middle graph) are followed by a muscle twitch (bottom graph).
Action potential from CNS
Muscle fiber
Motor end plate
Axon terminal
Recording electrodes
Muscle action potential
Motor Neuron Action Potential
+30
Neuron membrane potential in mV
-70
Time
Muscle Fiber Action Potential
+20
Muscle fiber membrane potential in mV
-80
2 msec
Time
Development of Tension during One Muscle Twitch
Latent period
Contraction phase
Relaxation phase
Tension
10–100 msec
Time
NAVIGATOR
Neuro-muscular junction (NMJ)
E-C coupling
Muscle twitch
FIGURE QUESTIONS
Movement of what ion(s) in what direction(s) creates
(a) the neuronal action potential?
(b) the muscle action potential?

Skeletal Muscle Contraction Requires a Steady Supply of ATP

The muscle fiber's use of ATP is a key feature of muscle physiology. Muscles require energy constantly: during contraction for crossbridge movement and release, during relaxation to pump Ca^{2+} back into the sarcoplasmic reticulum, and after E-C coupling to restore Na^+ and K^+ to the extracellular and intracellular compartments, respectively. Where do muscles get the ATP they need for this work?

The amount, or pool, of ATP stored in a muscle fiber at any one time is sufficient for only about eight twitches. As ATP is converted to ADP and P_i during contraction, the ATP pool must be replenished by transfer of energy from other high-energy phosphate bonds or by synthesis of ATP through the slower metabolic pathways of glycolysis and oxidative phosphorylation.

The backup energy source of muscles is **phosphocreatine,** a molecule whose high-energy phosphate bonds are created from creatine and ATP when muscles are at rest (**FIG. 12.12**). When muscles become active, such as during exercise, the high-energy phosphate group of phosphocreatine is quickly transferred to ADP, creating more ATP to power the muscles.

The enzyme that transfers the phosphate group from phosphocreatine to ADP is **creatine kinase (CK),** also known as *creatine phosphokinase* (CPK). Muscle cells contain large amounts of this enzyme. Consequently, elevated blood levels of creatine kinase usually indicate damage to skeletal or cardiac muscle. Because the

FIG. 12.12 Phosphocreatine

Resting muscle stores energy from ATP in the high-energy bonds of phosphocreatine. Working muscle then uses that stored energy.

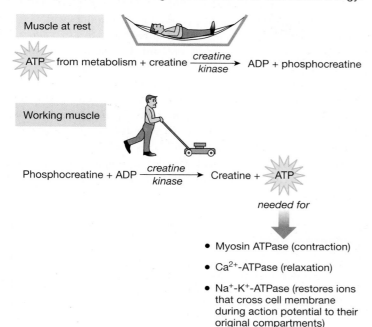

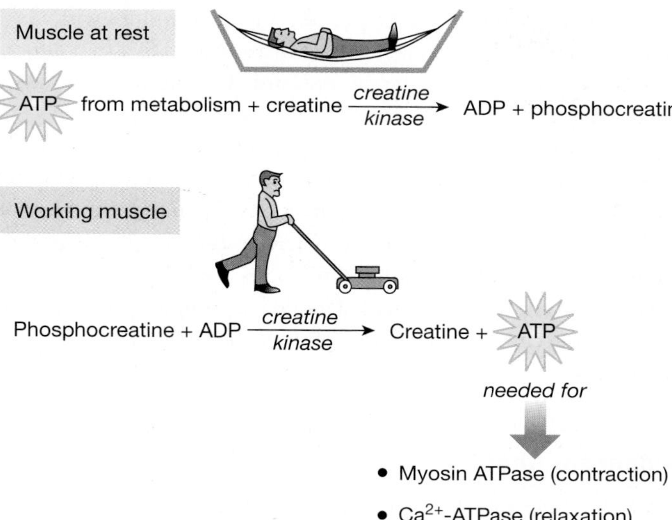
Muscle at rest
ATP from metabolism + creatine —creatine kinase→ ADP + phosphocreatine
Working muscle
Phosphocreatine + ADP —creatine kinase→ Creatine + ATP
needed for
• Myosin ATPase (contraction)
• Ca^{2+}-ATPase (relaxation)
• Na^+-K^+-ATPase (restores ions that cross cell membrane during action potential to their original compartments)

two muscle types contain different isozymes [p. 99], clinicians can distinguish cardiac tissue damage during a heart attack from skeletal muscle damage.

Energy stored in high-energy phosphate bonds is very limited, so muscle fibers must use metabolism of biomolecules to transfer energy from covalent bonds to ATP. Carbohydrates, particularly glucose, are the most rapid and efficient source of energy for ATP production. Glucose is metabolized through glycolysis to pyruvate [p. 106]. In the presence of adequate oxygen, pyruvate goes into the citric acid cycle, producing about 30 ATP for each molecule of glucose.

When oxygen concentrations fall during strenuous exercise, muscle fiber metabolism relies more on *anaerobic glycolysis.* In this pathway, glucose is metabolized to lactate with a yield of only 2 ATP per glucose [p. 109]. Anaerobic metabolism of glucose is a quicker source of ATP but produces many fewer ATP per glucose. When energy demands are greater than the amount of ATP that can be produced through anaerobic glucose metabolism, muscles can function for only a short time without fatiguing.

Muscle fibers also obtain energy from fatty acids, although this process always requires oxygen. During rest and light exercise, skeletal muscles burn fatty acids along with glucose, one reason that modest exercise programs of brisk walking are an effective way to reduce body fat. However, the metabolic process by which fatty acids are converted to acetyl CoA is relatively slow and cannot produce ATP rapidly enough to meet the energy needs of muscle fibers during strenuous exercise. Under these conditions, muscle fibers rely more on glucose.

Proteins normally are not a source of energy for muscle contraction. Most amino acids found in muscle fibers are used to synthesize proteins rather than to produce ATP.

Do muscles ever run out of ATP? You might think so if you have ever exercised to the point of fatigue, the point at which you feel that you cannot continue or your limbs refuse to obey commands from your brain. Most studies show, however, that even intense exercise uses only 30% of the ATP in a muscle fiber. The condition we call fatigue must come from other changes in the exercising muscle.

Concept Check

14. According to the convention for naming enzymes, what does the name creatine kinase tell you about this enzyme's function? [Hint: p. 101]

15. The reactions in Figure 12.12 show that creatine kinase catalyzes the creatine-phosphocreatine reaction in both directions. What then determines the direction that the reaction goes at any given moment? [Hint: p. 48]

Fatigue Has Multiple Causes

The physiological term **fatigue** describes a reversible condition in which an exercising muscle is no longer able to generate or sustain the expected power output. Fatigue is highly variable. It is influenced by the intensity and duration of the contractile activity, by whether the muscle fiber is using aerobic or anaerobic metabolism, by the composition of the muscle, and by the fitness level of the individual. The study of fatigue is complex, and research in this area is complicated by the fact that experiments are done under a wide range of conditions, from "skinned" (sarcolemma removed) single muscle fibers to exercising humans. Although many different factors have been *associated with* fatigue, the factors that *cause* fatigue are still uncertain.

Factors that have been proposed to play a role in fatigue are classified into **central fatigue** mechanisms, which arise in the central nervous system, and **peripheral fatigue** mechanisms, which arise anywhere between the neuromuscular junction and the contractile elements of the muscle (**FIG. 12.13**). Most experimental evidence suggests that muscle fatigue arises from excitation-contraction failure or changes in contraction force in the muscle fiber rather than from failure of control neurons or neuromuscular transmission.

Central fatigue includes subjective feelings of tiredness and a desire to cease activity. Several studies have shown that this psychological fatigue precedes physiological fatigue in the muscles and therefore may be a protective mechanism. Low pH from acid production during ATP hydrolysis is often mentioned as a possible cause of fatigue, and some evidence suggests that acidosis may influence the sensation of fatigue perceived by the brain. However, homeostatic mechanisms for pH balance maintain blood pH at normal levels until exertion is nearly maximal, so pH as a factor in central fatigue probably applies only in cases of maximal exertion.

FIG. 12.13 Muscle fatigue

Muscle fatigue has many possible causes but the strongest evidence supports failure of EC coupling and subsequent events. In recent years, research indicated that lactate accumulation is no longer a likely cause of fatigue.

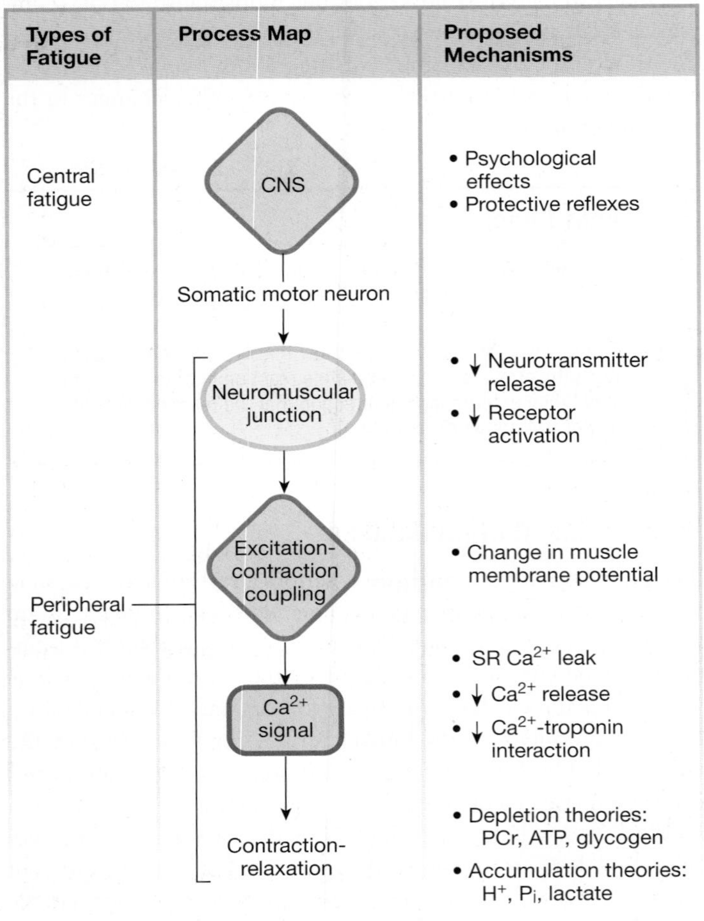

Types of Fatigue	Process Map	Proposed Mechanisms
Central fatigue	CNS ↓ Somatic motor neuron	• Psychological effects • Protective reflexes
Peripheral fatigue	Neuromuscular junction	• ↓ Neurotransmitter release • ↓ Receptor activation
	Excitation-contraction coupling	• Change in muscle membrane potential
	Ca^{2+} signal	• SR Ca^{2+} leak • ↓ Ca^{2+} release • ↓ Ca^{2+}-troponin interaction
	Contraction-relaxation	• Depletion theories: PCr, ATP, glycogen • Accumulation theories: H^+, P_i, lactate

Neural causes of fatigue could arise either from communication failure at the neuromuscular junction or from failure of the CNS command neurons. For example, if ACh is not synthesized in the axon terminal fast enough to keep up with neuron firing rate, neurotransmitter release at the synapse decreases. Consequently, the muscle end-plate potential fails to reach the threshold value needed to trigger a muscle fiber action potential, resulting in contraction failure. This type of fatigue is associated with some neuromuscular diseases, but it is probably not a factor in normal exercise.

Fatigue within the muscle fiber (peripheral fatigue) could occur in any of several sites. In extended submaximal exertion, fatigue is associated with the depletion of muscle glycogen stores. Because most studies show that lack of ATP is not a limiting factor, glycogen depletion may be affecting some other aspect of contraction, such as the release of Ca^{2+} from the sarcoplasmic reticulum.

The cause of fatigue in short-duration maximal exertion seems to be different. One theory is based on the increased levels of inorganic phosphate (P_i) produced when ATP and phosphocreatine are used for energy in the muscle fiber. Elevated cytoplasmic P_i may slow P_i release from myosin and thereby alter the power stroke (see Fig. 12.9 **4**).

Another theory suggests that elevated phosphate levels decrease Ca^{2+} release because the phosphate combines with Ca^{2+} to become calcium phosphate. Some investigators feel that alterations in Ca^{2+} release from the sarcoplasmic reticulum play a major role in fatigue.

Ion imbalances have also been implicated in fatigue. During maximal exercise, K^+ leaves the muscle fiber with each action potential, and as a result K^+ concentrations rise in the extracellular fluid of the t-tubules. The shift in K^+ alters the membrane potential of the muscle fiber. Changes in Na^+-K^+-ATPase activity may also be involved. In short, muscle fatigue is a complex phenomenon with multiple causes that interact with each other.

Concept Check

16. If K^+ concentration increases in the extracellular fluid surrounding a cell but does not change significantly in the cell's cytoplasm, the cell membrane (*depolarizes/hyperpolarizes*) and becomes (*more/less*) negative.

Skeletal Muscle Is Classified by Speed and Fatigue Resistance

Skeletal muscle fibers have traditionally been classified on the basis of their speed of contraction and their resistance to fatigue with repeated stimulation. But like so much in physiology, the more scientists learn, the more complicated the picture becomes. The current classification of muscle fiber types as type I or type II depends on the isoform of myosin expressed in the fiber.

Muscle fiber types are not fixed for life. Muscles have plasticity and can shift their type depending on their activity. The currently accepted muscle fiber types in humans include **slow-twitch fibers** (also called *ST* or *type I*), **fast-twitch oxidative-glycolytic fibers** (*FOG* or *type IIA*), and **fast-twitch glycolytic fibers** (*FG* or *type IIB/IIX*). Type IIX is found in humans; type IIB is found in other animals.

Fast-twitch muscle fibers (type II) develop tension two to three times faster than slow-twitch fibers (type I). The speed with which a muscle fiber contracts is determined by the isoform of myosin ATPase present in the fiber's thick filaments. Fast-twitch fibers split ATP more rapidly and can, therefore, complete multiple contractile cycles more rapidly than slow-twitch fibers. This speed translates into faster tension development in the fast-twitch fibers.

The duration of contraction also varies according to fiber type. Twitch duration is determined largely by how fast the sarcoplasmic reticulum removes Ca^{2+} from the cytosol. As cytosolic Ca^{2+} concentrations fall, Ca^{2+} unbinds from troponin, allowing tropomyosin to move into position to partially block the myosin-binding sites. With the power stroke inhibited in this way, the muscle fiber relaxes.

Fast-twitch fibers pump Ca^{2+} into their sarcoplasmic reticulum more rapidly than slow-twitch fibers do, so fast-twitch fibers have quicker twitches. The twitches in fast-twitch fibers last only about 7.5 msec, making these muscles useful for fine, quick movements, such as playing the piano. Contractions in slow-twitch muscle fibers may last more than 10 times as long. Fast-twitch fibers are used occasionally, but slow-twitch fibers are used almost constantly for maintaining posture, standing, or walking.

The second major difference between muscle fiber types is their ability to resist fatigue. Glycolytic fibers (fast-twitch type IIB/IIX) rely primarily on anaerobic glycolysis to produce ATP. However, the accumulation of H^+ from ATP hydrolysis contributes to acidosis, a condition implicated in the development of fatigue, as noted previously. As a result, glycolytic fibers fatigue more easily than do oxidative fibers, which do not depend on anaerobic metabolism.

Oxidative fibers rely primarily on oxidative phosphorylation [p. 108] for production of ATP—hence their descriptive name. These fibers, which include type I slow-twitch fibers and type IIA fast-twitch oxidative-glycolytic fibers, have more mitochondria (the site of enzymes for the citric acid cycle and oxidative phosphorylation) than glycolytic fibers do. They also have more blood vessels in their connective tissue to bring oxygen to the cells (**FIG. 12.14**).

The efficiency with which muscle fibers obtain oxygen is a factor in their preferred method of glucose metabolism. Oxygen in the blood must diffuse into the interior of muscle fibers in order to reach the mitochondria. This process is facilitated by the presence of **myoglobin**, a red oxygen-binding pigment with a high affinity for oxygen. This affinity allows myoglobin to act as a transfer molecule, bringing oxygen more rapidly to the interior of the fibers. Because oxidative fibers contain more myoglobin, oxygen diffusion is faster than in glycolytic fibers. Oxidative fibers are described as *red muscle* because large amounts of myoglobin give them their characteristic color.

In addition to myoglobin, oxidative fibers have smaller diameters, so the distance through which oxygen must diffuse before

RUNNING PROBLEM

Two forms of periodic paralysis exist. One form, called *hypokalemic periodic paralysis,* is characterized by decreased blood levels of K^+ during paralytic episodes. The other form, *hyperkalemic periodic paralysis* (hyperKPP), is characterized by either normal or increased blood levels of K^+ during episodes. Results of a blood test revealed that Paul has the hyperkalemic form.

Q3: *In people with hyperKPP, attacks may occur after a period of exercise (i.e., after a period of repeated muscle contractions). What ion is responsible for the repolarization phase of the muscle action potential, and in which direction does this ion move across the muscle fiber membrane? How might this be linked to hyperKPP?*

375
386
390
400
402
409

FIG. 12.14 Fast-twitch and slow-twitch muscles

Slow-twitch oxidative muscle has large amounts of red myoglobin, numerous mitochondria, and extensive capillary blood supply, in contrast to fast-twitch glycolytic muscle.

Slow-Twitch Oxidative Muscle Fibers. Note smaller diameter, darker color due to myoglobin. Fatigue-resistant.

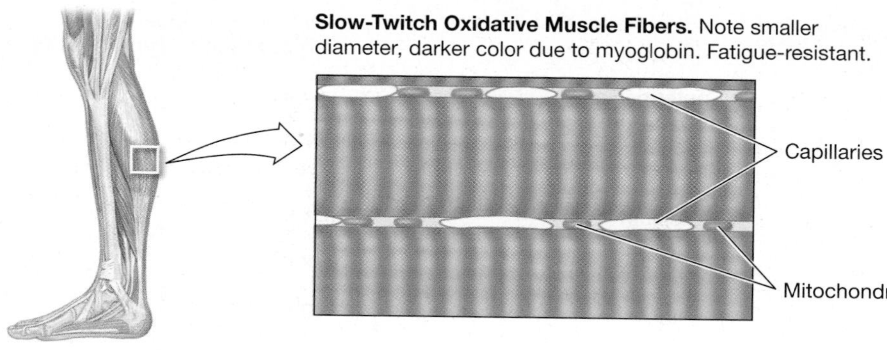
Capillaries
Mitochondria

Cross section of slow-twitch muscle fibers

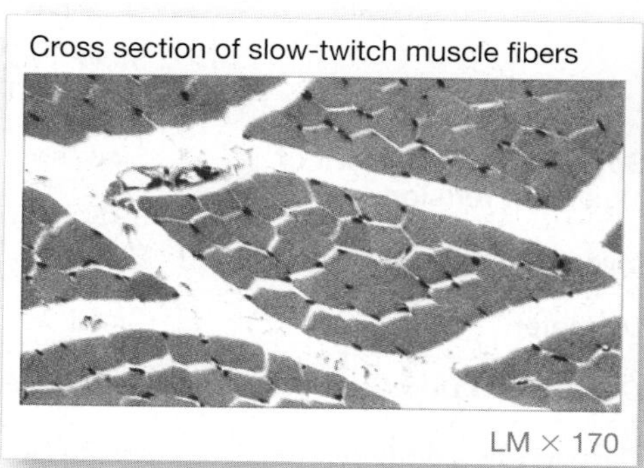
LM × 170

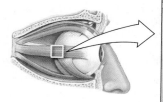

Fast-Twitch Glycolytic Muscle Fibers. Larger diameter, pale color. Easily fatigued.

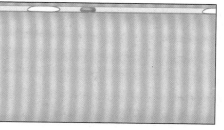

Cross section of fast-twitch muscle fibers

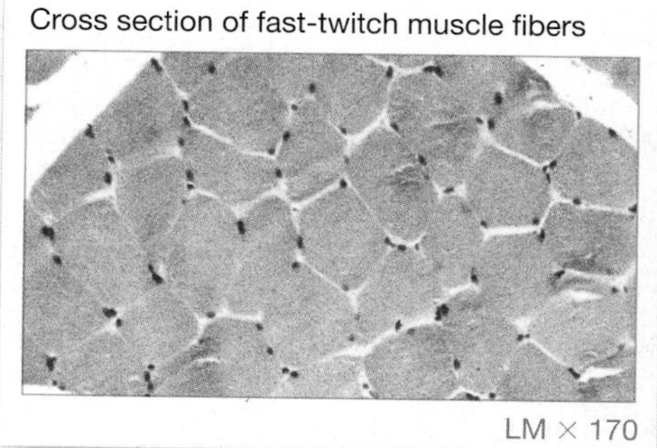
LM × 170

reaching the mitochondria is shorter. Because oxidative fibers have more myoglobin and more capillaries to bring blood to the cells and are smaller in diameter, they maintain a better supply of oxygen and are able to use oxidative phosphorylation for ATP production.

IIB/IIX glycolytic fibers, in contrast, are described as *white muscle* because of their lower myoglobin content. These muscle fibers are also larger in diameter than type I slow-twitch fibers. The combination of larger size, less myoglobin, and fewer blood vessels means that glycolytic fibers are more likely to run out of oxygen after repeated contractions. Glycolytic fibers therefore rely primarily on anaerobic glycolysis for ATP synthesis and fatigue most rapidly.

Type IIA fast-twitch oxidative-glycolytic fibers exhibit properties of both oxidative and glycolytic fibers. They are smaller than fast-twitch glycolytic fibers and use a combination of oxidative and glycolytic metabolism to produce ATP. Because of their intermediate size and the use of oxidative phosphorylation for ATP synthesis, type IIA fibers are more fatigue resistant than their IIB/IIX fast-twitch glycolytic cousins. Type IIA fibers, like type I slow-twitch fibers, are classified as red muscle because of their myoglobin content.

Human muscles are a mixture of fiber types, with the ratio of types varying from muscle to muscle and from one individual to another. For example, who would have more fast-twitch fibers in leg muscles, a marathon runner or a high-jumper? Characteristics of the three muscle fiber types are compared in **TABLE 12.2**.

Resting Fiber Length Affects Tension

In a muscle fiber, the tension developed during a twitch is a direct reflection of the length of individual sarcomeres before contraction begins (**FIG. 12.15**). Each sarcomere contracts with optimum force if it is at optimum length (neither too long nor too short) before the contraction begins. Fortunately, the normal resting length of skeletal muscles usually ensures that sarcomeres are at optimum length when they begin a contraction.

At the molecular level, sarcomere length reflects the overlap between the thick and thin filaments (Fig. 12.15). The sliding filament theory predicts that *the tension generated by a muscle fiber is directly proportional to the number of crossbridges formed between the thick and thin filaments.* If the fibers start a contraction at a very long sarcomere length, the thick and thin filaments barely overlap and form few crossbridges (Fig. 12.15e). This means that in the initial part of the contraction, the sliding filaments interact only minimally and therefore cannot generate much force.

At the optimum sarcomere length (Fig. 12.15c), the filaments begin contracting with numerous crossbridges between the thick and thin filaments, allowing the fiber to generate optimum force in that twitch. If the sarcomere is shorter than optimum length at the beginning of the contraction (Fig. 12.15b), the thick and thin fibers have too much overlap before the contraction begins. Consequently, the thick filaments can move the thin filaments only a short distance before the thin actin filaments from opposite ends of the sarcomere start to overlap. This overlap prevents crossbridge formation.

TABLE 12.2 Characteristics of Muscle Fiber Types

	Slow-Twitch Oxidative; Red Muscle (Type I)	Fast-Twitch Oxidative-Glycolytic; Red Muscle (Type IIA)	Fast-Twitch Glycolytic; White Muscle (Type IIB/IIX)
Speed of Development of Maximum Tension	Slowest	Intermediate	Fastest
Myosin ATPase Activity	Slow	Fast	Fast
Diameter	Small	Medium	Large
Contraction Duration	Longest	Short	Short
Ca^{2+}-ATPase Activity in SR	Moderate	High	High
Endurance	Fatigue resistant	Fatigue resistant	Easily fatigued
Use	Most used: posture	Standing, walking	Least used: jumping; quick, fine movements
Metabolism	Oxidative; aerobic	Glycolytic but becomes more oxidative with endurance training	Glycolytic; more anaerobic than fast-twitch oxidative-glycolytic type
Capillary Density	High	Medium	Low
Mitochondria	Numerous	Moderate	Few
Color	Dark red (myoglobin)	Red	Pale

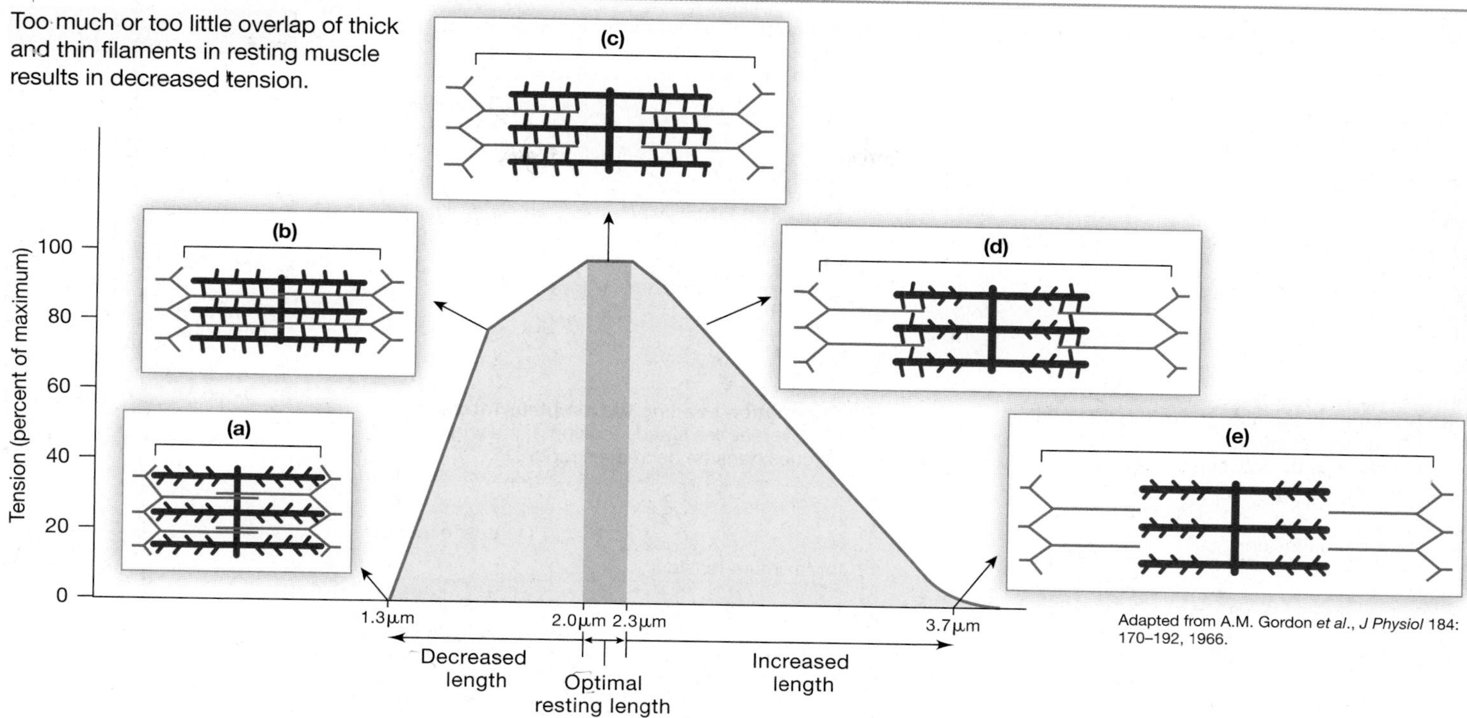
FIG. 12.15 Length-tension relationships
Too much or too little overlap of thick and thin filaments in resting muscle results in decreased tension.
Tension (percent of maximum)
100
80
60
40
20
0
(a)
(b)
(c)
(d)
(e)
1.3μm
2.0μm 2.3μm
3.7μm
Decreased length
Optimal resting length
Increased length
Adapted from A.M. Gordon et al., J Physiol 184: 170–192, 1966.

If the sarcomere is so short that the thick filaments run into the Z disks (Fig. 12.15a), myosin is unable to find new binding sites for crossbridge formation, and tension decreases rapidly. Thus, the development of single-twitch tension in a muscle fiber is a passive property that depends on filament overlap and sarcomere length.

Force of Contraction Increases with Summation

Although we have just seen that single-twitch tension is determined by the length of the sarcomere, it is important to note that a single twitch does not represent the maximum force that a muscle fiber can develop. The force generated by the contraction of a single muscle fiber can be increased by increasing the rate (frequency) at which muscle action potentials stimulate the muscle fiber.

A typical muscle action potential lasts between 1 and 3 msec, while the muscle contraction may last 100 msec (see Fig. 12.11). If repeated action potentials are separated by long intervals of time, the muscle fiber has time to relax completely between stimuli (**FIG. 12.16a**). If the interval of time between action potentials is shortened, the muscle fiber does not have time to relax completely between two stimuli, resulting in a more forceful contraction (Fig. 12.16b). This process is known as **summation** and is similar to the temporal summation of graded potentials that takes place in neurons [p. 261].

If action potentials continue to stimulate the muscle fiber repeatedly at short intervals (high frequency), relaxation between contractions diminishes until the muscle fiber achieves a state of maximal contraction known as **tetanus**. There are two types of tetanus. In *incomplete,* or *unfused, tetanus,* the stimulation rate of the muscle fiber is not at a maximum value, and consequently the fiber relaxes slightly between stimuli (Fig. 12.16c). In *complete,* or *fused, tetanus,* the stimulation rate is fast enough that the muscle fiber does not have time to relax. Instead, it reaches maximum tension and remains there (Fig. 12.16d).

Thus, it is possible to increase the tension developed in a single muscle fiber by changing the rate at which action potentials occur in the fiber. Muscle action potentials are initiated by the somatic motor neuron that controls the muscle fiber.

Concept Check

17. Summation in muscle fibers means that the ______ of the fiber increases with repeated action potentials.

18. Temporal summation in neurons means that the _______ of the neuron increases when two depolarizing stimuli occur close together in time.

A Motor Unit Is One Motor Neuron and Its Muscle Fibers

The basic unit of contraction in an intact skeletal muscle is a **motor unit**, composed of a group of muscle fibers that function together and the somatic motor neuron that controls them (**FIG. 12.17**). When the somatic motor neuron fires an action potential, all muscle fibers in the motor unit contract. Note that although one somatic motor neuron innervates multiple fibers, each muscle fiber is innervated by only a single neuron.

FIG. 12.16 Summation of contractions

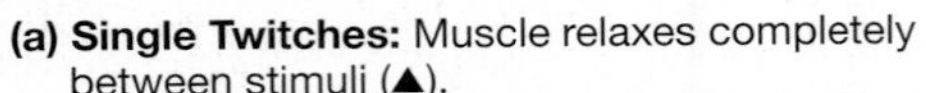

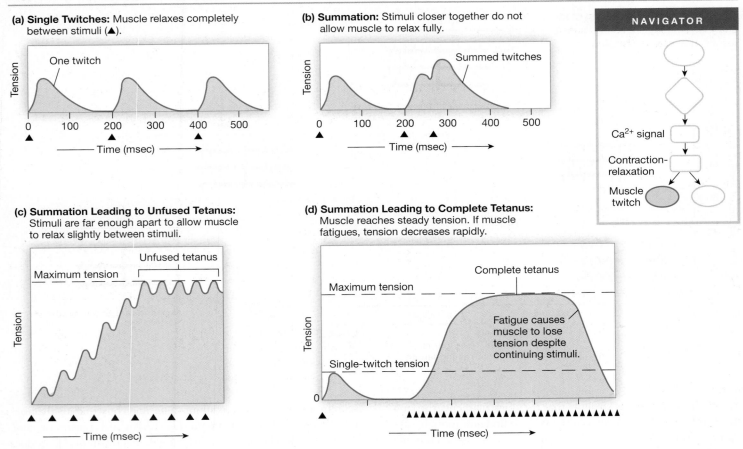

(a) Single Twitches: Muscle relaxes completely between stimuli (▲).

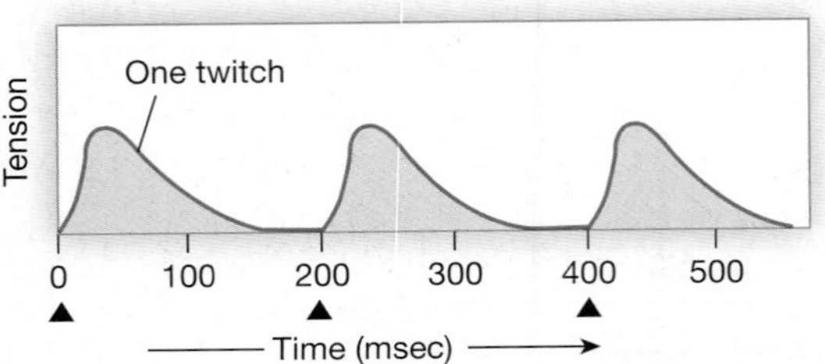
One twitch
Tension
0
100
200
300
400
500
Time (msec)

(b) Summation: Stimuli closer together do not allow muscle to relax fully.

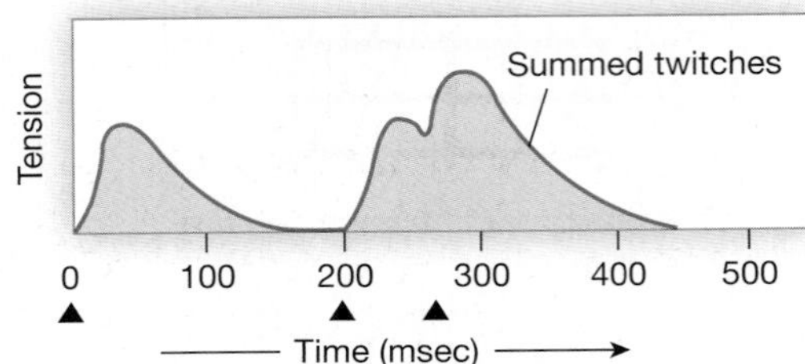
Summed twitches
Tension
0
100
200
300
400
500
Time (msec)

NAVIGATOR

Ca2+ signal

Contraction-relaxation

Muscle twitch

(c) Summation Leading to Unfused Tetanus: Stimuli are far enough apart to allow muscle to relax slightly between stimuli.

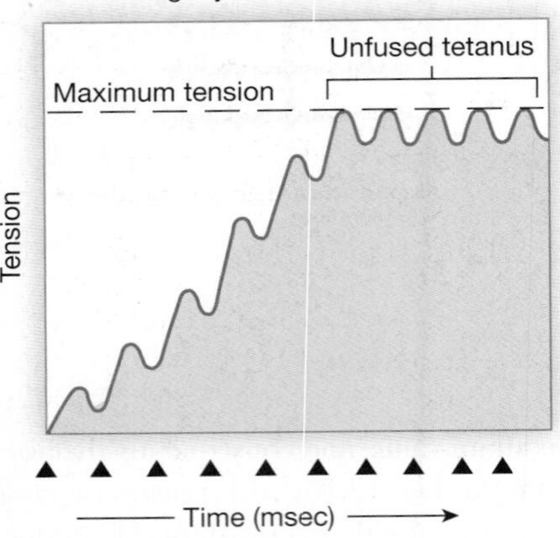
Unfused tetanus
Maximum tension
Tension
Time (msec)

(d) Summation Leading to Complete Tetanus: Muscle reaches steady tension. If muscle fatigues, tension decreases rapidly.

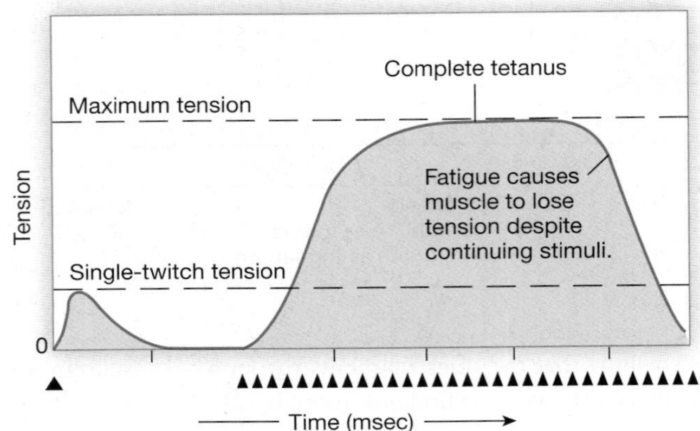
Complete tetanus
Maximum tension
Fatigue causes muscle to lose tension despite continuing stimuli.
Single-twitch tension
Tension
0
Time (msec)

The number of muscle fibers in a motor unit varies. In muscles used for fine motor actions, such as the *extraocular* muscles that move the eyes or the muscles of the hand, one motor unit contains as few as three to five muscle fibers. If one such motor unit is activated, only a few fibers contract, and the muscle response is quite small. If additional motor units are activated, the response increases by small increments because only a few more muscle fibers contract with the addition of each motor unit. This arrangement allows fine gradations of movement.

In muscles used for gross motor actions such as standing or walking, each motor unit may contain hundreds or even thousands of muscle fibers. The gastrocnemius muscle in the calf of the leg, for example, has about 2000 muscle fibers in each motor unit. Each time an additional motor unit is activated in these muscles, many more muscle fibers contract, and the muscle response jumps by correspondingly greater increments.

All muscle fibers in a single motor unit are of the same fiber type. For this reason there are fast-twitch motor units and slow-twitch motor units. Which kind of muscle fiber associates with a particular neuron appears to be a function of the neuron. During embryological development, each somatic motor neuron secretes a growth factor that directs the differentiation of all muscle fibers in its motor unit so that they develop into the same fiber type.

Intuitively, it would seem that people who inherit a predominance of one fiber type over another would excel in certain sports. They do, to some extent. Endurance athletes, such as distance runners and cross-country skiers, have a predominance of slow-twitch fibers, whereas sprinters, ice hockey players, and weight lifters tend to have larger percentages of fast-twitch fibers.

Inheritance is not the only determining factor for fiber composition in the body, however, because the metabolic characteristics of muscle fibers have some plasticity. With endurance training, the aerobic capacity of some fast-twitch fibers can be enhanced until they are almost as fatigue-resistant as slow-twitch fibers. Because the conversion occurs only in those muscles that are being trained, a neuromodulator chemical is probably involved. In addition, endurance training increases the number of capillaries and mitochondria in the muscle tissue, allowing more oxygen-carrying blood to reach the contracting muscle and contributing to the increased aerobic capacity of the muscle fibers.

Contraction Force Depends on the Types and Numbers of Motor Units

Within a skeletal muscle, each motor unit contracts in an all-or-none manner. How then can muscles create graded contractions of varying force and duration? The answer lies in the fact that muscles are composed of multiple motor units of different types (Fig. 12.17). This diversity allows the muscle to vary contraction by (1) changing the types of motor units that are active or

FIG. 12.17 Motor units

A motor unit consists of one motor neuron and all the muscle fibers it innervates. A muscle may have many motor units of different types.

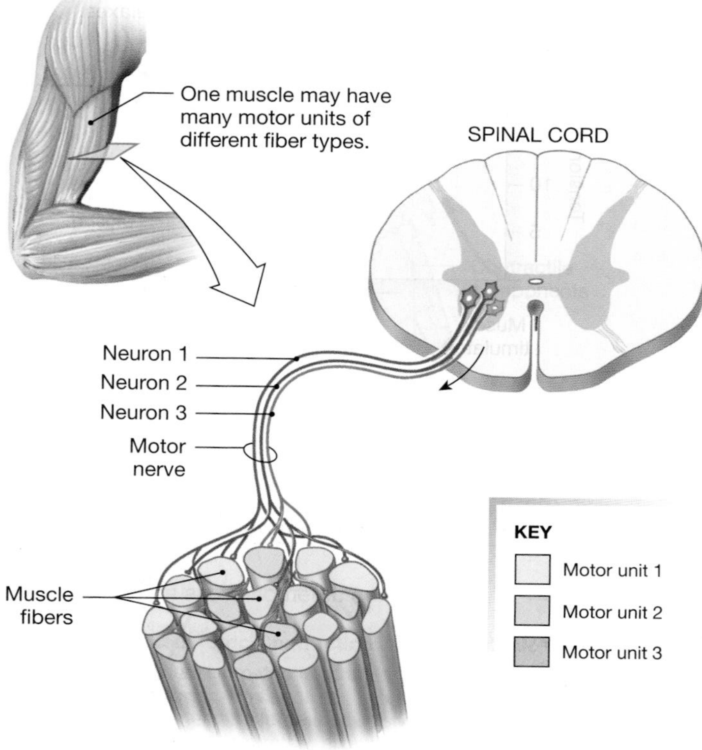
One muscle may have many motor units of different fiber types.
SPINAL CORD
Neuron 1
Neuron 2
Neuron 3
Motor nerve
Muscle fibers
KEY
Motor unit 1
Motor unit 2
Motor unit 3

(2) changing the number of motor units that are responding at any one time.

The force of contraction in a skeletal muscle can be increased by recruiting additional motor units. **Recruitment** is controlled by the nervous system and proceeds in a standardized sequence. A weak stimulus directed onto a pool of somatic motor neurons in the central nervous system activates only the neurons with the lowest thresholds [p. 239]. Studies have shown that these low-threshold neurons control fatigue-resistant slow-twitch fibers, which generate minimal force.

As the stimulus onto the motor neuron pool increases in strength, additional motor neurons with higher thresholds begin to fire. These neurons in turn stimulate motor units composed of fatigue-resistant, fast-twitch oxidative-glycolytic fibers. Because more motor units (and, thus, more muscle fibers) are participating in the contraction, greater force is generated in the muscle.

As the stimulus increases to even higher levels, somatic motor neurons with the highest thresholds begin to fire. These neurons stimulate motor units composed of glycolytic fast-twitch fibers. At this point, the muscle contraction is approaching its maximum force. Because of differences in myosin and crossbridge formation, fast-twitch fibers generate more force than slow-twitch fibers do. However, because fast-twitch fibers fatigue more rapidly, it is impossible to hold a muscle contraction at maximum force for an extended period of time. You can demonstrate this by clenching your fist as hard as you can: How long can you hold it before some of the muscle fibers begin to fatigue?

Sustained contractions in a muscle require a continuous train of action potentials from the central nervous system to the muscle. As you learned earlier, however, increasing the stimulation rate of a muscle fiber results in summation of its contractions. If the muscle fiber is easily fatigued, summation leads to fatigue and diminished tension (Fig. 12.16d).

One way the nervous system avoids fatigue in sustained contractions is by **asynchronous recruitment** of motor units. The nervous system modulates the firing rates of the motor neurons so that different motor units take turns maintaining muscle tension. The alternation of active motor units allows some of the motor units to rest between contractions, preventing fatigue.

Asynchronous recruitment prevents fatigue only in submaximal contractions, however. In high-tension, sustained contractions, the individual motor units may reach a state of unfused tetanus, in which the muscle fibers cycle between contraction and partial relaxation. In general, we do not notice this cycling because the different motor units in the muscle are contracting and relaxing at slightly different times. As a result, the contractions and relaxations of the motor units average out and appear to be one smooth contraction. But as different motor units fatigue, we are unable to maintain the same amount of tension in the muscle, and the force of the contraction gradually decreases.

Concept Check

19. Which type of runner would you expect to have more slow-twitch fibers, a sprinter or a marathoner?
20. What is the response of a muscle fiber to an increase in the firing rate of the somatic motor neuron?
21. How does the nervous system increase the force of contraction in a muscle composed of many motor units?

12.2 Mechanics Of Body Movement

Because one main role of skeletal muscles is to move the body, we now turn to the mechanics of body movement. The term *mechanics* refers to how muscles move loads and how the anatomical relationship between muscles and bones maximizes the work the muscles can do.

Isotonic Contractions Move Loads; Isometric Contractions Create Force without Movement

When we described the function of muscles earlier in this chapter, we noted that they can create force to generate movement but can also create force without generating movement. You can demonstrate both properties with a pair of heavy weights. Pick up one weight in each hand and then bend your elbows so that the weights touch your shoulders. You have just performed an **isotonic contraction** {*iso*, equal + *teinein*, to stretch}. Any

FIG. 12.26 ESSENTIALS Smooth Muscle Contraction and Relaxation

Smooth muscle contraction and relaxation are similar to those of skeletal muscle, but differ in several important ways: (1) Ca^{2+} comes from the ECF as well as the sarcoplasmic reticulum, (2) an action potential is not required for Ca^{2+} release, (3) there is no troponin, so Ca^{2+} initiates contraction through a cascade that includes phosphorylation of myosin light chains, and (4) an additional step in smooth muscle relaxation is dephosphorylation of myosin light chains by myosin phosphatase.

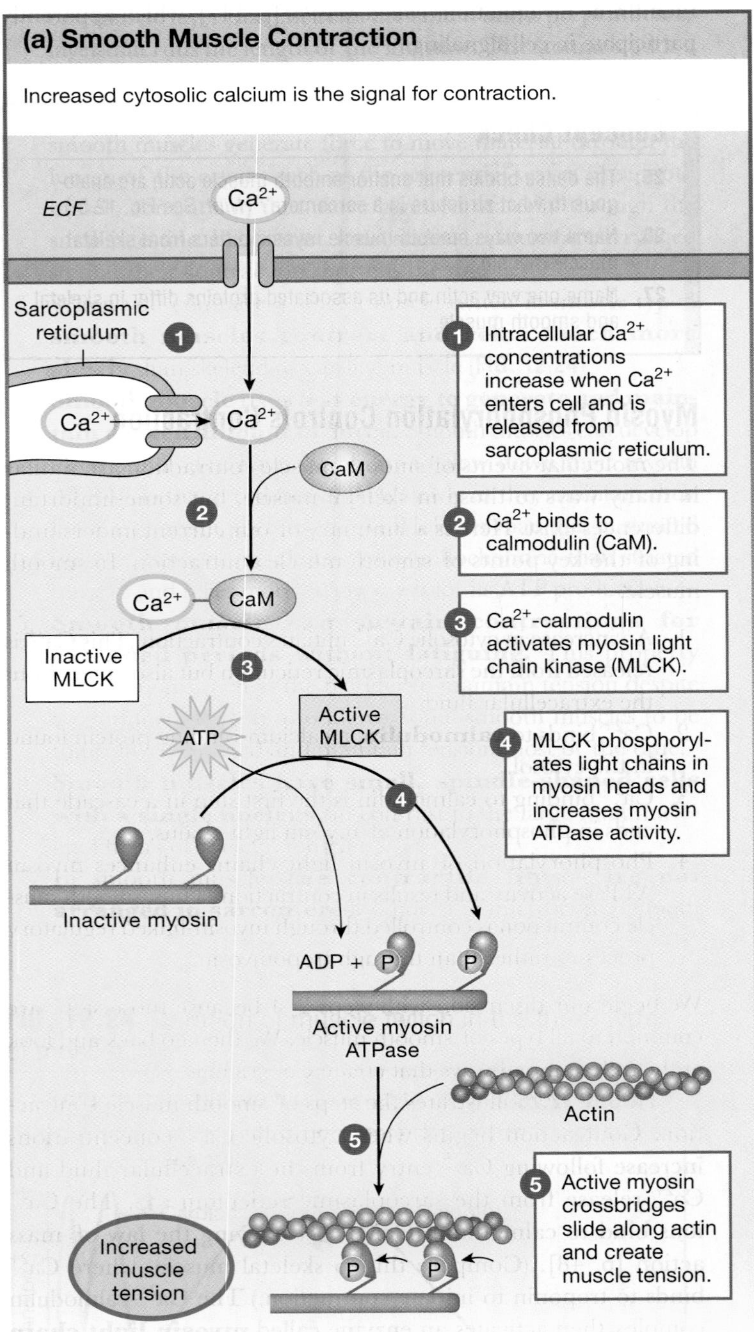

(a) Smooth Muscle Contraction
Increased cytosolic calcium is the signal for contraction.
ECF
Ca2+
Sarcoplasmic reticulum
1
Ca2+
Ca2+
CaM
2
Ca2+
CaM
Inactive MLCK
3
Active MLCK
ATP
4
Inactive myosin
ADP + P
P
Active myosin ATPase
Actin
5
P
P
Increased muscle tension
1 Intracellular Ca2+ concentrations increase when Ca2+ enters cell and is released from sarcoplasmic reticulum.
2 Ca2+ binds to calmodulin (CaM).
3 Ca2+-calmodulin activates myosin light chain kinase (MLCK).
4 MLCK phosphorylates light chains in myosin heads and increases myosin ATPase activity.
5 Active myosin crossbridges slide along actin and create muscle tension.

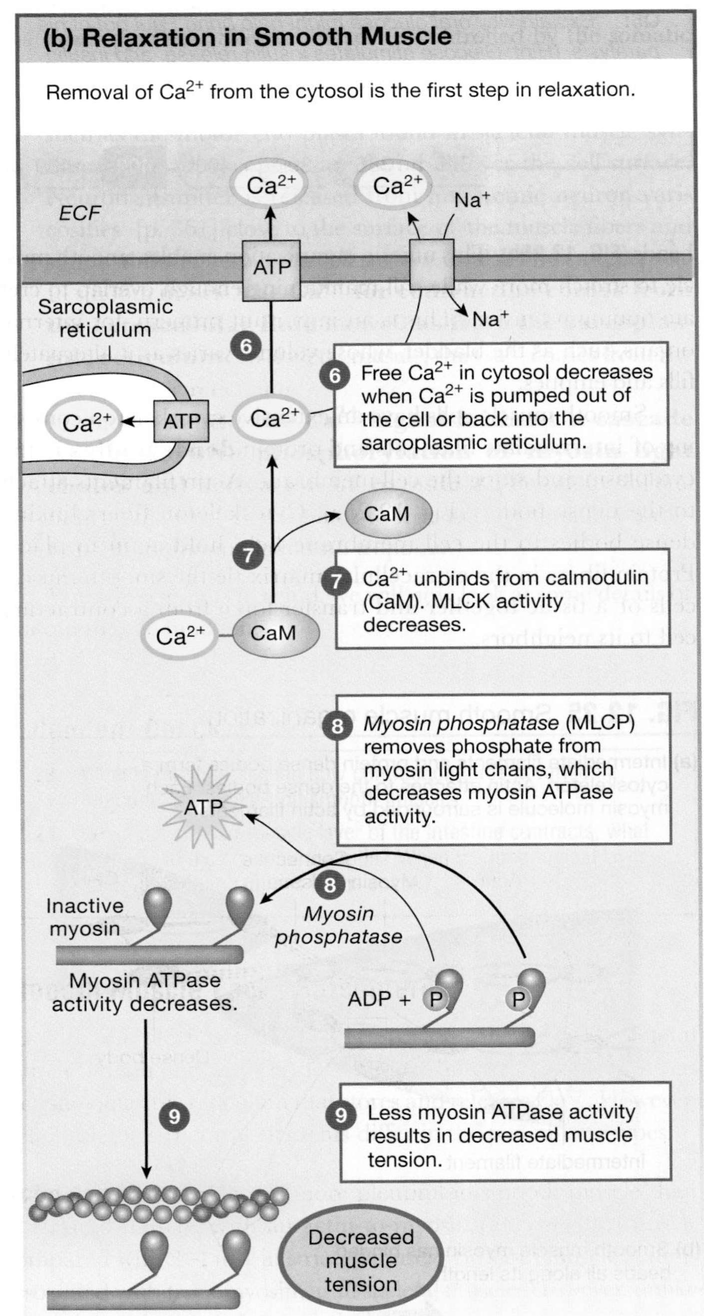

(b) Relaxation in Smooth Muscle
Removal of Ca2+ from the cytosol is the first step in relaxation.
Ca2+
Ca2+
Na+
ECF
ATP
Na+
Sarcoplasmic reticulum
6
Ca2+
ATP
Ca2+
CaM
7
Ca2+
CaM
ATP
8
Inactive myosin
Myosin phosphatase
Myosin ATPase activity decreases.
ADP + P
P
9
Decreased muscle tension
6 Free Ca2+ in cytosol decreases when Ca2+ is pumped out of the cell or back into the sarcoplasmic reticulum.
7 Ca2+ unbinds from calmodulin (CaM). MLCK activity decreases.
8 Myosin phosphatase (MLCP) removes phosphate from myosin light chains, which decreases myosin ATPase activity.
9 Less myosin ATPase activity results in decreased muscle tension.

KEY

MLCK = myosin light chain kinase

Phosphorylation of myosin enhances myosin ATPase activity. When myosin ATPase activity is high, actin binding and crossbridge cycling increase tension in the muscle **5**. The myosin ATPase isoform in smooth muscle is much slower than that in skeletal muscle, and this decreases the rate of crossbridge cycling.

Dephosphorylation of the myosin light chain by the enzyme **myosin light chain phosphatase (MLCP)** decreases myosin ATPase activity. Interestingly, dephosphorylation of myosin does not automatically result in relaxation. Under conditions that we do not fully understand, dephosphorylated myosin may remain in an isometric contraction called a **latch state**. This condition maintains tension in the muscle fiber while consuming minimal ATP. It is a significant factor in the ability of smooth muscle to sustain contraction without fatiguing.

Relaxation Because dephosphorylation of myosin does not automatically cause relaxation, it is the ratio of MLCK to MLCP activity that determines the contraction state of smooth muscle. MLCP is always active to some degree in smooth muscle, so the activity of MLCK is often the critical factor. As you learned, MLCK activity depends on Ca^{2+}-calmodulin.

Relaxation in a smooth muscle fiber is a multistep process (Fig. 12.26b). As in skeletal muscle, free Ca^{2+} is removed from the cytosol when Ca^{2+}-ATPase pumps it back into the sarcoplasmic reticulum. In addition, some Ca^{2+} is pumped out of the cell with the help of Ca^{2+}-ATPase and the Na^+-Ca^{2+} exchanger (NCX) [p. 144] **6**.

By the law of mass action, a decrease in free cytosolic Ca^{2+} causes Ca^{2+} to unbind from calmodulin **7**. In the absence of Ca^{2+}-calmodulin, myosin light chain kinase becomes inactivate. As MLCK becomes less active, myosin light chain phosphatase dephosphorylates myosin **8**. Myosin ATPase activity decreases **9**, and the muscle relaxes.

MLCP Controls Ca^{2+} Sensitivity

From the earlier discussion, it would appear that calcium and its regulation of MLCK activity is the primary factor responsible for control of smooth muscle contraction. But chemical signals such as neurotransmitters, hormones, and paracrine molecules alter smooth muscle Ca^{2+} **sensitivity** by modulating myosin light chain phosphatase (MLCP) activity. If MLCK and Ca^{2+}-calmodulin are constant but MLCP activity increases, the MLCK/MLCP ratio shifts so that MLCP dominates. Myosin ATPase dephosphorylates and contraction force decreases, even though the cytosolic Ca^{2+} concentration has not changed (**FIG. 12.27**). The contraction process is said to be *desensitized* to calcium—the calcium signal is less effective at causing a contraction. Conversely, signal molecules that *decrease* myosin light chain phosphatase activity make the cell *more sensitive* to Ca^{2+} and contraction force increases even though $[Ca^{2+}]$ has not changed.

Calcium Initiates Smooth Muscle Contraction

We now step back to look in detail at the processes that initiate smooth muscle contraction. Contraction can start with

FIG. 12.27 Phosphate-mediated Ca^{2+} sensitivity

Changes in phosphatase activity alter myosin's response to Ca^{2+}.

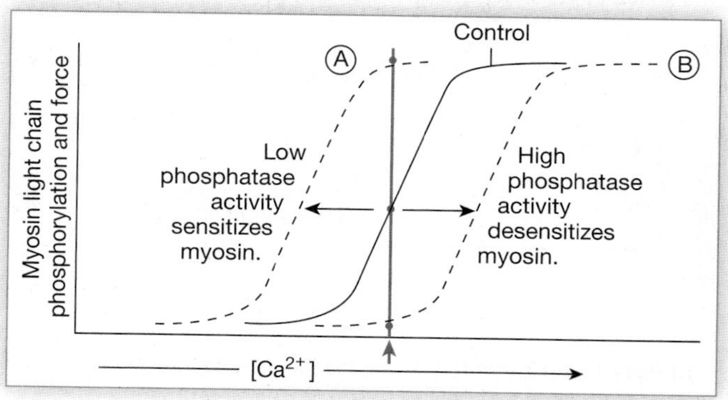

Control
A
B
Myosin light chain phosphorylation and force
Low phosphatase activity sensitizes myosin.
High phosphatase activity desensitizes myosin.
[Ca2+]

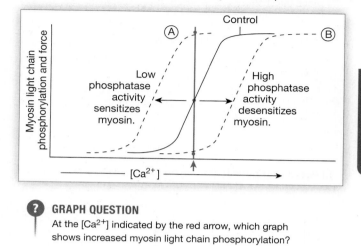

GRAPH QUESTION

At the $[Ca^{2+}]$ indicated by the red arrow, which graph shows increased myosin light chain phosphorylation?

electrical signals—changes in membrane potential—or chemical signals. Contraction caused by electrical signaling is termed *electromechanical coupling*. Contractions initiated by chemical signals without a significant change in membrane potential are called **pharmacomechanical coupling**. Chemical signals may also relax muscle tension without a change in membrane potential. **FIGURE 12.28** is a generalized summary of these pathways.

The Ca^{2+} to initiate contraction comes from two sources: the sarcoplasmic reticulum and the extracellular fluid (Fig. 12.26a). Variable amounts of Ca^{2+} can enter the cytosol from these sources, creating *graded contractions* whose force varies according to the strength of the Ca^{2+} signal.

Sarcoplasmic Ca^{2+} Release The smooth muscle's intracellular Ca^{2+} store is the sarcoplasmic reticulum (SR). SR Ca^{2+} release is mediated both by a ryanodine receptor (RyR) calcium release channel and by an **IP_3-receptor channel.** The RyR channel opens in response to Ca^{2+} entering the cell, a process known as **calcium-induced calcium release (CICR)**. You will learn more about CICR when you study cardiac muscle.

The IP_3 channels open when G protein-coupled receptors activate phospholipase C signal transduction pathways [p. 173]. *Inositol trisphosphate* (IP_3) is a second messenger created in that pathway. When IP_3 binds to the SR IP_3-receptor channel, the channel opens and Ca^{2+} flows out of the SR into the cytosol.

Smooth muscle cells have sufficient SR Ca^{2+} stores for contraction. However, because some Ca^{2+} is lost to the ECF through the membrane pumps, the cells must monitor their SR Ca^{2+} stores. When SR Ca^{2+} stores decrease, a protein sensor (*STIM1*) on the SR membrane interacts with **store-operated Ca^{2+} channels** on the cell membrane. These Ca^{2+} channels, made from the protein *Orai-1*, then open to allow more Ca^{2+} into the cell. The Ca^{2+}-ATPase pumps the cytosolic Ca^{2+} into the SR to replenish its stores.

FIG. 12.28 Membrane potentials vary in smooth muscle

(a) Slow wave potentials fire action potentials when they reach threshold.

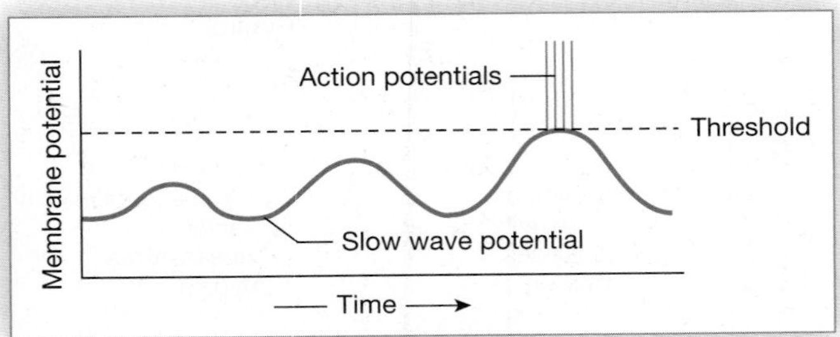
Membrane potential
Action potentials
Threshold
Slow wave potential
Time

(b) Pacemaker potentials always depolarize to threshold.

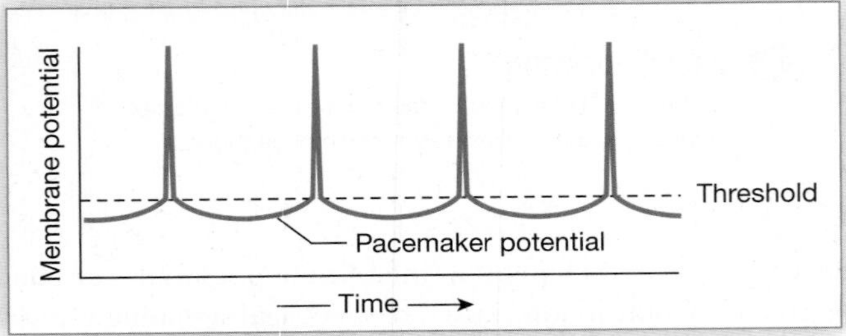
Membrane potential
Threshold
Pacemaker potential
Time

(c) Pharmacomechanical coupling occurs when chemical signals change muscle tension through signal transduction pathways with little or no change in membrane potential.

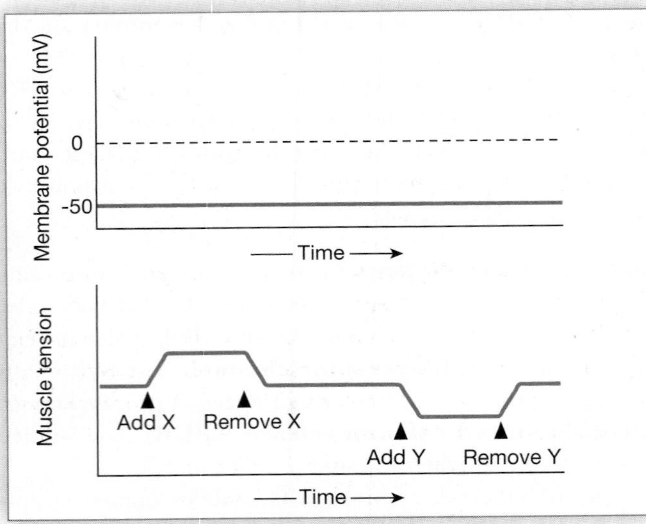
Membrane potential (mV)
0
-50
Time
Muscle tension
Add X
Remove X
Add Y
Remove Y
Time

Cell Membrane Ca^{2+} Entry Store-independent Ca^{2+} entry from the extracellular fluid takes place with the help of membrane channels that are voltage-gated, ligand-gated, or mechanically gated [p. 138].

1. Voltage-gated Ca^{2+} channels open in response to a depolarizing stimulus. Action potentials may be generated in the muscle cell or may enter from neighboring cells via gap junctions. Subthreshold graded potentials may open a few Ca^{2+} channels, allowing small amounts of Ca^{2+} into the cell. This cation entry depolarizes the cell and opens additional voltage-gated Ca^{2+} channels. Sometimes chemical signal molecules open cation channels, and the resulting depolarization opens the Ca^{2+} channels.
2. Ligand-gated Ca^{2+} channels are also known as *receptor-operated calcium channels* (ROCC). These channels open in response to ligand binding and allow enough Ca^{2+} into the cell to induce calcium release from the SR.
3. Stretch-activated channels: Some smooth muscle cells, such as those in blood vessels, contain stretch-activated channels that open when pressure or other force distorts the cell membrane. The exact process is still being worked out, but the cell depolarizes, opening neighboring voltage-gated Ca^{2+} channels. Because contraction in this instance originates from a property of the muscle fiber itself, it is known as a **myogenic contraction**. Myogenic contractions are common in blood vessels that maintain a certain amount of tone at all times.

Although stretch may initiate a contraction, some types of smooth muscle adapt if the muscle cells are stretched for an extended period of time. As the stretch stimulus continues, the Ca^{2+} channels begin to close in a time-dependent fashion. Then, as Ca^{2+} is pumped out of the cell, the muscle relaxes. This adaptation response explains why the bladder develops tension as it fills, then relaxes as it adjusts to the increased volume. (There is a limit to the amount of stretch the muscle can endure, however, and once a critical volume is reached, the urination reflex empties the bladder.)

Concept Check

28. Compare the following aspects of skeletal and smooth muscle contraction:
(a) signal for crossbridge activation
(b) source(s) of calcium for the Ca^{2+} signal
(c) signal that releases Ca^{2+} from the sarcoplasmic reticulum

29. What happens to contraction if a smooth muscle is placed in a saline bath from which all calcium has been removed?

30. Compare Ca^{2+} release channels in skeletal and smooth muscle sarcoplasmic reticulum.

Some Smooth Muscles Have Unstable Membrane Potentials

The role of membrane potentials in smooth muscle contraction is more complex than in skeletal muscle, where contraction always begins in response to an action potential. Smooth muscle exhibits a variety of electrical behaviors: it can hyperpolarize as well as depolarize. Hyperpolarization of the cell decreases the likelihood of contraction. Smooth muscle can also depolarize without firing action potentials. Contraction may take place after an action potential, after a subthreshold graded potential, or without any change in membrane potential.

Many types of smooth muscle display resting membrane potentials that vary between −40 and −80 mV. Cells that exhibit cyclic depolarization and repolarization of their membrane potential are said to have **slow wave potentials** (Fig. 12.28a). Sometimes, the cell simply cycles through a series of subthreshold slow waves. However, if the peak of the depolarization reaches threshold, action potentials fire, followed by contraction of the muscle.

Other types of smooth muscle with oscillating membrane potentials have regular depolarizations that always reach threshold and fire an action potential (Fig. 12.28b). These depolarizations are called **pacemaker potentials** because they create regular rhythms of contraction. Pacemaker potentials are found in some cardiac muscles as well as in smooth muscle. Both slow wave and pacemaker potentials are due to ion channels in the cell membrane that spontaneously open and close.

In pharmacomechanical coupling, the membrane potential of the muscle may not change at all. In the next section, we consider how this occurs.

Concept Check

31. How do pacemaker potentials differ from slow wave potentials?

32. When tetrodotoxin (TTX), a poison that blocks Na^+ channels, is applied to certain types of smooth muscle, it does not alter the spontaneous generation of action potentials. From this observation, what conclusion can you draw about the action potentials of these types of smooth muscle?

Chemical Signals Influence Smooth Muscle Activity

In this section, we look at how smooth muscle function is influenced by neurotransmitters, hormones, or paracrine signals. These chemical signals may be either excitatory or inhibitory, and they modulate contraction by second messenger action at the level of myosin as well as by influencing Ca^{2+} signals (**FIG. 12.29**). One of the interesting properties of smooth muscle is that signal transduction may cause muscle relaxation as well as contraction.

Autonomic Neurotransmitters and Hormones Many smooth muscles are under antagonistic control by both sympathetic and parasympathetic divisions of the autonomic nervous system. Other smooth muscles, such as those found in blood vessels, are under *tonic control* [p. 182] by only one of the two autonomic branches. In tonic control, the response is graded by increasing or decreasing the amount of neurotransmitter released onto the muscle.

A chemical signal can have different effects in different tissues, depending on the receptor type to which it binds [p. 179]. For this reason, it is important to specify the signal molecule and its receptor and subtype when describing the control of a tissue. For example, the sympathetic neurohormone epinephrine causes smooth muscle contraction when it binds to α-adrenergic receptors but relaxation when it binds to β_2-adrenergic receptors.

Most smooth muscle neurotransmitters and hormones bind to G protein-linked receptors. The second messenger pathways then determine the muscle response (**TBL. 12.3**). In general, IP_3 triggers contraction and cAMP promotes relaxation.

Pathways that increase IP_3 cause contraction several ways:

- IP_3 opens IP_3 channels on the SR to release Ca^{2+}.

FIG. 12.29 Control of smooth muscle contraction

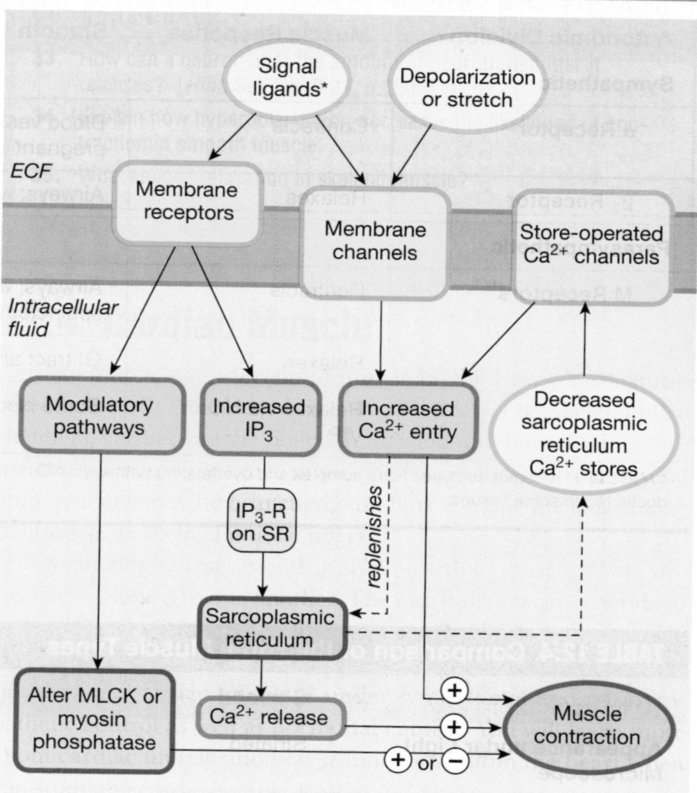
Signal ligands*
Depolarization or stretch
ECF
Membrane receptors
Membrane channels
Store-operated Ca2+ channels
Intracellular fluid
Modulatory pathways
Increased IP3
Increased Ca2+ entry
Decreased sarcoplasmic reticulum Ca2+ stores
IP3-R on SR
replenishes
Sarcoplasmic reticulum
Alter MLCK or myosin phosphatase
Ca2+ release
+ or −
Muscle contraction

KEY

IP_3-R = IP_3-activated receptor channel

*Ligands include norepinephrine, ACh, other neurotransmitters, hormones, and paracrine signals.

- Diacylglycerol (DAG), another product of the phospholipase C signal pathway, indirectly inhibits myosin phosphatase activity. Increasing the MLCK/MLCP ratio promotes crossbridge activity and muscle tension.

Signals that cause muscle relaxation work through the following mechanisms:

- Free cytosolic Ca^{2+} concentrations decrease when IP_3 channels are inhibited and the SR Ca^{2+}-ATPase is activated.
- K^+ leaking out of the cell hyperpolarizes it and decreases the likelihood of voltage-activated Ca^{2+} entry.
- Myosin phosphatase activity increases, which causes a decrease in muscle tension.

Paracrine Signals Locally released paracrine signal molecules can also alter smooth muscle contraction. For example, asthma is a condition in which smooth muscle of the airways constricts in response to histamine release. This constriction can be reversed by the administration of epinephrine, a neurohormone that relaxes

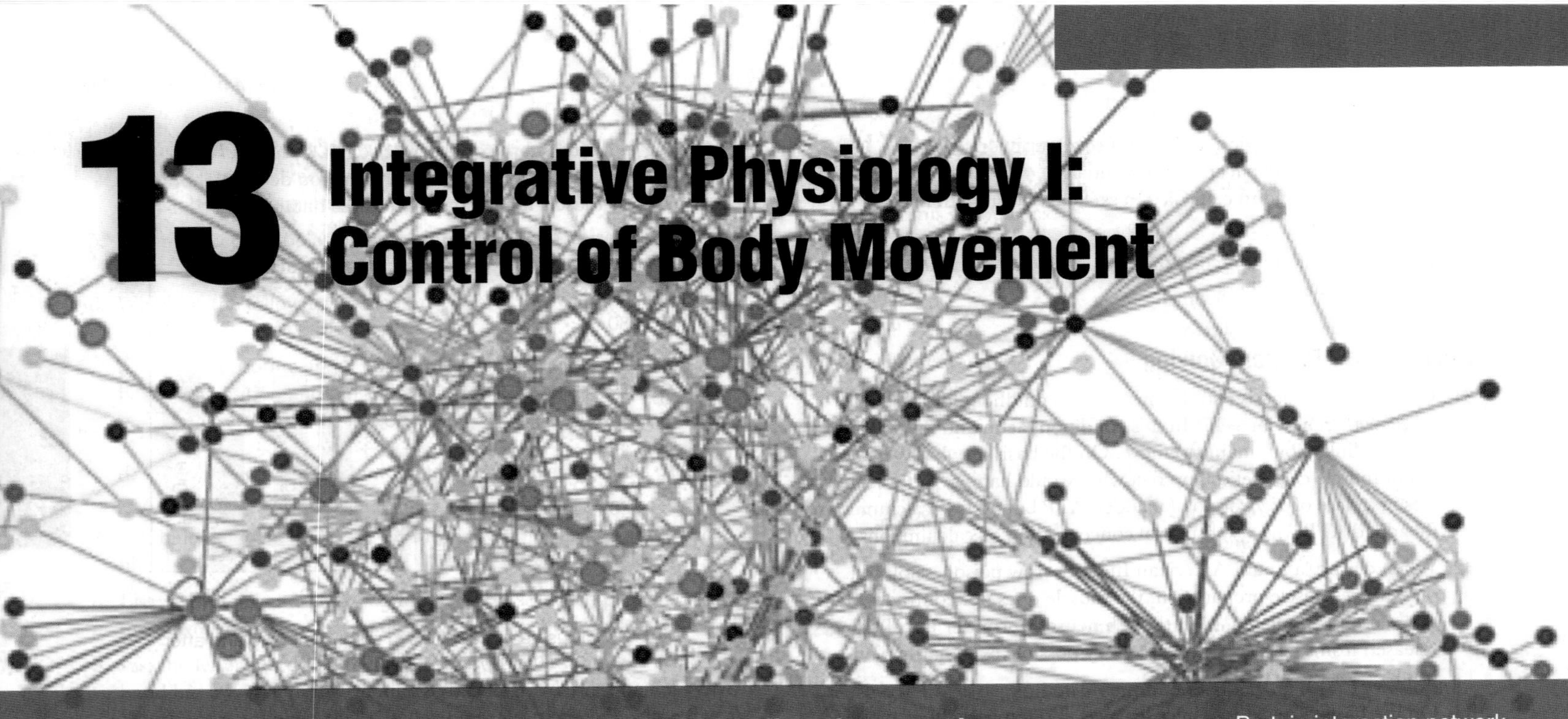

13 Integrative Physiology I: Control of Body Movement

Protein interaction network

Extracting signals directly from the brain to directly control robotic devices has been a science fiction theme that seems destined to become fact.

Dr. Eberhard E. Fetz, Rats Operate Robotic Arm via Brain Activity, Science News 156: 142, 8/28/1999

BACKGROUND BASICS

14	Reflex pathways
274	Central nervous system
261	Summation of action potentials
396	Isometric contraction
308	Sensory pathways and receptors
237	Graded potentials
182	Tonic control
82	Tendons

Think of a baseball pitcher standing on the mound. As he looks at the first batter, he receives sensory information from multiple sources: the sound of the crowd, the sight of the batter and the catcher, the smell of grass, the feel of the ball in his hand, and the alignment of his body as he begins his windup. Sensory receptors code this information and send it to the central nervous system (CNS), where it is integrated.

The pitcher acts consciously on some of the information: He decides to throw a fastball. But he processes other information at the subconscious level and acts on it without conscious thought. As he thinks about starting his motion, for instance, he shifts his weight to offset the impending movement of his arm. The integration of sensory information into an involuntary response is the hallmark of a *reflex* [p. 14].

RUNNING PROBLEM **Tetanus**

"She hasn't been able to talk to us. We're afraid she may have had a stroke." That is how her neighbors described 77-year-old Cecile Evans when they brought her to the emergency room. But when a neurological examination revealed no problems other than Mrs. Evans's inability to open her mouth plus stiffness in her neck, emergency room physician Dr. Ling began to consider other diagnoses. She noticed some scratches healing on Mrs. Evans's arms and legs and asked the neighbors if they knew what had caused them. "Oh, yes. She told us a few days ago that her dog jumped up and knocked her against the barbed wire fence." At that point, Dr. Ling realized she was probably dealing with her first case of tetanus.

415
417
421
423
428
428

CHAPTER 13

13.1 Neural Reflexes

All neural reflexes begin with a stimulus that activates a sensory receptor. The sensor sends information in the form of action potentials through sensory afferent neurons to the CNS [p. 186]. The CNS is the integrating center that evaluates all incoming information and selects an appropriate response. It then initiates action potentials in efferent neurons to direct the response of muscles and glands—the targets.

A key feature of many reflex pathways is *negative feedback* [p. 15]. Feedback signals from muscle and joint receptors keep the CNS continuously informed of changing body position. Some reflexes have a *feedforward* component that allows the body to anticipate a stimulus and begin the response [p. 17]. Bracing yourself in anticipation of a collision is an example of a feedforward response.

Neural Reflex Pathways Can Be Classified in Different Ways

Reflex pathways in the nervous system consist of chains or networks of neurons that link sensory receptors to muscles or glands. Neural reflexes can be classified in several ways (**TBL. 13.1**):

1. *By the efferent division of the nervous system that controls the response.* Reflexes that involve somatic motor neurons and skeletal muscles are known as **somatic reflexes**. Reflexes whose responses are controlled by autonomic neurons are called **autonomic reflexes**.
2. *By the CNS location where the reflex is integrated.* **Spinal reflexes** are integrated in the spinal cord. These reflexes may be modulated by higher input from the brain, but they can occur without that input. Reflexes integrated in the brain are called **cranial reflexes**.
3. *By whether the reflex is innate or learned.* Many reflexes are **innate**; in other words, we are born with them, and they are genetically determined. One example is the knee jerk, or patellar tendon reflex: When the patellar tendon at the lower edge of the kneecap is stretched with a tap from a reflex hammer, the lower leg kicks out. Other reflexes are acquired through experience (p. 298). The example of Pavlov's dogs salivating upon hearing a bell is the classic example of a **learned reflex**, also referred to as a **conditioned reflex**.
4. *By the number of neurons in the reflex pathway.* The simplest reflex is a **monosynaptic reflex**, named for the single synapse between the two neurons in the pathway: a sensory afferent neuron (often just called *a sensory afferent*) and an efferent somatic motor neuron (**FIG. 13.1a**). These two neurons synapse in the spinal cord, allowing a signal initiated at the receptor to go directly from the sensory neuron to the motor neuron. (The synapse between the somatic motor neuron and its muscle target is ignored.)

Most reflexes have three or more neurons in the pathway (and at least two synapses), leading to their designation as

TABLE 13.1 Classification of Neural Reflexes

Neural reflexes can be classified by:

1. **Efferent division that controls the effector**
   (a) Somatic motor neurons control skeletal muscles.
   (b) Autonomic neurons control smooth and cardiac muscle, glands, and adipose tissue.
2. **Integrating region within the central nervous system**
   (a) Spinal reflexes do not require input from the brain.
   (b) Cranial reflexes are integrated within the brain.
3. **Time at which the reflex develops**
   (a) Innate (inborn) reflexes are genetically determined.
   (b) Learned (conditioned) reflexes are acquired through experience.
4. **The number of neurons in the reflex pathway**
   (a) Monosynaptic reflexes have only two neurons: one afferent (sensory) and one efferent. Only somatic motor reflexes can be monosynaptic.
   (b) Polysynaptic reflexes include one or more interneurons between the afferent and efferent neurons. All autonomic reflexes are polysynaptic because they have three neurons: one afferent and two efferent.

FIG. 13.1 ESSENTIALS Neural Reflexes

Skeletal Muscle Reflexes

(a) A **monosynaptic reflex** has a single synapse between the afferent and efferent neurons.

Stimulus → Receptor → Sensory neuron → Spinal cord integrating center

Skeletal muscle

Somatic motor neuron

One synapse

Efferent neuron → Target cell → Response

(b) Polysynaptic reflexes have two or more synapses. This somatic motor reflex has both synapses in the CNS.

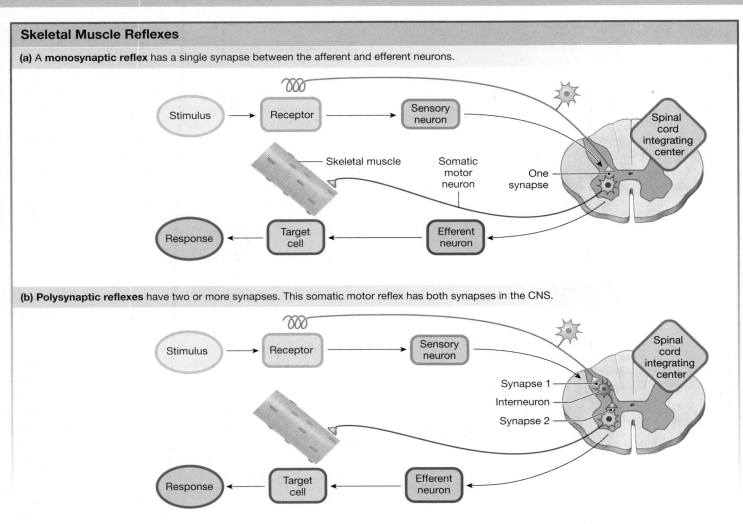

Stimulus → Receptor → Sensory neuron → Spinal cord integrating center

Synapse 1

Interneuron

Synapse 2

Efferent neuron → Target cell → Response

Autonomic Reflexes

(c) All autonomic reflexes are polysynaptic, with at least one synapse in the CNS and another in the autonomic ganglion.

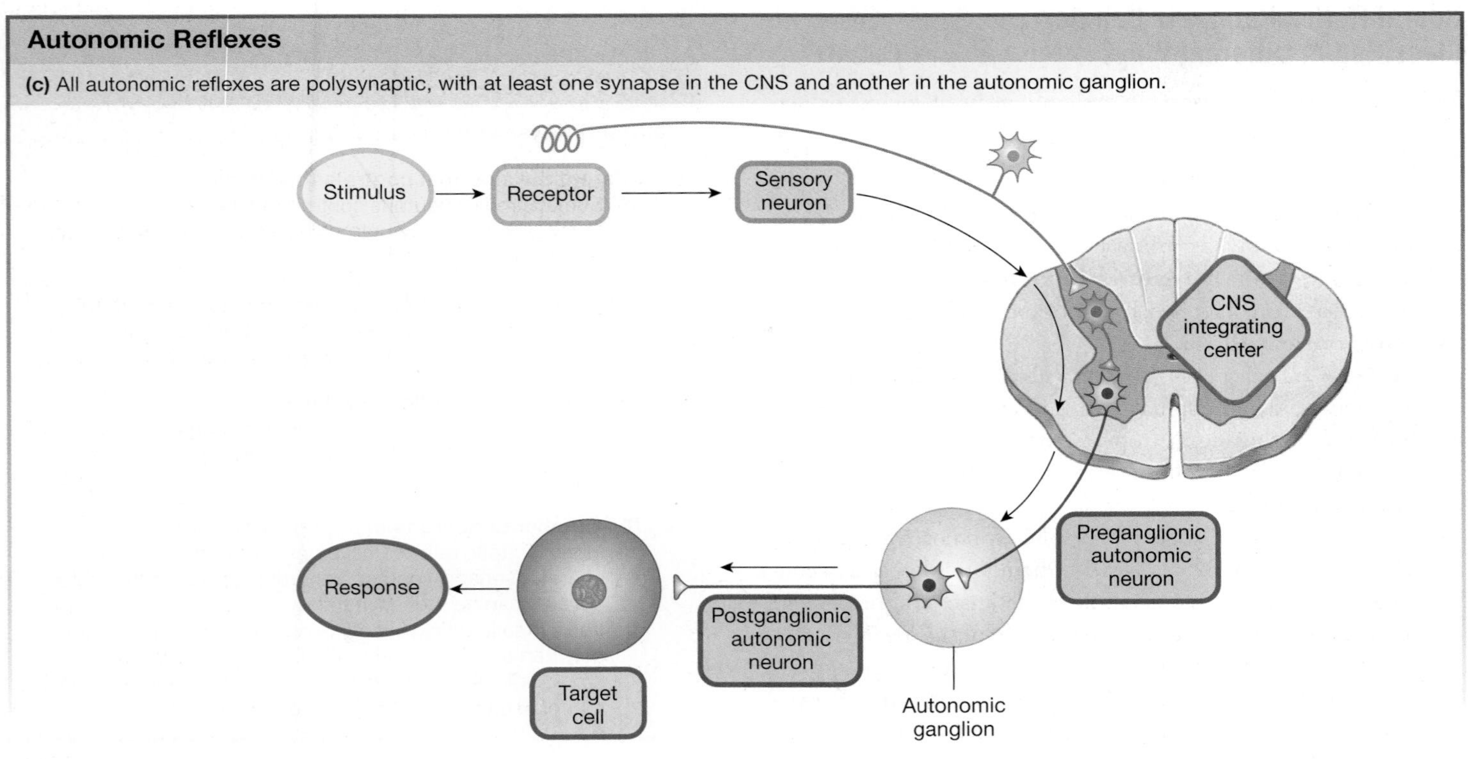
Stimulus
Receptor
Sensory neuron
CNS integrating center
Preganglionic autonomic neuron
Postganglionic autonomic neuron
Autonomic ganglion
Target cell
Response

RUNNING PROBLEM

Tetanus {*tetanus*, a muscle spasm}, also known as lockjaw, is a devastating disease caused by the bacterium *Clostridium tetani.* These bacteria are commonly found in soil and enter the human body through a cut or wound. As the bacteria reproduce in the tissues, they release a protein neurotoxin. This toxin, called *tetanospasmin,* is taken up by somatic motor neurons at the axon terminals. Tetanospasmin then travels along the axons until it reaches the nerve cell body in the spinal cord.

Q1:

a. *Tetanospasmin is a protein. By what process is it taken up into neurons? (Hint: p. 147)*

b. *By what process does it travel up the axon to the nerve cell body? (Hint: p. 228)*

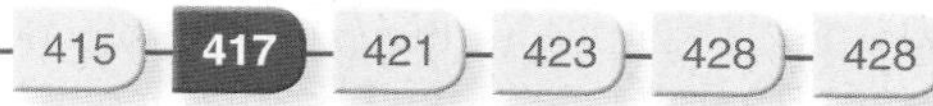
415
417
421
423
428
428

polysynaptic reflexes (Fig. 13.1b, c). Polysynaptic reflexes may be quite complex, with extensive branching in the CNS to form networks involving multiple interneurons. *Divergence* of pathways allows a single stimulus to affect multiple targets [p. 258]. *Convergence* integrates the input from multiple sources to modify the response. The modification in polysynaptic pathways may involve excitation or inhibition [p. 261].

13.2 Autonomic Reflexes

Autonomic reflexes are also known as *visceral reflexes* because they often involve the internal organs of the body. Some visceral reflexes, such as urination and defecation, are spinal reflexes that can take place without input from the brain. However, spinal reflexes are often modulated by excitatory or inhibitory signals from the brain, carried by descending tracts from higher brain centers.

For example, urination may be voluntarily initiated by conscious thought. Or it may be inhibited by emotion or a stressful situation, such as the presence of other people (a syndrome known as "bashful bladder"). Often, the higher control of a spinal reflex is a learned response. The toilet training we master as toddlers is an example of a learned reflex that the CNS uses to modulate the simple spinal reflex of urination.

Other autonomic reflexes are integrated in the brain, primarily in the hypothalamus, thalamus, and brain stem. These regions contain centers that coordinate body functions needed to maintain homeostasis, such as heart rate, blood pressure, breathing, eating, water balance, and maintenance of body temperature [see Fig. 11.3, p. 357]. The brain stem also contains the integrating centers for autonomic reflexes such as salivating, vomiting, sneezing, coughing, swallowing, and gagging.

An interesting type of autonomic reflex is the conversion of emotional stimuli into visceral responses. The limbic system [p. 288]—the site of primitive drives such as sex, fear, rage, aggression, and hunger—has been called the "visceral brain" because of its role in these emotionally driven reflexes. We speak of "gut feelings" and "butterflies in the stomach"—all transformations of emotion into somatic sensation and visceral function. Other emotion-linked autonomic reflexes include urination, defecation, blushing, blanching, and *piloerection,* in which tiny muscles in the hair follicles pull the shaft of the hair erect ("I was so scared my hair stood on end!").

Autonomic reflexes are all polysynaptic, with at least one synapse in the CNS between the sensory neuron and the preganglionic autonomic neuron, and an additional synapse in the ganglion between the preganglionic and postganglionic neurons (Fig. 13.1c).

Many autonomic reflexes are characterized by *tonic activity,* a continuous stream of action potentials that creates ongoing activity in the effector. For example, the tonic control of blood vessels is an example of a continuously active autonomic reflex [p. 182]. You will encounter many autonomic reflexes as you continue your study of the systems of the body.

Concept Check

1. List the general steps of a reflex pathway, including the anatomical structures in the nervous system that correspond to each step.
2. If a cell hyperpolarizes, does its membrane potential become more positive or more negative? Does the potential move closer to threshold or farther from threshold?

13.3 Skeletal Muscle Reflexes

Although we are not always aware of them, skeletal muscle reflexes are involved in almost everything we do. Receptors that sense changes in joint movements, muscle tension, and muscle length feed this information to the CNS, which responds in one of two ways. If muscle contraction is the appropriate response, the CNS activates somatic motor neurons to the muscle fibers. If a muscle needs to be relaxed to achieve the response, sensory input activates inhibitory interneurons in the CNS, and these interneurons *inhibit* activity in somatic motor neurons controlling the muscle.

Recall that excitation of somatic motor neurons always causes contraction in skeletal muscle [p. 368]. There is no inhibitory neuron that synapses on skeletal muscles to cause them to relax. Instead, *relaxation results from the absence of excitatory input by the somatic motor neuron.* Inhibition and excitation of somatic motor neurons and their associated skeletal muscles must occur at synapses within the CNS.

Skeletal muscle reflexes have the following components:

1. *Sensory receptors,* known as **proprioceptors**, are located in skeletal muscles, joint capsules, and ligaments. Proprioceptors monitor the position of our limbs in space, our movements, and the effort we exert in lifting objects. The input signal from proprioceptors goes to the CNS through sensory neurons.

   Three types of proprioceptors are found in the body: joint receptors, Golgi tendon organs, and muscle spindles.

CHAPTER 13

Joint receptors are found in the capsules and ligaments around joints in the body. They are stimulated by mechanical distortion that accompanies changes in the relative positioning of bones linked by flexible joints. Sensory information from joint receptors is integrated primarily in the cerebellum.

2. *The central nervous system* integrates the input signal using networks and pathways of *excitatory and inhibitory interneurons*. In a reflex, sensory information is integrated and acted on subconsciously. However, some sensory information may be integrated in the cerebral cortex and become perception, and some reflexes can be modulated by conscious input.
3. *Somatic motor neurons* carry the output signal. The somatic motor neurons that innervate skeletal muscle contractile fibers are called **alpha motor neurons** (**FIG. 13.2b**).
4. The effectors are contractile skeletal muscle fibers, also known as **extrafusal muscle fibers**. Action potentials in alpha motor neurons cause extrafusal fibers to contract.

Clinicians use reflexes to investigate the condition of the nervous system and the muscles.

For a reflex to be normal, there must be normal conduction through all neurons in the pathway, normal synaptic transmission at the neuromuscular junction, and normal muscle contraction. Any reflex that is absent, abnormally slow, or greater than normal (hyperactive) suggests the presence of a pathology. Interestingly, not all abnormal reflexes are caused by neuromuscular disorders. For example, slowed relaxation of the ankle flexion reflex suggests hypothyroidism. (The cellular mechanism linking low thyroid to slow reflexes is not known.)

Besides testing reflexes, clinicians assess **muscle tone**, which is resistance to stretch even when the muscle is relaxed and at rest. Muscle tone is the result of continuous (tonic) output by alpha motor neurons onto the extrafusal muscle fibers. The absence of muscle tone or increased muscle resistance to being stretched by an examiner (increased tone) usually indicates a problem with neural pathways.

In the next two sections we examine the function of Golgi tendon organs and muscle spindles, both interesting and unique receptors. These receptors lie inside skeletal muscles and sense changes in muscle length and tension. Their sensory output plays an important role in maintaining body position and movement.

Golgi Tendon Organs Respond to Muscle Tension

The **Golgi tendon organ (GTO)** is a type of receptor found at the junction of tendons and muscle fibers, placed in series with the muscle fibers (Fig. 13.2a). GTOs respond primarily to muscle tension created during the isometric phase of contraction and are relatively insensitive to muscle stretch.

Golgi tendon organs are composed of free nerve endings that wind between collagen fibers inside a connective tissue capsule (Fig. 13.2a). When a muscle contracts, its tendons act as a *series elastic element* during the isometric phase of the contraction [p. 396]. Muscle contraction pulls on collagen fibers within the GTO, pinching sensory endings of the afferent neurons and causing them to fire.

The classic view of Golgi tendon organs was that they were part of a protective reflex initiated by muscle contraction and ending with muscle relaxation. Research has now shown that the Golgi tendon organs primarily provide sensory information to CNS integrating centers. The sensory information from GTOs combines with feedback from muscle spindles and joint receptors to allow optimal motor control of posture and movement.

Muscle Spindles Respond to Muscle Stretch

Muscle spindles are stretch receptors that send information to the spinal cord and brain about muscle length and changes in muscle length. They are small, elongated structures scattered among and arranged parallel to the contractile extrafusal muscle fibers (Fig. 13.2b). With the exception of one muscle in the jaw, every skeletal muscle in the body has many muscle spindles. For example, a small muscle in the index finger of a newborn human has on average about 50 spindles.

Each muscle spindle consists of a connective tissue *capsule* that encloses a group of small muscle fibers known as **intrafusal fibers** {*intra-*, within + *fusus*, spindle}. Intrafusal muscle fibers are modified so that the ends are contractile but the central region lacks myofibrils (Fig. 13.2b). The noncontractile central region is wrapped by sensory nerve endings that are stimulated by stretch. The contractile ends of the intrafusal fibers have their own innervation from **gamma motor neurons**.

When a muscle is at its resting length, the central region of each muscle spindle is stretched enough to activate the sensory fibers (Fig. 13.2c). As a result, sensory neurons from the spindles are tonically active, sending a steady stream of action potentials to the spinal cord. The sensory neurons synapse directly on alpha motor neurons innervating the muscle in which the spindles lie, creating a monosynaptic reflex as shown in Figure 13.1a. The tonically active sensory neurons mean that tonically active alpha motor neurons are triggering muscle contraction. As a result, even a muscle at rest maintains a certain level of tension known as muscle tone.

Muscle spindles are anchored in parallel to the extrafusal muscle fibers. Any movement that increases muscle length also stretches the muscle spindles and causes their sensory fibers to fire more rapidly. Spindle and muscle stretch create a reflex contraction of the muscle to prevent damage from overstretching (**FIG. 13.3**). The reflex pathway in which muscle stretch initiates a contraction response is known as a **stretch reflex**.

An example of how muscle spindles work during a stretch reflex is shown in Figure 13.3. You can demonstrate this yourself with an unsuspecting friend. Have your friend stand with eyes closed, one arm extended with the elbow at 90°, and the hand palm up. Place a small book or other flat weight in the outstretched hand and watch the arm muscles contract to compensate for the added weight.

Now suddenly drop a heavier load, such as another book, onto your friend's hand. The added weight will send the hand downward, stretching the biceps muscle and activating its muscle

FIG. 13.2 ESSENTIALS Muscle Spindles and Golgi Tendon Organs

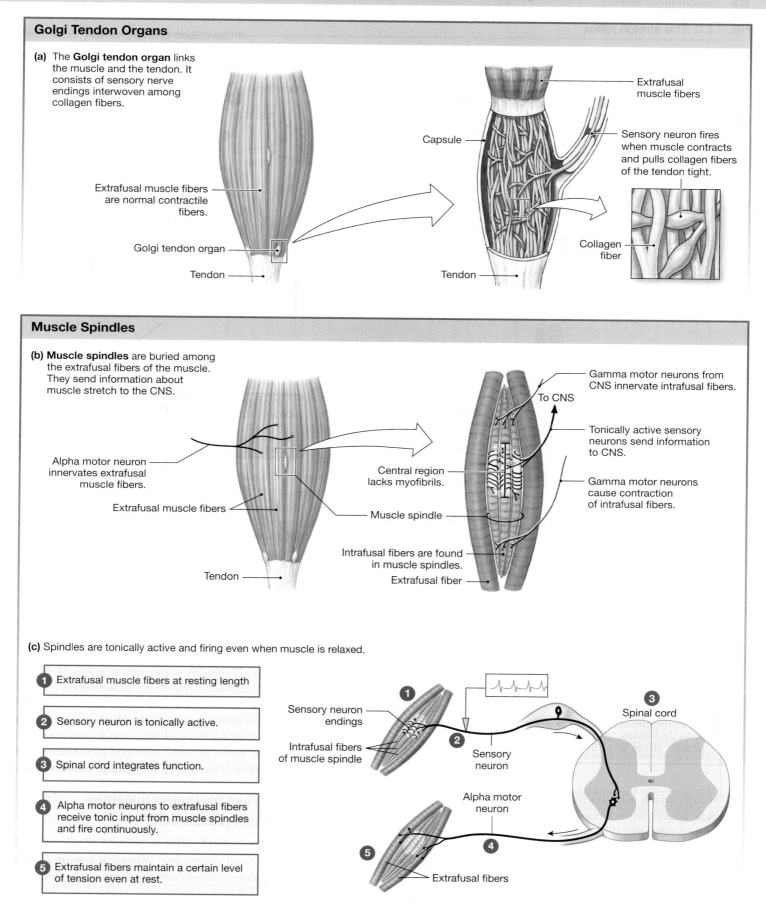

Golgi Tendon Organs

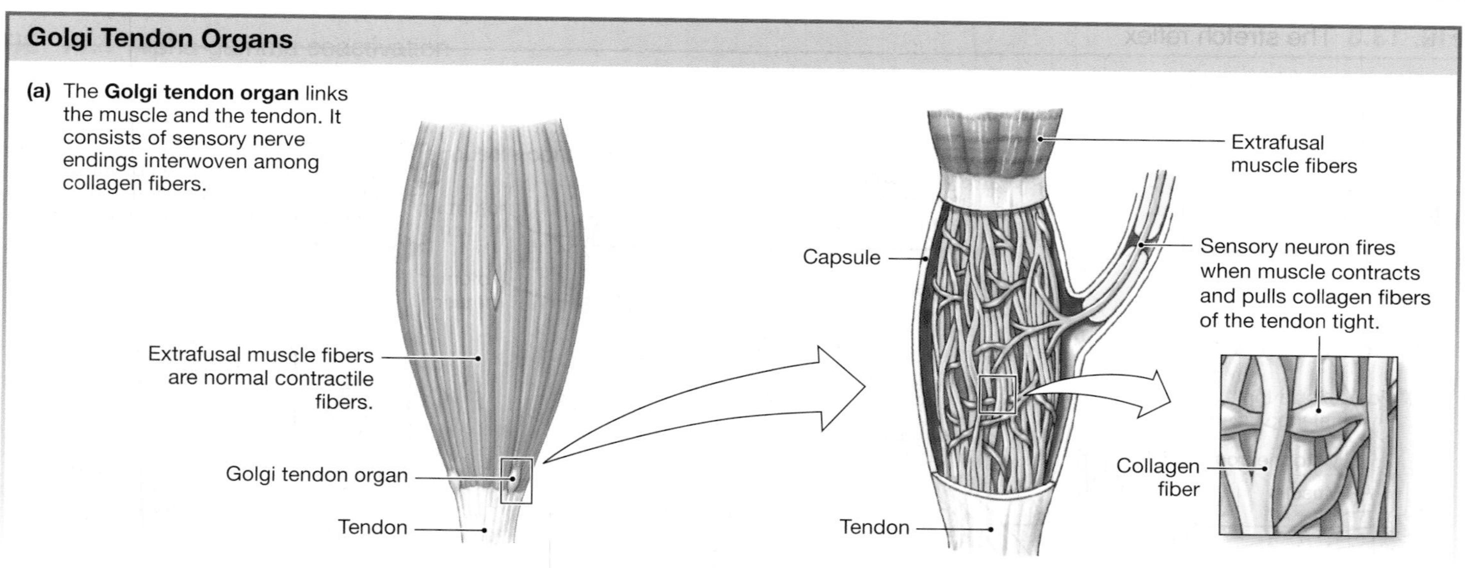
Extrafusal muscle fibers are normal contractile fibers.
Golgi tendon organ
Tendon
Capsule
Tendon
Extrafusal muscle fibers
Sensory neuron fires when muscle contracts and pulls collagen fibers of the tendon tight.
Collagen fiber

(a) The **Golgi tendon organ** links the muscle and the tendon. It consists of sensory nerve endings interwoven among collagen fibers.

Muscle Spindles

(b) **Muscle spindles** are buried among the extrafusal fibers of the muscle. They send information about muscle stretch to the CNS.

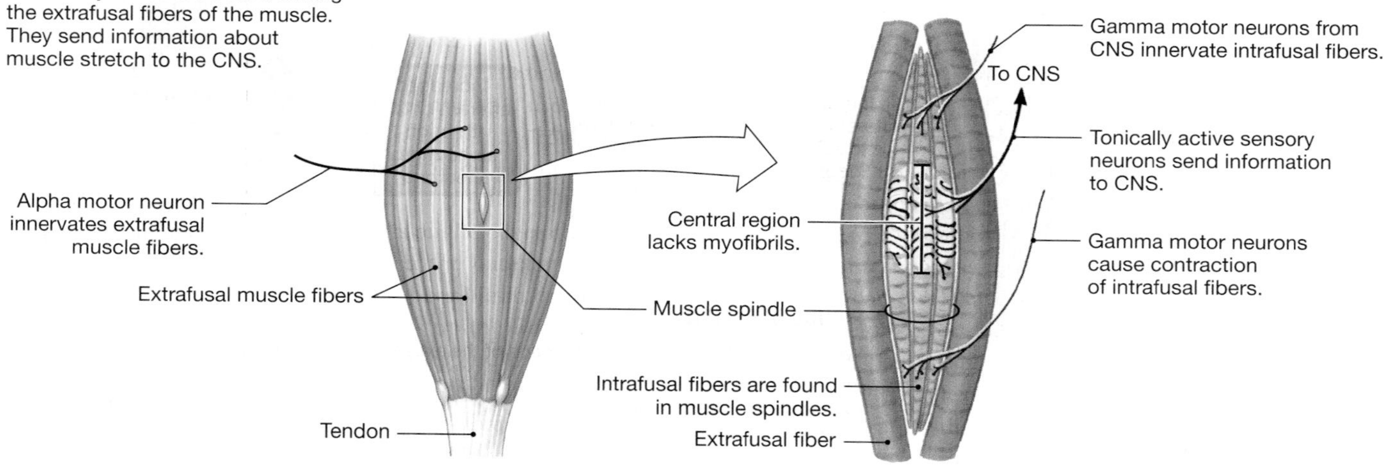

Alpha motor neuron innervates extrafusal muscle fibers.

Extrafusal muscle fibers

Tendon

Central region lacks myofibrils.

Muscle spindle

Intrafusal fibers are found in muscle spindles.

Extrafusal fiber

Gamma motor neurons from CNS innervate intrafusal fibers.

To CNS

Tonically active sensory neurons send information to CNS.

Gamma motor neurons cause contraction of intrafusal fibers.

(c) Spindles are tonically active and firing even when muscle is relaxed.

1. Extrafusal muscle fibers at resting length
2. Sensory neuron is tonically active.
3. Spinal cord integrates function.
4. Alpha motor neurons to extrafusal fibers receive tonic input from muscle spindles and fire continuously.
5. Extrafusal fibers maintain a certain level of tension even at rest.

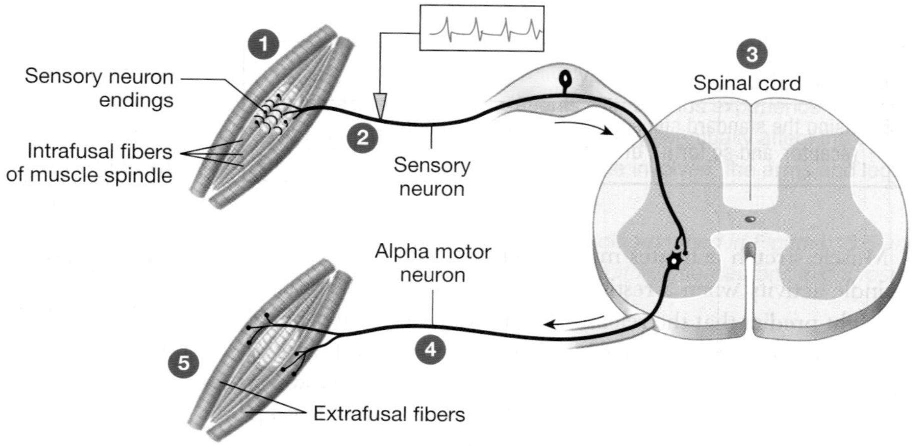

Sensory neuron endings

Intrafusal fibers of muscle spindle

Sensory neuron

Spinal cord

Alpha motor neuron

Extrafusal fibers

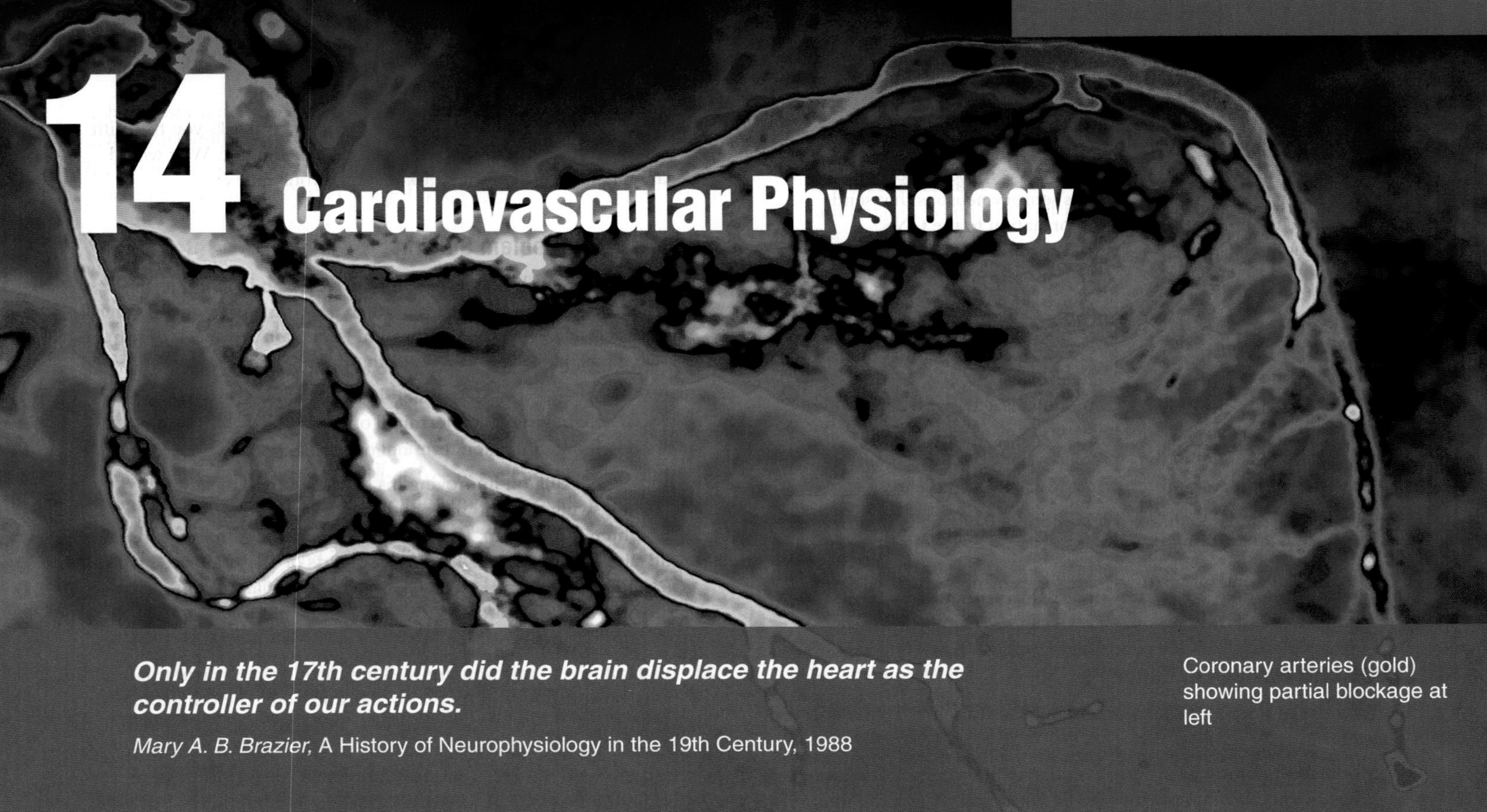

14 Cardiovascular Physiology

Only in the 17th century did the brain displace the heart as the controller of our actions.

Mary A. B. Brazier, A History of Neurophysiology in the 19th Century, 1988

Coronary arteries (gold) showing partial blockage at left

BACKGROUND BASICS

In the classic movie *Indiana Jones and the Temple of Doom*, the evil priest reaches into the chest of a sacrificial victim and pulls out his heart, still beating. This act was not dreamed up by some Hollywood scriptwriter—it was taken from rituals of the ancient Mayans, who documented this grisly practice in their carvings and paintings. The heart has been an object of fascination for centuries, but how can this workhorse muscle, which pumps 7200 liters of blood a day, keep beating outside the body? To answer that question, let's first consider the role of hearts in circulatory systems.

As life evolved, simple one-celled organisms began to band together, first into cooperative colonies and then into multicelled organisms. In most multicellular animals, only the surface layer of cells is in direct contact with the environment. This body plan presents a problem because diffusion slows exponentially as distance increases [p. 134]. Because of this, oxygen consumption in the interior cells of larger animals exceeds the rate at which oxygen can diffuse from the body surface.

One solution to overcome slow diffusion was the evolutionary development of circulatory systems that move fluid between the body's surface and its deepest parts. In simple animals, muscular activity creates fluid flow when the animal moves. More complex animals have muscular pumps called hearts to circulate internal fluid.

In the most efficient circulatory systems, the heart pumps blood through a closed system of vessels. This one-way circuit steers the blood along a specific route and ensures systematic distribution of gases, nutrients, signal molecules, and wastes. A circulatory system comprising a heart, blood vessels, and blood is known as a **cardiovascular system** {*kardia*, heart + *vasculum*, little vessel}.

Although the idea of a closed cardiovascular system that cycles blood in an endless loop seems intuitive to us today, it has not always been so. **Capillaries** {*capillus*, hair}, the microscopic vessels where blood exchanges material with the interstitial fluid, were not discovered until Marcello Malpighi, an Italian anatomist, observed them through a microscope in the middle of the seventeenth century. At that time, European medicine was still heavily influenced by the ancient belief that the cardiovascular system distributed both blood and air.

> **RUNNING PROBLEM** Myocardial Infarction
>
> The knock on Keesha's door came as she was settling down to study for her pathophysiology exam. In her door's viewer she saw Lisa Cooper, her 48-year-old neighbor. "Hi, Lisa. What's up?" As she opened the door, Lisa burst in. "I feel really awful. I've been nauseated all day and now I have this pain between my shoulder blades. Must have pulled something working out. You're a nursing student, right? Maybe you can help me?" Noticing Lisa's pale face, Keesha remembered, "Didn't your brother have some big medical problem recently?" "Yes, a heart attack three months ago. They put stents in and he's better now. But this isn't anything like that."
>
> 433 – 440 – 447 – 459 – 464 – 467 – 472

Blood was thought to be made in the liver and distributed throughout the body in the veins. Air went from the lungs to the heart, where it was digested and picked up "vital spirits." From the heart, air was distributed to the tissues through vessels called arteries. Anomalies—such as the fact that a cut artery squirted blood rather than air—were ingeniously explained by unseen links between arteries and veins that opened upon injury.

According to this model of the circulatory system, the tissues consumed all blood delivered to them, and the liver had to synthesize new blood continuously. It took the calculations of William Harvey (1578–1657), court physician to King Charles I of England, to show that the weight of blood pumped by the heart in a single hour exceeds the weight of the entire body! Once it became obvious that the liver could not make blood as rapidly as the heart pumped it, Harvey looked for an anatomical route that would allow the blood to recirculate rather than be consumed in the tissues. He showed that valves in the heart and veins created a one-way flow of blood, and that veins carried blood back to the heart, not out to the limbs. He also showed that blood entering the right side of the heart had to go to the lungs before it could go to the left side of the heart.

These studies created a furor among Harvey's contemporaries, leading Harvey to say in a huff that no one under the age of 40 could understand his conclusions. Ultimately, Harvey's work became the foundation of modern cardiovascular physiology. Today, we understand the structure of the cardiovascular system at microscopic and molecular levels that Harvey never dreamed existed. Yet some things have not changed. Even now, with our sophisticated technology, we are searching for "spirits" in the blood, although today we call them by names such as *hormone* and *cytokine*.

14.1 Overview of the Cardiovascular System

In the simplest terms, a cardiovascular system is a series of tubes (the blood vessels) filled with fluid (blood) and connected to a pump (the heart). Pressure generated in the heart propels blood through the system continuously. The blood picks up oxygen at the lungs and nutrients in the intestine and then delivers these substances to the body's cells while simultaneously removing cellular wastes and heat for excretion. In addition, the cardiovascular system plays an important role in cell-to-cell communication and in defending the body against foreign invaders. This chapter focuses on an overview of the cardiovascular system and on the heart as a pump. Later, you will learn about the properties of the blood vessels and the homeostatic controls that regulate blood flow and blood pressure.

The Cardiovascular System Transports Materials throughout the Body

The primary function of the cardiovascular system is the transport of materials to and from all parts of the body. Substances transported by the cardiovascular system can be divided into (1) nutrients, water, and gases that enter the body from the external environment, (2) materials that move from cell to cell within the body, and (3) wastes that the cells eliminate (**TBL. 14.1**).

coronary arteries, which nourish the heart muscle itself. Blood from these arteries flows into capillaries, then into the *coronary veins,* which empty directly into the right atrium at the *coronary sinus.*

Most nutrients absorbed in the intestine are routed directly to the liver, allowing that organ to process material before it is released into the general circulation. The two capillary beds of the digestive

TABLE 14.1 Transport in the Cardiovascular System		
Substance Moved	From	To

The Cardiovascular System Consists of the Heart, Blood Vessels, and Blood

The cardiovascular system is composed of the heart, the blood

FIG. 14.5 (Continued)

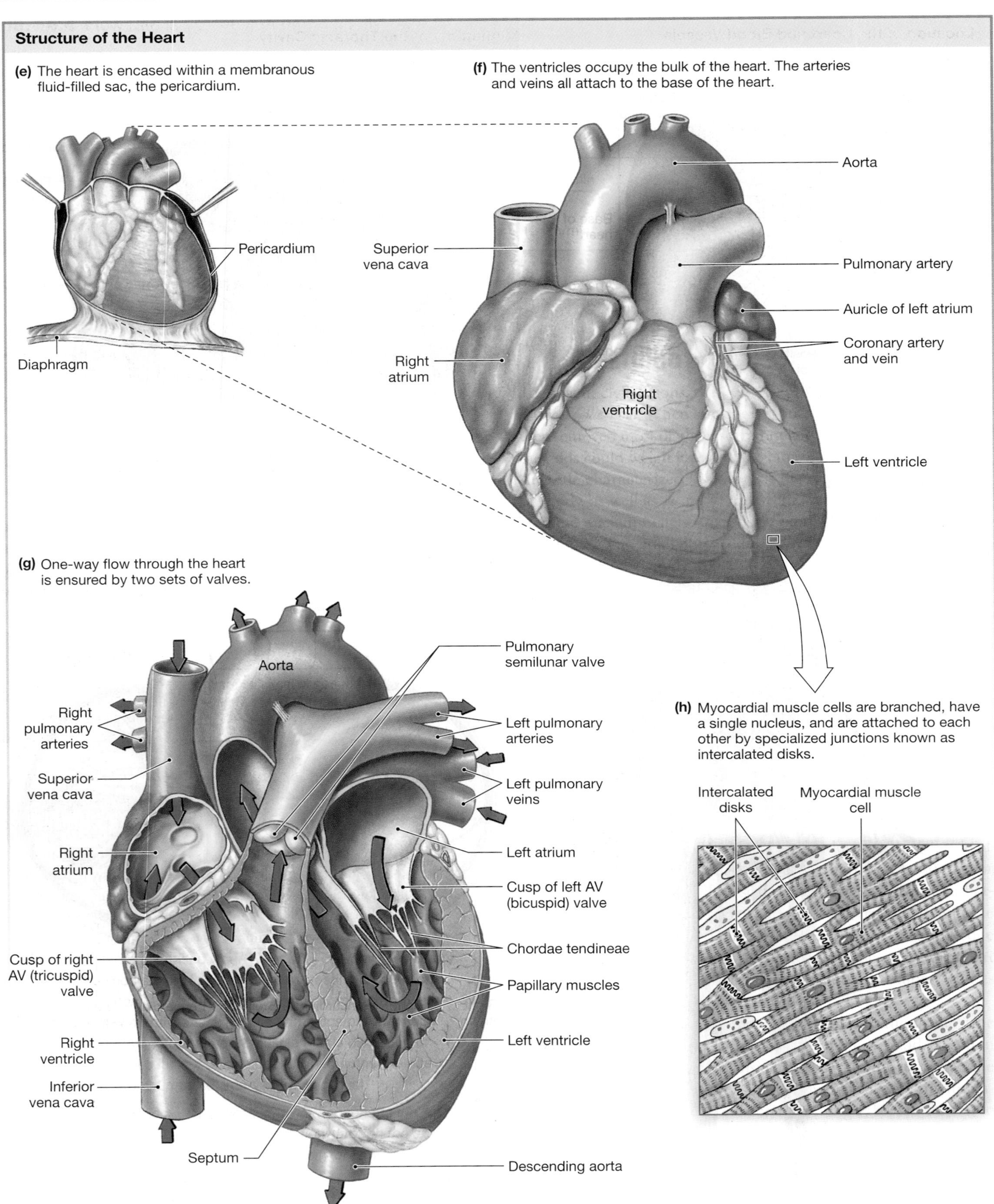
Structure of the Heart
(e) The heart is encased within a membranous fluid-filled sac, the pericardium.
Pericardium
Diaphragm
(f) The ventricles occupy the bulk of the heart. The arteries and veins all attach to the base of the heart.
Aorta
Superior vena cava
Pulmonary artery
Auricle of left atrium
Coronary artery and vein
Right atrium
Right ventricle
Left ventricle
(g) One-way flow through the heart is ensured by two sets of valves.
Aorta
Pulmonary semilunar valve
Right pulmonary arteries
Left pulmonary arteries
Superior vena cava
Left pulmonary veins
Right atrium
Left atrium
Cusp of left AV (bicuspid) valve
Chordae tendineae
Cusp of right AV (tricuspid) valve
Papillary muscles
Right ventricle
Left ventricle
Inferior vena cava
Septum
Descending aorta
(h) Myocardial muscle cells are branched, have a single nucleus, and are attached to each other by specialized junctions known as intercalated disks.
Intercalated disks
Myocardial muscle cell

TABLE 14.2 The Heart and Major Blood Vessels

Blue type indicates structures containing blood with lower oxygen content; red type indicates well-oxygenated blood.

	Receives Blood from	Sends Blood to
Heart		
Right atrium	Venae cavae	Right ventricle
Right ventricle	Right atrium	Lungs
Left atrium	Pulmonary veins	Left ventricle
Left ventricle	Left atrium	Body except for lungs
Vessels		
Venae cavae	Systemic veins	Right atrium
Pulmonary trunk (artery)	Right ventricle	Lungs
Pulmonary vein	Veins of the lungs	Left atrium
Aorta	Left ventricle	Systemic arteries

outer and inner layers of epithelium and connective tissue. Seen from the outside, the bulk of the heart is the thick muscular walls of the ventricles, the two lower chambers (Fig. 14.5f). The thinner-walled atria lie above the ventricles.

The major blood vessels all emerge from the base of the heart. The aorta and *pulmonary trunk* (artery) direct blood from the heart to the tissues and lungs, respectively. The venae cavae and pulmonary veins return blood to the heart (TBL. 14.2). When the heart is viewed from the front (anterior view), as in Figure 14.5f, the pulmonary veins are hidden behind the other major blood vessels.

The relationship between the atria and ventricles can be seen in a cross-sectional view of the heart (Fig. 14.5g). As noted earlier, the left and right sides of the heart are separated by a septum, so that blood on one side does not mix with blood on the other side. Although blood flow in the left heart is separated from flow in the right heart, the two sides contract in a coordinated fashion. First, the atria contract together, then the ventricles contract together.

Blood flows from veins into the atria and from there through one-way valves into the ventricles, the pumping chambers. Blood leaves the heart via the pulmonary trunk from the right ventricle and via the aorta from the left ventricle. A second set of valves guards the exits of the ventricles so that blood cannot flow back into the heart once it has been ejected.

Notice in Figure 14.5g that blood enters each ventricle at the top of the chamber but also leaves at the top. This is because during development, the tubular embryonic heart twists back on itself (FIG. 14.6b). This twisting puts the arteries (through which blood leaves) close to the top of the ventricles. Functionally, this means that the ventricles must contract from the bottom up so that blood is squeezed out of the top.

Four fibrous connective tissue rings surround the four heart valves (Fig. 14.5g). These rings form both the origin and insertion for the cardiac muscle, an arrangement that pulls the apex and base of the heart together when the ventricles contract. In addition, the fibrous connective tissue acts as an electrical insulator, blocking most transmission of electrical signals between the atria and the ventricles. This arrangement ensures that the electrical signals can be directed through a specialized conduction system to the apex of the heart for the bottom-to-top contraction.

Heart Valves Ensure One-Way Flow in the Heart

As the arrows in Figure 14.5g indicate, blood flows through the heart in one direction. Two sets of heart valves ensure this one-way flow: one set (the **atrioventricular valves**) between the atria and ventricles, and the second set (the **semilunar valves**, named for their crescent-moon shape) between the ventricles and the arteries. Although the two sets of valves are very different in structure, they serve the same function: preventing the backward flow of blood.

FIG. 14.6 In the embryo, the heart develops from a single tube

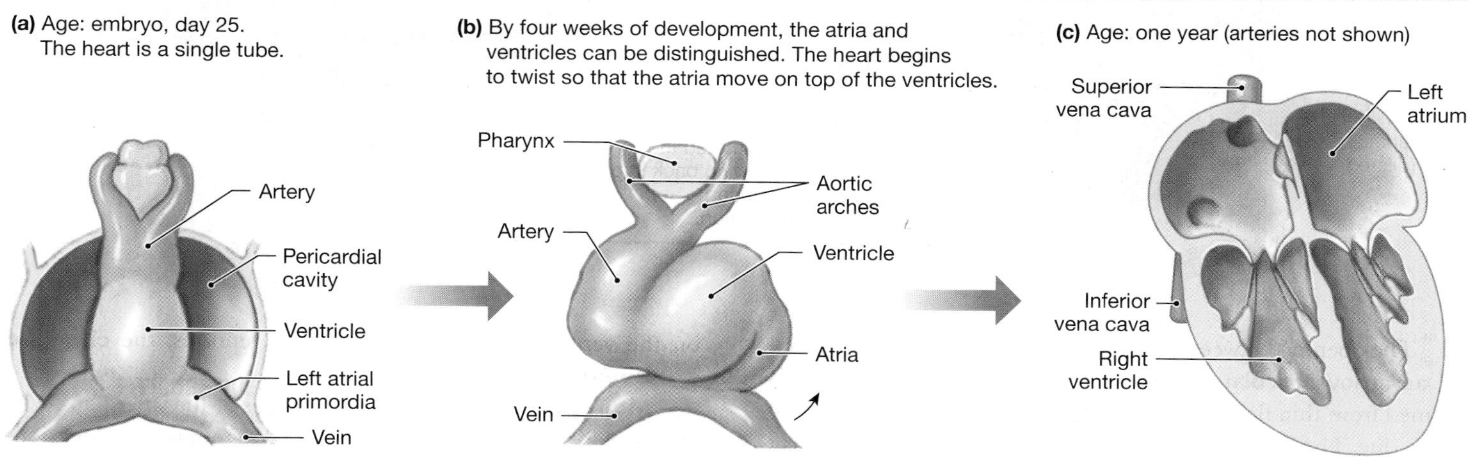
(a) Age: embryo, day 25.
The heart is a single tube.
Artery
Pericardial cavity
Ventricle
Left atrial primordia
Vein
(b) By four weeks of development, the atria and ventricles can be distinguished. The heart begins to twist so that the atria move on top of the ventricles.
Pharynx
Aortic arches
Artery
Ventricle
Atria
Vein
(c) Age: one year (arteries not shown)
Superior vena cava
Left atrium
Inferior vena cava
Right ventricle

There is considerable anatomic variability in terms of which regions of the heart are supplied by the different coronary artery branches. The two primary coronary arteries originate at the start of the aorta (the *root* of the aorta), just superior to the semilunar valve leaflets of the aortic valve. The *right coronary artery* (RCA) runs from the aorta around the right side of the heart in a groove (the *coronary sulcus*) between the right atrium and right ventricle. In the majority of people the RCA branches feed the right atrium, most of the right ventricle and some of the left ventricle, and the posterior ("back") portion of the interventricular septum.

The *left coronary artery* (LCA) leaves the left side of the aorta. It divides into two main branches: the *circumflex branch*, which continues around the left side of the heart to the posterior surface, and the *anterior interventricular branch*, which runs in a groove toward the apex of the heart. The anterior interventricular branch is often called the *LAD*, or left anterior descending artery, by clinicians. These branches of the left coronary artery supply blood to the left atrium, most of the left ventricle and interventricular septum, and some of the right ventricle.

Blood from the coronary circulation returns to the heart via three routes. Most venous blood (~85%) leaves the myocardium through *cardiac veins* that empty into the *coronary sinus* on the posterior aspect of the heart. Blood in the coronary sinus empties directly into the right atrium. Deep in the heart muscle, smaller blood channels empty their blood directly into the heart's chambers. In addition, a few small veins on the anterior portion of the right ventricle drain directly into the right atrium.

Venous blood in the coronary circulation is much lower in oxygen content than venous blood returning through the venae cavae. By one estimate, cardiac muscle consumes 70–80% of the oxygen delivered to it by the blood, more than twice the amount extracted by other cells in the body. During periods of increased activity, the heart uses almost all the oxygen brought to it by the coronary arteries. As a result, the only way to get more oxygen to exercising heart muscle is to increase the blood flow [discussed in Chapter 15]. Reduced myocardial blood flow from blockage of a coronary artery or excessively low blood pressure can damage or even kill the heart muscle.

Cardiac Muscle Cells Contract without Innervation

The bulk of the heart is composed of cardiac muscle cells, or myocardium. Most cardiac muscle is contractile, but about 1% of the myocardial cells are specialized to generate action potentials spontaneously. These cells account for a unique property of the heart: its ability to contract without any outside signal. As mentioned in the introduction to this chapter, records tell us of Spanish explorers in the New World witnessing human sacrifices in which hearts torn from the chests of living victims continued to beat for minutes. The heart can contract without a connection to other parts of the body because the signal for contraction is *myogenic*, originating within the heart muscle itself.

The signal for myocardial contraction comes not from the nervous system but from specialized myocardial cells known as **autorhythmic cells**. The autorhythmic cells are also called **pacemakers** because they set the rate of the heartbeat. Myocardial autorhythmic cells are anatomically distinct from contractile cells: autorhythmic cells are smaller and contain few contractile fibers. Because they do not have organized sarcomeres, autorhythmic cells do not contribute to the contractile force of the heart.

Contractile cells are typical striated muscle, however, with contractile fibers organized into sarcomeres [p. 380]. Cardiac muscle differs in significant ways from skeletal muscle and shares some properties with smooth muscle:

1. Cardiac muscle fibers are much smaller than skeletal muscle fibers and usually have a single nucleus per fiber.
2. Individual cardiac muscle cells branch and join neighboring cells end-to-end to create a complex network (Fig. 14.5h and FIG. 14.9b). The cell junctions, known as **intercalated disks** {*inter-*, between + *calare*, to proclaim}, consist of interdigitated membranes. Intercalated disks have two components: *desmosomes* and gap junctions [p. 73]. Desmosomes are strong connections that tie adjacent cells together, allowing force created in one cell to be transferred to the adjacent cell.

FIG. 14.9 Cardiac muscle

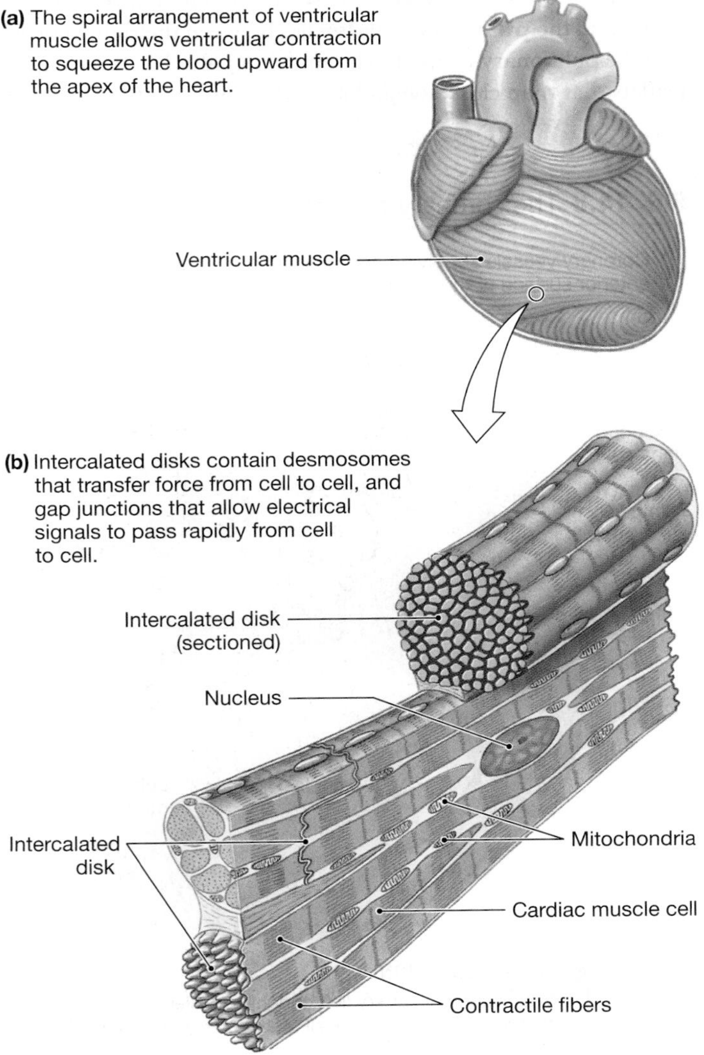
(a) The spiral arrangement of ventricular muscle allows ventricular contraction to squeeze the blood upward from the apex of the heart.
Ventricular muscle
(b) Intercalated disks contain desmosomes that transfer force from cell to cell, and gap junctions that allow electrical signals to pass rapidly from cell to cell.
Intercalated disk (sectioned)
Nucleus
Intercalated disk
Mitochondria
Cardiac muscle cell
Contractile fibers

3. **Gap junctions** in the intercalated disks electrically connect cardiac muscle cells to one another. They allow waves of depolarization to spread rapidly from cell to cell, so that all the heart muscle cells contract almost simultaneously. In this respect, cardiac muscle resembles single-unit smooth muscle.
4. The t-tubules of myocardial cells are larger than those of skeletal muscle, and they branch inside the myocardial cells.
5. Myocardial sarcoplasmic reticulum is smaller than that of skeletal muscle, reflecting the fact that cardiac muscle depends in part on extracellular Ca^{2+} to initiate contraction. In this respect, cardiac muscle resembles smooth muscle.
6. Mitochondria occupy about one-third the cell volume of a cardiac contractile fiber, a reflection of the high energy demand of these cells.

[See Tbl. 12.3, p. 408, for a summary comparison of the three muscle types.]

Calcium Entry Is a Feature of Cardiac EC Coupling

In skeletal muscle, acetylcholine from a somatic motor neuron causes a skeletal muscle action potential to begin excitation-contraction coupling (EC coupling) [p. 382]. In cardiac muscle, an action potential also initiates EC coupling, but the action potential originates spontaneously in the heart's pacemaker cells and spreads into the contractile cells through gap junctions. Other aspects of cardiac EC coupling are similar to processes you encountered in skeletal and smooth muscle contraction.

FIGURE 14.10 illustrates EC coupling and relaxation in cardiac muscle. An action potential that enters a contractile cell moves across the sarcolemma and into the t-tubules (1), where it opens voltage-gated L-type Ca^{2+} channels in the cell membrane (2). Ca^{2+} enters the cell through these channels, moving down its electrochemical gradient. Calcium entry opens *ryanodine receptor Ca^{2+} release channels* (RyR) in the sarcoplasmic reticulum (3). This process of EC coupling in cardiac muscle is also called **Ca^{2+}-induced Ca^{2+} release** (CICR). When the RyR channels open, stored Ca^{2+} flows out of the sarcoplasmic reticulum and into the cytosol (4), creating a Ca^{2+} "spark" that can be seen using special biochemical methods [p. 177]. Multiple sparks from different RyR channels sum to create a Ca^{2+} signal (5).

Calcium released from the sarcoplasmic reticulum provides about 90% of the Ca^{2+} needed for muscle contraction, with the remaining 10% entering the cell from the extracellular fluid. Calcium diffuses through the cytosol to the contractile elements, where the ions bind to troponin and initiate the cycle of crossbridge formation and movement (6). Contraction takes place by the same type of sliding filament movement that occurs in skeletal muscle [p. 382].

Relaxation in cardiac muscle is generally similar to that in skeletal muscle. As cytoplasmic Ca^{2+} concentrations decrease, Ca^{2+} unbinds from troponin, myosin releases actin, and the contractile filaments slide back to their relaxed position (7). As in skeletal muscle, Ca^{2+} is transported back into the sarcoplasmic reticulum with the help of a Ca^{2+}-ATPase (8). However, in cardiac muscle, Ca^{2+} is also removed from the cell via the *Na^+-Ca^{2+} exchanger* (NCX) (9). One Ca^{2+} moves out of the cell against its electrochemical gradient in exchange for 3 Na^+ entering the cell down their electrochemical gradient. Sodium that enters the cell during this transfer is removed by the Na^+-K^+-ATPase (10).

Cardiac Muscle Contraction Can Be Graded

A key property of cardiac muscle cells is the ability of a single muscle fiber to execute *graded contractions,* in which the fiber varies the amount of force it generates. (Recall that in skeletal muscle, contraction in a single fiber is all-or-none at any given fiber length.) The force generated by cardiac muscle is proportional to the number of crossbridges that are active. The number of active crossbridges is determined by how much Ca^{2+} is bound to troponin.

If cytosolic Ca^{2+} concentrations are low, some crossbridges are not activated and contraction force is small. If additional Ca^{2+} enters the cell from the extracellular fluid, more Ca^{2+} is released from the sarcoplasmic reticulum. This additional Ca^{2+} binds to troponin, enhancing the ability of myosin to form crossbridges with actin and creating additional force.

Another factor that affects the force of contraction in cardiac muscle is the sarcomere length at the beginning of contraction. In the intact heart, stretch on the individual fibers is a function of how much blood is in the chambers of the heart.

RUNNING PROBLEM

Lisa checked in at the emergency department and told the clerk she had back pain. Fortunately it was slow that night, so Lisa and Keesha were brought to a treatment room about 15 minutes later. Lisa's vital signs were taken, and she told the nurse about her symptoms. Shortly after, Dr. Dang, the emergency medicine resident physician on duty, came in. Dr. Dang did a quick history and physical examination. "I think you're probably right about a pulled muscle," he said. "I'm prescribing a muscle relaxant for you tonight. You can follow up with your regular doctor tomorrow. Any questions?" Lisa was relieved but Keesha asked, "Doctor, is there any chance this could be an MI? Lisa has a family history of heart disease, she smokes, and she's been told her cholesterol is too high, and she's not taking medicine for it. I know Lisa, and she doesn't want to let on, but she doesn't look right to me." Dr. Dang listened and thought for a minute. "You have a really good point. As you probably know, women often don't have chest pain with heart problems. Let's run some tests to make sure it's not that."

When blood supply to the heart muscle decreases too much, lack of oxygen can cause myocardial cells to die. Electrical conduction through the myocardium then must bypass the dead or dying cells.

Q2: *How do electrical signals pass from cell to cell in the myocardium?*

Q3: *What happens to contraction in a myocardial contractile cell if a wave of depolarization passing through the heart bypasses it?*

433 – 440 – **447** – 459 – 464 – 467 – 472

FIG. 14.10 EC coupling in cardiac muscle

This figure shows the cellular events leading to contraction and relaxation in a cardiac contractile cell.

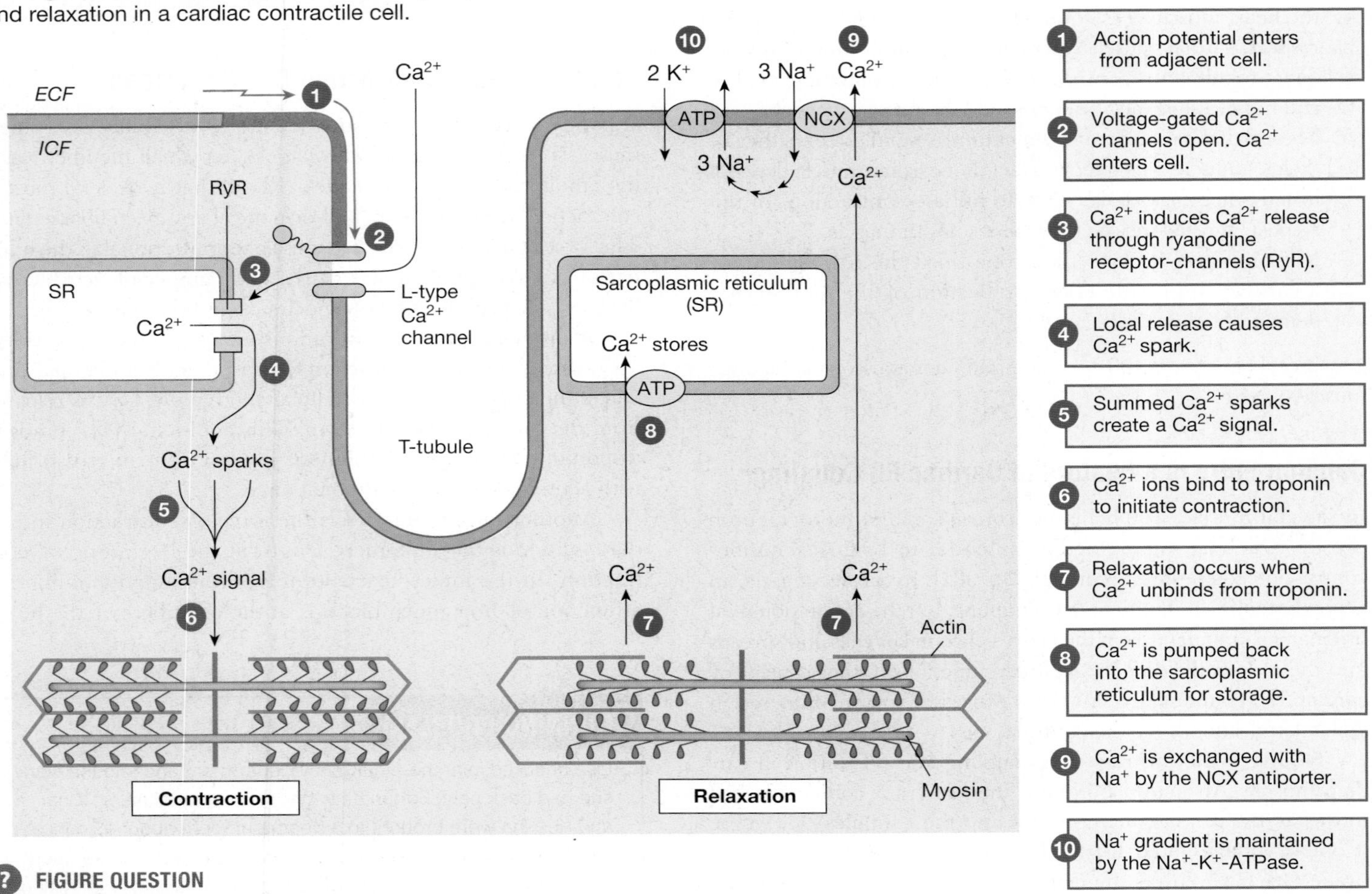

1. Action potential enters from adjacent cell.
2. Voltage-gated Ca^{2+} channels open. Ca^{2+} enters cell.
3. Ca^{2+} induces Ca^{2+} release through ryanodine receptor-channels (RyR).
4. Local release causes Ca^{2+} spark.
5. Summed Ca^{2+} sparks create a Ca^{2+} signal.
6. Ca^{2+} ions bind to troponin to initiate contraction.
7. Relaxation occurs when Ca^{2+} unbinds from troponin.
8. Ca^{2+} is pumped back into the sarcoplasmic reticulum for storage.
9. Ca^{2+} is exchanged with Na^+ by the NCX antiporter.
10. Na^+ gradient is maintained by the Na^+-K^+-ATPase.

FIGURE QUESTION

Using the numbered steps, compare the events shown to EC coupling in skeletal and smooth muscle [see Figs.12.10 and 12.26].

The relationship between force and ventricular volume is an important property of cardiac function, and we discuss it in detail later in this chapter.

Concept Check

10. Compare the receptors and channels involved in cardiac EC coupling to those found in skeletal muscle EC coupling. [*Hint:* p. 382]
11. If a myocardial contractile cell is placed in interstitial fluid and depolarized, the cell contracts. If Ca^{2+} is removed from the fluid surrounding the myocardial cell and the cell is depolarized, it does not contract. If the experiment is repeated with a skeletal muscle fiber, the skeletal muscle contracts when depolarized, whether or not Ca^{2+} is present in the surrounding fluid. What conclusion can you draw from the results of this experiment?
12. A drug that blocks all Ca^{2+} channels in the myocardial contractile cell membrane is placed in the solution around the cell. What happens to the force of contraction in that cell?

Myocardial Action Potentials Vary

Cardiac muscle, like skeletal muscle and neurons, is an excitable tissue with the ability to generate action potentials. Each of the two types of cardiac muscle cells has a distinctive action potential that will vary somewhat in shape depending on where in the heart it is recorded. In both autorhythmic and contractile myocardium, Ca^{2+} plays an important role in the action potential, in contrast to the action potentials of skeletal muscle and neurons, which depend solely on Na^+ and K^+ movement.

A point of confusion for many students as they study cardiac function is how a single ion channel, like a Ca^{2+} channel or a K^+ channel, can have so many different roles. The simple answer is that there is not just one channel type for each ion—there are families of ion channels, with multiple members in each family. By one estimate, there are at least 10 different types of K^+ channels that participate in myocardial action potentials. Each subtype of ion channel has slightly different properties that make it unique. In this book, you will learn about a few of the most important ion channel subtypes.

Myocardial Contractile Cells The action potentials of myocardial contractile cells are similar in several ways to those of neurons and skeletal muscle [p. 237]. The rapid depolarization phase of the action potential is the result of Na^+ entry, and the steep repolarization phase is due to K^+ leaving the cell (**FIG. 14.11**). The main difference between the action potential of the myocardial contractile cell and those of skeletal muscle fibers and neurons is that the myocardial cell has a longer action potential due to Ca^{2+} entry. Let's take a look at these longer action potentials. By convention, the action potential phases start with zero.

Phase 4: resting membrane potential. Myocardial contractile cells have a stable resting potential of about −90 mV.

Phase 0: depolarization. When a wave of depolarization moves into a contractile cell through gap junctions, the membrane potential becomes more positive. Voltage-gated Na^+ channels open, allowing Na^+ to enter the cell and rapidly depolarize it. The membrane potential reaches about +20 mV before the Na^+ channels close. These are double-gated Na^+ channels, similar to the voltage-gated Na^+ channels of the axon [p. 242].

Phase 1: initial repolarization. When the Na^+ channels close, the cell begins to repolarize as K^+ leaves through open K^+ channels.

Phase 2: the plateau. The initial repolarization is very brief. The action potential then flattens into a plateau as the result of two events: a decrease in K^+ permeability and an increase in Ca^{2+} permeability. Voltage-gated Ca^{2+} channels activated by depolarization have been slowly opening during phases 0 and 1. When they finally open, Ca^{2+} enters the cell. At the same time, some "fast" K^+ channels close. The combination of Ca^{2+} influx and decreased K^+ efflux causes the action potential to flatten out into a plateau.

Phase 3: rapid repolarization. The plateau ends when Ca^{2+} channels close and K^+ permeability increases once more. The "slow" K^+ channels responsible for this phase are similar to those in the neuron: They are activated by depolarization but are slow to open. When the slow K^+ channels open, K^+ exits rapidly, returning the cell to its resting potential (phase 4).

The influx of Ca^{2+} during phase 2 lengthens the total duration of a myocardial action potential. A typical action potential in a neuron or skeletal muscle fiber lasts between 1 and 5 msec. In a contractile myocardial cell, the action potential typically lasts 200 msec or more.

The longer myocardial action potential helps prevent the sustained contraction called *tetanus*. Prevention of tetanus in the heart is important because cardiac muscles must relax between contractions so the ventricles can fill with blood. To understand how a longer action potential prevents tetanus, let's compare the relationship between action potentials, refractory periods [p. 243], and contraction in skeletal and cardiac muscle cells (**FIG. 14.12**).

As you may recall, the *refractory period* is the time following an action potential during which a normal stimulus cannot trigger a second action potential. In cardiac muscle, the long action potential (red curve) means the refractory period (yellow background) and the contraction (blue curve) end almost simultaneously (Fig. 14.12a). By the time a second action potential can take place, the myocardial cell has almost completely relaxed. Consequently, no summation occurs (Fig. 14.12b).

In contrast, the skeletal muscle action potential and refractory period are ending just as contraction begins (Fig. 14.12c). For this reason, a second action potential fired immediately after the refractory period causes summation of the contractions (Fig. 14.12d). If a series of action potentials occurs in rapid succession, the sustained contraction known as tetanus results.

FIG. 14.11 Action potential of a cardiac contractile cell

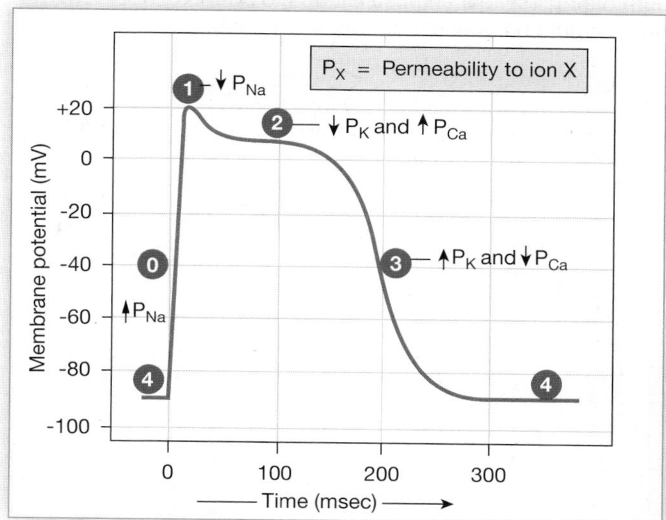

Px = Permeability to ion X
1 ↓PNa
2 ↓PK and ↑PCa
3 ↑PK and ↓PCa
0
↑PNa
4
4
Membrane potential (mV)
+20
0
-20
-40
-60
-80
-100
0
100
200
300
Time (msec)

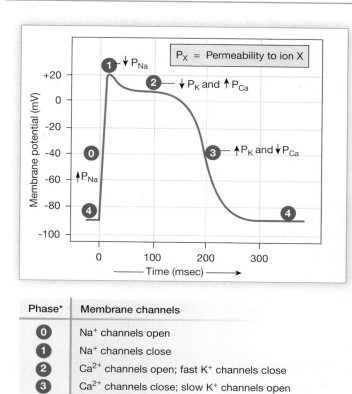

Phase*	Membrane channels
0	Na^+ channels open
1	Na^+ channels close
2	Ca^{2+} channels open; fast K^+ channels close
3	Ca^{2+} channels close; slow K^+ channels open
4	Resting potential

*The phase numbers are a convention.

FIGURE QUESTION

Compare ion movement during this action potential to ion movement of a neuron's action potential [Fig. 8.8].

Concept Check

13. Which ions moving in what directions cause the depolarization and repolarization phases of a neuronal action potential?
14. At the molecular level, what is happening during the refractory period in neurons and muscle fibers?
15. Lidocaine is a molecule that blocks the action of voltage-gated cardiac Na^+ channels. What happens to the action potential of a myocardial contractile cell if lidocaine is applied to the cell?

FIG. 14.12 Refractory periods and summation

Summation in skeletal muscle leads to tetanus, which would be fatal if it happened in the heart.

CARDIAC MUSCLE

(a) Cardiac muscle fiber: The refractory period lasts almost as long as the entire muscle twitch.

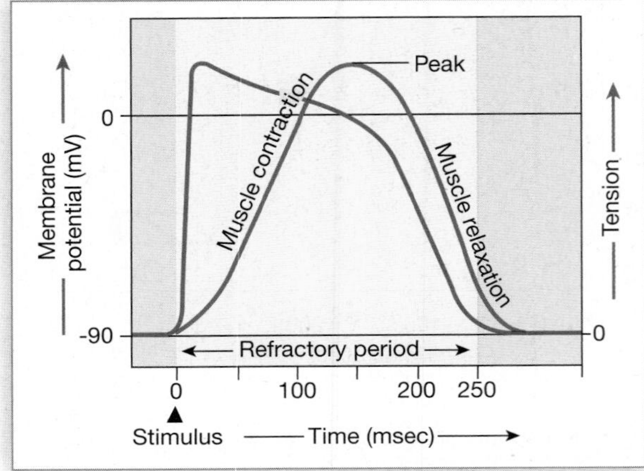

Peak
0
-90
Membrane potential (mV)
Muscle contraction
Muscle relaxation
Tension
0
Refractory period
0
100
200 250
Stimulus
Time (msec)

(b) Long refractory period in a cardiac muscle prevents tetanus.

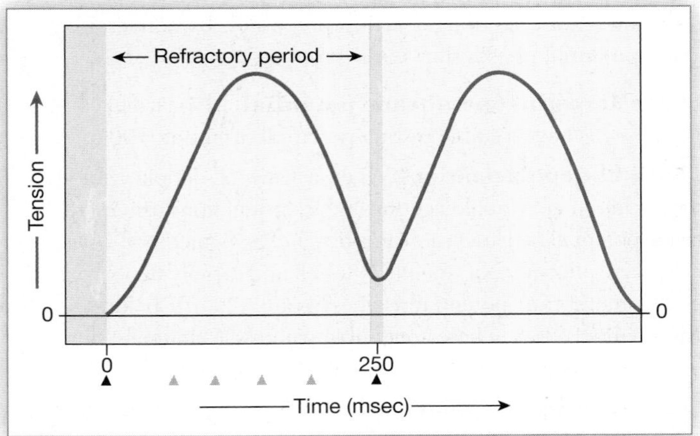

Refractory period
Tension
0
0
0
250
Time (msec)

SKELETAL MUSCLE

(c) Skeletal muscle fast-twitch fiber: The refractory period (yellow) is very short compared with the amount of time required for the development of tension.

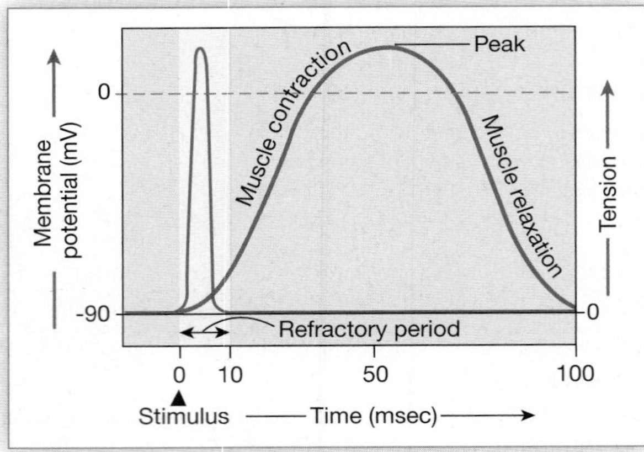

Peak
0
-90
Membrane potential (mV)
Muscle contraction
Muscle relaxation
Tension
0
Refractory period
0
10
50
100
Stimulus
Time (msec)

(d) Skeletal muscles that are stimulated repeatedly will exhibit summation and tetanus (action potentials not shown).

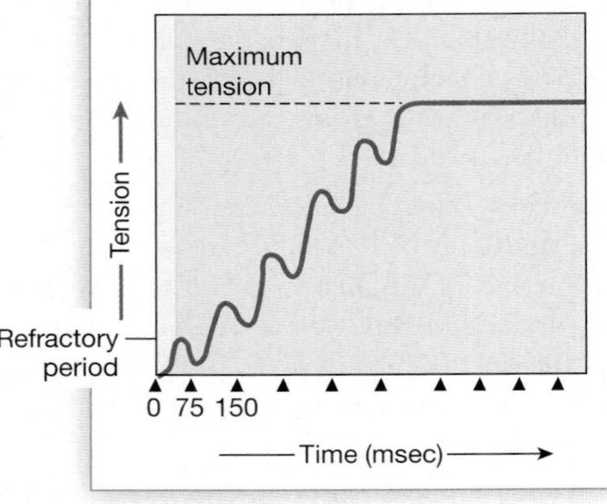

Maximum tension
Tension
Refractory period
0 75 150
Time (msec)

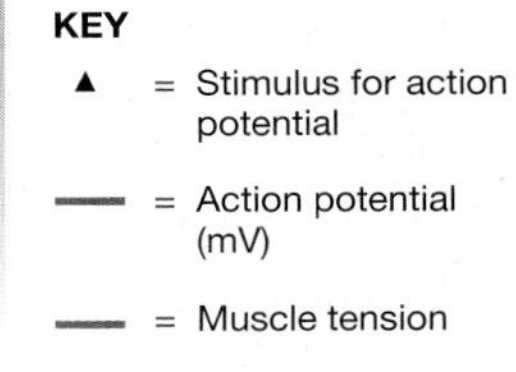

KEY
▲ = Stimulus for action potential
= Action potential (mV)
= Muscle tension

Myocardial Autorhythmic Cells What gives myocardial autorhythmic cells their unique ability to generate action potentials spontaneously in the absence of input from the nervous system? This ability results from their unstable membrane potential, which starts at −60 mV and slowly drifts upward toward threshold (**FIG. 14.13a**). This unstable membrane potential is called a **pacemaker potential** rather than a resting membrane potential because it never "rests" at a constant value. Whenever a pacemaker potential depolarizes to threshold, the autorhythmic cell fires an action potential.

What causes the membrane potential of these cells to be unstable? Our current understanding is that the autorhythmic cells contain channels that are different from the channels of other excitable tissues. When the cell membrane potential is −60 mV, **I_f channels** that are permeable to both K^+ and Na^+ open (Fig. 14.13c). These channels are called I_f channels because they allow current (I) to flow and because of their unusual properties. The researchers who first described the ion current through these channels initially did not understand its behavior and named it *funny* current—hence the subscript *f*. The I_f channels belong to the family of *HCN channels,* or *hyperpolarization-activated cyclic nucleotide-gated channels.* Other members of the HCN family are found in neurons.

When I_f channels open at negative membrane potentials, Na^+ influx exceeds K^+ efflux. (This is similar to what happens at the

FIG. 14.13 Action potentials in cardiac autorhythmic cells

Autorhythmic cells have unstable membrane potentials called pacemaker potentials.

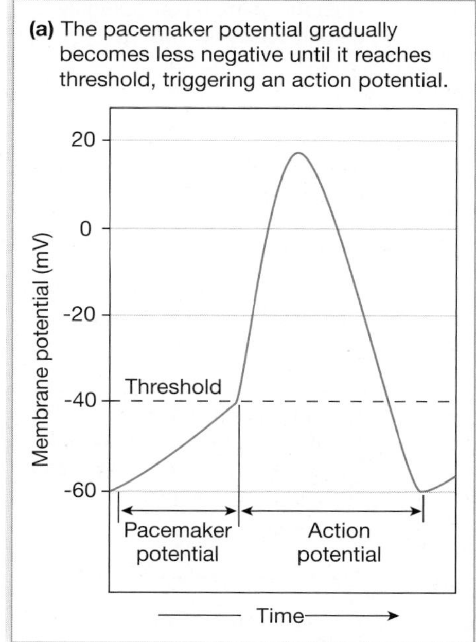
(a) The pacemaker potential gradually becomes less negative until it reaches threshold, triggering an action potential.
Membrane potential (mV)
20
0
-20
-40
-60
Threshold
Pacemaker potential
Action potential
Time

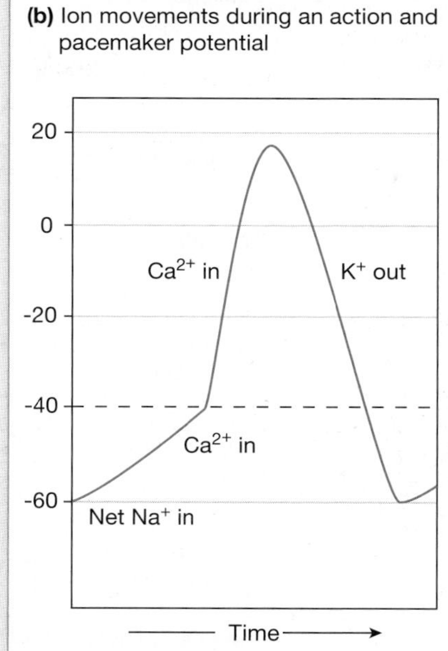
(b) Ion movements during an action and pacemaker potential
20
0
-20
-40
-60
Ca2+ in
K+ out
Ca2+ in
Net Na+ in
Time

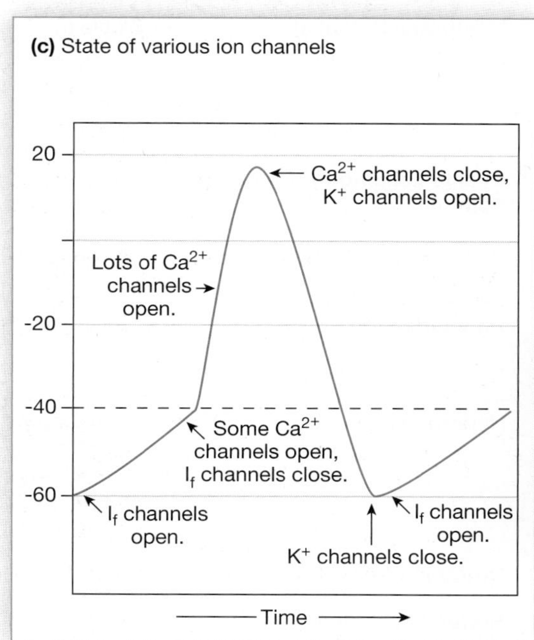
(c) State of various ion channels
20
-20
-40
-60
Ca2+ channels close, K+ channels open.
Lots of Ca2+ channels open.
Some Ca2+ channels open, If channels close.
If channels open.
If channels open.
K+ channels close.
Time

GRAPH QUESTIONS

1. Match the appropriate phases of the myocardial contractile cell action potential (Fig.14.11) to the pacemaker action potential above.
2. Which of the following would speed up the depolarization rate of the pacemaker potential?
   (a) increase in Ca^{2+} influx
   (b) increase in K^+ efflux
   (c) increase in Na^+ influx
   (d) none of these

neuromuscular junction when nonspecific cation channels open [p. 368].) The net influx of positive charge slowly depolarizes the autorhythmic cell (Fig. 14.13b). As the membrane potential becomes more positive, the I_f channels gradually close and one set of Ca^{2+} channels opens. The resulting influx of Ca^{2+} continues the depolarization, and the membrane potential moves steadily toward threshold.

When the membrane potential reaches threshold, a different subtype of voltage-gated Ca^{2+} channels opens. Calcium rushes into the cell, creating the steep depolarization phase of the action potential. Note that this process is different from that in other excitable cells, in which the depolarization phase is due to the opening of voltage-gated Na^+ channels.

When the Ca^{2+} channels close at the peak of the action potential, slow K^+ channels have opened (Fig. 14.13c). The repolarization phase of the autorhythmic action potential is due to the resultant efflux of K^+ (Fig. 14.13b). This phase is similar to repolarization in other types of excitable cells.

The speed with which pacemaker cells depolarize determines the rate at which the heart contracts (the heart rate). The interval between action potentials can be modified by altering the permeability of the autorhythmic cells to different ions, which in turn changes the duration of the pacemaker potential. This topic is discussed in detail at the end of the chapter.

TABLE 14.3 compares action potentials of the two types of myocardial muscle with those of skeletal muscle. Next we look at how action potentials of autorhythmic cells spread throughout the heart to coordinate contraction.

Concept Check

16. What does increasing K^+ permeability do to the membrane potential of the cell?
17. A new cardiac drug called *ivabradine* selectively blocks I_f channels in the heart. What effect would it have on heart rate and for what medical condition might it be used?
18. Do you think that the Ca^{2+} channels in autorhythmic cells are the same as the Ca^{2+} channels in contractile cells? Defend your answer.
19. What happens to the action potential of a myocardial autorhythmic cell if tetrodotoxin, which blocks voltage-gated Na^+ channels, is applied to the cell?
20. In an experiment, the *vagus nerve,* which carries parasympathetic signals to the heart, was cut. The investigators noticed that heart rate increased. What can you conclude about the vagal neurons that innervate the heart?

TABLE 14.3 Comparison of Action Potentials in Cardiac and Skeletal Muscle

	Skeletal Muscle	Contractile Myocardium	Autorhythmic Myocardium
Membrane Potential	Stable at −70-80 mV	Stable at −90 mV	Unstable pacemaker potential; usually starts at −60 mV
Events Leading to Threshold Potential	Net Na^+ entry through ACh-operated channels	Depolarization enters via gap junctions	Net Na^+ entry through I_f channels; reinforced by Ca^{2+} entry
Rising Phase of Action Potential	Na^+ entry	Na^+ entry	Ca^{2+} entry
Repolarization Phase	Rapid; caused by K^+ efflux	Extended plateau caused by Ca^{2+} entry; rapid phase caused by K^+ efflux	Rapid; caused by K^+ efflux
Hyperpolarization	Due to excessive K^+ efflux at high K^+ permeability. When K^+ channels close, leak of K^+ and Na^+ restores potential to resting state	None; resting potential is −90 mV, the equilibrium potential for K^+	Normally none; when repolarization hits −60 mV, the I_f channels open again. ACh can hyperpolarize the cell
Duration of Action Potential	Short: 1–2 msec	Extended: 200+ msec	Variable; generally 150+ msec
Refractory Period	Generally brief	Long because resetting of Na^+ channel gates delayed until end of action potential	Not significant in normal function

14.4 The Heart as a Pump

We now turn from single myocardial cells to the intact heart. How can one tiny noncontractile autorhythmic cell cause the entire heart to beat? And why do those doctors on TV shows shock patients with electric paddles when their hearts malfunction? You're about to learn the answers to these questions.

Electrical Signals Coordinate Contraction

A simple way to think of the heart is to imagine a group of people around a stalled car. One person can push on the car, but it's not likely to move very far unless everyone pushes together. In the same way, individual myocardial cells must depolarize and contract in a coordinated fashion if the heart is to create enough force to circulate the blood.

Electrical communication in the heart begins with an action potential in an autorhythmic cell. The depolarization spreads rapidly to adjacent cells through gap junctions in the intercalated disks (**FIG. 14.14**). The depolarization wave is followed by a wave of contraction that passes across the atria, then moves into the ventricles.

The depolarization begins in the **sinoatrial node (SA node)**, autorhythmic cells in the right atrium that serve as the main pacemaker of the heart (**FIG. 14.15**). The depolarization wave then spreads rapidly through a specialized conducting system of noncontractile autorhythmic fibers. A branched **internodal pathway** connects the SA node to the **atrioventricular node (AV node)**, a group of autorhythmic cells near the floor of the right atrium.

From the AV node, the depolarization moves into the ventricles. **Purkinje fibers**, specialized conducting cells of the ventricles, transmit electrical signals very rapidly down the **atrioventricular bundle (AV bundle)**, also called the **bundle of His** ("hiss"), in the ventricular septum. A short way down the septum, the AV bundle fibers divide into left and right **bundle branches**. The bundle branch fibers continue downward to the apex of the heart, where they divide into smaller Purkinje fibers that spread outward among the contractile cells. (Myocardial Purkinje fibers should not be confused with the brain neurons called Purkinje cells.)

The electrical signal for contraction begins when the SA node fires an action potential and the depolarization spreads to adjacent cells through gap junctions (Fig. 14.15 **1**). Electrical conduction is rapid through the internodal conducting pathways **2** but slower through the contractile cells of the atria **3**.

As action potentials spread across the atria, they encounter the fibrous skeleton of the heart at the junction of the atria and ventricles. This barricade prevents the transfer of electrical signals from the atria to the ventricles. Consequently, the AV node is the only pathway through which action potentials can reach the contractile fibers of the ventricles.

The electrical signal passes from the AV node through the AV bundle and bundle branches to the apex of the heart (Fig. 14.15 **4**). The Purkinje fibers transmit impulses very rapidly, with speeds

FIG. 14.14 Electrical conduction in myocardial cells

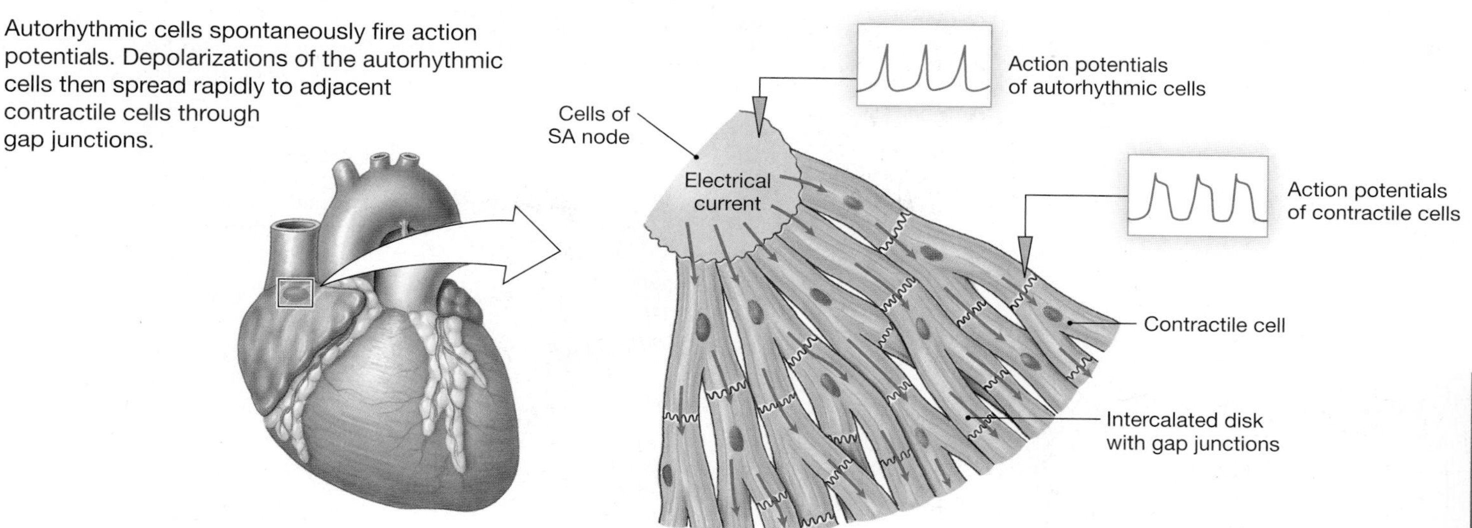
Autorhythmic cells spontaneously fire action potentials. Depolarizations of the autorhythmic cells then spread rapidly to adjacent contractile cells through gap junctions.
Cells of SA node
Electrical current
Action potentials of autorhythmic cells
Action potentials of contractile cells
Contractile cell
Intercalated disk with gap junctions

up to 4 m/sec, so that all contractile cells in the apex contract nearly simultaneously **5**.

Why is it necessary to direct the electrical signals through the AV node? Why not allow them to spread downward from the atria? The answer lies in the fact that blood is pumped out of the ventricles through openings at the top of the chambers (see Fig. 14.7a). If electrical signals from the atria were conducted directly into the ventricles, the ventricles would start contracting at the top. Then blood would be squeezed downward and would become trapped in the bottom of the ventricles (think of squeezing a toothpaste tube at the top). The apex-to-base contraction squeezes blood toward the arterial openings at the base of the heart.

The ejection of blood from the ventricles is aided by the spiral arrangement of the muscles in the walls (see Fig. 14.9a). As these muscles contract, they pull the apex and base of the heart closer together, squeezing blood out the openings at the top of the ventricles.

A second function of the AV node is to slow down the transmission of action potentials slightly. This delay allows the atria to complete their contraction before ventricular contraction begins. **AV node delay** is accomplished by slower conduction of signals through the nodal cells. Action potentials here move at only 1/20 the rate of action potentials in the atrial internodal pathway.

Pacemakers Set the Heart Rate

The cells of the SA node set the pace of the heartbeat. Other cells in the conducting system, such as the AV node and the Purkinje fibers, have unstable resting potentials and can also act as pacemakers under some conditions. However, because their rhythm is slower than that of the SA node, they do not usually have a chance to set the heartbeat. The Purkinje fibers, for example, can spontaneously fire action potentials, but their firing rate is very slow, between 25 and 40 beats per minute.

Why does the fastest pacemaker determine the pace of the heartbeat? Consider the following analogy. A group of people are playing "follow the leader" as they walk. Initially, everyone is walking at a different pace—some fast, some slow. When the game starts, everyone must match his or her pace to the pace of the person who is walking the fastest. The fastest person in the group is the SA node, walking at 70 steps per minute. Everyone else in the group (autorhythmic and contractile cells) sees that the SA node is fastest, and so they pick up their pace and follow the leader. In the heart, the cue to follow the leader is the electrical signal sent from the SA node to the other cells.

Now suppose the SA node gets tired and drops out of the group. The role of leader defaults to the next fastest person, the AV node, who is walking at a rate of 50 steps per minute. The group slows to match the pace of the AV node, but everyone is still following the fastest walker.

What happens if the group divides? Suppose that when they reach a corner, the AV node leader goes left but a renegade Purkinje fiber decides to go right. Those people who follow the AV node continue to walk at 50 steps per minute, but the people who follow the Purkinje fiber slow down to match his pace of 35 steps per minute. Now there are two leaders, each walking at a different pace.

In the heart, the SA node is the fastest pacemaker and normally sets the heart rate. If this node is damaged and cannot function, one of the slower pacemakers in the heart takes over. Heart rate then matches the rate of the new pacemaker. It is even possible for different parts of the heart to follow different pacemakers, just as the walking group split at the corner.

In a condition known as *complete heart block*, the conduction of electrical signals from the atria to the ventricles through the AV node is disrupted. The SA node fires at its rate of 70 beats per

FIG. 14.15 The conducting system of the heart

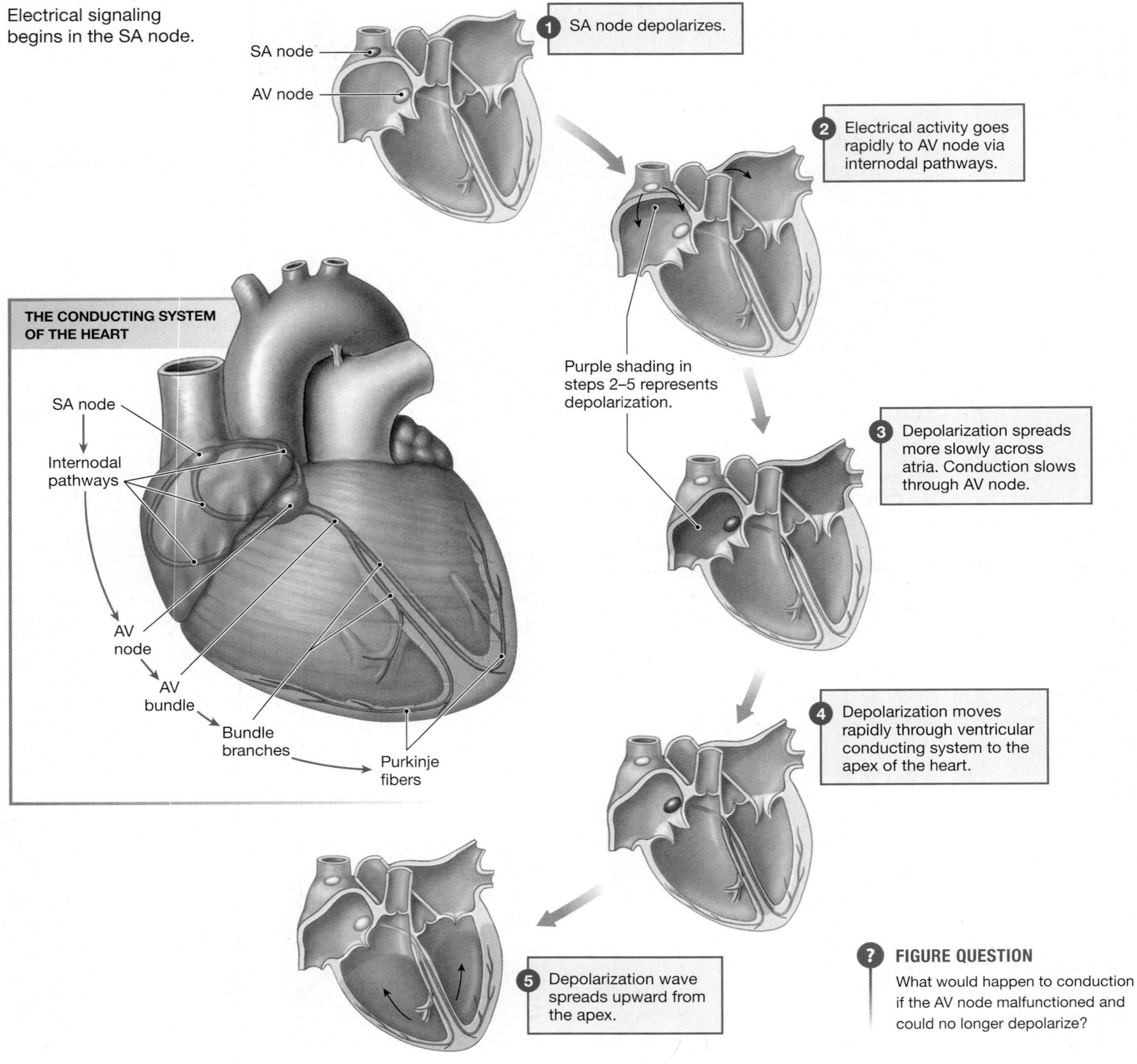
Electrical signaling begins in the SA node.
SA node
AV node
1 SA node depolarizes.
2 Electrical activity goes rapidly to AV node via internodal pathways.
Purple shading in steps 2–5 represents depolarization.
3 Depolarization spreads more slowly across atria. Conduction slows through AV node.
4 Depolarization moves rapidly through ventricular conducting system to the apex of the heart.
5 Depolarization wave spreads upward from the apex.
THE CONDUCTING SYSTEM OF THE HEART
SA node
Internodal pathways
AV node
AV bundle
Bundle branches
Purkinje fibers

FIGURE QUESTION

What would happen to conduction if the AV node malfunctioned and could no longer depolarize?

minute, but those signals never reach the ventricles. So the ventricles coordinate with their fastest pacemaker. Because ventricular autorhythmic cells discharge only about 35 times a minute, the rate at which the ventricles contract is much slower than the rate at which the atria contract. If ventricular contraction is too slow to maintain adequate blood flow, it may be necessary for the heart's rhythm to be set artificially by a surgically implanted mechanical pacemaker. These battery-powered devices artificially stimulate the heart at a predetermined rate.

Concept Check

21. Name two functions of the AV node. What is the purpose of AV node delay?
22. Where is the SA node located?
23. Occasionally an ectopic pacemaker {*ektopos*, out of place} develops in part of the heart's conducting system. What happens to heart rate if an ectopic atrial pacemaker depolarizes at a rate of 120 times per minute?

CLINICAL FOCUS

Fibrillation

Coordinated conduction of electrical signals through the heart's conducting system is essential for normal cardiac function. In extreme cases, the myocardial cells lose all coordination and contract in a disorganized manner, a condition known as *fibrillation* results. Atrial fibrillation is a common condition, often without symptoms, that can lead to serious consequences (such as stroke) if not treated. Ventricular fibrillation, on the other hand, is an immediately life-threatening emergency because without coordinated contraction of the muscle fibers, the ventricles cannot pump enough blood to supply adequate oxygen to the brain. One way to correct this problem is to administer an electrical shock to the heart. The shock creates a depolarization that triggers action potentials in all cells simultaneously, coordinating them again. You have probably seen this procedure on television hospital shows, when a doctor places flat paddles on the patient's chest and tells everyone to stand back ("Clear!") while the paddles pass an electrical current through the body.

The Electrocardiogram Reflects Electrical Activity

At the end of the nineteenth century, physiologists discovered that they could place electrodes on the skin's surface and record the electrical activity of the heart. It is possible to use surface electrodes to record internal electrical activity because salt solutions, such as our NaCl-based extracellular fluid, are good conductors of electricity. These recordings, called **electrocardiograms** (ECGs or, sometimes, EKGs—from the Greek word *kardia*, meaning *heart*) show the summed electrical activity generated by all cells of the heart (**FIG. 14.16a**).

The first human electrocardiogram was recorded in 1887, but the procedure was not refined for clinical use until the first years of the twentieth century. The father of the modern ECG was a Dutch physiologist named Walter Einthoven. He named the parts of the ECG as we know them today and created "Einthoven's triangle," a hypothetical triangle created around the heart when electrodes are placed on both arms and the left leg (Fig. 14.16b). The sides of the triangle are numbered to correspond with the three *leads* ("leeds"), or pairs of electrodes, used for a recording.

An ECG is recorded from one lead at a time. One electrode acts as the positive electrode of a lead, and a second electrode acts as the negative electrode of the lead. (The third electrode is inactive). For example, in lead I, the left arm electrode is designated as positive and the right arm electrode is designated as negative. When an electrical wave moving through the heart is directed toward the positive electrode, the ECG wave goes up from the baseline (Fig. 14.156d). If net charge movement through the heart is toward the negative electrode, the wave points downward.

An ECG is not the same as a single action potential (Fig. 14.16e). An action potential is one electrical event in a single cell, recorded using an intracellular electrode. The ECG is an extracellular recording that represents the sum of multiple action potentials taking place in many heart muscle cells. In addition, the amplitudes of action potential and ECG recordings are very different. A ventricular action potential has a voltage change of 110 mV, for example, but the ECG signal has an amplitude of only 1 mV by the time it reaches the surface of the body.

Waves of the ECG There are two major components of an ECG: waves and segments (Fig. 14.16f). *Waves* are the parts of the trace that go above or below the baseline. *Segments* are sections of baseline between two waves. *Intervals* are combinations of waves and segments. Different waves of the ECG reflect depolarization or repolarization of the atria and ventricles.

Three major waves can be seen on a normal ECG recorded from lead I (Fig. 14.16f). The first wave is the **P wave**, which corresponds to depolarization of the atria. The next trio of waves, the **QRS complex**, represents the progressive wave of ventricular depolarization. The Q wave is sometimes absent on normal ECGs. The final wave, the **T wave**, represents the repolarization of the ventricles. Atrial repolarization is not represented by a special wave but is incorporated into the QRS complex.

One thing many people find confusing is that you cannot tell if an ECG recording represents depolarization or repolarization simply by looking at the shape of the waves relative to the baseline. For example, the P wave represents atrial depolarization and the T wave represents ventricular repolarization, but both the P wave and the T wave are deflections above the baseline in lead I. This is very different from the intracellular recordings of neurons and muscle fibers, in which an upward deflection always represents depolarization [see Fig. 5.24, p. 156]. Remember that the direction of the ECG trace reflects only the direction of the current flow relative to the axis of the lead. Some waves even change direction in different leads.

The Cardiac Cycle Now let's follow an ECG through a single contraction-relaxation cycle, otherwise known as a **cardiac cycle** (**FIG. 14.17**). Because depolarization initiates muscle contraction, the *electrical events* (waves) of an ECG can be associated with contraction or relaxation (collectively referred to as the *mechanical events* in the heart). The mechanical events of the cardiac cycle lag slightly behind the electrical signals, just as the contraction of a single cardiac muscle cell follows its action potential (see Fig. 14.12a).

The cardiac cycle begins with both atria and ventricles at rest. The ECG begins with atrial depolarization. Atrial contraction starts during the latter part of the P wave and continues during the P-R segment. During the P-R segment, the electrical signal is slowing down as it passes through the AV node (AV node delay) and AV bundle.

Ventricular contraction begins just after the Q wave and continues through the T wave. The ventricles are repolarizing during the T wave, which is followed by ventricular relaxation. During the T-P segment the heart is electrically quiet.

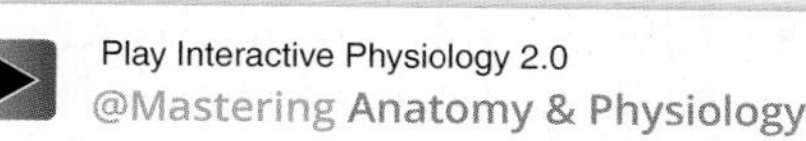
Play Interactive Physiology 2.0
@Mastering Anatomy & Physiology

CHAPTER 14

FIG. 14.16 ESSENTIALS The Electrocardiogram

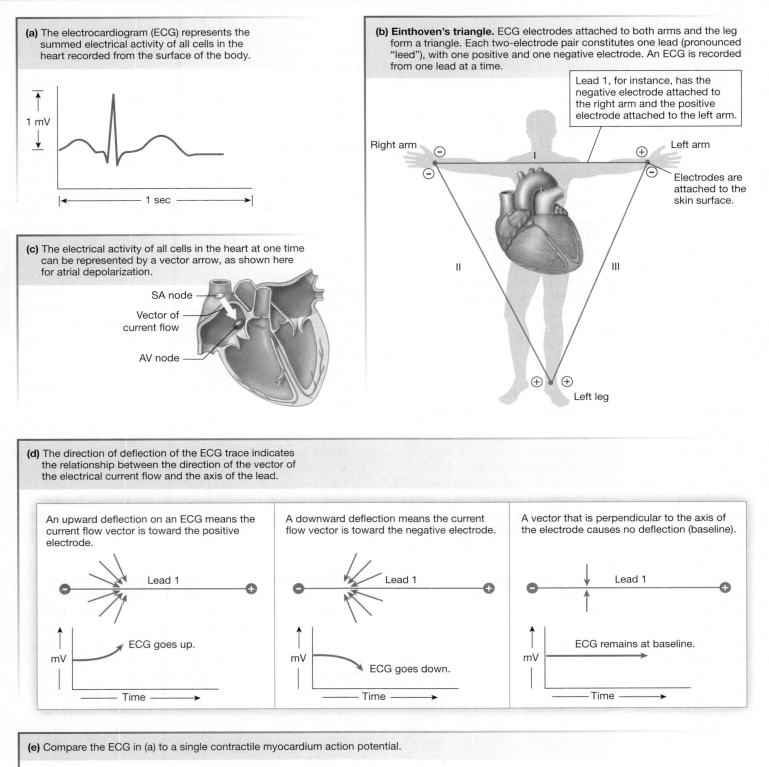

(a) The electrocardiogram (ECG) represents the summed electrical activity of all cells in the heart recorded from the surface of the body.

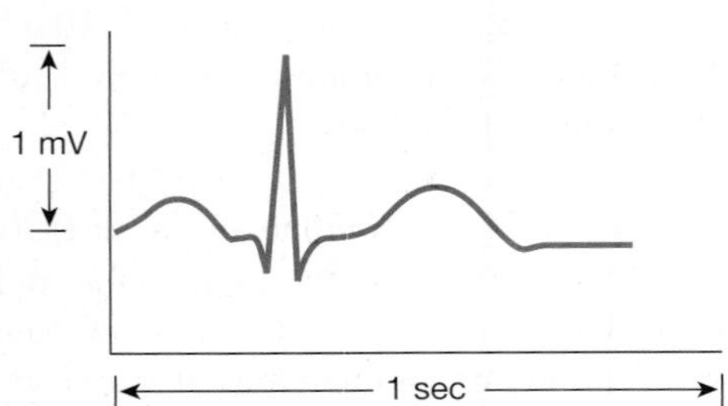

1 mV
1 sec

(b) Einthoven's triangle. ECG electrodes attached to both arms and the leg form a triangle. Each two-electrode pair constitutes one lead (pronounced "leed"), with one positive and one negative electrode. An ECG is recorded from one lead at a time.

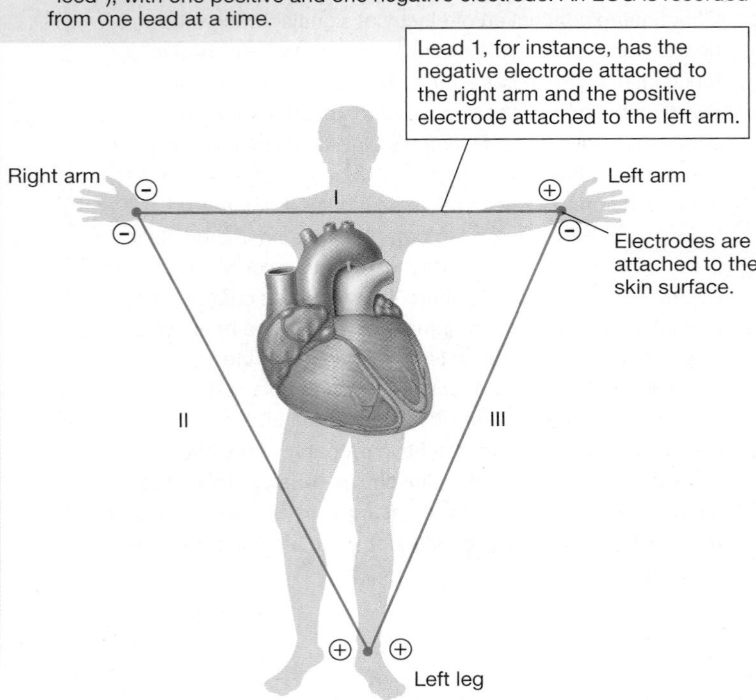

Lead 1, for instance, has the negative electrode attached to the right arm and the positive electrode attached to the left arm.
Right arm
Left arm
I
II
III
Electrodes are attached to the skin surface.
Left leg

(c) The electrical activity of all cells in the heart at one time can be represented by a vector arrow, as shown here for atrial depolarization.

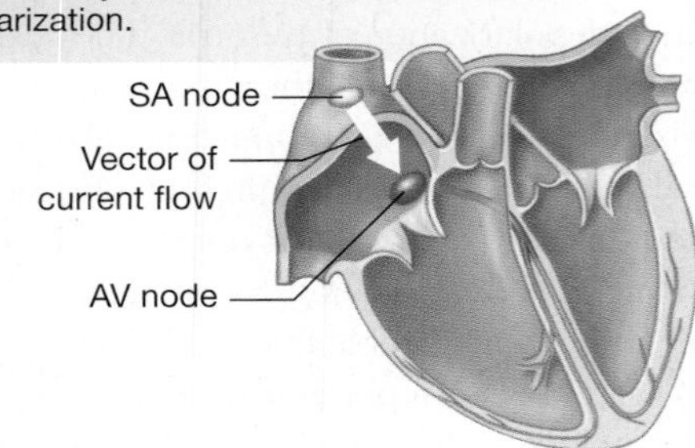

SA node
Vector of current flow
AV node

(d) The direction of deflection of the ECG trace indicates the relationship between the direction of the vector of the electrical current flow and the axis of the lead.

An upward deflection on an ECG means the current flow vector is toward the positive electrode.

Lead 1

ECG goes up.

mV

Time

A downward deflection means the current flow vector is toward the negative electrode.

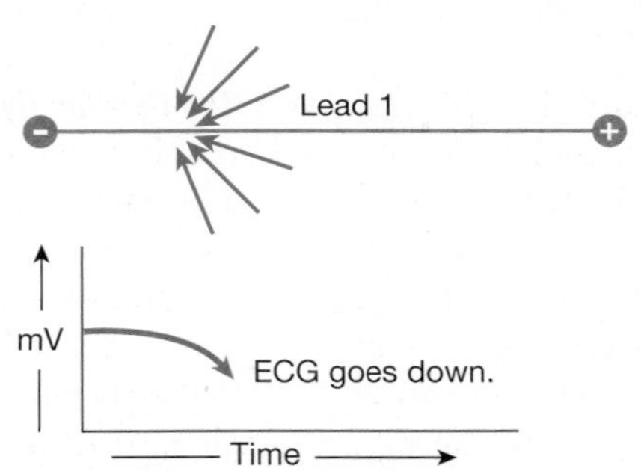

Lead 1
mV
ECG goes down.
Time

A vector that is perpendicular to the axis of the electrode causes no deflection (baseline).

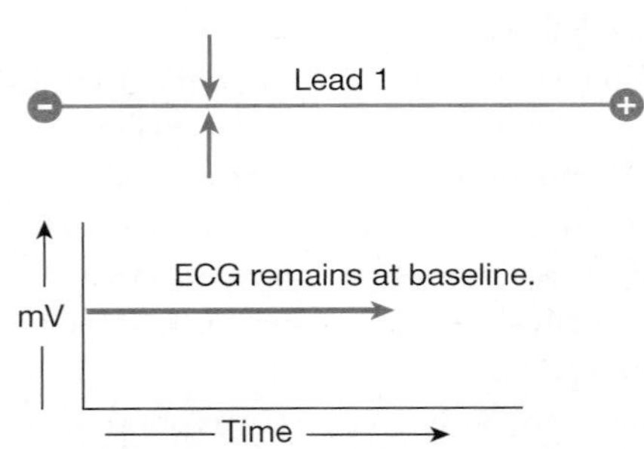

Lead 1
ECG remains at baseline.
mV
Time

(e) Compare the ECG in (a) to a single contractile myocardium action potential.

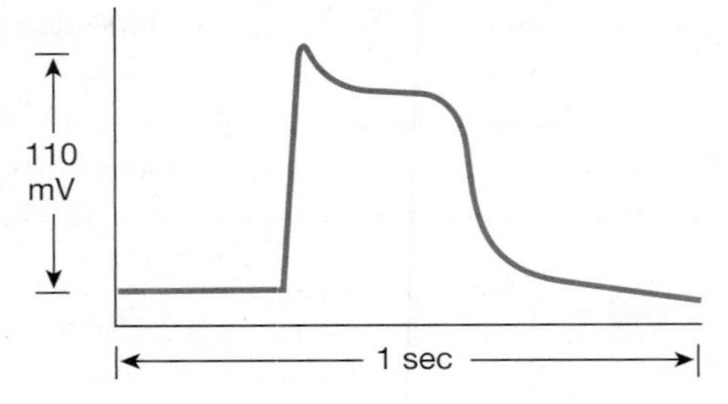

110 mV
1 sec

- The action potential of this ventricular cell is an intracellular recording made by placing one electrode inside the cell and a ground electrode outside the cell. [Fig. 5.23, p. 154]
- An upward deflection represents depolarization and a downward one represents repolarization.
- The action potential has much greater amplitude because it is being recorded close to the source of the signal.

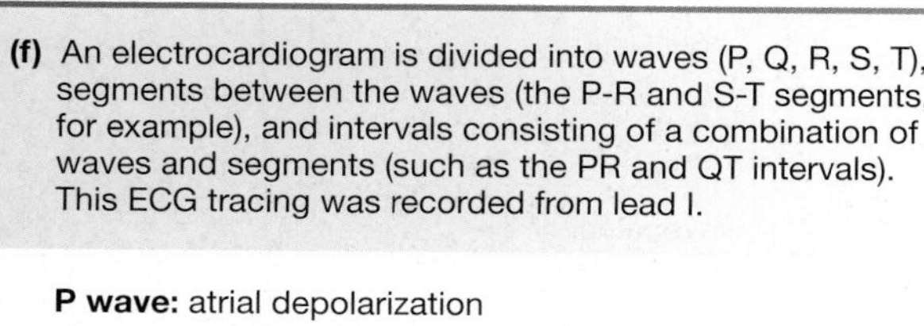

(f) An electrocardiogram is divided into waves (P, Q, R, S, T), segments between the waves (the P-R and S-T segments, for example), and intervals consisting of a combination of waves and segments (such as the PR and QT intervals). This ECG tracing was recorded from lead I.

P wave: atrial depolarization

P-R segment: conduction through AV node and AV bundle

QRS complex: ventricular depolarization

T wave: ventricular repolarization

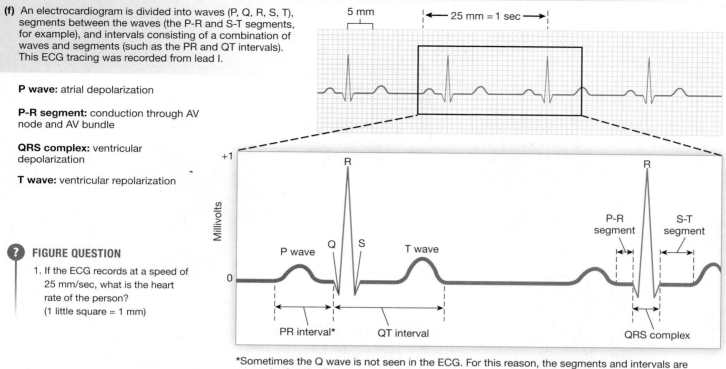

5 mm

25 mm = 1 sec

+1

0

Millivolts

R

R

P wave

Q

S

T wave

P-R segment

S-T segment

PR interval*

QT interval

QRS complex

*Sometimes the Q wave is not seen in the ECG. For this reason, the segments and intervals are named using the R wave but begin with the first wave of the QRS complex.

FIGURE QUESTION

1. If the ECG records at a speed of 25 mm/sec, what is the heart rate of the person? (1 little square = 1 mm)

(g) Normal and abnormal ECGs. All tracings represent 10-sec recordings.

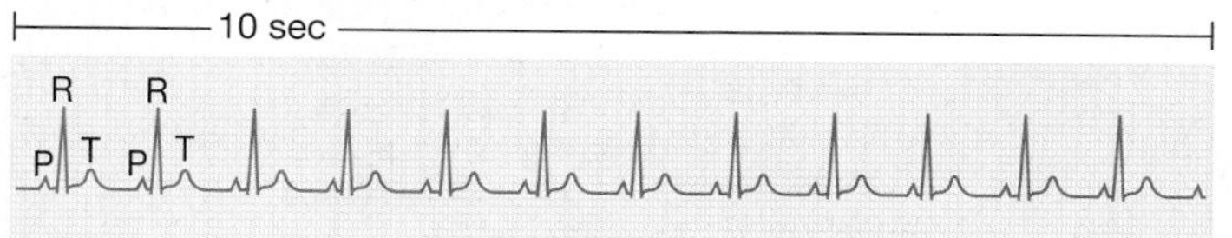

10 sec
R R
P T P T

(1) Normal ECG

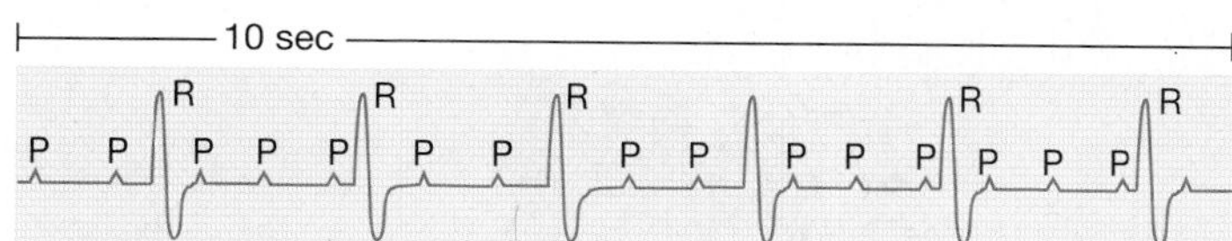

10 sec
R R R R R
P P P P P P P P P P P P P P P

(2) Third-degree block

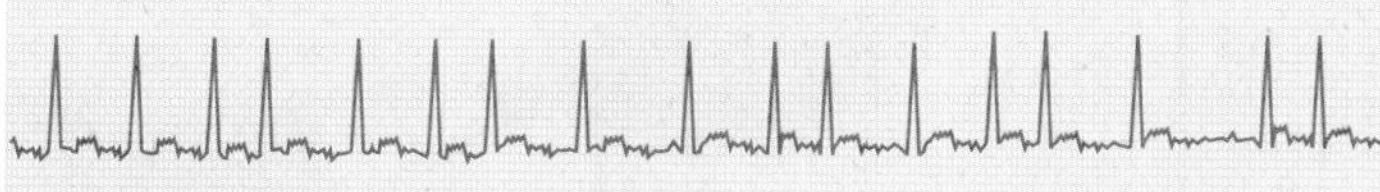

(3) Atrial fibrillation

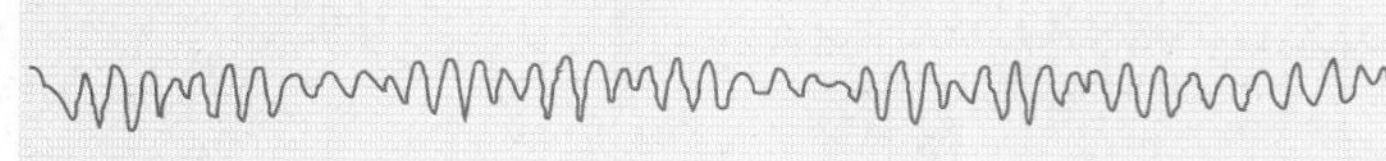

(4) Ventricular fibrillation

ECG Analysis: Questions to ask when analyzing ECG tracings.

1. What is the rate? Is it within the normal range of 60–100 beats per minute?
2. Is the rhythm regular?
3. Are all normal waves present in recognizable form?
4. Is there one QRS complex for each P wave? If yes, is the P-R segment constant in length? If there is not one QRS complex for each P wave, count the heart rate using the P waves, then count it according to the R waves. Are the rates the same? Which wave would agree with the pulse felt at the wrist?

FIGURE QUESTION

2. Three abnormal ECGs are shown above. Study them and see if you can relate the ECG changes to disruption of the normal electrical conduction pattern in the heart. Answer the ECG analysis questions for each trace.

FIGURE QUESTION

3. Identify the waves on the ECG in part (5). Look at the pattern of their occurrence and describe what has happened to electrical conduction in the heart.

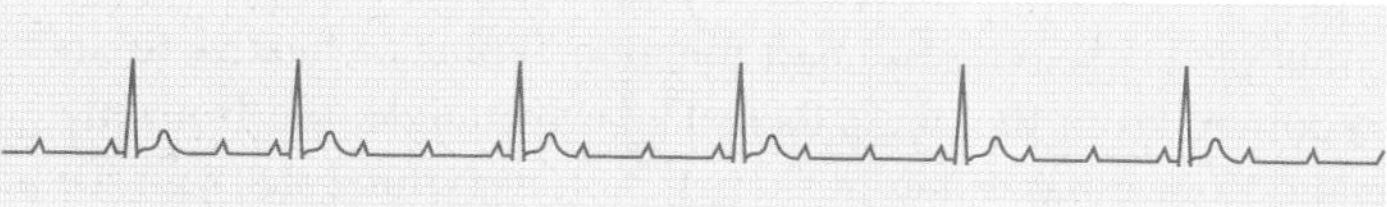

(5) Analyze this abnormal ECG.

FIG. 14.17 Correlation between an ECG and electrical events in the heart

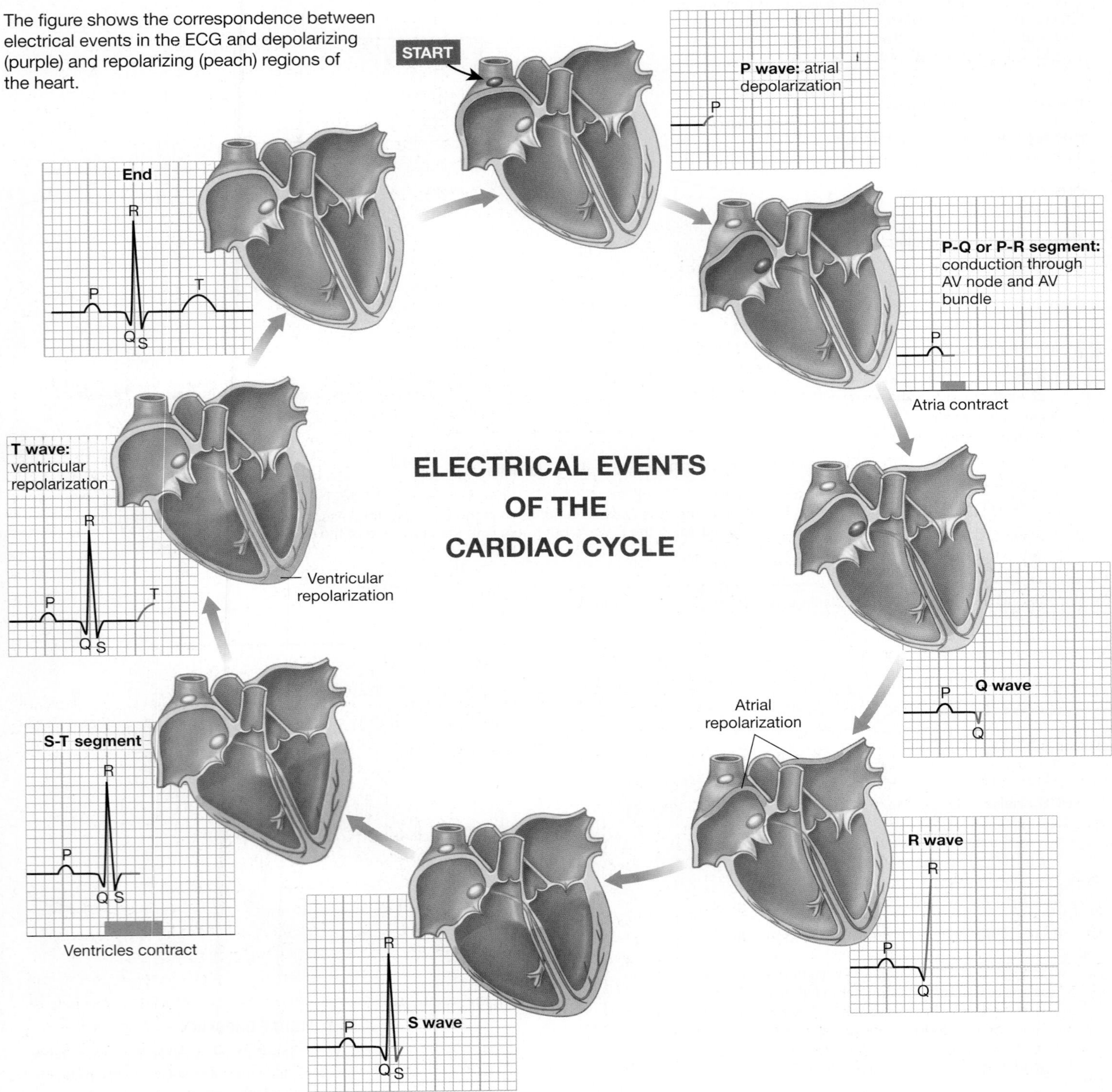
The figure shows the correspondence between electrical events in the ECG and depolarizing (purple) and repolarizing (peach) regions of the heart.
START
P wave: atrial depolarization
P
P-Q or P-R segment: conduction through AV node and AV bundle
Atria contract
ELECTRICAL EVENTS OF THE CARDIAC CYCLE
Q wave
Atrial repolarization
R wave
S wave
S-T segment
Ventricles contract
T wave: ventricular repolarization
Ventricular repolarization
End
R
T
Q
S

An important point to remember is that an ECG is an electrical "view" of a three-dimensional object. This is one reason we use multiple leads to assess heart function. Think of looking at an automobile. From the air, it looks like a rectangle, but from the side and front it has different shapes. Not everything that you see from the front of the car can be seen from its side, and vice versa. In the same way, the leads of an ECG provide different electrical "views" and give information about different regions of the heart.

A 12-lead ECG is now the standard for clinical use. It is recorded using various combinations of the three limb electrodes plus another six electrodes placed on the chest and trunk. The additional leads provide detailed information about electrical conduction in the heart. Electrocardiograms are important diagnostic tools in medicine because they are quick, painless, and noninvasive (that is, do not puncture the skin).

Interpretation of ECGs An ECG provides information on heart rate and rhythm, conduction velocity, and even the condition of tissues in the heart. Thus, although obtaining an ECG is simple, interpreting some of its subtleties can be quite complicated. The interpretation of an ECG begins with the following questions (Fig. 14.16g).

1. *What is the heart rate?* Heart rate is normally timed either from the beginning of one P wave to the beginning of the next P wave or from the peak of one R wave to the peak of the next R wave. A normal resting heart rate is 60–100 beats per minute, although trained athletes often have slower heart rates at rest. A faster-than-normal rate is known as *tachycardia,* and a slower-than-normal rate is called *bradycardia* {*tachys,* swift; *bradys,* slow}.
2. *Is the rhythm of the heartbeat regular (that is, occurs at regular intervals) or irregular?* An irregular rhythm, or *arrhythmia* {*a-*, without + rhythm}, can result from a benign extra beat or from more serious conditions such as atrial fibrillation, in which the SA node has lost control of the pacemaking.
3. *Are all normal waves present in recognizable form?* After determining heart rate and rhythm, the next step in analyzing an ECG is to look at the individual waves. To help your analysis, you might want to write the letters above the P, R, and T waves.
4. *Is there one QRS complex for each P wave? If yes, is the P-R segment constant in length?* If not, a problem with conduction of signals through the AV node may exist. In heart block (the conduction problem mentioned earlier), action potentials from the SA node sometimes fail to be transmitted through the AV node to the ventricles. In these conditions, one or more P waves may occur without initiating a QRS complex. In the most severe (third-degree) form of heart block, the atria depolarize regularly at one pace while the ventricles contract at a much slower pace (Fig. 14.16h).

Pathologies and ECGs The more difficult aspects of interpreting an ECG include looking for subtle changes, such as alterations in the shape, timing, or duration of various waves or segments. An experienced clinician can find signs pointing to changes in conduction velocity, enlargement of the heart, or tissue damage resulting from periods of *ischemia* (see Running Problem). An amazing number of conclusions can be drawn about heart function simply by looking at alterations in the heart's electrical activity as recorded on an ECG.

Cardiac arrhythmias are a family of cardiac pathologies that range from benign to those with potentially fatal consequences. Arrhythmias are electrical problems that arise during the generation or conduction of action potentials through the heart, and they can usually be seen on an ECG. Some arrhythmias are "dropped beats" that result when the ventricles do not get their usual signal to contract. Other arrhythmias, such as *premature ventricular contractions* (*PVCs*), are extra beats that occur when an autorhythmic cell other than the SA node jumps in and fires an action potential out of sequence.

One interesting heart condition that can be observed on an ECG is *long QT syndrome (LQTS),* named for the change in the QT interval. LQTS has several forms. Some are inherited *channelopathies,* in which mutations occur in myocardial Na^+ or K^+ channels [p. 236]. In another form of LQTS, the ion channels are normal but the protein *ankyrin-B* that anchors the channels to the cell membrane is defective.

Iatrogenic (physician-caused) forms of LQTS can occur as a side effect of taking certain medications. One well-publicized incident occurred in the 1990s when patients took a nonsedating antihistamine called terfenadine (Seldane) that binds to K^+ repolarization channels. After at least eight deaths were attributed to the drug, the U.S. Food and Drug Administration removed Seldane from the market.

The Heart Contracts and Relaxes during a Cardiac Cycle

Each cardiac cycle has two phases: **diastole**, the time during which cardiac muscle relaxes, and **systole**, the time during which the muscle contracts {*diastole,* dilation; *systole,* contraction}. Because the atria and ventricles do not contract and relax at the same time, we discuss atrial and ventricular events separately.

In thinking about blood flow during the cardiac cycle, remember that blood flows from an area of higher pressure to one of lower pressure, and that contraction increases pressure while relaxation decreases pressure. In this discussion, we divide the cardiac cycle into the five phases shown in **FIGURE 14.18a**:

1 **The heart at rest: atrial and ventricular diastole.** We enter the cardiac cycle at the brief moment during which both the atria and the ventricles are relaxing. The

RUNNING PROBLEM

Lisa's electrocardiogram showed a change called S-T depression. S-T segment depression on an ECG is often caused by decreased blood flow to a small area of the heart. Lisa's blood tests showed slightly elevated levels of cardiac troponin. "It's a good thing we checked," said Dr. Dang. "Ms. Cooper, you've had some damage to a small part of your heart from severe narrowing of a heart artery. This is not a complete blockage, which is much more dangerous, but we need to admit you to a cardiology team for further care. Good job taking the aspirin. It may have kept this from turning into a full-blown heart attack." Lisa's type of heart attack is called a *non-ST elevation MI*, or *non-STEMI*.

Q4: *Draw a normal ECG with two cardiac cycles. In a different color, draw how the ECG would change with S-T depression.*

Q5: *What is troponin, and why would elevated blood levels of troponin indicate heart damage? [Hint: p. 383]*

433 – 440 – 447 – 459 – 464 – 467 – 472

CLINICAL FOCUS

Gallops, Clicks, and Murmurs

The simplest direct assessment of heart function consists of listening to the heart through the chest wall, a process known as **auscultation** {*auscultare,* to listen to} that has been practiced since ancient times. In its simplest form, auscultation is done by placing an ear against the chest. Today, however, it is usually performed by listening through a stethoscope placed against the chest and the back. Normally, there are two audible heart sounds. The first ("lub") is associated with closure of the AV valves. The second ("dup") is associated with closure of the semilunar valves.
Two additional heart sounds can be recorded with very sensitive electronic stethoscopes. The third heart sound is caused by turbulent blood flow into the ventricles during ventricular filling, and the fourth sound is associated with turbulence during atrial contraction. In certain abnormal conditions, these latter two sounds may become audible through a regular stethoscope. They are called gallops because their timing puts them close to one of the normal heart sounds: "lub—dup-dup," or "lub-lub—dup." Other abnormal heart sounds include clicking, caused by abnormal movement of one of the valves, and murmurs, caused by the "whoosh" of blood leaking through an incompletely closed or excessively narrowed (*stenotic*) valve.

Pressure-Volume Curves Represent One Cardiac Cycle

Another way to describe the cardiac cycle is with a pressure-volume graph, shown in Figure 14.18b. This figure represents the changes in volume (x-axis) and pressure (y-axis) that occur during one cardiac cycle.

Recall that the flow of blood through the heart is governed by the same principle that governs the flow of all liquids and gases: Flow proceeds from areas of higher pressure to areas of lower pressure. When the heart contracts, the pressure increases and blood flows out of the heart into areas of lower pressure. Figure 14.18b represents pressure and volume changes in the left ventricle, which sends blood into the systemic circulation. The left side of the heart creates higher pressures than the right side, which sends blood through the shorter pulmonary circuit.

The cycle begins at point A. The ventricle has completed a contraction and contains the minimum amount of blood that it will hold during the cycle. It has relaxed, and its pressure is also at its minimum value. Blood is flowing into the atrium from the pulmonary veins.

Once pressure in the atrium exceeds pressure in the ventricle, the mitral valve between the atrium and ventricle opens (Fig. 14.18b, point A). Atrial blood now flows into the ventricle, increasing its volume (point A to point B). As blood flows in, the relaxing ventricle expands to accommodate the entering blood. Consequently, the volume of the ventricle increases, but the pressure in the ventricle goes up very little.

The last portion of ventricular filling is completed by atrial contraction (point A' to B). The ventricle now contains the maximum volume of blood that it will hold during this cardiac cycle, the end-diastolic volume or EDV (point B). In a 70-kg man at rest, end-diastolic volume is about 135 mL. However, EDV varies under different conditions. During periods of very high heart rate, for instance, when the ventricle does not have time to fill completely between beats, the end-diastolic value may be less than 135 mL.

When ventricular contraction begins, the mitral (AV) valve closes. With both the AV valve and the semilunar valve closed, blood in the ventricle has nowhere to go. Nevertheless, the ventricle continues to contract, causing the pressure in this chamber to increase rapidly during isovolumic contraction (B → C in Fig. 14.17b). Once ventricular pressure exceeds the pressure in the aorta, the aortic valve opens (point C). Pressure continues to increase as the ventricle contracts further, but ventricular volume decreases as blood is pushed out into the aorta (C → D).

The end-systolic volume or ESV (point D) is the minimum volume of blood the ventricle contains during one cycle. An average ESV value in a person at rest is 65 mL, meaning that nearly half of the 135 mL that was in the ventricle at the start of the contraction is still there at the end of the contraction.

At the end of each ventricular contraction, the ventricle begins to relax. As it does so, ventricular pressure decreases. Once pressure in the ventricle falls below aortic pressure, the semilunar valve closes, and the ventricle again becomes a sealed chamber. The remainder of relaxation occurs without a change in blood volume, and so this phase is called *isovolumic relaxation* (Fig. 14.18b, D → A). When ventricular pressure finally falls to the point at which atrial pressure exceeds ventricular pressure, the mitral valve opens and the cycle begins again.

The electrical and mechanical events of the cardiac cycle are summarized together in **FIGURE 14.19,** known as a Wiggers diagram after the physiologist who first created it.

Concept Check

27. In Figure 14.19, at what points in the cycle do EDV and ESV occur?

28. On the Wiggers diagram in Figure 14.19, match the following events to the lettered boxes:
(a) end-diastolic volume,
(b) aortic valve opens,
(c) mitral valve opens,
(d) aortic valve closes,
(e) mitral valve closes,
(f) end-systolic volume

29. Why does atrial pressure increase just to the right of point C in Figure 14.19? Why does it decrease during the initial part of ventricular systole, then increase? Why does it decrease to the right of point D?

30. Why does ventricular pressure shoot up suddenly at point C in Figure 14.19?

FIG. 14.19 The Wiggers diagram

This diagram follows left heart and aortic pressures, left ventricular volume, and the ECG through one cardiac cycle. The boxed letters refer to Concept Checks 28–30.

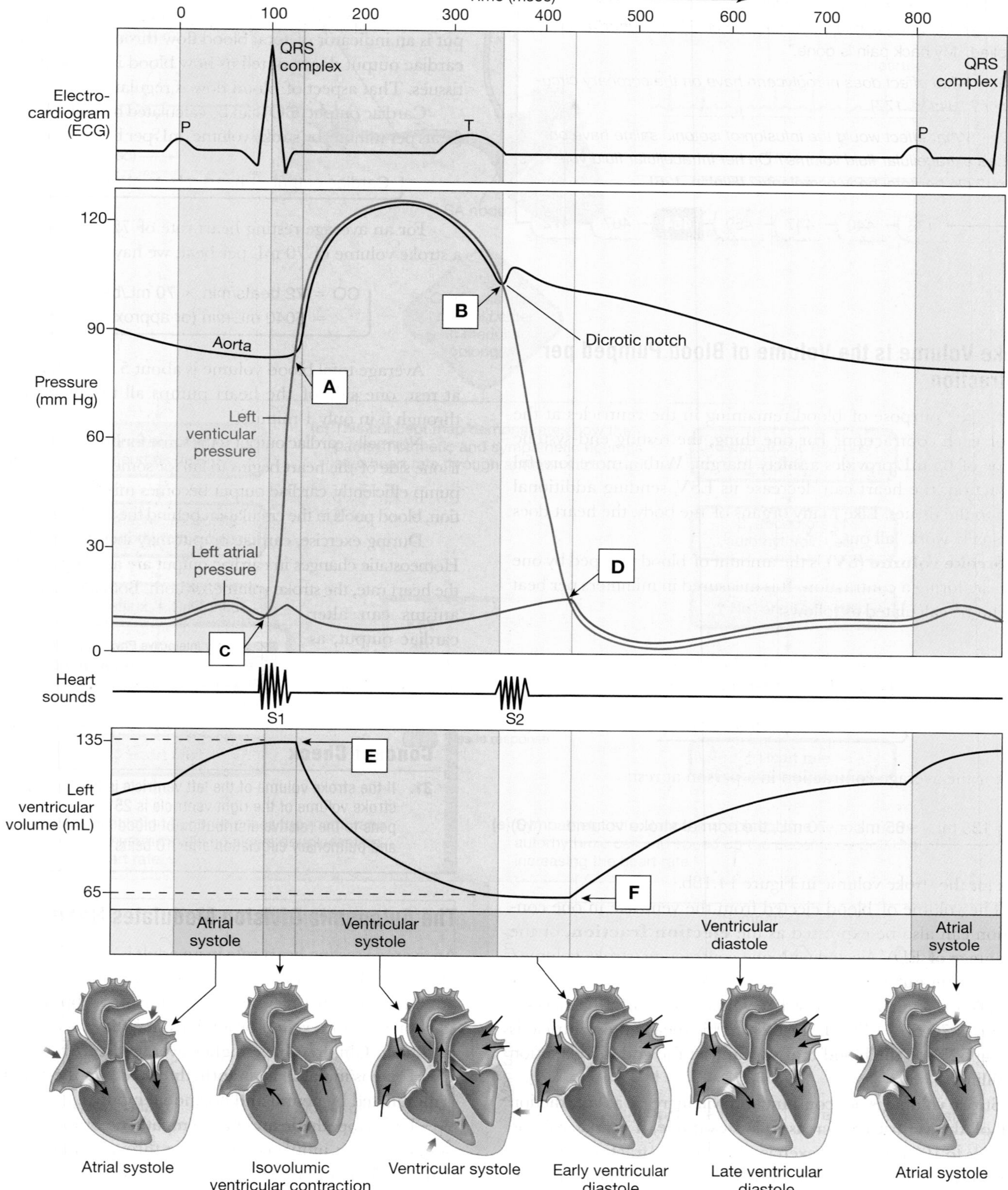
Time (msec)
0
100
200
300
400
500
600
700
800
Electro-cardiogram (ECG)
QRS complex
P
T
QRS complex
P
120
90
60
30
0
Pressure (mm Hg)
Aorta
Left venticular pressure
Left atrial pressure
Dicrotic notch
A
B
C
D
Heart sounds
S1
S2
135
65
Left ventricular volume (mL)
E
F
Atrial systole
Ventricular systole
Ventricular diastole
Atrial systole
Atrial systole
Isovolumic ventricular contraction
Ventricular systole
Early ventricular diastole
Late ventricular diastole
Atrial systole

FIG. 15.6 Systemic circulation pressures

Pressure waves created by ventricular contraction travel into the blood vessels. Pressure in the arterial side of the circulation cycles but the pressure waves diminish in amplitude with distance and disappear at the capillaries.

Pulse pressure = systolic pressure − diastolic pressure

Mean arterial pressure = diastolic pressure + 1/3 (pulse pressure)

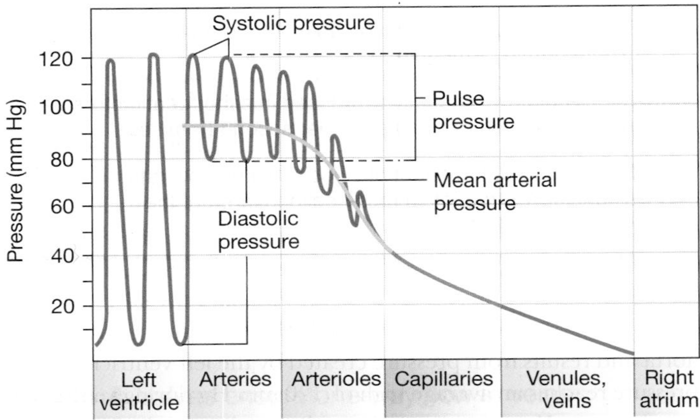
Systolic pressure
Pulse pressure
Mean arterial pressure
Diastolic pressure
Pressure (mm Hg)
120
100
80
60
40
20
Left ventricle
Arteries
Arterioles
Capillaries
Venules, veins
Right atrium

pressure in the ventricle falls to only a few mm Hg as the ventricle relaxes, but diastolic pressure in the large arteries remains relatively high. The high diastolic pressure in arteries reflects the ability of those vessels to capture and store energy in their elastic walls.

The rapid pressure increase that occurs when the left ventricle pushes blood into the aorta can be felt as a **pulse,** or pressure wave, transmitted through the fluid-filled arteries. The pressure wave travels about 10 times faster than the blood itself. Even so, a pulse felt in the arm is occurring slightly after the ventricular contraction that created the wave.

The amplitude of the pressure wave decreases over distance because of friction, and the wave finally disappears at the capillaries (Fig. 15.6). **Pulse pressure,** a measure of the strength of the pressure wave, is defined as systolic pressure minus diastolic pressure:

$$\text{Systolic pressure} - \text{diastolic pressure} = \text{pulse pressure} \quad (1)$$

For example, in the aorta:

$$120 \text{ mm Hg} - 80 \text{ mm Hg} = 40 \text{ mm Hg pressure} \quad (2)$$

By the time blood reaches the veins, pressure has decreased because of friction, and a pressure wave no longer exists. Venous blood flow is steady rather than *pulsatile* (in pulses), pushed along by the continuous movement of blood out of the capillaries.

Low-pressure blood in veins below the heart must flow "uphill," or against gravity, to return to the heart. Try holding your arm straight down without moving for several minutes and notice how the veins in the back of your hand begin to stand out as they fill with blood. (This effect may be more evident in older people, whose subcutaneous connective tissue has lost elasticity.) Then raise your hand so that gravity assists the venous flow and watch the bulging veins disappear.

Blood return to the heart, known as *venous return,* is aided by valves, the *respiratory pump,* and the *skeletal muscle pump.* When muscles such as those in the calf of the leg contract, they compress the veins, which forces blood upward past the valves. While your hand is hanging down, try clenching and unclenching your fist to see the effect muscle contraction has on distention of the veins.

Concept Check

1. Would you expect to find valves in the veins leading from the brain to the heart? Defend your answer.
2. If you check the pulse in a person's carotid artery and left wrist at the same time, would the pressure waves occur simultaneously? Explain.
3. Who has the higher pulse pressure, someone with blood pressure of 90/60 or someone with blood pressure of 130/95?

Arterial Blood Pressure Reflects the Driving Pressure for Blood Flow

Arterial blood pressure, or simply "blood pressure," reflects the driving pressure created by the pumping action of the heart. Because ventricular pressure is difficult to measure, it is customary to assume that arterial blood pressure reflects ventricular pressure. As you just learned, arterial pressure is pulsatile, so we use a single value—the **mean arterial pressure (MAP)**—to represent driving pressure. MAP is represented graphically in Figure 15.6.

Mean arterial pressure is estimated as diastolic pressure plus one-third of pulse pressure:

$$\text{MAP} = \text{diastolic P} + 1/3\,(\text{systolic P} - \text{diastolic P}) \quad (3)$$

For a person whose systolic pressure is 120 and diastolic pressure is 80:

$$\begin{aligned}\text{MAP} &= 80 \text{ mm Hg} + 1/3\,(120 - 80 \text{ mm Hg}) \\ &= 93 \text{ mm Hg}\end{aligned} \quad (4)$$

Mean arterial pressure is closer to diastolic pressure than to systolic pressure because diastole lasts twice as long as systole.

Abnormally high or low arterial blood pressure can be indicative of a problem in the cardiovascular system. If blood pressure falls too low (*hypotension*), the driving force for blood flow is unable to overcome opposition by gravity. In this instance, blood flow and oxygen supply to the brain are impaired, and the person may become dizzy or faint.

On the other hand, if blood pressure is chronically elevated (a condition known as *hypertension,* or high blood pressure), high pressure on the walls of blood vessels may cause weakened areas to rupture and bleed into the tissues. If a rupture occurs in the brain, it is called a *cerebral hemorrhage* and may cause the loss of neurological function commonly called a *stroke.* If a weakened area ruptures in a major artery, such as the descending aorta, rapid blood loss into the abdominal cavity causes blood pressure to fall below the critical minimum. Without prompt treatment, rupture of a major artery is fatal.

Concept Check

4. The formula given for calculating MAP applies to a typical resting heart rate of 60–80 beats/min. If heart rate increases, would the contribution of systolic pressure to mean arterial pressure decrease or increase, and would MAP decrease or increase?
5. Peter's systolic pressure is 112 mm Hg, and his diastolic pressure is 68 mm Hg (written 112/68). What is his pulse pressure? His mean arterial pressure?

Blood Pressure Is Estimated by Sphygmomanometry

We estimate arterial blood pressure in the radial artery of the arm using a *sphygmomanometer*, an instrument consisting of an inflatable cuff and a pressure gauge {*sphygmus*, pulse + {*manometer*, an *instrument for measuring pressure of a fluid*}. The cuff encircles the upper arm and is inflated until it exerts pressure higher than the systolic pressure driving arterial blood. When cuff pressure exceeds arterial pressure, blood flow into the lower arm stops (**FIG. 15.7a**).

Now pressure on the cuff is gradually released. When cuff pressure falls below systolic arterial blood pressure, blood begins to flow again. As blood squeezes through the still-compressed artery, a thumping noise called a **Korotkoff sound** can be heard with each pressure wave (Fig. 15.7b). Korotkoff sounds are caused by the turbulent flow of blood through the compression. Once the cuff pressure no longer compresses the artery, flow smooths out and the sounds disappear (Fig. 15.7c).

The pressure at which a Korotkoff sound is first heard represents the highest pressure in the artery and is recorded as the systolic pressure. The point at which the Korotkoff sounds disappear is the lowest pressure in the artery and is recorded as the diastolic pressure. By convention, blood pressure is written as systolic pressure over diastolic pressure.

For years, the "average" value for blood pressure has been stated as 120/80. Like many average physiological values, however, these numbers are subject to wide variability, both from one person to another and within a single individual from moment to moment. A systolic pressure that is consistently over 140 mm Hg at rest, or a diastolic pressure that is chronically over 90 mm Hg, is considered a sign of hypertension in an otherwise healthy person.*

CHAPTER 15

* 2014 Evidence-Based Guideline for the Management of High Blood Pressure in Adults *http://jamanetwork.com/journals/jama/fullarticle/1791497*

FIG. 15.7 Sphygmomanometry

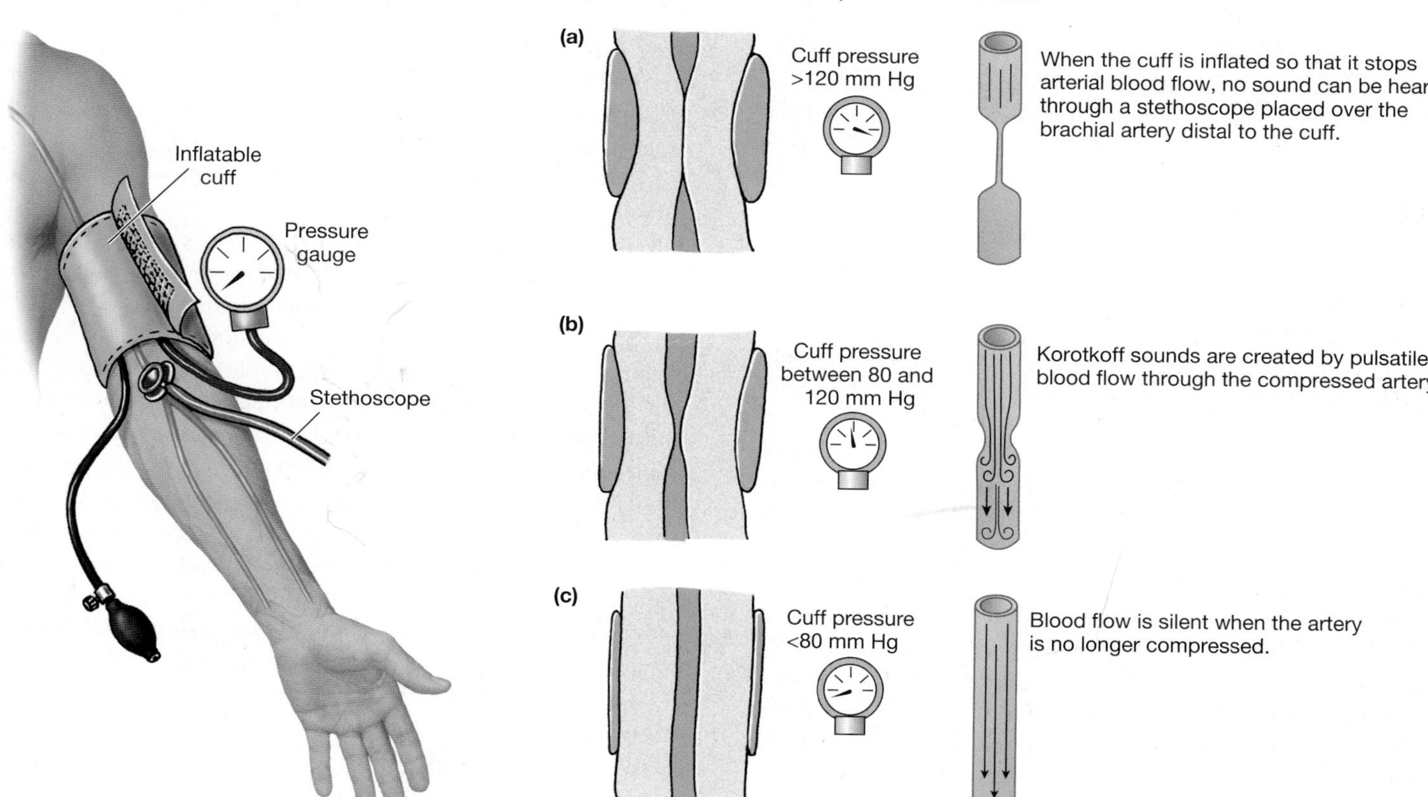
Arterial blood pressure is measured with a sphygmomanometer (an inflatable cuff plus a pressure gauge) and a stethoscope. The inflation pressure shown is for a person whose blood pressure is 120/80.
Inflatable cuff
Pressure gauge
Stethoscope
(a)
Cuff pressure >120 mm Hg
When the cuff is inflated so that it stops arterial blood flow, no sound can be heard through a stethoscope placed over the brachial artery distal to the cuff.
(b)
Cuff pressure between 80 and 120 mm Hg
Korotkoff sounds are created by pulsatile blood flow through the compressed artery.
(c)
Cuff pressure <80 mm Hg
Blood flow is silent when the artery is no longer compressed.

RUNNING PROBLEM

Kurt's second blood pressure reading is 158/98. Dr. Cortez asks him to take his blood pressure at home daily for two weeks and then return to the doctor's office. When Kurt comes back with his diary, the story is the same: his blood pressure continues to average 160/100. After running some tests, Dr. Cortez concludes that Kurt has high blood pressure or *hypertension,* like more than one out of every three adult Americans. If not controlled, hypertension can lead to heart failure, stroke, and kidney failure.

Q1: *Why are people with high blood pressure at greater risk for having a hemorrhagic (or bleeding) stroke?*

477 – **484** – 487 – 489 – 495 – 504

Cardiac Output and Peripheral Resistance Determine Mean Arterial Pressure

Mean arterial pressure is the driving force for blood flow, but what determines mean arterial pressure? Arterial pressure is a balance between blood flow into the arteries and blood flow out of the arteries. If flow in exceeds flow out, blood volume in the arteries increases, and mean arterial pressure increases. If flow out exceeds flow in, volume decreases and mean arterial pressure falls.

Blood flow into the aorta is equal to the cardiac output of the left ventricle. Blood flow out of the arteries is influenced primarily by **peripheral resistance,** defined as the resistance to flow offered by the arterioles (**FIG. 15.8a**). Mean arterial pressure then is proportional to cardiac output (CO) times resistance (R) of the arterioles:

$$\text{MAP} \propto \text{CO} \times R_{\text{arterioles}} \quad (5)$$

Let's consider how this works. If cardiac output increases, the heart pumps more blood into the arteries per unit time. If resistance to blood flow out of the arteries does not change, flow into the arteries is greater than flow out, blood volume in the arteries increases, and arterial blood pressure increases.

In another example, suppose cardiac output remains unchanged but peripheral resistance increases. Flow into arteries is unchanged, but flow out is decreased. Blood again accumulates in the arteries, and the arterial pressure again increases. Most cases of hypertension are believed to be caused by increased peripheral resistance without changes in cardiac output.

Two additional factors can influence arterial blood pressure: the distribution of blood in the systemic circulation and the total blood volume. The relative distribution of blood between the arterial and venous sides of the circulation can be an important factor in maintaining arterial blood pressure. Arteries are low-volume vessels that usually contain only about 11% of total blood volume at any one time. Veins, in contrast, are high-volume vessels that hold about 60% of the circulating blood volume at any one time.

The veins act as a *volume reservoir* for the circulatory system, holding blood that can be redistributed to the arteries if needed. If arterial blood pressure falls, increased sympathetic activity constricts veins, decreasing their holding capacity. Venous return sends blood to the heart, which according to the Frank–Starling law of the heart, pumps all the venous return out to the systemic side of the circulation [p. 466]. Thus, constriction of the veins redistributes blood to the arterial side of the circulation and raises mean arterial pressure.

Changes in Blood Volume Affect Blood Pressure

Although the volume of the blood in the circulation is usually relatively constant, changes in blood volume can affect arterial blood pressure (Fig. 15.8b). If blood volume increases, blood pressure increases. When blood volume decreases, blood pressure decreases.

To understand the relationship between blood volume and pressure, think of the circulatory system as an elastic balloon filled with water. If only a small amount of water is in the balloon, little pressure is exerted on the walls, and the balloon is soft and flabby. As more water is added to the balloon, more pressure is exerted on the elastic walls. If you fill a balloon close to the bursting point, you risk popping the balloon. The best way to reduce this pressure is to remove some of the water.

Small increases in blood volume occur throughout the day due to ingestion of food and liquids, but these increases usually do not create long-lasting changes in blood pressure because of homeostatic compensations. Adjustments for increased blood volume are primarily the responsibility of the kidneys. If blood volume increases, the kidneys restore normal volume by excreting excess water in the urine (**FIG. 15.9**).

Compensation for decreased blood volume is more difficult and requires an integrated response from the kidneys and the cardiovascular system. If blood volume decreases, *the kidneys cannot restore the lost fluid.* The kidneys can only *conserve* blood volume and thereby prevent further decreases in blood pressure.

The only way to restore lost fluid volume is through drinking or intravenous infusions. This is an example of mass balance: Volume lost to the external environment must be replaced from the external environment. Cardiovascular compensation for decreased blood volume includes vasoconstriction and increased sympathetic stimulation of the heart to increase cardiac output [Fig. 14.23, p. 471]. However, there are limits to the effectiveness of cardiovascular compensation—if fluid loss is too great, the body cannot maintain adequate blood pressure. Typical events that might cause significant changes in blood volume include dehydration, hemorrhage, and ingestion of a large quantity of fluid.

Figure 15.8b summarizes the four key factors that influence mean arterial blood pressure.

FIG. 15.8 ESSENTIALS Mean Arterial Blood Pressure

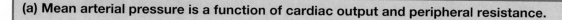

(a) Mean arterial pressure is a function of cardiac output and peripheral resistance.

Mean arterial pressure (MAP) is a function of cardiac output and resistance in the arterioles (peripheral resistance). MAP illustrates mass balance: the volume of blood in the arteries is determined by input (cardiac output) and flow out (altered by changing peripheral resistance). As arterial volume increases, pressure increases. In this model, the ventricle is represented by a syringe. The variable diameter of the arterioles is represented by adjustable screws.

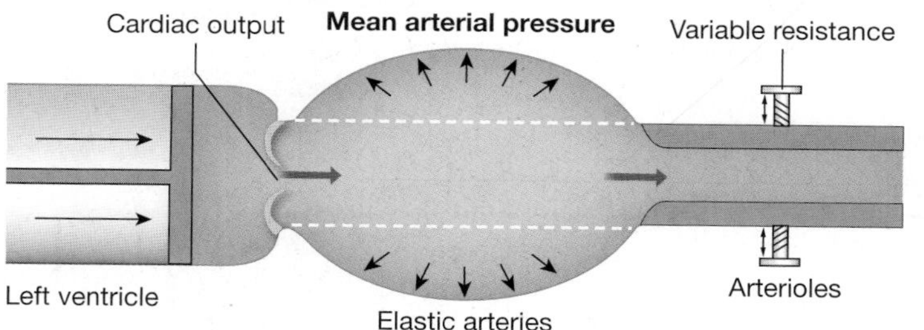

Cardiac output
Mean arterial pressure
Variable resistance
Left ventricle
Elastic arteries
Arterioles

Mean arterial pressure α cardiac output $\times$ resistance

FIGURE QUESTIONS

1. If arterioles constrict, what happens to blood flow out of the arteries? What happens to MAP?
2. If cardiac output decreases, what happens to arterial blood volume? What happens to MAP?
3. If veins constrict, what happens to blood volume in the veins? What happens to volume in the arteries and to MAP?

(b) Factors that influence mean arterial pressure

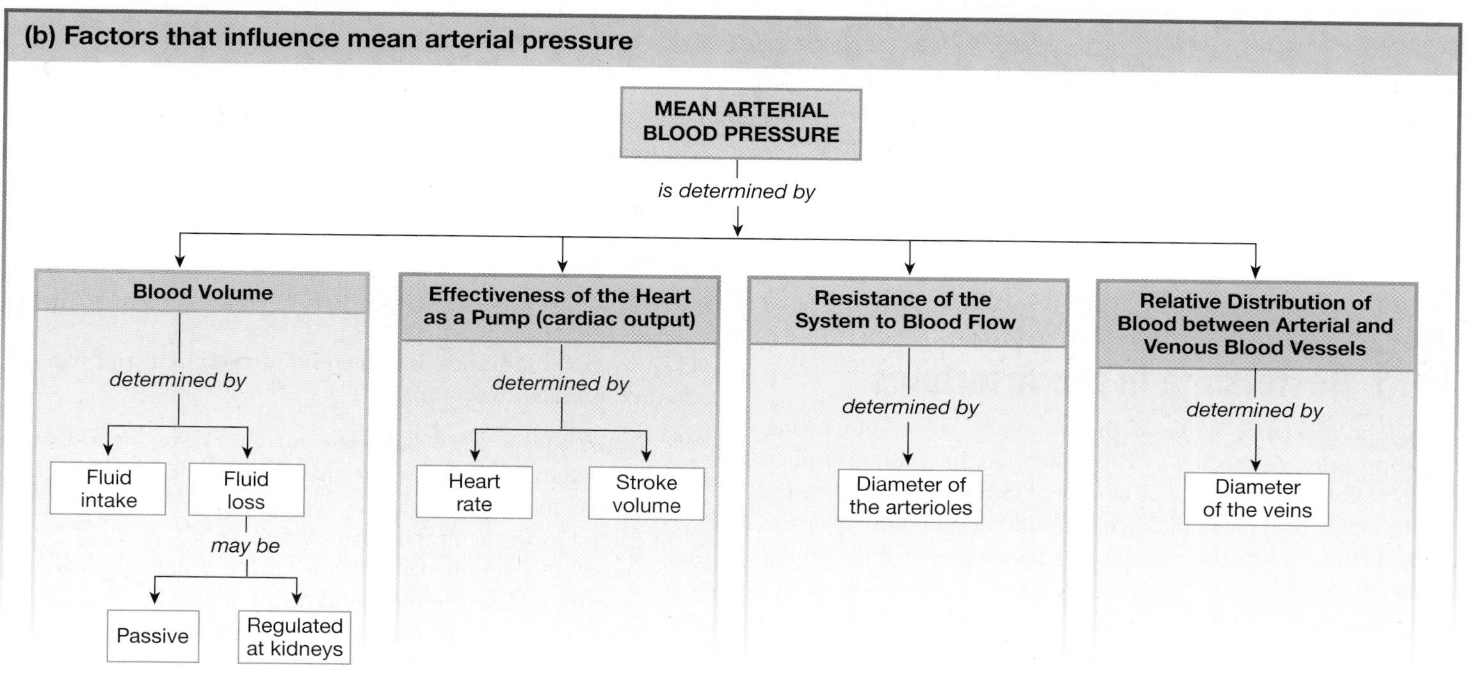

MEAN ARTERIAL BLOOD PRESSURE
is determined by
Blood Volume
determined by
Fluid intake
Fluid loss
may be
Passive
Regulated at kidneys
Effectiveness of the Heart as a Pump (cardiac output)
determined by
Heart rate
Stroke volume
Resistance of the System to Blood Flow
determined by
Diameter of the arterioles
Relative Distribution of Blood between Arterial and Venous Blood Vessels
determined by
Diameter of the veins

CLINICAL FOCUS

SHOCK

Shock is a broad term that refers to generalized, severe circulatory failure. Shock can arise from multiple causes: failure of the heart to maintain normal cardiac output (*cardiogenic shock*), decreased circulating blood volume (*hypovolemic shock*), failure of the sympathetic nervous system to maintain vascular tone (*neurogenic shock*), bacterial toxins (*septic shock*), and miscellaneous causes, such as the massive immune reactions that cause *anaphylactic shock*. No matter what the cause, the results are similar: low cardiac output and falling peripheral blood pressure. When tissue perfusion can no longer keep up with tissue oxygen demand, the cells begin to sustain damage from inadequate oxygen and from the buildup of metabolic wastes. Once this damage occurs, a positive feedback cycle begins. The shock becomes progressively worse until it becomes irreversible, and the patient dies. The management of shock includes administration of oxygen, fluids, and norepinephrine, which stimulates vasoconstriction and increases cardiac output. If the shock arises from a cause that is treatable, such as a bacterial infection, measures must also be taken to remove the precipitating cause.

FIG. 15.9 Compensation for increased blood volume

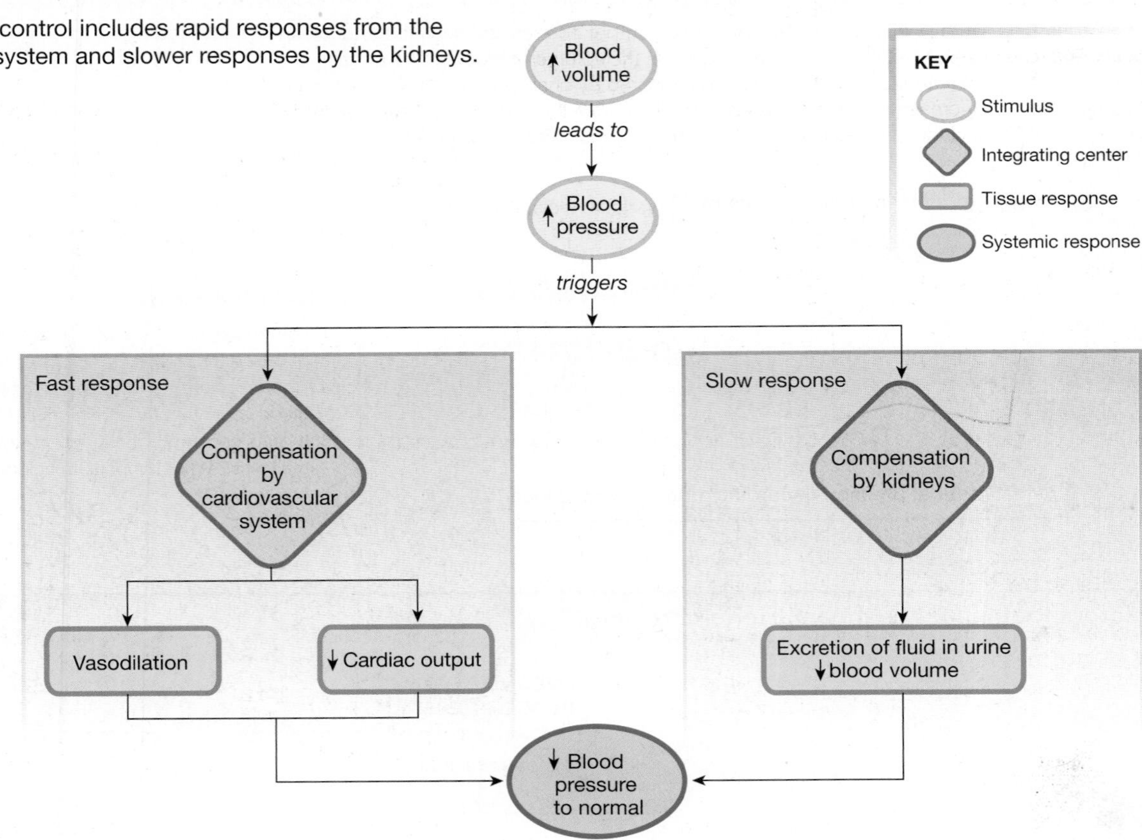
Blood pressure control includes rapid responses from the cardiovascular system and slower responses by the kidneys.
↑ Blood volume
leads to
↑ Blood pressure
triggers
Fast response
Compensation by cardiovascular system
Vasodilation
↓ Cardiac output
Slow response
Compensation by kidneys
Excretion of fluid in urine ↓ blood volume
↓ Blood pressure to normal
KEY
Stimulus
Integrating center
Tissue response
Systemic response

15.3 Resistance in the Arterioles

Peripheral resistance is one of the two main factors influencing blood pressure. According to Poiseuille's law [p. 438], resistance to blood flow (R) is directly proportional to the length of the tubing through which the fluid flows (L) and to the viscosity (η) of the fluid, and inversely proportional to the fourth power of the tubing radius (r):

$$R \propto L\eta/r^4 \quad (6)$$

Normally, the length of the systemic circulation and the blood's viscosity are relatively constant. That leaves only the radius of the blood vessels as the primary resistance to blood flow:

$$R \propto 1/r^4 \quad (7)$$

The arterioles are the main site of variable resistance in the systemic circulation and contribute more than 60% of the total resistance to flow in the system. Resistance in arterioles is variable because of the large amounts of smooth muscle in the arteriolar walls. When the smooth muscle contracts or relaxes, the radius of the arterioles changes.

Arteriolar resistance is influenced by both local and systemic control mechanisms:

1. *Local control of arteriolar resistance* matches tissue blood flow to the metabolic needs of the tissue. In the heart and skeletal muscle, these local controls often take precedence over reflex control by the central nervous system.
2. *Sympathetic reflexes* mediated by the CNS maintain mean arterial pressure and determine blood distribution to various tissues to meet homeostatic needs, such as temperature regulation.
3. *Hormones*—particularly those that regulate salt and water excretion by the kidneys—influence blood pressure by acting directly on the arterioles and by altering autonomic reflex control.

TABLE 15.2 lists significant chemicals that mediate arteriolar resistance by producing vasoconstriction or vasodilation. In the following sections, we look at some factors that influence blood flow at the tissue level.

Myogenic Autoregulation Adjusts Blood Flow

Vascular smooth muscle has the ability to regulate its own state of contraction, a process called **myogenic autoregulation.** In the absence of autoregulation, an increase in blood pressure increases

TABLE 15.2 Chemicals Mediating Vasoconstriction and Vasodilation

Chemical	Physiological Role	Source	Type
Vasoconstriction			
Norepinephrine (α-receptors)	Baroreceptor reflex	Sympathetic neurons	Neurotransmitter
Serotonin	Platelet aggregation, smooth muscle contraction	Neurons, digestive tract, platelets	Paracrine signal, neurotransmitter
Endothelin	Local control of blood flow	Vascular endothelium	Paracrine
Vasopressin	Increases blood pressure in hemorrhage	Posterior pituitary	Neurohormone
Angiotensin II	Increases blood pressure	Plasma hormone	Hormone
Vasodilation			
Epinephrine (β_2-receptors)	Increase blood flow to skeletal muscle, heart, liver	Adrenal medulla	Neurohormone
Acetylcholine	Erection reflex (indirectly through NO production)	Parasympathetic neurons	Neurotransmitter
Nitric oxide (NO)	Local control of blood flow	Endothelium	Paracrine signal
Bradykinin (via NO)	Increases blood flow	Multiple tissues	Paracrine signal
Adenosine	Increases blood flow to match metabolism	Hypoxic cells	Paracrine signal
$\downarrow O_2$, $\uparrow CO_2$, $\uparrow H^+$, $\uparrow K^+$	Increase blood flow to match metabolism	Cell metabolism	Paracrine molecule
Histamine	Increases blood flow	Mast cells	Paracrine signal
Natriuretic peptides (example: ANP)	Reduce blood pressure	Atrial myocardium, brain	Hormone, neurotransmitter
Vasoactive intestinal peptide	Digestive secretion, relax smooth muscle	Neurons	Neurotransmitter, neurohormone

CHAPTER 15

blood flow through an arteriole. However, when smooth muscle fibers in the wall of the arteriole stretch because of increased blood pressure, the arteriole constricts. This vasoconstriction increases the resistance offered by the arteriole, automatically decreasing blood flow through the vessel. With this simple and direct response to pressure, arterioles have limited ability to regulate their own blood flow.

How does myogenic autoregulation work at the cellular level? When vascular smooth muscle cells in arterioles are stretched, mechanically gated channels in the muscle membrane open. Cation entry depolarizes the cell. The depolarization opens voltage-gated Ca^{2+} channels, and Ca^{2+} flows into the cell down its electrochemical gradient. Calcium entering the cell combines with calmodulin and activates myosin light chain kinase [p. 403]. MLCK in turn increases myosin ATPase activity and crossbridge activity, resulting in contraction.

RUNNING PROBLEM

Most hypertension is *essential hypertension,* which means high blood pressure that cannot be attributed to any particular cause. "Since your blood pressure is only mildly elevated," Dr. Cortez tells Kurt, "let's see if we can control it with lifestyle changes and a diuretic. You need to reduce salt and fat in your diet, get some exercise, and lose some weight. The diuretic will help your kidneys get rid of excess fluid." "Looks like you're asking me to turn over a whole new leaf," says Kurt. "I'll try it."

Q2: *What is the rationale for reducing salt intake and taking a diuretic to control hypertension? (Hint: Salt causes water retention.)*

477 – 484 – **487** – 489 – 495 – 504

Paracrine Signals Influence Vascular Smooth Muscle

Local control is an important strategy by which individual tissues regulate their own blood supply. In a tissue, blood flow into individual capillaries can be regulated by the precapillary sphincters described earlier in the chapter. When these small bands of smooth muscle at metarteriole-capillary junctions constrict, they restrict blood flow into the capillaries (see Fig. 15.3). When the sphincters dilate, blood flow into the capillaries increases. This mechanism provides an additional site for local control of blood flow.

Local regulation also takes place by changing arteriolar resistance in a tissue. This is accomplished by paracrine molecules (including the gases O_2, CO_2, and NO) secreted by the vascular endothelium or by cells to which the arterioles are supplying blood (Tbl. 15.2).

The concentrations of many paracrine molecules change as cells become more or less metabolically active. For example, if aerobic metabolism increases, tissue O_2 levels decrease while CO_2 production goes up. Both low O_2 and high CO_2 dilate arterioles. This vasodilation increases blood flow into the tissue, bringing additional O_2 to meet the increased metabolic demand and removing waste CO_2 (**FIG. 15.10a**). The process in which an increase in blood flow accompanies an increase in metabolic activity is known as **active hyperemia** {*hyper-*, above normal + *(h)aimia*, blood}.

If blood flow to a tissue is occluded {*occludere*, to close up} for a few seconds to a few minutes, O_2 levels fall and metabolic paracrine signals such as CO_2 and H^+ accumulate in the interstitial fluid. Local *hypoxia* {*hypo-*, low + *oxia*, oxygen} causes endothelial cells to synthesize the vasodilator nitric oxide. When blood flow to the tissue resumes, the increased concentrations of NO, CO_2, and other paracrine molecules immediately trigger significant vasodilation. As the vasodilators are metabolized or washed away by the restored tissue blood flow, the radius of the arteriole gradually returns to normal. An increase in tissue blood flow following a period of low perfusion (blood flow) is known as **reactive hyperemia** (Fig. 15.10b).

Nitric oxide is probably best known for its role in the male erection reflex: Drugs used to treat erectile dysfunction prolong NO activity. Decreases in endogenous NO activity are suspected to play a role in other medical conditions, including hypertension and *preeclampsia*, the elevated blood pressure that sometimes occurs during pregnancy.

Not all vasoactive paracrine molecules reflect changes in metabolism. For example, *kinins* and *histamine* are potent vasodilators that play a role in inflammation. *Serotonin* (5-HT), previously mentioned as a CNS neurotransmitter [p. 253], is also a vasoconstricting signal molecule released by activated platelets. When damaged blood vessels activate platelets, the subsequent serotonin-mediated vasoconstriction helps slow blood loss.

FIG. 15.10 Hyperemia

Hyperemia is a locally mediated increase in blood flow.

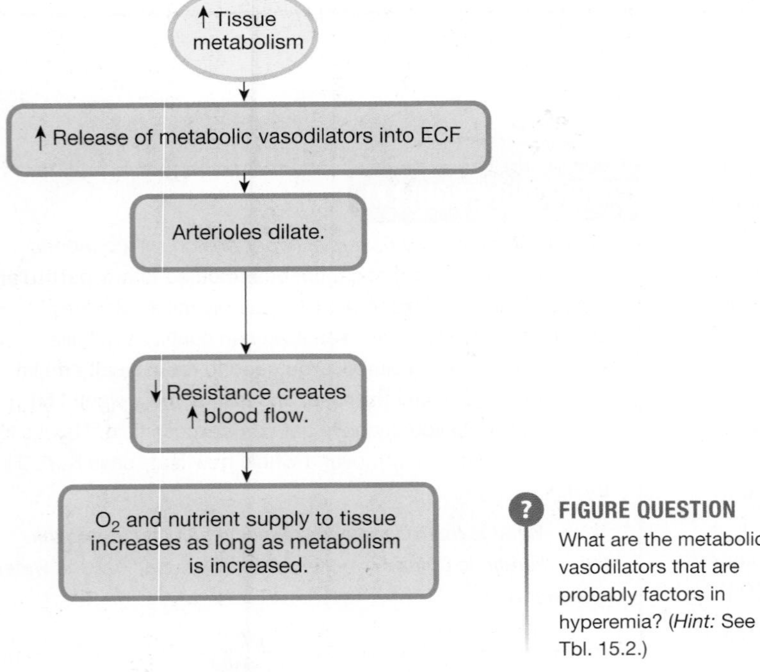
(a) Active hyperemia matches blood flow to increased metabolism.
↑ Tissue metabolism
↑ Release of metabolic vasodilators into ECF
Arterioles dilate.
↓ Resistance creates ↑ blood flow.
O_2 and nutrient supply to tissue increases as long as metabolism is increased.

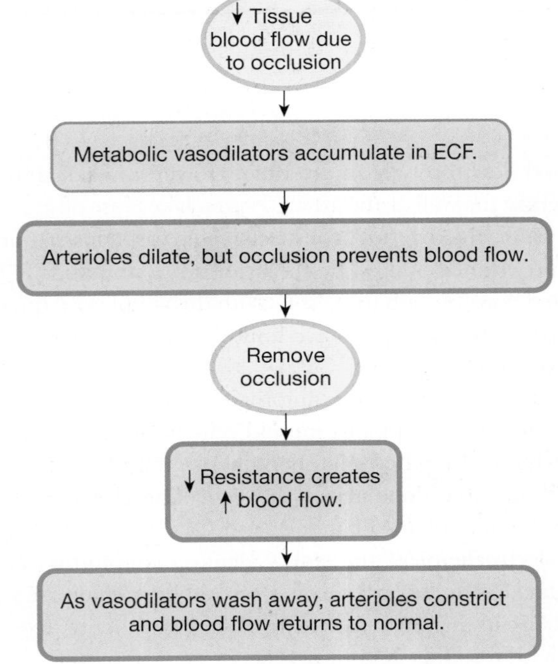
(b) Reactive hyperemia follows a period of decreased blood flow.
↓ Tissue blood flow due to occlusion
Metabolic vasodilators accumulate in ECF.
Arterioles dilate, but occlusion prevents blood flow.
Remove occlusion
↓ Resistance creates ↑ blood flow.
As vasodilators wash away, arterioles constrict and blood flow returns to normal.

FIGURE QUESTION
What are the metabolic vasodilators that are probably factors in hyperemia? (*Hint:* See Tbl. 15.2.)

Concept Check

6. Resistance to blood flow is determined *primarily* by which? (a) blood viscosity, (b) blood volume, (c) cardiac output, (d) blood vessel diameter, or (e) blood pressure gradient (ΔP)
7. The extracellular fluid concentration of K^+ increases in exercising skeletal muscles. What effect does this increase in K^+ have on blood flow in the muscles?

The Sympathetic Branch Controls Most Vascular Smooth Muscle

Smooth muscle contraction in arterioles is regulated by neural and hormonal signals in addition to locally produced paracrine molecules. Among the hormones with significant vasoactive properties are *atrial natriuretic peptide* and *angiotensin II* (ANG II). These hormones also have significant effects on the kidney's excretion of ions and water.

Most systemic arterioles are innervated by sympathetic neurons. A notable exception is arterioles involved in the erection reflex of the penis and clitoris. They are controlled indirectly by parasympathetic innervation. Acetylcholine from parasympathetic neurons causes paracrine release of nitric oxide, resulting in vasodilation.

Tonic discharge of norepinephrine from sympathetic neurons helps maintain tone of arterioles (**FIG. 15.11a**) [Fig. 6.15, p. 183]. Norepinephrine binding to α-receptors on vascular smooth muscle causes vasoconstriction. If sympathetic release of norepinephrine decreases, the arterioles dilate. If sympathetic stimulation increases, the arterioles constrict.

Epinephrine from the adrenal medulla travels through the blood and also binds with α-receptors, reinforcing vasoconstriction. However, α-receptors have a lower affinity for epinephrine and do not respond as strongly to it as they do to norepinephrine [p. 363]. In addition, epinephrine binds to β_2-receptors, found on vascular smooth muscle of heart, liver, and skeletal muscle arterioles. These receptors are not innervated and therefore respond primarily to circulating epinephrine. Activation of vascular β_2-receptors by epinephrine causes vasodilation [Fig. 6.14, p. 180].

One way to remember which tissues' arterioles have β_2-receptors is to think of a fight-or-flight response to a stressful event [p. 356]. This response includes a generalized increase in sympathetic activity, along with the release of epinephrine. Blood vessels that have β_2-receptors respond to epinephrine by vasodilating. Such β_2-mediated vasodilation enhances blood flow to the heart, skeletal muscles, and liver, tissues that are active during the fight-or-flight response. (The liver produces glucose for muscle contraction.) During fight or flight, increased sympathetic activity at arteriolar α-receptors causes vasoconstriction. The increase in resistance diverts blood from nonessential organs, such as the gastrointestinal tract, to the skeletal muscles, liver, and heart.

The map in Figure 15.11b summarizes the many factors that influence blood flow in the body. The pressure to drive blood flow is created by the pumping heart and captured by the arterial pressure reservoir, as reflected by the mean arterial pressure. Flow through the body as a whole is equal to the cardiac output, but flow to individual tissues can be altered by selectively changing resistance in a tissue's arterioles. In the next section, we consider the relationship between blood flow and arteriolar resistance.

RUNNING PROBLEM

After two months, Kurt returns to the doctor's office for a checkup. He has lost 5 pounds and is walking at least a mile daily, but his blood pressure has not changed. "I swear, I'm trying to do better," says Kurt, "but it's difficult." Because lifestyle changes and the diuretic have not lowered Kurt's blood pressure, Dr. Cortez adds an antihypertensive drug. "This drug, called an ACE inhibitor, blocks production of a chemical called angiotensin II, a powerful vasoconstrictor. This medication should bring your blood pressure back to a normal value."

Q3: *Why would blocking the action of a vasoconstrictor lower blood pressure?*

477 – 484 – 487 – 489 – 495 – 504

Concept Check

8. What happens when epinephrine combines with β_1-receptors in the heart? With β_2-receptors in the heart? (*Hint:* "in the heart" is vague. The heart has multiple tissue types. Which heart tissues possess the different types of β-receptors? [p. 469])
9. Skeletal muscle arterioles have both α- and β-receptors on their smooth muscle. Epinephrine can bind to both. Will these arterioles constrict or dilate in response to epinephrine? Explain.

15.4 Distribution of Blood to the Tissues

The body's ability to selectively alter blood flow to organs is an important aspect of cardiovascular regulation. The distribution of systemic blood varies according to the metabolic needs of individual organs and is governed by a combination of local control mechanisms and homeostatic reflexes. For example, skeletal muscles at rest receive about 20% of cardiac output. During exercise, when the muscles use more oxygen and nutrients, they receive as much as 85%.

Blood flow to individual organs is set to some degree by the number and size of arteries feeding the organ. **FIGURE 15.12** shows how blood is distributed to various organs when the body is at rest. Usually, more than two-thirds of the cardiac output is routed to the digestive tract, liver, muscles, and kidneys.

FIG. 15.11 ESSENTIALS Resistance and Flow

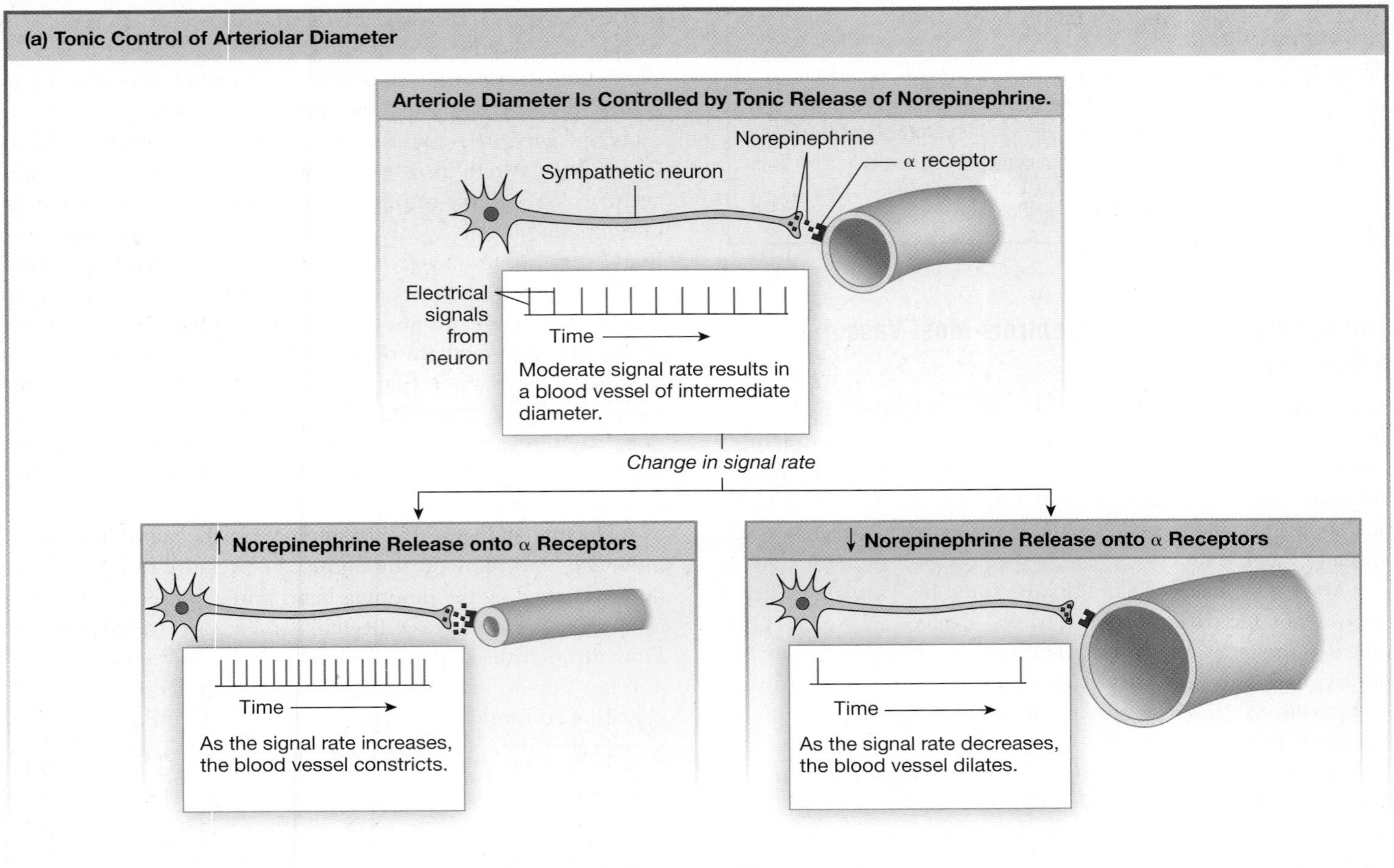

(a) Tonic Control of Arteriolar Diameter

Arteriole Diameter Is Controlled by Tonic Release of Norepinephrine.

Sympathetic neuron

Norepinephrine

α receptor

Electrical signals from neuron

Time →

Moderate signal rate results in a blood vessel of intermediate diameter.

Change in signal rate

↑ Norepinephrine Release onto α Receptors

Time →

As the signal rate increases, the blood vessel constricts.

↓ Norepinephrine Release onto α Receptors

Time →

As the signal rate decreases, the blood vessel dilates.

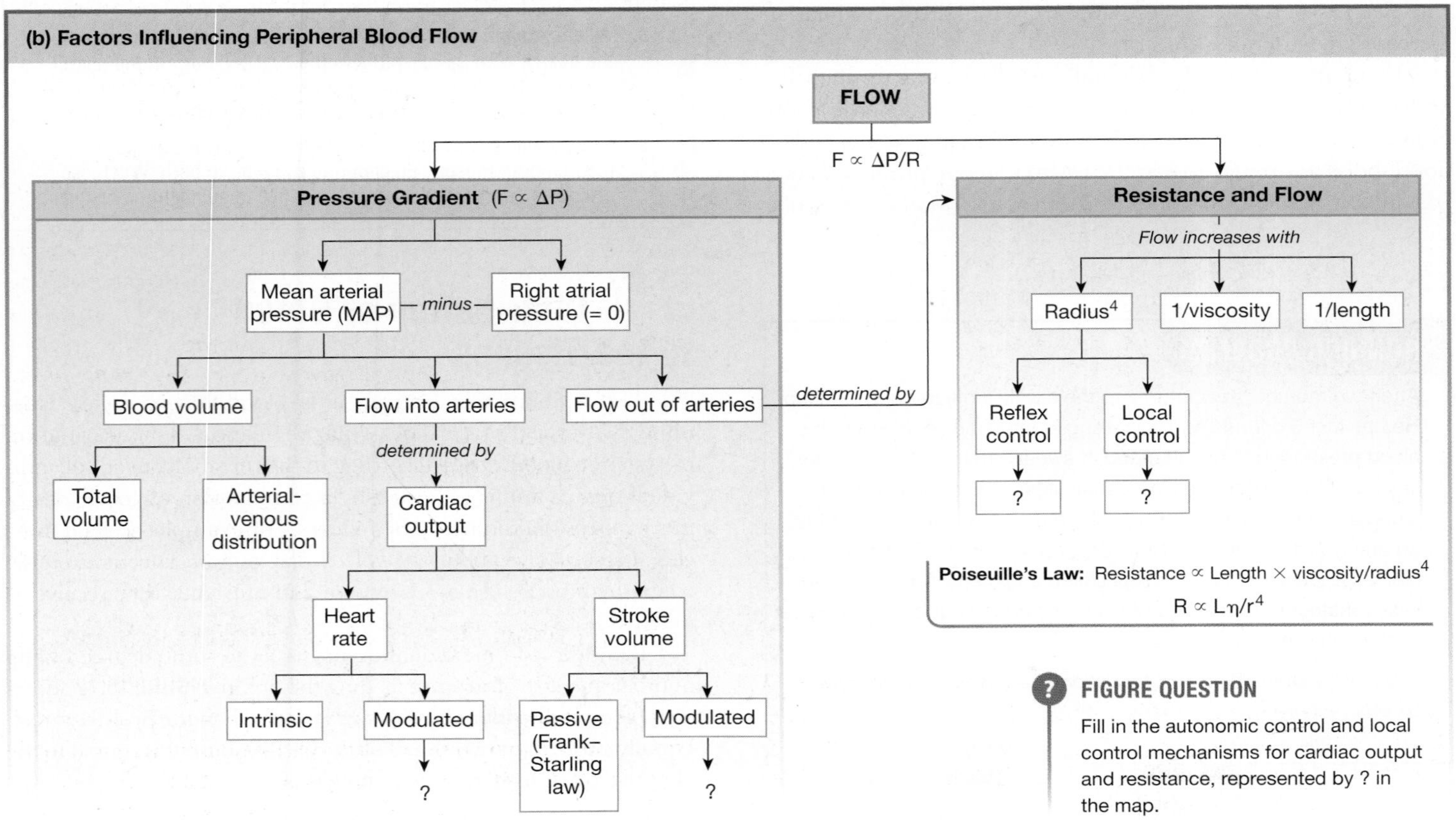

(b) Factors Influencing Peripheral Blood Flow

FLOW

$F \propto \Delta P/R$

Pressure Gradient ($F \propto \Delta P$)

Mean arterial pressure (MAP) *minus* Right atrial pressure (= 0)

- Blood volume
  - Total volume
  - Arterial-venous distribution
- Flow into arteries
  - *determined by* Cardiac output
    - Heart rate
      - Intrinsic
      - Modulated → ?
    - Stroke volume
      - Passive (Frank–Starling law)
      - Modulated → ?
- Flow out of arteries *determined by* → Resistance and Flow

Resistance and Flow

Flow increases with

- $Radius^4$
  - Reflex control → ?
  - Local control → ?
- 1/viscosity
- 1/length

Poiseuille's Law: Resistance ∝ Length × viscosity/$radius^4$

$$R \propto L\eta/r^4$$

FIGURE QUESTION

Fill in the autonomic control and local control mechanisms for cardiac output and resistance, represented by ? in the map.

FIG. 15.12 Distribution of blood in the body at rest

Blood flow to the major organs is represented in three ways: as a percentage of total flow, as volume per 100 grams of tissue per minute, and as an absolute rate of flow (in L/min).

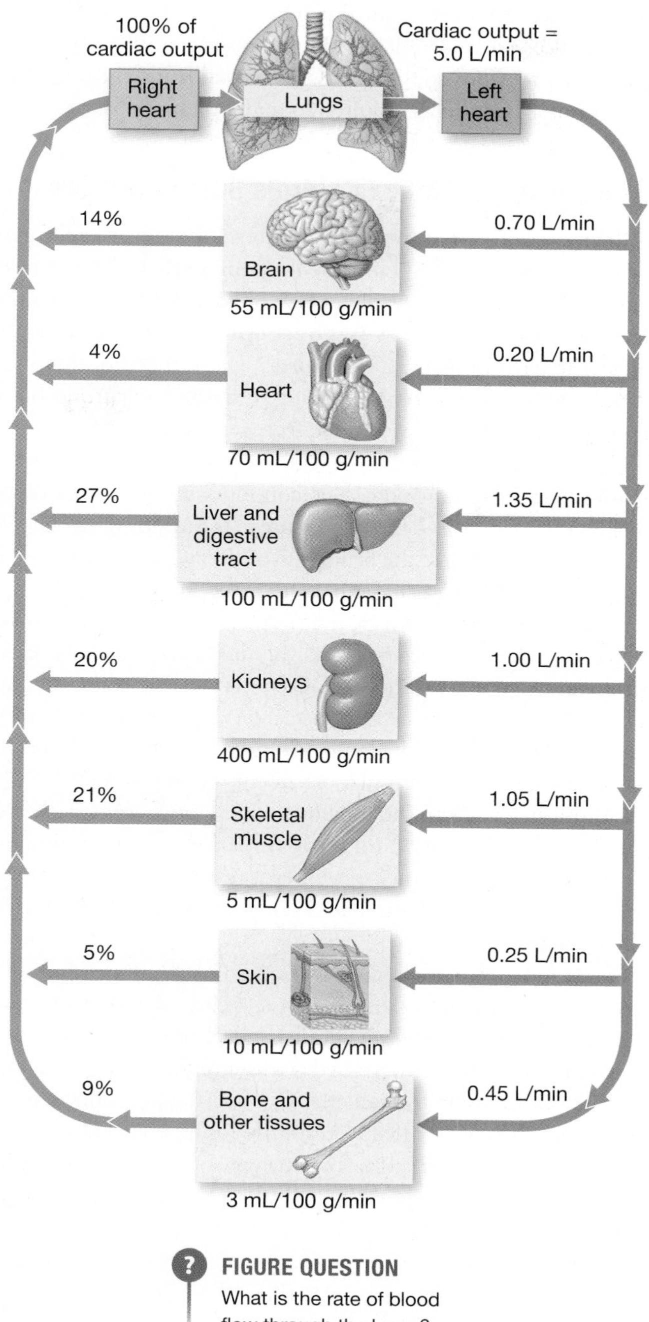
100% of cardiac output
Cardiac output = 5.0 L/min
Right heart
Lungs
Left heart
14%
0.70 L/min
Brain
55 mL/100 g/min
4%
0.20 L/min
Heart
70 mL/100 g/min
27%
1.35 L/min
Liver and digestive tract
100 mL/100 g/min
20%
1.00 L/min
Kidneys
400 mL/100 g/min
21%
1.05 L/min
Skeletal muscle
5 mL/100 g/min
5%
0.25 L/min
Skin
10 mL/100 g/min
9%
0.45 L/min
Bone and other tissues
3 mL/100 g/min

FIGURE QUESTION
What is the rate of blood flow through the lungs?

Variations in blood flow to individual tissues are possible because the arterioles in the body are arranged in parallel. In other words, all arterioles receive blood at the same time from the aorta (see Fig. 15.1). Total blood flow through *all* the arterioles of the body always equals the cardiac output.

However, the flow through individual arterioles in a branching system of arterioles depends on their resistance (R). The higher the resistance in an arteriole, the lower the blood flow through it. If an arteriole constricts and resistance increases, blood flow through that arteriole decreases (**FIG. 15.13**):

$$\text{Flow}_{\text{arteriole}} \propto 1/R_{\text{arteriole}} \quad (8)$$

If nothing else changes, the flow diverted away from the constricted arteriole then is distributed to the other arterioles, keeping total flow through the system constant. In other words, blood is diverted from high-resistance arterioles to lower-resistance arterioles. You might say that blood traveling through the arterioles takes the path of least resistance.

Concept Check

10. Use Fig. 15.12 to answer these questions. (a) Which tissue has the highest blood flow per unit weight? (b) Which tissue has the least blood flow, regardless of weight?

Arteriolar resistance in most tissues of the body is a balance between autonomic control by the brain and local control by tissue metabolites and paracrine signals. However, two critical organs, the brain and the heart, are so dependent on a steady supply of blood and oxygen that neural control of their arterioles plays only a small part in maintaining adequate perfusion. In these two organs, tissue metabolism is the primary factor that determines arteriolar resistance, as described in the next two sections.

Cerebral Blood Flow Stays Nearly Constant

Brain tissue is very sensitive to lack of oxygen and glucose, so loss of blood flow to the brain will cause unconsciousness in a matter of seconds, and brain damage within minutes. To avoid this, cerebral blood flow is relatively constant under normal circumstances. Increases in systemic blood pressure trigger myogenic responses that result in vasoconstriction. However, the primary factor that alters blood flow in the brain is tissue metabolism.

As discussed previously, accumulation of CO_2 around arterioles acts as a vasodilator. If metabolism in one region of the brain increases, oxygen consumption and CO_2 production both increase. The CO_2 vasodilates arterioles, increasing flow to the active region. Local increases in cerebral blood flow can be observed using imaging techniques such as PET scans and functional magnetic resonance imaging (fMRI) [p. 291].

Coronary Blood Flow Parallels the Work of the Heart

Heart muscle also demands a steady supply of oxygen and nutrients from the blood to provide ATP for contraction. In most resting tissues, only 25% of the oxygen that comes into the tissue is taken up by the cells, so venous blood leaving the tissues still has 75% of its

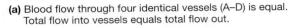

FIG. 15.13 Blood flow through individual blood vessels is determined by the vessel's resistance to flow

(a) Blood flow through four identical vessels (A–D) is equal. Total flow into vessels equals total flow out.

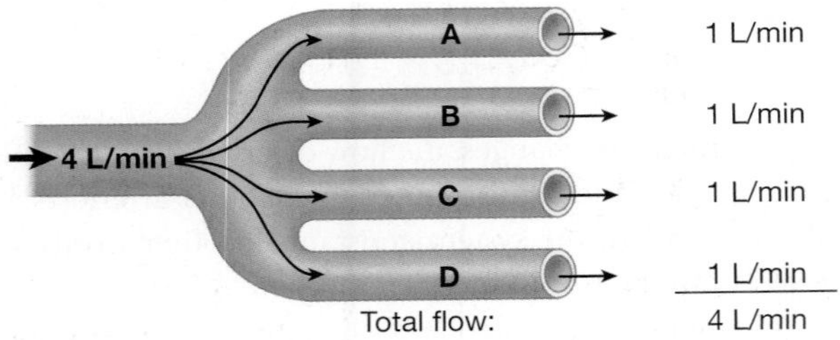
4 L/min
A
B
C
D
1 L/min
1 L/min
1 L/min
1 L/min
Total flow:
4 L/min

(b) When vessel B constricts, resistance of B increases and flow through B decreases. Flow diverted from B is divided among the lower-resistance vessels A, C, and D.

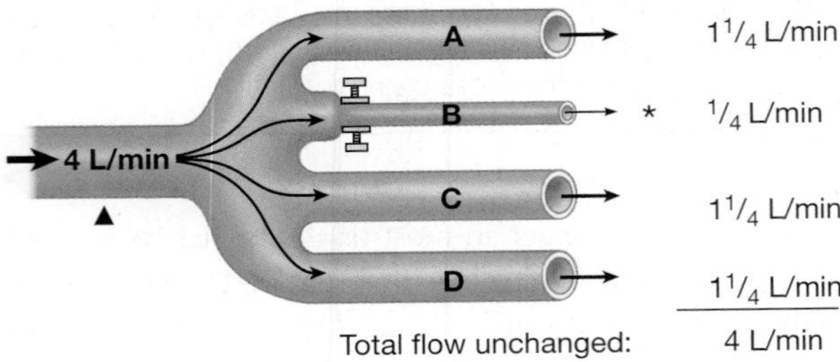
4 L/min
▲
A
B
C
D
*
$1\frac{1}{4}$ L/min
$\frac{1}{4}$ L/min
$1\frac{1}{4}$ L/min
$1\frac{1}{4}$ L/min
Total flow unchanged:
4 L/min

FIGURE QUESTION

1. You are monitoring blood pressure in the artery at the point indicated by ▲. What happens to arterial blood pressure when vessel B suddenly constricts?
2. Is pressure at the * end of B increased or decreased following constriction?

oxygen. In contrast, the myocardium, constantly working, extracts about 75% of the oxygen that comes to it, leaving little in reserve.

As the work of the heart increases, coronary blood flow must also increase to maintain the oxygen delivery. If oxygen consumption in heart muscle exceeds the rate at which oxygen is supplied by the blood, myocardial hypoxia results. In response to low tissue oxygen, the myocardial cells release the nucleotide **adenosine**. Adenosine dilates coronary arterioles to decrease their resistance and bring additional blood flow into the muscle.

15.5 Regulation of Cardiovascular Function

The central nervous system coordinates the reflex control of blood pressure and distribution of blood to the tissues. The main integrating center is in the medulla oblongata. Because of the complexity of the neural networks involved in cardiovascular control, we will simplify this discussion and refer to the CNS network as the **cardiovascular control center(CVCC)**.

The primary function of the cardiovascular control center is to ensure adequate blood flow to the brain and heart by maintaining sufficient mean arterial pressure. However, the CVCC also receives input from other parts of the brain, and it has the ability to alter function in one or two organs or tissues while leaving others unaffected. For example, thermoregulatory centers in the hypothalamus communicate with the CVCC to alter blood flow to the skin. Brain-gut communication following a meal increases blood flow to the intestinal tract. Reflex control of blood flow to selected tissues changes mean arterial pressure, so the CVCC is constantly monitoring MAP and adjusting its output as required to maintain homeostasis.

The Baroreceptor Reflex Controls Blood Pressure

The primary reflex pathway for homeostatic control of mean arterial blood pressure is the **baroreceptor reflex.** The components of the reflex are illustrated in **FIGURE 15.14a**. Stretch-sensitive mechanoreceptors known as **baroreceptors** are located in the walls of the carotid arteries and aorta, where they continuously monitor the pressure of blood flowing to the brain (carotid baroreceptors) and to the body (aortic baroreceptors).

The carotid and aortic baroreceptors are tonically active stretch receptors that fire action potentials continuously at normal blood pressures. When increased blood pressure in the arteries stretches the baroreceptor membrane, the firing rate of the receptor increases. If blood pressure falls, the firing rate of the receptor decreases.

If blood pressure changes, the frequency of action potentials traveling from the baroreceptors to the medullary cardiovascular control center changes. The CVCC integrates the sensory input and initiates an appropriate response. The response of the baroreceptor reflex is quite rapid: changes in cardiac output and peripheral resistance occur within two heartbeats of the stimulus.

Output signals from the cardiovascular control center are carried by both sympathetic and parasympathetic autonomic neurons. As you learned earlier, peripheral resistance is under tonic sympathetic control, with increased sympathetic discharge causing vasoconstriction.

Heart function is regulated by antagonistic control [p. 182]. Increased sympathetic activity increases heart rate, shortens conduction time through the AV node, and enhances the force of myocardial contraction. Increased parasympathetic activity slows heart rate but has only a small effect on ventricular contraction.

The baroreceptor reflex in response to increased blood pressure is mapped in Fig. 15.14b. Baroreceptors increase their firing rate as blood pressure increases, activating the medullary cardiovascular control center. In response, the cardiovascular control center increases parasympathetic activity and decreases sympathetic activity to slow down the heart and dilate arterioles.

When heart rate falls, cardiac output falls. In the vasculature, decreased sympathetic activity causes dilation of arterioles, lowering their resistance and allowing more blood to flow out of the arteries. Because mean arterial pressure is directly proportional to cardiac output and peripheral resistance ($MAP \propto CO \times R$), the combination of reduced cardiac output and decreased peripheral resistance lowers the mean arterial blood pressure.

It is important to remember that the baroreceptor reflex is functioning all the time, not just with dramatic disturbances in

FIG. 15.14 ESSENTIALS Cardiovascular Control

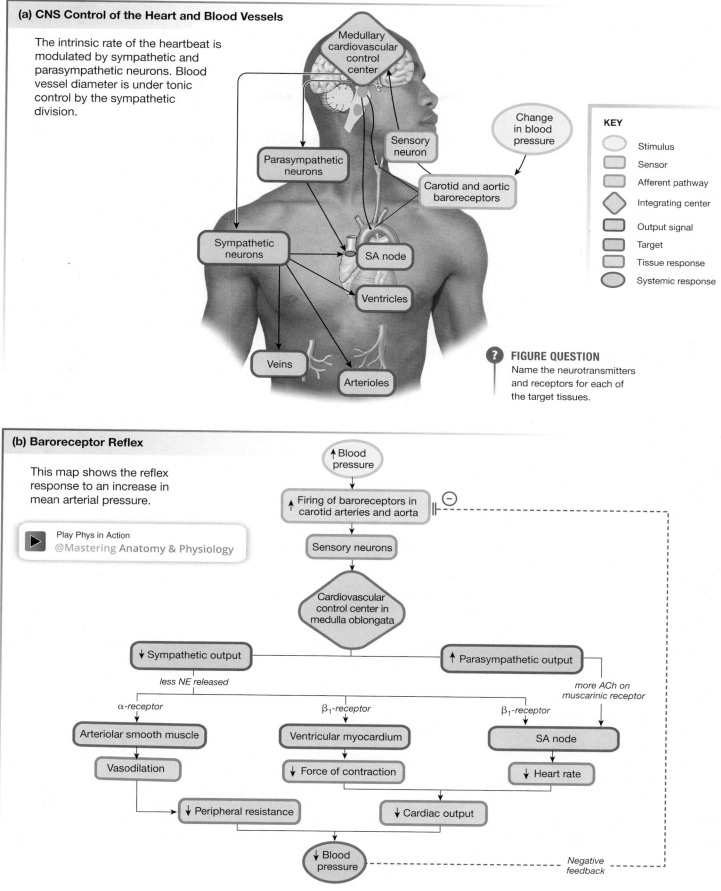

(a) CNS Control of the Heart and Blood Vessels

The intrinsic rate of the heartbeat is modulated by sympathetic and parasympathetic neurons. Blood vessel diameter is under tonic control by the sympathetic division.

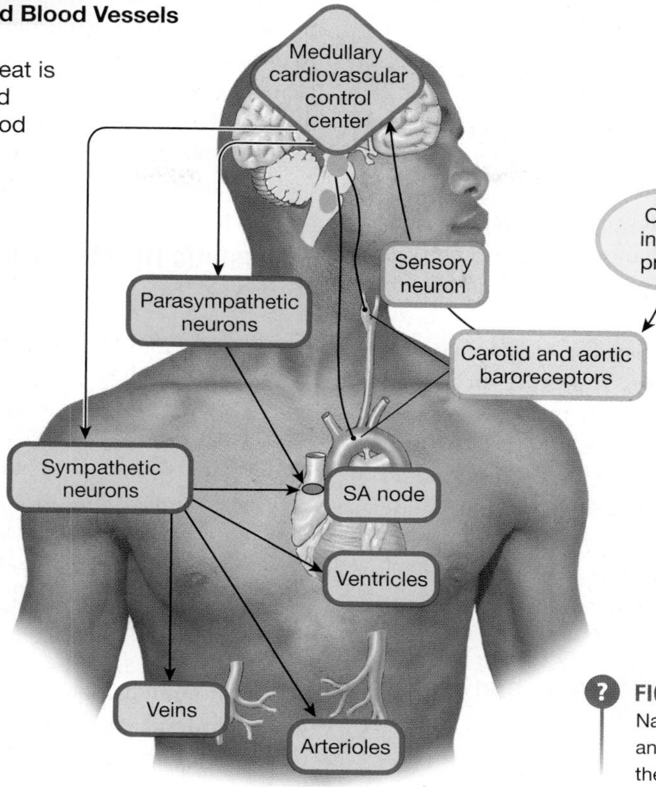

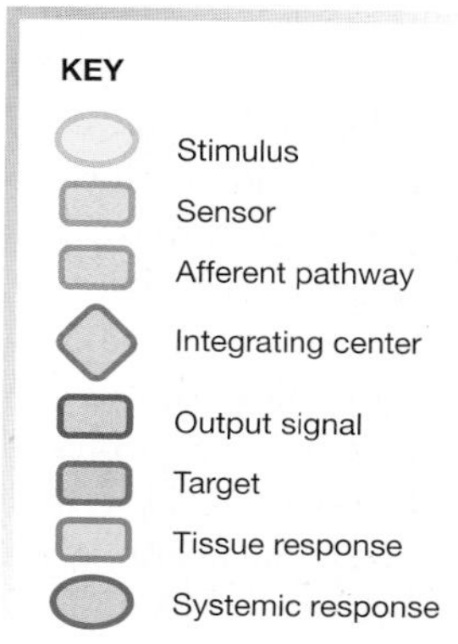

Medullary cardiovascular control center
Change in blood pressure
Sensory neuron
Parasympathetic neurons
Carotid and aortic baroreceptors
Sympathetic neurons
SA node
Ventricles
Veins
Arterioles
KEY
Stimulus
Sensor
Afferent pathway
Integrating center
Output signal
Target
Tissue response
Systemic response

FIGURE QUESTION
Name the neurotransmitters and receptors for each of the target tissues.

(b) Baroreceptor Reflex

This map shows the reflex response to an increase in mean arterial pressure.

Play Phys in Action
@Mastering Anatomy & Physiology

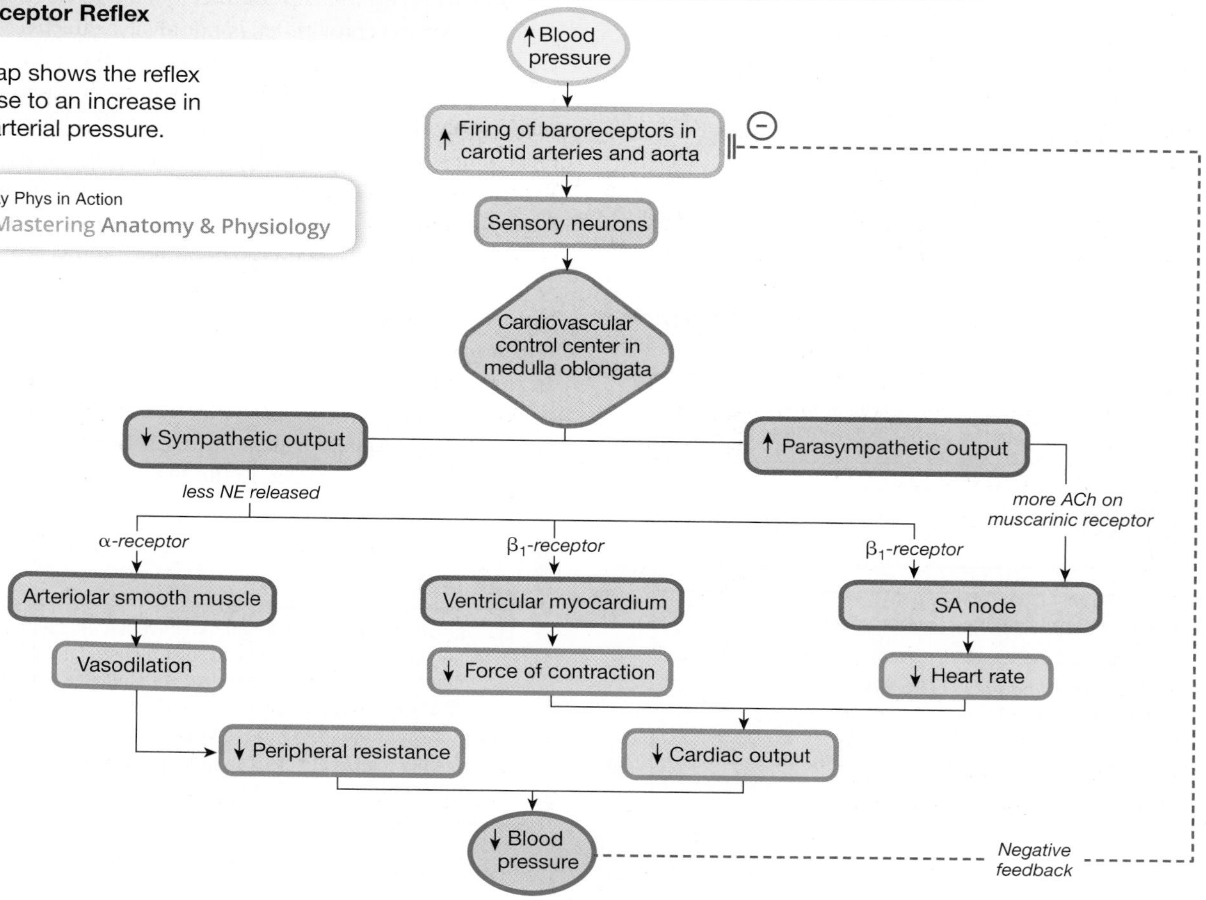

↑ Blood pressure
↑ Firing of baroreceptors in carotid arteries and aorta
Sensory neurons
Cardiovascular control center in medulla oblongata
↓ Sympathetic output
↑ Parasympathetic output
less NE released
more ACh on muscarinic receptor
α-receptor
β1-receptor
β1-receptor
Arteriolar smooth muscle
Ventricular myocardium
SA node
Vasodilation
↓ Force of contraction
↓ Heart rate
↓ Peripheral resistance
↓ Cardiac output
↓ Blood pressure
Negative feedback

blood pressure, and that it is not an all-or-none response. A change in blood pressure can result in a change in both cardiac output and peripheral resistance or a change in only one of the two variables. Let's look at an example.

For this example, we will use the schematic diagram in **FIGURE 15.15**, which combines the concepts introduced in Figures 15.8 and 15.13. In this model, there are four sets of variable resistance arterioles (A–D) whose diameters can be independently controlled by local or reflex control mechanisms. Baroreceptors in the arteries monitor mean arterial pressure and communicate with the medullary cardiovascular control center.

Suppose arteriole set A constricts because of local control mechanisms. Vasoconstriction increases resistance in A and decreases flow through A. Total peripheral resistance (TPR) across all the arterioles also increases. Using the relationship MAP $\propto$ CO $\times$ TPR, an increase in total resistance results in an increase in mean arterial pressure. The arterial baroreceptors sense the increase in MAP and activate the baroreceptor reflex.

Output from the cardiovascular control center can alter either cardiac output, arteriolar resistance, or both. In this instance, we can assume that blood flow in arteriole sets A–D now matches tissue needs and should remain constant. That means the only option left to decrease MAP is to decrease cardiac output. So efferent signals from the CVCC decrease cardiac output, which in turn brings mean arterial pressure down. Blood pressure homeostasis is restored. In this example, the output signal of the baroreceptor reflex altered cardiac output but did not change peripheral resistance.

FIG. 15.15 Integration of resistance changes and cardiac output

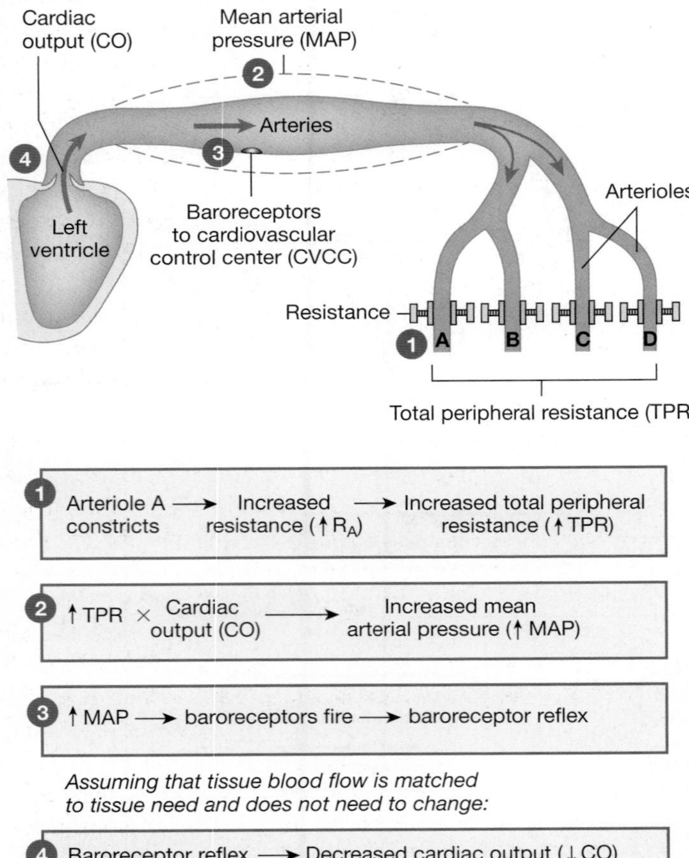
Cardiac output (CO)
Mean arterial pressure (MAP)
2
Arteries
3
4
Left ventricle
Baroreceptors to cardiovascular control center (CVCC)
Arterioles
Resistance
1
A
B
C
D
Total peripheral resistance (TPR)
1 Arteriole A constricts → Increased resistance (↑ R_A) → Increased total peripheral resistance (↑ TPR)
2 ↑ TPR × Cardiac output (CO) → Increased mean arterial pressure (↑ MAP)
3 ↑ MAP → baroreceptors fire → baroreceptor reflex
Assuming that tissue blood flow is matched to tissue need and does not need to change:
4 Baroreceptor reflex → Decreased cardiac output (↓ CO)
↑ TPR × ↓ CO = MAP restored to normal

Orthostatic Hypotension Triggers the Baroreceptor Reflex

The baroreceptor reflex functions every morning when you get out of bed. When you are lying flat, gravitational forces are distributed evenly up and down the length of your body, and blood is distributed evenly throughout the circulation. When you stand up, gravity causes blood to pool in the lower extremities.

This pooling creates an instantaneous decrease in venous return so that less blood is in the ventricles at the beginning of the next contraction. Cardiac output falls from 5 L/min to 3 L/min, causing arterial blood pressure to decrease. This decrease in blood pressure upon standing is known as *orthostatic hypotension* {*orthos*, upright + *statikos*, to stand}.

Orthostatic hypotension normally triggers the baroreceptor reflex. The result is increased cardiac output and increased peripheral resistance, which together increase mean arterial pressure and bring it back up to normal within two heartbeats. The skeletal muscle pump also contributes to the recovery by enhancing venous return when abdominal and leg muscles contract to maintain an upright position.

The baroreceptor reflex is not always effective, however. For example, during extended bed rest or in the zero-gravity conditions of space flights, blood from the lower extremities is distributed evenly throughout the body rather than pooled in the lower extremities. This even distribution raises arterial pressure, triggering the kidneys to excrete what the body perceives as excess fluid. Over the course of three days of bed rest or in space, excretion of water leads to a 12% decrease in blood volume.

When the person finally gets out of bed or returns to earth, gravity again causes blood to pool in the legs. Orthostatic hypotension occurs, and the baroreceptors attempt to compensate. In this instance, however, the cardiovascular system is unable to restore normal pressure because of the loss of blood volume. As a result, the individual may become light-headed or even faint from reduced delivery of oxygen to the brain.

Concept Check

11. Baroreceptors have stretch-sensitive ion channels in their cell membrane. Increased pressure stretches the receptor cell membrane, opens the channels, and initiates action potentials. What ion probably flows through these channels and in which direction (into or out of the cell)?
12. Use the map in Fig. 15.14b to map the reflex response to orthostatic hypotension.

Other Systems Influence Cardiovascular Function

Cardiovascular function can be modulated by input from peripheral receptors other than the baroreceptors. For example, arterial chemoreceptors activated by low blood oxygen levels increase cardiac output. The cardiovascular control center also has reciprocal communication with centers in the medulla that control breathing.

The integration of function between the respiratory and circulatory systems is adaptive. If tissues require more oxygen, it is supplied by the cardiovascular system working in tandem with the respiratory system. Consequently, increases in breathing rate are usually accompanied by increases in cardiac output.

Blood pressure is also subject to modulation by higher brain centers, such as the hypothalamus and cerebral cortex. The hypothalamus mediates vascular responses involved in body temperature regulation and for the fight-or-flight response. Learned and emotional responses may originate in the cerebral cortex and be expressed by cardiovascular responses such as blushing and fainting.

One such reflex is *vasovagal syncope,* which may be triggered in some people by the sight of blood or a hypodermic needle. (Recall Anthony's experience at the beginning of this chapter.) In this pathway, increased parasympathetic activity and decreased sympathetic activity slow heart rate and cause widespread vasodilation. Cardiac output and peripheral resistance both decrease, triggering a precipitous drop in blood pressure. With insufficient blood to the brain, the individual faints.

Regulation of blood pressure in the cardiovascular system is closely tied to regulation of body fluid balance by the kidneys. Certain hormones secreted from the heart act on the kidneys, while hormones secreted from the kidneys act on the heart and blood vessels. Together, the heart and kidneys play a major role in maintaining homeostasis of body fluids, an excellent example of the integration of organ system function.

RUNNING PROBLEM

Another few weeks go by, and Kurt again returns to Dr. Cortez for a checkup. Kurt's blood pressure is finally closer to the normal range and has been averaging 135/87. "But, Doc, can you give me something for this dry, hacking cough I've been having? I don't feel bad, but it's driving me nuts." Dr. Cortez explains that a dry cough is an occasional side effect of taking ACE inhibitors. "It is more of a nuisance than anything else, but let's change your medicine. I'd like to try you on a calcium channel blocker instead of the ACE inhibitor."

Q4: *How do calcium channel blockers lower blood pressure?*

477 – 484 – 487 – 489 – **495** – 504

Concept Check

13. In the classic movie *Jurassic Park,* Dr. Ian Malcolm must flee from the *T. rex.* Draw a reflex map showing the cardiovascular response to his fight-or-flight situation. Remember that fight-or-flight causes epinephrine secretion as well as output from the cardiovascular control center. (*Hints:* What is the stimulus? Fear is integrated in the limbic system.)

15.6 Exchange at the Capillaries

The transport of materials around the body is only part of the function of the cardiovascular system. Once blood reaches the capillaries, the plasma and the cells exchange materials across the thin capillary walls. Most cells are located within 0.1 mm of the nearest capillary, and diffusion over this short distance proceeds rapidly.

The capillary density in any given tissue is directly related to the metabolic activity of the tissue's cells. Tissues with a higher metabolic rate require more oxygen and nutrients. Those tissues have more capillaries per unit area. Subcutaneous tissue and cartilage have the lowest capillary density. Muscles and glands have the highest. By one estimate, the adult human body has about 50,000 miles of capillaries, with a total exchange surface area of more than 6300 m^2, nearly the surface area of two football fields.

Capillaries have the thinnest walls of all the blood vessels, composed of a single layer of flattened endothelial cells supported on a basal lamina (Fig. 15.2). The diameter of a capillary is barely that of a red blood cell, forcing the RBCs to squeeze through in single file. Cell junctions between the endothelial cells vary from tissue to tissue and help determine the "leakiness" of the capillary.

The most common capillaries are **continuous capillaries,** whose endothelial cells are joined to one another with leaky junctions (**FIG. 15.16a**). These capillaries are found in muscle, connective tissue, and neural tissue. The continuous capillaries of the brain have evolved to form the *blood-brain barrier,* with tight junctions that help protect neural tissue from toxins that may be present in the bloodstream [p. 279].

Fenestrated capillaries {*fenestra,* window} have large pores (*fenestrae* or *fenestrations*) that allow high volumes of fluid to pass rapidly between the plasma and interstitial fluid (Fig. 15.16b). These capillaries are found primarily in the kidney and the intestine, where they are associated with absorptive transporting epithelia.

Three tissues—the bone marrow, the liver, and the spleen—do not have typical capillaries. Instead they have modified vessels called **sinusoids** that are as much as five times wider than a capillary. The sinusoid endothelium has fenestrations, and there may be gaps between the cells as well. Sinusoids are found in locations where blood cells and plasma proteins need to cross the endothelium to enter the blood. [Fig. 16.4c, *Focus On: Bone Marrow,* shows blood cells leaving the bone marrow by squeezing between endothelial cells.] In the liver, the sinusoidal endothelium lacks a basal lamina, which allows even more free exchange between plasma and interstitial fluid.

Colloid osmotic pressure is *not* equivalent to the total osmotic pressure in a capillary. It is simply a measure of the osmotic pressure created by proteins. Because the capillary endothelium is freely permeable to ions and other solutes in the plasma and interstitial fluid, these other solutes do not contribute to the osmotic gradient.

Colloid osmotic pressure is higher in the plasma ($\pi_{cap} = 25$ mm Hg) than in the interstitial fluid ($\pi_{IF} = 0$ mm Hg). Therefore, the osmotic gradient favors water movement by osmosis from the interstitial fluid into the plasma, represented by the red arrows in **FIGURE 15.18b**. For the purposes of our discussion, colloid osmotic pressure is constant along the length of the capillary, at $\pi = 25$ mm Hg.

Capillary hydrostatic pressure (P_H), by contrast, decreases along the length of the capillary as energy is lost to friction. Average values for capillary hydrostatic pressure, shown in Fig. 15.18b, are 32 mm Hg at the arterial end of a capillary and 15 mm Hg at the venous end. The hydrostatic pressure of the interstitial fluid P_{IF} is very low, and so we consider it to be essentially zero. This means that water movement due to hydrostatic pressure is directed out of the capillary, as denoted by the blue arrows in Fig. 15.18b, with the pressure gradient decreasing from the arterial end to the venous end.

If we assume that the interstitial hydrostatic and colloid osmotic pressures are zero, as discussed earlier, then the net pressure driving fluid flow across the capillary is determined by the difference between the hydrostatic pressure P_H and the colloid osmotic pressure (π):

$$\text{Net pressure} = P_H - \pi \quad (9)$$

FIG. 15.18 Capillary fluid exchange

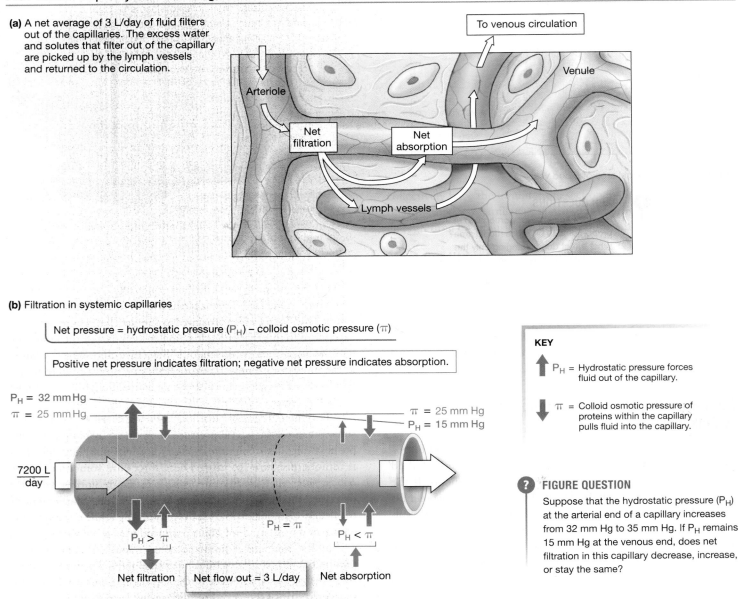

(a) A net average of 3 L/day of fluid filters out of the capillaries. The excess water and solutes that filter out of the capillary are picked up by the lymph vessels and returned to the circulation.

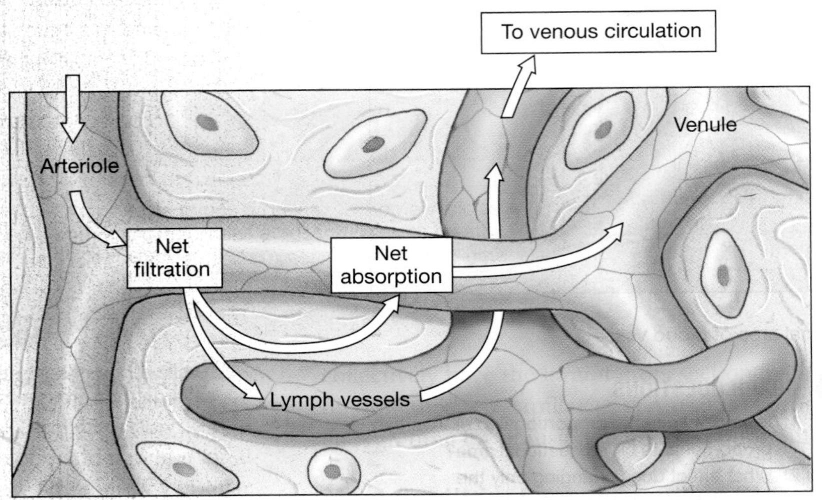

To venous circulation
Venule
Arteriole
Net filtration
Net absorption
Lymph vessels

(b) Filtration in systemic capillaries

Net pressure = hydrostatic pressure (P_H) – colloid osmotic pressure (π)

Positive net pressure indicates filtration; negative net pressure indicates absorption.

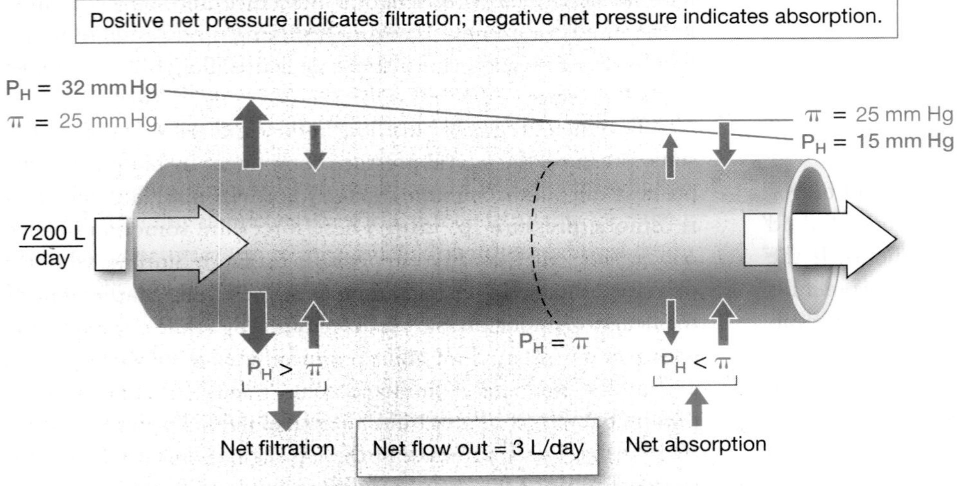

P_H = 32 mm Hg
π = 25 mm Hg
π = 25 mm Hg
P_H = 15 mm Hg
7200 L / day
$P_H > \pi$
$P_H = \pi$
$P_H < \pi$
Net filtration
Net flow out = 3 L/day
Net absorption

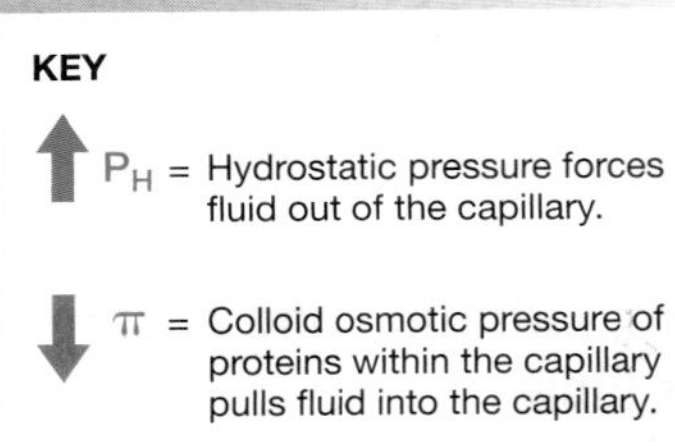

KEY
P_H = Hydrostatic pressure forces fluid out of the capillary.
π = Colloid osmotic pressure of proteins within the capillary pulls fluid into the capillary.

FIGURE QUESTION

Suppose that the hydrostatic pressure (P_H) at the arterial end of a capillary increases from 32 mm Hg to 35 mm Hg. If P_H remains 15 mm Hg at the venous end, does net filtration in this capillary decrease, increase, or stay the same?

A positive value for the net pressure indicates net filtration and a negative value indicates net absorption.

Using the hydrostatic and oncotic pressure values given in Fig. 15.18b, we can calculate the following values at the arterial end of a capillary:

$$\text{Net pressure} = P_H\ (32\text{ mm Hg}) - \pi\ (25\text{ mm Hg}) = 7\text{ mm Hg} \quad (10)$$

At the arterial end, P_H is greater than π, so the net pressure is 7 mm Hg of filtration pressure.

At the venous end, where capillary hydrostatic pressure is less:

$$\text{Net pressure}_{\text{venous end}} = (15\text{ mm Hg} - 25\text{ mm Hg}) = -10\text{ mm Hg} \quad (11)$$

At the venous end, π is greater than P_H. The net pressure is −10 mm Hg, favoring absorption. (A negative net pressure indicates absorption.)

Fluid movement down the length of a capillary is shown in Fig. 15.18b. There is net filtration at the arterial end and net absorption at the venous end. If the point at which filtration equals absorption occurred in the middle of the capillary, there would be no net movement of fluid. All volume that was filtered at the arterial end would be absorbed at the venous end. However, filtration is usually greater than absorption, resulting in bulk flow of fluid out of the capillary into the interstitial space.

By most estimates, that bulk flow amounts to about 3 liters per day, which is the equivalent of the entire plasma volume! If this filtered fluid could not be returned to the plasma, the blood would turn into a sludge of blood cells and proteins. Restoring fluid lost from the capillaries to the circulatory system is one of the functions of the lymphatic system, which we discuss next.

Concept Check

14. A man with liver disease loses the ability to synthesize plasma proteins. What happens to the colloid osmotic pressure of his blood? What happens to the balance between filtration and absorption in his capillaries?

15. Why did this discussion refer to the colloid osmotic pressure of the plasma rather than the osmolarity of the plasma?

15.7 The Lymphatic System

The vessels of the lymphatic system interact with three other physiological systems: the cardiovascular system, the digestive system, and the immune system. Functions of the lymphatic system include (1) returning fluid and proteins filtered out of the capillaries to the circulatory system, (2) picking up fat absorbed at the small intestine and transferring it to the circulatory system, and (3) serving as a filter to help capture and destroy foreign pathogens. In this discussion, we focus on the role of the lymphatic system in fluid transport.

The lymphatic system allows the one-way movement of interstitial fluid from the tissues into the circulation. Blind-end lymph vessels (*lymph capillaries*) lie close to all blood capillaries except those in the kidney and central nervous system (Fig. 15.18a). The smallest lymph vessels are composed of a single layer of flattened endothelium that is even thinner than the capillary endothelium.

The walls of these tiny lymph vessels are anchored to the surrounding connective tissue by fibers that hold the thin-walled vessels open. Large gaps between cells allow fluid, interstitial proteins, and particulate matter such as bacteria to be swept into the lymph vessels, also called *lymphatics*, by bulk flow. Once inside the lymphatics, this clear fluid is called simply **lymph.**

Lymph vessels in the tissues join one another to form larger lymphatic vessels that progressively increase in size (**FIG. 15.19**). These vessels have a system of semilunar valves, similar to valves in the venous circulation. The largest lymph ducts empty into the venous circulation just under the collarbones, where the left and right subclavian veins join the internal jugular veins. At intervals along the way, vessels enter **lymph nodes,** bean-shaped nodules of tissue with a fibrous outer capsule and an internal collection of immunologically active cells, including lymphocytes and macrophages.

The lymphatic system has no single pump like the heart. Lymph flow depends primarily on waves of contraction of smooth muscle in the walls of the larger lymph vessels. Flow is aided by contractile fibers in the endothelial cells, by the one-way valves, and by external compression created by skeletal muscles.

The skeletal muscle pump plays a significant role in lymph flow, as you know if you have ever injured a wrist or ankle. An immobilized limb frequently swells from the accumulation of fluid in the interstitial space, a condition known as **edema** {*oidema*, swelling}. Patients with edema in an injured limb are told to elevate the limb above the level of the heart so that gravity can assist lymph flow back to the blood.

An important reason for returning filtered fluid to the circulation is the recycling of plasma proteins. The body must maintain a low protein concentration in the interstitial fluid because colloid osmotic pressure is the only significant force that opposes capillary hydrostatic pressure. If proteins move from the plasma to the interstitial fluid, the osmotic pressure gradient that opposes filtration decreases. With less opposition to capillary hydrostatic pressure, additional fluid moves into the interstitial space.

Inflammation is an example of a situation in which the balance of colloid osmotic and hydrostatic pressures is disrupted. Histamine released in the inflammatory response makes capillary walls leakier and allows proteins to escape from the plasma into the interstitial fluid. The local swelling that accompanies a region of inflammation is an example of edema caused by redistribution of proteins from the plasma to the interstitial fluid.

CLINICAL FOCUS

Diabetes and Cardiovascular Disease

Having diabetes is one of the major risk factors for developing cardiovascular disease, and almost two-thirds of people with diabetes will die from cardiovascular problems. In diabetes, cells that cannot take up adequate glucose turn to fats and proteins for their energy. The body breaks down fat into fatty acids [p. 30] and dumps them into the blood. Plasma cholesterol levels are also elevated. When LDL-C remains in the blood, the excess is ingested by macrophages, starting a series of events that lead to atherosclerosis. Because of the pivotal role that LDL-C plays in atherosclerosis, many forms of therapy, ranging from dietary modification and exercise to drugs, are aimed at lowering LDL-C levels. Left untreated, blockage of small and medium-sized blood vessels in the lower extremities can lead to loss of sensation and *gangrene* (tissue death) in the feet. Atherosclerosis in larger vessels causes heart attacks and strokes. To learn more about diabetes and the increased risk of cardiovascular disease, visit the websites of the American Diabetes Association (*www.diabetes.org*) and the American Heart Association (*www.americanheart.org*).

Atherosclerosis Is an Inflammatory Process

Coronary heart disease accounts for the majority of cardiovascular disease deaths and is the single largest killer of Americans, both men and women. Let's look at the underlying cause of this disease: atherosclerosis.

The role of elevated blood cholesterol in the development of atherosclerosis is well established. Cholesterol, like other lipids, is not very soluble in aqueous solutions, such as the plasma. For this reason, when cholesterol in the diet is absorbed from the digestive tract, it combines with lipoproteins to make it more soluble. Clinicians generally are concerned with two of these lipoproteins: **high-density lipoprotein-cholesterol (HDL-C)** complexes and **low-density lipoprotein-cholesterol (LDL-C)** complexes. HDL-C is the more desirable form of blood cholesterol because high levels of HDL-C are associated with lower risk of heart attacks. (Memory aid: "H" in HDL stands for "healthy.")

LDL-C is sometimes called "bad" cholesterol because elevated plasma LDL-C levels are associated with coronary heart disease. (Remember this by associating "L" with "lethal.") Normal levels of LDL-C are not bad, however, because LDL is necessary for cholesterol transport into cells. LDL-C's binding site—a protein called **apoB**—combines with an LDL receptor found in clathrin-coated pits on the cell membrane. The receptor/LDL-C complex is then brought into the cell by endocytosis. The LDL receptor recycles to the cell membrane, and the endosome fuses with a lysosome. LDL-C's proteins are digested to amino acids, and the freed cholesterol is used to make cell membranes or steroid hormones.

Although LDL is needed for cellular uptake of cholesterol, excess levels of plasma LDL-C lead to atherosclerosis (**FIG. 15.21**). Endothelial cells lining the arteries transport LDL-C into the extracellular space so that it accumulates just under the intima **1**. There, white blood cells called *macrophages* ingest cholesterol and other lipids to become lipid-filled *foam cells* **2**. Cytokines released by the macrophages promote smooth muscle cell division **3**. This early-stage *lesion* {*laesio*, injury} is called a *fatty streak*.

As the condition progresses, the lipid core grows, and smooth muscle cells reproduce, forming bulging *plaques* that protrude into the lumen of the artery **4**. In the advanced stages of atherosclerosis, the plaques develop hard, calcified regions and fibrous collagen caps (**5** – **7**). The mechanism by which calcium carbonate is deposited is still being investigated.

Scientists once believed that the occlusion (blockage) of coronary blood vessels by large plaques that triggered blood clots was the primary cause of heart attacks, but that model has been revised. The new model indicates that blood clot formation on plaques is more dependent on the structure of a plaque than on its size. Atherosclerosis is now considered to be an inflammatory process in which macrophages release enzymes that convert stable plaques to vulnerable plaques **8**. *Stable plaques* have thick fibrous caps that separate the lipid core from the blood and do not activate platelets. *Vulnerable plaques* have thin fibrous caps that are more likely to rupture, exposing collagen and activating platelets that initiate a blood clot (*thrombus*) **9**.

If a clot blocks blood flow to the heart muscle, a heart attack, or *myocardial infarction*, results [see the Running Problem in Chapter 14]. Blocked blood flow in a coronary artery cuts off the oxygen supply to myocardial cells supplied by that artery. The oxygen-starved cells must then rely on anaerobic metabolism [p. 109], which produces lactate. As ATP production declines, the contractile cells are unable to pump Ca^{2+} out of the cell.

The unusually high Ca^{2+} concentration in the cytosol closes gap junctions in the damaged cells. Closure electrically isolates the damaged cells so that they no longer contract, and it forces action potentials to find an alternate route from cell to cell. If the damaged area of myocardium is large, the disruption can lead to an irregular heartbeat (*arrhythmia*) and potentially result in cardiac arrest or death.

Hypertension Represents a Failure of Homeostasis

One controllable risk factor for cardiovascular disease is hypertension—chronically elevated blood pressure, with systolic pressures greater than 140 mm Hg or diastolic pressures greater than 90 mm Hg. Hypertension is a common disease in the United States and is one of the most common reasons for visits to physicians and for the use of prescription drugs. High blood pressure is associated with increasing risk of CVD: the risk doubles for each 20/10 mm Hg increase in blood pressure over a baseline value of 115/75 (**FIG. 15.22**).

More than 90% of all patients with hypertension are considered to have *essential* (or *primary*) *hypertension,* with no clear-cut cause other than heredity. Cardiac output is usually normal in these people, and their elevated blood pressure appears to be associated with increased peripheral resistance. Some investigators have speculated that the increased resistance may be due to a lack of nitric oxide, the locally produced vasodilator formed by endothelial cells in the arterioles. In the remaining 5–10% of hypertensive cases, the cause is more apparent, and the hypertension is considered to

FIG. 15.21 The development of atherosclerotic plaques

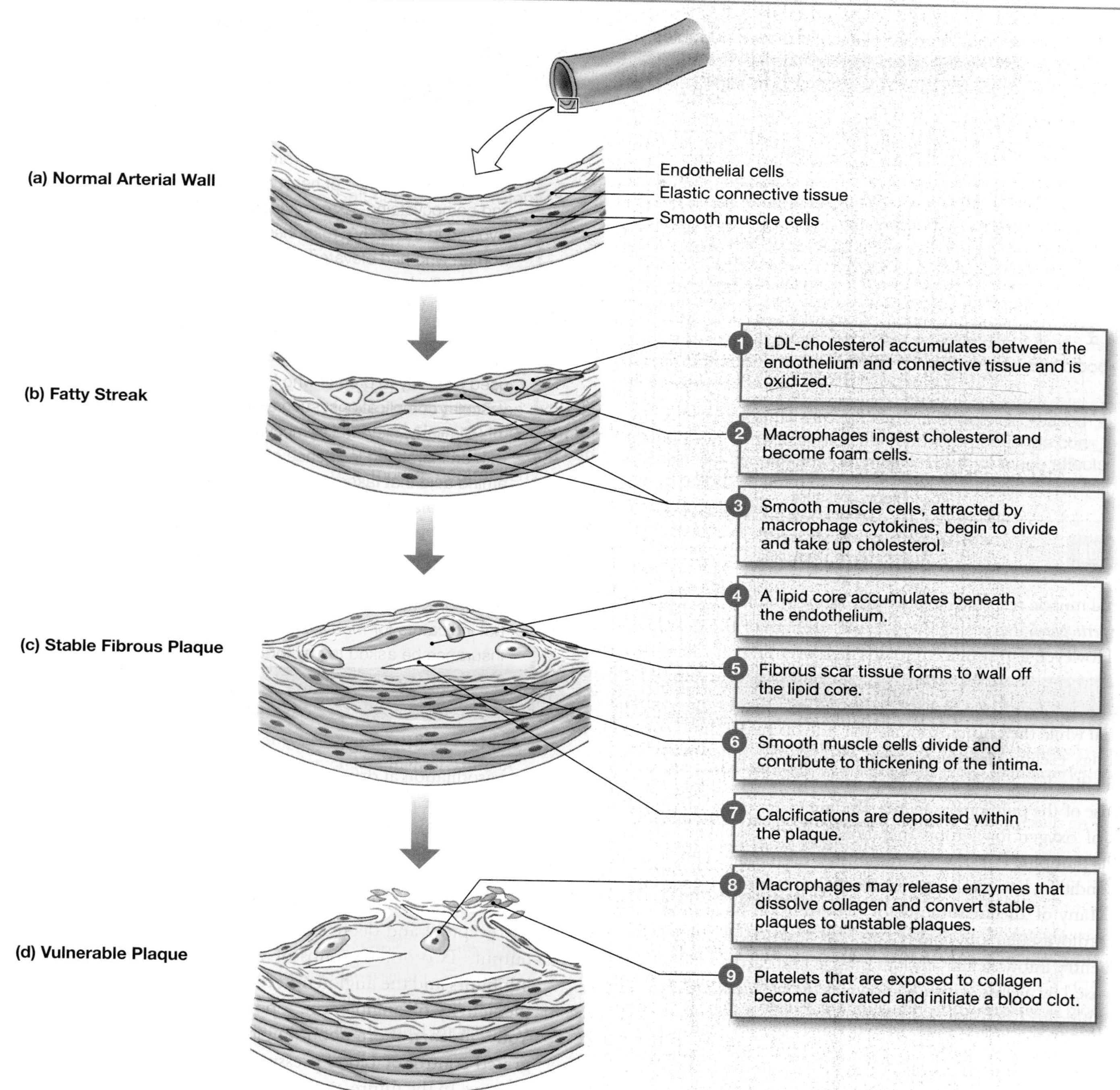

(a) Normal Arterial Wall
Endothelial cells
Elastic connective tissue
Smooth muscle cells
(b) Fatty Streak
1 LDL-cholesterol accumulates between the endothelium and connective tissue and is oxidized.
2 Macrophages ingest cholesterol and become foam cells.
3 Smooth muscle cells, attracted by macrophage cytokines, begin to divide and take up cholesterol.
(c) Stable Fibrous Plaque
4 A lipid core accumulates beneath the endothelium.
5 Fibrous scar tissue forms to wall off the lipid core.
6 Smooth muscle cells divide and contribute to thickening of the intima.
7 Calcifications are deposited within the plaque.
(d) Vulnerable Plaque
8 Macrophages may release enzymes that dissolve collagen and convert stable plaques to unstable plaques.
9 Platelets that are exposed to collagen become activated and initiate a blood clot.

be secondary to an underlying pathology. For instance, the cause might be an endocrine disorder that causes fluid retention.

A key feature of hypertension from all causes is adaptation of the carotid and aortic baroreceptors to higher pressure, with subsequent down-regulation of their activity. Without input from the baroreceptors, the cardiovascular control center interprets the high blood pressure as "normal," and no reflex reduction of pressure occurs.

Hypertension is a risk factor for atherosclerosis because high pressure in the arteries damages the endothelial lining of the vessels and promotes the formation of atherosclerotic plaques. In addition, high arterial blood pressure puts additional strain on the heart by increasing afterload [p. 470]. When resistance in the arterioles is high, the myocardium must work harder to push the blood into the arteries.

Amazingly, stroke volume in hypertensive patients remains constant up to a mean blood pressure of about 200 mm Hg, despite the increasing amount of work that the ventricle must perform as blood pressure increases. The cardiac muscle of the left ventricle responds to chronic high systemic resistance in the same way that

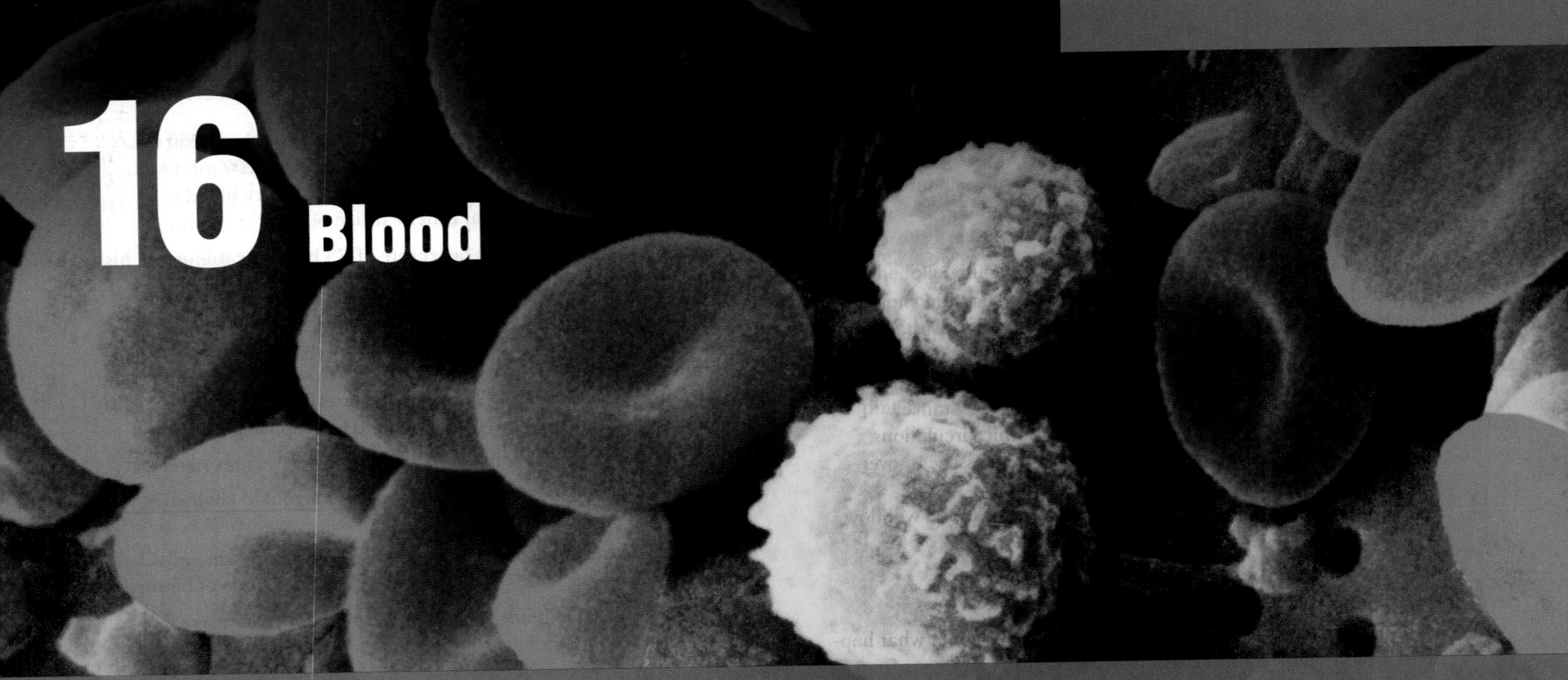

16 Blood

Who would have thought the old man to have had so much blood in him?

William Shakespeare, in *Macbeth*, V, i, 42

Red blood cells, white blood cells (yellow), and platelets (purple)

BACKGROUND BASICS

Blood, the fluid that circulates in the cardiovascular system, has occupied a prominent place throughout history as an almost mystical fluid. Humans undoubtedly had made the association between blood and life by the time they began to fashion tools and hunt animals. A wounded animal that lost blood would weaken and die if the blood loss was severe enough. The logical conclusion was that blood was necessary for existence. This observation eventually led to the term *lifeblood*, meaning anything essential for existence.

Ancient Chinese physicians linked blood to energy flow in the body. They wrote about the circulation of blood through the heart and blood vessels long before William Harvey described it in seventeenth-century Europe. In China, changes in blood flow were used as diagnostic clues to illness. Chinese physicians were expected to recognize some 50 variations in the pulse. Because blood was considered a vital fluid to be conserved and maintained, bleeding patients to cure disease was not a standard form of treatment.

In contrast, ancient Western civilizations came to believe that disease-causing evil spirits circulated in the blood. The way to remove these spirits was to remove the blood containing them. Because blood was recognized as an essential fluid, however, bloodletting had to be done judiciously. Veins were opened with knives or sharp instruments (*venesection*), or blood-sucking leeches were applied to the skin. In ancient India, people believed that leeches could distinguish between healthy and infected blood.

There is no written evidence that venesection was practiced in ancient Egypt, but the writings of Galen of Pergamum in the second century influenced Western medicine for nearly 2000 years. This early Greek physician advocated bleeding as treatment for many disorders. The location, timing, and frequency of the bleeding depended on the condition, and the physician was instructed to remove enough blood to bring the patient to the point of fainting. Over the years, this practice undoubtedly killed more people than it cured.

What is even more remarkable is the fact that as late as 1923, an American medical textbook advocated bleeding for treating certain infectious diseases, such as pneumonia! Now that we better understand the importance of blood in the immune response, it is doubtful that modern medicine will ever again turn to blood removal as a nonspecific means of treating disease. It is still used, however, for selected *hematological disorders* {*haima*, blood}.

RUNNING PROBLEM **Blood Doping in Athletes**

Athletes spend hundreds of hours training, trying to build their endurance. For Johann Muehlegg, a cross-country skier at the 2002 Salt Lake City Winter Olympics, it appeared that his training had paid off when he captured three gold medals. On the last day of the Games, however, Olympics officials expelled Muehlegg and stripped him of his gold medal in the 50-kilometer classical race. The reason? Muehlegg had tested positive for a performance-enhancing chemical that increased the oxygen-carrying capacity of his blood. Officials claimed Muehlegg's endurance in the grueling race was the result of blood doping, not training.

511 - 521 - 523 - 525 - 529

16.1 Plasma and the Cellular Elements of Blood

What is this remarkable fluid that flows through the circulatory system? Blood is a connective tissue composed of cellular elements suspended in an extensive fluid matrix called *plasma* [p. 82]. Plasma makes up one-fourth of the extracellular fluid, the internal environment that bathes cells and acts as a buffer between cells and the external environment. Blood is the circulating portion of the extracellular compartment, responsible for carrying material from one part of the body to another.

Total blood volume in a 70-kg man is equal to about 7% of his total body weight, or 0.07×70 kg $= 4.9$ kg. Thus, if we assume that 1 kg of blood occupies a volume of 1 liter, a 70-kg man has about 5 liters of blood. Of this volume, about 2 liters is composed of blood cells, while the remaining 3 liters is composed of plasma, the fluid portion of the blood. The 58-kg "Reference Woman" [p. 124] has about 4 liters total blood volume.

In this chapter, we present an overview of the components of blood and the functions of plasma, red blood cells, and platelets. You will learn more about hemoglobin when you study oxygen transport in the blood, and more about leukocytes and blood types when you study the immune system.

Plasma Is Extracellular Matrix

Plasma is the fluid matrix of the blood, within which cellular elements are suspended (**FIG. 16.1**). Water is the main component of plasma, accounting for about 92% of its weight. Proteins account for another 7%. The remaining 1% is dissolved organic molecules (amino acids, glucose, lipids, and nitrogenous wastes), ions (Na^+, K^+, Cl^-, H^+, Ca^{2+}, and HCO_3^-), trace elements and vitamins, and dissolved oxygen (O_2) and carbon dioxide (CO_2).

Plasma is identical in composition to interstitial fluid except for the presence of **plasma proteins. Albumins** are the most prevalent type of protein in the plasma, making up about 60% of the total. Albumins and nine other proteins—including *globulins*, the clotting protein *fibrinogen*, and the iron-transporting protein *transferrin*—make up more than 90% of all plasma proteins. The liver makes most plasma proteins and secretes them into the blood. Some globulins, known as *immunoglobulins* or *antibodies*, are synthesized and secreted by specialized blood cells rather than by the liver.

The presence of proteins in the plasma makes the osmotic pressure of the blood higher than that of the interstitial fluid. This osmotic gradient tends to pull water from the interstitial fluid into the capillaries and offset filtration out of the capillaries created by blood pressure [p. 497].

Plasma proteins participate in many functions, including blood clotting and defense against foreign invaders. In addition, they act as carriers for steroid hormones, cholesterol, drugs, and certain ions such as iron (Fe^{2+}). Finally, some plasma proteins act

CHAPTER 16

FIG. 16.1 Composition of blood

Blood consists of plasma and cellular elements.

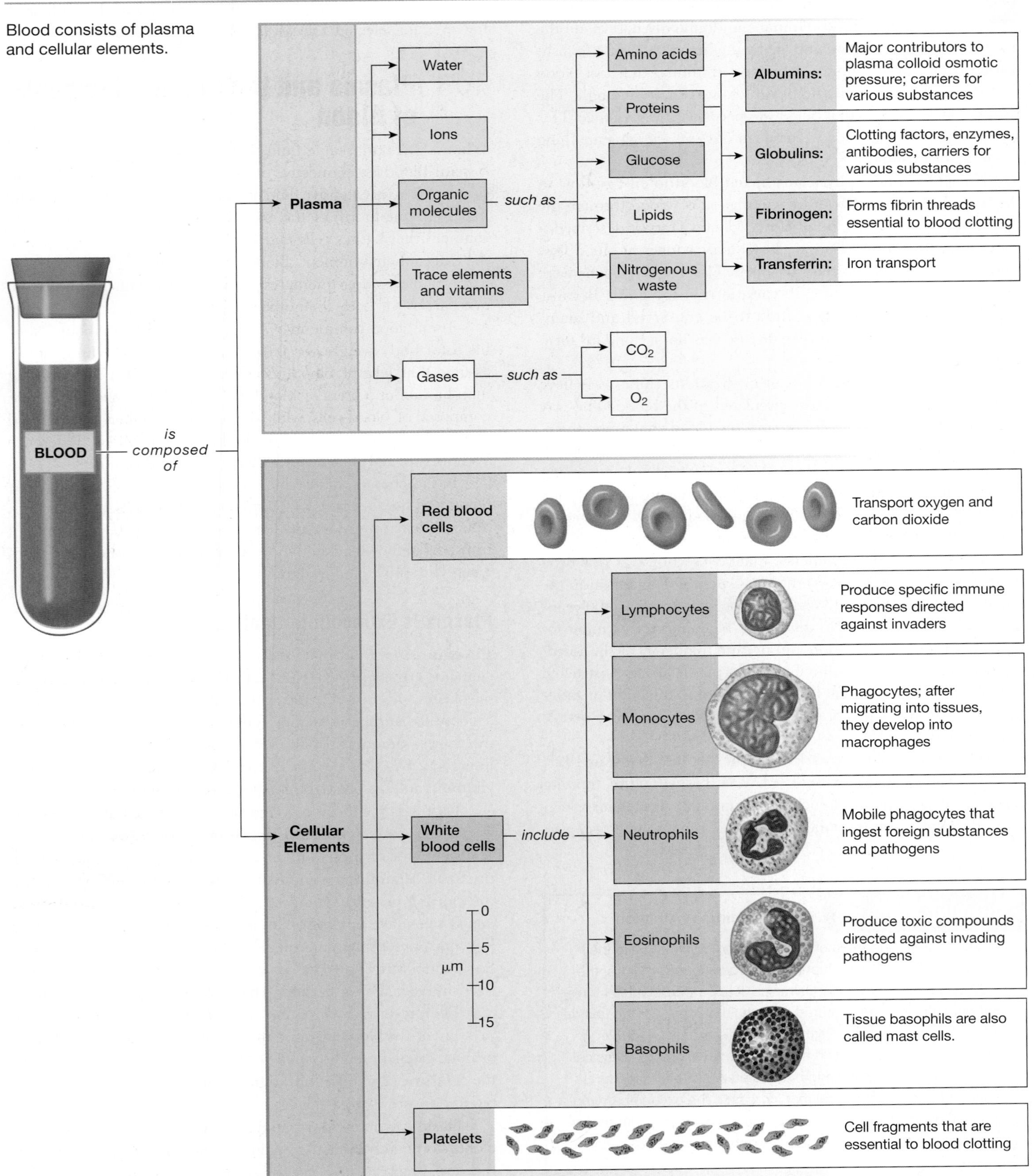

BLOOD
is composed of
Plasma
Water
Ions
Organic molecules
such as
Amino acids
Proteins
Glucose
Lipids
Nitrogenous waste
Albumins: Major contributors to plasma colloid osmotic pressure; carriers for various substances
Globulins: Clotting factors, enzymes, antibodies, carriers for various substances
Fibrinogen: Forms fibrin threads essential to blood clotting
Transferrin: Iron transport
Trace elements and vitamins
Gases
such as
CO_2
O_2
Cellular Elements
Red blood cells
Transport oxygen and carbon dioxide
White blood cells
include
Lymphocytes
Produce specific immune responses directed against invaders
Monocytes
Phagocytes; after migrating into tissues, they develop into macrophages
Neutrophils
Mobile phagocytes that ingest foreign substances and pathogens
Eosinophils
Produce toxic compounds directed against invading pathogens
Basophils
Tissue basophils are also called mast cells.
Platelets
Cell fragments that are essential to blood clotting
0
5
μm
10
15

as hormones or as extracellular enzymes. Functions of plasma proteins are summarized in Figure 16.1.

Cellular Elements Include RBCs, WBCs, and Platelets

Three main cellular elements are found in blood (Fig. 16.1): **red blood cells (RBCs),** also called **erythrocytes** {*erythros*, red}; **white blood cells (WBCs),** also called **leukocytes** {*leukos*, white}; and **platelets** or *thrombocytes* {*thrombo-*, lump, clot}. White blood cells are the only fully functional cells in the circulation. Red blood cells have lost their nuclei by the time they enter the bloodstream, and platelets, which also lack a nucleus, are cell fragments that have split off a relatively large parent cell known as a **megakaryocyte** {*mega*, extremely large + *karyon*, kernel + *-cyte*, cell}.

Red blood cells play a key role in transporting oxygen from lungs to tissues, and carbon dioxide from tissues to lungs. Platelets are instrumental in *coagulation*, the process by which blood clots prevent blood loss in damaged vessels. White blood cells play a key role in the body's immune responses, defending the body against foreign invaders, such as parasites, bacteria, and viruses. Most white blood cells circulate through the body in the blood, but their work is usually carried out in the tissues rather than in the circulatory system.

Blood contains five types of mature white blood cells: (1) **lymphocytes,** (2) **monocytes,** (3) **neutrophils,** (4) **eosinophils,** and (5) **basophils.** Monocytes that leave the circulation and enter the tissues develop into **macrophages.** Tissue basophils are called **mast cells.**

The types of white blood cells may be grouped according to common morphological or functional characteristics. Neutrophils, monocytes, and macrophages are collectively known as **phagocytes** because they can engulf and ingest foreign particles such as bacteria (phagocytosis) [p. 146]. Lymphocytes are sometimes called **immunocytes** because they are responsible for specific immune responses directed against invaders. Basophils, eosinophils, and neutrophils are called **granulocytes** because they contain cytoplasmic inclusions that give them a granular appearance.

Concept Check

1. Name the five types of leukocytes.
2. Why do we say that erythrocytes and platelets are not fully functional cells?
3. On the basis of what you have learned about the origin and role of plasma proteins, explain why patients with advanced liver degeneration frequently suffer from edema (p. 499).

16.2 Blood Cell Production

Where do these different blood cells come from? They are all descendants of a single precursor cell type known as the *pluripotent hematopoietic stem cell* (**FIG. 16.2**). This cell type is found primarily in **bone marrow,** a soft tissue that fills the hollow center of bones. Pluripotent stem cells have the remarkable ability to develop into many different cell types.

As they specialize, they narrow their possible fates. First, they become *uncommitted stem cells*, then *progenitor cells* that are committed to developing into one or perhaps two cell types. Progenitor cells differentiate into red blood cells, lymphocytes, other white blood cells, and megakaryocytes, the parent cells of platelets. It is estimated that only about one out of every 100,000 cells in the bone marrow is an uncommitted stem cell, making it difficult to isolate and study these cells.

In recent years, scientists have been working to isolate and grow uncommitted hematopoietic stem cells to use as replacements in patients whose own stem cells have been killed by cancer chemotherapy. Scientists obtain these stem cells from bone marrow or peripheral blood. Umbilical cord blood, collected at birth, has also been found to be a rich source of hematopoietic stem cells that can be used for transplants in patients with hematological diseases such as leukemia. Public and private cord blood banking programs are active in the United States and Europe, and the American National Marrow Donor Program Registry includes genetic marker information from banked cord blood to help patients find stem cell matches.

Blood Cells Are Produced in the Bone Marrow

Hematopoiesis {*haima*, blood + *poiesis*, formation}, the synthesis of blood cells, begins early in embryonic development and continues throughout a person's life. In about the third week of fetal development, specialized cells in the yolk sac of the embryo form clusters. Some of these cell clusters are destined to become the endothelial lining of blood vessels, while others become blood cells. The common embryological origin of the endothelium and blood cells perhaps explains why many cytokines that control hematopoiesis are released by the vascular endothelium.

As the embryo develops, blood cell production spreads from the yolk sac to the liver, spleen, and bone marrow. By birth, the liver and spleen no longer produce blood cells. Hematopoiesis continues in the marrow of all the bones of the skeleton until age five. As the child continues to age, the active regions of marrow decrease. In adults, the only areas producing blood cells are the pelvis, spine, ribs, cranium, and proximal ends of long bones.

Active bone marrow is red because it contains **hemoglobin,** the oxygen-binding protein of red blood cells. Inactive marrow is yellow because of an abundance of adipocytes (fat cells). (You can see the difference between red and yellow marrow the next time you look at bony cuts of meat in the grocery store.) Although blood synthesis in adults is limited, the liver, spleen, and inactive (yellow) regions of marrow can resume blood cell production in times of need.

In the regions of marrow that are actively producing blood cells, about 25% of the developing cells are red blood cells, while 75% are destined to become white blood cells. The life span of white blood cells is considerably shorter than that of red blood cells, and so WBCs must be replaced more frequently. For example, neutrophils have a 6-hour half-life, and the body must make more

FIG. 16.2 Hematopoiesis

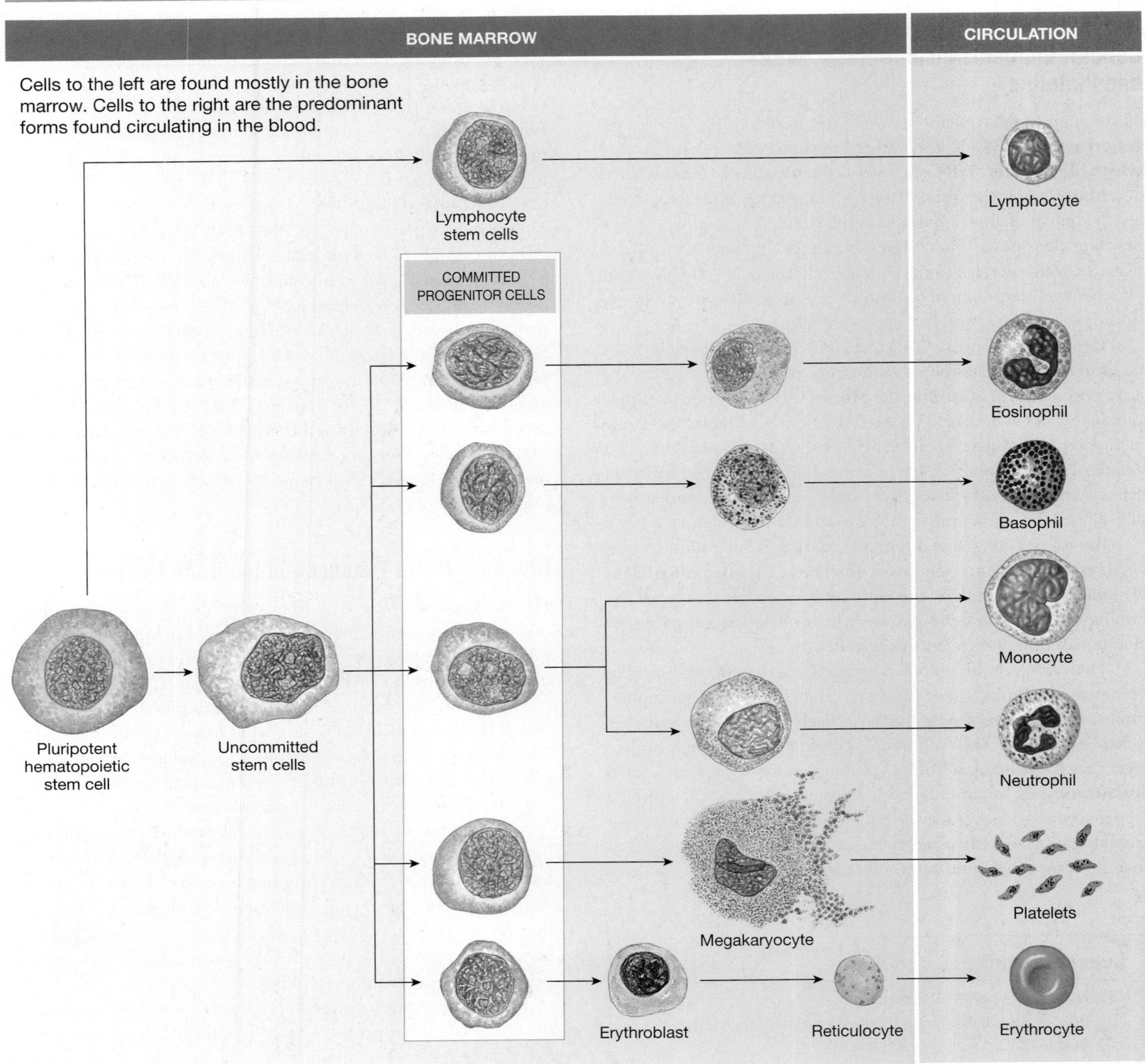
BONE MARROW
CIRCULATION
Cells to the left are found mostly in the bone marrow. Cells to the right are the predominant forms found circulating in the blood.
Lymphocyte stem cells
Lymphocyte
COMMITTED PROGENITOR CELLS
Eosinophil
Basophil
Monocyte
Neutrophil
Pluripotent hematopoietic stem cell
Uncommitted stem cells
Megakaryocyte
Platelets
Erythroblast
Reticulocyte
Erythrocyte

than *100 million* neutrophils each day to replace those that die. Red blood cells, on the other hand, live for nearly four months in the circulation.

Hematopoiesis Is Controlled by Cytokines

What controls the production and development of blood cells? The chemical factors known as cytokines are responsible. *Cytokines* are peptides or proteins released from one cell that affect the growth or activity of another cell [p. 167]. Newly discovered cytokines are often called *factors* and given a modifier that describes their actions: growth factor, differentiating factor, trophic (nourishing) factor.

Some of the best-known cytokines in hematopoiesis are the *colony-stimulating factors*, molecules made by endothelial cells and white blood cells. Others are the **interleukins** {*inter-*, between + *leuko*, white}, such as IL-3. The name *interleukin* was first given to cytokines released by one white blood cell to act on another white blood cell. Numbered interleukin names, such as interleukin-3, are given to cytokines once their amino acid sequences have been identified. Interleukins also play important roles in the immune system.

Another hematopoietic cytokine is *erythropoietin*, which controls red blood cell synthesis. Erythropoietin is usually called a hormone, but technically it fits the definition of a cytokine because it is made on demand rather than stored in vesicles like peptide hormones are.

TABLE 16.1 lists a few of the many cytokines linked to hematopoiesis. The role cytokines play in blood cell production is so complicated that one review on this topic was titled "Regulation of hematopoiesis in a sea of chemokine family members with a plethora of redundant activities"!* Because of the complexity of the subject, we give only an overview of the key hematopoietic cytokines.

Colony-Stimulating Factors Regulate Leukopoiesis

Colony-stimulating factors (CSFs) were identified and named for their ability to stimulate the growth of leukocyte colonies in culture. These cytokines, made by endothelial cells, marrow fibroblasts, and leukocytes, regulate leukocyte production and development, or **leukopoiesis.** CSFs induce both cell division (mitosis) and cell maturation in stem cells. Once a leukocyte matures, it loses its ability to undergo mitosis.

One fascinating aspect of leukopoiesis is that production of new leukocytes is regulated in part by existing white blood cells. This form of control allows leukocyte development to be very specific and tailored to the body's needs. When the body's defense system is called on to fight foreign invaders, both the absolute number of leukocytes and the relative proportions of the different types of leukocytes in the circulation change.

Clinicians often rely on a *differential white cell count* to help them arrive at a diagnosis (**FIG. 16.3**). For example, a person with a bacterial infection usually has a high total number of leukocytes in the blood, with an increased percentage that are neutrophils. Cytokines released by active leukocytes fighting the bacterial infection stimulate the production of additional neutrophils and monocytes. A person with a viral infection may have a high, normal, or low total white cell count but often shows an increase in the percentage of lymphocytes. The complex process by which leukocyte production is matched to need is still not completely understood and is an active area of research.

Scientists are working to create a model for the control of leukopoiesis so they can develop effective treatments for diseases characterized by either a lack or an excess of leukocytes. The *leukemias* are a group of diseases characterized by the abnormal growth and development of leukocytes. In *neutropenias* {*penia*, poverty}, patients have too few leukocytes and are unable to fight off bacterial and viral infections. Researchers hope to find better treatments for both leukemias and neutropenias by unlocking the secrets of how the body regulates cell growth and division.

*Broxmeyer H. E. and C. H. Kim, *Exp Hematol* 27(7): 1113–1123, 1999, July.

Thrombopoietin Regulates Platelet Production

Thrombopoietin (TPO) is a glycoprotein that regulates the growth and maturation of megakaryocytes, the parent cells of platelets. (Recall that *thrombocyte* is an alternative name for *platelet.*) TPO is produced primarily in the liver. This cytokine was first described in 1958, but its gene was not cloned until 1994. Within a year, genetically engineered TPO was widely available to researchers and physician-scientists hoping to use it to stimulate platelet production in patients with too few platelets, or *thrombocytopenia* {*penia*, poverty}. The first TPO drugs had to be recalled after patients developed adverse side effects but newer TPO agonists are currently in clinical use. Despite having these drugs, scientists still do not understand everything about the basic biology of thrombopoiesis, and research continues.

CHAPTER 16

Erythropoietin Regulates RBC Production

Red blood cell production (**erythropoiesis**) is controlled by the glycoprotein **erythropoietin (EPO),** assisted by several cytokines. Erythropoietin is made primarily in the kidneys of adults. The stimulus for EPO synthesis and release is *hypoxia*, low oxygen levels in the tissues. Hypoxia stimulates production of a transcription factor called *hypoxia-inducible factor 1* (HIF-1), which turns on the EPO gene to increase EPO synthesis. This pathway, like other endocrine pathways, helps the body maintain homeostasis. By stimulating the synthesis of red blood cells, EPO puts more hemoglobin into the circulation to carry oxygen.

The existence of a hormone controlling red blood cell production was first suggested in the 1950s, but two decades passed before scientists succeeded in purifying the substance. One reason for the delay is that EPO is made on demand and not stored, as classic peptide hormones are. It took scientists another nine years to identify the amino acid sequence of EPO and to isolate and clone the gene for it. However, an incredible leap was made after the EPO gene was isolated: only two years later, the hormone was produced by recombinant DNA technology and put into clinical use.

TABLE 16.1 Cytokines Involved in Hematopoiesis

Name	Sites of Production	Influences Growth or Differentiation of
Erythropoietin (EPO)	Kidney cells primarily	Red blood cells
Thrombopoietin (TPO)	Liver primarily	Megakaryocytes
Colony-stimulating factors, interleukins, stem cell factor	Endothelium and fibroblasts of bone marrow, leukocytes	All types of blood cells; mobilizes hematopoietic stem cells

17 Mechanics of Breathing

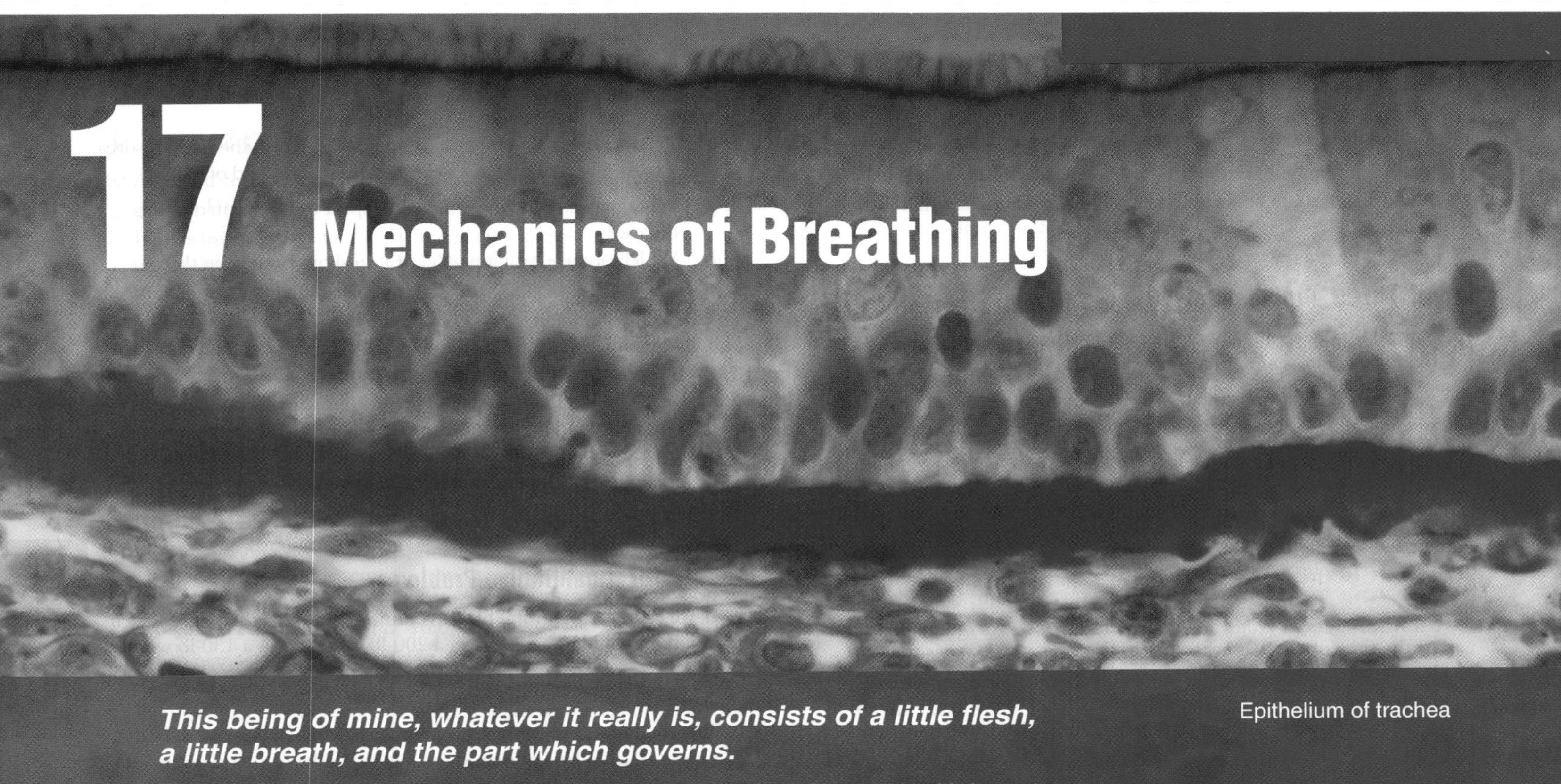

Epithelium of trachea

This being of mine, whatever it really is, consists of a little flesh, a little breath, and the part which governs.

Marcus Aurelius Antoninus (121–180 C.E.), Meditations *(c. 161–180 C.E.) New York: Dutton, 1937*

BACKGROUND BASICS

79 Ciliated and exchange epithelia
000 Pressure, volume, flow, and resistance
000 Pulmonary circulation
39 Surface tension
226 Autonomic and somatic motor neurons
439 Velocity of flow

Imagine covering the playing surface of a racquetball court (about 75 m^2) with thin plastic wrap, then crumpling up the wrap and stuffing it into a 3-liter soft drink bottle. Impossible? Maybe so, if you use plastic wrap and a drink bottle. But the lungs of a 70-kg man have a gas exchange surface the size of that plastic wrap, compressed into a volume that is less than that of the bottle. This tremendous surface area for gas exchange is needed to supply trillions of cells in the body with adequate amounts of oxygen.

Aerobic metabolism in cells depends on a steady supply of oxygen and nutrients from the environment, coupled with the removal of carbon dioxide. In very small aquatic animals, simple diffusion across the body surface meets these needs. Distance limits diffusion rate, however, so most multicelled animals require specialized respiratory organs associated with a circulatory system. Respiratory organs take a variety of forms, but all possess a large surface area compressed into a small space.

Besides needing a large exchange surface, humans and other terrestrial animals face an additional physiological challenge: dehydration. The exchange surface must be thin and moist to allow gases to pass from air into solution, and yet at the same time it must be protected from drying out as a result of exposure to air. Some terrestrial animals, such as the slug (a shell-less snail), meet the challenge of dehydration with behavioral adaptations that restrict them to humid environments and nighttime activities.

A more common solution is anatomical: an internalized respiratory epithelium. Human lungs are enclosed in the chest cavity to control their contact with the outside air. Internalization creates a humid environment for gas exchange with the blood and protects the delicate exchange surface from damage.

Internalized lungs create another challenge, however: how to move air between the atmosphere and an exchange surface deep within the body. Air flow requires a muscular pump to create pressure gradients. More complex respiratory systems therefore consist of two separate components: a muscle-driven pump and a thin, moist exchange surface. In humans, the pump is the musculoskeletal structure of the thorax. The lungs themselves consist of the exchange epithelium and associated blood vessels.

RUNNING PROBLEM **Emphysema**

You could hear the sound of her whistling, wheezing breathing preceding her down the hall. "Diagnosis: COPD," reads Edna Wilson's patient chart. COPD—chronic obstructive pulmonary disease—is the name given to diseases in which air exchange is impaired by narrowing of the lower airways. Most people with COPD have emphysema or chronic bronchitis or a combination of the two. Individuals in whom chronic bronchitis predominates are sometimes called "blue bloaters," owing to the bluish tinge of their skin (from low blood oxygen levels) and a tendency to be overweight. In contrast, patients with emphysema have been nicknamed "pink puffers." They tend to be thin, have normal (pink) skin coloration, and often breathe out through pursed lips, which helps open their airways. More than 15 million people in the United States have diagnosed COPD, and in 2011 COPD was the third leading cause of death. Most people can avoid the disease simply by not smoking cigarettes. Unfortunately, Edna has been a heavy smoker for 35 of her 47 years.

533 – 540 – 542 – 549 – 554 – 557

The four primary functions of the respiratory system are:

1. *Exchange of gases between the atmosphere and the blood.* The body brings in O_2 for distribution to the tissues and eliminates CO_2 waste produced by metabolism.
2. *Homeostatic regulation of body pH.* The lungs can alter body pH by selectively retaining or excreting CO_2.
3. *Protection from inhaled pathogens and irritating substances.* Like all other epithelia that contact the external environment, the respiratory epithelium is well supplied with defense mechanisms to trap and destroy potentially harmful substances before they can enter the body.
4. *Vocalization.* Air moving across the vocal cords creates vibrations used for speech, singing, and other forms of communication.

In addition to serving these functions, the respiratory system is also a significant source of water loss and heat loss from the body. These losses must be balanced using homeostatic compensations.

In this chapter, you will learn how the respiratory system carries out these functions by exchanging air between the environment and the interior air spaces of the lungs. This exchange is the *bulk flow* of air, and it follows many of the same principles that govern the bulk flow of blood through the cardiovascular system:

1. Flow takes place from regions of higher pressure to regions of lower pressure.
2. A muscular pump creates pressure gradients.
3. Resistance to air flow is influenced primarily by the diameter of the tubes through which the air is flowing.

Air and blood are both fluids. The primary difference between air flow in the respiratory system and blood flow in the circulatory system is that air is a less viscous, compressible mixture of gases while blood is a noncompressible liquid.

17.1 The Respiratory System

The word *respiration* has several meanings in physiology (**FIG. 17.1**). **Cellular respiration** refers to the intracellular reaction of oxygen with organic molecules to produce carbon dioxide, water, and energy in the form of ATP [p. 104]. **External respiration,** the topic of this chapter and the next, is the movement of gases between the environment and the body's cells. External respiration can be subdivided into four integrated processes, illustrated in Figure 17.1:

1. *The exchange of air between the atmosphere and the lungs.* This process is known as **ventilation,** or breathing. **Inspiration** (inhalation) is the movement of air into the lungs. **Expiration** (exhalation) is the movement of air out of the lungs. The mechanisms by which ventilation takes place are collectively called the *mechanics of breathing.*

FIG. 17.1 External respiration

The respiratory and circulatory systems coordinate the transfer of O_2 and CO_2 between the atmosphere and the cells.

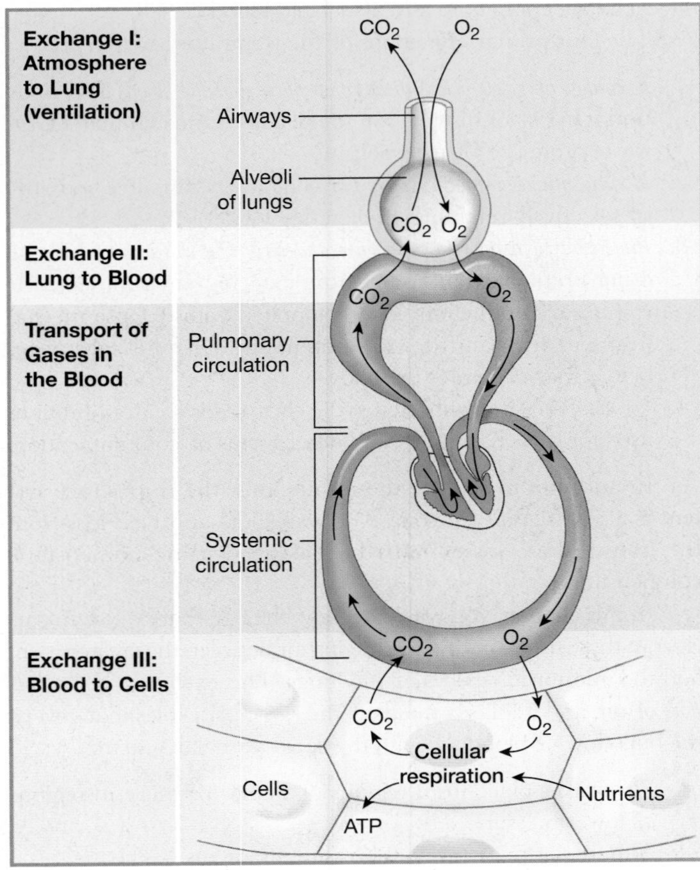
Exchange I: Atmosphere to Lung (ventilation)
Exchange II: Lung to Blood
Transport of Gases in the Blood
Exchange III: Blood to Cells
Airways
Alveoli of lungs
Pulmonary circulation
Systemic circulation
Cells
CO2
O2
Cellular respiration
Nutrients
ATP

2. *The exchange of O_2 and CO_2 between the lungs and the blood.*
3. *The transport of O_2 and CO_2 by the blood.*
4. *The exchange of gases between blood and the cells.*

External respiration requires coordination between the respiratory and cardiovascular systems. The **respiratory system** consists of structures involved in ventilation and gas exchange (**FIG. 17.2**):

1. The **conducting system** of passages, or **airways,** that lead from the external environment to the exchange surface of the lungs.
2. The **alveoli** (singular **alveolus**) {*alveus,* a concave vessel}, a series of interconnected sacs and their associated *pulmonary capillaries.* These structures form the exchange surface, where oxygen moves from inhaled air to the blood, and carbon dioxide moves from the blood to air that is about to be exhaled.
3. The bones and muscles of the thorax (chest cavity) and abdomen that assist in ventilation.

The respiratory system can be divided into two parts. The **upper respiratory tract** consists of the mouth, nasal cavity, pharynx, and larynx. The **lower respiratory tract** consists of the trachea, two primary bronchi {*bronchos,* windpipe; singular—bronchus}, their branches, and the lungs. The lower tract is also known as the *thoracic portion* of the respiratory system because it is enclosed in the thorax.

Bones and Muscles of the Thorax Surround the Lungs

The thorax is bounded by the bones of the spine and rib cage and their associated muscles. Together the bones and muscles are called the *thoracic cage.* The ribs and spine (the *chest wall*) form the sides and top of the cage. A dome-shaped sheet of skeletal muscle, the **diaphragm,** forms the floor (Fig. 17.2a).

Two sets of **intercostal muscles,** internal and external, connect the 12 pairs of ribs (Fig. 17.2c). Additional muscles, the **sternocleidomastoids** and the **scalenes,** run from the head and neck to the sternum and first two ribs.

Functionally, the thorax is a sealed container filled with three membranous bags, or sacs. One, the *pericardial sac,* contains the heart. The other two bags, the **pleural sacs,** each surround a lung {*pleura,* rib or side}. The esophagus and thoracic blood vessels and nerves pass between the pleural sacs (Fig. 17.2d).

Pleural Sacs Enclose the Lungs

The **lungs** (Fig. 17.2a, b) consist of light, spongy tissue whose volume is occupied mostly by air-filled spaces. These irregular cone-shaped organs nearly fill the thoracic cavity, with their bases resting on the curved diaphragm. Semi-rigid conducting airways—the bronchi—connect the lungs to the main airway, the trachea.

Each lung is surrounded by a double-walled pleural sac whose membranes line the inside of the thorax and cover the outer surface of the lungs (Fig. 17.2d, **FIG. 17.3**). Each *pleural membrane,* or **pleura,** contains several layers of elastic connective tissue and numerous capillaries. The opposing layers of pleural membrane are held together by a thin film of **pleural fluid** whose total volume is only about 25–30 mL in a 70-kg man. The result is similar to an air-filled balloon (the lung) surrounded by a water-filled balloon (the pleural sac). Most illustrations exaggerate the volume of the pleural fluid, but you can appreciate its thinness if you imagine spreading 25 mL of water evenly over the surface of a 3-liter soft drink bottle.

Pleural fluid serves several purposes. First, it creates a moist, slippery surface so that the opposing membranes can slide across one another as the lungs move within the thorax. Second, it holds the lungs tight against the thoracic wall. To visualize this arrangement, think of two panes of glass stuck together by a thin film of water. You can slide the panes back and forth across each other, but you cannot pull them apart because of the cohesiveness of the water [p. 39]. A similar fluid bond between the two pleural membranes makes the lungs "stick" to the thoracic cage and holds them stretched in a partially inflated state, even at rest.

FIG. 17.2 Anatomy Summary . . . The Lungs And Thoracic Cavity

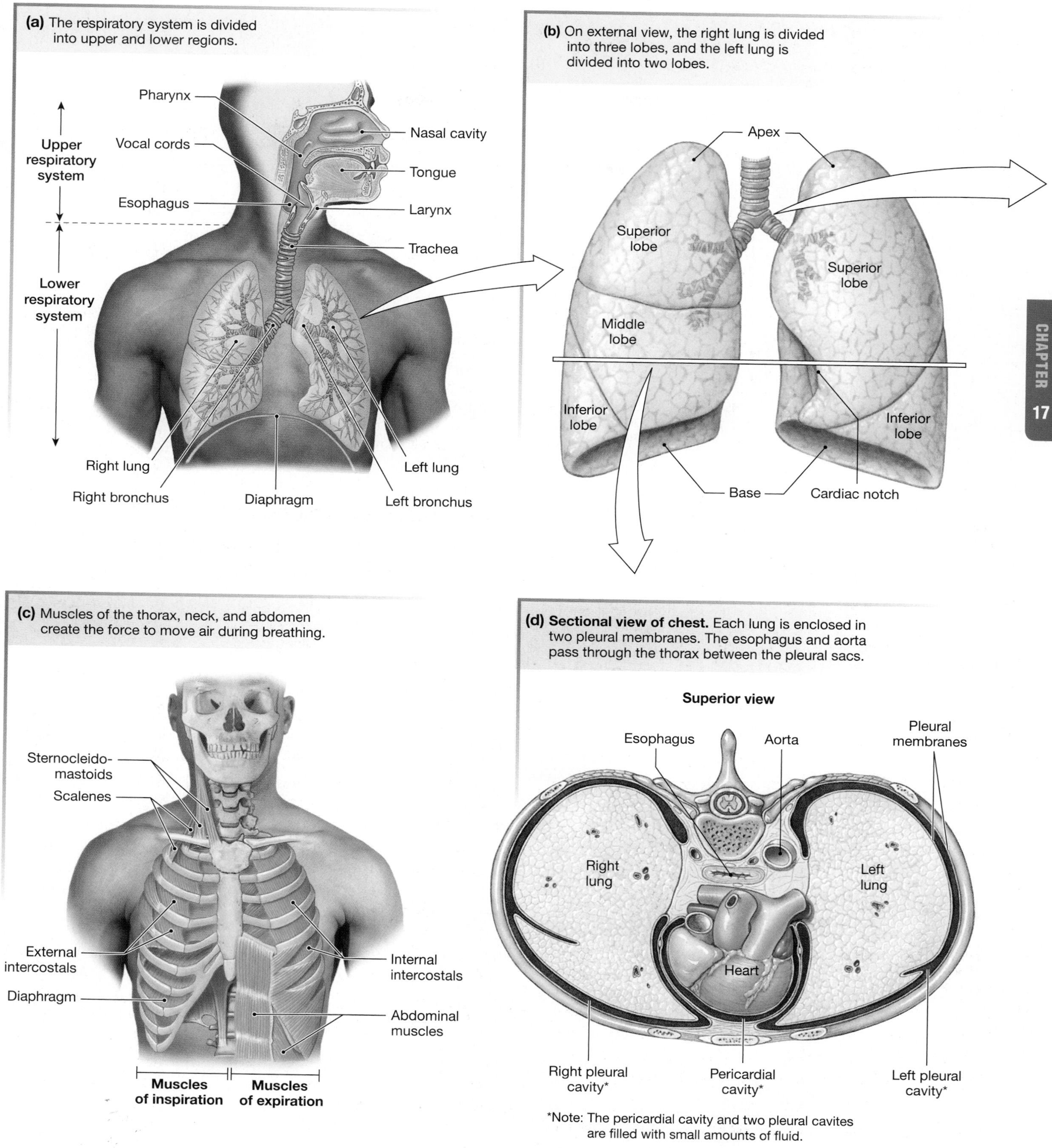

(a) The respiratory system is divided into upper and lower regions.

Pharynx
Nasal cavity
Upper respiratory system
Vocal cords
Tongue
Esophagus
Larynx
Trachea
Lower respiratory system
Right lung
Left lung
Right bronchus
Diaphragm
Left bronchus

(b) On external view, the right lung is divided into three lobes, and the left lung is divided into two lobes.

Apex
Superior lobe
Superior lobe
Middle lobe
Inferior lobe
Inferior lobe
Base
Cardiac notch

(c) Muscles of the thorax, neck, and abdomen create the force to move air during breathing.

Sternocleido-mastoids
Scalenes
External intercostals
Internal intercostals
Diaphragm
Abdominal muscles
Muscles of inspiration
Muscles of expiration

(d) Sectional view of chest. Each lung is enclosed in two pleural membranes. The esophagus and aorta pass through the thorax between the pleural sacs.

Superior view

Esophagus
Aorta
Pleural membranes
Right lung
Left lung
Heart
Right pleural cavity*
Pericardial cavity*
Left pleural cavity*

*Note: The pericardial cavity and two pleural cavites are filled with small amounts of fluid.

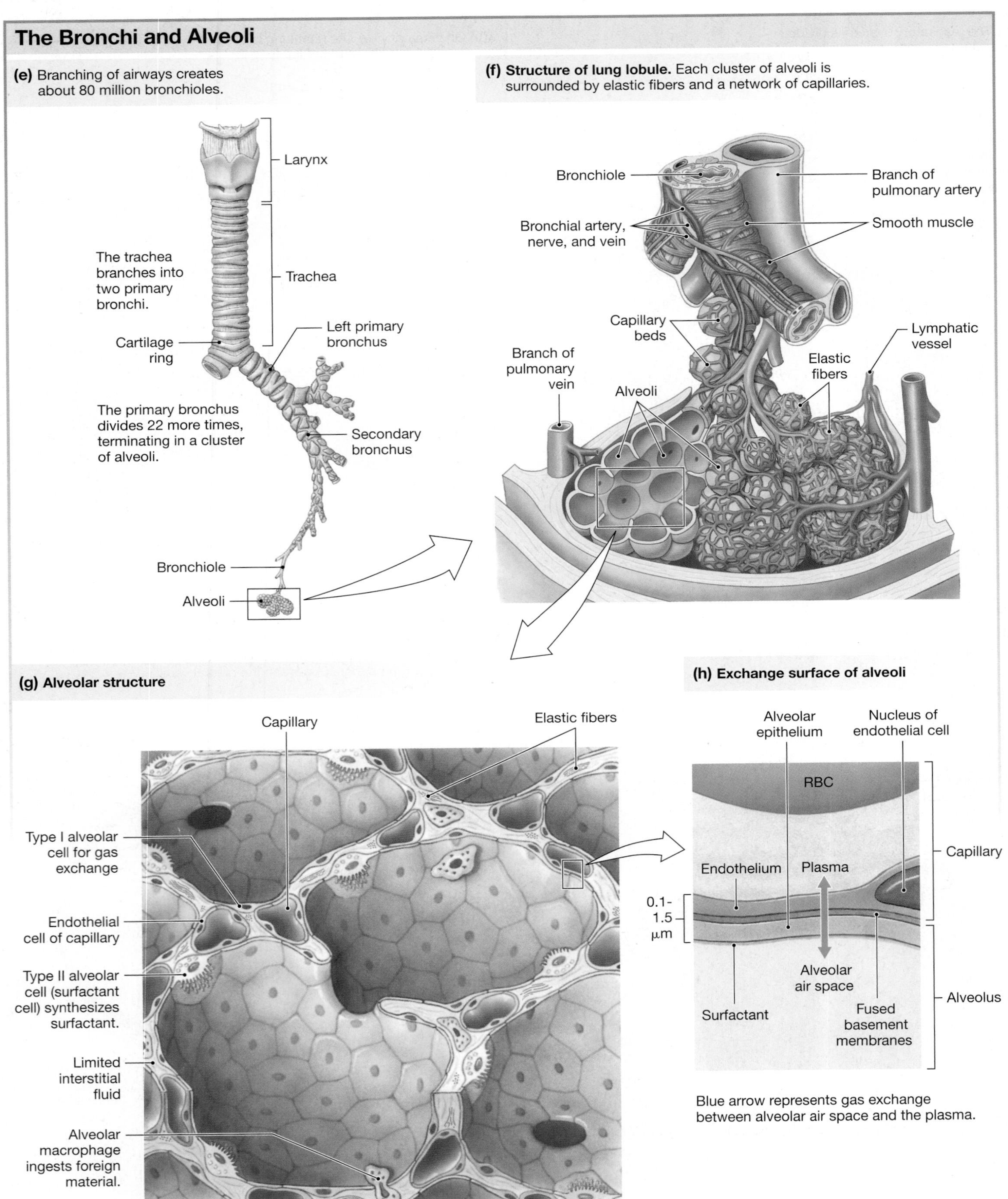
The Bronchi and Alveoli
(e) Branching of airways creates about 80 million bronchioles.
Larynx
Trachea
The trachea branches into two primary bronchi.
Cartilage ring
Left primary bronchus
The primary bronchus divides 22 more times, terminating in a cluster of alveoli.
Secondary bronchus
Bronchiole
Alveoli
(f) Structure of lung lobule. Each cluster of alveoli is surrounded by elastic fibers and a network of capillaries.
Bronchiole
Branch of pulmonary artery
Bronchial artery, nerve, and vein
Smooth muscle
Capillary beds
Lymphatic vessel
Branch of pulmonary vein
Elastic fibers
Alveoli
(g) Alveolar structure
Capillary
Elastic fibers
Type I alveolar cell for gas exchange
Endothelial cell of capillary
Type II alveolar cell (surfactant cell) synthesizes surfactant.
Limited interstitial fluid
Alveolar macrophage ingests foreign material.
(h) Exchange surface of alveoli
Alveolar epithelium
Nucleus of endothelial cell
RBC
Capillary
Endothelium
Plasma
0.1-1.5 µm
Alveolar air space
Surfactant
Fused basement membranes
Alveolus
Blue arrow represents gas exchange between alveolar air space and the plasma.

FIG. 17.3 The pleural sac

The pleural sac forms a double membrane surrounding the lung, similar to a fluid-filled balloon surrounding an air-filled balloon.

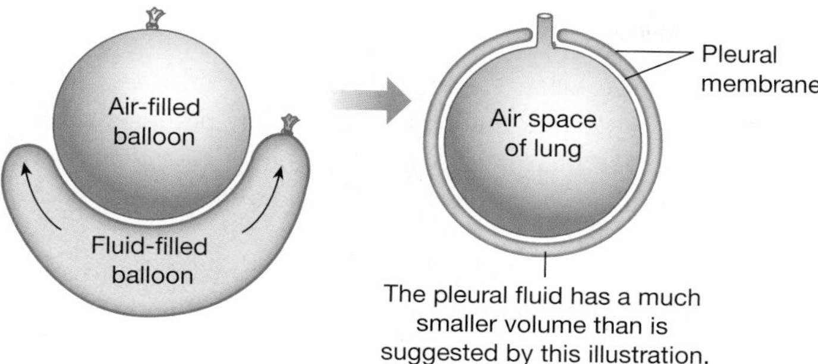

Air-filled balloon
Fluid-filled balloon
Air space of lung
Pleural membrane
The pleural fluid has a much smaller volume than is suggested by this illustration.

Airways Connect Lungs to the External Environment

Air enters the upper respiratory tract through the mouth and nose and passes into the **pharynx,** a common passageway for food, liquids, and air {*pharynx,* throat}. From the pharynx, air flows through the **larynx** into the **trachea,** or windpipe (Fig. 17.2a). The larynx contains the **vocal cords,** connective tissue bands that vibrate and tighten to create sound when air moves past them.

The trachea is a semiflexible tube held open by 15 to 20 C-shaped cartilage rings. It extends down into the thorax, where it branches (division 1) into a pair of **primary bronchi,** one *bronchus* to each lung (Fig. 17.2a). Within the lungs, the bronchi branch repeatedly (divisions 2–11) into progressively smaller bronchi (Fig. 17.2e). Like the trachea, the bronchi are semirigid tubes supported by cartilage.

Within the lungs, the smallest bronchi branch to become **bronchioles,** small collapsible passageways with walls of smooth muscle. The bronchioles continue branching (divisions 12–23) until the *respiratory bronchioles* form a transition between the airways and the exchange epithelium of the lung.

The diameter of the airways becomes progressively smaller from the trachea to the bronchioles, but as the individual airways get narrower, their numbers increase geometrically (**FIG. 17.4**). As a result, the total cross-sectional area increases with each division of the airways. Total cross-sectional area is lowest in the upper respiratory tract and greatest in the bronchioles, analogous to the increase in cross-sectional area that occurs from the aorta to the capillaries in the circulatory system [p. 496].

Velocity of air flow is inversely proportional to total cross-sectional area of the airways [p. 439]. This is similar to the velocity of blood flow through different parts of the circulatory system, and means that the velocity of air flow is greatest in the upper airways and slowest in the terminal bronchioles.

Concept Check

1. What is the difference between cellular respiration and external respiration?
2. Name the components of the upper respiratory tract and those of the lower respiratory tract.
3. Name the components (including muscles) of the thoracic cage. List the contents of the thorax.
4. Which air passages of the respiratory system are collapsible?

FIG. 17.4 Airway branching in the lower respiratory tract

Branching of the Airways

System Name	Name	Division	Diameter (mm)	How Many?	Cross-Sectional Area (cm^2)
Conducting system	Trachea	0	15–22	1	2.5
	Primary bronchi	1	10–15	2	↓
	Smaller bronchi ↓	2	1–10	4 ↓	
		3			
		4			
		5			
		6–11		1×10^4	
	Bronchioles	12–23	0.5–1	2×10^4 ↓	100 ↓
Exchange surface	Respiratory bronchioles			8×10^7	5×10^3
	Alveoli	24	0.3	$3–6 \times 10^8$	$>1 \times 10^6$

The Airways Warm, Humidify, and Filter Inspired Air

During breathing, the upper airways and the bronchi do more than simply serve as passageways for air. They play an important role in conditioning air before it reaches the alveoli. Conditioning has three components:

1. *Warming* air to body temperature (37 °C), so that core body temperature does not change and alveoli are not damaged by cold air;
2. *Adding water vapor* until the air reaches 100% humidity, so that the moist exchange epithelium does not dry out; and
3. *Filtering out foreign material,* so that viruses, bacteria, and inorganic particles do not reach the alveoli.

Inhaled air is warmed by the body's heat and moistened by water evaporating from the mucosal lining of the airways. Under normal circumstances, by the time air reaches the trachea, it has been conditioned to 100% humidity and 37 °C. Breathing through the mouth is not nearly as effective at warming and moistening air as breathing through the nose. If you exercise outdoors in very cold weather, you may be familiar with the ache in your chest that results from breathing cold air through your mouth.

Air is filtered both in the trachea and in the bronchi. These airways are lined with ciliated epithelium whose cilia are bathed in a watery saline layer (**FIG. 17.5**). The saline is produced by epithelial cells when Cl^- secreted into the lumen by apical anion channels draws Na^+ into the lumen through the paracellular pathway (Fig. 17.5c). Movement of solute from the ECF to the lumen creates an osmotic gradient, and water follows the ions into the airways. The *CFTR channel,* whose malfunction causes cystic fibrosis, is one of the anion channels found on the apical surface of this epithelium [p. 160].

A sticky layer of mucus floats over the cilia to trap most inhaled particles larger than 2 μm. The mucus layer is secreted by *goblet cells* in the epithelium (Fig. 17.5b). The cilia beat with an upward motion that moves the mucus continuously toward the pharynx, creating what is called the *mucociliary escalator.* Mucus contains *immunoglobulins* that can disable many pathogens. Once mucus reaches the pharynx, it can be spit out (*expectorated*) or swallowed. For swallowed mucus, stomach acid and enzymes destroy any remaining microorganisms.

Secretion of the watery saline layer beneath the mucus is essential for a functional mucociliary escalator. In the disease *cystic fibrosis,* for example, inadequate ion secretion decreases fluid movement in the airways. Without the saline layer, cilia become trapped in thick, sticky mucus and can no longer move. Mucus cannot be cleared, and bacteria colonize the airways, resulting in recurrent lung infections.

CLINICAL FOCUS

Congestive Heart Failure

When is a lung problem not a lung problem? The answer: when it's really a heart problem. *Congestive heart failure* (CHF) is an excellent example of the interrelationships among body systems and demonstrates how disruptions in one system can have a domino effect in the others. The primary symptoms of heart failure are shortness of breath (*dyspnea*), wheezing during breathing, and sometimes a productive cough that may be pinkish from the presence of blood. Congestive heart failure arises when the right heart is a more effective pump than the left heart. When blood accumulates in the pulmonary circulation, increased volume increases pulmonary blood pressure and capillary hydrostatic pressure. Capillary filtration exceeds the ability of the lymph system to drain interstitial fluid, resulting in pulmonary edema. Treatment of CHF includes increasing urinary output, which brings yet another organ system into the picture. By current estimates, about 5 million Americans suffer from CHF. To learn more about this condition, visit the American Heart Association web site (*www.americanheart.org*) or MedlinePlus, published by the National Institutes of Health (*www.nlm.nih.gov/medlineplus/heartfailure.html*).

Concept Check

5. Cigarette smoking paralyzes cilia in the airways and increases mucus production. Why would these effects cause smokers to develop a cough?

Alveoli Are the Site of Gas Exchange

The air-filled alveoli, clustered at the ends of terminal bronchioles, make up the bulk of lung tissue (Fig. 17.2f, g). Their primary function is the exchange of gases between themselves and the blood.

Each tiny alveolus is composed of a single layer of epithelium (Fig. 17.2g). Two types of epithelial cells are found in the alveoli. About 95% of the alveolar surface area is used for gas exchange and consists of the larger **type I alveolar cells.** These cells are very thin so that gases can diffuse rapidly through them (Fig. 17.2h). In much of the exchange area, a layer of basement membrane fuses the alveolar epithelium to the capillary endothelium. In the remaining area, only a small amount of interstitial fluid is present.

The smaller but thicker **type II alveolar cells** synthesize and secrete a chemical known as **surfactant.** Surfactant mixes with the thin fluid lining of the alveoli to aid lungs as they expand during breathing, as you will see later in this chapter. Type II cells also help minimize the amount of fluid present in the alveoli by transporting solutes, followed by water, out of the alveolar air space.

The thin walls of alveoli do not contain muscle because muscle fibers would block rapid gas exchange. As a result, lung tissue itself cannot contract. However, connective tissue between the alveolar epithelial cells contains many elastin and collagen fibers that create *elastic recoil* when lung tissue is stretched.

The close association of the alveoli with an extensive network of capillaries demonstrates the intimate link between the respiratory and cardiovascular systems. Blood vessels fill 80–90% of the space between alveoli, forming an almost continuous "sheet" of

FIG. 17.5 Airway epithelium

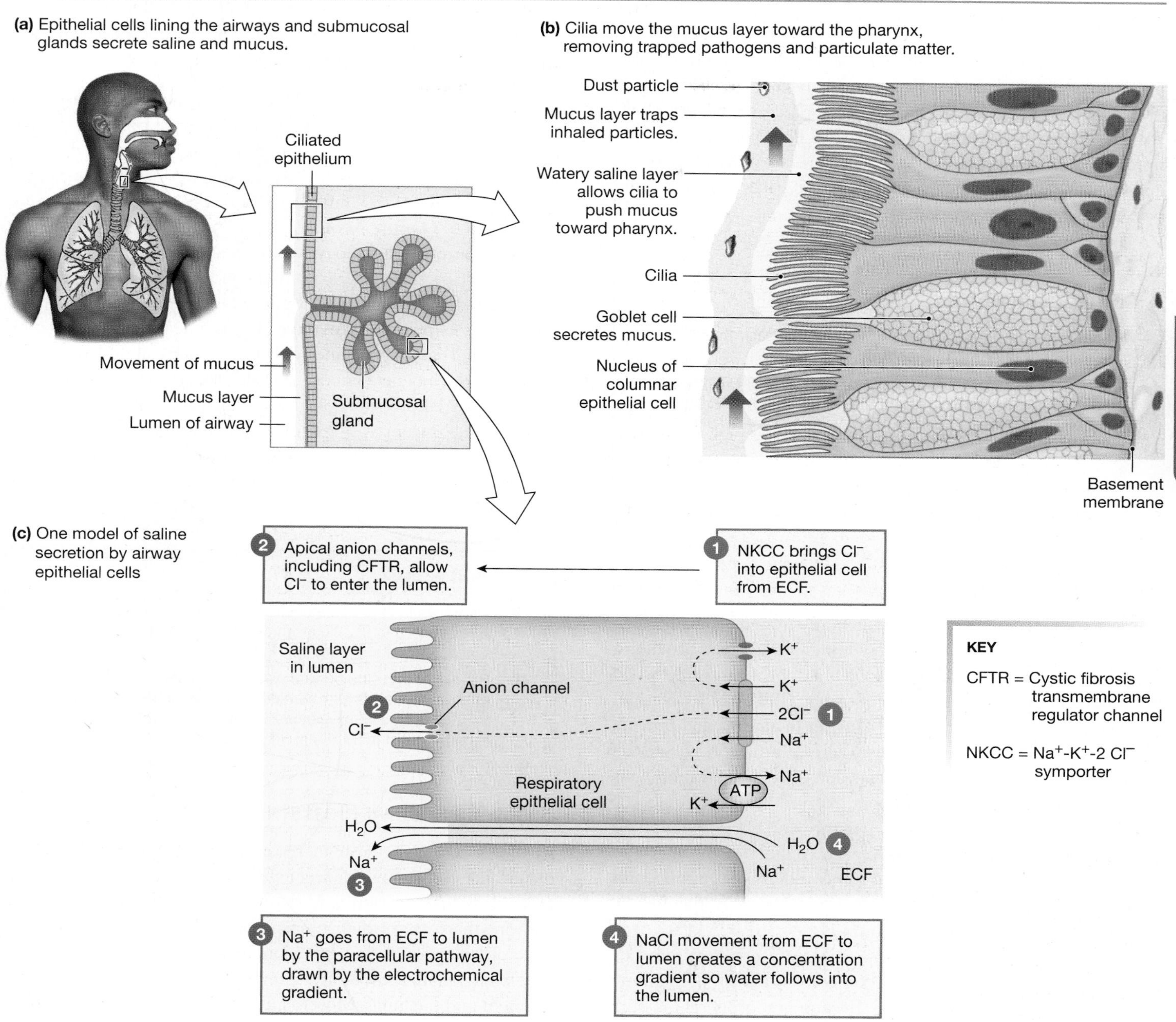
(a) Epithelial cells lining the airways and submucosal glands secrete saline and mucus.
Ciliated epithelium
Movement of mucus
Mucus layer
Lumen of airway
Submucosal gland
(b) Cilia move the mucus layer toward the pharynx, removing trapped pathogens and particulate matter.
Dust particle
Mucus layer traps inhaled particles.
Watery saline layer allows cilia to push mucus toward pharynx.
Cilia
Goblet cell secretes mucus.
Nucleus of columnar epithelial cell
Basement membrane
(c) One model of saline secretion by airway epithelial cells
2 Apical anion channels, including CFTR, allow Cl⁻ to enter the lumen.
1 NKCC brings Cl⁻ into epithelial cell from ECF.
Saline layer in lumen
Anion channel
K+
K+
2Cl⁻ 1
Na+
Na+
ATP
K+
Cl⁻
2
Respiratory epithelial cell
H2O
Na+
3
H2O 4
Na+
ECF
3 Na+ goes from ECF to lumen by the paracellular pathway, drawn by the electrochemical gradient.
4 NaCl movement from ECF to lumen creates a concentration gradient so water follows into the lumen.
KEY
CFTR = Cystic fibrosis transmembrane regulator channel
NKCC = Na+-K+-2 Cl⁻ symporter

blood in close contact with the air-filled alveoli. The proximity of capillary blood to alveolar air is essential for the rapid exchange of gases.

Pulmonary Circulation Is High-Flow, Low-Pressure

The pulmonary circulation begins with the pulmonary trunk, which receives low-oxygen blood from the right ventricle. The pulmonary trunk divides into two pulmonary arteries, one to each lung [Fig. 14.5, p. 442]. Oxygenated blood from the lungs returns to the left atrium via the pulmonary veins.

At any given moment, the pulmonary circulation contains about 0.5 liter of blood, or 10% of total blood volume. About 75 mL of this amount is found in the capillaries, where gas exchange takes place, with the remainder in pulmonary arteries and veins. The rate of blood flow through the lungs is much higher than the rate in other tissues [p. 491] because the lungs receive the entire cardiac output of the right ventricle: 5 L/min. This means that as much blood flows through the lungs in 1 minute as flows through the rest of the body in the same amount of time!

Despite the high flow rate, pulmonary blood pressure is low. Pulmonary arterial pressure averages 25/8 mm Hg, much lower than the average systemic pressure of 120/80 mm Hg. The right ventricle does not have to pump as forcefully to create blood flow through the lungs because resistance of the pulmonary circulation is low. This low resistance can be attributed to the shorter total length of pulmonary blood vessels and to the distensibility and large total cross-sectional area of pulmonary arterioles.

Normally, the net hydrostatic pressure filtering fluid out of a pulmonary capillary into the interstitial space is low because of low mean blood pressure [p. 498]. The lymphatic system efficiently removes filtered fluid, and lung interstitial fluid volume is usually minimal. As a result, the distance between the alveolar air space and the capillary endothelium is short, and gases diffuse rapidly between them.

Concept Check

6. Is blood flow through the pulmonary trunk greater than, less than, or equal to blood flow through the aorta?
7. A person has left ventricular failure but normal right ventricular function. As a result, blood pools in the pulmonary circulation, doubling pulmonary capillary hydrostatic pressure. What happens to net fluid flow across the walls of the pulmonary capillaries?
8. Calculate the mean pressure in a person whose pulmonary arterial pressure is 25/8 mm Hg. [p. 482]

17.2 Gas Laws

Respiratory air flow is very similar in many respects to blood flow in the cardiovascular system because both air and blood are fluids. Their primary difference is that blood is a noncompressible liquid but air is a compressible mixture of gases. **FIGURE 17.6** describes the laws that govern the behavior of gases in air and that provide the basis for the exchange of air between the atmosphere and the alveoli. We will consider the gas laws that govern the solubility of gases in solution when we talk about oxygen transport in blood.

In this book, we report blood pressure and environmental air pressure (**atmospheric pressure**) in millimeters of mercury (mm Hg). Respiratory physiologists sometimes report gas pressures in centimeters of water instead, where 1 mm Hg = 1.36 cm H_2O, or in kiloPascals (kPa), where 760 mm Hg = 101.325 kPa.

At sea level, normal atmospheric pressure is 760 mm Hg. However, in this book we follow the convention of designating atmospheric pressure as 0 mm Hg. Because atmospheric pressure varies with altitude and because very few people live exactly at sea level, this convention allows us to compare pressure differences that occur during ventilation without correcting for altitude.

RUNNING PROBLEM

Edna has not been able to stop smoking, and her COPD is a combination of emphysema and bronchitis. Patients with chronic bronchitis have excessive mucus production and exhibit general inflammation of the entire respiratory tract. The mucus narrows the airways and makes breathing difficult.

Q1: *What does narrowing of the airways do to airway resistance?* [Hint: *Poiseuille's law, p. 438*]

533 – 540 – 542 – 549 – 554 – 557

Air Is a Mixture of Gases

The atmosphere surrounding the earth is a mixture of gases and water vapor. **Dalton's law** states that the total pressure exerted by a mixture of gases is the sum of the pressures exerted by the individual gases (Fig. 17.6c). For example, in dry air at an atmospheric pressure of 760 mm Hg, 78% of the total pressure is due to N_2, 21% to O_2, and so on.

In respiratory physiology, we are concerned not only with total atmospheric pressure but also with the individual pressures of oxygen and carbon dioxide. The pressure of a single gas in a mixture is known as its **partial pressure (P_{gas})**. The pressure exerted by an individual gas is determined only by its relative abundance in the mixture and is independent of the molecular size or mass of the gas.

The partial pressures of gases in air vary slightly depending on how much water vapor is in the air because the pressure of water vapor "dilutes" the contribution of other gases to the total pressure. The table in Figure 17.6c compares the partial pressures of some gases in dry air and at 100% humidity.

Concept Check

9. If nitrogen is 78% of atmospheric air, what is the partial pressure of nitrogen (P_{N2}) in a sample of dry air that has an atmospheric pressure of 720 mm Hg?
10. The partial pressure of water vapor in inspired air is 47 mm Hg when inhaled air is fully humidified. If atmospheric pressure is 700 mm Hg and oxygen is 21% of the atmosphere at 0% humidity, what is the P_{O2} of fully humidified air?

Gases Move Down Pressure Gradients

Air flow occurs whenever there is a pressure gradient. Bulk flow of air, like blood flow, is directed from areas of higher pressure to areas of lower pressure. Meteorologists predict the weather by knowing that areas of high atmospheric pressure move in to replace areas of low pressure. In ventilation, bulk flow of air down pressure gradients explains how air is exchanged between the external environment and the lungs. Movement of the thorax during breathing creates alternating conditions of high and low pressure in the lungs.

Diffusion of gases down concentration (partial pressure) gradients applies to single gases. For example, oxygen moves from areas of higher oxygen partial pressure (P_{O_2}) to areas of lower oxygen partial pressure. Diffusion of individual gases is important in the exchange of oxygen and carbon dioxide between alveoli and blood and from blood to cells, as you will learn later.

Boyle's Law Describes Pressure-Volume Relationships

The pressure exerted by a gas or mixture of gases in a sealed container is created by the collisions of moving gas molecules with the walls of the container and with each other. If the size of the

FIG. 17.6 ESSENTIALS: Gas Laws

This figure summarizes the rules that govern the behavior of gases in air. These rules provide the basis for the exchange of air between the external environment and the alveoli.

(a) The Ideal Gas Equation

$PV = nRT$ Where P is pressure, V is volume, n is the moles of gas, T is absolute temperature, and R is the universal gas constant, 8.3145 j/mol × K

In the human body, we can assume that the number of moles and temperature are constant. Removing the constants leaves the following equation:

$V = 1/P$ This relationship says that if the volume of gas increases, the pressure decreases, and vice versa.

(b) Boyle's Law

Boyle's Law also expresses this inverse relationship between pressure and volume.

$P_1V_1 = P_2V_2$ For example, the container on the left is 1 L (V_1) and has a pressure of 100 mm Hg (P_1).

What happens to the pressure when the volume decreases to 0.5 L?

$$100 \text{ mm Hg} \times 1 \text{ L} = P_2 \times 0.5 \text{ L}$$

$$200 \text{ mm Hg} = P_2$$

The pressure has increased ×2.

The Ideal Gas Law and Boyle's Law apply to all gases or mixtures of gases.

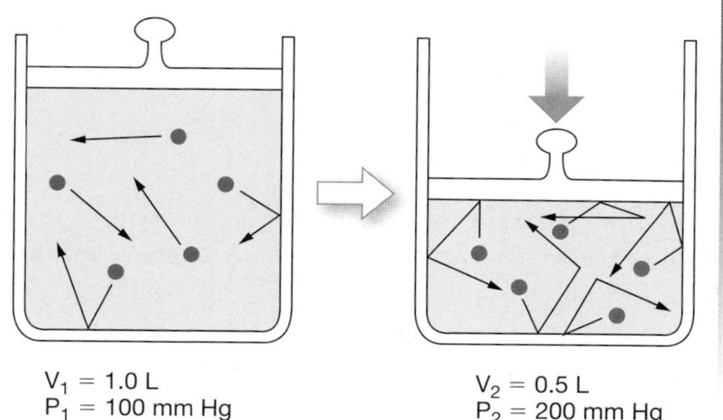

$V_1 = 1.0$ L
$P_1 = 100$ mm Hg
$V_2 = 0.5$ L
$P_2 = 200$ mm Hg

(c) Dalton's Law

Dalton's Law says that the total pressure of a mixture of gases is the sum of the pressures of the individual gases. The pressure of an individual gas in a mixture is known as the **partial pressure** of the gas (P_{gas}).

For example, at sea level, atmospheric pressure (P_{atm}) is 760 mm Hg, and oxygen is 21% of the atmosphere. What is the partial pressure of oxygen (P_{O_2})?

To find the partial pressure of any one gas in a sample of dry air, multiply the atmospheric pressure (P_{atm}) by the gas's relative contribution (%) to P_{atm}:

Partial pressure of a gas = P_{atm} × % of gas in atmosphere

$$P_{O_2} = 760 \text{ mm Hg} \times 21\% \text{ oxygen}$$

$$= 760 \text{ mm} \times 0.21 = 160 \text{ mm Hg}$$

The partial pressure of oxygen (P_{O_2}) in dry air at sea level is 160 mm Hg.

The pressure exerted by an individual gas is determined only by its relative abundance in the mixture and is independent of the molecular size or mass of the gas.

In humid air, water vapor "dilutes" the contribution of other gases to the total pressure.

Partial Pressures (P_{gas}) of Atmospheric Gases at 760 mm Hg

Gas and its percentage in air	P_{gas} in dry 25 °C air	P_{gas} in 25 °C air, 100% humidity	P_{gas} in 37 °C air, 100% humidity
O_2 21%	160 mm Hg	155 mm Hg	150 mm Hg
CO_2 0.03%	0.25 mm Hg	0.24 mm Hg	0.235 mm Hg
Water vapor	0 mm Hg	24 mm Hg	47 mm Hg

To calculate the partial pressure of a gas in humid air, you must first subtract the water vapor pressure from the total pressure. At 100% humidity and 25 °C, water vapor pressure (P_{H_2O}) is 24 mm Hg.

P_{gas} in humid air = (P_{atm} – P_{H_2O}) × % of gas

$$P_{O_2} = (760 - 24) \times 21\% = 155 \text{ mm Hg}$$

container is reduced, the collisions between the gas molecules and the walls become more frequent, and the pressure rises (Fig. 17.6b). This relationship between pressure and volume was first noted by Robert Boyle in the 1600s and can be expressed by the equation of **Boyle's law** of gases:

$$P_1V_1 = P_2V_2$$

where P represents pressure and V represents volume.

Boyle's law states that if the volume of a gas is reduced, the pressure increases. If the volume increases, the pressure decreases.

In the respiratory system, changes in the volume of the chest cavity during ventilation cause pressure gradients that create air flow. When chest volume increases, alveolar pressure falls, and air flows into the respiratory system. When the chest volume decreases, alveolar pressure increases, and air flows out into the atmosphere. This movement of air is bulk flow because the entire gas mixture is moving rather than merely one or two of the gases in the air.

The important gas laws for respiratory physiology are summarized in **TABLE 17.1**.

RUNNING PROBLEM

Edna's COPD began with chronic bronchitis and a morning cough that produced lots of mucus (*phlegm*). Cigarette smoke paralyzes the cilia that sweep debris and mucus out of the airways, and smoke irritation increases mucus production in the airway. Without functional cilia, mucus and debris pool in the airways, leading to a chronic cough. Eventually, smokers may begin to develop emphysema in addition to their bronchitis.

Q2: *Why do people with chronic bronchitis have a higher-than-normal rate of respiratory infections?*

533
540
542
549
554
557

17.3 Ventilation

The bulk flow exchange of air between the atmosphere and the alveoli is *ventilation*, or *breathing* (Fig. 17.1). A single **respiratory cycle** consists of an inspiration followed by an expiration.

Lung Volumes Change during Ventilation

Physiologists and clinicians assess a person's pulmonary function by measuring how much air the person moves during quiet breathing, then with maximum effort. These **pulmonary function tests** use a **spirometer,** an instrument that measures the volume of air moved with each breath (**FIG. 17.7a**). Most spirometers in clinical use today are small computerized machines rather than the traditional wet spirometer illustrated here. The wet spirometer, however, is easier to understand.

When a subject is attached to the traditional spirometer through a mouthpiece and the subject's nose is clipped closed, the subject's respiratory tract and the spirometer form a closed system. When the subject breathes in, air moves from the spirometer into the lungs, and the recording pen, which traces a graph on a rotating cylinder, moves up. When the subject exhales, air moves from the lungs back into the spirometer, and the pen moves down.

Lung Volumes The air moved during breathing can be divided into four lung volumes: (1) tidal volume, (2) inspiratory reserve volume, (3) expiratory reserve volume, and (4) residual volume. The numerical values used on the graph in Figure 17.7b represent average volumes for a 70-kg man. The volumes for women are typically less, as shown in Figure 17.7b. Lung volumes vary considerably with age, sex, height, and weight, so clinicians use algorithms based on those parameters to predict lung volumes. (An *algorithm* is an equation or series of steps used to solve a problem.) See question 36 at the end of this chapter for one of these algorithms.

Each of the following paragraphs begins with the instructions you would be given if you were being tested for these volumes.

"Breathe quietly." The volume of air that moves during a single inspiration or expiration is known as the **tidal volume (V_T).** Average tidal volume during quiet breathing is about 500 mL. (It is hard for subjects to breathe normally when they are thinking about their breathing, so the clinician may not give this instruction.)

"Now, at the end of a quiet inspiration, take in as much additional air as you possibly can." The additional volume you inspire above the tidal volume represents your **inspiratory reserve volume (IRV).** In a 70-kg man, this volume is about 3000 mL, a 6fold increase over the normal tidal volume.

"Now stop at the end of a normal exhalation, then exhale as much air as you possibly can." The amount of air forcefully exhaled after the end of a normal expiration is the **expiratory reserve volume (ERV),** which averages about 1100 mL.

The fourth volume cannot be measured directly. Even if you blow out as much air as you can, air still remains in the lungs

TABLE 17.1 Gas Laws

1. The total pressure of a mixture of gases is the sum of the pressures of the individual gases (Dalton's law).
2. To find the partial pressure of a gas (P_{gas}) in dry air, multiply the gas's contribution (%) times the atmospheric pressure (P_{atm}). $P_{gas} = P_{atm} \times \%$ gas
3. To calculate the partial pressure of a gas in humid air, you must first subtract water vapor pressure (P_{H_2O}) from the total pressure (P_{atm}). $P_{gas} = (P_{atm} - P_{H_2O}) \times \%$ gas
4. Gases, singly or in a mixture, move from areas of higher pressure to areas of lower pressure.
5. If the volume of a container of gas changes, the pressure of the gas will change in an inverse manner: $P_1V_1 = P_2V_2$ (Boyle's law).

FIG. 17.7 Pulmonary function tests

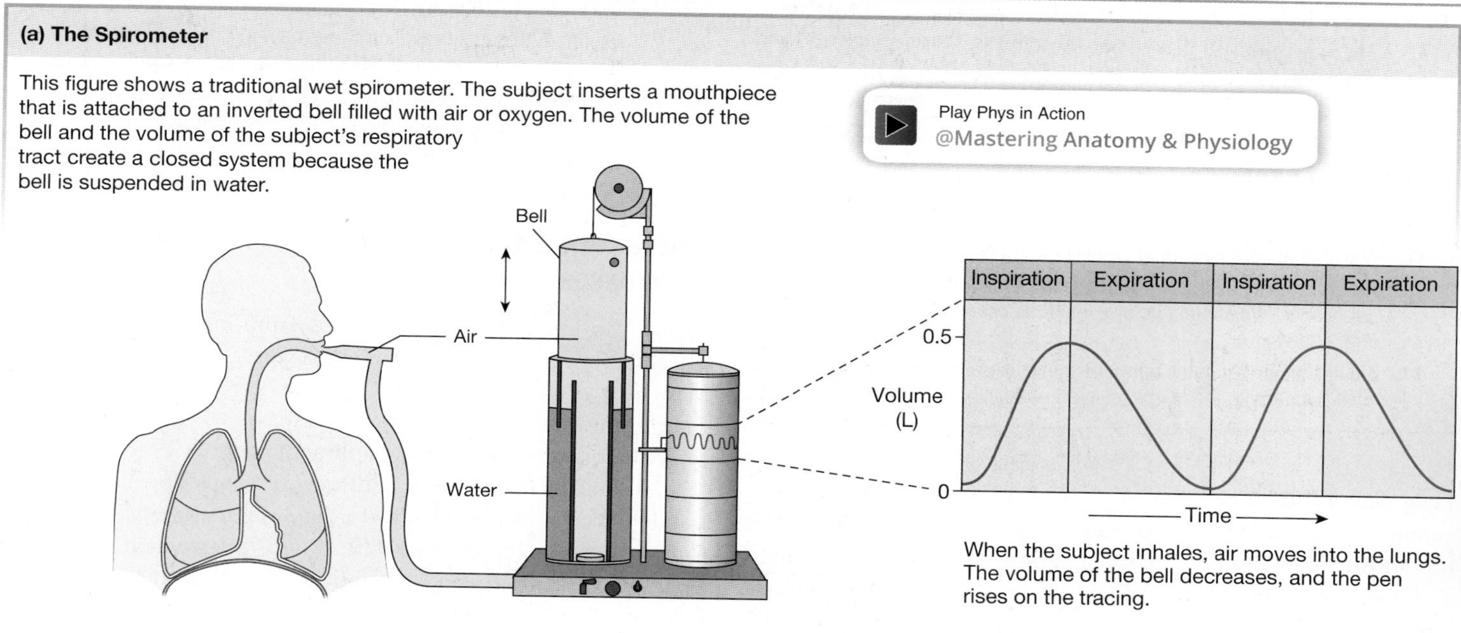
(a) The Spirometer
This figure shows a traditional wet spirometer. The subject inserts a mouthpiece that is attached to an inverted bell filled with air or oxygen. The volume of the bell and the volume of the subject's respiratory tract create a closed system because the bell is suspended in water.
Play Phys in Action
@Mastering Anatomy & Physiology
Bell
Air
Water
Inspiration
Expiration
Inspiration
Expiration
0.5
Volume (L)
0
Time
When the subject inhales, air moves into the lungs. The volume of the bell decreases, and the pen rises on the tracing.

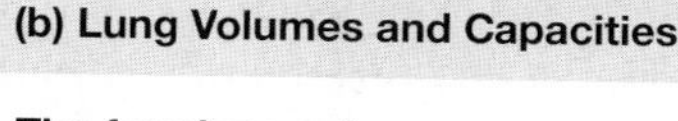

(b) Lung Volumes and Capacities

The four lung volumes:

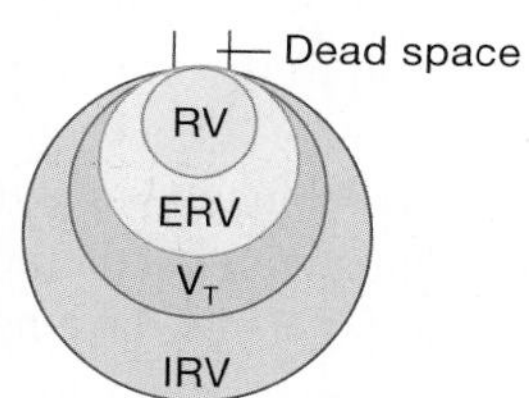
Dead space
RV
ERV
V_T
IRV

KEY

RV = Residual volume
ERV = Expiratory reserve volume
V_T = Tidal volume
IRV = Inspiratory reserve volume

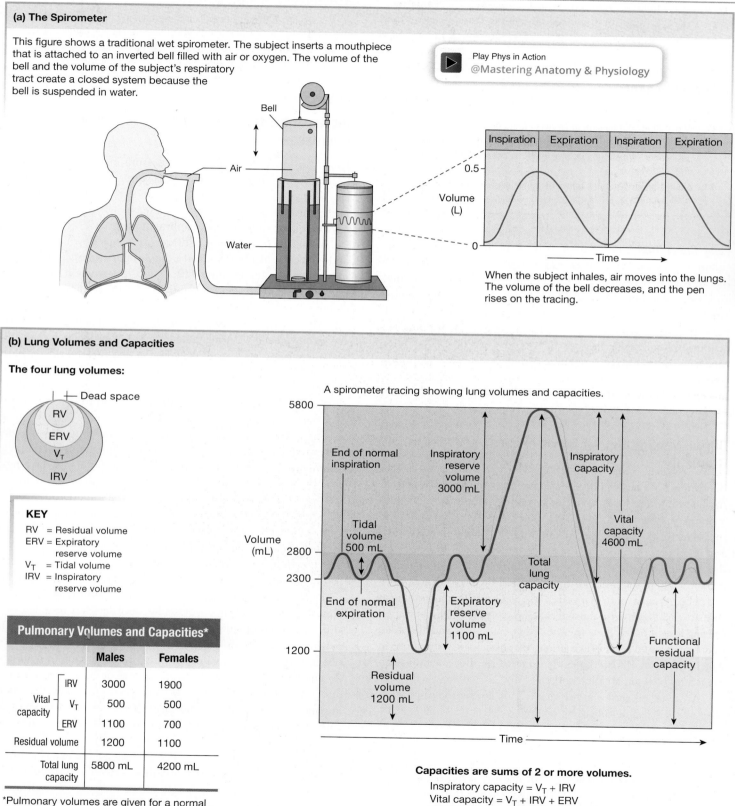

Pulmonary Volumes and Capacities*

		Males	Females
Vital capacity	IRV	3000	1900
	V_T	500	500
	ERV	1100	700
Residual volume		1200	1100
Total lung capacity		5800 mL	4200 mL

*Pulmonary volumes are given for a normal 70-kg man or a 50-kg woman, 28 years old.

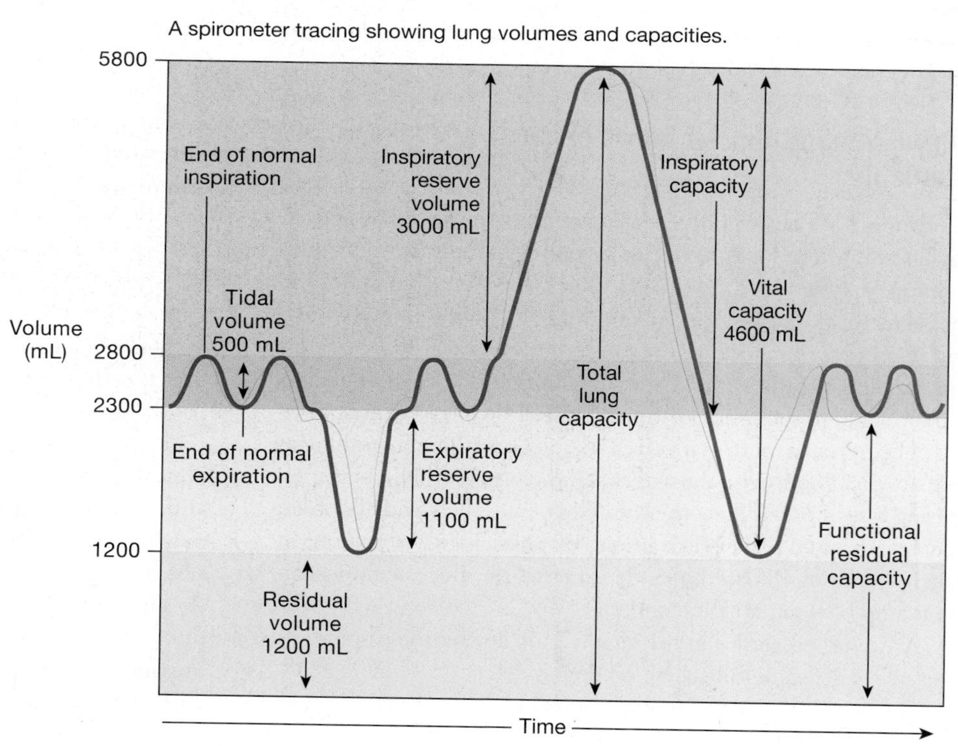
A spirometer tracing showing lung volumes and capacities.
5800
Volume (mL)
2800
2300
1200
End of normal inspiration
Tidal volume 500 mL
End of normal expiration
Inspiratory reserve volume 3000 mL
Expiratory reserve volume 1100 mL
Residual volume 1200 mL
Total lung capacity
Inspiratory capacity
Vital capacity 4600 mL
Functional residual capacity
Time

Capacities are sums of 2 or more volumes.

Inspiratory capacity = V_T + IRV
Vital capacity = V_T + IRV + ERV
Total lung capacity = V_T + IRV + ERV + RV
Functional residual capacity = ERV + RV

CHAPTER 17

and the airways. The volume of air in the respiratory system after maximal exhalation—about 1200 mL—is called the **residual volume (RV).** Most of this residual volume exists because the lungs are held stretched against the ribs by the pleural fluid.

Concept Check

11. How are lung volumes related to lung capacities?
12. Which lung volume cannot be measured directly?
13. If vital capacity decreases with age but total lung capacity does not change, which lung volume must be changing? In which direction?
14. As inhaled air becomes humidified passing down the airways, what happens to the P_{O_2} of the air?

Lung Capacities The sum of two or more lung volumes is called a *capacity*. The **vital capacity (VC)** is the sum of the inspiratory reserve volume, expiratory reserve volume, and tidal volume. Vital capacity represents the maximum amount of air that can be voluntarily moved into or out of the respiratory system with one breath. Vital capacity plus the residual volume yields the **total lung capacity (TLC)**. Other capacities of importance in pulmonary medicine include the **inspiratory capacity** (tidal volume + inspiratory reserve volume) and the **functional residual capacity** (expiratory reserve volume + residual volume).

During Ventilation, Air Flows because of Pressure Gradients

Breathing is an active process that requires muscle contraction. Air flows into the lungs because of pressure gradients created by a pump, just as blood flows because of the pumping action of the heart. In the respiratory system, muscles of the thoracic cage and diaphragm function as the pump because most lung tissue is thin exchange epithelium. When these muscles contract, the lungs expand, held to the inside of the chest wall by the pleural fluid.

The primary muscles involved in quiet breathing (breathing at rest) are the diaphragm, the external intercostals, and the scalenes. During forced breathing, other muscles of the chest and abdomen may be recruited to assist. Examples of physiological situations in which breathing is forced include exercise, playing a wind instrument, and blowing up a balloon.

As we noted earlier in the chapter, air flow in the respiratory tract obeys the same rule as blood flow:

$$\text{Flow} \propto \Delta P/R$$

This equation means that (1) air flows in response to a pressure gradient (ΔP) and (2) flow decreases as the resistance (R) of the system to flow increases. Before we discuss resistance, let's consider how the respiratory system creates a pressure gradient. The pressure-volume relationships of Boyle's law provide the basis for pulmonary ventilation.

Concept Check

15. Compare the direction of air movement during one respiratory cycle with the direction of blood flow during one cardiac cycle.
16. Explain the relationship between the lungs, the pleural membranes, the pleural fluid, and the thoracic cage.

Inspiration Occurs When Alveolar Pressure Decreases

For air to move into the alveoli, pressure inside the lungs must become lower than atmospheric pressure. According to Boyle's law, an increase in volume will create a decrease in pressure. During inspiration, thoracic volume increases when certain skeletal muscles of the rib cage and diaphragm contract.

When the diaphragm contracts, it drops down toward the abdomen. In quiet breathing, the diaphragm moves about 1.5 cm, increasing thoracic volume (**FIG. 17.8b**). Contraction of the diaphragm causes between 60% and 75% of the inspiratory volume change during normal quiet breathing.

Movement of the rib cage creates the remaining 25–40% of the volume change. During inhalation, the external intercostal and scalene muscles (see Fig. 17.2c) contract and pull the ribs upward and out (Fig. 17.8b). Rib movement during inspiration has been likened to a pump handle lifting up and away from the pump (the ribs moving up and away from the spine) and to the movement of a bucket handle as it lifts away from the side of a bucket (ribs moving outward in a lateral direction). The combination of these two movements broadens the rib cage in all directions. As thoracic volume increases, pressure decreases, and air flows into the lungs.

For many years, quiet breathing was attributed solely to the action of the diaphragm and the external intercostal muscles. It was thought that the scalenes and sternocleidomastoid muscles were active only during deep breathing. In recent years, however, studies have changed our understanding of how these accessory muscles contribute to quiet breathing.

If an individual's scalenes are paralyzed, inspiration is achieved primarily by contraction of the diaphragm. Observation of patients with neuromuscular disorders has revealed that although the contracting diaphragm increases thoracic volume by moving toward the abdominal cavity, it also tends to pull the lower ribs inward, working against inspiration. In normal individuals, we know that the lower ribs move up and out during inspiration rather than inward. The fact that there is no up-and-out rib motion in patients with paralyzed scalenes tells us that normally the scalenes must be contributing to inspiration by lifting the sternum and upper ribs.

New evidence also downplays the role of the external intercostal muscles during quiet breathing. However, the external intercostals play an increasingly important role as respiratory activity increases. Because the exact contribution of external intercostals and scalenes varies depending on the type of breathing, we group these muscles together and simply call them the *inspiratory muscles*.

FIG. 17.8 Movement of the thoracic cage and diaphragm during breathing

(a) At rest: Diaphragm is relaxed.

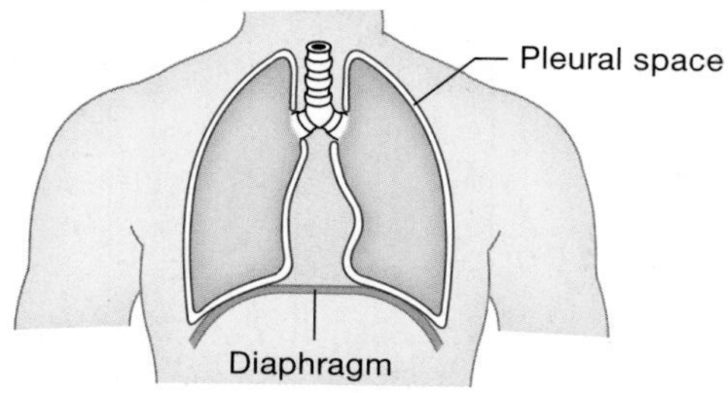
Pleural space
Diaphragm

(b) Inspiration: Thoracic volume increases.

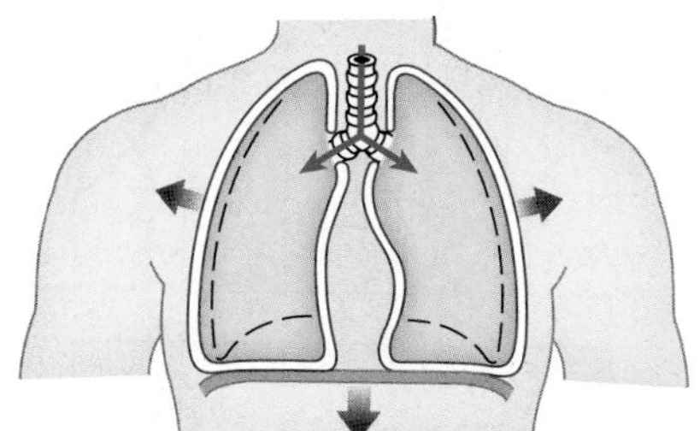
Diaphragm contracts and flattens.

(c) Expiration: Diaphragm relaxes, thoracic volume decreases.

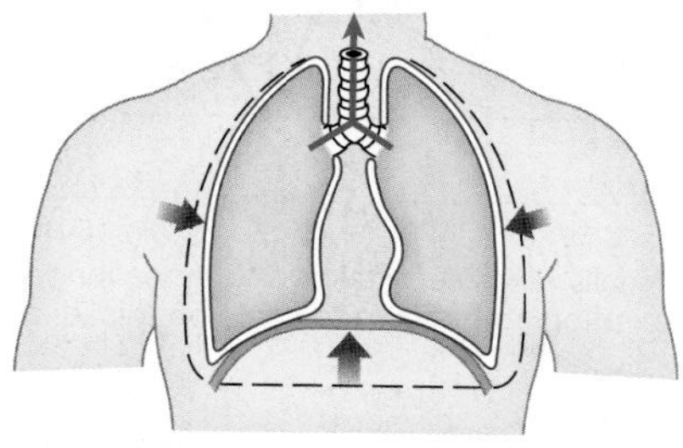

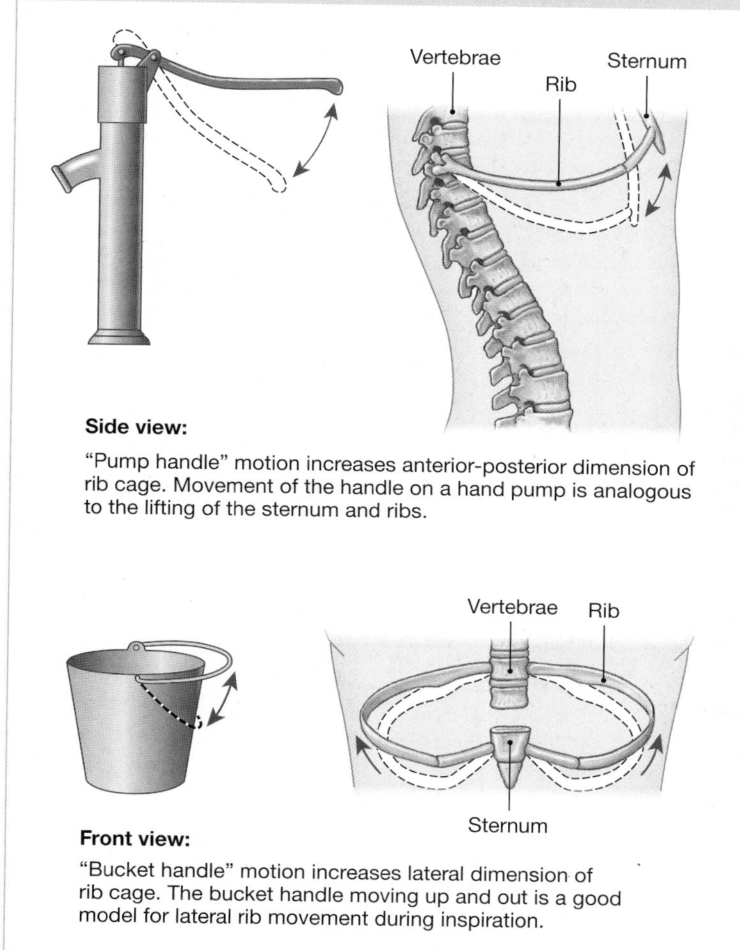
During inspiration, the dimensions of the thoracic cavity increase.
Vertebrae
Rib
Sternum
Side view:
"Pump handle" motion increases anterior-posterior dimension of rib cage. Movement of the handle on a hand pump is analogous to the lifting of the sternum and ribs.
Vertebrae
Rib
Sternum
Front view:
"Bucket handle" motion increases lateral dimension of rib cage. The bucket handle moving up and out is a good model for lateral rib movement during inspiration.

Now let's see how alveolar pressure (P_A) changes during a single inspiration. Follow the graphs in **FIGURE 17.9** as you read through the process. Remember that atmospheric pressure is assigned a value of 0 mm Hg. Negative numbers designate sub-atmospheric pressures, and positive numbers denote higher-than-atmospheric pressures.

Time 0 In the brief pause between breaths, alveolar pressure is equal to atmospheric pressure (0 mm Hg at point A_1). When pressures are equal, there is no air flow.

Time 0–2 sec: Inspiration As inspiration begins, inspiratory muscles contract, and thoracic volume increases. With the increase in volume, alveolar pressure falls about 1 mm Hg below atmospheric pressure (-1 mm Hg, point A_2), and air flows into the alveoli (point C_1 to point C_2). Because the thoracic volume changes faster than air can flow, alveolar pressure reaches its lowest value about halfway through inspiration (point A_2).

As air continues to flow into the alveoli, pressure increases until the thoracic cage stops expanding, just before the end of inspiration. Air movement continues for a fraction of a second longer, until pressure inside the lungs equalizes with atmospheric pressure (point A_3). At the end of inspiration, lung volume is at its maximum for the respiratory cycle (point C_2), and alveolar pressure is equal to atmospheric pressure.

You can demonstrate this phenomenon by taking a deep breath and stopping the movement of your chest at the end of inspiration. (Do not "hold your breath" because doing so closes the opening of the pharynx and prevents air flow.) If you do this correctly, you notice that air flow stops after you freeze the inspiratory movement. This exercise shows that at the end of inspiration, alveolar pressure is equal to atmospheric pressure.

FIG. 17.9 Pressure changes during quiet breathing

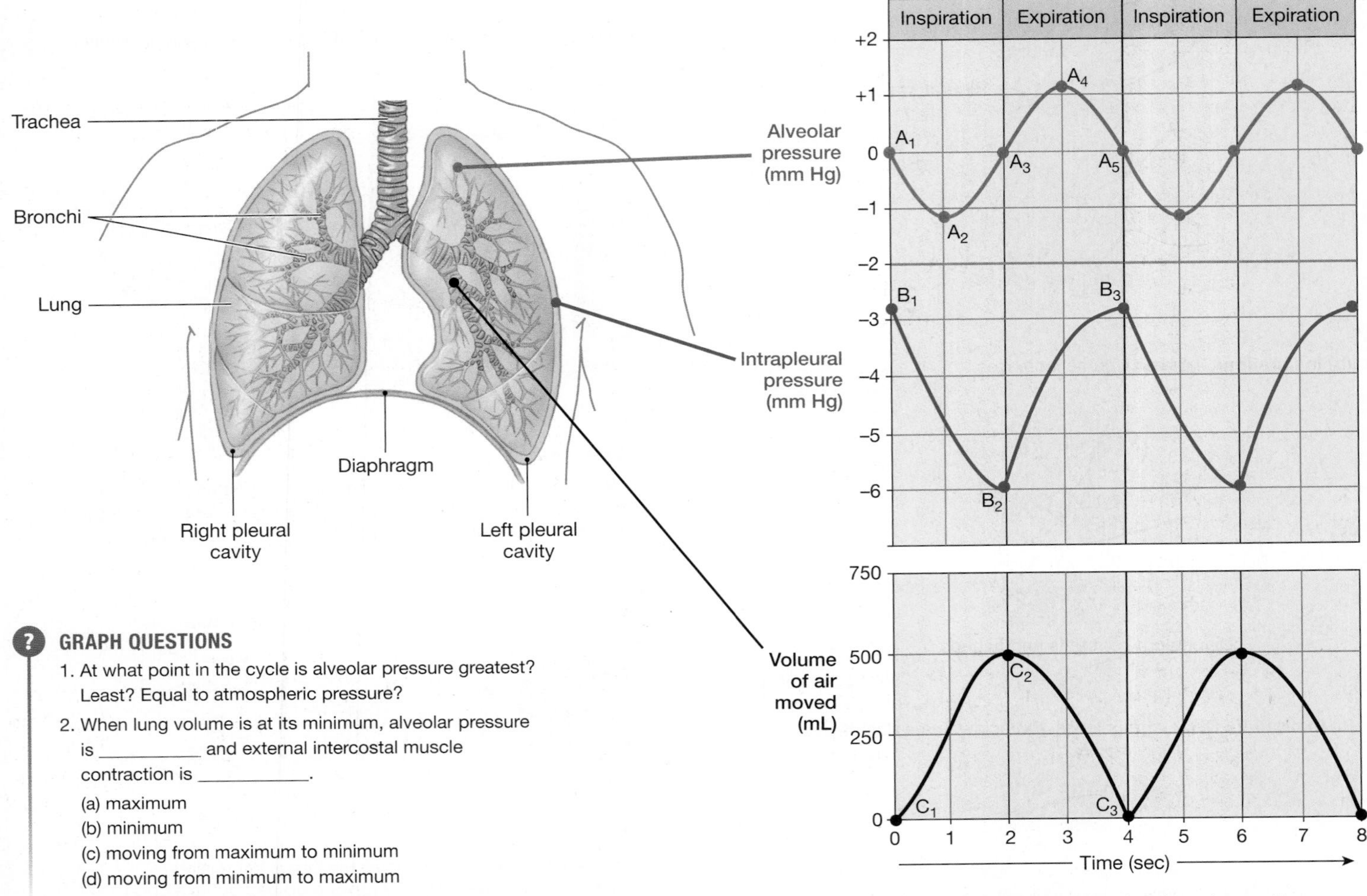
Trachea
Bronchi
Lung
Diaphragm
Right pleural cavity
Left pleural cavity
Alveolar pressure (mm Hg)
Intrapleural pressure (mm Hg)
Volume of air moved (mL)
Inspiration
Expiration
Inspiration
Expiration
+2
+1
0
−1
−2
−3
−4
−5
−6
A1
A2
A3
A4
A5
B1
B2
B3
750
500
250
0
C1
C2
C3
0 1 2 3 4 5 6 7 8
Time (sec)

Normally, expiration takes 2–3 times longer than inspiration (not shown to scale on this idealized graph).

GRAPH QUESTIONS

1. At what point in the cycle is alveolar pressure greatest? Least? Equal to atmospheric pressure?
2. When lung volume is at its minimum, alveolar pressure is ___________ and external intercostal muscle contraction is ___________.
   (a) maximum
   (b) minimum
   (c) moving from maximum to minimum
   (d) moving from minimum to maximum
3. What is this person's ventilation rate?

Expiration Occurs When Alveolar Pressure Increases

At the end of inspiration, impulses from somatic motor neurons to the inspiratory muscles cease, and the muscles relax. Elastic recoil of the lungs and thoracic cage returns the diaphragm and rib cage to their original relaxed positions, just as a stretched elastic waistband recoils when released. Because expiration during quiet breathing involves passive elastic recoil rather than active muscle contraction, it is called passive expiration.

Time 2–4 sec: Expiration As lung and thoracic volumes decrease during expiration, air pressure in the lungs increases, reaching a maximum of about 1 mm Hg above atmospheric pressure (Fig. 17.9, point A_4). Alveolar pressure is now higher than atmospheric pressure, so air flow reverses and air moves out of the lungs.

Time 4 sec. At the end of expiration, air movement ceases when alveolar pressure is again equal to atmospheric pressure (point A_5). Lung volume reaches its minimum for the respiratory cycle (point C_3). At this point, the respiratory cycle has ended and is ready to begin again with the next breath.

The pressure differences shown in Figure 17.9 apply to quiet breathing. During exercise or forced heavy breathing, these values become proportionately larger. **Active expiration** occurs during voluntary exhalations and when ventilation exceeds 30–40 breaths per minute. (Normal resting ventilation rate is 12–20 breaths per minute for an adult.) Active expiration uses the internal intercostal muscles and the abdominal muscles (see Fig. 17.2c), which are not used during inspiration. These muscles are collectively called the *expiratory muscles.*

The internal intercostal muscles line the inside of the rib cage. When they contract, they pull the ribs inward, reducing the volume of the thoracic cavity. To feel this action, place your hands on your rib cage. Forcefully blow as much air out of your lungs as you can, noting the movement of your hands as you do so.

The internal and external intercostals function as antagonistic muscle groups [p. 376] to alter the position and volume of the rib cage during ventilation. The diaphragm, however, has no antagonistic muscles. Instead, abdominal muscles contract during active expiration to supplement the activity of the internal intercostals.

Abdominal contraction pulls the lower rib cage inward and decreases abdominal volume, actions that displace the intestines and liver upward. The displaced viscera push the diaphragm up into the thoracic cavity and passively decrease chest volume even more. The action of abdominal muscles during forced expiration is why aerobics instructors tell you to blow air out as you lift your head and shoulders during abdominal "crunches." The active process of blowing air out helps contract the abdominals, the muscles you are trying to strengthen.

Any neuromuscular disease that weakens skeletal muscles or damages their motor neurons can adversely affect ventilation. With decreased ventilation, less fresh air enters the lungs. In addition, loss of the ability to cough increases the risk of pneumonia and other infections. Examples of diseases that affect the motor control of ventilation include *myasthenia gravis* [p. 253], an illness in which acetylcholine receptors of the motor end plates of skeletal muscles are destroyed, and *polio* (poliomyelitis), a viral illness that paralyzes skeletal muscles.

Concept Check

17. Scarlett O'Hara is trying to squeeze herself into a corset with an 18-inch waist. Will she be more successful by taking a deep breath and holding it or by blowing all the air out of her lungs? Why?
18. Why would loss of the ability to cough increase the risk of respiratory infections? (*Hint:* What does coughing do to mucus in the airways?)

Intrapleural Pressure Changes during Ventilation

Ventilation requires that the lungs, which are unable to expand and contract on their own, move in association with the expansion and relaxation of the thorax. As we noted earlier in this chapter, the lungs are enclosed in the fluid-filled pleural sac. The surface of the lungs is covered by the *visceral pleura,* and the portion of the sac that lines the thoracic cavity is called the *parietal pleura* {*paries,* wall}. Cohesive forces of the intrapleural fluid cause the stretchable lung to adhere to the thoracic cage. When the thoracic cage moves during breathing, the lungs move with it.

Subatmospheric Intrapleural Pressure The intrapleural pressure in the fluid between the pleural membranes is normally subatmospheric. This subatmospheric pressure arises during fetal development, when the thoracic cage with its associated pleural membrane grows more rapidly than the lung with its associated pleural membrane. The two pleural membranes are held together by the pleural fluid bond, so the elastic lungs are forced to stretch to conform to the larger volume of the thoracic cavity. At the same time, however, elastic recoil of the lungs creates an inwardly directed force that tries to pull the lungs away from the chest wall (**FIG. 17.10a**). The combination of the outward pull of the thoracic cage and inward recoil of the elastic lungs creates a subatmospheric intrapleural pressure of about −3 mm Hg.

You can create a similar situation by half-filling a syringe with water, removing all air, and capping the end with a plugged-up needle or three-way valve. At this point, the pressure inside the barrel is equal to atmospheric pressure. Now hold the syringe barrel (the chest wall) in one hand while you try to withdraw the plunger

FIG. 17.10 Subatmospheric pressure in the pleural cavity helps keep the lungs inflated

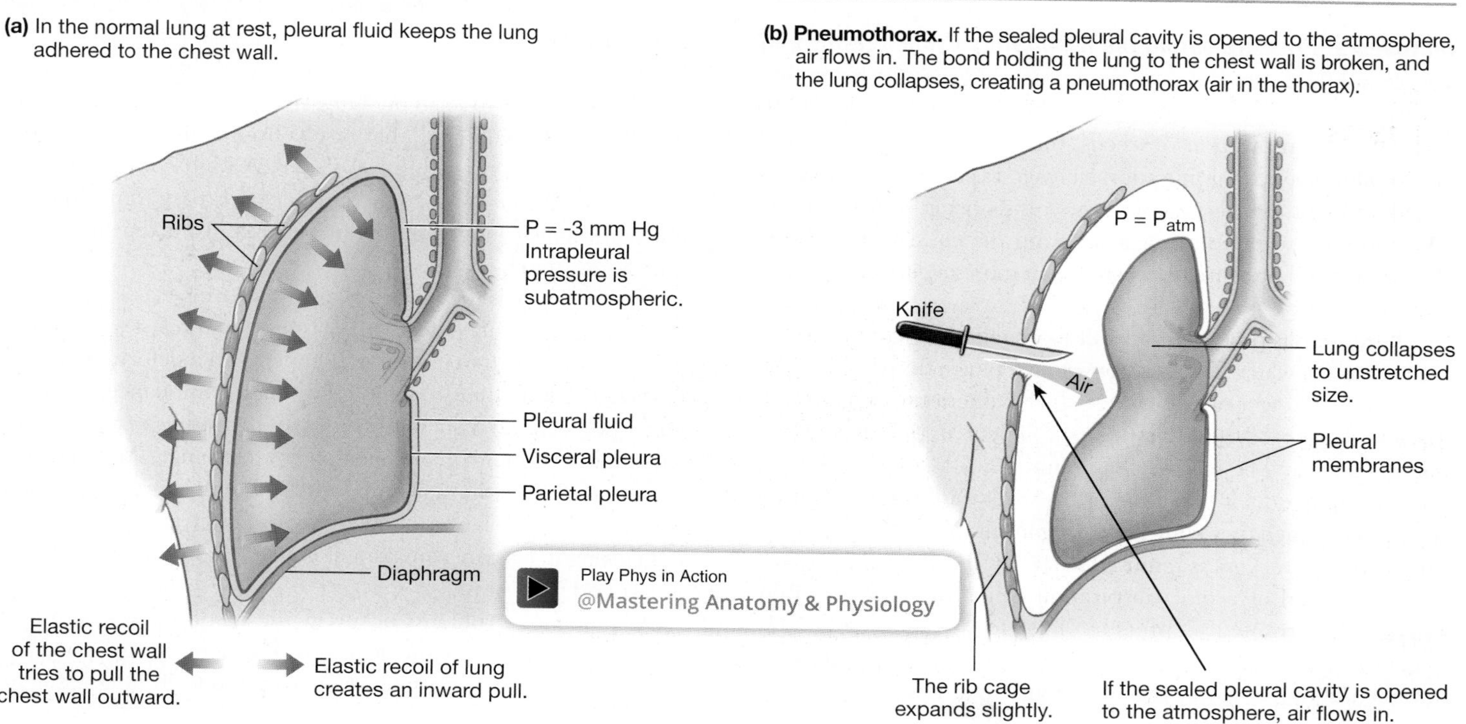
(a) In the normal lung at rest, pleural fluid keeps the lung adhered to the chest wall.
Ribs
P = -3 mm Hg Intrapleural pressure is subatmospheric.
Pleural fluid
Visceral pleura
Parietal pleura
Diaphragm
Elastic recoil of the chest wall tries to pull the chest wall outward.
Elastic recoil of lung creates an inward pull.
Play Phys in Action
@Mastering Anatomy & Physiology
(b) Pneumothorax. If the sealed pleural cavity is opened to the atmosphere, air flows in. The bond holding the lung to the chest wall is broken, and the lung collapses, creating a pneumothorax (air in the thorax).
P = Patm
Knife
Air
Lung collapses to unstretched size.
Pleural membranes
The rib cage expands slightly.
If the sealed pleural cavity is opened to the atmosphere, air flows in.

(the elastic lung pulling away from the chest wall). As you pull on the plunger, the volume inside the barrel increases very slightly, but the cohesive forces between the water molecules cause the water to resist expansion. The pressure inside the barrel, which was initially equal to atmospheric pressure, decreases slightly as you pull on the plunger. If you release the plunger, it snaps back to its resting position, restoring atmospheric pressure inside the syringe. This experiment demonstrates the cohesive forces of water: water will resist being "stretched."

Pneumothorax Now suppose an opening is made between the sealed pleural cavity with its subatmospheric pressure and the atmosphere. In real life, this might happen with a knife thrust between the ribs, a broken rib that punctures the pleural membrane, or any other event that breaks the seal of the pleural cavity. Air moves down pressure gradients, so opening the pleural cavity to the atmosphere allows air to flow into the cavity, just as air enters a vacuum-packed can when you break the seal with a can opener.

Air entering the pleural cavity breaks the fluid bond holding the lung to the chest wall. The chest wall expands outward while the elastic lung collapses to an unstretched state, like a deflated balloon (Fig. 17.10b). This condition, called **pneumothorax** {*pneuma*, air + *thorax*, chest}, results in a collapsed lung that is unable to function normally. Pneumothorax can be due to trauma but can also occur spontaneously when a congenital *bleb*, a weakened section of lung tissue, ruptures, allowing air from inside the lung to enter the pleural cavity.

Correction of a pneumothorax has two components: removing as much air from the pleural cavity as possible with a suction pump, and sealing the hole to prevent more air from entering. Any air remaining in the cavity is gradually absorbed into the blood, restoring the pleural fluid bond and reinflating the lung.

Intrapleural Pressure during the Respiratory Cycle Pressures in the pleural fluid vary during a respiratory cycle. At the beginning of inspiration, intrapleural pressure is about −3 mm Hg (Fig. 17.9, point B_1). As inspiration proceeds, the pleural membranes and lungs follow the expanding thoracic cage because of the pleural fluid bond, but the elastic lung tissue resists being stretched. The lungs attempt to pull farther away from the chest wall, causing the intrapleural pressure to become even more negative (Fig. 17.9, point B_2).

Because this process is difficult to visualize, let's return to the analogy of the water-filled syringe with the plugged-up needle. You can pull the plunger out a small distance without much effort, but the cohesiveness of the water makes it difficult to pull the plunger out any farther. The increased amount of work you do trying to pull the plunger out is paralleled by the work your inspiratory muscles must do when they contract during inspiration. The bigger the breath, the more work is required to stretch the elastic lung.

By the end of a quiet inspiration, when the lungs are fully expanded, intrapleural pressure falls to around −6 mm Hg (Fig. 17.9, point B_2). During exercise or other powerful inspirations, intrapleural pressure may reach −8 mm Hg or lower.

During expiration, the thoracic cage returns to its resting position. The lungs are released from their stretched position, and the intrapleural pressure returns to its normal value of about −3 mm Hg (point B_3). Notice that intrapleural pressure never equilibrates with atmospheric pressure because the pleural cavity is a sealed compartment.

Concept Check

19. A person has periodic spastic contractions of the diaphragm, otherwise known as hiccups. What happens to intrapleural and alveolar pressures when a person hiccups?
20. A stabbing victim is brought to the emergency room with a knife wound between the ribs on the left side of his chest. What has probably happened to his left lung? To his right lung? Why does the left side of his rib cage seem larger than the right side?

Lung Compliance and Elastance May Change in Disease States

Pressure gradients required for air flow are created by the work of skeletal muscle contraction. Normally, about 3–5% of the body's energy expenditure is used for quiet breathing. During exercise, the energy required for breathing increases substantially. The two factors that have the greatest influence on the amount of work needed for breathing are the stretchability of the lungs and the resistance of the airways to air flow.

Adequate ventilation depends on the ability of the lungs to expand normally. Most of the work of breathing goes into overcoming the resistance of the elastic lungs and the thoracic cage to stretching. Clinically, the ability of the lung to stretch is called **compliance.**

Compliance refers to the amount of force that must be exerted on a body to deform it. In the lung, we can express compliance as the change of volume (V) that results from a given force or pressure (P) exerted on the lung: $\Delta V/\Delta P$. A high-compliance lung stretches easily, just as a compliant person is easy to persuade. A low-compliance lung requires more force from the inspiratory muscles to stretch it.

Compliance is the reciprocal of **elastance** (elastic recoil), the ability to resist being deformed. Elastance also refers to the ability of a body to return to its original shape when a deforming force is removed. A lung that stretches easily (high compliance) has probably lost its elastic tissue and will not return to its resting volume when the stretching force is released (low elastance). You may have experienced something like this with old gym shorts. After many washings the elastic waistband is easy to stretch (high compliance) but lacking in elastance, making it impossible for the shorts to stay up around your waist.

Analogous problems occur in the respiratory system. For example, as noted in the Running Problem, emphysema is a disease in which elastin fibers normally found in lung tissue are destroyed. Destruction of elastin results in lungs that exhibit high compliance and stretch easily during inspiration. However, these lungs also have decreased elastance, so they do not recoil to their resting position during expiration.

To understand the importance of elastic recoil to expiration, think of an inflated balloon and an inflated plastic bag. The balloon is similar to the normal lung. Its elastic walls squeeze on the air inside the balloon, thereby increasing the internal air pressure. When the neck of the balloon is opened to the atmosphere, elastic recoil causes air to flow out of the balloon.

The inflated plastic bag, on the other hand, is like the lung of an individual with emphysema. It has high compliance and is easily inflated, but it has little elastic recoil. If the inflated plastic bag is opened to the atmosphere, most of the air remains inside the bag. To get the air out of the bag, you must squeeze it with your hands. Patients with emphysema contract their expiratory muscles (active expiration) to force out air that is not leaving from elastic recoil.

RUNNING PROBLEM

Emphysema is characterized by a loss of *elastin*, the elastic fibers that help the alveoli recoil during expiration. Elastin is destroyed by *elastase,* an enzyme released by alveolar macrophages, which work overtime in smokers to rid the lungs of irritants. People with emphysema have more difficulty exhaling than inhaling. Their alveoli have lost elastic recoil, which makes expiration—normally a passive process—require conscious effort.

Q3: *Name the muscles that patients with emphysema use to exhale actively.*

533 - 540 - 542 - 549 - 554 - 557

Surfactant Decreases the Work of Breathing

For years, physiologists assumed that elastin and other elastic fibers were the primary source of resistance to stretch in the lung. However, studies comparing the work required to expand air-filled and saline-filled lungs showed that air-filled lungs are much harder to inflate. From this result, researchers concluded that lung tissue itself contributes less to resistance than once thought. Some other property of the normal air-filled lung, a property not present in the saline-filled lung, must create most of the resistance to stretch.

This property is the *surface tension* [p. 39] created by the thin fluid layer between the alveolar cells and the air. At any air-fluid interface, the surface of the fluid is under tension, like a thin membrane being stretched. When the fluid is water, surface tension arises because of the hydrogen bonds between water molecules. The water molecules on the fluid's surface are attracted to other water molecules beside and beneath them but are not attracted to gases in the air at the air-fluid interface.

Alveolar surface tension is similar to the surface tension that exists in a spherical bubble, even though alveoli are not perfect spheres. The surface tension created by the thin film of fluid is directed toward the center of the bubble and creates pressure in the interior of the bubble. The **law of LaPlace** is an expression of this pressure. It states that the pressure (P) inside a bubble formed by a fluid film is a function of two factors: the surface tension of the fluid (T) and the radius of the bubble (r). This relationship is expressed by the equation

$$P = 2T/r$$

Notice in **FIGURE 17.11a** that if two bubbles have different diameters but are formed by fluids with the same surface tension, the pressure inside the smaller bubble is greater than that inside the larger bubble.

How does this apply to the lung? In physiology, we can equate the bubble to a fluid-lined alveolus (although alveoli are

FIG. 17.11 Law of LaPlace

(a) The two bubbles shown have the same surface tension (T). According to the Law of LaPlace, pressure is greater in the smaller bubble.

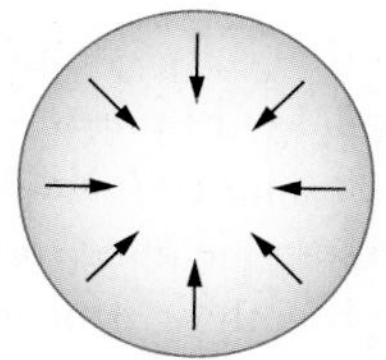
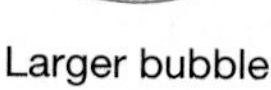

Larger bubble

r = 2
T = 3
P = (2 × 3)/2
P = 3

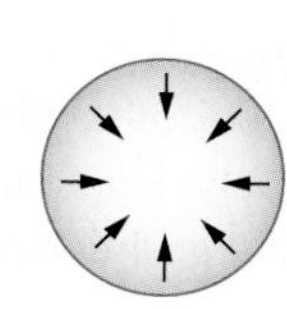

Smaller bubble

r = 1
T = 3
P = (2 × 3)/1
P = 6

Law of LaPlace

$P = 2T/r$

P = pressure
T = surface tension
r = radius

According to the law of LaPlace, if two bubbles have the same surface tension, the smaller bubble will have higher pressure.

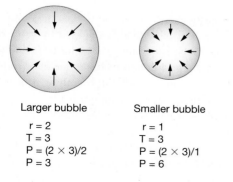

(b) Surfactant (●) reduces surface tension (T). In the lungs, smaller alveoli have more surfactant, which equalizes the pressure between large and small alveoli.

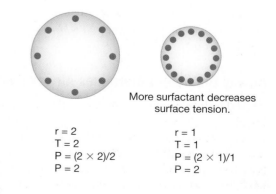

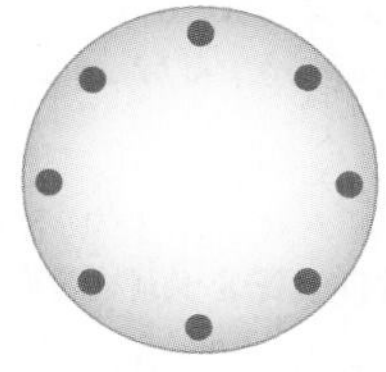

r = 2
T = 2
P = (2 × 2)/2
P = 2

More surfactant decreases surface tension.

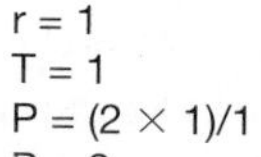

r = 1
T = 1
P = (2 × 1)/1
P = 2

not perfect spheres). The fluid lining all the alveoli creates surface tension. If the surface tension (T) of the fluid were the same in small and large alveoli, small alveoli would have higher inwardly directed pressure than larger alveoli, and increased resistance to stretch. As a result, more work would be needed to expand smaller alveoli.

Normally, however, our lungs secrete a surfactant that reduces surface tension. Surfactants ("**surf**ace **act**ive **a**ge**nts**") are molecules that disrupt cohesive forces between water molecules by substituting themselves for water at the surface. For example, that product you add to your dishwasher to aid in the rinse cycle is a surfactant that keeps the rinse water from beading up on the dishes (and forming spots when the water beads dry). In the lungs, surfactant decreases surface tension of the alveolar fluid and thereby decreases resistance of the lung to stretch.

Surfactant is more concentrated in smaller alveoli, making their surface tension less than that in larger alveoli (Fig. 17.11b). Lower surface tension helps equalize the pressure among alveoli of different sizes and makes it easier to inflate the smaller alveoli. With lower surface tension, the work needed to expand the alveoli with each breath is greatly reduced. Human surfactant is a mixture containing proteins and phospholipids, such as *dipalmitoylphosphatidylcholine,* which are secreted into the alveolar air space by type II alveolar cells (see Fig. 17.2g).

Normally, surfactant synthesis begins about the 25th week of fetal development under the influence of various hormones. Production usually reaches adequate levels by the 34th week (about 6 weeks before normal delivery). Babies who are born prematurely without adequate concentrations of surfactant in their alveoli develop *newborn respiratory distress syndrome* (NRDS). In addition to having "stiff" (low-compliance) lungs, NRDS babies also have alveoli that collapse each time they exhale. These infants must use a tremendous amount of energy to expand their collapsed lungs with each breath. Unless treatment is initiated rapidly, about 50% of these infants die. Until about 1970, all physicians could do for NRDS babies was administer oxygen.

Today, however, the prognosis for NRDS babies is much better. Before birth, amniotic fluid can be sampled to assess whether or not the fetal lungs are producing adequate amounts of surfactant. If they are not, and if delivery cannot be delayed, the mother is given a glucocorticoid drug that helps the baby's alveolar type II cells mature. If premature babies develop NRDS, they can be treated with aerosol administration of artificial surfactant. The current treatment also includes artificial ventilation that forces air into the lungs (*positive-pressure ventilation*) and keeps the alveoli open.

Airway Diameter Determines Airway Resistance

The other factor besides compliance that influences the work of breathing is the resistance of the respiratory system to air flow. Resistance in the respiratory system is similar in many ways to resistance in the cardiovascular system [p. 438]. Three parameters contribute to resistance (R): the system's length (L), the viscosity of the substance flowing through the system (η), and the radius of the tubes in the system (r). As with flow in the cardiovascular system, Poiseuille's law relates these factors to one another:

$$R \propto L\eta/r^4$$

Because the length of the respiratory system is constant, we can ignore L in the equation. The viscosity of air is almost constant, although you may have noticed that it feels harder to breathe in a sauna filled with steam than in a room with normal humidity. Water droplets in the steam increase the viscosity of the steamy air, thereby increasing its resistance to flow. Viscosity also changes slightly with atmospheric pressure, decreasing as pressure decreases. A person at high altitude may feel less resistance to air flow than a person at sea level. Despite these exceptions, viscosity plays a very small role in resistance to air flow.

Length and viscosity are essentially constant for the respiratory system. As a result, the radius (or diameter) of the airways becomes the primary determinant of airway resistance. Normally, however, the work needed to overcome resistance of the airways to air flow is much less than the work needed to overcome the resistance of the lungs and thoracic cage to stretch.

Nearly 90% of airway resistance normally can be attributed to the trachea and bronchi, rigid structures with the smallest total cross-sectional area. Because these structures are supported by cartilage and bone, their diameters normally do not change, and their resistance to air flow is constant. However, accumulation of mucus from allergies or infections can dramatically increase resistance. If you have ever tried breathing through your nose when you have a cold, you can appreciate how the narrowing of an upper airway limits air flow!

The bronchioles normally do not contribute significantly to airway resistance because their total cross-sectional area is about 2000 times that of the trachea. Because the bronchioles are collapsible tubes, however, a decrease in their diameter can suddenly turn them into a significant source of airway resistance. **Bronchoconstriction** increases resistance to air flow and decreases the amount of fresh air that reaches the alveoli.

Bronchioles, like arterioles, are subject to reflex control by the nervous system and by hormones. However, most minute-to-minute changes in bronchiolar diameter occur in response to paracrine signals. Carbon dioxide in the airways is the primary paracrine molecule that affects bronchiolar diameter. Increased CO_2 in expired air relaxes bronchiolar smooth muscle and causes **bronchodilation.**

Histamine is a paracrine signal molecule that acts as a powerful bronchoconstrictor. This chemical is released by *mast cells* [p. 513] in response to either tissue damage or allergic reactions. In severe allergic reactions, large amounts of histamine may lead to widespread bronchoconstriction and difficult breathing. Immediate medical treatment in these patients is imperative.

The primary neural control of bronchioles comes from parasympathetic neurons that cause bronchoconstriction, a reflex designed to protect the lower respiratory tract from inhaled irritants. There is no significant sympathetic innervation of the bronchioles in humans. However, smooth muscle in the bronchioles is well supplied with β_2-receptors that respond to circulating epinephrine. Stimulation of β_2-receptors relaxes airway smooth muscle and results in bronchodilation. This reflex is used therapeutically in the treatment of asthma and various allergic reactions characterized by histamine release and bronchoconstriction. **TABLE 17.2** summarizes the factors that alter airway resistance.

Concept Check

21. In a normal person, which contributes more to the work of breathing: airway resistance or lung and chest wall elastance?
22. Coal miners who spend years inhaling fine coal dust have much of their alveolar surface area covered with scarlike tissue. What happens to their lung compliance as a result?
23. How does the work required for breathing change when surfactant is not present in the lungs?
24. A cancerous lung tumor has grown into the walls of a group of bronchioles, narrowing their lumens. What has happened to the resistance to air flow in these bronchioles?
25. Name the neurotransmitter and receptor for parasympathetic bronchoconstriction.

Rate and Depth of Breathing Determine the Efficiency of Breathing

You may recall that the efficiency of the heart is measured by the cardiac output, which is calculated by multiplying heart rate by stroke volume. Likewise, we can estimate the effectiveness of ventilation by calculating **total pulmonary ventilation,** the volume of air moved into and out of the lungs each minute (**FIG. 17.12a**). Total pulmonary ventilation, also known as the *minute volume,* is calculated as follows:

$$\text{Total pulmonary ventilation} = \text{ventilation rate} \times \text{tidal volume}$$

The normal ventilation rate for an adult is 12–20 breaths (br) per minute. Using the average tidal volume (500 mL) and the slowest ventilation rate, we get:

$$\text{Total pulmonary ventilation} = 12 \text{ br/min} \times 500 \text{ mL/br} = 6000 \text{ mL/min} = 6 \text{ L/min}$$

Total pulmonary ventilation represents the physical movement of air into and out of the respiratory tract, but is it a good indicator of how much fresh air reaches the alveolar exchange surface? Not necessarily.

Some air that enters the respiratory system does not reach the alveoli because part of every breath remains in the conducting airways, such as the trachea and bronchi. Because the conducting airways do not exchange gases with the blood, they are known as the **anatomic dead space.** Anatomic dead space averages about 150 mL.

To illustrate the difference between the total volume of air that enters the airways and the volume of fresh air that reaches the alveoli, let's consider a typical breath that moves 500 mL of air during a respiratory cycle (Fig. 17.12b).

1. Start at the end of inspiration **1**: Lung volume is maximal at 2700 mL, and fresh air from the atmosphere fills the 150 mL of the upper airways (the dead space).
2. Now exhale **2**: the tidal volume of 500 mL leaves the body. However, the first portion of exhaled air is the 150 mL of fresh air from the dead space, followed by 350 mL of "stale" air from the alveoli. So even though 500 mL of low-oxygen air exited the alveoli, only 350 mL of that volume left the body. The remaining 150 mL of alveolar air stays in the dead space.
3. At the end of expiration **3**, lung volume is at its minimum and low-oxygen air from the alveoli fills the anatomic dead space.
4. With the next inspiration **4**, another 500 mL of fresh air enters the airways. The first air to enter the alveoli is the 150 mL of stale air that was in the anatomic dead space. The remaining 350 mL of air to go into the alveoli is fresh air. The last 150 mL of inspired fresh air remains in the dead space and never reaches the alveoli.

Thus, although 500 mL of air enters the alveoli with each breath, only 350 mL of that volume is fresh air. The volume of fresh air entering the alveoli equals the tidal volume minus the dead space volume: $V_T - V_{DS}$

CHAPTER 17

TABLE 17.2 Factors That Affect Airway Resistance

Factor	Affected by	Mediated by
Length of the system	Constant; not a factor	
Viscosity of air	Usually constant; humidity and altitude may alter slightly	
Diameter of airways		
Upper airways	Physical obstruction	Mucus and other factors
Bronchioles	Bronchoconstriction	Parasympathetic neurons (muscarinic receptors), histamine, leukotrienes
	Bronchodilation	Carbon dioxide, epinephrine (β_2-receptors)

FIG. 17.12 ESSENTIALS Ventilation

(a) Total pulmonary ventilation is greater than alveolar ventilation because of dead space.

Total pulmonary ventilation:

Total pulmonary ventilation = ventilation rate × tidal volume (V_T)

For example: 12 breaths/min × 500 mL breath = 6000 mL/min

Alveolar ventilation:

Alveolar ventilation is a better indication of how much fresh air reaches the alveoli. Fresh air remaining in the dead space does not get to the alveoli.

Alveolar ventilation = ventilation rate × (V_T – dead space volume V_D)

If dead space is 150 mL: 12 breaths/min × (500 – 150 mL) = 4200 mL/min

(b) Because the conducting airways do not exchange gases with the blood, they are known as anatomic dead space.

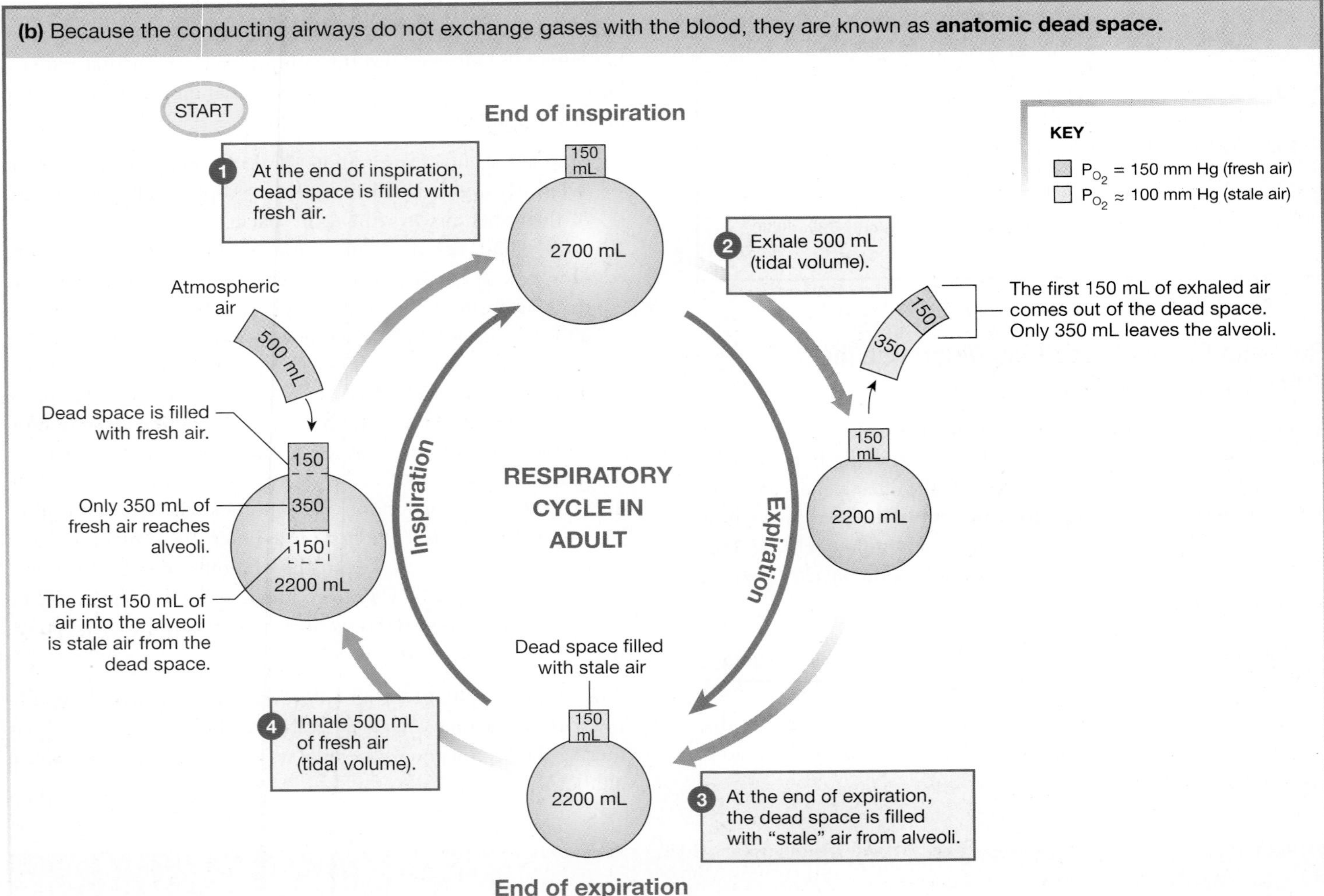

START
End of inspiration
1 At the end of inspiration, dead space is filled with fresh air.
150 mL
2700 mL
2 Exhale 500 mL (tidal volume).
KEY
P_{O_2} = 150 mm Hg (fresh air)
P_{O_2} ≈ 100 mm Hg (stale air)
The first 150 mL of exhaled air comes out of the dead space. Only 350 mL leaves the alveoli.
150
350
150 mL
2200 mL
Atmospheric air
500 mL
Dead space is filled with fresh air.
150
350
150
2200 mL
Only 350 mL of fresh air reaches alveoli.
The first 150 mL of air into the alveoli is stale air from the dead space.
Inspiration
RESPIRATORY CYCLE IN ADULT
Expiration
Dead space filled with stale air
150 mL
2200 mL
4 Inhale 500 mL of fresh air (tidal volume).
3 At the end of expiration, the dead space is filled with "stale" air from alveoli.
End of expiration

FIGURE QUESTION

Complete this table showing the effects of breathing pattern on alveolar ventilation. Assume dead space volume is 150 mL. Which pattern is the most efficient?

Tidal Volume (mL)	Ventilation Rate (breaths/min)	Total Pulmonary Ventilation (mL/min)	Fresh Air to Alveoli (mL)	Alveolar Ventilation (mL/min)
500 (normal)	12 (normal)	6000	350	4200
300 (shallow)	20 (rapid)			
750 (deep)	8 (slow)			

Because a significant portion of inspired air never reaches an exchange surface, a more accurate indicator of ventilation efficiency is **alveolar ventilation,** the volume of fresh air that reaches the alveoli each minute. Alveolar ventilation is calculated by multiplying ventilation rate by the volume of fresh air that reaches the alveoli:

$$\text{Alveolar ventilation} = \text{ventilation rate} \times (\text{tidal volume} - \text{dead space volume})$$

Using the same ventilation rate and tidal volume as before, and a dead space of 150 mL, then

$$\text{Alveolar ventilation} = 12 \text{ br/min} \times (500 - 150 \text{ mL/br}) = 4200 \text{ mL/min}$$

Thus, at 12 breaths per minute, the alveolar ventilation is 4.2 L/min. Although 6 L/min of fresh air enters the respiratory system, only 4.2 L of fresh air reaches the alveoli.

Alveolar ventilation can be drastically affected by changes in the rate or depth of breathing, as you can calculate using the figure question in Figure 17.12. **Maximum voluntary ventilation,** which involves breathing as deeply and quickly as possible, may increase total pulmonary ventilation to as much as 170 L/min. **TABLE 17.3** describes various patterns of ventilation, and **TABLE 17.4** gives normal ventilation values.

Alveolar Gas Composition Varies Little during Normal Breathing

The P_{O_2} and P_{CO_2} in the alveoli change surprisingly little during normal quiet breathing. Alveolar P_{O_2} is fairly constant at 100 mm Hg, and alveolar P_{CO_2} stays close to 40 mm Hg.

Intuitively, you might think that P_{O_2} would increase when fresh air first enters the alveoli, then decrease steadily as oxygen leaves to enter the blood. Instead, we find only very small swings in P_{O_2}. Why? The reasons are that (1) the amount of oxygen that enters the alveoli with each breath is roughly equal to the amount of oxygen that enters the blood, and (2) the amount of fresh air that enters the lungs with each breath is only a little more than 10% of the total lung volume at the end of inspiration.

You can see this in Figure 17.12b. In this example, at the end of inspiration **4** only 350 mL out of the total volume of 2700 mL is higher-oxygen fresh air. This comes to about 13% of the total lung volume.

Although alveolar gases do not change much with quiet breathing, changes in alveolar ventilation can significantly affect the amount of fresh air and oxygen that reach the alveoli. **FIGURE 17.13** shows how partial pressures P_{O_2} and P_{CO_2} in the alveoli vary with increased alveolar ventilation (*hyperventilation*) and decreased hypoventilation (*hypoventilation*).

As alveolar ventilation increases during hyperventilation, alveolar P_{O_2} increases and alveolar P_{CO_2} falls. During hypoventilation, when less fresh air enters the alveoli, alveolar P_{O_2} decreases and alveolar P_{CO_2} increases. Carbon dioxide concentrations in the blood are closely linked to the body's pH, and you will learn later how the body uses changes in ventilation to help maintain pH homeostasis.

Ventilation and Alveolar Blood Flow Are Matched

Moving oxygen from the atmosphere to the alveolar exchange surface is only the first step in external respiration. Next, gas exchange must occur across the alveolar-capillary interface.

TABLE 17.4 Normal Ventilation Values in Pulmonary Medicine

Total pulmonary ventilation	6 L/min
Total alveolar ventilation	4.2 L/min
Maximum voluntary ventilation	125–170 L/min
Respiration rate	12–20 breaths/min

TABLE 17.3 Types and Patterns of Ventilation

Name	Description	Examples
Eupnea	Normal quiet breathing	
Hyperpnea	Increased respiratory rate and/or volume in response to increased metabolism	Exercise
Hyperventilation	Increased respiratory rate and/or volume without increased metabolism	Emotional hyperventilation; blowing up a balloon
Hypoventilation	Decreased alveolar ventilation	Shallow breathing; asthma; restrictive lung disease
Tachypnea	Rapid breathing; usually increased respiratory rate with decreased depth	Panting
Dyspnea	Difficulty breathing (a subjective feeling sometimes described as "air hunger")	Various pathologies or hard exercise
Apnea	Cessation of breathing	Voluntary breath-holding; depression of CNS control centers

Finally, blood flow (*perfusion*) past the alveoli must be high enough to pick up the available oxygen. Matching the ventilation rate into groups of alveoli with blood flow past those alveoli is a two-part process involving local regulation of both air flow and blood flow.

Alterations in pulmonary blood flow depend almost exclusively on properties of the capillaries and on such local factors as the concentrations of oxygen and carbon dioxide in the lung tissue. Capillaries in the lungs are unusual because they are collapsible. If the pressure of blood flowing through the capillaries falls below a certain point, the capillaries close off, diverting blood to pulmonary capillary beds in which blood pressure is higher.

In a person at rest, some capillary beds in the apex (top) of the lung are closed off because of low hydrostatic pressure. Capillary beds at the base of the lung have higher hydrostatic pressure because of gravity and thus remain open. Consequently, blood flow is diverted toward the base of the lung. During exercise, when blood pressure rises, the closed apical capillary beds open, ensuring that the increased cardiac output can be fully oxygenated as it passes through the lungs. The ability of the lungs to recruit additional capillary beds during exercise is an example of the reserve capacity of the body.

At the local level, the body attempts to match air flow and blood flow in each section of the lung by regulating the diameters of the arterioles and bronchioles. Bronchiolar diameter is mediated primarily by CO_2 levels in exhaled air passing through them (**FIG. 17.14**). An increase in the P_{CO_2} of expired air causes bronchioles to dilate. A decrease in the P_{CO_2} of expired air causes bronchioles to constrict.

Although there is some autonomic innervation of pulmonary arterioles, there is apparently little neural control of pulmonary blood flow. The resistance of pulmonary arterioles to blood flow is regulated primarily by the oxygen content of the interstitial fluid around the arteriole. If ventilation of alveoli in one area of the lung is diminished, as shown in Figure 17.14b, the P_{O_2} in that area decreases, and the arterioles respond by constricting, as shown in Figure 17.14c. This local vasoconstriction is adaptive because it diverts blood away from the under-ventilated region to better-ventilated parts of the lung.

Note that constriction of pulmonary arterioles in response to low P_{O_2} is the opposite of what occurs in the systemic circulation [p. 488]. In the systemic circulation, a decrease in the P_{O_2} of a tissue causes local arterioles to dilate, delivering more oxygen-carrying blood to those tissues that are consuming oxygen. In the lungs, blood is picking up oxygen, so it does not make sense to send more blood to an area with low tissue P_{O_2} due to poor ventilation.

Another important point must be noted here. Local control mechanisms are not effective regulators of air and blood flow under all circumstances. If blood flow is blocked in one pulmonary artery, or if air flow is blocked at the level of the larger airways, local responses that shunt air or blood to other parts of the lung are ineffective because in these cases not enough of the lung has normal ventilation or perfusion.

FIG. 17.13 Alveolar gases

As alveolar ventilation increases, alveolar P_{O_2} increases and P_{CO_2} decreases. The opposite occurs as alveolar ventilation decreases.

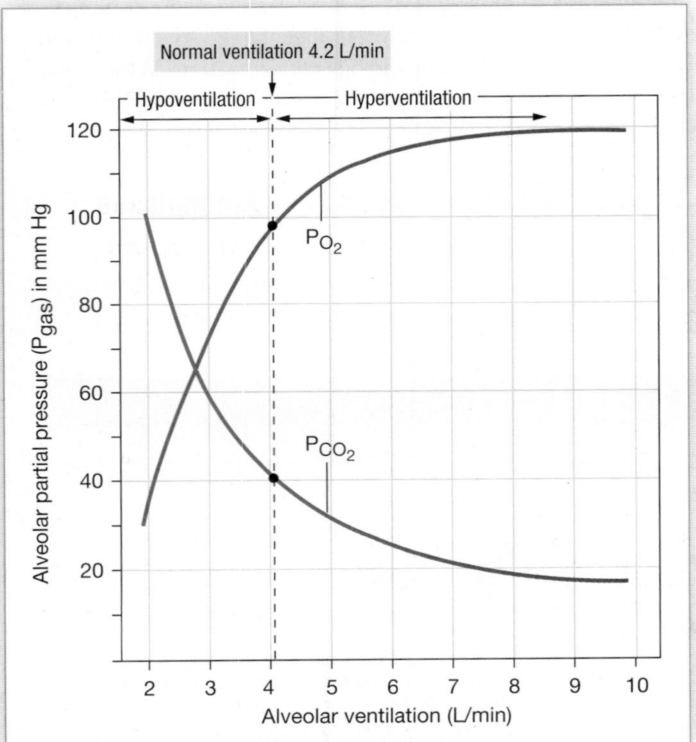
Normal ventilation 4.2 L/min
Hypoventilation
Hyperventilation
120
100
80
60
40
20
P_{O_2}
P_{CO_2}
Alveolar partial pressure (P_{gas}) in mm Hg
2 3 4 5 6 7 8 9 10
Alveolar ventilation (L/min)

GRAPH QUESTION

What are the maximum alveolar P_{O_2} and minimum P_{CO_2} shown in this graph?

Play Phys in Action
@Mastering Anatomy & Physiology

RUNNING PROBLEM

Edna has been experiencing shortness of breath while exercising, so her physician runs some tests, including measuring Edna's lung volumes with spirometry. Part of the test is a forced expiration. With her lungs filled to their maximum with air, Edna is told to blow out as fast and as forcefully as she can. The volume of air that Edna expels in the first second of the test (the *forced expiratory volume in one second*, or FEV_1) is lower than normal because in COPD, airway resistance is increased, slowing the flow of air. Another test the physician orders is a complete blood count (CBC). The results of this test show that Edna has higher-than-normal red blood cell count and hematocrit [p. 517].

Q4: *When Edna fills her lungs maximally, the volume of air in her lungs is known as the ______________ capacity. When she exhales all the air she can, the volume of air left in her lungs is the ______________.*

Q5: *Why are Edna's RBC count and hematocrit increased?* (Hint: *Because of Edna's COPD, her arterial* P_{O_2} *is low.*)

533 – 540 – 542 – 549 – **554** – 557

FIG. 17.14 Local control mechanisms attempt to match ventilation and perfusion

(a) Normally, perfusion of blood past alveoli is matched to alveolar ventilation to maximize gas exchange.

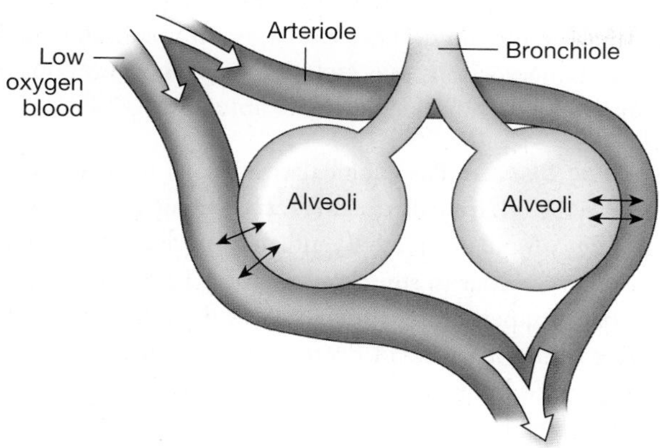

Arteriole
Bronchiole
Low oxygen blood
Alveoli
Alveoli

(b) Ventilation-Perfusion Mismatch Caused by Under-Ventilated Alveoli

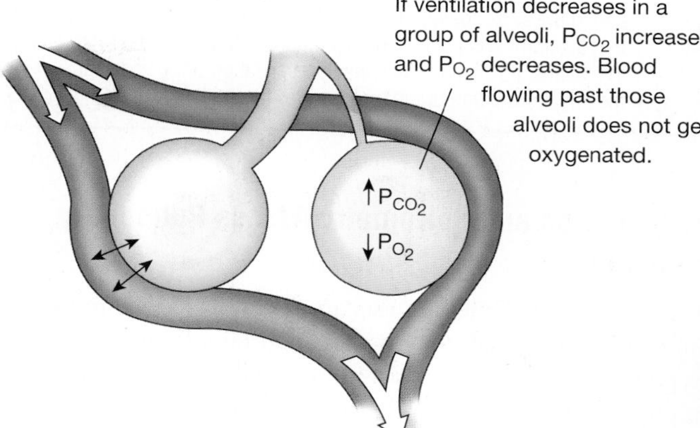

If ventilation decreases in a group of alveoli, P_{CO_2} increases and P_{O_2} decreases. Blood flowing past those alveoli does not get oxygenated.
↑P_{CO_2}
↓P_{O_2}

(c) Local control mechanisms try to keep ventilation and perfusion matched.

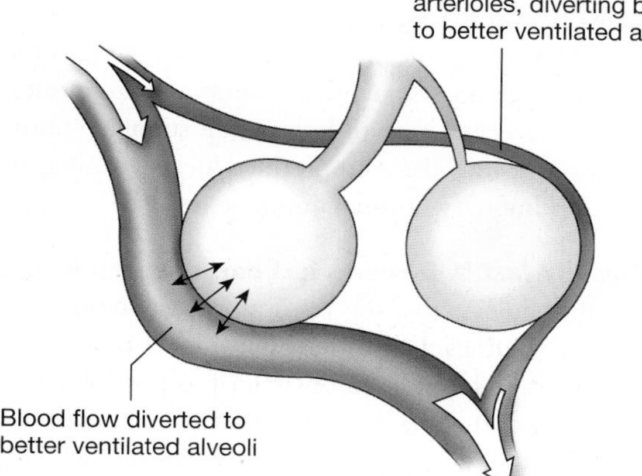

Decreased tissue P_{O_2} around underventilated alveoli constricts their arterioles, diverting blood to better ventilated alveoli.
Blood flow diverted to better ventilated alveoli

(d) Bronchiole diameter is mediated primarily by CO_2 levels in exhaled air passing through them.

Local Control of Arterioles and Bronchioles by Oxygen and Carbon Dioxide

Gas Composition	Bronchioles	Pulmonary Arteries	Systemic Arteries
P_{CO_2} increases	Dilate	(Constrict)*	Dilate
P_{CO_2} decreases	Constrict	(Dilate)	Constrict
P_{O_2} increases	(Constrict)	(Dilate)	Constrict
P_{O_2} decreases	(Dilate)	Constrict	Dilate

*Parentheses indicate weak responses.

FIGURE QUESTIONS

A blood clot prevents gas exchange in a group of alveoli.

1. What happens to tissue and alveolar gases?
2. What do bronchioles and arterioles do in response?

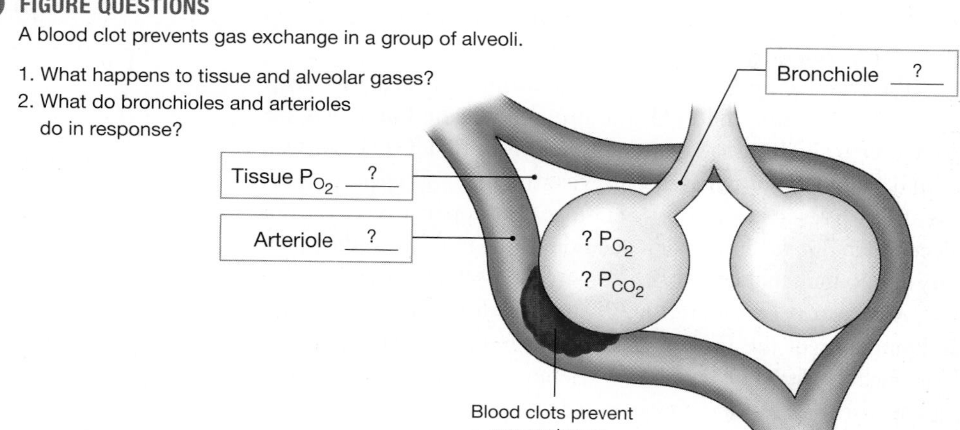

Bronchiole ?
Tissue P_{O_2} ?
Arteriole ?
? P_{O_2}
? P_{CO_2}
Blood clots prevent gas exchange.

Concept Check

26. If a lung tumor decreases blood flow in one small section of the lung to a minimum, what happens to P_{O_2} in the alveoli in that section and in the surrounding interstitial fluid? What happens to P_{CO_2} in that section? What is the compensatory response of the bronchioles in the affected section? Will the compensation bring ventilation in the affected section of the lung back to normal? Explain.

Auscultation and Spirometry Assess Pulmonary Function

Most pulmonary function tests are relatively simple to perform. Auscultation of breath sounds is an important diagnostic technique in pulmonary medicine, just as auscultation of heart sounds is an important technique in cardiovascular diagnosis [p. 462]. Breath sounds are more complicated to interpret than heart sounds, however, because breath sounds have a wider range of normal variation.

Normally, breath sounds are distributed evenly over the lungs and resemble a quiet "whoosh" made by flowing air. When air flow is reduced, such as in pneumothorax, breath sounds may be either diminished or absent. Abnormal sounds include various squeaks, pops, wheezes, and bubbling sounds caused by fluid and secretions in the airways or alveoli. Inflammation of the pleural membrane results in a crackling or grating sound known as a *friction rub*. It is caused by swollen, inflamed pleural membranes rubbing against each other, and it disappears when fluid again separates them.

Auscultation and pulmonary function tests, described earlier, are non-invasive tests that allow quick assessment of lung function. They can also be used to differentiate between two major types of lung disease, obstructive and restrictive.

Obstructive Lung Disease Diseases in which air flow is diminished because of increased airway resistance are known as **obstructive lung diseases.** When patients with obstructive lower airway diseases are asked to exhale forcefully, air whistling through the narrowed airways creates a wheezing sound that can be heard even without a stethoscope. Depending on the severity of the disease, the bronchioles may even collapse and close off before a forced expiration is completed, reducing both the amount and rate of air flow as measured by a spirometer.

Obstructive lung diseases include asthma, obstructive sleep apnea, emphysema, and chronic bronchitis. The latter two are sometimes called *chronic obstructive pulmonary disease* (COPD) because of their ongoing, or chronic, nature. *Obstructive sleep apnea* {*apnoia*, breathless} results from obstruction of the upper airway during sleep, often due to abnormal relaxation of the muscles of the pharynx and tongue that increases airway resistance during inspiration.

Asthma is considered an obstructive lung disease because inflammation of the airways, often associated with allergies, results in bronchoconstriction and airway edema. Asthma can be triggered by exercise (*exercise-induced asthma*) or by rapid changes in the temperature or humidity of inspired air. Asthmatic patients complain of "air hunger" and difficulty breathing, or *dyspnea*. The severity of asthma attacks ranges from mild to life threatening. Studies of asthma at the cellular level show that a variety of chemical signals may be responsible for inducing asthmatic bronchoconstriction. Among these are acetylcholine, histamine, *substance P* (a neuropeptide), and leukotrienes secreted by mast cells, macrophages, and eosinophils. *Leukotrienes* are lipidlike bronchoconstrictors that are released during the inflammatory response. Asthma is treated with inhaled and oral medications that include β_2-adrenergic agonists, anti-inflammatory drugs, and leukotriene antagonists.

Restrictive Lung Disease Pathological conditions in which lung compliance is reduced are called **restrictive lung diseases.** A decrease in lung compliance affects ventilation because respiratory muscles must work harder to stretch a stiff lung. In restrictive lung disease the energy expenditure can far exceed the normal work of breathing. Two common causes of decreased compliance are inelastic scar tissue formed in *fibrotic lung diseases*, and inadequate alveolar production of surfactant, the chemical that facilitates lung expansion.

Pulmonary **fibrosis** is characterized by the development of stiff, fibrous scar tissue that restricts lung inflation. In *idiopathic* pulmonary fibrosis {*idios*, one's own}, the cause is unknown. Other forms of fibrotic lung disease result from chronic inhalation of fine particulate matter, such as asbestos and silicon, that escapes the mucus lining the airways and reaches the alveoli. Wandering alveolar macrophages (see Fig. 17.2g) then ingest the inhaled particulate matter. If the particles are organic, the macrophages can digest them with lysosomal enzymes. However, if the particles cannot be digested or if they accumulate in large numbers, an inflammatory process ensues. The macrophages then secrete growth factors that stimulate fibroblasts in the lung's connective tissue to produce inelastic collagen, stiffening the tissue. Pulmonary fibrosis cannot be reversed.

Forced Vital Capacity Test A **forced vital capacity** test with a spirometer allows the clinician to assess respiratory system function as well as static lung volumes. In this test, the subject takes in as much air as possible, then blows it all out as fast as possible. The spirometer measures both the total volume of air exhaled (the vital capacity) and how fast that air leaves the airways (**FIG. 17.15**). The

FIG. 17.15 The forced vital capacity test

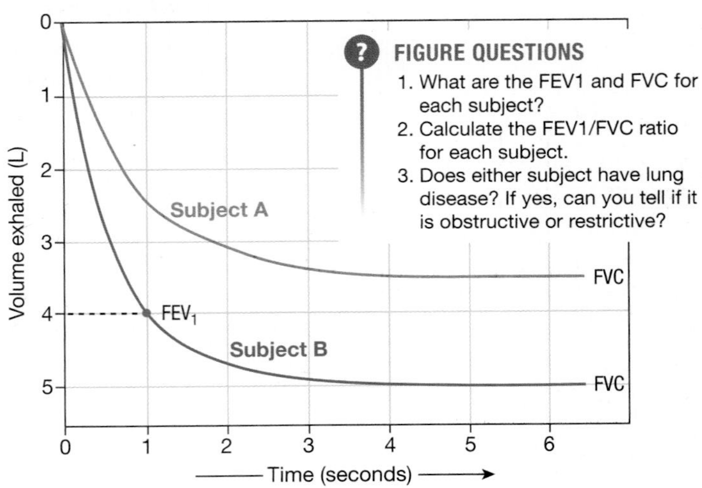
FIGURE QUESTIONS
1. What are the FEV1 and FVC for each subject?
2. Calculate the FEV1/FVC ratio for each subject.
3. Does either subject have lung disease? If yes, can you tell if it is obstructive or restrictive?
Volume exhaled (L)
0
1
2
3
4
5
Subject A
Subject B
FEV1
FVC
FVC
0
1
2
3
4
5
6
Time (seconds)

volume that leaves the airways in the first second of expiration is a measurement known as the **FEV_1**, or **forced expiratory volume in 1 second**. FEV_1 decreases in both obstructive and restrictive lung diseases, and it also decreases with age as muscles weaken and the lungs become less elastic.

The forced vital capacity test can be used to distinguish between restrictive and obstructive lung disease through calculation of the **FEV_1/FVC ratio**. In both types of lung disease, the FEV1 and FVC are decreased. In restrictive lung disease, where the lung tissue has stiffened, FEV1 and FVC decrease by about the same proportion, leaving the FEV_1/FVC ratio unchanged. In obstructive lung diseases, the loss of elastance and narrowed airways cause smaller airways to collapse during forced expiration, decreasing the FEV1 more than the FVC. This results in a lower FEV_1/FVC ratio than normal. Normally the FEV_1/FVC is 80% or more, indicating that most of the air exhaled in a forced vital capacity test leaves the airways in that first second of expiration.

Concept Check

27. Restrictive lung diseases decrease lung compliance. How will inspiratory reserve volume change in patients with a restrictive lung disease?
28. Chronic obstructive lung disease causes patients to lose the ability to exhale fully. How does residual volume change in these patients?

This completes our discussion of the mechanics of ventilation. Next, we shift focus from the bulk flow of air to the diffusion and transport of oxygen and carbon dioxide as they travel between the air spaces of the alveoli and the cells of the body.

CHAPTER 17

RUNNING PROBLEM **CONCLUSION** Emphysema

Edna leaves the office with prescriptions for a mucus-thinning drug, a bronchodilator, and anti-inflammatory drugs to keep her airways as open as possible. She has agreed to try to stop smoking once more and also has a prescription and brochures for that. Unfortunately, the lung changes that take place with COPD are not reversible, and Edna will require treatment for the rest of her life. According to the American Lung Association (*www.lung.org*), COPD costs are about $50 billion per year in direct medical costs and indirect costs such as lost wages.

In this running problem, you learned about chronic obstructive pulmonary disease. Now check your understanding of the physiology in the problem by comparing your answers with those in the following table.

Question	Facts	Integration and Analysis
Q1: *What does narrowing of the airways do to the resistance airways offer to air flow?*	The relationship between tube radius and resistance is the same for air flow as for blood flow: as radius decreases, resistance increases [p. 438].	When resistance increases, the body must use more energy to create air flow.
Q2: *Why do people with chronic bronchitis have a higher-than-normal rate of respiratory infections?*	Cigarette smoke paralyzes the cilia that sweep debris and mucus out of the airways. Without the action of cilia, mucus and trapped particles pool in the airways.	Bacteria trapped in the mucus can multiply and cause respiratory infections.
Q3: *Name the muscles that patients with emphysema use to exhale actively.*	Normal passive expiration depends on elastic recoil of muscles and elastic tissue in the lungs.	Forceful expiration involves the internal intercostal muscles and the abdominal muscles.
Q4: *When Edna fills her lungs maximally, the volume of air in her lungs is known as the ___________ capacity. When she exhales all the air she can, the volume of air left in her lungs is the ____________.*	The maximum volume of air in the lungs is the *total lung capacity*. Air left in the lungs after maximal exhalation is the *residual volume*.	N/A
Q5: *Why are Edna's RBC count and hematocrit increased?*	Because of Edna's COPD, her arterial P_{O_2} is low. The major stimulus for red blood cell synthesis is hypoxia.	Low arterial oxygen levels trigger EPO release, which increases the synthesis of red blood cells [p. 515]. More RBCs provide more binding sites for oxygen transport.

533 – 540 – 542 – 549 – 554 – 557

CHAPTER SUMMARY

Air flow into and out of the lungs is another example of the principle of *mass flow*. Like blood flow, air flow is bulk flow that requires a pump to create a pressure gradient and that encounters resistance, primarily from changes in the diameter of the tubes through which it flows. The *mechanical properties* of the pleural sacs and elastic recoil in the chest wall and lung tissue are essential for normal ventilation.

1. Aerobic metabolism in living cells consumes oxygen and produces carbon dioxide. (p. 533)
2. Gas exchange requires a large, thin, moist exchange surface; a pump to move air; and a circulatory system to transport gases to the cells. (p. 533)
3. Respiratory system functions include gas exchange, pH regulation, vocalization, and protection from foreign substances. (p. 533)

17.1 The Respiratory System

4. **Cellular respiration** refers to cellular metabolism that consumes oxygen. **External respiration** is the exchange of gases between the atmosphere and cells of the body. It includes ventilation, gas exchange at the lung and cells, and transport of gases in the blood. **Ventilation** is the movement of air into and out of the lungs. (p. 533; Fig. 17.1)
5. The **respiratory system** consists of anatomical structures involved in ventilation and gas exchange. (p. 534)
6. The **upper respiratory tract** includes the mouth, nasal cavity, **pharynx,** and **larynx.** The **lower respiratory tract** includes the **trachea, bronchi, bronchioles,** and exchange surfaces of the **alveoli.** (p. 534; Fig. 17.2a)
7. The thoracic cage is bounded by the ribs, spine, and **diaphragm.** Two sets of **intercostal muscles** connect the ribs. (p. 534; Fig. 17.2c)
8. Each **lung** is contained within a double-membrane **pleural sac** that contains a small quantity of **pleural fluid.** (p. 534; Figs. 17.2d, 17.3)
9. The two **primary bronchi** enter the lungs. Each primary bronchus divides into progressively smaller bronchi and finally into collapsible **bronchioles.** (p. 537; Figs. 17.2e, 17.4)
10. The upper respiratory system filters, warms, and humidifies inhaled air. (p. 538)
11. The alveoli consist mostly of thin-walled **type I alveolar cells** for gas exchange. **Type II alveolar cells** produce surfactant. A network of capillaries surrounds each alveolus. (p. 538; Fig. 17.2f, g)
12. Blood flow through the lungs equals cardiac output. Resistance to blood flow in the pulmonary circulation is low. Pulmonary arterial pressure averages 25/8 mm Hg. (p. 539)

17.2 Gas Laws

13. **Dalton's law** states that the total pressure of a mixture of gases is the sum of the pressures of the individual gases in the mixture. **Partial pressure** is the pressure contributed by a single gas in a mixture. (p. 540; Fig. 17.6)
14. Bulk flow of air occurs down pressure gradients, as does the movement of any individual gas making up the air. (p. 540)
15. **Boyle's law** states that as the volume available to a gas increases, the gas pressure decreases. The body creates pressure gradients by changing thoracic volume. (p. 542; Fig. 17.6b)

17.3 Ventilation

16. A single **respiratory cycle** consists of one inspiration followed by one expiration. (p. 542)
17. **Tidal volume** is the amount of air taken in during a single normal inspiration. **Vital capacity** is tidal volume plus **expiratory** and **inspiratory reserve volumes.** Air volume in the lungs at the end of maximal expiration is the **residual volume.** (p. 542; Fig. 17.7b)
18. Air flow in the respiratory system is directly proportional to the pressure gradient, and inversely related to the resistance to air flow offered by the airways. (p. 544)
19. During **inspiration, alveolar pressure** decreases, and air flows into the lungs. Inspiration requires contraction of the inspiratory muscles and the diaphragm. (p. 544; Fig. 17.9)
20. **Expiration** is usually passive, resulting from elastic recoil of the lungs. (p. 546)
21. **Active expiration** requires contraction of the internal intercostal and abdominal muscles. (p. 546)
22. **Intrapleural pressures** are subatmospheric because the pleural cavity is a sealed compartment. (p. 547; Figs. 17.9, 17.10)
23. **Compliance** is a measure of the ease with which the chest wall and lungs expand. Loss of compliance increases the work of breathing. **Elastance** is the ability of a lung to resist stretching or to return to its unstretched state. (p. 548)
24. **Surfactant** decreases surface tension in the fluid lining the alveoli. Reduced surface tension prevents smaller alveoli from collapsing and also makes it easier to inflate the lungs. (p. 549; Fig. 17.11)
25. The diameter of the bronchioles determines how much resistance they offer to air flow. (p. 550)
26. Increased CO_2 in expired air dilates bronchioles. Parasympathetic neurons cause **bronchoconstriction** in response to irritant stimuli. There is no significant sympathetic innervation of bronchioles, but epinephrine causes **bronchodilation.** (p. 550; Tbl. 17.2)
27. **Total pulmonary ventilation** = ventilation rate × tidal volume. **Alveolar ventilation** = ventilation rate × (tidal volume − dead space volume). (p. 551; Fig. 17.12a)
28. Alveolar gas composition changes very little during a normal respiratory cycle. **Hyperventilation** increases alveolar P_{O_2} and decreases alveolar P_{CO_2}. **Hypoventilation** has the opposite effect. (p. 553; Fig. 17.13)
29. Local mechanisms match air flow and blood flow around the alveoli. Increased levels of CO_2 dilate bronchioles, and decreased O_2 constricts pulmonary arterioles. (p. 554; Fig. 17.14)
30. **Restrictive lung diseases** are characterized by loss of compliance. **Obstructive lung disease** has decreased airflow. (p. 556)

REVIEW QUESTIONS

In addition to working through these questions and checking your answers on p. A-21, review the Learning Outcomes at the beginning of this chapter.

Level One Reviewing Facts and Terms

1. List four functions of the respiratory system.
2. Give two definitions for the word *respiration*.
3. Which sets of muscles are used for normal quiet inspiration? For normal, quiet expiration? For active expiration? What kind(s) of muscles are the different respiratory muscles (skeletal, cardiac, or smooth)?
4. Give two functions of pleural fluid.
5. Name the anatomical structures that an oxygen molecule passes on its way from the atmosphere to the blood.
6. Diagram the structure of an alveolus, and state the function of each part. How are capillaries associated with an alveolus?
7. Trace the path of the pulmonary circulation. About how much blood is found here at any given moment? What is a typical arterial blood pressure for the pulmonary circuit, and how does this pressure compare with that of the systemic circulation?
8. What happens to inspired air as it is conditioned during its passage through the airways?
9. During inspiration, most of the thoracic volume change is the result of movement of the ________.
10. Describe the changes in alveolar and intrapleural pressure during one respiratory cycle.
11. Refer to the spirogram in the following figure:
    a. Label tidal volume (V_T), inspiratory and expiratory reserve volumes (IRV and ERV), residual volume (RV), vital capacity (VC), total lung capacity (TLC).
    b. What is the value of each of these volumes and capacities?
    c. What is this person's ventilation rate?

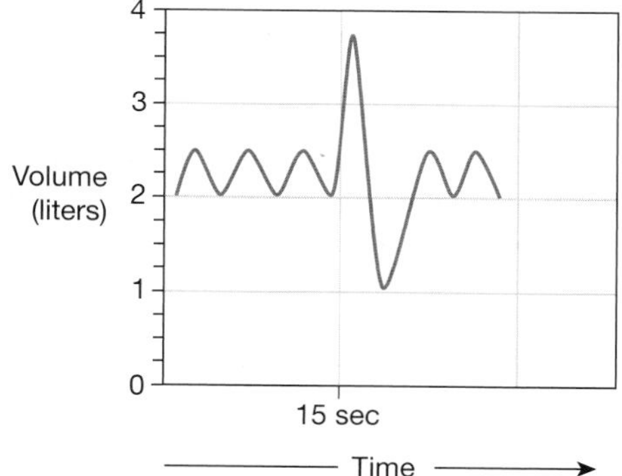
Volume (liters)
4
3
2
1
0
15 sec
Time

12. Of the three factors that contribute to the resistance of air flow through a tube, which plays the largest role in changing resistance in the human respiratory system?
13. Match the following items with their correct effect on the bronchioles:

a. histamine	1. bronchoconstriction
b. epinephrine	2. bronchodilation
c. acetylcholine	3. no effect
d. increased P_{CO_2}	

14. What is the function of surfactants in general? In the respiratory system?
15. If a person increases her tidal volume, what would happen to her alveolar P_{O_2}?

Level Two Reviewing Concepts

16. Compare and contrast the terms in each of the following sets:
    a. compliance and elastance
    b. inspiration, expiration, and ventilation
    c. intrapleural pressure and alveolar pressure
    d. total pulmonary ventilation and alveolar ventilation
    e. type I and type II alveolar cells
    f. pulmonary circulation and systemic circulation
17. List the major paracrines and neurotransmitters that cause bronchoconstriction and bronchodilation. What receptors do they act through? (muscarinic, nicotinic, α, β_1, β_2)
18. Compile the following terms into a map of ventilation. Use up arrows, down arrows, greater than symbols (>), and less than symbols (<) as modifiers. You may add other terms.

• abdominal muscles	• inspiratory muscles
• air flow	• internal intercostals
• contract	• P_A
• diaphragm	• P_{atm}
• expiratory muscles	• $P_{intrapleural}$
• external intercostals	• quiet breathing
• forced breathing	• relax
• in, out, from, to	• scalenes

19. Decide whether each of the following parameters will increase, decrease, or not change in the situations given.
    a. airway resistance with bronchodilation
    b. intrapleural pressure during inspiration
    c. air flow with bronchoconstriction
    d. bronchiolar diameter with increased P_{CO_2}
    e. tidal volume with decreased compliance
    f. alveolar pressure during expiration
20. Define the following terms: pneumothorax, spirometer, auscultation, hypoventilation, bronchoconstriction, minute volume, partial pressure of a gas.
21. The cartoon coyote is blowing up a balloon in another attempt to catch the roadrunner. He first breathes in as much air as he can, then blows out all he can into the balloon.
    a. The volume of air in the balloon is equal to the ________ of the coyote's lungs. This volume can be measured directly by measuring the balloon volume or by adding which respiratory volumes together?
    b. In 10 years, when the coyote is still chasing the roadrunner, will he still be able to put as much air into the balloon in one breath? Explain.

18 Gas Exchange and Transport

Lung tissue

The successful ascent of Everest without supplementary oxygen is one of the great sagas of the 20th century.

John B. West, Climbing with O's, *NOVA Online* (www.pbs.org)

BACKGROUND BASICS

The book *Into Thin Air* by Jon Krakauer chronicles an ill-fated trek to the top of Mt. Everest. To reach the summit of Mt. Everest, climbers must pass through the "death zone" located at about 8000 meters (over 26,000 ft.). Of the thousands of people who have attempted the summit, only about 2000 have been successful, and more than 185 have died. What are the physiological challenges of climbing Mt. Everest (8850 m or 29,035 ft.), and why did it take so many years before humans successfully reached the top? The lack of oxygen at high altitude is part of the answer.

The mechanics of breathing include the events that create bulk flow of air into and out of the lungs. In this chapter, we focus on the two gases most significant to human physiology, oxygen and carbon dioxide, and look at how they move between alveolar air spaces and the cells of the body. The process can be divided into two components: the exchange of gases between compartments, which requires diffusion across cell membranes, and the transport of gases in the blood. **FIGURE 18.1** presents an overview of the topics that we cover in this chapter.

If the diffusion of gases between alveoli and blood is significantly impaired, or if oxygen transport in the blood is inadequate, **hypoxia** (a state of too little oxygen) results. Hypoxia frequently (but not always!) goes hand in hand with **hypercapnia,** elevated concentrations of carbon dioxide. These two conditions are clinical signs, not diseases, and clinicians must gather additional information to pinpoint their cause. **TABLE 18.1** lists several types of hypoxia and some typical causes.

To avoid hypoxia and hypercapnia, the body uses sensors that monitor arterial blood composition. These sensors respond to three regulated variables:

1. *Oxygen.* Arterial oxygen delivery to the cells must be adequate to support aerobic respiration and ATP production.
2. *Carbon dioxide* (CO_2) is produced as a waste product during the citric acid cycle [p. 107]. Excretion of CO_2 by the lungs is important for two reasons: high levels of CO_2 are a central nervous system depressant, and elevated CO_2 causes a state of acidosis (low pH) through the following reaction:

$$CO_2 + H_2O \rightleftharpoons H^+ + HCO_3^-$$

3. *pH.* Maintaining pH homeostasis is critical to prevent denaturation of proteins [p. 51]. The respiratory system monitors plasma pH and uses changes in ventilation to alter pH. We discuss this process later along with renal contributions to pH homeostasis.

RUNNING PROBLEM **High Altitude**

In 1981, a group of 20 physiologists, physicians, and climbers, supported by 42 Sherpa assistants, formed the American Medical Research Expedition to Mt. Everest. The purpose of the expedition was to study human physiology at extreme altitudes, starting with the base camp at 5400 m (18,000 ft) and continuing on to the summit at 8850 m (over 29,000 ft). From the work of these scientists and others, we now have a good picture of the physiology of high-altitude acclimatization [p. 18].

563 – 565 – 569 – 573 – 577 – 582 – 583

FIG. 18.1 Pulmonary gas exchange and transport

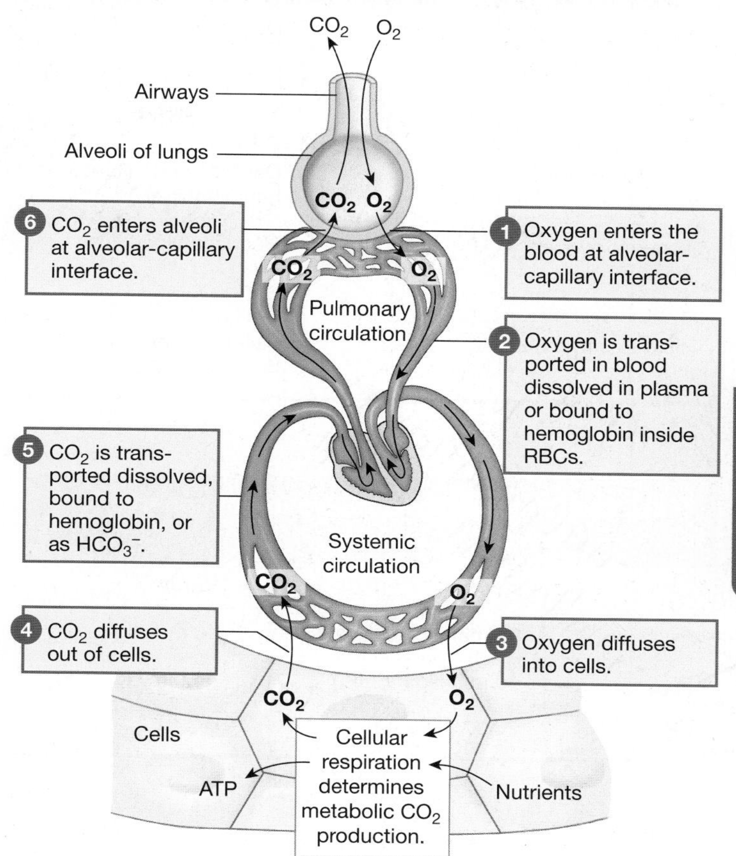
CO2 O2
Airways
Alveoli of lungs
CO2 O2
6 CO2 enters alveoli at alveolar-capillary interface.
1 Oxygen enters the blood at alveolar-capillary interface.
CO2
O2
Pulmonary circulation
2 Oxygen is transported in blood dissolved in plasma or bound to hemoglobin inside RBCs.
5 CO2 is transported dissolved, bound to hemoglobin, or as HCO3−.
Systemic circulation
CO2
O2
4 CO2 diffuses out of cells.
3 Oxygen diffuses into cells.
CO2
O2
Cells
Cellular respiration determines metabolic CO2 production.
ATP
Nutrients

The normal values for these three parameters are given in **TABLE 18.2**. In this chapter, we will consider the mechanisms by which oxygen and CO_2 move from the lungs to the cells and back again.

18.1 Gas Exchange in the Lungs and Tissues

Breathing is the bulk flow of air into and out of the lungs. Once air reaches the alveoli, individual gases such as oxygen and CO_2 diffuse from the alveolar air space into the blood. Recall that diffusion is movement of a molecule from a region of higher concentration to one of lower concentration [p. 134].

When we think of concentrations of solutions, units such as moles/liter and milliosmoles/liter come to mind. However, respiratory physiologists commonly express plasma gas concentrations in partial pressures to establish whether there is a concentration gradient between the alveoli and the blood. Gases move from regions of higher partial pressure to regions of lower partial pressure.

TABLE 18.1 Classification of Hypoxias

Type	Definition	Typical Causes
Hypoxic hypoxia	Low arterial P_{O_2}	High altitude; alveolar hypoventilation; decreased lung diffusion capacity; abnormal ventilation-perfusion ratio
Anemic hypoxia	Decreased total amount of O_2 bound to hemoglobin	Blood loss; anemia (low [Hb] or altered HbO_2 binding); carbon monoxide poisoning
Ischemic hypoxia	Reduced blood flow	Heart failure (whole-body hypoxia); shock (peripheral hypoxia); thrombosis (hypoxia in a single organ)
Histotoxic hypoxia	Failure of cells to use O_2 because cells have been poisoned	Cyanide and other metabolic poisons

TABLE 18.2 Normal Blood Values in Pulmonary Medicine

	Arterial	Venous
P_{O_2}	95 mm Hg (85–100)	40 mm Hg
P_{CO_2}	40 mm Hg (35–45)	46 mm Hg
pH	7.4 (7.38–7.42)	7.37

FIGURE 18.2 shows the partial pressures of oxygen and CO_2 in air, the alveoli, and inside the body. Normal alveolar P_{O_2} at sea level is about 100 mm Hg. The P_{O_2} of "deoxygenated" venous blood arriving at the lungs is about 40 mm Hg. Oxygen therefore diffuses down its partial pressure (concentration) gradient from the alveoli into the capillaries. Diffusion goes to equilibrium, and the P_{O_2} of arterial blood leaving the lungs is the same as in the alveoli: 100 mm Hg.

When arterial blood reaches tissue capillaries, the gradient is reversed. Cells are continuously using oxygen for oxidative phosphorylation [p. 108]. In the cells of a person at rest, intracellular P_{O_2} averages 40 mm Hg. Arterial blood arriving at the cells has a P_{O_2} of 100 mm Hg. Because P_{O_2} is lower in the cells, oxygen diffuses down its partial pressure gradient from plasma into cells. Once again, diffusion goes to equilibrium. As a result, venous blood has the same P_{O_2} as the cells it just passed.

Conversely, P_{CO_2} is higher in tissues than in systemic capillary blood because of CO_2 production during metabolism (Fig. 18.2). Cellular P_{CO_2} in a person at rest is about 46 mm Hg, compared to an arterial plasma P_{CO_2} of 40 mm Hg. The gradient causes CO_2 to diffuse out of cells into the capillaries. Diffusion goes to equilibrium, and systemic venous blood averages a P_{CO_2} of 46 mm Hg.

At the pulmonary capillaries, the process reverses. Venous blood bringing waste CO_2 from the cells has a P_{CO_2} of 46 mm Hg. Alveolar P_{CO_2} is 40 mm Hg. Because P_{CO_2} is higher in the plasma, CO_2 moves from the capillaries into the alveoli. By the time blood leaves the alveoli, it has a P_{CO_2} of 40 mm Hg, identical to the P_{CO_2} of the alveoli.

In the sections that follow, we will consider some of the other factors that affect the transfer of gases between the alveoli and the body's cells.

Concept Check

1. Cellular metabolism review: which of the following three metabolic pathways—glycolysis, the citric acid cycle, and the electron transport system—is *directly* associated with (a) O_2 consumption and with (b) CO_2 production?
2. Why doesn't the movement of oxygen from the alveoli to the plasma decrease the P_{O_2} of the alveoli? [*Hint:* p. 553]
3. If nitrogen is 78% of atmospheric air, what is the partial pressure of this gas when the dry atmospheric pressure is 720 mm Hg?

Lower Alveolar P_{O_2} Decreases Oxygen Uptake

Many variables influence the efficiency of alveolar gas exchange and determine whether arterial blood gases are normal (**FIG. 18.3a**). First, adequate oxygen must reach the alveoli. A decrease in alveolar P_{O_2} means that less oxygen is available to enter the blood. There can also be problems with the transfer of gases between the alveoli and pulmonary capillaries. Finally, blood flow, or *perfusion*, of the alveoli must be adequate [p. 554]. If something impairs blood flow to the lung, then the body is unable to acquire the oxygen it needs. Let's look in more detail at these factors.

There are two possible causes of low alveolar P_{O_2}: either (1) the inspired air has low oxygen content or (2) alveolar ventilation [p. 553] is inadequate.

Composition of the Inspired Air The first requirement for adequate oxygen delivery to the tissues is adequate oxygen intake from the atmosphere. The main factor that affects atmospheric oxygen content is altitude. The partial pressure of oxygen in air decreases along with total atmospheric pressure as you move from sea level (where normal atmospheric pressure is 760 mm Hg) to higher altitudes.

FIG. 18.2 Gases diffuse down concentration gradients

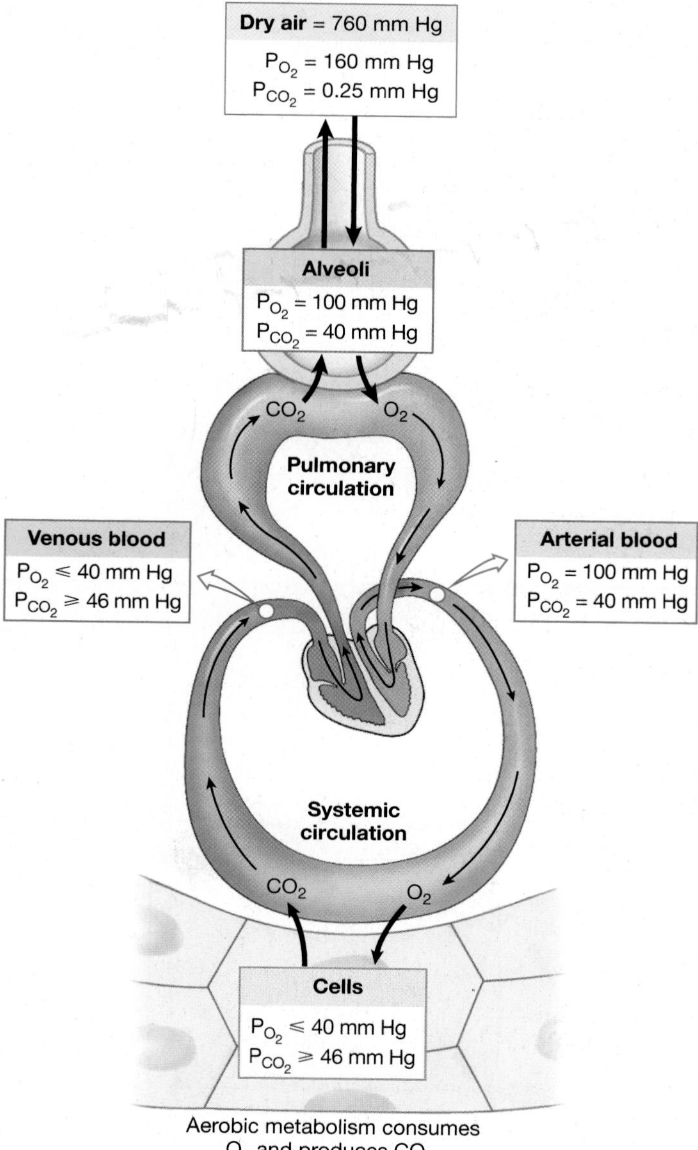
Dry air = 760 mm Hg
P_{O_2} = 160 mm Hg
P_{CO_2} = 0.25 mm Hg
Alveoli
P_{O_2} = 100 mm Hg
P_{CO_2} = 40 mm Hg
CO_2
O_2
Pulmonary circulation
Venous blood
$P_{O_2} \leq 40$ mm Hg
$P_{CO_2} \geq 46$ mm Hg
Arterial blood
P_{O_2} = 100 mm Hg
P_{CO_2} = 40 mm Hg
Systemic circulation
CO_2
O_2
Cells
$P_{O_2} \leq 40$ mm Hg
$P_{CO_2} \geq 46$ mm Hg
Aerobic metabolism consumes O_2 and produces CO_2.

For example, Denver, 1609 m above sea level, has an atmospheric pressure of about 628 mm Hg. The P_{O_2} of dry air in Denver is 132 mm Hg, down from 160 mm Hg at sea level. For fully humidified atmospheric air reaching the alveoli, the P_{O_2} is even lower: P_{atm} 628 mm Hg − P_{H_2O} 47 mm Hg = 581 mm Hg × 21% = P_{O_2} of 122 mm Hg, down from 150 mm Hg at sea level. Notice that water vapor pressure at 100% humidity is the same no matter what the altitude, making its contribution to total pressure in the lungs more important as you go higher.

Alveolar Ventilation Unless a person is traveling, altitude remains constant. If the composition of inspired air is normal but alveolar P_{O_2} is low, then the problem must lie with alveolar ventilation. Low alveolar ventilation is also known as *hypoventilation* and is characterized by lower-than-normal volumes of fresh air entering the alveoli. Pathological changes that can result in alveolar hypoventilation (Fig. 18.3c) include decreased lung compliance [p. 548], increased airway resistance [p. 550], or central nervous system (CNS) depression that slows ventilation rate and decreases depth. Common causes of CNS depression in young people include alcohol poisoning and drug overdoses.

RUNNING PROBLEM

Hypoxia is the primary problem that people experience when ascending to high altitude. High altitude is considered anything above 1500 m (5000 ft), but most pathological responses to altitude occur above 2500 m (about 8000 ft). By one estimate, 25% of people arriving at 2500 m will experience some form of altitude sickness.

Q1: *If water vapor contributes 47 mm Hg to the pressure of fully humidified air, what is the P_{O_2} of inspired air reaching the alveoli at 2500 m, where dry atmospheric pressure is 542 mm Hg? How does this value for P_{O_2} compare with that of fully humidified air at sea level? [p. 541]*

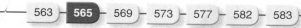

563 – 565 – 569 – 573 – 577 – 582 – 583

Concept Check

4. At the summit of Mt. Everest, an altitude of 8850 m, atmospheric pressure is only 250 mm Hg. What is the P_{O_2} of dry atmospheric air atop Everest? If water vapor added to inhaled air at the summit has a partial pressure of 47 mm Hg, what is the P_{O_2} of the inhaled air when it reaches the alveoli?

Diffusion Problems Cause Hypoxia

If hypoxia is not caused by hypoventilation, then the problem usually lies with some aspect of gas exchange between alveoli and blood. In these situations, alveolar P_{O_2} may be normal, but the P_{O_2} of arterial blood leaving the lungs is low. The transfer of oxygen from alveoli to blood requires diffusion across the barrier created by type I alveolar cells and the capillary endothelium (Fig. 18.3b).

The exchange of oxygen and carbon dioxide across this diffusion barrier obeys the same rules as simple diffusion across a membrane [p. 134]. The diffusion rate is directly proportional to the available surface area, the concentration gradient of the gas, and the permeability of the barrier:

$$\text{Diffusion rate} \propto \text{surface area} \times \text{concentration gradient} \times \text{barrier permeability}$$

From the general rules for diffusion, we can add a fourth factor: *diffusion distance*. Diffusion is inversely proportional to the square of the distance or, in simpler terms—diffusion is most rapid over short distances [p. 134]:

$$\text{Diffusion rate} \propto 1/\text{distance}^2$$

FIG. 18.3 ESSENTIALS Factors Affecting Gas Exchange in the Alveoli

(a) Alveolar gas exchange

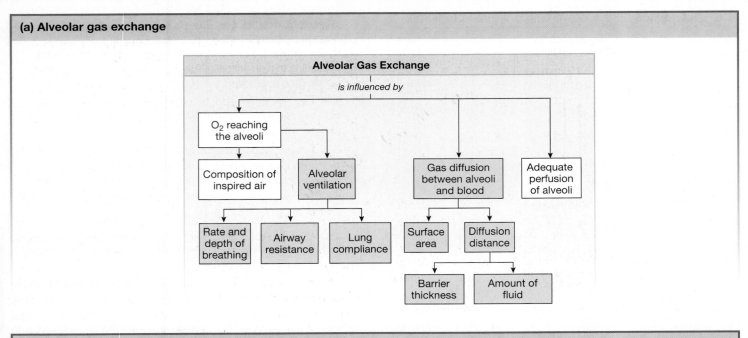

Alveolar Gas Exchange

is influenced by

- O_2 reaching the alveoli
  - Composition of inspired air
  - Alveolar ventilation
    - Rate and depth of breathing
    - Airway resistance
    - Lung compliance
- Gas diffusion between alveoli and blood
  - Surface area
  - Diffusion distance
    - Barrier thickness
    - Amount of fluid
- Adequate perfusion of alveoli

(b) Cells form a diffusion barrier between lung and blood.

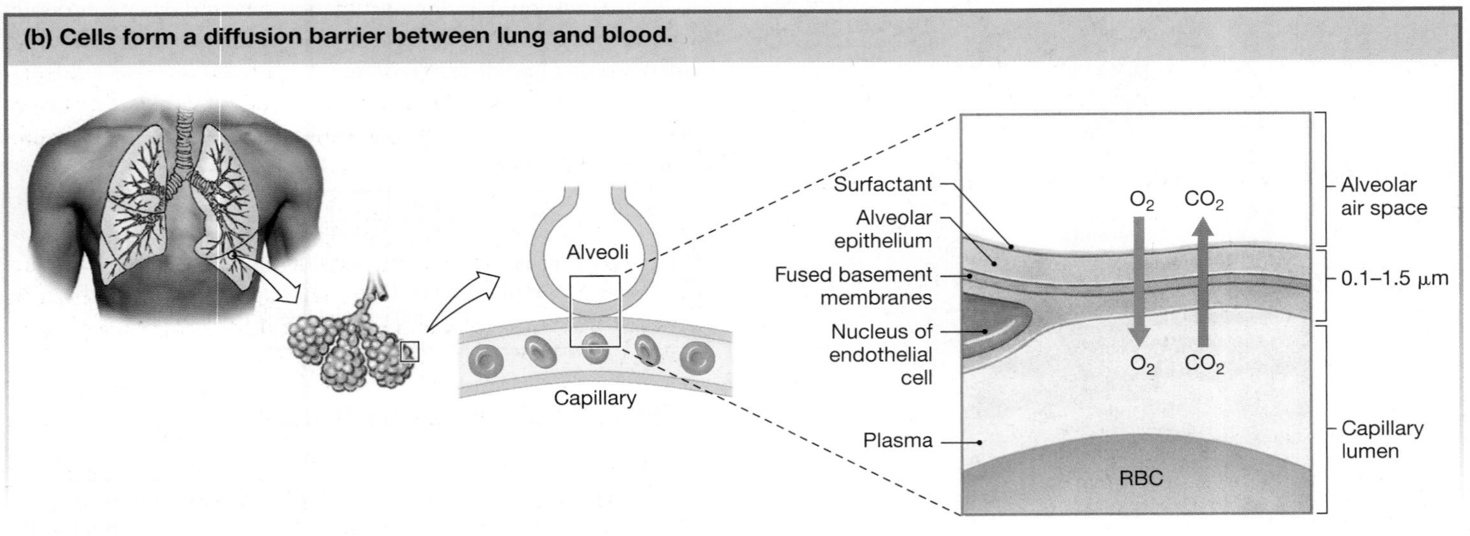

Alveoli
Capillary
Surfactant
Alveolar epithelium
Fused basement membranes
Nucleus of endothelial cell
Plasma
O_2
CO_2
O_2
CO_2
RBC
Alveolar air space
0.1–1.5 μm
Capillary lumen

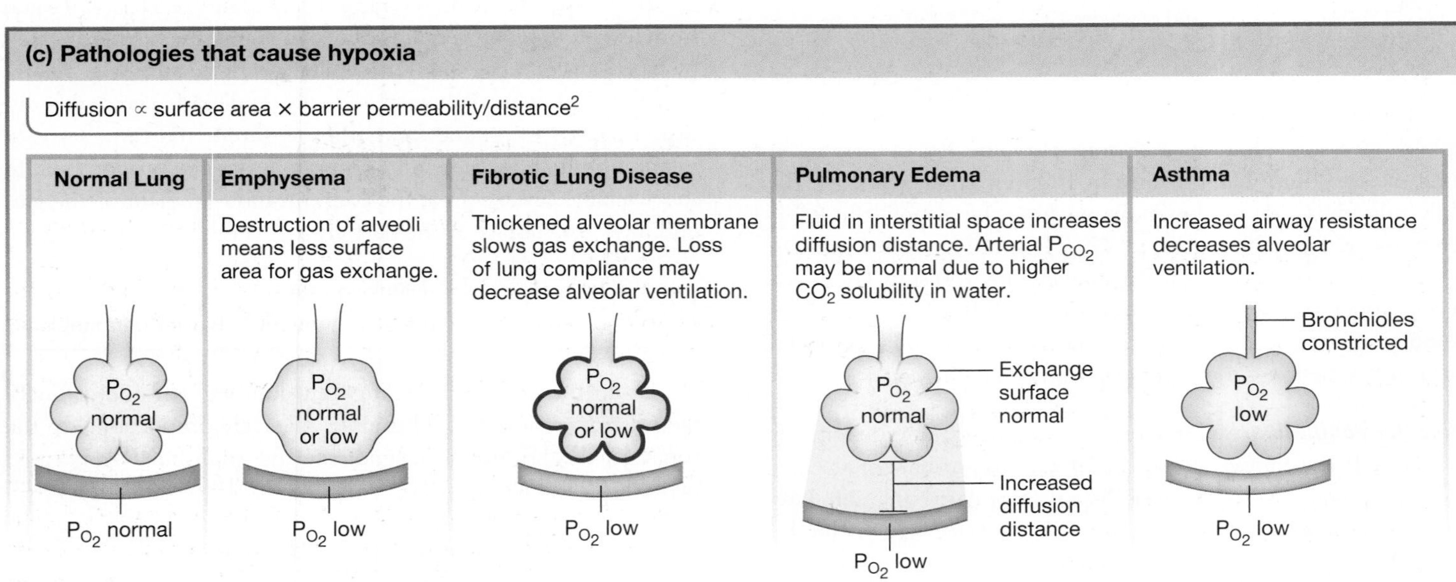

(c) Pathologies that cause hypoxia
Diffusion ∝ surface area × barrier permeability/distance^2
Normal Lung
P_{O_2} normal
P_{O_2} normal
Emphysema
Destruction of alveoli means less surface area for gas exchange.
P_{O_2} normal or low
P_{O_2} low
Fibrotic Lung Disease
Thickened alveolar membrane slows gas exchange. Loss of lung compliance may decrease alveolar ventilation.
P_{O_2} normal or low
P_{O_2} low
Pulmonary Edema
Fluid in interstitial space increases diffusion distance. Arterial P_{CO_2} may be normal due to higher CO_2 solubility in water.
P_{O_2} normal
Exchange surface normal
Increased diffusion distance
P_{O_2} low
Asthma
Increased airway resistance decreases alveolar ventilation.
Bronchioles constricted
P_{O_2} low
P_{O_2} low

Under most circumstances, diffusion distance, surface area, and barrier permeability in the body are constants and are maximized to facilitate diffusion. Gas exchange in the lungs is rapid, blood flow through pulmonary capillaries is slow, and diffusion reaches equilibrium in less than 1 second. This leaves the concentration gradient between alveoli and blood as the primary factor affecting gas exchange in healthy people.

The factors of surface area, diffusion distance, and membrane permeability do come into play with various diseases. Pathological changes that adversely affect gas exchange include (1) a decrease in the amount of alveolar surface area available for gas exchange, (2) an increase in the thickness of the alveolar-capillary exchange barrier, and (3) an increase in the diffusion distance between the alveolar air space and the blood.

Surface Area Physical loss of alveolar surface area can have devastating effects in *emphysema,* a degenerative lung disease most often caused by cigarette smoking (Fig. 18.3c). The irritating effect of smoke chemicals and tar in the alveoli activates alveolar macrophages that release *elastase* and other proteolytic enzymes. These enzymes destroy the elastic fibers of the lung [p. 538] and induce apoptosis of cells, breaking down the walls of the alveoli. The result is a high-compliance/low-elastic recoil lung with fewer and larger alveoli and less surface area for gas exchange.

Diffusion Barrier Permeability Pathological changes in the alveolar-capillary diffusion barrier may alter its properties and slow gas exchange. For example, in fibrotic lung diseases, scar tissue thickens the alveolar wall (Fig. 18.3c). Diffusion of gases through this scar tissue is much slower than normal. However, because the lungs have a built-in reserve capacity, one-third of the exchange epithelium must be incapacitated before arterial P_{O_2} falls significantly.

BIOTECHNOLOGY

The Pulse Oximeter

One important clinical indicator of the effectiveness of gas exchange in the lungs is the concentration of oxygen in arterial blood. Obtaining an arterial blood sample is difficult for the clinician and painful for the patient because it means finding an accessible artery. (Most blood is drawn from superficial veins rather than from arteries, which lie deeper within the body.) Over the years, however, scientists have developed instruments that quickly and painlessly measure blood oxygen levels through the surface of the skin on a finger or earlobe. One such instrument, the *pulse oximeter*, clips onto the skin and in seconds gives a digital reading of arterial hemoglobin saturation. The oximeter works by measuring light absorbance of the tissue's hemoglobin at two wavelengths. Another instrument, the *transcutaneous oxygen sensor,* measures dissolved oxygen using a variant of traditional gas-measuring electrodes. Both methods have limitations but are popular because they provide a rapid, noninvasive means of estimating arterial oxygen content.

Diffusion Distance Normally, the pulmonary diffusion distance is small because the alveolar and endothelial cells are thin and there is little or no interstitial fluid between the two cell layers (Fig. 18.3b). However, in certain pathological states, excess fluid increases the diffusion distance between the alveolar air space and the blood. Fluid accumulation may occur inside the alveoli or in the interstitial compartment between the alveolar epithelium and the capillary.

In **pulmonary edema,** accumulation of interstitial fluid increases the diffusion distance and slows gas exchange (Fig. 18.3c). Normally, only small amounts of interstitial fluid are present in the lungs, the result of low pulmonary blood pressure and effective lymph drainage. However, if pulmonary blood pressure rises for some reason, such as left ventricular failure or mitral valve dysfunction, the normal filtration/reabsorption balance at the capillary is disrupted [Fig. 15.18, p. 498].

When capillary hydrostatic pressure increases, more fluid filters out of the capillary. If filtration increases too much, the lymphatics are unable to remove all the fluid, and excess accumulates in the pulmonary interstitial space, creating pulmonary edema. In severe cases, if edema exceeds the tissue's ability to retain it, fluid leaks from the interstitial space into the alveolar air space, flooding the alveoli. Normally the inside of the alveoli is a moist surface lined by a very thin (about 2–5 μm) layer of fluid with surfactant (see Fig. 18.3b). With alveolar flooding, this fluid layer can become much thicker and seriously impair gas exchange. Alveolar flooding can also occur with leakage when alveolar epithelium is damaged, such as from inflammation or inhaling toxic gases. If hypoxia due to alveolar fluid accumulation is severe and cannot be corrected by oxygen therapy, the condition may be called *adult respiratory distress syndrome* or ARDS.

Concept Check

5. Why would left ventricular failure or mitral valve dysfunction cause elevated pulmonary blood pressure?
6. If alveolar ventilation increases, what happens to arterial P_{O_2}? To arterial P_{CO_2}? To venous P_{O_2} and P_{CO_2}? Explain your answers.

Gas Solubility Affects Diffusion

A final factor that can affect gas exchange in the alveoli is the solubility of the gas. The movement of gas molecules from air into a liquid is directly proportional to three factors: (1) the pressure gradient of the gas, (2) the solubility of the gas in the liquid, and (3) temperature. Because temperature is relatively constant in mammals, we can ignore its contribution in this discussion.

When a gas is placed in contact with water and there is a pressure gradient, gas molecules move from one phase to the other. If gas pressure is higher in the water than in the gaseous phase, then gas molecules leave the water. If gas pressure is higher in the gaseous phase than in water, then the gas dissolves into the water.

For example, consider a container of water exposed to air with a P_{O_2} of 100 mm Hg (**FIG. 18.4a**). Initially, the water has no

FIG. 18.6 Mass balance and the Fick equation

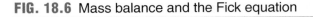

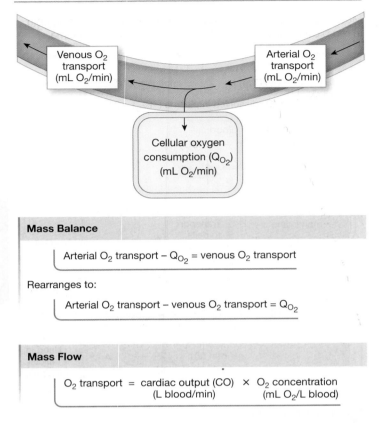

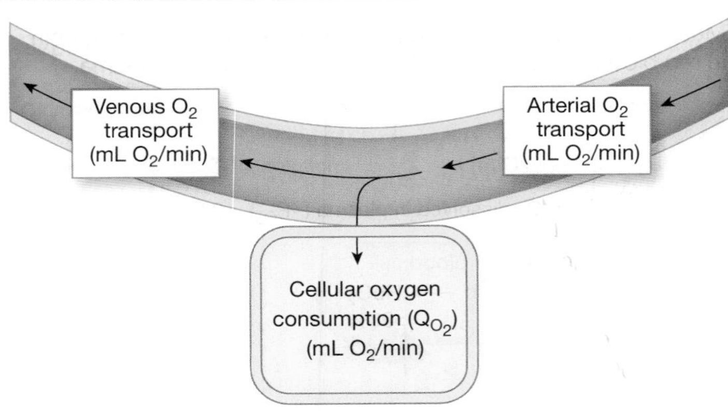
Venous O_2 transport (mL O_2/min)
Arterial O_2 transport (mL O_2/min)
Cellular oxygen consumption (Q_{O_2}) (mL O_2/min)

Mass Balance

Arterial O_2 transport – Q_{O_2} = venous O_2 transport

Rearranges to:

Arterial O_2 transport – venous O_2 transport = Q_{O_2}

Mass Flow

O_2 transport = cardiac output (CO) (L blood/min) × O_2 concentration (mL O_2/L blood)

Fick Equation

Substitute the mass flow equation for O_2 transport in the mass balance equation:

(CO × Arterial $[O_2]$) – (CO × Venous $[O_2]$) = Q_{O_2}

Using algebra (AB) – (AC) = A(B – C):

CO × (Arterial $[O_2]$ – Venous $[O_2]$) = Q_{O_2}

FIGURE QUESTION

During exercise, a man consumes 1.8 L of oxygen per minute. His arterial oxygen content is 190 mL O_2/L blood, and the oxygen content of his venous blood is 134 mL O_2/L blood. What is his cardiac output?

four globular protein chains (*globins*), each centered around an iron-containing *heme* group [p. 517]. The central iron atom of each heme group can bind reversibly with one oxygen molecule. The iron-oxygen interaction is a weak bond that can be easily broken without altering either the hemoglobin or the oxygen.

With four heme groups per hemoglobin molecule, one hemoglobin molecule has the potential to bind four oxygen molecules. Hemoglobin bound to oxygen is known as **oxyhemoglobin,** abbreviated HbO_2. It would be most accurate to show the number of oxygen molecules carried on each hemoglobin molecule—$Hb(O_2)_{1-4}$—but we use the simpler abbreviation of HbO_2 because the number of bound oxygen molecules varies from one hemoglobin molecule to another.

Oxygen Binding Obeys the Law of Mass Action

The hemoglobin binding reaction $Hb + O_2 \rightleftarrows HbO_2$ obeys the law of mass action [p. 48]. As the concentration of free O_2 increases, more oxygen binds to hemoglobin and the equation shifts to the right, producing more HbO_2. If the concentration of O_2 decreases, the equation shifts to the left. Hemoglobin releases oxygen, and the amount of oxyhemoglobin decreases.

In the blood, the free oxygen available to bind to hemoglobin is dissolved oxygen, indicated by the P_{O_2} of plasma (Fig. 18.5). In pulmonary capillaries, oxygen from the alveoli first dissolves in plasma. Dissolved O_2 then diffuses into the red blood cells, where it can bind to hemoglobin. The hemoglobin acts like a sponge, soaking up oxygen from the plasma until the reaction $Hb + O_2 \rightleftarrows HbO_2$ reaches equilibrium.

The transfer of oxygen from alveolar air to plasma to red blood cells and onto hemoglobin occurs so rapidly that blood in the pulmonary capillaries normally picks up as much oxygen as the P_{O_2} of the plasma and the number of red blood cells allow.

Once arterial blood reaches the tissues, the exchange process that took place in the lungs reverses. Dissolved oxygen diffuses out of systemic capillaries into cells, which have a lower P_{O_2}. This decreases plasma P_{O_2} and disturbs the equilibrium of the oxygen-hemoglobin binding reaction by removing O_2 from the left side of the equation. The equilibrium shifts to the left according to the law of mass action, causing hemoglobin molecules release their oxygen stores (bottom half of Fig. 18.5).

Like oxygen loading at the lungs, transferring oxygen to the body's cells takes place very rapidly and goes to equilibrium. The

RUNNING PROBLEM

Acute mountain sickness is the mildest illness caused by altitude hypoxia. The primary symptom is a headache that may be accompanied by dizziness, nausea, fatigue, or confusion. More severe illnesses are *high-altitude pulmonary edema* (HAPE) and high-altitude cerebral edema. HAPE is the major cause of death from altitude sickness. It is characterized by high pulmonary arterial pressure, extreme shortness of breath, and sometimes a productive cough yielding a pink, frothy fluid. Treatment is immediate relocation to lower altitude and administration of oxygen.

Q2: *Why would someone with HAPE be short of breath?*

Q3: *Based on what you learned about the mechanisms for matching ventilation and perfusion in the lung [p. 554], can you explain why patients with HAPE have elevated pulmonary arterial blood pressure?*

563 – 565 – **570** – 573 – 579 – 581 – 583

P_{O_2} of the cells determines how much oxygen is unloaded from hemoglobin. As cells increase their metabolic activity, their P_{O_2} decreases, and hemoglobin releases more oxygen to them.

Hemoglobin Transports Most Oxygen to the Tissues

We must have adequate amounts of hemoglobin in our blood or we cannot survive. To understand why, consider the following example.

Assume that a person's total oxygen consumption at rest is about 250 mL O_2/min and cardiac output is 5 L blood/min. How much oxygen must blood contain to meet this demand?

$$250 \text{ mL } O_2\text{/min consumed} = 5 \text{ L blood/min} \times ? \text{ mL } O_2\text{/L blood}$$

To meet the cells' needs for oxygen, the 5 L of blood/min coming to the tissues would need to contain at least 250 mL O_2. This calculates to a blood oxygen concentration of 50 mL O_2/L blood.

The low solubility of oxygen in plasma means that only 3 mL of O_2 can dissolve in the plasma fraction of each liter of arterial blood (**FIG. 18.7a**). The dissolved oxygen delivered to the cells in plasma therefore is

$$3 \text{ mL } O_2\text{/L blood} \times 5 \text{ L blood/min} = 15 \text{ mL } O_2\text{/min}$$

The cells require at least 250 mL O_2/min, so the small amount of oxygen that dissolves in plasma cannot meet the needs of the tissues at rest.

Now let's consider the difference in oxygen delivery if hemoglobin is available. At normal hemoglobin levels, red blood cells carry about 197 mL O_2/L blood (Fig. 18.7b).

$$\begin{aligned}\text{Total blood } O_2 \text{ content} &= \text{dissolved } O_2 + O_2 \text{ bound to Hb}\\ &= 3 \text{ mL } O_2\text{/L blood} + 197 \text{ mL } HbO_2\text{/L blood}\\ &= 200 \text{ mL } O_2\text{/L blood}\end{aligned}$$

If cardiac output remains 5 L/min, total oxygen delivery to cells is 1000 mL/min with hemoglobin present:

$$200 \text{ mL } O_2\text{/L blood} \times 5 \text{ L blood/min} = 1000 \text{ mL } O_2\text{/min}$$

This is four times the oxygen consumption needed by the tissues at rest. The extra O_2 serves as a reserve for times when oxygen demand increases, such as with exercise.

CHAPTER 18

P_{O_2} Determines Oxygen-Hb Binding

The amount of oxygen that binds to hemoglobin depends on two factors: (1) the P_{O_2} in the plasma surrounding the red blood cells and (2) the number of potential Hb binding sites available in the

FIG. 18.7 Hemoglobin increases oxygen transport

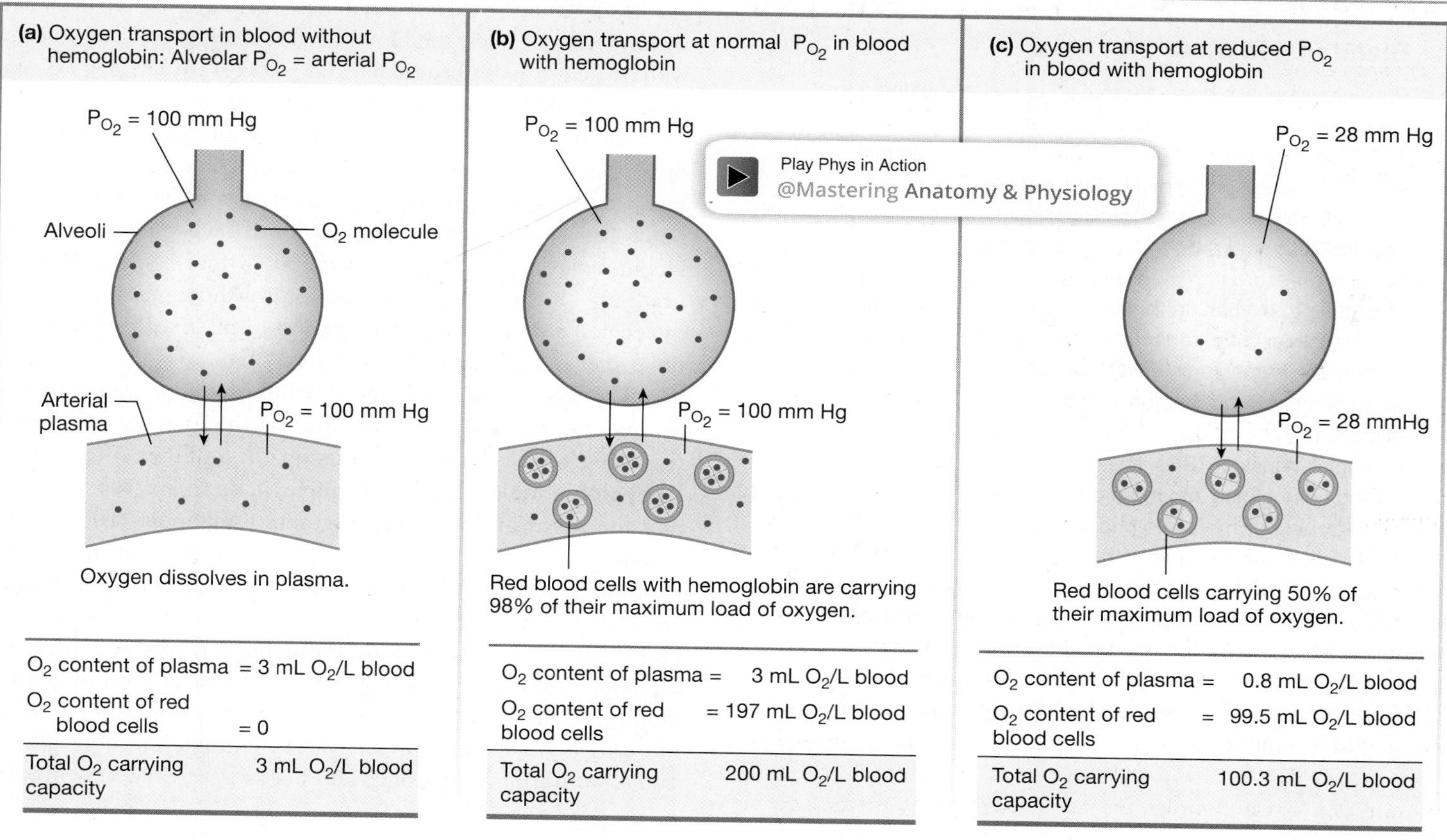
(a) Oxygen transport in blood without hemoglobin: Alveolar P_{O_2} = arterial P_{O_2}
P_{O_2} = 100 mm Hg
Alveoli
O_2 molecule
Arterial plasma
P_{O_2} = 100 mm Hg
Oxygen dissolves in plasma.
O_2 content of plasma = 3 mL O_2/L blood
O_2 content of red blood cells = 0
Total O_2 carrying capacity 3 mL O_2/L blood
(b) Oxygen transport at normal P_{O_2} in blood with hemoglobin
P_{O_2} = 100 mm Hg
Play Phys in Action
@Mastering Anatomy & Physiology
P_{O_2} = 100 mm Hg
Red blood cells with hemoglobin are carrying 98% of their maximum load of oxygen.
O_2 content of plasma = 3 mL O_2/L blood
O_2 content of red blood cells = 197 mL O_2/L blood
Total O_2 carrying capacity 200 mL O_2/L blood
(c) Oxygen transport at reduced P_{O_2} in blood with hemoglobin
P_{O_2} = 28 mm Hg
P_{O_2} = 28 mmHg
Red blood cells carrying 50% of their maximum load of oxygen.
O_2 content of plasma = 0.8 mL O_2/L blood
O_2 content of red blood cells = 99.5 mL O_2/L blood
Total O_2 carrying capacity 100.3 mL O_2/L blood

red blood cells (FIG. 18.8). Plasma P_{O_2} is the primary factor determining what percentage of the available hemoglobin binding sites are occupied by oxygen, known as *the percent saturation of hemoglobin.* As you learned in previous sections, arterial P_{O_2} is established by (1) the composition of inspired air, (2) the alveolar ventilation rate, and (3) the efficiency of gas exchange from alveoli to blood. Figure 18.7c shows what happens to O_2 transport when P_{O_2} decreases.

The total number of oxygen-binding sites depends on the number of hemoglobin molecules in red blood cells. Clinically, this number can be estimated either by counting the red blood cells and quantifying the amount of hemoglobin per red blood cell (*mean corpuscular hemoglobin*) or by determining the blood hemoglobin content (g Hb/dL whole blood). Any pathological condition that decreases the amount of hemoglobin in red blood cells or the number of red blood cells will adversely affect the blood's oxygen-transporting capacity.

People who have lost large amounts of blood need to replace hemoglobin for oxygen transport. A blood transfusion is the ideal replacement for blood loss, but in emergencies this is not always possible. Saline infusions can replace lost blood volume, but saline, like plasma, cannot transport sufficient quantities of oxygen to support cellular respiration. Faced with this problem, researchers are currently testing artificial oxygen carriers to replace hemoglobin. In times of large-scale disasters, these hemoglobin substitutes could eliminate the need to identify a patient's blood type before giving transfusions.

EMERGING CONCEPTS

Blood Substitutes

Physiologists have been attempting to find a substitute for blood ever since 1878, when an intrepid physician named T. Gaillard Thomas transfused a patient with whole milk in place of blood. (It helped but the patient died anyway.) Although milk seems an unlikely replacement for blood, it has two important properties: proteins to provide colloid osmotic pressure and molecules (emulsified lipids) capable of binding to oxygen. In the development of hemoglobin substitutes, oxygen transport is the most difficult property to mimic. A hemoglobin solution would seem to be the obvious answer, but hemoglobin that is not compartmentalized in red blood cells behaves differently than native hemoglobin inside RBCs. Investigators have tried polymerizing hemoglobin into more stable molecules, loading hemoglobin polymers into phospholipid liposomes [p. 63], or combining hemoglobin with other compounds. Unfortunately, the products developed to date have adverse side effects. To learn more about this ongoing research, read "Evaluating the Safety and Efficacy of Hemoglobin-based Blood Substitutes (FDA)" *(www.fda.gov/biologicsbloodvaccines/scienceresearch/biologicsresearchareas/ucm127061.htm).*

FIG. 18.8 Factors controlling oxygen-hemoglobin binding

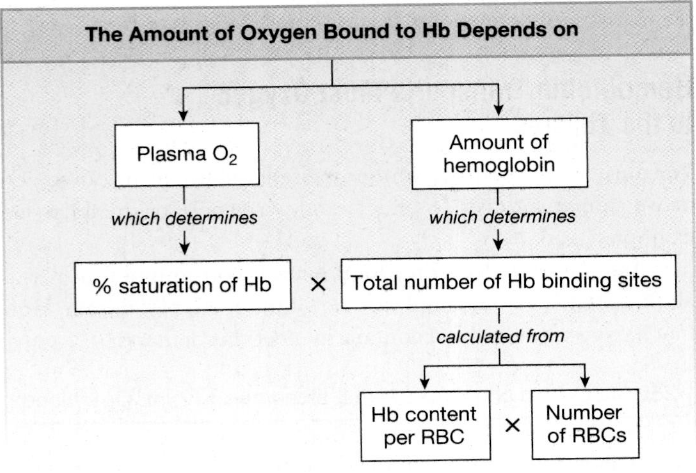
The Amount of Oxygen Bound to Hb Depends on
Plasma O_2
Amount of hemoglobin
which determines
which determines
% saturation of Hb
×
Total number of Hb binding sites
calculated from
Hb content per RBC
×
Number of RBCs

Oxygen Binding Is Expressed as a Percentage

As you just learned, the amount of oxygen bound to hemoglobin at any given P_{O_2} is expressed as the **percent saturation of hemoglobin,** where

$$(\text{Amount of } O_2 \text{ bound/maximum that could be bound}) \times 100 = \text{percent saturation of hemoglobin}$$

If all binding sites on all hemoglobin molecules are occupied by oxygen molecules, the blood is 100% oxygenated, or *saturated* with oxygen. If half the available binding sites are carrying oxygen, the hemoglobin is 50% saturated, and so on.

The relationship between plasma P_{O_2} and percent saturation of hemoglobin can be explained with the following analogy. The hemoglobin molecules carrying oxygen are like students moving books from an old library to a new one. Each student (a hemoglobin molecule) can carry a maximum of four books (100% saturation). The librarian in charge controls how many books (O_2 molecules) each student will carry, just as plasma P_{O_2} determines the percent saturation of hemoglobin.

The total number of books being carried depends on the number of available students, just as the amount of oxygen delivered to the tissues depends on the number of available hemoglobin molecules. For example, if there are 100 students, and the librarian gives each of them four books (100% saturation), then 400 books are carried to the new library. If the librarian gives three books to each student (decreased plasma P_{O_2}), then only 300 books go to the new library, even though each student could carry four. (Students carrying only three of a possible four books correspond to 75% saturation of hemoglobin.) If the librarian is handing out four books per student but only 50 students show up (fewer hemoglobin molecules at 100% saturation), then only 200 books get to the new library, even though the students are taking the maximum number of books they can carry.

The physical relationship between P_{O_2} and how much oxygen binds to hemoglobin can be studied *in vitro.* Researchers expose samples of hemoglobin to various P_{O_2} levels and quantitatively determine the amount of oxygen that binds. **Oxyhemoglobin saturation curves,** such as the ones shown in **FIGURE 18.9**, are the result of these *in vitro* binding studies. (These curves are also called *dissociation curves.*)

The shape of the HbO_2 saturation curve reflects the properties of the hemoglobin molecule and its affinity for oxygen. If you look at the curve, you find that at normal alveolar and arterial P_{O_2} (100 mm Hg), 98% of the hemoglobin is bound to oxygen (Fig. 18.9a). In other words, as blood passes through the lungs under normal conditions, hemoglobin picks up nearly the maximum amount of oxygen that it can carry.

Notice that the curve is nearly flat at P_{O_2} levels higher than 100 mm Hg (that is, the slope approaches zero). At P_{O_2} above 100 mm Hg, even large changes in P_{O_2} cause only minor changes in percent saturation. In fact, hemoglobin is not 100% saturated until the P_{O_2} reaches nearly 650 mm Hg, a partial pressure far higher than anything we encounter in everyday life.

The flattening of the saturation curve at higher P_{O_2} also means that alveolar P_{O_2} can fall a good bit below 100 mm Hg without significantly lowering hemoglobin saturation. As long as P_{O_2} in the alveoli (and thus in the pulmonary capillaries) stays above 60 mm Hg, hemoglobin is more than 90% saturated and maintains near-normal levels of oxygen transport. However, once P_{O_2} falls below 60 mm Hg, the curve becomes steeper. The steep slope means that a small decrease in P_{O_2} causes a relatively large release of oxygen.

For example, if P_{O_2} falls from 100 mm Hg to 60 mm Hg, the percent saturation of hemoglobin goes from 98 to about 90%, a decrease of 8%. This is equivalent to a saturation change of 2% for each 10 mm Hg change. If P_{O_2} falls further, from 60 to 40 mm Hg, the percent saturation goes from 90 to 75%, a decrease of 7.5% for each 10 mm Hg. In the 40–20 mm Hg range, the curve is even steeper. Hemoglobin saturation declines from 75 to 35%, a change of 20% for each 10 mm Hg change.

What is the physiological significance of the shape of the saturation curve? In blood leaving systemic capillaries with a P_{O_2} of 40 mm Hg (an average value for venous blood in a person at rest), hemoglobin is still 75% saturated. This means that at the cells, blood releases only one-fourth of the oxygen it is capable of carrying. The oxygen that remains bound serves as a reservoir that cells can draw on if metabolism increases.

When metabolically active tissues use additional oxygen, their cellular P_{O_2} decreases, and hemoglobin releases additional O_2 at the cells. At a P_{O_2} of 20 mm Hg (an average value for exercising muscle), hemoglobin saturation falls to about 35%. With this 20 mm Hg decrease in P_{O_2} (40 mm Hg to 20 mm Hg), hemoglobin releases an additional 40% of the oxygen it is capable of carrying. This is another example of the built-in reserve capacity of the body.

RUNNING PROBLEM

In most people arriving at high altitude, normal physiological responses kick in to help acclimatize the body to the chronic hypoxia. Within two hours of arrival, hypoxia triggers the release of erythropoietin from the kidneys and liver. This hormone stimulates red blood cell production, and as a result, new erythrocytes appear in the blood within days.

Q4: *How does adding erythrocytes to the blood help a person acclimatize to high altitude?*

Q5: *What does adding erythrocytes to the blood do to the viscosity of the blood? What effect will that change in viscosity have on blood flow?*

563 – 565 – 570 – **573** – 577 – 581 – 583

Several Factors Affect O_2-Hb Binding

Any factor that changes the conformation of the hemoglobin protein may affect its ability to bind oxygen. In humans, physiological changes in plasma pH, temperature, and P_{CO_2} all alter the oxygen-binding affinity of hemoglobin. Changes in binding affinity are reflected by changes in the shape of the HbO_2 saturation curve.

Decreased pH, increased temperature, or increased P_{CO_2}, decrease the affinity of hemoglobin for oxygen and shift the oxygen-hemoglobin saturation curve to the right (Fig. 18.9c–e). When these factors change in the opposite direction, binding affinity increases, and the curve shifts to the left. Notice that when the curve shifts in either direction, the changes are much more pronounced in the steep part of the curve. Physiologically, this means that oxygen binding at the lungs (in the 90–100 mm Hg P_{O_2} range) is not greatly affected, but oxygen delivery at the tissues (in the 20–40 mm Hg range) is significantly altered.

Let's examine one situation, the affinity shift that takes place when pH decreases from 7.4 (normal) to 7.2 (more acidic). (The normal range for blood pH is 7.38–7.42, but a pH of 7.2 is compatible with life.) Look at the graph in Figure 18.9c.

At a P_{O_2} of 40 mm Hg (equivalent to a resting cell) and pH of 7.4, hemoglobin is about 75% saturated. At the same P_{O_2}, if the pH falls to 7.2, the percent saturation decreases to about 62%. This means that hemoglobin molecules release 13% more oxygen at pH 7.2 than they do at pH 7.4.

When does the body undergo shifts in blood pH? One situation is with maximal exertion that pushes cells into anaerobic metabolism. Anaerobic metabolism in exercising muscle fibers releases H^+ into the cytoplasm and extracellular fluid. As H^+ concentrations increase, pH falls, the affinity of hemoglobin for oxygen decreases, and the HbO_2 saturation curve shifts to the right. More oxygen is released at the tissues as the blood becomes more acidic (pH decreases). A shift in the hemoglobin saturation curve that results from a change in pH is called the **Bohr effect.**

An additional factor that affects oxygen-hemoglobin binding is **2,3-bisphosphoglycerate** (2,3-BPG; previously called

FIG. 18.9 ESSENTIALS Oxygen-Hemoglobin Binding Curves

Binding Properties of Adult and Fetal Hemoglobin

(a) The oxyhemoglobin saturation curve is determined *in vitro* in the laboratory.

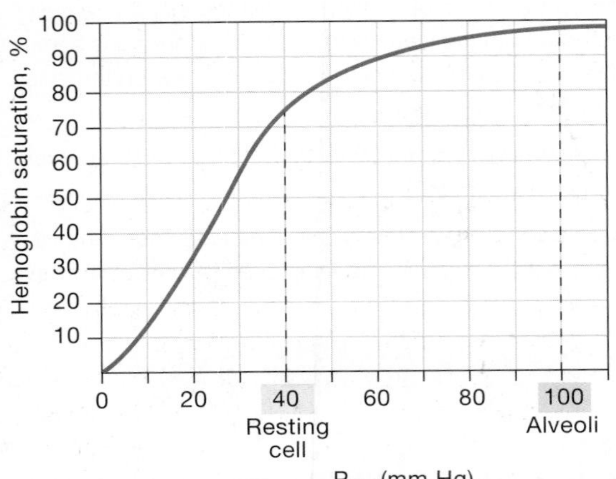

Hemoglobin saturation, %
100
90
80
70
60
50
40
30
20
10
0
20
40
Resting cell
60
80
100
Alveoli
P_{O_2} (mm Hg)

(b) Maternal and fetal hemoglobin have different oxygen-binding properties.

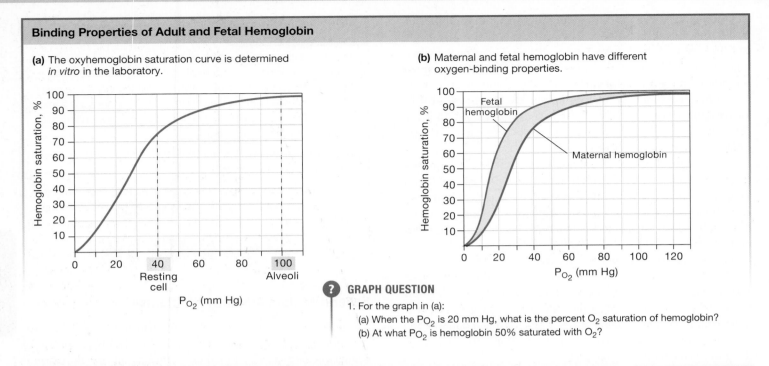

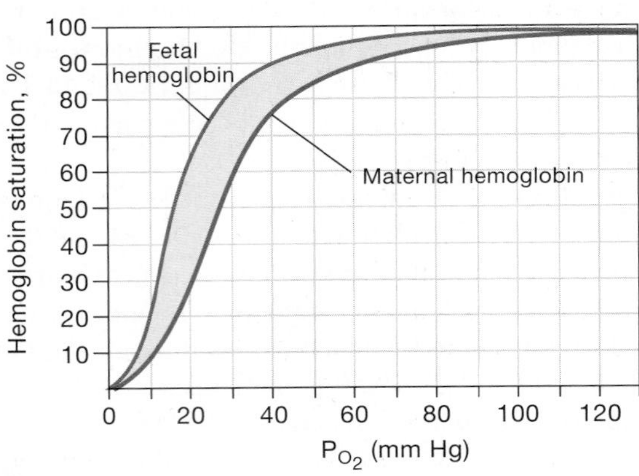

Hemoglobin saturation, %
Fetal hemoglobin
Maternal hemoglobin
100
90
80
70
60
50
40
30
20
10
0
20
40
60
80
100
120
P_{O_2} (mm Hg)

GRAPH QUESTION

1. For the graph in (a):
   (a) When the P_{O_2} is 20 mm Hg, what is the percent O_2 saturation of hemoglobin?
   (b) At what P_{O_2} is hemoglobin 50% saturated with O_2?

Physical Factors Alter Hemoglobin's Affinity for Oxygen

(c) Effect of pH

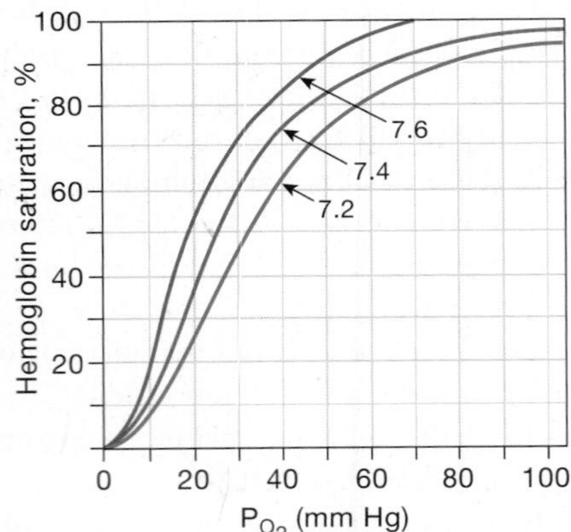

Hemoglobin saturation, %
100
80
60
40
20
7.6
7.4
7.2
0
20
40
60
80
100
P_{O_2} (mm Hg)

(d) Effect of temperature

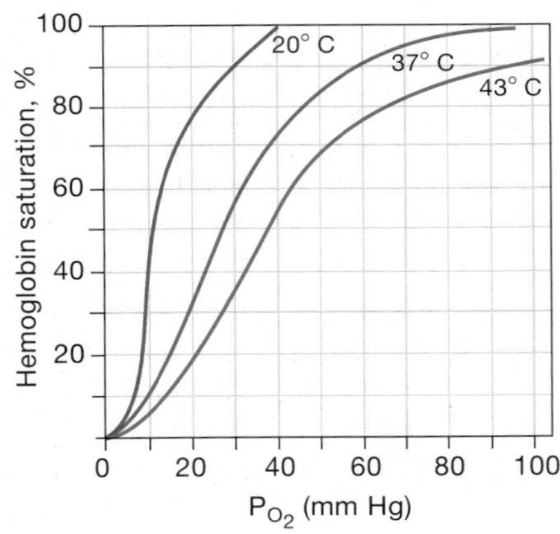

Hemoglobin saturation, %
100
80
60
40
20
20° C
37° C
43° C
0
20
40
60
80
100
P_{O_2} (mm Hg)

(e) Effect of P_{CO_2}

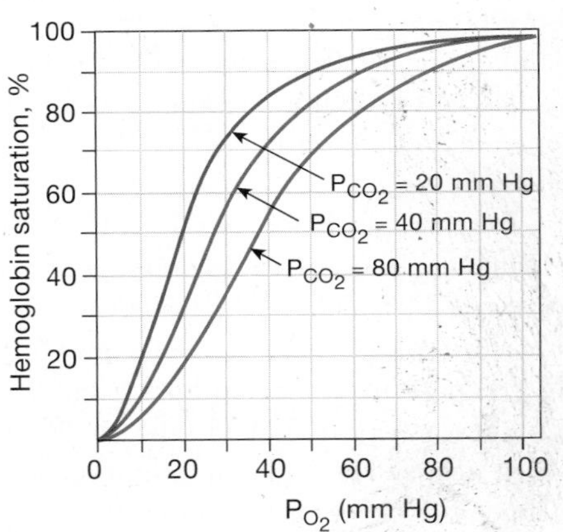

Hemoglobin saturation, %
100
80
60
40
20
P_{CO_2} = 20 mm Hg
P_{CO_2} = 40 mm Hg
P_{CO_2} = 80 mm Hg
0
20
40
60
80
100
P_{O_2} (mm Hg)

(f) Effect of the metabolic compound 2,3-BPG

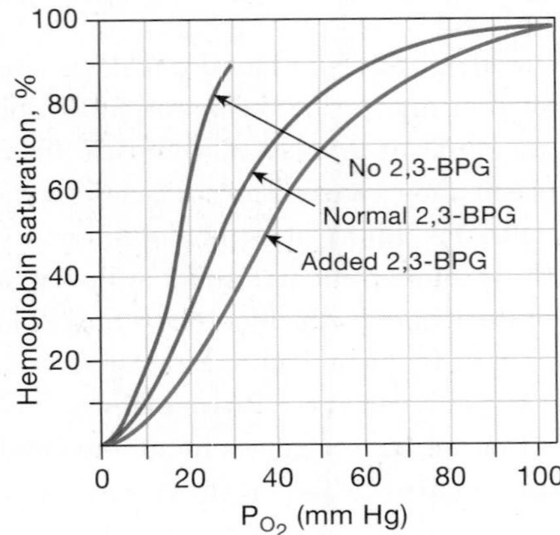

Hemoglobin saturation, %
100
80
60
40
20
No 2,3-BPG
Normal 2,3-BPG
Added 2,3-BPG
0
20
40
60
80
100
P_{O_2} (mm Hg)

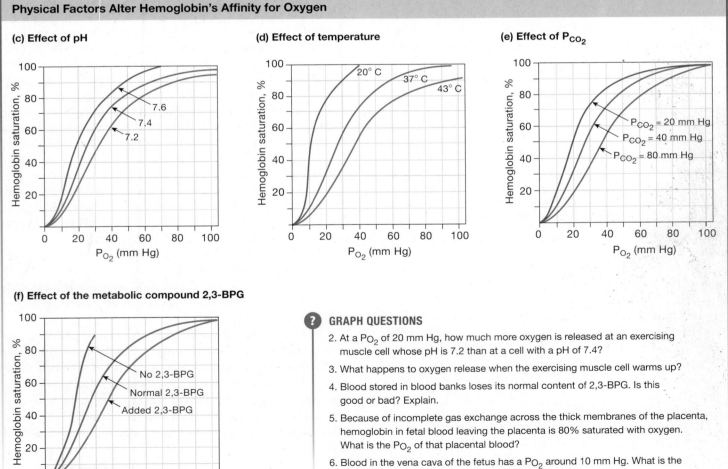

GRAPH QUESTIONS

2. At a P_{O_2} of 20 mm Hg, how much more oxygen is released at an exercising muscle cell whose pH is 7.2 than at a cell with a pH of 7.4?
3. What happens to oxygen release when the exercising muscle cell warms up?
4. Blood stored in blood banks loses its normal content of 2,3-BPG. Is this good or bad? Explain.
5. Because of incomplete gas exchange across the thick membranes of the placenta, hemoglobin in fetal blood leaving the placenta is 80% saturated with oxygen. What is the P_{O_2} of that placental blood?
6. Blood in the vena cava of the fetus has a P_{O_2} around 10 mm Hg. What is the percent O_2 saturation of maternal hemoglobin at the same P_{O_2}?

2,3-diphosphoglycerate or 2,3-DPG), a compound made from an intermediate of the glycolysis pathway. **Chronic hypoxia** (extended periods of low oxygen) triggers an increase in 2,3-BPG production in red blood cells. Increased levels of 2,3-BPG lower the binding affinity of hemoglobin and shift the HbO_2 saturation curve to the right (Fig. 18.9f). Ascent to high altitude and anemia are two situations that increase 2,3-BPG production.

Changes in hemoglobin's structure also change its oxygen-binding affinity. For example, *fetal hemoglobin* (HbF) has two gamma protein chains in place of the two beta chains found in adult hemoglobin. The presence of gamma chains enhances the ability of fetal hemoglobin to bind oxygen in the low-oxygen environment of the placenta. The altered binding affinity is reflected by the different shape of the fetal HbO_2 saturation curve (Fig. 18.9b). At any given placental P_{O_2}, oxygen released by maternal hemoglobin is picked up by the higher-affinity fetal hemoglobin for delivery to the developing fetus. Shortly after birth, fetal hemoglobin is replaced with the adult form as new red blood cells are made.

FIGURE 18.10 summarizes all the factors that influence the total oxygen content of arterial blood.

Concept Check

9. Can a person breathing 100% oxygen at sea level achieve 100% saturation of her hemoglobin?
10. What effect does hyperventilation have on the percent saturation of arterial hemoglobin? [*Hint:* Fig. 17.13, p. 554]
11. A muscle that is actively contracting may have a cellular P_{O_2} of 25 mm Hg. What happens to oxygen binding to hemoglobin at this low P_{O_2}? What is the P_{O_2} of the venous blood leaving the active muscle?

Carbon Dioxide Is Transported in Three Ways

Gas transport in the blood includes carbon dioxide removal from the cells as well as oxygen delivery to cells. Carbon dioxide is a by-product of cellular respiration [p. 104] and is potentially toxic if not excreted (removed from the body). Elevated P_{CO_2} (*hypercapnia*) causes the pH disturbance known as *acidosis*. Extremes of pH interfere with hydrogen bonding of molecules and can denature proteins [p. 51]. Abnormally high P_{CO_2} levels also depress central nervous system function, causing confusion, coma, or even death. For these reasons, CO_2 must be removed, making CO_2 homeostasis an important function of the lungs.

Carbon dioxide is more soluble in body fluids than oxygen is, but the cells produce far more CO_2 than can dissolve in the plasma. Only about 7% of the CO_2 carried by venous blood is dissolved in the plasma. The other 93% diffuses into red blood cells, where 23% binds to hemoglobin ($HbCO_2$) while the remaining 70% is converted to bicarbonate ion (HCO_3^-), as explained next. **FIGURE 18.11** summarizes these three mechanisms of carbon dioxide transport in the blood.

CO_2 and Bicarbonate Ions Most of the CO_2 that enters the blood is transported to the lungs as bicarbonate ions (HCO_3^-) dissolved in the plasma. The conversion of CO_2 to HCO_3^- serves two purposes: (1) it provides an additional way to transport CO_2 from cells to lungs, and (2) HCO_3^- is available to act as a buffer for metabolic acids [p. 41], thereby helping stabilize the body's pH.

How does CO_2 turn into HCO_3^-? The rapid production of HCO_3^- depends on the presence of **carbonic anhydrase (CA),** an enzyme found concentrated in red blood cells. Let's see how this happens. Dissolved CO_2 in the plasma diffuses into red blood

FIG. 18.10 Arterial oxygen

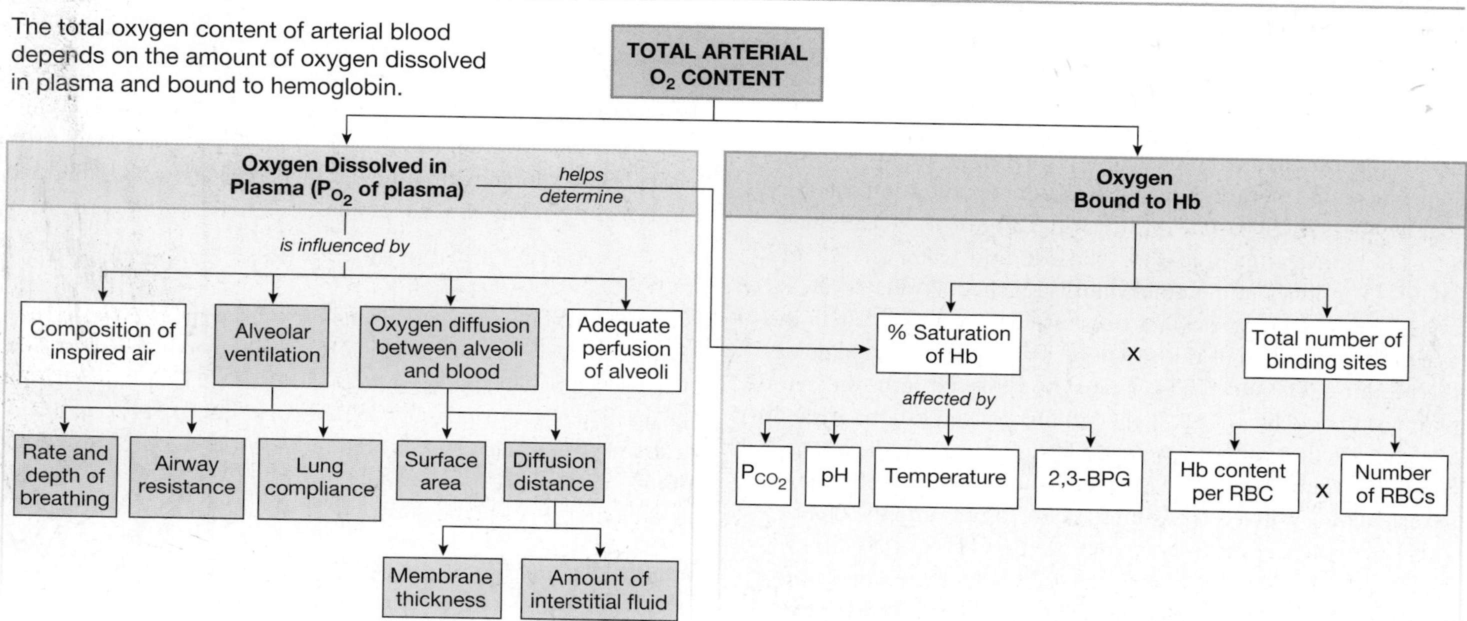

The total oxygen content of arterial blood depends on the amount of oxygen dissolved in plasma and bound to hemoglobin.
TOTAL ARTERIAL O2 CONTENT
Oxygen Dissolved in Plasma (PO2 of plasma)
helps determine
Oxygen Bound to Hb
is influenced by
Composition of inspired air
Alveolar ventilation
Oxygen diffusion between alveoli and blood
Adequate perfusion of alveoli
% Saturation of Hb
X
Total number of binding sites
affected by
Rate and depth of breathing
Airway resistance
Lung compliance
Surface area
Diffusion distance
PCO2
pH
Temperature
2,3-BPG
Hb content per RBC
X
Number of RBCs
Membrane thickness
Amount of interstitial fluid

CHAPTER 18

FIG. 18.11 Carbon dioxide transport

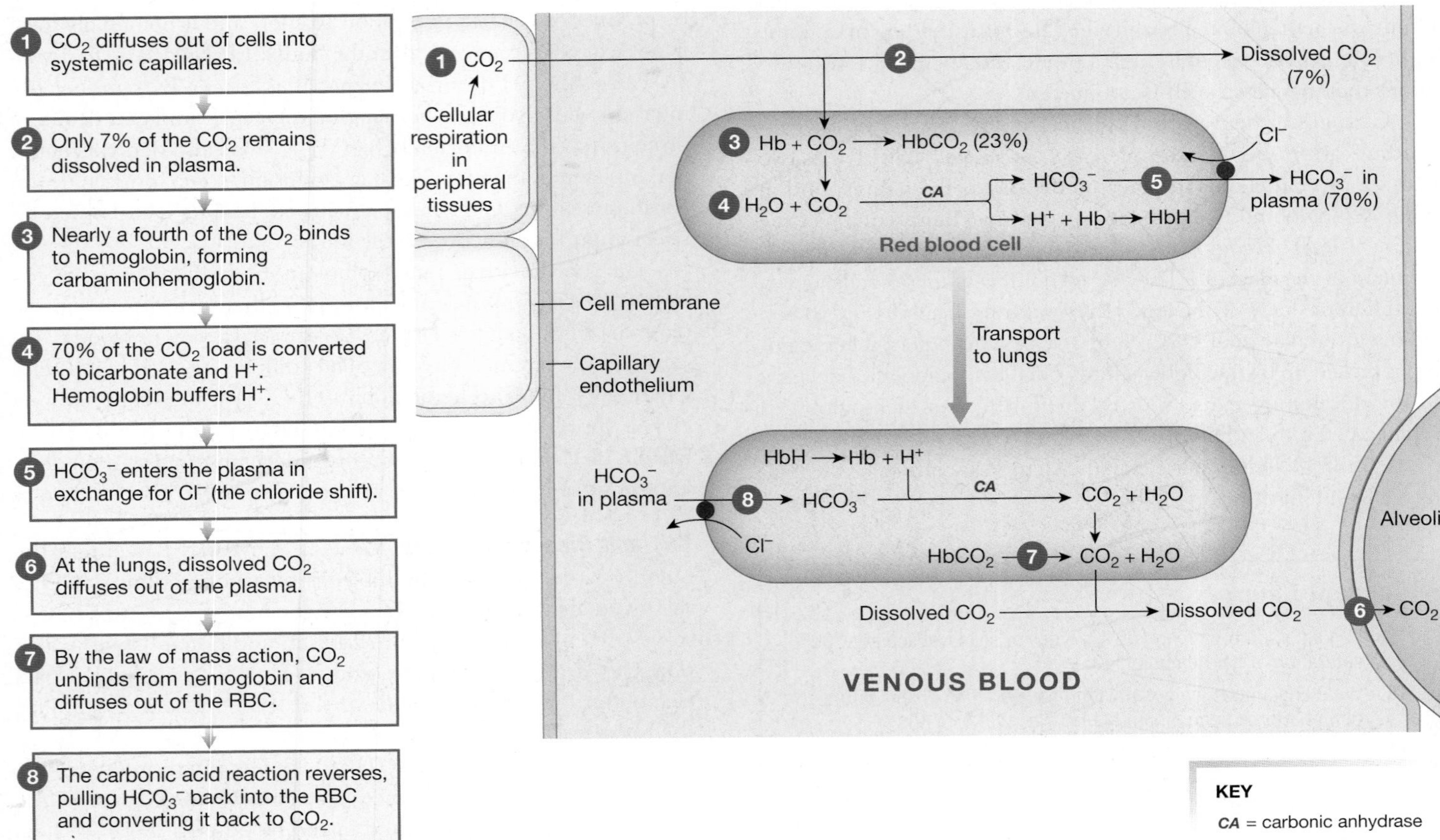
Most CO_2 in the blood has been converted to bicarbonate ion, HCO_3^-.
1 CO_2 diffuses out of cells into systemic capillaries.
2 Only 7% of the CO_2 remains dissolved in plasma.
3 Nearly a fourth of the CO_2 binds to hemoglobin, forming carbaminohemoglobin.
4 70% of the CO_2 load is converted to bicarbonate and H^+. Hemoglobin buffers H^+.
5 HCO_3^- enters the plasma in exchange for Cl^- (the chloride shift).
6 At the lungs, dissolved CO_2 diffuses out of the plasma.
7 By the law of mass action, CO_2 unbinds from hemoglobin and diffuses out of the RBC.
8 The carbonic acid reaction reverses, pulling HCO_3^- back into the RBC and converting it back to CO_2.
1 CO_2
Cellular respiration in peripheral tissues
2
Dissolved CO_2 (7%)
3 Hb + CO_2 → $HbCO_2$ (23%)
4 H_2O + CO_2
CA
HCO_3^-
5
H^+ + Hb → HbH
Cl^-
HCO_3^- in plasma (70%)
Red blood cell
Cell membrane
Capillary endothelium
Transport to lungs
HbH → Hb + H^+
HCO_3^- in plasma
8 HCO_3^-
CA
CO_2 + H_2O
Cl^-
$HbCO_2$ 7 CO_2 + H_2O
Alveoli
Dissolved CO_2
Dissolved CO_2
6 CO_2
VENOUS BLOOD
KEY
CA = carbonic anhydrase

cells, where it can react with water in the presence of carbonic anhydrase to form a hydrogen ion and a bicarbonate ion:

$$CO_2 + H_2O \overset{\text{carbonic anhydrase}}{\rightleftharpoons} H^+ + HCO_3^-$$

This reaction is reversible and obeys the law of mass action. We used to believe that carbonic acid was formed as an intermediate step, but studies of carbonic anhydrase function now indicate that the enzyme combines OH^- directly with CO_2 to form bicarbonate.

The conversion of carbon dioxide and water to H^+ and HCO_3^- continues until equilibrium is reached. (Water is always in excess in the body, so water concentration plays no role in the dynamic equilibrium of this reaction.) To keep the reaction going, the products (H^+ and HCO_3^-) must be removed from the cytoplasm of the red blood cell. If the product concentrations are kept low, the reaction cannot reach equilibrium. Carbon dioxide will continue to move out of plasma into the red blood cells, which in turn allows more CO_2 to diffuse out of tissues into the blood.

Two separate mechanisms remove free H^+ and HCO_3^-. In the first, bicarbonate leaves the red blood cell on an antiport protein [p. 140]. This transport process, known as the **chloride shift,** exchanges HCO_3^- for Cl^-. The anion exchange maintains the cell's electrical neutrality. The transfer of HCO_3^- into the plasma makes this buffer available to moderate pH changes caused by the production of metabolic acids. Bicarbonate is the most important extracellular buffer in the body.

Hemoglobin and H⁺ The second mechanism for keeping product concentrations low removes free H^+ from the red blood cell cytoplasm. Hemoglobin within the red blood cell acts as a buffer and binds hydrogen ions in the reaction

$$H^+ + Hb \rightleftharpoons HbH$$

Hemoglobin's buffering of H^+ is an important step that prevents large changes in the body's pH. If blood P_{CO_2} is elevated much above normal, the hemoglobin buffer cannot soak up all the H^+ produced from the reaction of CO_2 and water. In those cases, excess H^+ accumulates in the plasma, causing the condition known as **respiratory acidosis.** You will learn more about the role of the respiratory system in maintaining pH homeostasis when you study acid-base balance.

Hemoglobin and CO_2 Most carbon dioxide that enters red blood cells is converted to bicarbonate ions, but about 23% of the CO_2 in venous blood binds directly to hemoglobin. At the tissues, when

RUNNING PROBLEM

The usual homeostatic response to high-altitude hypoxia is hyperventilation, which begins on arrival. Hyperventilation enhances alveolar ventilation, but this may not help elevate arterial P_{O_2} levels significantly when atmospheric P_{O_2} is low. However, hyperventilation does lower plasma P_{CO_2}.

Q6: *What happens to plasma pH during hyperventilation? (Hint: Apply the law of mass action to figure out what happens to the balance between CO_2 and $H^+ + HCO_3^-$).*

Q7: *How does this change in pH affect oxygen binding at the lungs when P_{O_2} is decreased? How does it affect unloading of oxygen at the cells?*

563 – 565 – 570 – 573 – 577 – 581 – 583

oxygen leaves its binding sites on the hemoglobin molecule, CO_2 binds with free hemoglobin at exposed amino groups ($-NH_2$), forming **carbaminohemoglobin:**

$$CO_2 + Hb \rightleftharpoons HbCO_2 \text{ (carbaminohemoglobin)}$$

The presence of CO_2 and H^+ facilitates formation of carbaminohemoglobin because both of these factors decrease hemoglobin's binding affinity for oxygen (see Fig. 18.9c and e).

CO_2 Removal at the Lungs When venous blood reaches the lungs, the processes that took place in the systemic capillaries reverse (bottom portion of Fig. 18.11). The P_{CO_2} of the alveoli is lower than that of venous blood in the pulmonary capillaries. In response to this gradient, CO_2 diffuses out of plasma into the alveoli, and the plasma P_{CO_2} begins to fall.

The decrease in plasma P_{CO_2} allows dissolved CO_2 to diffuse out of the red blood cells. As CO_2 levels in the red blood cells decrease, the equilibrium of the CO_2-HCO_3^- reaction is disturbed, shifting toward production of more CO_2:

$$H^+ + HCO_3^- \rightarrow CO_2 + H_2O$$

H^+ unbinds from hemoglobin molecules and HCO_3^- moves back into the red blood cells when the chloride shift reverses. The HCO_3^- and newly released H^+ are converted back into water and CO_2. This newly made CO_2 is then free to diffuse out of the red blood cell into the plasma and from there into the alveoli.

FIGURE 18.12 shows the combined transport of CO_2 and O_2 in the blood. At the alveoli, O_2 diffuses down its pressure gradient, moving from the alveoli into the plasma and then from the plasma into the red blood cells. Hemoglobin binds to O_2, increasing the amount of oxygen that can be transported to the cells.

At the cells, the process reverses. Because P_{O_2} is lower in cells than in the arterial blood, O_2 diffuses from the plasma into the cells. The decrease in plasma P_{O_2} causes hemoglobin to release O_2, making additional oxygen available to enter cells.

FIG. 18.12 Summary of O_2 and CO_2 exchange and transport

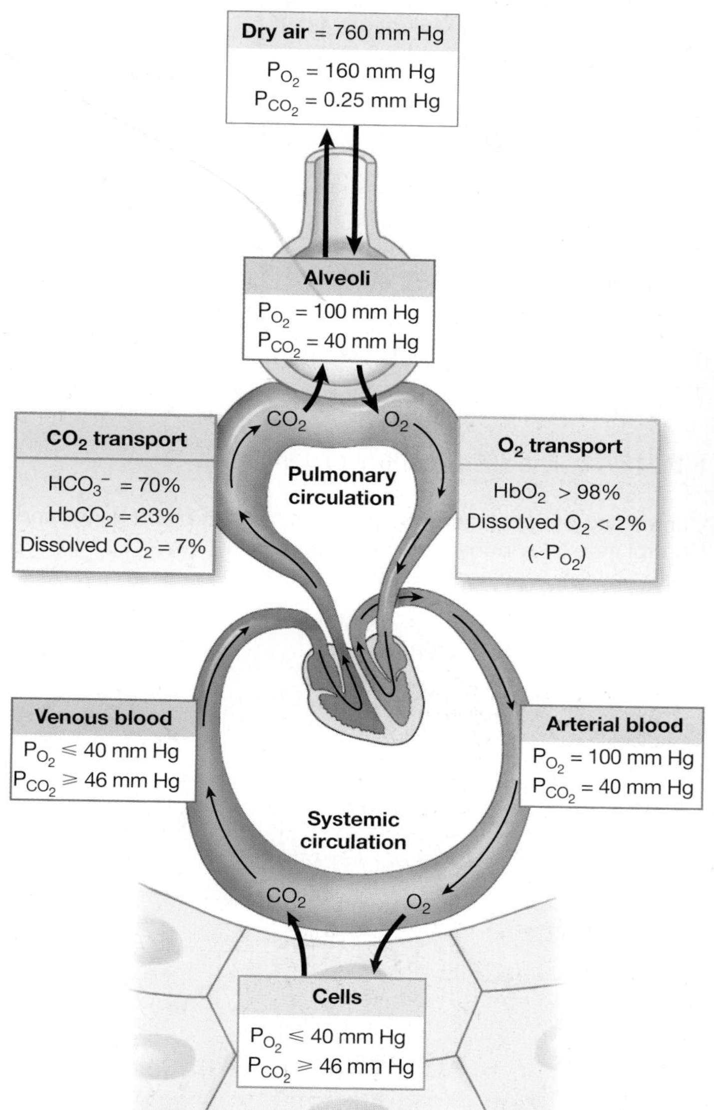

Dry air = 760 mm Hg
P_{O_2} = 160 mm Hg
P_{CO_2} = 0.25 mm Hg
Alveoli
P_{O_2} = 100 mm Hg
P_{CO_2} = 40 mm Hg
CO_2
O_2
CO_2 transport
HCO_3^- = 70%
$HbCO_2$ = 23%
Dissolved CO_2 = 7%
Pulmonary circulation
O_2 transport
HbO_2 > 98%
Dissolved O_2 < 2%
(~P_{O_2})
Venous blood
P_{O_2} ≤ 40 mm Hg
P_{CO_2} ≥ 46 mm Hg
Arterial blood
P_{O_2} = 100 mm Hg
P_{CO_2} = 40 mm Hg
Systemic circulation
CO_2
O_2
Cells
P_{O_2} ≤ 40 mm Hg
P_{CO_2} ≥ 46 mm Hg

Carbon dioxide from aerobic metabolism simultaneously leaves cells and enters the blood, dissolving in the plasma. From there, CO_2 enters red blood cells, where most is converted to HCO_3^- and H^+. The HCO_3^- is returned to the plasma in exchange for a Cl^- while the H^+ binds to hemoglobin. A fraction of the CO_2 that enters red blood cells binds directly to hemoglobin. At the lungs, the process reverses as CO_2 diffuses out of the pulmonary capillaries and into the alveoli.

To understand fully how the respiratory system coordinates delivery of oxygen to the lungs with transport of oxygen in the circulation, we now consider the central nervous system control of ventilation and the factors that influence it.

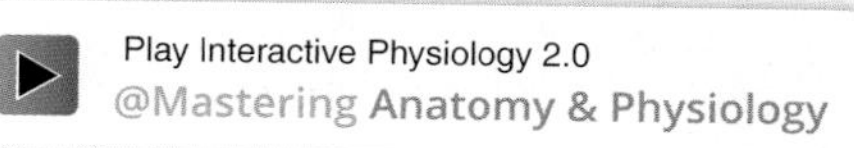

Play Interactive Physiology 2.0
@Mastering Anatomy & Physiology

Concept Check

12. How would an obstruction of the airways affect alveolar ventilation, arterial P_{CO_2}, and the body's pH?

18.3 Regulation of Ventilation

Breathing is a rhythmic process that usually occurs without conscious thought or awareness. In that respect, it resembles the rhythmic beating of the heart. However, skeletal muscles, unlike autorhythmic cardiac muscles, are not able to contract spontaneously. Instead, skeletal muscle contraction must be initiated by somatic motor neurons, which in turn are controlled by the central nervous system.

In the respiratory system, contraction of the diaphragm and other muscles is initiated by a spontaneously firing network of neurons in the brain stem (**FIG. 18.13**). Breathing occurs automatically throughout a person's life but can also be controlled voluntarily, up to a point. Complicated synaptic interactions between neurons in the network create the rhythmic cycles of inspiration and expiration, influenced continuously by sensory input, especially that from chemoreceptors for CO_2, O_2, and H^+. Ventilation pattern depends in large part on the levels of those three substances in the arterial blood and extracellular fluid.

The neural control of breathing is one of the few "black boxes" left in systems-level physiology. As you have learned, the "facts" presented in a textbook like this are really just our latest models of how the body works [p. 19]. Of all the models presented in this book, the model for neural control of breathing is the one

FIG. 18.13 The reflex control of ventilation

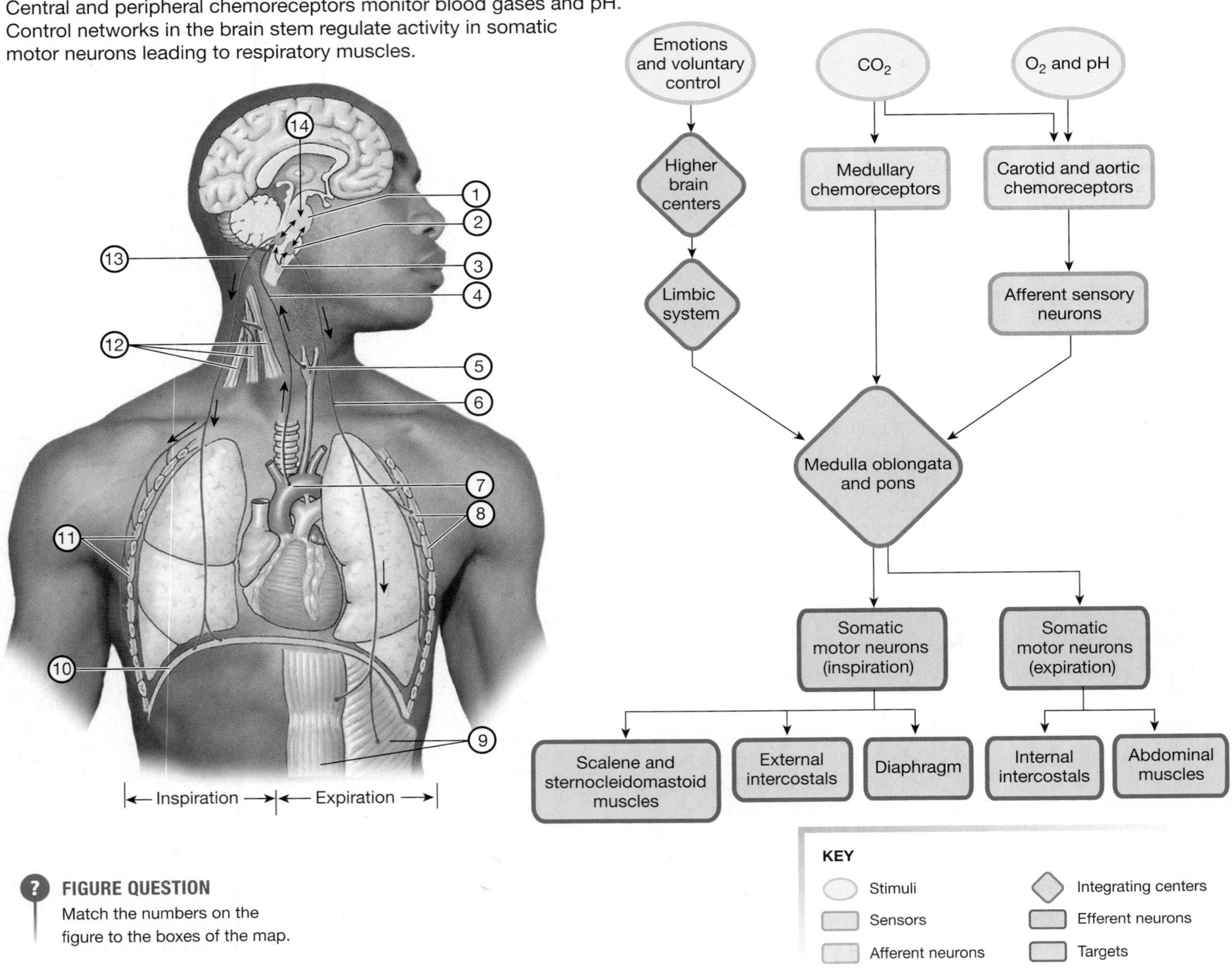
Central and peripheral chemoreceptors monitor blood gases and pH. Control networks in the brain stem regulate activity in somatic motor neurons leading to respiratory muscles.
1
2
3
4
5
6
7
8
9
10
11
12
13
14
Inspiration
Expiration
Emotions and voluntary control
CO2
O2 and pH
Higher brain centers
Medullary chemoreceptors
Carotid and aortic chemoreceptors
Limbic system
Afferent sensory neurons
Medulla oblongata and pons
Somatic motor neurons (inspiration)
Somatic motor neurons (expiration)
Scalene and sternocleidomastoid muscles
External intercostals
Diaphragm
Internal intercostals
Abdominal muscles
KEY
Stimuli
Sensors
Afferent neurons
Integrating centers
Efferent neurons
Targets

FIGURE QUESTION

Match the numbers on the figure to the boxes of the map.

that has changed the most in the past 20 years. We know the major regions of the brain stem that are involved, but the details of the neural networks involved remain elusive. The brain stem network that controls breathing behaves like a *central pattern generator* [p. 424], with intrinsic rhythmic activity that probably arises from *pacemaker neurons* with unstable membrane potentials. Tonic input from CO_2-sensitive and other chemoreceptors adds to the complexity.

Some of our understanding of how ventilation is controlled has come from observing patients with brain damage. Other information has come from animal experiments in which neural connections between major parts of the brain stem are severed, or sections of brain are studied in isolation. Research on CNS respiratory control is difficult because of the complexity of the neural networks and their anatomical locations. In recent years scientists have developed better techniques for studying the system.

The details that follow represent a contemporary model for the control of ventilation. Although some parts of the model are well supported with experimental evidence, other aspects are still under investigation. This model states that:

1. Respiratory neurons in the medulla control inspiratory and expiratory muscles.
2. Neurons in the pons integrate sensory information and interact with medullary neurons to influence ventilation.
3. The rhythmic pattern of breathing arises from a brainstem neural network with spontaneously discharging neurons.
4. Ventilation is subject to continuous modulation by various chemoreceptor- and mechanoreceptor-linked reflexes and by higher brain centers.

Neurons in the Medulla Control Breathing

Classic descriptions of how the brain controls ventilation divided the brain stem into various control centers. More recent descriptions, however, are less specific about assigning function to particular "centers" and instead look at complex interactions between neurons in a network. We know that respiratory neurons are concentrated bilaterally in two areas of the medulla oblongata. **FIGURE 18.14** shows these areas on the left side of the brain stem. One area called the **nucleus tractus solitarius (NTS)** contains the **dorsal respiratory group (DRG)** of neurons that control mostly muscles of inspiration. Output from the DRG goes via the **phrenic nerves** to the diaphragm and via the **intercostal nerves** to the intercostal muscles. In addition, the NTS receives sensory information from peripheral chemo- and mechanoreceptors through the *vagus* and *glossopharyngeal nerves* (cranial nerves X and IX).

Respiratory neurons in the pons receive sensory information from the DRG and in turn influence the initiation and termination of inspiration. The **pontine respiratory groups** (previously called the pneumotaxic center) and other pontine neurons provide tonic input to the medullary networks to help coordinate a smooth respiratory rhythm.

The **ventral respiratory group (VRG)** of the medulla has multiple regions with different functions. One area known

FIG. 18.14 Neural networks in the brain stem control ventilation

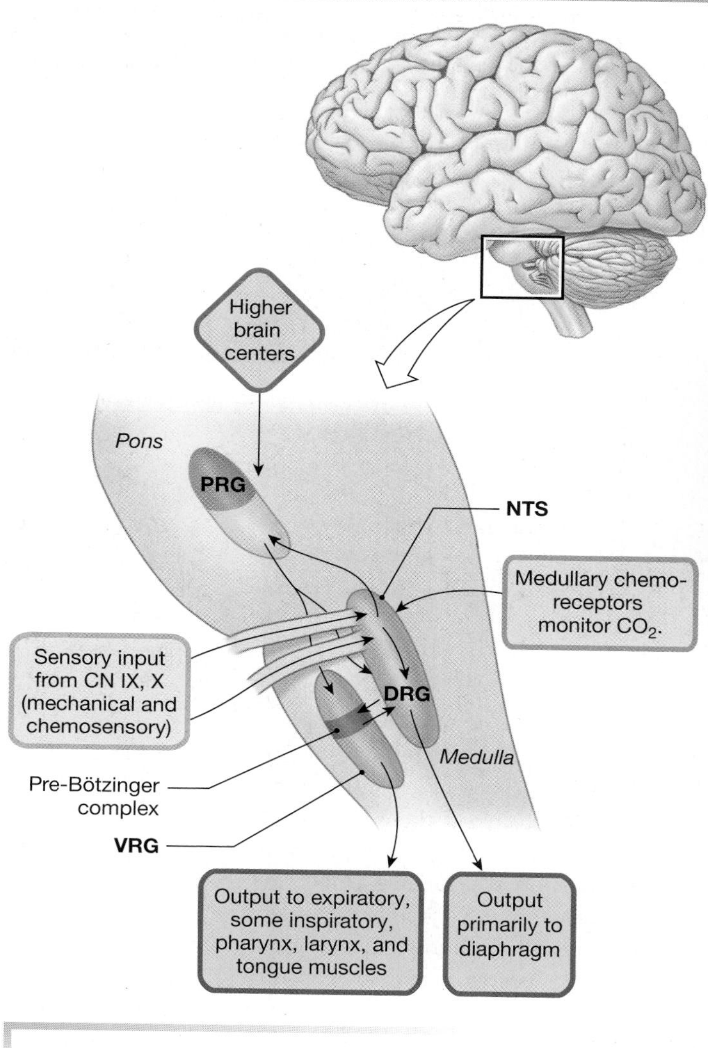
Higher brain centers
Pons
PRG
NTS
Medullary chemoreceptors monitor CO_2.
Sensory input from CN IX, X (mechanical and chemosensory)
DRG
Medulla
Pre-Bötzinger complex
VRG
Output to expiratory, some inspiratory, pharynx, larynx, and tongue muscles
Output primarily to diaphragm
KEY
PRG = Pontine respiratory group
DRG = Dorsal respiratory group
VRG = Ventral respiratory group
NTS = Nucleus tractus solitarius

as the **pre-Bötzinger complex** contains spontaneously firing neurons that may act as the basic pacemaker for the respiratory rhythm. Other areas control muscles used for active expiration or for greater-than-normal inspiration, such as occurs during vigorous exercise. In addition, nerve fibers from the VRG innervate muscles of the larynx, pharynx, and tongue to keep the upper airways open during breathing. Inappropriate relaxation of these muscles during sleep contributes to *obstructive sleep apnea*, a sleeping disorder associated with snoring and excessive daytime sleepiness.

The integrated action of the respiratory control networks can be seen by monitoring electrical activity in the phrenic nerve and other motor nerves (**FIG. 18.15**). During quiet breathing, a pacemaker initiates each cycle, and inspiratory neurons gradually increase stimulation of the inspiratory muscles. This increase is

FIG. 18.15 Neural activity during quiet breathing

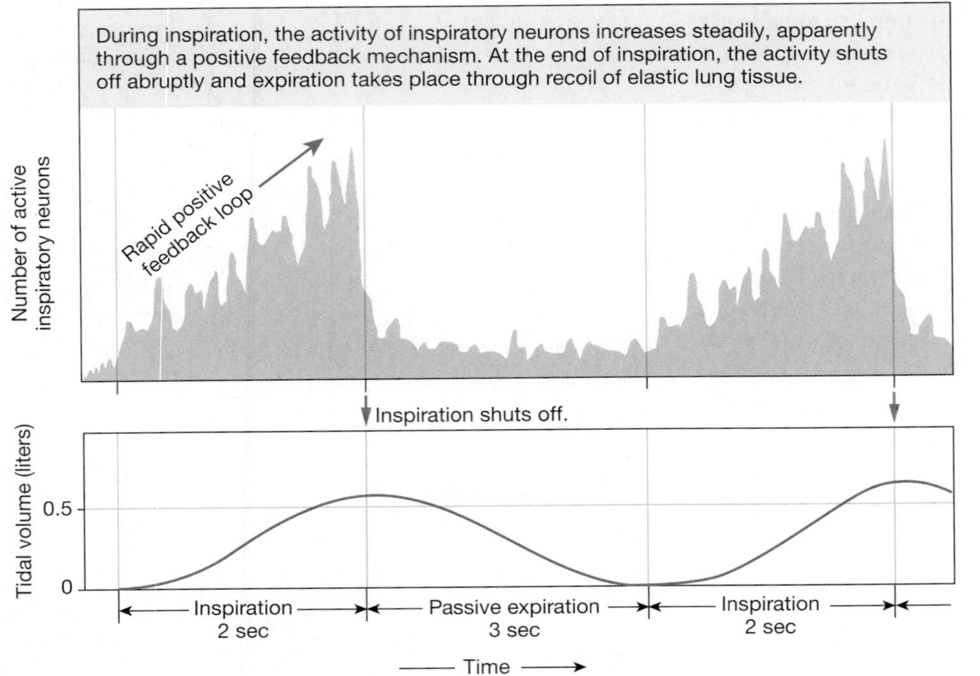
During inspiration, the activity of inspiratory neurons increases steadily, apparently through a positive feedback mechanism. At the end of inspiration, the activity shuts off abruptly and expiration takes place through recoil of elastic lung tissue.
Number of active inspiratory neurons
Rapid positive feedback loop
Inspiration shuts off.
Tidal volume (liters)
0.5
0
Inspiration 2 sec
Passive expiration 3 sec
Inspiration 2 sec
Time

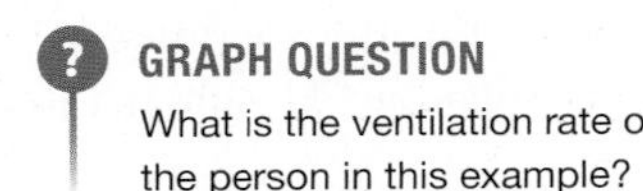
GRAPH QUESTION
What is the ventilation rate of the person in this example?

sometimes called *ramping* because of the shape of the graph of inspiratory neuron activity. A few inspiratory neurons fire to begin the ramp. The firing of these neurons recruits other inspiratory neurons to fire in an apparent positive feedback loop. As more neurons fire, more skeletal muscle fibers are recruited. The rib cage expands smoothly as the diaphragm contracts.

At the end of inspiration, the inspiratory neurons abruptly stop firing, and the respiratory muscles relax. Over the next few seconds, passive expiration occurs because of elastic recoil of the inspiratory muscles and elastic lung tissue. However, some motor neuron activity can be observed during passive expiration, suggesting that perhaps muscles in the upper airways contract to slow the flow of air out of the respiratory system.

Many neurons of the VRG remain inactive during quiet respiration. They function primarily during forced breathing, when inspiratory movements are exaggerated, and during active expiration. In forced breathing, increased activity of inspiratory neurons stimulates accessory muscles, such as the sternocleidomastoids. Contraction of these accessory muscles enhances expansion of the thorax by raising the sternum and upper ribs.

With active expiration, expiratory neurons from the VRG activate the internal intercostal and abdominal muscles. There seems to be some communication between inspiratory and expiratory neurons, as inspiratory neurons are inhibited during active expiration.

CO_2, Oxygen, and pH Influence Ventilation

Sensory input from central and peripheral chemoreceptors modifies the rhythmicity of the control network to help maintain blood gas homeostasis. Carbon dioxide is the primary stimulus for changes in ventilation. Oxygen and plasma pH play lesser roles.

The chemoreceptors for oxygen and carbon dioxide are strategically associated with the arterial circulation. If too little oxygen is present in arterial blood destined for the brain and other tissues, the rate and depth of breathing increase. If the rate of CO_2 production by the cells exceeds the rate of CO_2 removal by the lungs, arterial P_{CO_2} increases, and ventilation is intensified to match CO_2 removal to production. These homeostatic reflexes operate constantly, keeping arterial P_{O_2} and P_{CO_2} within a narrow range.

Peripheral chemoreceptors outside the CNS sense changes in the P_{O_2}, pH, and P_{CO_2} of the plasma (Fig. 18.13). The **carotid bodies** in the carotid arteries are the primary peripheral chemoreceptors. They are located close to the baroreceptors involved in reflex control of blood pressure [p. 492]. **Central chemoreceptors** in the brain respond to changes in the concentration of CO_2 in the cerebrospinal fluid. The primary central receptors lie on the ventral surface of the medulla, close to neurons involved in respiratory control.

Peripheral Chemoreceptors When specialized *type 1* or **glomus cells** {*glomus*, a ball-shaped mass} in the carotid bodies are activated by a decrease in P_{O_2} or pH or by an increase in P_{CO_2}, they trigger a reflex increase in ventilation. Under most normal circumstances, oxygen is not an important factor in modulating ventilation because arterial P_{O_2} must fall to less than 60 mm Hg before ventilation is stimulated. This large decrease in P_{O_2} is equivalent to ascending to an altitude of 3000 m. (For reference, Denver is located at an altitude of 1609 m.) However, any condition that reduces plasma pH or increases P_{CO_2} will activate the carotid and aortic glomus cells and increase ventilation.

The details of glomus cell function remain to be worked out, but the current model suggests that hypoxia in the cell closes oxygen-sensitive K^+ channels, depolarizing the cell. From that point, the basic mechanism by which these chemoreceptors respond is similar to the mechanism you learned for insulin release by pancreatic beta cells [p. 158] or taste transduction in taste buds [p. 325].

In all three examples, a stimulus inactivates K^+ channels and depolarizes the receptor cell (**FIG. 18.16**). Depolarization opens voltage-gated Ca^{2+} channels, and Ca^{2+} entry causes exocytosis of neurotransmitter onto the sensory neuron. In the carotid bodies, neurotransmitters initiate action potentials in sensory neurons leading to the brain stem respiratory networks, signaling them to increase ventilation.

Arterial oxygen concentrations do not play a role in the everyday regulation of ventilation because the peripheral chemoreceptors respond only to dramatic changes in arterial P_{O_2}. However, unusual physiological conditions, such as ascending to high altitude, and pathological conditions, such as chronic obstructive pulmonary disease (COPD), heart failure, and obstructive sleep apnea, seem to alter function of the carotid bodies. As a result, plasticity of the signaling between carotid bodies and the brain has become a recent focus of biomedical research.

FIG. 18.16 Carotid body cells respond to P_{O_2} below 60 mm Hg

The carotid body oxygen sensor releases neurotransmitter when P_{O_2} decreases.

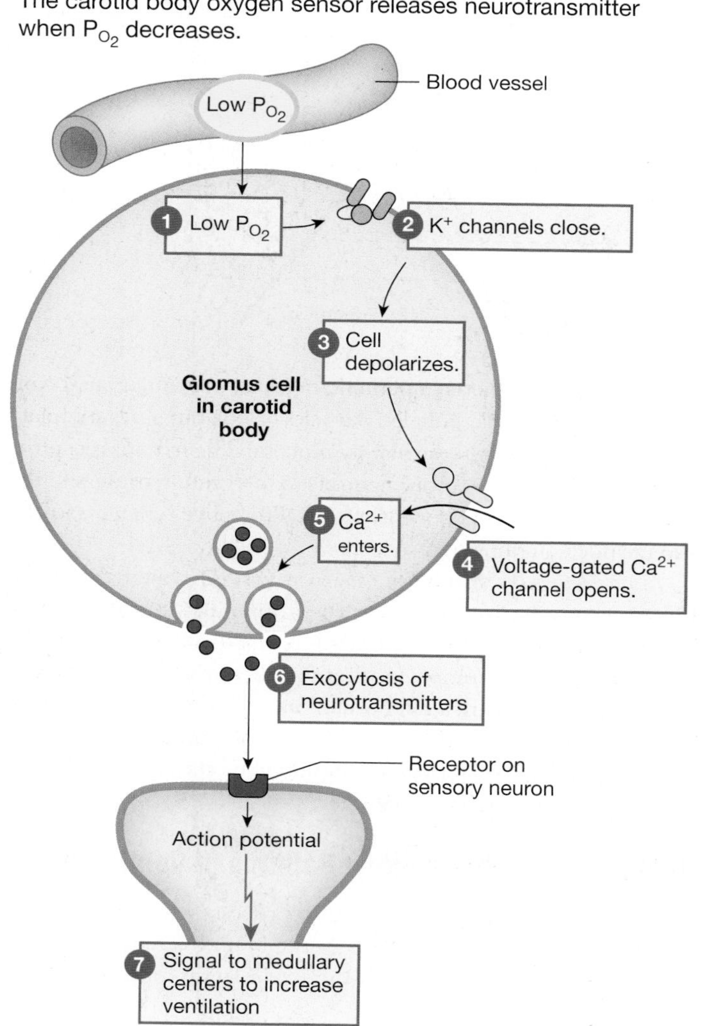

Blood vessel
Low P_{O_2}
1 Low P_{O_2}
2 K^+ channels close.
3 Cell depolarizes.
Glomus cell in carotid body
5 Ca^{2+} enters.
4 Voltage-gated Ca^{2+} channel opens.
6 Exocytosis of neurotransmitters
Receptor on sensory neuron
Action potential
7 Signal to medullary centers to increase ventilation

Central Chemoreceptors The most important chemical controller of ventilation is carbon dioxide, mediated both through the peripheral chemoreceptors just discussed and through central chemoreceptors located in the medulla (**FIG. 18.17**). These receptors set the respiratory pace, providing continuous input into the control network. When arterial P_{CO_2} increases, CO_2 crosses the blood-brain barrier and activates the central chemoreceptors. These receptors signal the control network to increase the rate and depth of ventilation, thereby enhancing alveolar ventilation and removing CO_2 from the blood.

Although we say that the central chemoreceptors monitor CO_2, they actually respond to pH changes in the cerebrospinal fluid (CSF). Carbon dioxide that diffuses across the blood-brain barrier into the CSF is converted to bicarbonate and H^+. Experiments indicate that the H^+ produced by this reaction is what initiates the chemoreceptor reflex, rather than the increased level of CO_2.

Note, however, that pH changes in the plasma *do not* usually influence the central chemoreceptors directly. Although plasma P_{CO_2} enters the CSF readily, plasma H^+ crosses the blood-brain barrier very slowly and therefore has little direct effect on the central chemoreceptors.

When plasma P_{CO_2} increases, the chemoreceptors initially respond strongly by increasing ventilation. However, if P_{CO_2} remains elevated for several days, ventilation falls back toward normal rates as the chemoreceptor response adapts. The adaptation appears to be due to increased CSF bicarbonate concentrations that buffer the H^+. The mechanism by which bicarbonate increases is not clear.

Even though the central chemoreceptor response adapts to chronically high P_{CO_2}, the response of peripheral chemoreceptor to low arterial P_{O_2} remains intact over time. In some situations, low P_{O_2} becomes the primary chemical stimulus for ventilation. For example, patients with severe chronic lung disease, such as COPD, have chronic hypercapnia and hypoxia. Their arterial P_{CO_2} may rise to 50–55 mm Hg (normal is 35–45) while their P_{O_2} falls to 45–50 mm Hg (normal 75–100). Because these levels are chronic, the chemoreceptor response adapts to the elevated P_{CO_2}.

RUNNING PROBLEM

The hyperventilation response to hypoxia creates a peculiar breathing pattern called *periodic breathing*, in which the person goes through a 10–15-second period of breath-holding followed by a short period of hyperventilation. Periodic breathing occurs most often during sleep.

Q8: *Based on your understanding of how the body controls ventilation, why do you think periodic breathing occurs most often during sleep?*

563 – 565 – 570 – 573 – 577 – **581** – 583

FIG. 18.17 Chemoreceptor response

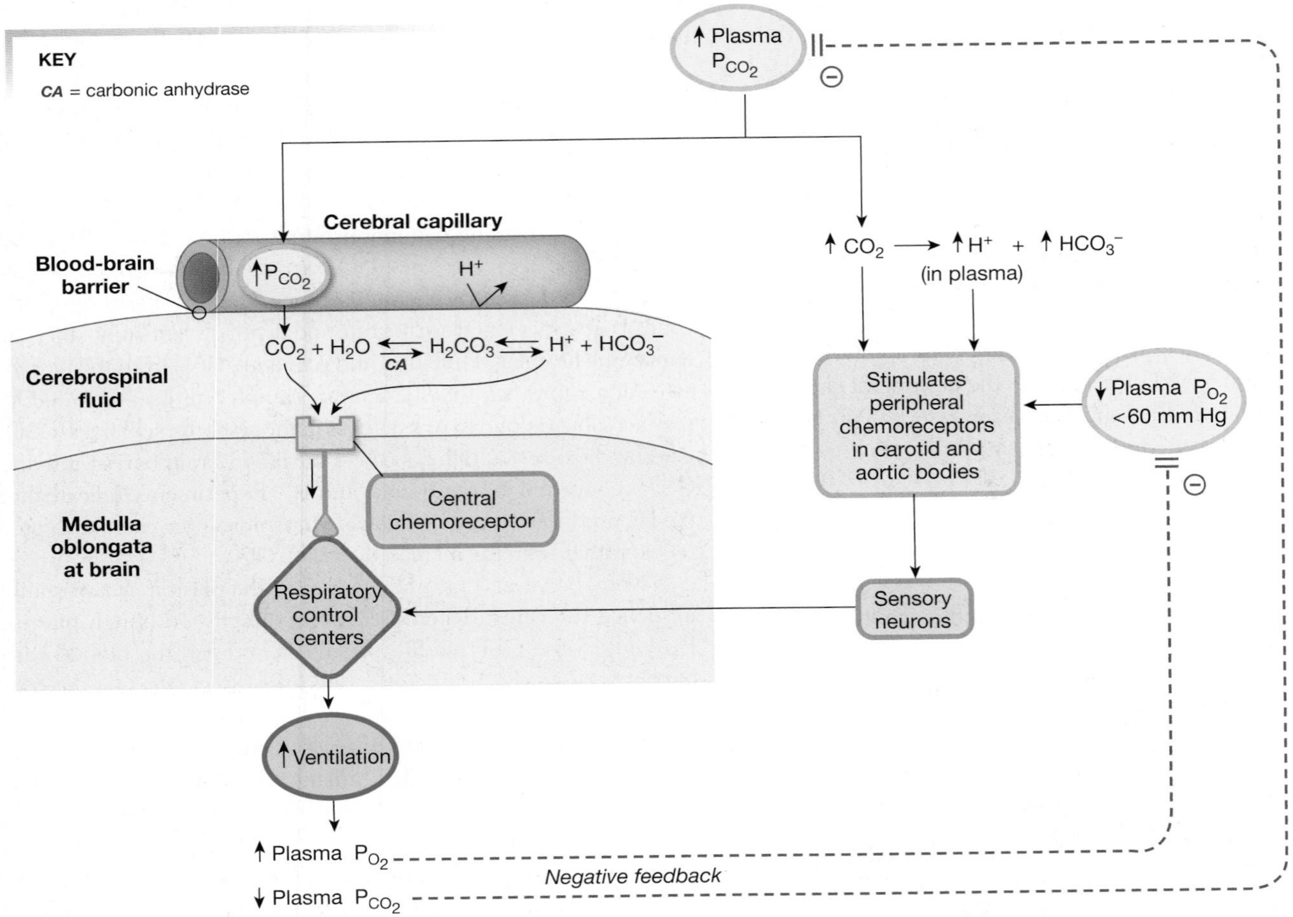
Central chemoreceptors monitor CO_2 in cerebrospinal fluid.
Carotid and aortic chemoreceptors monitor CO_2, O_2, and H^+.
KEY
CA = carbonic anhydrase
↑ Plasma P_{CO_2}
Cerebral capillary
Blood-brain barrier
↑P_{CO_2}
H^+
$CO_2 + H_2O \rightleftharpoons H_2CO_3 \rightleftharpoons H^+ + HCO_3^-$
CA
Cerebrospinal fluid
Central chemoreceptor
Medulla oblongata at brain
Respiratory control centers
↑Ventilation
↑ CO_2 → ↑H^+ + ↑HCO_3^-
(in plasma)
Stimulates peripheral chemoreceptors in carotid and aortic bodies
↓Plasma P_{O_2} <60 mm Hg
Sensory neurons
↑ Plasma P_{O_2}
↓ Plasma P_{CO_2}
Negative feedback

Most of the chemical stimulus for ventilation in this situation then comes from low P_{O_2}, sensed by the carotid body chemoreceptors. If these patients are given too much oxygen, they may stop breathing because their chemical stimulus for ventilation is eliminated.

The central chemoreceptors respond to decreases in arterial P_{CO_2} as well as to increases. If alveolar P_{CO_2} falls, as it does during hyperventilation, plasma P_{CO_2} and cerebrospinal fluid P_{CO_2} follow suit. As a result, central chemoreceptor activity declines, and the control network slows the ventilation rate. When ventilation decreases, carbon dioxide begins to accumulate in alveoli and the plasma. Eventually, the arterial P_{CO_2} rises above the threshold level for the chemoreceptors. At that point, the receptors fire, and the control network again increases ventilation.

Protective Reflexes Guard the Lungs

In addition to the chemoreceptor reflexes that help regulate ventilation, the body has protective reflexes that respond to physical injury or irritation of the respiratory tract and to overinflation of the lungs. The major protective reflex is *bronchoconstriction*, mediated through parasympathetic neurons that innervate bronchiolar smooth muscle. Inhaled particles or noxious gases stimulate **irritant receptors** in the airway mucosa. The irritant receptors send signals through sensory neurons to integrating centers in the CNS that trigger bronchoconstriction. Protective reflex responses also include coughing and sneezing.

The *Hering-Breuer inflation reflex* was first described in the late 1800s in anesthetized dogs. In these animals, if tidal volume exceeded a certain volume, stretch receptors in the lung signaled the brain stem to terminate inspiration. However, this reflex is difficult to demonstrate in adult humans and does not operate during quiet breathing and mild exertion. Studies on human infants, however, suggest that the Hering-Breuer inflation reflex may play a role in limiting their ventilation volumes.

Higher Brain Centers Affect Patterns of Ventilation

Conscious and unconscious thought processes also affect respiratory activities. Higher centers in the hypothalamus and cerebrum can alter the activity of the brain stem control network to change ventilation rate and depth. Voluntary control of ventilation falls into this category. Higher brain center control is not a *requirement* for ventilation, however. Even if the brain stem above the pons is severely damaged, essentially normal respiratory cycles continue.

Respiration can also be affected by stimulation of portions of the limbic system. For this reason, emotional and autonomic activities such as fear and excitement may affect the pace and depth of respiration. In some of these situations, the neural pathway goes directly to the somatic motor neurons, bypassing the control network in the brain stem.

Although we can temporarily alter our respiratory performance, we cannot override the chemoreceptor reflexes. Holding your breath is a good example. You can hold your breath voluntarily only until elevated P_{CO_2} in the blood and cerebrospinal fluid activates the chemoreceptor reflex, forcing you to inhale.

Small children having temper tantrums sometimes attempt to manipulate parents by threatening to hold their breath until they die. However, the chemoreceptor reflexes make it impossible for the children to carry out that threat. Extremely strong-willed children can continue holding their breath until they turn blue and pass out from hypoxia, but once they are unconscious, normal breathing automatically resumes.

Breathing is intimately linked to cardiovascular function. The integrating centers for both functions are located in the brain stem, and interneurons project between the two networks, allowing signaling back and forth. The cardiovascular, respiratory, and renal systems all work together to maintain fluid and acid-base homeostasis, as you will see.

RUNNING PROBLEM **CONCLUSION** High Altitude

On May 29, 1953, Edmund Hillary and Tenzing Norgay of the British Everest Expedition were the first humans to reach the summit of Mt. Everest. They carried supplemental oxygen with them, as it was believed that this feat was impossible without it. In 1978, however, Reinhold Messner and Peter Habeler achieved the "impossible." On May 8, they struggled to the summit using sheer willpower and no extra oxygen. In Messner's words, "I am nothing more than a single narrow gasping lung, floating over the mists and summits." Learn more about these Everest expeditions by doing a Google search for Hillary Everest or Messner Everest.

To learn more about different types of mountain sickness, see "High altitude medicine," *Am Fam Physician* 1998 Apr. 15 (*www.aafp.org/afp/980415ap/harris.html*) and "Altitude Illness" in the Centers for Disease Control and Prevention book *CDC Health Information for International Travel 2014* (*www.nc.cdc.gov/travel/yellowbook/2014/chapter-2-the-pre-travel-consultation/altitude-illness*).

In this running problem, you learned about normal and abnormal responses to high altitude. Check your understanding of the physiology behind this respiratory challenge by comparing your answers with the information in the following table.

Question	Facts	Integration and Analysis
Q1: *What is the P_{O_2} of inspired air reaching the alveoli when dry atmospheric pressure is 542 mm Hg? How does this value for P_{O_2} compare with the P_{O_2} value for fully humidified air at sea level?*	Water vapor contributes a partial pressure of 47 mm Hg to fully humidified air. Oxygen is 21% of dry air. Normal atmospheric pressure at sea level is 760 mm Hg.	Correction for water vapor: $542 - 47 = 495$ mm Hg $\times$ 21% $=$ 104 mm Hg P_{O_2}. In humidified air at sea level, $P_{O_2} = 150$ mm Hg.
Q2: *Why would someone with HAPE be short of breath?*	Pulmonary edema increases the diffusion distance for oxygen.	Slower oxygen diffusion means less oxygen reaching the blood, which worsens the normal hypoxia of altitude.
Q3: *Based on mechanisms for matching ventilation and perfusion in the lung, why do patients with HAPE have elevated pulmonary arterial blood pressure?*	Low oxygen levels constrict pulmonary arterioles.	Constriction of pulmonary arterioles causes blood to collect in the pulmonary arteries behind the constriction. This increases pulmonary arterial blood pressure.
Q4: *How does adding erythrocytes to the blood help a person acclimatize to high altitude?*	98% of arterial oxygen is carried bound to hemoglobin.	Additional hemoglobin increases the total oxygen-carrying capacity of the blood.
Q5: *What does adding erythrocytes to the blood do to the viscosity of the blood? What effect will that change in viscosity have on blood flow?*	Adding cells increases blood viscosity.	According to Poiseuille's law, increased viscosity increases resistance to flow, so blood flow will decrease.

Continued

25. You are a physiologist on a space flight to a distant planet. You find intelligent humanoid creatures inhabiting the planet, and they willingly submit to your tests. Some of the data you have collected are described below.

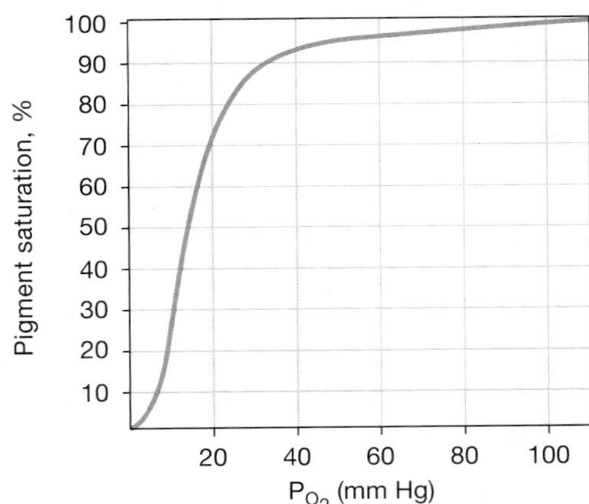
Pigment saturation, %
100
90
80
70
60
50
40
30
20
10
20
40
60
80
100
P_{O_2} (mm Hg)

The graph above shows the oxygen saturation curve for the oxygen-carrying molecule in the blood of the humanoid named Bzork. Bzork's normal alveolar P_{O_2} is 85 mm Hg. His normal cell P_{O_2} is 20 mm Hg, but it drops to 10 mm Hg with exercise.

a. What is the percent saturation for Bzork's oxygen-carrying molecule in blood at the alveoli? In blood at an exercising cell?

b. Based on the graph, what conclusions can you draw about Bzork's oxygen requirements during normal activity and during exercise?

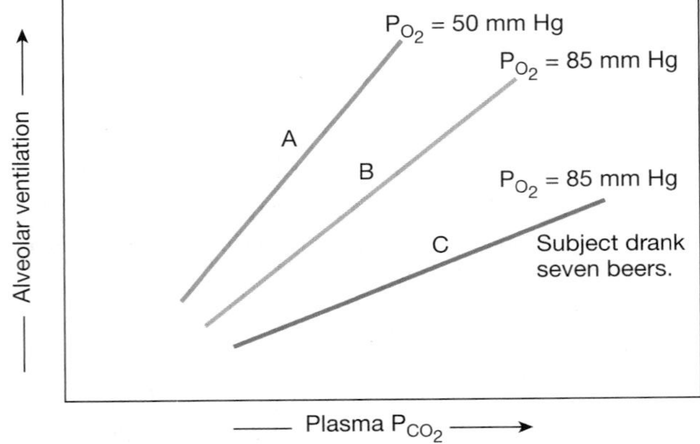
P_{O_2} = 50 mm Hg
P_{O_2} = 85 mm Hg
A
B
P_{O_2} = 85 mm Hg
C
Subject drank seven beers.
Alveolar ventilation
Plasma P_{CO_2}

26. The next experiment on Bzork involves his ventilatory response to different conditions. The data from that experiment are graphed below. Interpret the results of experiments A and C.

27. The alveolar epithelium is an absorptive epithelium and is able to transport ions from the fluid lining of alveoli into the interstitial space, creating an osmotic gradient for water to follow. Draw an alveolar epithelium and label apical and basolateral surfaces, the airspace, and interstitial fluid. Arrange the following proteins on the cell membrane so that the epithelium absorbs sodium and water: aquaporins, Na^+-K^+-ATPase, epithelial Na^+ channel (ENaC). (*Remember:* Na^+ concentrations are higher in the ECF than in the ICF.)

Level Four Quantitative Problems

28. You are given the following information on a patient.

    Blood volume = 5.2 liters
    Hematocrit = 47%
    Hemoglobin concentration = 12 g/dL whole blood
    Total amount of oxygen carried in blood = 1015 mL
    Arterial plasma = 100 mm Hg

    You know that when plasma P_{O_2} is 100 mm Hg, plasma contains 0.3 mL O_2/dL, and that hemoglobin is 98% saturated. Each hemoglobin molecule can bind to a maximum of four molecules of oxygen. Using this information, calculate the maximum oxygen-carrying capacity of hemoglobin (100% saturated). Units will be mL O_2/g Hb.

29. Adolph Fick, the nineteenth-century physiologist who derived Fick's law of diffusion, also developed the Fick equation that relates oxygen consumption, cardiac output, and blood oxygen content:

    O_2 consumption =
    cardiac output × (arterial oxygen content − venous oxygen content)

    A person has a cardiac output of 4.5 L/min, an arterial oxygen content of 105 mL O_2/L blood, and a vena cava oxygen content of 50 mL O_2/L blood. What is this person's oxygen consumption?

30. Describe what happens to the oxygen-hemoglobin saturation curve in Figure 18.9a when blood hemoglobin falls from 15 g/dL blood to 10 g/dL blood.

Answers to Concept Checks, Figure and Graph Questions, and end-of-chapter Review Questions can be found in Appendix A [p. A-1].

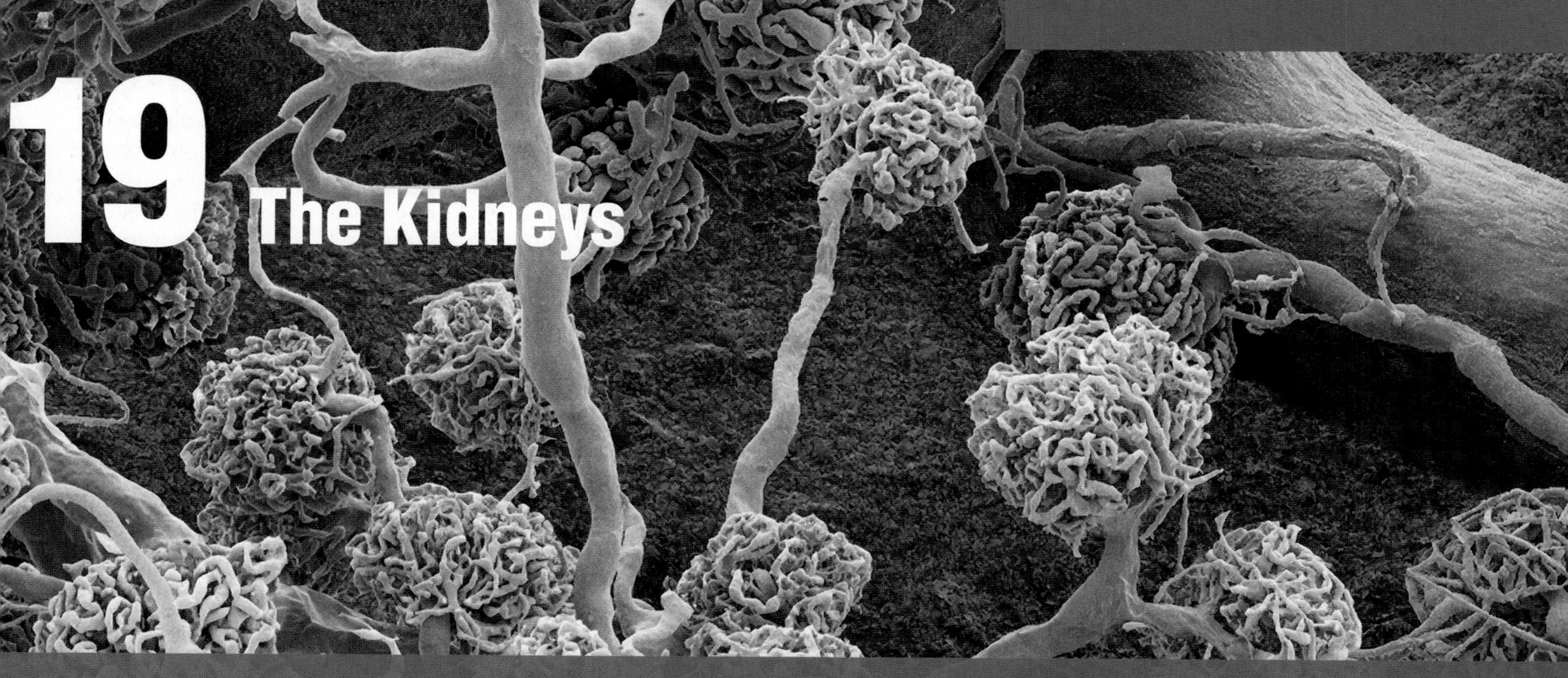

19 The Kidneys

Plasma undergoes modification to urine in the nephron.

Arthur Grollman, in Clinical Physiology: The Functional Pathology of Disease, *1957*

Renal glomeruli and blood vessels

BACKGROUND BASICS

About 100 C.E., Aretaeus the Cappadocian wrote, "Diabetes is a wonderful affection, not very frequent among men, being a melting down of the flesh and limbs into urine . . . . The patients never stop making water {urinating}, but the flow is incessant, as if from the opening of aqueducts."* Physicians have known since ancient times that **urine**, the fluid waste produced by the kidneys, reflects the functioning of the body. To aid them in their diagnosis of illness, they even carried special flasks for the collection and inspection of patients' urine.

The first step in examining a urine sample is to determine its color. Is it dark yellow (concentrated), pale straw (dilute), red (indicating the presence of blood), or black (indicating the presence of hemoglobin metabolites)? One form of malaria was called *blackwater fever* because metabolized hemoglobin from the abnormal breakdown of red blood cells turned victims' urine black or dark red.

Physicians also inspected urine samples for clarity, froth (indicating abnormal presence of proteins), smell, and even taste. Physicians who did not want to taste the urine themselves would allow their students the "privilege" of tasting it for them. A physician without students might expose insects to the urine and study their reaction.

Probably the most famous example of using urine for diagnosis was the taste test for diabetes mellitus, historically known as the *honey-urine disease*. Diabetes is an endocrine disorder characterized by the presence of glucose in the urine. The urine of diabetics tasted sweet and attracted insects, making the diagnosis clear.

Today, we have much more sophisticated tests for glucose in the urine, but the first step of a *urinalysis* is still to examine the color, clarity, and odor of the urine. In this chapter, you will learn why we can tell so much about how the body is functioning by what is present in the urine.

19.1 Functions of the Kidneys

If you ask people on the street, "What is the most important function of the kidney?" they are likely to say, "The removal of wastes." Actually, the most important function of the kidney is the homeostatic regulation of the water and ion content of the blood, also called *salt and water balance* or *fluid and electrolyte balance*. Waste removal is important, but disturbances in blood volume or ion levels cause serious medical problems before the accumulation of metabolic wastes reaches toxic levels.

The kidneys maintain normal blood concentrations of ions and water by balancing intake of those substances with their excretion in the urine, obeying the principle of *mass balance* [p. 10]. We can divide kidney function into six general areas:

1. **Regulation of extracellular fluid volume and blood pressure.** When extracellular fluid volume decreases, blood pressure also decreases [p. 484]. If ECF volume and blood pressure fall too low, the body cannot maintain adequate blood flow to the brain and other essential organs. The kidneys work in an integrated fashion with the cardiovascular system to ensure that blood pressure and tissue perfusion remain within an acceptable range.
2. **Regulation of osmolarity.** The body integrates kidney function with behavioral drives, such as thirst, to maintain blood osmolarity at a value close to 290 mOsM. We examine the reflex pathways for regulation of ECF volume and osmolarity later.
3. **Maintenance of ion balance.** The kidneys keep concentrations of key ions within a normal range by balancing dietary intake with urinary loss. Sodium (Na^+) is the major ion involved in the regulation of extracellular fluid volume and osmolarity. Potassium (K^+) and calcium (Ca^{2+}) concentrations are also closely regulated.
4. **Homeostatic regulation of pH.** The pH of plasma is normally kept within a narrow range. If extracellular fluid becomes too acidic, the kidneys remove H^+ and conserve bicarbonate ions (HCO_3^-) which act as a buffer [p. 41]. Conversely, when extracellular fluid becomes too alkaline, the kidneys remove HCO_3^- and conserve H^+. The kidneys play a significant role in pH homeostasis, but they do not correct pH disturbances as rapidly as the lungs do.
5. **Excretion of wastes.** The kidneys remove metabolic waste products and *xenobiotics*, or foreign substances, such as drugs and environmental toxins. Metabolic wastes include *creatinine* from muscle metabolism [p. 388] and the nitrogenous wastes *urea* and *uric acid*. A metabolite of hemoglobin called *urobilinogen* gives urine its characteristic yellow color. Hormones are another endogenous substance the kidneys clear from the blood. Examples of foreign substances that the kidneys actively remove include the artificial sweetener *saccharin* and the anion *benzoate*, part of the preservative *potassium benzoate*, which you ingest each time you drink a diet soft drink.
6. **Production of hormones.** Although the kidneys are not endocrine glands, they play important roles in three endocrine pathways. Kidney cells synthesize *erythropoietin*, the cytokine/hormone that regulates red blood cell synthesis [p. 515]. They also release *renin*, an enzyme that regulates the production of hormones involved in sodium balance and blood pressure homeostasis. Finally, renal enzymes help convert vitamin D_3 into a hormone that regulates Ca^{2+} balance.

* *The Extant Works of Aretaeus the Cappadocian*. Edited and translated by Adams, F. London, 1856.

RUNNING PROBLEM **Gout**

Michael, 43, had spent the past two days on the sofa, suffering from a relentless throbbing pain in his left big toe. When the pain began, Michael thought he had a mild sprain or perhaps the beginnings of arthritis. Then the pain intensified, and the toe joint became hot and red. Michael finally hobbled into his doctor's office, feeling a little silly about his problem. On hearing his symptoms and looking at the toe, the doctor seemed to know instantly what was wrong. "Looks to me like you have gout," said Dr. Garcia.

588 – 589 – 607 – 607 – 611 – 613

The kidneys, like many other organs in the body, have a tremendous reserve capacity. By most estimates, you must lose nearly three-fourths of your kidney function before homeostasis begins to be affected. Many people function perfectly normally with only one kidney, including the one person in 1000 born with only one kidney (the other fails to develop during gestation) or those people who donate a kidney for transplantation.

Concept Check

1. Ion regulation is a key feature of kidney function. What happens to the resting membrane potential of a neuron if extracellular K^+ levels decrease? [p. 249]
2. What happens to the force of cardiac contraction if plasma Ca^{2+} levels decrease substantially? [p. 466]

19.2 Anatomy of the Urinary System

The **urinary system** is composed of the kidneys and accessory structures (**FIG. 19.1a**). The study of kidney function is called **renal physiology**, from the Latin word *renes*, meaning "kidneys."

The Urinary System Consists of Kidneys, Ureters, Bladder, and Urethra

Let's begin by following the route a drop of water takes on its way from plasma to excretion in the urine. Urine production begins when water and solutes move from plasma into the hollow tubules (*nephrons*) that make up the bulk of the paired **kidneys**. These tubules modify the composition of the fluid as it passes through. The modified fluid, now called *urine*, leaves the kidney and passes into a smooth muscle tube called a **ureter**. There are two ureters, one leading from each kidney to the **urinary bladder**. The bladder expands and fills with urine until, in a reflex called *micturition* or urination, the bladder contracts and expels urine through a single tube, the **urethra**.

The urethra in males exits the body through the shaft of the penis. In females, the urethral opening is found anterior to the openings of the vagina and anus. Because of the shorter length of the female urethra and its proximity to bacteria leaving the large intestine, women are more prone than men to develop bacterial infections of the bladder and kidneys, or **urinary tract infections (UTIs)**.

The most common cause of UTIs is the bacterium *Escherichia coli*, a normal inhabitant of the human large intestine. *E. coli* is not harmful while restricted to the lumen of the large intestine, but it is pathogenic {*patho-*, disease + *-genic*, causing} if it gets into the urethra. The most common symptoms of a UTI are pain or burning during urination and increased frequency of urination. A urine sample from a patient with a UTI often contains many red and white blood cells, neither of which is commonly found in normal urine. UTIs are treated with antibiotics.

The Kidneys The kidneys are the site of urine formation. They lie on either side of the spine at the level of the 11th and 12th ribs, just above the waist (Fig. 19.1b). Although they are below the diaphragm, they are technically outside the abdominal cavity, sandwiched between the membranous **peritoneum**, which lines the abdomen, and the bones and muscles of the back. Because of their location behind the peritoneal cavity, the kidneys are sometimes described as being *retroperitoneal* {*retro-*, behind}.

The concave surface of each kidney faces the spine. The renal blood vessels, nerves, lymphatics, and ureters all emerge from this surface. **Renal arteries**, which branch off the abdominal aorta, supply blood to the kidneys. **Renal veins** carry blood from the kidneys to the inferior vena cava.

At any given time, the kidneys receive 20–25% of the cardiac output, even though they constitute only 0.4% of total body weight (4.5–6 ounces each). This high rate of blood flow through the kidneys is critical to renal function.

The Nephron Is the Functional Unit of the Kidney

A cross section through a kidney shows that the interior is arranged in two layers: an outer **cortex** and inner **medulla** (Fig. 19.1c). The layers are formed by the organized arrangement of microscopic tubules called **nephrons**. About 80% of the nephrons in a kidney are almost completely contained within the cortex (*cortical* nephrons), but the other 20%—called *juxtamedullary* nephrons {*juxta-*, beside}—dip down into the medulla (Fig. 19.1f, h).

The nephron is the functional unit of the kidney. (A *functional unit* is the smallest structure that can perform all the functions of an organ.) Each of the 1 million nephrons in a kidney is divided into sections (Fig. 19.1i), and each section is closely associated with specialized blood vessels (Fig. 19.1g, h).

Vascular Elements of the Kidney Blood enters the kidney through the renal artery before flowing into smaller arteries and then into arterioles in the cortex (Fig. 19.1d, e). At this point, the arrangement of blood vessels forms a *portal system*, one of three in the body [p. 436]. Recall that a portal system consists of two capillary beds in series (one after the other).

RUNNING PROBLEM

Gout is a metabolic disease characterized by high blood concentrations of uric acid (*hyperuricemia*). If uric acid concentrations reach a critical level (6.8 mg/dL), monosodium urate precipitates out of solution and forms crystals in peripheral joints, particularly in the feet, ankles, and knees. These crystals trigger an inflammatory reaction and cause periodic attacks of excruciating pain. Uric acid crystals may also form kidney stones in the *renal pelvis* (Fig. 19.1c).

Q1: *Trace the route followed by these kidney stones when they are excreted.*

Q2: *Name the anion formed when uric acid dissociates.*

588 – 589 – 607 – 607 – 611 – 613

FIG. 19.1 Anatomy summary . . . The Urinary System

Overview of the Urinary System

(a) Urinary system

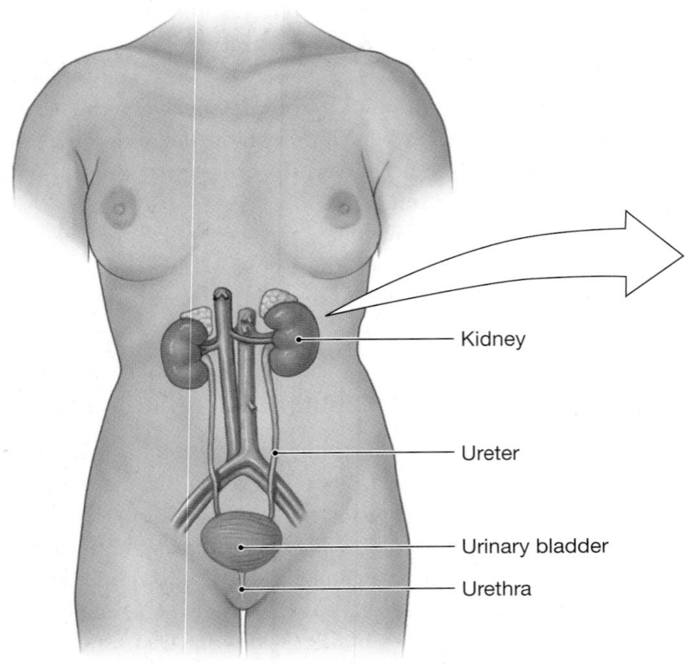

Kidney
Ureter
Urinary bladder
Urethra

(b) The kidneys are located retroperitoneally at the level of the lower ribs.

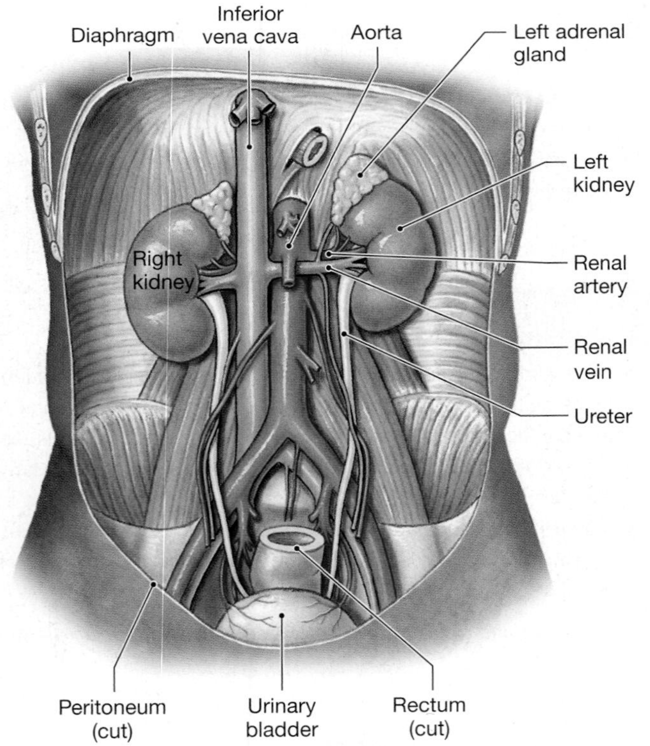

Diaphragm
Inferior vena cava
Aorta
Left adrenal gland
Left kidney
Right kidney
Renal artery
Renal vein
Ureter
Peritoneum (cut)
Urinary bladder
Rectum (cut)

Structure of the Kidney

(c) In cross section, the kidney is divided into an outer cortex and an inner medulla. Urine leaving the nephrons flows into the renal pelvis prior to passing through the ureter into the bladder.

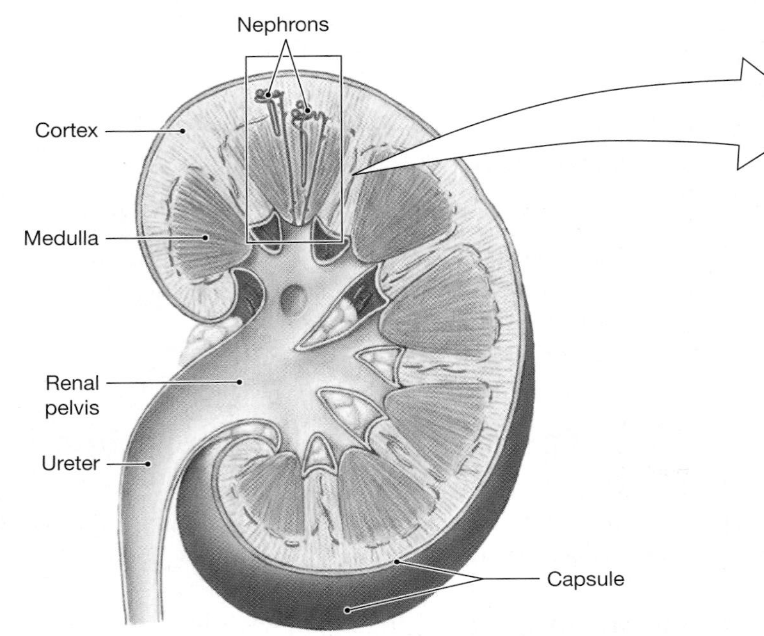

Nephrons
Cortex
Medulla
Renal pelvis
Ureter
Capsule

(d) Renal arteries take blood to the cortex. Afferent arterioles and glomeruli are all found in the cortex.

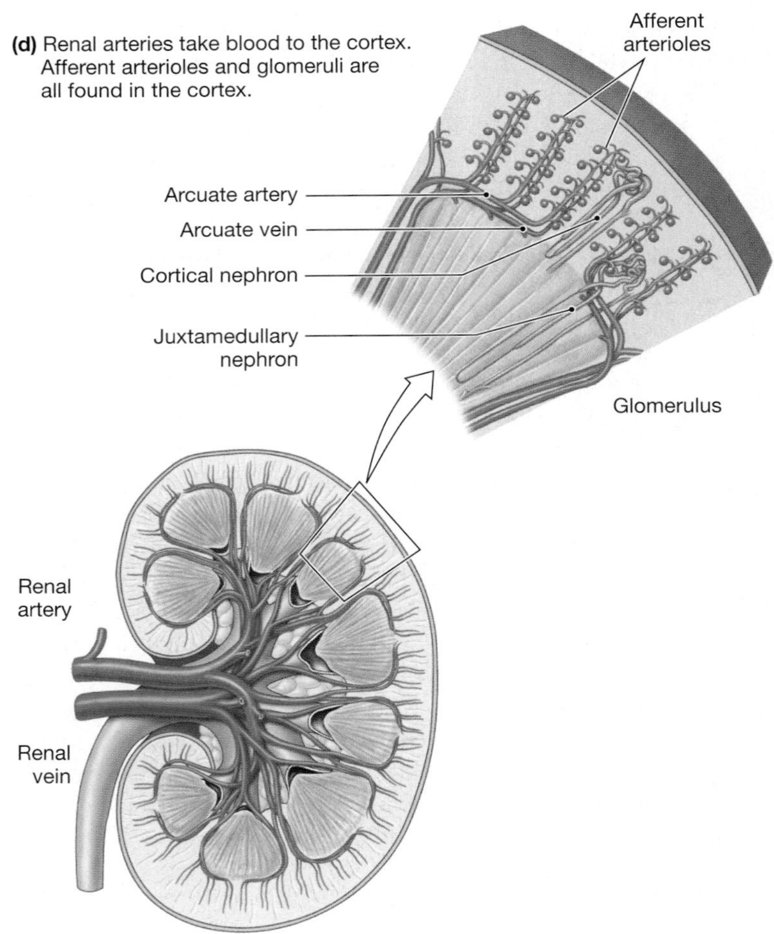

Afferent arterioles
Arcuate artery
Arcuate vein
Cortical nephron
Juxtamedullary nephron
Glomerulus
Renal artery
Renal vein

Structure of the Nephron

(e) Some nephrons dip deep into the medulla.

Arterioles

Nephrons

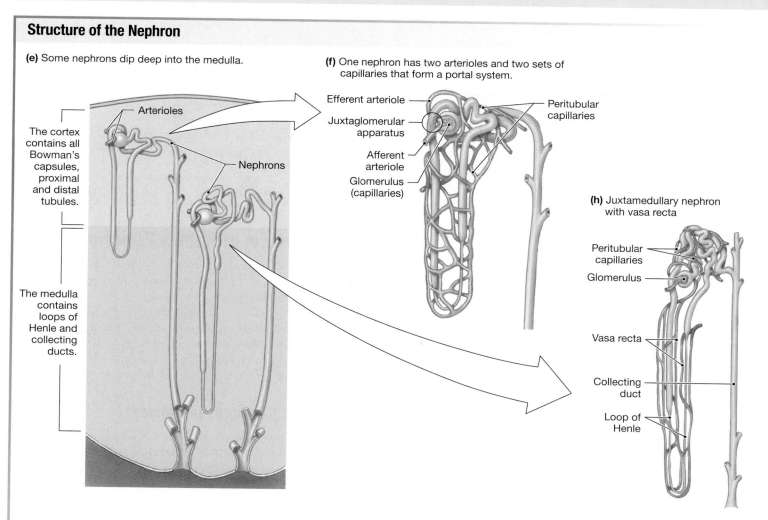

The cortex contains all Bowman's capsules, proximal and distal tubules.

The medulla contains loops of Henle and collecting ducts.

(f) One nephron has two arterioles and two sets of capillaries that form a portal system.

Efferent arteriole

Juxtaglomerular apparatus

Afferent arteriole

Glomerulus (capillaries)

Peritubular capillaries

(h) Juxtamedullary nephron with vasa recta

Peritubular capillaries

Glomerulus

Vasa recta

Collecting duct

Loop of Henle

(g) Parts of a nephron. In this view, the nephron has been untwisted so that flow goes left to right. Compare with the nephrons in (f).

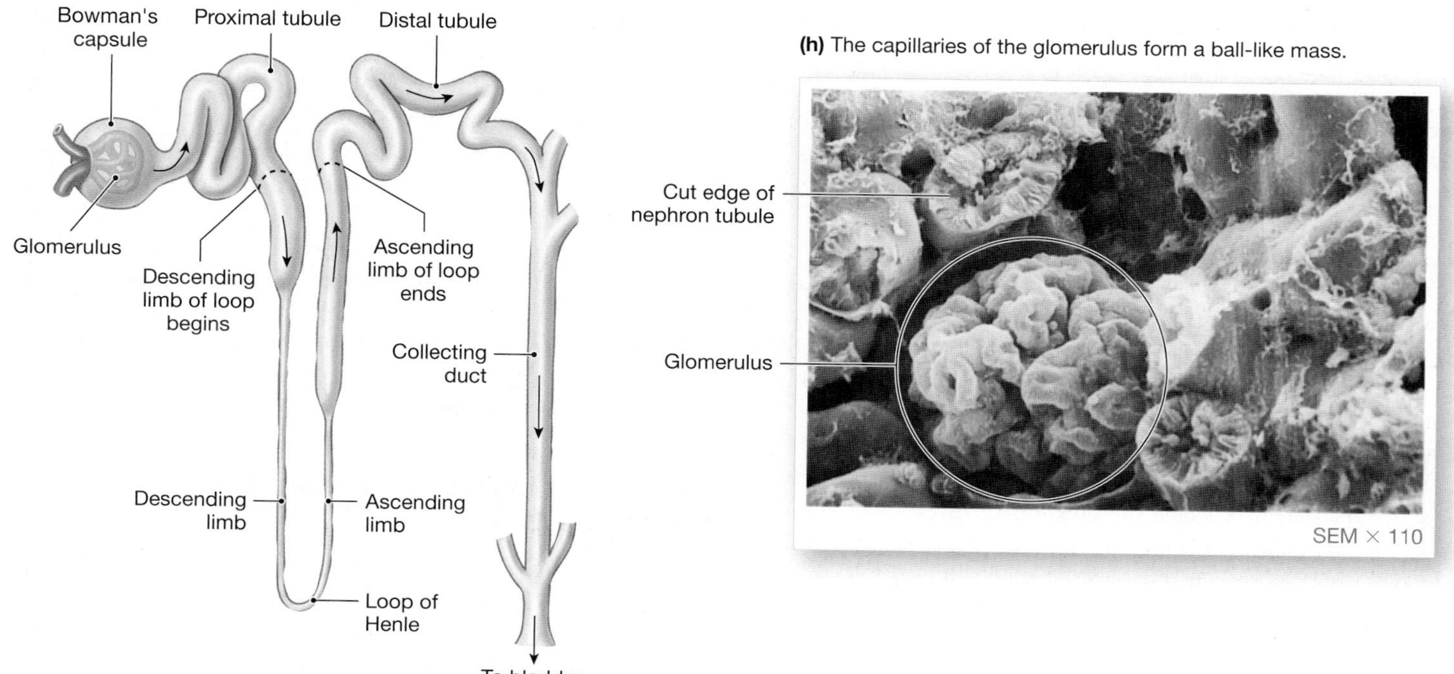
Bowman's capsule
Proximal tubule
Distal tubule
Glomerulus
Descending limb of loop begins
Ascending limb of loop ends
Collecting duct
Descending limb
Ascending limb
Loop of Henle
To bladder
Cut edge of nephron tubule
Glomerulus
SEM × 110

(h) The capillaries of the glomerulus form a ball-like mass.

In the renal portal system, blood flows from renal arteries into an **afferent arteriole**. From the afferent arteriole it goes into the first capillary bed, a ball-like network known as the **glomerulus** {*glomus,* a ball-shaped mass; plural *glomeruli*} (Fig. 19.1g, j). Blood leaving the glomerulus flows into an **efferent arteriole**, then into the second set of capillaries, the **peritubular capillaries** {*peri-*, around} that surround the tubule (Fig. 19.1g). In juxtamedullary nephrons, the long peritubular capillaries that dip into the medulla are called the **vasa recta** (Fig. 19.1h). Finally, peritubular capillaries converge to form venules and small veins, sending blood out of the kidney through the renal vein.

The function of the renal portal system is to filter fluid out of the blood and into the lumen of the nephron at the glomerular capillaries, then to *reabsorb* fluid from the tubule lumen back into the blood at the peritubular capillaries. The forces behind fluid movement in the renal portal system are similar to those that govern filtration of water and molecules out of systemic capillaries in other tissues, as we will describe shortly.

Concept Check

3. If net filtration out of glomerular capillaries occurs, then you know that capillary hydrostatic pressure must be (*greater than/less than/equal to*) capillary colloid osmotic pressure. [p. 497]
4. If net reabsorption into peritubular capillaries occurs, then capillary hydrostatic pressure must be (*greater than/less than/equal to*) the capillary colloid osmotic pressure.

Tubular Elements of the Kidney The kidney tubule consists of a single layer of epithelia cells connected together near their apical surface. The apical surfaces are folded into *microvilli* [p. 69] or other area-increasing folds, and the basal side of the polarized epithelium [p. 149] rests on a basement membrane. The cell-cell junctions are mostly tight but some have selective permeability for ions.

The nephron begins with a hollow, ball-like structure called **Bowman's capsule** that surrounds the glomerulus (Fig. 19.1i). The endothelium of the glomerulus is fused to the epithelium of Bowman's capsule so that fluid filtering out of the capillaries passes directly into the lumen of the tubule. The combination of glomerulus and Bowman's capsule is called the **renal corpuscle**.

From Bowman's capsule, filtered fluid flows into the **proximal tubule** {*proximal,* close or near}, then into the **loop of Henle**, a hairpin-shaped segment that dips down toward the medulla and then back up. The loop of Henle is divided into two limbs, a thin **descending limb** and an **ascending limb** with thin and thick segments. The fluid then passes into the **distal tubule** {*distal,* distant or far}. The distal tubules of up to eight nephrons drain into a single larger tube called the **collecting duct**. (The distal tubule and its collecting duct together form the **distal nephron**.) Collecting ducts pass from the cortex through the medulla and drain into the **renal pelvis** (Fig. 19.1c). From the renal pelvis, the filtered and modified fluid, now called urine, flows into the ureter on its way to excretion.

Notice in Figure 19.1g how the nephron twists and folds back on itself so that the final part of the ascending limb of the loop of Henle passes between the afferent and efferent arterioles. This region is known as the **juxtaglomerular apparatus**. The proximity of the ascending limb and the arterioles allows paracrine communication between the two structures, a key feature of kidney autoregulation. Because the twisted configuration of the nephron makes it difficult to follow fluid flow, we unfold the nephron in many of the remaining figures in this chapter so that fluid flows from left to right across the figure, as in Figure 19.1i.

Play Interactive Physiology 2.0
@Mastering Anatomy & Physiology

19.3 Overview of Kidney Function

Imagine drinking a 12-ounce soft drink every three minutes around the clock: By the end of 24 hours, you would have consumed the equivalent of 90 two-liter bottles. The thought of putting 180 liters of liquid into your intestinal tract is staggering, but that is how much plasma passes into the nephrons every day! Because the average volume of urine leaving the kidneys is only 1.5 L/day, more than 99% of the fluid that enters nephrons must find its way back into the blood, or the body would rapidly dehydrate.

Kidneys Filter, Reabsorb, and Secrete

Three basic processes take place in the nephron: filtration, reabsorption, and secretion (**FIG. 19.2**). **Filtration** is the movement of fluid from blood into the lumen of the nephron. Filtration takes place only in the renal corpuscle, where the walls of glomerular capillaries and Bowman's capsule are modified to allow bulk flow of fluid.

Once the filtered fluid, called *filtrate,* passes into the lumen of the nephron, it becomes part of the body's external environment, just as substances in the lumen of the intestinal tract are part of the external environment [Fig. 1.2, p. 4]. For this reason, anything that filters into the nephron is destined for **excretion**, removal in the urine, unless it is reabsorbed into the body.

After filtrate leaves Bowman's capsule, it is modified by reabsorption and secretion. **Reabsorption** is the process of moving substances in the filtrate from the lumen of the tubule back into the blood flowing through peritubular capillaries. **Secretion** selectively removes molecules from the blood and adds them to the filtrate in the tubule lumen. Although secretion and glomerular filtration both move substances from blood into the tubule, secretion is a more selective process that usually uses membrane proteins to move molecules across the tubule epithelium.

The Nephron Modifies Fluid Volume and Osmolarity

Now let's follow some filtrate through the nephron to learn what happens to it in the various segments (Fig. 19.2). The 180 liters of fluid that filters into Bowman's capsule each day are almost identical in composition to plasma and nearly isosmotic—about

FIG. 19.2 ESSENTIALS Nephron Function

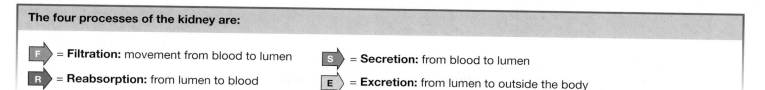

The four processes of the kidney are:

 = **Filtration:** movement from blood to lumen

R = **Reabsorption:** from lumen to blood

 = **Secretion:** from blood to lumen

 = **Excretion:** from lumen to outside the body

This model nephron has been untwisted so that fluid flows left to right.

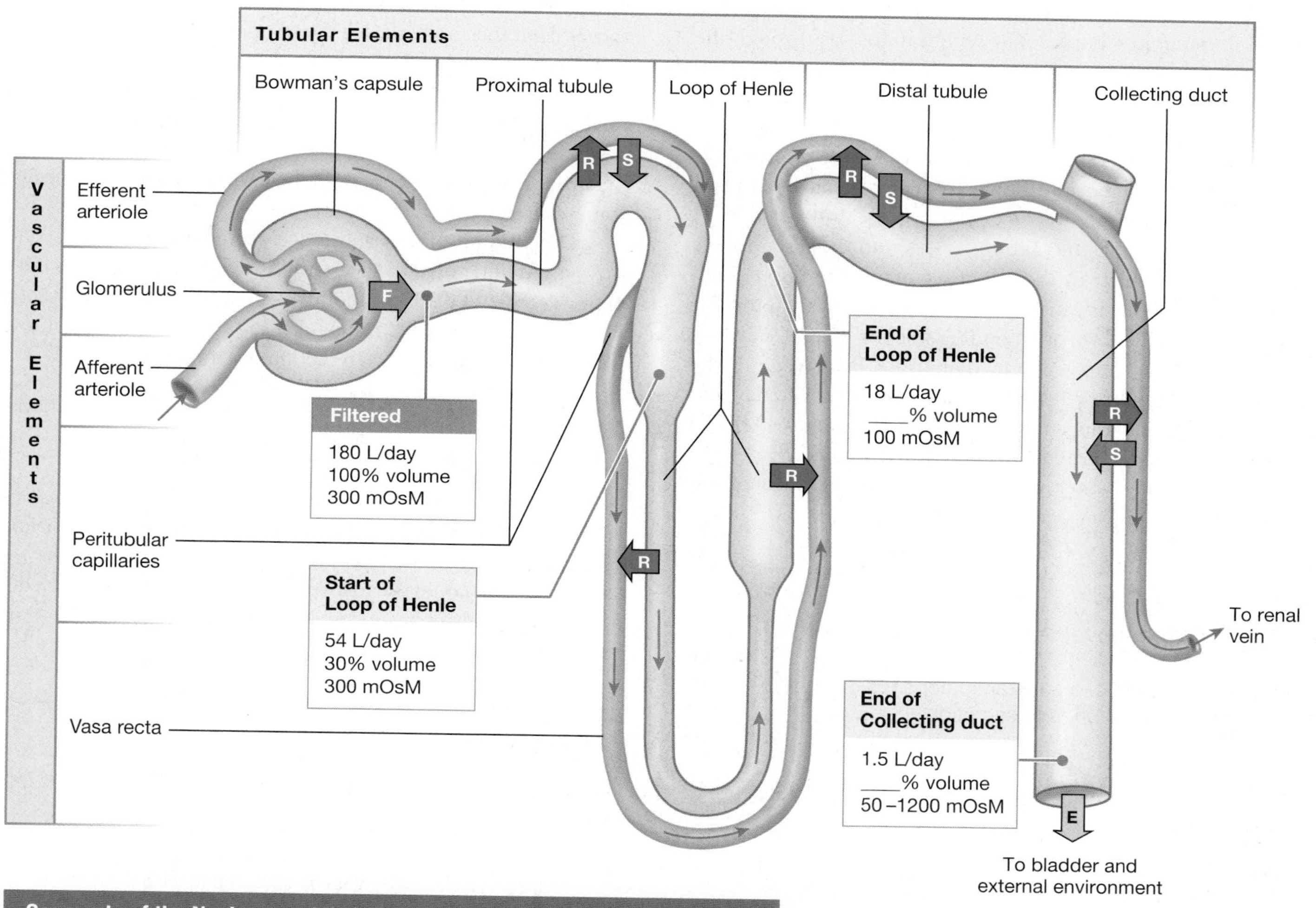

Tubular Elements
Bowman's capsule
Proximal tubule
Loop of Henle
Distal tubule
Collecting duct
Vascular Elements
Efferent arteriole
Glomerulus
Afferent arteriole
Peritubular capillaries
Vasa recta
F
R
S
E
Filtered
180 L/day
100% volume
300 mOsM
Start of Loop of Henle
54 L/day
30% volume
300 mOsM
End of Loop of Henle
18 L/day
____% volume
100 mOsM
End of Collecting duct
1.5 L/day
____% volume
50–1200 mOsM
To renal vein
To bladder and external environment

Segments of the Nephron and their Functions	
Segment of Nephron	**Processes**
Renal corpuscle (glomerulus + Bowman's capsule)	Filtration of mostly protein-free plasma from the capillaries into the capsule
Proximal tubule	Isosmotic reabsorption of organic nutrients, ions, and water. Secretion of metabolites and xenobiotic molecules such as penicillin.
Loop of Henle	Reabsorption of ions in excess of water to create dilute fluid in the lumen. Countercurrent arrangements contributes to concentrated interstitial fluid in the renal medulla [see Chapter 20].
Distal nephron (distal tubule + collecting duct)	Regulated reabsorption of ions and water for salt and water balance and pH homeostasis.

FIGURE QUESTIONS

1. In which segments of the nephron do the following processes take place:
   (a) filtration
   (b) reabsorption
   (c) secretion
   (d) excretion
2. Calculate the percentage of filtered volume that leaves
   (a) the loop of Henle
   (b) the collecting duct

300 mOsM [p. 125]. As this filtrate flows through the proximal tubule, about 70% of its volume is reabsorbed, leaving 54 liters in the lumen.

Reabsorption occurs when proximal tubule cells transport solutes out of the lumen, and water follows by osmosis. Filtrate leaving the proximal tubule has the same osmolarity as filtrate that entered. For this reason, we say that the primary function of the proximal tubule is the isosmotic reabsorption of solutes and water.

Filtrate leaving the proximal tubule passes into the loop of Henle, the primary site for creating dilute urine [by a process discussed in detail in Chapter 20]. As filtrate passes through the loop, proportionately more solute is reabsorbed than water, and the filtrate becomes hyposmotic relative to the plasma. By the time filtrate flows out of the loop, it averages 100 mOsM, and its volume has fallen from 54 L/day to about 18 L/day. Most of the volume originally filtered into Bowman's capsule has been reabsorbed into the capillaries.

From the loop of Henle, filtrate passes into the distal tubule and the collecting duct. In these two segments, the fine regulation of salt and water balance takes place under the control of several hormones. Reabsorption and (to a lesser extent) secretion determine the final composition of the filtrate. By the end of the collecting duct, the filtrate has a volume of about 1.5 L/day and an osmolarity that can range from 50 mOsM to 1200 mOsM. The final volume and osmolarity of urine depend on the body's need to conserve or excrete water and solute. Hormonal control of salt and water balance is discussed in the next chapter.

FIG. 19.3 Solute movement through the nephron

The urinary excretion of a substance depends on its filtration, reabsorption, and secretion.

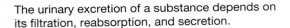

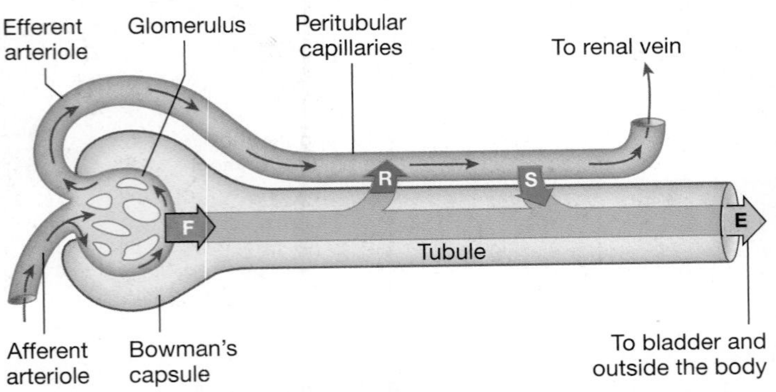
Efferent arteriole
Glomerulus
Peritubular capillaries
To renal vein
R
S
F
E
Tubule
Afferent arteriole
Bowman's capsule
To bladder and outside the body

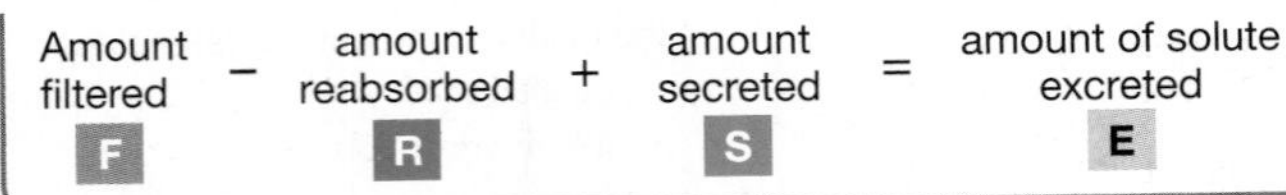
Amount filtered F − amount reabsorbed R + amount secreted S = amount of solute excreted E

FIGURE QUESTION

A person filters 720 millimoles of K^+ in a day and secretes 43 millimoles. She excretes 79 millimoles in her urine. What happened to the rest of the K^+ and how much was it?

A word of caution here: It is very easy to confuse *secretion* with *excretion*. Try to remember the origins of the two prefixes. *Se-* means *apart*, as in to separate something from its source. In the nephron, secreted solutes are moved from plasma to tubule lumen. *Ex-* means *out*, or *away*, as in out of or away from the body. Excretion refers to the removal of a substance from the body. Besides the kidneys, other organs that carry out excretory processes include the lungs (CO_2) and intestines (undigested food, bilirubin).

Figure 19.2 summarizes filtration, reabsorption, secretion, and excretion. Filtration takes place in the renal corpuscle as fluid moves from the capillaries of the glomerulus into Bowman's capsule. Reabsorption and secretion occur along the remainder of the tubule, transferring materials between the lumen and the peritubular capillaries. The quantity and composition of the substances being reabsorbed and secreted vary in different segments of the nephron. Filtrate that remains in the lumen at the end of the nephron is excreted as urine.

The amount of any substance excreted in the urine reflects the net handling of that substance during its passage through the nephron (**FIG. 19.3**). The amount excreted is equal to the amount filtered into the tubule, minus any amount reabsorbed into the blood, plus any amount secreted into the tubule lumen:

$$\text{Amount excreted} = \text{amount filtered} - \text{amount reabsorbed} + \text{amount secreted}$$

This equation is a useful way to think about *renal handling* of solutes. Note, however, that not every substance in the plasma is filtered. And substances that are filtered may or may not be reabsorbed or secreted. In the following sections, we look in more detail at the important processes of filtration, reabsorption, secretion, and excretion.

Concept Check

5. Name one way in which filtration and secretion are alike. Name one way in which they differ.
6. A water molecule enters the renal corpuscle from the blood and ends up in the urine. Name all the anatomical structures that the molecule passes through on its trip to the outside world.
7. What would happen to the body if filtration continued at a normal rate but reabsorption dropped to half the normal rate?

19.4 Filtration

The filtration of plasma into the kidney tubule is the first step in urine formation. This relatively nonspecific process creates a filtrate whose composition is like that of plasma minus most of the plasma proteins. Under normal conditions, blood cells remain in the capillary, so that the filtrate is composed of water and dissolved solutes.

When you visualize plasma filtering out of the glomerular capillaries, it is easy to imagine that all the plasma in the capillary moves into Bowman's capsule. However, filtration of all the plasma would leave behind a sludge of blood cells and proteins that could not flow out of the glomerulus. Instead, only about one-fifth of the plasma that flows through the kidneys filters into the nephrons. The remaining four-fifths of the plasma, along with most plasma proteins and blood cells, flows into the peritubular capillaries (**FIG. 19.4**). The percentage of renal plasma flow that filters into the tubule is called the **filtration fraction**.

The Renal Corpuscle Contains Filtration Barriers

Filtration takes place in the renal corpuscle (**FIG. 19.5**), which consists of the glomerular capillaries surrounded by Bowman's capsule. Substances leaving the plasma must pass through three *filtration barriers* before entering the tubule lumen: the glomerular capillary endothelium, a basement membrane sandwiched in the middle, and the epithelium of Bowman's capsule (Fig. 19.5d).

The first filtration barrier is the capillary endothelium. Glomerular capillaries are *fenestrated capillaries* [p. 495] with large pores (*fenestra*) that allow most components of the plasma to filter through the endothelium. The luminal surface of the capillary and the pores are lined with a lattice-like layer of glycoproteins called the *glycocalyx*. The negatively charged proteins of the glycocalyx help repel negatively charged plasma proteins, and the endothelial pores are small enough to prevent blood cells from leaving the capillary.

The second filtration barrier is the basement membrane, the acellular layer of extracellular matrix that separates the capillary endothelium from the epithelium of Bowman's capsule (Fig. 19.5d). The basement membrane consists of negatively charged glycoproteins, collagen, and other proteins, and it too acts like a coarse sieve, excluding most plasma proteins from the fluid that filters through it.

The third filtration barrier is the epithelium of Bowman's capsule, consisting of specialized cells called **podocytes** {*podos*, foot} that surround each glomerular capillary (Fig. 19.5c). Podocytes have long cytoplasmic extensions called **foot processes** that extend from the main cell body (Fig. 19.5a, b). Foot processes wrap around the glomerular capillaries and interlace with one another, leaving narrow **filtration slits** closed by a *slit diaphragm*. The slit diaphragm contains several structural unique proteins, including *nephrin* and *podocin*, that seem to form a two-layer sieve. The slit diaphragm proteins were discovered by investigators looking for the gene mutations responsible for two congenital kidney diseases. In these diseases, where nephrin or podocin are absent or abnormal, the filtration barrier does not function properly and proteins leak across the barrier into the urine.

Glomerular **mesangial cells** lie between and around the glomerular capillaries, creating a support structure for the tuft of capillaries (Fig. 19.5c). Mesangial cells influence filtration by altering the surface area of the filtration slits. In addition, mesangial cells secrete cytokines associated with immune and inflammatory processes. Disruptions of mesangial cell function have been linked to several disease processes in the kidney.

EMERGING CONCEPTS

Diabetes: Diabetic Nephropathy

End-stage renal failure, in which kidney function has deteriorated beyond recovery, is a life-threatening complication in 30–40% of people with type 1 diabetes and in 10–20% of those with type 2 diabetes. As with many other complications of diabetes, the exact causes of renal failure are not clear. *Diabetic nephropathy* usually begins with an increase in glomerular filtration. This is followed by the appearance of proteins in the urine (*proteinuria*), an indication that the normal filtration barrier has been altered. In later stages, filtration rates decline. This stage is associated with thickening of the glomerular basal lamina and changes in podocytes and mesangial cells. Abnormal growth of mesangial cells compresses the glomerular capillaries and impedes blood flow, contributing to the decrease in glomerular filtration. At this point, patients must have their kidney function supplemented by dialysis, and eventually they may need a kidney transplant.

FIG. 19.4 The filtration fraction

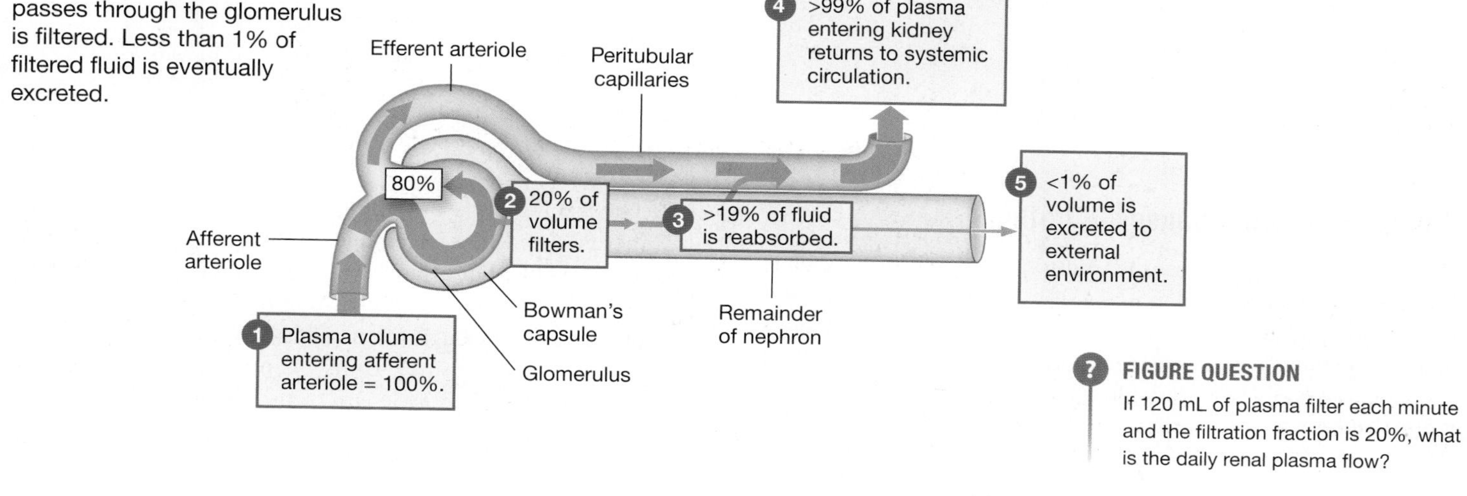
Only 20% of the plasma that passes through the glomerulus is filtered. Less than 1% of filtered fluid is eventually excreted.
Efferent arteriole
Peritubular capillaries
4 >99% of plasma entering kidney returns to systemic circulation.
80%
2 20% of volume filters.
3 >19% of fluid is reabsorbed.
5 <1% of volume is excreted to external environment.
Afferent arteriole
Bowman's capsule
Remainder of nephron
1 Plasma volume entering afferent arteriole = 100%.
Glomerulus

FIGURE QUESTION

If 120 mL of plasma filter each minute and the filtration fraction is 20%, what is the daily renal plasma flow?

FIG. 19.5 The renal corpuscle

(a) The epithelium around glomerular capillaries is modified into podocytes.

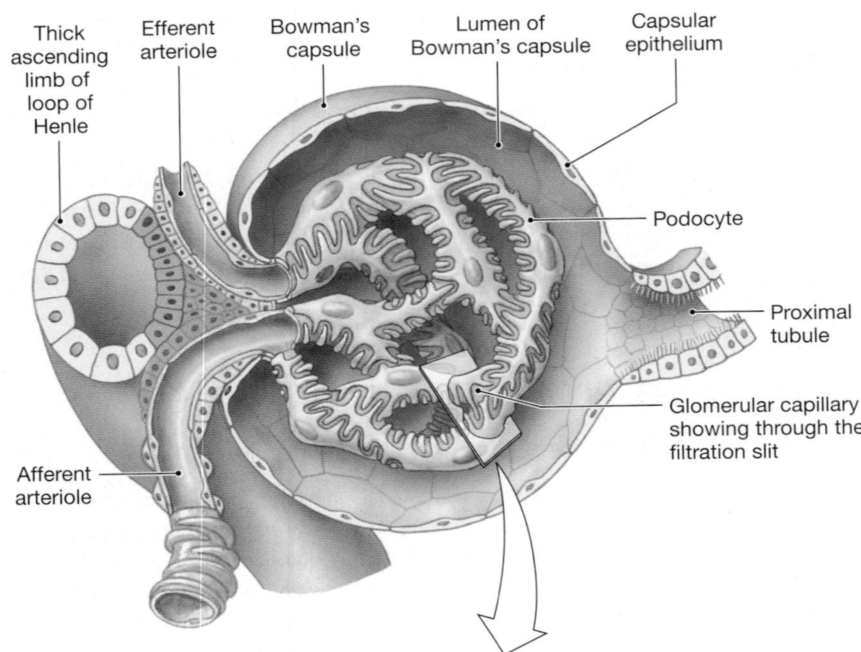
Thick ascending limb of loop of Henle
Efferent arteriole
Bowman's capsule
Lumen of Bowman's capsule
Capsular epithelium
Podocyte
Proximal tubule
Glomerular capillary showing through the filtration slit
Afferent arteriole

(b) Micrograph showing podocyte foot processes around glomerular capillary.

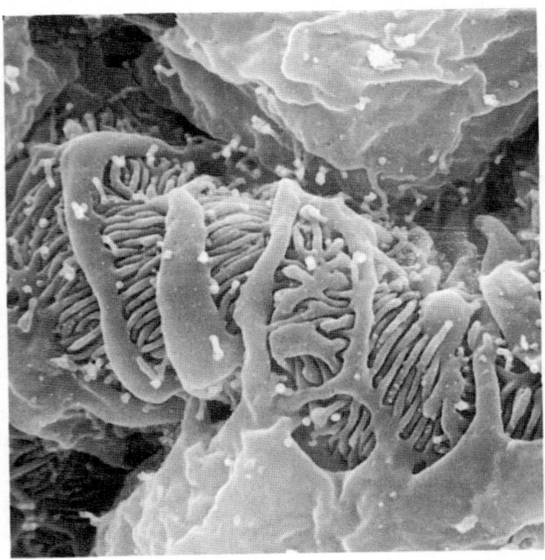

(c) Podocyte foot processes surround each capillary, leaving slits through which filtration takes place. Mesangial cells between the capillaries contract to alter blood flow.

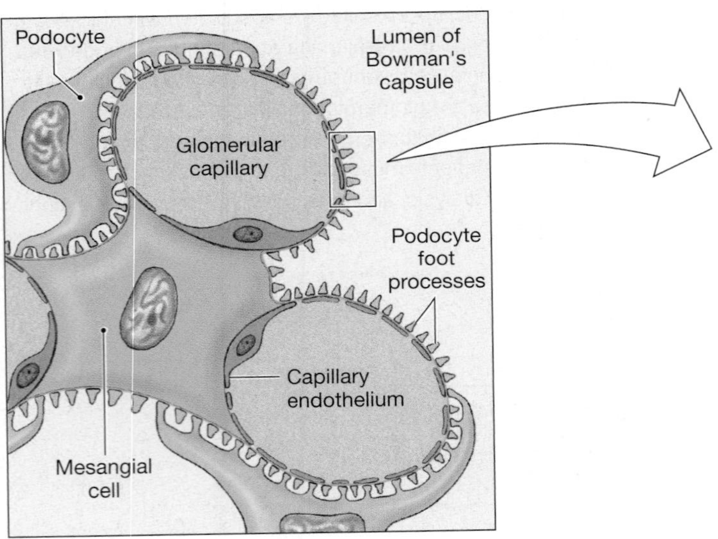
Podocyte
Lumen of Bowman's capsule
Glomerular capillary
Podocyte foot processes
Capillary endothelium
Mesangial cell

(d) The glomerular capillary endothelium, basement membrane, and podocytes create a three-layer filtration barrier. Filtered substances pass through endothelial pores and filtration slits.

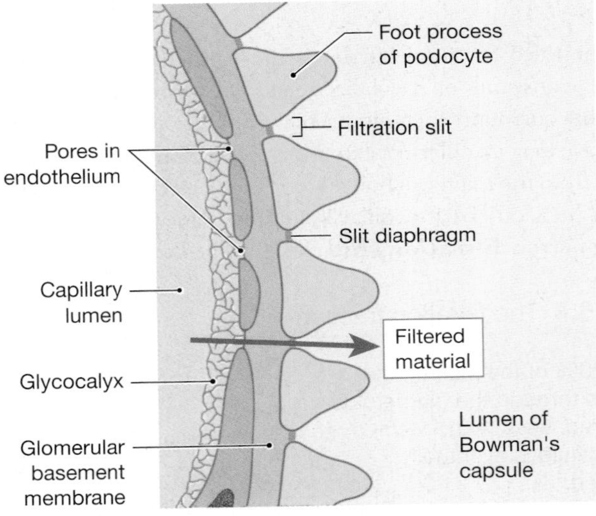
Foot process of podocyte
Filtration slit
Pores in endothelium
Slit diaphragm
Capillary lumen
Filtered material
Glycocalyx
Lumen of Bowman's capsule
Glomerular basement membrane

Capillary Pressure Causes Filtration

What drives filtration across the walls of the glomerular capillaries? The process is similar in many ways to filtration of fluid out of systemic capillaries [p. 497]. The three pressures that influence glomerular filtration—capillary blood pressure, capillary colloid osmotic pressure, and capsule fluid pressure—are summarized in **FIGURE 19.6a**.

1. The *hydrostatic pressure* (P_H) of blood flowing through the glomerular capillaries forces fluid through the leaky endothelium. Capillary blood pressure averages 55 mm Hg and favors filtration into Bowman's capsule. Although pressure decreases as blood moves through the capillaries, it remains higher than the opposing pressures. Consequently, filtration takes place along nearly the entire length of the glomerular capillaries.

FIG. 19.6 ESSENTIALS Glomerula Filtration Rate

Filtration pressure depends on hydrostatic pressure, and is opposed by colloid osmotic pressure and capsule fluid pressure.

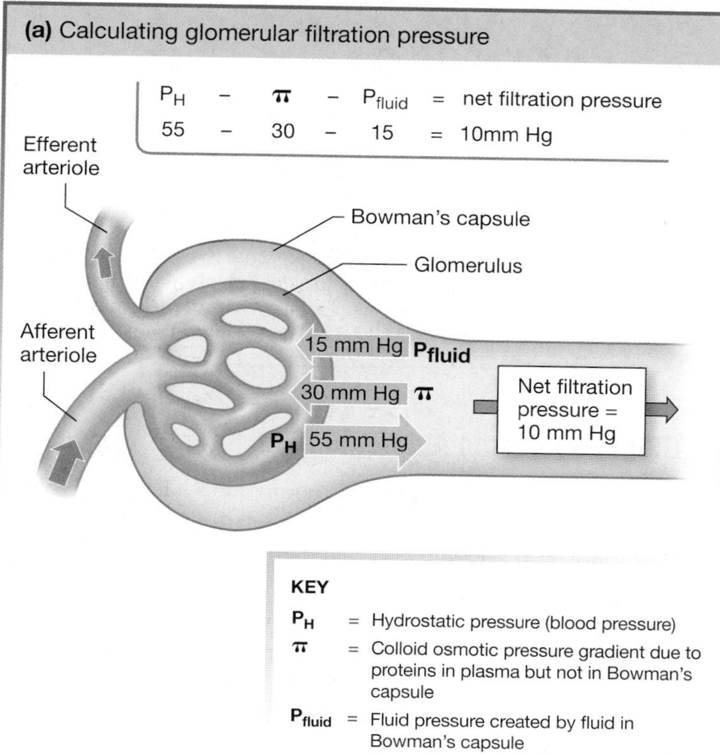

(a) Calculating glomerular filtration pressure
P_H − π − P_{fluid} = net filtration pressure
55 − 30 − 15 = 10mm Hg
Efferent arteriole
Bowman's capsule
Glomerulus
Afferent arteriole
15 mm Hg P_{fluid}
30 mm Hg π
P_H 55 mm Hg
Net filtration pressure = 10 mm Hg
KEY
P_H = Hydrostatic pressure (blood pressure)
π = Colloid osmotic pressure gradient due to proteins in plasma but not in Bowman's capsule
P_{fluid} = Fluid pressure created by fluid in Bowman's capsule

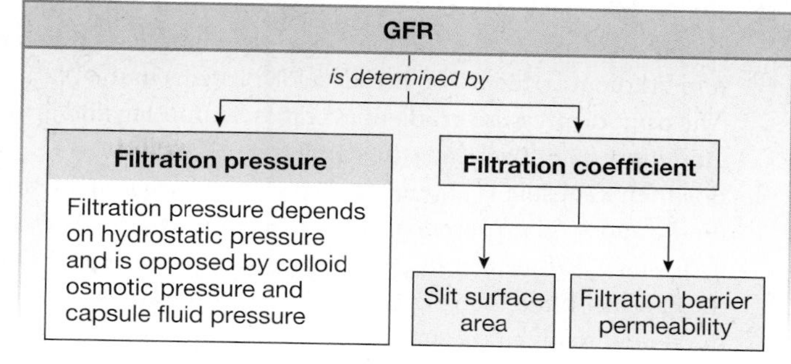

GFR
is determined by
Filtration pressure
Filtration pressure depends on hydrostatic pressure and is opposed by colloid osmotic pressure and capsule fluid pressure
Filtration coefficient
Slit surface area
Filtration barrier permeability

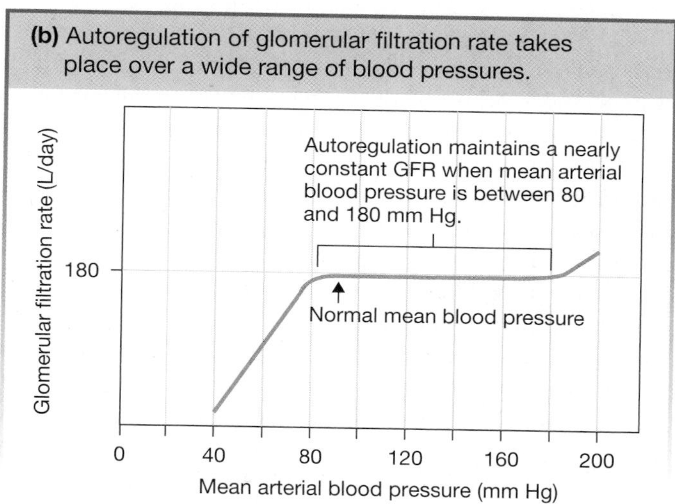

(b) Autoregulation of glomerular filtration rate takes place over a wide range of blood pressures.
Glomerular filtration rate (L/day)
180
Autoregulation maintains a nearly constant GFR when mean arterial blood pressure is between 80 and 180 mm Hg.
Normal mean blood pressure
0
40
80
120
160
200
Mean arterial blood pressure (mm Hg)

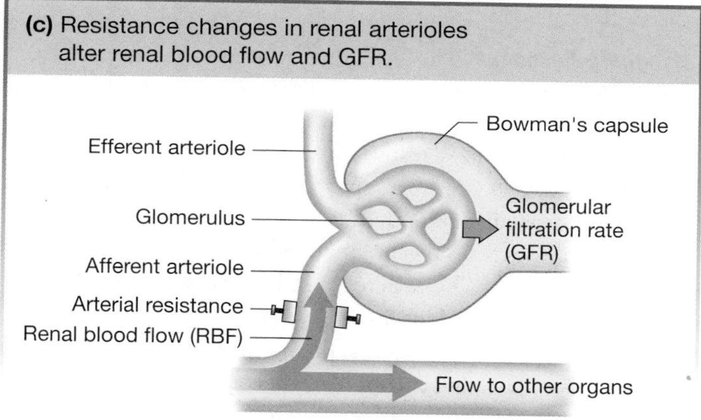

(c) Resistance changes in renal arterioles alter renal blood flow and GFR.
Efferent arteriole
Bowman's capsule
Glomerulus
Glomerular filtration rate (GFR)
Afferent arteriole
Arterial resistance
Renal blood flow (RBF)
Flow to other organs

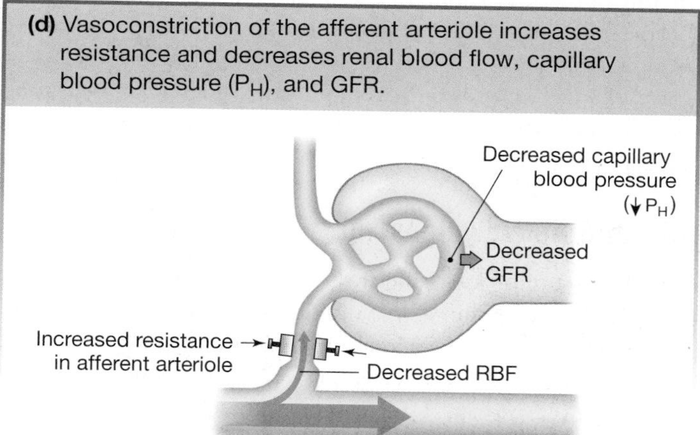

(d) Vasoconstriction of the afferent arteriole increases resistance and decreases renal blood flow, capillary blood pressure (P_H), and GFR.
Decreased capillary blood pressure (↓P_H)
Decreased GFR
Increased resistance in afferent arteriole
Decreased RBF

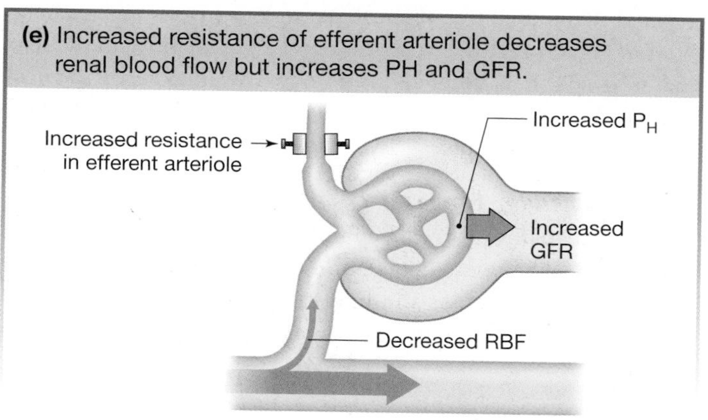

(e) Increased resistance of efferent arteriole decreases renal blood flow but increases PH and GFR.
Increased resistance in efferent arteriole
Increased P_H
Increased GFR
Decreased RBF

FIGURE QUESTION

What happens to capillary blood pressure, GFR, and RBF when the afferent arteriole dilates?

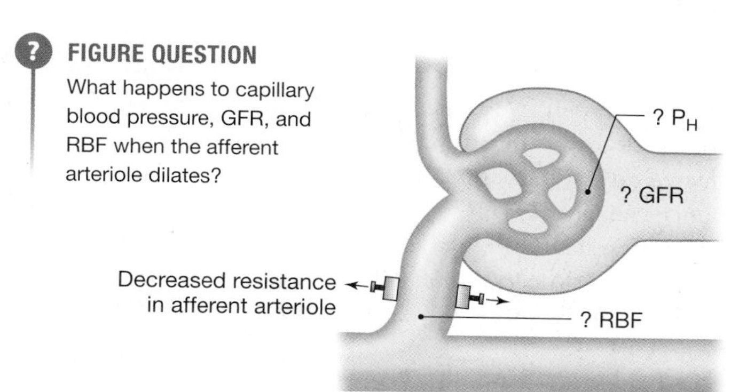

? P_H
? GFR
Decreased resistance in afferent arteriole
? RBF

2. The *colloid osmotic pressure* (π) inside glomerular capillaries is higher than that of the fluid in Bowman's capsule. This pressure gradient is due to the presence of proteins in the plasma. The osmotic pressure gradient averages 30 mm Hg and favors fluid movement back into the capillaries.
3. Bowman's capsule is an enclosed space (unlike the interstitial fluid), and so the presence of fluid in the capsule creates a hydrostatic *fluid pressure* (P_{fluid}) that opposes fluid movement into the capsule. Fluid filtering out of the capillaries must displace the fluid already in the capsule lumen. Hydrostatic fluid pressure in the capsule averages 15 mm Hg, opposing filtration.

The net driving force is 10 mm Hg in the direction favoring filtration. This pressure may not seem very high, but when combined with the very leaky nature of the fenestrated glomerular capillaries, it results in rapid fluid filtration into the tubules.

The volume of fluid that filters into Bowman's capsule per unit time is the **glomerular filtration rate (GFR)**. Average GFR is 125 mL/min, or 180 L/day, an incredible rate considering that total plasma volume is only about 3 liters. This rate means that the kidneys filter the entire plasma volume 60 times a day, or 2.5 times every hour. If most of the filtrate were not reabsorbed during its passage through the nephron, we would run out of plasma in only 24 minutes of filtration!

GFR is influenced by two factors: the net filtration pressure just described and the filtration coefficient. Filtration pressure is determined primarily by *renal blood flow* and blood pressure. The **filtration coefficient** has two components: (1) the surface area of the glomerular capillaries available for filtration, and (2) the permeability of the filtration slits. In this respect, glomerular filtration is similar to gas exchange at the alveoli, where the rate of gas exchange depends on partial pressure differences, the surface area of the alveoli, and the permeability of the alveolar-capillary diffusion barrier [p. 565].

GFR Is Relatively Constant

Blood pressure provides the hydrostatic pressure that drives glomerular filtration. Therefore, it might seem reasonable to assume that if blood pressure increased, GFR would increase, and if blood pressure fell, GFR would decrease. That is not usually the case, however. Instead, GFR is remarkably constant over a wide range of blood pressures. As long as mean arterial blood pressure remains between 80 mm Hg and 180 mm Hg, GFR averages 180 L/day (Fig. 19.6b).

GFR is controlled primarily by regulation of blood flow through the renal arterioles. If the overall resistance of the renal arterioles increases, renal blood flow decreases, and blood is diverted to other organs [Fig. 15.13, p. 492]. The effect of increased resistance on GFR, however, depends on *where* the resistance change takes place.

If resistance increases in the *afferent* arteriole (Fig. 19.6d), hydrostatic pressure decreases on the glomerular side of the constriction. This translates into a decrease in GFR. If resistance increases in the *efferent* arteriole, blood "dams up" in front of the constriction, and hydrostatic pressure in the glomerular capillaries increases (Fig. 19.6e). Increased glomerular pressure increases GFR. The opposite changes occur with decreased resistance in the afferent or efferent arterioles. Most regulation occurs at the afferent arteriole.

Concept Check

8. Why is the osmotic pressure of plasma in efferent arterioles higher than that in afferent arterioles?
9. If a hypertensive person's blood pressure is 143/107 mm Hg, and mean arterial pressure = diastolic pressure + 1/3 the pulse pressure, what is this person's mean arterial pressure? What is this person's GFR according to Figure 19.6b?

GFR Is Subject to Autoregulation

Autoregulation of glomerular filtration rate is a local control process in which the kidney maintains a relatively constant GFR in the face of normal fluctuations in blood pressure. One important function of GFR autoregulation is to protect the filtration barriers from high blood pressures that might damage them. We do not completely understand the autoregulation process, but several mechanisms are at work. The **myogenic response** is the intrinsic ability of vascular smooth muscle to respond to pressure changes. **Tubuloglomerular feedback** is a paracrine signaling mechanism through which changes in fluid flow through the loop of Henle influence GFR.

Myogenic Response The myogenic response of afferent arterioles is similar to autoregulation in other systemic arterioles. When smooth muscle in the arteriole wall stretches because of increased blood pressure, stretch-sensitive ion channels open, and the muscle cells depolarize. Depolarization opens voltage-gated Ca^{2+} channels, and the vascular smooth muscle contracts [p. 405]. Vasoconstriction increases resistance to flow, and so blood flow through the arteriole diminishes. The decrease in blood flow decreases filtration pressure in the glomerulus.

If blood pressure decreases, the tonic level of arteriolar contraction disappears, and the arteriole becomes maximally dilated. However, vasodilation is not as effective at maintaining GFR as vasoconstriction because normally the afferent arteriole is fairly relaxed. Consequently, when mean blood pressure drops below 80 mm Hg, GFR decreases. This decrease is adaptive in the sense that if less plasma is filtered, the potential for fluid loss in the urine is decreased. In other words, a decrease in GFR helps the body conserve blood volume.

Tubuloglomerular Feedback Tubuloglomerular feedback is a local control pathway in which fluid flow through the tubule influences GFR. The twisted configuration of the nephron, as shown in **FIGURE 19.7a**, causes the final portion of the thick ascending limb of the loop of Henle to pass between the afferent and

FIG. 19.7 The juxtaglomerular apparatus

The juxtaglomerular apparatus consists of macula densa and granular cells. Paracrine signaling between the nephron and afferent arteriole influences GFR.

(a) The nephron loops back on itself so that the ascending limb of the loop of Henle passes between the afferent and efferent arterioles.

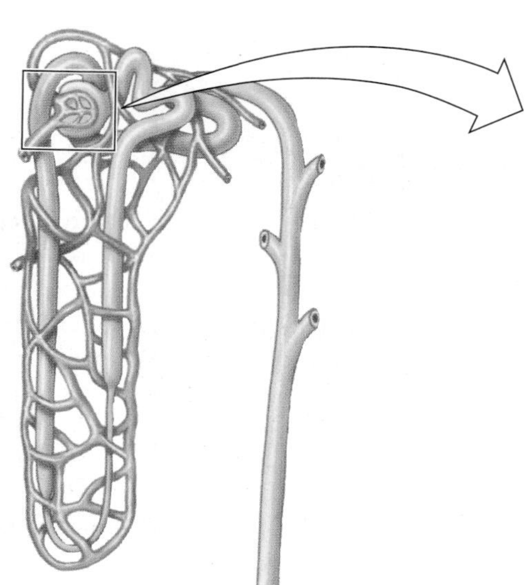

(b) The macula densa cells sense distal tubule flow and release paracrine signals that affect afferent and efferent arteriole diameter.

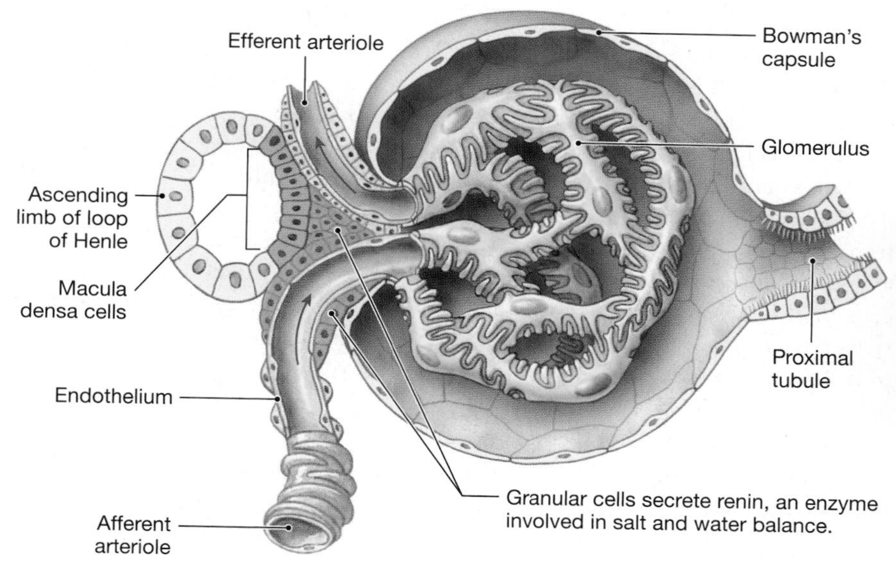
Efferent arteriole
Bowman's capsule
Glomerulus
Ascending limb of loop of Henle
Macula densa cells
Endothelium
Proximal tubule
Granular cells secrete renin, an enzyme involved in salt and water balance.
Afferent arteriole

(c) Tubuloglomerular feedback helps GFR autoregulation.

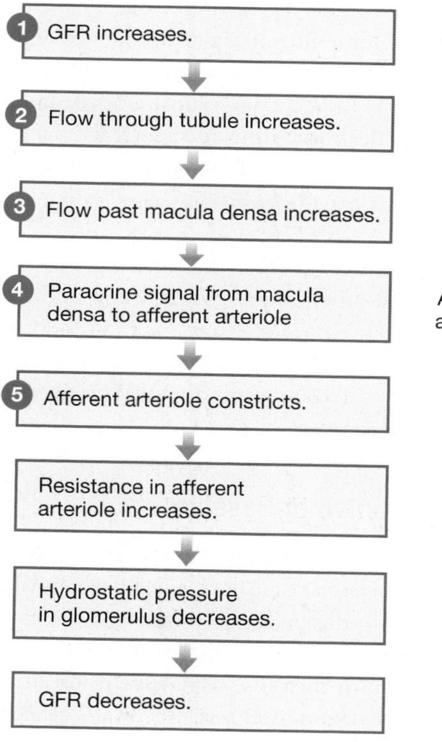
1 GFR increases.
2 Flow through tubule increases.
3 Flow past macula densa increases.
4 Paracrine signal from macula densa to afferent arteriole
5 Afferent arteriole constricts.
Resistance in afferent arteriole increases.
Hydrostatic pressure in glomerulus decreases.
GFR decreases.

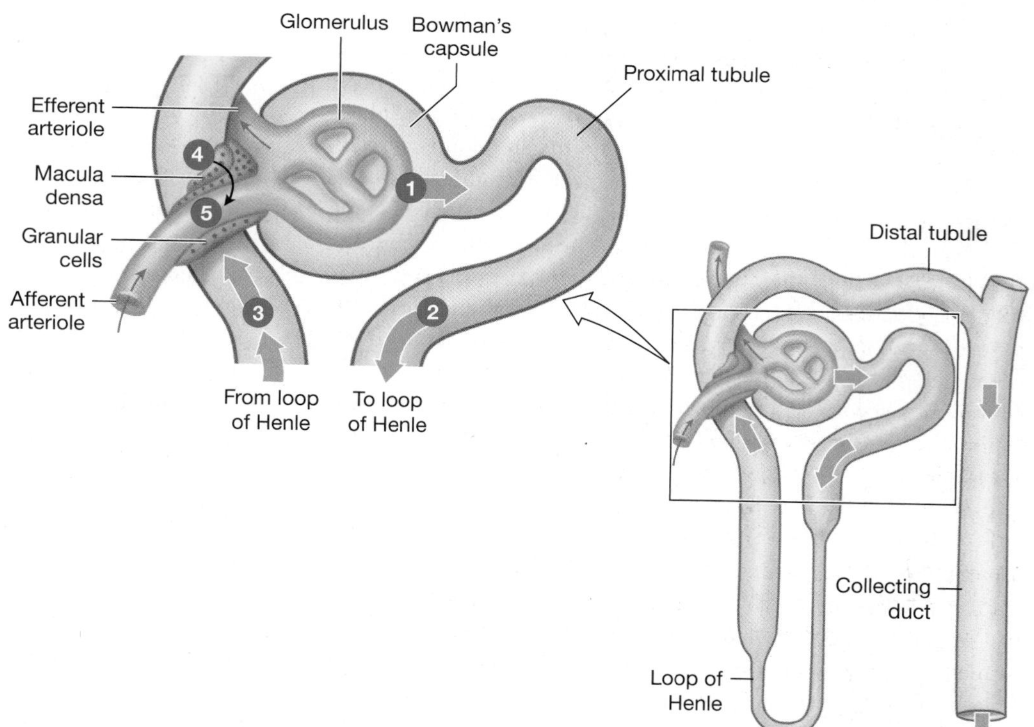
Glomerulus
Bowman's capsule
Proximal tubule
Efferent arteriole
Macula densa
Granular cells
Afferent arteriole
From loop of Henle
To loop of Henle
Distal tubule
Collecting duct
Loop of Henle

efferent arterioles. The tubule and arteriolar walls are modified in the regions where they contact each other and together form the **juxtaglomerular apparatus**.

The modified portion of the tubule epithelium is a plaque of cells called the **macula densa** (Fig. 19.7b). The adjacent wall of the afferent arteriole has specialized smooth muscle cells called **granular cells** (also known as *juxtaglomerular cells* or *JG cells*). The granular cells secrete *renin,* an enzyme involved in salt and water balance [Chapter 20] as well as paracrine signal molecules. When NaCl delivery past the macula densa increases as a result of increased GFR, the macula densa cells send a paracrine message to the neighboring afferent arteriole (Fig. 19.7c). The afferent arteriole constricts, increasing resistance and decreasing GFR.

Experimental evidence indicates that the macula densa cells transport NaCl, and that increases in salt transport initiate tubuloglomerular feedback. Flow can also be sensed in renal tubule cells by the *primary cilia* [p. 68], which are located on the apical surface facing the lumen. The renal primary cilia are known to act as flow sensors as well as signal transducers for normal development.

Paracrine signaling between the macula densa and the afferent arteriole is complex, and the details are still being worked out. Experiments show that multiple paracrine signals, including ATP, adenosine, and nitric oxide, pass from the macula densa to the arteriole as part of tubuloglomerular feedback.

Hormones and Autonomic Neurons Also Influence GFR

Although local mechanisms within the kidney attempt to maintain a constant GFR, the importance of the kidneys in systemic blood pressure homeostasis means that integrating centers outside the kidney can override local controls. Hormones and the autonomic nervous system alter glomerular filtration rate two ways: by changing resistance in the arterioles and by altering the filtration coefficient.

Neural control of GFR is mediated by sympathetic neurons that innervate both the afferent and efferent arterioles. Sympathetic innervation of α-receptors on vascular smooth muscle causes vasoconstriction [p. 489]. If sympathetic activity is moderate, there is little effect on GFR. If systemic blood pressure drops sharply, however, as occurs with hemorrhage or severe dehydration, sympathetically induced vasoconstriction of the arterioles decreases GFR and renal blood flow. This is an adaptive response that helps conserve fluid volume.

A variety of hormones also influence arteriolar resistance. Among the most important are *angiotensin II,* a potent vasoconstrictor, and *prostaglandins* [p. 178] that are vasodilators. These same hormones may also affect the filtration coefficient by acting on podocytes or mesangial cells. Podocytes change the size of the glomerular filtration slits. If the slits widen, more surface area is available for filtration, and GFR increases. We still have much to learn about these processes, and physiologists are actively investigating them.

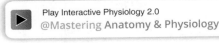

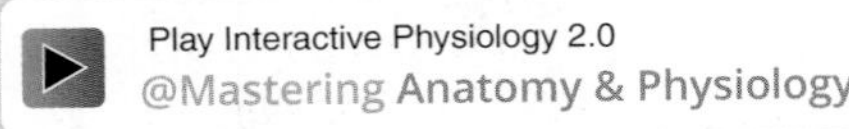
Play Interactive Physiology 2.0
@Mastering Anatomy & Physiology

Concept Check

10. If systemic blood pressure remains constant but the afferent arteriole of a nephron constricts, what happens to renal blood flow and GFR in that nephron?
11. A person with cirrhosis of the liver has lower-than-normal levels of plasma proteins and consequently a higher-than-normal GFR. Explain why a decrease in plasma protein concentration causes an increase in GFR.

19.5 Reabsorption

Each day, 180 liters of filtered fluid pass from the glomerular capillaries into the tubules, yet only about 1.5 liters are excreted in the urine. Thus, more than 99% of the fluid entering the tubules must be reabsorbed into the blood as filtrate moves through the nephrons. Most of this reabsorption takes place in the proximal tubule, with a smaller amount of reabsorption in the distal segments of the nephrons. Regulated reabsorption in the distal nephron allows the kidneys to return ions and water to the plasma selectively—as needed to maintain homeostasis.

One question you might be asking is, "Why bother to filter 180 L/day and then reabsorb 99% of it? Why not simply filter and excrete the 1% that needs to be eliminated?" There are two reasons. First, many foreign substances are filtered into the tubule but not reabsorbed into the blood. Once a substance filters into Bowman's capsule, it is no longer part of the body's internal environment. The lumen of the nephron is external environment, and anything in the filtrate is destined to leave the body in the urine unless reabsorbed. The high daily filtration rate of unwanted molecules helps clear such substances from the plasma very rapidly. Many desirable nutrients, such as glucose and citric acid cycle intermediates, also filter in large amounts, but the proximal tubule has transporters that very efficiently reabsorb them.

The second reason for a high GFR is that filtering ions and water into the tubule simplifies their regulation. If water and ions reach the distal nephron and are not needed to maintain homeostasis, they pass into the urine. With a high GFR, this excretion can occur quite rapidly. However, if the ions and water are needed, they are reabsorbed and return to the blood.

Reabsorption May Be Active or Passive

Reabsorption of water and solutes from the tubule lumen to the extracellular fluid depends on active transport. The filtrate flowing out of Bowman's capsule into the proximal tubule has the same solute concentrations as extracellular fluid. To move solute out of the lumen, the tubule cells must therefore use active transport to create concentration or electrochemical gradients. Water osmotically follows solutes as they are reabsorbed.

FIGURE 19.8a is an overview of renal reabsorption. Active transport of Na^+ from the tubule lumen to the extracellular fluid

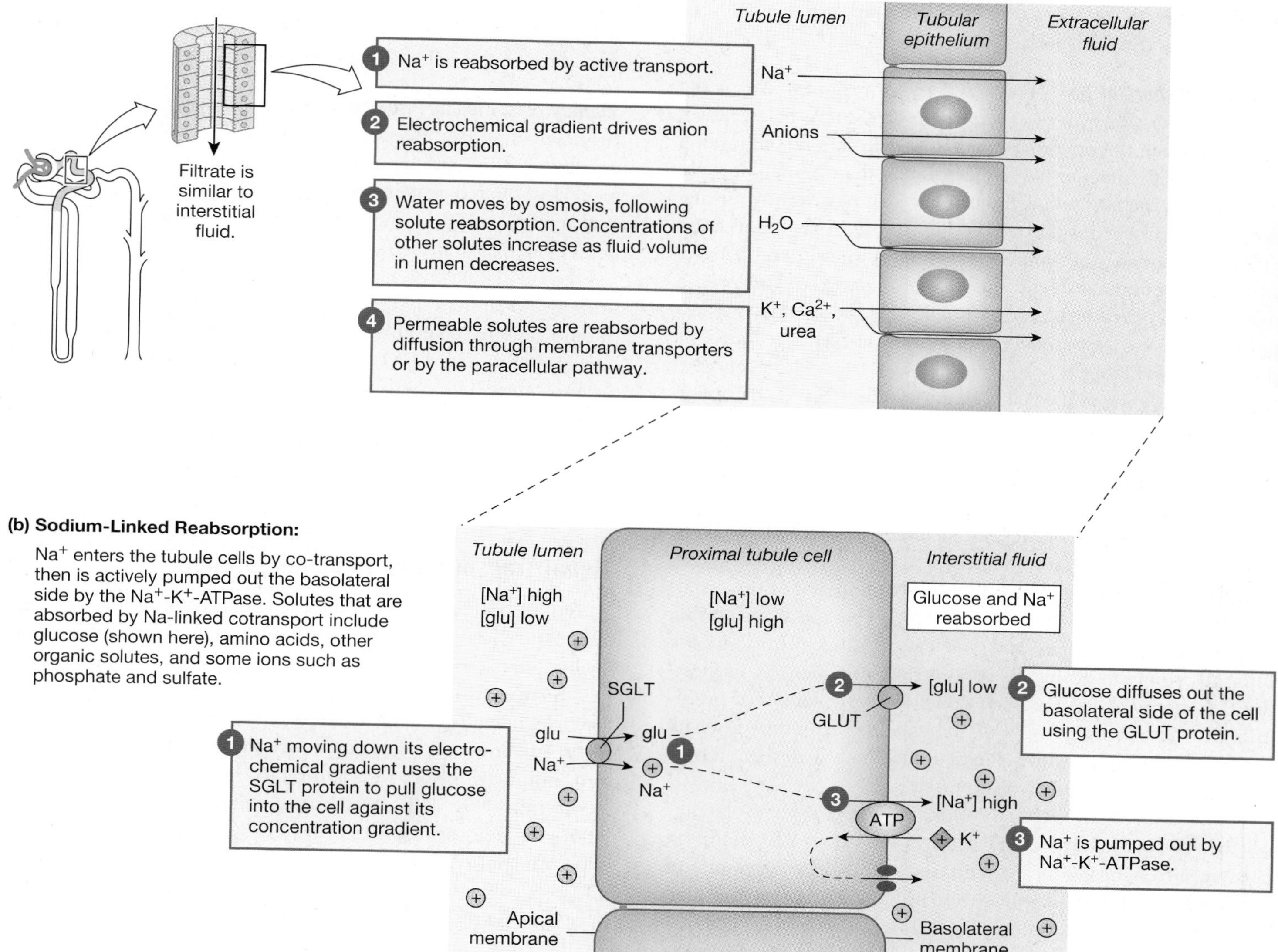
FIG. 19.8 Reabsorption
(a) Principles Governing the Tubular Reabsorption of Solutes
Some solutes and water move into and then out of epithelial cells (transcellular or epithelial transport); other solutes move through junctions between epithelial cells (the paracellular pathway). Membrane transporters are not shown in this illustration.
Filtrate is similar to interstitial fluid.
Tubule lumen
Tubular epithelium
Extracellular fluid
1 Na+ is reabsorbed by active transport.
2 Electrochemical gradient drives anion reabsorption.
3 Water moves by osmosis, following solute reabsorption. Concentrations of other solutes increase as fluid volume in lumen decreases.
4 Permeable solutes are reabsorbed by diffusion through membrane transporters or by the paracellular pathway.
Na+
Anions
H2O
K+, Ca2+, urea
(b) Sodium-Linked Reabsorption:
Na+ enters the tubule cells by co-transport, then is actively pumped out the basolateral side by the Na+-K+-ATPase. Solutes that are absorbed by Na-linked cotransport include glucose (shown here), amino acids, other organic solutes, and some ions such as phosphate and sulfate.
Tubule lumen
Proximal tubule cell
Interstitial fluid
[Na+] high
[glu] low
[Na+] low
[glu] high
Glucose and Na+ reabsorbed
SGLT
GLUT
glu
Na+
[glu] low
[Na+] high
ATP
K+
1 Na+ moving down its electrochemical gradient uses the SGLT protein to pull glucose into the cell against its concentration gradient.
2 Glucose diffuses out the basolateral side of the cell using the GLUT protein.
3 Na+ is pumped out by Na+-K+-ATPase.
Apical membrane
Basolateral membrane

creates a transepithelial electrical gradient in which the lumen is more negative than the ECF. Anions then follow the positively charged Na^+ out of the lumen. The removal of Na^+ and anions from lumen to ECF dilutes the luminal fluid and increases the concentration of the ECF, so water leaves the tubule by osmosis.

The loss of volume from the lumen increases the concentration of solutes (including K^+, Ca^{2+}, and urea) left behind in the filtrate: The same amount of solute in a smaller volume equals higher solute concentration. Once luminal solute concentrations are higher than solute concentrations in the extracellular fluid, the solutes diffuse out of the lumen if the epithelium of the tubule is permeable to them.

Reabsorption involves both transepithelial transport and paracellular transport. In **transepithelial transport** (also called *transcellular transport*), substances cross the apical and basolateral membranes of the tubule epithelial cell [p. 149] to reach the interstitial fluid. In the **paracellular pathway**, substances pass through the cell-cell junction between two adjacent cells. Which route a solute takes depends on the permeability of the epithelial junctions and on the electrochemical gradient for the solute.

For solutes that move by transepithelial transport, their concentration or electrochemical gradients determine their transport mechanisms. Solutes moving down their gradient use open leak channels or facilitated diffusion carriers to cross the cell membrane. Molecules that need to be pushed against their gradient are moved by either primary or indirect (usually secondary) active transport. Sodium is directly or indirectly involved in many instances of both passive and active transport.

Active Transport of Na^+ The active reabsorption of Na^+ is the primary driving force for most renal reabsorption. As noted earlier, filtrate entering the proximal tubule is similar in ion composition to plasma, with a higher Na^+ concentration than is found in cells. Thus, Na^+ in the filtrate can enter tubule cells passively by moving down its electrochemical gradient (Fig. 19.8b). Apical movement of Na^+ uses a variety of symport and antiport transport proteins [p. 140]. In the proximal tubule, the *Na^+-H^+ exchanger* (NHE) plays a major role in Na^+. Once inside a tubule cell, Na^+ is actively transported out across the basolateral membrane in exchange for K^+ by the Na^+-K^+-ATPase. A basolateral K^+ leak channel prevents K^+ from accumulating in the cell. The end result is Na^+ reabsorption across the tubule epithelium.

Secondary Active Transport: Symport with Na^+ Sodium-linked secondary active transport in the nephron is responsible for the reabsorption of many substances, including glucose, amino acids, ions, and various organic metabolites. Figure 19.8c shows one example: Na^+-dependent glucose reabsorption across the proximal tubule epithelium [Fig. 5.21, p. 150]. The apical membrane contains the Na^+-*glucose cotransporter* (SGLT) that brings glucose into the cytoplasm against its concentration gradient by harnessing the energy of Na^+ moving down its electrochemical gradient. On the basolateral side of the cell, Na^+ is pumped out by the Na^+-K^+-ATPase, while glucose diffuses out with the aid of a facilitated diffusion GLUT transporter.

The same basic pattern holds for many other molecules absorbed by Na^+-dependent transport: an apical symport protein and a basolateral facilitated diffusion carrier or ion exchanger. Other molecules that are reabsorbed by similar mechanisms include amino acids, lactate, citric acid cycle intermediates such as citrate and α-ketoglutarate (αKG), and ions such as phosphate and sulfate. A few of the transporters use H^+ in place of Na^+.

Passive Reabsorption: Urea The nitrogenous waste product urea has no active transporters in the proximal tubule but can move through the epithelial junctions by diffusion if there is a urea concentration gradient. Initially, urea concentrations in the filtrate and extracellular fluid are equal. However, the active transport of Na^+ and other solutes out of the proximal tubule lumen creates a urea concentration gradient by the following process.

When Na^+ and other solutes are reabsorbed from the proximal tubule, the transfer of osmotically active particles makes the extracellular fluid more concentrated than the filtrate remaining in the lumen (see Fig. 19.8a). In response to the osmotic gradient, water moves by osmosis across the epithelium. Up to this point, no urea molecules have moved out of the lumen because there has been no urea concentration gradient.

When water is reabsorbed, the concentration of urea in the lumen increases—the same amount of urea is contained in a smaller volume. Once a concentration gradient for urea exists, urea moves out of the lumen into the extracellular fluid by transport through the cells or by the paracellular pathway.

Endocytosis: Plasma Proteins Filtration of plasma at the glomerulus normally leaves most plasma proteins in the blood, but some smaller proteins and peptides can pass through the filtration barrier. Most filtered proteins are removed from filtrate in the proximal tubule, with the result that normally only trace amounts of protein appear in urine.

Small as they are, filtered proteins are too large to be reabsorbed by carriers or through channels. Most enter proximal tubule cells by receptor-mediated endocytosis [p. 147] at the apical membrane. Once in the cells, the proteins are digested in lysosomes. The resulting amino acids are transported across the basolateral membrane and absorbed into the blood. The renal digestion of small filtered proteins is actually a significant method by which peptide signal molecules can be removed from the circulation.

Renal Transport Can Reach Saturation

Most transport in the nephron uses membrane proteins and exhibits the three characteristics of mediated transport: saturation, specificity, and competition [p. 46].

Saturation refers to the maximum rate of transport that occurs when all available carriers are occupied by (are saturated with) substrate. At substrate concentrations below the saturation point, transport rate is directly related to substrate concentration (**FIG. 19.9**). At substrate concentrations equal to or above the saturation point, transport occurs at a maximum rate. The transport rate at saturation is the **transport maximum (T_m)** [p. 51].

Glucose reabsorption in the nephron is an excellent example of the consequences of saturation. At normal plasma glucose concentrations, all glucose that enters the nephron is reabsorbed before it reaches the end of the proximal tubule. The tubule epithelium is well supplied with carriers to capture glucose as the filtrate flows past.

But what happens if blood glucose concentrations become excessive, as they do in diabetes mellitus? In that case, glucose is filtered faster than the carriers can reabsorb it. The carriers become saturated and are unable to reabsorb all the glucose that flows through the tubule. As a result, some glucose escapes reabsorption and is excreted in the urine.

Consider the following analogy. Assume that the carriers are like seats on a train at Disney World. Instead of boarding the stationary train from a stationary platform, passengers step

FIG. 19.9 Saturation of mediated transport

The transport rate of a substance is proportional to the plasma concentration of the substance, up to the point at which transporters become saturated. Once saturation occurs, transport rate reaches a maximum. The plasma concentration of substrate at which the transport maximum occurs is called the renal threshold.

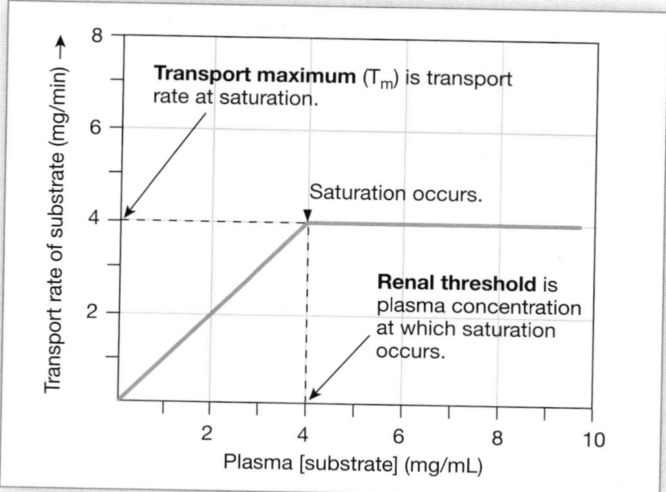

Transport rate of substrate (mg/min) →
Plasma [substrate] (mg/mL)
Transport maximum (T_m) is transport rate at saturation.
Saturation occurs.
Renal threshold is plasma concentration at which saturation occurs.
8
6
4
2
2
4
6
8
10

GRAPH QUESTION

What is the transport rate at the following plasma substrate concentrations: 3 mg/mL, 5 mg/mL, 8 mg/mL? At what plasma substrate concentration is the transport rate 2 mg/min?

onto a moving sidewalk that rolls them past the train. As the passengers see an open seat, they grab it. However, if more people are allowed onto the moving sidewalk than there are seats in the train, some people will not find seats. And because the sidewalk is moving people past the train toward an exit, they cannot wait for the next train. Instead, they end up being transported out the exit.

Glucose molecules entering Bowman's capsule in the filtrate are like passengers stepping onto the moving sidewalk. To be reabsorbed, each glucose molecule must bind to a transporter as the filtrate flows through the proximal tubule. If only a few glucose molecules enter the tubule at a time, each one can find a free transporter and be reabsorbed, just as a small number of people on the moving sidewalk all find seats on the train. However, if glucose molecules filter into the tubule faster than the glucose carriers can transport them, some glucose remains in the lumen and is excreted in the urine.

FIGURE 19.10 is a graphic representation of glucose handling by the kidney. Figure 19.10a shows that the filtration rate of glucose from plasma into Bowman's capsule is proportional to the plasma concentration of glucose. Because filtration does not exhibit saturation, the graph continues infinitely in a straight line: The filtrate glucose concentration is always equal to the plasma glucose concentration.

Figure 19.10b plots the reabsorption rate of glucose in the proximal tubule against the plasma concentration of glucose. Reabsorption exhibits a maximum transport rate (T_m) when the carriers reach saturation. Notice that normal plasma glucose concentrations are well below the saturation point.

Figure 19.10c plots the excretion rate of glucose in relation to the plasma concentration of glucose. Remember that excretion equals filtration minus reabsorption ($E = F - R$). When plasma glucose concentrations are low enough that 100% of the filtered glucose is reabsorbed, no glucose is excreted. Once the carriers reach saturation, glucose excretion begins. The plasma concentration at which glucose first appears in the urine is called the **renal threshold** for glucose.

Figure 19.10d is a composite graph that compares filtration, reabsorption, and excretion of glucose. Recall from our earlier discussion that

Amount excreted = amount filtered − amount reabsorbed + amount secreted

For glucose, which is not secreted, the equation can be rewritten as

Glucose excreted = glucose filtered − glucose reabsorbed

BIOTECHNOLOGY

Artificial Kidneys

Many people with severe renal disease depend on *dialysis,* a medical procedure that either supplements or completely replaces their kidney function. Imagine trying to make a machine or develop a procedure that performs the functions of the kidney. What would it have to do? Dialysis is based on diffusion through a semipermeable membrane. Solutes and water pass from a patient's extracellular fluid across the membrane into a dialysis fluid. *Hemodialysis* routes blood from the arm past a membrane in an external dialysis machine. This technique requires attachment to the machine for 3–5 hours three days a week and is used for more severe cases of renal failure. *Peritoneal dialysis* is also called *continuous ambulatory peritoneal dialysis* (CAPD) because it takes place while the patient moves about during daily activities. In CAPD, dialysis fluid is injected into the peritoneal cavity, where it accumulates waste products from the blood for 4–6 hours before being drained out. For more information about dialysis, see the website for the National Institute of Diabetes and Digestive and Kidney Diseases (*www.niddk.nih.gov*) and search for *dialysis.*

FIG. 19.10 Glucose handling by the nephron

(a) Filtration

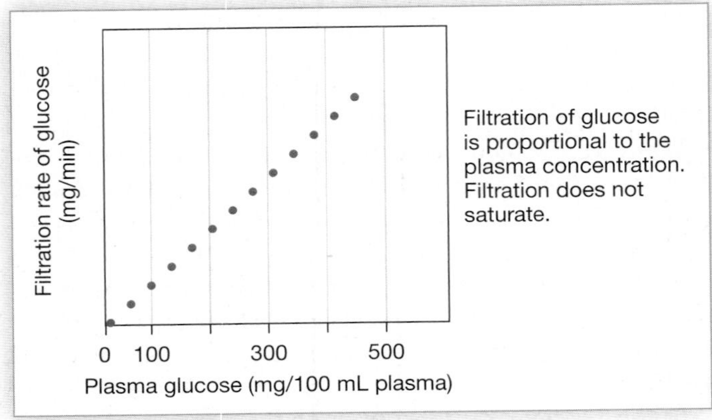
Filtration rate of glucose (mg/min)
Filtration of glucose is proportional to the plasma concentration. Filtration does not saturate.
0
100
300
500
Plasma glucose (mg/100 mL plasma)

(b) Reabsorption

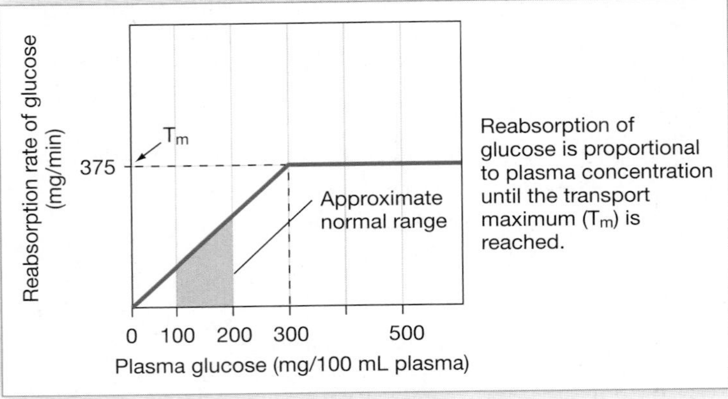
Reabsorption rate of glucose (mg/min)
375
Tm
Approximate normal range
Reabsorption of glucose is proportional to plasma concentration until the transport maximum (Tm) is reached.
0
100
200
300
500
Plasma glucose (mg/100 mL plasma)

(c) Excretion = Filtration – Reabsorption

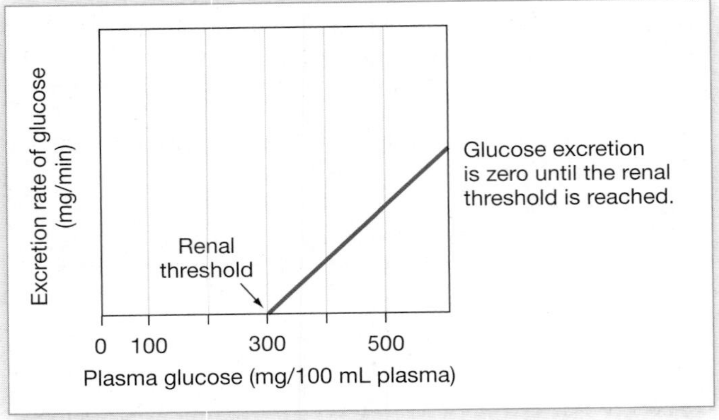
Excretion rate of glucose (mg/min)
Glucose excretion is zero until the renal threshold is reached.
Renal threshold
0
100
300
500
Plasma glucose (mg/100 mL plasma)

(d) Composite graph shows the relationship between filtration, reabsorption, and excretion of glucose.

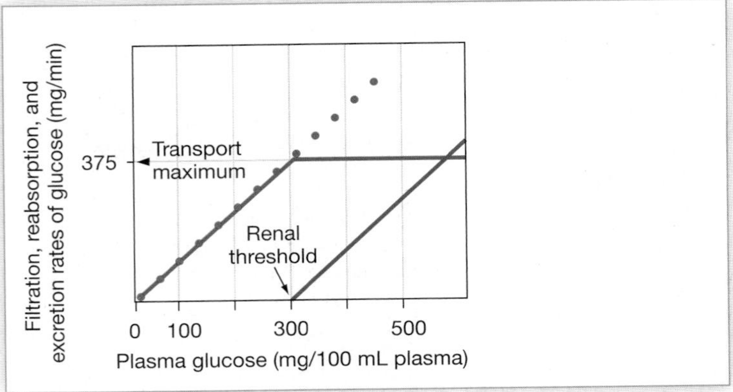
Filtration, reabsorption, and excretion rates of glucose (mg/min)
375
Transport maximum
Renal threshold
0
100
300
500
Plasma glucose (mg/100 mL plasma)

Under normal conditions, all filtered glucose is reabsorbed. In other words, filtration is equal to reabsorption.

Notice in Figure 19.10d that the lines representing filtration and reabsorption are identical up to the plasma glucose concentration that equals the renal threshold. If filtration equals reabsorption, the algebraic difference between the two is zero, and there is no excretion. Once the renal threshold is reached, filtration begins to exceed reabsorption. Notice on the graph that the filtration and reabsorption lines diverge at this point. The difference between the filtration line and the reabsorption line represents the excretion rate:

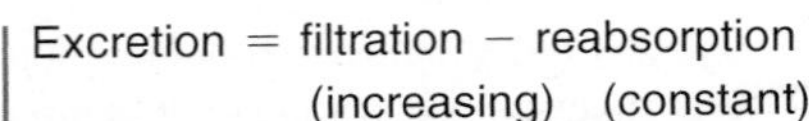
Excretion = filtration – reabsorption
(increasing) (constant)

Excretion of glucose in the urine is called **glucosuria** or **glycosuria** {*-uria*, in the urine} and usually indicates an elevated blood glucose concentration. Rarely, glucose appears in the urine even though the blood glucose concentrations are normal. This situation is due to a genetic disorder in which the nephron does not make enough carriers.

Peritubular Capillary Pressures Favor Reabsorption

The reabsorption we have just discussed refers to the movement of solutes and water from the tubule lumen to the interstitial fluid. How does that reabsorbed fluid then get into the capillary? The answer is that the hydrostatic pressure that exists along the entire length of the peritubular capillaries is less than the colloid osmotic pressure, so the net pressure gradient favors reabsorption (**FIG. 19.11**).

The peritubular capillaries have an average hydrostatic pressure of 10 mm Hg (in contrast to the glomerular capillaries, where hydrostatic pressure averages 55 mm Hg). Colloid osmotic pressure, which favors movement of fluid into the capillaries, is 30 mm Hg. As a result, the pressure gradient in peritubular capillaries is 20 mm Hg, favoring the absorption of fluid into the capillaries. Fluid that is reabsorbed passes from the peritubular capillaries to the venous circulation and returns to the heart.

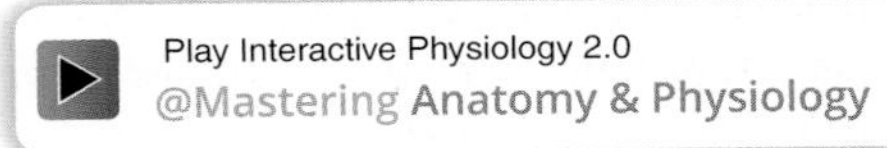
Play Interactive Physiology 2.0
@Mastering Anatomy & Physiology

Instructors: A version of this Try It! Activity can be assigned in @Mastering Anatomy & Physiology

TRY IT! Glycosuria in Diabetes Mellitus

The discovery and production of insulin for the treatment of diabetes was one of the miracles of modern science. In the late 1800s Oskar Minkowski removed the pancreas of a dog as part of an experiment. A technician in Minkowski's laboratory noticed that the once-housebroken animal began urinating on the floor of the lab. The dog's urine contained glucose, which was not there before its pancreas was removed. Minkowski knew that patients with diabetes mellitus also suffer *glycosuria* and *polyuria* (increased urine flow), and he made the connection between loss of pancreatic function and diabetes. Several decades later Frederick Banting and Charles Best in Toronto would go on to isolate insulin from the pancreas and use it to successfully treat patients with type 1 diabetes mellitus. The graph below shows the effect of insulin administration on glucose excretion in one of their first patients, L.T., a 14-year-old boy with type I diabetes.

Effect of insulin injections on the excretion of glucose in the urine of L.T. Arrows indicate when injections were given.

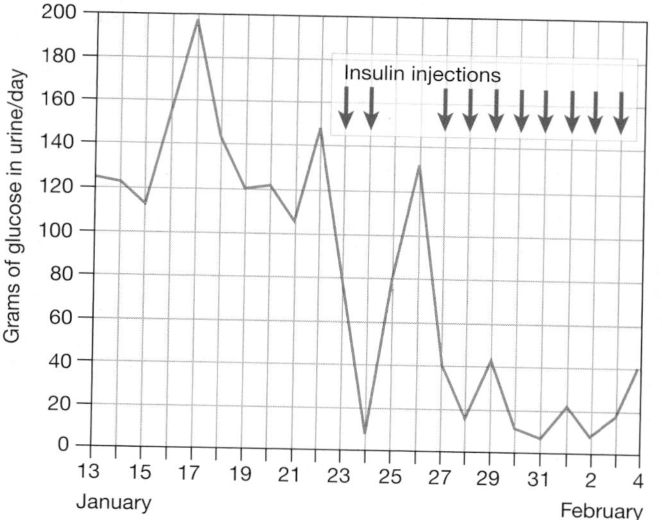
200
180
160
140
120
100
80
60
40
20
0
Grams of glucose in urine/day
Insulin injections
13
15
17
19
21
23
25
27
29
31
2
4
January
February

Data from Banting et al., 1922.*

GRAPH QUESTIONS

- Summarize the graph in words and draw a conclusion about the effect of insulin injections on glucose excretion.
- How much glucose is excreted in the urine when blood glucose levels are normal? What causes increased glucose excretion in diabetes when blood glucose concentrations are elevated?
- Explain teleologically why reabsorbing glucose in the kidney would be advantageous to a person.

*Pancreatic Extracts in the Treatment of Diabetes Mellitus. *Can Med Assoc J* 12, 141–146 (1922). https://www.ncbi.nlm.nih.gov/pmc/articles/PMC1524425/pdf/canmedaj00414-0043.pdf

19.6 Secretion

Secretion is the transfer of molecules from extracellular fluid into the lumen of the nephron (see Fig. 19.2). Secretion, like reabsorption, depends mostly on membrane transport systems. The secretion of K^+ and H^+ by the distal nephron is important in the homeostatic regulation of those ions. In addition, many organic compounds are secreted. These compounds include both metabolites produced in the body and substances brought into the body, or *xenobiotics*.

Secretion enables the nephron to enhance excretion of a substance. If a substance is filtered and not reabsorbed, it is excreted very efficiently. If, however, the substance is filtered into the tubule, not reabsorbed, *and* then more of it is secreted into the tubule from the peritubular capillaries, excretion is even more efficient.

Secretion is an active process because it requires moving substrates against their concentration gradients. Most organic compounds are secreted across the proximal tubule epithelium into the lumen by indirect active transport. Let's look at how the tubule handles the secretion of organic anions (**FIG. 19.12**).

The transporters responsible for organic solute excretion have broad specificity. For example, the **organic anion transporter (OAT)** family, shown in this figure, is able to transport a wide variety of endogenous and exogenous anions, ranging from bile salts to benzoate used as a preservative in soft drinks, salicylate from aspirin, and the artificial sweetener saccharine. Secretion of organic anions on the OAT is an example of *tertiary* active transport, where the use of energy from ATP is two steps removed from the OAT.

Let's see how this works. In the first step of the process, which is direct active transport, the proximal tubule cell uses ATP to maintain the low intracellular concentration of Na^+. In the second step, the Na^+ gradient is then used to concentrate a *dicarboxylate* inside the tubule cell, using a **Na^+-dicarboxylate cotransporter** called the **NaDC**. The NaDC is found on both apical and basolateral membranes in the proximal tubule.

Dicarboxylates are the anion form of dicarboxylic acids, which have two carboxyl (−COOH) groups. Most of the citric acid cycle intermediates, such as citrate, oxaloacetate, and α-ketoglutarate (αKG), are dicarboxylates. Figure 19.12 shows αKG as the dicarboxylate.

The concentration of dicarboxylate inside the tubule cell drives the third step of organic anion secretion. OAT transporters

CHAPTER 19

FIG. 19.11 Reabsorption in peritubular capillaries

Lower hydrostatic pressure in peritubular capillaries results in net reabsorption of interstitial fluid.

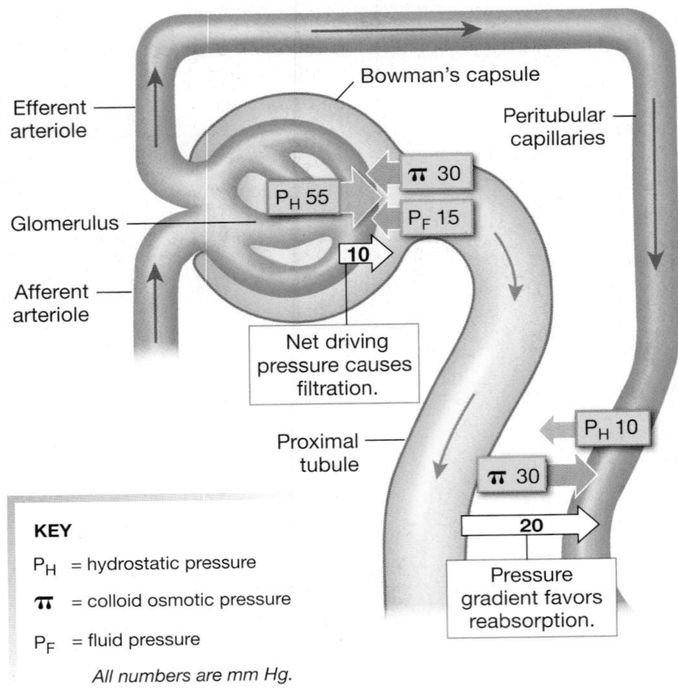

Efferent arteriole
Bowman's capsule
Peritubular capillaries
Glomerulus
Afferent arteriole
π 30
P_H 55
P_F 15
10
Net driving pressure causes filtration.
Proximal tubule
P_H 10
π 30
20
Pressure gradient favors reabsorption.
KEY
P_H = hydrostatic pressure
π = colloid osmotic pressure
P_F = fluid pressure
All numbers are mm Hg.

FIGURE QUESTION

Why is the hydrostatic pressure so much lower in the peritubular capillaries than in the glomerulus?

OAT1, OAT2, and OAT3 are indirect active transporters that use a dicarboxylate or other organic anion moving out of the cell down their concentration gradient to move an organic anion into the cell against its gradient. In the final step, once the organic anion is concentrated inside the tubule cell, an apical OAT4 sends the organic anion into the lumen in exchange for a dicarboxylate anion. The net result is reabsorption of desirable metabolic intermediates in exchange for the secretion of unwanted organic anions.

Competition Decreases Penicillin Secretion

The broad specificity of the organic anion transporters means that different substrates must compete for the transporter binding sites [p. 48]. An interesting and important example of an organic molecule secreted by the OAT is the antibiotic *penicillin*. Many people today take antibiotics for granted, but until the early decades of the twentieth century, infections were a leading cause of death.

In 1928, Alexander Fleming discovered a substance in the bread mold *Penicillium* that retarded the growth of bacteria. But the antibiotic was difficult to isolate, so it did not become available for clinical use until the late 1930s. During World War II, penicillin made a major difference in the number of deaths and amputations caused by infected wounds. The only means of producing penicillin, however, was to isolate it from bread mold, and supplies were limited.

Demand for the drug was heightened by the fact that kidney tubules secrete penicillin. Renal secretion is so efficient at clearing foreign molecules from the blood that within three to four hours after a dose of penicillin has been administered, about 80% has been excreted in the urine. During the war, the drug was in such short supply that it was common procedure to collect the urine from patients being treated with penicillin so that the antibiotic could be isolated and reused.

This solution was not satisfactory, however, and so researchers looked for a way to slow penicillin secretion. They hoped to find a molecule that could compete with penicillin for the organic anion transporter responsible for secretion. That way, when presented with both drugs, the OAT carrier would bind

FIG. 19.12 Organic anion secretion

Proximal tubule secretion of organic anions by the organic anion transporter (OAT) is an example of tertiary active transport.

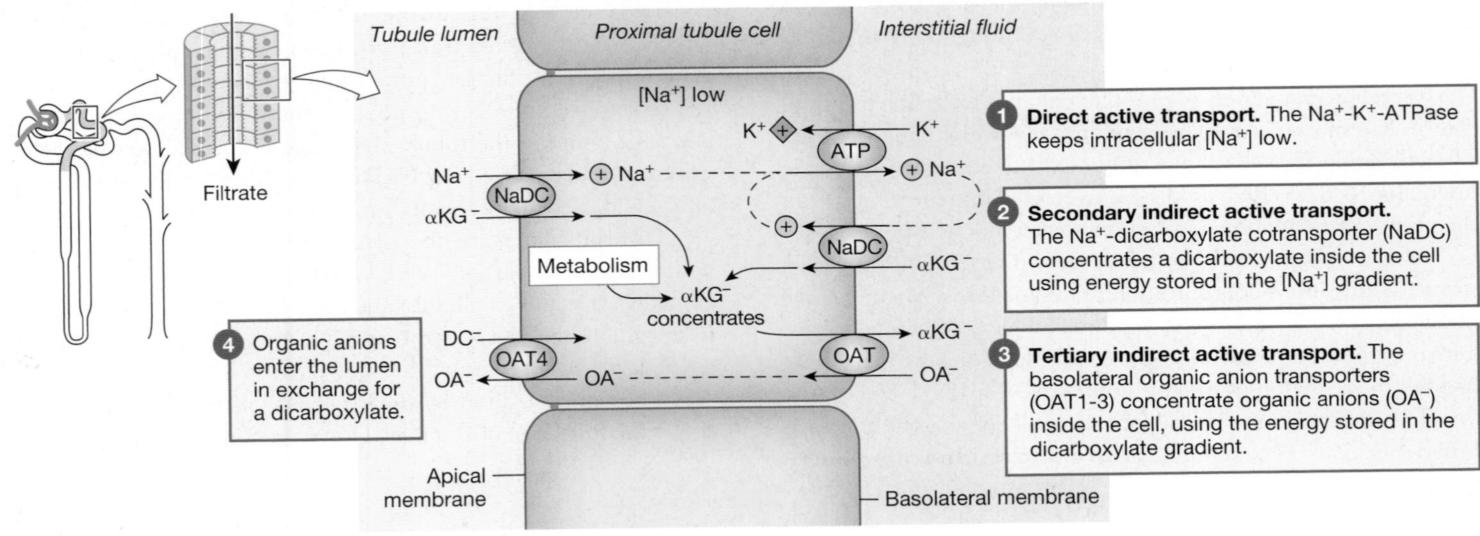

Filtrate
Tubule lumen
Proximal tubule cell
Interstitial fluid
[Na+] low
K+
ATP
Na+
NaDC
αKG−
Metabolism
αKG− concentrates
DC−
OAT4
OA−
OAT
1 Direct active transport. The Na+-K+-ATPase keeps intracellular [Na+] low.
2 Secondary indirect active transport. The Na+-dicarboxylate cotransporter (NaDC) concentrates a dicarboxylate inside the cell using energy stored in the [Na+] gradient.
3 Tertiary indirect active transport. The basolateral organic anion transporters (OAT1-3) concentrate organic anions (OA−) inside the cell, using the energy stored in the dicarboxylate gradient.
4 Organic anions enter the lumen in exchange for a dicarboxylate.
Apical membrane
Basolateral membrane

RUNNING PROBLEM

Uric acid, the molecule that causes gout, is a normal product of purine metabolism. Increased uric acid production may be associated with cell and tissue breakdown, or it may occur as a result of inherited enzyme defects. Plasma *urate*, the anionic form of uric acid, filters freely into Bowman's capsule but is almost totally reabsorbed in the first part of the proximal tubule. The middle section of the proximal tubule then secretes about half of the reabsorbed urate back into the lumen. Finally, the terminal section of the proximal tubule again reabsorbs some of the urate. The end result is net secretion.

Q3: *Purines are part of which category of biomolecules? Using that information, explain why uric acid levels in the blood go up when cell breakdown increases.*

Q4: *Based on what you have learned about uric acid and urate, predict two ways a person may develop hyperuricemia.*

588 – 589 – **607** – 607 – 611 – 613

preferentially to the competitor and secrete it, leaving penicillin behind in the blood.

A synthetic compound named *probenecid* was the answer. When probenecid is administered concurrently with penicillin, the transporter removes probenecid preferentially, prolonging the activity of penicillin in the body. Once mass-produced synthetic penicillin became available and supply was no longer a problem, the medical use of probenecid declined.

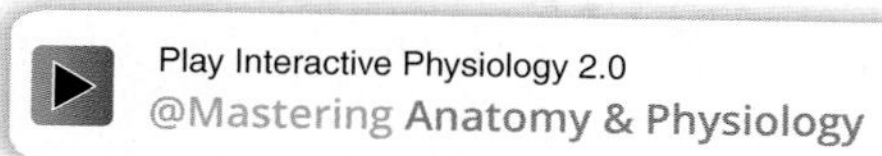

Play Interactive Physiology 2.0 @Mastering Anatomy & Physiology

19.7 Excretion

Urine output is the result of all the processes that take place in the kidney. By the time fluid reaches the end of the nephron, it bears little resemblance to the filtrate that started in Bowman's capsule. Glucose, amino acids, and useful metabolites are gone, having been reabsorbed into the blood, and organic wastes are more concentrated. The concentrations of ions and water in the urine are highly variable, depending on the state of the body.

Although excretion tells us what the body is eliminating, excretion by itself cannot tell us the details of renal function. Recall that for any substance,

$$\text{Excretion} = \text{filtration} - \text{reabsorption} + \text{secretion}$$

Simply looking at the excretion rate of a substance tells us nothing about the tubular transport mechanisms for that substance, collectively described as **renal handling** of the substance. The excretion rate of a substance depends on (1) the filtration rate of the substance and (2) whether the substance is reabsorbed, secreted, or both, as it passes through the tubule. **FIGURE 19.13** summarizes the renal handling of some common substances.

Renal handling of a substance and GFR are often of clinical interest. For example, clinicians use information about a person's glomerular filtration rate as an indicator of overall kidney function. And pharmaceutical companies developing drugs must provide the Food and Drug Administration with complete information on how the human kidney handles each new compound.

But how can investigators dealing with living humans assess filtration, reabsorption, and secretion at the level of the individual nephron? They have no way to do this directly: The kidneys are not easily accessible and the nephrons are microscopic. Scientists therefore had to develop a technique that would allow them to assess renal function using only analysis of the urine and blood. To do this, they apply the concept of clearance.

Clearance Is a Noninvasive Way to Measure GFR

Clearance of a solute is the rate at which that solute disappears from the body by excretion or by metabolism[p. 12]. The general equation for clearance is:

$$\text{Clearance of } X = \frac{\text{excretion rate of } X \text{ (mg/min)}}{[X]_{\text{plasma}} \text{ (mg/mL plasma)}}$$

where clearance is mL plasma cleared of X per minute. Notice that the units for clearance are mL plasma and time. Substance X does not appear anywhere in the clearance units.

For any solute that is filtered by the nephron, clearance is expressed as the volume of plasma passing through the kidneys that has been totally cleared of that solute in a given period of time. Because this is such an indirect way to think of excretion (how much blood has been cleared of X rather than how much X has been excreted), clearance is often a difficult concept to grasp.

RUNNING PROBLEM

Michael found it amazing that a metabolic problem could lead to pain in his big toe. "How do we treat gout?" he asked. Dr. Garcia explained that the treatment includes anti-inflammatory agents, lots of water, and avoidance of alcohol, which can trigger gout attacks. "In addition, I would like to put you on a uricosuric agent, like probenecid, which will enhance renal excretion of urate," replied Dr. Garcia. "By enhancing excretion, we can reduce uric acid levels in your blood and thus provide relief." Michael agreed to try these measures.

Q5: *Urate is reabsorbed by some proximal tubule cells and secreted by others using membrane transporters, one on the apical membrane and one on the basolateral membrane. Could the same transporters be used by cells that reabsorb urate and cells that secrete it? Defend your reasoning.*

Q6: *Uricosuric agents, like urate, are organic acids. Given this fact, explain how uricosuric agents might enhance excretion of urate.*

588 – 589 – 607 – **607** – 611 – 613

FIG. 19.13 Renal handling of common substances

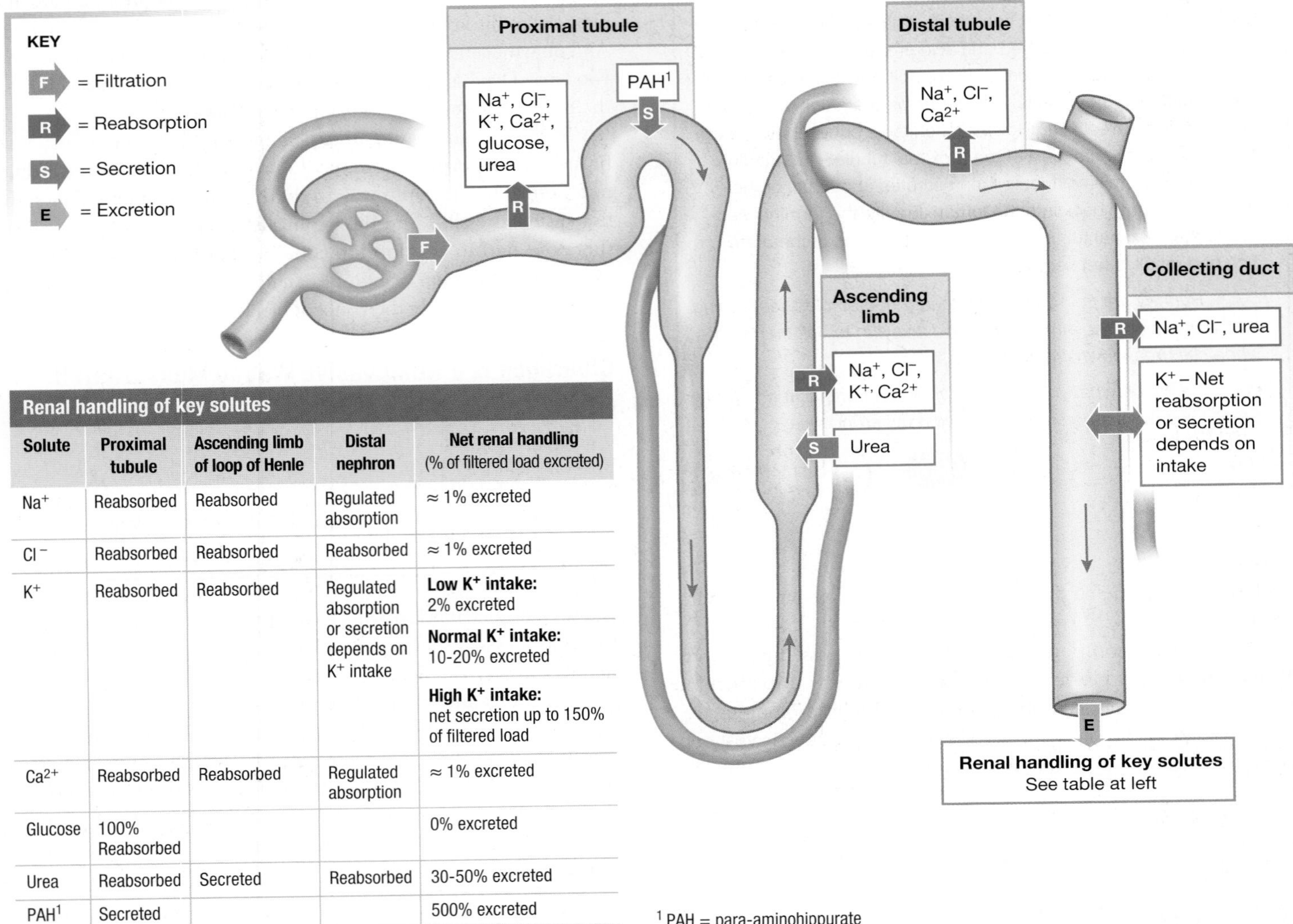
KEY
F = Filtration
R = Reabsorption
S = Secretion
E = Excretion
Proximal tubule
Na+, Cl−, K+, Ca2+, glucose, urea
PAH1
Distal tubule
Na+, Cl−, Ca2+
Ascending limb
Na+, Cl−, K+, Ca2+
Urea
Collecting duct
Na+, Cl−, urea
K+ – Net reabsorption or secretion depends on intake
Renal handling of key solutes
See table at left

Renal handling of key solutes

Solute	Proximal tubule	Ascending limb of loop of Henle	Distal nephron	Net renal handling (% of filtered load excreted)
Na^+	Reabsorbed	Reabsorbed	Regulated absorption	≈ 1% excreted
Cl^-	Reabsorbed	Reabsorbed	Reabsorbed	≈ 1% excreted
K^+	Reabsorbed	Reabsorbed	Regulated absorption or secretion depends on K^+ intake	**Low K^+ intake:** 2% excreted **Normal K^+ intake:** 10-20% excreted **High K^+ intake:** net secretion up to 150% of filtered load
Ca^{2+}	Reabsorbed	Reabsorbed	Regulated absorption	≈ 1% excreted
Glucose	100% Reabsorbed			0% excreted
Urea	Reabsorbed	Secreted	Reabsorbed	30-50% excreted
PAH[1]	Secreted			500% excreted

[1] PAH = para-aminohippurate

Before we jump into the mathematical expression of clearance, let's look at an example that shows how clearance relates to kidney function. For our example, we use **inulin**, a polysaccharide isolated from the tuberous roots of a variety of plants. (Inulin is not the same as *insulin,* the protein hormone that regulates glucose metabolism.) Scientists discovered from experiments with isolated nephrons that inulin injected into the plasma filters freely into the nephron. As it passes through the kidney tubule, inulin is neither reabsorbed nor secreted. In other words, 100% of the inulin that filters into the tubule is excreted.

How does this relate to clearance? To answer this question, take a look at **FIGURE 19.14**, which assumes that 100% of a filtered volume of plasma is reabsorbed. (This is not too far off the actual value, which is more than 99%.) In Figure 19.14a, inulin has been injected so that its plasma concentration is 4 inulin molecules per 100 mL plasma. If GFR is 100 mL plasma filtered per minute, we can calculate the filtration rate, or *filtered load,* of inulin using the equation

$$\text{Filtered load of } X = [X]_{\text{plasma}} \times \text{GFR}$$

$$\begin{aligned}\text{Filtered load of inulin} &= (4 \text{ inulin}/100 \text{ mL plasma}) \\ &\quad \times 100 \text{ mL plasma filtered/min} \\ &= 4 \text{ inulin/min filtered}\end{aligned}$$

As the filtered inulin and the filtered plasma pass along the nephron, all the plasma is reabsorbed, but all the inulin remains in the tubule. The reabsorbed plasma contains no inulin, so we say it has been totally *cleared* of inulin. The *inulin clearance,* therefore, is 100 mL of plasma cleared/min. At the same time, the excretion rate of inulin is 4 inulin molecules excreted per minute.

What good is this information? For one thing, we can use it to calculate the glomerular filtration rate. Notice from Figure 19.14a that inulin clearance (100 mL plasma cleared/min) is equal to the GFR (100 mL plasma filtered/min). Thus, *for any substance that is freely filtered but neither reabsorbed nor secreted, its clearance is equal to GFR.*

Now let's show mathematically that inulin clearance is equal to GFR. We already know that

$$\text{Filtered load of } X = [X]_{plasma} \times \text{GFR} \quad (1)$$

We also know that 100% of the inulin that filters into the tubule is excreted. In other words:

$$\text{Filtered load of inulin} = \text{excretion rate of inulin} \quad (2)$$

Because of this equality, we can substitute excretion rate for filtered load in equation (1) by using algebra (if A = B and A = C, then B = C):

$$\text{Excretion rate of inulin} = [\text{inulin}]_{plasma} \times \text{GFR} \quad (3)$$

This equation can be rearranged to read

$$\text{GFR} = \frac{\text{excretion rate of inulin}}{[\text{inulin}]_{plasma}} \quad (4)$$

It turns out that the right side of this equation is identical to the clearance equation for inulin. Thus the general equation for the clearance of any substance X (mL plasma cleared/min) is

$$\text{Clearance of } X = \frac{\text{excretion rate of } X \text{ (mg/min)}}{[X]_{plasma}\text{(mg/mL plasma)}} \quad (5)$$

For inulin:

$$\text{Inulin clearance} = \frac{\text{excretion rate of inulin}}{[\text{inulin}]_{plasma}} \quad (6)$$

The right sides of equations (4) and (6) are identical, so by using algebra again, we can say that:

$$\text{GFR} = \text{inulin clearance} \quad (7)$$

So why is this important? For one thing, you have just learned how we can measure GFR in a living human by taking only blood and urine samples. Try the example in Concept Check 12 to see if you understand the preceding discussion. **TABLE 19.1** is a summary table of equations you will find useful for renal physiology.

Inulin is not practical for routine clinical applications because it does not occur naturally in the body and must be administered by continuous intravenous infusion. As a result, inulin use is restricted to research. Unfortunately, no substance that occurs naturally in the human body is handled by the kidney exactly the way inulin is handled.

TABLE 19.1 Useful Equations in Renal Physiology

- Excretion = Filtration − Reabsorption + Secretion
- Filtration rate of $X = [X]_{plasma} \times \text{GFR}$
- Excretion rate of $X = \text{urine flow} \times [X]_{urine}$
- Clearance of $X = \dfrac{\text{excretion rate of } X \text{ (mg/min)}}{[X]_{plasma} \text{ (mg/mL plasma)}}$
- When $[X]_{plasma}$ = renal threshold for X, then reabsorption of X = transport maximum for X.

In clinical settings, physicians use creatinine to estimate GFR. **Creatinine** is a breakdown product of phosphocreatine, an energy-storage compound found primarily in muscles [p. 388]. It is constantly produced by the body and need not be administered. Normally, the production and breakdown rates of phosphocreatine are relatively constant, and the plasma concentration of creatinine does not vary much.

Although creatinine is always present in the plasma and is easy to measure, it is not the perfect molecule for estimating GFR because a small amount is secreted into the urine. However, the amount secreted is small enough that, in most people, *creatinine clearance* is routinely used to estimate GFR.

Concept Check

12. If plasma creatinine = 1.8 mg/100 mL plasma, urine creatinine = 1.5 mg/mL urine, and urine volume is 1100 mL in 24 hours, what is the creatinine clearance? What is GFR?

Clearance Helps Us Determine Renal Handling

Once we know a person's GFR, we can determine how the kidney handles any solute by measuring the solute's plasma concentration and its excretion rate. If we assume that the solute is freely filtered at the glomerulus, we know from equation (1) that

$$\text{Filtered load of } X = [X]_{plasma} \times \text{GFR}$$

By comparing the filtered load of the solute with its excretion rate, we can tell how the nephron handled that substance (Fig. 19.14). For example, if less of the substance appears in the urine than was filtered, net reabsorption occurred (excreted = filtered − reabsorbed). If more of the substance appears in the urine than was filtered, there must have been net secretion of the substance into the lumen (excreted = filtered + secreted). If the same amount of the substance is filtered and excreted, then the substance is handled

FIG. 19.14 ESSENTIALS Renal Clearance

These figures show the relationship between clearance and excretion. Each figure represents the events taking place in one minute. For simplicity, 100% of the filtered volume is assumed to be reabsorbed.

Play Phys in Action
@Mastering Anatomy & Physiology

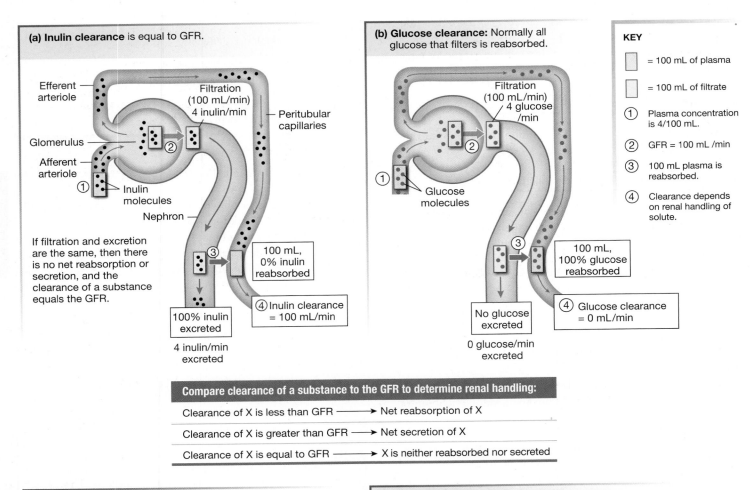

(a) Inulin clearance is equal to GFR.

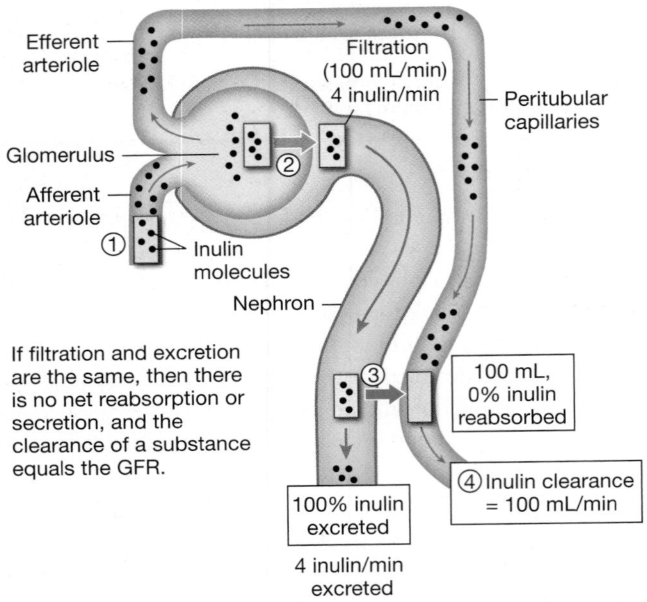

Efferent arteriole
Filtration (100 mL/min) 4 inulin/min
Peritubular capillaries
Glomerulus
Afferent arteriole
① Inulin molecules
Nephron

If filtration and excretion are the same, then there is no net reabsorption or secretion, and the clearance of a substance equals the GFR.

②
③ 100 mL, 0% inulin reabsorbed
100% inulin excreted
④ Inulin clearance = 100 mL/min

4 inulin/min excreted

(b) Glucose clearance: Normally all glucose that filters is reabsorbed.

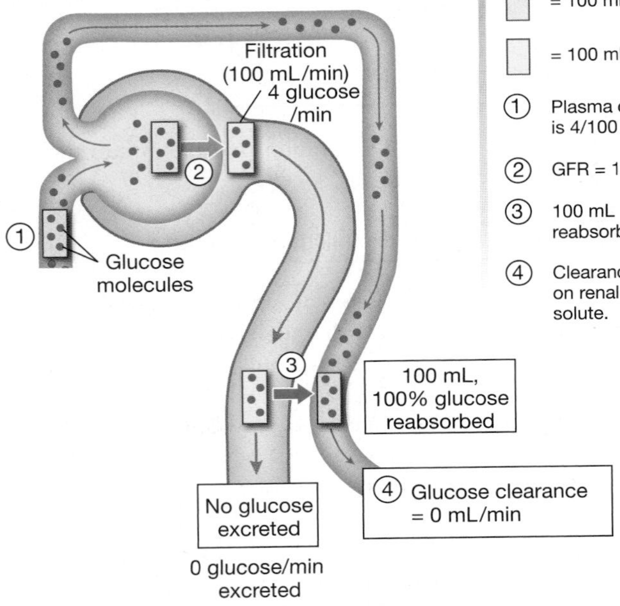

Filtration (100 mL/min) 4 glucose /min
①
Glucose molecules
②
③ 100 mL, 100% glucose reabsorbed
No glucose excreted
④ Glucose clearance = 0 mL/min

0 glucose/min excreted

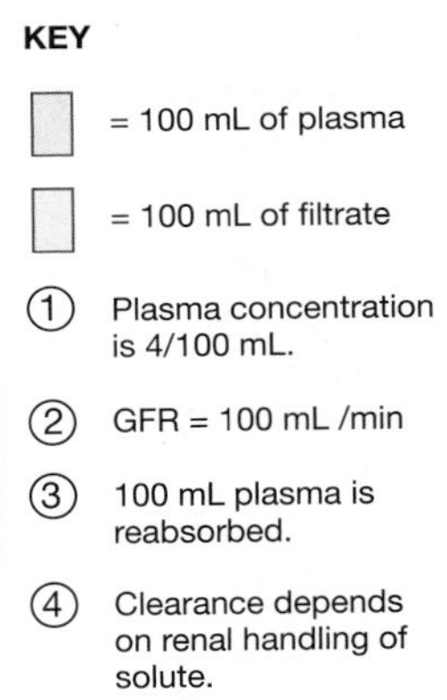

KEY

- ☐ = 100 mL of plasma
- ☐ = 100 mL of filtrate
- ① Plasma concentration is 4/100 mL.
- ② GFR = 100 mL /min
- ③ 100 mL plasma is reabsorbed.
- ④ Clearance depends on renal handling of solute.

Compare clearance of a substance to the GFR to determine renal handling:	
Clearance of X is less than GFR →	Net reabsorption of X
Clearance of X is greater than GFR →	Net secretion of X
Clearance of X is equal to GFR →	X is neither reabsorbed nor secreted

(c) Urea clearance is an example of net reabsorption. If filtration is greater than excretion, there is net reabsorption.

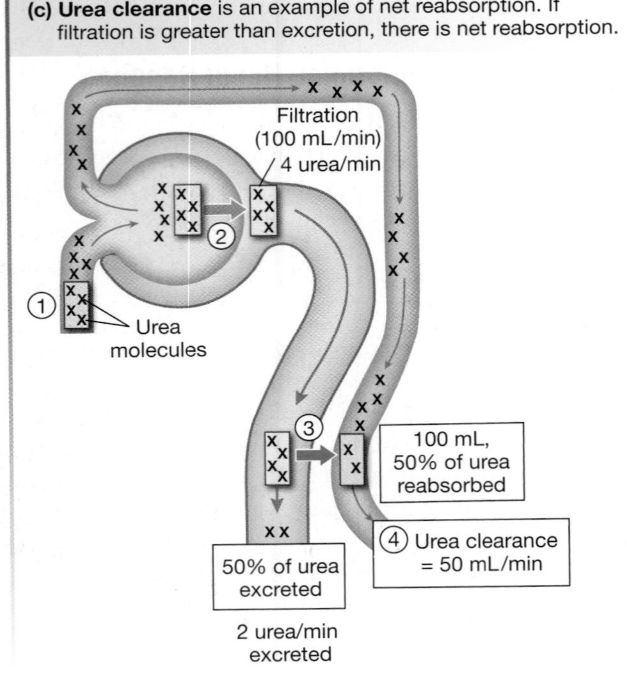

Filtration
(100 mL/min)
4 urea/min
①
Urea
molecules
②
③
100 mL,
50% of urea
reabsorbed
50% of urea
excreted
④ Urea clearance
= 50 mL/min
2 urea/min
excreted

(d) Penicillin clearance is an example of net secretion. If excretion is greater than filtration, there is net secretion.

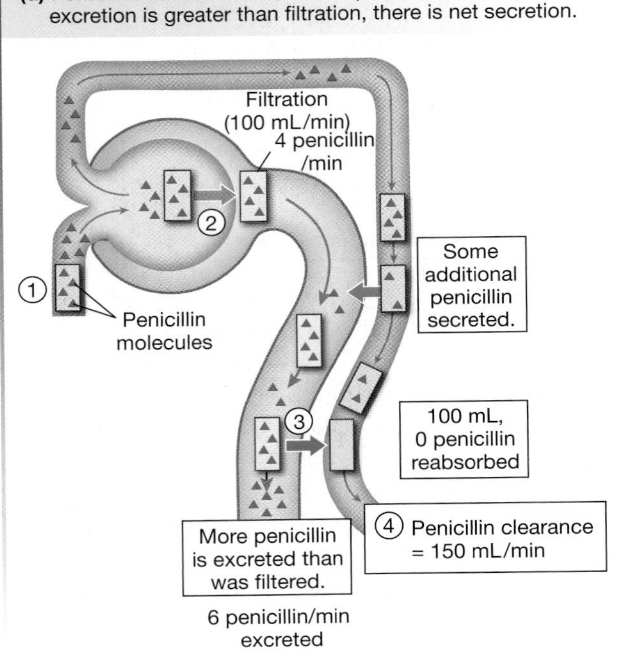

Filtration
(100 mL/min)
4 penicillin
/min
①
Penicillin
molecules
②
Some
additional
penicillin
secreted.
③
100 mL,
0 penicillin
reabsorbed
More penicillin
is excreted than
was filtered.
④ Penicillin clearance
= 150 mL/min
6 penicillin/min
excreted

like inulin—neither reabsorbed nor secreted. Let's look at some examples.

Suppose that glucose is present in the plasma at 100 mg glucose/dL plasma, and GFR is calculated from creatinine clearance to be 125 mL plasma/min. For these values, equation (1) tells us that

$$\text{Filtered load of glucose} = (100 \text{ mg glucose}/100 \text{ mL plasma}) \times 125 \text{ mL plasma/min}$$
$$\text{Filtered load of glucose} = 125 \text{ mg glucose/min}$$

There is no glucose in this person's urine, however: glucose excretion is zero. Because glucose was filtered at a rate of 125 mg/min but excreted at a rate of 0 mg/min, it must have been totally reabsorbed.

Clearance values can also be used to determine how the nephron handles a filtered solute. In this method, researchers calculate creatinine or inulin clearance, then compare the clearance of the solute being investigated with the creatinine or inulin clearance. If clearance of the solute is less than the inulin clearance, the solute has been reabsorbed. If the clearance of the solute is higher than the inulin clearance, additional solute has been secreted into the urine. More plasma was cleared of the solute than was filtered, so the additional solute must have been removed from the plasma by secretion.

Figure 19.14 shows filtration, excretion, and clearance of three molecules: glucose, urea, and penicillin. All solutes have the same concentration in the blood entering the glomerulus: 4 molecules/100 mL plasma. GFR is 100 mL/min, and we assume for simplicity that the entire 100 mL of plasma filtered into the tubule is reabsorbed.

For any solute, its clearance reflects how the kidney tubule handles it. For example, 100% of the glucose that filters is reabsorbed, and glucose clearance is zero (Fig. 19.14b). On the other hand, urea is partially reabsorbed: Four molecules filter, but two are reabsorbed and two are excreted (Fig. 19.14c). Consequently, urea clearance is 50 mL plasma per minute. Urea and glucose clearance are both less than the inulin clearance of 100 mL/min, which tells you that urea and glucose have been reabsorbed.

As you learned earlier, penicillin is filtered, not reabsorbed, and additional penicillin molecules are secreted from plasma in the peritubular capillaries. In Figure 19.14d, four penicillin are filtered, but six are excreted. An extra 50 mL of plasma have been cleared of penicillin in addition to the original 100 mL that were filtered. The penicillin clearance therefore is 150 mL plasma cleared per minute. Penicillin clearance is greater than the inulin clearance of 100 mL/min, which tells you that net secretion of penicillin occurs.

Note that a comparison of clearance values tells you only the *net* handling of the solute. It does not tell you if a molecule is both reabsorbed and secreted. For example, nearly all K^+ filtered is reabsorbed in the proximal tubule and loop of Henle, and then a small amount is secreted back into the tubule lumen at the distal nephron. On the basis of K^+ clearance, it appears that only reabsorption occurred.

Clearance calculations are relatively simple because all you need to know are the urine excretion rates and the plasma concentrations for any solute of interest, and both values are easily obtained. If you also know either inulin or creatinine clearance, then you can determine the renal handling of any compound.

RUNNING PROBLEM

Three weeks later, Michael was back in Dr. Garcia's office. The anti-inflammatory drugs and probenecid had eliminated the pain in his toe, but last night he had gone to the hospital with a very painful kidney stone. "We'll have to wait until the analysis comes back," said Dr. Garcia, "but I will guess that it is a uric acid stone. Did you drink as much water as I told you to?" Sheepishly, Michael admitted that he had good intentions but could never find the time at work to drink much water. "You have to drink enough water while on this drug to produce 3 liters or more of urine a day. That's more than 3 quarts. Otherwise, you may end up with another uric acid kidney stone." Michael remembered how painful the kidney stone was and agreed that this time he would follow instructions to the letter.

Q7: *Explain why not drinking enough water while taking uricosuric agents may cause uric acid crystals to form kidney stones in the urinary tract.*

588 – 589 – 607 – 607 – **611** – 613

PAH Clearance and Renal Plasma Flow One organic anion, *para*-aminohippurate or **PAH**, has become a useful tool in physiological research because in one pass through the kidneys, PAH is completely cleared from the plasma. In other words, PAH clearance is equal to the renal plasma flow. Let's see how this works.

Let's assume PAH has been administered by IV so that its plasma concentration is 100 mg PAH/100 mL of plasma. If 100 mL of plasma/min enter the glomerulus, 20 mL of that plasma will filter into the tubule (20%, the filtration fraction), along with 20 mg of PAH. Once PAH is in the nephron lumen, it cannot be reabsorbed. The remaining 80 mL of plasma + 80 mg PAH pass into the peritubular capillaries. As the blood flows past the proximal tubule, the tubule cells "recognize" PAH as a xenobiotic and use the organic anion transport system, shown in Figure 19.12, to extract PAH from the plasma and secrete it into the tubule lumen. By the end of the nephron, essentially all of the 20 mL of filtered plasma have been reabsorbed and the 80 mg of PAH in the peritubular capillaries secreted into the tubule lumen. All 100 mL of plasma/min that entered the kidney were cleared of PAH. Thus, PAH clearance is equal to renal plasma flow.

19.8 Micturition

Once filtrate leaves the collecting ducts, it can no longer be modified, and its composition does not change. The filtrate, now called urine, flows into the renal pelvis and then down the *ureter* with the help of rhythmic smooth muscle contractions that spurt urine into the bladder. The bladder is a hollow organ whose walls contain well-developed layers of smooth muscle. In the bladder, urine is stored until released in the process known as urination, voiding, or more formally, **micturition** (*micturire*, to desire to urinate).

The bladder can expand to hold a volume of about 500 mL. The neck of the bladder is continuous with the *urethra*, a single tube through which urine passes to reach the external environment. The opening between the bladder and urethra is closed by two rings of muscle called *sphincters* (**FIG. 19.14a**).

The **internal sphincter** is a continuation of the bladder wall and consists of smooth muscle. Its normal tone keeps it contracted. The **external sphincter** is a ring of skeletal muscle controlled by somatic motor neurons. Tonic stimulation from the central nervous system maintains contraction of the external sphincter except during urination.

Micturition is a simple spinal reflex that is subject to both conscious and unconscious control from higher brain centers. As the bladder fills with urine and its walls expand, stretch receptors send signals via sensory neurons to the spinal cord (**FIG. 19.14b**). There the information is integrated and transferred to two sets of neurons. The stimulus of a full bladder excites parasympathetic neurons leading to the smooth muscle in the bladder wall. The smooth muscle contracts, increasing the pressure on the bladder contents. Simultaneously, somatic motor neurons leading to the external sphincter are inhibited.

Contraction of the bladder occurs in a wave that pushes urine downward toward the urethra. Pressure exerted by the urine forces the internal sphincter open while the external sphincter relaxes. Urine passes into the urethra and out of the body, aided by gravity.

This simple micturition reflex occurs primarily in infants who have not yet been toilet trained. A person who has been toilet trained acquires a learned reflex that keeps the micturition reflex inhibited until she or he consciously desires to urinate. The learned reflex involves additional sensory fibers in the bladder that signal the degree of fullness. Centers in the brain stem and cerebral cortex receive that information and override the basic micturition reflex

FIG. 19.15 Micturition

Micturition is a spinal reflex subject to higher brain control.

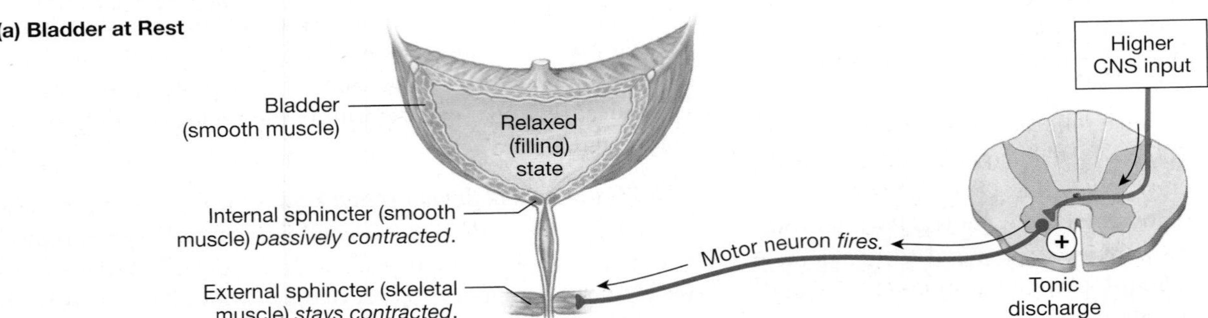
(a) Bladder at Rest
Bladder (smooth muscle)
Relaxed (filling) state
Internal sphincter (smooth muscle) *passively contracted*.
External sphincter (skeletal muscle) *stays contracted*.
Motor neuron *fires*.
Higher CNS input
+
Tonic discharge

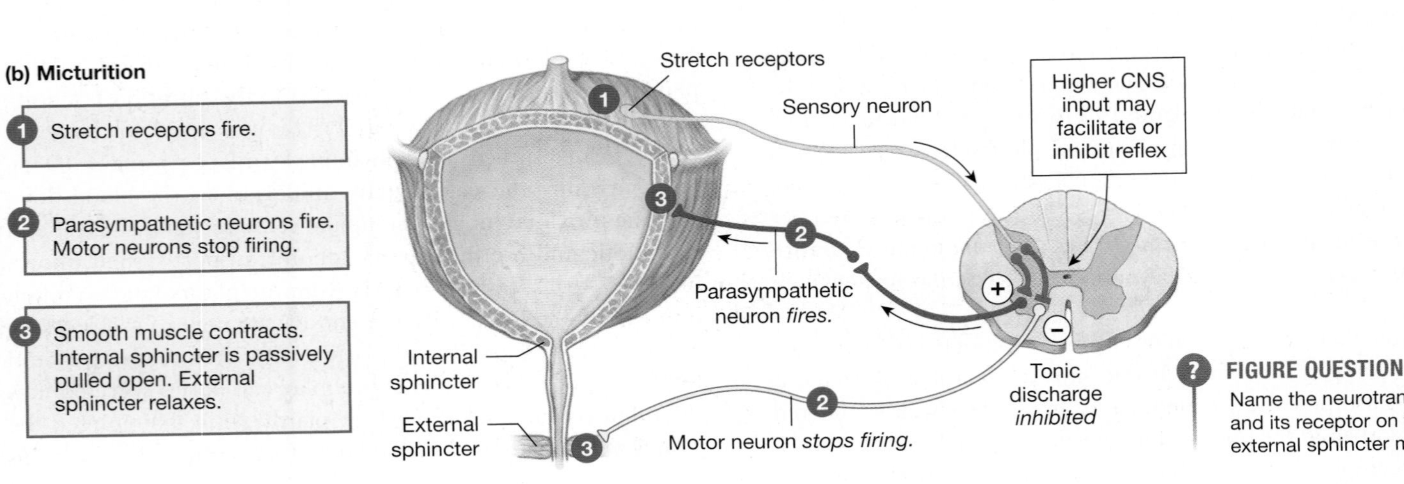
(b) Micturition
1 Stretch receptors fire.
2 Parasympathetic neurons fire. Motor neurons stop firing.
3 Smooth muscle contracts. Internal sphincter is passively pulled open. External sphincter relaxes.
Stretch receptors
1
Sensory neuron
Higher CNS input may facilitate or inhibit reflex
3
2
Parasympathetic neuron *fires*.
+
–
Internal sphincter
External sphincter
3
2
Motor neuron *stops firing*.
Tonic discharge *inhibited*

FIGURE QUESTION

Name the neurotransmitter and its receptor on the external sphincter muscle.

by directly inhibiting the parasympathetic fibers and by reinforcing contraction of the external sphincter. When an appropriate time to urinate arrives, those same centers remove the inhibition and facilitate the reflex by inhibiting contraction of the external sphincter.

In addition to conscious control of urination, various subconscious factors can affect the micturition reflex. "Bashful bladder" is a condition in which a person is unable to urinate in the presence of other people despite the conscious intent to do so. The sound of running water facilitates micturition and is often used to help patients urinate if the urethra is irritated from insertion of a *catheter*, a tube inserted into the bladder to drain it passively.

RUNNING PROBLEM **CONCLUSION** Gout

In this running problem, you learned that gout, which often presents as a debilitating pain in the big toe, is a metabolic problem whose cause and treatment can be linked to kidney function. Urate handling by the kidney is a complex process because urate is both secreted and reabsorbed in different segments of the proximal tubule. Scientists have now identified three different but related transport proteins that are involved in the process: the *organic anion transporter* (OAT), which exchanges anions in an electrically neutral exchange; *urate transporter* 1 (URAT1), which is also an anion exchanger but with high specificity for urate; and *urate transporter* (UAT), an electrogenic uniport urate transporter. The arrangement of these transport proteins on the polarized cell membrane determines whether the cell reabsorbs or secretes urate.

Gout is one of the oldest known diseases and for many years was considered a "rich man's" disease caused by too much rich food and drink. Thomas Jefferson and Benjamin Franklin both suffered from gout. To learn more about its causes, symptoms, and treatments, go to the Mayo Clinic's health information pages (*www.mayoclinic.com*) and search for *gout*. Check your understanding of this running problem by comparing your answers against the information in the summary table.

Question	Facts	Integration and Analysis
Q1: *Trace the route followed by kidney stones when they are excreted.*	Kidney stones often form in the renal pelvis.	From the renal pelvis, a stone passes down the ureter, into the urinary bladder, then into the urethra and out of the body.
Q2: *Name the anion formed when uric acid dissociates.*	The suffix *–ate* is used to identify the anion of organic acids.	The anion of uric acid is urate.
Q3: *Purines are part of which category of biomolecules? Using that information, explain why uric acid levels in the blood go up when cell breakdown increases.*	Purines include adenine and guanine, which are components of DNA, RNA, and ATP [p. 34]. When a cell dies, nuclear DNA and other chemical components are broken down.	Degradation of the cell's DNA, RNA, and ATP increases purine production, which in turn increases uric acid production.
Q4: *Based on what you have learned about uric acid and urate, predict two ways a person may develop hyperuricemia.*	Hyperuricemia is a disturbance of mass balance. Uric acid is made from purines. Urate is filtered by the kidneys with net secretion.	Hyperuricemia results either from overproduction of uric acid or from a defect in the renal excretion of urate.
Q5: *Could the same transporters be used by cells that reabsorb urate and cells that secrete it? Defend your reasoning.*	Some transporters move substrates in one direction only but others are reversible. Assume one urate transporter brings urate into the cell and another takes it out.	You could use the same two transporters if you reverse their positions on the apical and basolateral membranes. Cells reabsorbing urate would bring it in on the apical side and move it out on the basolateral. Cells secreting urate would reverse this pattern.
Q6: *Uricosuric agents, like urate, are organic acids. With that information, explain how uricosuric agents might enhance excretion of urate.*	Mediated transport exhibits competition, in which related molecules compete for one transporter. Usually, one molecule binds preferentially and therefore inhibits transport of the second molecule [p. 49].	Uricosuric agents are organic anions, so they may compete with urate for the proximal tubule organic anion transporter. Preferential binding of the uricosuric agents would block urate access to the OAT, leaving urate in the lumen and increasing its excretion.

Continued

RUNNING PROBLEM **CONCLUSION** *Continued*

Question	Facts	Integration and Analysis
Q7: *Explain why not drinking enough water while taking uricosuric agents may cause uric acid stones to form in the urinary tract.*	Uric acid stones form when uric acid concentrations exceed a critical level and crystals precipitate.	If a person drinks large volumes of water, the excess water will be excreted by the kidneys. Large amounts of water dilute the urine, thereby preventing the high concentrations of uric acid needed for stone formation.

588 – 589 – 607 – 607 – 611 – 613

CHAPTER SUMMARY

The urinary system, like the lungs, uses the principle of *mass balance* to maintain homeostasis. The components of urine are constantly changing and reflect the kidney's functions of regulating ions and water and removing wastes. One of the body's three *portal systems*—each of which includes two capillary beds—is found in the kidney. Filtration occurs in the first capillary bed and reabsorption in the second. The *pressure-flow-resistance* relationship you encountered in the cardiovascular and pulmonary systems also plays a role in glomerular filtration and urinary excretion. *Compartmentation* is illustrated by the movement of water and solutes between the internal and external environments as filtrate is modified along the nephron. Reabsorption and secretion of solutes depend on *molecular interactions* and on the *movement of molecules across membranes* of the tubule cells.

19.1 Functions of the Kidneys

1. The kidneys regulate extracellular fluid volume, blood pressure, and osmolarity; maintain ion balance; regulate pH; excrete wastes and foreign substances; and participate in endocrine pathways. (p. 588)

19.2 Anatomy of the Urinary System

2. The **urinary system** is composed of two kidneys, two ureters, a bladder, and a urethra. (p. 590; Fig. 19.1a)
3. Each **kidney** has about 1 million microscopic **nephrons.** In cross section, a kidney is arranged into an outer **cortex** and inner **medulla.** (p. 590; Fig. 19.1c)
4. Renal blood flow goes from **afferent arteriole** to **glomerulus** to **efferent arteriole** to **peritubular capillaries.** The **vasa recta** capillaries dip into the medulla. (p. 591; Fig. 19.1g, h, j)
5. Fluid filters from the glomerulus into **Bowman's capsule.** From there, it flows through the **proximal tubule, loop of Henle, distal tubule,** and **collecting duct,** then drains into the **renal pelvis. Urine** flows through the **ureter** to the **urinary bladder.** (p. 591; Fig. 19.1b, c, i)

19.3 Overview of Kidney Function

6. **Filtration** is the movement of fluid from plasma into Bowman's capsule. **Reabsorption** is the movement of filtered materials from tubule to blood. **Secretion** is the movement of selected molecules from blood to tubule. (p. 593; Fig. 19.2)
7. Average urine volume is 1.5 L/day. Osmolarity varies between 50 and 1200 mOsM. (p. 593; Fig. 19.2)
8. The amount of a solute excreted equals the amount filtered minus the amount reabsorbed plus the amount secreted. (p. 594; Fig. 19.3)

19.4 Filtration

9. One-fifth of renal plasma flow filters into the tubule lumen. The percentage of total plasma volume that filters is called the **filtration fraction.** (p. 595; Fig. 19.4)
10. Bowman's capsule epithelium has specialized cells called **podocytes** that wrap around the glomerular capillaries and create **filtration slits. Mesangial cells** support the glomerular capillaries. (p. 596; Fig. 19.5a, c)
11. Filtered solutes must pass first through glomerular capillary endothelium, then through a basement membrane, and finally through podocyte filtration slits before reaching the lumen of Bowman's capsule. (p. 596; Fig. 19.5d)
12. Filtration allows most components of plasma to enter the tubule but excludes blood cells and almost all plasma proteins. (p. 596)
13. Hydrostatic pressure in glomerular capillaries averages 55 mm Hg, favoring filtration. Opposing filtration are colloid osmotic pressure of 30 mm Hg and hydrostatic capsule fluid pressure averaging 15 mm Hg. The net driving force is 10 mm Hg, favoring filtration. (p. 597; Fig. 19.6)
14. The **glomerular filtration rate (GFR)** is the amount of fluid that filters into Bowman's capsule per unit time. Average GFR is 125 mL/min, or 180 L/day. (p. 598)

15. Hydrostatic pressure in glomerular capillaries can be altered by changing resistance in the afferent and efferent arterioles. (p. 597; Fig. 19.6c–e)
16. Autoregulation of glomerular filtration is accomplished by a **myogenic response** of vascular smooth muscle in response to pressure changes and by **tubuloglomerular feedback.** When fluid flow through the distal tubule increases, the **macula densa** cells send a paracrine signal to the afferent arteriole, which constricts. (p. 599; Fig. 19.7c)
17. Reflex control of GFR is mediated through systemic signals, such as hormones, and through the autonomic nervous system. (p. 600)

19.5 Reabsorption

18. Most reabsorption takes place in the proximal tubule. Finely regulated reabsorption takes place in the more distal segments of the nephron. (p. 600)
19. The active transport of Na^+ and other solutes creates concentration gradients for passive reabsorption of urea and other solutes. (p. 601; Fig. 19.8a)
20. Most reabsorption involves transepithelial transport, but some solutes and water are reabsorbed by the paracellular pathway. (p. 601)
21. Glucose, amino acids, ions, and various organic metabolites are reabsorbed by Na^+-linked secondary active transport. (p. 602; Fig. 19.8c)
22. Most renal transport is mediated by membrane proteins and exhibits saturation, specificity, and competition. The **transport maximum T_m** is the transport rate at saturation. (p. 602; Fig. 19.9)
23. The **renal threshold** is the plasma concentration at which a substance first appears in the urine. (p. 603; Fig. 19.9)
24. Peritubular capillaries reabsorb fluid along their entire length. (p. 604; Fig. 19.11)

19.6 Secretion

25. Secretion enhances excretion by removing solutes from the peritubular capillaries. K^+, H^+, and a variety of organic compounds are secreted. (p. 606; Fig. 19.12)
26. Molecules that compete for renal carriers slow the secretion of a molecule. (p. 606)

19.7 Excretion

27. The excretion rate of a solute depends on (1) its filtered load and (2) whether it is reabsorbed or secreted as it passes through the nephron. (p. 607)
28. **Clearance** describes how many milliliters of plasma passing through the kidneys have been totally cleared of a solute in a given period of time. (p. 607)
29. **Inulin** clearance is equal to GFR. In clinical settings, **creatinine** is used to measure GFR. (p. 612; Fig. 19.13)
30. Clearance can be used to determine how the nephron handles a solute filtered into it. (p. 612; Fig. 19.13)

19.8 Micturition

31. The external sphincter of the bladder is skeletal muscle that is tonically contracted except during urination. (p. 612; Fig. 19.14)
32. Micturition is a simple spinal reflex subject to conscious and unconscious control. (p. 612)
33. Parasympathetic neurons cause contraction of the smooth muscle in the bladder wall. Somatic motor neurons leading to the external sphincter are simultaneously inhibited. (p. 612)

REVIEW QUESTIONS

In addition to working through these questions and checking your answers on p. A-24, review the Learning Outcomes at the beginning of this chapter.

Level One Reviewing Facts and Terms

1. List and explain the significance of the five characteristics of urine that can be found by physical examination.
2. List and explain the six major kidney functions.
3. At any given time, what percentage of cardiac output goes to the kidneys?
4. List the major structures of the urinary system in their anatomical sequence, from the kidneys to the urine leaving the body. Describe the function of each structure.
5. Arrange the following structures in the order that a drop of water entering the nephron would encounter them:
   a. afferent arteriole
   b. Bowman's capsule
   c. collecting duct
   d. distal tubule
   e. glomerulus
   f. loop of Henle
   g. proximal tubule
   h. renal pelvis
6. Name the three filtration barriers that solutes must cross as they move from plasma to the lumen of Bowman's capsule. What components of blood are usually excluded by these layers?
7. What force(s) promote(s) glomerular filtration? What force(s) oppose(s) it? What is meant by the term *net driving force*?
8. What does the abbreviation GFR stand for? What is a typical numerical value for GFR in milliliters per minute? In liters per day?
9. Identify the following structures, then explain their significance in renal physiology:
   a. juxtaglomerular apparatus
   b. macula densa
   c. mesangial cells
   d. podocytes
   e. sphincters in the bladder
   f. renal cortex
10. In which segment of the nephron does most reabsorption take place? When a molecule or ion is reabsorbed from the lumen of the nephron, where does it go? If a solute is filtered and not reabsorbed from the tubule, where does it go?

20 Integrative Physiology II: Fluid and Electrolyte Balance

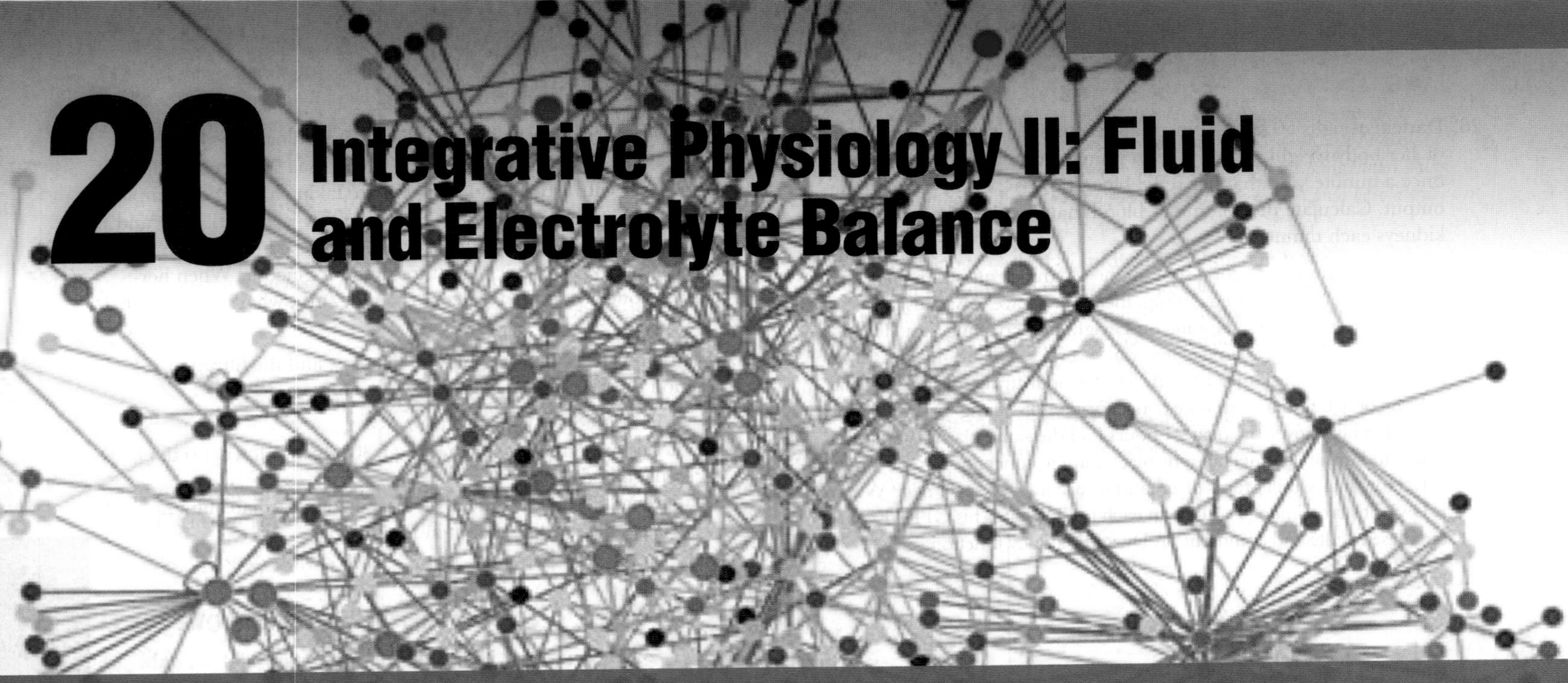

Protein interaction network

At a 10% loss of body fluid, the patient will show signs of confusion, distress, and hallucinations and at 20%, death will occur.

Poul Astrup, in Salt and Water in Culture and Medicine, *1993*

BACKGROUND BASICS

The American businesswoman in Tokyo finished her workout and stopped at the snack bar of the fitness club to ask for a sports drink. The attendant handed her a bottle labeled *Pocari Sweat®*. Although the thought of drinking sweat is not very appealing, the physiological basis for the name is sound.

During exercise, the body secretes sweat, a dilute solution of water and ions, particularly Na^+, K^+, and Cl^-. To maintain homeostasis, the body must replace any substances it has lost to the external environment. For this reason, the replacement fluid a person consumes after exercise should resemble sweat.

In this chapter, we explore how humans maintain salt and water balance, also known as fluid and electrolyte balance. The homeostatic control mechanisms for fluid and electrolyte balance in the body are aimed at maintaining four parameters: fluid volume, osmolarity, the concentrations of individual ions, and pH.

20.1 Fluid and Electrolyte Homeostasis

The human body is in a state of constant flux. Over the course of a day, we ingest about 2 liters of fluid that contains 6–15 grams of NaCl. In addition, we take in varying amounts of other electrolytes, including K^+, H^+, Ca^{2+}, HCO_3^-, and phosphate ions (HPO_4^{2-}). The body's task is to maintain *mass balance* [p. 10]: What comes in must be excreted if the body does not need it.

The body has several routes for excreting ions and water. The kidneys are the primary route for water loss and for removal of many ions. Under normal conditions, small amounts of both water and ions are lost in the feces and sweat as well. In addition, the lungs lose water and help remove H^+ and HCO_3^- by excreting CO_2.

Although physiological mechanisms that maintain fluid and electrolyte balance are important, behavioral mechanisms also play an essential role. *Thirst* is critical because drinking is the only normal way to replace lost water. *Salt appetite* is a behavior that leads people and animals to seek and ingest salt (sodium chloride, NaCl).

Why are we concerned with homeostasis of these substances? Water and Na^+ are associated with extracellular fluid volume and osmolarity. Disturbances in K^+ balance can cause serious problems with cardiac and muscle function by disrupting the membrane potential of excitable cells. Ca^{2+} is involved in a variety of body processes, from exocytosis and muscle contraction to bone formation and blood clotting, and H^+ and HCO_3^- are the ions whose balance determines body pH.

ECF Osmolarity Affects Cell Volume

Why is maintaining osmolarity so important to the body? The answer lies in the fact that water crosses most cell membranes freely. If the osmolarity of the extracellular fluid changes, water moves into or out of cells and changes intracellular volume. If extracellular fluid (ECF) osmolarity decreases as a result of excess water intake, water moves into the cells and they swell. If ECF osmolarity increases as a result of salt intake, water moves out of the cells and they shrink. Cell volume is so important that many cells have independent mechanisms for maintaining it.

For example, renal tubule cells in the medulla of the kidney are constantly exposed to high extracellular fluid osmolarity, yet these cells maintain normal cell volume. They do so by synthesizing organic solutes as needed to make their intracellular osmolarity match that of the medullary interstitial fluid. The organic solutes used to raise intracellular osmolarity include sugar alcohols and certain amino acids. Other cells in the body regulate their volume by changing their ionic composition.

In a few instances, changes in cell volume are believed to act as signals that initiate certain cellular responses. For example, swelling of liver cells activates protein and glycogen synthesis, and shrinkage of these cells causes protein and glycogen breakdown. In many cases, however, inappropriate changes in cell volume—either shrinking or swelling—impair cell function. The brain, encased in the rigid skull, is particularly vulnerable to damage from swelling. In general, maintenance of ECF osmolarity within a normal range is essential to maintain cell volume homeostasis.

Multiple Systems Integrate Fluid and Electrolyte Balance

The process of fluid and electrolyte balance is truly integrative because it involves the respiratory and cardiovascular systems in addition to renal and behavioral responses. Adjustments made by the lungs and cardiovascular system are primarily under neural control and can be made quite rapidly. Homeostatic compensation by the kidneys occurs more slowly because the kidneys are primarily under endocrine and neuroendocrine control. For example, small changes in blood pressure that result from increases or decreases in blood volume are quickly corrected by the cardiovascular control centers in the brain [p. 492]. If volume changes are persistent or of large magnitude, the kidneys step in to help maintain homeostasis.

FIGURE 20.1 summarizes the integrated response of the body to changes in blood volume and blood pressure. Signals from carotid and aortic baroreceptors and atrial volume receptors initiate a quick neural response mediated through the cardiovascular control center and a slower response elicited from the kidneys. In addition, low blood pressure stimulates thirst. In both situations,

RUNNING PROBLEM Hyponatremia

Lauren was competing in her first ironman distance triathlon, a 140.6-mile race consisting of 2.4 miles of swimming, 112 miles of cycling, and 26.2 miles of running. At mile 22 of the run, approximately 16 hours after starting the race, she collapsed. On being admitted to the medical tent, Lauren complained of nausea, a headache, and general fatigue. The medical staff noted that Lauren's face and clothing were covered in white crystals. When they weighed her and compared that value with her prerace weight recorded at registration, they realized Lauren had gained 2 kg during the race.

617 – 619 – 628 – 634 – 639 – 648

CHAPTER 20

FIG. 20.3 The kidneys conserve volume

Kidneys cannot restore lost volume. They only conserve fluid.

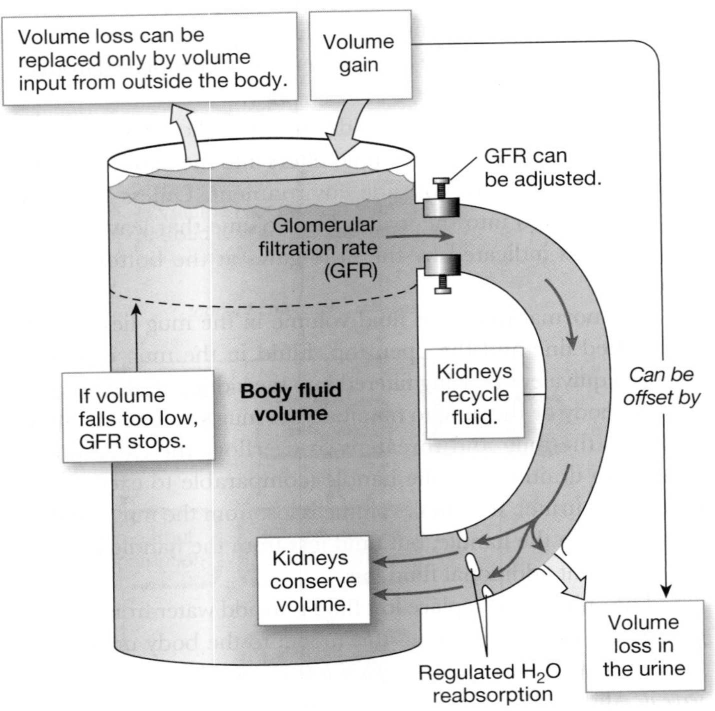
Volume loss can be replaced only by volume input from outside the body.
Volume gain
GFR can be adjusted.
Glomerular filtration rate (GFR)
If volume falls too low, GFR stops.
Body fluid volume
Kidneys recycle fluid.
Can be offset by
Kidneys conserve volume.
Regulated H_2O reabsorption
Volume loss in the urine

create concentrated urine. How can the kidney reabsorb water without first reabsorbing solute? At one time, scientists speculated that water was actively transported on carriers, just as Na^+ and other ions are. However, once scientists developed micropuncture techniques for sampling fluid inside kidney tubules, they discovered that water is reabsorbed by osmosis through water pores (*aquaporins*).

The mechanism for absorbing water without solute turned out to be simple: make the collecting duct cells and interstitial fluid surrounding them more concentrated than the fluid flowing into the tubule. Then, if the tubule cells have water pores, water can be absorbed from the lumen without first reabsorbing solute.

This is indeed the situation in the kidney. Through an unusual arrangement of blood vessels and renal tubules, which we discuss later, the renal medulla maintains a high osmotic concentration in its cells and interstitial fluid. This high *medullary interstitial osmolarity* allows urine to be concentrated as it flows through the collecting duct.

Let's follow some filtered fluid through a nephron to see where these changes in osmolarity take place (**FIG. 20.4**). The renal cortex has an interstitial osmolarity of about 300 mOsM. Reabsorption in the proximal tubule is isosmotic [p. 594], and filtrate entering the loop of Henle has an osmolarity of about 300 mOsM (Fig. 20.4 **1**).

As the nephrons dip into the medulla, the interstitial osmolarity steadily increases until it reaches about 1200 mOsM where the collecting ducts empty into the renal pelvis [Fig. 19.1c, p. 591]. Fluid passing through the descending limb of the loop loses water to the interstitium. Tubule fluid at the bottom of the loop will be of the same osmolarity as in the medulla.

In the ascending limb, the permeability of the tubule wall changes. The cells in the thick portion of the ascending limb of the loop have apical surfaces (facing the tubule lumen) that are impermeable to water. These cells do transport ions out of the tubule lumen (Fig. 20.4 **2**), but in this part of the nephron, solute movement is not followed by water movement. The reabsorption of solute without water decreases the concentration of the tubule fluid. Fluid leaving the loop of Henle therefore is hyposmotic, with an osmolarity of around 100 mOsM. The loop of Henle is the primary site where the kidney creates hyposmotic fluid.

Once hyposmotic fluid leaves the loop of Henle, it passes into the distal nephron. Here the water permeability of the tubule cells is variable and under hormonal control (Fig. 20.4 **3**). When the apical membrane of distal nephron cells is not permeable to water, water cannot leave the tubule, and the filtrate remains dilute. A small amount of additional solute can be reabsorbed as fluid passes along the collecting duct, making the filtrate even more dilute. When this happens, the concentration of urine can be as low as 50 mOsM (Fig. 20.4 **4**).

On the other hand, if the body needs to conserve water by reabsorbing it, the tubule epithelium in the distal nephron must become permeable to water. Under hormonal control, the cells insert water pores into their apical membranes. Once water can enter the epithelial cells, osmosis draws water out of the less-concentrated lumen and into the more concentrated interstitial fluid. At maximal water permeability, removal of water from the tubule leaves behind concentrated urine with an osmolarity that can be as high as 1200 mOsM (Fig. 20.4 **4**).

Water reabsorption in the kidneys conserves water and can decrease body osmolarity to some degree when coupled with excretion of solute in the urine. But remember that the kidney's homeostatic mechanisms can do nothing to restore lost fluid volume. Only the ingestion or infusion of water can replace water that has been lost.

RUNNING PROBLEM

The medical staff was concerned with Lauren's large weight increase during the race. They asked her to recall what she ate and drank during the race. Lauren reported that to avoid getting dehydrated in the warm weather, she had drunk large quantities of water in addition to sports gel and sports drinks containing carbohydrates and electrolytes.

Q1: *Name the two major body fluid compartments and give the major ions in each compartment.*

Q2: *Based on Lauren's history, give a reason for why her weight increased during the race.*

617 – **619** – 628 – 634 – 639 – 648

FIG. 20.4 Osmolarity changes as fluid flow through the nephron

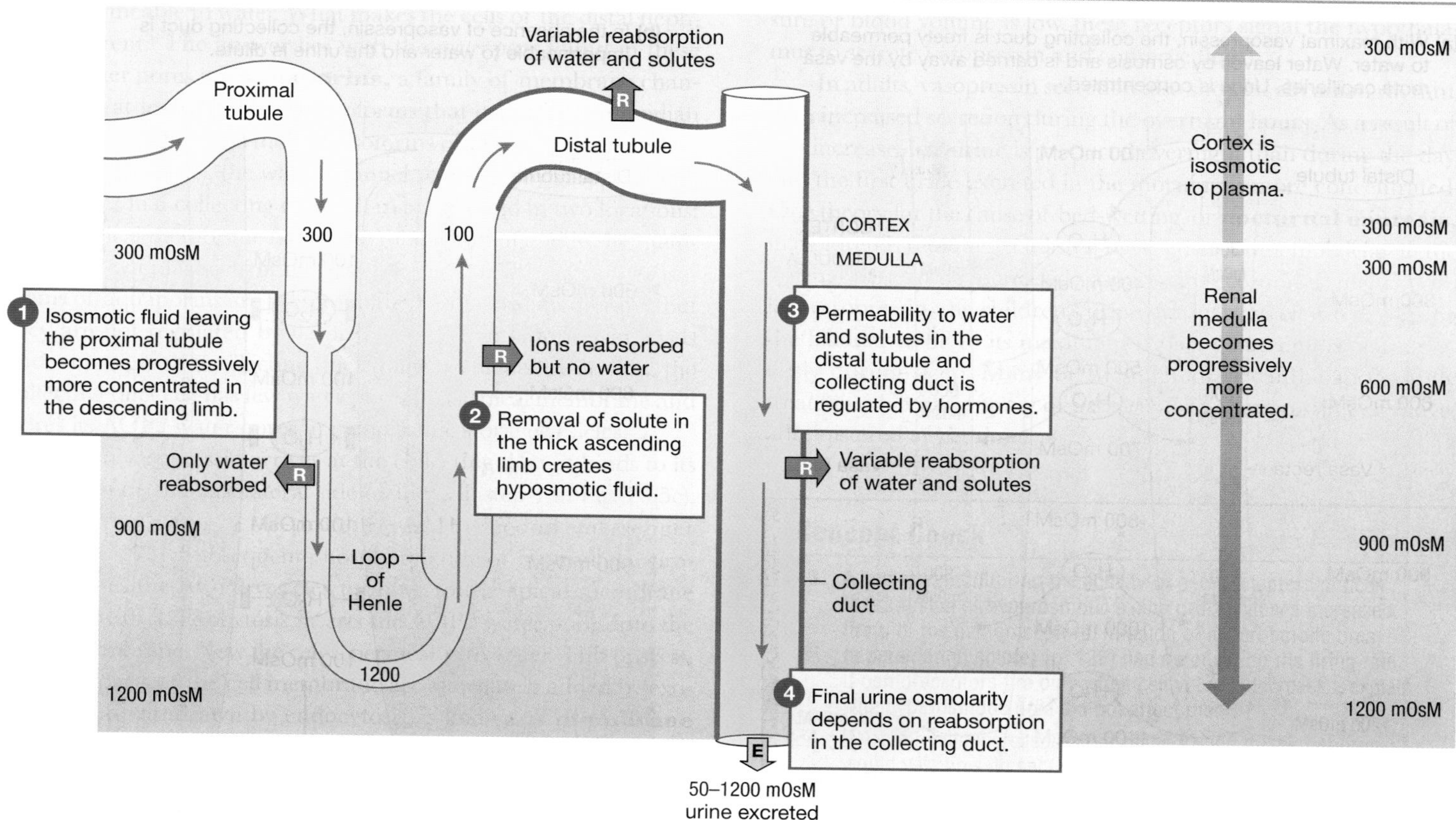
Variable reabsorption of water and solutes
Proximal tubule
Distal tubule
300
100
300 mOsM
1 Isosmotic fluid leaving the proximal tubule becomes progressively more concentrated in the descending limb.
Ions reabsorbed but no water
2 Removal of solute in the thick ascending limb creates hyposmotic fluid.
Only water reabsorbed
900 mOsM
Loop of Henle
1200
1200 mOsM
CORTEX
MEDULLA
3 Permeability to water and solutes in the distal tubule and collecting duct is regulated by hormones.
Variable reabsorption of water and solutes
Collecting duct
4 Final urine osmolarity depends on reabsorption in the collecting duct.
50–1200 mOsM urine excreted
300 mOsM
Cortex is isosmotic to plasma.
300 mOsM
300 mOsM
Renal medulla becomes progressively more concentrated.
600 mOsM
900 mOsM
1200 mOsM
R
E

CLINICAL FOCUS

Diabetes: Osmotic Diuresis

The primary sign of diabetes mellitus is an elevated blood glucose concentration. In untreated diabetics, if blood glucose levels exceed the renal threshold for glucose reabsorption [p. 603], glucose is excreted in the urine. This may not seem like a big deal, but any additional solute that remains in the lumen forces additional water to be excreted, causing *osmotic diuresis*. Suppose, for example, that the nephrons must excrete 300 milliosmoles of NaCl. If the urine is maximally concentrated at 1200 mOsM, the NaCl is excreted in a volume of 0.25 L. However, if the NaCl is joined by 300 milliosmoles of glucose that must be excreted, the volume of urine doubles, to 0.5 L. Osmotic diuresis in untreated diabetics (primarily type 1) causes *polyuria* (excessive urination) and *polydipsia* (excessive thirst) {*dipsios,* thirsty} as a result of dehydration and high plasma osmolarity.

Vasopressin Controls Water Reabsorption

How do the collecting duct cells alter their permeability to water? The process involves adding or removing water pores in the apical membrane under the direction of the posterior pituitary hormone **vasopressin** [p. 207]. In most mammals, the nine-amino-acid peptide contains the amino acid arginine, so vasopressin is called *arginine vasopressin* or AVP. Because vasopressin causes the body to retain water, its alternate name is *antidiuretic hormone* (ADH).

When vasopressin acts on target cells, the collecting duct epithelium becomes permeable to water, allowing water to move out of the lumen (**FIG. 20.5a**). The water moves by osmosis because osmolarity of tubule cells and the medullary interstitial fluid is higher than osmolarity of fluid in the tubule. In the absence of vasopressin, the collecting duct is impermeable to water (Fig. 20.5b). Although a concentration gradient is present across the epithelium, water remains in the tubule, producing dilute urine.

The water permeability of the collecting duct is not an all-or-none phenomenon, as the previous paragraph might suggest. Permeability is variable, depending on how much vasopressin is present. The graded effect of vasopressin allows the body to match urine concentration closely to the body's needs: the more vasopressin is present, the more water is reabsorbed.

One point that is sometimes difficult to remember is that this is not a static system, where the filtrate sits passively in the lumen waiting for solutes and water to be reabsorbed. The collecting duct, like other segments of the nephron, is a flow-through system. If the apical membrane has low water permeability, most of the water in filtrate will pass through the tubule unabsorbed and end up in the urine.

RUNNING PROBLEM

The medical staff analyzed Lauren's blood for electrolyte concentrations. Her serum Na^+ concentration was 124 mEq/L. The normal range is 135–145 mEq/L. Lauren's diagnosis was **hyponatremia** {*hypo-*, below + *natri-*, sodium + *-emia*, blood}, defined as a serum Na^+ concentration below 135 mEq/L. Hyponatremia induced by the consumption of large quantities of low-sodium or sodium-free fluid, which is what happened in Lauren's case, is sometimes called *dilutional hyponatremia.*

Q3: *Which body fluid compartment is being diluted in dilutional hyponatremia?*

Q4: *One way to estimate body osmolarity is to double the plasma Na^+ concentration. Estimate Lauren's osmolarity and explain what effect the dilutional hyponatremia has on her cells.*

Q5: *In dilutional hyponatremia, the medical personnel are most concerned about which organ or tissue?*

617 – 619 – 628 – 634 – 639 – 648

Aldosterone Controls Sodium Balance

The regulation of blood Na^+ levels takes place through one of the most complicated endocrine pathways of the body. The reabsorption of Na^+ in the distal tubules and collecting ducts of the kidney is regulated by the steroid hormone **aldosterone**: the more aldosterone, the more Na^+ reabsorption. Because one target of aldosterone is increased activity of the Na^+-K^+-ATPase, aldosterone also causes K^+ secretion (**FIG. 20.9**).

Aldosterone is a steroid hormone synthesized in the adrenal cortex, the outer portion of the adrenal gland that sits atop each kidney [p. 200]. Like other steroid hormones, aldosterone is secreted into the blood and transported on a protein carrier to its target.

The primary site of aldosterone action is the last third of the distal tubule and the portion of the collecting duct that runs through the kidney cortex (the *cortical collecting duct*). The primary target of aldosterone is **principal cells (P cells)** (Fig. 20.9b), the main cell type found in the distal nephron epithelium. Principal cells are arranged much like other polarized transporting epithelial cells, with Na^+-K^+-ATPase, pumps on the basolateral membrane and various channels and transporters on the apical membrane [p. 77]. In principal cells, the apical membranes contain leak channels for Na^+ (called ENaC, for **e**pithelial **Na^+** **c**hannel) and for K^+ (called ROMK, for **r**enal **o**uter **m**edulla **K^+** channel).

Aldosterone enters P cells by simple diffusion. Once inside, aldosterone combines with a cytoplasmic receptor (Fig. 20.9b 1). In the early response phase, apical Na^+ and K^+ channels increase their open time and existing channels are inserted into the apical membrane. As intracellular Na^+ levels rise from apical entry, the Na^+-K^+-ATPase, pump speeds up transporting cytoplasmic Na^+ into the ECF and bringing K^+ from the ECF into the P cell. The net result is a rapid increase in Na^+ reabsorption and K^+ secretion that does not require the synthesis of new channel or ATPase proteins. In the slower phase of aldosterone action, newly synthesized channels and pumps are inserted into epithelial cell membranes (Fig. 20.9b).

Note that Na^+ and water reabsorption are separately regulated in the distal nephron. Water does not automatically follow Na^+ reabsorption: Vasopressin must be present to make the distal nephron epithelium permeable to water. In contrast, Na^+ reabsorption in the proximal tubule is automatically followed by water reabsorption because the proximal tubule epithelium is always freely permeable to water.

Concept Check

8. In Figure 20.9b, what forces cause Na^+ and K^+ to cross the apical membrane?
9. If a person experiences hyperkalemia, what happens to resting membrane potential and the excitability of neurons and the myocardium?
10. Laboratory values for ions may be reported as mg/L, mmol/L, or mEq/L. If normal plasma Na^+ is 140 mmol/L, what is that concentration expressed as mEq/L? [Fig. 2.7, p. 42].

Low Blood Pressure Stimulates Aldosterone Secretion

What controls physiological aldosterone secretion from the adrenal cortex? There are two primary stimuli: increased extracellular K^+ concentration and decreased blood pressure (Fig. 20.9a). Elevated K^+ concentrations act directly on the adrenal cortex in a reflex that protects the body from hyperkalemia. Decreased blood pressure initiates a complex pathway that results in release of a hormone, **angiotensin II,** that stimulates aldosterone secretion in most situations.

Two additional factors modulate aldosterone release in pathological states: An increase in ECF osmolarity acts directly on adrenal cortex cells to inhibit aldosterone secretion during severe dehydration, and an abnormally large (10–20 mEq/L) decrease in plasma Na^+ can directly stimulate aldosterone secretion.

The Renin-Angiotensin Pathway Angiotensin II (ANG II) is the usual signal controlling aldosterone release from the adrenal cortex. ANG II is one component of the **renin-angiotensin system (RAS),** a complex, multistep pathway for maintaining blood pressure. The RAS pathway begins when juxtaglomerular granular cells in the afferent arterioles of a nephron [p. 600] secrete an enzyme called **renin** (**FIG. 20.10**). Renin converts an inactive plasma protein, **angiotensinogen,** into **angiotensin I (ANG I).** (The suffix *-ogen* indicates an inactive precursor.) When ANG I in the blood encounters an enzyme called **angiotensin-converting enzyme (ACE)**, ANG I is converted into ANG II.

This conversion was originally thought to take place only in the lungs, but ACE is now known to occur on the endothelium of blood vessels throughout the body. When ANG II in the blood reaches the adrenal gland, it causes synthesis and release of aldosterone. Finally, at the distal nephron, aldosterone initiates the intracellular reactions that cause the tubule to reabsorb Na^+.

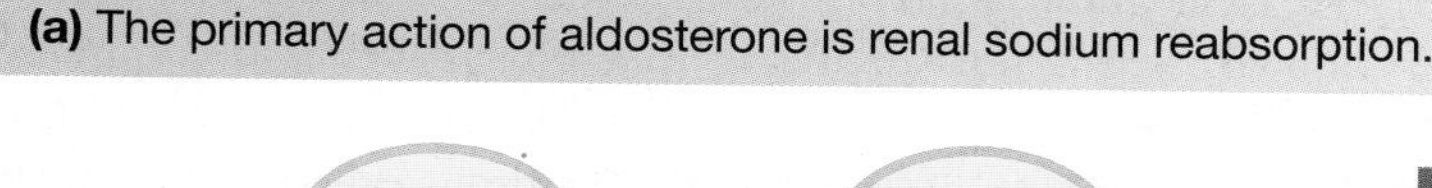

FIG. 20.9 ESSENTIALS: Aldosterone

(a) The primary action of aldosterone is renal sodium reabsorption.

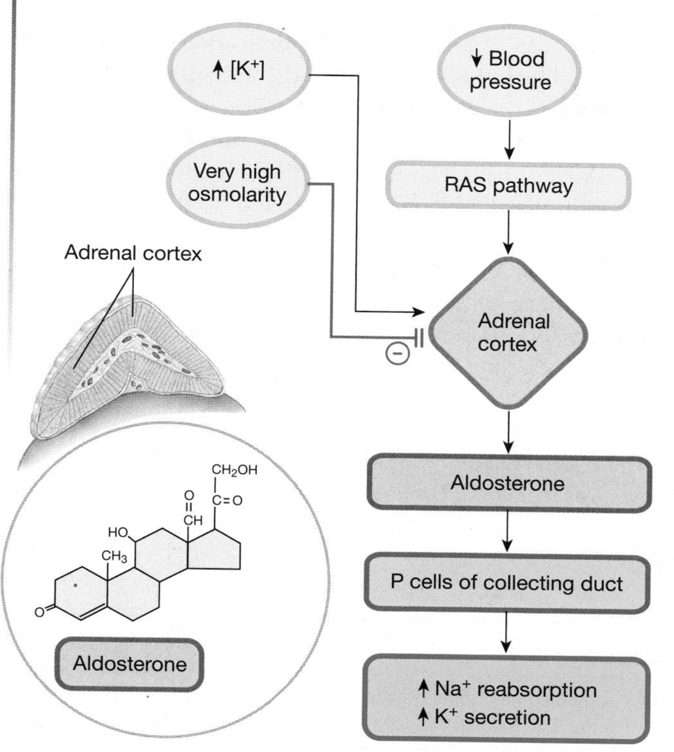

↑ [K+]
↓ Blood pressure
Very high osmolarity
RAS pathway
Adrenal cortex
Adrenal cortex
Aldosterone
P cells of collecting duct
↑ Na+ reabsorption
↑ K+ secretion
CH2OH
C=O
CH
HO
CH3
O
Aldosterone

ALDOSTERONE	
Origin	Adrenal cortex, zona glomerulosa
Chemical Nature	Steroid
Biosynthesis	Made on demand
Transport in the Circulation	50-70% bound to plasma protein
Half-Life	15 min
Factors Affecting Release	↓ Blood pressure (via renin) ↑ K^+ (hyperkalemia) Natriuretic peptides inhibit release
Target Cells or Tissues	Renal collecting duct - principal cells
Receptor	Cytosolic mineralocorticoid (MR) receptor
Tissue Action	Increases Na^+ reabsorption and K^+ secretion
Action at Cellular-Molecular Level	Synthesis of new ion channels (ENaC and ROMK) and pumps (Na^+-K^+-ATPase); increased activity of existing channels and pumps.

(b) Aldosterone acts on principal cells.

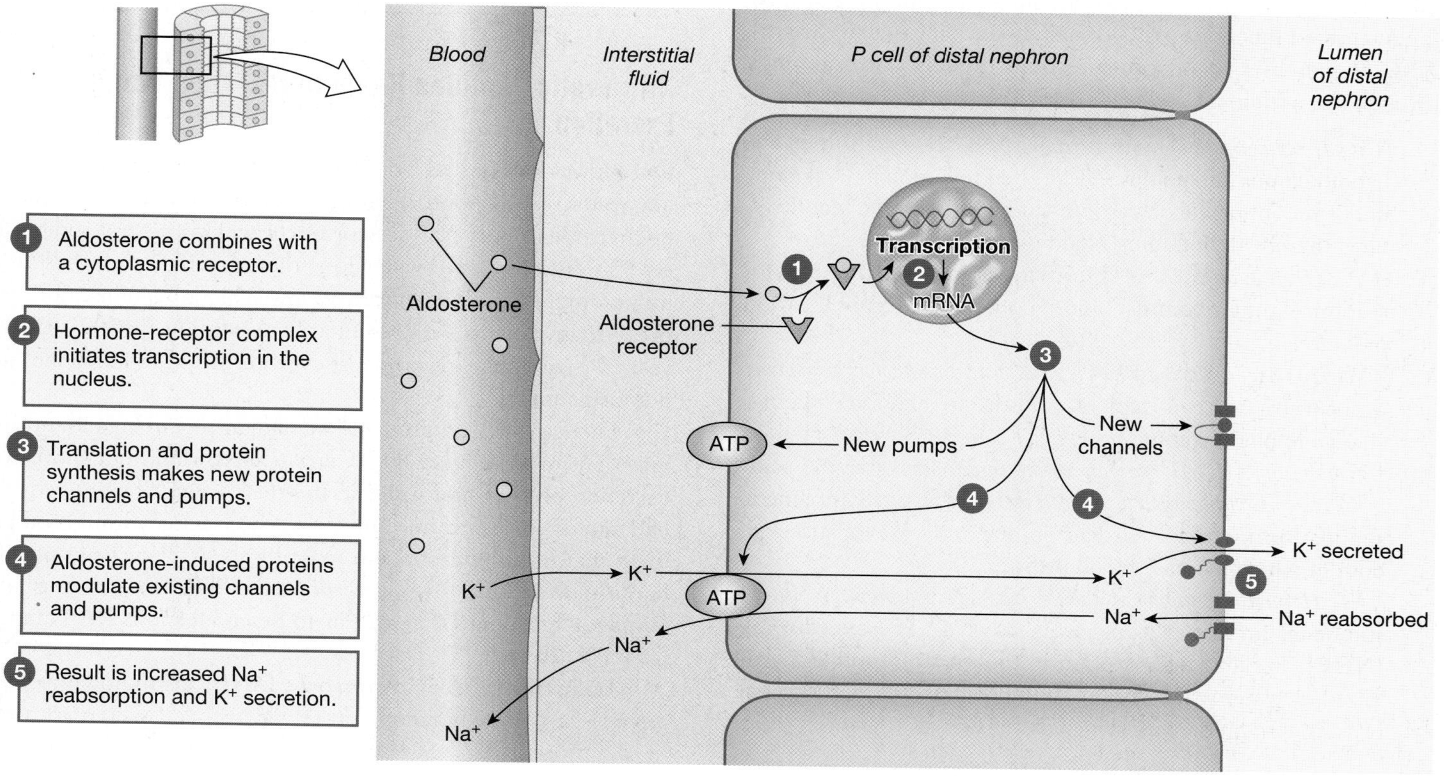

1. Aldosterone combines with a cytoplasmic receptor.
2. Hormone-receptor complex initiates transcription in the nucleus.
3. Translation and protein synthesis makes new protein channels and pumps.
4. Aldosterone-induced proteins modulate existing channels and pumps.
5. Result is increased Na^+ reabsorption and K^+ secretion.

Blood
Interstitial fluid
P cell of distal nephron
Lumen of distal nephron
Aldosterone
Aldosterone receptor
Transcription
mRNA
ATP
New pumps
New channels
K^+
K^+ secreted
ATP
Na^+
Na^+ reabsorbed

The stimuli that begin the RAS pathway are all related either directly or indirectly to low blood pressure (**FIG. 20.10**):

1. The *granular cells* are directly sensitive to blood pressure. They respond to low blood pressure in renal arterioles by secreting renin.
2. *Sympathetic neurons,* activated by the cardiovascular control center when blood pressure decreases, terminate on the granular cells and stimulate renin secretion.
3. *Paracrine feedback*—from the macula densa in the distal tubule to the granular cells—stimulates renin release [p. 600]. When fluid flow through the distal tubule is relatively high, the macula densa cells release paracrine signals that inhibit renin release. When fluid flow in the distal tubule decreases, macula densa cells signal the granular cells to secrete renin.

Sodium reabsorption does not directly raise low blood pressure, but retention of Na^+ increases osmolarity, which stimulates thirst. Fluid intake when the person drinks more water increases ECF volume (see Fig. 20.8). When blood volume increases, blood pressure also increases.

The effects of the RAS pathway are not limited to aldosterone release, however. Angiotensin II is a remarkable hormone with additional effects directed at raising blood pressure. These actions make ANG II an important hormone in its own right, not merely an intermediate step in the aldosterone control pathway.

ANG II Has Many Effects

Angiotensin II has significant effects on fluid balance and blood pressure beyond stimulating aldosterone secretion, underscoring the integrated functions of the renal and cardiovascular systems. ANG II increases blood pressure both directly and indirectly through five additional pathways (Fig. 20.10):

1. *ANG II increases vasopressin secretion.* ANG II receptors in the hypothalamus initiate this reflex. Fluid retention in the kidney under the influence of vasopressin helps conserve blood volume, thereby maintaining blood pressure.
2. *ANG II stimulates thirst.* Fluid ingestion is a behavioral response that expands blood volume and raises blood pressure.
3. *ANG II is one of the most potent vasoconstrictors known in humans.* Vasoconstriction causes blood pressure to increase without a change in blood volume.
4. *Activation of ANG II receptors in the cardiovascular control center increases sympathetic output to the heart and blood vessels.* Sympathetic stimulation increases cardiac output and vasoconstriction, both of which increase blood pressure.
5. *ANG II increases proximal tubule Na^+ reabsorption.* ANG II stimulates an apical transporter, the **Na^+-H^+ exchanger (NHE).** Sodium reabsorption in the proximal tubule is followed by water reabsorption, so the net effect is reabsorption of isosmotic fluid, conserving volume.

Once these blood pressure–raising effects of ANG II became known, it was not surprising that pharmaceutical companies started looking for drugs to block ANG II. Their research produced a new class of antihypertensive drugs called *ACE inhibitors.* These drugs block the ACE-mediated conversion of ANG I to ANG II, thereby helping to relax blood vessels and lower blood pressure. Less ANG II also means less aldosterone release, a decrease in Na^+ reabsorption and, ultimately, a decrease in ECF volume. All these responses contribute to lowering blood pressure.

However, the ACE inhibitors have side effects in some patients. ACE inactivates a cytokine called *bradykinin.* When ACE is inhibited by drugs, bradykinin levels increase, and in some patients this creates a dry, hacking cough. One solution was the development of drugs called *angiotensin receptor blockers* (ARBs), which block the blood pressure–raising effects of ANG II at target cells by binding to *AT_1 receptors.* Recently another new class of drugs, *direct renin inhibitors,* was approved. Direct renin inhibitors decrease the plasma activity of renin, which in turn blocks production of ANG I and inhibits the entire RAS pathway.

Concept Check

11. A man comes to the doctor with high blood pressure. Tests show that he also has elevated plasma renin levels and atherosclerotic plaques that have nearly blocked blood flow through his renal arteries. How does decreased blood flow in his renal arteries cause elevated renin levels?
12. Map the pathways through which elevated renin causes high blood pressure in the man mentioned in Concept Check 11.
13. Why is it more efficient to put ACE in the pulmonary vasculature than in the systemic vasculature?

Natriuretic Peptides Promote Na^+ and Water Excretion

Once it was known that aldosterone and vasopressin increase Na^+ and water reabsorption, scientists speculated that other hormones might cause urinary Na^+ loss, or **natriuresis** {*natrium,* sodium + *ourein,* to urinate} and water loss (diuresis). If found, these hormones might be used clinically to lower blood volume and blood pressure in patients with essential hypertension [p. 502]. During years of searching, however, evidence for the other hormones was not forthcoming.

Then, in 1981, a group of Canadian researchers found that injections of homogenized rat atria caused rapid but short-lived excretion of Na^+ and water in the rats' urine. They hoped they had found the missing hormone, one whose activity would complement that of aldosterone and vasopressin. As it turned out, they had discovered the first *natriuretic peptide* (NP), one member of a family of hormones that appear to be endogenous RAS antagonists (**FIG. 20.11**).

Atrial natriuretic peptide (**ANP;** also known as *atriopeptin*) is a peptide hormone produced in specialized myocardial

FIG. 20.10 ESSENTIALS The Renin-Angiotensin System (RAS)

This map outlines the control of aldosterone secretion as well as the blood pressure–raising effects of ANG II. The pathway begins when decreased blood pressure stimulates renin secretion.

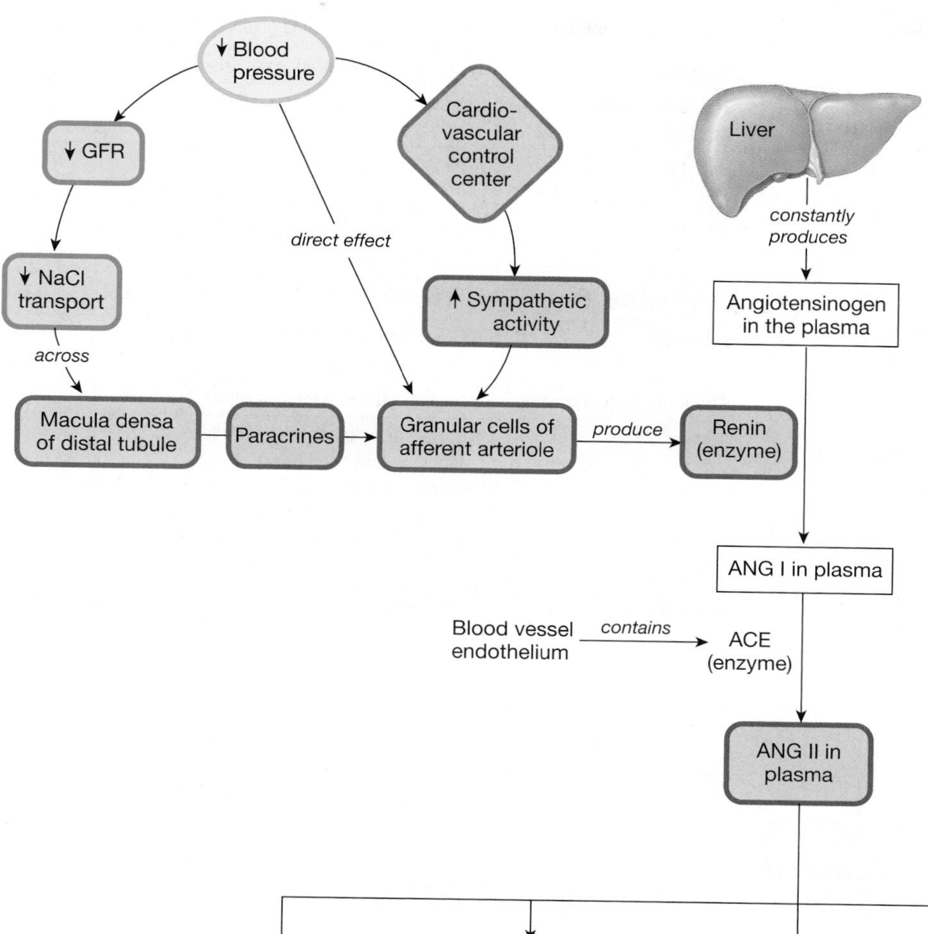

↓ Blood pressure
↓ GFR
Cardio-vascular control center
Liver
constantly produces
direct effect
↓ NaCl transport
↑ Sympathetic activity
Angiotensinogen in the plasma
across
Macula densa of distal tubule
Paracrines
Granular cells of afferent arteriole
produce
Renin (enzyme)
ANG I in plasma
Blood vessel endothelium
contains
ACE (enzyme)
ANG II in plasma

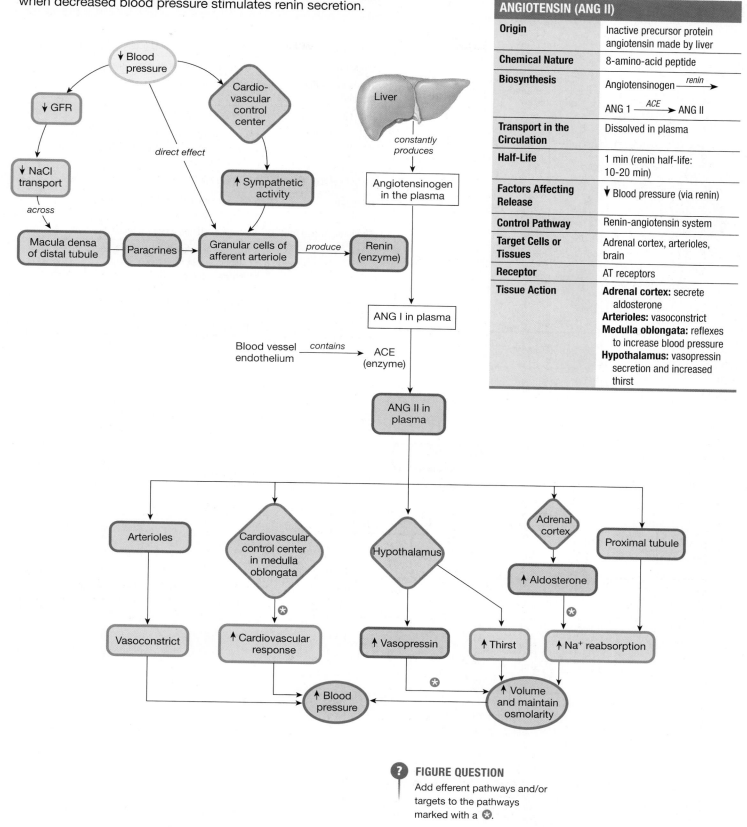

ANGIOTENSIN (ANG II)	
Origin	Inactive precursor protein angiotensin made by liver
Chemical Nature	8-amino-acid peptide
Biosynthesis	Angiotensinogen —*renin*→ ANG 1 —*ACE*→ ANG II
Transport in the Circulation	Dissolved in plasma
Half-Life	1 min (renin half-life: 10-20 min)
Factors Affecting Release	↓ Blood pressure (via renin)
Control Pathway	Renin-angiotensin system
Target Cells or Tissues	Adrenal cortex, arterioles, brain
Receptor	AT receptors
Tissue Action	**Adrenal cortex:** secrete aldosterone **Arterioles:** vasoconstrict **Medulla oblongata:** reflexes to increase blood pressure **Hypothalamus:** vasopressin secretion and increased thirst

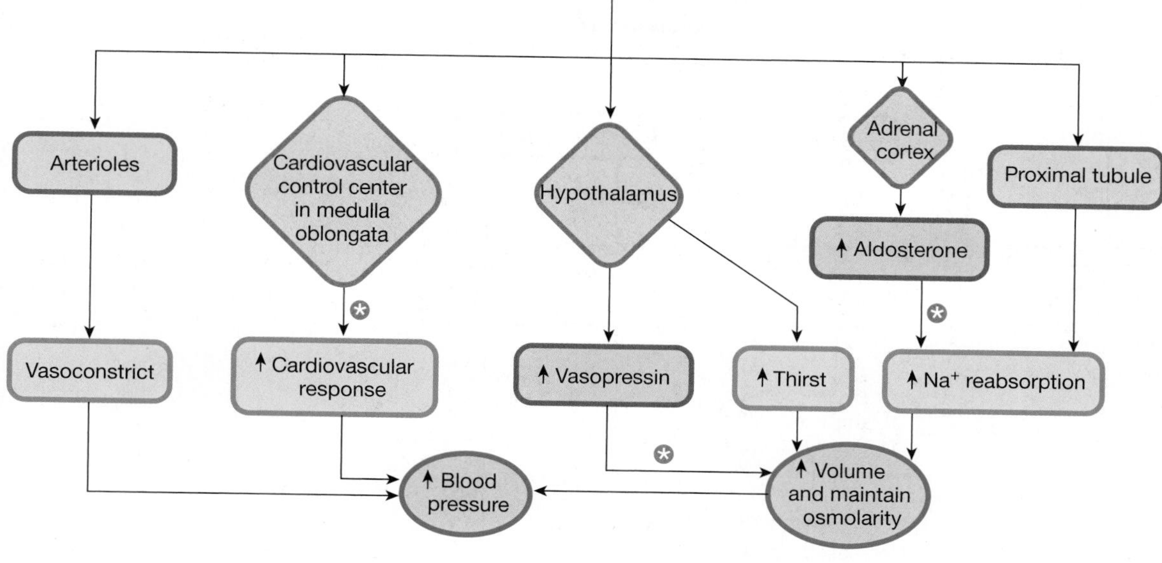

Arterioles
Cardiovascular control center in medulla oblongata
Hypothalamus
Adrenal cortex
Proximal tubule
↑ Aldosterone
Vasoconstrict
↑ Cardiovascular response
↑ Vasopressin
↑ Thirst
↑ Na⁺ reabsorption
↑ Blood pressure
↑ Volume and maintain osmolarity

FIGURE QUESTION

Add efferent pathways and/or targets to the pathways marked with a ✪.

FIG. 20.11 ESSENTIALS Natriuretic Peptides

Atrial natriuretic peptide (ANP) promotes salt and water excretion. Brain natriuretic peptide (BNP) is a clinical marker for heart failure.

NATRIURETIC PEPTIDES (ANP, BNP)	
Origin	Myocardial cells
Chemical Nature	Peptides. ANP: 28 amino acids, BNP: 32 amino acids
Biosynthesis	Typical peptide. Stored in secretory cells
Transport in the Circulation	Dissolved in plasma
Half-Life	ANP: 2-3 min, BNP: 12 min
Factors Affecting Release	↑ Myocardial stretch. ANP: atrial stretch due to increased blood volume. BNP: ventricular stretch in heart failure
Target Cells or Tissues	ANP: kidney, brain, adrenal cortex primarily
Receptor	NPR receptors. Guanylyl cyclase-linked receptor-enzymes
Systemic Action of ANP	Increase salt and water excretion
Tissue Action	**Afferent arterioles:** vasodilate to increase GFR; inhibit renin secretion **Nephron:** decrease Na^+ and water reabsorption **Adrenal cortex:** inhibit aldosterone secretion **Medulla oblongata:** reflexes to decrease blood pressure **Hypothalamus:** inhibit vasopressin secretion

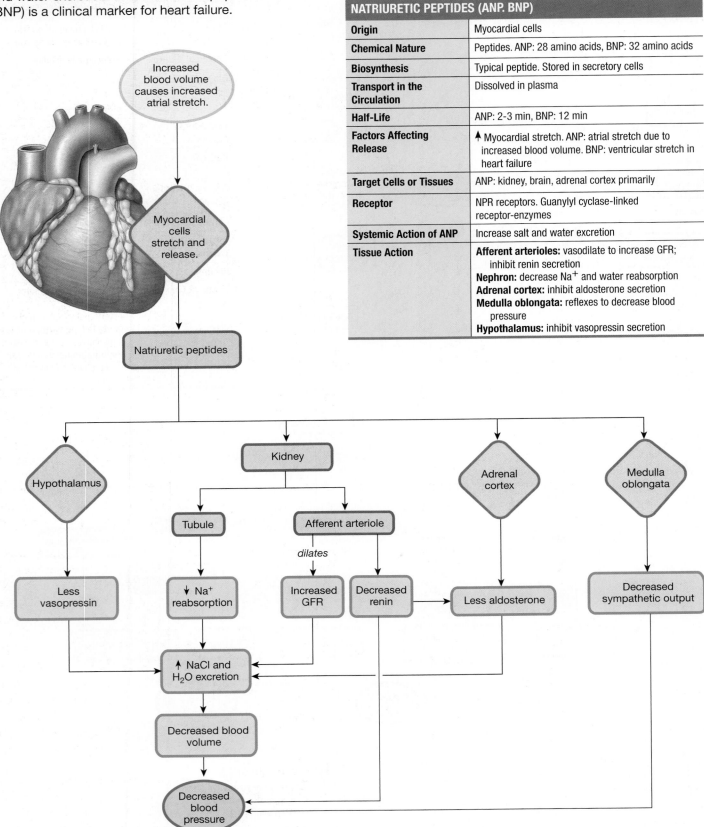

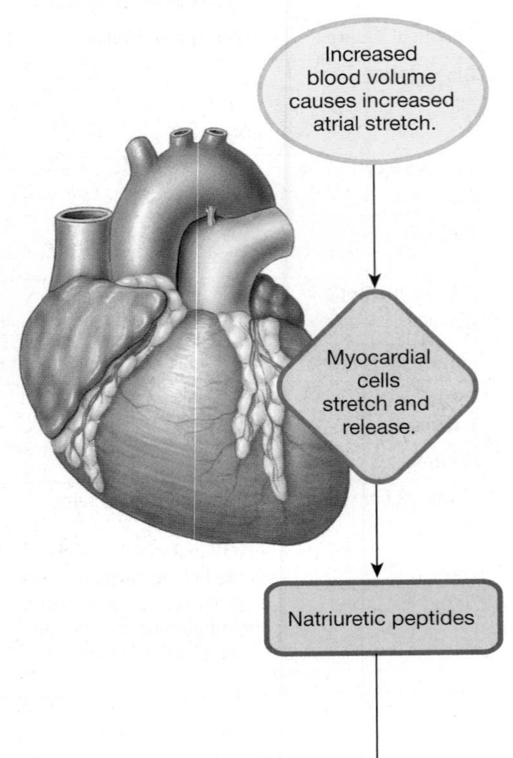

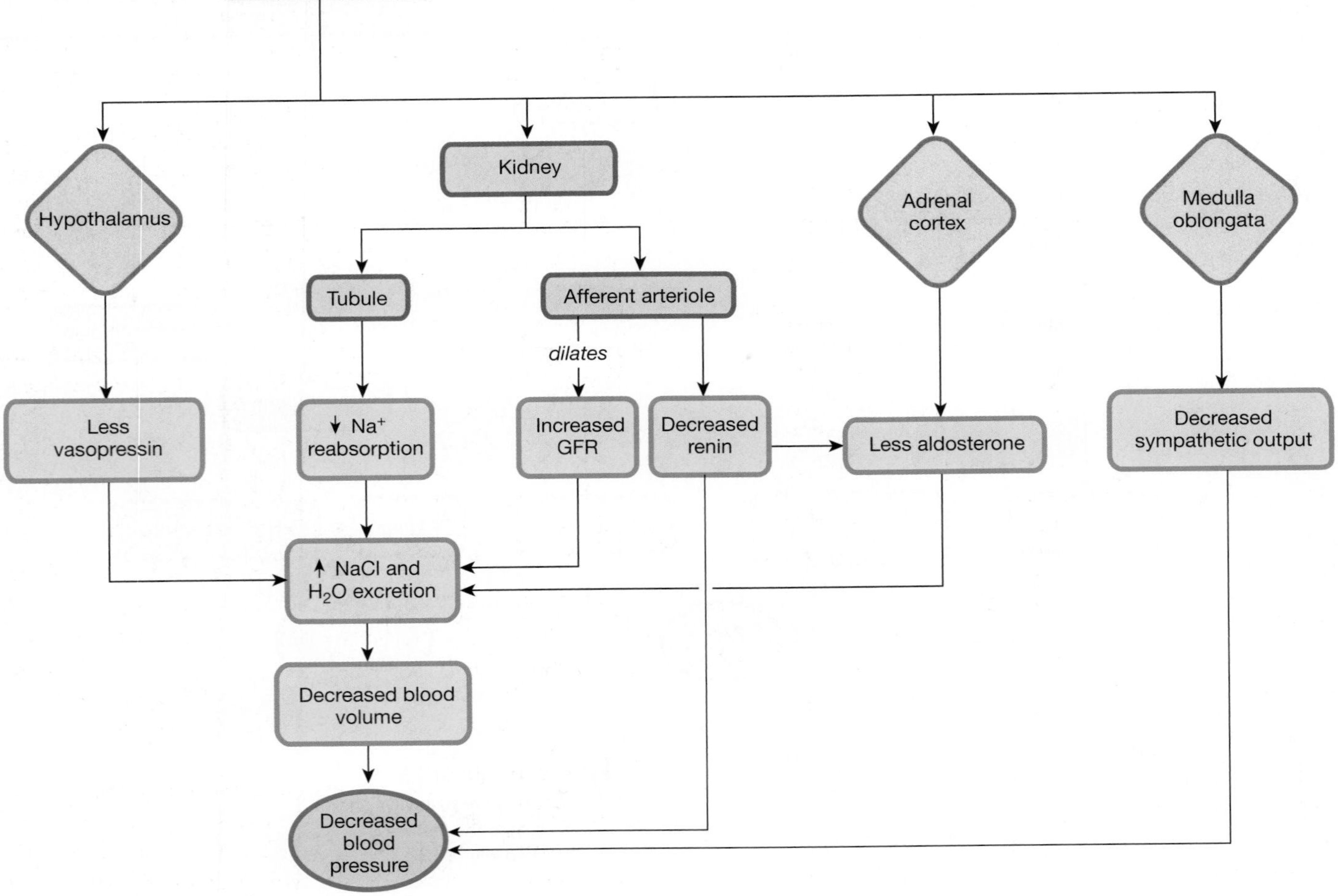

Increased blood volume causes increased atrial stretch.
Myocardial cells stretch and release.
Natriuretic peptides
Hypothalamus
Kidney
Adrenal cortex
Medulla oblongata
Tubule
Afferent arteriole
dilates
Less vasopressin
↓ Na^+ reabsorption
Increased GFR
Decreased renin
Less aldosterone
Decreased sympathetic output
↑ NaCl and H_2O excretion
Decreased blood volume
Decreased blood pressure

cells primarily in the atria of the heart. ANP is synthesized as part of a large prohormone that is cleaved into several active hormone fragments [p. 199]. A related hormone, **brain natriuretic peptide (BNP)**, is synthesized by ventricular myocardial cells and certain brain neurons. Natriuretic peptides are released by the heart when myocardial cells stretch more than normal. The natriuretic peptides bind to membrane receptor-enzymes that work through a cGMP second messenger system.

ANP is the more important signal molecule in normal physiology. ANP and its co-secreted natriuretic peptides are released when increased blood volume causes increased atrial stretch. At the systemic level, ANP enhances Na^+ and water excretion to decrease blood volume. ANP acts at multiple sites. In the kidney it increases GFR by dilating the afferent arterioles, and it directly decreases Na^+ reabsorption in the collecting duct.

Natriuretic peptides also act indirectly to increase Na^+ and water excretion by suppressing the release of renin, aldosterone, and vasopressin (Fig. 20.11), actions that reinforce the natriuretic-diuretic effect. In addition, natriuretic peptides act directly on the cardiovascular control center of the medulla to lower blood pressure.

BNP is now recognized as an important biological marker for heart failure because production of this substance increases with ventricular dilation and increased ventricular pressure. Hospital emergency departments now use BNP levels to distinguish *dyspnea* (difficulty breathing) in heart failure from other causes. BNP levels are also used as an independent predictor of heart failure and sudden death from cardiac arrhythmias.

20.4 Potassium Balance

Aldosterone (but not other factors in the RAS pathway) plays a critical role in potassium homeostasis. Only about 2% of the body's K^+ load is in the ECF, but regulatory mechanisms keep plasma K^+ concentrations within a narrow range (3.5–5 mEq/L). Under normal conditions, mass balance matches K^+ excretion to K^+ ingestion.

Renal handling of K^+ is complicated because K^+ is both reabsorbed and secreted. The net handling depends on the body load and the need to maintain K^+ homeostasis. K^+ is reabsorbed in the proximal tubule and ascending limb of the loop and may be secreted in the collecting duct [Fig. 19.13, p. 609]. With normal plasma concentrations of K^+, reabsorption is less than filtration, resulting in excretion of 10–20% of the filtered load of K^+. If plasma K^+ concentrations fall, more K^+ is reabsorbed and excretion can be as little as 2% of the filtered load. In other words, reabsorption of filtered K^+ is maximal when the body needs to conserve K^+.

On the other hand, if K^+ intake exceeds excretion and plasma K^+ goes up, homeostatic mechanisms kick in to get rid of the excess K^+. Hyperkalemia acts directly on adrenal cortex cells to promote secretion of aldosterone (Fig. 20.9a). Aldosterone acting on distal-nephron P cells keeps the cells' apical ion channels open longer and speeds up the Na^+-K^+-ATPase pump, increasing renal excretion of K^+. Under the influence of aldosterone, the kidney shifts to net secretion of K^+, with as much as 150% of the filtered load being excreted.

The regulation of body potassium levels is essential in maintaining a state of well-being. Changes in extracellular K^+ concentration affect the resting membrane potential of all cells [Fig. 8.17, p. 250]. If plasma (and ECF) K^+ concentrations decrease (*hypokalemia*), the concentration gradient between the cell and the ECF becomes larger, more K^+ leaves the cell, and the resting membrane potential becomes more negative. If ECF K^+ concentrations increase (*hyperkalemia*), the concentration gradient decreases and more K^+ remains in the cell, depolarizing it. (Remember that when plasma K^+ concentrations change, anions such as Cl^- are also added to or subtracted from the ECF in a 1:1 ratio, maintaining overall electrical neutrality.)

Because of the effect of plasma K^+ on excitable tissues, such as the heart, clinicians are always concerned about keeping plasma K^+ within its normal range. If K^+ falls below 3 mEq/L or rises above 6 mEq/L, the excitable tissues of muscle and nerve begin to show altered function. For example, hypokalemia causes muscle weakness because it is more difficult for hyperpolarized neurons and muscles to fire action potentials. The danger in this condition lies in the failure of respiratory muscles and the heart. Fortunately, skeletal muscle weakness is usually significant enough to lead patients to seek treatment before cardiac problems occur. Mild hypokalemia may be corrected by oral intake of K^+ supplements and K^+-rich foods, such as orange juice and bananas.

Hyperkalemia is a more dangerous potassium disturbance because in this case depolarization of excitable tissues makes them more excitable initially. Subsequently, the cells are unable to repolarize fully and actually become *less* excitable. In this state, they have action potentials that are either smaller than normal or nonexistent. Cardiac muscle excitability affected by changes in plasma K^+ can lead to life-threatening cardiac arrhythmias.

Disturbances in K^+ balance may result from kidney disease, eating disorders, loss of K^+ in diarrhea, or the use of certain types of diuretics that prevent the kidneys from fully reabsorbing K^+. Inappropriate correction of dehydration can also create K^+ imbalance. Consider a golfer playing a round of golf when the temperature was above 100 °F. He was aware of the risk of dehydration, so he drank lots of water to replace fluid lost through sweating. The replacement of lost sweat with pure water kept his ECF volume normal but dropped his total blood osmolarity and his K^+ and Na^+ concentrations. He was unable to finish the round of golf because of muscle weakness, and he required medical attention that included ion replacement therapy. A more suitable replacement fluid would have been one of the sports drinks that include salt and K^+.

Potassium balance is also closely tied to acid-base balance, as you will learn in the final section of this chapter. Correction of a pH disturbance requires close attention to plasma K^+ levels. Similarly, correction of K^+ imbalance may alter body pH.

RUNNING PROBLEM

During exercise in the heat, sweating rate and sweat composition are quite variable among athletes and depend partly on how acclimatized the individual is to the heat. Sweat fluid losses can range from less than 0.6 L/h to more than 2.5 L/h, and sweat Na^+ concentrations can range from less than 20 mEq/L to more than 90 mEq/L. The white salt crystals noted on Lauren's face and clothing suggest that she is a "salty sweater" who probably lost a large amount of salt during the race. Follow-up testing revealed that Lauren's sweat Na^+ concentration was 70 mEq/L.

Q6: *Assuming a sweating rate of 1 L/hr, how much Na^+ did Lauren lose during the 16-hour race?*

Q7: *Total body water for a 60-kg female is approximately 30 L, and her ECF volume is 10 L. Based on the information given in the problem so far, calculate how much fluid Lauren probably ingested during the race.*

617
619
628
634
639
648

20.5 Behavioral Mechanisms in Salt and Water Balance

Although neural, neuroendocrine, and endocrine reflexes play key roles in salt and water homeostasis, behavioral responses are critical in restoring the normal state, especially when ECF volume decreases or osmolarity increases. Drinking water is normally the only way to restore lost water, and eating salt is the only way to raise the body's Na^+ content. Both behaviors are essential for normal salt and water balance. Clinicians must recognize the absence of these behaviors in patients who are unconscious or otherwise unable to obey behavioral urges, and must adjust treatment accordingly. The study of the biological basis for behaviors, including drinking and eating, is a field known as *physiological psychology*.

Drinking Replaces Fluid Loss

Thirst is one of the most powerful urges known in humans. In 1952, the Swedish physiologist Bengt Andersson showed that stimulating certain regions of the hypothalamus triggered drinking behavior. This discovery led to the identification of hypothalamic osmoreceptors that initiate drinking when body osmolarity rises above 280 mOsM. This is an example of a behavior initiated by an internal stimulus.

It is interesting to note that although increased osmolarity triggers thirst, the act of drinking is sufficient to relieve thirst. The ingested water need not be absorbed for thirst to be quenched. As-yet-unidentified receptors in the mouth and pharynx (*oropharynx receptors*) respond to cold water by decreasing thirst and decreasing vasopressin release even though plasma osmolarity remains high. This oropharynx reflex is one reason surgery patients are allowed to suck on ice chips: the ice alleviates their thirst without putting significant amounts of fluid into the digestive system.

A similar reflex exists in camels. When led to water, they drink just enough to replenish their water deficit. Oropharynx receptors apparently act as a feedforward "metering" system that helps prevent wide swings in osmolarity by matching water intake to water need.

In humans, cultural rituals complicate the thirst reflex. For example, we may drink during social events, whether or not we are thirsty. As a result, our bodies must be capable of eliminating fluid ingested in excess of our physiological needs.

Low Na^+ Stimulates Salt Appetite

Thirst is not the only urge associated with fluid balance. **Salt appetite** is a craving for salty foods that occurs when plasma Na^+ concentrations drop. It can be observed in deer and cattle attracted to salt blocks or naturally occurring salt licks. In humans, salt appetite is linked to aldosterone and angiotensin, hormones that regulate Na^+ balance. The centers for salt appetite are in the hypothalamus close to the center for thirst.

Avoidance Behaviors Help Prevent Dehydration

Other behaviors play a role in fluid balance by preventing or promoting dehydration. Desert animals avoid the heat of the day and become active only at night, when environmental temperatures fall and humidity rises. Humans, especially now that we have air conditioning, are not always so wise.

The midday nap, or *siesta,* is a cultural adaptation in tropical countries that keeps people indoors during the hottest part of the day, thereby helping prevent dehydration and overheating. In the United States, we have abandoned this civilized custom and are active continuously during daylight hours, even when the temperature soars during summer in the South and Southwest. Fortunately, our homeostatic mechanisms usually keep us out of trouble.

Concept Check

14. Incorporate the thirst reflex into Figure 20.8.

20.6 Integrated Control of Volume, Osmolarity, and Blood Pressure

The body uses an integrated response to correct disruptions of salt and water balance. The cardiovascular system responds to changes in blood pressure and blood volume, and the kidneys respond to changes in blood volume or osmolarity. Maintaining homeostasis throughout the day is a continuous process in which the amounts of salt and water in the body shift, according to whether you just drank a soft drink or sweated through an aerobics class.

In that respect, maintaining fluid balance is like driving a car down the highway and making small adjustments to keep the car in the center of the lane. However, just as exciting movies feature wild car chases, not sedate driving, the exciting part of fluid homeostasis is the body's response to crisis situations, such as severe dehydration or hemorrhage. In this section, we examine more extreme challenges to salt and water balance.

Osmolarity and Volume Can Change Independently

Normally, volume and osmolarity are maintained within an acceptable range through homeostatic control pathways. Under some circumstances, however, fluid loss exceeds fluid gain or vice versa, and the body goes out of balance. Common pathways for fluid loss include excessive sweating, vomiting, diarrhea, and hemorrhage. All these situations may require medical intervention. In contrast, fluid gain is seldom a medical emergency, unless it is addition of water that decreases osmolarity below an acceptable range, as described in this chapter's Running Problem.

Volume and osmolarity of the ECF can each have three possible states: normal, increased, or decreased. The relation of volume and osmolarity changes can be represented by the matrix in **FIGURE 20.12**. The center box represents the normal state, and the surrounding boxes represent the most common examples of the variations from normal.

In all cases, the appropriate homeostatic compensation for the change acts according to the principle of mass balance: Whatever fluid and solute were added to the body must be removed, or whatever was lost must be replaced. However, perfect compensation is not always possible. Let's begin at the upper right corner of Figure 20.12 and move right to left across each row.

1. *Increased volume, increased osmolarity.* A state of increased volume and increased osmolarity might occur if you ate salty food and drank liquids at the same time, such as popcorn and a soft drink at the movies. The net result could be ingestion of hypertonic saline that increases ECF volume and osmolarity. The appropriate homeostatic response is excretion of hypertonic urine. For homeostasis to be maintained, the osmolarity and volume of the urinary output must match the salt and water input from the popcorn and soft drink.
2. *Increased volume, no change in osmolarity.* Moving one cell to the left in the top row, we see that if the proportion of salt and water in ingested food is equivalent to an isotonic NaCl solution, volume increases but osmolarity does not change. The appropriate response is excretion of isotonic urine whose volume equals that of the ingested fluid.

FIG. 20.12 Disturbances in volume and osmolarity

Volume	Osmolarity: Decrease	Osmolarity: No change	Osmolarity: Increase
Increase	Drinking large amount of water	Ingestion of isotonic saline	Ingestion of hypertonic saline
No change	Replacement of sweat loss with plain water	Normal volume and osmolarity	Eating salt without drinking water
Decrease	Incomplete compensation for dehydration	Hemorrhage	Dehydration (e.g., sweat loss or diarrhea)

3. *Increased volume, decreased osmolarity.* This situation would occur if you drank pure water without ingesting any solute. The goal here would be to excrete very dilute urine to maximize water loss while conserving salts. However, because our kidneys cannot excrete pure water, there is always some loss of solute in the urine. In this situation, urinary output cannot exactly match input, and so compensation is imperfect.
4. *No change in volume, increased osmolarity.* This disturbance (middle row, right cell) might occur if you ate salted popcorn without drinking anything. The ingestion of salt without water increases ECF osmolarity and causes some water to shift from cells to the ECF. The homeostatic response is intense thirst, which prompts ingestion of water to dilute the added solute. The kidneys help by creating highly concentrated urine of minimal volume, conserving water while removing excess NaCl. Once water is ingested, the disturbance becomes that described in situation 1 or situation 2.
5. *No change in volume, decreased osmolarity.* This scenario (middle row, left cell) might occur when a person who is dehydrated replaces lost fluid with pure water, like the golfer described earlier. The decreased volume resulting from the dehydration is corrected, but the replacement fluid has no solutes to replace those lost. Consequently, a new imbalance is created.

This situation led to the development of electrolyte-containing sports beverages. If people working out in hot weather replace lost sweat with pure water, they may restore volume but run the risk of diluting plasma K^+ and Na^+ concentrations to dangerously low levels (*dilutional hypokalemia* and *hyponatremia*, respectively).

6. *Decreased volume, increased osmolarity.* Dehydration is a common cause of this disturbance (bottom row, right cell). Dehydration has multiple causes. During prolonged heavy exercise, water loss from the lungs can double while sweat loss may increase from 0.1 liter to as much as 5 liters! Because the fluid secreted by sweat glands is hyposmotic, the fluid left behind in the body becomes hyperosmotic.

Diarrhea {*diarhein*, to flow through}, excessively watery feces, is a pathological condition involving major water and solute loss, this time from the digestive tract. In both sweating and diarrhea, if too much fluid is lost from the circulatory system, blood volume decreases to the point that the heart can no longer pump blood effectively to the brain. In addition, cell shrinkage caused by increased osmolarity disrupts cell function.

7. *Decreased volume, no change in osmolarity.* This situation (bottom row, middle cell) occurs with hemorrhage. Blood loss represents the loss of isosmotic fluid from the extracellular compartment, similar to scooping a cup of seawater out of a large bucketful. If a blood transfusion is not immediately available, the best replacement solution is one that is isosmotic and remains in the ECF, such as isotonic NaCl.
8. *Decreased volume, decreased osmolarity.* This situation (bottom row, left cell) might also result from incomplete compensation of dehydration, but it is uncommon.

CHAPTER 20

Dehydration Triggers Homeostatic Responses

To understand the body's integrated response to changes in volume and osmolarity, you must first have a clear idea of which pathways become active in response to various stimuli. **TABLE 20.1** is a summary of the many pathways involved in the homeostasis of salt and water balance. For details of individual pathways, refer to the figures cited in Table 20.1.

The homeostatic response to severe dehydration is an excellent example of how the body works to maintain blood volume and cell volume in the face of decreased volume and increased osmolarity. It also illustrates the role of neural and endocrine integrating centers. In severe dehydration, the adrenal cortex receives two opposing signals. One says, "Secrete aldosterone"; the other says, "Do not secrete aldosterone." The body has multiple mechanisms for dealing with diminished blood volume, but high ECF osmolarity causes cells to shrink and presents a more immediate threat to well-being. Thus, faced with decreased volume and increased osmolarity, the adrenal cortex does not secrete aldosterone. (If secreted, aldosterone would cause Na^+ reabsorption, which could worsen the already-high osmolarity associated with dehydration.)

In severe dehydration, compensatory mechanisms are aimed at restoring normal blood pressure, ECF volume, and osmolarity by (1) conserving fluid to prevent additional loss, (2) triggering cardiovascular reflexes to increase blood pressure, and (3) stimulating thirst so that normal fluid volume and osmolarity can be restored. **FIGURE 20.13** maps the interwoven nature of these responses. This figure is complex and intimidating at first glance, so let's discuss it step by step.

At the top of the map (in yellow) are the two stimuli caused by dehydration: decreased blood volume/pressure, and increased osmolarity. Decreased ECF volume causes decreased blood pressure. Blood pressure acts both directly and as a stimulus for several reflex pathways that are mediated through the carotid and aortic baroreceptors and the pressure-sensitive granular cells. Decreased volume is sensed by the atrial volume receptors.

1. *The carotid and aortic baroreceptors signal the cardiovascular control center (CVCC) to raise blood pressure.* Sympathetic output from the CVCC increases while parasympathetic output decreases.
   a. Heart rate goes up as control of the SA node shifts from predominantly parasympathetic to sympathetic.
   b. The force of ventricular contraction also increases under sympathetic stimulation. The increased force of contraction combines with increased heart rate to increase cardiac output.
   c. Simultaneously, sympathetic input causes arteriolar vasoconstriction, increasing peripheral resistance.
   d. Sympathetic vasoconstriction of afferent arterioles in the kidneys decreases GFR, helping conserve fluid.
   e. Increased sympathetic activity at the granular cells of the kidneys increases renin secretion.
2. *Decreased peripheral blood pressure directly decreases GFR.* A lower GFR conserves ECF volume by filtering less fluid into the nephron.
3. *Paracrine feedback causes the granular cells to release renin.* Lower GFR decreases fluid flow past the macula densa. This triggers renin release.
4. *Granular cells respond to decreased blood pressure by releasing renin.* The combination of decreased blood pressure, increased sympathetic input onto granular cells, and signals from the macula densa stimulates renin release and ensures increased production of ANG II.
5. *Decreased blood pressure, decreased blood volume, increased osmolarity, and increased ANG II production all stimulate vasopressin and the thirst centers of the hypothalamus.*

The redundancy in the control pathways ensures that all four main compensatory mechanisms are activated: cardiovascular responses, ANG II, vasopressin, and thirst.

1. *Cardiovascular responses* combine increased cardiac output and increased peripheral resistance to raise blood pressure. Note, however, that this increase in blood pressure does *not necessarily* mean that blood pressure returns to normal. If dehydration is severe, compensation may be incomplete, and blood pressure may remain below normal.
2. *Angiotensin II* has a variety of effects aimed at raising blood pressure, including stimulation of thirst, vasopressin release, direct vasoconstriction, and reinforcement of cardiovascular control center output. ANG II also reaches the adrenal cortex and attempts to stimulate aldosterone release. In severe dehydration, however, Na^+ reabsorption worsens the already high osmolarity. Consequently, high osmolarity at the adrenal cortex directly inhibits aldosterone release, blocking the action of ANG II. The RAS pathway in dehydration produces the beneficial blood pressure–enhancing effects of ANG II while avoiding the detrimental effects of Na^+ reabsorption. This is a beautiful example of integrated function.
3. *Vasopressin* increases the water permeability of the renal collecting ducts, allowing water reabsorption to conserve fluid. Without fluid replacement, however, vasopressin cannot bring volume and osmolarity back to normal.
4. *Oral (or intravenous) intake of water* in response to thirst is the only mechanism for replacing lost fluid volume and for restoring ECF osmolarity to normal.

The net result of all four mechanisms is (1) restoration of volume by water conservation and fluid intake, (2) maintenance of blood pressure through increased blood volume, increased cardiac output, and vasoconstriction, and (3) restoration of normal osmolarity by decreased Na^+ reabsorption and increased water reabsorption and intake.

Using the pathways listed in Table 20.1 and Figure 20.13 as a model, try to create reflex maps for the seven other disturbances of volume and osmolarity shown in Figure 20.12.

TABLE 20.1 Reponses Triggered by Changes in Volume, Blood Pressure, and Osmolarity

Stimulus	Organ or Tissue Involved	Response(s)	Figure(s)
Decreased Blood Pressure/Volume			
Direct effects			
	Granular cells	Renin secretion	20.10
	Glomerulus	Decreased GFR	19.6, 20.10
Reflexes			
Carotid and aortic baroreceptors	Cardiovascular control center	Increased sympathetic output, decreased parasympathetic output	15.14b, 20.10
Carotid and aortic baroreceptors	Hypothalamus	Thirst stimulation	20.1a
Carotid and aortic baroreceptors	Hypothalamus	Vasopressin secretion	20.6
Atrial volume receptors	Hypothalamus	Thirst stimulation	20.1a
Atrial volume receptors	Hypothalamus	Vasopressin secretion	20.6
Increased Blood Pressure			
Direct effects			
	Glomerulus	Increased GFR (transient)	19.6, 19.7
	Myocardial cells	Natriuretic peptide secretion	20.11
Reflexes			
Carotid and aortic baroreceptors	Cardiovascular control center	Decreased sympathetic output, increased parasympathetic output	15.14b
Carotid and aortic baroreceptors	Hypothalamus	Thirst inhibition	
Carotid and aortic baroreceptors	Hypothalamus	Vasopressin inhibition	
Atrial volume receptors	Hypothalamus	Thirst inhibition	
Atrial volume receptors	Hypothalamus	Vasopressin inhibition	
Increased Osmolarity			
Direct effects			
Pathological dehydration	Adrenal cortex	Decreased aldosterone secretion	20.13
Reflexes			
Osmoreceptors	Hypothalamus	Thirst stimulation	20.8
Osmoreceptors	Hypothalamus	Vasopressin secretion	20.6
Decreased Osmolarity			
Direct effects			
Pathological hyponatremia	Adrenal cortex	Increased aldosterone secretion	
Reflexes			
Osmoreceptors	Hypothalamus	Decreased vasopressin secretion	

CHAPTER 20

FIG. 20.13 Homeostatic compensation for severe dehydration

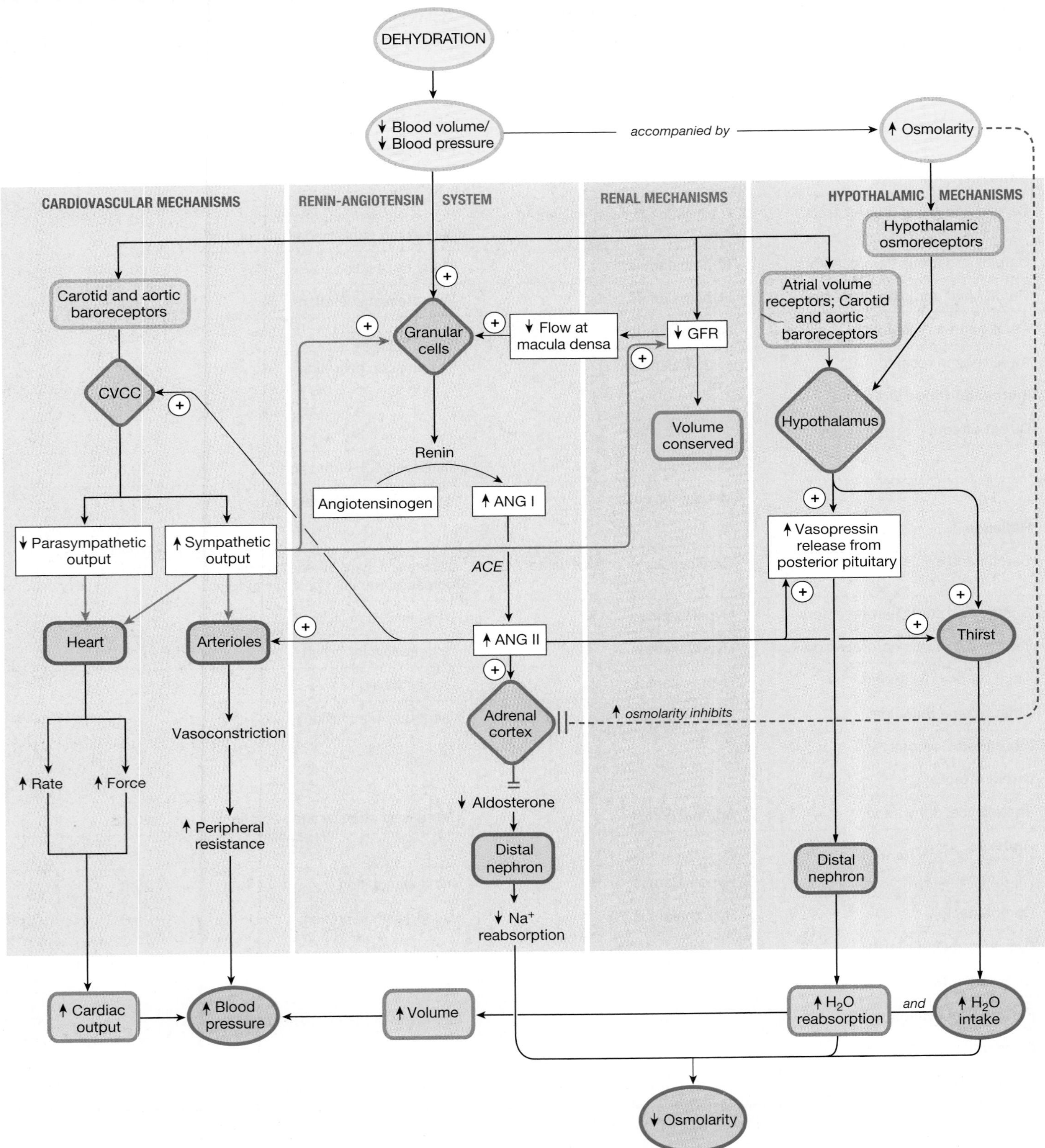
DEHYDRATION
↓ Blood volume/
↓ Blood pressure
accompanied by
↑ Osmolarity
CARDIOVASCULAR MECHANISMS
RENIN-ANGIOTENSIN SYSTEM
RENAL MECHANISMS
HYPOTHALAMIC MECHANISMS
Hypothalamic osmoreceptors
Carotid and aortic baroreceptors
Granular cells
↓ Flow at macula densa
↓ GFR
Atrial volume receptors; Carotid and aortic baroreceptors
CVCC
Volume conserved
Hypothalamus
Renin
Angiotensinogen
↑ ANG I
↓ Parasympathetic output
↑ Sympathetic output
ACE
↑ Vasopressin release from posterior pituitary
Heart
Arterioles
↑ ANG II
Thirst
Adrenal cortex
↑ osmolarity inhibits
Vasoconstriction
↑ Rate
↑ Force
↓ Aldosterone
↑ Peripheral resistance
Distal nephron
Distal nephron
↓ Na+ reabsorption
↑ Cardiac output
↑ Blood pressure
↑ Volume
↑ H2O reabsorption
and
↑ H2O intake
↓ Osmolarity

Kidneys Assist in Blood Pressure Homeostasis

The link between blood volume and blood pressure, as mapped in Figure 20.1, makes the kidneys key players in the body's work to maintain blood pressure homeostasis. As you learned previously, central control of blood pressure resides in medullary centers responding to signals from the carotid and aortic baroreceptors [Fig. 15.14, p. 493]. If blood pressure falls, increased sympathetic output to the vasculature causes vasoconstriction. In the kidney, sympathetic innervation vasocontricts both afferent and efferent arterioles, decreasing renal blood flow and conserving fluid. Sympathetic innervation to granular cells triggers renin release, which begins the cascade of events that include ANG II and aldosterone. And finally, sympathetic innervation of the proximal tubule decreases Na^+ reabsorption, apparently by modulating Na^+-linked transporters.

Many of the drugs prescribed for hypertension act on pathways linked to the kidney. *Diuretics* decrease ion reabsorption in the ascending limb of the loop of Henle ("loop diuretics") or in the distal convoluted tubule (thiazide diuretics), causing additional water excretion along with the unreabsorbed ions. Drugs that target the renin-angiotensin-aldosterone pathway include *ACE inhibitors* and *angiotensin II receptor blockers* (ARBs). Decreasing activity of the RAAS pathway removes the blood pressure–raising effects of ANG II in addition to promoting excretion of Na^+ and water.

Renal dysfunction can also *cause* problems with blood pressure. In a small percentage (<5%) of people with high blood pressure, the cause of hypertension is narrowing (*stenosis*) of the renal artery. Decreasing blood flow to the kidney triggers a series of compensatory events that result in elevated blood pressure. High blood pressure caused this way is called *renovascular hypertension*.

Concept Check

15. Map the pathway that begins with renal artery stenosis and ends with hypertension. (*Hint:* It involves the RAAS pathway.)

Endocrine Problems Disrupt Fluid Balance

Normally the body's hormones help maintain homeostasis, but they can also contribute to disruptions of homeostasis. Endocrine pathologies fall into three categories [p. 214]: too much hormone (hypersecretion), too little hormone (hyposecretion), and abnormal tissue responsiveness. The latter is a broad category that includes problems with receptors as well as disruptions in the normal pathways for the hormone's action.

Hypersecretion of aldosterone and pathologies of vasopressin secretion and action are the most common endocrine issues affecting fluid and electrolyte balance. Hypersecretion of aldosterone (*hyperaldosteronism*) may be primary, originating in the adrenal cortex, or secondary, from excess secretion of ANG II, as might occur in renovascular hypertension described in the previous section. Tumors that secrete aldosterone or ANG II are common causes of hyperaldosteronism. Disorders of aldosterone secretion are complicated by the interconnected biochemical pathways for steroid hormones in the adrenal cortex [p. 200] that link aldosterone production with cortisol and sex steroid synthesis.

Insufficient vasopressin activity in the nephron leads to inability of the kidney to reabsorb water and make concentrated urine. The large volumes of dilute urine produced in the absence of adequate vasopressin activity create a condition called **diabetes insipidus**. Diabetes refers to the excessive flow of urine, and *insipid* is "without taste," meaning the dilute nature of the urine. Diabetes insipidus (DI) can be caused by lack of vasopressin secretion (neurogenic DI) or by faculty vasopressin receptors in the kidney tubule (nephrogenic DI).

Vasopressin can also be over-secreted, a condition known as **SIADH**, which stands for ***s**yndrome of **i**nappropriate **a**nti **d**iuretic **h**ormone secretion.* (Vasopressin is also called ADH.) SIADH has multiple causes, including tumors that secrete the hormone, some lung diseases, and a variety of central nervous system disorders.

CHAPTER 20

20.7 Acid-Base Balance

Acid-base balance (also called pH homeostasis) is one of the essential functions of the body. The pH of a solution is a measure of its H^+ concentration [p. 41]. The H^+ concentration of normal arterial plasma sample is 0.00004 mEq/L, minute compared with the concentrations of other ions. (For example, the plasma concentration of Na^+ is about 135 mEq/L.)

Because the body's H^+ concentration is so low, it is commonly expressed on a logarithmic pH scale of 0–14, in which a pH of 7.0 is neutral (neither acidic nor basic). If the pH of a solution is below 7.0, the H^+ concentration is greater than 1×10^{-7} M and the solution is considered acidic. If the pH is above 7.0, the H^+ concentration is lower than 1×10^{-7} M and the solution is considered alkaline (basic).

The normal pH of the body is 7.40, slightly alkaline. A change of 1 pH unit represents a 10-fold change in H^+ concentration. [To review the concept of pH, see Fig. 2.9, p. 45. To review logarithms, see Appendix B.]

pH Changes Can Denature Proteins

The normal pH range of plasma is 7.38–7.42. Extracellular pH usually reflects intracellular pH, and vice versa. Because monitoring intracellular conditions is difficult, plasma values are used clinically as an indicator of ECF and whole body pH. Body fluids that are "outside" the body's internal environment, such as those in the lumen of the gastrointestinal tract or kidney tubule, can have a pH that far exceeds the normal range. Acidic secretions in the stomach, for instance, may create a gastric pH as low as 1. The pH of urine varies between 4.5 and 8.5, depending on the body's need to excrete H^+ or HCO_3^-.

The concentration of H^+ in the body is closely regulated. Intracellular proteins, such as enzymes and membrane channels, are particularly sensitive to pH because the function of these proteins depends on their three-dimensional shape. Changes in H^+ concentration alter the tertiary structure of proteins by interacting with hydrogen bonds in the molecules, disrupting the proteins' three-dimensional structures and activities [p. 51].

FIG. 20.14 pH balance in the body

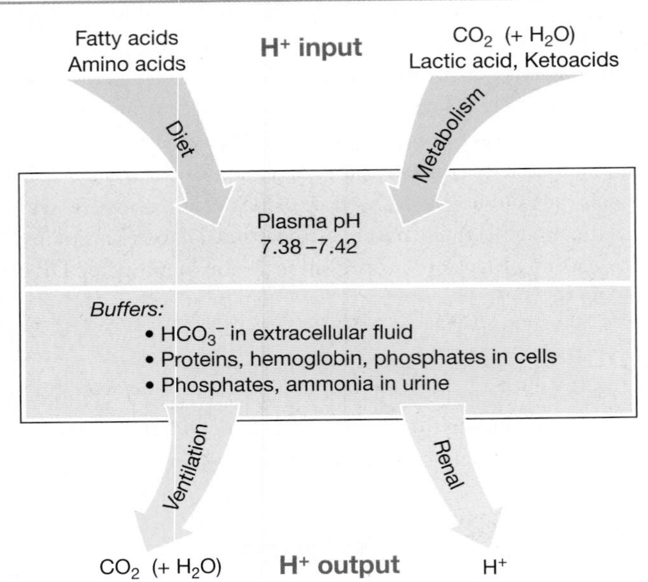
Fatty acids
Amino acids
H+ input
CO2 (+ H2O)
Lactic acid, Ketoacids
Diet
Metabolism
Plasma pH
7.38–7.42
Buffers:
• HCO3− in extracellular fluid
• Proteins, hemoglobin, phosphates in cells
• Phosphates, ammonia in urine
Ventilation
Renal
CO2 (+ H2O)
H+ output
H+

Abnormal pH may significantly affect the activity of the nervous system. If pH is too low—the condition known as **acidosis**—neurons become less excitable, and CNS depression results. Patients become confused and disoriented, then slip into a coma. If CNS depression progresses, the respiratory centers cease to function, causing death.

If pH is too high—the condition known as **alkalosis**—neurons become hyperexcitable, firing action potentials at the slightest signal. This condition shows up first as sensory changes, such as numbness or tingling, then as muscle twitches. If alkalosis is severe, muscle twitches turn into sustained contractions (*tetanus*) that paralyze respiratory muscles.

Disturbances of acid-base balance are associated with disturbances in K^+ balance. This is partly due to a renal transporter that moves K^+ and H^+ ions in an antiport fashion. In acidosis, the kidneys excrete H^+ and reabsorb K^+ using an *H^+-K^+-ATPase*. In alkalosis, the kidneys reabsorb H^+ and excrete K^+. Potassium imbalance usually shows up as disturbances in excitable tissues, especially the heart.

Acids and Bases in the Body Come from Many Sources

In day-to-day functioning, the body is challenged by intake and production of acids more than bases. Hydrogen ions come from both food and internal metabolism. Maintaining mass balance requires that acid intake and production be balanced by acid excretion. Hydrogen balance in the body is summarized in **FIGURE 20.14**.

Acid Input Many metabolic intermediates and foods are organic acids that ionize and contribute H^+ to body fluids.* Examples of organic acids include amino acids, fatty acids, intermediates in the citric acid cycle, and lactate produced by anaerobic metabolism. Metabolic production of organic acids each day generates a significant amount of H^+ that must be excreted to maintain mass balance.

Under extraordinary circumstances, metabolic organic acid production can increase significantly and create a crisis. For example, severe anaerobic conditions, such as circulatory collapse, produce so much lactate that normal homeostatic mechanisms cannot keep pace, resulting in a state of *lactic acidosis*. In diabetes mellitus, abnormal metabolism of fats and amino acids creates strong acids known as **ketoacids.** These acids cause a state of metabolic acidosis known as *ketoacidosis*.

The biggest source of acid on a daily basis is the production of CO_2 during aerobic respiration. Carbon dioxide is not an acid because it does not contain any hydrogen atoms. However, CO_2 from respiration combines with water to form H^+ and bicarbonate ion, HCO_3^-.

$$CO_2 + H_2O \rightleftharpoons H^+ + HCO_3^-$$

This reaction takes place in all cells and in the plasma, but at a slow rate. However, in certain cells of the body, the reaction proceeds very rapidly because of the presence of large amounts of *carbonic anhydrase* [p. 575]. This enzyme catalyzes the conversion of CO_2 and H_2O to H^+ and HCO_3^-.

The production of H^+ from CO_2 and H_2O is the single biggest source of acid input under normal conditions. By some estimates, CO_2 from resting metabolism produces 12,500 mEq of H^+ each day. If this amount of acid were placed in a volume of water equal to the plasma volume, it would create an H^+ concentration of 4167 mEq/L, over one hundred million (10^8) times as concentrated as the normal plasma H^+ concentration of 0.00004 mEq/L!

These numbers show that CO_2 from aerobic respiration has the potential to affect pH in the body dramatically. Fortunately, homeostatic mechanisms normally prevent CO_2 from accumulating in the body.

Base Input Acid-base physiology focuses on acids for good reasons. First, our diet and metabolism have few significant sources of bases. Some fruits and vegetables contain anions that metabolize to HCO_3^-, but the influence of these foods is far outweighed by the contribution of acidic fruits, amino acids, and fatty acids. Second, acid-base disturbances due to excess acid are more common than those due to excess base. For these reasons, the body uses far more resources removing excess acid.

pH Homeostasis Depends on Buffers, Lungs, and Kidneys

How does the body cope with minute-to-minute changes in pH? There are three mechanisms: (1) buffers, (2) ventilation, and (3) renal regulation of H^+ and HCO_3^-. Buffers are the first line of defense, always present and waiting to prevent wide swings in pH. Ventilation, the second line of defense, is a rapid, reflexively controlled response that can take care of 75% of most pH disturbances. The final line of defense lies with the kidneys. They are slower than buffers or the lungs but are very effective at coping with any remaining pH disturbance under normal conditions. Usually these three

*The anion forms of many organic acids end with the suffix–ate, such as pyruvate and lactate.

mechanisms help the body balance acid so effectively that normal body pH varies only slightly. Let's take a closer look at each of them.

Buffer Systems Include Proteins, Phosphate Ions, and HCO_3^-

A buffer is a molecule that moderates but does not prevent changes in pH by combining with or releasing H^+ [p. 41]. In the absence of buffers, the addition of acid to a solution causes a sharp change in pH. In the presence of a buffer, the pH change is moderated or may even be unnoticeable. Because acid production is the major challenge to pH homeostasis, most physiological buffers combine with H^+.

Buffers are found both within cells and in the plasma. Intracellular buffers include cellular proteins, phosphate ions (HPO_4^{2-}), and hemoglobin. Hemoglobin in red blood cells buffers the H^+ produced by the reaction of CO_2 with H_2O [Fig. 18.11, p. 576].

Each H^+ ion buffered by hemoglobin leaves a matching bicarbonate ion inside the red blood cell. This HCO_3^- can then leave the red blood cell in exchange for plasma Cl^-, the *chloride shift* [p. 576].

The large amounts of plasma bicarbonate produced from metabolic CO_2 create the most important extracellular buffer system of the body. Plasma HCO_3^- concentration averages 24 mEq/L, which is approximately 600,000 times as concentrated as plasma H^+. Although H^+ and HCO_3^- are created in a 1:1 ratio from CO_2 and H_2O, intracellular buffering of H^+ by hemoglobin is a major reason the two ions do not appear in the plasma in the same concentration. The HCO_3^- in plasma is then available to buffer H^+ from nonrespiratory sources, such as metabolism.

The relationship between CO_2, HCO_3^-, and H^+ in the plasma is expressed by the equation we just looked at:

$$CO_2 + H_2O \rightleftharpoons H^+ + HCO_3^- \quad (1)$$

According to the law of mass action, any change in the amount of CO_2, H^+, or HCO_3^- in the reaction solution causes the reaction to shift until a new equilibrium is reached. (Water is always in excess in the body and does not contribute to the reaction equilibrium.) For example, if CO_2 increases (red), the equation shifts to the right, creating one additional H^+ and one additional HCO_3^- from each CO_2 and water:

$$\uparrow CO_2 + H_2O \rightleftharpoons \uparrow H^+ + \uparrow HCO_3^- \quad (2)$$

Once a new equilibrium is reached, both H^+ and HCO_3^- levels have increased. The addition of H^+ makes the solution more acidic and therefore lowers its pH.

Note that in reaction (2), it does not matter that a bicarbonate molecule has also been produced. HCO_3^- acts as a buffer only when it binds to H^+ and becomes carbonic acid. When the reaction is at equilibrium, as shown here, HCO_3^- will not combine with H^+.

Now suppose H^+ is added to the plasma from some metabolic source, such as lactic acid:

$$\text{organic acid} \rightleftharpoons \text{organic anion}^- + H^+ \quad (3)$$

Adding H^+ (red) disturbs the equilibrium state of the CO_2-HCO_3^--H^+ reaction. By the law of mass action [p. 48], adding a molecule to the right side of the equilibrium will send the equation to the left:

$$CO_2 + H_2O \leftarrow \uparrow\uparrow H^+ + HCO_3^- \quad (4)$$

Now plasma HCO_3^- can act as a buffer and combine with some of the added H^+. The reaction shifts to the left, converting some of the added H^+ and bicarbonate buffer to CO_2 and H_2O. When the equation comes back to equilibrium, H^+ is still elevated, but not as much as it was initially. The concentration of HCO_3^- is decreased because some bicarbonate has been used as a buffer. CO_2 and H_2O have both increased. At equilibrium, the reaction looks like this:

$$\uparrow CO_2 + \uparrow H_2O \rightleftharpoons \uparrow H^+ + \downarrow HCO_3^- \quad (5)$$

The law of mass action is a useful way to think about the relationship between changes in the concentrations of H^+, HCO_3^-, and CO_2 as long as you remember certain caveats. First, a change in HCO_3^- concentration (as indicated in reaction 2) may not show up clinically as a HCO_3^- concentration outside the normal range. This is because HCO_3^- is 600,000 times more concentrated in the plasma than H^+ is. If both H^+ and HCO_3^- are added to the plasma, you may observe changes in pH but not in HCO_3^- concentration because so much bicarbonate was present initially. Both H^+ and HCO_3^- experience an *absolute* increase in concentration, but because so many HCO_3^- were in the plasma to begin with, the *relative increase* in HCO_3^- goes unnoticed.

As an analogy, think of two football teams playing in a stadium packed with 80,000 fans. If 10 more players (H^+) run out onto the field, everyone notices. But if 10 people (HCO_3^-) come into the stands at the same time, no one pays any attention because there were already so many people watching the game that 10 more make no significant difference.

The relationship between pH, HCO_3^- concentration in mM, and dissolved CO_2 concentration is expressed mathematically by the **Henderson-Hasselbalch equation.** One variant of the

RUNNING PROBLEM

The human body attempts to maintain fluid and sodium balance via several hormonal mechanisms. During exercise sessions, increased sympathetic output causes increased production of aldosterone and vasopressin, which promote the retention of Na^+ and water by the kidneys.

Q8: *What would you expect to happen to vasopressin and aldosterone production in response to dilutional hyponatremia?*

617 – 619 – 628 – 634 – **639** – 648

equation that is more useful in clinical medicine uses P_{CO_2} instead of dissolved CO_2 concentration:

$$pH = 6.1 + \log \frac{[HCO_3^-]}{0.03 \times P_{CO_2}}$$

With this equation, if you know a patient's P_{CO_2} and plasma bicarbonate concentration, you can predict the plasma pH.

The second qualification for the law of mass action is that when the reaction shifts to the left and increases plasma CO_2, a nearly instantaneous increase in ventilation takes place in a normal person. If extra CO_2 is ventilated off, arterial P_{CO_2} may remain normal or even fall below normal as a result of hyperventilation.

Ventilation Can Compensate for pH Disturbances

The increase in ventilation just described is a *respiratory compensation* for acidosis. Ventilation and acid-base status are intimately linked, as shown by the equation

$$CO_2 + H_2O \rightleftharpoons H^+ + HCO_3^-$$

Changes in ventilation can correct disturbances in acid-base balance, but they can also cause them. Because of the dynamic equilibrium between CO_2 and H^+, any change in plasma P_{CO_2} affects both H^+ and HCO_3^- content of the blood.

Hypoventilation For example, if a person hypoventilates and P_{CO_2} increases (red), the equation shifts to the right. More carbonic acid is formed, and H^+ goes up, creating a more *acidotic* state:

$$\uparrow CO_2 + H_2O \rightleftharpoons \uparrow H^+ + \uparrow HCO_3^- \quad (6)$$

Hyperventilation On the other hand, if a person hyperventilates, blowing off CO_2 and thereby decreasing the plasma P_{CO_2} (red), the equation shifts to the left, which means that H^+ combines with HCO_3^- and becomes $CO_2 + H_2O$, thereby decreasing the H^+ concentration. Lower H^+ means an increase in pH:

$$\downarrow CO_2 + H_2O \rightleftharpoons \downarrow H^+ + \downarrow HCO_3^- \quad (7)$$

In these two examples, you can see that a change in P_{CO_2} affects the H^+ concentration and therefore the pH of the plasma.

Ventilation Reflexes The body uses ventilation as a homeostatic method for adjusting pH only if a stimulus associated with pH triggers the reflex response. Two stimuli can do so: H^+ and CO_2.

Ventilation is affected directly by plasma H^+ levels primarily through carotid body chemoreceptors (**FIG. 20.15**). These

FIG. 20.15 Respiratory compensation for metabolic acidosis

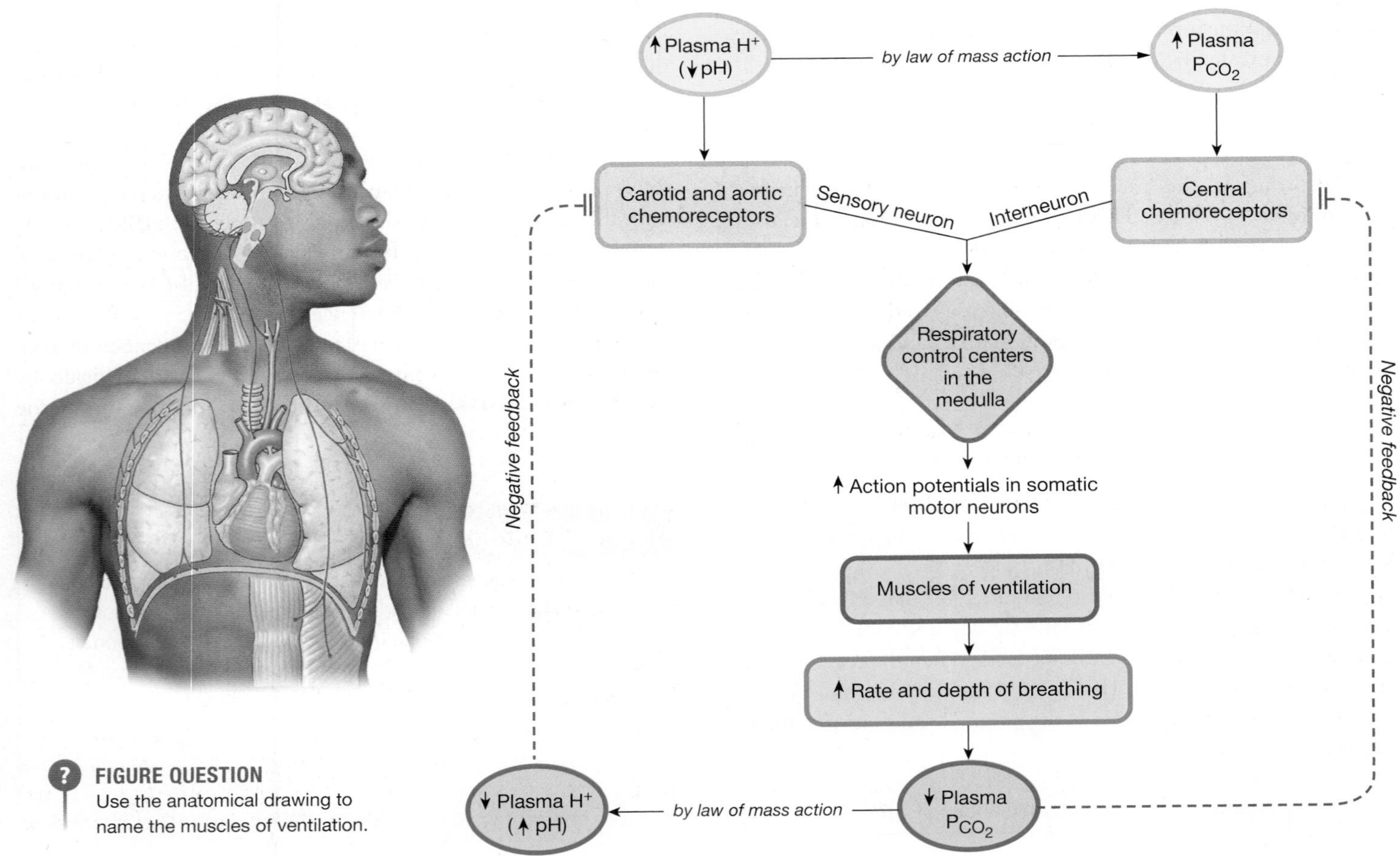
↑ Plasma H+ (↓pH)
by law of mass action
↑ Plasma PCO2
Carotid and aortic chemoreceptors
Sensory neuron
Interneuron
Central chemoreceptors
Respiratory control centers in the medulla
↑ Action potentials in somatic motor neurons
Muscles of ventilation
↑ Rate and depth of breathing
↓ Plasma H+ (↑ pH)
by law of mass action
↓ Plasma PCO2
Negative feedback
Negative feedback
FIGURE QUESTION
Use the anatomical drawing to name the muscles of ventilation.

chemoreceptors are located in the carotid arteries along with oxygen sensors and blood pressure sensors [p. 580]. An increase in plasma H^+ stimulates the chemoreceptors, which in turn signal the medullary respiratory control centers to increase ventilation. Increased alveolar ventilation allows the lungs to excrete more CO_2 and convert H^+ to $CO_2 + H_2O$.

The central chemoreceptors of the medulla oblongata cannot respond directly to changes in plasma pH because H^+ does not cross the blood-brain barrier. However, changes in pH change P_{CO_2}, and CO_2 stimulates the central chemoreceptors [Fig. 18.17, p. 582]. Dual control of ventilation through the central and peripheral chemoreceptors helps the body respond rapidly to changes in either pH or plasma CO_2.

Concept Check

16. In equation 6, the amount of HCO_3^- present is increased at equilibrium. Why doesn't this HCO_3^- act as a buffer and prevent acidosis from occurring?

Kidneys Use Ammonia and Phosphate Buffers

The kidneys take care of the 25% of compensation that the lungs cannot handle. They alter pH two ways: (1) directly, by excreting or reabsorbing H^+ and (2) indirectly, by changing the rate at which HCO_3^- buffer is reabsorbed or excreted.

In acidosis, the kidney secretes H^+ into the tubule lumen using direct and indirect active transport (**FIG. 20.16**). Ammonia from amino acids and phosphate ions (HPO_4^{2-}) in the kidney act as buffers, trapping large amounts of H^+ as NH_4^+ and $H_2PO_4^-$. These buffers allow more H^+ to be excreted. Phosphate ions are present in filtrate and combine with H^+ secreted into the nephron lumen:

$$HPO_4^{2-} + H^+ \rightleftharpoons H_2PO_4^-$$

Even with these buffers, urine can become quite acidic, down to a pH of about 4.5. While H^+ is being excreted, the kidneys make new HCO_3^- from CO_2 and H_2O. The HCO_3^- is reabsorbed into the blood to act as a buffer and increase pH.

In alkalosis, the kidney reverses the general process just described for acidosis, excreting HCO_3^- and reabsorbing H^+ in an effort to bring pH back into the normal range. Renal compensations are slower than respiratory compensations, and their effect on pH may not be noticed for 24–48 hours. However, once activated, renal compensations effectively handle all but severe acid-base disturbances.

The cellular mechanisms for renal handling of H^+ and HCO_3^- resemble transport processes in other epithelia. However, these mechanisms involve some membrane transporters that you have not encountered before:

1. The apical *Na^+-H^+ exchanger* (NHE) is an indirect active transporter that brings Na^+ into the epithelial cell in exchange for moving H^+ against its concentration gradient into the lumen. This transporter also plays a role in proximal tubule Na^+ reabsorption.

FIG. 20.16 Overview of renal compensation for acidosis

The kidney secretes H^+, which is buffered in the urine by ammonia and phosphate ions. It reabsorbs bicarbonate to act as an extracellular buffer.

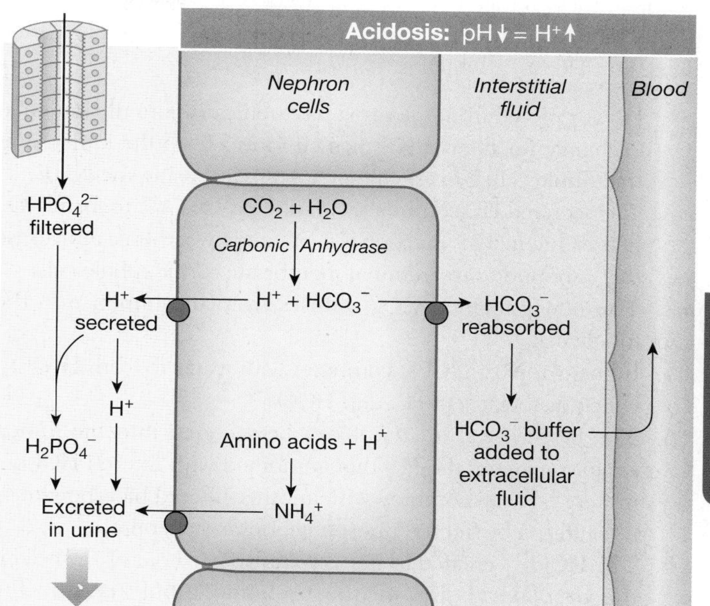
Acidosis: pH↓ = H+↑
Nephron cells
Interstitial fluid
Blood
HPO4 2− filtered
CO2 + H2O
Carbonic Anhydrase
H+ secreted
H+ + HCO3−
HCO3− reabsorbed
H+
H2PO4−
Amino acids + H+
HCO3− buffer added to extracellular fluid
Excreted in urine
NH4+

The transporters shown here are generic membrane proteins. For specific transporters involved, see Figure 20.17.

2. The basolateral **Na^+-HCO_3^- symporter** moves Na^+ and HCO_3^- out of the epithelial cell and into the interstitial fluid. This indirect active transporter couples the energy of HCO_3^- diffusing down its concentration gradient to the uphill movement of Na^+ from the cell to the ECF.
3. The **H^+-ATPase** uses energy from ATP to acidify the urine, pushing H^+ against its concentration gradient into the lumen of the distal nephron. The H^+-ATPase is also called the *proton pump*.
4. The **H^+-K^+-ATPase** puts H^+ into the urine in exchange for reabsorbed K^+. This exchange contributes to the potassium imbalance that sometimes accompanies acid-base disturbances.
5. An **Na^+-NH_4^+ antiporter** moves NH_4^+ from the cell to the lumen in exchange for Na^+.

In addition to these transporters, the renal tubule also uses the ubiquitous Na^+-K^+-ATPase and the same HCO_3^--Cl^- antiport protein that is responsible for the chloride shift in red blood cells.

The Proximal Tubule Secretes H^+ and Reabsorbs HCO_3^-

The amount of bicarbonate ion the kidneys filter each day is equivalent to the bicarbonate in a pound of baking soda ($NaHCO_3$)! Most of this HCO_3^- must be reabsorbed to maintain the body's buffer capacity. The proximal tubule reabsorbs most filtered HCO_3^- by indirect methods because there is no apical membrane transporter to bring HCO_3^- into the tubule cell.

FIGURE 20.17 shows the two pathways by which bicarbonate is reabsorbed in the proximal tubule. (The numbers in the following lists correspond to the steps shown in the figure.) By following this illustration, you will see how the transporters listed in the previous section function together.

The first pathway converts filtered HCO_3^- into CO_2, then back into HCO_3^-, which is reabsorbed:

1. H^+ is secreted from the proximal tubule cell into the lumen in exchange for filtered Na^+, which moves from the lumen into the tubule cell. This exchange takes place using the NHE.
2. The secreted H^+ combines with filtered HCO_3^- to form CO_2 in the lumen. This reaction is facilitated by carbonic anhydrase that is bound to the luminal membrane of the tubule cells.
3. The newly formed CO_2 diffuses from the lumen into the tubule cell.
4. In the cytoplasm, CO_2 combines with water to form H_3CO_3, which dissociates to H^+ and HCO_3^-.
5. The H^+ created in step 4 can be secreted into the lumen again, replacing the H^+ that combined with filtered HCO_3^- in step 2. It can combine with another filtered bicarbonate or be buffered by filtered phosphate ion and excreted.
6. The HCO_3^- created in step 3 is transported out of the cell on the basolateral side of the proximal tubule cell by the HCO_3^--Na^+ symporter.

The net result of this process is reabsorption of filtered Na^+ and HCO_3^-, and secretion of H^+.

A second way to reabsorb bicarbonate and excrete H^+ comes from metabolism of the amino acid glutamine:

7. Glutamine in the proximal tubule cell is metabolized to α-ketoglutarate (α-KG) and two amino groups ($-NH_2$). The amino groups become ammonia (NH_3), and the ammonia buffers H^+ to become ammonium ion NH_4^+. The ammonium ion NH_4^+ is transported into the lumen in exchange for Na^+. The α-ketoglutarate molecule is metabolized further to HCO_3^-, which is transported into the blood along with Na^+.

The net result of both pathways shown in Figure 20.17 is secretion of acid (H^+) and reabsorption of buffer in the form of sodium bicarbonate—baking soda, $NaHCO_3$.

The Distal Nephron Controls Acid Excretion

The distal nephron plays a significant role in the fine regulation of acid-base balance. Specialized cells called **intercalated cells (I cells)** interspersed among the principal cells are primarily responsible for acid-base regulation.

Intercalated cells are characterized by high concentrations of carbonic anhydrase in their cytoplasm. This enzyme allows them to rapidly convert CO_2 and water into H^+ and HCO_3^-. The H^+

FIG. 20.17 The proximal tubule reabsorbs filtered bicarbonate

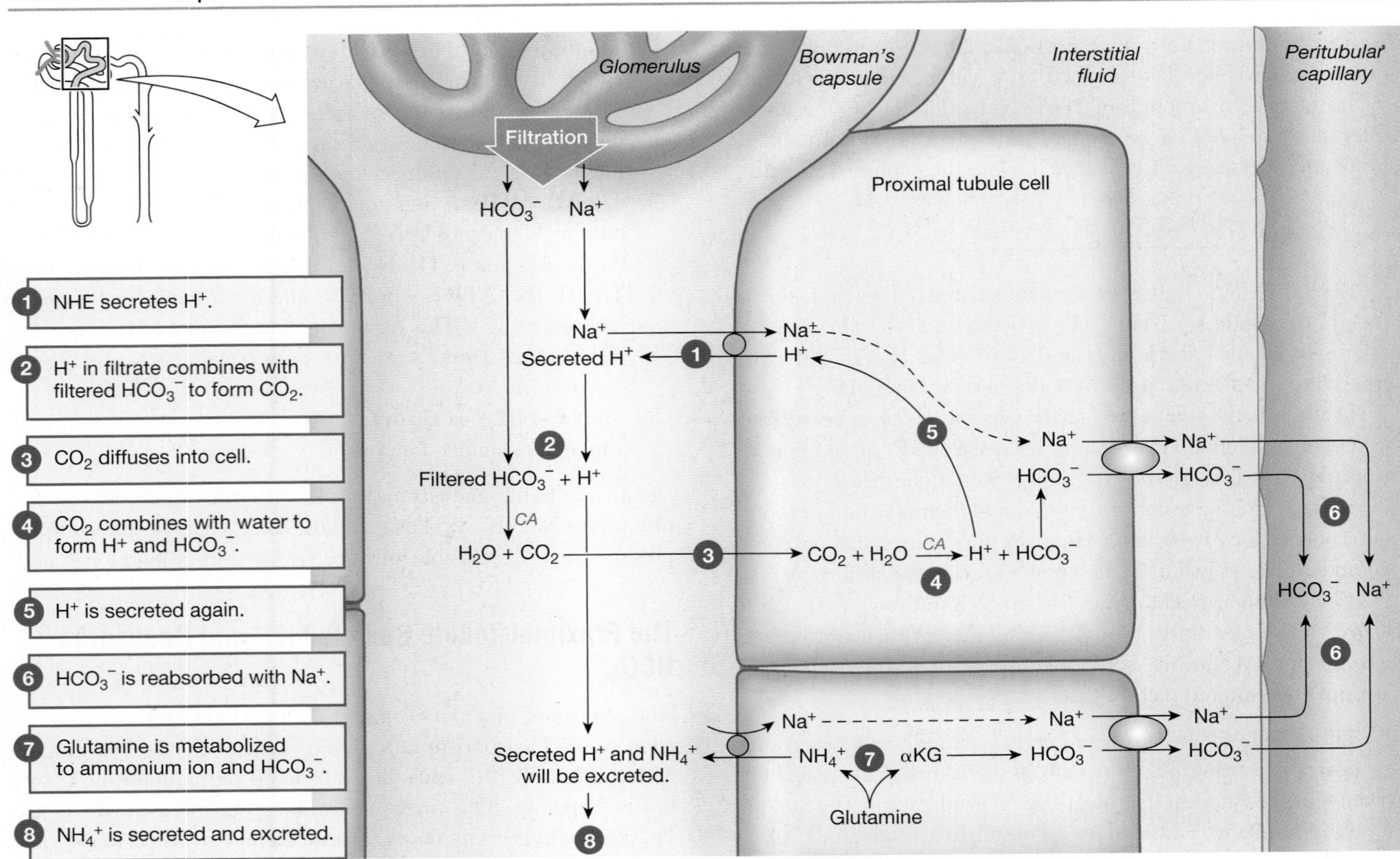
Glomerulus
Bowman's capsule
Interstitial fluid
Peritubular capillary
Filtration
Proximal tubule cell
HCO3⁻ Na+
1 NHE secretes H+.
2 H+ in filtrate combines with filtered HCO3⁻ to form CO2.
3 CO2 diffuses into cell.
4 CO2 combines with water to form H+ and HCO3⁻.
5 H+ is secreted again.
6 HCO3⁻ is reabsorbed with Na+.
7 Glutamine is metabolized to ammonium ion and HCO3⁻.
8 NH4+ is secreted and excreted.
Na+
Secreted H+
Filtered HCO3⁻ + H+
CA
H2O + CO2
CO2 + H2O → H+ + HCO3⁻
HCO3⁻ Na+
NH4+ αKG
Glutamine
Secreted H+ and NH4+ will be excreted.

ions are pumped out of the intercalated cell either by the H^+-ATPase or by the H^+-K^+-ATPase. Bicarbonate leaves the cell by means of the HCO_3^--Cl^- antiport exchanger.

There are two types of intercalated cells, and their transporters are found on different faces of the epithelial cell. During periods of acidosis, type A intercalated cells secrete H^+ and reabsorb bicarbonate. During periods of alkalosis, type B intercalated cells secrete HCO_3^- and reabsorb H^+.

FIGURE 20.18a shows how type A intercalated cells work during acidosis, secreting H^+ and reabsorbing HCO_3^-. The process is similar to H^+ secretion in the proximal tubule except for the specific H^+ transporters. The distal nephron uses apical H^+-ATPase and H^+-K^+-ATPase rather than the Na^+-H^+ antiport protein found in the proximal tubule.

During alkalosis, when the H^+ concentration of the body is too low, H^+ is reabsorbed and HCO_3^- buffer is excreted in the urine (Fig. 20.18b). Once again, the ions are formed from H_2O and CO_2. Hydrogen ions are reabsorbed by transport into the ECF on the basolateral side of the cell, and HCO_3^- is secreted into the lumen. The polarity of the two types of I cells is reversed, with the same transport processes taking place, but on the opposite sides of the cell.

The H^+-K^+-ATPase of the distal nephron helps create parallel disturbances of acid-base balance and K^+ balance. In acidosis, when plasma H^+ is high, the kidney secretes H^+ and reabsorbs K^+. For this reason, acidosis is often accompanied by hyperkalemia. (Other non-renal events also contribute to elevated ECF K^+ concentrations in acidosis.) The reverse is true for alkalosis, when blood H^+ levels are low. The mechanism that allows the distal nephron to reabsorb H^+ simultaneously causes it to secrete K^+, with the result that alkalosis goes hand in hand with hypokalemia.

Concept Check

17. Why is ATP required for H^+ secretion by the H^+-K^+ transporter but not for the Na^+-H^+ exchanger?
18. In hypokalemia, the intercalated cells of the distal nephron reabsorb K^+ from the tubule lumen. What happens to blood pH as a result?

Acid-Base Disturbances May Be Respiratory or Metabolic

The three compensatory mechanisms (buffers, ventilation, and renal excretion) take care of most variations in plasma pH. But under some circumstances, the production or loss of H^+ or HCO_3^- is so extreme that compensatory mechanisms fail to maintain pH homeostasis. In these states, the pH of the blood moves out of the normal range of 7.38–7.42. If the body fails to keep pH between 7.00 and 7.70, acidosis or alkalosis can be fatal (**FIG. 20.19**).

Acid-base problems are classified both by the direction of the pH change (acidosis or alkalosis) and by the underlying cause

FIG. 20.18 Intercalated cells function in acid-base disturbances

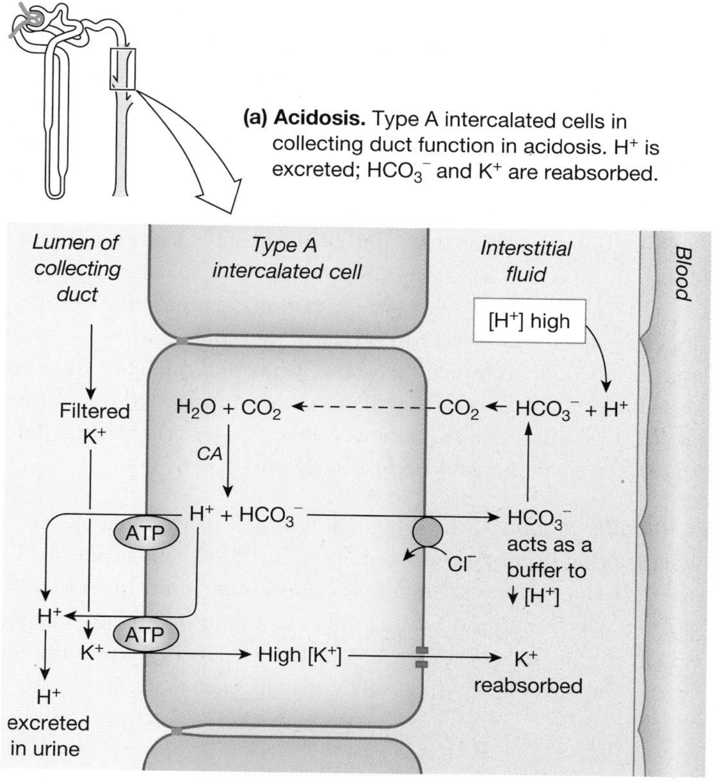

(a) Acidosis. Type A intercalated cells in collecting duct function in acidosis. H⁺ is excreted; HCO₃⁻ and K⁺ are reabsorbed.
Lumen of collecting duct
Type A intercalated cell
Interstitial fluid
Blood
[H⁺] high
Filtered K⁺
H₂O + CO₂
CO₂
HCO₃⁻ + H⁺
CA
H⁺ + HCO₃⁻
ATP
HCO₃⁻ acts as a buffer to [H⁺]
Cl⁻
H⁺
ATP
K⁺
High [K⁺]
K⁺ reabsorbed
H⁺ excreted in urine

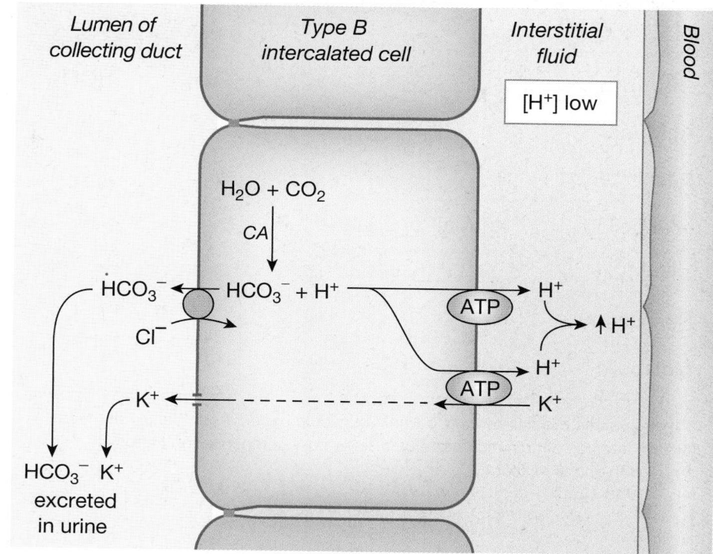

(b) Alkalosis. Type B intercalated cells in collecting duct function in alkalosis. HCO₃⁻ and K⁺ are excreted; H⁺ is reabsorbed.
Lumen of collecting duct
Type B intercalated cell
Interstitial fluid
Blood
[H⁺] low
H₂O + CO₂
CA
HCO₃⁻
HCO₃⁻ + H⁺
ATP
H⁺
H⁺
Cl⁻
H⁺
ATP
K⁺
K⁺
HCO₃⁻ K⁺ excreted in urine

FIG. 20.19 Acid-base disturbances may be incompletely compensated

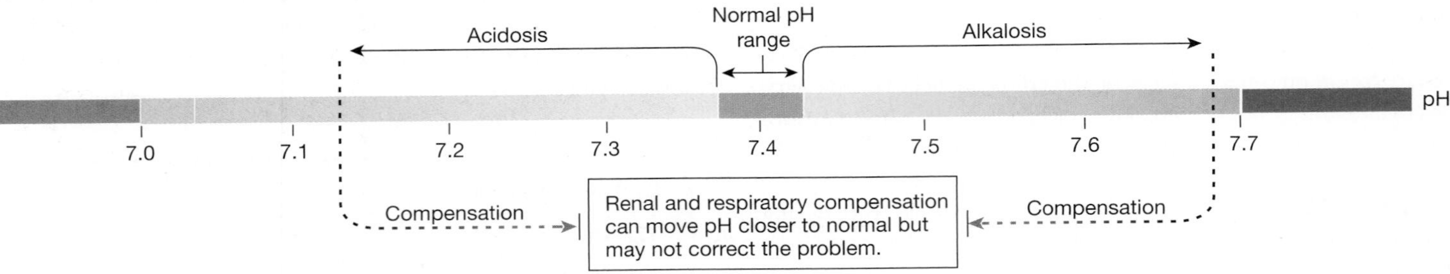
Acidosis
Normal pH range
Alkalosis
pH
7.0
7.1
7.2
7.3
7.4
7.5
7.6
7.7
Compensation
Renal and respiratory compensation can move pH closer to normal but may not correct the problem.
Compensation

(metabolic or respiratory). Changes in P_{CO_2} resulting from hyperventilation or hypoventilation cause pH to shift. These disturbances are said to be of respiratory origin. If the pH problem arises from acids or bases of non-CO_2 origin, the problem is said to be a metabolic problem.

Note that by the time an acid-base disturbance becomes evident as a change in plasma pH, the body's buffers are ineffectual. The loss of buffering ability leaves the body with only two options: respiratory compensation or renal compensation. And if the problem is of respiratory origin, only one homeostatic compensation is available—the kidneys. If the problem is of metabolic origin, both respiratory and renal mechanisms can compensate. Compensation can bring the pH back closer to normal but may not correct the disturbance completely (Fig. 20.19).

The combination of an initial pH disturbance and the resultant compensatory changes is one factor that makes analysis of acid-base disorders in the clinical setting so difficult. In this book, we concentrate on simple scenarios with a single underlying cause. Changes that occur in the four simple acid-base disturbances are listed in **TABLE 20.2**.

TABLE 20.2 Plasma P_{CO_2}, Ions, and pH in Acid-Base Disturbances

Disturbance	P_{CO_2}	H^+	pH	HCO_3^-
Acidosis				
Respiratory	↑	↑	↓	↑
Metabolic	Normal* or ↓	↑	↓	↓
Alkalosis				
Respiratory	↓	↓	↑	↓
Metabolic	Normal* or ↑	↓	↑	↑

*These values are different from what you would expect from the law of mass action because almost instantaneous respiratory compensation keeps P_{CO_2} from changing significantly.

Respiratory Acidosis A state of respiratory acidosis occurs when alveolar hypoventilation results in CO_2 retention and elevated plasma P_{CO_2}. Some situations in which this occurs are respiratory depression due to drugs (including alcohol), increased airway resistance in asthma, impaired gas exchange in fibrosis or severe pneumonia, and muscle weakness in muscular dystrophy and other muscle diseases. The most common cause of respiratory acidosis is *chronic obstructive pulmonary disease* (COPD), which includes emphysema. In emphysema, inadequate alveolar ventilation is compounded by loss of alveolar exchange area.

No matter what the cause of respiratory acidosis, plasma CO_2 levels increase (red), leading to elevated H^+ and HCO_3^-:

$$\uparrow CO_2 + H_2O \rightarrow \uparrow H^+ + \uparrow HCO_3^- \quad (8)$$

The hallmark of respiratory acidosis is decreased pH with elevated bicarbonate levels (Tbl. 20.2). Because the problem is of respiratory origin, the body cannot carry out respiratory compensation. (However, depending on the problem, mechanical ventilation can sometimes be used to assist breathing.)

Any compensation for respiratory acidosis must occur through renal mechanisms that excrete H^+ and reabsorb HCO_3^-. The excretion of H^+ raises plasma pH. Reabsorption of HCO_3^- provides additional buffer that combines with H^+, lowering the H^+ concentration and therefore raising the pH.

In chronic obstructive pulmonary disease, renal compensation mechanisms for acidosis can moderate the pH change, but they may not be able to return the pH to its normal range. If you look at pH and HCO_3^- levels in patients with compensated respiratory acidosis, you find that both those values are closer to normal than they were when the acidosis was at its worst.

Metabolic Acidosis Metabolic acidosis is a disturbance of mass balance that occurs when the dietary and metabolic input of H^+ exceeds H^+ excretion. Metabolic causes of acidosis include *lactic acidosis*, which is a result of anaerobic metabolism, and *ketoacidosis*, which results from excessive breakdown of fats or certain amino acids. The metabolic pathway that produces ketoacids is associated with type 1 diabetes mellitus and with low-carbohydrate diets, like the Atkins diet [Chapter 22]. Ingested substances that cause metabolic acidosis include methanol, aspirin, and ethylene glycol (antifreeze).

Metabolic acidosis is expressed by the equation

$$\uparrow CO_2 + H_2O \leftarrow \uparrow H^+ + \downarrow HCO_3^- \quad (9)$$

Hydrogen ion concentration increases (red) because of the H^+ contributed by organic acids. This increase shifts the equilibrium represented in the equation to the left, increasing CO_2 levels and using up HCO_3^- buffer.

Metabolic acidosis can also occur if the body loses HCO_3^-. The most common cause of bicarbonate loss is diarrhea, during which HCO_3^- is lost from the intestines. The pancreas produces HCO_3^- from CO_2 and H_2O by a mechanism similar to the renal mechanism illustrated in Figure 20.16. The H^+ made at the same time is released into the blood. Normally, the HCO_3^- is released into the small intestine, then reabsorbed into the blood, buffering the H^+. However, if a person is experiencing diarrhea, HCO_3^- is not reabsorbed, and a state of acidosis may result.

Whether HCO_3^- concentration is elevated or decreased is an important criterion for distinguishing metabolic acidosis from respiratory acidosis (Tbl. 20.2).

You would think from looking at equation 9 that metabolic acidosis would be accompanied by elevated P_{CO_2}. However, unless the individual also has a lung disease, respiratory compensation takes place almost instantaneously. Both elevated CO_2 and elevated H^+ stimulate ventilation through the pathways described earlier. As a result, P_{CO_2} decreases to normal or even below-normal levels due to hyperventilation.

Uncompensated metabolic acidosis is rarely seen clinically. Actually, a common sign of metabolic acidosis is hyperventilation, evidence of respiratory compensation occurring in response to the acidosis.

The renal compensations discussed for respiratory acidosis also take place in metabolic acidosis: secretion of H^+ and reabsorption of HCO_3^-. Renal compensations take several days to reach full effectiveness, and so they are not usually seen in recent-onset (acute) disturbances.

Respiratory Alkalosis States of alkalosis are much less common than acidotic conditions. Respiratory alkalosis occurs as a result of hyperventilation, when alveolar ventilation increases without a matching increase in metabolic CO_2 production. Consequently, plasma P_{CO_2} falls (red), and alkalosis results when the equation shifts to the left:

$$\downarrow CO_2 + H_2O \leftarrow \downarrow H^+ + \downarrow HCO_3^- \quad (10)$$

The decrease in CO_2 shifts the equilibrium to the left, and both plasma H^+ and plasma HCO_3^- decrease. Low plasma HCO_3^- levels in alkalosis indicate a respiratory disorder.

The primary clinical cause of respiratory alkalosis is excessive artificial ventilation. Fortunately, this condition is easily corrected by adjusting the ventilator. The most common physiological cause of respiratory alkalosis is hysterical hyperventilation caused by anxiety. When this is the cause, the neurological symptoms caused by alkalosis can be partially reversed by having the patient breathe into a paper bag. In doing so, the patient rebreathes exhaled CO_2, a process that raises arterial P_{CO_2} and corrects the problem.

Because this alkalosis has respiratory cause, the only compensation available to the body is renal. Filtered bicarbonate is not reabsorbed in the proximal tubule and is secreted in the distal nephron. The combination of HCO_3^- excretion and H^+ reabsorption in the distal nephron decreases the body's HCO_3^- buffer load and increases its H^+, both of which help correct the alkalosis.

Metabolic Alkalosis Metabolic alkalosis has two common causes: excessive vomiting of acidic stomach contents and excessive ingestion of bicarbonate-containing antacids. In both cases, the resulting alkalosis reduces H^+ concentration (red):

$$\downarrow CO_2 + H_2O \rightarrow \downarrow H^+ + \uparrow HCO_3^- \quad (11)$$

The decrease in H^+ shifts the equilibrium to the right, meaning that carbon dioxide (P_{CO_2}) decreases and HCO_3^- goes up.

Just as in metabolic acidosis, respiratory compensation for metabolic alkalosis is rapid. The increase in pH and decrease in P_{CO_2} depress ventilation. Hypoventilation means the body retains CO_2, raising the P_{CO_2} and creating more H^+ and HCO_3^-. This respiratory compensation helps correct the pH problem but elevates HCO_3^- levels even more. However, this respiratory compensation is limited because hypoventilation also causes hypoxia. Once the arterial P_{O_2} drops below 60 mm Hg, hypoventilation ceases.

The renal response to metabolic alkalosis is the same as that for respiratory alkalosis: HCO_3^- is excreted and H^+ is reabsorbed.

This chapter has used fluid balance and acid-base balance to illustrate functional integration in the cardiovascular, respiratory, and renal systems. Changes in body fluid volume, reflected by changes in blood pressure, trigger both cardiovascular and renal homeostatic responses. Disturbances of acid-base balance are met with compensatory responses from both the respiratory and renal systems. Because of the interwoven responsibilities of these three systems, a disturbance in one system is likely to cause disturbances in the other two. Recognition of this fact is an important aspect of treatment for many clinical conditions.

RUNNING PROBLEM CONCLUSION Hyponatremia

In acute cases of dilutional hyponatremia such as Lauren's, the treatment goal is to correct the body's depleted Na^+ load and raise the plasma osmolarity to reduce cerebral swelling. The physicians in the emergency medical tent started a slow intravenous drip of 3% saline and restricted Lauren's oral fluid intake. Over the course of several hours, the combination of Na^+ intake and excretion of dilute urine returned Lauren's plasma Na^+ to normal levels.

Hyponatremia has numerous causes, including inappropriate secretion of antidiuretic hormone. To learn more about medical causes of hyponatremia, see "Clinical Practice Guideline on Diagnosis and Treatment of Hyponatraemia," *Nephrol. Dial. Transplant.* (2014) 29 (suppl 2): i1–i39. Available at *www.eje-online.org/content/170/3/G1.full.pdf. doi: 10.1093/ndt/gfu040* (free pdf; note the British spelling of *hyponatremia*).

Question	Facts	Integration and Analysis
Q1: *Name the two major body fluid compartments and give the major ions in each compartment.*	The major compartments are the intracellular fluid (ICF) and extracellular fluid (ECF) compartments. The primary ICF ion is K^+, and the major ECF ions are Na^+ and Cl^-.	N/A
Q2: *Based on Lauren's history, give a reason for why her weight increased during the race.*	Lauren reported drinking lots of water and sports drinks. One liter of pure water has a mass of 1 kg.	Lauren's fluid intake was greater than her fluid loss from sweating. A 2-kg increase in body weight means she drank an excess of about 2 L.
Q3: *Which body fluid compartment is being diluted in dilutional hyponatremia?*	Ingested water distributes itself throughout the ECF and ICF. Sodium is one of the major extracellular cations.	Lauren consumed a large amount of Na-free fluid and therefore diluted her Na^+ stores. However, the body compartments are in osmotic equilibrium so both ECF and ICF have lower osmolarities.
Q4: *One way to estimate osmolarity is to double the plasma Na^+ concentration. Estimate Lauren's osmolarity and explain what effect the dilutional hyponatremia has on her cells.*	Lauren's plasma Na^+ is 124 mEq/L. For Na^+, 1 mEq = 1 milliosmole. Doubling this value tells you that Lauren's estimated plasma osmolarity is 248 mOsM. Water distributes to maintain osmotic equilibrium.	At the start of the race, Lauren's cells were at 280 mOsM. The water she ingested distributed to maintain osmotic equilibrium, so water entered the ICF from the ECF, resulting in cell swelling.
Q5: *In dilutional hyponatremia, the medical personnel are most concerned about which organ or tissue?*	All cells in Lauren's body swell as a result of excess water ingestion. The brain is encased in the rigid skull.	The bony skull restricts the swelling of brain tissue, causing neurological symptoms, including confusion, headache, and loss of coordination. With lower Na^+ concentrations, death can result.
Q6: *Assuming a sweating rate of 1 L/hr, how much Na^+ did Lauren lose during the 16-hour race?*	1 L sweat lost/hr $\times$ 16 hr $\times$ 70 mEq Na^+/L sweat = 1120 mEq Na^+ lost during the 16-hour race.	N/A
Q7: *Total body water for a 60-kg female is approximately 30 L, and her ECF volume is 10 L. Based on the information given so far, how much fluid did Lauren ingest during the race?*	From the sweating rate given in question 6, you know that Lauren lost 16 L of sweat during the race. You also know that she gained 2 kg in weight. One liter of water weighs 1 kg.	Lauren must have ingested at least 18 L of fluid. You have no information on other routes of fluid loss, such as urine and insensible water lost during breathing.
Q8: *What would you expect to happen to vasopressin and aldosterone production in response to dilutional hyponatremia?*	Vasopressin secretion is inhibited by a decrease in osmolarity. The usual stimuli for renin or aldosterone release are low blood pressure and hyperkalemia.	Vasopressin secretion decreases with hyponatremia. The usual stimuli for aldosterone secretion are absent, but a pathological decrease in plasma Na^+ of 10 mEq/L can stimulate the adrenal cortex to secrete aldosterone. Thus, Lauren's plasma Na^+ may be low enough to increase her aldosterone secretion.

This problem was developed by Matt Pahnke while he was a kinesiology graduate student at the University of Texas.

617 – 619 – 628 – 634 – 639 – 648

CHAPTER SUMMARY

Homeostasis of body fluid volume, electrolytes, and pH follows the principle of *mass balance:* To maintain constant amount of a substance in the body, any intake or production must be offset by metabolism or excretion. The *control systems* that regulate these parameters are among the most complicated reflexes of the body because of the overlapping functions of the kidneys, lungs, and cardiovascular system. At the cellular level, however, the *movement of molecules across membranes* follows familiar patterns, as transfer of water and solutes from one *compartment* to another depends on osmosis, diffusion, and protein-mediated transport.

20.1 Fluid and Electrolyte Homeostasis

1. The renal, respiratory, and cardiovascular systems control fluid and electrolyte balance. Behaviors such as drinking also play an important role. (p. 617; Fig. 20.1)
2. Pulmonary and cardiovascular compensations are more rapid than renal compensation. (p. 617)

20.2 Water Balance

3. Most water intake comes from food and drink. The largest water loss is 1.5 liters/day in urine. Smaller amounts are lost in feces, by evaporation from skin, and in exhaled humidified air. (p. 618; Fig. 20.2)
4. Renal water reabsorption conserves water but cannot restore water lost from the body. (p. 619; Fig. 20.3)
5. To produce dilute urine, the nephron must reabsorb solute without reabsorbing water. To concentrate urine, the nephron must reabsorb water without reabsorbing solute. (p. 620)
6. Filtrate leaving the ascending limb of the loop of Henle is dilute. The final concentration of urine depends on the water permeability of the collecting duct. (p. 620; Fig. 20.4)
7. The hypothalamic hormone **vasopressin** controls collecting duct permeability to water in a graded fashion. When vasopressin is absent, water permeability is nearly zero. (p. 620; Fig. 20.5a, b)
8. Vasopressin causes distal nephron cells to insert **aquaporin** water pores in their apical membrane. (p. 621; Fig. 20.5c)
9. An increase in ECF osmolarity or a decrease in blood pressure stimulates vasopressin release from the posterior pituitary. Osmolarity is monitored by hypothalamic **osmoreceptors**. Blood pressure and blood volume are sensed by receptors in the carotid and aortic bodies, and in the atria, respectively. (p. 623; Fig. 20.6)
10. The loop of Henle is a **countercurrent multiplier** that creates high osmolarity in the medullary interstitial fluid by actively transporting Na^+, Cl^-, and K^+ out of the nephron. This high medullary osmolarity is necessary for formation of concentrated urine as filtrate flows through the collecting duct. (p. 623; Fig. 20.7)
11. The **vasa recta** capillaries form a countercurrent exchanger that carries away water leaving the tubule so that the water does not dilute the medullary interstitium. (p. 625; Fig. 20.7)
12. Urea contributes to the high osmolarity in the renal medulla. (p. 627)

20.3 Sodium Balance and ECF Volume

13. The total amount of Na^+ in the body is a primary determinant of ECF volume. (p. 627; Fig. 20.8)
14. The steroid hormone **aldosterone** increases Na^+ reabsorption and K^+ secretion. (p. 628; Fig. 20.9a)
15. Aldosterone acts on **principal cells (P cells)** of the distal nephron. This hormone enhances Na^+-K^+-ATPase activity and increases open time of Na^+ and K^+ leak channels. It also stimulates the synthesis of new pumps and channels. (p. 628; Fig. 20.9b)
16. Aldosterone secretion can be controlled directly at the adrenal cortex. Increased ECF K^+ stimulates aldosterone secretion, but very high ECF osmolarity inhibits it. (p. 628; Fig. 20.9)
17. Aldosterone secretion is also stimulated by **angiotensin II.** In response to signals associated with low blood pressure, granular cells in the kidney secrete **renin,** which converts **angiotensinogen** in the blood to **angiotensin I. Angiotensin-converting enzyme (ACE)** converts ANG I to ANG II. (p. 628; Fig. 20.10)
18. The signals for the release of renin are related either directly or indirectly to low blood pressure. (p. 630; Fig. 20.10)
19. ANG II has additional effects that raise blood pressure, including increased vasopressin secretion, stimulation of thirst, vasoconstriction, and activation of the cardiovascular control center. (p. 630; Fig. 20.10)
20. **Atrial natriuretic peptide (ANP)** and **brain natriuretic peptide (BNP)** enhance Na^+ excretion and urinary water loss by increasing GFR, inhibiting tubular reabsorption of NaCl, and inhibiting the release of renin, aldosterone, and vasopressin. (p. 630; Fig. 20.11)

20.4 Potassium Balance

21. Potassium homeostasis keeps plasma K^+ concentrations in a narrow range. **Hyperkalemia** and **hypokalemia** cause problems with excitable tissues, especially the heart. (p. 633)

20.5 Behavioral Mechanisms in Salt and Water Balance

22. Thirst is triggered by hypothalamic osmoreceptors and relieved by drinking. (p. 634)
23. **Salt appetite** is triggered by aldosterone and angiotensin. (p. 634)

20.6 Integrated Control of Volume, Osmolarity, and Blood Pressure

24. Homeostatic compensations for changes in salt and water balance follow the law of mass balance. Fluid and solute added to the body must be removed, and fluid and solute lost from the body must be replaced. However, perfect compensation is not always possible. (p. 634; Tbl. 20.1)

20.7 Acid-Base Balance

25. The body's pH is closely regulated because pH affects intracellular proteins, such as enzymes and membrane channels. (p. 639)
26. Acid intake from foods and acid production by the body's metabolic processes are the biggest challenge to body pH. The most significant source of acid is CO_2 from respiration, which combines with water in the presence of **carbonic anhydrase** to form H^+ and HCO_3^-. (p. 640; Fig. 20.14)
27. The body copes with changes in pH by using buffers, ventilation, and renal secretion or reabsorption of H^+ and HCO_3^-. (p. 640; Fig. 20.14)

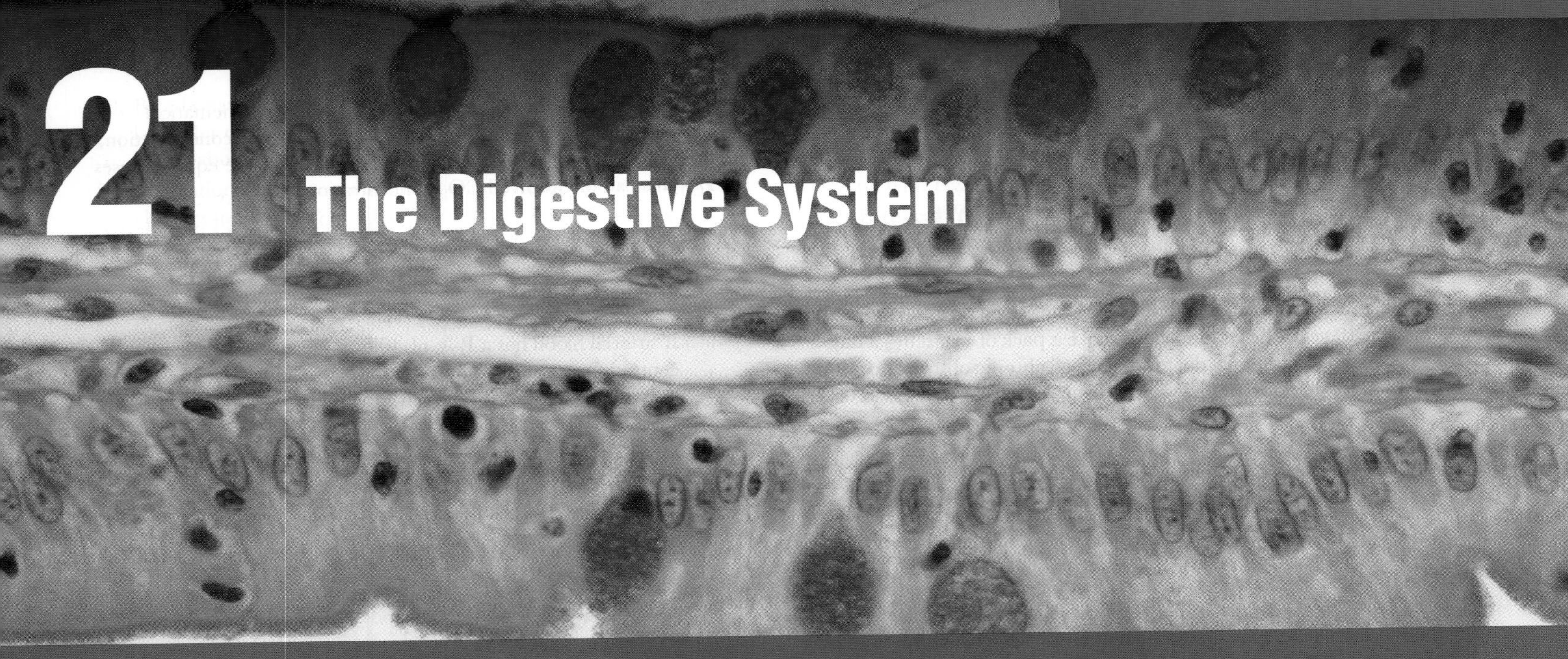

21 The Digestive System

Give me a good digestion, Lord, and also something to digest.

Anonymous, *A Pilgrim's Grace*

Intestinal epithelium with goblet cells (red)

BACKGROUND BASICS

A shotgun wound to the stomach seems an unlikely beginning to the scientific study of digestive processes. But in 1822, at Fort Mackinac, a young Canadian trapper named Alexis St. Martin narrowly escaped death when a gun discharged 3 feet from him, tearing open his chest and abdomen and leaving a hole in his stomach wall. U.S. Army surgeon William Beaumont attended to St. Martin and nursed him back to health over the next two years.

The gaping wound over the stomach failed to heal properly, leaving a *fistula*, or opening, into the lumen. St. Martin was destitute and unable to care for himself, so Beaumont "retained St. Martin in his family for the special purpose of making physiological experiments." In a legal document, St. Martin even agreed to "obey, suffer, and comply with all reasonable and proper experiments of the said William [Beaumont] in relation to . . . the exhibiting . . . of his said stomach and the power and properties . . . and states of the contents thereof."

Beaumont's observations on digestion and on the state of St. Martin's stomach under various conditions created a sensation. In 1832, just before Beaumont's observations were published, the nature of gastric juice {*gaster*, stomach} and digestion in the stomach was a subject of much debate. Beaumont's careful observations went far toward solving the mystery. Like physicians of old who tasted urine when making a diagnosis, Beaumont tasted the mucous lining of the stomach and the gastric juices. He described them both as "saltish," but mucus was not at all acidic, and gastric fluid was very acidic. Beaumont collected copious amounts of gastric fluid through the fistula, and in controlled experiments he confirmed that gastric fluid digested meat, using a combination of hydrochloric acid and another active factor now known to be the enzyme pepsin.

These observations and others about motility and digestion in the stomach became the foundation of what we know about digestive physiology. Although research today is conducted more at the cellular and molecular level, researchers still create surgical fistulas in experimental animals to observe and sample the contents of the digestive tract.

Why is the digestive system—also referred to as the **gastrointestinal system** {*intestinus*, internal}—of such great interest? The reason is that gastrointestinal diseases today account for nearly 10% of the money spent on health care. Many of these conditions, such as heartburn, indigestion, gas, and constipation, are troublesome rather than major health risks, but their significance should not be underestimated. Go into any drugstore and look at the number of over-the-counter medications for digestive disorders to get a feel for the impact digestive diseases have on our society. In this chapter, we examine the gastrointestinal system and the remarkable way in which it transforms the food we eat into nutrients for the body's use.

RUNNING PROBLEM **Cholera in India**

Anish was looking for a way to spend his summer break, so he decided to volunteer with a medical mission team going to staff a clinic in an impoverished rural region of India. At the briefing meeting, Anish and the other volunteers were warned that with the onset of the rainy season in August, they would be seeing patients with cholera, an acute diarrheal disease caused by the bacterium *Vibrio cholera*. Toxins from the cholera bacterium cause vomiting and massive volumes of watery diarrhea in people who consume contaminated food or water. Unless treated promptly, cholera can be fatal.

653 – 657 – 670 – 674 – 681 – 686

21.1 Anatomy of the Digestive System

The digestive system begins with the oral cavity (mouth and pharynx), which serves as a receptacle for food (**FIG. 21.1a**). Swallowed food enters the **gastrointestinal tract (GI tract)** consisting of esophagus, stomach, small intestine, and large intestine. The portion of the GI tract running from the stomach to the anus is also called the **gut**.

Digestion, the chemical and mechanical breakdown of food, takes place primarily in the lumen of the gut. Along the way, secretions are added to ingested food by secretory epithelial cells and by *accessory glandular organs* that include salivary glands, the liver, the gallbladder, and the pancreas. The soupy mixture of food and secretions is known as **chyme**.

The GI tract is a long tube with muscular walls lined by secretory and transporting epithelium [p. 75]. At intervals along the tract, rings of muscle function as *sphincters* to separate the tube into segments with distinct functions. Food moves through the tract propelled by waves of muscle contraction.

The products of digestion are absorbed across the intestinal epithelium and pass into the interstitial fluid. From there, they move into the blood or lymph for distribution throughout the body. Any waste remaining in the lumen at the end of the GI tract leaves the body through an opening called the *anus*.

Because the digestive system opens to the outside world, the tract lumen and its contents are actually part of the external environment. (Think of a hole passing through the center of a bead.) [Fig. 1.2, p. 4] This allows an amazing variety of bacteria to live in the lumen, particularly in the large intestine. The arrangement is usually described as a *commensal* relationship, in which the bacteria benefit from having a home and food supply, while the human body is not affected. However, we are discovering ways in which the body does benefit from its bacterial companions. The relationship between humans and their bacterial *microbiome* is a hot topic in physiology today, and you will learn more about it at the end of the chapter.

The Digestive System Is a Tube

In the oral cavity, the first stages of digestion begin with chewing and the secretion of saliva by three pairs of **salivary glands**: *sublingual glands* under the tongue, *submandibular glands* under the mandible (jawbone), and *parotid glands* lying near the hinge of the jaw (Fig. 21.1b). Swallowed food passes into the **esophagus,** a narrow tube that travels through the thorax to the abdomen (Fig. 21.1a).

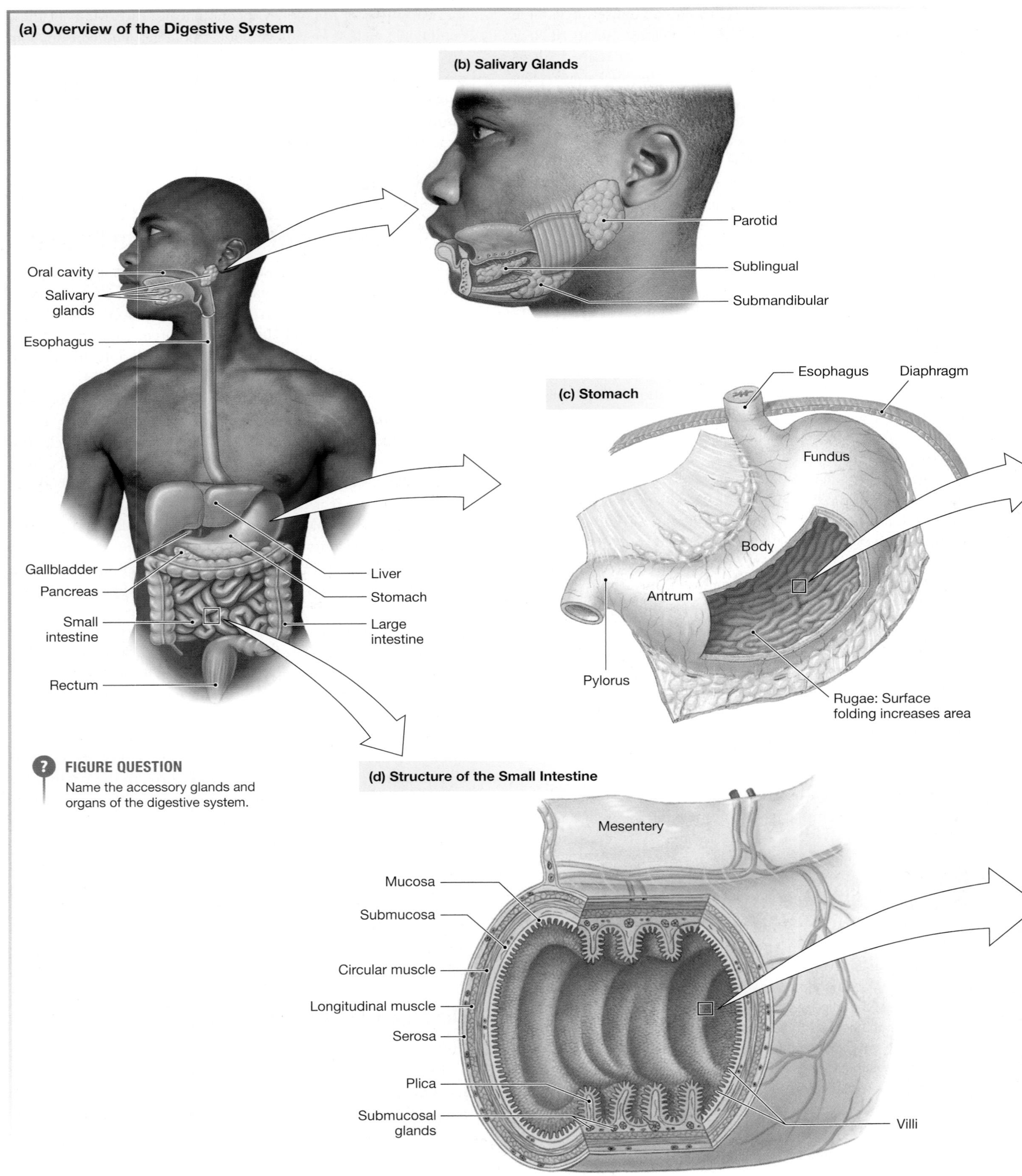
(a) Overview of the Digestive System
Oral cavity
Salivary glands
Esophagus
Gallbladder
Pancreas
Small intestine
Rectum
Liver
Stomach
Large intestine
(b) Salivary Glands
Parotid
Sublingual
Submandibular
(c) Stomach
Esophagus
Diaphragm
Fundus
Body
Antrum
Pylorus
Rugae: Surface folding increases area
(d) Structure of the Small Intestine
Mesentery
Mucosa
Submucosa
Circular muscle
Longitudinal muscle
Serosa
Plica
Submucosal glands
Villi

FIGURE QUESTION

Name the accessory glands and organs of the digestive system.

(e) Sectional View of the Stomach

In the stomach, surface area is increased by invaginations called gastric glands.

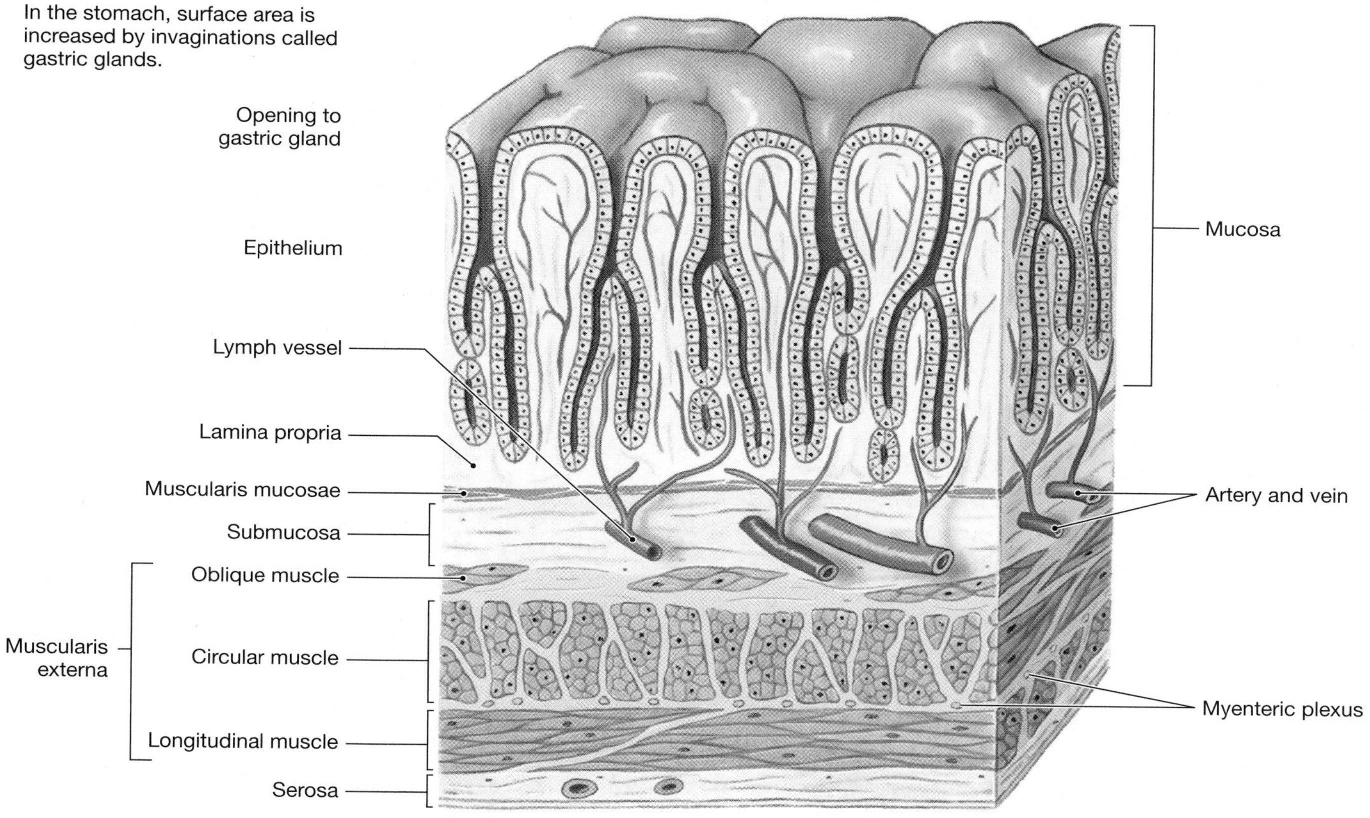

Opening to gastric gland
Epithelium
Lymph vessel
Lamina propria
Muscularis mucosae
Submucosa
Oblique muscle
Muscularis externa
Circular muscle
Longitudinal muscle
Serosa
Mucosa
Artery and vein
Myenteric plexus

(f) Sectional View of the Small Intestine

Intestinal surface area is enhanced by fingerlike villi and invaginations called crypts.

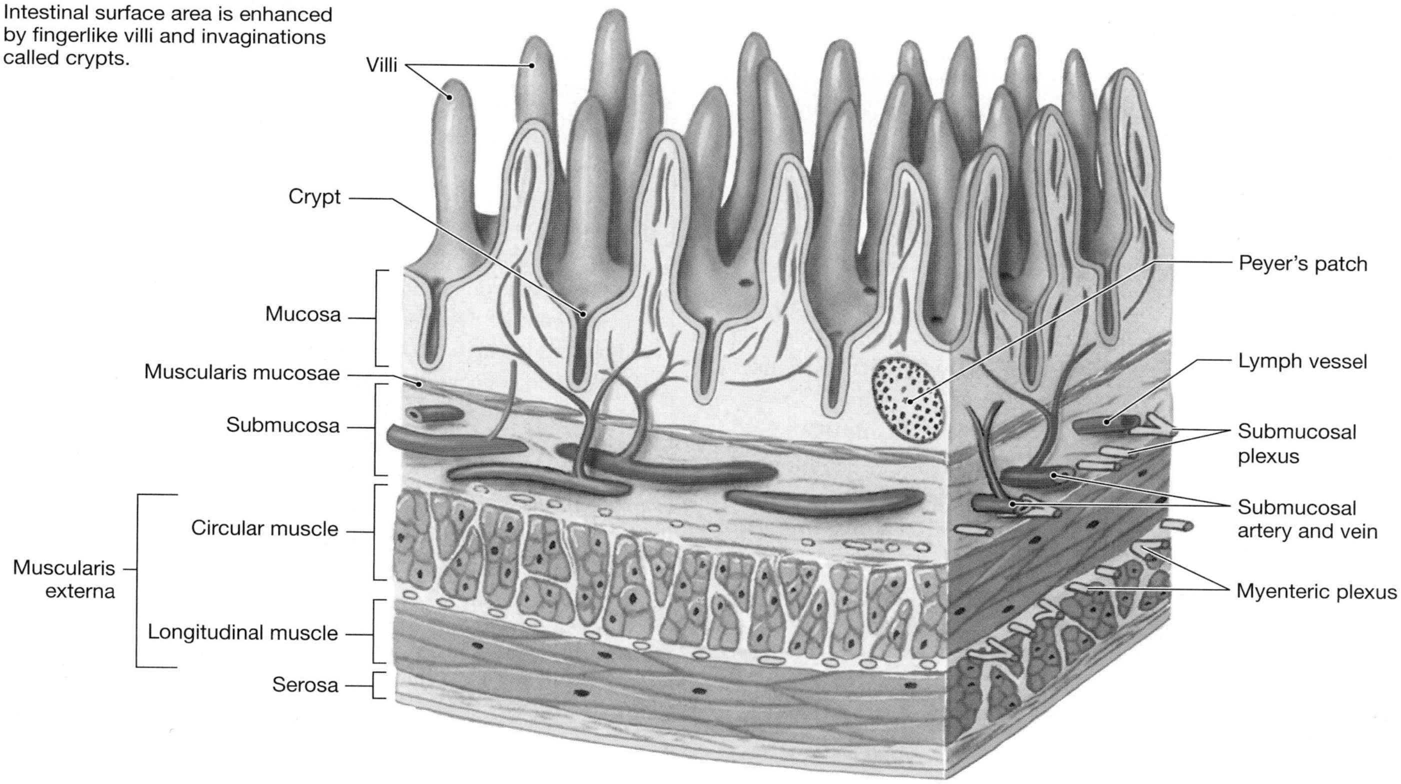

Villi
Crypt
Mucosa
Muscularis mucosae
Submucosa
Muscularis externa
Circular muscle
Longitudinal muscle
Serosa
Peyer's patch
Lymph vessel
Submucosal plexus
Submucosal artery and vein
Myenteric plexus

The esophageal walls are skeletal muscle initially but transition to smooth muscle about two-thirds of the way down the length. Just below the diaphragm, the esophagus ends at the **stomach,** a baglike organ that can hold as much as 2 liters of food and fluid when fully (if uncomfortably) expanded.

The stomach has three sections: the upper **fundus,** the central **body,** and the lower **antrum** (Fig. 21.1c). The stomach continues digestion that began in the mouth by mixing food with acid and enzymes to create chyme. The **pylorus** {gatekeeper} or opening between the stomach and the **small intestine** is guarded by the **pyloric valve.** This thickened band of smooth muscle relaxes to allow only small amounts of chyme into the small intestine at any one time.

The stomach acts as an intermediary between the behavioral act of eating and the physiological events of digestion and absorption in the intestine. Integrated signals and feedback loops between the intestine and stomach regulate the rate at which chyme enters the duodenum. This ensures that the intestine is not overwhelmed with more than it can digest and absorb.

Most digestion takes place in the small intestine, which has three sections: the **duodenum** (the first 25 cm), **jejunum,** and **ileum** (the latter two together are about 260 cm long). Digestion is carried out by intestinal enzymes, aided by exocrine secretions from two accessory glandular organs: the pancreas and the liver (Fig. 21.1a). Secretions from these two organs enter the initial section of the duodenum through ducts. A tonically contracted sphincter (the *sphincter of Oddi*) keeps pancreatic fluid and bile from entering the small intestine except during a meal.

Digestion finishes in the small intestine, and nearly all digested nutrients and secreted fluids are absorbed there, leaving about 1.5 liters of chyme per day to pass into the **large intestine** (Fig. 21.1a). In the **colon**—the proximal section of the large intestine—watery chyme becomes semisolid **feces** {*faeces*, dregs} as water and electrolytes are absorbed out of the chyme and into the extracellular fluid (ECF).

When feces are propelled into the terminal section of the large intestine, known as the **rectum,** distension of the rectal wall triggers a *defecation reflex*. Feces leave the GI tract through the **anus,** with its **external anal sphincter** of skeletal muscle, which is under voluntary control.

In a living person, the digestive system from mouth to anus is about 450 cm (nearly 15 ft.) long! Of this length, 395 cm (about 13 ft.) consists of the large and small intestines. Try to imagine 13 ft. of rope ranging from 1 to 3 inches in diameter all coiled up inside your abdomen from the belly button down. The tight arrangement of the abdominal organs helps explain why you feel the need to loosen your belt after consuming a large meal.

Measurements of intestinal length made during autopsies are nearly double those given here because after death, the longitudinal muscles of the intestinal tract relax. This relaxation accounts for the wide variation in intestinal length you may encounter in different references.

The GI Tract Wall Has Four Layers

The basic structure of the gastrointestinal wall is similar in the stomach and intestines, although variations exist from one section of the GI tract to another (Fig. 21.1e, f). The gut wall is crumpled into folds to increase its surface area. These folds are called *rugae* in the stomach and *plicae* in the small intestine. The intestinal mucosa also projects into the lumen in small fingerlike extensions known as **villi** (Fig. 21.1f). Additional surface area is added by tubular invaginations of the surface that extend down into the supporting connective tissue. These invaginations are called **gastric glands** in the stomach and **crypts** in the intestine. Some of the deepest invaginations form secretory **submucosal glands** that open into the lumen through ducts.

The gut wall consists of four layers: (1) an inner *mucosa* facing the lumen, (2) a layer known as the *submucosa*, (3) layers of smooth muscle known collectively as the *muscularis externa*, and (4) a covering of connective tissue called the *serosa*.

Mucosa The **mucosa,** the inner lining of the gastrointestinal tract, has three layers: a single layer of **mucosal epithelium** facing the lumen; the **lamina propria,** subepithelial connective tissue that holds the epithelium in place; and the **muscularis mucosae,** a thin layer of smooth muscle. Several structural modifications increase the amount of mucosal surface area to enhance absorption.

1. *The mucosal epithelium* is the most variable feature of the GI tract, changing from section to section. The cells of the mucosa include transporting epithelial cells (called *enterocytes* in the small intestine), endocrine and exocrine secretory cells, and stem cells. At the *mucosal* (apical) surface of the epithelium [p. 77], cells secrete ions, enzymes, mucus, and paracrine

RUNNING PROBLEM

Cholera is *endemic* in parts of India, meaning that it occurs on a regular basis. A volunteer trained by the U.S. Centers for Disease Control and Prevention (CDC) spoke to the medical mission team about what to expect when treating patients and also about proper precautions for themselves. He explained that the oral cholera vaccine they had taken would protect against the O1 strains but the area they would be visiting also had the newer O139 strain not covered by the vaccine. Once in India, everything went well at the clinic initially. Then about five days into his trip, Anish had several bouts of copious and watery diarrhea. When he developed dizziness and a rapid heartbeat, he visited the medical officer for the team. There, he was diagnosed with dehydration from cholera-induced diarrhea.

Q1: *Given Anish's watery diarrhea, what would you expect his ECF volume to be?*

Q2: *Why was Anish experiencing a rapid heartbeat?*

653 – **657** – 670 – 674 – 681 – 686

molecules into the lumen. On the *serosal* (basolateral) surface of the epithelium, substances being absorbed from the lumen and molecules secreted by epithelial cells enter the ECF.

The cell-to-cell junctions that tie GI epithelial cells together vary [p. 73]. In the stomach and colon, the junctions form a tight barrier so that little can pass between the cells. In the small intestine, junctions are not as tight. This intestinal epithelium is considered "leaky" because some water and some solutes can be absorbed between the cells (the *paracellular pathway*) instead of through them. We now know that these junctions have plasticity and that their "tightness" and selectivity can be regulated to some extent.

GI *stem cells* are rapidly dividing, undifferentiated cells that continuously produce new epithelium in the crypts and gastric glands. As stem cells divide, the newly formed cells are pushed toward the luminal surface of the epithelium. The average life span of a GI epithelial cell is only a few days, a good indicator of the rough life such cells lead. As with other types of epithelium, the rapid turnover and cell division rate in the GI tract make these organs susceptible to developing cancers. In 2013, cancers of the colon and rectum (colorectal cancer) were the second leading cause of cancer deaths in the United States. The death rate has been steadily falling, however, due to more screening examinations and better treatments.

2. The *lamina propria* is subepithelial connective tissue that contains nerve fibers and small blood and lymph vessels. Absorbed nutrients pass into the blood and lymph here (Fig. 21.1e). This layer also contains wandering immune cells, such as macrophages and lymphocytes, patrolling for invaders that enter through breaks in the epithelium.

   In the intestine, collections of lymphoid tissue adjoining the epithelium form small *nodules* and larger **Peyer's patches** that create visible bumps in the mucosa (Fig. 21.1f). These lymphoid aggregations are a major part of the **gut-associated lymphoid tissue (GALT)**.
3. The *muscularis mucosae*, a thin layer of smooth muscle, separates the lamina propria from the submucosa. Contraction of muscles in this layer alters the effective surface area for absorption by moving the villi back and forth, like the waving tentacles of a sea anemone.

Submucosa The **submucosa** is the middle layer of the gut wall. It is composed of connective tissue with larger blood and lymph vessels running through it (Fig. 21.1e, f). The submucosa also contains the **submucosal plexus** {*plexus*, interwoven}, one of the two major nerve networks of the **enteric nervous system** [p. 226]. The submucosal plexus (also called *Meissner's plexus*) innervates cells in the epithelial layer as well as smooth muscle of the muscularis mucosae.

Muscularis Externa The outer wall of the gastrointestinal tract, the **muscularis externa,** consists primarily of two layers of smooth muscle: an inner circular layer and an outer longitudinal layer (Fig. 21.1d, f). Contraction of the circular layer decreases the diameter of the lumen. Contraction of the longitudinal layer shortens the tube. The stomach has an incomplete third layer of oblique muscle between the circular muscles and the submucosa (Fig. 21.1e).

The second nerve network of the enteric nervous system, the **myenteric plexus** {*myo-*, muscle + *enteron*, intestine}, lies between the longitudinal and circular muscle layers. The myenteric plexus (also called *Auerbach's plexus*) controls and coordinates the motor activity of the muscularis externa.

Serosa The outer covering of the entire digestive tract, the **serosa**, is a connective tissue membrane that is a continuation of the **peritoneal membrane** (*peritoneum*) lining the abdominal cavity [p. 59]. The peritoneum also forms sheets of **mesentery** that hold the intestines in place so that they do not become tangled as they move.

The next section is a brief look at the four processes of secretion, digestion, absorption, and motility. Gastrointestinal physiology is a rapidly expanding field, and this textbook does not attempt to be all inclusive. Instead, it focuses on selected broad aspects of digestive physiology.

Concept Check

1. Is the lumen of the digestive tract on the apical or basolateral side of the intestinal epithelium? On the serosal or mucosal side?
2. Name the four layers of the GI tract wall, starting at the lumen and moving out.
3. Name the structures a piece of food passes through as it travels from mouth to anus.

21.2 Digestive Function and Processes

The primary function of the digestive system is to move nutrients, water, and electrolytes from the external environment into the body's internal environment. To accomplish this, the system uses four basic processes: digestion, absorption, secretion, and motility (**FIG. 21.2**). **Digestion** is the chemical and mechanical breakdown of foods into smaller units that can be taken across the intestinal epithelium into the body. **Absorption** is the movement of substances from the lumen of the GI tract to the extracellular fluid. **Secretion** in the GI tract has two meanings. It can mean the movement of water and ions from the ECF to the digestive tract lumen (the opposite of absorption), but it can also mean the release of substances synthesized by GI epithelial cells into either the lumen or the ECF. **Motility** {*movere*, move + *tillis*, characterized by} is movement of material in the GI tract as a result of muscle contraction.

Although it might seem simple to digest and absorb food, the digestive system faces three significant challenges:

1. *Avoiding autodigestion.* The food we eat is mostly in the form of macromolecules, such as proteins and complex carbohydrates, so our digestive systems must secrete powerful enzymes

FIG. 21.2 Four processes of the digestive system

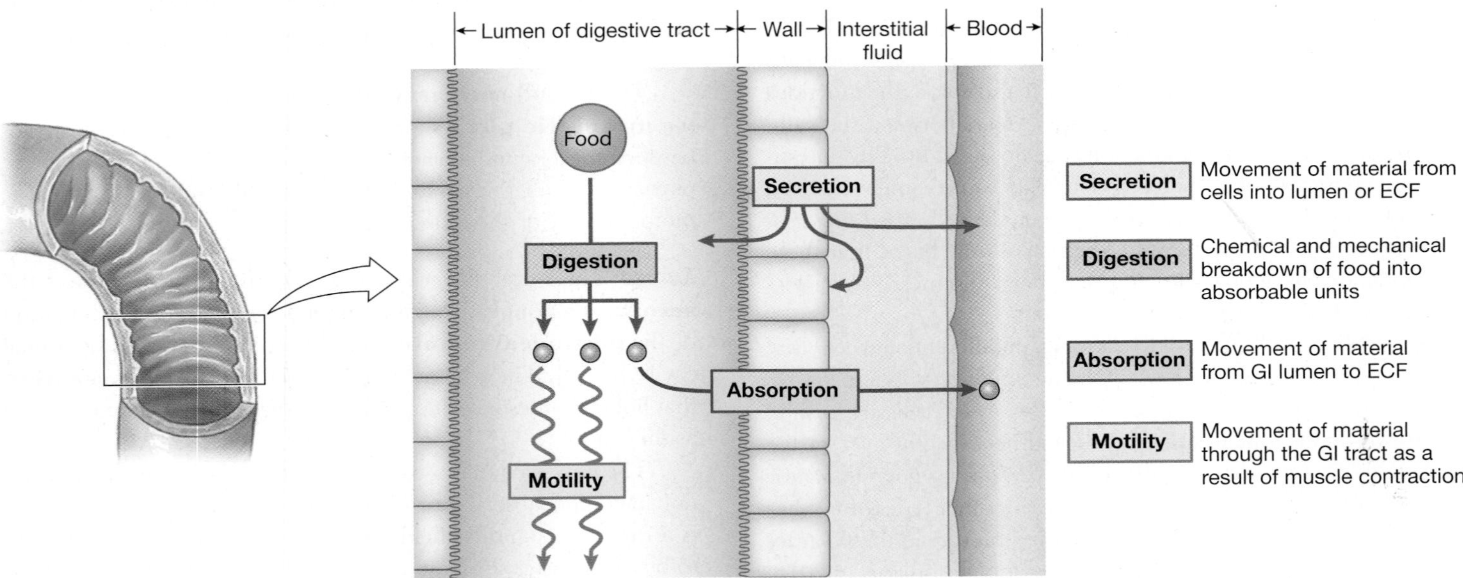
Lumen of digestive tract
Wall
Interstitial fluid
Blood
Food
Secretion
Digestion
Absorption
Motility
Secretion
Movement of material from cells into lumen or ECF
Digestion
Chemical and mechanical breakdown of food into absorbable units
Absorption
Movement of material from GI lumen to ECF
Motility
Movement of material through the GI tract as a result of muscle contraction

to digest food into molecules that are small enough to be absorbed into the body. At the same time, however, these enzymes must not digest the cells of the GI tract itself (*autodigestion*). If protective mechanisms against autodigestion fail, raw patches known as *peptic ulcers* {*peptos*, digested} develop on the walls of the GI tract.

2. *Mass balance.* Another challenge the digestive system faces daily is maintaining mass balance by matching fluid input with output (**FIG. 21.3**). People ingest about 2 liters of fluid a day. In addition, exocrine glands and cells secrete 7 liters or so of enzymes, mucus, electrolytes, and water into the lumen of the GI tract. That volume of secreted fluid is the equivalent of one-sixth of the body's total body water (42 liters), or more than twice the plasma volume of 3 liters. If the secreted fluid could not be reabsorbed, the body would rapidly dehydrate.
3. Normally intestinal reabsorption is very efficient, and only about 100 mL of fluid is lost in the feces. However, vomiting and diarrhea (excessively watery feces) can become an emergency when GI secretions are lost to the environment instead of being reabsorbed. In severe cases, this fluid loss can deplete extracellular fluid volume to the point that the circulatory system is unable to maintain adequate blood pressure.
4. *Defense.* A final challenge the digestive system faces is protecting the body from foreign invaders. It is counterintuitive, but the largest area of contact between the body's internal environment and the outside world is in the lumen of the digestive system. As a result, the GI tract, with a total surface area about the size of a tennis court, faces daily conflict between the need to absorb water and nutrients, and the need to keep bacteria, viruses, and other pathogens from entering the body. To this end, the transporting epithelium of the GI tract is assisted by an array of physiological defense mechanisms, including mucus, digestive enzymes, acid, and the largest collection of lymphoid tissue in the body, the *gut-associated lymphoid tissue* (GALT). By one estimate, 80% of all lymphocytes [p. 513] in the body are found in the small intestine.

The human body meets these sometimes conflicting physiological challenges by coordinating motility and secretion to maximize digestion and absorption.

We Secrete More Fluid than We Ingest

In a typical day, 9 liters of fluid pass through the lumen of an adult's gastrointestinal tract—equal to the contents of three 3-liter soft drink bottles! Only about 2 liters of that volume enter the GI system through the mouth. The remaining 7 liters of fluid come from body water secreted along with ions, enzymes, and mucus (see Fig. 21.3). The ions are transported from the ECF into the lumen. Water then follows the osmotic gradient created by this transfer of solutes from one side of the epithelium to the other. Water moves through the epithelial cells via channels or through leaky junctions between cells (the paracellular pathway).

Gastrointestinal epithelial cells, like those in the kidney, are *polarized* [p. 149], with distinct apical and basolateral membranes. Each cell surface contains membrane proteins for solute and water movement, many of them similar to those of the renal tubule. The arrangement of transport proteins on the apical and basolateral membranes determines the direction of movement of solutes and water across the epithelium.

Digestive Enzymes Digestive enzymes are secreted either by exocrine glands (salivary glands and the pancreas) or by epithelial cells in the stomach and small intestine. Enzymes are proteins, which means that they are synthesized on the rough endoplasmic reticulum, packaged by the Golgi complex into secretory vesicles, and then stored in the cell until needed. On demand, they are released

FIG. 21.3 Mass balance in the digestive system

To maintain homeostasis, the volume of fluid entering the GI tract by intake or secretion must equal the volume leaving the lumen.

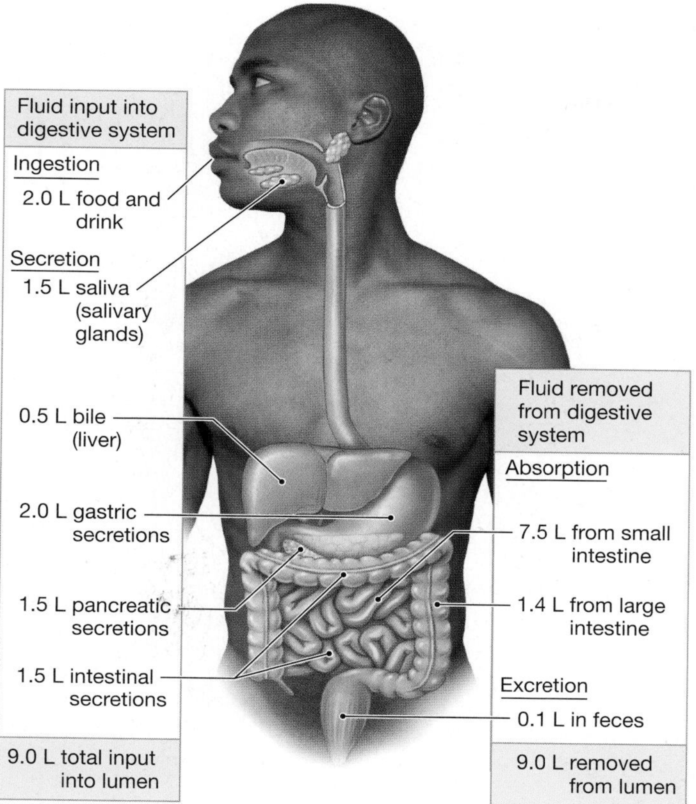
Fluid input into digestive system
Ingestion
2.0 L food and drink
Secretion
1.5 L saliva (salivary glands)
0.5 L bile (liver)
2.0 L gastric secretions
1.5 L pancreatic secretions
1.5 L intestinal secretions
9.0 L total input into lumen
Fluid removed from digestive system
Absorption
7.5 L from small intestine
1.4 L from large intestine
Excretion
0.1 L in feces
9.0 L removed from lumen

by exocytosis [p. 147]. Many intestinal enzymes remain bound to the apical membranes of intestinal cells, anchored by transmembrane protein "stalks" or lipid anchors [p. 64].

Some digestive enzymes are secreted in an inactive *proenzyme* form known collectively as *zymogens* [p. 99]. Zymogens must be activated in the GI lumen before they can carry out digestion. Synthesizing the enzymes in a nonfunctional form allows them to be stockpiled in the cells that make them without damaging those cells. Zymogen names often have the suffix *-ogen* added to the enzyme name, such as *pepsinogen.*

Mucus *Mucus* is a viscous secretion composed primarily of glycoproteins collectively called **mucins.** The primary functions of mucus are to form a protective coating over the GI mucosa and to lubricate the contents of the gut. Mucus is made in specialized exocrine cell called *mucous cells* in the stomach and salivary glands, and *goblet cells* in the intestine [Fig. 3.10, p. 78]. Goblet cells make up between 10% and 24% of the intestinal cell population.

The signals for mucus release include parasympathetic innervation, a variety of neuropeptides found in the enteric nervous system, and cytokines from immunocytes. Parasitic infections and inflammatory processes in the gut also cause substantial increases in mucus secretion as the body attempts to fortify its protective barrier.

Concept Check

4. Define digestion. What is the difference between digestion and metabolism [p. 102]?
5. Why is the digestive system associated with the largest collection of lymphoid tissue in the body?
6. Draw a cell showing (1) an enzyme in a cytoplasmic secretory vesicle, (2) exocytosis of the vesicle, and (3) the enzyme remaining bound to the surface membrane of the cell rather than floating away.

Digestion and Absorption Make Food Usable

Most GI secretions facilitate digestion. The GI system digests macromolecules into absorbable units using a combination of mechanical and chemical breakdown. Chewing and churning create smaller pieces of food with more surface area exposed to digestive enzymes. The pH at which different digestive enzymes function best [p. 100] reflects the location where they are most active. For example, enzymes that act in the stomach work well at acidic pH, and those that are secreted into the small intestine work best at alkaline pH.

Most absorption takes place in the small intestine, with additional absorption of water and ions in the large intestine. Absorption, like secretion, uses many of the same transport proteins as the kidney tubule. Once absorbed, nutrients enter the blood or the lymphatic circulation.

Motility: GI Smooth Muscle Contracts Spontaneously

Motility in the gastrointestinal tract serves two purposes: moving food from the mouth to the anus and mechanically mixing food to break it into uniformly small particles. This mixing maximizes exposure of the particles to digestive enzymes by increasing particle surface area. Gastrointestinal motility is determined by the properties of the GI smooth muscle and is modified by chemical input from nerves, hormones, and paracrine signals.

Most of the gastrointestinal tract is composed of single-unit smooth muscle, with groups of cells electrically connected by gap junctions [p. 73] to create contracting segments. Different regions exhibit different types of contraction. **Tonic contractions** are sustained for minutes or hours. They occur in some smooth muscle sphincters and in the anterior portion of the stomach. **Phasic contractions,** with contraction-relaxation cycles lasting only a few seconds, occur in the posterior region of the stomach and in the small intestine.

Cycles of smooth muscle contraction and relaxation are associated with cycles of depolarization and repolarization known as **slow wave potentials** or simply *slow waves* (**FIG. 21.4a**). Current research indicates that slow waves originate in a network of cells called the **interstitial cells of Cajal** (named after the Spanish neuroanatomist Santiago Ramón y Cajal), or ICCs. These modified smooth muscle cells lie between smooth muscle layers and the intrinsic nerve plexuses, and they may act as an intermediary between the neurons and smooth muscle.

It appears that ICCs function as the pacemakers for slow wave activity in different regions of the GI tract, just as cells of the cardiac

FIG. 21.4 ESSENTIALS Gastrointestinal Motility

(a) Slow waves are spontaneous depolarizations in GI smooth muscle.

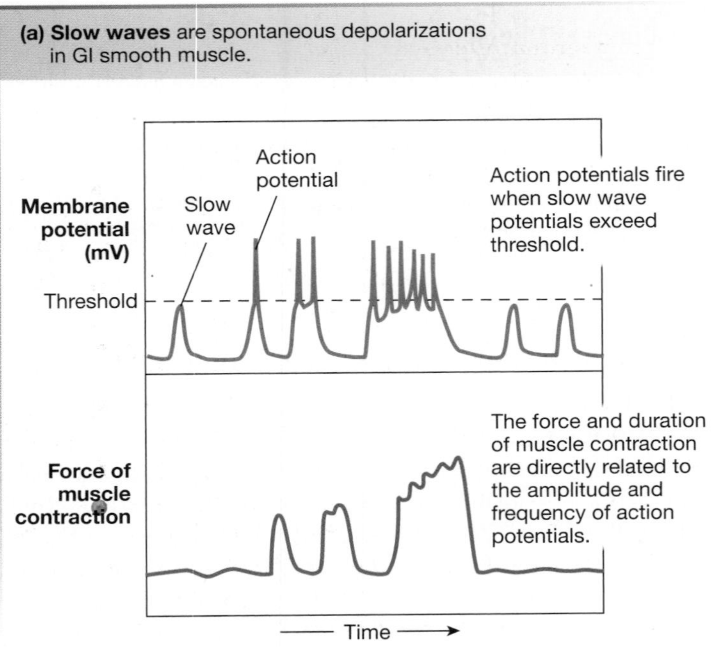

Membrane potential (mV)
Slow wave
Action potential
Action potentials fire when slow wave potentials exceed threshold.
Threshold
Force of muscle contraction
The force and duration of muscle contraction are directly related to the amplitude and frequency of action potentials.
Time

FIGURE QUESTION

Why do the peaks of the contraction waves occur after the peaks of the action potentials?

(b) The migrating motor complex (MMC) is a series of contractions that begin in the empty stomach and end in the large intestine.

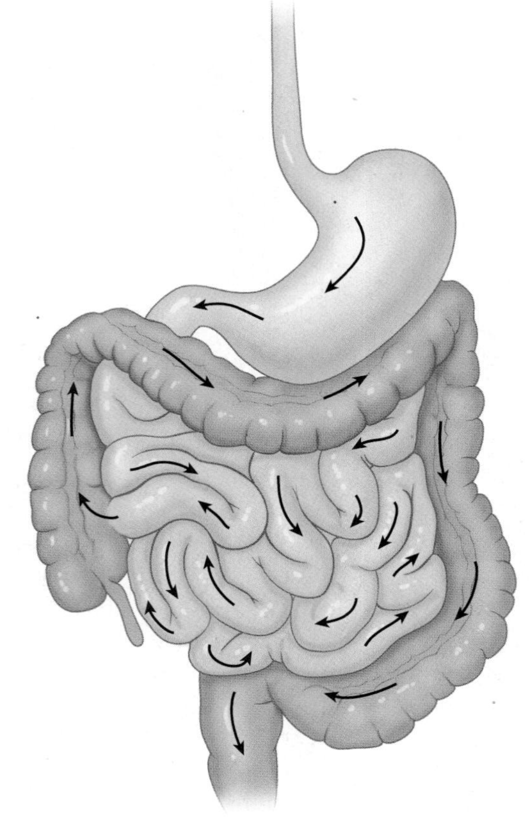

(c) Peristaltic contractions are responsible for forward movement.

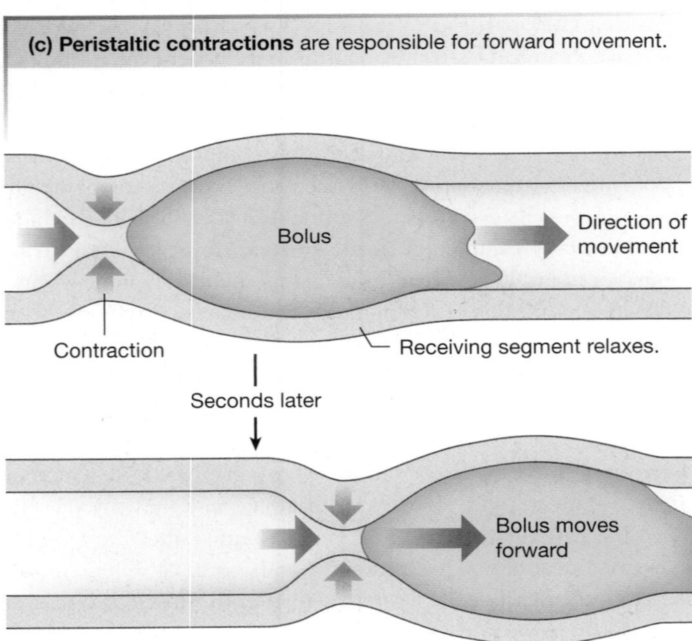

Bolus
Direction of movement
Contraction
Receiving segment relaxes.
Seconds later
Bolus moves forward

(d) Segmental contractions are responsible for mixing.

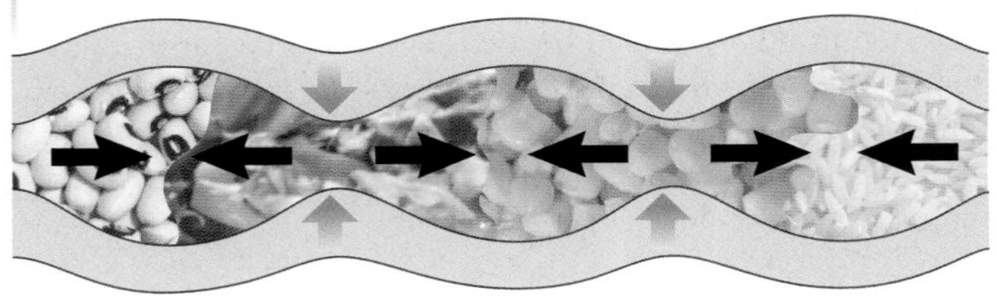

Alternate segments contract, and there is little or no net forward movement.

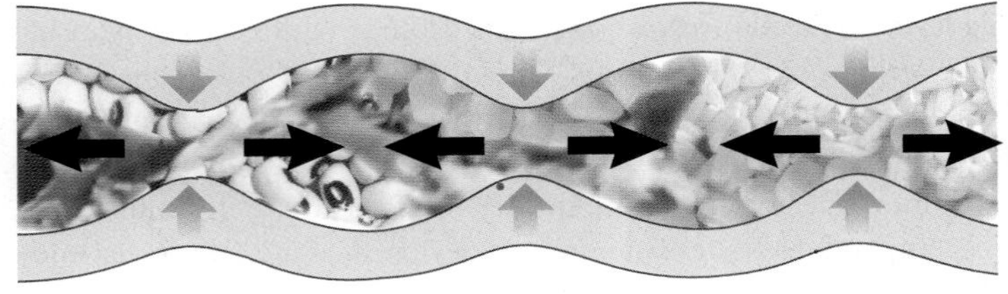

conduction system act as pacemakers for the heart [p. 450]. Slow wave potentials differ from myocardial pacemaker potentials in that the GI waves occur at a much slower frequency (3–12 waves/min GI versus 60–90 waves/min myocardial). In addition, slow wave frequency varies by region of the digestive tract, ranging from 3 waves/min in the stomach to 12 waves/min in the duodenum.

Slow waves that begin spontaneously in ICCs spread to adjacent smooth muscle layers through gap junctions. Just as in the cardiac conducting system, the fastest pacemaker in a group of ICCs sets the pace for the entire group [p. 453]. The observation that ICCs seem to coordinate GI motility now has researchers working to establish a link between ICCs and functional bowel disorders, such as irritable bowel syndrome and chronic constipation.

One difference between slow waves and cardiac pacemaker potentials is that slow waves do not reach threshold with each cycle, and a slow wave that does not reach threshold will not cause muscle contraction. When a slow wave potential does reach threshold, voltage-gated Ca^{2+} channels in the muscle fiber open, Ca^{2+} enters, and the cell fires one or more action potentials. The depolarization phase of the slow wave action potential, like that in myocardial autorhythmic cells, is the result of Ca^{2+} entry into the cell. In addition, Ca^{2+} entry initiates muscle contraction [p. 405].

Contraction of smooth muscle, like that of cardiac muscle, is graded according to the amount of Ca^{2+} that enters the fiber. The longer the duration of the slow wave, the more action potentials fire, and the greater the contraction force in the muscle. The likelihood of a slow wave firing an action potential depends primarily on input from the enteric nervous system.

GI Smooth Muscle Exhibits Different Patterns of Contraction

Muscle contractions in the gastrointestinal tract occur in three patterns that bring about different types of movement within the tract. Between meals, when the tract is largely empty, a series of contractions begins in the stomach and passes slowly from section to section, each series taking about 90 minutes to reach the large intestine. This pattern, known as the **migrating motor complex**, is a "housekeeping" function that sweeps food remnants and bacteria out of the upper GI tract and into the large intestine (Fig. 21.4b).

Muscle contractions during and following a meal fall into one of two other patterns (Fig. 21.4). **Peristalsis** {*peri-*, surrounding + *stalsis*, contraction} is progressive waves of contraction that move from one section of the GI tract to the next, just like the human "waves" that ripple around a football stadium or basketball arena. In peristalsis, circular muscles contract just behind a mass, or **bolus**, of food (Fig. 21.4c). This contraction pushes the bolus forward into a *receiving segment*, where the circular muscles are relaxed. The receiving segment then contracts, continuing the forward movement.

Peristaltic contractions push a bolus forward at speeds between 2 and 25 cm/sec. Peristalsis in the esophagus propels material from pharynx to stomach. Peristalsis contributes to food mixing in the stomach but in normal digestion, intestinal peristaltic waves are limited to short distances.

In **segmental contractions**, short (1–5 cm) segments of intestine alternately contract and relax (Fig. 21.4d). In the contracting segments, circular muscles contract while longitudinal muscles relax. These contractions may occur randomly along the intestine or at regular intervals. Alternating segmental contractions churn the intestinal contents, mixing them and keeping them in contact with the absorptive epithelium. When segments contract sequentially, in an oral-to-aboral direction {*ab-*, away}, intestinal contents are propelled short distances.

Motility disorders are among the more common gastrointestinal problems. They range from esophageal spasms and delayed gastric (stomach) emptying to constipation and diarrhea. *Irritable bowel syndrome* is a chronic functional disorder characterized by altered bowel habits and abdominal pain.

Around 20 years ago researchers studying diarrhea caused by pathogenic bacterial discovered that the bacterial toxin was binding to a previously unknown receptor-enzyme [p. 169] on the luminal side of intestinal epithelia cells. The receptor, called *guanylate cyclase-C* (GC-C), normally helps regulate fluid secretion in the intestine under the control of two gut peptides, **uroguanylin** and **guanylin**. When the receptors are overactivated by bacterial toxin, fluid secretion becomes excessive, causing diarrhea. After this observation researchers wondered if a drug that activated the GC-C receptor could be used to treat chronic constipation, which is characterized by dry, hard stools. The result of their work was a GC-C agonist called *plecanatide* that was recently approved by the U.S. Food and Drug Administration for treating constipation.

Concept Check

7. What is the difference between absorption and secretion?
8. How do fats absorbed into the lymphatic system get into the general circulation for distribution to cells? [*Hint:* p. 499]
9. Why are some sphincters of the digestive system tonically contracted?

CLINICAL FOCUS

Diabetes: Delayed Gastric Emptying

Diabetes mellitus has an impact on almost every organ system, and the digestive tract is not exempt. One problem that plagues more than a third of all people with diabetes is *gastroparesis*, also called delayed gastric emptying. In these patients, the migrating motor complex is absent between meals, and the stomach empties very slowly after meals. Many patients suffer nausea and vomiting as a result. The causes of diabetic gastroparesis are unclear, but recent studies of animal models and human patients show loss or dysfunction of the interstitial cells of Cajal, which serve as pacemakers and as a link between GI smooth muscle and the enteric and autonomic nervous systems. Adopting the cardiac model of an external pacemaker, scientists are now testing an implantable gastric pacemaker to promote gastric motility in diabetic patients with severe gastroparesis.

FIG. 21.8 Cephalic and gastric phase reflexes

The sight, smell, and taste of food initiate long reflexes that prepare the stomach for the arrival of food.

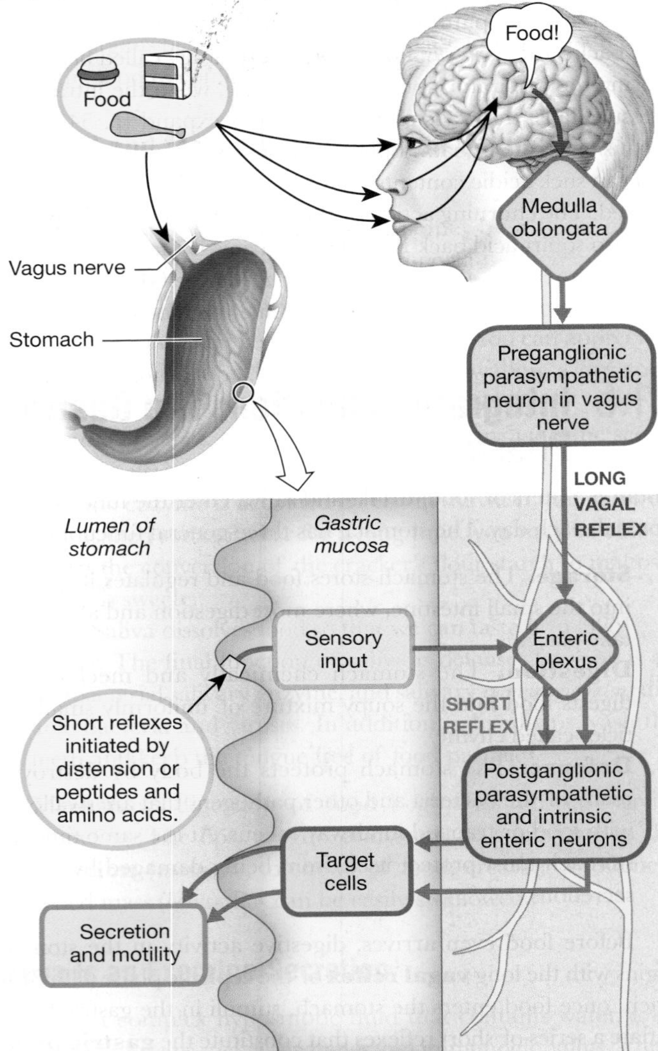
Food
Food!
Medulla oblongata
Vagus nerve
Stomach
Preganglionic parasympathetic neuron in vagus nerve
LONG VAGAL REFLEX
Lumen of stomach
Gastric mucosa
Sensory input
Enteric plexus
SHORT REFLEX
Short reflexes initiated by distension or peptides and amino acids.
Postganglionic parasympathetic and intrinsic enteric neurons
Target cells
Secretion and motility

amounts of unabsorbed chyme would pass into the large intestine. The epithelium of the large intestine is not designed for large-scale nutrient absorption, so most of the chyme would become feces, resulting in diarrhea. This "dumping syndrome" is one of the less pleasant side effects of surgery that removes portions of either the stomach or small intestine.

While the upper stomach is quietly holding food, the lower stomach is busy with digestion. In the distal half of the stomach, a series of peristaltic waves pushes the food down toward the pylorus, mixing food with acid and digestive enzymes. As large food particles are digested to the more uniform texture of chyme, each contractile wave squirts a small amount of chyme through the pylorus into the duodenum. Enhanced gastric motility during a meal is primarily under neural control and is stimulated by distension of the stomach.

Gastric Secretions Protect and Digest

The lumen of the stomach is lined with mucus-producing epithelium punctuated by the openings of *gastric pits*. The pits lead to **gastric glands** deep within the mucosal layer (see Fig. 21.1e). Multiple cell types within the glands produce gastric acid (HCl), enzymes, hormones, and paracrine molecules. The various secretions of gastric mucosa cells, their stimuli for release, and their functions are summarized in **FIGURE 21.9** and described next.

Gastrin Secretion **G cells,** found deep in the gastric glands, secrete the hormone **gastrin** into the blood. In short reflexes, gastrin release is stimulated by the presence of amino acids and peptides in the stomach and by distension of the stomach. Coffee (even if decaffeinated) also stimulates gastrin release—one reason people with excess acid secretion syndromes are advised to avoid coffee.

Gastrin release is also triggered by neural reflexes. Short reflexes are mediated by an ENS neurotransmitter called **gastrin-releasing peptide (GRP)**. In cephalic reflexes, parasympathetic neurons from the vagus nerve stimulate G cells to release gastrin into the blood.

Gastrin's primary action is to promote acid release. It does this directly by acting on parietal cells and indirectly by stimulating histamine release.

Acid Secretion **Parietal cells** deep in the gastric glands secrete **gastric acid** (HCl) into the lumen of the stomach. Acid secretion in the stomach averages 1–3 liters per day and can create a luminal pH as low as 1. The cytoplasmic pH of the parietal cells is about 7.2, which means the cells are pumping H^+ against a gradient that can be 1.5 million times more concentrated in the lumen.

Gastric acid has multiple functions:

- Acid in the stomach lumen causes release and activation of pepsin, an enzyme that digests proteins.
- Acid triggers somatostatin release from D cells. Somatostatin is discussed later in the section on paracrine signals.
- HCl *denatures* proteins by breaking disulfide and hydrogen bonds that hold the protein in its tertiary structure [p. 32]. Unfolding protein chains make the peptide bonds between amino acids more accessible to digestion by pepsin.
- Gastric acid helps kill bacteria and other ingested microorganisms.
- Acid inactivates salivary amylase, stopping carbohydrate digestion that began in the mouth.

The parietal cell pathway for acid secretion is depicted in Figure 21.9c. The process begins when H^+ from water inside the parietal cell is pumped into the stomach lumen by an H^+-K^+-ATPase in exchange for K^+ entering the cell. Cl^- then follows the electrical gradient created by H^+ by moving through open chloride channels. The net result is secretion of HCl by the cell.

By learning the cellular mechanism of parietal cell acid secretion, scientists were able to develop a new class of drugs to treat oversecretion of gastric acid. These drugs, known as *proton pump inhibitors* (PPIs), block activity of the H^+-K^+-ATPase. Generic

FIG. 21.9 ESSENTIALS Gastric Secretions

(a) Secretory Cells of the Gastric Mucosa

Opening of gastric gland

Gastric Mucosa / Cell Types	Substance Secreted	Function of Secretion	Stimulus for Release
Mucous surface cell	Mucus	Physical barrier between lumen and epithelium	Tonic secretion; irritation of mucosa
Mucous neck cell	Bicarbonate	Buffers gastric acid to prevent damage to epithelium	Secreted with mucus
Parietal cells	Gastric acid (HCl)	Activates pepsin; kills bacteria	Acetylcholine, gastrin, histamine
	Intrinsic factor	Complexes with vitamin B12 to permit absorption	
Enterochromaffin-like cell	Histamine	Stimulates gastric acid secretion	Acetylcholine, gastrin
Chief cells	Pepsin(ogen)	Digests proteins	Acetylcholine, acid secretion
	Gastric lipase	Digests fats	
D cells	Somatostatin	Inhibits gastric acid secretion	Acid in the stomach
G cells	Gastrin	Stimulates gastric acid secretion	Acetylcholine, peptides, and amino acids

(b) Mucus-Bicarbonate Barrier

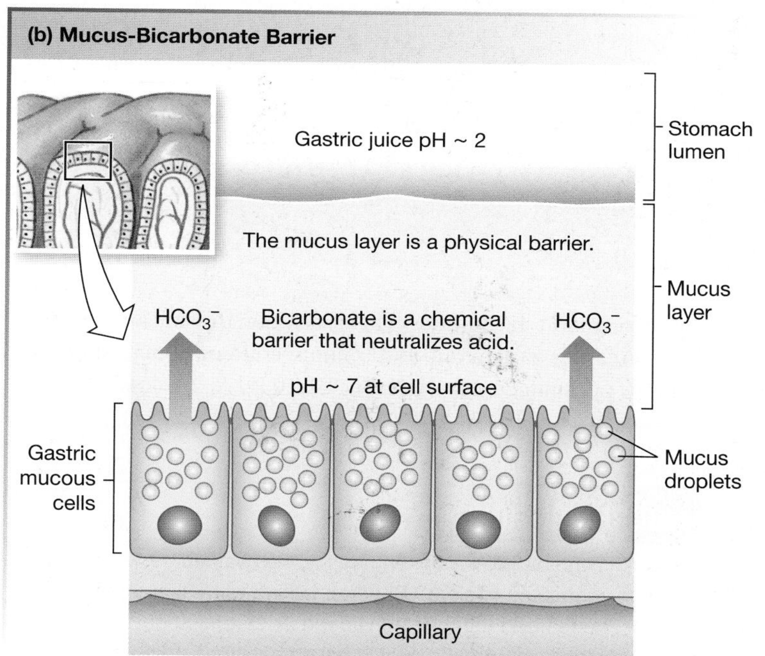

Gastric juice pH ~ 2
Stomach lumen
The mucus layer is a physical barrier.
Mucus layer
HCO_3^-
Bicarbonate is a chemical barrier that neutralizes acid.
HCO_3^-
pH ~ 7 at cell surface
Gastric mucous cells
Mucus droplets
Capillary

(c) Acid Secretion in the Stomach

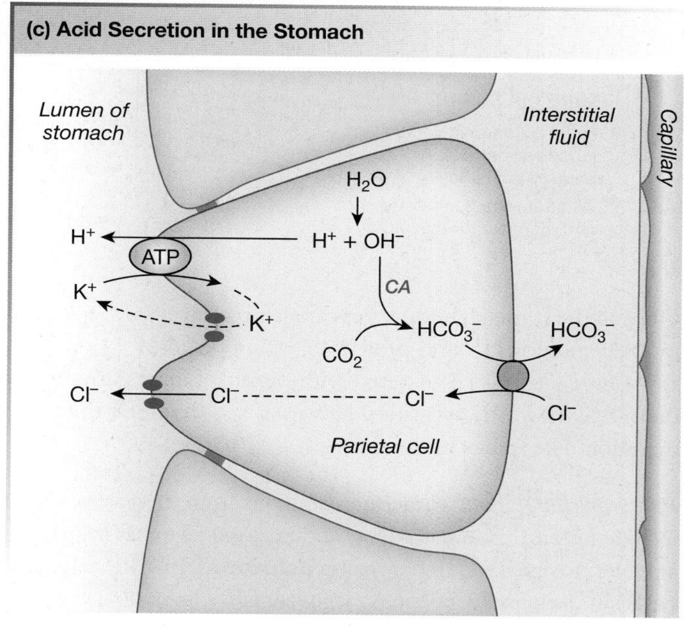

Lumen of stomach
Interstitial fluid
Capillary
H_2O
H^+
ATP
$H^+ + OH^-$
K^+
CA
K^+
HCO_3^-
HCO_3^-
CO_2
Cl^-
Cl^-
Cl^-
Cl^-
Parietal cell

versions of some PPIs (omeprazole, for example) are available over the counter in the United States.

While acid is being secreted into the lumen, bicarbonate made from CO_2 and the OH^- from water is absorbed into the blood. The buffering action of HCO_3^- makes blood leaving the stomach less acidic, creating an *alkaline tide* that can be measured as a meal is being digested.

Enzyme Secretion The stomach produces two enzymes: pepsin and a gastric lipase. **Pepsin** carries out the initial digestion of proteins. It is particularly effective on collagen and therefore plays an important role in digesting meat.

Pepsin is secreted as the inactive enzyme *pepsinogen* by **chief cells** in the gastric glands. Acid stimulates pepsinogen release from chief cells through a short reflex mediated in the ENS (**FIG. 21.10**).

FIG. 21.10 Integration of cephalic and gastric phase secretion

The cephalic phase is initiated by the sight, smell, sound, or thought of food or by the presence of food in the mouth. The gastric phase is initiated by the arrival of food in the stomach.

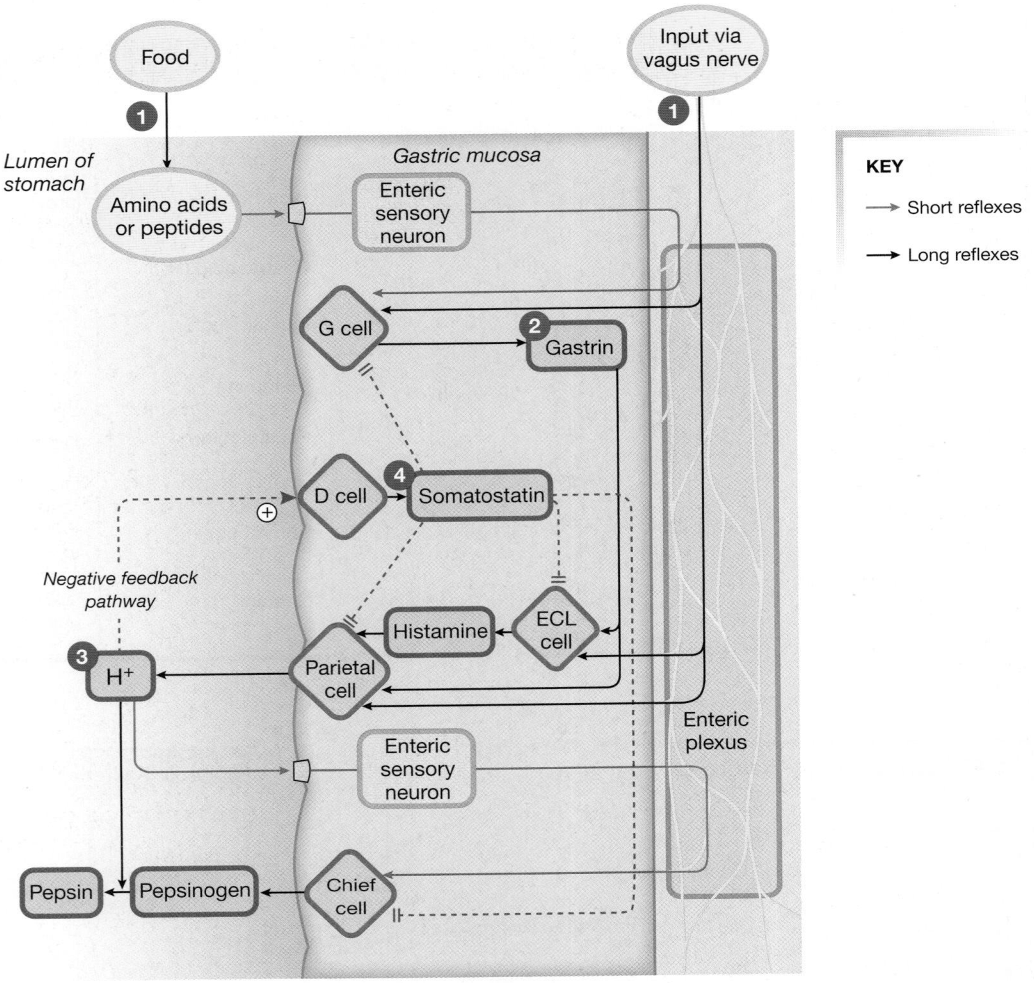

1. Food or cephalic reflexes initiate gastric secretion of gastrin, histamine, and acid.
2. Gastrin stimulates acid secretion by direct action on parietal cells or indirectly through histamine.
3. Acid stimulates short reflex secretion of pepsinogen.
4. Somatostatin release by H^+ is the negative feedback signal that modulates acid and pepsin release.

Food
Input via vagus nerve
Lumen of stomach
Gastric mucosa
Amino acids or peptides
Enteric sensory neuron
G cell
Gastrin
D cell
Somatostatin
Negative feedback pathway
Histamine
ECL cell
H^+
Parietal cell
Enteric sensory neuron
Enteric plexus
Pepsin
Pepsinogen
Chief cell

KEY
- Short reflexes
- Long reflexes

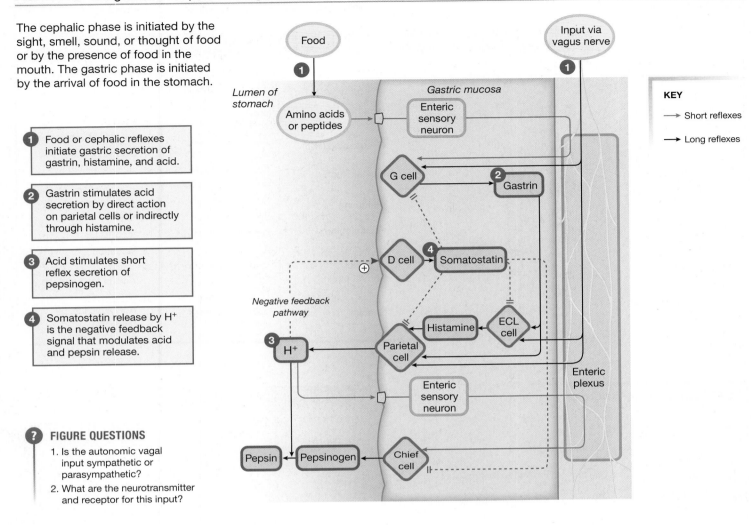

FIGURE QUESTIONS

1. Is the autonomic vagal input sympathetic or parasympathetic?
2. What are the neurotransmitter and receptor for this input?

Once in the stomach lumen, pepsinogen is cleaved to active pepsin by the action of H^+, and protein digestion begins.

Gastric lipase is co-secreted with pepsin. Lipases are enzymes that break down triglycerides. However, less than one-third of fat digestion takes place in the stomach.

Paracrine Secretions Paracrine secretions from the gastric mucosa include histamine, somatostatin, and intrinsic factor. **Histamine** is a paracrine signal secreted by **enterochromaffin-like cells (ECL cells)** in response to gastrin or acetylcholine stimulation. Histamine diffuses to its target, the parietal cells, and stimulates acid secretion by combining with *H_2 receptors* on parietal cells (Fig. 21.10). *H_2 receptor antagonists* (cimetidine and ranitidine, for example) that block histamine action are a second class of drugs used to treat acid hypersecretion.

Intrinsic factor is a protein secreted by the same gastric parietal cells that secrete acid. In the lumen of the stomach intrinsic factor complexes with vitamin B_{12}, a step that is needed for the vitamin's absorption in the intestine.

Somatostatin (SS), also known as hypothalamic growth hormone-inhibiting hormone, is secreted by **D cells** in the stomach. Somatostatin is the primary negative feedback signal for gastric phase secretion. It shuts down acid secretion directly and indirectly by decreasing gastrin and histamine secretion. Somatostatin also inhibits pepsinogen secretion (Fig. 21.10).

RUNNING PROBLEM

Anish, who had always been healthy, was baffled. How could he have contracted cholera? But after talking with other team members, he realized that he hadn't been as careful about consuming only bottled water as he should have been. One of the doctors noticed that Anish's medical history form listed Nexium® (esomeprazole) among his current medications. "You know, taking Nexium might have also contributed to your contracting cholera."

Q3: *Esomeprazole is a proton pump inhibitor (PPI). For what symptom or condition might Anish have been taking this drug?*

Q4: *Why might taking a protein pump inhibitor like esomeprazole have increased Anish's chances of contracting cholera?*

653 – 657 – 670 – 674 – 681 – 686

The Stomach Balances Digestion and Defense

Under normal conditions, the gastric mucosa protects itself from autodigestion by acid and enzymes with a mucus-bicarbonate barrier. **Mucous cells** on the luminal surface and in the neck of gastric glands secrete both substances. The mucus forms a physical barrier, and the bicarbonate creates a chemical buffer barrier underlying the mucus (Fig. 21.9b).

Researchers using microelectrodes have shown that the bicarbonate layer just above the cell surface in the stomach has a pH that is close to 7, even when the pH in the lumen is highly acidic at pH 2. Mucus secretion is increased when the stomach is irritated, such as by the ingestion of aspirin (acetylsalicylic acid) or alcohol.

Even the protective mucus-bicarbonate barrier can fail at times. In *Zollinger-Ellison syndrome*, patients secrete excessive levels of gastrin, usually from gastrin-secreting tumors in the pancreas. As a result, hyperacidity in the stomach overwhelms the normal protective mechanisms and causes a peptic ulcer. In peptic ulcers, acid and pepsin destroy the mucosa, creating holes that extend into the submucosa and muscularis of the stomach and duodenum. *Acid reflux* into the esophagus can erode the mucosal layer there as well.

Excess acid secretion is an uncommon cause of peptic ulcers. By far the most common causes are nonsteroidal anti-inflammatory drugs (NSAIDs), such as aspirin, and *Helicobacter pylori*, a bacterium that creates inflammation of the gastric mucosa.

For many years the primary therapy for excess acid secretion, or *dyspepsia*, was the ingestion of *antacids*, agents that neutralize acid in the gastric lumen. But as molecular biologists discovered the mechanism for acid secretion by parietal cells, the potential for new therapies became obvious. Today, we have two classes of drugs to fight hyperacidity: the H_2 receptor antagonists and proton pump inhibitors that block the H^+-K^+-ATPase.

21.6 Integrated Function: The Intestinal Phase

Once chyme passes into the small intestine, the **intestinal phase** of digestion begins. Chyme entering the small intestine has undergone relatively little chemical digestion, so its entry must be controlled to avoid overwhelming the small intestine. Motility in the small intestine is also controlled. Intestinal contents are slowly propelled forward by a combination of segmental and peristaltic contractions. These actions mix chyme with enzymes and they expose digested nutrients to the mucosal epithelium for absorption. Forward movement of chyme through the intestine must be slow enough to allow digestion and absorption to go to completion. Parasympathetic innervation and the GI hormones gastrin and CCK promote intestinal motility; sympathetic innervation inhibits it.

About 5.5 liters of food, fluid, and secretions enter the small intestine each day, and about 3.5 liters of hepatic, pancreatic, and intestinal secretions are added there, making a total input of 9 liters into the lumen (see Fig. 21.3). All but about 1.5 liters of this volume is absorbed in the small intestine, mostly in the duodenum and jejunum.

The anatomy of the small intestine facilitates secretion, digestion, and absorption by maximizing surface area (**FIGS. 21.11** and 21.1f). At the macroscopic level, the surface of the lumen is sculpted into fingerlike villi and deep crypts. Most absorption takes place along the villi while fluid and hormone secretion and cell renewal from stem cells occurs in the crypts. On a microscopic level the apical surface of the enterocytes is modified into microvilli whose surfaces are covered with membrane-bound enzymes and a *glycocalyx* coat [p. 64]. The surface of the intestinal epithelium is called the **brush border** from the bristle-like appearance of the microvilli.

Most nutrients absorbed across the intestinal epithelium move into capillaries in the villi for distribution through the circulatory system. The exception is digested fats, most of which pass into lacteals of the lymphatic system. Venous blood from the digestive tract does not go directly back to the heart. Instead, it passes into the *hepatic portal system* [p. 435]. This specialized region of the circulation has two sets of capillary beds: one that picks up absorbed nutrients at the intestine, and another that delivers the nutrients directly to the liver (**FIG. 21.12**).

The delivery of absorbed materials directly to the liver underscores the importance of that organ as a biological filter.

FIG. 21.11 The villus and a crypt in the small intestine

Villi and crypts increase the effective surface area of the small intestine. Stem cells in the crypts produce new epithelial cells to replace those that die or are damaged. Most absorption occurs along the villi. Most fluid secretion occurs in the crypts.

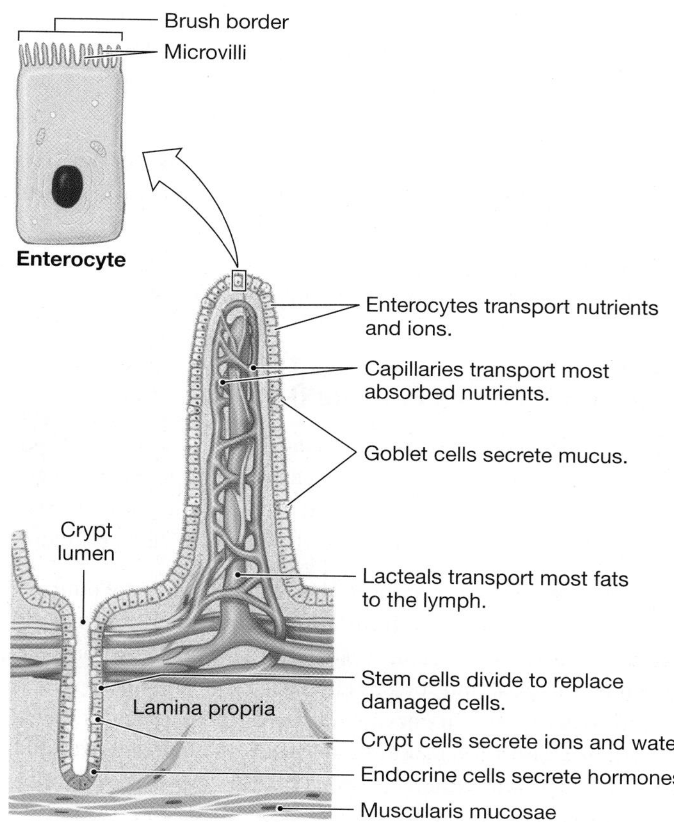
Brush border
Microvilli
Enterocyte
Enterocytes transport nutrients and ions.
Capillaries transport most absorbed nutrients.
Goblet cells secrete mucus.
Crypt lumen
Lacteals transport most fats to the lymph.
Lamina propria
Stem cells divide to replace damaged cells.
Crypt cells secrete ions and water.
Endocrine cells secrete hormones.
Muscularis mucosae

FIG. 21.12 The hepatic portal system

Most nutrients absorbed by the intestine pass through the liver, which serves as a filter that can remove potentially harmful xenobiotics before they get into the systemic circulation.

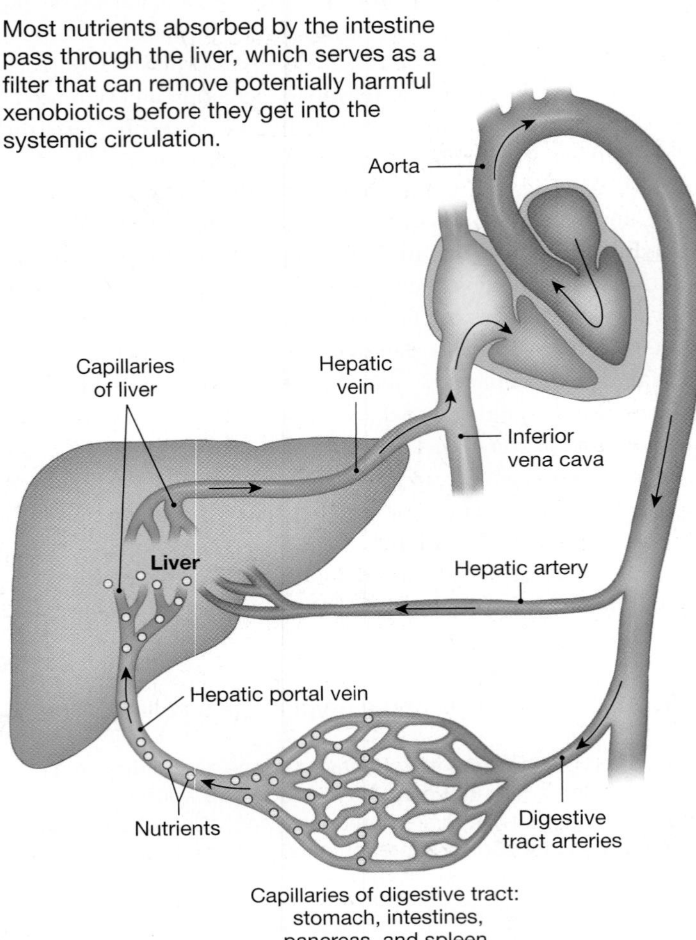
Aorta
Capillaries of liver
Hepatic vein
Inferior vena cava
Liver
Hepatic artery
Hepatic portal vein
Nutrients
Digestive tract arteries
Capillaries of digestive tract: stomach, intestines, pancreas, and spleen

Hepatocytes contain a variety of enzymes, such as the *cytochrome P450* isozymes, that metabolize drugs and xenobiotics and clear them from the bloodstream before they reach the systemic circulation. Hepatic clearance is one reason a drug administered orally must often be given in higher doses than the same drug administered by IV infusion.

Intestinal Secretions Promote Digestion

Each day, the liver, pancreas, and intestine produce more than 3 liters of secretions whose contents are necessary for completing the digestion of ingested nutrients. The added secretions include digestive enzymes, bile, bicarbonate, mucus, and an isotonic NaCl solution.

1. *Digestive enzymes* are produced by the intestinal epithelium and the exocrine pancreas. Intestinal brush border enzymes are anchored to the luminal cell membranes and are not swept out of the small intestine as chyme is propelled forward. The control pathways for enzyme release vary but include a variety of neural, hormonal, and paracrine signals. Usually, stimulation of parasympathetic neurons in the vagus nerve enhances enzyme secretion.
2. *Bile* made in the liver and released from the gall bladder is a nonenzymatic solution that facilitates the digestion of fats.
3. *Bicarbonate secretion* into the small intestine neutralizes the highly acidic chyme that enters from the stomach. Most bicarbonate comes from the pancreas and is released in response to neural stimuli and secretin.
4. *Mucus* from intestinal goblet cells protects the epithelium and lubricates the gut's contents.
5. An *isotonic NaCl solution* mixes with mucus to help lubricate the contents of the gut.

Isotonic NaCl Secretion Crypt cells in the small intestine and colon secrete an isotonic NaCl solution in a process similar to the initial step of salivation (**FIG. 21.13**). Chloride from the ECF enters cells via NKCC transporters, then exits into the lumen via an apical gated Cl^- channel known as the **cystic fibrosis transmembrane conductance regulator,** or **CFTR channel**. Movement of negatively charged Cl^- into the lumen draws Na^+ down the electrical gradient through leaky cell junctions. Water follows Na^+ along the osmotic gradient created by redistribution of NaCl. The result is secretion of isotonic saline solution.

The Pancreas Secretes Enzymes and Bicarbonate

The pancreas is an organ that contains both types of secretory epithelium: endocrine and exocrine [p. 79]. Endocrine secretions come from clusters of cells called *islets* and include the hormones insulin and glucagon (**FIG. 21.14**). Exocrine secretions include digestive enzymes and a watery solution of sodium bicarbonate, $NaHCO_3$.

FIG. 21.13 Isotonic NaCl secretion

Intestinal and colonic crypt cells and salivary gland acini secrete isotonic NaCl solutions.

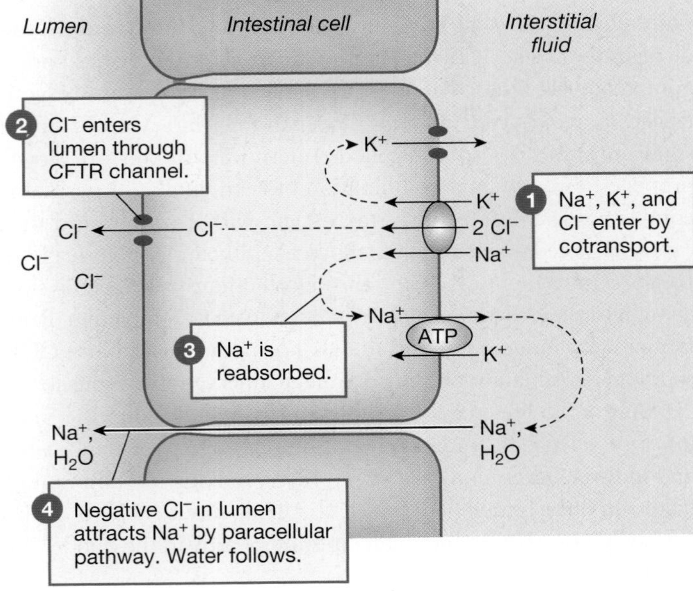
Lumen
Intestinal cell
Interstitial fluid
2 Cl⁻ enters lumen through CFTR channel.
K⁺
K⁺
2 Cl⁻
Na⁺
1 Na⁺, K⁺, and Cl⁻ enter by cotransport.
Cl⁻
Cl⁻
Cl⁻
Cl⁻
Na⁺
ATP
K⁺
3 Na⁺ is reabsorbed.
Na⁺, H₂O
Na⁺, H₂O
4 Negative Cl⁻ in lumen attracts Na⁺ by paracellular pathway. Water follows.

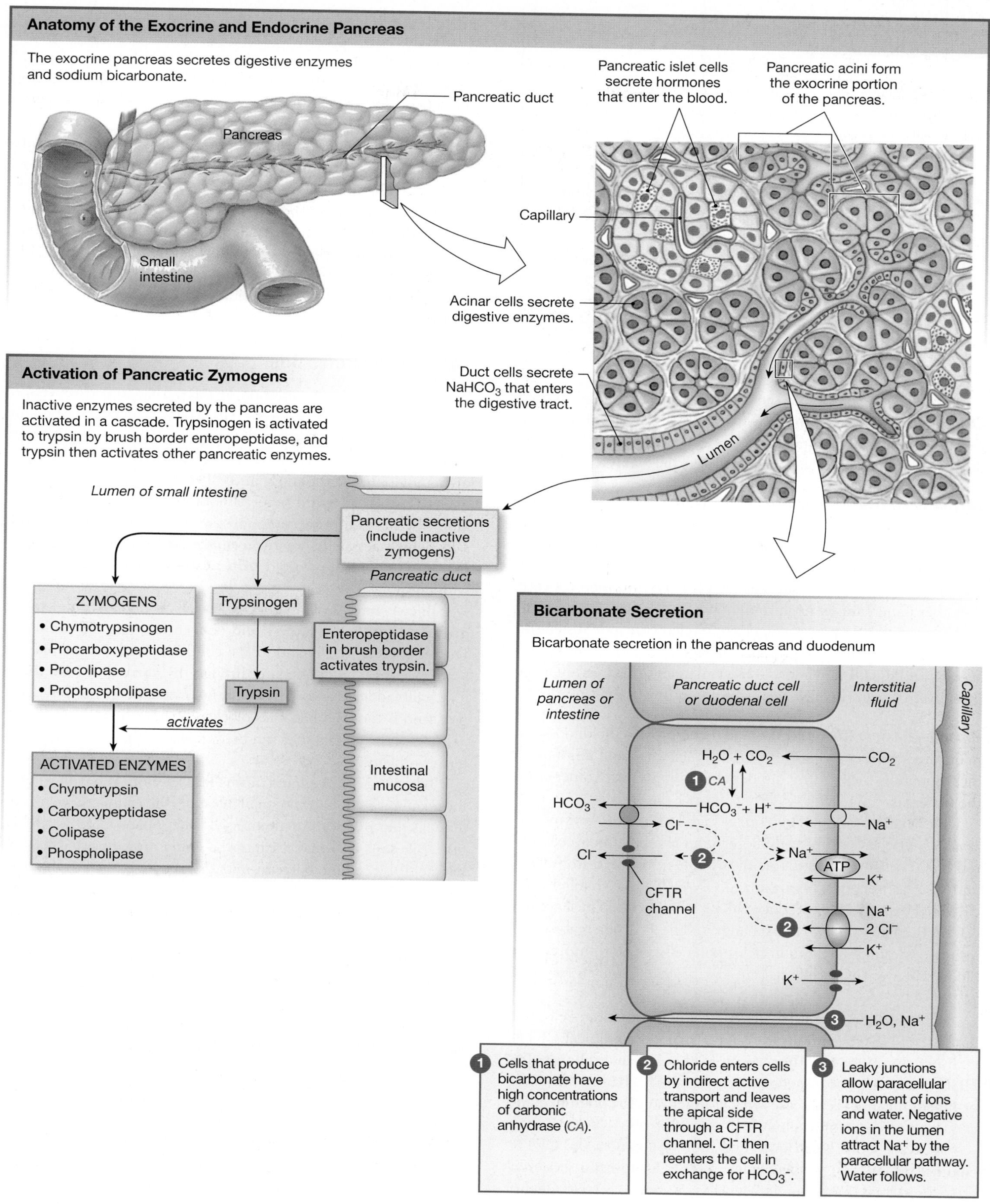
Anatomy of the Exocrine and Endocrine Pancreas
The exocrine pancreas secretes digestive enzymes and sodium bicarbonate.
Pancreatic duct
Pancreas
Small intestine
Pancreatic islet cells secrete hormones that enter the blood.
Pancreatic acini form the exocrine portion of the pancreas.
Capillary
Acinar cells secrete digestive enzymes.
Duct cells secrete $NaHCO_3$ that enters the digestive tract.
Lumen
Activation of Pancreatic Zymogens
Inactive enzymes secreted by the pancreas are activated in a cascade. Trypsinogen is activated to trypsin by brush border enteropeptidase, and trypsin then activates other pancreatic enzymes.
Lumen of small intestine
Pancreatic secretions (include inactive zymogens)
Pancreatic duct
ZYMOGENS
• Chymotrypsinogen
• Procarboxypeptidase
• Procolipase
• Prophospholipase
Trypsinogen
Enteropeptidase in brush border activates trypsin.
Trypsin
activates
ACTIVATED ENZYMES
• Chymotrypsin
• Carboxypeptidase
• Colipase
• Phospholipase
Intestinal mucosa
Bicarbonate Secretion
Bicarbonate secretion in the pancreas and duodenum
Lumen of pancreas or intestine
Pancreatic duct cell or duodenal cell
Interstitial fluid
Capillary
$H_2O + CO_2$
CO_2
1 CA
HCO_3^-
$HCO_3^- + H^+$
Cl^-
Na^+
Cl^-
2
Na^+
ATP
K^+
CFTR channel
Na^+
2 $2\ Cl^-$
K^+
K^+
3 H_2O, Na^+
1 Cells that produce bicarbonate have high concentrations of carbonic anhydrase (CA).
2 Chloride enters cells by indirect active transport and leaves the apical side through a CFTR channel. Cl^- then reenters the cell in exchange for HCO_3^-.
3 Leaky junctions allow paracellular movement of ions and water. Negative ions in the lumen attract Na^+ by the paracellular pathway. Water follows.

RUNNING PROBLEM

A hallmark of *Vibrio cholerae* infection is profuse, isosmotic diarrhea sometimes said to resemble "rice water." The toxin secreted by *Vibrio cholerae* is a protein complex with six subunits. Cholera toxin binds to intestinal cells, and the A subunit is taken into the enterocytes by endocytosis. Once inside the enterocyte, the toxin turns on adenylyl cyclase, which then produces cAMP continuously. Because the CFTR channel of the enterocyte is a cAMP-gated channel, the effect of cholera toxin is to open the CFTR channels and keep them open.

Q5: *Why would continuously open enterocyte CFTR channels cause secretory diarrhea and dehydration in humans?*

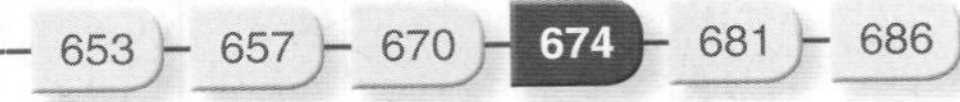
653
657
670
674
681
686

The exocrine portion of the pancreas consists of lobules called *acini*, similar to those of the salivary glands. Ducts from the acini empty into the duodenum (Fig. 21.14a). The acinar cells secrete digestive enzymes, and the duct cells secrete the $NaHCO_3$ solution.

Enzyme Secretion Most pancreatic enzymes are secreted as zymogens that must be activated upon arrival in the intestine. This activation process is a cascade that begins when brush border **enteropeptidase** (previously called *enterokinase*) converts inactive trypsinogen to active trypsin (Fig. 21.14b). Trypsin then converts the other pancreatic zymogens to their active forms.

The signals for pancreatic enzyme release include distension of the small intestine, the presence of food in the intestine, neural signals, and the GI hormone CCK. Pancreatic enzymes enter the intestine in a watery fluid that also contains bicarbonate.

Bicarbonate Secretion Bicarbonate secretion into the duodenum neutralizes acid entering from the stomach. A small amount of bicarbonate is secreted by duodenal cells, but most comes from the pancreas.

Bicarbonate production requires high levels of the enzyme *carbonic anhydrase*, levels similar to those found in renal tubule cells and red blood cells [pp. 575, 646]. Bicarbonate produced from CO_2 and water is secreted by an apical Cl^--HCO_3^- exchanger (Fig. 21.14c). Hydrogen ions produced along with bicarbonate leave the cell on basolateral Na^+-H^+ exchangers. The H^+ thus reabsorbed into the intestinal circulation helps balance HCO_3^- put into the blood when parietal cells secrete H^+ into the stomach (see Fig. 21.9c).

The chloride for bicarbonate exchange enters the cell on a basolateral NKCC cotransporter and leaves via an apical CFTR channel. Luminal Cl^- then reenters the cell in exchange for HCO_3^- entering the lumen. Defects in CFTR channel structure or function cause the disease *cystic fibrosis*, and disruption of pancreatic secretion is one hallmark of cystic fibrosis.

In cystic fibrosis, an inherited mutation causes the CFTR channel protein to be defective or absent. As a result, secretion of Cl^- and fluid ceases but goblet cells continue to secrete mucus, resulting in thickened mucus. In the digestive system, the thick mucus clogs small pancreatic ducts and prevents digestive enzyme secretion into the intestine. In airways of the respiratory system, where the CFTR channel is also found, failure to secrete fluid clogs the mucociliary escalator [Fig. 17.5c, p. 539] with thick mucus, leading to recurrent lung infections.

In both the pancreas and intestinal crypts, sodium and water secretion is a passive process, driven by electrochemical and osmotic gradients. The movement of negative ions from the ECF to the lumen creates a lumen-negative electrical gradient that attracts Na^+. Sodium moves down its electrochemical gradient through leaky junctions between the cells. The transfer of Na^+ and HCO_3^- from ECF into the lumen creates an osmotic gradient, and water follows by osmosis. The net result is secretion of a watery sodium bicarbonate solution.

The Liver Secretes Bile

Bile is a nonenzymatic solution secreted from **hepatocytes,** or liver cells (see *Focus On: The Liver*, **FIG. 21.15**). The key components of bile are (1) **bile salts,** which facilitate enzymatic fat digestion, (2) *bile pigments*, such as bilirubin, which are the waste products of hemoglobin degradation, and (3) *cholesterol*, which is excreted in the feces. Drugs and other xenobiotics are cleared from the blood by hepatic processing and are also excreted in bile. Bile salts, which act as detergents to make fats soluble during digestion, are made from steroid **bile acids** combined (conjugated) with amino acids.

Bile secreted by hepatocytes travels in hepatic ducts to the **gallbladder,** which stores and concentrates the bile solution. During a meal that includes fats, contraction of the gallbladder sends bile into the duodenum through the **common bile duct.** The gallbladder is an organ that is not essential for normal digestion, and if the duct becomes blocked by hard deposits known as gallstones, the gallbladder can be removed without creating long-term problems.

Bile salts are not altered during fat digestion. When they reach the terminal section of the small intestine (the ileum), they encounter cells that reabsorb them and send them back into the circulation. Bile salts that make it into the colon are converted back to bile acids by colonic bacteria and also recycled back to the liver. Through the hepatic portal vein, bile salts return to the liver, where the hepatocytes take them back up and resecrete them. The recirculation of bile salts is essential to fat digestion because the body's pool of bile salts must cycle from two to five times for each meal. Bilirubin and other wastes secreted in bile cannot be reabsorbed and are excreted in the feces.

Most Digestion Occurs in the Small Intestine

The intestinal, pancreatic, and hepatic secretion of enzymes and bile is essential for normal digestive function. Although a significant amount of mechanical digestion takes place in the mouth and stomach, chemical digestion of food there is limited to a small amount of starch breakdown and incomplete protein digestion in the stomach. When chyme enters the small intestine, protein

FIG.21.15 FOCUS ON . . . The Liver

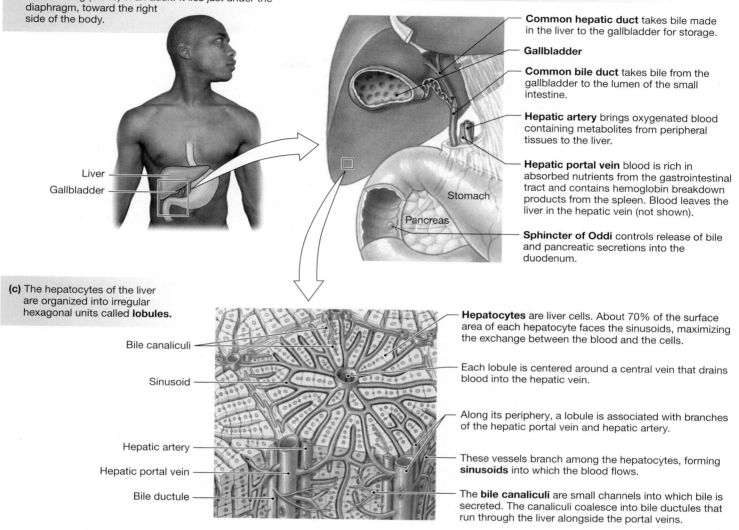

(a) The liver is the largest of the internal organs, weighing about 1.5 kg (3.3 lb) in an adult. It lies just under the diaphragm, toward the right side of the body.

Liver

Gallbladder

(b) Gallbladder and bile ducts

Common hepatic duct takes bile made in the liver to the gallbladder for storage.

Gallbladder

Common bile duct takes bile from the gallbladder to the lumen of the small intestine.

Hepatic artery brings oxygenated blood containing metabolites from peripheral tissues to the liver.

Hepatic portal vein blood is rich in absorbed nutrients from the gastrointestinal tract and contains hemoglobin breakdown products from the spleen. Blood leaves the liver in the hepatic vein (not shown).

Sphincter of Oddi controls release of bile and pancreatic secretions into the duodenum.

Stomach

Pancreas

(c) The hepatocytes of the liver are organized into irregular hexagonal units called **lobules.**

Bile canaliculi

Sinusoid

Hepatic artery

Hepatic portal vein

Bile ductule

Hepatocytes are liver cells. About 70% of the surface area of each hepatocyte faces the sinusoids, maximizing the exchange between the blood and the cells.

Each lobule is centered around a central vein that drains blood into the hepatic vein.

Along its periphery, a lobule is associated with branches of the hepatic portal vein and hepatic artery.

These vessels branch among the hepatocytes, forming **sinusoids** into which the blood flows.

The **bile canaliculi** are small channels into which bile is secreted. The canaliculi coalesce into bile ductules that run through the liver alongside the portal veins.

(d) Blood entering the liver brings nutrients and foreign substances from the digestive tract, bilirubin from hemoglobin breakdown, and metabolites from peripheral tissues of the body. In turn, the liver excretes some of these in the bile and stores or metabolizes others. Some of the liver's products are wastes to be excreted by the kidney; others are essential nutrients, such as glucose. In addition, the liver synthesizes an assortment of plasma proteins.

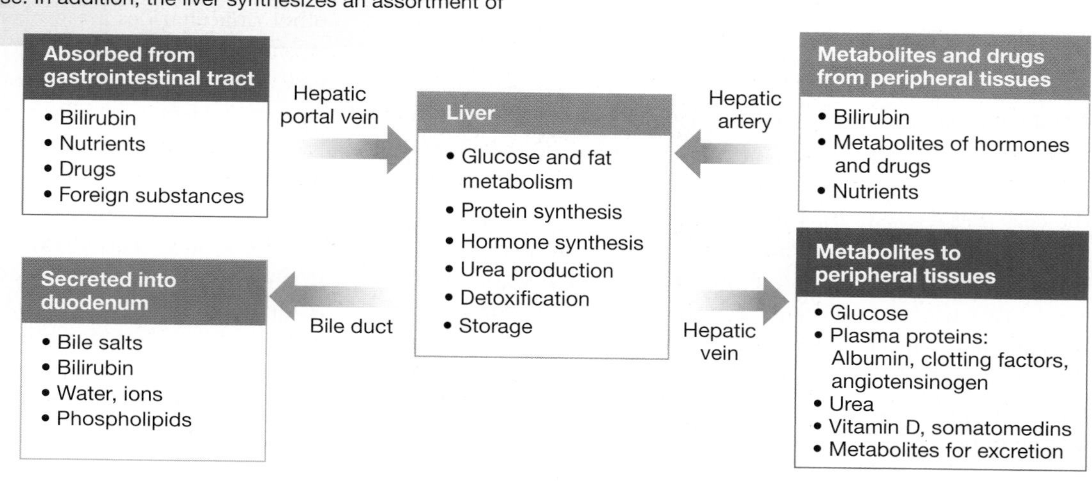
Absorbed from gastrointestinal tract
• Bilirubin
• Nutrients
• Drugs
• Foreign substances
Hepatic portal vein
Liver
• Glucose and fat metabolism
• Protein synthesis
• Hormone synthesis
• Urea production
• Detoxification
• Storage
Hepatic artery
Metabolites and drugs from peripheral tissues
• Bilirubin
• Metabolites of hormones and drugs
• Nutrients
Secreted into duodenum
• Bile salts
• Bilirubin
• Water, ions
• Phospholipids
Bile duct
Hepatic vein
Metabolites to peripheral tissues
• Glucose
• Plasma proteins: Albumin, clotting factors, angiotensinogen
• Urea
• Vitamin D, somatomedins
• Metabolites for excretion

digestion stops when pepsin is inactivated at the higher intestinal pH. Pancreatic and brush border enzymes then finish digestion of peptides, carbohydrates, and fats into smaller molecules that can be absorbed.

Bile Salts Facilitate Fat Digestion

Fats and related molecules in the Western diet include triglycerides, cholesterol, phospholipids, long-chain fatty acids, and the fat-soluble vitamins [Fig. 2.1, p. 30]. Nearly 90% of our fat calories come from triglycerides because they are the primary form of lipid in both plants and animals.

Fat digestion is complicated by the fact that most lipids are not particularly water soluble. As a result, the aqueous chyme leaving the stomach contains a coarse emulsion of large fat droplets, which have less surface area than smaller particles. To increase the surface area available for enzymatic fat digestion, the liver secretes bile salts into the small intestine (**FIG. 21.16a**). Bile salts help break down the coarse emulsion into smaller, more stable particles.

Bile salts, like phospholipids of cell membranes, are *amphipathic* {*amphi-*, on both sides + *pathos*, experience}, meaning that they have both a hydrophobic region and a hydrophilic region. The hydrophobic regions of bile salts associate with the surface of lipid droplets while the polar side chains interact with water, creating a stable emulsion of small, water-soluble fat droplets (Fig. 21.16a). You can see a similar emulsion when you shake a bottle of salad dressing to combine the oil and aqueous layers.

Enzymatic fat digestion is carried out by **lipases,** enzymes that remove two fatty acids from each triglyceride molecule. The result is one monoglyceride and two free fatty acids (Fig. 21.16c). The bile salt coating of the intestinal emulsion complicates digestion, however, because lipase is unable to penetrate the bile salts. For this reason, fat digestion also requires **colipase,** a protein cofactor secreted by the pancreas. Colipase displaces some bile salts, allowing lipase access to fats inside the bile salt coating.

Phospholipids are digested by pancreatic *phospholipase*. Free cholesterol is not digested and is absorbed intact.

As enzymatic and mechanical digestion proceed, fatty acids, bile salts, mono- and diglycerides, phospholipids, and cholesterol coalesce to form small disk-shaped **micelles** (Fig. 21.16b) [p. 63]. Micelles then enter the unstirred aqueous layer at the edge of the brush border.

Fat Absorption Lipophilic fats such as fatty acids and monoglycerides are absorbed primarily by simple diffusion. They move out of their micelles and diffuse across the enterocyte membrane into the cells (Fig. 21.16d). Initially scientists believed that cholesterol also diffused across the enterocyte membrane, but the discovery of a drug called *ezetimibe* that inhibits cholesterol absorption suggested that transport proteins were involved. Experiments now indicate that some cholesterol is transported across the brush border membrane on specific, energy-dependent membrane transporters, including one named *NPC1L1*, the protein that is inhibited by ezetimibe.

Once monoglycerides and fatty acids are inside the enterocytes, they move to the smooth endoplasmic reticulum, where they recombine into triglycerides (Fig. 21.16d). The triglycerides then join cholesterol and proteins to form large droplets called **chylomicrons**. Because of their size, chylomicrons must be packaged into secretory vesicles by the Golgi. The chylomicrons then leave the cell by exocytosis.

The large size of chylomicrons also prevents them from crossing the basement membrane of capillaries (Fig. 21.16d). Instead, chylomicrons are absorbed into *lacteals*, the lymph vessels of the villi. Chylomicrons pass through the lymphatic system and finally enter the venous blood just before it flows into the right side of the heart [p. 499].

Some shorter fatty acids (10 or fewer carbons) are not assembled into chylomicrons. These fatty acids can therefore cross the capillary basement membrane and go directly into the blood.

Concept Check

12. Do bile salts digest triglycerides into monoglycerides and free fatty acids?
13. Bile acids are reabsorbed in the distal intestine by an apical sodium-dependent bile acid transporter (ASBT) and a basolateral organic anion transporter (OAT). Draw one enterocyte. Label the lumen, ECF, and basolateral and apical sides. Diagram bile acid reabsorption as described.
14. Explain how pH can be used to predict the location where a particular digestive enzyme might be most active.

Carbohydrates Are Absorbed as Monosaccharides

About half the calories the average American ingests are in the form of carbohydrates, mainly *starch* and *sucrose* (table sugar). Other dietary carbohydrates include the glucose polymers *glycogen* and *cellulose*, disaccharides such as *lactose* (milk sugar) and *maltose*, and the monosaccharides *glucose* and *fructose* [Fig. 2.2, p. 31]. The enzyme *amylase* breaks long glucose polymers into smaller glucose chains and into the disaccharide maltose (**FIG. 21.17a**).

Starch digestion starts in the mouth with salivary amylase but that enzyme is denatured in the acidic stomach. Pancreatic amylase then resumes digestion of starch into maltose. Maltose and other disaccharides are broken down by intestinal brush-border enzymes known as **disaccharidases** (maltase, sucrase, and lactase). The absorbable end products of carbohydrate digestion are glucose, galactose, and fructose.

Because intestinal carbohydrate absorption is restricted to monosaccharides, all larger carbohydrates must be digested if they are to be used by the body. The complex carbohydrates we can digest are starch and glycogen. We are unable to digest cellulose because we lack the necessary enzymes. As a result, the cellulose in plant matter becomes what is known as dietary *fiber* or *roughage* and is excreted undigested. Similarly, *sucralose* (Splenda®), the artificial sweetener made from sucrose, cannot be digested because chlorine atoms substituted for three hydroxyl groups block enzymatic digestion of the sugar derivative.

FIG. 21.16 ESSENTIALS Digestion and Absorption of Fats

Most lipids are hydrophobic and must be emulsified to facilitate digestion in the aqueous environment of the intestine.

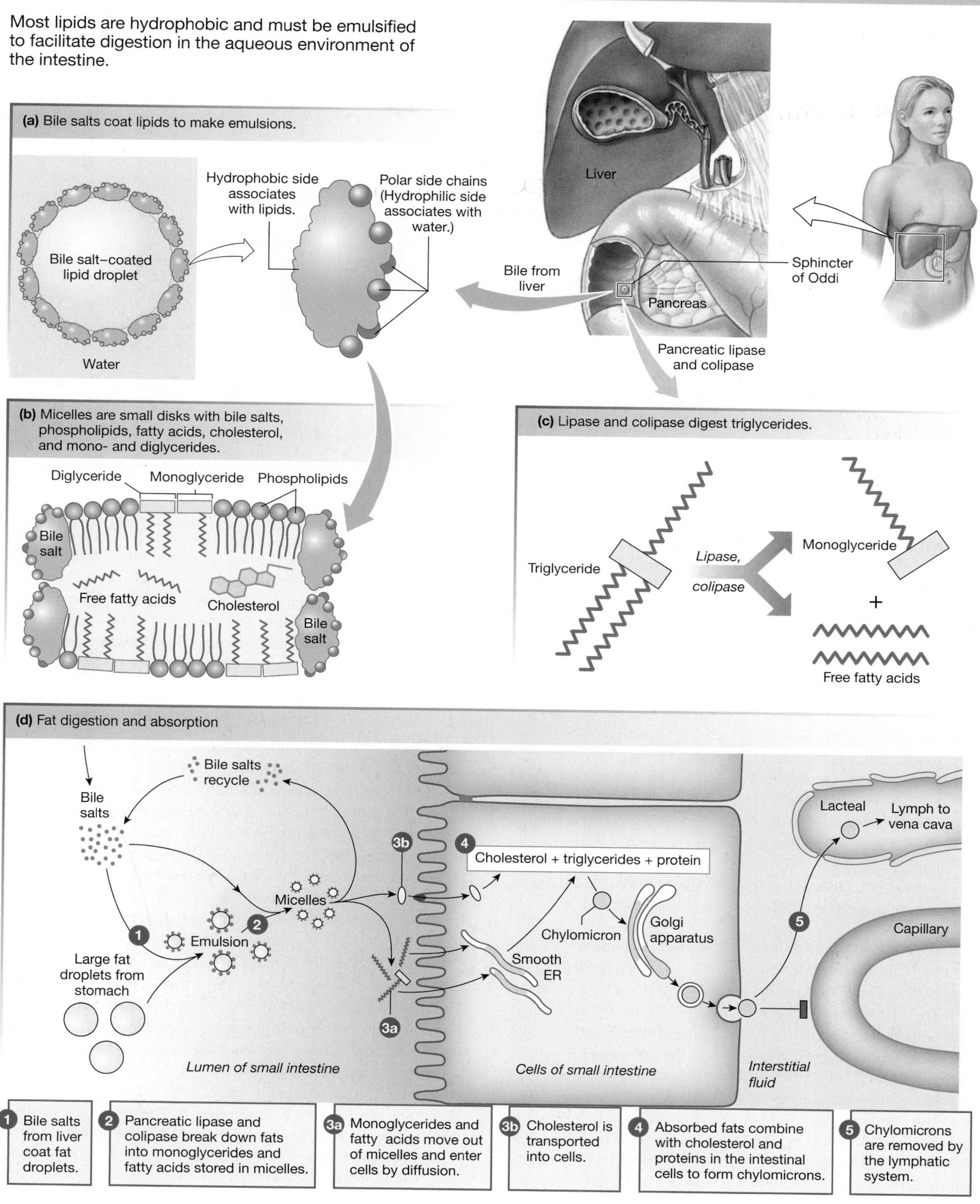

(a) Bile salts coat lipids to make emulsions.
Bile salt–coated lipid droplet
Water
Hydrophobic side associates with lipids.
Polar side chains (Hydrophilic side associates with water.)
Liver
Bile from liver
Sphincter of Oddi
Pancreas
Pancreatic lipase and colipase
(b) Micelles are small disks with bile salts, phospholipids, fatty acids, cholesterol, and mono- and diglycerides.
Diglyceride
Monoglyceride
Phospholipids
Bile salt
Free fatty acids
Cholesterol
Bile salt
(c) Lipase and colipase digest triglycerides.
Triglyceride
Lipase, colipase
Monoglyceride
+
Free fatty acids
(d) Fat digestion and absorption
Bile salts recycle
Bile salts
Micelles
Emulsion
Large fat droplets from stomach
Cholesterol + triglycerides + protein
Chylomicron
Golgi apparatus
Smooth ER
Lacteal
Lymph to vena cava
Capillary
Lumen of small intestine
Cells of small intestine
Interstitial fluid
1 Bile salts from liver coat fat droplets.
2 Pancreatic lipase and colipase break down fats into monoglycerides and fatty acids stored in micelles.
3a Monoglycerides and fatty acids move out of micelles and enter cells by diffusion.
3b Cholesterol is transported into cells.
4 Absorbed fats combine with cholesterol and proteins in the intestinal cells to form chylomicrons.
5 Chylomicrons are removed by the lymphatic system.

FIG. 21.17 ESSENTIALS Digestion and Absorption of Carbohydrates

Most carbohydrates in our diets are disaccharides and complex carbohydrates. Cellulose is not digestible. All other carbohydrates must be digested to monosaccharides before they can be absorbed.

(a) Carbohydrates break down into monosaccharides.

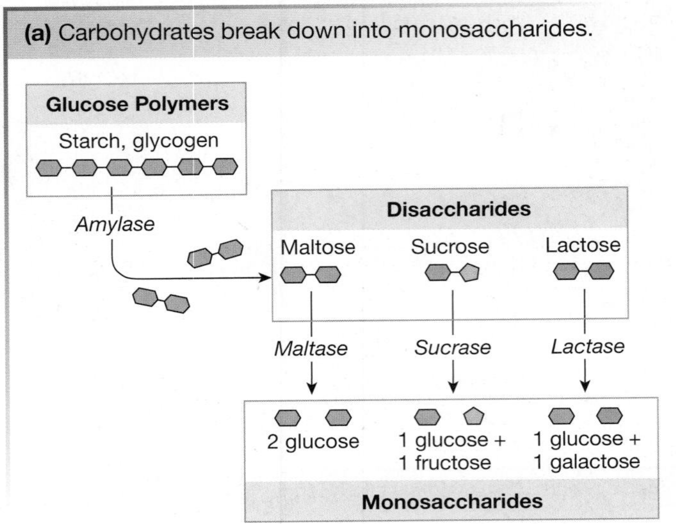

Glucose Polymers
Starch, glycogen
Amylase
Disaccharides
Maltose
Sucrose
Lactose
Maltase
Sucrase
Lactase
2 glucose
1 glucose + 1 fructose
1 glucose + 1 galactose
Monosaccharides

(b) Carbohydrate absorption in the small intestine

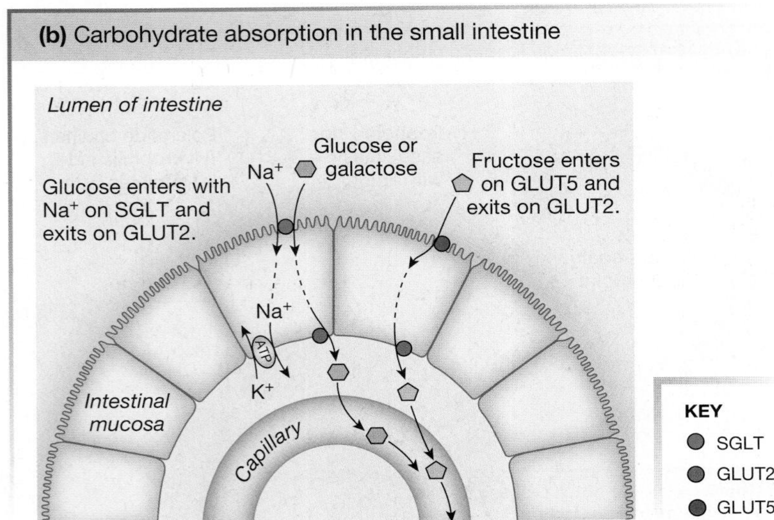

Lumen of intestine
Glucose enters with Na+ on SGLT and exits on GLUT2.
Na+
Glucose or galactose
Fructose enters on GLUT5 and exits on GLUT2.
Na+
ATP
K+
Intestinal mucosa
Capillary
KEY
SGLT
GLUT2
GLUT5

Lactose Intolerance Lactose, or milk sugar, is a disaccharide composed of glucose and galactose. Ingested lactose must be digested accomplished by the intestinal brush border enzyme *lactase* before it can be absorbed. Generally, lactase is found only in juvenile mammals, except in some humans of European descent. Those people inherit a dominant gene that allows them to produce lactase after childhood. Scientists believe the lactase gene provided a selective advantage to their ancestors, who developed a culture in which milk and milk products played an important role. In cultures where dairy products are not part of the diet after weaning, most adults lack the gene and synthesize less intestinal lactase.

Decreased lactase activity is associated with a condition known as *lactose intolerance*. If a person with lactose intolerance drinks milk or eats dairy products, diarrhea may result. In addition, bacteria in the large intestine ferment lactose to gas and organic acids, leading to bloating and *flatulence* (intestinal gas). The simplest treatment for lactose intolerance is to remove milk products from the diet, although milk predigested with lactase is available.

Carbohydrate Absorption Intestinal glucose and galactose absorption uses transporters identical to those found in the renal proximal tubule: the apical Na^+-glucose SGLT symporter and the basolateral GLUT2 transporter (Fig. 21.17b). These transporters move galactose as well as glucose. Fructose absorption, however, is not Na^+-dependent. Fructose moves across the apical membrane by facilitated diffusion on the GLUT5 transporter and across the basolateral membrane by GLUT2 [p. 144].

How are enterocytes able to keep intracellular glucose concentrations high so that facilitated diffusion moves glucose into the extracellular space? In most cells, glucose is the major metabolic substrate for aerobic respiration and is immediately phosphorylated when it enters the cell [p. 141]. However, the metabolism of enterocytes (and proximal tubule cells) apparently differs from that of most other cells. These transporting epithelial cells do not use glucose as their preferred energy source. Current studies indicate that these cells use the amino acid glutamine as their main source of energy, thus allowing absorbed glucose to pass unchanged into the bloodstream.

Proteins Are Digested into Small Peptides and Amino Acids

Unlike carbohydrates, which are ingested in forms ranging from simple to complex, most ingested proteins are polypeptides or larger [Fig. 2.3, p. 32]. Not all proteins are equally digestible by humans, however. Plant proteins are the least digestible. Among the most digestible is egg protein, 85–90% of which is in a form that can be digested and absorbed. Surprisingly, between 30% and 60% of the protein found in the intestinal lumen comes not from ingested food but from the sloughing of dead cells and from protein secretions such as enzymes and mucus.

The enzymes for protein digestion are classified into two broad groups: endopeptidases and exopeptidases (**FIG. 21.18B**). **Endopeptidases**, more commonly called **proteases**, attack peptide bonds in the interior of the amino acid chain and break a long peptide chain into smaller fragments. Proteases are secreted as inactive *proenzymes* (zymogens) from epithelial cells in the stomach, intestine, and pancreas. They are activated once they reach the GI tract lumen. Examples of proteases include **pepsin** secreted in the stomach, and **trypsin** and **chymotrypsin** secreted by the pancreas.

Exopeptidases release single amino acids from peptides by chopping them off the ends, one at a time. *Aminopeptidases* act on

FIG. 21.18 ESSENTIALS Digestion and Absorption of Proteins

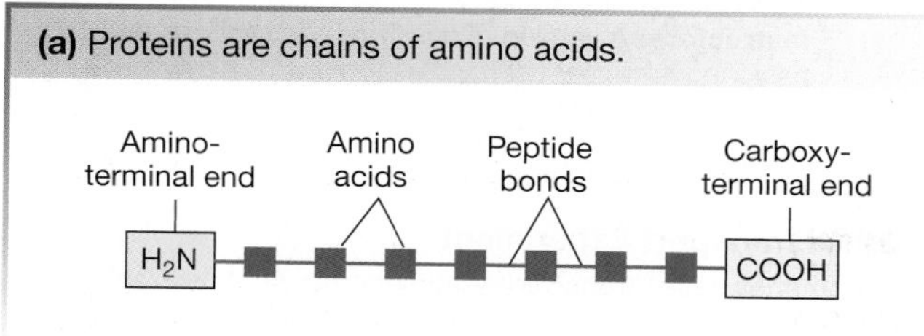
(a) Proteins are chains of amino acids.
Amino-terminal end
Amino acids
Peptide bonds
Carboxy-terminal end
H2N
COOH

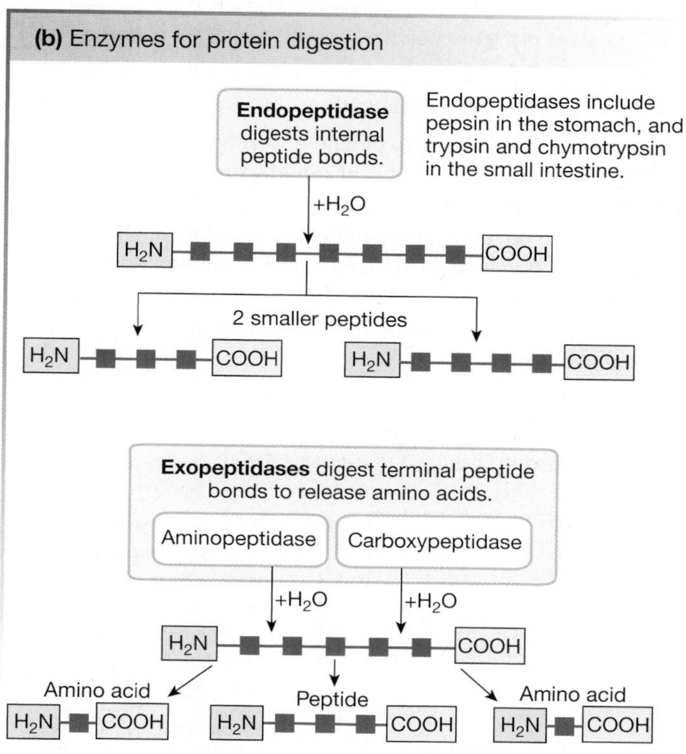
(b) Enzymes for protein digestion
Endopeptidase digests internal peptide bonds.
Endopeptidases include pepsin in the stomach, and trypsin and chymotrypsin in the small intestine.
+H2O
H2N
COOH
2 smaller peptides
Exopeptidases digest terminal peptide bonds to release amino acids.
Aminopeptidase
Carboxypeptidase
+H2O
+H2O
Amino acid
Peptide
Amino acid

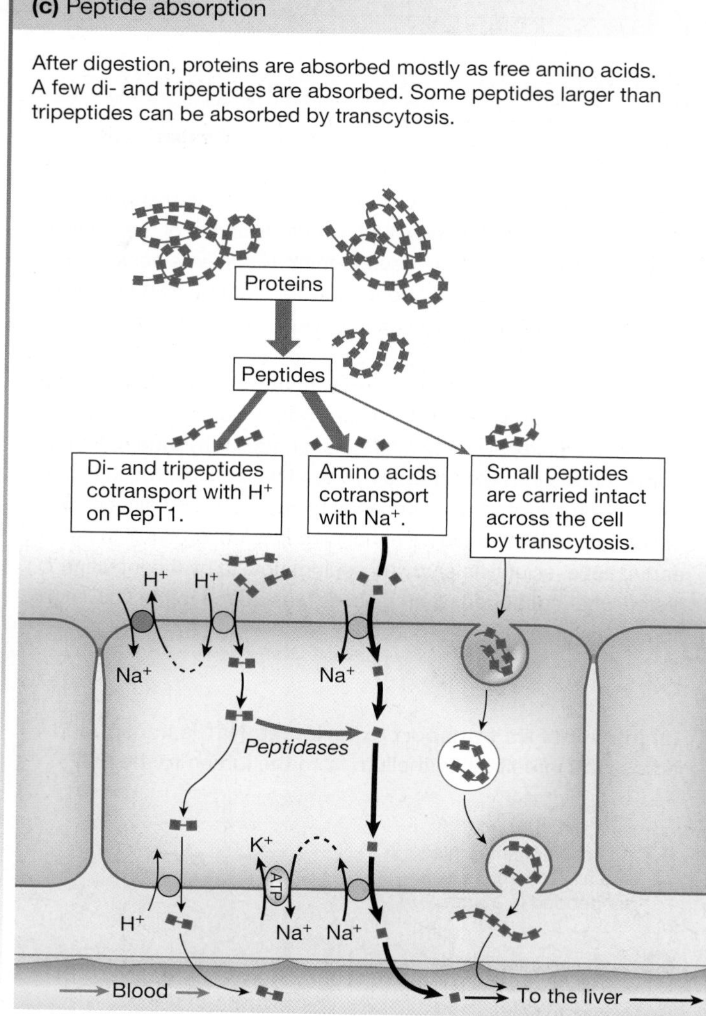
(c) Peptide absorption
After digestion, proteins are absorbed mostly as free amino acids. A few di- and tripeptides are absorbed. Some peptides larger than tripeptides can be absorbed by transcytosis.
Proteins
Peptides
Di- and tripeptides cotransport with H+ on PepT1.
Amino acids cotransport with Na+.
Small peptides are carried intact across the cell by transcytosis.
H+
Na+
Peptidases
K+
ATP
Blood
To the liver

the amino-terminal end of the protein; *carboxypeptidases* act at the carboxy-terminal end. The most important digestive exopeptidases are two isozymes of carboxypeptidase secreted by the pancreas. Aminopeptidases play a lesser role in digestion.

> **Concept Check**
>
> **15.** What activates pepsinogen, trypsinogen, and chymotrypsinogen?

Peptide Absorption The primary products of protein digestion are free amino acids, dipeptides, and tripeptides, all of which can be absorbed. Amino acid structure is so variable that multiple amino acid transport systems occur in the intestine. Most free amino acids are carried by Na^+-dependent cotransport proteins similar to those in the proximal tubule of the kidney (Fig. 21.18b). A few amino acid transporters are H^+-dependent.

Dipeptides and tripeptides are carried into enterocytes on the oligopeptide transporter *PepT1* that uses H^+-dependent cotransport (Fig. 21.18c). Once inside the epithelial cell, these *oligopeptides* {*oligos*, little} have two possible fates. Most are digested by cytoplasmic peptidases into individual amino acids, which are then transported across the basolateral membrane and into the circulation. Those oligopeptides that are not digested are transported intact across the basolateral membrane on an H^+-dependent exchanger. The transport system that moves oligopeptides also is responsible for intestinal uptake of certain drugs, including some antibiotics, angiotensin-converting enzyme inhibitors, and thrombin inhibitors.

Some Larger Peptides Can Be Absorbed Intact

Some peptides larger than three amino acids are absorbed by transcytosis [p. 151] after binding to membrane receptors on the luminal surface of the intestine. The discovery that ingested proteins can be absorbed as small peptides has implications in medicine because these peptides may act as *antigens*, substances that stimulate antibody formation and result in allergic reactions. Consequently, the intestinal absorption of peptides may be a significant factor in the development of food allergies and food intolerances.

Instructors: A version of this Try It! Activity can be assigned in Mastering Anatomy & Physiology

TRY IT! Oral Rehydration Therapy

The World Health Organization (WHO) estimates that diarrhea is the second largest killer of children under five, with nearly 2 billion cases of diarrhea annually. Cholera, a water-borne diarrheal disease caused by the bacteria *Vibrio cholera*, is one of these killers. Clinicians and researchers used to think *V. cholerae* destroyed the intestinal mucosa, so early treatments for cholera included restricting oral intake of fluids to give the bowel a chance to "rest" and recover.

In the late 1950s, however, researchers showed that the intestinal mucosa of patients with cholera was not damaged. In this case, could fluids given by mouth adequately rehydrate patients? Some reports indicated that dehydrated patients improved faster when given carrot soup or bananas by mouth. Could glucose from the carrots and bananas be enhancing fluid absorption? In the early 1960s, scientists studying sodium uptake by the intestine tested what happened when glucose was added to the NaCl solution in the lumen.

Intestinal Na Transport Experiment

The graph shows Na^+ transport across the intestine without glucose and after glucose was added to the lumen of the intestinal sac.

- What effect did adding glucose have on Na^+ transport?
- How would this graph differ if glucose was added to the serosal (i.e., ECF) instead of the mucosal (i.e., luminal) side of the epithelium?
- What did adding glucose to a saline solution in the lumen do to water absorption by the intestine?

Using the basic principles of fluid absorption in the intestine led to the creation of **oral rehydration therapy**: packets of NaCl and glucose that can be added to water and administered to patients. This simple, inexpensive, and life-saving intervention is used in many cases of cholera and in other diarrheal diseases. Oral rehydration powders have saved so many lives that these powders are included in WHO's list of essential medicines.

(a) Intestinal Na transport experiment: Na^+ is transported across the intestinal epithelium from the lumen to the ECF.

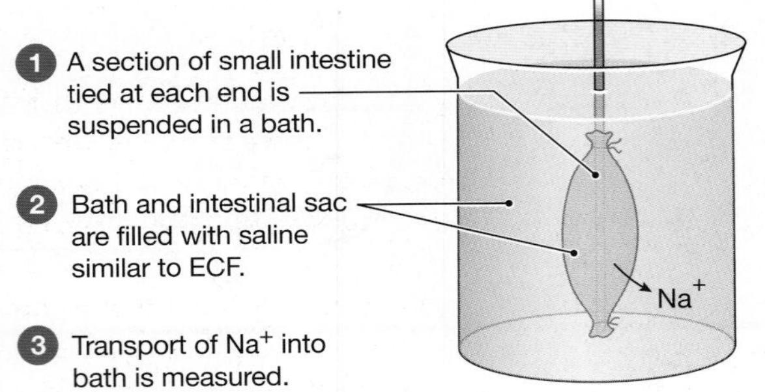
1 A section of small intestine tied at each end is suspended in a bath.
2 Bath and intestinal sac are filled with saline similar to ECF.
3 Transport of Na+ into bath is measured.
Na+

(b) Graph 1: *Data from Schultz and Zalusky, 1963.*

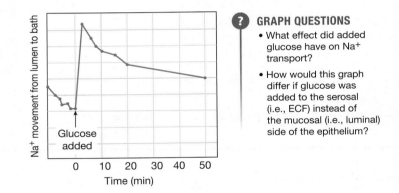

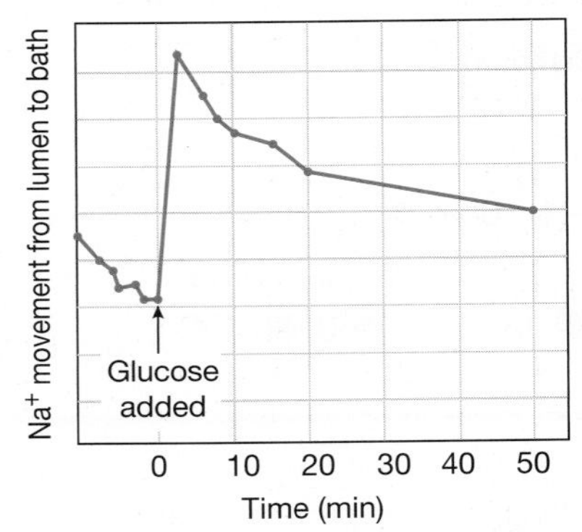
Na+ movement from lumen to bath
Glucose added
0
10
20
30
40
50
Time (min)

GRAPH QUESTIONS

- What effect did added glucose have on Na^+ transport?
- How would this graph differ if glucose was added to the serosal (i.e., ECF) instead of the mucosal (i.e., luminal) side of the epithelium?

Schultz S. G. & Zalusky R. The interaction between active sodium transport and active sugar transport in the isolated rabbit ileum. *Biochim Biophys Acta*, 71, 503–505 (1963).

In newborns, peptide absorption takes place primarily in intestinal crypt cells (Fig. 21.11). At birth, intestinal villi are very small, so the crypts are well exposed to the luminal contents. As the villi grow and the crypts have less access to chyme, the high peptide absorption rates present at birth decline steadily. If parents delay feeding the infant allergy-inducing peptides, the gut has a chance to mature, lessening the likelihood of antibody formation.

One of the most common antigens responsible for food allergies is gluten, a component of wheat. The incidence of childhood gluten allergies has decreased since the 1970s, when parents were cautioned not to feed infants gluten-based cereals until they were several months old.

In another medical application, pharmaceutical companies have developed indigestible peptide drugs that can be given orally instead of by injection. Probably the best-known example is *DDAVP* (1-deamino-8-D-arginine vasopressin), the synthetic analog of vasopressin. If the natural hormone vasopressin is ingested, it is digested rather than absorbed intact. By changing the structure of the hormone slightly, scientists created a synthetic peptide that has the same activity but is absorbed without being digested.

Nucleic Acids Are Digested into Bases and Monosaccharides

The nucleic acid polymers DNA and RNA are only a very small part of most diets. They are digested by pancreatic and intestinal enzymes, first into their component nucleotides and then into nitrogenous bases and monosaccharides [Fig. 2.4, p. 34]. The bases are absorbed by active transport, and the monosaccharides are absorbed by facilitated diffusion and secondary active transport, as other simple sugars are.

The Intestine Absorbs Vitamins and Minerals

In general, the fat-soluble vitamins (A, D, E, and K) are absorbed in the small intestine along with fats—one reason that health professionals are concerned about excessive consumption of "fake fats," such as Olestra, that are not absorbed. The same concern exists with orlistat (Alli®), a lipase inhibitor used for weight loss. Users of these weight-loss aids are advised to take a daily multivitamin to avoid vitamin deficiencies.

The water-soluble vitamins (C and most B vitamins) are absorbed by mediated transport. The major exception is **vitamin B_{12}**, also known as *cobalamin* because it contains the element cobalt. We obtain most of our dietary supply of B_{12} from seafood, meat, and milk products. The intestinal transporter for B_{12} is found only in the ileum and recognizes B_{12} only when the vitamin is complexed with a protein called **intrinsic factor,** secreted by the same gastric parietal cells that secrete acid.

One concern about extended use of drugs that inhibit gastric acid secretion, such as the proton-pump inhibitors discussed earlier, is that they may cause decreased absorption of vitamin B_{12}. In the complete absence of intrinsic factor, severe vitamin B_{12} deficiency causes the condition known as *pernicious anemia*. In this state, red blood cell synthesis (*erythropoiesis*), which depends on vitamin B_{12}, is severely diminished. Lack of intrinsic factor cannot be remedied directly, but patients with pernicious anemia can be given vitamin B_{12} shots.

RUNNING PROBLEM

Rehydrating people with cholera is the key to their survival. Most patients who develop cholera can be treated successfully with oral rehydration solutions. However, in about 5% of patients, the dehydration caused by cholera-induced diarrhea can be severe. If left untreated, these patients can die from circulatory collapse as soon as 18 hours after infection. Because Anish's blood pressure was so low, the medical personnel decided that he needed intravenous (IV) fluids to restore his volume.

Q6: *Which type of IV solution would you select for Anish, and why? Your choices are normal (isotonic) saline, half-normal saline, and 5% dextrose in water (D-5-W).*

653 – 657 – 670 – 674 – 681 – 686

Iron and Calcium Mineral absorption usually occurs by active transport. Iron and calcium are two of the few substances whose intestinal absorption is regulated. For both minerals, a decrease in body concentrations of the mineral leads to enhanced uptake at the intestine.

Dietary iron is ingested as heme iron [p. 517] in meat and as ionized iron in some plant products. Heme iron is absorbed by an apical transporter on the enterocyte (**FIG. 21.19a**). Ionized iron Fe^{2+} is actively absorbed by apical cotransport with H^+ on a protein called the *divalent metal transporter 1* (DMT1). Inside the cell, enzymes convert heme iron to Fe^{2+} and both pools of ionized iron leave the cell on a transporter called *ferroportin*.

Iron uptake by the body is regulated by a peptide hormone called *hepcidin*. When body stores of iron are high, the liver secretes hepcidin, which binds to ferroportin. Hepcidin binding causes the enterocyte to destroy the ferroportin transporter, which results in decreased iron uptake across the intestine.

Most Ca^{2+} absorption in the gut occurs by passive, unregulated movement through paracellular pathways (Fig. 21.19b). Hormonally regulated transepithelial Ca^{2+} transport takes place in the duodenum. Calcium enters the enterocyte through apical Ca^{2+} channels and is actively transported across the basolateral membrane by either a Ca^{2+}-ATPase or by the Na^+-Ca^{2+} antiporter. Calcium absorption is regulated by vitamin D_3, discussed in Chapter 23.

The Intestine Absorbs Ions and Water

Most water absorption takes place in the small intestine, with an additional 0.5 liter per day absorbed in the colon. The absorption of nutrients moves solute from the lumen of the intestine to the ECF, creating an osmotic gradient that allows water to follow.

Ion absorption into the body also creates the osmotic gradients needed for water movement. Enterocytes in the small intestine and **colonocytes**, the epithelial cells on the luminal surface of the colon, absorb Na^+ using three membrane proteins (Fig. 21.19c): apical Na^+ channels such as ENaC, a Na^+-Cl^- symporter, and the Na^+-H^+ exchanger (NHE). In the small intestine, a significant fraction of Na^+ absorption also takes place through Na^+-dependent organic solute uptake, such as the SGLT and Na^+-amino acid transporters.

On the basolateral side of both enterocytes and colonocytes, the primary transporter for Na^+ is Na^+-K^+-ATPase. Chloride uptake uses an apical Cl^--HCO_3^- exchanger and a basolateral Cl channel to move across the cells. Potassium and water absorption in the intestine occur primarily by the paracellular pathway.

Regulation of the Intestinal Phase

The regulation of intestinal digestion and absorption comes primarily from signals that control motility and secretion. Sensors in the intestine trigger neural and endocrine reflexes that feed

FIG. 21.19 Ion and water absorption

(a) Iron absorption

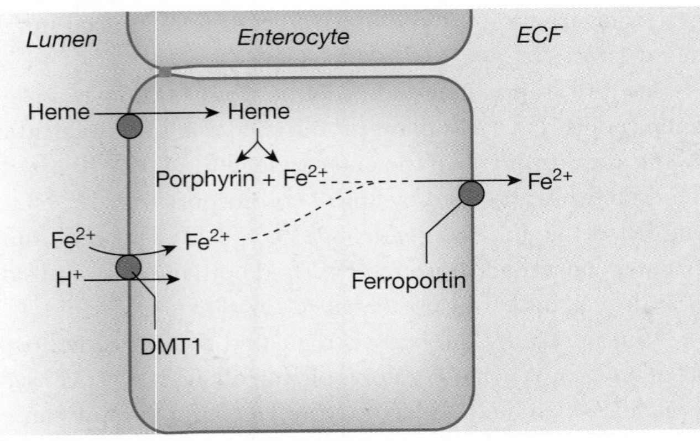
Lumen
Enterocyte
ECF
Heme
Heme
Porphyrin + Fe^{2+}
Fe^{2+}
Fe^{2+}
Fe^{2+}
H^+
Ferroportin
DMT1

(b) Calcium absorption

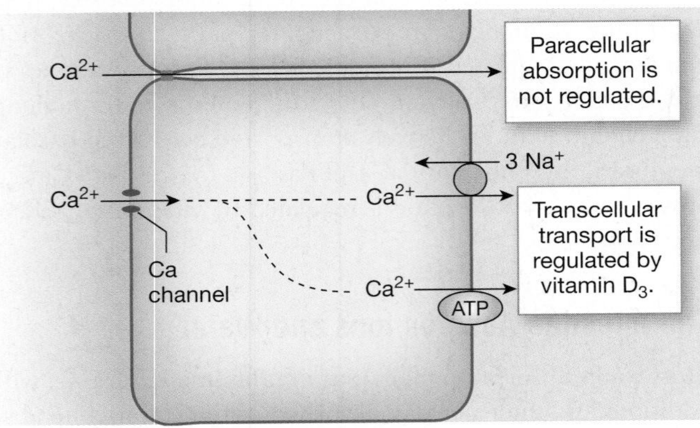
Ca^{2+}
Paracellular absorption is not regulated.
3 Na^+
Ca^{2+}
Ca^{2+}
Transcellular transport is regulated by vitamin D_3.
Ca channel
Ca^{2+}
ATP

(c) Na^+, K^+, Cl^-, and water absorption

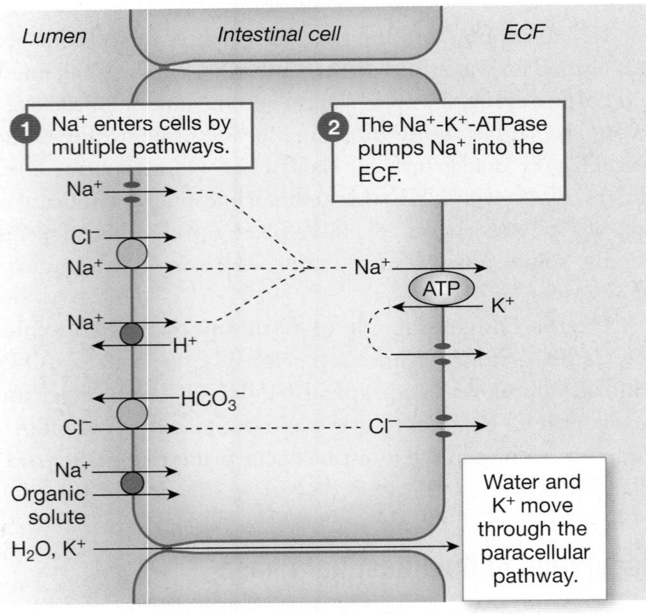
Lumen
Intestinal cell
ECF
1 Na^+ enters cells by multiple pathways.
2 The Na^+-K^+-ATPase pumps Na^+ into the ECF.
Na^+
Cl^-
Na^+
Na^+
ATP
K^+
Na^+
H^+
HCO_3^-
Cl^-
Cl^-
Na^+
Organic solute
H_2O, K^+
Water and K^+ move through the paracellular pathway.

back to regulate the delivery rate of chyme from the stomach, and feed forward to promote digestion, motility, and utilization of nutrients.

Scientists have known for years that the GI tract has the ability to sense and respond specifically and differentially to the composition of a meal. Fats and proteins do not stimulate the same endocrine and exocrine responses as a meal of pure carbohydrate. But how does the gut "know" what is in a meal? Traditional sensory receptors, such as osmoreceptors and stretch receptors, are not tuned to respond to biomolecules. New research indicates that epithelial cells in the gut, especially some of the endocrine cells, express the same G protein-coupled receptors and the taste-linked G protein gustducin as taste buds [p. 325]. Researchers using knockout mice and cultured cell lines are now trying to establish the functional link between gut "taste receptors" and physiological responses to food.

The efferent control signals from the gut to the stomach and pancreas are both neural and hormonal (**FIG. 21.20**):

1. Chyme entering the intestine activates *the enteric nervous system*, which then decreases gastric motility and secretion and slows gastric emptying. In addition, three hormones reinforce the "decrease motility" signal: secretin, cholecystokinin (CCK), and gastric inhibitory peptide (GIP) (see Tbl. 21.1).
2. *Secretin* is released by the presence of acidic chyme in the duodenum. Secretin inhibits acid production and decreases gastric motility. In addition, secretin stimulates production of pancreatic bicarbonate to neutralize the acidic chyme that has entered the intestine.
3. *CCK* is secreted into the bloodstream if a meal contains fats. CCK also slows gastric motility and acid secretion. Because fat digestion proceeds more slowly than either protein or carbohydrate digestion, it is crucial that the stomach allow only small amounts of fat into the intestine at one time.
4. The *incretin hormones* GIP and glucagon-like peptide-1 (GLP-1) are released if the meal contains carbohydrates. Both hormones feed forward to promote insulin release by the endocrine pancreas, allowing cells to prepare for glucose that is about to be absorbed. They also slow the entry of food into the intestine by decreasing gastric motility and acid secretion.
5. The mixture of acid, enzymes, and digested food in chyme usually forms a hyperosmotic solution. Sensors in the intestine wall are sensitive to the osmolarity of the entering chyme. When stimulated by high osmolarity, the sensors inhibit gastric emptying in a reflex mediated by some unknown blood-borne substance.

The Large Intestine Concentrates Waste

By the end of the ileum, only about 1.5 liters of unabsorbed chyme remain. The colon absorbs most of this volume so that normally only about 0.1 liter of water is lost daily in feces. Chyme enters the large intestine through the **ileocecal valve**. This is a tonically

FIG. 21.20 Integration of gastric and intestinal phases

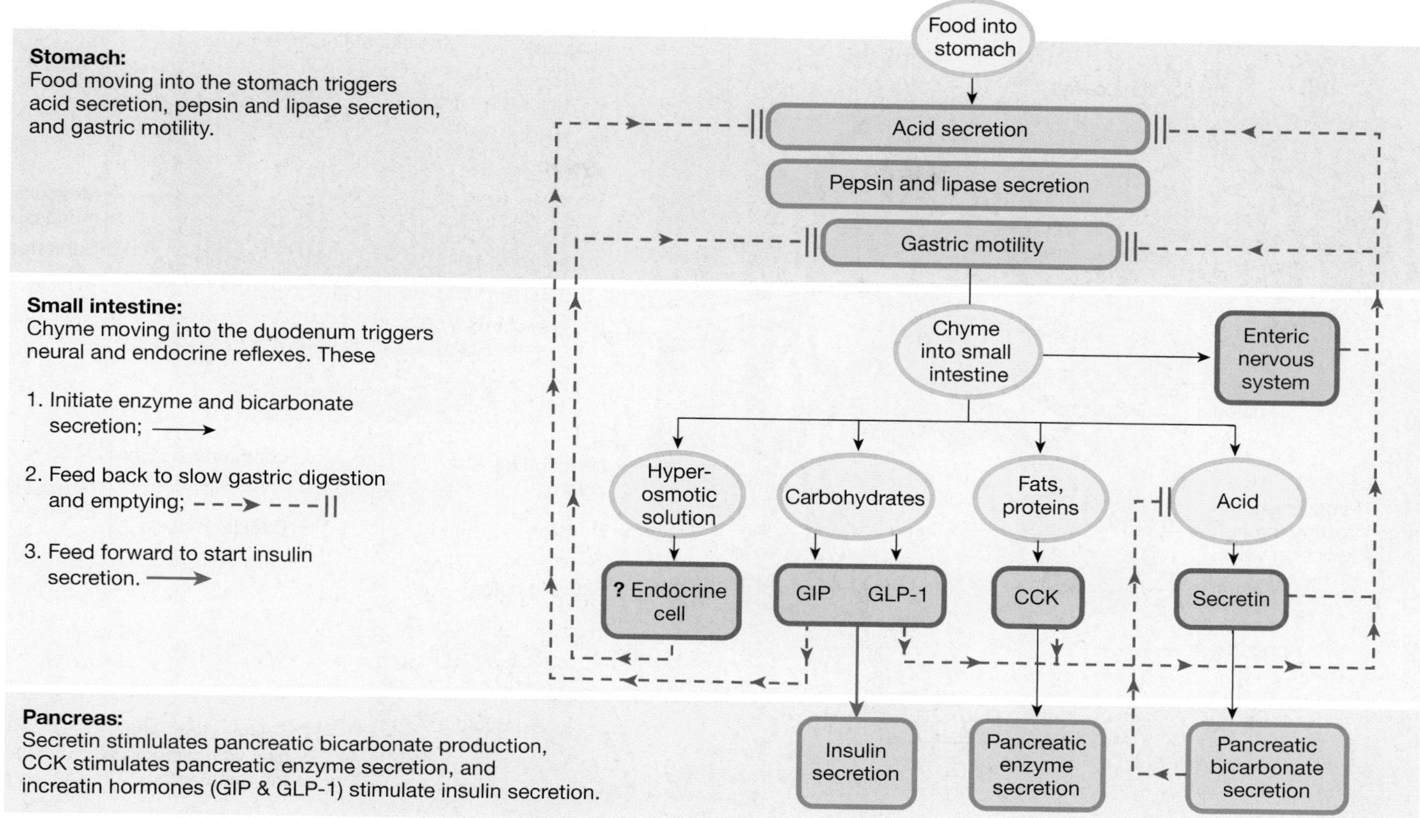
Stomach:
Food moving into the stomach triggers acid secretion, pepsin and lipase secretion, and gastric motility.
Small intestine:
Chyme moving into the duodenum triggers neural and endocrine reflexes. These
1. Initiate enzyme and bicarbonate secretion;
2. Feed back to slow gastric digestion and emptying;
3. Feed forward to start insulin secretion.
Pancreas:
Secretin stimlulates pancreatic bicarbonate production,
CCK stimulates pancreatic enzyme secretion, and
increatin hormones (GIP & GLP-1) stimulate insulin secretion.
Food into stomach
Acid secretion
Pepsin and lipase secretion
Gastric motility
Chyme into small intestine
Enteric nervous system
Hyper-osmotic solution
Carbohydrates
Fats, proteins
Acid
? Endocrine cell
GIP
GLP-1
CCK
Secretin
Insulin secretion
Pancreatic enzyme secretion
Pancreatic bicarbonate secretion

contracted region of muscularis that narrows the opening between the ileum and the **cecum**, the initial section of the large intestine (**FIG. 21.21**). The ileocecal valve relaxes each time a peristaltic wave reaches it. It also relaxes when food leaves the stomach as part of the *gastroileal reflex*.

The large intestine has seven regions. The cecum is a dead-end pouch with the *appendix*, a small fingerlike projection, at its ventral end. Material moves from the cecum upward through the **ascending colon**, horizontally across the body through the **transverse colon**, then down through the **descending colon** and **sigmoid colon** {*sigmoeides*, shaped like a sigma, Σ}. The rectum is the short (about 12 cm) terminal section of the large intestine. It is separated from the external environment by the anus, an opening closed by two sphincters, an internal smooth muscle sphincter and an external skeletal muscle sphincter.

The wall of the colon differs from that of the small intestine in that the muscularis of the large intestine has an inner circular layer but a discontinuous longitudinal muscle layer concentrated into three bands called the **tenia coli**. Contractions of the tenia pull the wall into bulging pockets called **haustra** {*haustrum*, bucket or scoop}.

The mucosa of the colon has two regions, like that of the small intestine. The luminal surface lacks villi and appears smooth. It is composed of colonocytes and mucus-secreting goblet cells. The crypts contain stem cells that divide to produce new epithelium, as well as goblet cells, endocrine cells, and maturing colonocytes.

Motility in the Large Intestine Chyme that enters the colon continues to be mixed by segmental contractions. Forward movement is minimal during mixing contractions and depends primarily on a unique colonic contraction known as **mass movement**. A wave of contraction decreases the diameter of a segment of colon and sends a substantial bolus of material forward. These contractions occur 3–4 times a day and are associated with eating and distension of the stomach through the *gastrocolic reflex*. Mass movement is responsible for the sudden distension of the rectum that triggers defecation.

The **defecation reflex** removes undigested feces from the body. Defecation resembles urination in that it is a spinal reflex triggered by distension of the organ wall. The movement of fecal material into the normally empty rectum triggers the reflex. Smooth muscle of the **internal anal sphincter** relaxes, and peristaltic contractions in the rectum push material toward the anus. At the same time, the external anal sphincter, which is under voluntary control, is consciously relaxed if the situation is appropriate. Defecation is often aided by conscious abdominal contractions and forced expiratory movements against a closed glottis (the *Valsalva maneuver*).

FIG. 21.21 Anatomy of the large intestine

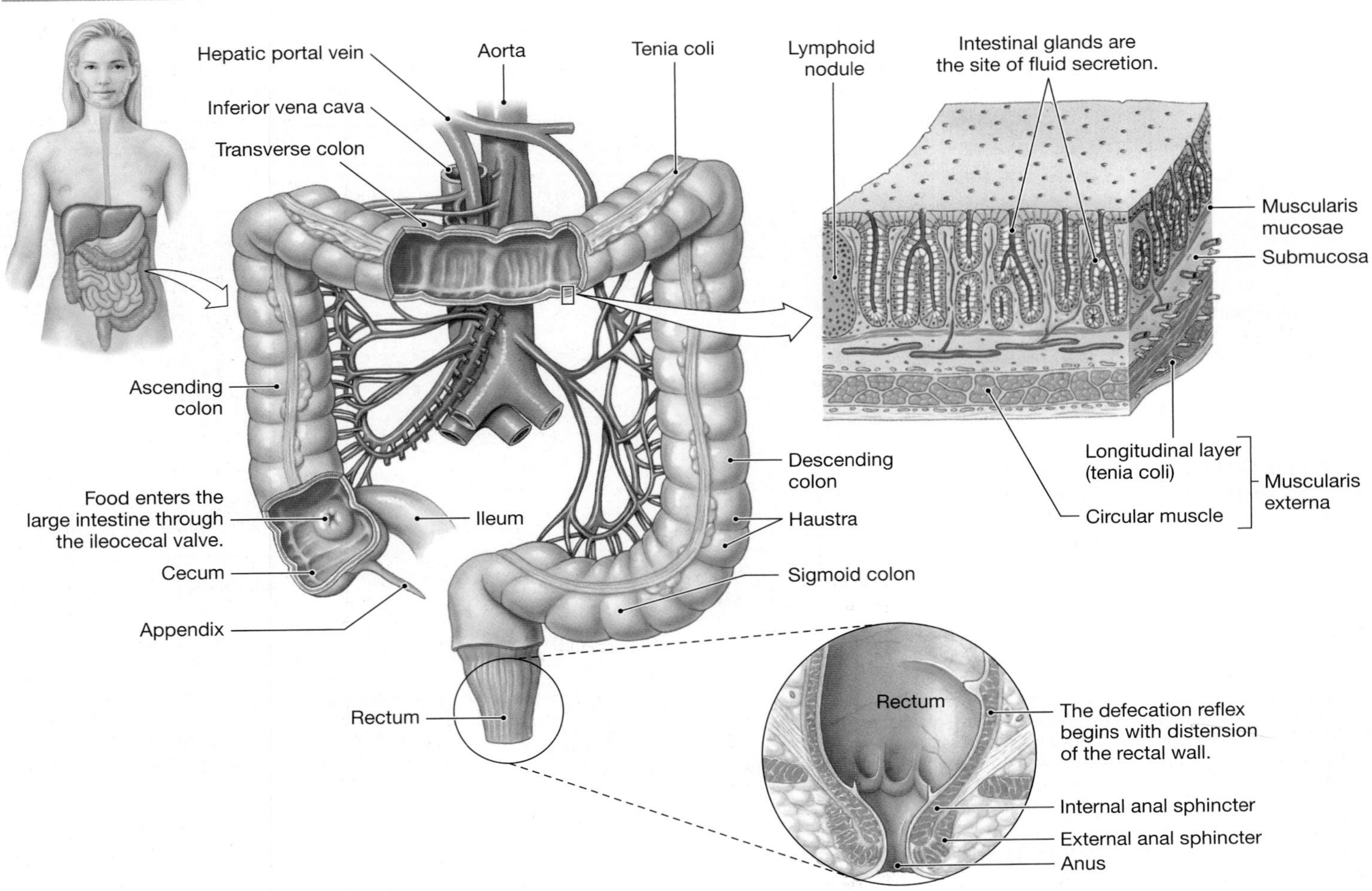
Hepatic portal vein
Aorta
Tenia coli
Lymphoid nodule
Intestinal glands are the site of fluid secretion.
Inferior vena cava
Transverse colon
Muscularis mucosae
Submucosa
Ascending colon
Longitudinal layer (tenia coli)
Circular muscle
Muscularis externa
Descending colon
Food enters the large intestine through the ileocecal valve.
Ileum
Haustra
Cecum
Sigmoid colon
Appendix
Rectum
Rectum
The defecation reflex begins with distension of the rectal wall.
Internal anal sphincter
External anal sphincter
Anus

Defecation, like urination, is subject to emotional influence. Stress may increase intestinal motility and cause psychosomatic diarrhea in some individuals but may decrease motility and cause *constipation* in others. When feces are retained in the colon, either through consciously ignoring a defecation reflex or through decreased motility, continued water absorption creates hard, dry feces that are difficult to expel. One treatment for constipation is glycerin suppositories, small bullet-shaped wads that are inserted through the anus into the rectum. Glycerin attracts water and helps soften the feces to promote defecation.

Digestion and Absorption in the Large Intestine According to the traditional view of the large intestine, no significant digestion of organic molecules takes place there. However, in recent years, this view has been revised. We now know that the numerous bacteria inhabiting the colon break down significant amounts of undigested complex carbohydrates and proteins through fermentation. The end products include lactate and short-chain fatty acids, such as butyric acid. Several of these products are lipophilic and can be absorbed by simple diffusion. The fatty acids, for example, are used by colonocytes as their preferred energy substrate.

Colonic bacteria also produce significant amounts of absorbable vitamins, especially vitamin K. Intestinal gases, such as hydrogen sulfide, that escape from the gastrointestinal tract are a less useful product. Some starchy foods, such as beans, are notorious for their tendency to produce intestinal gas (**flatus**).

Diarrhea Can Cause Dehydration

Diarrhea is a pathological state in which intestinal secretion of fluid is not balanced by absorption, resulting in watery stools. Diarrhea occurs if normal intestinal water absorption mechanisms are disrupted or if there are unabsorbed osmotically active solutes that "hold" water in the lumen. Substances that cause *osmotic diarrhea* include undigested lactose and sorbitol, a sugar alcohol from plants. Sorbitol is used as an "artificial" sweetener in some chewing gums and in foods made for people with diabetes. Another unabsorbed solute that can cause osmotic diarrhea, intestinal cramping, and gas is Olestra, the "fake fat" made from vegetable oil and sugar.

In clinical settings, patients who need to have their bowels cleaned out before surgery or other procedures are often given 4 liters of an isotonic solution of polyethylene glycol and electrolytes

EMERGING CONCEPTS

The Human Microbiome Project

Did you realize that the average human body has many more bacteria living on and in it than it has cells? And that most of these bacteria reside in the gut? Scientists have known for decades about intestinal bacteria and the problems they cause when they leave the external environment of the gut lumen and enter the body proper. Bacterial infections are common if your appendix ruptures or if trauma, such as a stab wound, punctures the wall of the intestine. At the same time, our continued health depends on absorption of vitamins and other nutrients from bacterial metabolism. The relationship between our *microbiota* (the bacteria that inhabit our bodies) and our health has become a topic of research studies in recent years, and data are being collected by an international collaboration known as the Human Microbiome Project (*http://ihmpdcc.org/*). Do foods advertised as "probiotics" really do anything? Can bacteria influence whether we gain weight or not? Do they affect fetal development and our susceptibility to disease? We will be learning more about the answers to these questions in the years to come.

to drink. Because polyethylene glycol cannot be absorbed, a large volume of unabsorbed solution passes into the colon, where it triggers copious diarrhea that removes all solid waste from the GI tract.

Secretory diarrheas occur when bacterial toxins, such as cholera toxin from *Vibrio cholerae* and *Escherichia coli* enterotoxin, enhance colonic Cl^- and fluid secretion (see Fig. 21.13). When excessive fluid secretion is coupled with increased motility, diarrhea results. Secretory diarrhea in response to intestinal infection can be viewed as adaptive because it helps flush pathogens out of the lumen. However, it also has the potential to cause dehydration if fluid loss is excessive.

The World Health Organization estimates that in developing countries, 4 million people die from diarrhea each year. In the United States, diarrhea in children causes about 200,000 hospitalizations a year. Oral replacement fluids for treatment of diarrheal salt and water loss can prevent the *morbidity* (illness) and *mortality* (death) associated with diarrhea. Oral rehydration solutions usually contain glucose or sucrose as well as Na^+, K^+, and Cl^- because the inclusion of a sugar enhances Na^+ absorption. If dehydration is severe, intravenous fluid therapy may be necessary.

Concept Check

16. In secretory diarrhea, epithelial cells in the intestinal villi may be damaged or may slough off. In these cases, would it be better to use an oral rehydration solution containing glucose or one containing sucrose? Explain your reasoning.

21.7 Immune Functions of the GI Tract

As you learned at the beginning of the chapter, the GI tract is the largest immune organ in the body. Its luminal surface is continuously exposed to disease-causing organisms, and the immune cells of the GALT must prevent these pathogens from entering the body through delicate absorptive tissues. The first lines of defense are the enzymes and immunoglobulins in saliva and the highly acidic environment of the stomach. If pathogens or toxic materials make it into the small intestine, sensory receptors and the immune cells of the GALT respond. Two common responses are diarrhea, just described, and vomiting.

M Cells Sample Gut Contents

The immune system of the intestinal mucosa consists of immune cells scattered throughout the mucosa, clusters of immune cells in Peyer's patches (see Fig. 21.1f), and specialized epithelial cells called **M cells** (*microfold cells*) that overlie the Peyer's patches. The M cells provide information about the contents of the lumen to the immune cells of the GALT.

The microvilli of M cells are fewer in number and more widely spaced than in the typical intestinal cell. The apical surface of M cells contains clathrin-coated pits [p. 147] with embedded membrane receptors. When antigens bind to these receptors, the M cell uses endocytosis and transcytosis to transport them to its basolateral membrane, where they are released into the interstitial fluid. Macrophages and lymphocytes [p. 513] are waiting in the extracellular compartment for the M cell to present them with antigens.

If the antigens are substances that threaten the body, the immune cells swing into action. They secrete cytokines to attract additional immune cells that can attack the invaders and cytokines that trigger an inflammatory response. A third response to cytokines is increased secretion of Cl^-, fluid, and mucus to flush the invaders from the GI tract.

In *inflammatory bowel diseases* (such as ulcerative colitis and Crohn's disease), the immune response is triggered inappropriately by the normal contents of the gut. One apparently successful experimental therapy for these diseases involves blocking the action of cytokines released by the gut-associated lymphoid tissues.

How certain pathogenic bacteria cross the barrier created by the intestinal epithelium has puzzled scientists for years. The discovery of M cells may provide the answer. It appears that some bacteria, such as *Salmonella* and *Shigella*, have evolved surface molecules that bind to M cell receptors. The M cells then obligingly transport the bacteria across the epithelial barrier and deposit them inside the body, where the immune system immediately reacts. Both bacteria cause diarrhea, and *Salmonella* also causes fever and vomiting.

Vomiting Is a Protective Reflex

Vomiting, or *emesis*, the forceful expulsion of gastric and duodenal contents from the mouth, is a protective reflex that removes toxic materials from the GI tract before they can be absorbed. However, excessive or prolonged vomiting, with its loss of gastric acid, can cause metabolic alkalosis [p. 649].

The vomiting reflex is coordinated through a vomiting center in the medulla. The reflex begins with stimulation of sensory receptors and is often (but not always) accompanied by nausea. A variety of stimuli from all over the body can trigger vomiting. They include chemicals in the blood, such as cytokines and certain drugs; pain; disturbed equilibrium, such as occurs in a moving car or rocking boat, and emotional stress. Tickling the back of the pharynx can also induce vomiting.

Efferent signals from the vomiting center initiate a wave of reverse peristalsis that begins in the small intestine and moves upward. The motility wave is aided by abdominal contraction that increases intra-abdominal pressure. The stomach relaxes so that the increased pressure forces gastric and intestinal contents back into the esophagus and out of the mouth.

During vomiting, respiration is inhibited. The epiglottis and soft palate close off the trachea and nasopharynx to prevent the *vomitus* from being inhaled (*aspirated*). Should acid or small food particles get into the airways, they could damage the respiratory system and cause *aspiration pneumonia*.

RUNNING PROBLEM **CONCLUSION** Cholera in India

Vibrio cholerae, the bacterium that causes cholera, was first identified in India in the 1800s. It has caused seven worldwide epidemics in the years since. About 75% of people who become infected with *V. cholerae* have no symptoms, but the remaining 25% develop potentially fatal *secretory diarrhea*. The gut immune systems in most people overcome the infection within about a week. But until that happens, even asymptomatic people shed the bacteria in their feces, which contributes to the spread of the disease. To learn more about cholera, see the CDC (*www.cdc.gov*) and WHO (*www.who.int*) websites. Now check your understanding of this running problem by comparing your answers to the information in the following summary table.

Question	Facts	Integration and Analysis
Q1: *What would you expect Anish's ECF volume to be?*	Most fluid in diarrhea has been secreted from the body into the lumen of the GI tract.	Loss of fluid from the body would decrease ECF volume.
Q2: *Why was Anish experiencing a rapid heartbeat?*	Loss of ECF volume with the diarrhea decreased Anish's blood pressure.	Decreased blood pressure triggered a baroreceptor reflex [p. 492]. Increased sympathetic and decreased parasympathetic output to the SA node resulted in a faster heart rate.
Q3: *Esomeprazole is a proton pump inhibitor (PPI). For what symptom or condition might Anish have been taking this drug?*	"Proton pump" is another name for an ATP-dependent H^+ transporter. Stomach acid is secreted by H^+-K^+-ATPase.	A proton pump inhibitor would decrease stomach acid, so Anish may have been taking the PPI for heartburn or gastroesophageal reflex disorder (GERD).
Q4: *Why might taking a protein pump inhibitor like esomeprazole have increased Anish's chances of contracting cholera?*	Proton pump inhibitors decrease the acidity in the stomach. Low gastric pH is one of the body's defense mechanisms.	In a less acidic stomach environment, more cholera bacteria might survive passage through the stomach to the small intestine, where they could infect the enterocytes.
Q5: *Why would continuously open enterocyte CFTR channels cause secretory diarrhea and dehydration in humans?*	Chloride leaves enterocytes by the CFTR channel. Na^+ and water follow by the paracellular pathway. See Figure 21.13.	A continuously open CFTR channel means increased secretion of NaCl and water into the lumen, which leads to watery diarrhea. The salt and water come from the ECF, and their loss causes dehydration.
Q6: *Which type of IV solution would you select for Anish and why? Your choices are normal (isotonic) saline, half-normal saline, and 5% dextrose in water (D-5-W).*	Chloride secretion by enterocytes causes Na^+ and water to follow, with the net result being secretion of isotonic fluid. The replacement fluid should match the fluid loss as closely as possible.	Normal saline (isosmotic) approximates the fluid lost in cholera diarrhea. Half-normal saline would dilute the body's osmolarity. D-5-W is not acceptable because it is equivalent to giving pure water and would not replace the lost NaCl.

This problem was developed by Claire Conroy when she was an undergraduate Nutritional Sciences/Pre-Physical Therapy student at the University of Texas at Austin.

653 - 657 - 670 - 674 - 681 - 686

CHAPTER SUMMARY

The digestive system, like the renal system, plays a key role in *mass balance* in the body. Most material that enters the system, whether by mouth or by secretion, is absorbed before it reaches the end of the GI tract. In pathologies such as diarrhea, in which absorption and secretion are unbalanced, the loss of material through the GI tract can seriously disrupt *homeostasis*. Absorption and secretion in the GI tract provide numerous examples of *movement across membranes*, and most transport processes follow patterns you have encountered in the kidney and other systems. Finally, regulation of GI tract function illustrates the complex interactions that take place between endocrine and neural *control systems* and the immune system.

21.1 Anatomy of the Digestive System

1. Food entering the digestive system passes through the mouth, pharynx, **esophagus, stomach (fundus, body, antrum), small intestine (duodenum, jejunum, ileum), large intestine (colon, rectum)**, and **anus**. (p. 654; Fig. 21.1a)
2. The **salivary glands, pancreas**, and **liver** add exocrine secretions containing enzymes and mucus to the lumen. (p. 654; Fig. 21.1a)
3. **Chyme** is a soupy substance created as ingested food is broken down by mechanical and chemical digestion. (p. 653)
4. The wall of the GI tract consists of four layers: mucosa, submucosa, muscle layers, and serosa. (p. 654; Fig. 21.1d)
5. The **mucosa** faces the lumen and consists of epithelium, the **lamina propria**, and the **muscularis mucosae**. The lamina propria contains immune cells. Small **villi** and invaginations increase the surface area. (p. 655; Fig. 21.1 e, f)
6. The **submucosa** contains blood vessels and lymph vessels and the **submucosal plexus** of the **enteric nervous system**. (p. 655; Fig. 21.1f)
7. The **muscularis externa** consists of a layer of circular muscle and a layer of longitudinal muscle. The **myenteric plexus** lies between these two muscle layers. (p. 655; Fig. 21.1e, f)
8. The **serosa** is the outer connective tissue layer that is a continuation of the peritoneal membrane. (p. 654; Fig. 21.1d)

21.2 Digestive Function and Processes

9. The GI tract moves nutrients, water, and electrolytes from the external environment to the internal environment. (p. 657)
10. **Digestion** is chemical and mechanical breakdown of foods into absorbable units. **Absorption** is transfer of substances from the lumen of the GI tract to the ECF. **Motility** is movement of material through the GI tract. **Secretion** is the transfer of fluid and electrolytes from ECF to lumen or the release of substances from cells. (p. 658; Fig. 21.2)
11. About 2 L of fluid per day enter the GI tract through the mouth. Another 7 L of water, ions, and proteins are secreted by the body. To maintain mass balance, nearly all of this volume is reabsorbed. (p. 659; Fig. 21.3)
12. Many digestive enzymes are secreted as inactive zymogens to prevent autodigestion. (p. 659)
13. For defense from invaders, the GI tract contains the largest collection of lymphoid tissue in the body, the **gut-associated lymphoid tissue (GALT)**. (p. 658)
14. GI smooth muscle cells depolarize spontaneously and are electrically connected by gap junctions. Some segments of the gut are **tonically contracted**, but others exhibit **phasic contractions**. (p. 659)
15. Intestinal smooth muscle exhibits spontaneous **slow wave potentials** that originate in the **interstitial cells of Cajal**. (p. 659)
16. When a slow wave reaches threshold, it fires action potentials and the muscle contracts. (p. 660; Fig. 21.4a)
17. Between meals, the **migrating motor complex** moves food remnants from the upper GI tract to the lower regions. (p. 660; Fig. 21.4b)
18. **Peristaltic contractions** are progressive waves of contraction that occur mainly in the esophagus. **Segmental contractions** are mixing contractions. (p. 660; Fig. 21.4c, d)

21.3 Regulation of GI Function

19. The **enteric nervous system** can integrate information without input from the CNS. **Intrinsic neurons** lie completely within the ENS. (p. 662)
20. **Short reflexes** originate in the ENS and are integrated there. **Long reflexes** may originate in the ENS or outside it but are integrated in the CNS. (p. 662; Fig. 21.5)
21. Generally, parasympathetic innervation is excitatory for GI function, and sympathetic innervation is inhibitory. (p. 663)
22. GI peptides excite or inhibit motility and secretion. Most stimuli for GI peptide secretion arise from the ingestion of food. (p. 663)
23. GI hormones are divided into the gastrin family (**gastrin, cholecystokinin**), secretin family (**secretin, gastric inhibitory peptide, glucagon-like peptide-1**), and hormones that do not fit into either of those two families (**motilin**). (p. 664; Tbl. 21.1)

21.4 Integrated Function: The Cephalic Phase

24. The sight, smell, or taste of food initiates GI reflexes in the **cephalic phase** of digestion. (p. 665; Fig. 21.8)
25. Mechanical digestion begins with chewing, or **mastication**. Saliva moistens and lubricates food. Salivary amylase digests carbohydrates. (p. 666)
26. **Saliva** is an exocrine secretion that contains water, ions, mucus, and proteins. Salivation is under autonomic control. (p. 666)
27. Swallowing, or **deglutition**, is a reflex integrated by a medullary center. (p. 667; Fig. 21.7)

21.5 Integrated Function: The Gastric Phase

28. The stomach stores food, begins protein and fat digestion, and protects the body from swallowed pathogens. (p. 667)
29. The stomach secretes mucus and bicarbonate from **mucous cells**, pepsinogen from **chief cells**, somatostatin from **D cells**, histamine from **ECL cells**, and gastrin from **G cells**. (pp. 669–670; Fig. 21.9a, b; Fig. 21.10)
30. **Parietal cells** in gastric glands secrete hydrochloric acid. (p. 669; Fig. 21.9c)
31. Gastric function is integrated with the cephalic and intestinal phases of digestion. (pp. 668, 683; Fig. 21.8, Fig. 21.20)

21.6 Integrated Function: The Intestinal Phase

32. Most nutrient absorption takes place in the small intestine. The large intestine absorbs water and ions. (p. 671)
33. Most absorbed nutrients go directly to the liver via the hepatic portal system before entering the systemic circulation. (p. 671; Fig. 21.12)
34. Intestinal enzymes are part of the **brush border. Goblet cells** secrete mucus. (p. 671)
35. Intestinal cells secrete an isotonic NaCl solution using the **CFTR chloride channel**. Water and Na^+ follow Cl^- down osmotic and electrochemical gradients. (p. 672; Fig. 21.13)
36. The pancreas secretes a watery $NaHCO_3$ solution from duct cells and inactive digestive enzymes from the acini. (p. 673; Fig. 21.14)
37. Bile made by **hepatocytes** contains **bile salts**, bilirubin, and cholesterol. Bile is stored and concentrated in the gallbladder (p. 675; Fig. 21.15)
38. Fat digestion is facilitated by bile salts. As digestion proceeds, fat droplets form **micelles**. (p. 678; Fig. 21.16)
39. Fat digestion requires the enzyme **lipase** and the cofactor **colipase**. (p. 678; Fig. 21.16)
40. Fat absorption occurs primarily by simple diffusion. Cholesterol is actively transported. (p. 677; Fig. 21.16)
41. **Chylomicrons**, made of monoglycerides, fatty acids, cholesterol, and proteins, are absorbed into the lymph. (p. 677; Fig. 21.16)
42. **Amylase** digests starch to maltose. **Disaccharidases** digest disaccharides to monosaccharides. (p. 678; Fig. 21.17)
43. Glucose absorption uses the SGLT Na^+-glucose symporter and GLUT2 transporter. Fructose uses the GLUT5 and GLUT2 transporters. (p. 678; Fig. 21.17)
44. **Endopeptidases** (also called proteases) break proteins into smaller peptides. **Exopeptidases** remove amino acids from peptides. (p. 679; Fig. 21.18)
45. Amino acids are absorbed via Na^+- or H^+-dependent cotransport. Dipeptides and tripeptides are absorbed via H^+-dependent cotransport. Some larger peptides are absorbed intact via transcytosis. (p. 679; Fig. 21.18)
46. Nucleic acids are digested and absorbed as nitrogenous bases and monosaccharides. (p. 681)
47. Fat-soluble vitamins are absorbed along with fats. Water-soluble vitamins are absorbed by mediated transport. Vitamin B_{12} absorption requires **intrinsic factor** secreted by the stomach. (p. 683)
48. Mineral absorption usually occurs via active transport. Some calcium moves by the paracellular pathway. Ions and water move by the paracellular pathway as well as by membrane proteins. (p. 682; Fig. 21.19)
49. Acid in the intestine, CCK, and secretin delay gastric emptying. (p. 683; Fig. 21.20)
50. Undigested material in the colon moves forward by **mass movement**. The **defecation reflex** is triggered by sudden distension of the rectum. (p. 685; Fig. 21.21)
51. Colonic bacteria use fermentation to digest organic material. (p. 684)
52. Cells of the colon can both absorb and secrete fluid. Excessive fluid secretion or decreased absorption causes diarrhea. (p. 685)

21.7 Immune Functions of the GI Tract

53. Protective mechanisms of the GI tract include acid and mucus production, vomiting, and diarrhea. (p. 685)
54. **M cells** sample gut contents and present antigens to cells of the GALT. (p. 685)
55. **Vomiting** is a protective reflex integrated in the medulla. (p. 686)

REVIEW QUESTIONS

In addition to working through these questions and checking your answers on p. A-27, review the Learning Outcomes at the beginning of this chapter.

Level One Reviewing Facts and Terms

1. Match each of the following descriptions with the appropriate term(s):

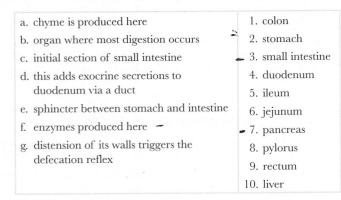

a. chyme is produced here	1. colon
b. organ where most digestion occurs	2. stomach
c. initial section of small intestine	3. small intestine
d. this adds exocrine secretions to duodenum via a duct	4. duodenum
e. sphincter between stomach and intestine	5. ileum
f. enzymes produced here	6. jejunum
g. distension of its walls triggers the defecation reflex	7. pancreas
	8. pylorus
	9. rectum
	10. liver

2. For most nutrients, which two processes are not regulated? Which two are continuously regulated? Why do you think these differences exist? Defend your answer.
3. Define the four basic processes of the digestive system and give an example of each.
4. List the four layers of the GI tract walls. What type of tissue predominates in each layer?
5. Describe the functional types of epithelium lining the stomach and intestines.
6. What are Peyer's patches? M cells of the intestine?
7. What purposes does motility serve in the gastrointestinal tract? Which types of tissue contribute to gut motility? Which types of contraction do the tissues undergo?
8. What is a zymogen? What is a proenzyme? List two examples of each.
9. Match each of the following cells with the product(s) it secretes. Items may be used more than once.

a. parietal cells	1. enzymes
b. goblet cells	2. histamine
c. brush border cells	3. mucus
d. pancreatic cells	4. pepsinogen
e. D cells	5. gastrin
f. ECL cells	6. somatostatin
g. chief cells	7. HCO_3^-
h. G cells	8. HCl
	9. intrinsic factor

10. How does each of the following factors affect digestion? Briefly explain how and where each factor exerts its effects.
    a. emulsification
    b. neural activity
    c. low pH
    d. size of food particles
11. Most digested nutrients are absorbed into the __________ of the __________ system, delivering nutrients to the __________ (organ). However, digested fats go into the __________ system because intestinal capillaries have a(n) __________ around them that most lipids are unable to cross.
12. What is the enteric nervous system, and what is its function?
13. What are short reflexes? What types of responses do they regulate? What is meant by the term *long reflex*?
14. What role do paracrines play in digestion? Give specific examples.

Level Two Reviewing Concepts

15. **Map 1:** List the three major groups of biomolecules across the top of a large piece of paper. Down the left side of the paper write *mouth, stomach, small intestine*. For each biomolecule in each location, fill in the enzymes that digest the biomolecule, the products of digestion for each enzyme, and the location and mechanisms by which these products are absorbed.

    Map 2: Create a diagram or map using the following terms related to iron absorption:

• DMT1	• heme iron
• endocytosis	• hepcidin
• enterocyte	• Fe^{2+}
• ferroportin	• liver

16. Define, compare, and contrast the following pairs or sets of terms:
    a. mastication, deglutition
    b. microvilli, villi
    c. peristalsis, segmental contractions, migrating motor complex, mass movements
    d. chyme, feces
    e. short reflexes, long reflexes
    f. submucosal plexus, myenteric plexus, vagus nerve
    g. cephalic, gastric, and intestinal phases of digestion
17. a. Diagram the cellular mechanisms by which Na^+, K^+, and Cl^- are reabsorbed from the intestine.
    b. Diagram the cellular mechanisms by which H^+ and HCO_3^- are secreted into the lumen.
18. Compare the enteric nervous system with the cephalic brain. Give some specific examples of neurotransmitters, neuromodulators, and supporting cells in the two.
19. List and briefly describe the actions of the members of each of the three groups of GI hormones.
20. Explain how H_2 receptor antagonists and proton pump inhibitors decrease gastric acid secretion.

Level Three Problem Solving

21. In the disease state called *hemochromatosis*, the hormone hepcidin is either absent or not functional. Use your understanding of iron homeostasis to predict what would happen to intestinal iron uptake and plasma levels of iron in this disease.
22. Erica's baby, Justin, has had a severe bout of diarrhea and is now dehydrated. Is his blood more likely to be acidotic or alkalotic? Why?
23. Mary Littlefeather arrives in her physician's office complaining of severe, steady pain in the upper right quadrant of her abdomen. The pain began shortly after she ate a meal of fried chicken, French fries, and peas. Lab tests and an ultrasound reveal the presence of gallstones in the common bile duct running from the liver, gallbladder, and pancreas into the small intestine.
    a. Why was Mary's pain precipitated by the meal she ate?
    b. Which of the following processes will be affected by the gallstones: micelle formation in the intestine, carbohydrate digestion in the intestine, and protein absorption in the intestine. Explain your reasoning.
24. Using what you have learned about epithelial transport, draw a picture of the salivary duct cells and lumen. Arrange the following membrane channels and transporters on the apical and basolateral membranes so that the duct cell absorbs Na^+ and secretes K^+: ENaC, Na^+-K^+-ATPase, and K^+ leak channel. With neural stimulation, the flow rate of saliva can increase from 0.4 mL/min to 2 mL/min. What do you think happens to the Na^+ and K^+ content of saliva at the higher flow rate?

CHAPTER 21

Level Four Quantitative Problems

25. Intestinal transport of the amino acid analog MIT (monoiodotyrosine) can be studied using the "everted sac" preparation. A length of intestine is turned inside out, filled with a solution containing MIT, tied at both ends, and then placed in a bath containing nutrients, salts, and an equal concentration of MIT. Changes in the concentration of MIT are monitored in the bath (mucosal or apical side of the inverted intestine), in the intestinal cells (tissue), and within the sac (serosal or basolateral side of the intestine) over a 240-minute period. The results are displayed in the graph shown here. (Data from Nathans *et al., Biochimica et Biophysica Acta 41:* 271–282, 1960.)
    a. Based on the data shown, is the transepithelial transport of MIT a passive process or an active process?
    b. Which way does MIT move: (1) apical to tissue to basolateral, or (2) basolateral to tissue to apical? Is this movement absorption or secretion?
    c. Is transport across the apical membrane active or passive? Explain your reasoning.
    d. Is transport across the basolateral membrane active or passive? Explain your reasoning.

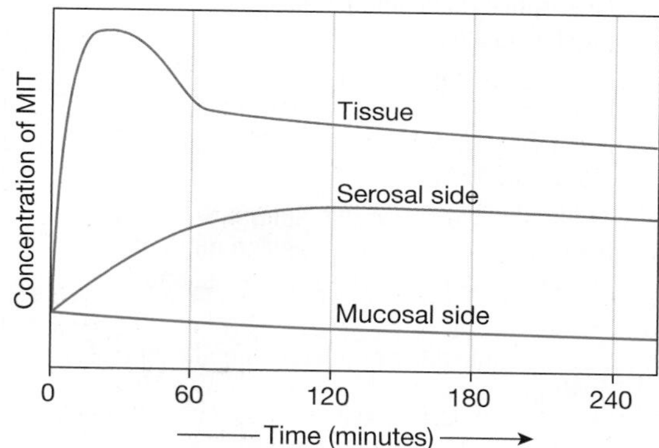
Concentration of MIT
Tissue
Serosal side
Mucosal side
0
60
120
180
240
Time (minutes)

Answers to Concept Checks, Figure and Graph Questions, and end-of-chapter Review Questions can be found in Appendix A [p. A-1].

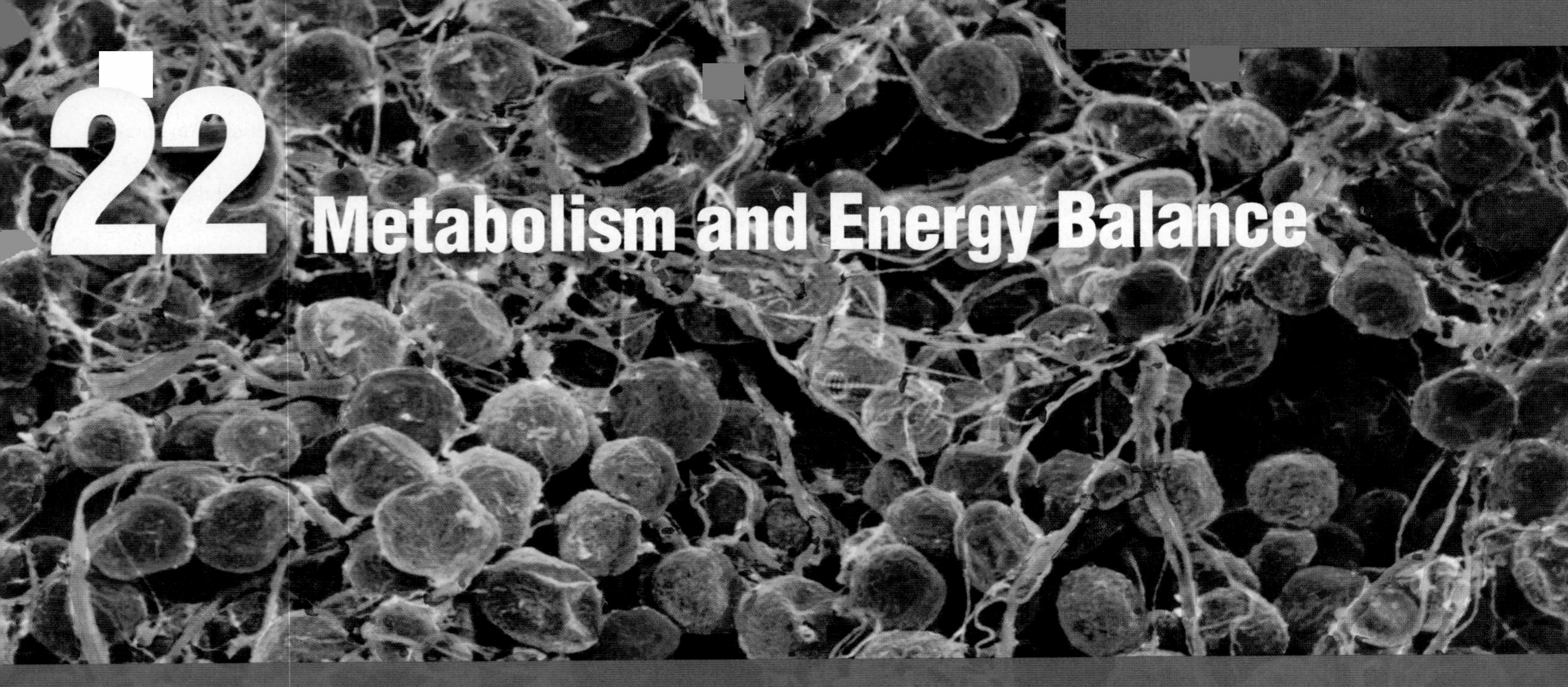

22 Metabolism and Energy Balance

Adipocytes and connective tissue

Probably no single abnormality has contributed more to our knowledge of the intermediary metabolism of animals than has the disease known as diabetes.

Helen R. Downes, The Chemistry of Living Cells, *1955*

BACKGROUND BASICS

Magazine covers at grocery store checkout stands reveal a lot about Americans. Headlines screaming "Lose 10 pounds in a week without dieting" or "CCK: the hormone that makes you thin" vie for attention with glossy photographs of fat-laden, high-calorie desserts dripping with chocolate and whipped cream. As one magazine article put it, we are a nation obsessed with staying trim and with eating—two mutually exclusive occupations. But what determines when, what, and how much we eat? The factors influencing food intake are an area of intense research because the act of eating is the main point at which our bodies exert control over energy input.

22.1 Appetite and Satiety

The control of food intake is a complex process. The digestive system does not regulate energy intake, so we must depend on behavioral mechanisms, such as hunger and **satiety** {*satis*, enough}, to tell us when and how much to eat. Psychological and social aspects of eating, such as parents who say "Clean your plate," complicate the physiological control of food intake. As a result, we still do not fully understand what governs when, what, and how much we eat. What follows is an overview of this increasingly complex and constantly changing field.

Our simplest model for behavioral regulation of food intake is based on two hypothalamic centers [Fig. 11.3, p. 357] whose output signals guide behavior and create sensations of hunger and fullness. One region is a **feeding center** that is tonically active, the other a **satiety center** that stops food intake by inhibiting the feeding center. Animals whose feeding centers are destroyed cease to eat. If the satiety centers are destroyed, animals overeat and become obese {*obesus*, plump or fat}. A more complex model divides feeding behaviors into food seeking and food consumption, governed by separate neural circuits.

The control of food intake is complicated and not well understood. Studies using transgenic and knockout mice show that the hypothalamic centers are part of a complicated neural network secreting and responding to a variety of chemical messengers. Higher brain centers, including the cerebral cortex and limbic system, provide input to the hypothalamus. Some of the chemical signals that influence food intake and satiety include neuropeptides, "brain-gut" hormones secreted by the GI tract, and chemical signals secreted by adipose tissue called *adipocytokines* or **adipokines** for short.

There are two classic theories for regulation of food intake: the glucostatic theory and the lipostatic theory. Current evidence indicates that these two classic theories are too simple to be the only models, however. The **glucostatic theory** states that glucose metabolism by hypothalamic centers regulates food intake. When blood glucose concentrations decrease, the satiety center is suppressed, and the feeding center is dominant. When glucose metabolism increases, the satiety center inhibits the feeding center.

The **lipostatic theory** of energy balance proposes that a signal from the body's fat stores to the brain modulates eating behavior so that the body maintains a particular weight. If fat stores are increased, eating decreases. In times of starvation, eating increases. Obesity results from disruption of this pathway.

The 1994 discovery of **leptin** {*leptos*, thin}, a protein hormone synthesized in adipocytes, provided evidence for the lipostatic theory. Leptin acts as a negative-feedback signal between adipose tissue and the brain. As fat stores increase, adipose cells secrete more leptin, and food intake decreases.

Leptin is synthesized in adipocytes under control of the *obese (ob) gene*. Mice that lack the *ob* gene (and therefore lack leptin) become obese, as do mice with defective leptin receptors. However, these findings did not translate well to humans, as only a small percentage of obese humans are leptin deficient. The majority of them have *elevated* leptin levels and are considered *leptin-resistant*. In other words, the problem is abnormal tissue responsiveness rather than too little hormone [p. 214].

Leptin is only part of the story. Another key signal molecule is **neuropeptide Y (NPY),** a brain neurotransmitter that seems to be the stimulus for food intake. In normal-weight animals, leptin inhibits NPY in a negative feedback pathway (**FIG. 22.1**). Other neuropeptides, hormones, and adipokines also influence NPY secretion, leptin release by adipocytes, and the hypothalamic centers controlling food intake.

For example, the peptide **ghrelin** is secreted by the stomach during fasting and increases hunger when infused into human subjects. Other peptides, such as the hormones CCK and GLP-1, are released by the gut during a meal and help decrease hunger. Many of these appetite-regulating peptides have functions in addition to control of food intake. Ghrelin promotes release of growth hormone, for instance. The brain peptides called *orexins* appear to play a role in wakefulness and sleep. Some of the major signal molecules being studied are listed in the table in Figure 22.1. Our understanding of how these factors interact is incomplete, and many scientists are studying this topic because of the relationship between food intake and the obesity problem in the United States.

Appetite and eating are also influenced by sensory input through the nervous system. The simple acts of swallowing and chewing food help create a sensation of fullness. The sight, smell, and taste of food can either stimulate or suppress appetite.

RUNNING PROBLEM **Eating Disorders**

Sara and Nicole had been best friends growing up but had not seen each other for several semesters. When Sara heard that Nicole was in the hospital, she went to visit but wasn't prepared for what she saw. Nicole, who had always been thin, now was so emaciated that Sara hardly recognized her. Nicole had been taken to the emergency room when she fainted in a ballet class. When Dr. Ayani saw Nicole, her concern about her weight was as great as her concern about her broken wrist. At 5′6″, Nicole weighed only 95 pounds (normal healthy range for that height is 118–155 pounds). Dr. Ayani suspected an eating disorder, probably anorexia nervosa, and she ordered blood tests to confirm her suspicion.

693 – 695 – 696 – 699 – 718 – 724

FIG. 22.1 Complex chemical signaling controls food intake

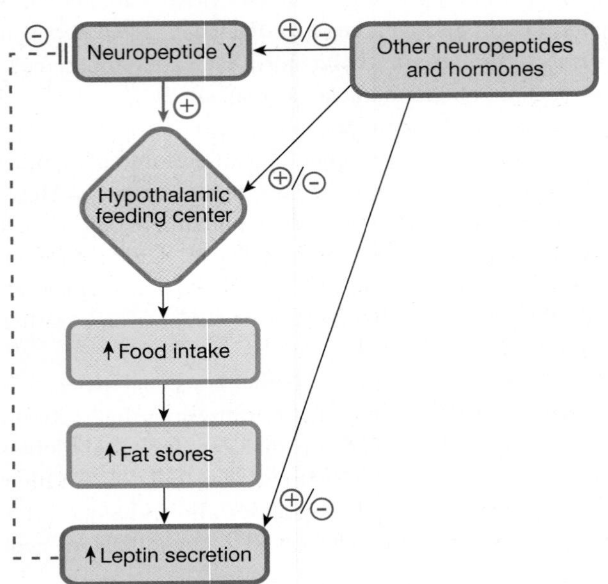
Neuropeptide Y
Other neuropeptides and hormones
Hypothalamic feeding center
↑ Food intake
↑ Fat stores
↑ Leptin secretion

Key Peptides that Modulate Food Intake	
Secretion	**Source**
Increase Food Intake	
Ghrelin	Stomach
NPY and Agouti-related protein (AgRP)	Co-expressed in hypothalamus
Orexins (hypocretins)	Hypothalamus
Decrease Food Intake	
CCK	Small intestine, neurons
Glucagon-like peptide-1 (GLP-1)	Intestines
PYY	Intestines
Leptin	Adipose cells
Corticotropin-releasing hormone (CRH)	Hypothalamus
α-Melanocyte-stimulating hormone (α-MSH)	Hypothalamus
CART (cocaine-and-amphetamine-regulated transcript) and POMC (pro-opiomelanocortin)	Co-expressed in hypothalamus

In one interesting study researchers tried to determine whether chocolate craving is attributable to psychological factors or to physiological stimuli, such as chemicals in the chocolate.* Subjects were given either dark chocolate, white chocolate (which contains none of the pharmacological agents of cocoa), cocoa capsules, or placebo capsules. The researchers found that white chocolate was the best substitute for the real thing, which suggests that taste and aroma play a significant role in satisfying chocolate craving.

*W. Michener and P. Rozin. Pharmacological versus sensory factors in the satiation of chocolate craving. *Physiol Behav* 56(3): 419–422, 1994.

BIOTECHNOLOGY

Discovering Peptides: Research in Reverse

In the early days of molecular biology, scientists collected tissues that contained active peptides, isolated and purified the peptides, and then analyzed their amino acid sequence. Now, in the era of proteomics, investigators are using reverse techniques to discover new proteins in the body. One group of researchers, for example, found the hunger-inducing *orexin* (or *hypocretin*) peptides by isolating mRNA expressed in a particular region of the hypothalamus. The investigators then used this mRNA to create the amino acid sequence of the prepropeptide. At the same time, a different group of scientists also discovered orexin peptides by working backwards from an "orphan" G protein-coupled receptor [p. 173] to find its peptide ligand. The endogenous hunger hormone *ghrelin* was discovered by a similar method. Pharmacologists testing synthetic peptides to stimulate the release of growth hormone found that their peptides were binding to a previously unknown receptor, and from that receptor they discovered ghrelin.

Psychological factors, such as stress, can also play a significant role in regulating food intake. In another study,** researchers found that subjects who imagined eating 30 M&M's one at a time ate fewer real M&M's than subjects who thought about eating only 3 M&M's. A repeat of the experiment using cheese cubes instead of M&M's had the same result.

The eating disorder *anorexia nervosa* has both psychological and physiological components, which complicates its treatment. And the concept of *appetite* is closely linked to the psychology of eating, which may explain why dieters who crave smooth, cold ice cream cannot be satisfied by a crunchy carrot stick. A considerable amount of money is directed to research on eating behaviors.

Concept Check

1. Explain the roles of the satiety and feeding centers. Where are they located?
2. Studies show that most obese humans have elevated leptin levels in their blood. Based on your understanding of endocrine disorders [p. 214], propose some reasons why leptin is not decreasing food intake in these people.

22.2 Energy Balance

Once food has been digested and absorbed, the body's chemical reactions—known collectively as metabolism—determine what happens to the nutrients in the food. Are they destined to burn off as heat? Become muscle? Or turn into extra pounds that make it difficult to zip up blue jeans? In this section, we examine energy balance in the body.

**C. K. Morewedge *et al.* Thought for food: Imagined consumption reduces actual consumption. *Science* 330: 1530–1533, 2010.

RUNNING PROBLEM

Sara and Nicole first met in ballet class at age 11. They were both serious about dance and worked hard to maintain the perfect thin ballet-dancer's body. Both girls occasionally took diet pills or laxatives, and it was almost a contest to see who could eat the least food. Sara was always impressed with Nicole's willpower, but then again, Nicole was a perfectionist. Two years ago, Sara found herself less interested in dance and went away to college. Nicole was accepted into a prestigious ballet company and remained focused on her dancing—and on her body image. Nicole's regimented diet became stricter, and if she felt she'd eaten too much, she would make herself throw up. Her weight kept dropping, but each time she looked in the mirror, she saw a fat girl looking back.

Q1: *If you measured Nicole's leptin level, what would you expect to find?*

Q2: *Would you expect Nicole to have elevated or depressed levels of neuropeptide Y?*

693 – **695** – 696 – 699 – 718 – 724

Energy Input Equals Energy Output

The *first law of thermodynamics* [p. 95] states that the total amount of energy in the universe is constant. By extension, this statement means that all energy that goes into a biological system, such as the human body, can be accounted for (**FIG. 22.2**). In the body, most stored energy is contained in the chemical bonds of molecules.

We can apply the concept of mass balance to energy balance: changes to the body's energy stores result from the difference between the energy put into the body and the energy used.

$$\text{Total body energy} = \text{energy stored} + \text{energy intake} - \text{energy output} \quad (1)$$

Energy intake for humans consists of energy in the nutrients we eat, digest, and absorb. *Energy output* is a combination of work performed and energy returned to the environment as heat:

$$\text{Energy output} = \text{work} + \text{heat} \quad (2)$$

In the human body, at least half the energy released in chemical reactions is lost to the environment as unregulated "waste" heat.

The work in equation (2) takes one of three forms [p. 94]:

1. **Transport work** moves molecules from one side of a membrane to the other. Transport processes bring materials into and out of the body and transfer them between compartments.
2. **Mechanical work** uses intracellular fibers and filaments to create movement. This form of work includes external work, such as movement created by skeletal muscle contraction, and internal work, such as the movement of cytoplasmic vesicles and the pumping of the heart.
3. **Chemical work** is used for growth, maintenance, and storage of information and energy. Chemical work in the body can be subdivided into synthesis and storage. Storage includes both *short-term energy storage* in high-energy phosphate compounds such as ATP and *long-term energy storage* in the chemical bonds of glycogen and fat.

Most of this energy-consuming work in the body is not under conscious control. The only way people can voluntarily increase energy *output* is through body movement, such as walking and exercise. People can control their energy *intake*, however, by watching what they eat.

Although energy balance is a very simple concept, it is a difficult one for people to accept. Behavior modifications, such as eating less and exercising more, are among the most frequent instructions healthcare professionals give to patients. These instructions are also among the most difficult for people to follow, and patient compliance is low.

Oxygen Consumption Reflects Energy Use

To compile an energy balance sheet for the human body, we must estimate both the energy content of food (energy intake) and the energy expenditure from heat loss and various types of work (energy output). The most direct way to measure the energy content of food is by **direct calorimetry**. In this procedure, food is burned in an instrument called a *bomb calorimeter*, and the heat released is trapped and measured.

The heat released is a direct measure of the energy content of the burned food and is usually measured in kilocalories (kcal). One **kilocalorie** (kcal) is the amount of heat needed to raise the

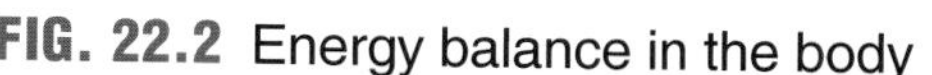

FIG. 22.2 Energy balance in the body

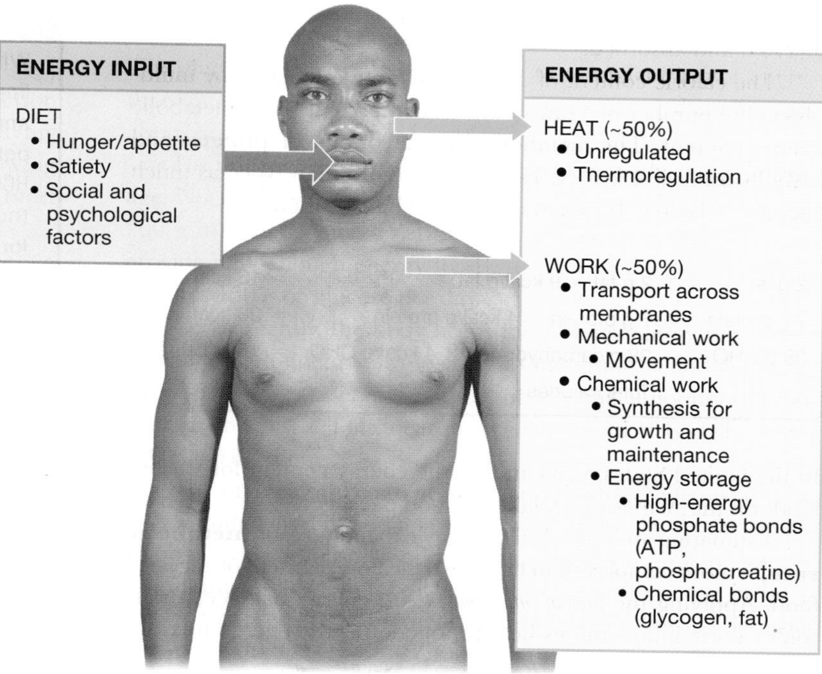
ENERGY INPUT
DIET
• Hunger/appetite
• Satiety
• Social and psychological factors
ENERGY OUTPUT
HEAT (~50%)
• Unregulated
• Thermoregulation
WORK (~50%)
• Transport across membranes
• Mechanical work
• Movement
• Chemical work
• Synthesis for growth and maintenance
• Energy storage
• High-energy phosphate bonds (ATP, phosphocreatine)
• Chemical bonds (glycogen, fat)

Concept Check

3. Name seven factors that can influence a person's metabolic rate.
4. Why does the body store most of its extra energy in fat and not in glycogen?
5. Complete and balance the following equation for aerobic metabolism of one glucose molecule: $C_6H_{12}O_6 + O_2 \rightarrow ? + ?$
6. What is the RQ for the balanced equation in Concept Check 5?

22.3 Metabolism

Metabolism is the sum of all chemical reactions in the body. The reactions making up these pathways (1) extract energy from nutrients, (2) use energy for work, and (3) store excess energy so that it can be used later. Metabolic pathways that synthesize large molecules from smaller ones are called **anabolic pathways** {*ana-*, completion + *metabole*, change}. Those that break large molecules into smaller ones are called **catabolic pathways** {*cata-*, down or back}.

The classification of a pathway is its net result, not what happens in any individual step of the pathway. For example, in the first step of *glycolysis* [p. 106], glucose gains a phosphate to become a larger molecule, glucose 6-phosphate. This single reaction is anabolic, but by the end of glycolysis, the initial 6-carbon glucose molecule has been converted to two 3-carbon pyruvate molecules. The breakdown of one glucose to two pyruvate makes glycolysis a catabolic pathway.

In the human body, we divide metabolism into two states. The period of time following a meal, when the products of digestion are being absorbed, used, and stored, is called the **fed state** or the **absorptive state**. This is an anabolic state in which the energy of nutrient biomolecules is transferred to high-energy compounds or stored in the chemical bonds of other molecules.

Once nutrients from a recent meal are no longer in the bloodstream and available for use by the tissues, the body enters what is called the **fasted state** or the **postabsorptive state**. As the pool of available nutrients in the blood decreases, the body taps into its stored reserves. The fasted state is catabolic because cells break down large molecules into smaller molecules. The energy released by breaking chemical bonds of large molecules is used to do work.

Ingested Energy May Be Used or Stored

The biomolecules we ingest are destined to meet one of three fates:

1. **Energy.** Biomolecules can be metabolized immediately, with the energy released from broken chemical bonds trapped in ATP, phosphocreatine, and other high-energy compounds. This energy can then be used to do mechanical work.
2. **Synthesis.** Biomolecules entering cells can be used to synthesize basic components needed for growth and maintenance of cells and tissues.
3. **Storage.** If the amount of food ingested exceeds the body's requirements for energy and synthesis, the excess energy goes into storage in the bonds of glycogen and fat. Storage makes energy available for times of fasting.

The fate of an absorbed biomolecule depends on whether it is a carbohydrate, protein, or fat. **FIGURE 22.3** is a schematic diagram that follows these biomolecules from the diet into the three *nutrient pools* of the body: the free fatty acid pool, the glucose pool, and the amino acid pool. **Nutrient pools** are nutrients that are available for immediate use. They are located primarily in the plasma.

Free fatty acids form the primary pool of fats in the blood. They can be used as an energy source by many tissues but are also easily stored as fat (*triglycerides*) in adipose tissue.

Carbohydrates are absorbed mostly as glucose. Plasma glucose concentration is the most closely regulated of the three nutrient pools because glucose is the only fuel the brain can metabolize, except in times of starvation. Notice the locations of the exit "pipes" on the glucose pool in Figure 22.3. If the glucose pool falls below a certain level, only the brain has access to glucose. This conservation measure ensures that the brain has an adequate energy supply. Just as the circulatory system gives priority to supplying oxygen to the brain, metabolism also gives priority to the brain.

If the body's glucose pool is within the normal range, most tissues use glucose for their energy source. Excess glucose goes into storage as glycogen. The synthesis of glycogen from glucose is known as **glycogenesis**. Glycogen stores are limited, however, and additional excess glucose is converted to fat by **lipogenesis**.

If plasma glucose concentrations decrease, the body converts glycogen to glucose through **glycogenolysis**. The body maintains plasma glucose concentrations within a narrow range by balancing oxidative metabolism, glycogenesis, glycogenolysis, and lipogenesis.

If homeostasis fails and plasma glucose exceeds a critical level, as occurs in diabetes mellitus, excess glucose is excreted in the urine. Glucose excretion takes place only when the *renal threshold* for glucose reabsorption is exceeded [p. 603].

The amino acid pool of the body is used primarily for protein synthesis. However, if glucose intake is low, amino acids can be converted into glucose through the pathways known as **gluconeogenesis**. This word literally means "the birth (*genesis*) of new (*neo*) glucose" and refers to the synthesis of glucose from a noncarbohydrate precursor.

Amino acids are the main source for glucose through the gluconeogenesis pathways, but glycerol from triglycerides can also be used. Both gluconeogenesis and glycogenolysis are important backup sources for glucose during periods of fasting.

Enzymes Control the Direction of Metabolism

How does the body control the shift of metabolism between the fed state and the fasted state? One key feature of metabolic regulation is the use of different enzymes to catalyze forward and reverse reactions. This dual control, sometimes called *push-pull control*, allows close regulation of a reaction's direction.

FIG. 22.3 Nutrient pools and metabolism

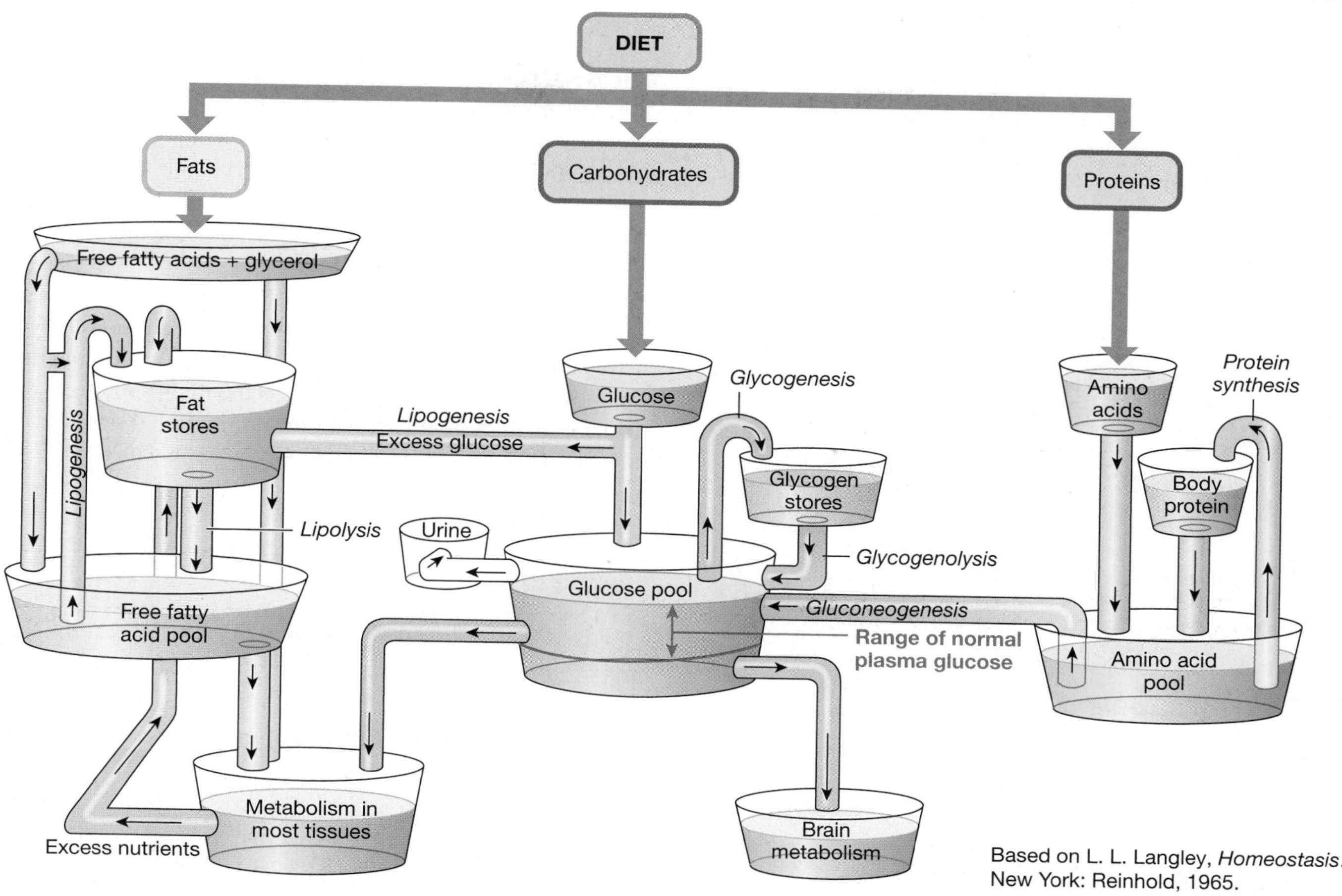

DIET
Fats
Carbohydrates
Proteins
Free fatty acids + glycerol
Fat stores
Lipogenesis
Lipogenesis
Excess glucose
Glucose
Glycogenesis
Amino acids
Protein synthesis
Body protein
Glycogen stores
Lipolysis
Urine
Glycogenolysis
Glucose pool
Gluconeogenesis
Free fatty acid pool
Range of normal plasma glucose
Amino acid pool
Metabolism in most tissues
Excess nutrients
Brain metabolism

Based on L. L. Langley, *Homeostasis*. New York: Reinhold, 1965.

RUNNING PROBLEM

When Nicole was admitted to the hospital, her blood pressure was 80/50, and her pulse was a weak and irregular 90 beats per minute. She weighed less than 85% of the minimal healthy weight for a woman of her height and age. She had an intense fear of gaining weight, even though she was underweight. Her menstrual periods were irregular, she had just suffered a fractured wrist from a fall that normally shouldn't have caused a fracture, and her hair was thinning. When Dr. Ayani questioned Nicole, she admitted that she had been feeling weak during dance rehearsals and had been having difficulty concentrating at times.

Q6: *Based on what you know about heart rate and blood pressure, speculate on why Nicole has low blood pressure with a rapid pulse.*

Q7: *Would you expect Nicole's renin and aldosterone levels to be normal, elevated, or depressed? How might these levels relate to her K^+ disturbance?*

Q8: *Give some possible reasons Nicole had been feeling weak during dance rehearsals.*

693 – 695 – 696 – 699 – 718 – 724

FIGURE 22.4 shows how push-pull control can regulate the flow of nutrients through metabolic pathways. In Figure 22.4a, enzyme 1 catalyzes the reaction A → B, and enzyme 2 catalyzes the reverse reaction, B → A. When the activity of the two enzymes is roughly equal, as soon as A is converted into B, B is converted back into A. Turnover of the two substrates is rapid, but there is no net production of either A or B.

To alter the net direction of the reaction, the enzyme activity must change. Enzymes are proteins that bind ligands, so their activity can be modulated [p. 49]. Most modulation of metabolic enzymes is controlled by hormones.

Figure 22.4b represents the series of reactions through which glucose becomes glycogen. During the fed state, the pancreatic hormone insulin stimulates the enzymes promoting glycogenesis and inhibits the enzymes for glycogenolysis. The net result is glycogen synthesis from glucose.

The reverse pattern is shown in Figure 22.4c. In the fasted state, *glucagon*, another pancreatic hormone, is dominant. Glucagon stimulates the enzymes of glycogenolysis while inhibiting the enzymes for glycogenesis. The net result is glucose synthesis from glycogen.

FIG. 22.4 Push-pull control

In push-pull control, different enzymes catalyze forward and reverse reactions.

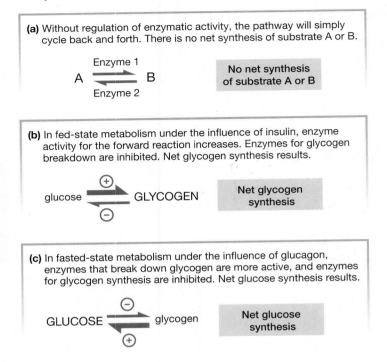

(a) Without regulation of enzymatic activity, the pathway will simply cycle back and forth. There is no net synthesis of substrate A or B.

Enzyme 1
A ⇌ B
Enzyme 2

No net synthesis of substrate A or B

(b) In fed-state metabolism under the influence of insulin, enzyme activity for the forward reaction increases. Enzymes for glycogen breakdown are inhibited. Net glycogen synthesis results.

+
glucose
GLYCOGEN
−

Net glycogen synthesis

(c) In fasted-state metabolism under the influence of glucagon, enzymes that break down glycogen are more active, and enzymes for glycogen synthesis are inhibited. Net glucose synthesis results.

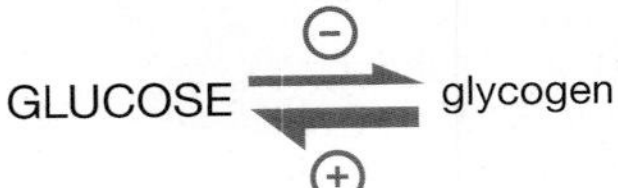
−
GLUCOSE
glycogen
+

Net glucose synthesis

22.4 Fed-State Metabolism

The fed state following ingestion of nutrients is anabolic: absorbed nutrients are being used for energy, synthesis, and storage. The table at the bottom of **FIGURE 22.5** summarizes the fates of carbohydrates, proteins, and fats in the fed state. In the sections that follow, we examine some of these pathways.

Carbohydrates Make ATP

The most important biochemical pathways for energy production are summarized in Figure 22.5. This figure does not include all of the metabolic intermediates in each pathway [see Chapter 4, p. 107 for detailed pathways]. Instead, it emphasizes the points at which different pathways intersect, because these intersections are often key points at which metabolism is controlled.

Glucose is the primary substrate for ATP production. Glucose absorbed from the intestine enters the hepatic portal vein and is taken directly to the liver, where about 30% of all ingested glucose is metabolized. The remaining 70% continues in the bloodstream for distribution to the brain, muscles, and other organs and tissues.

Glucose moves from interstitial fluid into cells by membrane GLUT transporters [p. 141]. Most glucose absorbed from a meal goes immediately into glycolysis and the citric acid cycle to make ATP. Some glucose is used by the liver for lipoprotein synthesis. Glucose that is not required for energy and synthesis is stored either as glycogen or fat. The body's ability to store glycogen is limited, so most excess glucose is converted to triglycerides and stored in adipose tissue.

Glucose Storage Glycogen, a large polysaccharide, is the main storage form of glucose in the body. Glycogen is a glucose polymer, created by linking many individual glucose molecules together into a branching chain [see Fig. 2.2, p. 31]. A single glycogen particle in the cytoplasm may contain as many as 55,000 linked glucose molecules! Glycogen granules occur as insoluble inclusions in the cytosol of cells [p. 31].

Glycogen is found in all cells of the body, but the liver and skeletal muscles contain especially high concentrations. Glycogen in skeletal muscles provides a ready energy source for muscle contraction. Glycogen in the liver acts as the main source of glucose for the body in periods between meals (the fasted state). It is estimated that the liver keeps about a 4-hour supply of glucose stored as glycogen.

Concept Check

7. Are GLUT transporters active or passive transporters?

Amino Acids Make Proteins

Most amino acids absorbed from a meal go to the tissues for protein synthesis [p. 112]. Like glucose, amino acids are taken first to the liver by the hepatic portal system. The liver uses them to synthesize lipoproteins and plasma proteins, such as albumin, clotting factors, and angiotensinogen.

Amino acids not taken up by the liver are used by cells to create structural or functional proteins, such as cytoskeletal elements, enzymes, and hormones. Amino acids are also incorporated into nonprotein molecules, such as amine hormones and neurotransmitters.

If glucose intake is low, amino acids can be used for energy, as described in the next section on fasted-state metabolism. However, if more protein is ingested than is needed for synthesis and energy expenditures, excess amino acids are converted to fat. Some bodybuilders spend large amounts of money on amino-acid supplements advertised to build bigger muscles. But these amino acids do not automatically go into protein synthesis. When amino acid intake exceeds the body's need for protein synthesis, excess amino acids are burned for energy or stored as fat.

Fats Store Energy

Most ingested fats are assembled inside intestinal epithelial cells into lipoprotein and lipid complexes called *chylomicrons* [p. 678]. Chylomicrons leave the intestine and enter the venous circulation via the lymphatic vessels (**FIG. 22.6a**). Chylomicrons consist of cholesterol, triglycerides, phospholipids, and lipid-binding proteins called **apoproteins,** or *apolipoproteins* {*apo-*, derived from}. Once

FIG. 22.5 ESSENTIALS Biochemical Pathways For Energy Production

(a) Summary of Biochemical Pathways for Energy Production

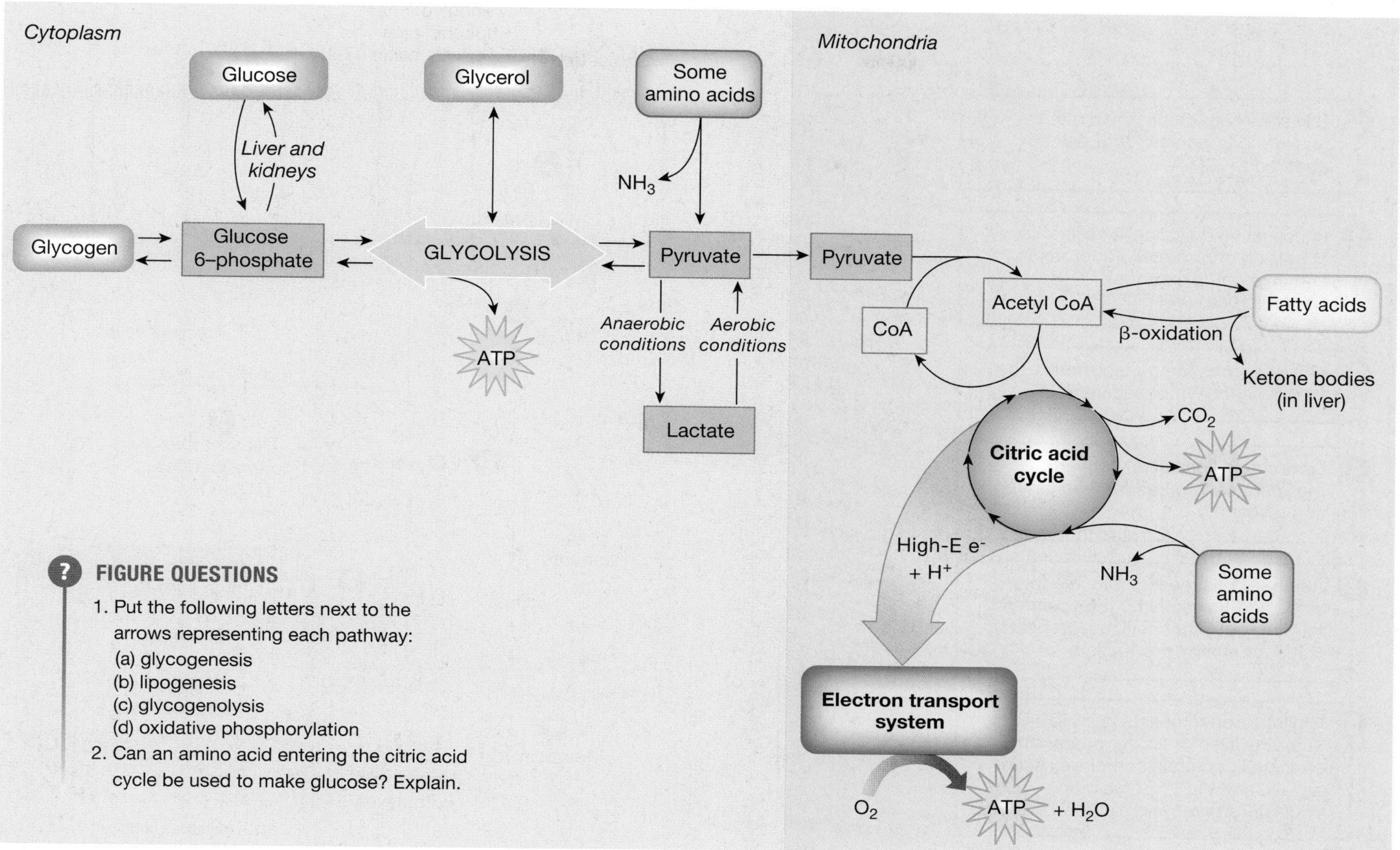

Cytoplasm
Mitochondria
Glucose
Glycerol
Some amino acids
Liver and kidneys
NH_3
Glycogen
Glucose 6–phosphate
GLYCOLYSIS
Pyruvate
Pyruvate
ATP
Anaerobic conditions
Aerobic conditions
Lactate
CoA
Acetyl CoA
Fatty acids
β-oxidation
Ketone bodies (in liver)
CO_2
Citric acid cycle
ATP
High-E e⁻ + H⁺
NH_3
Some amino acids
Electron transport system
O_2
ATP
+ H_2O

FIGURE QUESTIONS

1. Put the following letters next to the arrows representing each pathway:
   (a) glycogenesis
   (b) lipogenesis
   (c) glycogenolysis
   (d) oxidative phosphorylation
2. Can an amino acid entering the citric acid cycle be used to make glucose? Explain.

(b) Fates of Nutrients in Fed-State and Fasted-State Metabolism

Nutrient	Absorbed as	Fed-State Metabolism	Fasted-State Metabolism
Carbohydrates	Glucose primarily; also fructose and galactose	• Used immediately for energy through aerobic pathways* (glycolysis and citric acid cycle) • Used for lipoprotein synthesis in liver • Stored as glycogen in liver and muscle (glycogenesis) • Excess converted to fat and stored in adipose tissue (lipogenesis)	• Glycogen polymers broken down (glycogenolysis) to glucose in liver and kidney or to glucose 6-phosphate for use in glycolysis
Proteins	Amino acids primarily plus some small peptides	• Most amino acids go to tissues for protein synthesis* in liver to intermediates for aerobic metabolism (deamination) • Excess converted to fat and stored in adipose tissue (lipogenesis)	• Proteins broken down into amino acids • Amino acids deaminated in liver for ATP production or used to make glucose (gluconeogenesis)
Fats	Fatty acids, triglycerides and cholesterol	• Stored as triglycerides primarily in the liver and adipose tissue* (lipogenesis) • Cholesterol used for steroid synthesis or as a membrane component • Fatty acids used for lipoprotein and eicosanoid synthesis	• Triglycerides broken down into fatty acids and glycerol (lipolysis) • Fatty acids used for ATP production through aerobic pathways (β-oxidation)

*Primary fate

FIG. 22.6 ESSENTIALS Fat Synthesis

(a) Transport and Fate of Dietary Fats

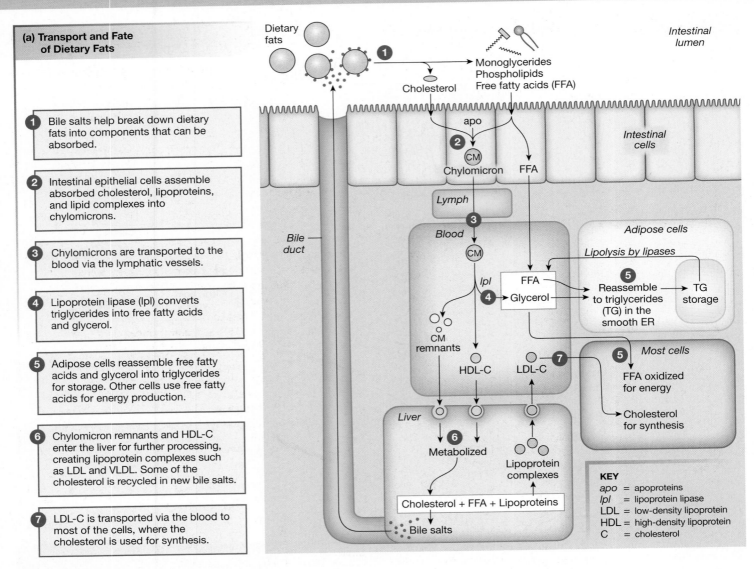

1. Bile salts help break down dietary fats into components that can be absorbed.
2. Intestinal epithelial cells assemble absorbed cholesterol, lipoproteins, and lipid complexes into chylomicrons.
3. Chylomicrons are transported to the blood via the lymphatic vessels.
4. Lipoprotein lipase (lpl) converts triglycerides into free fatty acids and glycerol.
5. Adipose cells reassemble free fatty acids and glycerol into triglycerides for storage. Other cells use free fatty acids for energy production.
6. Chylomicron remnants and HDL-C enter the liver for further processing, creating lipoprotein complexes such as LDL and VLDL. Some of the cholesterol is recycled in new bile salts.
7. LDL-C is transported via the blood to most of the cells, where the cholesterol is used for synthesis.

Dietary fats
Intestinal lumen
Monoglycerides
Phospholipids
Free fatty acids (FFA)
Cholesterol
apo
CM
Chylomicron
FFA
Intestinal cells
Lymph
Bile duct
Blood
Adipose cells
Lipolysis by lipases
lpl
FFA
Glycerol
Reassemble to triglycerides (TG) in the smooth ER
TG storage
CM remnants
HDL-C
LDL-C
Most cells
FFA oxidized for energy
Cholesterol for synthesis
Liver
Metabolized
Lipoprotein complexes
Cholesterol + FFA + Lipoproteins
Bile salts

KEY
apo = apoproteins
lpl = lipoprotein lipase
LDL = low-density lipoprotein
HDL = high-density lipoprotein
C = cholesterol

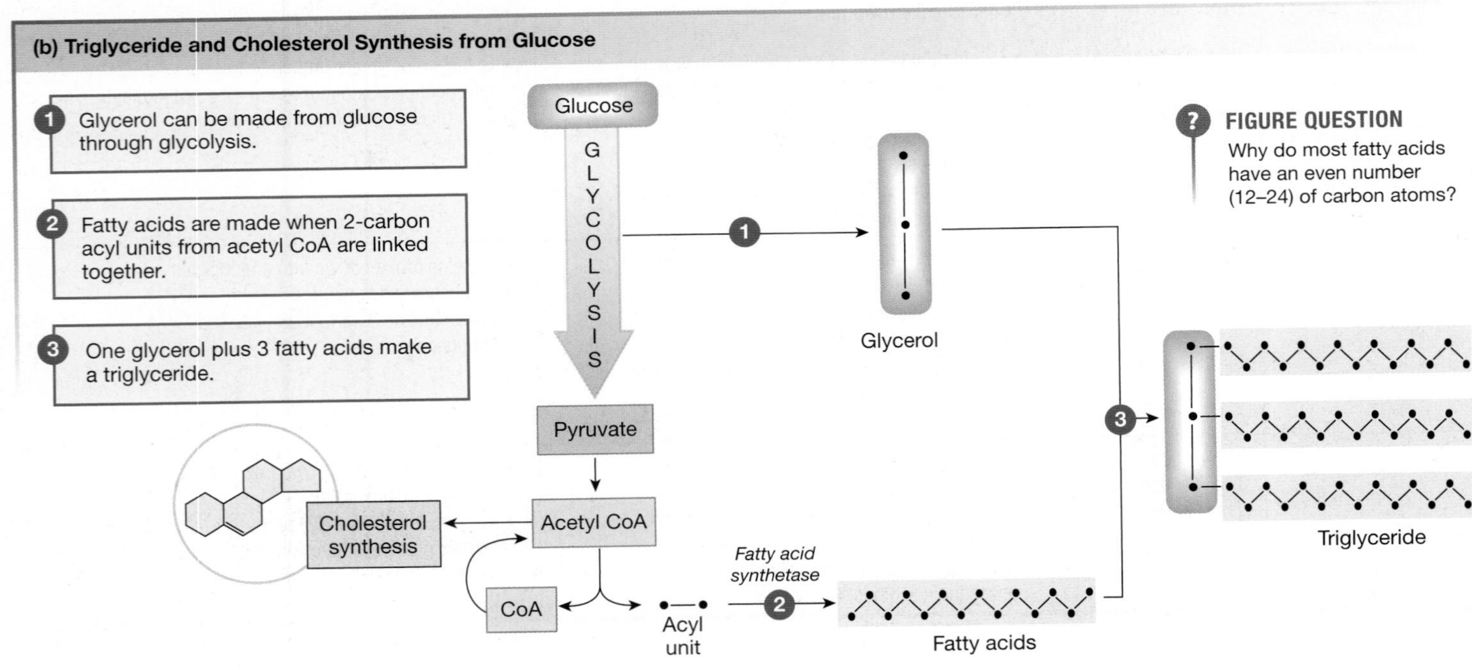

(b) Triglyceride and Cholesterol Synthesis from Glucose
1 Glycerol can be made from glucose through glycolysis.
2 Fatty acids are made when 2-carbon acyl units from acetyl CoA are linked together.
3 One glycerol plus 3 fatty acids make a triglyceride.
Glucose
GLYCOLYSIS
Pyruvate
Acetyl CoA
Cholesterol synthesis
CoA
Acyl unit
Fatty acid synthetase
Fatty acids
Glycerol
Triglyceride
FIGURE QUESTION
Why do most fatty acids have an even number (12–24) of carbon atoms?

these lipid complexes begin to circulate through the blood, the enzyme **lipoprotein lipase** (lpl) bound to the capillary endothelium of muscles and adipose tissue converts the triglycerides to free fatty acids and glycerol. These molecules may be used for energy by most cells or reassembled into triglycerides for storage in adipose tissue.

Chylomicron remnants that remain in the circulation are taken up and metabolized by the liver (Fig. 22.6a). Cholesterol from the remnants joins the liver's pool of lipids. If cholesterol is in excess, some may be converted to bile salts and excreted in the bile. The remaining cholesterol is added to newly synthesized cholesterol and fatty acids, and packaged into lipoprotein complexes for secretion into the blood.

The lipoprotein complexes that reenter the blood contain varying amounts of triglycerides, phospholipids, cholesterol, and apoproteins. The more protein a complex contains, the heavier it is, with plasma lipoprotein complexes ranging from *very-low-density lipoprotein* (VLDL) to *high-density lipoprotein* (HDL). The combination of lipids with proteins makes cholesterol more soluble in plasma, but the complexes are unable to diffuse through cell membranes. Instead, they must be brought into cells by *receptor-mediated endocytosis* [p. 147]. The apoproteins in the complexes have specific membrane receptors in different tissues.

Most lipoprotein in the blood is *low-density lipoprotein-cholesterol* (LDL-C) [p. 147]. LDL-C is sometimes known as the "lethal cholesterol" because elevated concentrations of plasma LDL-C are associated with the development of atherosclerosis [p. 501]. LDL-C complexes contain *apoprotein B* (apoB), which combines with receptors that bring LDL-C into most cells of the body. Several inherited forms of *hypercholesterolemia* (elevated plasma cholesterol levels) have been linked to defective forms of apoB. These abnormal apoproteins may help explain the accelerated development of atherosclerosis in people with hypercholesterolemia.

The second most common lipoprotein in the blood is *high-density lipoprotein-cholesterol* (HDL-C). HDL-C is sometimes called the "healthy cholesterol" because HDL is the lipoprotein involved in cholesterol transport out of the plasma. HDL-C contains *apoprotein A* (apoA), which facilitates cholesterol uptake by the liver and other tissues.

Lipid Synthesis Most people get sufficient cholesterol from animal products in the diet, but cholesterol is such an important molecule that the body will synthesize it if the diet is deficient. Even vegetarians who eat no animal products (vegans) have substantial amounts of cholesterol in their cells. The body can make cholesterol from acetyl CoA through a series of reactions. Once the ring structure of cholesterol is synthesized, it is a fairly simple matter for the cell to change cholesterol into hormones and other steroids.

Other fats needed for cell structure and function, such as phospholipids, can also be made from nonlipid precursors during the fed state. Lipids are so diverse that generalizing about their synthesis is difficult. Enzymes in the smooth endoplasmic reticulum and cytosol of cells are responsible for most lipid synthesis. For example, the phosphorylation steps that convert triglycerides into phospholipids take place in the smooth endoplasmic reticulum (ER).

CLINICAL FOCUS

Antioxidants Protect the Body

One of the daily rituals of childhood is taking your chewable vitamin. Many vitamins are coenzymes [p. 100] and necessary in small amounts for metabolic reactions. Some, such as vitamins C and E, also act as antioxidants. We hear antioxidants promoted everywhere—in cosmetics as well as foods. Many antioxidants occur naturally in fruits and vegetables. **Antioxidants** are molecules that prevent damage to our cells by taking or giving up an electron without becoming free radicals. So what are free radicals, and why are they bad?

Free radicals are unstable molecules with an unpaired electron. Electrons in an atom are most stable in pairs, so free radicals, with their one unpaired electron, look for an electron they can "steal" from another atom. This creates a chain reaction of free radical production that can disrupt normal cell function. Free radicals are thought to contribute to aging and to the development of certain diseases, such as some cancers.

Free radicals are created during normal metabolism or through exposure to radiation, both natural, such as from the sun, and manufactured, such as from microwave ovens. One common free radical is the *superoxide* ion, $\cdot O_2^-$. The body constantly manufactures this free radical during normal metabolism whenever a neutral oxygen molecule (O_2) gains an extra electron (represented by placing a · in front of the molecule). Antioxidants can absorb the extra electron from superoxide, thereby stopping the destructive chain of free radical formation.

Triglyceride synthesis from excess ingested glucose and protein is an important part of fed-state metabolism. Figure 22.6b shows some pathways for triglyceride synthesis. Glycerol can be made from glucose or from glycolysis intermediates (Fig. 22.6b). Fatty acids are made from acetyl CoA when a cytosolic enzyme called *fatty acid synthetase* links the 2-carbon acyl groups into carbon chains. This process also requires hydrogens and high-energy electrons from NADPH. The combination of glycerol and fatty acids into triglycerides takes place in the smooth endoplasmic reticulum.

Plasma Cholesterol Predicts Heart Disease

Of the nutrients in the plasma, lipids and glucose receive the most attention from health professionals. Abnormal glucose metabolism is the hallmark of diabetes mellitus, described later in this chapter. Abnormal plasma lipids are used as predictors of atherosclerosis and coronary heart disease (CHD) [p. 501].

Tests to measure blood lipids and assess cardiovascular risk range from simple but less accurate finger-stick blood samples to expensive tests on venous blood that look at all sizes of lipoproteins, from VLDL through HDL. As more epidemiological and treatment data are gathered, experts continue to redefine desirable lipid values. The U.S. National Cholesterol Education Panel issued guidelines in 2004 that are still in use today. Emphasis over the years has shifted from concern about total cholesterol levels (<200 mg/dL of plasma is recommended) to the absolute amounts and relative proportions of the various subtypes. You can calculate your risk of developing cardiovascular disease based on lipid levels with online risk calculators, such as the one provided by the American College of Cardiology and American Heart Association (*www.cvriskcalculator.com/*).

Some studies indicate that elevated LDL-C is the single largest cholesterol risk factor for CHD (**FIG. 22.7**). Oxidized LDL-C in the circulation is taken up by macrophages and leads to the development of atherosclerotic plaques inside blood vessels [Fig. 15.21, p. 503]. Higher concentrations of LDL-C increase a person's risk for CHD.

The HDL-C level in plasma has also been used to predict risk of developing atherosclerosis. Like high LDL-C, low HDL-C (<40 mg/dL) is associated with a higher risk of developing CHD. More recently, health professionals have started looking at the *non-HDL cholesterol value* (total cholesterol minus HDL-C) as perhaps a better indicator of CHD risk.

Lifestyle modifications (improved diet, smoking cessation, and exercise) can be very effective in improving lipid profiles but can be difficult for patients to implement and sustain. All the pharmacological therapies developed to treat elevated cholesterol target some aspect of cholesterol absorption and metabolism, underscoring the principle of mass balance in cholesterol homeostasis. Decreasing cholesterol uptake or synthesis and increasing cholesterol clearance by metabolism or excretion are effective mechanisms for reducing the amount of cholesterol in the body.

The drugs known as *bile acid sequestrants* {*sequestrare*, to put in the hands of a trustee} bind to bile acids in the intestinal lumen, preventing their reabsorption and increasing cholesterol excretion [p. 676]. In the absence of recycled bile salts, the liver increases bile acid synthesis from cholesterol, thereby decreasing plasma cholesterol when hepatic LDL receptors import cholesterol from the blood. A drug-free way to accomplish the same result is to eat more soluble fiber—indigestible fiber found in oatmeal, for example. Soluble fiber traps bile salts and increases their excretion in the feces.

FIG. 22.7 The relationship between LDL-C and risk of developing heart disease

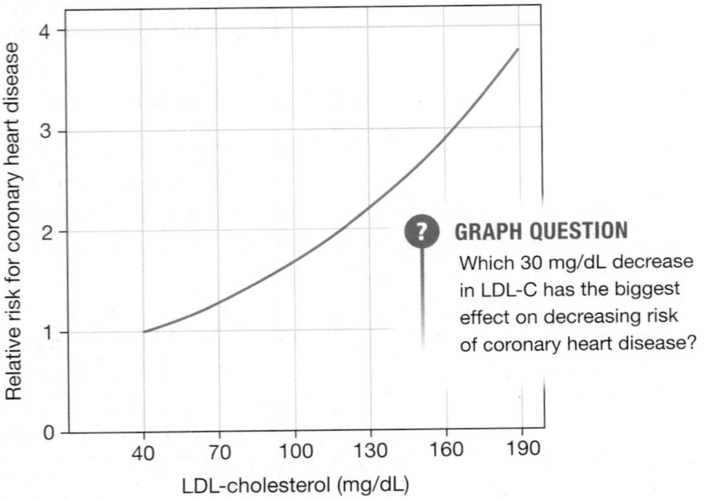
Relative risk for coronary heart disease
0
1
2
3
4
40
70
100
130
160
190
LDL-cholesterol (mg/dL)
GRAPH QUESTION
Which 30 mg/dL decrease in LDL-C has the biggest effect on decreasing risk of coronary heart disease?

Data taken from Grundy *et al.*, *Circulation* 110: 227–239, 2004 July 13.

Another lipid-lowering drug, *ezetimibe*, inhibits intestinal cholesterol transport [p. 678]. A different strategy for reducing intestinal cholesterol uptake is adding plant *sterols* (steroid-alcohols) and *stanols* (saturated sterols) to the diet by eating nuts, seeds, cereals, and vegetable oils. Sterols and stanols displace cholesterol in chylomicrons, decreasing cholesterol absorption.

The remaining lipid-lowering drugs all affect cholesterol metabolism in the liver. The drugs called *statins* inhibit the enzyme *HMG CoA reductase*, which mediates cholesterol synthesis in hepatocytes. The *fibrates*, which stimulate a transcription factor called *PPAR*α (pronounced *p-par-alpha*), and *niacin* (vitamin B_3, nicotinic acid) decrease LDL-C and increase HDL-C by mechanisms that are not well understood.

Concept Check

8. What is a dL (as in mg/dL)?
9. Use your understanding of digestive physiology to predict the common side effects of taking bile acid sequestrants and ezetimibe.

22.5 Fasted-State Metabolism

Once all nutrients from a meal have been digested, absorbed, and distributed to various cells, plasma concentrations of glucose begin to fall. This is the signal for the body to shift from fed (absorptive) state to fasted (postabsorptive) state metabolism. Metabolism is under the control of hormones whose goal is to maintain blood glucose homeostasis so that the brain and neurons have adequate fuel.

Glucose homeostasis is achieved through catabolic pathways that convert glycogen, proteins, and fats into intermediates that can be used to make either glucose or ATP. Using proteins or fats for ATP production spares plasma glucose for use by the brain. **FIGURE 22.8** summarizes the catabolic pathways of the fasted state in different organs.

Glycogen Converts to Glucose

The most readily available source of glucose for plasma glucose homeostasis is the body's glycogen store, most of which is in the liver (Fig. 22.8). Liver glycogen can provide enough glucose to meet 4–5 hours of the body's energy needs.

In glycogenolysis, glycogen is broken down to glucose or glucose 6-phosphate (**FIG. 22.9**). Most glycogen is converted directly to glucose 6-phosphate in a reaction that splits a glucose molecule from the polymer with the help of an inorganic phosphate from the cytosol. Only about 10% of stored glycogen is hydrolyzed to plain glucose molecules.

In the fasted state, skeletal muscle glycogen can be metabolized to glucose, but not directly (Fig. 22.8). Muscle cells, like most other cells, lack the enzyme that makes glucose from glucose 6-phosphate. Consequently, glucose 6-phosphate produced from glycogenolysis in skeletal muscle is metabolized to either pyruvate (aerobic conditions) or lactate (anaerobic conditions). Pyruvate and lactate are then transported to the liver, which uses them to make glucose via gluconeogenesis.

FIG. 22.8 Fasted-state metabolism

Fasted-state metabolism must maintain plasma glucose homeostasis for the brain.

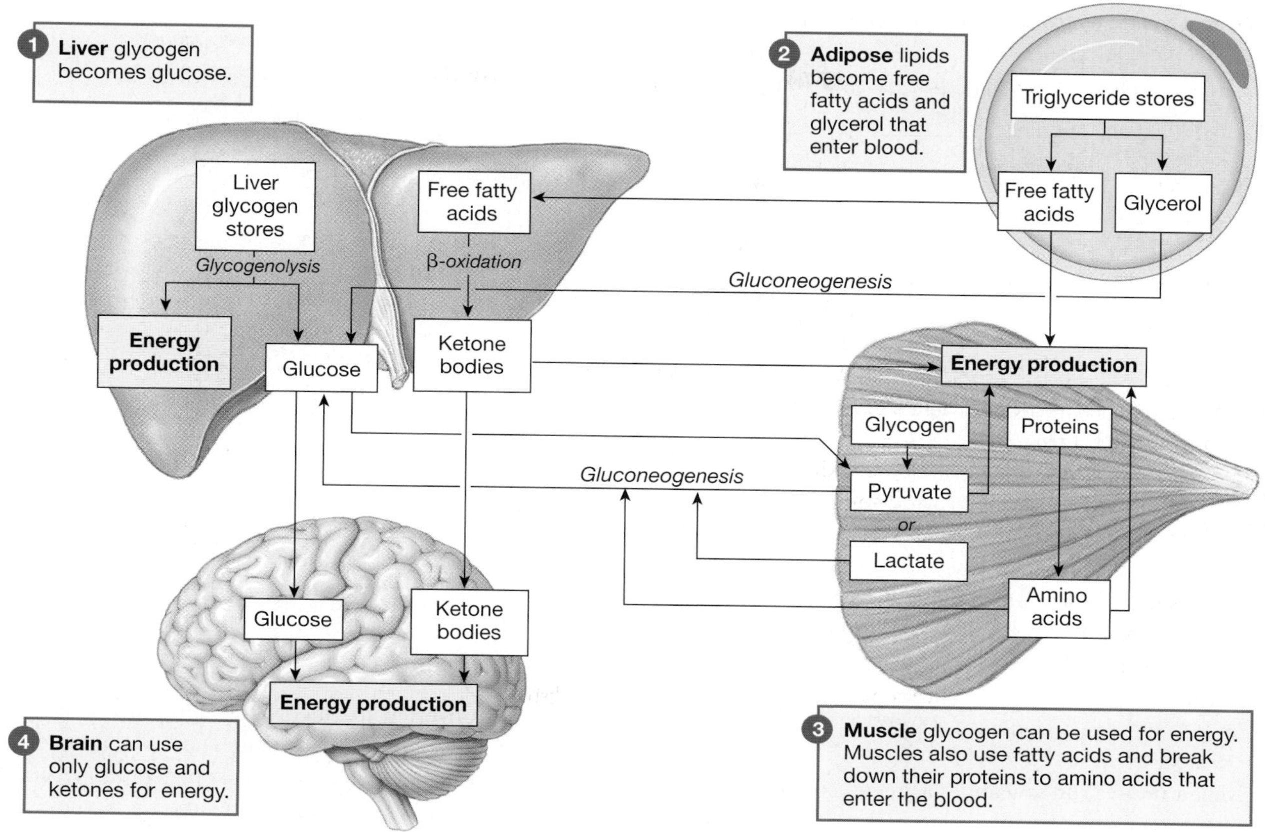
1 Liver glycogen becomes glucose.
2 Adipose lipids become free fatty acids and glycerol that enter blood.
Triglyceride stores
Free fatty acids
Glycerol
Liver glycogen stores
Free fatty acids
Glycogenolysis
β-oxidation
Gluconeogenesis
Energy production
Glucose
Ketone bodies
Energy production
Glycogen
Proteins
Gluconeogenesis
Pyruvate
or
Lactate
Amino acids
Glucose
Ketone bodies
Energy production
4 Brain can use only glucose and ketones for energy.
3 Muscle glycogen can be used for energy. Muscles also use fatty acids and break down their proteins to amino acids that enter the blood.

FIG. 22.9 Glycogenolysis

Glycogen can be converted directly to glucose 6-phosphate by the addition of phosphate. Glycogen that is broken down first to glucose and then phosphorylated "costs" the cell an extra ATP.

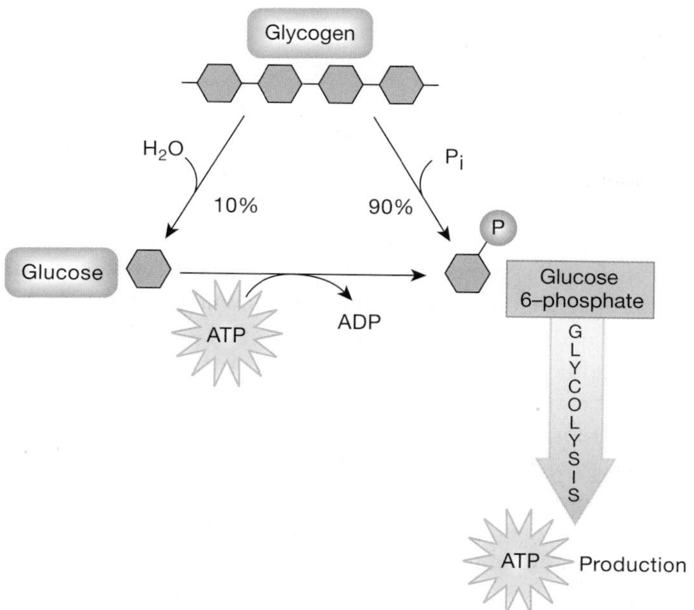
Glycogen
H_2O
P_i
10%
90%
P
Glucose
Glucose 6-phosphate
ATP
ADP
GLYCOLYSIS
ATP
Production

Proteins Can Be Used to Make ATP

The free amino acid pool of the body normally provides substrate for ATP production during the fasted state. If the fasted state goes on for an extended time, however, muscle proteins can be broken down to provide energy. The first step in protein catabolism is the digestion of a protein into smaller polypeptides by enzymes called *proteases* (*endopeptidases*) [p. 680]. Then enzymes known as *exopeptidases* break the peptide bonds at the ends of the smaller polypeptide, freeing individual amino acids.

Amino acids can be converted into intermediates that enter either glycolysis or the citric acid cycle (Fig. 22.5). The first step in this conversion is **deamination**, the removal of the amino group from the amino acid (**FIG. 22.10a**). Deamination creates an ammonia molecule and an organic acid. Some of the organic acids created this way are pyruvate, acetyl CoA, and several intermediates of the citric acid cycle. The organic acids can then enter the pathways of aerobic metabolism for ATP production.

The *ammonia* (NH_3) molecules created during deamination rapidly pick up hydrogen ions (H^+) to become *ammonium ions* (NH_4^+), as indicated in Figure 22.10b. Both ammonia and ammonium ions are toxic, however, so liver cells rapidly convert them into **urea** (CH_4N_2O). Urea is the main nitrogenous waste of the body and is excreted by the kidneys.

FIG. 22.10 Amino acid catabolism

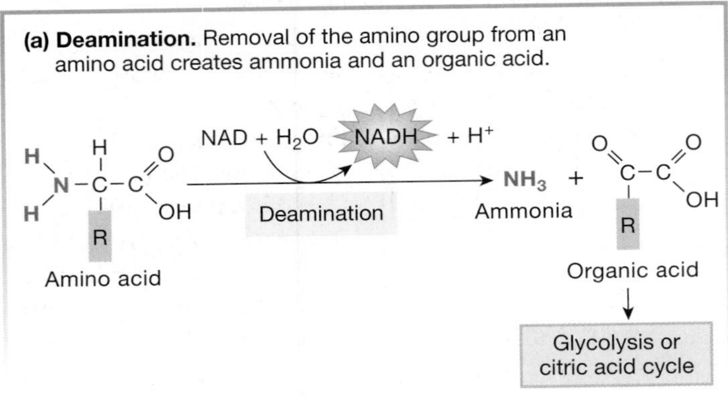

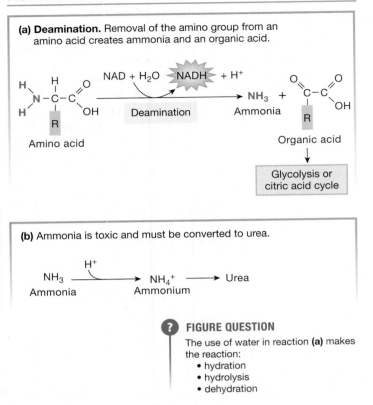

(a) Deamination. Removal of the amino group from an amino acid creates ammonia and an organic acid.
NAD + H_2O
NADH
+ H^+
Deamination
NH_3 +
Ammonia
Amino acid
Organic acid
Glycolysis or citric acid cycle

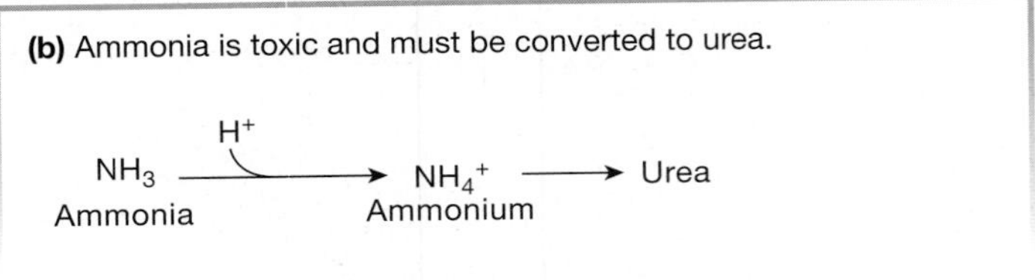

(b) Ammonia is toxic and must be converted to urea.
H^+
NH_3
Ammonia
NH_4^+
Ammonium
Urea

FIGURE QUESTION

The use of water in reaction **(a)** makes the reaction:
- hydration
- hydrolysis
- dehydration

FIG. 22.11 Gluconeogenesis

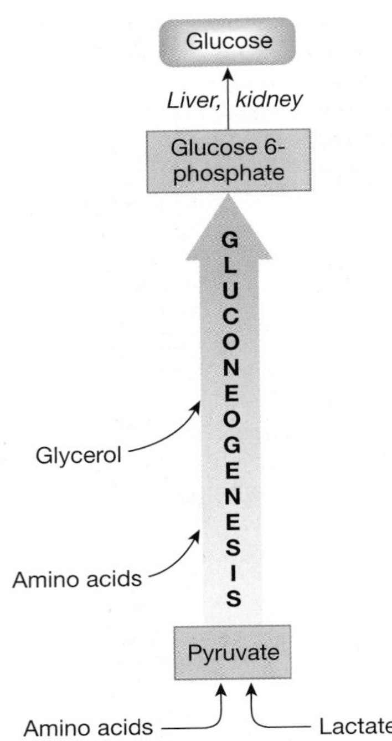

Glucose
Liver, kidney
Glucose 6-phosphate
GLUCONEOGENESIS
Glycerol
Amino acids
Pyruvate
Amino acids
Lactate

If glycogen stores run low and plasma glucose homeostasis is threatened, proteins can also be used to make glucose. In the liver, amino acids or pyruvate made from amino acids enter the glycolysis pathway (**FIG. 22.11**). The pathway then runs backward to create glucose 6-phosphate and glucose (gluconeogenesis).

Lipids Store More Energy than Glucose or Protein

Lipids are the primary fuel-storage molecule of the body because they have higher energy content than proteins or carbohydrates. When the fasted-state body needs to use stored energy, *lipases* break down triglycerides into glycerol and fatty acids through a series of reactions collectively known as **lipolysis** (**FIG. 22.12**). Glycerol feeds into glycolysis about halfway through the pathway **2**. It then goes through the same reactions as glucose from that point on.

The long carbon chains of fatty acids are more difficult to turn into ATP. Most fatty acids must first be transported from the cytosol into the mitochondrial matrix (Fig. 22.12). There they are slowly disassembled as 2-carbon units are chopped off the end of the chain, one unit at a time, in a process called **beta-oxidation** (β-oxidation).

In most cells, the 2-carbon units from fatty acids are converted into acetyl CoA, whose 2-carbon acyl units feed directly into the citric acid cycle (Fig. 22.12 **4**). Because many acetyl CoA molecules can be produced from a single fatty acid, lipids contain 9 kcal of stored energy per gram, compared with 4 kcal per gram for proteins and carbohydrates.

In the fasted state, adipose tissue releases fatty acids and glycerol into the blood (Fig. 22.8 **2**). Glycerol goes to the liver and can be converted to glucose through gluconeogenesis. The fatty acids are taken up by many tissues and used for energy production.

In the liver, if fatty acid breakdown produces acetyl CoA faster than the citric acid cycle can metabolize it, the excess acyl units become **ketone bodies** (often called simply *ketones* in physiology and medicine). Ketone bodies become a significant source of energy for the brain in cases of prolonged starvation, when glucose is low (Fig. 22.8). Ketone bodies enter the blood, creating a state of *ketosis*. The breath of people in ketosis has a fruity odor caused by acetone, a volatile ketone whose odor you might recognize from nail polish remover.

Low-carbohydrate diets, such as the Atkins and South Beach diet plans, are *ketogenic* because most of their calories come from fats. The diets are very low in carbohydrate and high in fat and protein, which shifts metabolism to β-oxidation of fats and production of ketone bodies. People who go on these diets are often pleased by initial rapid weight loss, but it is due to glycogen breakdown and water loss, not body fat reduction.

Ketogenic diets have been used successfully to treat children who have epilepsy that is not responding fully to drug therapy. For reasons that we do not understand, maintaining a state of ketosis in these children decreases the incidence of seizures. But ketogenic diets can also be dangerous. Ketone bodies such as *acetoacetic acid* and β-*hydroxybutyric acid* are strong metabolic acids that can seriously disrupt the body's pH balance and lead to a state of *ketoacidosis*, a metabolic acidosis [p. 648]. Among the other risks associated with ketogenic diets are dehydration, electrolyte loss, inadequate intake of calcium and vitamins, gout, and kidney problems.

With this summary of metabolic pathways as background, we now turn to the endocrine and neural regulation of metabolism.

Concept Check

10. What is the difference between glycogenesis and gluconeogenesis?
11. When amino acids are used for energy, which pathways in Figure 22.5 do they follow?
12. Cholesterol is soluble in lipids, so why does plasma cholesterol need the help of a membrane transporter to enter cells?

22.6 Homeostatic Control of Metabolism

The endocrine system has primary responsibility for metabolic regulation, although the nervous system also plays a role, particularly in terms of governing food intake. Many hormones are involved in long-term regulation of metabolism but hour-to-hour regulation depends primarily on the ratio of insulin to glucagon, two hormones secreted by endocrine cells of the pancreas. Both hormones have short half-lives and must be continuously secreted if they are to have a sustained effect.

The Pancreas Secretes Insulin and Glucagon

In 1869, the German anatomist Paul Langerhans described small clusters of cells—now known as the **islets of Langerhans**—scattered throughout the body of the pancreas (**FIG. 22.13b**). Most pancreatic tissue is devoted to the production and exocrine secretion of digestive enzymes and bicarbonate [Fig. 21.14, p. 675] but Langerhans had found the pancreas's endocrine cells, which make up less than 2% of the total mass. The islets of Langerhans contain four distinct cell types, each associated with secretion of one or more peptide hormones.

Nearly three-quarters of the islet cells are **beta cells**, which produce *insulin* and a peptide called *amylin*. Another 20% are **alpha cells**, which secrete **glucagon**. Most of the remaining cells are *somatostatin*-secreting **D cells**. A few rare cells called *PP cells* (or *F cells*) produce *pancreatic polypeptide*.

Like all endocrine glands, the islets are closely associated with capillaries into which the hormones are released. Both sympathetic and parasympathetic neurons terminate on the islets, providing a means by which the nervous system can influence metabolism.

The Insulin-to-Glucagon Ratio Regulates Metabolism

As noted earlier, insulin and glucagon act in antagonistic fashion to keep plasma glucose concentrations within an acceptable range. Both hormones are present in the blood most of the time. The ratio of the two hormones determines which hormone dominates.

In the fed state, when the body is absorbing nutrients, insulin dominates, and the body undergoes net anabolism (**FIG. 22.14a**). Ingested glucose is used for energy production, and any excess is stored as glycogen or fat. Amino acids go primarily to protein synthesis.

In the fasted state, metabolic regulation prevents low plasma glucose concentrations (*hypoglycemia*). When glucagon predominates, the liver uses glycogen and nonglucose intermediates to synthesize glucose for release into the blood (Fig. 22.14b).

Figure 22.14c shows plasma glucose, glucagon, and insulin concentrations before and after a meal. In a person with normal metabolism, fasting plasma glucose is maintained around 90 mg/dL of plasma, insulin secretion is low, and plasma glucagon levels are relatively high. After absorption of nutrients from a meal, plasma glucose rises. The increase in blood glucose inhibits glucagon secretion and stimulates insulin release. Insulin in turn promotes glucose transfer into cells. As a result, plasma glucose concentrations begin to fall back toward fasting levels shortly after each meal. Insulin secretion decreases along with the glucose, and glucagon secretion slowly begins to increase.

FIG. 22.12 Lipolysis

Triglycerides can be metabolized for ATP production.

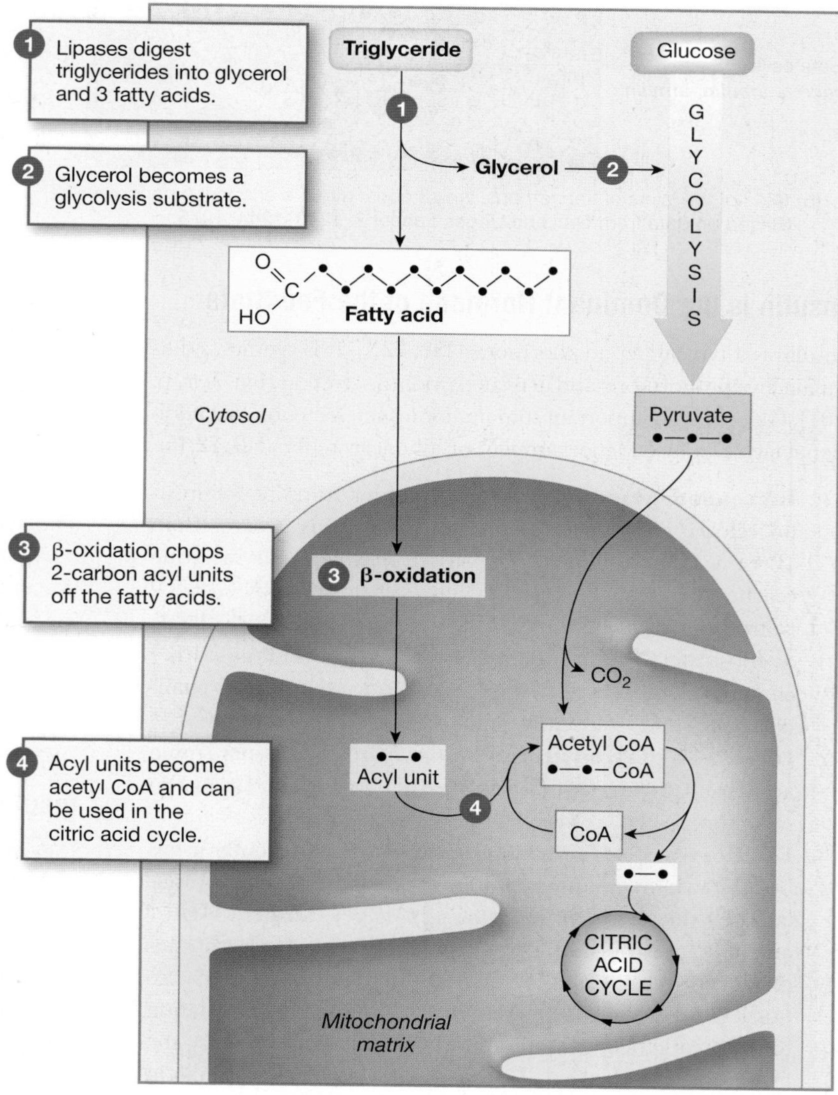

1 Lipases digest triglycerides into glycerol and 3 fatty acids.
2 Glycerol becomes a glycolysis substrate.
3 β-oxidation chops 2-carbon acyl units off the fatty acids.
4 Acyl units become acetyl CoA and can be used in the citric acid cycle.
Triglyceride
Glucose
Glycerol
GLYCOLYSIS
O
C
HO
Fatty acid
Cytosol
Pyruvate
β-oxidation
CO2
Acetyl CoA
CoA
Acyl unit
CoA
CITRIC ACID CYCLE
Mitochondrial matrix

FIG. 22.13 The pancreas

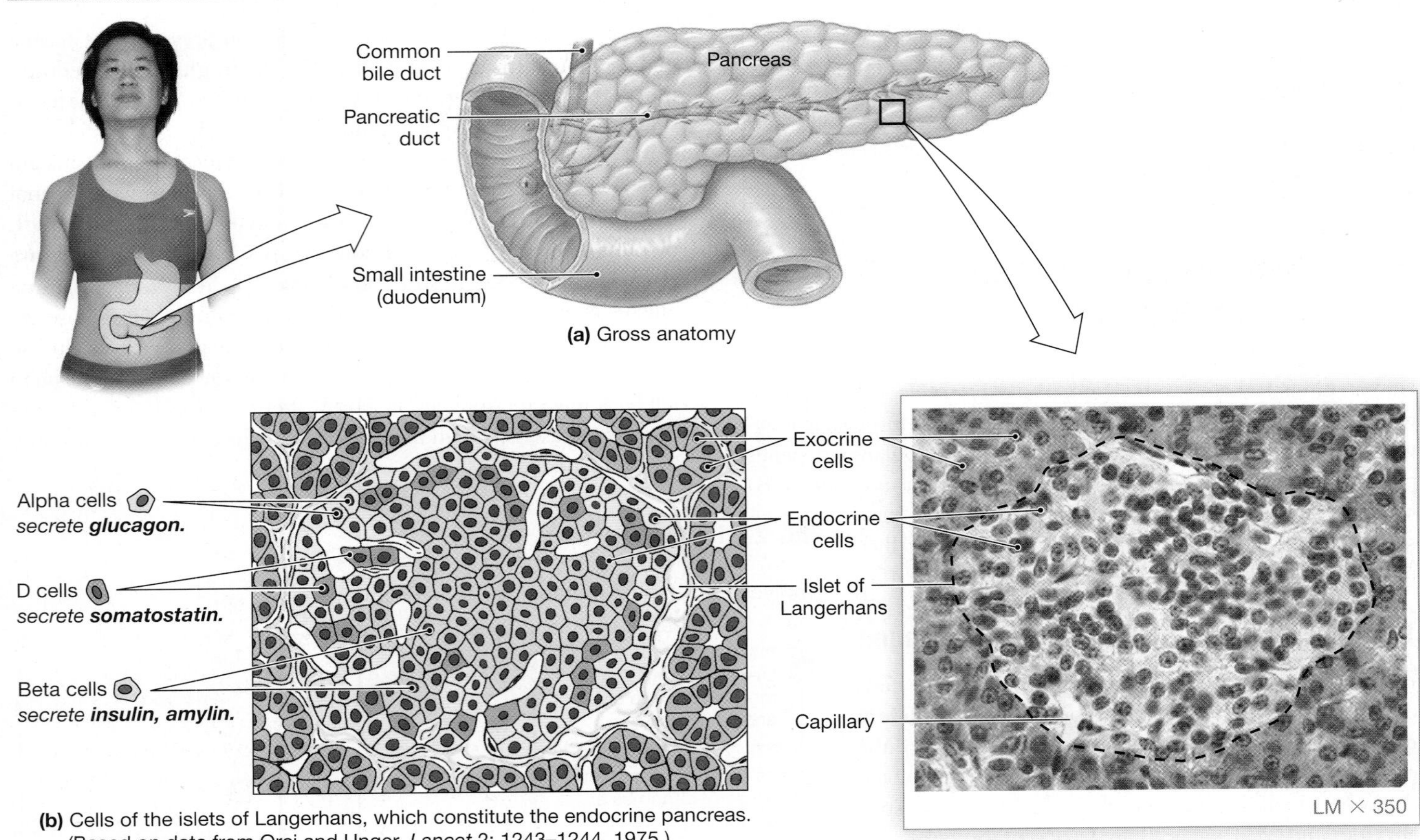
Common bile duct
Pancreas
Pancreatic duct
Small intestine (duodenum)
(a) Gross anatomy
Alpha cells secrete glucagon.
D cells secrete somatostatin.
Beta cells secrete insulin, amylin.
Exocrine cells
Endocrine cells
Islet of Langerhans
Capillary
LM × 350

(b) Cells of the islets of Langerhans, which constitute the endocrine pancreas. (Based on data from Orci and Unger, *Lancet* 2: 1243–1244, 1975.)

Insulin Is the Dominant Hormone of the Fed State

Insulin is a typical peptide hormone (**TBL. 22.1**). It is synthesized as an inactive prohormone and activated prior to secretion [Fig. 7.3c, p. 201]. Glucose is an important stimulus for insulin secretion, but additional factors may enhance, amplify, or inhibit secretion (**FIG. 22.15**).

1. **Increased plasma glucose.** A major stimulus for insulin release is plasma glucose concentrations greater than 100 mg/dL. Glucose absorbed from the small intestine reaches pancreatic beta cells, where it is taken up by GLUT2 transporters [Fig. 5.26b, p. 158]. With more glucose available as substrate, ATP production increases and ATP-gated K^+ channels close. The cell depolarizes, voltage-gated Ca^{2+} channels open, and Ca^{2+} entry initiates exocytosis of insulin.
2. **Increased plasma amino acids.** Increased plasma amino acid concentrations following a meal also trigger insulin secretion.
3. **Feedforward effects of GI hormones.** Recently it has been shown that as much as 50% of insulin secretion is stimulated by the hormone *glucagon-like peptide-1* (GLP-1). GLP-1 and GIP (gastric inhibitory peptide) are *incretin* hormones produced by cells of the ileum and jejunum in response to nutrient ingestion. The incretins travel through the circulation to pancreatic beta cells and may reach them even before the first glucose is absorbed. The anticipatory release of insulin in response to these hormones prevents a sudden surge in plasma glucose concentrations when the meal is absorbed. Other GI hormones, such as CCK and gastrin, amplify insulin secretion.
4. **Parasympathetic activity.** Parasympathetic activity to the GI tract and pancreas increases during and following a meal. Parasympathetic input to beta cells stimulates insulin secretion.
5. **Sympathetic activity.** Insulin secretion is inhibited by sympathetic neurons. In times of stress, sympathetic input to the endocrine pancreas increases, reinforced by catecholamine release from the adrenal medulla (**TBL. 22.2**). Epinephrine and norepinephrine inhibit insulin secretion and switch metabolism to gluconeogenesis to provide extra fuel for the nervous system and skeletal muscles.

Insulin Promotes Anabolism

Like other peptide hormones, insulin combines with a membrane receptor on its target cells (**FIG. 22.16**). The insulin receptor has *tyrosine kinase* activity, which initiates complex intracellular cascades whose details are still not completely understood. The activated insulin receptor phosphorylates proteins that include a group known as the **insulin-receptor substrates (IRS)**. These proteins act through complicated pathways to influence transport and cellular metabolism. The enzymes that regulate metabolic

FIG. 22.14 Insulin and glucagon

Metabolism is controlled by the insulin:glucagon ratio.

(a) Fed state: Insulin dominates

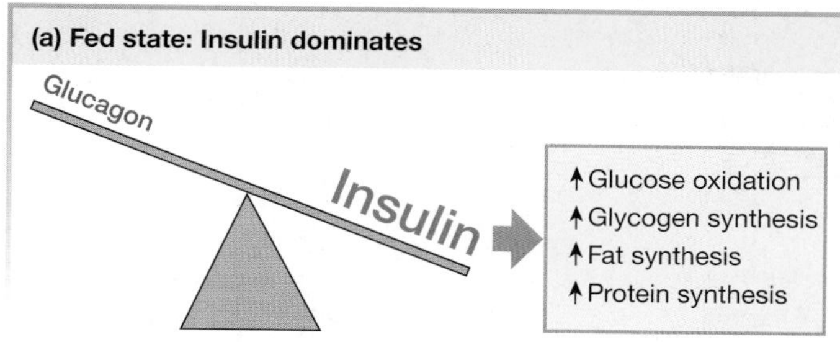
Glucagon
Insulin
↑Glucose oxidation
↑Glycogen synthesis
↑Fat synthesis
↑Protein synthesis

(b) Fasted state: Glucagon dominates

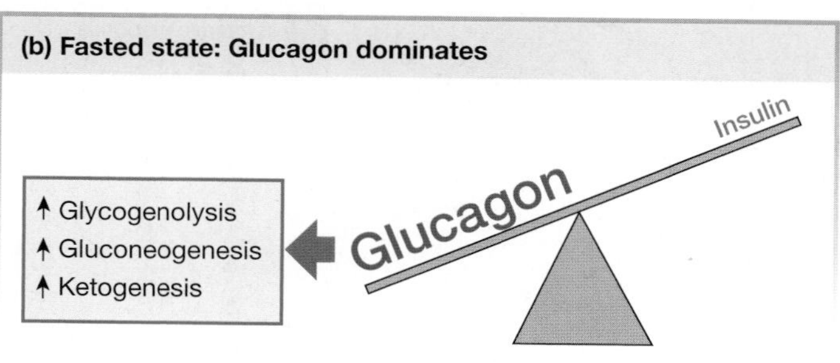
Insulin
Glucagon
↑Glycogenolysis
↑Gluconeogenesis
↑Ketogenesis

(c) Glucose, insulin and glucagon levels before and after a meal

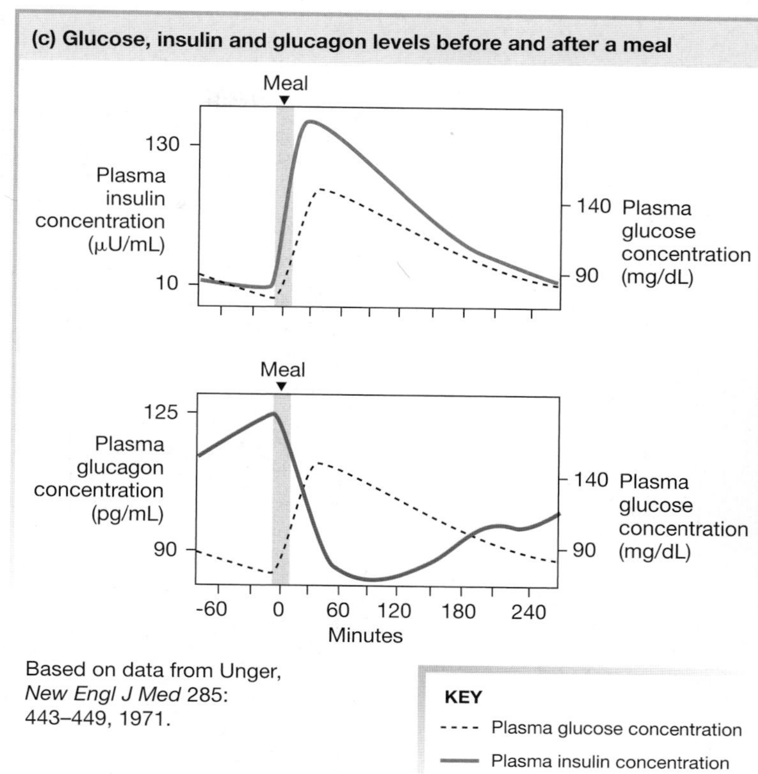
Meal
130
10
Plasma insulin concentration (μU/mL)
140
90
Plasma glucose concentration (mg/dL)
Meal
125
90
Plasma glucagon concentration (pg/mL)
140
90
Plasma glucose concentration (mg/dL)
-60
0
60
120
180
240
Minutes

Based on data from Unger, *New Engl J Med* 285: 443–449, 1971.

KEY
- - - - Plasma glucose concentration
—— Plasma insulin concentration
—— Plasma glucagon concentration

TABLE 22.1 Insulin	
Cell of origin	Beta cells of pancreas
Chemical nature	51-amino acid peptide
Biosynthesis	Typical peptide
Transport in the circulation	Dissolved in plasma
Half-life	5 minutes
Factors affecting release	Plasma [glucose] > 100 mg/dL; ↑ blood amino acids; GLP-1 (feedforward reflex). Parasympathetic activity amplifies. Sympathetic activity inhibits.
Target cells or tissues	Liver, muscle, and adipose tissue primarily; brain, kidney, and intestine not insulin dependent
Target receptor	Membrane receptor with tyrosine kinase activity; pathways with insulin-receptor substrates
Whole body or tissue action	↓ Plasma [glucose] by ↑ transport into cells or ↑ metabolic use of glucose
Action at cellular level	↑ Glycogen synthesis; ↑ aerobic metabolism of glucose; ↑ protein and triglyceride synthesis
Action at molecular level	Inserts GLUT4 transporters in muscle and adipose cells; alters enzyme activity. Complex signal transduction pathways
Feedback regulation	↓ Plasma [glucose] shuts off insulin release
Other information	Growth hormone and cortisol are antagonistic

FIG. 22.15 Insulin in the fed state

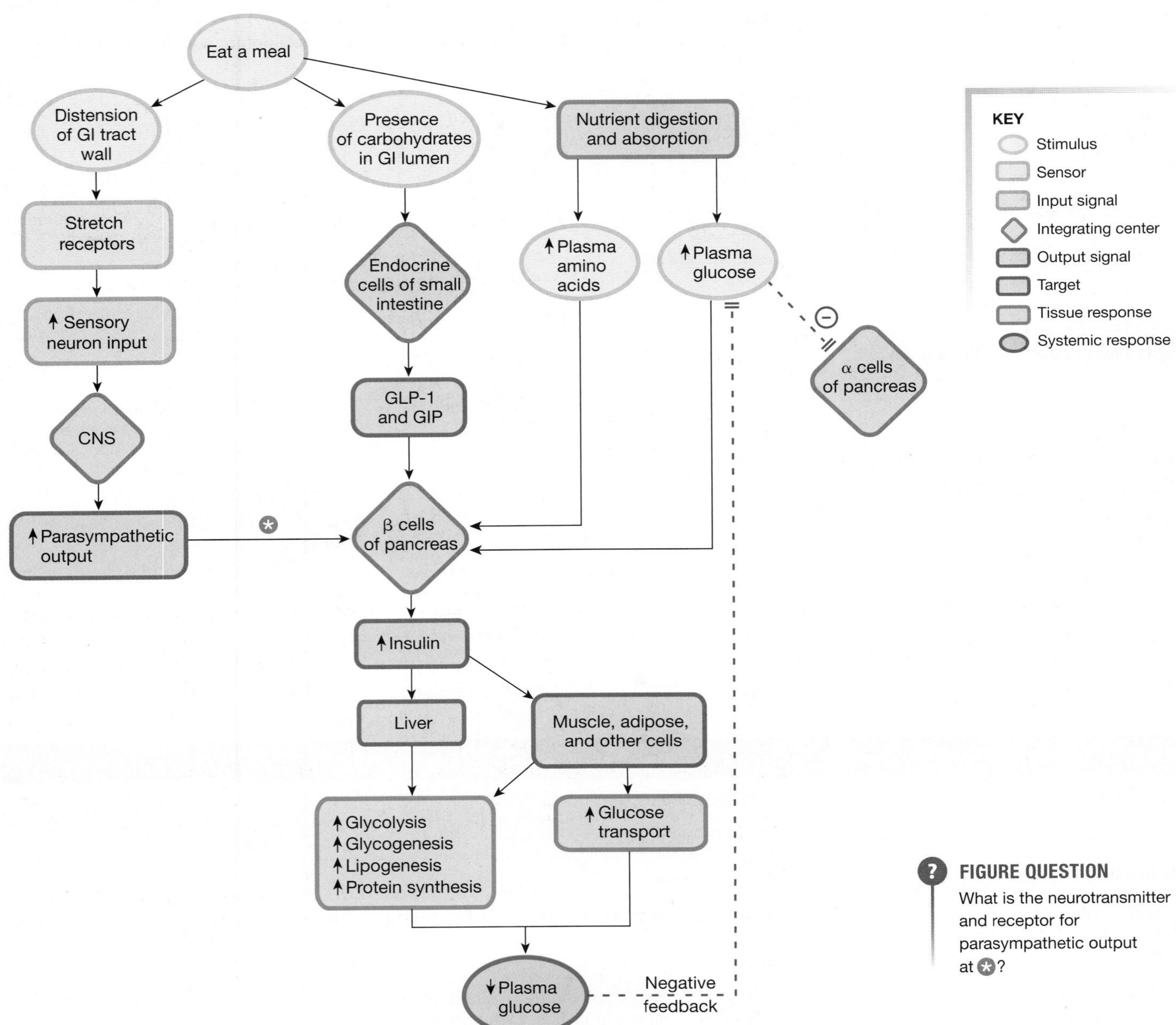
Eat a meal
Distension of GI tract wall
Presence of carbohydrates in GI lumen
Nutrient digestion and absorption
Stretch receptors
↑ Sensory neuron input
CNS
↑ Parasympathetic output
Endocrine cells of small intestine
GLP-1 and GIP
↑ Plasma amino acids
↑ Plasma glucose
α cells of pancreas
β cells of pancreas
↑ Insulin
Liver
Muscle, adipose, and other cells
↑ Glycolysis
↑ Glycogenesis
↑ Lipogenesis
↑ Protein synthesis
↑ Glucose transport
↓ Plasma glucose
Negative feedback
KEY
Stimulus
Sensor
Input signal
Integrating center
Output signal
Target
Tissue response
Systemic response

FIGURE QUESTION

What is the neurotransmitter and receptor for parasympathetic output at ✪?

pathways may be inhibited or activated directly, or their synthesis may be influenced indirectly through transcription factors.

The primary target tissues for insulin are the liver, adipose tissue, and skeletal muscles (Fig. 22.15). The usual target cell response is increased glucose metabolism. In some target tissues, insulin also regulates the GLUT transporters. Other tissues, including the brain and transporting epithelia of the kidney and intestine, are insulin-independent—they do not require insulin for glucose uptake and metabolism.

Insulin lowers plasma glucose in the following four ways:

1. **Insulin increases glucose transport into most, but not all, insulin-sensitive cells.** Adipose tissue and resting skeletal muscle require insulin for glucose uptake (**FIG. 22.17a**). Without insulin, their GLUT4 transporters are withdrawn from the membrane and stored in cytoplasmic vesicles—another example of membrane recycling. When insulin binds to the receptor and activates it, the resulting signal transduction cascade causes the vesicles to move to the cell membrane and insert the GLUT4 transporters by exocytosis (Fig. 22.17b). The cells then take up glucose from the interstitial fluid by facilitated diffusion.

   Interestingly, *exercising* skeletal muscle is not dependent on insulin activity for its glucose uptake. When muscles contract, GLUT4 transporters are inserted into the membrane even in the absence of insulin, and glucose uptake increases. The intracellular signal for this is complex but appears to involve Ca^{2+} as well as a variety of intracellular proteins.

TABLE 22.2 Adrenal Catecholamines (Epinephrine and Norepinephrine)

Origin	Adrenal medulla (90% epinephrine and 10% norepinephrine)
Chemical nature	Amines made from tyrosine
Biosynthesis	Typical peptide
Transport in the circulation	Some bound to sulfate
Half-life	2 minutes
Factors affecting release	Primarily fight-or-flight reaction through CNS and autonomic nervous system; hypoglycemia
Target cells or tissues	Mainly neurons, pancreatic endocrine cells, heart, blood vessels, adipose tissue
Target receptor	G protein-coupled receptors; α and β subtypes
Second messenger	cAMP for α_2 and all β receptors; IP_3 for α_1 receptors
Whole body or tissue action	↑ Plasma [glucose]; activate fight-or-flight and stress reactions; ↑ glucagon and ↓ insulin secretion
Onset and duration of action	Rapid and brief

Glucose transport in liver cells (*hepatocytes*) is not *directly* insulin dependent but is influenced by the presence or absence of insulin. Hepatocytes have GLUT2 transporters that are always present in the cell membrane. In the fasted state, when insulin levels are low, glucose moves *out* of the liver and into the blood to help maintain glucose homeostasis. In this process (Fig. 22.17c), hepatocytes are converting glycogen stores and amino acids to glucose. Newly formed glucose moves down its concentration gradient out of the cell using the GLUT2 facilitated diffusion transporters. If GLUT transporters were pulled from the hepatocyte membrane during the fasted state, as they are in muscle and adipose tissue, glucose would have no way to leave the hepatocytes.

In the fed state (Fig. 22.17d), insulin activates *hexokinase*, the enzyme that phosphorylates glucose to glucose 6-phosphate. This phosphorylation reaction keeps free intracellular glucose concentrations low relative to plasma concentrations [Fig. 5.13, p. 141]. Now glucose diffuses into the hepatocyte on the GLUT2 transporter operating in the reverse direction. So insulin does increase hepatic glucose uptake but not by direct action on glucose transporters.

2. **Insulin enhances cellular utilization and storage of glucose.** Insulin activates enzymes for glucose utilization (*glycolysis*), and for glycogen synthesis (*glycogenesis*). Insulin simultaneously inhibits enzymes for glycogen breakdown (*glycogenolysis*), glucose synthesis (*gluconeogenesis*), and fat breakdown (*lipolysis*) to ensure that metabolism moves in the anabolic direction. If more glucose has been ingested than is needed for energy and synthesis, the excess is made into glycogen or fatty acids.
3. **Insulin enhances utilization of amino acids.** Insulin activates enzymes for protein synthesis and inhibits enzymes that promote protein breakdown. If a meal includes protein, amino acids in the ingested food are used for protein synthesis by both the liver and muscle. Excess amino acids are converted to fatty acids.
4. **Insulin promotes fat synthesis.** Insulin inhibits β-oxidation of fatty acids and promotes conversion of excess glucose or amino acids into triglycerides (*lipogenesis*). Excess triglyceride is stored as lipid droplets in adipose tissue.

In summary, insulin is an anabolic hormone because it promotes glycogen, protein, and fat synthesis. When insulin is absent or deficient, cells go into catabolic metabolism.

Concept Check

13. What are the primary target tissues for insulin?
14. Why are glucose metabolism and glucose transport independent of insulin in renal and intestinal epithelium and in neurons?
15. What is the advantage to the body of inhibiting insulin release during a sympathetically mediated fight-or-flight response?

Glucagon Is Dominant in the Fasted State

Glucagon, secreted by pancreatic alpha cells, is generally antagonistic to insulin in its effects on metabolism (**TBL. 22.3**). When plasma glucose concentrations decrease after a meal, insulin secretion slows, and the effects of glucagon on tissue metabolism take

FIG. 22.16 Insulin's cellular mechanism of action

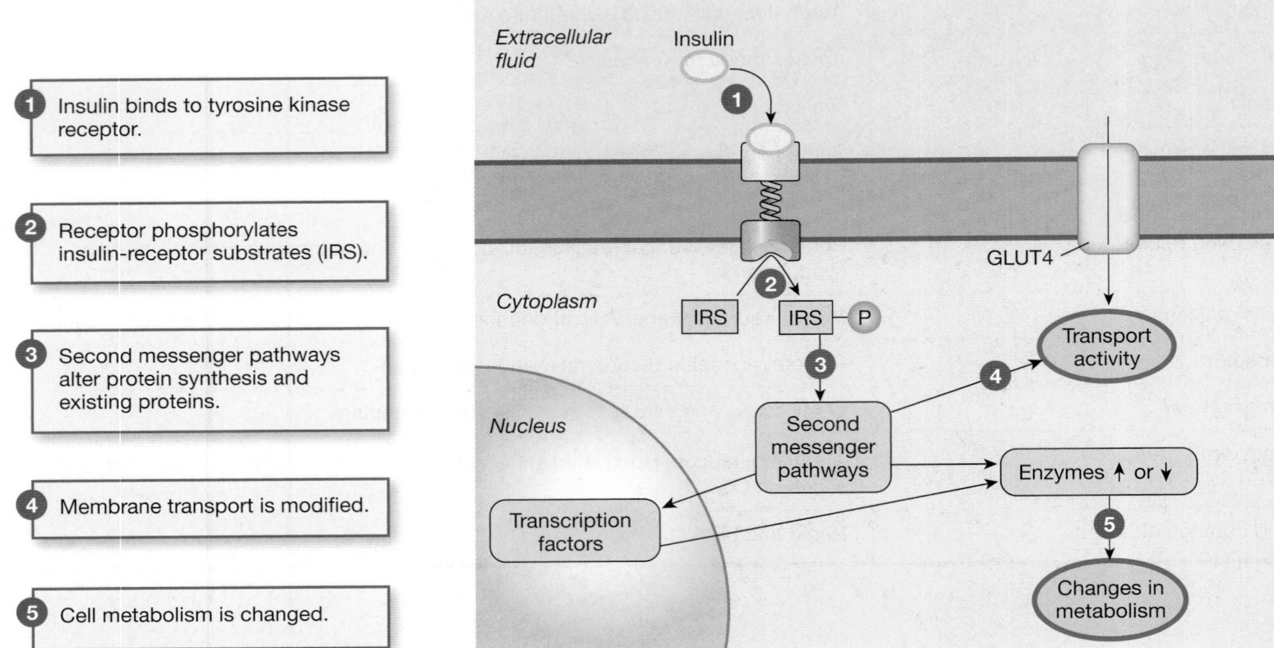
1 Insulin binds to tyrosine kinase receptor.
2 Receptor phosphorylates insulin-receptor substrates (IRS).
3 Second messenger pathways alter protein synthesis and existing proteins.
4 Membrane transport is modified.
5 Cell metabolism is changed.
Extracellular fluid
Insulin
GLUT4
Cytoplasm
IRS
IRS
P
Transport activity
Nucleus
Second messenger pathways
Enzymes ↑ or ↓
Transcription factors
Changes in metabolism

on greater significance (Fig. 22.14c). As noted earlier, it is the ratio of insulin to glucagon that determines the direction of metabolism rather than an absolute amount of either hormone.

The function of glucagon is to prevent hypoglycemia, so the primary stimulus for glucagon release is low plasma glucose concentration (**FIG. 22.18**). When plasma glucose falls below 100 mg/dL, glucagon secretion rises dramatically. At glucose concentrations above 100 mg/dL, when insulin is being secreted, glucagon secretion is inhibited and remains at a low but relatively constant level. The strong relationship between insulin secretion and glucagon inhibition has led to speculation that alpha cells are regulated by some factor linked to insulin rather than by plasma glucose concentrations directly.

The liver is glucagon's primary target tissue (Fig. 22.18). Glucagon stimulates glycogenolysis and gluconeogenesis to increase glucose output. It is estimated that during an overnight fast, 75% of the glucose produced by the liver comes from glycogen stores, and the remaining 25% from gluconeogenesis.

Glucagon release is also stimulated by plasma amino acids. This pathway prevents hypoglycemia after ingestion of a pure protein meal. Let's see how hypoglycemia might occur in the absence of glucagon.

If a meal contains protein but no carbohydrate, amino acids absorbed from the food cause insulin secretion. Even though no glucose has been absorbed, insulin-stimulated glucose uptake increases, and plasma glucose concentrations fall. Unless something counteracts this process, the brain's fuel supply is threatened by hypoglycemia.

Co-secretion of glucagon in this situation prevents hypoglycemia by stimulating hepatic glucose output. As a result, although only amino acids were ingested, both glucose and amino acids are made available to peripheral tissues.

Diabetes Mellitus Is a Family of Diseases

The most common pathology of the pancreatic endocrine system is the family of metabolic disorders known as **diabetes mellitus (DM)**. Diabetes is characterized by abnormally elevated plasma glucose concentrations (*hyperglycemia*) resulting from inadequate insulin secretion, abnormal target cell responsiveness [p. 214], or both. Chronic hyperglycemia and its associated metabolic abnormalities cause the many complications of diabetes, including damage to blood vessels, eyes, kidneys, and the nervous system.

Diabetes is reaching epidemic proportions in the United States. In 2014, the U.S. Centers for Disease Control and Prevention estimated that over 29 million people in the United States (9.3% of the population) have diabetes and that more than a quarter of these people were unaware that they have the disease. Another 86 million, or 37% of the population, have prediabetes. Prediabetes is a condition that will likely become diabetes if those people do not alter their eating and exercise habits. Experts attribute the cause of the epidemic to our sedentary lifestyle, ample food, and overweight and obesity, which affect more than 50% of the population.

Diabetes has been known to affect humans since ancient times, and written accounts of the disorder highlight the calamitous consequences of insulin deficiency. Aretaeus the Cappadocian (81–138 C.E.) wrote of the "wonderful"* nature of this disease that consists of the "melting down of the flesh . . . into urine,"

*In the sense of "causing wonder."

FIG. 22.17 Glucose transport in fed and fasted states

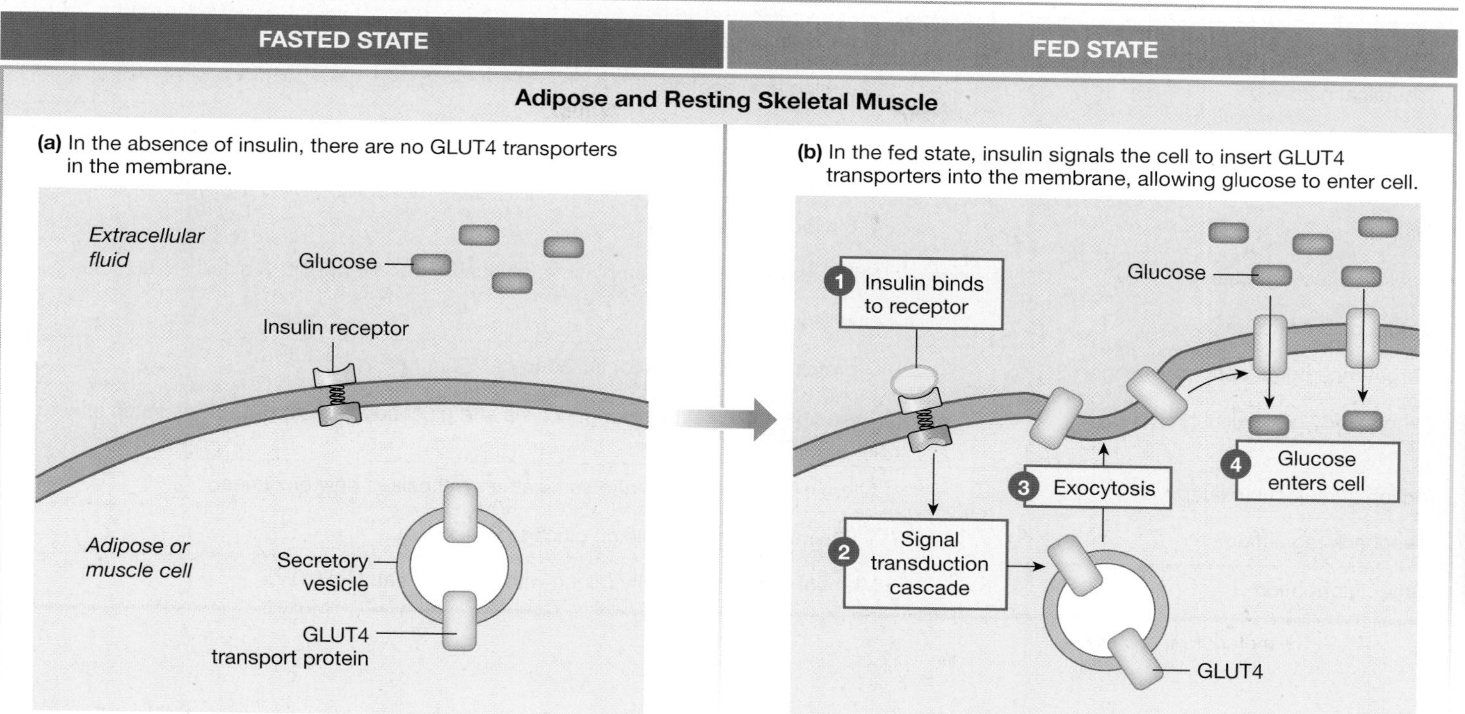
FASTED STATE
FED STATE
Adipose and Resting Skeletal Muscle
(a) In the absence of insulin, there are no GLUT4 transporters in the membrane.
Extracellular fluid
Glucose
Insulin receptor
Adipose or muscle cell
Secretory vesicle
GLUT4 transport protein
(b) In the fed state, insulin signals the cell to insert GLUT4 transporters into the membrane, allowing glucose to enter cell.
1 Insulin binds to receptor
Glucose
2 Signal transduction cascade
3 Exocytosis
4 Glucose enters cell
GLUT4

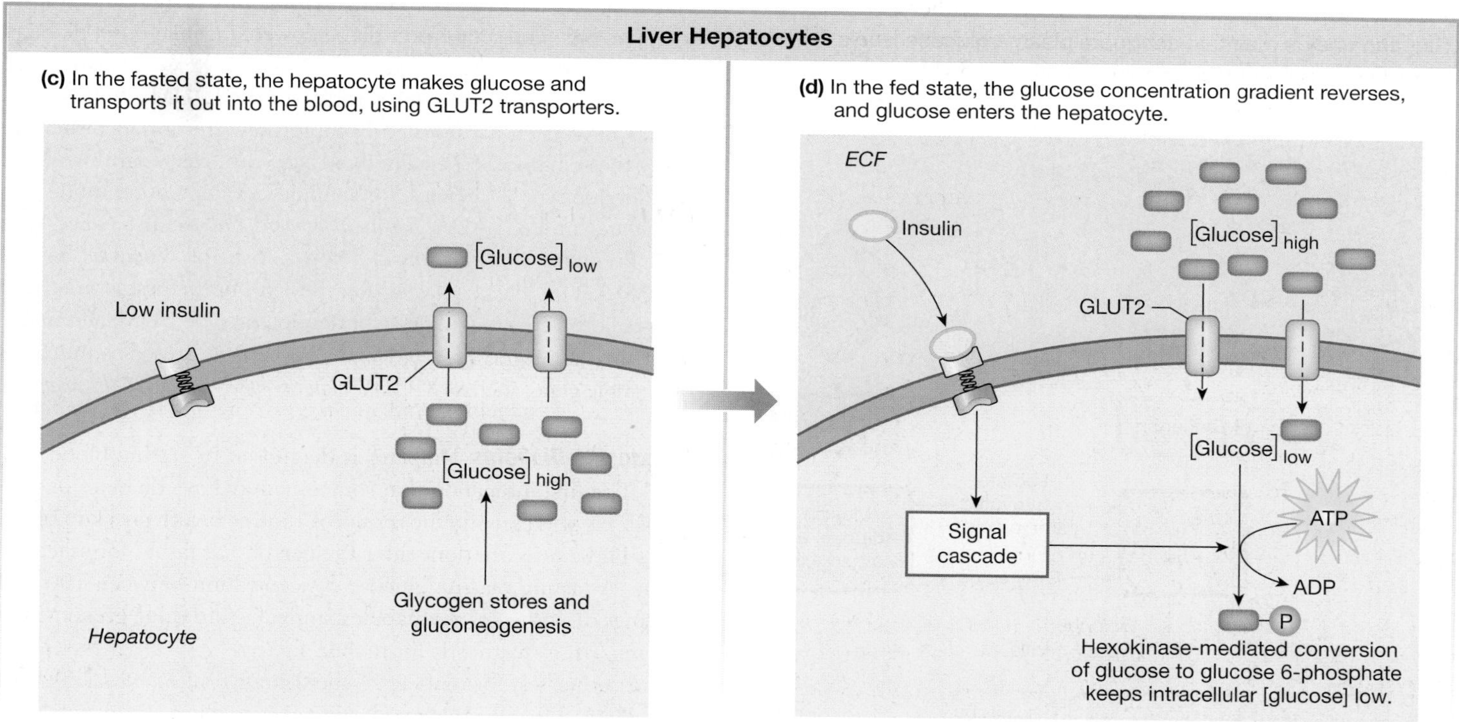
Liver Hepatocytes
(c) In the fasted state, the hepatocyte makes glucose and transports it out into the blood, using GLUT2 transporters.
[Glucose] low
Low insulin
GLUT2
[Glucose] high
Glycogen stores and gluconeogenesis
Hepatocyte
(d) In the fed state, the glucose concentration gradient reverses, and glucose enters the hepatocyte.
ECF
Insulin
[Glucose] high
GLUT2
[Glucose] low
ATP
Signal cascade
ADP
P
Hexokinase-mediated conversion of glucose to glucose 6-phosphate keeps intracellular [glucose] low.

accompanied by terrible thirst that cannot be quenched. The copious production of glucose-laden urine gave the disease its name. *Diabetes* refers to the flow of fluid through a siphon, and *mellitus* comes from the word for honey. In the Middle Ages, diabetes was known as "the pissing evil."

The severe type of diabetes described by Aretaeus is **type 1 diabetes mellitus**. It is a condition of insulin deficiency resulting from beta cell destruction. Type 1 diabetes is most commonly an *autoimmune disease* in which the body fails to recognize the beta cells as "self" and destroys them with antibodies and white blood cells.

The other major variant of diabetes is **type 2 diabetes mellitus**. This type of diabetes is also known as *insulin-resistant diabetes* because in most patients, insulin levels in the blood are normal or

TABLE 22.3 Glucagon

Cell of origin	Alpha cells of pancreas
Chemical nature	29-amino acid peptide
Biosynthesis	Typical peptide
Transport in the circulation	Dissolved in plasma
Half-life	4–6 minutes
Factors affecting release	Enhanced secretion when plasma [glucose] <65–70 mg/dL; ↑ blood amino acids
Target cells or tissues	Liver primarily
Target receptor/second messenger	G protein-coupled receptor linked to cAMP
Whole body or tissue action	↑ Plasma [glucose] by glycogenolysis and gluconeogenesis; ↑ lipolysis leads to ketogenesis in liver
Action at molecular level	Alters existing enzymes and stimulates synthesis of new enzymes
Feedback regulation	↑ Plasma [glucose] shuts off glucagon secretion
Other information	Member of secretin family (along with VIP, GIP, and GLP-1)

FIG. 22.18 Endocrine response to hypoglycemia

Glucagon helps maintain adequate plasma glucose levels by promoting glycogenolysis and gluconeogenesis.

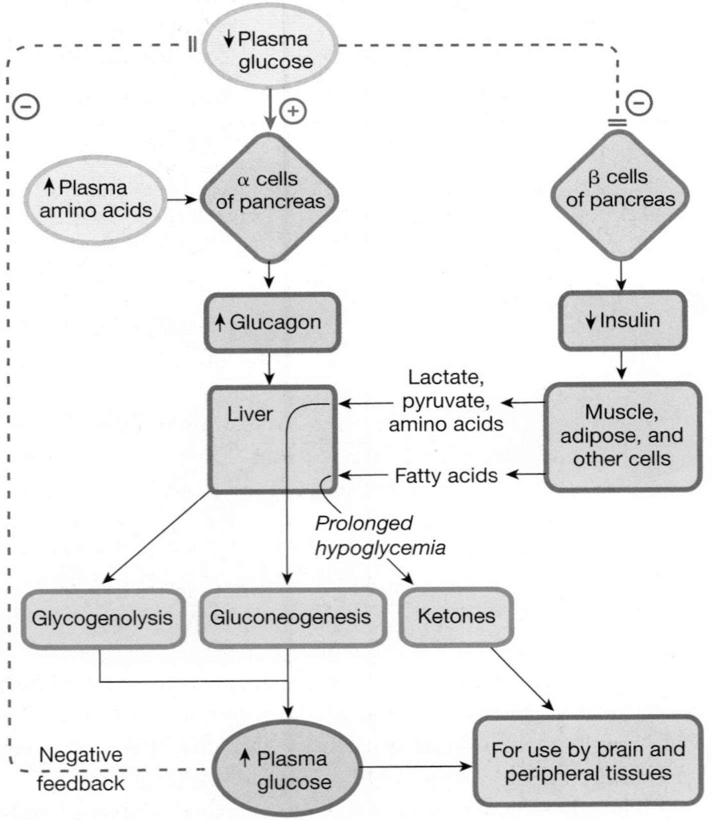
↓Plasma glucose
↑Plasma amino acids
α cells of pancreas
β cells of pancreas
↑Glucagon
↓Insulin
Liver
Lactate, pyruvate, amino acids
Muscle, adipose, and other cells
Fatty acids
Prolonged hypoglycemia
Glycogenolysis
Gluconeogenesis
Ketones
Negative feedback
↑ Plasma glucose
For use by brain and peripheral tissues

even elevated initially. Later in the disease process, many type 2 diabetics become insulin deficient and require insulin injections.

The old names for type 1 diabetes were juvenile-onset diabetes and insulin-dependent diabetes (IDDM); type 2 was called adult-onset and non-insulin-dependent diabetes (NIDDM). Both sets of names have been abandoned because they are not accurate. We know that children, especially those who are overweight or obese, can develop type 2 diabetes; some adults develop autoimmune diabetes later in life (LADA, *latent autoimmune diabetes*; also called type 1.5). Pregnant women may develop a form of *gestational diabetes* (GDM) that usually goes away after they give birth; these women and their children are at higher risk for developing type 2 DM later in life. There are also some inherited forms of diabetes caused by mutation of a single gene, such as MODY, *maturity-onset diabetes of the young*.

Diagnosing Diabetes Diabetes is diagnosed by testing blood glucose. The first indication that someone may have diabetes or prediabetes is often seen in the results of routine health checkup blood tests. These tests are done after the person has fasted for at least 8 hours. A fasting plasma glucose concentration between 100 and 125 mg/dL indicates prediabetes, and a fasting value greater than 125 mg/dL is diagnostic for diabetes.

Another test for diabetes is the 2-hour oral *glucose tolerance test* (**FIG. 22.19**). First, the person's fasting plasma glucose concentration is determined (time 0). Then the person consumes a special drink containing 75 g of glucose dissolved in water. Plasma glucose is measured every 30 minutes for 2 hours.

Normal individuals show a slight increase in plasma glucose concentration soon after drinking glucose, but the level rapidly returns to normal with insulin secretion. In diabetic patients, however, fasting plasma glucose concentrations are above normal and go even higher as glucose is absorbed into the body. In diabetes, plasma glucose remains elevated above 200 mg/dL after 2 hours. This slow response indicates that cells are not taking up and metabolizing glucose normally. Prediabetic patients show an intermediate response, with 2-hour plasma glucose concentrations in the 140–199 mg/dL range.

Elevated fasting glucose levels and abnormal oral glucose tolerance tests simply show that the body's response to an ingested glucose load is not normal. The tests cannot distinguish between problems with insulin synthesis, insulin release, or the responsiveness of target tissues to insulin.

> **Concept Check**
>
> **16.** For the oral glucose tolerance test, could you use equal amounts (in grams) of table sugar and expect the same results? Explain.

FIG. 22.19 Diagnosing diabetes

(a) Normal and Abnormal Glucose Tolerance Tests

At time 0, the subject consumes a drink containing 75 g anhydrous glucose dissolved in water.

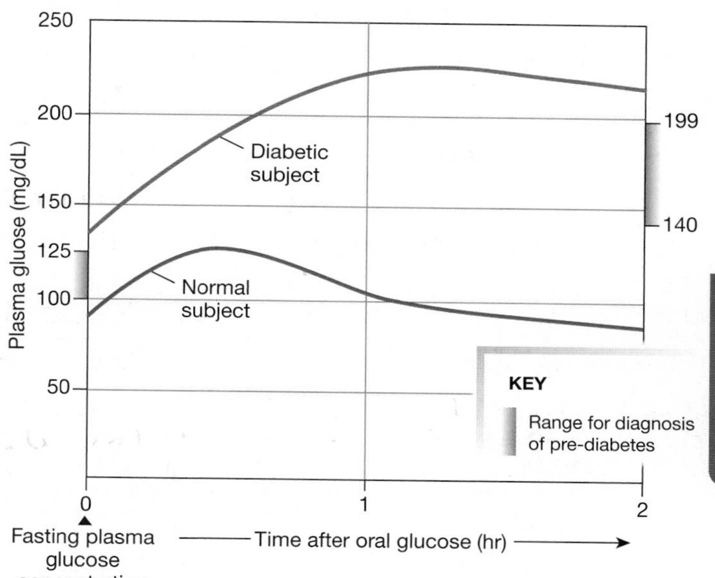
250
200
150
125
100
50
199
140
Plasma gluose (mg/dL)
Diabetic subject
Normal subject
KEY
Range for diagnosis of pre-diabetes
0
1
2
Fasting plasma glucose concentration
Time after oral glucose (hr)

(b) Diagnostic Criteria for Diabetes

Condition	Fasting Blood Glucose	After 2-Hour Oral Glucose Tolerance Test
Normal	<100 mg/dl (<5.6 mM)	<140 mg/dL (<7.8 mM)
Pre-diabetes	100-125 mg/dL (5.6–6.9 mM)	140-199 mg/dL (7.8-11 mM)
Diabetes	>125 mg/dL (>6.9 mM)	>199 mg/dL (>11 mM)

Type 1 Diabetics Are Prone to Ketoacidosis

Type 1 diabetes is a complex disorder whose onset in genetically susceptible individuals is sometimes preceded by a viral infection. Many type 1 diabetics develop their disease in childhood, giving rise to the old name *juvenile-onset diabetes*. About 10% of all diabetics have type 1 diabetes.

Because individuals with type 1 diabetes are insulin deficient, the only treatment is insulin injections. Until the arrival of genetic engineering, most pharmaceutical insulin came from swine, cow, and sheep pancreases. However, once the gene for human insulin was cloned, biotechnology companies began to manufacture artificial human insulin for therapeutic use. In addition, scientists are developing techniques for implanting encapsulated beta cells in the body, in the hope that individuals with type 1 diabetes will no longer need to rely on regular insulin injections.

The events that follow ingestion of food in an insulin-deficient diabetic create a picture of what happens to metabolism in the absence of insulin (**FIG. 22.20**). They also show the integrative nature of physiology because the problems that arise from abnormal metabolism affect nearly every organ system of the body.

Following a meal, nutrient absorption by the intestine proceeds normally because this process is insulin independent. However, nutrient uptake from the blood and cellular metabolism in many tissues are insulin dependent and therefore severely diminished in the absence of insulin. Lacking nutrients to metabolize, cells go into fasted-state metabolism:

1. *Protein metabolism.* Without glucose for energy or amino acids for protein synthesis, muscles break down their proteins to provide a substrate for ATP production. Amino acids are also converted to pyruvate and lactate, which leave the muscles and are transported to the liver.
2. *Fat metabolism.* Adipose tissue in fasted-state metabolism breaks down its fat stores. Fatty acids enter the blood for transport to the liver. The liver uses β-oxidation to break down fatty acids. However, this organ is limited in its ability to send fatty acids through the citric acid cycle, and the excess fatty acids are converted to ketones.

   Ketone bodies reenter the circulation and can be used by other tissues (such as muscle and brain) for ATP synthesis. (The breakdown of muscle and adipose tissue in the absence of insulin leads to tissue loss and the "melting down of the flesh" described by Aretaeus.) However, ketones are also metabolic acids, creating a state of ketoacidosis (see point 7).
3. *Glucose metabolism.* In the absence of insulin, glucose remains in the blood, causing hyperglycemia. The liver, unable to metabolize this glucose, initiates fasted-state pathways of glycogenolysis and gluconeogenesis. These pathways produce *additional* glucose from glycogen, amino acids, and glycerol. When the liver dumps this glucose into the blood, hyperglycemia worsens.

Diabetic hyperglycemia will increase the osmolarity of the blood and create a *hyperglycemic hyperosmolar state.* Plasma

FIG. 22.20 Acute pathophysiology of type 1 diabetes mellitus

Untreated type 1 diabetes is marked by tissue breakdown, glucosuria, polyuria, polydipsia, polyphagia, and metabolic ketoacidosis.

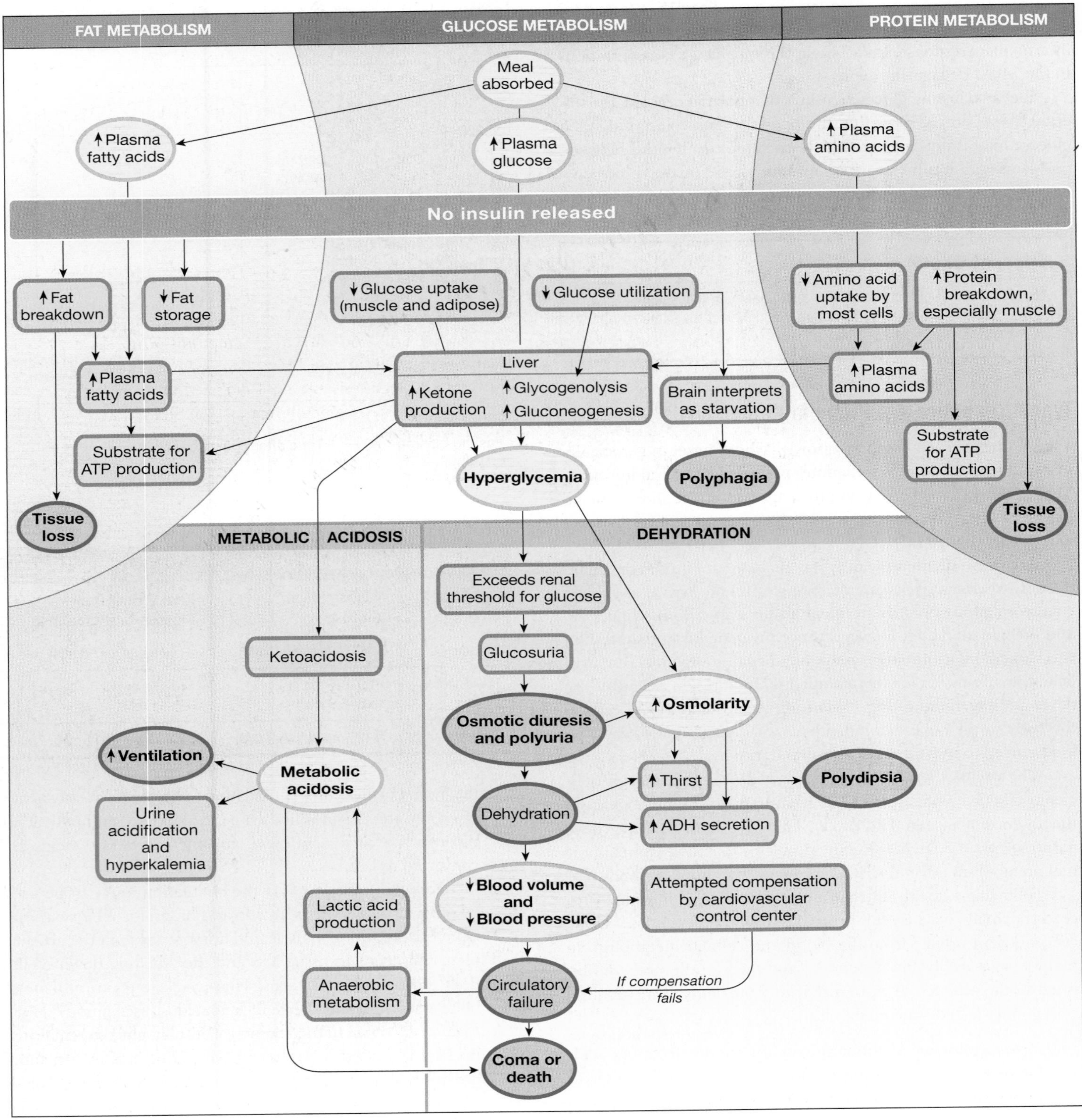
FAT METABOLISM
GLUCOSE METABOLISM
PROTEIN METABOLISM
Meal absorbed
↑ Plasma fatty acids
↑ Plasma glucose
↑ Plasma amino acids
No insulin released
↑ Fat breakdown
↓ Fat storage
↓ Glucose uptake (muscle and adipose)
↓ Glucose utilization
↓ Amino acid uptake by most cells
↑ Protein breakdown, especially muscle
↑ Plasma fatty acids
Liver
↑ Ketone production
↑ Glycogenolysis
↑ Gluconeogenesis
Brain interprets as starvation
↑ Plasma amino acids
Substrate for ATP production
Hyperglycemia
Polyphagia
Substrate for ATP production
Tissue loss
Tissue loss
METABOLIC ACIDOSIS
DEHYDRATION
Exceeds renal threshold for glucose
Ketoacidosis
Glucosuria
↑ Osmolarity
Osmotic diuresis and polyuria
↑ Ventilation
Metabolic acidosis
↑ Thirst
Polydipsia
Urine acidification and hyperkalemia
Dehydration
↑ ADH secretion
↓ Blood volume and ↓ Blood pressure
Attempted compensation by cardiovascular control center
Lactic acid production
Anaerobic metabolism
Circulatory failure
If compensation fails
Coma or death

glucose may be as high as 600–1200 mg/dL and total osmolarity ranges from 330–380 mOsM. The high osmolarity will trigger vasopressin (ADH) secretion and thirst in an effort to conserve water and return osmolarity back to the normal range [p. 626].

4. *Brain metabolism.* Tissues that are not insulin dependent, such as most neurons in the brain, carry on metabolism as usual. However, neurons in the brain's satiety center *are* insulin sensitive. Therefore, in the absence of insulin, the satiety center is unable to take up plasma glucose. The center perceives the absence of intracellular glucose as starvation and allows the feeding center to increase food intake. The result is *polyphagia* (excessive eating), a classic symptom associated with untreated type 1 diabetes mellitus.
5. *Osmotic diuresis and polyuria.* If the hyperglycemia of diabetes causes plasma glucose concentrations to exceed the *renal threshold* for glucose, glucose reabsorption in the proximal tubule of the kidney becomes saturated [p. 603]. As a result, some filtered glucose is not reabsorbed, and it is excreted in the urine (*glucosuria*).

   The presence of additional solute in the collecting duct lumen causes less water to be reabsorbed and more to be excreted [see Chapter 20, question 33, p. 653]. This creates large volumes of urine (*polyuria*) and, if unchecked, results in dehydration. The loss of water in the urine due to unreabsorbed solutes is known as **osmotic diuresis**.
6. *Dehydration.* Dehydration caused by osmotic diuresis leads to decreased circulating blood volume and decreased blood pressure. Low blood pressure triggers homeostatic mechanisms for raising blood pressure, including secretion of vasopressin, thirst that causes constant drinking (*polydipsia*), and cardiovascular compensations [Fig. 20.13, p. 640].
7. *Metabolic acidosis.* Metabolic acidosis in diabetes has two potential sources: anaerobic metabolism and ketone body production. The primary cause of metabolic acidosis in type 1 diabetics is the production of acidic ketone bodies by the liver. Patients in *diabetic ketoacidosis* (DKA) exhibit the signs of metabolic acidosis: increased ventilation, acidification of the urine, and hyperkalemia [p. 648].

Tissues may also go into anaerobic glycolysis (which creates lactate) if low blood pressure decreases blood flow to the point that oxygen delivery to peripheral tissues becomes inadequate. Lactate leaves the cells and enters the blood, contributing to a state of metabolic acidosis. If untreated, the combination of ketoacidosis and hypoxia from circulatory collapse can cause coma and even death. The treatment for a patient in diabetic ketoacidosis is insulin replacement, accompanied by fluid and electrolyte therapy to replenish lost volume and ions.

Type 2 Diabetics Often Have Elevated Insulin Levels

Type 2 diabetics account for 90% of all diabetics. A significant genetic predisposition to develop the disease exists among certain ethnic groups. For example, about 25% of Hispanics over age 45 have diabetes. The disease is more common in people over the age of 40, but there is growing concern about the increased diagnosis of type 2 diabetes in children and adolescents. About 80% of type 2 diabetics are obese.

A common hallmark of type 2 diabetes is **insulin resistance**, demonstrated by the delayed response to an ingested glucose load seen in the 2-hour oral glucose tolerance test. Some type 2 diabetics have both resistance to insulin action and decreased insulin secretion. Others have normal-to-high insulin secretion but decreased target cell responsiveness.

In addition, although type 2 diabetics are hyperglycemic, they often have elevated glucagon levels as well. This seems contradictory until you realize that the pancreatic alpha cells, like muscle and adipose cells, require insulin for glucose uptake. This means that in diabetes, the alpha cells do not take up glucose, which prompts them to secrete glucagon. Glucagon then contributes to hyperglycemia by promoting glycogenolysis and gluconeogenesis.

In type 2 diabetes, the acute symptoms of the disease are not nearly as severe as in type 1 because insulin is usually present, and the cells, although resistant to insulin's action, are able to carry out a certain amount of glucose metabolism. The liver, for example, does not have to turn to ketone production. As a result, ketosis is uncommon in type 2 diabetes.

Nevertheless, overall metabolism is not normal in type 2 diabetes, and patients with this condition develop a variety of physiological problems because of abnormal glucose and fat metabolism. Complications of type 2 diabetes include atherosclerosis, neurological changes, renal failure, and blindness from diabetic retinopathy. As many as 70% of type 2 diabetics die from cardiovascular disease.

Because many people with type 2 diabetes are asymptomatic when diagnosed, they can be very difficult to treat. People who come in for their yearly checkup feeling fine, only to be told that they have diabetes, can be reluctant to make dramatic lifestyle changes when they do not feel sick. Unfortunately, by the time diabetic symptoms appear, damage to tissues and organs is well under way. Patient compliance at that point can slow the progress of the disease but cannot reverse the pathological changes. The goal of treatment is to correct hyperglycemia to prevent the complications described earlier.

The first therapy recommended for most type 2 diabetics and prediabetics, and for those individuals at high risk of developing the disease, is to exercise and lose weight. For some patients, simply losing weight eliminates their insulin resistance. Exercise decreases hyperglycemia because exercising skeletal muscle does not require insulin for glucose uptake.

Drugs used to treat type 2 diabetes may (1) stimulate beta-cell secretion of insulin, (2) slow the digestion or absorption of carbohydrates in the intestine, (3) inhibit hepatic glucose output, (4) make target tissues more responsive to insulin, or (5) promote glucose excretion in the urine (**TBL. 22.4**). Many of the newest antidiabetic drugs mimic endogenous hormones. For example, *pramlintide* is an analog of **amylin,** a peptide hormone that is co-secreted with insulin. Amylin helps regulate glucose homeostasis following a meal by slowing digestion and absorption of carbohydrates.

TABLE 22.4 Drugs for Treating Diabetes

Drug Class	Effect	Mechanism of Action
Sulfonylureas and meglitinides	Stimulate insulin secretion	Close beta cell K_{ATP} channels and depolarize the cell
α-Glucosidase inhibitors	Decrease intestinal glucose uptake	Block intestinal enzymes that digest complex carbohydrates
Sodium-glucose cotransporter 2 (SGLT2) inhibitors (e.g., dapagliflozin)	Increase glucose excretion in urine	Inhibit glucose reabsorption in the renal proximal tubule
Biguanides (metformin)	Reduce plasma glucose by decreasing hepatic gluconeogenesis	Increases activity of AMPK, AMP-dependent protein kinase
PPAR activators ("glitazones")	Increase gene transcription for proteins that promote glucose utilization and fatty acid metabolism	Activate PPARγ, a nuclear receptor for genes involved in metabolism
Amylin analogs (pramlintide)	Reduce plasma glucose	Delay gastric emptying, suppress glucagon secretion, and promote satiety
Incretin (GLP-1) analogs (exendin-4)	Reduce plasma glucose and induce weight loss	Stimulate insulin secretion, reduce glucagon secretion, delay gastric emptying, and promote satiety
DPP4 inhibitors (sitagliptin)	Increase insulin secretion and decrease gastric emptying	Inhibit dipeptidyl peptidase-4, which breaks down GLP-1 and GIP

Amylin also decreases food intake by a central effect on appetite, and it decreases secretion of glucagon.

Other hormone-based therapies approved by the FDA are incretin *mimetics* (agonists). *Exendin-4* (Byetta®) is a GLP-1 mimetic derived from a compound found in the venomous saliva of gila monsters. Exendin-4 has four primary effects: It increases insulin production, decreases production of glucagon, slows digestion, and increases satiety. It has also been associated with weight loss.

In normal physiology, the combined actions of amylin, GIP, and GLP-1 create a self-regulating cycle for glucose absorption and fed-state glucose metabolism. Glucose in the intestine following a meal causes feedforward release of GIP and GLP-1 (Fig. 22.15). The two incretins travel through the circulation to the pancreas, where they initiate insulin and amylin secretion. Amylin then acts on the GI tract to slow the rate at which food enters the intestine, while insulin acts on target tissues to promote glucose uptake and utilization.

Metabolic Syndrome Links Diabetes and Cardiovascular Disease

Clinicians have known for years that people who are overweight are prone to develop type 2 diabetes, atherosclerosis, and high blood pressure. The combination of these three conditions has been formalized into a diagnosis called **metabolic syndrome**, which highlights the integrative nature of metabolic pathways. People with metabolic syndrome meet at least three of the following five criteria: central (visceral) obesity, blood pressure ≥130/85 mm Hg, fasting plasma glucose ≥110 mg/dL, elevated fasting plasma triglyceride levels, and low plasma HDL-C levels. *Central obesity* is defined as a waist circumference >40″ in men and >35″ in women. Women who have an apple-shaped body (widest at the waist) are more prone to developing metabolic syndrome than women who have a pear-shaped body (widest at the hips).

The association between obesity, diabetes, and cardiovascular disease illustrates the fundamental disturbances in cellular metabolism that occur with obesity. One common mechanism known to play a role in both glucose metabolism and lipid metabolism involves the family of nuclear receptors called *peroxisome*

RUNNING PROBLEM

The medical staff and Nicole's family decided to talk to her about her weight and eating habits. The disorder Dr. Ayani suspects, **anorexia nervosa (AN)**, can have serious physiological consequences. As a result, AN has the highest death rate of any psychiatric disorder, and the mortality rate of young women ages 15–24 with AN is 12 times greater than that of the general population. The most common causes of death are cardiac arrest, electrolyte imbalance, and suicide. AN reportedly affects as many as 3% of females in industrialized nations at some point in their lifetime. (While 90% of AN cases are female, the number of male cases is increasing.) Successful treatment of AN includes providing nutrition, psychotherapy, and family therapy. Current research is investigating the usefulness of a ghrelin agonist and other brain peptides in treating anorexia.

Q9: *Why might a ghrelin agonist help in cases of anorexia?*

693 – 695 – 696 – 699 – **718** – 724

proliferator-activated receptors (PPARs, pronounced *p-pars*). Lipids and lipid-derived molecules bind to PPARs, which then turn on a variety of genes. The PPAR subtype called PPARγ (*p-par-gamma*) has been linked to adipocyte differentiation, type 2 diabetes, and foam cells, the endothelial macrophages that have ingested oxidized cholesterol. PPARα, mentioned earlier in the discussion of cholesterol metabolism, is important in hepatic cholesterol metabolism. The PPARs may be important clues to the link between obesity, type 2 diabetes, and atherosclerosis that has evaded scientists for so long.

Multiple Hormones Influence Metabolism

The long-term regulation of metabolism is much more complicated than we are able to present here, and it is still not completely understood. Many of the neuropeptides and hypothalamic neurons mentioned in the discussion of hunger and satiety also have metabolic effects. Hormones such as thyroid hormone, hormones of the cortisol pathway [p. 212], growth hormone, and epinephrine modulate metabolic pathways both directly and indirectly through their influence on insulin secretion.

Concept Check

17. Why must insulin be administered as a shot and not as an oral pill?
18. Patients admitted to the hospital with acute diabetic ketoacidosis and dehydration are given insulin and fluids that contain K^+ and other ions. The acidosis is usually accompanied by hyperkalemia, so why is K^+ included in the rehydration fluids? (*Hint:* Dehydrated patients may have a high *concentration* of K^+, but their total body fluid volume is low.)
19. In 2006, the FDA approved sitagliptin (Januvia®), a DPP4 inhibitor. This drug blocks action of the enzyme *dipeptidyl peptidase-4*, which breaks down GLP-1 and GIP. Explain how sitagliptin is helpful in treating diabetes.
20. One of the newest drugs for treating diabetes increases urinary excretion of glucose. It does so by inhibiting the Na^+-glucose cotransporter (SGLT) that allows glucose reabsorption in the proximal tubule. What are some potential side effects of increasing urinary glucose excretion? [*Hint:* p. 621]

For example, in times of stress, cortisol and circulating epinephrine levels increase. Sympathetic influence on the endocrine pancreas decreases insulin secretion and enhances glucagon secretion. The combined metabolic effects of insulin, cortisol, and glucagon are *synergistic*, or more than additive [Fig. 7.12, p. 213], and blood glucose concentrations rise sharply. When this happens in diabetics who are stressed, they may need to increase their medications to keep their blood glucose levels under control.

22.7 Regulation of Body Temperature

Type 2 diabetes is an excellent example of the link between body weight and metabolism. The development of obesity may be linked to the efficiency with which the body converts food energy into cell and tissue components. According to one theory, people who are more efficient in transferring energy from food to fat are the ones who put on weight. In contrast, people who are less metabolically efficient can eat the same number of calories and not gain weight because more food energy is released as heat. Much of what we know about the regulation of energy balance comes from studies on body temperature regulation.

Body Temperature Balances Heat Production, Gain, and Loss

Temperature regulation in humans is linked to metabolic heat production (*thermogenesis*). Humans are **homeothermic** animals, which means our bodies regulate internal temperature within a relatively narrow range. Average body temperature is 37 °C (98.6 °F), with a normal range of 35.5–37.7 °C (96–99.9 °F).

These values are subject to considerable variation, both among individuals and throughout the day in a single individual. The site at which temperature is measured also makes a difference because core body temperature may be higher than temperature at the skin surface. Oral temperatures are about 0.5 °C lower than rectal temperatures.

Several factors affect body temperature in a given individual. Body temperature increases with exercise or after a meal (because of *diet-induced thermogenesis*). Temperature also cycles throughout the day: The lowest (basal) body temperature occurs in the early morning, and the highest occurs in the early evening. Women of reproductive age also exhibit a monthly temperature cycle: Basal body temperatures are about 0.5 °C higher in the second half of the menstrual cycle (after ovulation) than before ovulation.

Heat Gain and Loss Are Balanced Temperature balance in the body, like energy balance, depends on a dynamic equilibrium between input and output (**FIG. 22.21**). Heat input has two components: *internal heat production*, which includes heat from normal metabolism and heat released during muscle contraction, and *external heat input* from the environment through either *radiation* or *conduction*.

All objects with a temperature above absolute zero give off radiant energy (radiation) with infrared or visible wavelengths. This energy can be absorbed by other objects and constitutes **radiant heat gain** for those objects. You absorb radiant energy each time you sit in the sun or in front of a fire. **Conductive heat gain** is the transfer of heat between objects that are in contact with each other, such as the skin and a heating pad or the skin and hot water.

We lose heat from the body in four ways: conduction, radiation, convection, and evaporation. **Conductive heat loss** is the loss of body heat to a cooler object that is touching the body, such as an icepack or a cold stone bench. **Radiant heat loss** from the human body is estimated to account for nearly half the heat lost from a person at rest in a normal room. *Thermography* is a diagnostic imaging technique that measures radiant heat loss. Some cancerous tumors can be visually identified because they have higher metabolic activity and give off more heat than surrounding tissues.

Sweat glands are made of transporting epithelium. Cells deep in the gland secrete an isotonic solution similar to interstitial fluid. As the fluid travels through the duct to the skin, NaCl is reabsorbed, resulting in hypotonic sweat. A typical value for sweat production is 1.5 L/hr. With acclimatization to hot weather, some people sweat at rates of 4–6 L/hr. However, they can maintain this high rate only for short periods unless they are drinking to replace lost fluid volume. Sweat production is regulated by cholinergic sympathetic neurons.

Cooling by evaporative heat loss depends on the evaporation of water from sweat on the skin's surface. Because water evaporates rapidly in dry environments but slowly or not at all in humid ones, the body's ability to withstand high temperatures is directly related to the relative humidity of the air. Meteorologists report the combination of heat and humidity as the *heat index* or *humidex*. Air moving across a sweaty skin surface enhances evaporation even with high humidity, which is one reason fans are useful in hot weather.

Movement and Metabolism Produce Heat

Heat production by the body falls into two broad categories: (1) unregulated heat production from voluntary muscle contraction and normal metabolic pathways, and (2) regulated heat production for maintaining temperature homeostasis in low environmental temperatures. Regulated heat production is further divided into shivering thermogenesis and nonshivering thermogenesis (Fig. 22.21).

In **shivering thermogenesis**, the body uses shivering (rhythmic tremors caused by skeletal muscle contraction) to generate heat. Signals from the hypothalamic thermoregulatory center initiate these skeletal muscle tremors. Shivering muscle generates five to six times as much heat as resting muscle. Shivering can be partially suppressed by voluntary control.

Nonshivering thermogenesis is metabolic heat production by means other than shivering. In laboratory animals such as the rat, cold exposure significantly increases heat production in *brown adipose tissue* (BAT), also known as brown fat [p. 82]. The mechanism for brown fat heat production is *mitochondrial uncoupling*, induced by BAT-produced *uncoupling protein 1* (UCP1).

In mitochondrial uncoupling, energy flowing through the electron transport system [p. 108] is released as heat rather than being trapped in ATP. Mitochondrial uncoupling in response to cold is promoted by thyroid hormones and by increased sympathetic activity on β_3 adrenergic receptors in brown fat.

The importance of nonshivering thermogenesis in adult humans is becoming a topic of increasing interest. Humans are born with significant amounts of brown fat, found primarily in the *interscapular* area between the shoulder blades. In newborns, nonshivering thermogenesis in this brown fat contributes significantly to raising and maintaining body temperature. We used to believe that as children age, white fat replaced most brown fat. Recently, however, imaging studies used for cancer diagnosis showed that adult humans still have active brown fat. Scientists are now investigating whether increasing brown fat activity might be one way to help people burn calories as heat instead of storing them as fat.

The body's responses to high and low temperatures are summarized in **FIGURE 22.23**. In cold environments, the body tries to reduce heat loss while increasing internal heat production. In hot temperatures, the opposite is true. Notice from Figure 22.23 that voluntary behavioral responses play a significant role in temperature regulation. We reduce activity during hot weather, thereby decreasing muscle heat production. In cold weather, we put on extra clothing, tuck our hands in our armpits, or curl up in a ball to slow heat loss.

FIG. 22.23 Homeostatic responses to environmental extremes

High environmental temperature

Maximize Heat Loss

- Vasodilation of cutaneous blood vessels
- Increased sweating
- Behavioral responses
  - Use of fans to increase convective heat loss
  - Immersion in water to increase conductive heat loss
  - Staying out of sun to prevent radiant heat gain
  - Removing clothes

Minimize Heat Production

- Diminished food intake to lessen obligatory heat production
- Behavioral responses
  - Decreased physical activity

Low environmental temperature

Minimize Heat Loss

- Vasoconstriction of cutaneous blood vessels
- Lack of sweating
- Behavioral responses
  - Adding layers of protective clothing
  - Curling up to minimize exposed surface
  - Standing near heat source to enhance radiant heat gain

Maximize Heat Production

- Shivering thermogenesis
- Nonshivering thermogenesis
- Behavioral responses
  - Increased voluntary activity (stamping feet, etc.)

The Body's Thermostat Can Be Reset

Variations in body temperature regulation can be either physiological or pathological. Examples of physiological variation include the circadian rhythm of body temperature mentioned earlier, menstrual cycle variations, postmenopausal hot flashes, and fever. These processes share a common mechanism: resetting of the hypothalamic thermostat.

Hot flashes appear to be transient decreases in the thermostat's setpoint caused by the absence of estrogen. When the setpoint is lower, a room temperature that had previously been comfortable suddenly feels too hot. This discomfort triggers the usual thermoregulatory responses to heat, including sweating and cutaneous vasodilation, which leads to flushing of the skin.

For many years, *fever* was thought to be a pathological response to infection, but it is now considered part of the body's normal immune response. Toxins from bacteria and other pathogens trigger the release of chemicals known as **pyrogens** {*pyr*, fire} from various immunocytes. Pyrogens are fever-producing cytokines that also have many other effects.

Experimentally, some interleukins (IL-1, IL-6), some interferons, and tumor necrosis factor have all been shown to induce fever. They do so by resetting the hypothalamic thermostat to a higher setpoint. Normal room temperature feels too cold, and the patient begins to shiver, creating additional heat. Pyrogens may also increase nonshivering thermogenesis, causing body temperature to rise.

The adaptive significance of fever is still unclear, but it seems to enhance the activity of white blood cells involved in the immune response. For this reason, some people question whether patients with a fever should be given aspirin and other fever-reducing drugs simply for the sake of comfort. High fever can be dangerous, however, as a fever of 41 °C (106 °F) for more than a brief period causes brain damage.

Pathological conditions in which body temperature strays outside the normal range include different states of *hyperthermia* and *hypothermia*. Heat exhaustion and heat stroke are the most common forms of **hyperthermia**, a condition in which body temperature rises to abnormally high values. **Heat exhaustion** is marked by severe dehydration and core body temperatures of 37.5–39 °C (99.5–102.2 °F). Patients may experience muscle cramps, nausea, and headache. They are usually pale and sweating profusely. Heat exhaustion often occurs in people who are physically active in hot, humid climates to which they are not acclimatized. It also occurs in the elderly, whose ability to thermoregulate is diminished.

Heat stroke is a more severe form of hyperthermia, with higher core body temperatures. The skin is usually flushed and dry. Immediate and rapid cooling of these patients is important, as enzymes and other proteins begin to denature at temperatures above 41 °C (106 °F). Mortality in heat stroke is nearly 50%.

Malignant hyperthermia, in which body temperature becomes abnormally elevated, is a genetically linked condition [see the running problem in Chapter 25]. A defective Ca^{2+} channel in skeletal muscle releases too much Ca^{2+} into the cytoplasm. As cell transporters work to move the Ca^{2+} back into mitochondria and the sarcoplasmic reticulum, the heat released from ATP hydrolysis substantially raises body temperature. Some investigators have suggested that a mild version of this process plays a role in nonshivering thermogenesis in mammals.

Hypothermia, a condition in which body temperature falls abnormally low, is also a dangerous condition. As core body temperature falls, enzymatic reactions slow, and the person loses consciousness. When metabolism slows, oxygen consumption also decreases.

Victims of drowning in cold water can sometimes be revived without brain damage if they have gone into a state of hypothermia. This observation led to the development of induced hypothermia for certain surgical procedures, such as heart surgery. The patient is cooled to 21–24 °C (70–75 °F) so that tissue oxygen demand can be met by artificial oxygenation of the blood as it passes through a bypass pump. After surgery is complete, the patient is gradually rewarmed.

Concept Check

23. Why must a water bed be heated to allow a person to sleep on it comfortably?

24. Will a person who is exercising outside overheat faster when the air humidity is low or when it is high?

RUNNING PROBLEM **CONCLUSION** Eating Disorders

Nicole finally agreed to undergo counseling and enter a treatment program for anorexia nervosa. She was lucky–her wrist would heal, and her medical complications could have been much worse. After seeing Nicole and discussing her anorexia, Sara realized that she also needed to see a counselor. Even though she was no longer dancing, Sara still used diet pills, diuretics, and laxatives when she became uncomfortable with her weight, and she had started bingeing and purging–eating much more than normal when she was stressed and then forcing herself to vomit to avoid gaining any weight. These are the behavioral patterns of bulimia nervosa (BN), a condition that is as serious as AN and that affects an estimated 4% of females. Its physiological effects and treatments are similar to those for AN.

To learn more about anorexia and bulimia, and for help finding a support group, see the National Association of Anorexia Nervosa and Associated Disorders website at *www.anad.org* or *www.nationaleatingdisorders.org*.

Question	Facts	Integration and Analysis
Q1: *If you measured Nicole's leptin level, what would you expect to find?*	Leptin is a hormone secreted by adipose tissue.	Nicole has little adipose tissue, so she would have a low leptin level.
Q2: *Would you expect Nicole to have elevated or depressed levels of neuropeptide Y?*	NPY is inhibited by leptin. NPY stimulates feeding centers.	Because her leptin level is low, you might predict that NPY would be elevated and feeding stimulated. However, the feeding center is affected by other factors besides NPY (Fig. 22.1). Brain studies of anorexic patients show high levels of CRH, which opposes NPY and depresses feeding.
Q3: *What is Nicole's K^+ disturbance called? What effect does it have on the resting membrane potential of her cells?*	Nicole's K^+ is 2.5 mEq/L, and normal is 3.5–5 mEq/L.	Low plasma K^+ is called hypokalemia. Hypokalemia causes the membrane potential to hyperpolarize [p. 249].
Q4: *Why does Dr. Ayani want to monitor Nicole's cardiac function?*	Cardiac muscle is an excitable tissue whose activity depends on changes in membrane potential.	Hypokalemia can alter the membrane potential of cardiac autorhythmic and contractile cells and cause a potentially fatal cardiac arrhythmia.
Q5: *Based on her clinical values, what is Nicole's acid-base status?*	Nicole's pH is 7.52, and her plasma HCO_3^- is elevated at 40 mEq/L.	Normal pH is 7.38–7.42, so she is in alkalosis. Her elevated HCO_3^- indicates a metabolic alkalosis. The cause is probably induced vomiting and loss of HCl from her stomach.
Q6: *Based on what you know about heart rate and blood pressure, speculate on why Nicole has low blood pressure with a rapid pulse.*	Her blood pressure is 80/50 (low), and her pulse is 90 (high).	Normally, increasing the heart rate would increase blood pressure. In this case, the increased pulse is a compensatory attempt to raise her low blood pressure. The low blood pressure probably results from dehydration.
Q7: *Would you expect Nicole's renin and aldosterone levels to be normal, elevated, or depressed? How might these levels relate to her K^+ disturbance?*	All the primary stimuli for renin secretion are associated with low blood pressure. Renin begins the RAAS pathway that stimulates aldosterone secretion.	Because Nicole's blood pressure is low, you would expect elevated renin and aldosterone levels. Aldosterone promotes renal K^+ secretion, which would lower her body load of K^+. She probably also has low dietary K^+ intake, which contributes to her hypokalemia.
Q8: *Give some possible reasons Nicole had been feeling weak during dance rehearsals.*	In fasted-state metabolism, the body breaks down skeletal muscle.	Loss of skeletal muscle proteins, hypokalemia, and possibly hypoglycemia could all be causes of Nicole's weakness.
Q9: *Why might a ghrelin agonist help in cases of anorexia?*	Ghrelin stimulates the feeding center.	A ghrelin agonist might stimulate the feeding center and help Nicole want to eat.

693 – 695 – 696 – 699 – 718 – 724

CHAPTER SUMMARY

Energy balance in the body means that the body's energy intake equals its energy output. The same balance principle applies to metabolism and body temperature. The amount of nutrient in each of the body's nutrient pools depends on intake and output. Glucose *homeostasis* is one of the most important goals of regulated metabolism, for without adequate glucose, the brain is unable to function. Flow of material through the biochemical pathways of metabolism depends on the *molecular interactions* of substrates and enzymes.

22.1 Appetite and Satiety

1. The hypothalamus contains a tonically active **feeding center** and a **satiety center** that inhibits the feeding center. (p. 693)
2. Blood glucose concentrations (the **glucostatic theory**) and body fat content (the **lipostatic theory**) influence food intake. (p. 693)
3. Food intake is influenced by a variety of peptides, including **leptin**, **neuropeptide Y**, and **ghrelin**. (p. 693; Fig. 22.1)

22.2 Energy Balance

4. To maintain a constant amount of energy in the body, energy intake must equal energy output. (p. 695; Fig. 22.2)
5. The body uses energy for transport, movement, and chemical work. About half of energy used is released as heat. (p. 695)
6. **Direct calorimetry** measures the energy content of food. (p. 695)
7. The body's **oxygen consumption rate** is the most common method of estimating energy expenditure. (p. 695)
8. The **respiratory quotient (RQ)** or **respiratory exchange ratio (RER)** is the ratio of CO_2 produced to O_2 consumed. RQ varies with diet. (p. 696)
9. **Basal metabolic rate (BMR)** is an individual's lowest metabolic rate. Metabolic rate (kcal/day) $=$ L O_2 consumed/day $\times$ kcal/L O_2. (p. 697)
10. **Diet-induced thermogenesis** is an increase in heat production after eating. (p. 697)
11. Glycogen and fat are the two primary forms of energy storage in the human body. (p. 697)

22.3 Metabolism

12. **Metabolism** is all the chemical reactions that extract, use, or store energy. (p. 698; Figs. 22.3, 22.5)
13. **Anabolic pathways** synthesize small molecules into larger ones. **Catabolic pathways** break large molecules into smaller ones. (p. 698)
14. Metabolism is divided into the **fed** (absorptive) **state** and **fasted** (postabsorptive) **state**. The fed state is anabolic; the fasted state is catabolic. (p. 698)

22.4 Fed-State Metabolism

15. **Glycogenesis** is glycogen synthesis. (p. 700; Fig. 22.5)
16. Ingested fats enter the circulation as chylomicrons. **Lipoprotein lipase** removes triglycerides, leaving chylomicron remnants to be taken up and metabolized by the liver. (p. 700; Fig. 22.6)
17. The liver secretes lipoprotein complexes, such as LDL-C. **Apoproteins A** and **B** are the ligands for receptor-mediated endocytosis of lipoprotein complexes. (p. 700; Fig. 22.6)
18. Elevated blood LDL-C and low blood HDL-C are risk factors for coronary heart disease. Therapies for lowering cholesterol decrease cholesterol uptake or synthesis or increase cholesterol clearance. (p. 703)

22.5 Fasted-State Metabolism

19. The function of fasted-state metabolism is to maintain adequate plasma glucose concentrations because glucose is normally the only fuel that the brain can metabolize. (p. 704; Fig. 22.8)
20. **Glycogenolysis** is glycogen breakdown. **Gluconeogenesis** is glucose synthesis from noncarbohydrate precursors, especially amino acids. (p. 705; Figs. 22.9, 22.11)
21. In the fasted state, the liver produces glucose from glycogen and amino acids. **Beta oxidation** of fatty acids forms acidic **ketone bodies**. (p. 706; Fig. 22.8)

22.6 Homeostatic Control of Metabolism

22. Hour-to-hour metabolic regulation depends on the ratio of insulin to glucagon. Insulin dominates the fed state and decreases plasma glucose. Glucagon dominates the fasted state and increases plasma glucose. (p. 707; Fig. 22.14)
23. The **islets of Langerhans** secrete insulin and amylin from beta cells, glucagon from alpha cells, and somatostatin from D cells. (p. 707; Fig. 22.13)
24. Increased plasma glucose and amino acid levels stimulate insulin secretion. GI hormones and parasympathetic input amplify it. Sympathetic signals inhibit insulin secretion. (p. 708; Fig. 22.15)
25. Insulin binds to a tyrosine kinase receptor and activates multiple **insulin-receptor substrates**. (p. 708; Fig. 22.16)
26. Major insulin target tissues are the liver, adipose tissue, and skeletal muscles. Some tissues are insulin independent. (p. 710)
27. Insulin increases glucose transport into muscle and adipose tissue, as well as glucose utilization and storage of glucose and fat. (p. 710; Fig. 22.17)
28. **Glucagon** stimulates glycogenolysis and gluconeogenesis. (p. 711; Fig. 22.18)
29. **Diabetes mellitus** is a family of disorders marked by abnormal secretion or activity of insulin that causes hyperglycemia. In **type 1 diabetes**, pancreatic beta cells are destroyed by antibodies. In **type 2 diabetes**, target tissues fail to respond normally to insulin. (p. 712)
30. Type 1 diabetes is marked by catabolism of muscle and adipose tissue, glucosuria, polyuria, and metabolic ketoacidosis. Type 2 diabetics have less acute symptoms. In both types, complications include atherosclerosis, neurological changes, and problems with the eyes and kidneys. (p. 712; Fig. 22.20)
31. **Metabolic syndrome** is a condition in which people have central obesity, elevated fasting glucose levels, and elevated lipids. These people are at high risk for developing cardiovascular disease. (p. 718)

22.7 Regulation of Body Temperature

32. Body temperature homeostasis is controlled by the hypothalamus. (p. 720)
33. Heat loss from the body takes place by radiation, conduction, convection, and evaporation. Heat loss is promoted by cutaneous vasodilation and sweating. (p. 719; Figs. 22.21, 22.22)
34. Heat is generated by **shivering thermogenesis** and by **nonshivering thermogenesis**. (p. 719; Fig. 22.21)

REVIEW QUESTIONS

In addition to working through these questions and checking your answers on p. A-29, review the Learning Outcomes at the beginning of this chapter.

Level One Reviewing Facts and Terms

1. Define metabolic, anabolic, and catabolic pathways.
2. List and briefly explain the three forms of biological work.
3. Define a kilocalorie. What is direct calorimetry?
4. What is the respiratory quotient (RQ)? What is a typical RQ value for an American diet?
5. Define basal metabolic rate (BMR). Under what conditions is it measured? Why does the average BMR differ in adult males and females? List at least four factors other than sex that may affect BMR in humans.
6. What are the three general fates of biomolecules in the body?
7. What are the main differences between metabolism in the absorptive and postabsorptive states?
8. What is a nutrient pool? What are the three primary nutrient pools of the body?
9. What is the primary goal of fasted-state metabolism?
10. In what forms is excess energy stored in the body?
11. What are the three possible fates for ingested proteins? For ingested fats?
12. Name the two hormones that regulate glucose metabolism, and explain what effect each hormone has on blood glucose concentrations.
13. Which noncarbohydrate molecules can be made into glucose? What are the pathways called through which these molecules are converted to glucose?
14. Under what circumstances are ketone bodies formed? From what biomolecule are ketone bodies formed? How are they used by the body, and why is their formation potentially dangerous?
15. Name two stimuli that increase insulin secretion, and one stimulus that inhibits insulin secretion.
16. What are the two types of diabetes mellitus? How do their causes and basic symptoms differ?
17. What factors release glucagon? What organ is the primary target of glucagon? What effect(s) do(es) glucagon produce?
18. Define the following terms and explain their physiological significance:
    (a) lipoprotein lipase
    (b) amylin
    (c) ghrelin
    (d) neuropeptide Y
    (e) apoprotein
    (f) leptin
    (g) osmotic diuresis
    (h) insulin resistance
19. What effect does insulin have on:
    (a) glycolysis
    (b) gluconeogenesis
    (c) glycogenesis
    (d) lipogenesis
    (e) protein synthesis

Level Two Reviewing Concepts

20. **Map:** Draw a map that compares the fed state and the fasted state. For each state, compare metabolism in skeletal muscles, the brain, adipose tissue, and the liver. Indicate which hormones are active in each stage and at what points they exert their influence.
21. Examine the graphs of insulin and glucagon secretion in Figure 22.14c. Why have some researchers concluded that the ratio of these two hormones determines whether glucose is stored or removed from storage?
22. Define, compare, and contrast or relate the terms in each of the following sets:
    (a) glucose, glycogenolysis, glycogenesis, gluconeogenesis, glucagon, glycolysis
    (b) shivering thermogenesis, nonshivering thermogenesis, diet-induced thermogenesis
    (c) lipoproteins, chylomicrons, cholesterol, HDL-C, LDL-C, apoproteins
    (d) direct and indirect calorimetry
    (e) conductive heat loss, radiant heat loss, convective heat loss, evaporative heat loss
    (f) absorptive and postabsorptive states
23. Describe (or map) the physiological events that lead to the following signs or symptoms in a type 1 diabetic:
    (a) hyperglycemia
    (b) glucosuria
    (c) polyuria
    (d) ketosis
    (e) dehydration
    (f) severe thirst
24. Both insulin and glucagon are released following ingestion of a protein meal that raises plasma amino acid levels. Why is the secretion of both hormones necessary?
25. Explain the current theory of the control of food intake. Use the following terms in your explanation: hypothalamus, feeding center, satiety center, appetite, leptin, NPY, neuropeptides.
26. Compare human thermoregulation in hot environments and cold environments.

Level Three Problem Solving

27. Scott is a bodybuilder who consumes large amounts of amino acid supplements in the belief that they will increase his muscle mass. He believes that the amino acids he consumes are stored in his body until he needs them. Is Scott correct? Explain.
28. Draw and label a graph showing the effect of insulin secretion on plasma glucose concentration.
29. The signal molecules involved in active cutaneous vasodilation are unclear but it is known that sympathetic cholinergic neurons are involved. In one experiment scientists used botulinum toxin [p. 399] to block release of chemicals from the sympathetic axon terminal.* When they did this, the vasodilation response disap-

*D. L. Kellogg Jr. *et al.* Cutaneous active vasodilation in humans is mediated by cholinergic nerve cotransmission. *Circ Res* 77: 1222–1228, 1995.

peared. In the next experiment, they applied atropine, a muscarinic receptor antagonist and observed that some but not all of the vasodilation response disappeared. What conclusion could they draw from these two experiments?

30. One of the debates in fluid therapy for diabetic ketoacidosis (DKA) is whether to administer bicarbonate (bicarb). Although it is generally accepted that bicarb should be given if the patient's blood pH is <7.1 (life-threatening), most authorities do not give bicarb otherwise. One reason for not administering bicarb relates to the oxygen-binding capacity of hemoglobin. In DKA, patients have low levels of 2,3-BPG [p. 573]. When acidosis is corrected rapidly, 2,3-BPG is much slower to recover and may take 24 or more hours to return to normal.

    Draw and label a graph of the normal oxygen-dissociation curve [p. 573]. *Briefly explain and draw lines on the same graph to show:*

    (a) what happens to oxygen release during DKA as a result of acidosis and low 2,3-BPG levels.
    (b) what happens to oxygen release when the metabolic acidosis is rapidly corrected with bicarbonate.

Level Four Quantitative Problems

31. One way to estimate obesity is to calculate a person's body mass index (BMI). A body mass index greater than 30 is considered a sign of obesity. To calculate BMI, divide body weight in kilograms by the square of height in meters: kg/m^2. (To convert weight from pounds to kilograms, use the conversion factor 1 kg/2.2 lb. To convert height from inches to meters, use the factor 1 m/39.24 in.)

    (a) Anita is 5′1″ tall and weighs 101 lb. What is her BMI? Is this in the normal range?
    (b) Calculate your own BMI. Is it in the normal range?

32. What is the calorie content of a serving of spaghetti and meatballs that contains 6 g fat, 30 g carbohydrate, and 8 g protein? What percentage of the calories comes from fat?

Answers to Concept Checks, Figure and Graph Questions, and end-of-chapter Review Questions can be found in Appendix A [p. A-1].

23 Endocrine Control of Growth and Metabolism

Spongy bone

Disorders of hormone action will be more common causes of endocrinopathy than states of hormone deficiency and excess combined.

Jean D. Wilson, "Endocrinology: Survival as a discipline in the 21st century?" Annu Rev Physiol *62: 947–950, 2000*

BACKGROUND BASICS

73 Extracellular matrix
102 Intermediary metabolism
199 Peptide hormones
200 Steroid hormones
205 Control pathways for pituitary hormones
213 Permissiveness
214 Primary and secondary endocrine pathologies
364 Adrenal medulla

In 1998, Mark McGwire made news when he hit 70 home runs, surpassing the single-season home run record Roger Maris established in 1961. McGwire also created a firestorm of controversy when he admitted to taking *androstenedione*, a performance-enhancing steroid prohormone banned by the International Olympic Committee and other groups but not by professional baseball. As a result of the controversy, Congress passed the Anabolic Steroids Act of 2004, which made androstenedione and some other steroid prohormones controlled substances available only by prescription.

What is this prohormone, and why is it so controversial? You will learn more about androstenedione in this chapter as we discuss the hormones that play a role in long-term regulation of metabolism and growth. In individuals with normal metabolism, these hormones can be difficult to study because their effects are subtle and their interactions with one another complex. As a result, much of what we know about endocrinology comes from studying pathological conditions in which a hormone is either oversecreted or undersecreted. In recent years, however, advances in molecular biology and the use of transgenic animal models have enabled scientists to learn more about hormone action at the cellular level.

23.1 Review Of Endocrine Principles

Before we delve into the different hormones, let's do a quick review of some basic principles and patterns of endocrinology.

1. **The hypothalamic-pituitary control system** [p. 207]. Several of the hormones described in this chapter are controlled by hypothalamic and anterior pituitary (*adenohypophyseal*) trophic hormones.
2. **Feedback patterns** [p. 15]. The negative feedback signal for simple endocrine pathways is the systemic response to the hormone. For example, insulin secretion shuts off when blood glucose concentrations decrease. In complex pathways using the hypothalamic-pituitary control system, the feedback signal may be the hormone itself. In pathological states, endocrine cells may not respond appropriately to feedback signals.
3. **Hormone receptors** [p. 197]. Hormone receptors may be on the cell surface or inside the cell.
4. **Cellular responses** [p. 200]. In general, hormone target cells respond by altering existing proteins or by making new proteins. The historical distinctions between the actions of peptide and steroid hormones no longer apply. Some steroid hormones exert rapid, nongenomic effects, and some peptide hormones alter transcription and translation.
5. **Modulation of target cell response** [p. 180]. The amount of active hormone available to the cell and the number and activity of target cell receptors determine the magnitude of target cell response. Cells may up-regulate or down-regulate their receptors to alter their response. Cells that do not have hormone receptors are nonresponsive.
6. **Endocrine pathologies** [p. 214]. Endocrine pathologies result from (a) excess hormone secretion, (b) inadequate hormone secretion, and (c) abnormal target cell response to the hormone. It now appears that failure of the target cell to respond appropriately to its hormone is a major cause of endocrine disorders.

RUNNING PROBLEM Hyperparathyroidism

"Broken bones, kidney stones, abdominal groans, and psychic moans." Medical students memorize this saying when they learn about hyperparathyroidism, a disease in which parathyroid glands (see Fig. 23.12, p. 745) work overtime and produce excess parathyroid hormone (PTH). Dr. Spinks suddenly recalls the saying as she examines Prof. Magruder, who has arrived at the Emergency Room in pain from a kidney stone lodged in his ureter. When questioned about his symptoms, Prof. Magruder also mentions pain in his shin bones, muscle weakness, stomach upset, and a vague feeling of depression. "I thought it was all just the stress of getting my book published," he says. To Dr. Spinks, however, Prof. Magruder's combination of symptoms sounds suspiciously like he might be suffering from hyperparathyroidism.

729 – 731 – 734 – 744 – 750 – 750

In the following sections, we first examine adrenal corticosteroids and thyroid hormones, two groups of hormones that influence long-term metabolism. We then consider the endocrine control of growth.

23.2 Adrenal Glucocorticoids

The paired adrenal glands sit on top of the kidneys like little caps (**FIG. 23.1**). Each adrenal gland, like the pituitary gland, is two embryologically distinct tissues that merged during development. This complex organ secretes multiple hormones, both neurohormones and classic hormones. The *adrenal medulla* occupies a little over a quarter of the inner mass and is composed of modified sympathetic ganglia that secrete catecholamines (mostly epinephrine) to mediate rapid responses in fight-or-flight situations [p. 356]. The *adrenal cortex* forms the outer three-quarters of the gland and secretes a variety of steroid hormones.

The Adrenal Cortex Secretes Steroid Hormones

The adrenal cortex secretes three major types of steroid hormones: aldosterone (sometimes called a *mineralocorticoid* because of its effect on the minerals sodium and potassium) [p. 629], glucocorticoids, and sex hormones. Histologically, the adrenal cortex is divided into three layers, or zones (Fig. 23.1a). The outer *zona glomerulosa* secretes only aldosterone. The inner *zona reticularis* secretes mostly *androgens*, the sex hormones dominant in men. The middle *zona fasciculata* secretes mostly **glucocorticoids**, named for their ability to increase plasma glucose concentrations. **Cortisol** is the main glucocorticoid secreted by the adrenal cortex.

CHAPTER 23

FIG. 23.1 The adrenal gland

(a) The paired adrenal glands sit on top of the kidneys. Each region secretes different hormones.

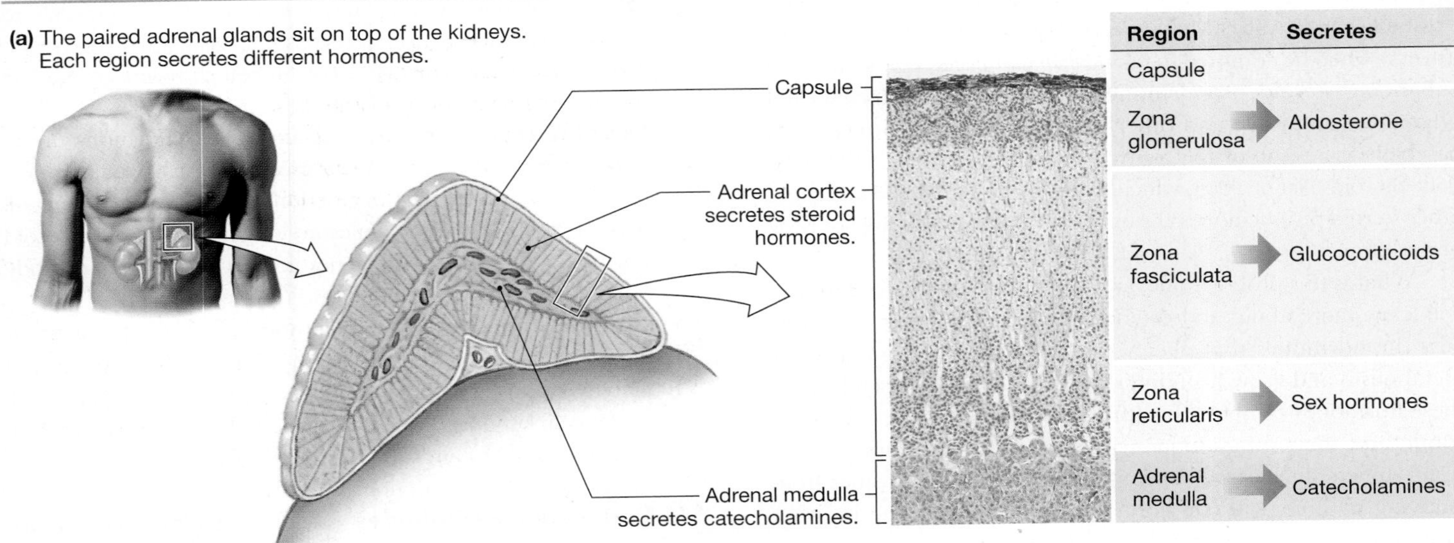
Capsule
Adrenal cortex secretes steroid hormones.
Adrenal medulla secretes catecholamines.
Region
Secretes
Capsule
Zona glomerulosa
Aldosterone
Zona fasciculata
Glucocorticoids
Zona reticularis
Sex hormones
Adrenal medulla
Catecholamines

(b) Synthesis pathways for steroid hormones

All steroid hormones are synthesized from cholesterol. The gray boxes represent intermediate compounds whose names have been omitted for simplicity. Each step is catalyzed by an enzyme, but only two enzymes are shown in the figure.

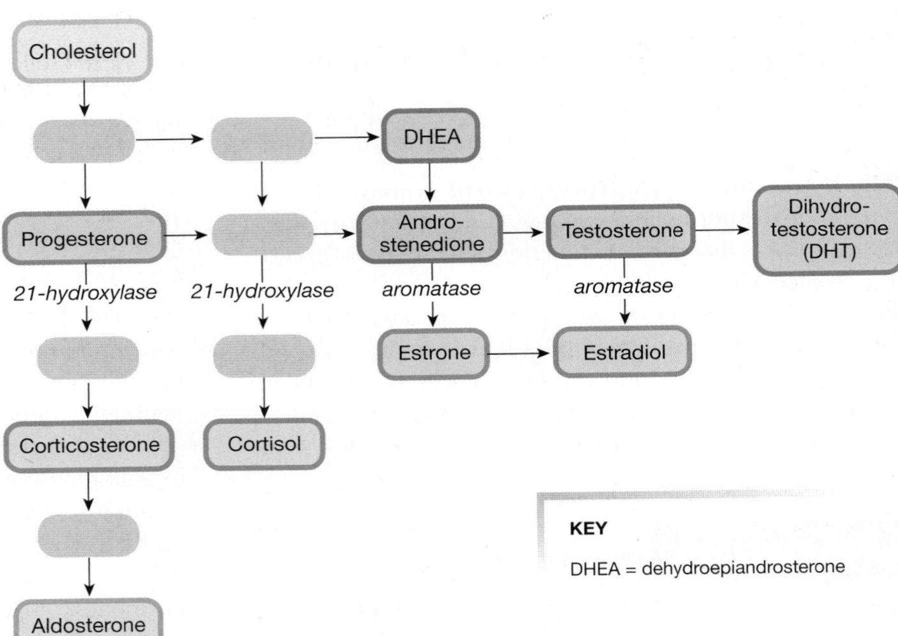
Cholesterol
DHEA
Progesterone
Andro-stenedione
Testosterone
Dihydro-testosterone (DHT)
21-hydroxylase
21-hydroxylase
aromatase
aromatase
Estrone
Estradiol
Corticosterone
Cortisol
Aldosterone
KEY
DHEA = dehydroepiandrosterone

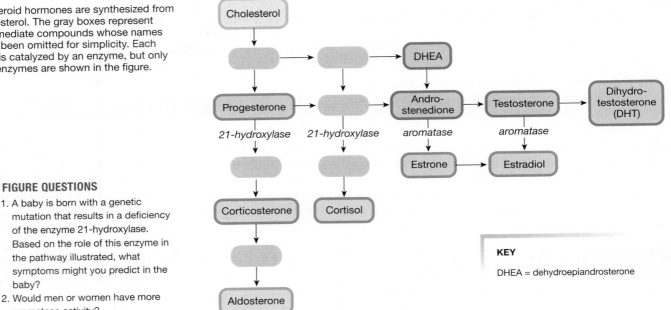

FIGURE QUESTIONS

1. A baby is born with a genetic mutation that results in a deficiency of the enzyme 21-hydroxylase. Based on the role of this enzyme in the pathway illustrated, what symptoms might you predict in the baby?
2. Would men or women have more aromatase activity?

The generalized synthesis pathway for steroid hormones is shown in Figure. 23.1b. All steroid hormones begin with cholesterol, which is modified by multiple enzymes to end up as aldosterone, glucocorticoids, or sex steroids (androgens as well as *estrogens* and *progesterone*, the dominant sex hormones in females). The pathways are the same in the adrenal cortex, gonads, and placenta, but what differs from tissue to tissue is the distribution of enzymes that catalyze the different reactions. For example, the enzyme that makes aldosterone is found in only one of the three adrenal cortex zones.

This chapter opened with the story of baseball player Mark McGwire and his controversial use of the supplement androstenedione. Figure 23.1b shows that this prohormone is one intermediate in the synthesis of testosterone and dihydrotestosterone. One androstenedione precursor, *dehydroepiandrosterone* (DHEA), is used as a dietary supplement. In the United States, purchase of DHEA is not regulated, despite the fact that this substance is metabolically converted to androstenedione and testosterone, both controlled substances whose use is widely banned by sports associations.

The close structural similarity among steroid hormones means that the binding sites on their receptors are also similar, leading to *crossover effects* when one steroid binds to the receptor for a related molecule. For example, *mineralocorticoid receptors* (MRs) for aldosterone are found in the distal nephron. MRs also bind and respond to cortisol, which may be 100 times more concentrated in the blood than aldosterone. What is to keep cortisol from binding to an MR and influencing Na^+ and K^+ excretion? It turns out that renal tubule cells with MRs have an enzyme (*11β-hydroxysteroid dehydrogenase*) that converts cortisol to a less active form with low specificity for the MR. By inactivating cortisol, renal cells normally prevent crossover effects from cortisol. However, crossover activity and the structural similarities of steroid hormones mean that in many endocrine disorders, patients may experience symptoms related to more than one hormone.

Concept Check

1. Name the two parts of the adrenal gland and the major hormones secreted by each part.
2. For what hormones is androstenedione a prohormone? (See Fig. 23.1b.) Why might this prohormone give an athlete an advantage?

Cortisol Secretion Is Controlled by ACTH

The control pathway for cortisol secretion is known as the *hypothalamic-pituitary-adrenal (HPA) pathway* (**FIG. 23.2a**). The HPA pathway begins with hypothalamic **corticotropin-releasing hormone (CRH)**, which is secreted into the hypothalamic-hypophyseal portal system and transported to the anterior pituitary. CRH stimulates release of **adrenocorticotropic hormone (ACTH)** (or *corticotropin*) from the anterior pituitary. ACTH in turn acts on the adrenal cortex to promote synthesis and release of cortisol. Cortisol then acts as a negative feedback signal, inhibiting ACTH and CRH secretion.

Cortisol secretion is continuous and has a strong diurnal rhythm (Fig. 23.2c). Secretion normally peaks in the morning and diminishes during the night. Cortisol secretion also increases with stress.

Cortisol is a typical steroid hormone and is synthesized on demand [Fig. 7.5, p. 203]. Once synthesized, it diffuses out of adrenal cells into the plasma, where most of it is transported by a carrier protein, *corticosteroid-binding globulin* (CBG or *transcortin*). Unbound hormone is free to diffuse into target cells.

All nucleated cells of the body have cytoplasmic glucocorticoid receptors. The hormone-receptor complex enters the nucleus, binds to DNA, and alters gene expression, transcription, and translation. In general, a tissue's response to glucocorticoid hormones is not evident for 60–90 minutes. However, cortisol's negative feedback effect on ACTH secretion occurs within minutes.

RUNNING PROBLEM

Hyperparathyroidism causes breakdown of bone and the release of calcium phosphate into the blood. Elevated plasma Ca^{2+} can affect the function of excitable tissues, such as muscles and neurons. Surprisingly, however, most people with hyperparathyroidism have no symptoms. The condition is usually discovered during blood work performed for a routine health evaluation.

Q1: *What role does Ca^{2+} play in the normal functioning of muscles and neurons?*

Q2: *What is the technical term for "elevated levels of calcium in the blood"? (Use your knowledge of word roots to construct this term.)*

729
731
734
744
750
750

Cortisol Is Essential for Life

Adrenal glucocorticoids are sometimes called the body's stress hormones because of their role in the mediation of long-term stress. Adrenal catecholamines, particularly epinephrine, are responsible for rapid metabolic responses needed in fight-or-flight situations.

Cortisol is essential for life. Animals whose adrenal glands have been removed die if exposed to any significant environmental stress. The most important metabolic effect of cortisol is its protective effect against *hypoglycemia*. When blood glucose decreases, the normal response is secretion of pancreatic glucagon, which promotes gluconeogenesis and glycogen breakdown [p. 711]. In the absence of cortisol, however, glucagon is unable to respond adequately to a hypoglycemic challenge. Because cortisol is required for full glucagon and catecholamine activity, it is said to have a *permissive effect* on those hormones [p. 213].

Cortisol receptors are found in every tissue of the body, but for many targets we do not fully understand the physiological actions of cortisol. However, we can speculate on these actions based on tissue responses to high levels (*pharmacological doses*) of cortisol administered for therapeutic reasons or associated with hypersecretion.

All the metabolic effects of cortisol are directed at preventing hypoglycemia. Overall, cortisol is catabolic (Fig. 23.2a, b).

1. **Cortisol promotes gluconeogenesis** in the liver. Some glucose produced in the liver is released into the blood, and the rest is stored as glycogen. As a result, cortisol increases blood glucose concentrations.
2. **Cortisol causes the breakdown of skeletal muscle proteins** to provide a substrate for gluconeogenesis.
3. **Cortisol enhances lipolysis** so that fatty acids are available to peripheral tissues for energy use. The glycerol from fatty acids can be used for gluconeogenesis.
4. **Cortisol suppresses the immune system** through multiple pathways. This effect is discussed in more detail next.
5. **Cortisol causes negative calcium balance.** Cortisol decreases intestinal Ca^{2+} absorption and increases renal Ca^{2+}

FIG. 23.2 The hypothalamic-pituitary-adrenal (HPA) pathway

(a) The control of cortisol secretion

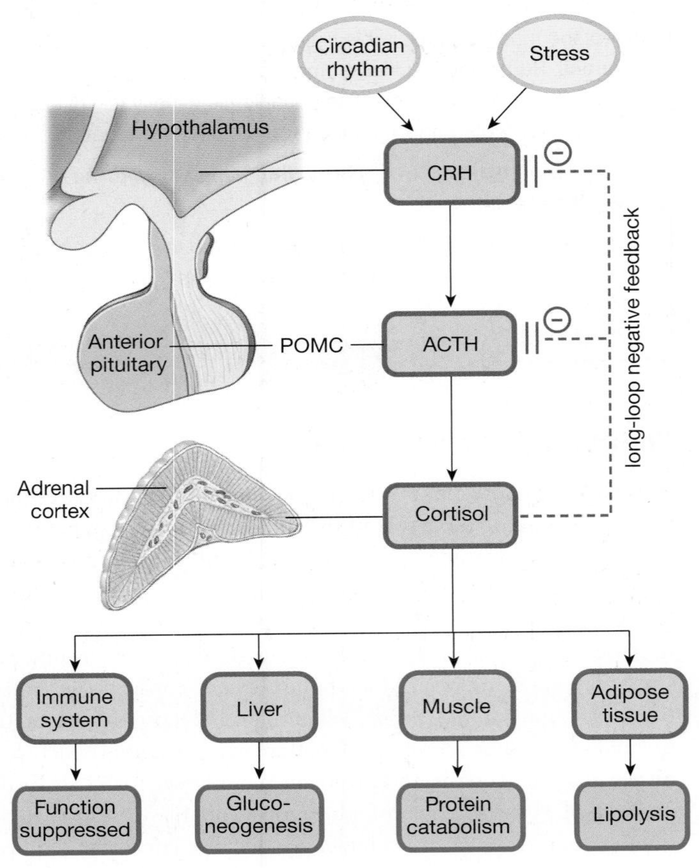
Circadian rhythm
Stress
Hypothalamus
CRH
Anterior pituitary
POMC
ACTH
long-loop negative feedback
Adrenal cortex
Cortisol
Immune system
Liver
Muscle
Adipose tissue
Function suppressed
Gluco-neogenesis
Protein catabolism
Lipolysis

(b) Cortisol	
Origin	Adrenal cortex
Chemical Nature	Steroid
Biosynthesis	From cholesterol; made on demand; not stored
Transport in the Circulation	On corticosteroid-binding globulin (made in liver)
Half-Life	60-90 min
Factors Affecting Release	Circadian rhythm of tonic secretion; stress enhances release
Control Pathway	CRH (hypothalamus) → ACTH (anterior pituitary) → cortisol (adrenal cortex)
Target Cells or Tissues	Most tissues
Target Receptor	Intracellular
Whole Body or Tissue Reaction	↑ Plasma (glucose); ↓ immune activity; permissive for glucagon and catecholamines
Action at Cellular Level	↑ Gluconeogenesis and glycogenolysis; ↑ protein catabolism. Blocks cytokine production by immune cells
Action at Molecular Level	Initiates transcription, translation, and new protein synthesis
Feedback Regulation	Negative feedback to anterior pituitary and hypothalamus

(c) The circadian rhythm of cortisol secretion

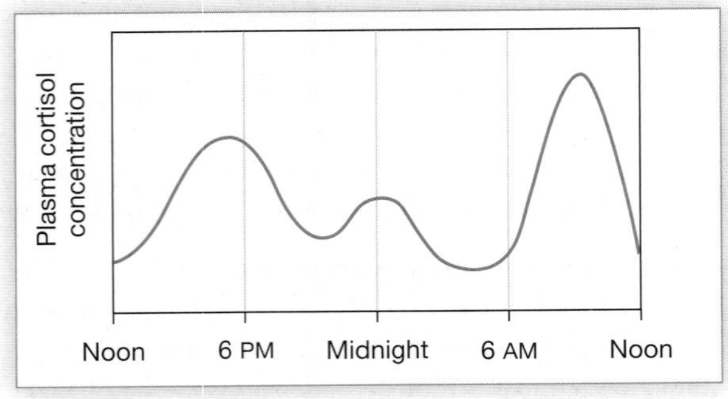
Plasma cortisol concentration
Noon
6 PM
Midnight
6 AM
Noon

FIGURE QUESTION

What do the following abbreviations stand for? ACTH, CRH, MSH

(d) Post-translational processing of POMC creates a variety of active peptides.

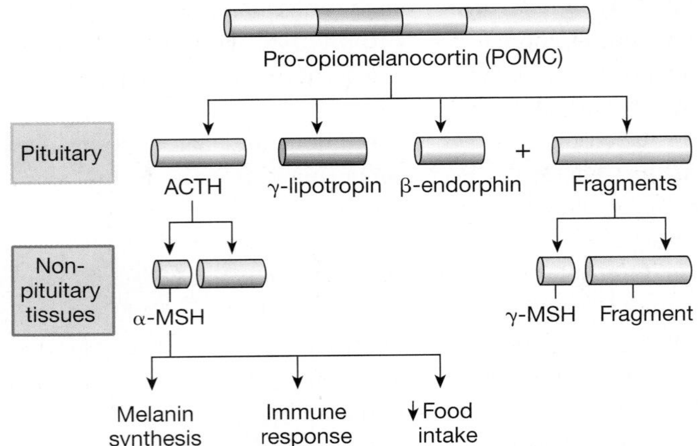
Pro-opiomelanocortin (POMC)
Pituitary
ACTH
γ-lipotropin
β-endorphin
+
Fragments
Non-pituitary tissues
α-MSH
γ-MSH
Fragment
Melanin synthesis
Immune response
↓ Food intake

excretion, resulting in net Ca^{2+} loss from the body. In addition, cortisol is catabolic in bone tissue, causing net breakdown of calcified bone matrix. As a result, people who take therapeutic cortisol for extended periods have a higher-than-normal incidence of broken bones.

6. **Cortisol influences brain function.** States of cortisol excess or deficiency cause mood changes as well as memory and learning alterations. Some of these effects may be mediated by hormones in the cortisol release pathway, such as CRH. We discuss this effect of cortisol in more detail next.

Concept Check

3. What do the abbreviations HPA and CBG stand for? If there is an alternate name for each term, what is it?
4. You are mountain-biking in Canada and encounter a bear, which chases you up a tree. Is your stress response mediated by cortisol? Explain.
5. The illegal use of anabolic steroids by bodybuilders and athletes periodically receives much attention. Do these illegal steroids include cortisol? Explain.

Cortisol Is a Useful Therapeutic Drug

Cortisol suppresses the immune system by preventing cytokine release and antibody production by white blood cells. It also inhibits the inflammatory response by decreasing leukocyte mobility and migration. These *immunosuppressant effects* of cortisol make it a useful drug for treating a variety of conditions, including bee stings, poison ivy, and pollen allergies. Cortisol also helps prevent rejection of transplanted organs. However, glucocorticoids also have potentially serious side effects because of their metabolic actions. Once *nonsteroidal anti-inflammatory drugs* (NSAIDs), such as ibuprofen, were developed, the use of glucocorticoids for treating minor inflammatory problems was discontinued.

Exogenous administration of glucocorticoids has a negative feedback effect on the anterior pituitary and may shut down ACTH production [Fig. 7.13, p. 214]. Without ACTH stimulation, the adrenal cells that produce cortisol atrophy. For this reason, it is essential that patients taking steroids taper their dose gradually, giving the pituitary and adrenal glands a chance to recover, rather than stopping the drug abruptly.

Cortisol Pathologies Result from Too Much or Too Little Hormone

The most common HPA pathologies result from hormone deficiency and hormone excess. Abnormal tissue responsiveness is an uncommon cause of adrenal steroid disorders.

Hypercortisolism Excess cortisol in the body is called **hypercortisolism**. It can arise from hormone-secreting tumors or from exogenous administration of the hormone. Cortisol therapy with high doses for more than a week has the potential to cause hypercortisolism—also known as *Cushing's syndrome*, after Dr. Harvey Cushing, who first described the condition in 1932.

Most signs of hypercortisolism can be predicted from the normal actions of the hormone. Excess gluconeogenesis causes hyperglycemia, which mimics diabetes. Muscle protein breakdown and lipolysis cause tissue wasting. Paradoxically, excess cortisol deposits extra fat in the trunk and face, perhaps in part because of increased appetite and food intake. The classic appearance of patients with hypercortisolism is thin arms and legs, obesity in the trunk, and a "moon face" with plump cheeks (**FIG. 23.3**). CNS effects of too much cortisol include initial mood elevation followed by depression, as well as difficulty with learning and memory.

Hypercortisolism has three common causes:

1. **An adrenal tumor that autonomously secretes cortisol.** These tumors are not under the control of pituitary ACTH. This condition is an instance of *primary hypercortisolism* [p. 215].
2. **A pituitary tumor that autonomously secretes ACTH.** Excess ACTH prompts the adrenal gland to oversecrete cortisol (*secondary hypercortisolism*). The tumor does not respond to negative feedback. This condition is also called *Cushing's disease* because it was the actual disease described by Dr. Cushing. (Hypercortisolism from any cause is called Cushing's *syndrome*.)
3. **Iatrogenic (physician-caused) hypercortisolism** occurs secondary to cortisol therapy for some other condition.

Hypocortisolism Hyposecretion pathologies are far less common than Cushing's syndrome. *Adrenal insufficiency*, commonly known as **Addison's disease**, is hyposecretion of all adrenal steroid hormones, usually following autoimmune destruction of the adrenal cortex. Hereditary defects in the enzymes needed for adrenal steroid production cause several related syndromes collectively known as *congenital adrenal hyperplasia* (see the question in Fig. 23.1). In some of these inherited disorders, excess androgens are secreted because substrate that cannot be made into cortisol or aldosterone is converted to androgens. In newborn girls, excess androgens cause masculinization of the external genitalia, a condition called *adrenogenital syndrome*.

Concept Check

6. For primary, secondary, and iatrogenic hypercortisolism, indicate whether the ACTH level is normal, higher than normal, or lower than normal.
7. Would someone with Addison's disease have normal, low, or high levels of ACTH in the blood?

FIG. 23.3 Hypercortisolism (Cushing's syndrome)

(a) Moon face. A "moon face" with red cheeks is typical in this condition.

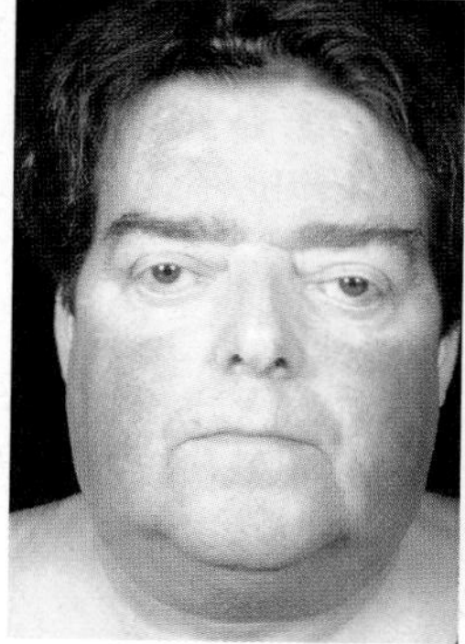

(b) Abdominal fat with striations. Fat also deposits in the trunk. The dark striations result from protein breakdown in the skin.

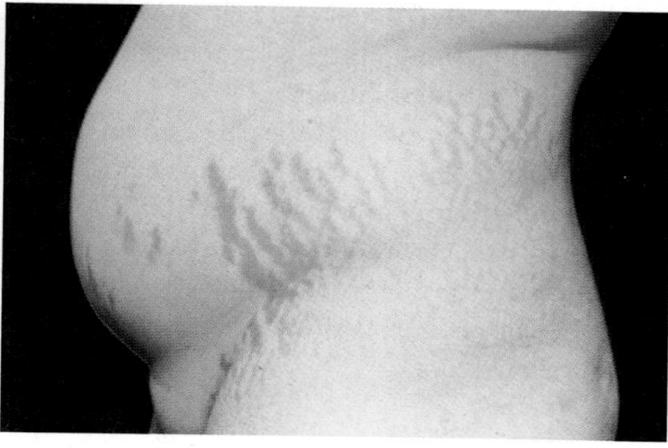

CRH and ACTH Have Additional Physiological Functions

In recent years, research interest has shifted away from glucocorticoids to CRH and ACTH, the trophic hormones of the HPA pathway. Both peptides are now known to belong to larger families of related molecules, with multiple receptor types found in numerous tissues. Experiments with knockout mice lacking a particular receptor have revealed some of the physiological functions of peptides related to CRH and ACTH.

Two interesting findings of this research are that cytokines secreted by the immune system can stimulate the HPA pathway and that immune cells have receptors for ACTH and CRH. The association between stress and immune function appears to be mediated through CRH and ACTH, and this association provides one explanation for mind–body interactions in which mental state influences physiological function.

CRH Family The CRH family includes CRH and a related brain neuropeptide called *urocortin*. In addition to its involvement in inflammation and the immune response, CRH is known to decrease food intake [Fig. 22.1, p. 694] and has been associated with signals that mark the onset of labor in pregnant women. Additional evidence links CRH to anxiety, depression, and other mood disorders.

POMC and Melanocortins CRH acting on the anterior pituitary stimulates secretion of ACTH. ACTH is synthesized from a large glycoprotein called **pro-opiomelanocortin** (POMC, pronounced *pom-see*). POMC undergoes posttranslational processing to produce a variety of biologically active peptides in addition to ACTH (Fig. 23.2d). In the pituitary, POMC products include **β-endorphin**, an endogenous opioid that binds to receptors that block pain perception [p. 254].

Processing of POMC in nonpituitary tissues creates additional peptides, such as *melanocyte-stimulating hormone* (MSH). α-**MSH** is produced in the brain, where it inhibits food intake, and in the skin, where it acts on **melanocytes**. Melanocytes contain pigments called **melanins** that influence skin color in humans and coat color in mice.

The MSH hormones plus ACTH have been given the family name **melanocortins**. Five *melanocortin receptors* (MCRs) have been identified. **MC2R** responds only to ACTH and is the adrenal cortex receptor. **MC1R** is found in skin melanocytes and responds equally to α-MSH and ACTH. When ACTH is elevated in Addison's disease, the action of ACTH on MC1R leads to increased melanin production and the apparent "tan," or skin darkening, characteristic of this disorder.

Much of what we have learned about MCRs started with research on the *agouti* mouse, a strain that resulted from a spontaneous mutation first described in 1905. *Agouti* mice with one mutated gene overproduce *agouti protein*, an antagonist to the MC1R melanocortin receptor. MC1R controls melanin synthesis in hair, so blocking its pathway causes mice to develop a characteristic yellow coat.

Of more interest to physiologists, however, is the fact that *agouti* mice overeat and develop adult-onset obesity, hyperglycemia, and insulin resistance—in other words, these mice are a model for obesity-related type 2 diabetes. In 1997 scientists identified *agouti-related protein* (AgRP) in hypothalamic neurons related to feeding behaviors. Neurons in the same region express **MC4R**, melanocortin receptors that depress feeding behavior. Our current model shows that AgRP is a MC4R receptor antagonist. High levels of AgRP inactivate MC4R, removing the inhibition of feeding, so the animal overeats and becomes obese.

POMC-producing neurons in the hypothalamus also affect food intake and energy balance. Hypothalamic neurons apparently release α-MSH made from POMC. α-MSH is an agonist of MC4R, so α-MSH decreases food intake when it activates MC4R. Recent investigations suggest that the action of nicotine on POMC neurons explains why smoking decreases food intake. Other research suggests that these POMC neurons respond to changes in blood glucose and possibly participate in the glucostatic control mechanism influencing food intake [p. 693]. The link between melanocortin receptors, eating behavior, and diabetes has opened up a new area of research on treatments to prevent type 2 diabetes.

Concept Check

8. Can you think of a situation where it might be the advantage for the body to co-secrete ACTH and β-endorphin?

23.3 Thyroid Hormones

The thyroid gland is a butterfly-shaped gland that lies across the trachea at the base of the throat, just below the larynx (**FIG. 23.4a**). It is one of the larger endocrine glands, weighing 15–20 g. The thyroid gland has two distinct endocrine cell types: *C* ("clear") *cells*, which secrete a calcium-regulating hormone called *calcitonin*, and *follicular cells*, which secrete thyroid hormone. We discuss calcitonin later, with calcium homeostasis.

Thyroid Hormones Contain Iodine

Thyroid hormones, like glucocorticoids, have long-term effects on metabolism. Unlike glucocorticoids, however, thyroid hormones are not essential for life: adults can live, although not comfortably, without thyroid hormone or a thyroid gland. Thyroid hormones are essential for normal growth and development in children, however, and infants born with thyroid deficiency will be developmentally delayed unless treated promptly. Because of the importance of thyroid hormones in children, the United States and Canada test all newborns for thyroid deficiency.

Thyroid hormones are amines derived from the amino acid tyrosine, and they are unusual because they contain the element iodine (Fig. 23.4c). Currently, thyroid hormones are the only

FIG. 23.4 Thyroid hormone synthesis

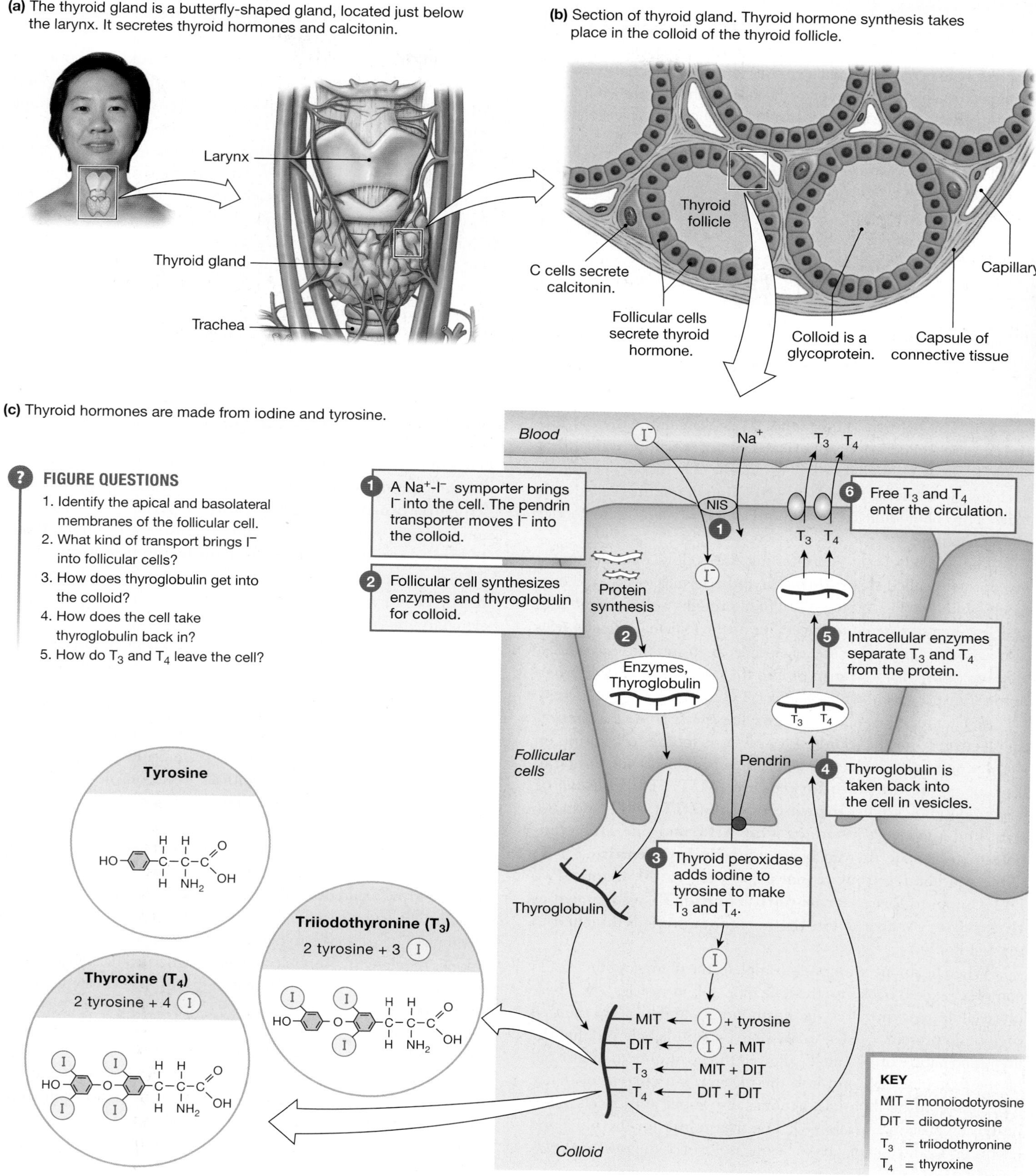

(a) The thyroid gland is a butterfly-shaped gland, located just below the larynx. It secretes thyroid hormones and calcitonin.
Larynx
Thyroid gland
Trachea
(b) Section of thyroid gland. Thyroid hormone synthesis takes place in the colloid of the thyroid follicle.
Thyroid follicle
C cells secrete calcitonin.
Follicular cells secrete thyroid hormone.
Colloid is a glycoprotein.
Capsule of connective tissue
Capillary
(c) Thyroid hormones are made from iodine and tyrosine.
Blood
Na+
T3 T4
NIS
1 A Na+-I− symporter brings I− into the cell. The pendrin transporter moves I− into the colloid.
2 Follicular cell synthesizes enzymes and thyroglobulin for colloid.
Protein synthesis
Enzymes, Thyroglobulin
6 Free T3 and T4 enter the circulation.
5 Intracellular enzymes separate T3 and T4 from the protein.
4 Thyroglobulin is taken back into the cell in vesicles.
Follicular cells
Pendrin
3 Thyroid peroxidase adds iodine to tyrosine to make T3 and T4.
Thyroglobulin
MIT ← I + tyrosine
DIT ← I + MIT
T3 ← MIT + DIT
T4 ← DIT + DIT
Colloid
KEY
MIT = monoiodotyrosine
DIT = diiodotyrosine
T3 = triiodothyronine
T4 = thyroxine
Tyrosine
Triiodothyronine (T3)
2 tyrosine + 3 I
Thyroxine (T4)
2 tyrosine + 4 I

FIGURE QUESTIONS

1. Identify the apical and basolateral membranes of the follicular cell.
2. What kind of transport brings I^- into follicular cells?
3. How does thyroglobulin get into the colloid?
4. How does the cell take thyroglobulin back in?
5. How do T_3 and T_4 leave the cell?

RUNNING PROBLEM

Elevated blood Ca^{2+} leads to high Ca^{2+} concentrations in the kidney filtrate. Calcium-based kidney stones occur when calcium phosphate or calcium oxalate crystals form and aggregate with organic material in the lumen of the kidney tubule. Once Prof. Magruder's kidney stone passes into the urine, Dr. Spinks sends it for a chemical analysis.

Q3: *Only free Ca^{2+} in the blood filters into Bowman's capsule at the nephron. A significant portion of plasma Ca^{2+} cannot be filtered. Use what you have learned about filtration at the glomerulus to speculate on why some plasma Ca^{2+} cannot filter [p. 598].*

729
731
734
744
750
750

known use for iodine in the body, although a few other tissues also concentrate this mineral.

Synthesis of thyroid hormones takes place in the thyroid **follicles** (also called *acini*), spherical structures whose walls are a single layer of epithelial cells (Fig. 23.4b). The hollow center of each follicle is filled with a sticky glycoprotein mixture called **colloid**. The colloid holds a 2–3 month supply of thyroid hormones at any one time.

The follicular cells surrounding the colloid manufacture a glycoprotein called **thyroglobulin** and enzymes for thyroid hormone synthesis (Fig. 23.4c 1). These proteins are packaged into vesicles, then secreted into the center of the follicle. Follicular cells also actively concentrate dietary iodide, I^-, using the *sodium-iodide symporter* (NIS) 2. I^- transport into the colloid is mediated by an anion transporter known as *pendrin* (SLC26A4).

As I^- enters the colloid, the enzyme *thyroid peroxidase* removes an electron from the iodide ion and adds iodine to tyrosine on the thyroglobulin molecule 3. The addition of one iodine to tyrosine creates **monoiodotyrosine (MIT)**. The addition of a second iodine creates **diiodotyrosine (DIT)**. MIT and DIT then undergo *coupling reactions*. One MIT and one DIT combine to create the thyroid hormone **triiodothyronine**, or $\mathbf{T_3}$ (Note the change from *tyrosine* to *thyronine* in the name.) Two DIT couple to form **tetraiodothyronine** ($\mathbf{T_4}$, also known as **thyroxine**). At this point, the hormones are still attached to thyroglobulin.

When hormone synthesis is complete, the thyroglobulin–T_3/T_4 complex is taken back into the follicular cells in vesicles 4. There intracellular enzymes free the hormones T_3 and T_4 from the thyroglobulin protein 5. For many years, scientists believed that the lipophilic nature of T_3 and T_4 allowed the hormones to diffuse out of the follicular cells and into the plasma, but current evidence indicates that the thyroid hormones also move across cell membranes by protein carriers 6. The transporter for thyroid gland export of T_3 and T_4 appears to be one isoform of the *monocarboxylate transporter* (MCT8).

T_3 and T_4 have limited solubility in plasma because they are lipophilic molecules. As a result, thyroid hormones bind to plasma proteins, such as **thyroid-binding globulin (TBG)**. Most thyroid hormone in the plasma is in the form of T_4.

Target tissue uptake transporters for thyroid hormones vary from tissue to tissue. They include the monocarboxylate transporters MCT8 and MCT10 as well as one member of the organic anion transporter (OAT) family.

For years it was thought that T_4 was the active hormone but we now know that T_3 is three to five times more active biologically, and that T_3 is the active hormone in target cells. Target cells make about 85% of active T_3 by using enzymes called **deiodinases** to remove an iodine from T_4. Target tissue activation of the hormone adds another layer of control because individual target tissues can alter their exposure to active thyroid hormone by regulating their tissue deiodinase synthesis.

Thyroid receptors, with multiple isoforms, are in the nucleus of target cells. Hormone binding initiates transcription, translation, and synthesis of new proteins.

TSH Controls the Thyroid Gland

The control of thyroid hormone secretion follows the typical hypothalamic-pituitary-peripheral endocrine gland pattern (**FIG. 23.5**). **Thyrotropin-releasing hormone (TRH)** from the hypothalamus controls secretion of the anterior pituitary hormone **thyrotropin**, also known as **thyroid-stimulating hormone (TSH)**. TSH in turn acts on the thyroid gland to promote hormone synthesis. The thyroid hormones normally act as a negative feedback signal to prevent oversecretion.

The main function of thyroid hormones in adults is to provide substrates for oxidative metabolism. Thyroid hormones are thermogenic [p. 720] and increase oxygen consumption in most tissues. The molecular mechanism is unclear but is at least partly related to increased Na^+-K^+-ATPase activity and effects on the electron transport system. Thyroid hormones also interact with other hormones in a complex and tissue-specific fashion to modulate protein, carbohydrate, and fat metabolism.

In children, thyroid hormones are necessary for full expression of growth hormone, which means thyroid function is essential for normal growth and development, especially of the nervous system. In the first few years after birth, myelin and synapse formation requires T_3 and T_4. Cytological studies suggest that thyroid hormones regulate microtubule assembly, which is an essential part of neuronal growth. Thyroid hormone is also necessary for proper bone growth.

The actions of thyroid hormones are most observable in people who secrete too much or too little hormone. Physiological effects that are subtle in people with normal hormone secretion often become exaggerated in patients with endocrine disorders. Patients with thyroid excess or deficiency may experience decreased tolerance to heat or cold and mood disturbances, in addition to other symptoms.

FIG. 23.5 Thyroid hormones

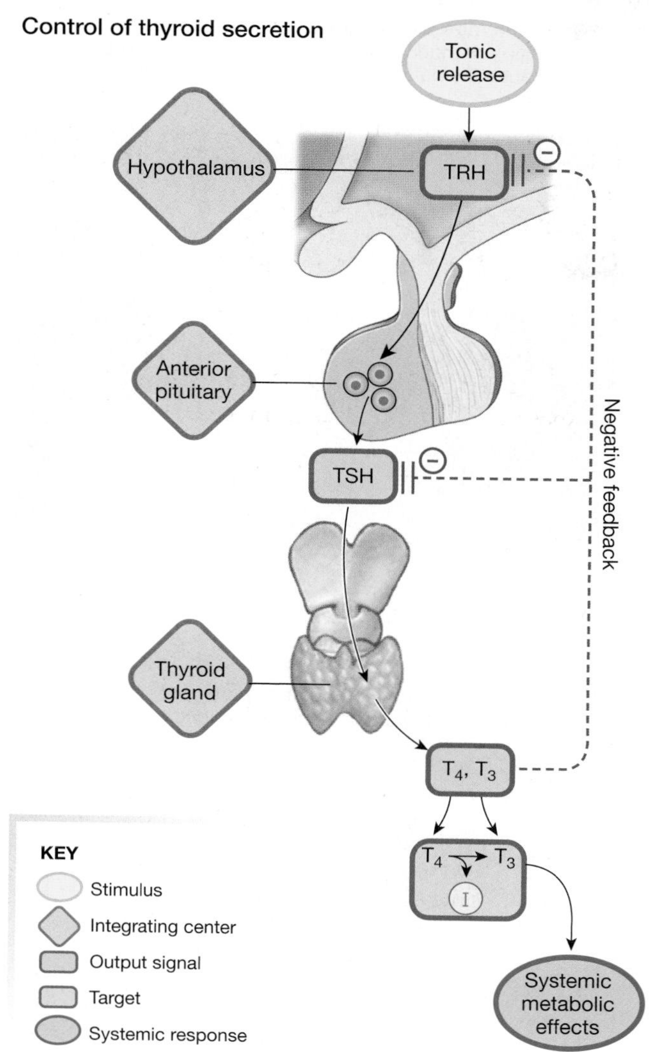
Control of thyroid secretion
Tonic release
Hypothalamus
TRH
Anterior pituitary
TSH
Thyroid gland
T_4, T_3
T_4 → T_3
I
Negative feedback
Systemic metabolic effects
KEY
Stimulus
Integrating center
Output signal
Target
Systemic response

Thyroid Hormones	
Cell of Origin	Thyroid follicle cells
Chemical Nature	Iodinated amine
Biosynthesis	From iodine and tyrosine. Formed and stored on thyroglobulin in follicle colloid.
Transport in the Circulation	Bound to thyroxine-binding globulin and albumins
Half-Life	6-7 days for thyroxine (T_4); about 1 day for triiodothyronine (T_3)
Factors Affecting Release	Tonic release
Control Pathway	TRH (hypothalamus) → TSH (anterior pituitary) → $T_3 + T_4$ (thyroid) → T_4 deiodinates in tissues to form more T_3
Target Cells or Tissues	Most cells of the body
Target Receptor	Nuclear receptor
Whole Body or Tissue Reaction	↑ Oxygen consumption (thermogenesis). Protein catabolism in adults but anabolism in children. Normal development of nervous system.
Action at Cellular Level	Increases activity of metabolic enzymes and $Na^+ - K^+$ – ATPase
Action at Molecular Level	Production of new enzymes
Feedback Regulation	Free T_3 and T_4 have negative feedback on anterior pituitary and hypothalamus.

Thyroid Pathologies Affect Quality of Life

Problems with thyroid hormone secretion can arise either in the thyroid gland or along the control pathway depicted in Figure 23.5. The trophic action of TSH on the thyroid gland causes enlargement, or *hypertrophy*, of follicular cells. In pathological conditions with elevated TSH levels, the thyroid gland will enlarge, a condition known as a **goiter**. A large goiter can weigh hundreds of grams and almost encircle the neck (**FIG. 23.6a**).

Goiters are the result of excess TSH stimulation of the thyroid gland. Simply knowing that someone has a goiter does not tell you what the pathology is, however. Let's see how both hypothyroidism and hyperthyroidism can be associated with goiter.

Hyperthyroidism A person whose thyroid gland secretes too much hormone suffers from **hyperthyroidism**. Excess thyroid hormone causes changes in metabolism, the nervous system, and the heart.

1. Hyperthyroidism increases oxygen consumption and metabolic heat production. Because of the internal heat generated, these patients have warm, sweaty skin and may complain of being intolerant of heat.
2. Excess thyroid hormone increases muscle protein catabolism and may cause muscle weakness. Patients often report weight loss.
3. The effects of excess thyroid hormone on the nervous system include hyperexcitable reflexes and psychological disturbances ranging from irritability and insomnia to psychosis. The mechanism for psychological disturbances is unclear, but morphological changes in the hippocampus and effects on β-adrenergic receptors have been suggested.
4. Thyroid hormones are known to influence β-adrenergic receptors in the heart, and these effects are exaggerated with hypersecretion. A common sign of hyperthyroidism is rapid heartbeat and increased force of contraction due to up-regulation of β_1-receptors on the myocardium [p. 469].

The most common cause of hyperthyroidism is *Graves' disease* (**FIG. 23.7a**). In this condition, the body produces antibodies

FIG. 23.6 Signs of thyroid pathologies

(a) Goiter. Excessive stimulation of the thyroid gland by TSH causes the gland to enlarge (goiter).

(b) Myxedema. In hypothyroid individuals, mucopolysaccharide deposits beneath the skin may cause bags under the eyes.

(c) Exophthalmos. In hyperthyroid states, excessive deposition of mucopolysaccharides in the bony orbit may cause the bulging eyeball called exophthalmos.

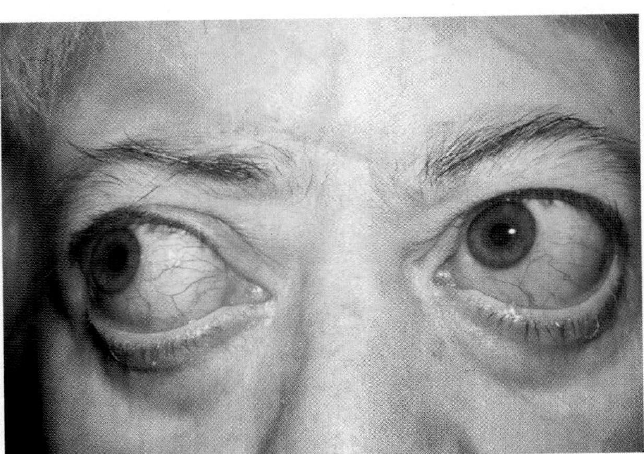

FIG. 23.7 Thyroid pathologies

(a) Hyperthyroidism due to Graves' disease. In Graves' disease, thyroid-stimulating immune proteins (TSI) bind to thyroid gland TSH receptors and cause the gland to hypertrophy.

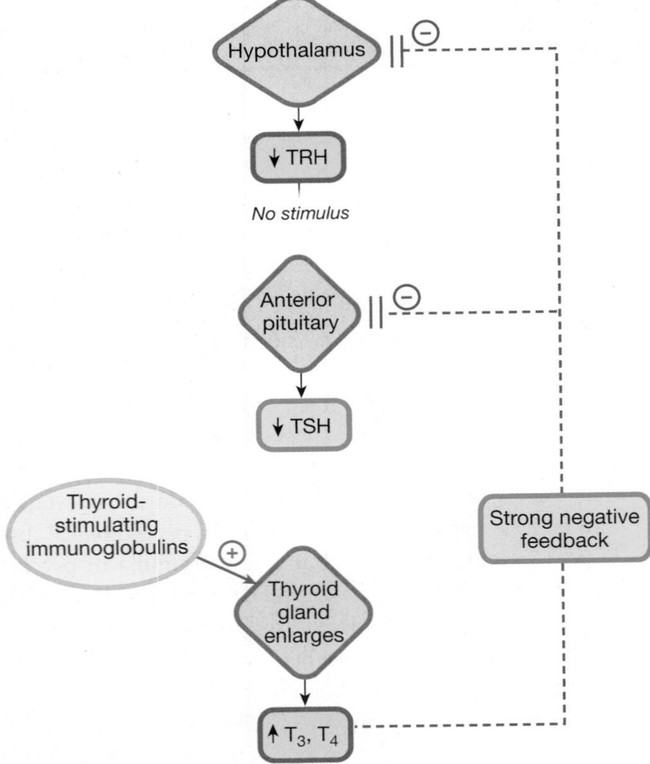
Hypothalamus
⊖
↓ TRH
No stimulus
Anterior pituitary
⊖
↓ TSH
Thyroid-stimulating immunoglobulins
⊕
Thyroid gland enlarges
↑ T_3, T_4
Strong negative feedback

(b) Hypothyroidism due to low iodine. In hypothyroidism caused by iodine deficiency, absence of negative feedback increases TSH secretion and results in goiter.

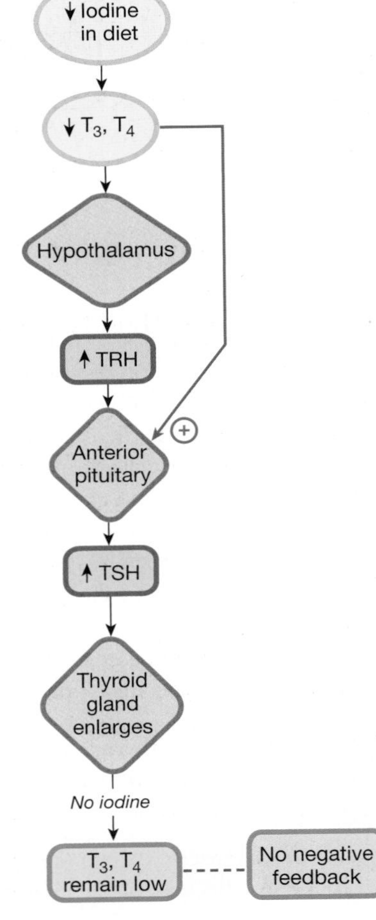
↓ Iodine in diet
↓ T_3, T_4
Hypothalamus
↑ TRH
⊕
Anterior pituitary
↑ TSH
Thyroid gland enlarges
No iodine
T_3, T_4 remain low
No negative feedback

FIGURE QUESTION

Draw the pathway for a person with a pituitary tumor that is oversecreting TSH. Would this person be hypothyroid or hyperthyroid? Would this person have a goiter?

called **thyroid-stimulating immunoglobulins (TSI)**. These antibodies mimic the action of TSH by combining with and activating TSH receptors on the thyroid gland. The result is goiter, hypersecretion of T_3 and T_4 and symptoms of hormone excess.

Negative feedback by the high levels of T_3 and T_4 shuts down the body's TRH and TSH secretion but does nothing to block the TSH-like activity of TSI on the thyroid gland. Graves' disease is often accompanied by **exophthalmos** (Fig. 23.6c), a bug-eyed appearance caused by immune-mediated enlargement of muscles and tissue in the eye socket. The English comic Marty Feldman was known for his wild-eyed appearance caused by exophthalmos.

Thyroid gland tumors are another cause of primary hyperthyroidism. Secondary hyperthyroidism will occur with pituitary tumors secreting TSH.

Hypothyroidism Hyposecretion of thyroid hormones affects the same systems altered by hyperthyroidism.

1. Decreased thyroid hormone secretion slows metabolic rate and oxygen consumption. Patients become intolerant of cold because they are generating less internal heat.
2. Hypothyroidism decreases protein synthesis. In adults, this causes brittle nails, thinning hair, and dry, thin skin. Hypothyroidism also causes accumulation of *mucopolysaccharides* under the skin. These molecules attract water and cause the puffy appearance of *myxedema* (Fig. 23.6b). Hypothyroid children have slow bone and tissue growth and are shorter than normal for their age.
3. Nervous system changes in adults include slowed reflexes, slow speech and thought processes, and feelings of fatigue. Deficient thyroid hormone secretion in infancy causes **cretinism**, a condition marked by decreased mental capacity.
4. The primary cardiovascular change in hypothyroidism is *bradycardia* (slow heart rate).

Primary hypothyroidism is most commonly caused by a lack of iodine in the diet. Without iodine, the thyroid gland cannot make thyroid hormones (Fig. 23.7b). Low levels of T_3 and T_4 in the blood mean no negative feedback to the hypothalamus and anterior pituitary. In the absence of negative feedback, TSH secretion rises dramatically, and TSH stimulation enlarges the thyroid gland (goiter). Despite hypertrophy, the gland cannot obtain iodine to make hormone, so the patient remains hypothyroid. These patients exhibit the previously described signs of hypothyroidism. The goiter shown in the photograph of Figure. 23.6a is probably due to iodine deficiency.

Therapy for thyroid disorders depends on the cause of the problem. Hypothyroidism from lack of iodine in the diet can be treated by supplementation, such as iodized salt. Hypothyroidism from other causes is treated with oral thyroid hormone. Hyperthyroidism can be treated by surgical removal of all or part of the gland, by destruction of thyroid cells with radioactive iodine, or by drugs that block either hormone synthesis (thiourea drugs) or peripheral conversion of T_4 to T_3 (propylthiouracil).

Concept Check

9. A woman who had her thyroid gland removed because of cancer was given pills containing only T_4. Why was this less active form of the hormone an effective treatment for her hypothyroidism?
10. Why would excessive production of thyroid hormone, which uncouples mitochondrial ATP production and proton transport [p. 108], cause a person to become intolerant of heat?

23.4 Growth Hormone

Growth in human beings is a continuous process that begins before birth. However, growth rates in children are not steady, with the first two years of life and the adolescent years marked by spurts of rapid growth and development. In adults, bone growth ceases, but soft tissue growth, as reflected by body mass, can continue to increase.

Normal growth before adulthood is a complex process that depends on a number of factors:

1. **Growth hormone and other hormones.** Without adequate amounts of growth hormone, children simply fail to grow. Thyroid hormones, insulin, and the sex hormones at puberty also play both direct and permissive roles. A deficiency in any one of these hormones leads to abnormal growth and development.
2. **An adequate diet** that includes protein, sufficient energy (caloric intake), vitamins, and minerals. Many amino acids can be manufactured in the body from other precursors, but essential amino acids must come from dietary sources. Among the minerals, calcium in particular is needed for proper bone formation.
3. **Absence of chronic stress.** Cortisol from the adrenal cortex is released in times of stress and has significant catabolic effects that inhibit growth. Children who are subjected to stressful environments may exhibit a condition known as *failure to thrive* that is marked by abnormally slow growth.
4. **Genetics.** Each human's potential adult size is genetically determined at conception.

Growth Hormone Is Anabolic

Growth hormone (GH or somatotropin [p. 209]) is released throughout life, although its biggest role is in children. Peak GH secretion occurs during the teenage years. The stimuli for growth hormone release are complex and not completely understood, but they include circulating nutrients, stress, and other hormones such as ghrelin interacting with a daily rhythm of secretion (**FIG. 23.8**).

The stimuli for GH secretion are integrated in the hypothalamus, which secretes two neuropeptides into the hypothalamic-hypophyseal portal system: **growth hormone-releasing hormone (GHRH)** and *growth hormone-inhibiting hormone*, better known as *somatostatin* (SS). On a daily basis, pulses of GHRH from the hypothalamus stimulate GH release. In adults, the largest pulse of GH release occurs in the first two hours of sleep. It is speculated

FIG. 23.8 Growth hormone

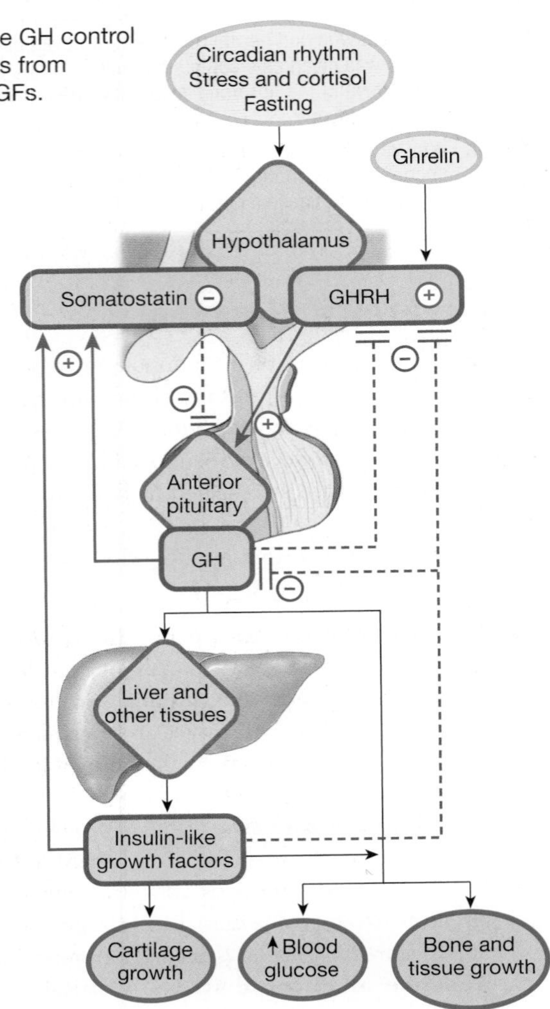
Feedback in the GH control pathway comes from both GH and IGFs.
Circadian rhythm
Stress and cortisol
Fasting
Ghrelin
Hypothalamus
Somatostatin
GHRH
Anterior pituitary
GH
Liver and other tissues
Insulin-like growth factors
Cartilage growth
Blood glucose
Bone and tissue growth

Growth Hormone (hGH)	
Origin	Anterior pituitary
Chemical Nature	191-amino acid peptide; several closely related forms
Biosynthesis	Typical peptide
Transport in the Circulation	Half is dissolved in plasma, half is bound to a binding protein whose structure is identical to that of the GH receptor
Half-Life	18 minutes
Factors Affecting Release	Circadian rhythm of tonic secretion; influenced by circulating nutrients, stress, and other hormones in a complex fashion
Control Pathway	GHRH, somatostatin (hypothalamus) → growth hormone (anterior pituitary)
Target Cells or Tissues	Trophic on liver for insulin-like growth factor production; also acts directly on many cells
Target Receptor	Membrane receptor with tyrosine kinase activity
Whole Body or Tissue Reaction (with IGFs)	Bone and cartilage growth; soft tissue growth; ↑ plasma glucose
Action at Cellular Level	Receptor linked to kinases that phosphorylate proteins to initiate transcription

that GHRH has sleep-inducing properties, but the role of GH in sleep cycles is unclear.

GH is secreted by cells in the anterior pituitary. It is a typical peptide hormone in most respects, except that nearly half the GH in blood is bound to a plasma **growth hormone-binding protein**. The binding protein protects plasma GH from being filtered into the urine and extends its half-life by 12 minutes. Researchers have hypothesized that genetic determination of binding protein concentration plays a role in determining adult height.

The target tissues for GH include both endocrine and nonendocrine cells (Fig. 23.8). GH acts as a trophic hormone to stimulate secretion of **insulin-like growth factors (IGFs)** (formerly called *somatomedins*) from the liver and other tissues. IGFs act in concert with growth hormone to stimulate growth in the following ways:

1. GH and IGFs promote protein synthesis. This is particularly true in skeletal muscle.
2. GH causes lipolysis, decreased glucose uptake by muscle, and gluconeogenesis in the liver. Collectively these actions increase blood glucose concentrations.
3. Both GH and IGFs act on bones to increase bone growth. Only IGFs stimulate cartilage synthesis directly.

Feedback Control of GH Secretion The homeostatic feedback loops that are typical for hormones of the anterior pituitary [Fig. 7.11a, p. 212] become much more complex in the GH pathway, where there are two hypothalamic hormones (GHRH and somatostatin) acting on the pituitary, plus two systemic signals (GH and IGFs). In the classic pattern, GH feeds back to inhibit GHRH (Fig. 23.8). However, GH also promotes release of somatostatin to reinforce shutting off GH secretion. IGFs have the same effect as GH on the hypothalamic hormones but in addition, IGFs directly inhibit GH secretion.

Concept Check

11. Which pituitary hormone in addition to GH has two hypothalamic factors that regulate its release?

Growth Hormone Is Essential for Normal Growth

The disorders that reflect the actions of growth hormone are most obvious in children. Severe growth hormone deficiency in childhood leads to **dwarfism**, which can result from a problem either with growth hormone synthesis or with defective GH receptors. Unfortunately, only primate growth hormone is active in humans, and prior to 1985, donated human pituitaries harvested at autopsy were the only source of growth hormone. Fortunately, severe growth hormone deficiency is relatively rare.

At the opposite extreme, oversecretion of growth hormone in children leads to **giantism**. Once bone growth stops in late adolescence, growth hormone cannot further increase height. GH and IGFs will continue to act on cartilage and soft tissues, however. Adults with excessive secretion of growth hormone develop a condition known as **acromegaly**, characterized by lengthening of the jaw, coarsening of facial features, and growth of hands and feet (**FIG. 23.9**). André the Giant, a French wrestler who also had a role in the classic movie *The Princess Bride*, exhibited signs of both giantism (he grew to 7′4″ tall) and acromegaly before his death at age 47.

FIG. 23.9 Acromegaly

Excess growth hormone secretion in adults causes acromegaly, with lengthening of the jaw, coarsening of the features, and growth in hands and feet. Compare these signs in André the Giant (top) to the features of his co-stars in *The Princess Bride*, Mandy Patinkin (middle) and Wallace Shawn (bottom).

Genetically Engineered hGH Raises Ethical Questions

When genetically engineered human growth hormone (hGH) became available in the mid-1980s, the medical profession was faced with a dilemma. Obviously, the hormone should be used to treat children who would otherwise be dwarfs, but what about children with only partial GH deficiency or genetically short children with normal GH secretion levels? This question is complicated by the difficulty of accurately identifying children with partial growth hormone deficiency. And what about children whose parents simply want them to be taller? Should these healthy children be given the hormone?

In 2003, the U.S. Food and Drug Administration approved use of a recombinant human growth hormone for treating children with non-GH-deficient short stature, defined as being more than 2.25 standard deviations below the mean height for their age and sex. (This means children in the bottom 1% of their age-sex group.) In clinical trials, daily injections of the drug for two years resulted in an average height increase of 1.3″ (3.3 cm). According to a 2006 analysis in a pediatric medicine journal, the cost for this treatment was more than $52,000 per inch of height gained. There are potential side effects from taking any drug, and parents must be made aware that hGH therapy has the potential to create psychological problems in children if the results are less than optimum.

The newest ethical issue with recombinant hGH is its use in adults without documented GH deficiency. As with children, adults who have hyposecretion of GH are candidates for hGH therapy. However, articles promoting GH as the fountain of youth have created a significant market for hGH injections, often purchased without a prescription. Side effects of GH administration reported during hGH studies include glucose intolerance [p. 714] and *pancreatitis* (inflammation of the pancreas). Long-term risks associated with hGH treatment are unknown.

23.5 Tissue and Bone Growth

Growth can be divided into two general areas: soft tissue growth and linear bone growth. In children, bone growth is usually assessed by measuring height, and tissue growth by measuring weight. Multiple hormones have direct or permissive effects on growth. In addition, we are just beginning to understand how paracrine growth factors interact with classic hormones to influence tissue development and differentiation.

Tissue Growth Requires Hormones and Paracrine Factors

Soft tissue growth requires adequate amounts of growth hormone, thyroid hormone, and insulin. Growth hormone and IGFs are required for tissue protein synthesis and cell division. Under the influence of these hormones, cells undergo both *hypertrophy* (increased cell size) and **hyperplasia** (increased cell number).

Thyroid hormones play a permissive role in growth and contribute directly to nervous system development. At the target tissue level, thyroid hormone interacts synergistically with growth hormone in protein synthesis and nervous system development. Children with untreated hypothyroidism (cretinism) do not grow to normal height even if they secrete normal amounts of growth hormone.

Insulin supports tissue growth by stimulating protein synthesis and providing energy in the form of glucose. Because insulin is permissive for growth hormone, insulin-deficient children fail to grow normally even though they may have normal concentrations of growth and thyroid hormones.

Bone Growth Requires Adequate Dietary Calcium

Bone growth, like soft tissue development, requires the proper hormones and adequate amounts of protein and calcium. Bones have extensive calcified extracellular matrix, formed when calcium phosphate crystals precipitate and attach to a collagenous lattice support. Although bone looks "dead," spaces in the collagen-calcium matrix are occupied by living cells. Blood vessels running through adjacent channels supply the cells with oxygen and nutrients (**FIG. 23.10**).

Bones of the skeleton come in different shapes and sizes but they generally have two layers: an outer layer of dense **compact bone** and an inner layer of spongy or **trabecular bone**. Compact bone provides strength and is thickest where support is needed (such as in the long bones of the legs) or where muscles attach. Trabecular bone is less sturdy and has open, cell-filled spaces between struts of calcified lattice. In some bones, a central cavity is filled with marrow [Fig. 16.4, p. 518].

Although the large amount of inorganic matrix in bone makes some people think of it as nonliving, bone is a dynamic tissue, constantly being formed and broken down. Bone formation occurs when specialized cells called **osteoblasts** synthesize and deposit matrix. The *resorption* or breakdown of bone takes place when a different set of cells, the **osteoclasts**, secrete acid that dissolves calcified matrix.

Bone diameter increases when matrix deposits on the surface of the bone. Linear growth of long bones in children and adolescents occurs in specialized bands of cartilage called **epiphyseal plates**, located at each end of the *diaphysis* or bone shaft (Fig. 23.10b). The side of the plate closer to the end (*epiphysis*) of the bone contains continuously dividing columns of **chondrocytes**, the collagen-producing cells of cartilage. These chrondrocytes lay down new cartilage and lengthen the bone. At the same time, older chondrocytes closer to the diaphysis die, leaving spaces that osteoblasts invade. The osteoblasts secrete calcium phosphate and a protein mixture called *osteoid* on top of the cartilage base. The combination of calcium phosphate and osteoid creates new bone. When osteoblasts complete their work, they revert to a less active form known as **osteocytes**.

When matrix is added at the ends of long bones, the shaft lengthens. This bone growth continues as long as the epiphyseal plate is active. In adolescents, sex hormones eventually inactivate the epiphyseal plate. Because the epiphyseal plates of various bones close in a regular, ordered sequence, X-rays that show which plates are open and which have closed can be used to calculate a child's "bone age."

CLINICAL FOCUS

New Growth Charts

When you were growing up, did your family mark your growth each year on a special wall chart? Monitoring growth is an important part of health care for children and adolescents, particularly as we see a growing problem with childhood obesity in the United States. In 2000, the U.S. Centers for Disease Control and Prevention (CDC) issued new growth charts for the first time since 1977. In 2006, they recommended that clinicians use an international chart from the World Health Organization for children under two years of age. The old charts were based on 1929–1979 data from mostly bottle-fed, middle-class white children. We now know that breast-fed babies grow more rapidly than bottle-fed infants in the first two months, then more slowly for the remainder of the first year. We also have data showing that babies in lower socioeconomic groups grow more slowly. The new charts take these differences into account and also include body mass index (BMI) information up to age 20. To see the new charts and learn more about monitoring growth in infants and children, visit the CDC website at *www.cdc.gov/growthcharts*.

Linear bone growth ceases after adolescence, but bones undergo continual remodeling throughout life. Most bone turnover in adults takes place in spongy bone, such as that found in vertebra of the spine. The vertebral body has a thin outer layer of compact bone and a large central area of spongy bone, making it one of the most active regions of bone remodeling.

Bone mass in the body is an example of mass balance. In children, bone deposition exceeds bone resorption, and bone mass increases. In young adults up to about age 30, deposition and resorption are balanced. From age 30 on, resorption begins to exceed deposition, with concurrent loss of bone from the skeleton. We discuss bone loss and osteoporosis in more detail at the end of this chapter.

Control of Bone Growth Growth of long bone is under the influence of growth hormone and the insulin-like growth factors. In the absence of these hormones, normal bone growth does not occur. Long bone growth is also influenced by steroid sex hormones. The growth spurt of adolescent boys used to be attributed solely to increased androgen production but it now appears that estrogens play a significant role in pubertal bone growth in both sexes.

One nonendocrine factor that plays an important role in bone mass is mechanical stress on the bone. High-impact exercise, such as running, helps build bone, but non-weight-bearing exercise such as swimming will not. Osteocytes and chondrocytes act as mechanosensors and are able to transduce mechanical

FIG. 23.10 ESSENTIALS Bone Growth

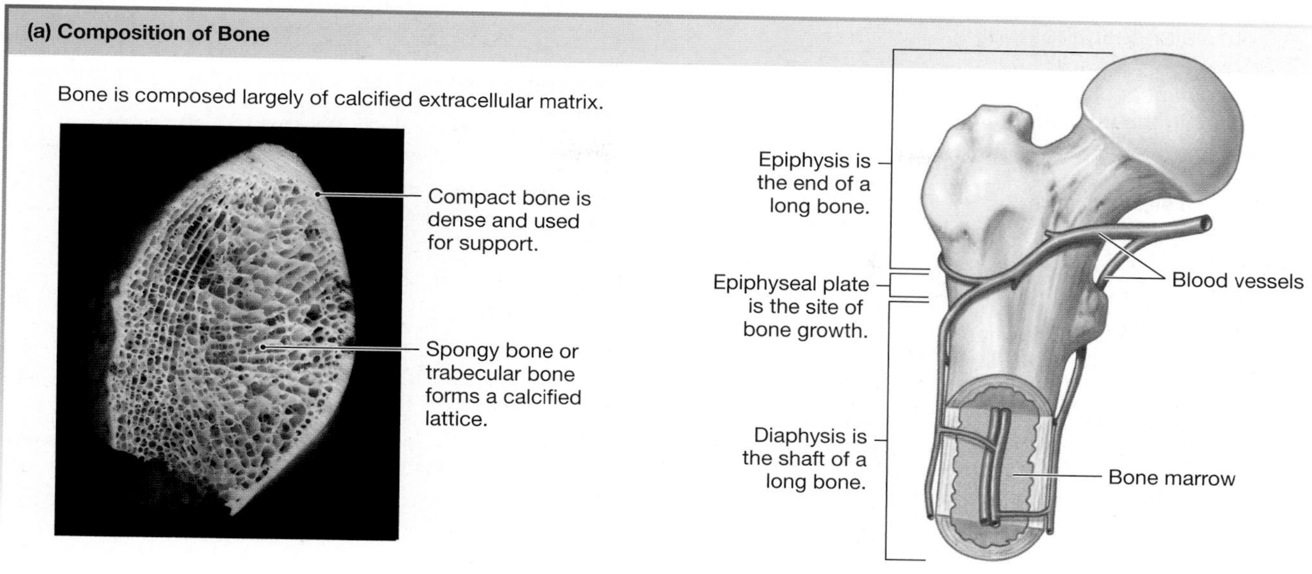
(a) Composition of Bone
Bone is composed largely of calcified extracellular matrix.
Compact bone is dense and used for support.
Spongy bone or trabecular bone forms a calcified lattice.
Epiphysis is the end of a long bone.
Epiphyseal plate is the site of bone growth.
Diaphysis is the shaft of a long bone.
Blood vessels
Bone marrow

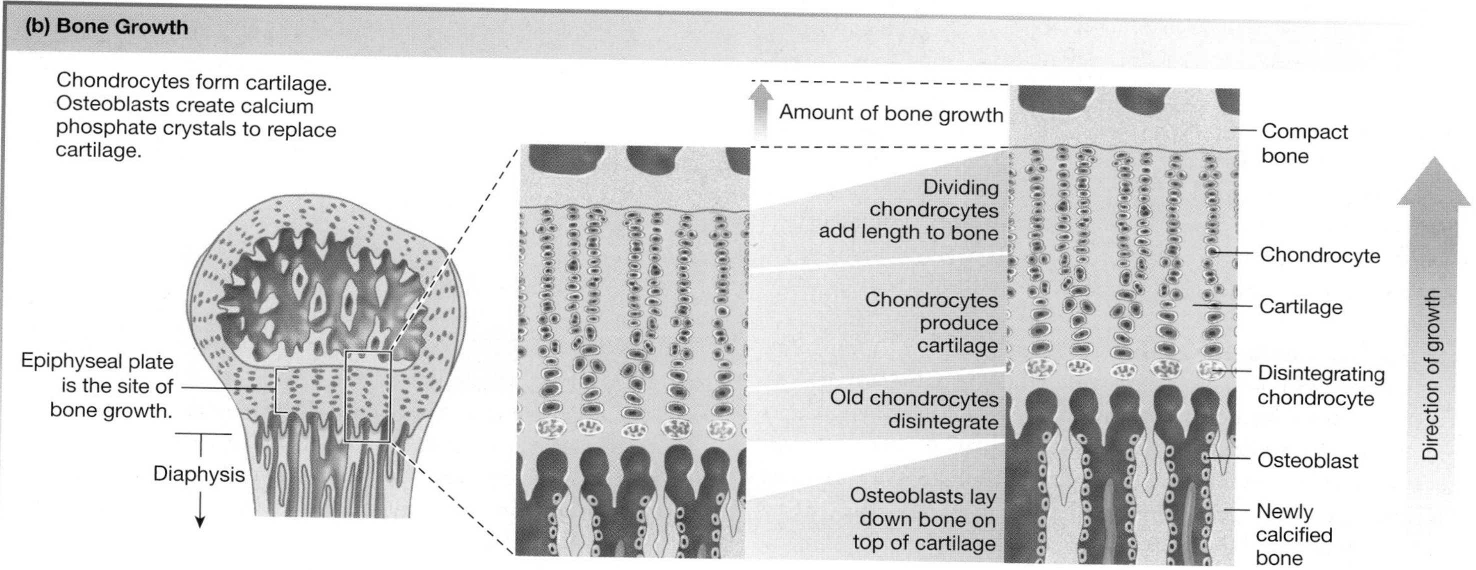
(b) Bone Growth
Chondrocytes form cartilage. Osteoblasts create calcium phosphate crystals to replace cartilage.
Epiphyseal plate is the site of bone growth.
Diaphysis
Amount of bone growth
Dividing chondrocytes add length to bone
Chondrocytes produce cartilage
Old chondrocytes disintegrate
Osteoblasts lay down bone on top of cartilage
Compact bone
Chondrocyte
Cartilage
Disintegrating chondrocyte
Osteoblast
Newly calcified bone
Direction of growth

stimuli into intracellular signals to lay down new bone. It appears that the primary cilia [p. 68] of these cells are the sensory structures responding to mechanical stress, and that signaling to other bone cells takes place through gap junctions connecting the cells. Evidence supporting this hypothesis includes skeletal deformities observed in genetic conditions where primary cilia malfunction (*ciliopathies*).

Concept Check

12. Which hormones are essential for normal growth and development?

13. Why don't adults with growth hormone hypersecretion grow taller?

23.6 Calcium Balance

Most calcium in the body—99%, or nearly 2.5 pounds—is found in the bones. This pool is relatively stable, however, so it is the body's small fraction of non-bone calcium that is most critical to physiological functioning (**FIG. 23.11**). As you have learned, Ca^{2+} has several physiological functions:

1. **Ca^{2+} is an important signal molecule.** The movement of Ca^{2+} from one body compartment to another creates Ca^{2+} signals. Calcium entering the cytoplasm initiates exocytosis of synaptic and secretory vesicles, contraction in muscle fibers, or altered activity of enzymes and transporters. Removal of Ca^{2+} from the cytoplasm requires active transport.
2. **Ca^{2+} is part of the intercellular cement that holds cells together at tight junctions.**

FIG. 23.11 Calcium balance in the body

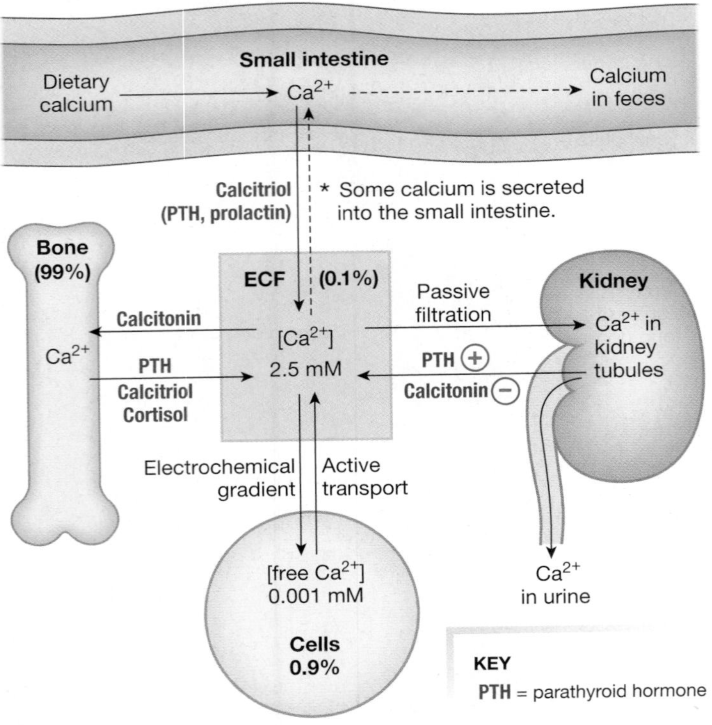
To maintain calcium balance, dietary intake should equal Ca2+ loss in the urine and feces.
Small intestine
Dietary calcium
Ca2+
Calcium in feces
Calcitriol (PTH, prolactin)
* Some calcium is secreted into the small intestine.
Bone (99%)
Ca2+
ECF (0.1%)
Calcitonin
PTH
Calcitriol
Cortisol
[Ca2+] 2.5 mM
Passive filtration
PTH +
Calcitonin −
Kidney
Ca2+ in kidney tubules
Electrochemical gradient
Active transport
[free Ca2+] 0.001 mM
Cells 0.9%
Ca2+ in urine
KEY
PTH = parathyroid hormone

Functions of Calcium in the Body	
Location	**Function**
Extracellular matrix Ca^{2+} (99%)	• Calcified matrix of bone and teeth
Extracellular fluid Ca^{2+} (0.1%)	• Neurotransmitter release at synapse • Role in myocardial and smooth muscle contraction • Cofactor in coagulation cascade • "Cement" for tight junctions • Influences excitability of neurons
Intracellular Ca^{2+} (0.9%)	• Muscle contraction • Signal in second messenger pathways

3. **Ca^{2+} is a cofactor in the coagulation cascade** [p. 523]. Although Ca^{2+} is essential for blood coagulation, body Ca^{2+} concentrations never decrease to the point at which coagulation is inhibited. However, removal of Ca^{2+} from a blood sample will prevent the specimen from clotting in the test tube.
4. **Plasma Ca^{2+} concentrations affect the excitability of neurons.** This function of Ca^{2+} has not been introduced before in this text, but it is the function that is most obvious in Ca^{2+}-related disorders. If plasma Ca^{2+} falls too low (**hypocalcemia**), neuronal permeability to Na^+ increases, neurons depolarize, and the nervous system becomes hyperexcitable. In its most extreme form, hypocalcemia causes sustained contraction (*tetany*) of the respiratory muscles, resulting in asphyxiation. **Hypercalcemia** has the opposite effect, depressing neuromuscular activity. The alterations in Na^+ permeability occur when a G protein-coupled *calcium-sensing receptor* (CaSR) alters regulatory proteins that control gating of a Na^+ leak channel in neurons (the NALCN channel).

Plasma Calcium Is Closely Regulated

Because calcium is critical to so many physiological functions, the body's plasma Ca^{2+} concentration is very closely regulated. Calcium homeostasis follows the principle of mass balance:

Total body calcium = intake − output

RUNNING PROBLEM

Prof. Magruder's blood work reveals that his Ca^{2+} level is 12.3 mg/dL plasma (normal is 8.5–10.5 mg/dL). These results support the suspected diagnosis of hyperparathyroidism. "Do you take vitamin D or use a lot of antacids?" Dr. Spinks asks. "Those could raise your blood calcium." Prof. Magruder denies using either substance. "Well, you'll need one more test before we can say conclusively that you have hyperparathyroidism," Dr. Spinks says.

Q4: *What one test could definitively prove that Prof. Magruder has hyperparathyroidism?*

729 – 731 – 734 – **744** – 750 – 750

1. **Total body Ca^{2+}** is all the calcium in the body, distributed among three compartments (Fig. 23.11):
   (a) *Extracellular fluid.* Ionized Ca^{2+} is concentrated in the ECF. In the plasma, nearly half the Ca^{2+} is bound to plasma proteins and other molecules. The unbound Ca^{2+} is free to diffuse across membranes through open Ca^{2+} channels. Total plasma Ca^{2+} concentration is about 2.5 mM.
   (b) *Intracellular Ca^{2+}.* The concentration of free Ca^{2+} in the cytosol is about 0.001 mM. In addition, Ca^{2+} is

concentrated inside mitochondria and the sarcoplasmic reticulum. Electrochemical gradients favor movement of Ca^{2+} into the cytosol when Ca^{2+} channels open.

(c) *Extracellular matrix (bone)*. Bone is the largest Ca^{2+} reservoir in the body, with most bone Ca^{2+} in the form of calcium phosphate crystals called **hydroxyapatite**, $Ca_{10}(PO_4)_6(OH)_2$. Bone Ca^{2+} forms a reservoir that can be tapped to maintain plasma Ca^{2+} homeostasis. Usually only a small fraction of bone Ca^{2+} is ionized and readily exchangeable, and this pool remains in equilibrium with Ca^{2+} in the interstitial fluid.

2. **Intake** is the Ca^{2+} ingested in the diet and absorbed in the small intestine. Only about one-third of ingested Ca^{2+} is absorbed, and unlike organic nutrients, Ca^{2+} absorption is hormonally regulated. Many people do not eat enough Ca^{2+}-containing foods, however, and intake may not match output.

   Intestinal calcium absorption takes place both between the cells (paracellular transport) and through the cells (**FIG. 23.12A**). In transcellular transport, Ca^{2+} enters the enterocyte through apical Ca^{2+} channels (*ECaC, TRPV6*). Once inside the cell, Ca^{2+} binds to a protein called *calbindin* that helps keep free intracellular $[Ca^{2+}]$ low. This is necessary because of the role of free Ca^{2+} as an intracellular signal molecule. On the basolateral side of the cell, Ca^{2+} exits through basolateral Ca^{2+}-ATPase or Na^+-Ca^{2+} exchangers (NCX). Transcellular absorption is hormonally regulated; paracellular absorption is unregulated.

3. **Output**, or Ca^{2+} loss from the body, occurs primarily through the kidneys, with a small amount excreted in feces. Ionized Ca^{2+} is freely filtered at the glomerulus. Most (90%) of the filtered Ca^{2+} is reabsorbed through paracellular pathways in the proximal tubule and ascending limb of the loop of Henle. Hormonally regulated reabsorption takes place in the distal nephron and uses the same transporters found in the intestine (Fig. 23.12b).

Concept Check

14. What does hypercalcemia do to neuronal membrane potential, and why does that effect depress neuromuscular excitability?
15. Describe the movement of Ca^{2+} from lumen of the nephron or intestine to the ECF as active, passive, facilitated diffusion, and so on.

CHAPTER 23

FIG. 23.12 Intestinal and renal transport of Ca^{2+}

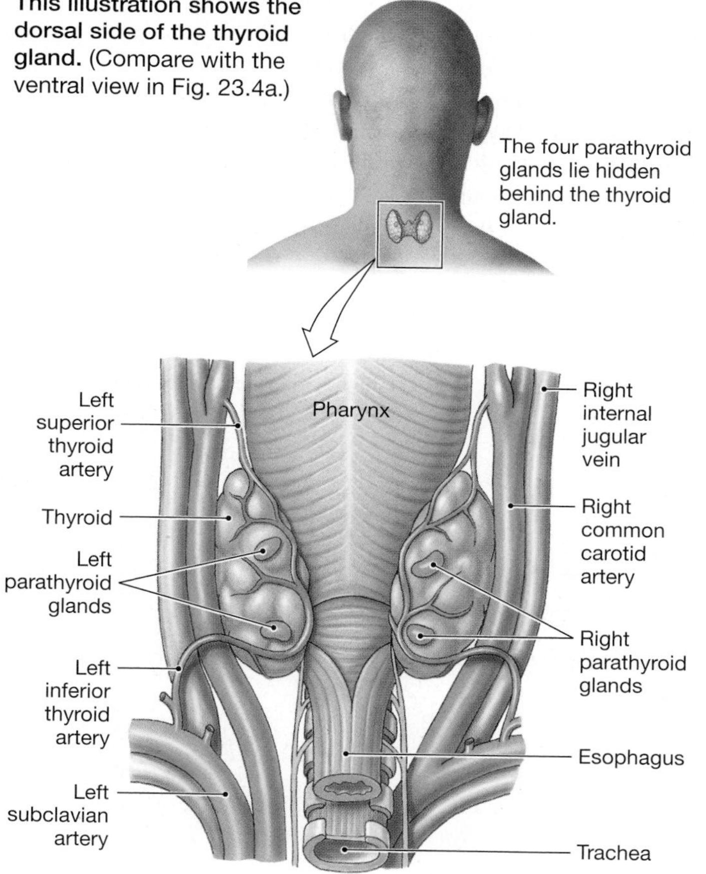
This illustration shows the dorsal side of the thyroid gland. (Compare with the ventral view in Fig. 23.4a.)
The four parathyroid glands lie hidden behind the thyroid gland.
Left superior thyroid artery
Pharynx
Right internal jugular vein
Thyroid
Right common carotid artery
Left parathyroid glands
Right parathyroid glands
Left inferior thyroid artery
Esophagus
Left subclavian artery
Trachea

Parathyroid Hormone (PTH)	
Origin	Parathyroid glands
Chemical Nature	84-amino acid peptide
Biosynthesis	Continuous production, little stored
Transport in the Circulation	Dissolved in plasma
Half-Life	Less than 20 minutes
Factors Affecting Release	↓ Plasma Ca^{2+}
Target Cells or Tissues	Kidney, bone, intestine
Target Receptor	Membrane receptor acts via cAMP
Whole Body or Tissue Reaction	↑ Plasma Ca^{2+}
Action at Cellular Level	↑Vitamin D synthesis; ↑ renal reabsorption of Ca^{2+}; ↑ bone resorption
Action at Molecular Level	Rapidly alters Ca^{2+} transport but also initiates protein synthesis in osteoclasts
Onset of Action	2–3 hours for bone, with increased osteoclast activity requiring 1–2 hours. 1–2 days for intestinal absorption. Within minutes for renal transport
Feedback Regulation	Negative feedback by ↑ plasma Ca^{2+}
Other Information	Osteoclasts have no PTH receptors and are regulated by PTH-induced paracrines. PTH is essential for life. Absence causes hypocalcemic tetany.

Three Hormones Control Calcium Balance

Three hormones regulate the movement of Ca^{2+} between bone, kidney, and intestine: parathyroid hormone, calcitriol (vitamin D_3), and calcitonin (Fig. 23.11). Of these, parathyroid hormone and calcitriol are the most important in adult humans.

The **parathyroid glands**, which secrete parathyroid hormone {*para-*, alongside of}, were discovered in the 1890s by physiologists studying the role of the thyroid gland. These scientists noticed that if they removed all of the thyroid gland from dogs and cats, the animals died in a few days. In contrast, rabbits died only if the little parathyroid "glandules" alongside the thyroid were removed. The scientists then looked for parathyroid glands in dogs and cats and found them tucked away behind the larger thyroid gland. If the parathyroid glands were left behind when the thyroid was surgically removed, the animals lived.

The scientists concluded that the parathyroid glands contained a substance that was essential for life, although the thyroid gland did not. That essential substance was parathyroid hormone. The absence of parathyroid hormones causes hypocalcemic tetany and respiratory paralysis, as mentioned in the section on functions of calcium.

Parathyroid Hormone Four small parathyroid glands lie on the dorsal surface of the thyroid gland (**FIG. 23.13**). They secrete **parathyroid hormone (PTH)** (also called *parathormone*), a peptide whose main function is to increase plasma Ca^{2+} concentrations. The stimulus for PTH release is a decrease in plasma Ca^{2+} monitored by a cell membrane **Ca^{2+}-sensing receptor (CaSR)**. The CaSR, a G protein-coupled receptor, was the first membrane receptor identified whose ligand was an ion rather than an organic molecule.

PTH acts on bone, kidney, and intestine to increase plasma Ca^{2+} concentrations (Fig. 23.12). Increased plasma Ca^{2+} acts as negative feedback and shuts off PTH secretion. Parathyroid hormone raises plasma Ca^{2+} in three ways:

1. **PTH mobilizes calcium from bone.** The complex control of bone remodeling is discussed in the next section.
2. **PTH enhances renal reabsorption of calcium.** As we mentioned previously, regulated Ca^{2+} reabsorption takes place in the distal nephron. PTH simultaneously enhances renal excretion of phosphate by reducing its reabsorption. The opposing effects of PTH on calcium and phosphate are needed to keep their combined concentrations below a critical level. If the concentrations exceed that level, calcium phosphate crystals form and precipitate out of solution. High concentrations of calcium phosphate in the urine are one cause of kidney stones. We discuss additional aspects of phosphate homeostasis later.
3. **PTH indirectly increases intestinal absorption of calcium** through its influence on vitamin D_3, a process described next.

Calcitriol Intestinal absorption of calcium is enhanced by the action of a hormone known as **1,25-dihydroxycholecalciferol** or 1,25$(OH)_2D_3$, also known as **calcitriol** or **vitamin D_3** (**FIG. 23.14**). The body makes calcitriol from vitamin D that has been obtained through diet or made in the skin by the action of sunlight on precursors made from acetyl CoA. People who live above 37 degrees of latitude north or below 37 degrees south do not get enough sunlight to make adequate vitamin D except in the summer, and they should consider taking vitamin supplements.

Vitamin D is modified in two steps—first in the liver, then in the kidneys—to make vitamin D_3 or calcitriol. Calcitriol is the primary hormone responsible for enhancing Ca^{2+} uptake from the small intestine. In addition, calcitriol facilitates renal reabsorption of Ca^{2+} and helps mobilize Ca^{2+} out of bone.

The production of calcitriol is regulated at the kidney by the action of PTH. Decreased plasma Ca^{2+} increases PTH secretion, which stimulates calcitriol synthesis. Intestinal and renal absorption of Ca^{2+} raises blood Ca^{2+}, turning off PTH in a negative feedback loop that decreases calcitriol synthesis.

FIG. 23.13 Parathyroid glands and parathyroid hormone (PTH)

(a) Intestinal absorption is regulated by vitamin D_3.

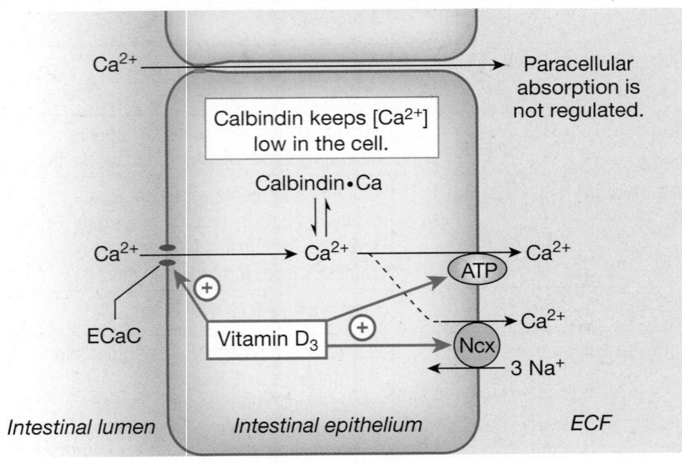
Ca2+
Paracellular absorption is not regulated.
Calbindin keeps [Ca2+] low in the cell.
Calbindin•Ca
Ca2+
Ca2+
Ca2+
ATP
+
Ca2+
ECaC
Vitamin D3
+
Ncx
3 Na+
Intestinal lumen
Intestinal epithelium
ECF

(b) Renal reabsorption in distal tubule is regulated by PTH and vitamin D_3.

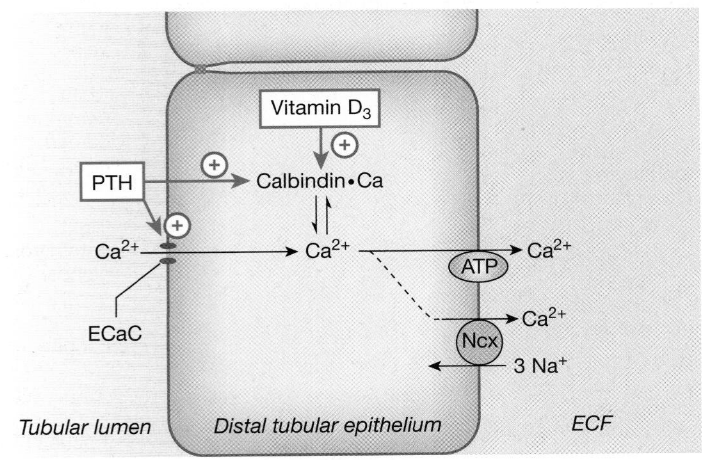
Vitamin D3
+
PTH
+
Calbindin•Ca
+
Ca2+
Ca2+
ATP
Ca2+
Ca2+
ECaC
Ncx
3 Na+
Tubular lumen
Distal tubular epithelium
ECF

FIG. 23.14 Endocrine control of calcium balance

PTH works with calcitriol to promote bone resorption, intestinal Ca^{2+} absorption, and distal nephron Ca^{2+} reabsorption, all of which tend to elevate plasma Ca^{2+} concentrations.

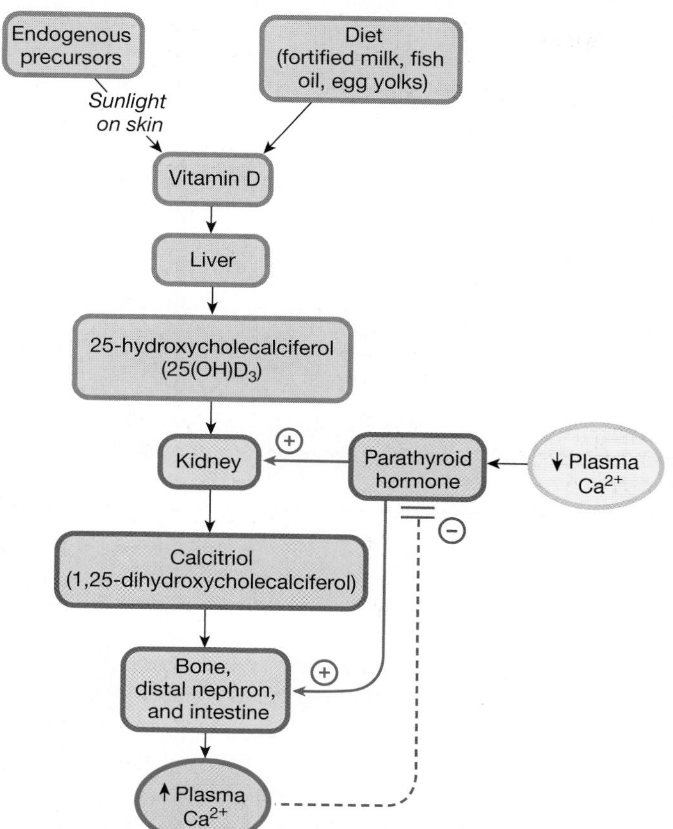
Endogenous precursors
Diet (fortified milk, fish oil, egg yolks)
Sunlight on skin
Vitamin D
Liver
25-hydroxycholecalciferol (25(OH)D3)
Kidney
Parathyroid hormone
↓ Plasma Ca2+
Calcitriol (1,25-dihydroxycholecalciferol)
Bone, distal nephron, and intestine
↑ Plasma Ca2+

Vitamin D_3 (calcitrol, 1,25-dihydroxycholecalciferol)	
Origin	Complex biosynthesis; see below
Chemical Nature	Steroid
Biosynthesis	Vitamin D formed by sunlight on precursor molecules or ingested in food; converted in two steps (liver and kidney) to $1,25(OH)_2D_3$
Transport in the Circulation	Bound to plasma protein
Stimulus for Synthesis	↓ Ca^{2+}. Indirectly via PTH. Prolactin also stimulates synthesis.
Target Cells or Tissues	Kidney, bone, intestine
Target Receptor	Nuclear
Whole Body or Tissue Reaction	↑ Plasma Ca^{2+}
Action at Molecular Level	Stimulates production of calbindin, a Ca^{2+}–binding protein, and of CaSR in parathyroid gland. Associated with intestinal transport by unknown mechanism
Feedback Regulation	↑ Plasma Ca^{2+} shuts off PTH secretion

Prolactin, the hormone responsible for milk production in breast-feeding (lactating) women, also stimulates calcitriol synthesis. This action ensures maximal absorption of Ca^{2+} from the diet at a time when metabolic demands for calcium are high.

Calcitonin The third hormone involved with calcium metabolism is **calcitonin**, a peptide produced by the C cells of the thyroid gland (**TBL. 23.1**). Its actions are opposite to those of parathyroid hormone. Calcitonin is released when plasma Ca^{2+} increases. Experiments in animals have shown that calcitonin decreases bone resorption and increases renal calcium excretion.

Calcitonin apparently plays only a minor role in daily calcium balance in adult humans. Patients whose thyroid glands have been removed show no disturbance in calcium balance, and people with thyroid tumors that secrete large amounts of calcitonin also show no ill effects.

Calcitonin has been used medically to treat patients with *Paget's disease*, a genetically linked condition in which osteoclasts are overactive and bone is weakened by resorption. Calcitonin in these patients stabilizes the abnormal bone loss, leading scientists to speculate that this hormone is most important during childhood growth, when net bone deposition is needed, and during pregnancy and lactation, when the mother's body must supply calcium for both herself and her child.

Multiple Factors Control Bone Remodeling

Bone mass in adults is determined by the relative activity of bone-forming osteoblasts and bone-dissolving osteoclasts, which is turn are controlled by an alphabet soup of hormones, cytokines, and their receptors. If bone resorption is greater than bone deposition, bone mass is lost, leading first to *osteopenia* {*penia*, poverty} and then to *osteoporosis*, which is described later. This section looks at the mechanisms by which normal adult bone mass is maintained.

Osteoblasts Osteoblasts are responsible for production of the calcified matrix of bone (**FIG. 23.15a**). Under the influence of PTH and vitamin D_3, osteoblasts secrete proteins onto the surface of the bone: enzymes, collagen fibers, *osteocalcin*, and *osteonectin*. They also secrete *proteoglycans* [p. 73], the typical glycoprotein of extracellular matrix. The mixture of collagen and other proteins is called **osteoid**.

At the same time, osteoblasts concentrate calcium and phosphate compounds into vesicles, then release the contents into the

TABLE 23.1 Calcitonin

Cell of Origin	C Cells of Thyroid Gland (parafollicular cells)
Chemical nature	32-amino acid peptide
Biosynthesis	Typical peptide
Transport in the circulation	Dissolved in plasma
Half-life	<10 minutes
Factors affecting release	↑ plasma [Ca^{2+}]
Target cells or tissues	Bone and kidney
Target receptor	G protein-coupled membrane receptor
Whole body or tissue action	Prevents bone resorption. Enhances kidney excretion
Action at molecular level	Signal transduction pathways appear to vary during cell cycle
Other information	Experimentally decreases plasma Ca^{2+} but has little apparent physiological effect in adult humans. Possible effect on skeletal development; possible protection of bone Ca^{2+} stores during pregnancy and lactation

extracellular space. The secreted enzymes free Ca^{2+} and PO_4^- from the compounds, resulting in high concentrations of the ions that precipitate into hydroxyapatite crystals. The crystals interacting with osteoid become the mineralized matrix of bone.

Osteoclasts Osteoclasts are the bone-dissolving cells. They are large, mobile, multinucleate cells derived from the same hematopoietic stem cells as macrophages [p. 514]. Mature osteoclasts attach to a section of matrix with tight junctions around their edges, much like a suction cup (Fig. 23.15b).

The central region of the osteoclast secretes hydrochloric acid with the aid of carbonic anhydrase, a chloride channel, and an H^+-ATPase. Osteoclasts also secrete *protease* enzymes that work at low pH. The combination of acid and enzymes dissolves the calcified hydroxyapatite matrix and its collagen support. Ca^{2+} from hydroxyapatite becomes part of the ionized Ca^{2+} pool and can enter the blood. Increased bone resorption by osteoclasts takes about 12 hours to become measurable.

Control of Bone Remodeling Curiously, although osteoclasts are responsible for dissolving the calcified matrix and would be logical targets for PTH trying to raise plasma Ca^{2+}, they do not have PTH receptors. Instead, PTH effects are mediated through a collection of paracrine molecules, including *osteoprotegerin* (OPG) and an osteoclast differentiation factor called *RANKL*. The details are illustrated in Figure. 23.15a.

In bone, PTH receptors are found on the osteoblasts. When PTH activates osteoblasts, they secrete factors that regulate differentiation and activity of the osteoclasts. The primary signal molecule from osteoblasts is called **RANKL**. RANKL binds to **RANK** on osteoclast precursors and mature osteoclasts. (RANK = *receptor for activation of nuclear factor* κβ.) Activated RANK receptors increase acid secretion by mature osteoclasts and promote formation of new osteoclasts from precursor cells.

At the same time that osteoblasts secrete RANKL, they also secrete a molecule, **osteoprotegerin (OPG)**, that binds to RANKL before it can combine with RANK. By adjusting the ratio of RANKL and OPG, osteoblasts are able to control osteoclast activity.

Calcium and Phosphate Homeostasis Are Linked

Phosphate homeostasis is closely linked to calcium homeostasis. Phosphate is the second key ingredient in the hydroxyapatite of bone, $Ca_{10}(PO_4)_6(OH)_2$, and most phosphate in the body is found in bone. However, phosphates have other significant physiological roles, including energy transfer and storage in high-energy phosphate bonds, and activation or deactivation of enzymes, transporters, and ion channels through phosphorylation and dephosphorylation. Phosphates also form part of the DNA and RNA backbone.

Phosphate homeostasis parallels that of Ca^{2+}. Phosphate is absorbed in the intestines, filtered and reabsorbed in the kidneys, and divided between bone, ECF, and intracellular compartments. Vitamin D_3 enhances intestinal absorption of phosphate. Renal excretion is affected by both PTH (which promotes phosphate excretion) and vitamin D_3 (which promotes phosphate reabsorption).

CONCEPT CHECK

16. Name two compounds that store energy in high-energy phosphate bonds.
17. What are the differences between a kinase, a phosphatase, and a phosphorylase?

FIG. 23.15 ESSENTIALS Bone Remodeling

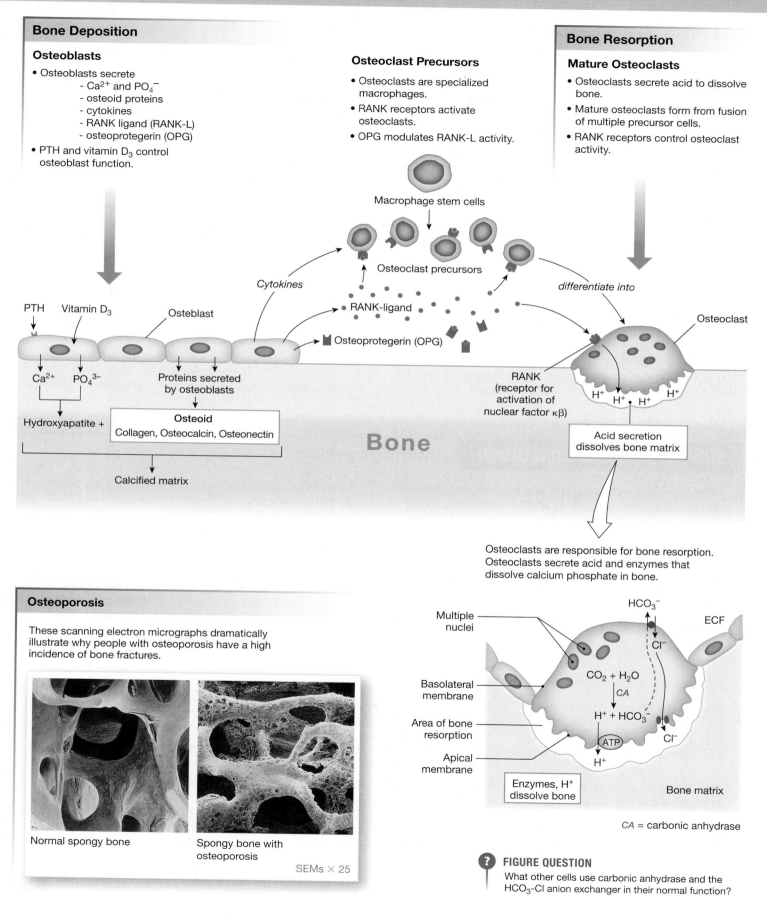

Bone Deposition

Osteoblasts

- Osteoblasts secrete
  - Ca^{2+} and PO_4^-
  - osteoid proteins
  - cytokines
  - RANK ligand (RANK-L)
  - osteoprotegerin (OPG)
- PTH and vitamin D_3 control osteoblast function.

Osteoclast Precursors

- Osteoclasts are specialized macrophages.
- RANK receptors activate osteoclasts.
- OPG modulates RANK-L activity.

Bone Resorption

Mature Osteoclasts

- Osteoclasts secrete acid to dissolve bone.
- Mature osteoclasts form from fusion of multiple precursor cells.
- RANK receptors control osteoclast activity.

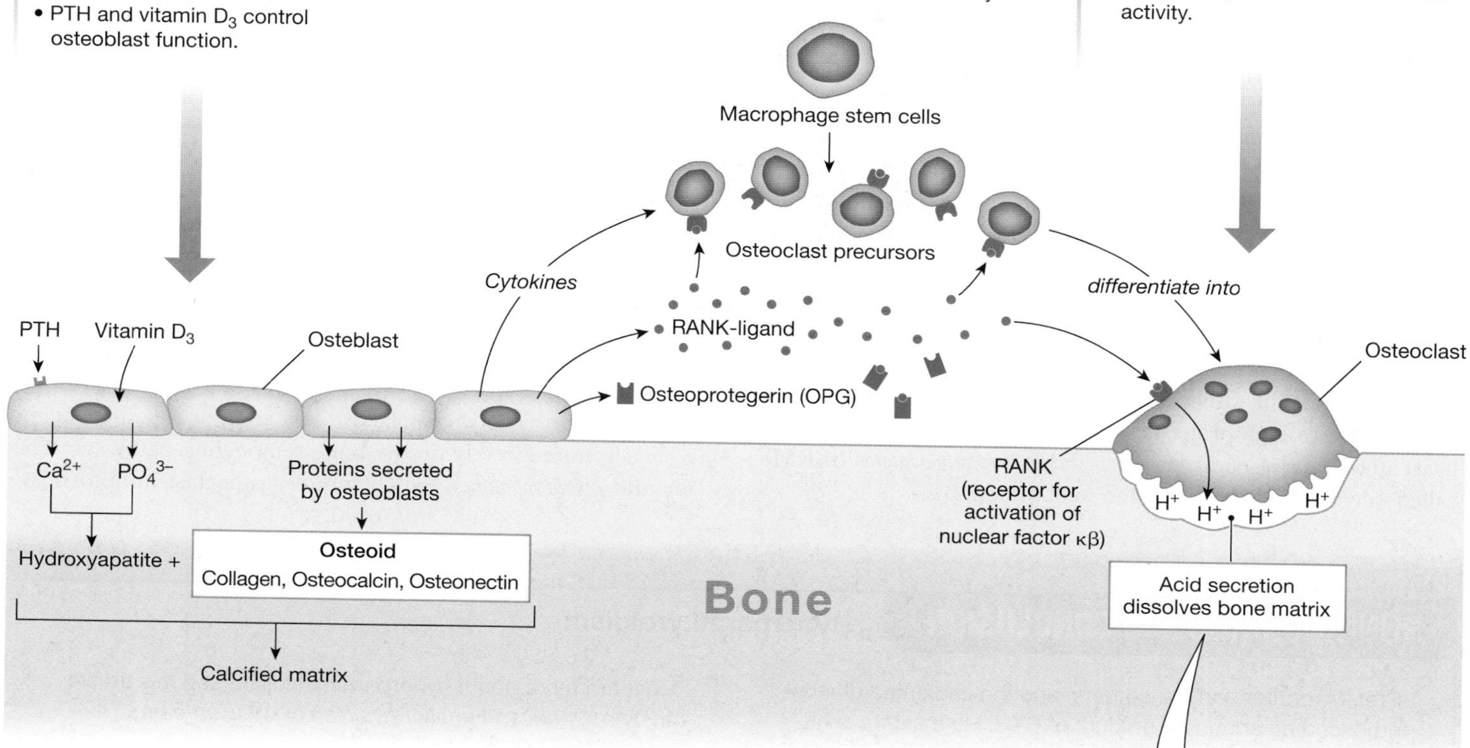

Macrophage stem cells

Osteoclast precursors

Cytokines

RANK-ligand

differentiate into

PTH

Vitamin D_3

Osteblast

Osteoprotegerin (OPG)

Osteoclast

Ca^{2+} PO_4^{3-}

Proteins secreted by osteoblasts

Hydroxyapatite +

Osteoid
Collagen, Osteocalcin, Osteonectin

Calcified matrix

RANK (receptor for activation of nuclear factor κβ)

H^+ H^+ H^+ H^+

Acid secretion dissolves bone matrix

Bone

Osteoclasts are responsible for bone resorption. Osteoclasts secrete acid and enzymes that dissolve calcium phosphate in bone.

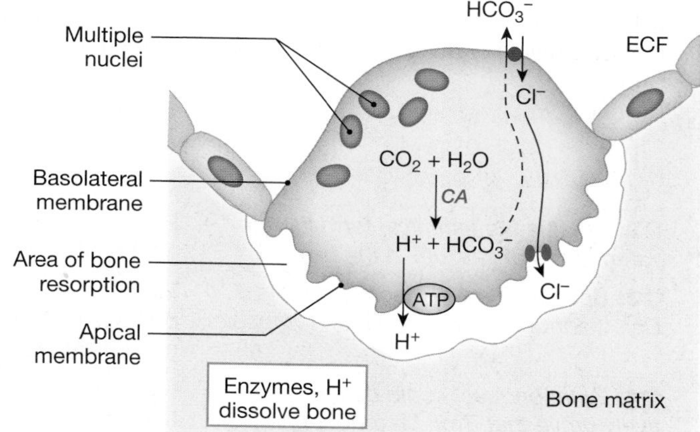

HCO_3^-
ECF
Multiple nuclei
Cl^-
$CO_2 + H_2O$
Basolateral membrane
CA
$H^+ + HCO_3^-$
Area of bone resorption
ATP
Cl^-
Apical membrane
H^+
Enzymes, H^+ dissolve bone
Bone matrix

CA = carbonic anhydrase

Osteoporosis

These scanning electron micrographs dramatically illustrate why people with osteoporosis have a high incidence of bone fractures.

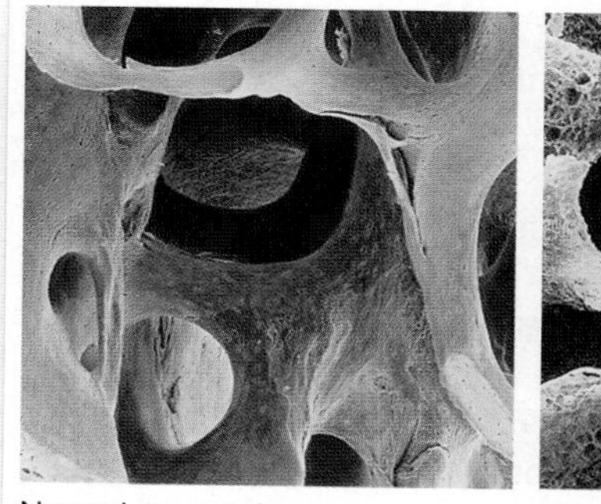

Normal spongy bone

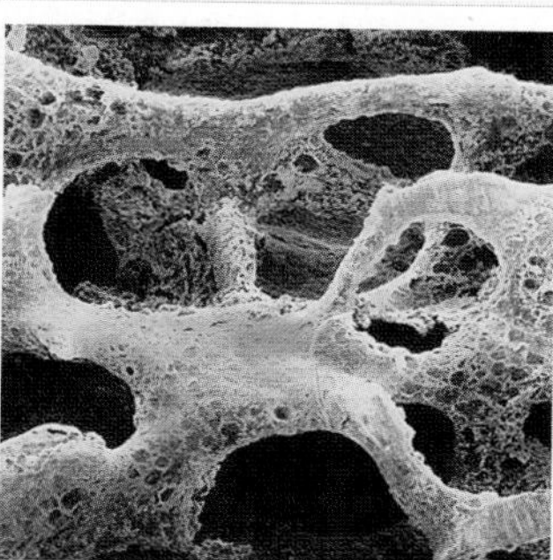

Spongy bone with osteoporosis

SEMs × 25

FIGURE QUESTION

What other cells use carbonic anhydrase and the HCO_3-Cl anion exchanger in their normal function?

Osteoporosis Is a Disease of Bone Loss

One of the best-known pathologies of bone function is **osteoporosis**, a metabolic disorder in which bone resorption exceeds bone deposition. The result is fragile, weakened bones that are more easily fractured (Fig. 23.10c). Most bone resorption takes place in spongy trabecular bone, particularly in the vertebrae, hips, and wrists.

Osteoporosis is most common in women after menopause, when estrogen concentrations fall. However, older men also develop osteoporosis. Bone loss and small fractures and compression in the spinal column lead to *kyphosis* {hump-back}, the stooped, hunchback appearance that is characteristic of advanced osteoporosis in the elderly. Osteoporosis is a complex disease with genetic and environmental components. Risk factors include small, thin body type; postmenopausal age; smoking; and low dietary Ca^{2+} intake.

For many years estrogen or estrogen/progesterone *hormone replacement therapy* (HRT) was recommended to prevent osteoporosis. However, estrogen therapy alone increases the risk of endometrial and possibly other cancers, and some studies suggest that combined estrogen/progesterone HRT might increase risk of heart attacks and strokes. A *selective estrogen receptor modulator* (SERM) called raloxifene has been used to treat osteoporosis.

RUNNING PROBLEM

The results of Prof. Magruder's last test confirm that he has hyperparathyroidism. He goes on a low-calcium diet, avoiding milk, cheese, and other dairy products, but several months later he returns to the emergency room with another painful kidney stone. Dr. Spinks sends him to an endocrinologist, who recommends surgical removal of the overactive parathyroid glands. "We can't tell which of the parathyroid glands is most active," the specialist says, "and we'd like to leave you with some parathyroid hormone of your own. So I will take out all four glands, but we'll reimplant two of them in the muscle of your forearm. In many patients, the implanted glands secrete just enough PTH to maintain calcium homeostasis. And if they secrete too much PTH, it is much easier to take them out of your arm than do major surgery on your neck again."

Q5: *Why can't Prof. Magruder simply take replacement PTH by mouth?* (Hint: *PTH is a peptide hormone.*)

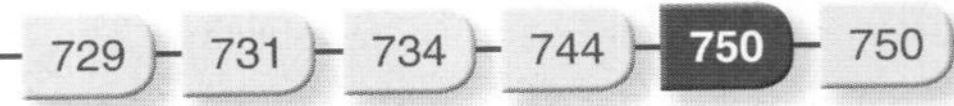
729
731
734
744
750
750

The most effective drugs for preventing or treating osteoporosis act more directly on the bone remodeling pathways. They include *bisphosphonates*, which induce osteoclast apoptosis and

RUNNING PROBLEM CONCLUSION Hyperparathyroidism

Prof. Magruder had the surgery, and the implanted glands produced an adequate amount of PTH. He must have his plasma Ca^{2+} levels checked regularly for the rest of his life to ensure that the glands continue to function adequately.

To learn more about hyperparathyroidism, see this article in *The American Family Physician* at *www.aafp.org/afp/20040115/333.html.*

Question	Facts	Integration and Analysis
Q1: *What role does Ca^{2+} play in the normal functioning of muscles and neurons?*	Calcium triggers neurotransmitter release [p. 254] and uncovers myosin-binding sites on muscle actin filaments [p. 383].	Muscle weakness in hyperparathyroidism is the opposite of what you would predict from knowing the role of Ca^{2+} in muscles and neurons. However, Ca^{2+} also affects the Na^+ permeability of neurons, and this effect leads to muscle weakness and CNS effects.
Q2: *What is the technical term for "elevated levels of calcium in the blood"?*	Prefix for elevated levels: *hyper-*. Suffix for "in the blood": *-emia.*	Hypercalcemia
Q3: *Speculate on why some plasma Ca^{2+} cannot filter into Bowman's capsule.*	Filtration at the glomerulus is a selective process that excludes blood cells and most plasma proteins [p. 598].	A significant amount of plasma Ca^{2+} is bound to plasma proteins and therefore cannot filter.
Q4: *What one test could definitively prove that Prof. Magruder has hyperparathyroidism?*	Hyperparathyroidism is excessive secretion of PTH.	A test for PTH concentration in the blood would confirm the diagnosis of hyperparathyroidism.
Q5: *Why can't Prof. Magruder simply take replacement PTH by mouth?*	PTH is a peptide hormone. Ingested peptides are digested by proteolytic enzymes.	PTH taken orally will be digested and not absorbed intact. Consequently, it will not be effective.

729 – 731 – 734 – 744 – 750 – 750

suppress bone resorption, and *teriparatide*, a PTH derivative, which stimulates formation of new bone. Teriparatide consists of the first 34 amino acids of the 84-amino acid PTH molecule and must be injected rather than taken orally. Currently clinical studies are investigating whether some combination of bisphosphonates and teriparatide is more effective in combating osteoporosis than either drug alone. In late 2010, a RANKL inhibitor named *denosumab* was approved for use as treatment for conditions with excessive bone loss.

Fractures from osteoporosis are a significant concern in the elderly population because the mortality rate from a fractured hip is about 50% in the first year after the fracture. To avoid osteoporosis in later years, young women need to maintain adequate dietary calcium intake and perform weight-bearing exercises, such as running or aerobics, which increase bone density. Loss of bone mass begins by age 30, long before people think they are at risk, and many women suffer from low bone mass (*osteopenia*) before they are aware of a problem. Bone mass testing can help with early diagnosis of osteopenia.

CHAPTER SUMMARY

Endocrinology is based on the physiological principles of *homeostasis* and *control systems*. Each hormone has stimuli that initiate its secretion, and feedback signals that modulate its release. *Molecular interactions* and *communication across membranes* are also essential to hormone activity. In many instances, such as calcium and phosphate homeostasis, the principle of *mass balance* is the focus of homeostatic regulation.

23.1 Review of Endocrine Principles

1. Basic components of endocrine pathways include hormone receptors, feedback loops, and cellular responses. (p. 729)

23.2 Adrenal Glucocorticoids

2. The **adrenal cortex** secretes **glucocorticoids**, sex steroids, and aldosterone. (p. 729; Fig. 23.1)
3. **Cortisol** secretion is controlled by hypothalamic **CRH** and **ACTH** from the pituitary. Cortisol is the feedback signal. Cortisol is a typical steroid hormone in its synthesis, secretion, transport, and action. (p. 729; Fig. 23.2)
4. Cortisol is catabolic and essential for life. It promotes gluconeogenesis, breakdown of skeletal muscle proteins and adipose tissue, Ca^{2+} excretion, and suppression of the immune system. (p. 731)
5. **Hypercortisolism** usually results from a tumor or therapeutic administration of the hormone. **Addison's disease** is hyposecretion of all adrenal steroids. (p. 733)
6. CRH and the **melanocortins** have physiological actions in addition to cortisol release. (p. 734; Fig. 23.2d)

23.3 Thyroid Hormones

7. The thyroid **follicle** has a hollow center filled with **colloid** containing **thyroglobulin** and enzymes. (p. 736; Fig. 23.4b)
8. Thyroid hormones are made from tyrosine and iodine. **Tetraiodothyronine** (thyroxine, T_4) is converted in target tissues to the more active hormone **triiodothyronine** (T_3). (p. 736; Fig. 23.4)
9. Thyroid hormones are not essential for life, but they influence metabolic rate as well as protein, carbohydrate, and fat metabolism. (p. 736)
10. Thyroid hormone secretion is controlled by **thyrotropin thyroid-stimulating hormone** (TSH) and **thyrotropin-releasing hormone (TRH)**. (p. 736; Fig. 23.5)

23.4 Growth Hormone

11. Normal growth requires growth hormone, thyroid hormones, insulin, and sex hormones at puberty. Growth also requires adequate diet and absence of chronic stress. (p. 739)
12. Growth hormone is secreted by the anterior pituitary and stimulates secretion of **insulin-like growth factors (IGFs)** from the liver and other tissues. These hormones promote bone and soft tissue growth. (p. 740; Fig. 23.8)
13. Secretion of growth hormone is controlled by **growth hormone-releasing hormone (GHRH)** and growth hormone-inhibiting hormone (**somatostatin**). (p. 741; Fig. 23.8)

23.5 Tissue and Bone Growth

14. **Bone** is composed of **hydroxyapatite** crystals attached to a collagenous support. Bone is a dynamic tissue with living cells. (p. 742)
15. **Osteoblasts** synthesize bone. Long bone growth occurs at **epiphyseal plates**, where **chondrocytes** produce cartilage. (p. 742; Fig. 23.10)

23.6 Calcium Balance

16. Calcium acts as an intracellular signal for second messenger pathways, exocytosis, and muscle contraction. It also plays a role in cell junctions, coagulation, and neural function. (p. 743)
17. Ca^{2+} homeostasis balances dietary intake, urinary output, and distribution of Ca^{2+} among bone, cells, and the ECF. (p. 744; Fig. 23.11)
18. Decreased plasma Ca^{2+} stimulates **parathyroid hormone (PTH)** secretion by the **parathyroid glands**. (p. 747; Fig. 23.13)
19. PTH promotes Ca^{2+} resorption from bone, enhances renal Ca^{2+} reabsorption, and increases intestinal Ca^{2+} absorption through its effect on **calcitriol**. (p. 747; Fig. 23.12)
20. Calcitonin from the thyroid gland plays only a minor role in daily calcium balance in adult humans. (p. 747; Tbl. 23.1)
21. Bone mass homeostasis is a balance between the activity of osteoblasts and osteoclasts, mediated by hormones, RANKL, and osteoprotegerin. (p. 747; Fig. 23.15)

24 The Immune System

Virus particles

Although, at first sight, the immune system may appear to be autonomous, it is connected by innumerable structural and functional bridges with the nervous system and the endocrine system, so as to constitute a multisystem.

Branislav D. Jankovic, Neuroimmunomodulation: The State of the Art, *1994*

BACKGROUND BASICS

"Laughter is the best medicine." This old saying seemed silly for many years, when our attention was focused on the complex interactions between the cells and chemicals of the immune system. But even as our list of immune chemicals grows, scientists are recognizing that the immune system is only one part of a complex communication network that includes the nervous and endocrine systems. The brain-immune connection has even been given its own name: *psychoneuroimmunology*.

The connections between the immune system and other body systems are becoming more important in our understanding of physiology. For example, we have learned that bacteria are not always harmful, and that the commensal bacteria permitted by our immune system to live inside our gut play an important role in metabolism. Inflammation created by the immune system is a factor in many disease states, including atherosclerosis [p. 501]. And on the flip side, the brain can assist or derail the immune system, depending on our psychological state. In this chapter we examine the key processes by which the immune system responds to and reduces challenges to homeostasis.

RUNNING PROBLEM **HPV: To Vaccinate or Not?**

Rebecca and her daughter Lizzie arrived at the doctor's office for Lizzie's annual back-to-school checkup. Lizzie is 12 years old and will be in sixth grade in the fall. Halfway through the appointment, Dr. Paul asked Rebecca if Lizzie had started the series of vaccinations to protect her against HPV, the human papillomavirus. "Isn't HPV the virus that causes cervical cancer?" Rebecca asked. When the doctor confirmed that it was, Rebecca said, "No, Lizzie doesn't need that yet. She's only 12 and she's a good girl." "Let's talk about it," Dr. Paul suggested. "The American Academy of Pediatrics, the American Cancer Society, The American Congress of Obstetricians and Gynecologists, and the Centers for Disease Control and Prevention all recommend this vaccine for girls and boys, and now is the time to start it."

755 – 762 – 771 – 773 – 780 – 782

24.1 Overview

The immune system is a physiological system whose primary job is to protect the body from damage. The body's ability to protect itself is known as **immunity**, from the Latin word *immunis*, meaning *exempt*. The immune system carries out its functions by distinguishing "self"—the body's normal cells—from "nonself." Nonself includes viruses, bacteria, parasites, allergens, and other disease-causing **pathogens** {*pathos*, suffering, disease} in addition to any of our own cells that have become defective and threaten to do harm, such as become cancer.

The body's first line of defense against external pathogens includes physical, chemical, and mechanical barriers, such as skin, tears, mucus, and stomach acid (**FIG. 24.1**). These protective barriers attempt to keep pathogens from entering the extracellular fluid but if pathogens evade them, the body initiates an immediate internal immune response with four basic steps:

1. *detection* and *identification* of the pathogen,
2. *communication* with other immune cells to rally an organized response,
3. *recruitment* of assistance and *coordination* of the response among all participants, and
4. *destruction* or *suppression* of the pathogen.

Substances that trigger the body's immune response are called *immunogens*. Immunogens that react with products of the immune response are known as **antigens**.

The internal immune response is carried out by *leukocytes* [p. 513], and is heavily dependent on cell-to-cell communication. Chemical communication includes substances released by damaged or dying cells as well as *cytokines*, protein signal molecules released by one cell that affect the growth or activity of another cell [p. 167]. The immune system is also the primary user of *contact-dependent signaling* that occurs when surface receptors on one cell recognize and bind to surface receptors on another cell [p. 165].

The internal immune response can be divided into two phases: a rapid *innate response* and a slower *adaptive response*. **Innate immunity** is present from birth {*innatus*, inborn} and is the body's **immediate immune response** to invasion. The innate immune response is not specific to any one pathogen, so it begins within minutes to hours. (Ever get a mosquito bite?) **Inflammation**, visible on the skin as a red, warm, swollen area, is a classic sign of innate immunity. An innate immune response to a pathogen is not remembered by the immune system and must be triggered anew with each exposure.

The cells responsible for the rapid innate response are circulating and stationary leukocytes that are genetically programmed to respond to a broad range of material that they identify as foreign. For example, one molecular signal that is both unique and common to a group of pathogenic bacteria is a component of the bacterial cell wall. When certain types of leukocytes called *phagocytes* identify the bacterium as a pathogen, they ingest it via *phagocytosis* [p. 146] and digest it. Some types of phagocytes then display bits of digested pathogen on their cell surface to attract cells involved in the adaptive immune response. The cells that display pathogen this way are called **antigen-presenting cells** or **APCs**.

Adaptive immunity (also called *acquired immunity*) is directed at particular invaders and is the body's **specific immune response**. One characteristic of adaptive immunity is that the steps needed to launch a specific immune response following first exposure to a pathogen may take days to weeks. However, upon reexposure certain immune cells called *memory cells* "remember" their prior exposure to the pathogen and react more rapidly.

Adaptive immunity can be divided into cell-mediated immunity and *antibody-mediated* immunity. *Cell-mediated immunity* requires contact-dependent signaling between an immune cell and receptors on its target cell. *Antibody-mediated immunity*, also known as *humoral immunity*, uses **antibodies**, proteins secreted by immune cells, to carry out the immune response. Antibodies bind to foreign substances to disable them or make them more visible to the cells of the immune system. (The term *humoral*, referring to the blood,

FIG. 24.2 ANATOMY SUMMARY The Immune System

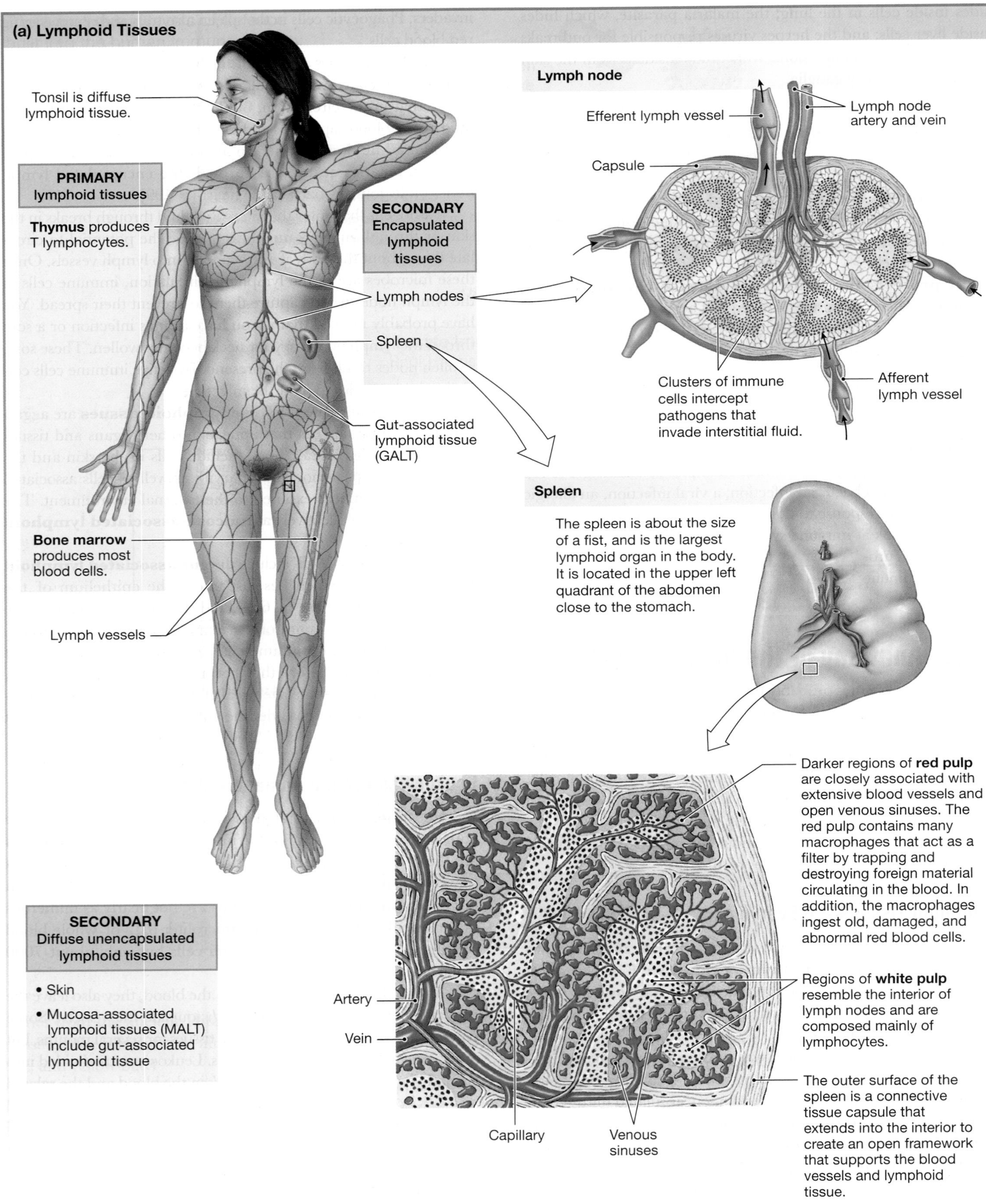

(a) Lymphoid Tissues
Tonsil is diffuse lymphoid tissue.
PRIMARY lymphoid tissues
Thymus produces T lymphocytes.
Bone marrow produces most blood cells.
Lymph vessels
SECONDARY Encapsulated lymphoid tissues
Lymph nodes
Spleen
Gut-associated lymphoid tissue (GALT)
SECONDARY Diffuse unencapsulated lymphoid tissues
• Skin
• Mucosa-associated lymphoid tissues (MALT) include gut-associated lymphoid tissue
Lymph node
Efferent lymph vessel
Lymph node artery and vein
Capsule
Clusters of immune cells intercept pathogens that invade interstitial fluid.
Afferent lymph vessel
Spleen
The spleen is about the size of a fist, and is the largest lymphoid organ in the body. It is located in the upper left quadrant of the abdomen close to the stomach.
Darker regions of red pulp are closely associated with extensive blood vessels and open venous sinuses. The red pulp contains many macrophages that act as a filter by trapping and destroying foreign material circulating in the blood. In addition, the macrophages ingest old, damaged, and abnormal red blood cells.
Regions of white pulp resemble the interior of lymph nodes and are composed mainly of lymphocytes.
The outer surface of the spleen is a connective tissue capsule that extends into the interior to create an open framework that supports the blood vessels and lymphoid tissue.
Artery
Vein
Capillary
Venous sinuses

Leukocytes can be distinguished from one another in stained tissue samples by the shape and size of the nucleus, the staining characteristics of the cytoplasm and cytoplasmic inclusions, and the regularity of the cell border.

Immune Cell Names The terminology associated with immune cells can be very confusing. Some cell types have several variants, and others have been given multiple names for historical reasons (Fig. 24.2c). One morphological group of leukocytes is the **granulocytes**, white blood cells whose cytoplasm contains prominent granules. Granulocytes include the basophils, eosinophils, and neutrophils. The cell names come from the staining properties of the granules. Basophil granules stain dark blue with basic (alkaline) dye, and eosinophil granules stain dark pink with the acidic dye *eosin* {Eos, Greek goddess of the dawn}. Neutrophil granules do not stain darkly with standard blood stains and are therefore "neutral." In all three types of granulocytes, the activated leukocyte releases its granule contents by exocytosis, a process known as *degranulation*.

Two functional groups of leukocytes are the phagocytes and the antigen-presenting cells. *Phagocytes* ingest material from the extracellular fluid using a large vesicle [p. 146] and include neutrophils, macrophages, dendritic cells. *Antigen-presenting cells*, or APCs, have the ability to display bits of antigen on their surface as a signal to other immune cells. The primary APCs are the macrophages and dendritic cells.

Basophils and Mast Cells **Basophils** are rare in the circulation but are easily recognized in a stained blood smear by the large, dark blue granules in their cytoplasm. They are related to the **mast cells** of tissues. Mast cells are concentrated in the connective tissue of skin, lungs, and the gastrointestinal tract: locations where they are ideally situated to intercept pathogens. Both basophils and mast cells release chemicals that contribute to inflammation and the innate immune response.

Eosinophils **Eosinophils** are easily recognized by the bright pink-staining granules in their cytoplasm. Normally, few eosinophils are found in the peripheral circulation. Most functioning eosinophils are found in the digestive tract, lungs, urinary and genital epithelia, and connective tissue of the skin. These locations reflect their role in defense against parasitic invaders. Eosinophils attach to large antibody-coated parasites, such as the blood fluke *Schistosoma*, and release substances from their granules that damage or kill the parasites. Eosinophils also participate in allergic reactions, where they contribute to inflammation and tissue damage by releasing toxic enzymes and oxidative substances.

Neutrophils **Neutrophils** are phagocytic cells that typically ingest and kill 5–20 bacteria during their short, programmed life span of one or two days. They are the most abundant white blood cells (50–70% of the total) and are most easily identified by a segmented nucleus made up of three to five lobes connected by thin strands of nuclear material (Fig. 24.2c). Because of the segmented nucleus, neutrophils are also called *polymorphonuclear leukocytes* ("*polys*") and *"segs."* Immature neutrophils, occasionally found in the circulation, can be identified by their horseshoe-shaped nucleus. These immature neutrophils go by the nicknames of *"bands"* and *"stabs."*

Neutrophils, like other blood cells, are formed in the bone marrow and released into the circulation. Most neutrophils remain in the blood but can leave the circulation if attracted to an extravascular site of damage or infection. In addition to ingesting bacteria and foreign particles, neutrophils release a variety of cytokines, including fever-causing *pyrogens* [p. 723] and chemical mediators of the inflammatory response.

Monocytes and Macrophages **Monocytes** are the precursor cells of tissue macrophages. Monocytes are not very common in the blood (1–6% of all white blood cells). By some estimates, they spend only eight hours there in transit from the bone marrow to their permanent positions in the tissues.

Once in the tissues, monocytes enlarge and differentiate into phagocytic **macrophages**. Some tissue macrophages patrol the tissues, creeping along by amoeboid motion. Others find a location and remain fixed in place. In either case, macrophages are the primary scavengers within tissues. They are larger and more effective than neutrophils, ingesting up to 100 bacteria during their life span. Macrophages also remove larger particles, such as old red blood cells and dead neutrophils.

The terminology associated with macrophages has changed over the history of histology and immunology. For many years, tissue macrophages were known as the *reticuloendothelial system* and were not associated with white blood cells. To confuse matters, the cells were named when they were first described in different tissues, before they were all identified as macrophages. For this reason, *histiocytes* in skin, *Kupffer cells* in the liver, *osteoclasts* in bone, *microglia* in the brain, and *reticuloendothelial cells* in the spleen are all names for specialized macrophages. The new name for the reticuloendothelial system is the **mononuclear phagocyte system**, a term that refers both to macrophages in the tissues and to their parent monocytes circulating in the blood.

Dendritic Cells **Dendritic cells** are macrophage relatives characterized by long, thin processes that resemble the dendrites of neurons. Dendritic cells are found in the skin (where they are called *Langerhans cells*) and in various organs. They play a key role in linking innate and adaptive immune responses by displaying bits of foreign antigen that they have ingested and processed.

Lymphocytes **Lymphocytes** and their derivative **plasma cells** are the key cells that mediate the specific adaptive immune response of the body. By one estimate, the adult body contains a trillion lymphocytes at any one time. Only 5% of these are found in the circulation, where they constitute 20–35% of all white blood cells. Most lymphocytes are found in lymphoid tissues, where they are especially likely to encounter invaders. Although lymphocytes all look alike under the microscope, there are three major sub-types with significant differences in function and specificity, as you will learn later.

Play Interactive Physiology 2.0 @Mastering Anatomy & Physiology

CHAPTER 24

24.3 Development of Immune Cells

All white blood cells begin development, or *hematopoiesis*, in the bone marrow (**FIG. 24.3**) under the influence of cytokines called *colony-stimulating factors* and *interleukins* [p. 514]. During embryonic development, one set of immature lymphocyte precursor cells, the **T lymphocytes (T cells)**, migrates from the bone marrow to the thymus gland, where they mature (**FIG. 24.4**). Another group known as **B lymphocytes (B cells)** remains in the bone marrow. (Memory aid: bone starts with B and thymus starts with T.) **Natural killer** or **NK cells** form a third category of lymphocytes. They are thought to develop in bone marrow as well as in other tissues.

Activated B lymphocytes differentiate primarily into specialized plasma cells that secrete antibodies. The word *antibody* describes what the molecules do: work against foreign bodies (antigens). Antibodies are also called **immunoglobulins**, and this alternative name describes what the molecules are: globular proteins that participate in the adaptive immune response. *T lymphocytes* and NK cells play important roles in defense against intracellular pathogens, such as viruses. Recently an NK cell-related lymphocyte, the *innate lymphoid cell* (ILC), was described but little is known about it at this time and we will not discuss it further.

FIG. 24.3 Development of immune cells

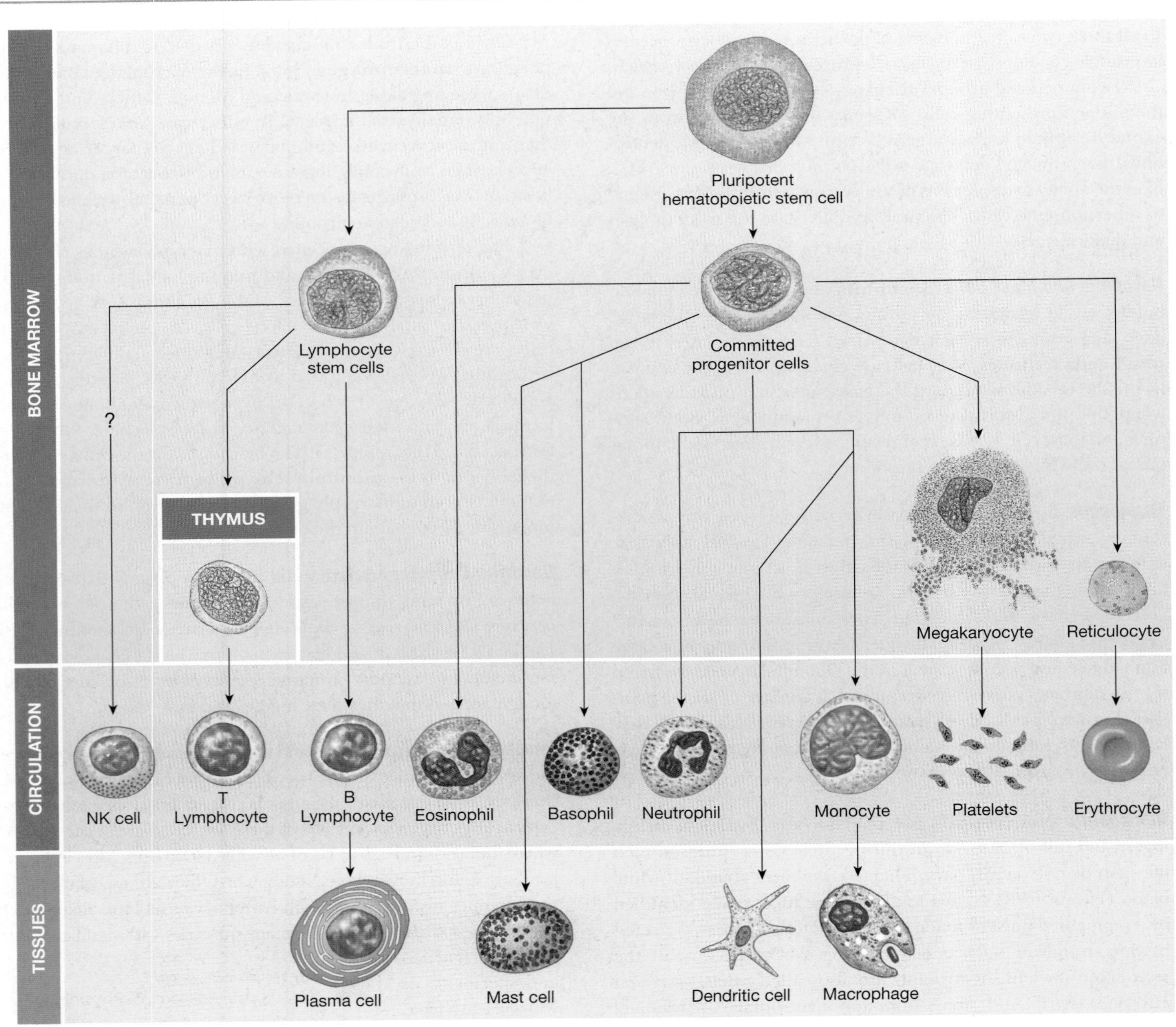
BONE MARROW
CIRCULATION
TISSUES
Pluripotent hematopoietic stem cell
Lymphocyte stem cells
Committed progenitor cells
?
THYMUS
Megakaryocyte
Reticulocyte
NK cell
T Lymphocyte
B Lymphocyte
Eosinophil
Basophil
Neutrophil
Monocyte
Platelets
Erythrocyte
Plasma cell
Mast cell
Dendritic cell
Macrophage

FIG. 24.4 FOCUS ON . . . The Thymus Gland

The thymus gland is a two-lobed organ located in the thorax just above the heart.

The thymus gland reaches its greatest size during adolescence. Then it shrinks and is largely replaced by adipose tissue as a person ages.

During development in the thymus, those cells that would be self-reactive are eliminated. Those that do not react with "self" tissues multiply to form clones.

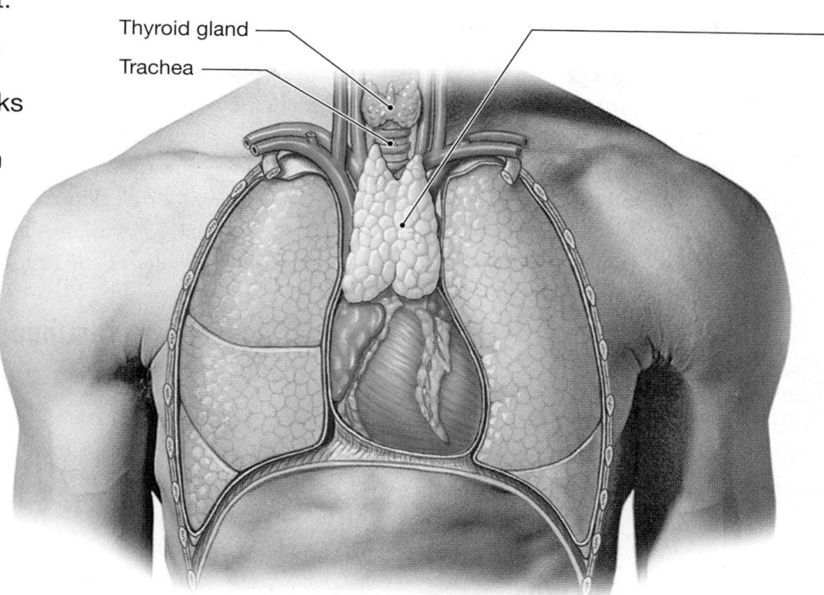

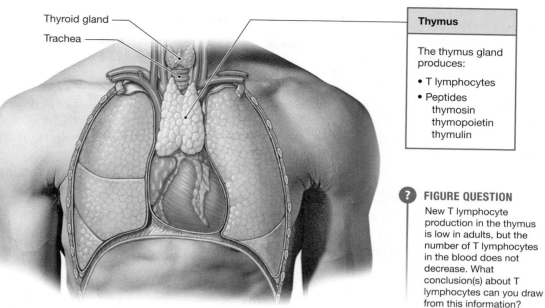

Thyroid gland
Trachea

Thymus

The thymus gland produces:

- T lymphocytes
- Peptides
  thymosin
  thymopoietin
  thymulin

FIGURE QUESTION

New T lymphocyte production in the thymus is low in adults, but the number of T lymphocytes in the blood does not decrease. What conclusion(s) about T lymphocytes can you draw from this information?

Lymphocytes Mediate the Adaptive Immune Response

Lymphocytes are the primary effector cell for the antigen-specific responses of adaptive immunity. On a microscopic level, all lymphocytes look alike. At the molecular level, however, the different cell types can be distinguished from one another by their membrane receptors. Each B and T lymphocyte binds only one particular antigen. All lymphocytes that bind that particular antigen form a group known as a **clone** {*klon*, a twig}.

If every pathogen that enters the body needs a dedicated type of lymphocyte, there must be millions of different types of lymphocytes ready to combat millions of different pathogens. But how can the body store both the number and variety of lymphocytes needed for adequate defense? As it turns out, the immune system keeps only a few cells of each lymphocyte clone on hand. If a pathogen appears, the clone whose cell receptors match the pathogen quickly reproduces to provide the additional cells needed, a process described in more detail later.

The Immune System Must Recognize "Self"

Before attempting any other function, the immune system must accomplish one task first – distinguish its own cells from foreign

FIG. 24.5 Development of self-tolerance

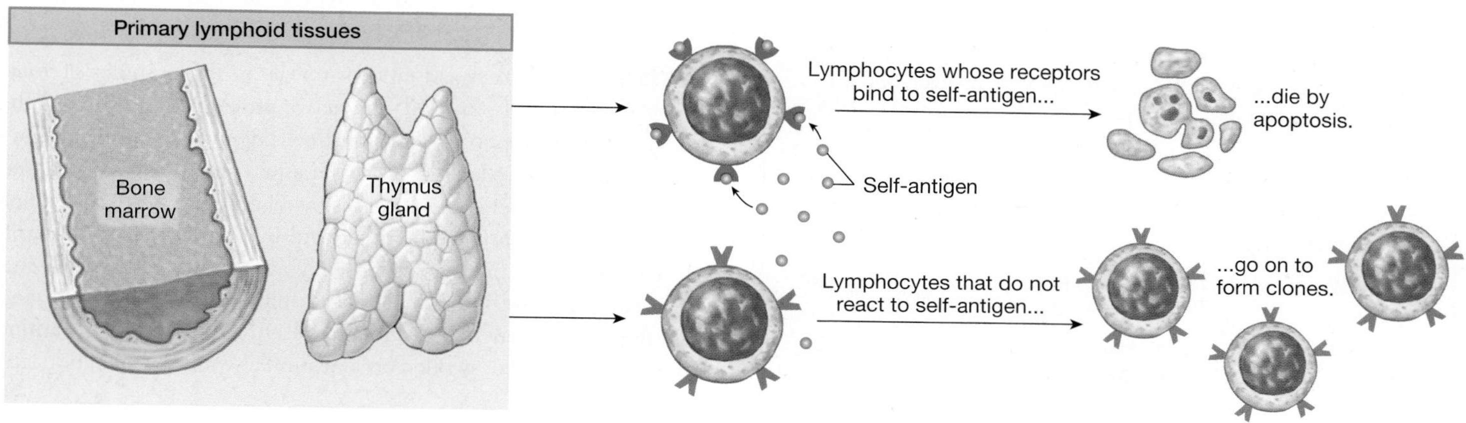

During embryonic development, lymphocytes insert their receptors into the membrane.
Primary lymphoid tissues
Bone marrow
Thymus gland
Lymphocytes whose receptors bind to self-antigen...
...die by apoptosis.
Self-antigen
Lymphocytes that do not react to self-antigen...
...go on to form clones.

cells. The lack of immune response by lymphocytes to cells of the body is known as **self-tolerance**, and it begins during embryonic development. The specificity of lymphocyte clones resides in the proteins that become cell surface receptors or antibodies. During development, a complex genetic mechanism rearranges the sequence of amino acids in the receptor proteins to form millions of different combinations. Some of these combinations will not bind to the body's own components, or *self-antigens*, so those cells that do not react with self-antigens go on to form clones (**FIG. 24.5**).

On the other hand, many lymphocytes develop with receptor proteins that will recognize self-antigens. It would be damaging to the body if these lymphocytes persisted because they would create an immune response against the body—an *autoimmune response*. So when these self-reactive lymphocytes combine with self-antigen in the primary lymphoid tissues, they are immediately targeted for destruction (*negative selection*) by apoptosis [p. 84]. The removal of clones of self-reactive lymphocytes is known as *clonal deletion*.

Early Pathogen Exposure Strengthens Immunity

Over the past century, as developed countries eradicated many parasitic, viral, and bacterial diseases, they have seen an increase in autoimmune and allergic diseases, such as asthma, food allergies, and irritable bowel syndrome (IBS). One hypothesis proposed to explain this observation is the **hygiene hypothesis**, which says that challenging the immune system early in life strengthens it, and that a too-clean environment contributes to a weakened immune system. The hygiene hypothesis comes from epidemiological data, and there are only a limited number of controlled animal studies to support it. One group of scientists has proposed that it is not lack of exposure to microbes that is at fault but lack of diversity in the human microbiome that is to blame.

RUNNING PROBLEM

Dr. Paul explained to Rebecca that there are over 100 known types of human papillomavirus. HPV is a virus inserts its DNA into the DNA of its host cell. HPV can live only inside the thin, flat epithelial cells known as *squamous epithelium* {*squama,* a scale}. Squamous epithelium is found on the surface of the skin and in mucous membrane, such as those lining the mouth and genital tract. HPV can cause both ordinary warts on the skin and the growths known as genital warts. More importantly, HPV infections cause nearly all cases of cervical cancer. (The *cervix* is the neck of the uterus that opens into the vagina.) Two high-risk strains of HPV, designated types 16 and 18, account for about 70% of cervical cancer cases.

Q1: *How might you test a person to see if she is infected with HPV?*

755
762
771
773
780
782

24.4 Molecules of the Innate Immune Response

Now that you have been introduced to the cellular effectors of the immune system, we will take a look at the some of the molecules that participate in immune responses. The immune system is distinguished by its extensive use of chemical signaling. Detection, identification, communication, recruitment, coordination, and the attack on the invader all depend on membrane receptors and signal molecules such as cytokines and antibodies. In this section we consider the molecules of the nonspecific innate immune response.

Many Molecules of the Innate Immune Response Are Always Present

The rapid response of innate immunity depends heavily on molecules that are always present in the extracellular fluid. Other chemicals are secreted in response to an immune challenge, or in some cases, produced by the pathogen. These molecules may activate leukocytes or serve one of several other roles. *Chemotaxins* are signal molecules that attract leukocytes to help fight the infection. **Opsonins** {*opsonin,* to buy provisions} are molecules that coat foreign particles to make them visible "food" for phagocytic leukocytes. Some cytokines act as *pyrogens* [p. 723] that raise body temperature by altering the hypothalamic setpoint [p. 720]. A few molecules involved in innate immunity are described here; others will be discussed later. A summary of the major innate immune response molecules can be found in **TABLE 24.1**.

Acute-Phase Proteins In the time immediately following an injury or pathogen invasion (the *acute phase*), the body responds by increasing the concentration of various plasma proteins. Some of these proteins, produced mostly by the liver, are given the general name of **acute-phase proteins**. They include molecules that act as opsonins by coating pathogens; enzyme inhibitors that help prevent tissue damage; and *C-reactive protein* (CRP).

Normally, the levels of acute-phase proteins decline to normal as the immune response proceeds, but in *chronic inflammatory diseases*, such as rheumatoid arthritis, elevated levels of acute-phase proteins may persist. Increased levels of CRP, for example, are also associated atherosclerosis and increased risk of coronary heart disease [p. 501].

Histamine Histamine is found primarily in the granules of mast cells and basophils, and it is the active molecule that initiates the inflammatory response when mast cells degranulate. Histamine's actions bring more leukocytes to the injury site to kill bacteria and remove cellular debris. It does this by dilating blood vessels, which increases blood flow to the area, and by opening pores in capillaries. Making capillaries more leaky also allows plasma proteins to escape into the interstitial space, pulling water with them and leading to tissue edema (swelling) [p. 499]. The result of histamine release is a hot, red, swollen area around a wound or infection site.

TABLE 24.1 Chemicals of the Innate Immune Response

FUNCTIONAL CLASSES
Chemotaxins: Molecules that attract phagocytes to a site of infection
Opsonins: Proteins that coat pathogens so that phagocytes recognize and ingest them
Pyrogens: Fever-producing substances
SPECIFIC CHEMICALS AND THEIR FUNCTIONS
Acute phase proteins: Liver proteins that act as opsonins and that enhance the inflammatory response
Bradykinin: Stimulates pain receptors; vasodilator
Complement: Plasma and cell membrane proteins that act as opsonins, cytolytic agents, and mediators of inflammation
C-reactive protein: Opsonin that activates complement cascade
Granzymes: Cytotoxic enzymes that initiate apoptosis
Heparin: An anticoagulant
Histamine: Vasodilator and bronchoconstrictor released by mast cells and basophils
Interferons (IFN): Cytokines that inhibit viral replication and modulate the immune response
Interleukins (IL): Cytokines secreted by leukocytes to act primarily on other leukocytes; IL-1 mediates inflammatory response and induces fever
Kinins: Plasma proteins that activate to form bradykinin
Lysozyme: An extracellular enzyme that attacks bacteria
Membrane attack complex: A membrane pore protein made in the complement cascade
Perforin: A membrane pore protein that allows granzymes to enter the cell; made by NK and cytotoxic T cells
Superoxide anion (O_2^-): Powerful oxidant in phagocyte lysosomes
Tumor necrosis factor (TNF): Cytokines that promote inflammation and that can cause cells to self-destruct through apoptosis

Complement Proteins **Complement** is a collective term for a group of more than 25 plasma proteins and cell membrane proteins. The *complement cascade* is similar to the blood coagulation cascade. The complement proteins are secreted in inactive forms that are activated as the cascade proceeds. Intermediates of the complement cascade act as opsonins, chemical attractants for leukocytes, and agents that cause mast cell degranulation.

The complement cascade terminates with the formation of **membrane attack complex**, a group of lipid-soluble proteins that insert themselves into the cell membranes of pathogens and virus-infected cells and form giant pores (**FIG. 24.6**). These pores allow water and ions to enter the pathogen cells. As a result, the cells swell and lyse.

FIG. 24.6 Membrane attack complex creates pores in pathogens

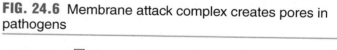

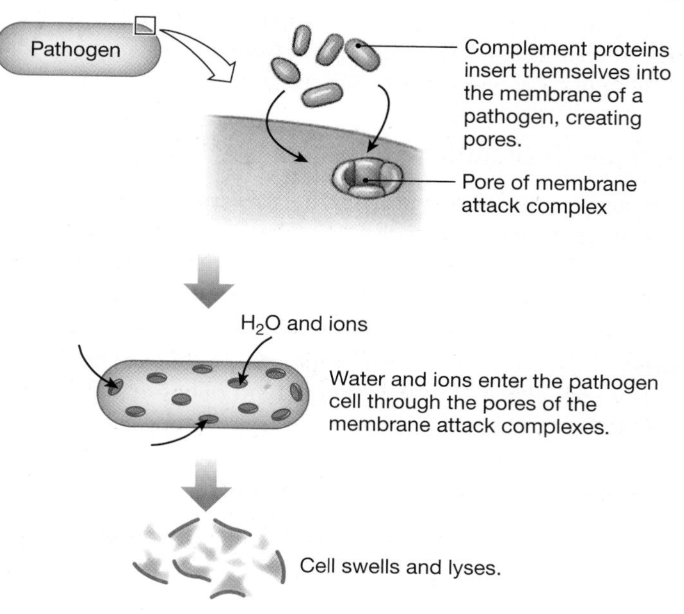
Pathogen
Complement proteins insert themselves into the membrane of a pathogen, creating pores.
Pore of membrane attack complex
H_2O and ions
Water and ions enter the pathogen cell through the pores of the membrane attack complexes.
Cell swells and lyses.

24.5 Antigen Presentation and Recognition Molecules

Adaptive immunity requires the presentation of pathogen antigen to immune cells so they can respond to destroy or suppress that particular pathogen. Three groups of molecules participate in this process (**TBL. 24.2**) Antigen presentation is the job of membrane

TABLE 24.2 Molecules of the Adaptive Immune Response

Major histocompatibility complex (MHC): Membrane protein complexes that display antigen fragments on the cell surface. All cells have MHC-I. Antigen-presenting leukocytes have MHC-II.
Antibodies (immunoglobulins, gamma globulins): Proteins secreted by B lymphocytes that fight specific invaders. Antibodies inserted into the cell membrane of B cells act as surface receptors.
T-cell receptors: T lymphocyte receptors that recognize and bind antigen presented by MHC receptors

proteins known as MHC, or **major histocompatibility complexes.** Recognition of antigen is the responsibility of B and T lymphocyte receptors.

Major Histocompatibility Complexes, MHC The **major histocompatibility complexes** are a family of membrane protein complexes encoded by a specific set of genes. Every nucleated cell of the body has **MHC class I molecules** on its membrane. **MHC class II molecules** are found primarily on antigen-presenting cells (APCs), including macrophages and dendritic cells.

MHC proteins combine with peptide fragments of antigens that have been digested within the cell. The MHC-antigen complex is then inserted into the cell membrane so that the antigen is visible on the extracellular surface. Contact-dependent signaling occurs when the MHC-antigen complex interacts with a T-cell immune receptor. Free antigen in the extracellular fluid cannot bind to unoccupied MHC receptors on the cell surface.

MHC proteins were named when they were discovered to play a role in rejecting foreign tissue following organ or transplants. All MHC proteins are related, but they vary from person to person because of the huge number of potential MHC variants people can inherit from their parents. There are so many variants that it is unlikely that any two people other than identical twins inherit exactly the same set. Major histocompatibility complexes are one reason tissues cannot be transplanted from one person to another without first establishing compatibility.

Antigen-Recognition Molecules

The antigen-recognition proteins that mediate adaptive immunity are highly specific and can distinguish between different pathogens. B lymphocytes make **antibodies**, proteins that bind antigens and make them more visible to the immune system. T lymphocytes have antigen-specific membrane proteins known as **T cell receptors**. T-cell receptors are not antibodies, although the proteins are closely related. T cell receptors bind only to MHC-antigen complexes on the surface of an antigen-presenting cell.

B Lymphocytes Produce Antibodies

Antibodies were among the first aspects of the immune system to be discovered. Most antibodies are found in the blood, where they make up about 20% of the plasma proteins in a healthy individual. These antibodies are most effective against extracellular pathogens (such as bacteria), some parasites, antigenic macromolecules, and viruses that have not yet invaded their host cells. Because antibodies are not toxic, they cannot destroy antigens. Their primary role is to help the immune system react to specific antigens.

Antibody Proteins The basic antibody molecule has four polypeptide chains linked into a Y shape (**FIG. 24.7a**). The arms contain the *antigen-binding sites* (Fig. 24.7b). The stem of the Y-shaped molecule, the **Fc region**, determines the class to which the antibody belongs. A *hinge region* between the arms and the stem allows flexible positioning of the arms as the antibody binds to the antigen.

FIG. 24.7 Antibodies

(a) Antibody Structure

An antibody molecule is composed of two identical light chains and two identical heavy chains, linked by disulfide bonds.

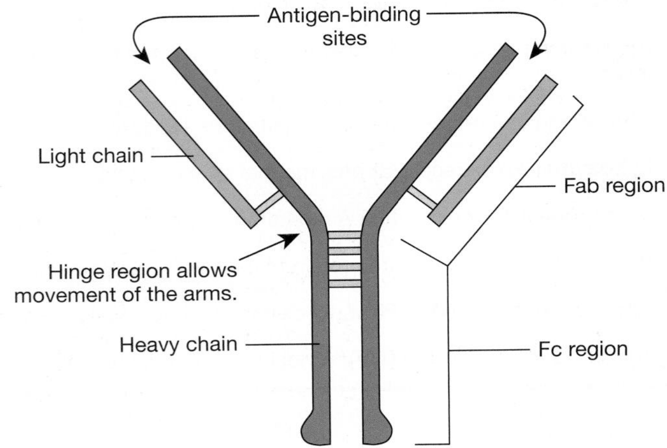
Antigen-binding sites
Light chain
Fab region
Hinge region allows movement of the arms.
Heavy chain
Fc region

(b) Antigen Binding

Antibodies have antigen-binding sites on the arms.

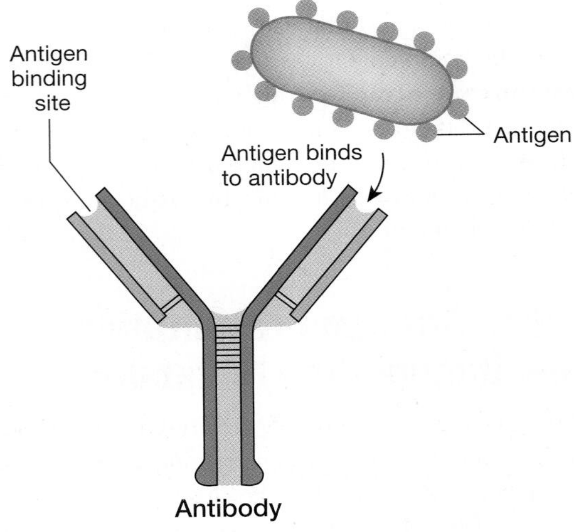
Antigen binding site
Antigen
Antigen binds to antibody
Antibody

Each arm of the Y, or **Fab region**, consists of one **light chain** and one **heavy chain with one antigen-binding site**. In any given antibody molecule, the two light chains are identical and the two heavy chains are identical. However, the chains vary widely among different antibodies, giving the antibody its specificity. Each B lymphocyte clone produces a unique antibody.

Classes of Antibodies Traditional antibody names—agglutinins, precipitins, hemolysins, and so on—indicated what they do. Today, however, antibodies or *immunoglobulins* (Ig) are divided into five general classes: IgG, IgA, IgE, IgM, and IgD (pronounced *eye-gee-[letter]*). Antibodies collectively are known as **gamma globulins**.

IgGs make up 75% of plasma antibody in adults. Some maternal IgGs cross the placental membrane and give infants immunity in the first few months of life. Some IgGs activate complement.

IgA antibodies are found in external secretions, such as saliva, tears, intestinal and bronchial mucus, and breast milk, where they bind to pathogens and flag them for phagocytosis if they reach the internal environment.

IgEs target parasites and are associated with allergic responses. When mast cell receptors bind with IgEs and antigen, the mast cells degranulate and release chemical mediators, such as histamine.

IgM antibodies are associated with early immune responses to a pathogen. IgMs strongly activate complement. They are also the antibodies that react to antigens found on red blood cells.

IgD antibody proteins appear as receptors on the surface of B lymphocytes and help activate the B cells.

Two classes of immunoglobulins (IgM and IgA) are secreted as polymers: IgM is made up of five Y-shaped antibody molecules, and IgA has from one to four antibody molecules.

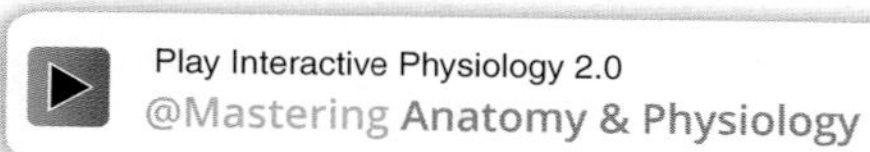
Play Interactive Physiology 2.0
@Mastering Anatomy & Physiology

Concept Check

1. Because antibodies are proteins, they are too large to cross cell membranes on transport proteins or through channels. How then do IgAs and other antibodies become part of external secretions such as saliva, tears, and mucus?

24.6 Pathogens of the Human Body

Before we begin our discussion of integrated immune function, let's review the major pathogens of the human body. In the United States, our most prevalent infectious diseases are of viral and bacterial origin. Common multicellular parasites include hookworms and tapeworms. Almost any exogenous molecule or cell has the potential to elicit an immune response. Pollens, chemicals, and foreign bodies are examples of non-infectious substances to which the body may react.

Worldwide, parasites are a significant public health concern. For example, malaria, a pathogenic protozoan whose life cycle alternates between human and mosquito hosts, was estimated to infect more than 212 million people in 2015, with over 400,000 deaths that year. Many microbes and parasitic organisms, such as the malaria protozoan and the *Zika virus*, are introduced into the body by biting insects.

Some pathogens enter the body through the digestive tract, brought in by contaminated food and water. Others, such as the fungi that cause valley fever and histoplasmosis, are inhaled. A few, such as the blood fluke *Schistosoma*, burrow through the host's skin. Once in the body, microbes and parasites may enter host cells in an effort to evade the immune response, or they may remain in the extracellular compartment.

Bacteria and Viruses Require Different Defense Mechanisms

Bacteria and viruses differ from each other in several ways (**TBL. 24.3**). Because of these differences, the body has evolved a variety of immune responses.

1. **Structure.** Bacteria are cells, with a cell membrane that is usually surrounded by a cell wall. Some *encapsulated* bacteria produce an additional protective glycoprotein outer layer known as a *capsule*. Viruses are not cells. They consist of nucleic acid (DNA or RNA) enclosed in a coat of viral proteins called a *capsid*. Some viruses add an *envelope* of phospholipid and protein made from the host's cell membrane and incorporate viral proteins into the envelope.
2. **Living conditions and reproduction.** Most bacteria can survive and reproduce outside a host if they have the required

TABLE 24.3 Differences between Bacteria and Viruses

	Bacteria	Viruses
Structure	Cells. No organelles. Usually surrounded by cell wall. Some have an additional capsule.	Not cells. Nucleic acid core enclosed in protein capsid. Some have external envelope.
Living conditions	Most can survive and reproduce outside a host.	Parasitic. Must have a host cell to reproduce.
Genetic material	Singular circular DNA chromosome.	May be DNA or RNA
Susceptibility to drugs	Most can be killed or suppressed by antibiotics	Cannot be killed with antibiotics. Some can be suppressed with antiviral drugs.

nutrients, temperature, pH, and so on. Viruses *must* use the intracellular machinery of a host cell to replicate. Eliminating pathogens in intracellular and extracellular compartments of the body requires different defense mechanisms.

3. **Susceptibility to drugs.** Most bacteria can be killed by the drugs we call **antibiotics**. These drugs act directly on bacteria and destroy them or inhibit their growth. Viruses cannot be killed by antibiotics. A few viral infections can be treated with *antiviral drugs,* which target specific stages of viral replication.

Viruses Can Only Replicate inside Host Cells

The replication cycle of a virus begins when the virus invades the host cell. Once inside and free of the capsid, the virus's nucleic acid takes over the host cell's resources to make new virus particles that can infect other cells. Some viruses kill the host cell. Some, like *Herpes simplex type 1* and varicella-zoster virus, which cause cold sores and chicken pox, respectively, "hide out" in the host cell and replicate only sporadically. Other viruses incorporate their DNA into the host cell DNA. Viruses with this characteristic include *human immunodeficiency virus* (HIV) and **oncogenic viruses**, which cause cancer.

24.7 The Immune Response

The human body has two lines of defense. Physical and chemical barriers, such as skin, first try to keep pathogens out of the body's internal environment. If this first line of defense fails, then the internal **immune response** takes over. The internal immune response begins with the rapid but nonspecific innate responses, followed more slowly by the adaptive response to specific antigens.

Barriers Are the Body's First Line of Defense

The body's first line of defense is to exclude pathogens by physical, mechanical, and chemical barriers (**FIG. 24.8**). Physical barriers of the body include the skin [p. 86], the protective mucous linings of the gastrointestinal and genitourinary tracts, and the ciliated epithelium of the respiratory tract. The digestive and respiratory systems are most vulnerable to microbial invasion because these regions have extensive areas of thin epithelium in direct contact with the external environment. In women, the reproductive tract is also vulnerable, but to a lesser degree. The opening to the uterus is normally sealed by a plug of mucus that keeps bacteria in the vagina from ascending into the uterine cavity.

Secretions from exocrine glands and mechanical removal of pathogens assist the physical barriers. In the respiratory system, inhaled particulate matter is trapped by mucus lining the upper respiratory system. The mucus is then transported upward on the *mucociliary escalator* to be expelled or swallowed [p. 538], or it may be coughed out. Swallowed pathogens may be disabled or killed by the acidity of the stomach. Respiratory tract secretions, saliva, and tears contain **lysozyme**, an enzyme with antibacterial activity. Lysozyme attacks cell wall components of unencapsulated bacteria and breaks them down. However, it cannot digest the capsules of encapsulated bacteria. In the intestines, the presence of pathogens may trigger increased motility and secretory diarrhea [p. 687].

FIG. 24.8 Physical, mechanical, and chemical barriers

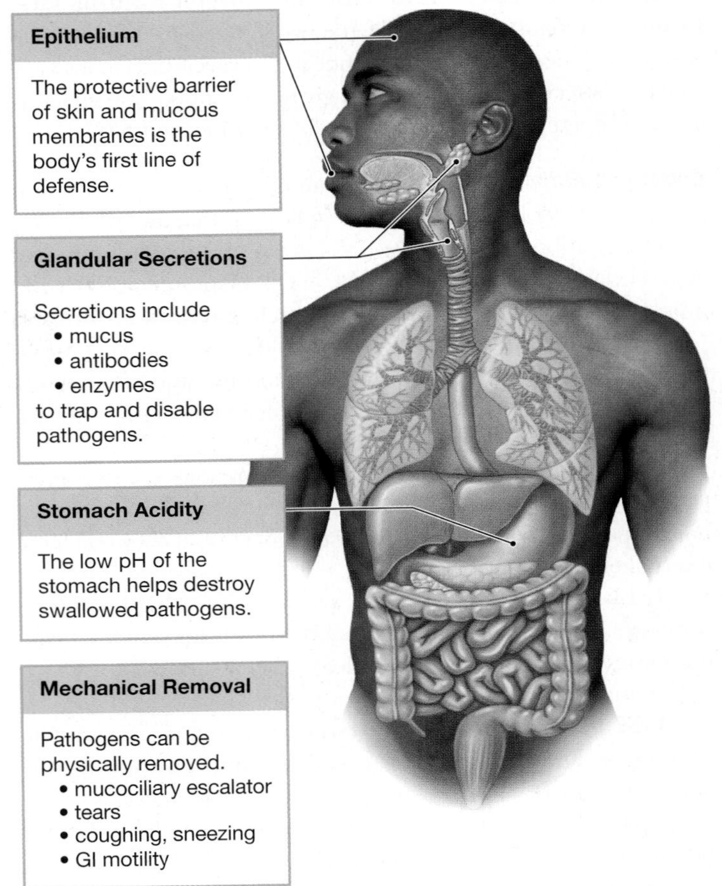

Epithelium
The protective barrier of skin and mucous membranes is the body's first line of defense.
Glandular Secretions
Secretions include
• mucus
• antibodies
• enzymes
to trap and disable pathogens.
Stomach Acidity
The low pH of the stomach helps destroy swallowed pathogens.
Mechanical Removal
Pathogens can be physically removed.
• mucociliary escalator
• tears
• coughing, sneezing
• GI motility

Innate Immunity Provides Nonspecific Responses

Pathogens that get past the physical barriers of skin and mucous membranes trigger the innate immune response. Innate immunity either clears the infection or contains it until the adaptive immune response is active. A key element of the innate immune response is its broad specificity. Some immune cells recognize classes of molecules that are unique to microorganisms, generally called *pathogen-associated molecular patterns,* or PAMPs. PAMPs bind to leukocyte *pattern recognition receptors* (PRRs) and activate responses that attempt to kill or ingest the invader.

First, patrolling and stationary leukocytes are attracted to areas of invasion by chemical signals known as **chemotaxins**. Chemotaxins include bacterial toxins or cell wall components that act as PAMPs. Products of tissue injury, such as fibrin and collagen fragments, may also indicate a location that needs defending. Endogenous "danger signals" are sometimes called DAMPSs, for *danger-associated molecular patterns*. Once on site, activated leukocytes secrete their own chemotaxic cytokines to bring additional leukocytes to the infection site.

Tissue macrophages and neutrophils are the primary phagocytic cells responsible for the initial defense. Circulating leukocytes leave the blood (*extravasation*) by squeezing through pores in the capillary endothelium. If an area of infection attracts a large number of leukocytes, the material we call **pus** may form. This thick, whitish to greenish substance is a collection of living and dead neutrophils and macrophages, along with tissue fluid, cell debris, and other remnants of the immune process.

Phagocytosis Phagocytes are a functional group of white blood cells that engulf and ingest their targets by phagocytosis [p. 146]. The primary phagocytes of the innate response are the neutrophils and macrophages. Dendritic cells, which link innate and adaptive immunity, are also phagocytic.

Phagocytosis is a receptor-mediated event, which ensures that only unwanted particles are ingested. The phagocyte receptors recognize many different types of foreign particles, both organic and inorganic, leading the cells to ingest everything from unencapsulated bacteria and cell fragments to carbon and asbestos particles. They will even ingest tiny polystyrene beads, providing one way that scientists in the laboratory analyze phagocytic activity.

In the simplest phagocytic reactions, surface molecules on the pathogen (PAMPs) act as ligands that bind directly to pattern recognition receptors on the phagocyte membrane (**FIG. 24.9**). In a sequence reminiscent of a zipper closing 1, the ligands and receptors combine sequentially, so that the phagocyte surrounds the unwanted foreign particle. The process is aided by actin filaments that push arms of the phagocytic cell around the invader.

The ingested particle ends up in a cytoplasmic vesicle called a **phagosome** (Fig. 24.9 2). Phagosomes fuse with intracellular lysosomes [p. 71], which contain enzymes and oxidizing agents, such as hydrogen peroxide (H_2O_2) and the superoxide anion (O_2^-). These powerful chemicals then digest organic pathogens 3. Inorganic particles, such as asbestos and carbon encountered by macrophages in the lung, cannot be digested and remain inside the cell.

Phagocytes cannot instantly recognize all foreign substances, however, because some pathogens lack markers that react with pattern recognition receptors. For example, certain bacteria have evolved a polysaccharide capsule that masks their surface markers from the host immune system. These encapsulated bacteria are not as quickly recognized by phagocytes and consequently are more pathogenic because they can grow unchecked until the immune system finally recognizes them and makes antibodies against them.

NK Cells Kill Virus-Infected Cells One class of lymphocyte—natural killer (NK) cells—participates in the innate response against viral infections. NK cells act more rapidly than other lymphocytes, responding within hours of a primary viral infection. NK cells are programmed to recognize virus-infected cells and induce them to commit suicide by *apoptosis* [p. 84] before the virus can replicate. Complete elimination of the virus requires activation of a specific immune response.

NK cells target virus-infected cells by looking for cells without MHC class I proteins on their surface. Some viruses try to evade the human immune system by blocking the host cell's synthesis of MHC proteins. Without MCH protein, the host cell cannot display viral antigen on its surface, which allows the virus to hide undetected inside the cell. But NK cells do not need viral antigen to activate them. Instead they are programmed to find and attack cells displaying low concentrations of MHC-I.

FIG. 24.9 Phagocytosis

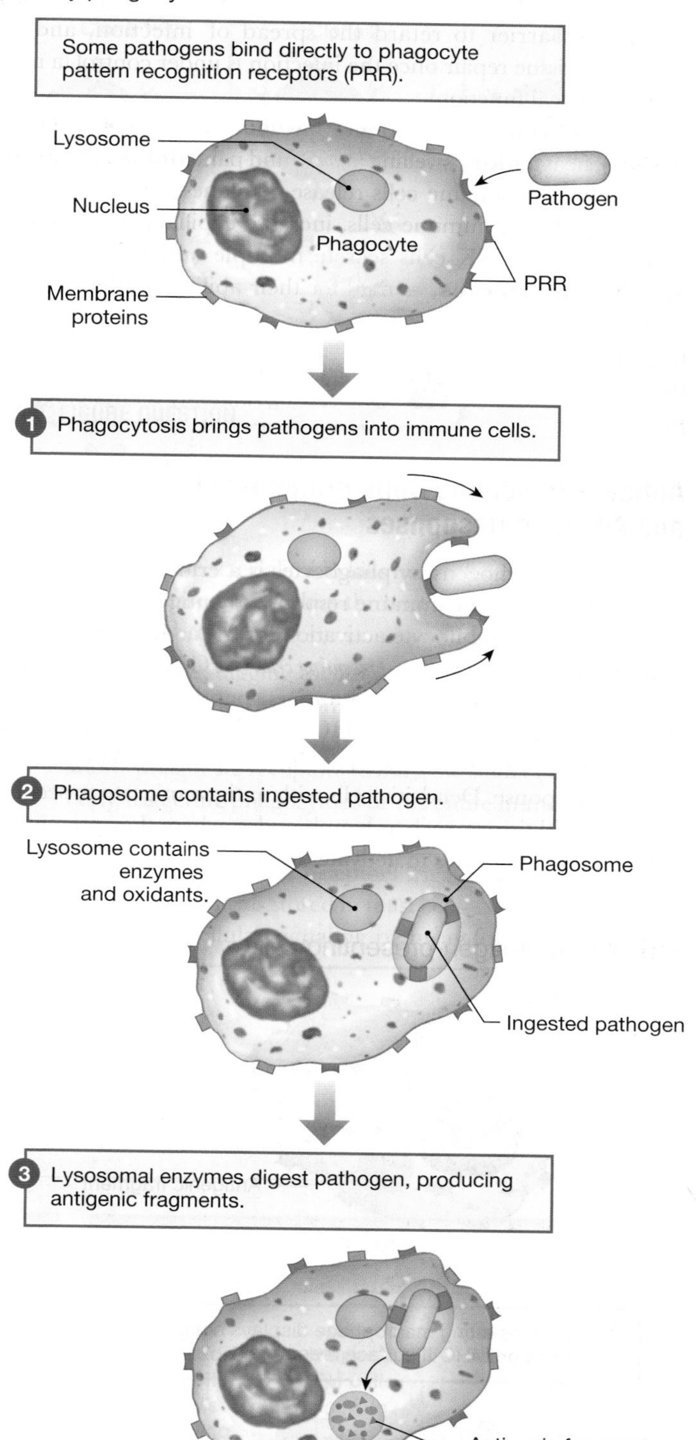
Macrophages, neutrophils, and dendritic cells are the primary phagocytes.
Some pathogens bind directly to phagocyte pattern recognition receptors (PRR).
Lysosome
Nucleus
Phagocyte
Membrane proteins
Pathogen
PRR
1 Phagocytosis brings pathogens into immune cells.
2 Phagosome contains ingested pathogen.
Lysosome contains enzymes and oxidants.
Phagosome
Ingested pathogen
3 Lysosomal enzymes digest pathogen, producing antigenic fragments.
Antigenic fragment

FIG. 24.16 ESSENTIALS Immune Responses to Extracellular Bacteria

Bacterial infections cause inflammation and trigger specific immune responses.

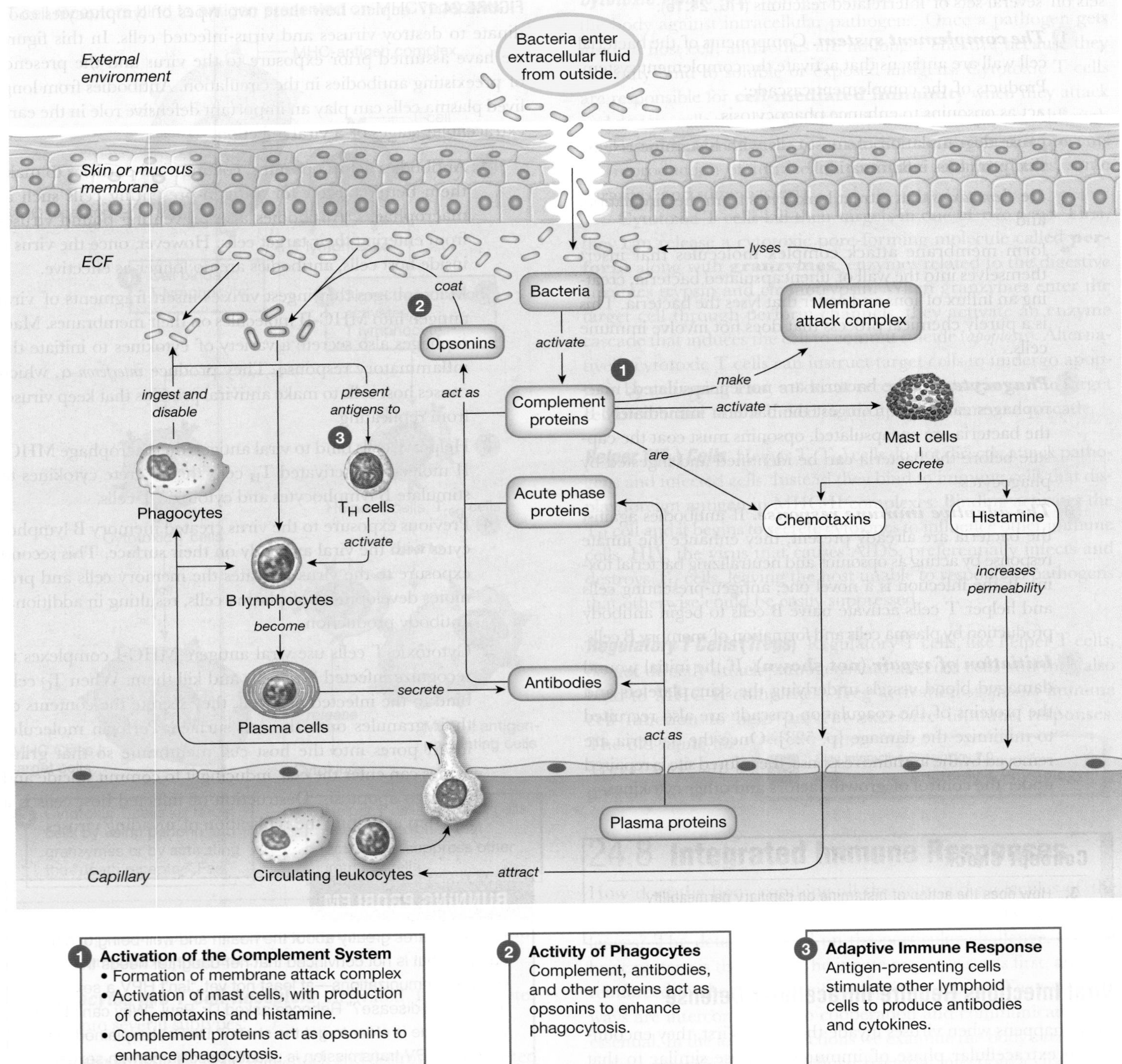

Bacteria enter extracellular fluid from outside.
External environment
Skin or mucous membrane
ECF
coat
2
Bacteria
lyses
Membrane attack complex
Opsonins
activate
1
make
activate
ingest and disable
present antigens to
act as
Complement proteins
Mast cells
3
are
secrete
Phagocytes
T_H cells
Acute phase proteins
Chemotaxins
Histamine
activate
increases permeability
B lymphocytes
become
secrete
Antibodies
Plasma cells
act as
Plasma proteins
Capillary
Circulating leukocytes
attract
1 Activation of the Complement System
• Formation of membrane attack complex
• Activation of mast cells, with production of chemotaxins and histamine.
• Complement proteins act as opsonins to enhance phagocytosis.
2 Activity of Phagocytes
Complement, antibodies, and other proteins act as opsonins to enhance phagocytosis.
3 Adaptive Immune Response
Antigen-presenting cells stimulate other lymphoid cells to produce antibodies and cytokines.

FIG. 24.17 ESSENTIALS Immune Responses to Viruses

This figure assumes prior exposure to the virus and preexisting antibodies.

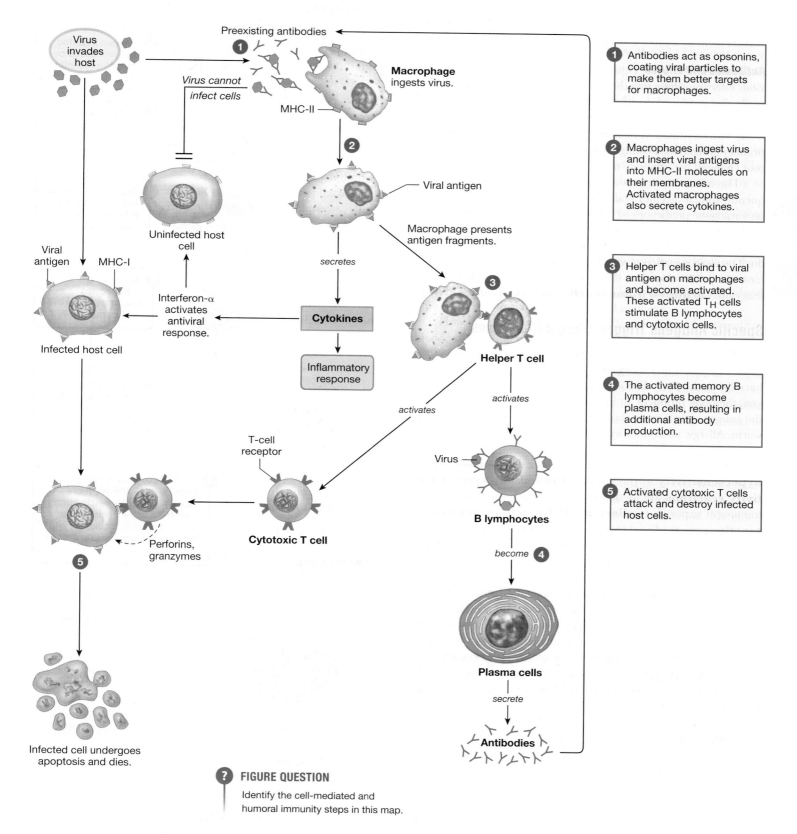

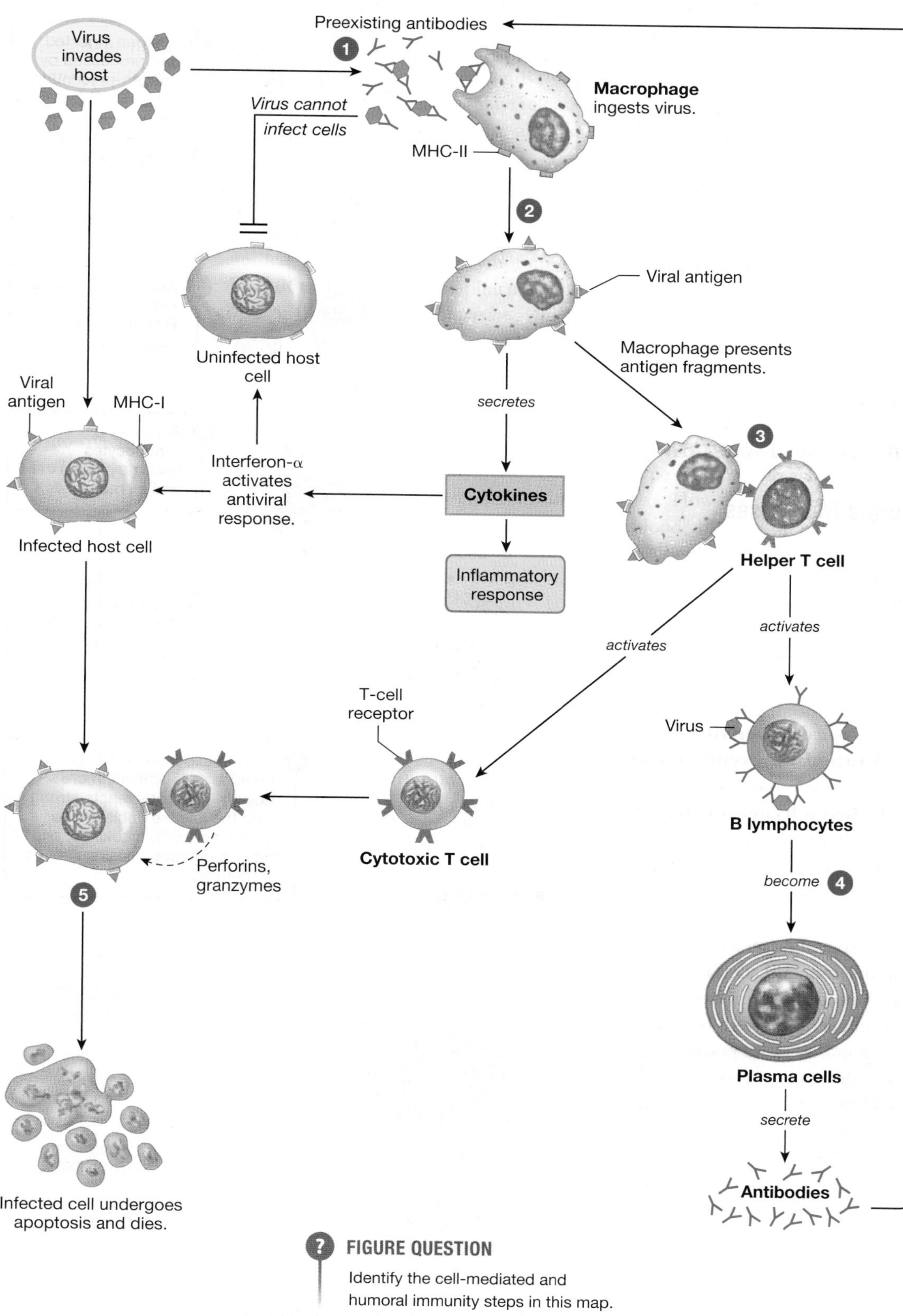

Virus invades host
Preexisting antibodies
Virus cannot infect cells
Macrophage ingests virus.
MHC-II
Viral antigen
Uninfected host cell
Macrophage presents antigen fragments.
Viral antigen
MHC-I
secretes
Interferon-α activates antiviral response.
Cytokines
Infected host cell
Inflammatory response
Helper T cell
activates
activates
T-cell receptor
Virus
Perforins, granzymes
Cytotoxic T cell
B lymphocytes
become
Plasma cells
secrete
Antibodies
Infected cell undergoes apoptosis and dies.

1. Antibodies act as opsonins, coating viral particles to make them better targets for macrophages.

2. Macrophages ingest virus and insert viral antigens into MHC-II molecules on their membranes. Activated macrophages also secrete cytokines.

3. Helper T cells bind to viral antigen on macrophages and become activated. These activated T_H cells stimulate B lymphocytes and cytotoxic cells.

4. The activated memory B lymphocytes become plasma cells, resulting in additional antibody production.

5. Activated cytotoxic T cells attack and destroy infected host cells.

FIGURE QUESTION

Identify the cell-mediated and humoral immunity steps in this map.

FIG. 24.19 ABO blood groups

(a) Characteristics of the ABO Blood Groups

Blood Type	Antigen on Red Blood Cell	Antibodies in Plasma
O	RBC No A or B antigens	"Anti-A" and "anti-B"
A	A antigens	"Anti-B"
B	B antigens	"Anti-A"
AB	A and B antigens	None to A or B

(b) A Mixture of Type O and Type A Blood

When red blood cells with group A antigens on their membranes are mixed with plasma containing antibodies to group A, the antibodies cause the blood cells to clump, or agglutinate.

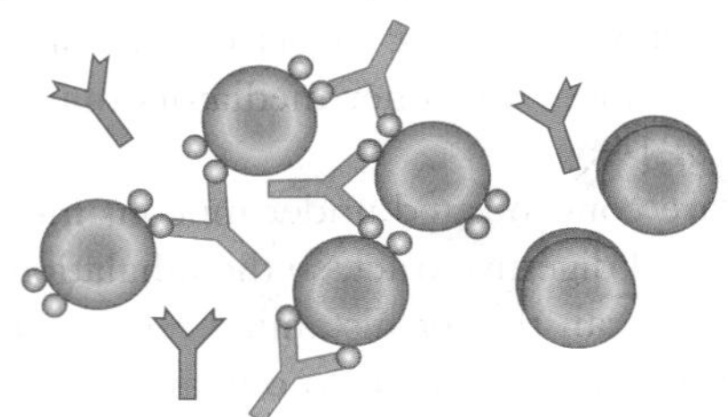

FIGURE QUESTION

Each person inherits one allele for ABO blood groups from each parent. A and B are dominant to O but equal if they occur together (blood type AB). Fill in the table showing combinations of inherited alleles. In the shaded blocks, show the blood type that would be expressed.

		Mother		
		A	B	O
Father	A	A A A		
	B			
	O			

people describe themselves as "Rh-positive" or "Rh-negative." Someone who lacks the Rh D antigen (i.e., is Rh-negative) and is exposed to the D antigen on foreign red blood cells will make antibodies to Rh D antigen.

Production of anti-D antibodies may occur during pregnancy if an Rh-negative mother is pregnant with an Rh-positive fetus. Leakage of the placental barrier can allow fetal blood into the mother's circulation, and the mother's immune system will react to the foreign D antigen. Exposure can also occur during delivery of an Rh-positive child.

Antibodies are transported across the placenta from mother to fetus so that a newborn will have some immunity at birth. If a mother has anti-D antibodies, they will be transferred to the fetus. If the fetus is Rh D-positive, the anti-D antibodies will attack the fetus's red blood cells and destroy them. This destruction causes *hemolytic disease of the newborn* (HDN), also known as *erythroblastosis fetalis*. HDN can also occur if there are other blood type incompatibilities, such as an ABO mismatch.

Rh D incompatibility can be fatal, but treatments have been developed to help prevent HDN. These treatments involve the use of manufactured antibodies that bind to fetal D antigen or bind to the mother's anti-D antibodies.

Concept Check

4. A person with AB blood type is transfused with type O red blood cells. What happens and why?
5. A person with O blood type is transfused with type A blood. What happens? Why?

24.9 Immune System Pathologies

Sometimes, the body's immune system fails to perform its normal functions. Pathologies of the immune system generally fall into one of three categories: incorrect responses, overactive responses, or lack of response.

1. ***Incorrect responses.*** If mechanisms for distinguishing self from nonself fail and the immune system attacks the body's normal cells, an *autoimmune disease* results.
2. ***Overactive responses.*** Allergies and *hypersensitivity reactions* are conditions in which the immune system creates a response that is out of proportion to the threat posed by the antigen. In extreme cases, the systemic effects of hypersensitivity reactions can be life threatening.

3. ***Lack of response.*** **Immunodeficiency diseases** arise when some component of the immune system fails to work properly. *Primary immunodeficiency* is a family of genetically inherited disorders that range from mild to severe. *Acquired immunodeficiencies* may occur as a result of infection, such as *acquired immunodeficiency syndrome* (AIDS) caused by the *human immunodeficiency virus* (HIV). Acquired immunodeficiencies may also arise as a side effect of drug or radiation therapies, such as those used to treat cancer.

Autoimmune Disease Results from Antibodies against Self-Antigen

When self-tolerance fails, the body makes antibodies against its own components through T cell–activated B lymphocytes. The body's attack on its own cells leads to **autoimmune diseases** {*auto-*, self}. The antibodies produced in autoimmune diseases are specific against a particular antigen and are usually restricted to a particular organ or tissue type. **TABLE 24.5** lists some common autoimmune diseases of humans.

Why does self-tolerance suddenly fail? We still do not know. We have learned, however, that autoimmune diseases often begin in association with an infection. One potential trigger for autoimmune diseases is foreign antigens that are similar to human antigens. When the body makes antibodies to the foreign antigen, those antibodies have enough cross-reactivity with human tissues to do some damage.

One example of an autoimmune disease is type 1 diabetes mellitus, in which the body makes *islet cell antibodies* that destroy pancreatic beta cells but leave other endocrine cells untouched [p. 708]. Another autoimmune condition is the hyperthyroidism of Graves' disease [p. 195]. The body makes *thyroid-stimulating immunoglobulins* that mimic thyroid-stimulating hormone and cause the thyroid gland to oversecrete hormone.

Severe autoimmune diseases are sometimes treated by administering glucocorticosteroids, such as cortisol and its derivatives. Glucocorticoids depress immune system function by suppressing bone marrow production of leukocytes and decreasing the activity of circulating immune cells. Although helpful in suppressing symptoms of the autoimmune disease, large doses of steroids may trigger the same signs and symptoms as hormone oversecretion, causing patients to develop the central obesity and moon face characteristic of Cushing's syndrome [p. 733].

Immune Surveillance Removes Abnormal Cells

The diseases we call *cancer* result from abnormal cells that multiply uncontrollably, crowding out normal cells and disrupting function. What is the role of the human immune system in protecting the body against cancer? Scientists believe that cancer cells form on a regular basis.

The **immune surveillance** hypothesis proposes that these abnormal cells are detected by the immune system and destroyed before they can spread. Some evidence supports that hypothesis, but it does not explain why so many cancerous tumors develop every year, or why people with depressed immune function do not develop cancerous tumors in all types of tissues.

Immune surveillance does appear to recognize and control some virus-associated tumors. In addition, some types of endogenous tumors lack surface MHC antigens. Their absence allows NK cells to recognize these cells as abnormal and destroy them, just as with virus-infected cells discussed earlier. One active area of cancer research is investigating ways to activate the immune system to fight cancerous tumor cells.

24.10 Neuro-Endocrine-Immune Interactions

One of the most fascinating and rapidly developing areas in physiology involves the link between the mind and the body. Although serious scientists scoffed at the topic for many years, the interaction of emotions and somatic illnesses has been described for centuries. The Old Testament says, "A merry heart doeth good like a medicine, but a broken spirit drieth the bones" (Proverbs 17:22, King James Version). Most societies have tales of people who lost the will to live and subsequently died without obvious illness, or of people given up for dead who made remarkable recoveries.

Today, the medical field is beginning to recognize the reality of the mind-body connection. Psychosomatic illnesses and the placebo effect have been accepted for many years. Why should we not investigate the possibility that cancer patients can enhance

TABLE 24.5 Some Common Autoimmune Diseases in Humans

Disease	Antibodies Produced Against
Graves' disease (hyperthyroidism)	TSH receptor on thyroid cells
Insulin-dependent diabetes mellitus	Pancreatic beta cell antigens
Multiple sclerosis	Myelin of CNS neurons
Myasthenia gravis	Acetylcholine receptor of motor endplate
Rheumatoid arthritis	Collagen
Systemic lupus erythematosus	Intracellular nucleic acid protein complexes (antinuclear antibodies)
Guillain-Barré syndrome (acute inflammatory demyelinating polyneuropathy)	Myelin of peripheral nerves

13. Blood flow through exercising muscle increases dramatically when skeletal muscle arterioles dilate. Arterioles in other tissues constrict. (p. 791; Fig. 25.7)
14. Decreased tissue O_2 and glucose or increased muscle temperature, CO_2, and acid act as paracrine signals and cause local vasodilation. (p. 792)
15. Mean arterial blood pressure increases slightly as exercise intensity increases. The baroreceptors that control blood pressure change their setpoints during exercise. (p. 792; Fig. 25.8)

25.4 Feedforward Responses to Exercise

16. When exercise begins, feedforward responses prevent significant disruption of homeostasis. (p. 793)

25.5 Temperature Regulation during Exercise

17. Heat released during exercise is dissipated by sweating and increased cutaneous blood flow. (p. 794)

25.6 Exercise and Health

18. Physical activity can help prevent or decrease the risk of developing high blood pressure, strokes, and type 2 diabetes mellitus. (p. 794)
19. Studies suggest that serotonin release during exercise may help alleviate depression. (p. 796)

REVIEW QUESTIONS

In addition to working through these questions and checking your answers on p. A-33, review the Learning Outcomes at the beginning of this chapter.

Level One Reviewing Facts and Terms

1. Name the two muscle compounds that store energy in the form of high-energy phosphate bonds.
2. The most efficient ATP production is through *aerobic/anaerobic* pathways. When these pathways are being used, then *glucose/fatty acids/both/neither* can be metabolized to provide ATP.
3. What are the differences between aerobic and anaerobic metabolism?
4. List three sources of glucose that can be metabolized to ATP, either directly or indirectly.
5. List four hormones that promote the conversion of triglycerides into fatty acids. What effects do these hormones have on plasma glucose levels?
6. What is meant by the term oxygen deficit, and how is it related to excessive postexercise oxygen consumption?
7. What organ system is the limiting factor for maximal exertion?
8. In endurance events, body temperature can reach 40–42 °C. What is normal body temperature? Which two thermoregulatory mechanisms are triggered by this change in temperature during exercise?

Level Two Reviewing Concepts

9. Concept map: Map the metabolic, cardiovascular, and respiratory changes that occur during exercise. Include the signals to and from the nervous system, and show what specific areas signal and coordinate the exercise response.
10. What causes insulin secretion to decrease during exercise, and why is this decrease adaptive?
11. State two advantages and two disadvantages of anaerobic glycolysis.
12. Compare and contrast each of the terms in the following sets of terms, especially as they relate to exercise:
    a. ATP, ADP, PCr
    b. myoglobin, hemoglobin
13. Match the following brain areas with the response(s) that each controls. Brain areas may control one response, more than one response, or none at all. Some responses may be associated with more than one brain area.

a. pons	1. changes in cardiac output
b. medulla oblongata	2. vasoconstriction
c. midbrain	3. exercise hyperventilation
d. motor cortex	4. increased stroke volume
e. hypothalamus	5. increased heart rate
f. cerebellum	6. coordination of skeletal muscle movement
g. brain not involved (i.e., local control)	

14. Specify whether each of the following parameters stays the same, increases, or decreases when a person becomes better conditioned for athletic activities:
    a. heart rate during exercise
    b. resting heart rate
    c. cardiac output during exercise
    d. resting cardiac output
    e. breathing rate during exercise
    f. blood flow to muscles during exercise
    g. blood pressure during exercise
    h. total peripheral resistance during exercise
15. Why doesn't increased venous return during exercise overstretch the heart muscle?
16. Diagram the three theories that explain why the normal baroreceptor reflex is absent during exercise.

17. List and briefly discuss the benefits of a lifestyle that includes regular exercise.
18. Explain how exercise decreases blood glucose in type 2 diabetes mellitus.

Level Three Problem Solving

19. You have decided to manufacture a new sports drink that will help athletes, from football players to gymnasts. List at least four different ingredients you would include in your drink, and indicate why each is important for the athlete.

Level Four Quantitative Problems

20. You are a well-conditioned athlete. At rest, your heart rate is 60 beats/minute and your stroke volume is 70 mL/beat. What is your cardiac output? At one point during exercise, your heart rate increases to 120 beats/min. Does your cardiac output increase proportionately? Explain.
21. The following graph shows left ventricular pressure-volume curves in one individual. Curve A is the person sitting at rest. Curve B shows the person's cardiac response to mild exercise on a stationary bicycle. Curve C shows the cardiac response during maximum intensity cycling.
    a. Calculate the stroke volume for each of the curves.
    b. Given the following cardiac outputs (CO), calculate the heart rates for each condition.
    $CO_A = 6$ L/min, $CO_B = 10.5$ L/min, $CO_C = 19$ L/min

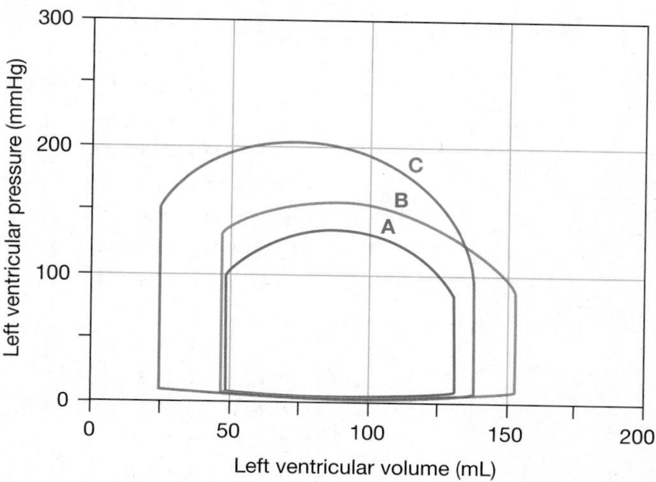

Left ventricular pressure (mmHg)
300
200
100
0
C
B
A
0
50
100
150
200
Left ventricular volume (mL)

Data from G. D. Plotnick, *et al.* *Am J Physiol* 251: H1101–H1105, 1986.

    c. Which exercise curve shows an increase in stroke volume due primarily to increased contractility? Which exercise curve shows an increase in stroke volume due primarily to increased venous return?
    d. Mechanistically, why did the end-diastolic volume in curve C fall back toward the resting value?

Answers to Concept Checks, Figure and Graph Questions, and end-of-chapter Review Questions can be found in Appendix A [p. A-1].

26 Reproduction and Development

Ovum with sperm

Teenagers who received some type of comprehensive sex education were 60 percent less likely to get pregnant or get someone else pregnant.

Amanda Peterson Beadle, *Teen Pregnancies Highest in States with Abstinence-Only Policies*, April 10, 2012. This material was published by ThinkProgress. *http://thinkprogress.org/health/2012/04/10/461402/teen-pregnancy-sex-education/.*

BACKGROUND BASICS

Imagine growing up as a girl, then at the age of 12 or so, finding that your voice is deepening and your genitals are developing into those of a man. This scenario actually happens to a small number of men who have a condition known as *pseudohermaphroditism* {*pseudes*, false + *hermaphrodites*, the dual-sex offspring of Hermes and Aphrodite}. These men have the internal sex organs of a male but inherit a gene that causes a deficiency in one of the male hormones. Consequently, they are born with external genitalia that appear feminine, and they are raised as girls. At **puberty** {*pubertas*, adulthood}, the period when a person makes the transition from being nonreproductive to being reproductive, individuals with pseudohermaphroditism begin to secrete more male hormones. As a result, they develop some, but not all, of the characteristics of men. Not surprisingly, a conflict arises: Should these individuals change gender or remain female? Most choose to change and continue life as men.

Reproduction is one area of physiology in which we humans like to think of ourselves as significantly advanced over other animals. We mate for pleasure as well as procreation, and women are always sexually receptive (i.e., not only during fertile periods). But just how different are we?

Like many other terrestrial animals, humans have internal fertilization that allows motile flagellated sperm to remain in an aqueous environment. To facilitate the process, we have mating and courtship rituals, as do other animals. Development is also internal, within the uterus, which protects the growing embryo from dehydration and cushions it in a layer of fluid.

Humans are *sexually dimorphic* {*di-*, two + *morphos*, form}, meaning that males and females are physically distinct. This distinction is sometimes blurred by dress and hairstyle, but these are cultural acquisitions. Everyone agrees that male and female humans are physically dimorphic, but we are still debating just how behaviorally and psychologically dimorphic we are.

Sex hormones play a significant role in the behavior of other mammals, acting on adults as well as influencing the brain of the developing embryo. Their role in humans is more controversial. Human fetuses are exposed to sex hormones while in the uterus, but it is unclear how much influence these hormones have on behavior later in life. Does the preference of little girls for dolls and of little boys for toy guns have a biological basis or a cultural basis? We have no answer yet, but growing evidence suggests that at least part of our brain structure is influenced by sex hormones before we ever leave the womb.

In this chapter, we address the biology of human reproduction and development. The topic is an incredibly complex one, with numerous hormones, cytokines, and paracrine signal molecules interacting in constantly shifting interplay. In recent decades, reproductive research has raised many new questions. Because of the complexity of the topic, this chapter presents only a generalized overview.

We begin our discussion with gametes that fuse to form the fertilized egg, or **zygote**. As the zygote begins to divide (2-cell stage, 4-cell stage, etc.), it becomes first an **embryo** (weeks 0–8 of development), then a **fetus** (eight weeks until birth).

RUNNING PROBLEM **Infertility**

Kate and Jon have just about everything to make them happy: successful careers, a loving marriage, a comfortable home. But one thing is missing: After seven years of marriage, they have been unable to have a child. Today, Kate and Jon have their first appointment with Dr. Baker, an infertility specialist. "Finding the cause of your infertility is going to require some painstaking detective work," Dr. Baker explains. She begins her workup of Kate and Jon by asking detailed questions about their reproductive histories. Based on the answers to these questions, she will then order tests to pinpoint the problem.

801 – 811 – 820 – 823 – 827 – 834

26.1 Sex Determination

The male and female sex organs consist of three sets of structures: the gonads, the internal genitalia, and the external genitalia. **Gonads** {*gonos*, seed} are the organs that produce **gametes** {*gamein*, to marry}, the eggs and sperm that unite to form new individuals. The male gonads are the **testes** (singular *testis*), which produce **sperm** (*spermatozoa*). The female gonads are the **ovaries**, which produce eggs, or **ova** (singular *ovum*). The undifferentiated gonadal cells destined to produce eggs and sperm are called **germ cells**. The **internal genitalia** consist of accessory glands and ducts that connect the gonads with the outside environment. The **external genitalia** include all external reproductive structures.

Sexual development is programmed in the human genome. Each nucleated cell of the body except eggs and sperm contains 46 chromosomes. This set of chromosomes is called the *diploid number* because the chromosomes occur in pairs: 22 matched, or *homologous*, pairs of **autosomes** plus one pair of **sex chromosomes** (**FIG. 26.1a**). The amount of DNA in a diploid cell is written as 2n, indicating duplicated chromosomes.

The 22 pairs of autosomal chromosomes in our cells direct development of the human body form and of variable characteristics such as hair color and blood type. The two sex chromosomes, designated as either X or Y, contain genes that direct development of internal and external sex organs. The X chromosome is larger than the Y chromosome and includes many genes that are missing from the Y chromosome.

Eggs and sperm are *haploid* (1n) cells with 23 chromosomes, one from each of the 22 matched pairs plus one sex chromosome. When egg and sperm unite, the resulting zygote then contains a unique set of 46 chromosomes, with one chromosome of each matched pair coming from the mother and the other from the father.

Concept Check

1. Name the male and female gonads and gametes.

Sex Chromosomes Determine Genetic Sex

The sex chromosomes a person inherits determine the genetic sex of that individual. Genetic females are XX, and genetic males are XY (Fig. 26.1b). Females inherit one X chromosome from each parent. Males inherit a Y chromosome from the father and an X chromosome from the mother. The Y chromosome is essential for development of the male reproductive organs.

If sex chromosomes are abnormally distributed at fertilization, the presence or absence of a Y chromosome determines whether development proceeds along male or female lines. The presence of a Y chromosome means the embryo will become male, even if the zygote also has multiple X chromosomes. For instance, an XXY zygote will become male. A zygote that inherits only a Y chromosome (YO) will die because the larger X chromosome contains essential genes that are missing from the Y chromosome.

In the absence of a Y chromosome, an embryo will develop into a female. For this reason, a zygote that gets only one X chromosome (XO; Turner syndrome) will develop into a female. Two X chromosomes are needed for normal female reproductive function, however.

Once the ovaries develop in a female fetus, one X chromosome in each cell of her body is inactivated and condenses into a clump of nuclear chromatin known as a *Barr body*. (Barr bodies in females can be seen in stained cheek epithelium.) The selection of the X chromosome that becomes inactive during development is random: Some cells will have an active maternal X chromosome and others have an active paternal X chromosome. Because inactivation occurs early in development—before cell division is complete—all cells of a given tissue will usually have the same active X chromosome, either maternal or paternal.

Sexual Differentiation Occurs Early in Development

The sex of an early embryo is difficult to determine because reproductive structures do not begin to differentiate until the seventh week of development. Before differentiation, the embryonic tissues are considered *bipotential* because they cannot be morphologically identified as male or female.

The bipotential gonad has an outer cortex and an inner medulla (**FIG. 26.2**). Under the influence of the appropriate developmental signal (described later), the medulla will develop into a testis. In the absence of that signal, the cortex will differentiate into ovarian tissue.

The bipotential internal genitalia consist of two pairs of accessory ducts: **Wolffian ducts** (*mesonephric ducts*) derived from the embryonic kidney, and **Müllerian ducts** (*paramesonephric ducts*). As development proceeds along either male or female lines, one pair of ducts develops while the other degenerates (Fig. 26.2b).

The bipotential external genitalia consist of a *genital tubercle, urethral folds, urethral groove,* and *labioscrotal swellings* (Fig. 26.2a). These structures differentiate into the male and female reproductive structures as development progresses.

FIG. 26.1 Human chromosomes

(a) Humans have 23 pairs of chromosomes: 22 pairs of autosomes and one pair of sex chromosomes. X and Y chromosomes (lower right) mean that these chromosomes came from a male. The autosomes are arranged in homologous pairs in this figure.

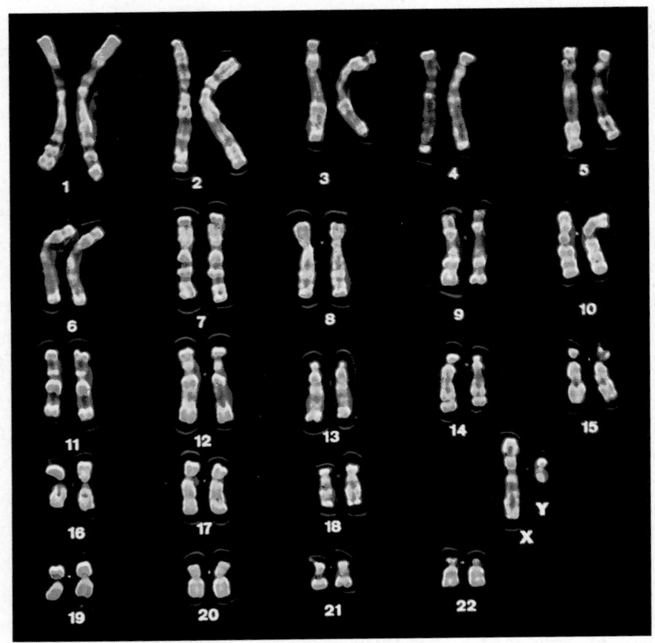
1
2
3
4
5
6
7
8
9
10
11
12
13
14
15
16
17
18
X
Y
19
20
21
22

(b) X and Y chromosomes determine sex. Each egg produced by a female (XX) has an X chromosome. Sperm produced by a male (XY) have either an X chromosome or a Y chromosome.

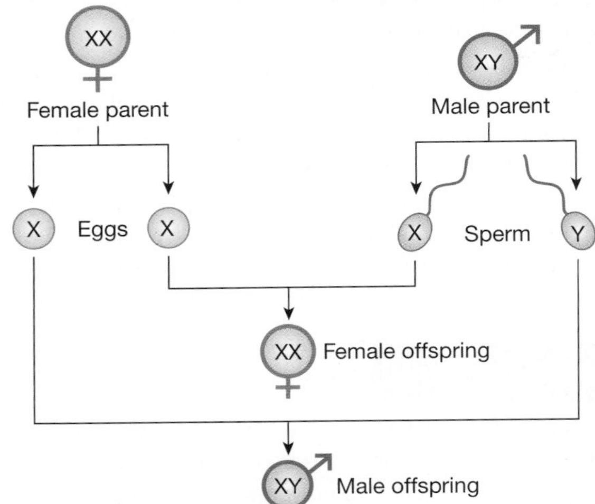
XX
Female parent
XY
Male parent
X
Eggs
X
X
Sperm
Y
XX
Female offspring
XY
Male offspring

What directs some single-cell zygotes to become males, and others to become females? Sex determination depends on the presence or absence of the ***s**ex-determining **r**egion of the **Y** chromosome,* or **SRY gene**. In the presence of a functional SRY gene, the bipotential gonads develop into testes. In the absence of the SRY gene and under the direction of multiple female-specific genes, the gonads develop into ovaries.

FIG. 26.2 ESSENTIALS Sexual Development in the Human Embryo

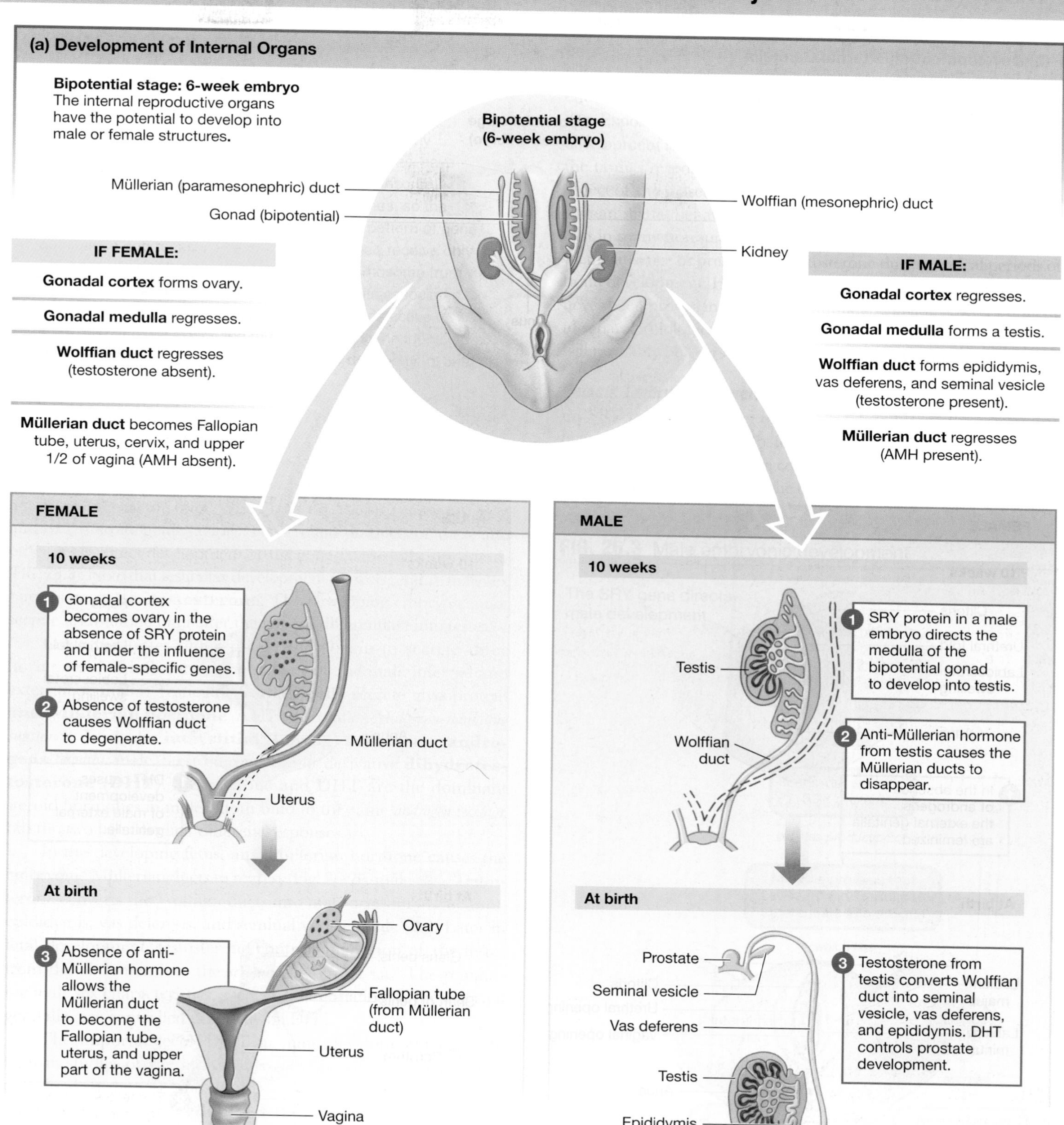

(a) Development of Internal Organs
Bipotential stage: 6-week embryo
The internal reproductive organs have the potential to develop into male or female structures.
Bipotential stage (6-week embryo)
Müllerian (paramesonephric) duct
Gonad (bipotential)
Wolffian (mesonephric) duct
Kidney
IF FEMALE:
Gonadal cortex forms ovary.
Gonadal medulla regresses.
Wolffian duct regresses (testosterone absent).
Müllerian duct becomes Fallopian tube, uterus, cervix, and upper 1/2 of vagina (AMH absent).
IF MALE:
Gonadal cortex regresses.
Gonadal medulla forms a testis.
Wolffian duct forms epididymis, vas deferens, and seminal vesicle (testosterone present).
Müllerian duct regresses (AMH present).
FEMALE
10 weeks
1 Gonadal cortex becomes ovary in the absence of SRY protein and under the influence of female-specific genes.
2 Absence of testosterone causes Wolffian duct to degenerate.
Müllerian duct
Uterus
At birth
Ovary
3 Absence of anti-Müllerian hormone allows the Müllerian duct to become the Fallopian tube, uterus, and upper part of the vagina.
Fallopian tube (from Müllerian duct)
Uterus
Vagina
MALE
10 weeks
Testis
1 SRY protein in a male embryo directs the medulla of the bipotential gonad to develop into testis.
Wolffian duct
2 Anti-Müllerian hormone from testis causes the Müllerian ducts to disappear.
At birth
Prostate
Seminal vesicle
Vas deferens
Testis
Epididymis
3 Testosterone from testis converts Wolffian duct into seminal vesicle, vas deferens, and epididymis. DHT controls prostate development.

In the absence of testicular AMH, the Müllerian ducts develop into the upper portion of the **vagina**, the **uterus**, and the **Fallopian tubes**, named after the anatomist Fallopius, who first described them (Fig. 26.2a female 3). Fallopian tubes are also called *uterine tubes* or *oviducts*. Without testosterone, the Wolffian ducts degenerate (Fig. 26.2a female 2). Without DHT, the external genitalia take on female characteristics.

Concept Check

2. Where in a target cell would you expect to find receptors for androgens? Where would you expect to find receptors for AMH?
3. Why was King Henry VIII of England wrong to blame his wives when they were unable to produce a male heir to the throne?
4. Which sex will a zygote become if it inherits only one X chromosome (XO)?
5. If the testes are removed from an early male embryo, why does it develop a uterus and Fallopian tubes rather than the normal male accessory structures? Will the embryo have male or female external genitalia? Explain.

FIG. 26.4 Synthesis pathways for steroid hormones

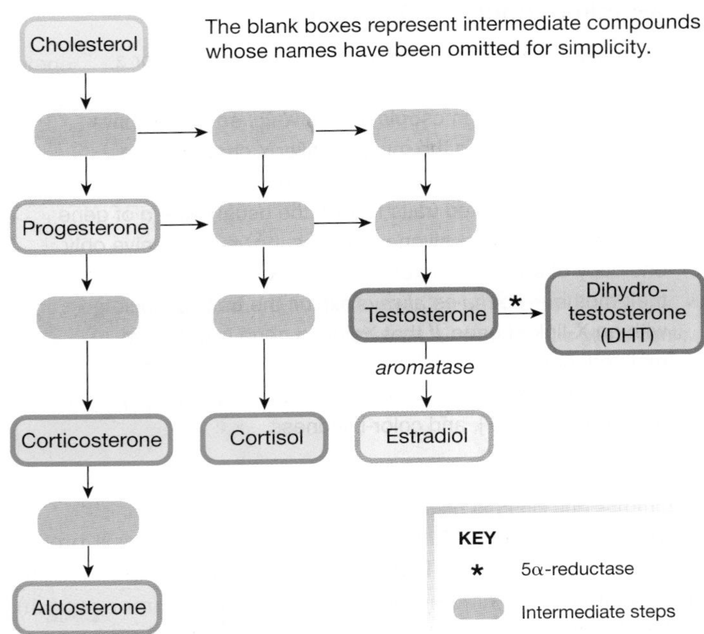
Cholesterol
The blank boxes represent intermediate compounds whose names have been omitted for simplicity.
Progesterone
Testosterone
Dihydro-testosterone (DHT)
aromatase
Corticosterone
Cortisol
Estradiol
Aldosterone
KEY
* 5α-reductase
Intermediate steps

26.2 Basic Patterns of Reproduction

The testis and ovary both produce hormones and gametes, and they share other similarities, as might be expected of organs having the same origin. However, male and female gametes are very different from each other. Eggs are some of the largest cells in the body [Fig. 3.1, p. 60]. They are nonmotile and must be moved through the reproductive tract on currents created by smooth muscle contraction or the beating of cilia. Sperm, in contrast, are quite small. They are the only flagellated cells in the body and are highly motile so that they can swim up the female reproductive tract in their search for an egg to fertilize.

The timing of gamete production, or **gametogenesis**, is also very different in males and females. Most evidence indicates that women are born with all the eggs, or **oocytes**, they will ever have. During the reproductive years, eggs mature in a cyclic pattern and are released from the ovaries roughly once a month. After about 40 years, female reproductive cycles cease (*menopause*).

Men, in contrast, manufacture sperm continuously from the time they reach reproductive maturity. Sperm production and testosterone secretion diminish with age but do not cease as women's reproductive cycles do.

CLINICAL FOCUS

Determining Sex

The first question new parents typically ask about their child is, "Is it a boy or a girl?" Sometimes the answer is not obvious because in approximately 1 in 3000 births, the sex of the child cannot easily be determined. Multiple criteria might be used to establish an individual's sex: genetic, chromosomal, gonadal, morphological, or even psychological characteristics. For example, presence of a Y chromosome with a functional SRY gene could be one criterion for "maleness." However, it is possible for an infant to have a Y chromosome and not appear to be male because of a defect in some aspect of development. Traditionally, sex determination has been based on appearance of the external genitalia at birth, but the idea that individuals should be allowed to choose their sex when they become old enough is gaining ground. The sex a person considers himself or herself to be is called the person's *gender identity.* Learn and read more about the current criteria used to decide a child's sex in cases of ambiguous genitalia in the American Academy of Pediatrics policy statement "Evaluation of the Newborn with Developmental Anomalies of the External Genitalia," *Pediatrics* 106(1): 138–142, 2000 (July).

Gametogenesis Begins in Utero

FIGURE 26.5 compares the male and female patterns of gametogenesis. In both sexes, germ cells in the embryonic gonads first undergo a series of mitotic divisions to increase their numbers 1. After that, the germ cells are ready to undergo **meiosis**, the cell division process that forms gametes.

In the first step of meiosis 2, the germ cell's DNA (2n) replicates until each chromosome is duplicated (46 chromosomes duplicated = 92 chromosomes). The cell, now called a **primary spermatocyte** or **primary oocyte**, contains twice the normal amount of DNA (4n). However, cell and chromosomal division do not take place as they do in mitosis. Instead, each duplicated chromosome forms two identical **sister chromatids**, linked together at a region known as the **centromere**. The primary gametes are then ready to undergo meiotic divisions to create haploid cells.

In the **first meiotic division** 3, one primary gamete divides into two *secondary gametes* (**secondary spermatocyte** or **secondary oocyte**). Each secondary gamete gets one copy of each duplicated autosome plus one sex chromosome. In the **second meiotic division** 4, the sister chromatids separate. In males, the cells split during the second meiotic division, resulting in two haploid (1n) sperm from each secondary spermatocyte.

In females, the second meiotic division creates one egg and one small cell called a *polar body.* What happens after that depends on whether or not the egg is fertilized.

The timing of mitotic and meiotic divisions is very different in males and females. Let's take a closer look at gametogenesis in each sex.

Male Gametogenesis At birth, the testes of a newborn boy have not progressed beyond mitosis and contain only immature germ cells (Fig. 26.5 1). After birth, the gonads become *quiescent* (relatively inactive) until puberty, the period in the early teen years when the gonads mature.

At puberty, germ cell mitosis resumes. From that point onward, the germ cells, known as **spermatogonia** (singular *spermatogonium*), have two possible fates. Some continue to undergo mitosis throughout the male's reproductive life. Others are destined to start meiosis and become primary spermatocytes 2.

Each primary spermatocyte creates four sperm. In the first meiotic division 3, a primary spermatocyte (4n) divides into two secondary spermatocytes. In the second meiotic division 4, each secondary spermatocyte divides into two spermatids. Each spermatid has 23 single chromosomes, the haploid number (1n) characteristic of a gamete. The spermatids then mature into sperm.

Female Gametogenesis In the embryonic ovary, germ cells are called **oögonia** (singular *oögonium*) (Fig. 26.5 1). Oögonia complete mitosis and the DNA duplication stage of meiosis by the fifth month of fetal development, resulting in **primary oocytes** (4n) 2. At birth each ovary contains about half a million primary oocytes. The best evidence indicates that at this time, germ cell mitosis ceases and no additional oocytes can be formed.

In the ovary, meiosis does not resume until puberty 3. If a primary oocyte develops, it divides into two cells, a large **egg (secondary oocyte)** and a tiny **first polar body**. Despite the size difference, the egg and polar body each contain 23 duplicated chromosomes. The first polar body disintegrates.

If the secondary oocyte is selected for ovulation, the second meiotic division takes place just before the egg is released from the ovary 4. The sister chromatids separate but now meiosis pauses again. The final step of meiosis, in which sister chromatids go to separate cells, does not take place unless the egg is fertilized.

The ovary releases the mature egg during a process known as **ovulation**. If the egg is not fertilized, meiosis never goes to completion, and the egg disintegrates 5. If fertilization by a sperm occurs, the final step of meiosis takes place 6. Half the sister chromatids remain in the fertilized egg (zygote), while the other half are released in a **second polar body** (1n). The second polar body, like the first, degenerates. As a result of meiosis, each primary oocyte gives rise to only one egg.

Gametogenesis in both males and females is under the control of hormones from the brain and from endocrine cells in the gonads. Some of these hormones are identical in males and females, but others are different.

Concept Check

6. The gametes in a newborn male are at what stage of development? Is it the same in a newborn female?
7. Compare the amount of DNA in the first polar body with the amount of DNA in the second polar body.
8. How many gametes are formed from one primary oocyte? From one primary spermatocyte?

The Brain Directs Reproduction

The reproductive system has some of the most complex control pathways of the body, in which multiple hormones interact in an ever-changing fashion. The endocrine pathways that regulate reproduction begin with secretion of peptide hormones by the hypothalamus and anterior pituitary. These trophic hormones control gonadal secretion of the steroid sex hormones, including androgens, and the so-called female sex hormones **estrogen** and **progesterone**.

These sex steroids are closely related to one another and arise from the same steroid precursors (Fig. 26.4). Both sexes produce all three groups of hormones, but androgens predominate in males, and estrogens and progesterone are dominant in females.

In men, most testosterone is secreted by the testes, but about 5% comes from the adrenal cortex. Testosterone is converted in peripheral tissues to its more potent derivative DHT. Some of the physiological effects attributed to testosterone are actually the result of DHT activity.

Males synthesize some estrogens, but the feminizing effects of these compounds are usually not obvious in males. Both testes and ovaries contain the enzyme **aromatase**, which converts androgens to estrogens, the female sex hormones. A small amount of estrogen is also made in peripheral tissues.

In women, the ovary produces estrogens (particularly **estradiol** and *estrone*) and *progestins,* particularly progesterone. The ovary and the adrenal cortex produce small amounts of androgens.

Control Pathways The hormonal control of reproduction in both sexes follows the basic hypothalamus-anterior pituitary-peripheral gland pattern (**FIG. 26.6**). **Gonadotropin-releasing hormone (GnRH)*** from the hypothalamus controls secretion of two anterior pituitary **gonadotropins: follicle-stimulating hormone (FSH)** and **luteinizing hormone (LH)**. FSH and LH in turn act on the gonads. FSH, along with steroid sex hormones, is

*GnRH is sometimes called *luteinizing hormone releasing hormone* (*LHRH*) because it was first thought to have its primary effect on LH.

The combination of estrogen and progesterone exerts negative feedback on the hypothalamus and anterior pituitary (Fig. 26.12c). Gonadotropin secretion, further suppressed by luteal inhibin production, remains shut down throughout most of the luteal phase.

Under the influence of progesterone, the endometrium continues its preparation for pregnancy and becomes a secretory structure. Endometrial glands coil, and additional blood vessels grow into the connective tissue layer. Endometrial cells deposit lipids and glycogen in their cytoplasm. These deposits will provide nourishment for a developing embryo while the **placenta**, the fetal-maternal connection, is developing.

Progesterone also causes cervical mucus to thicken. Thicker mucus creates a plug that blocks the cervical opening, preventing bacteria as well as sperm from entering the uterus.

One interesting effect of progesterone is its thermogenic ability. During the luteal phase of an ovulatory cycle, a woman's basal body temperature, taken immediately upon awakening and before getting out of bed, jumps 0.3–0.5 °F and remains elevated until

FIG. 26.12 Hormonal control of the menstrual cycle

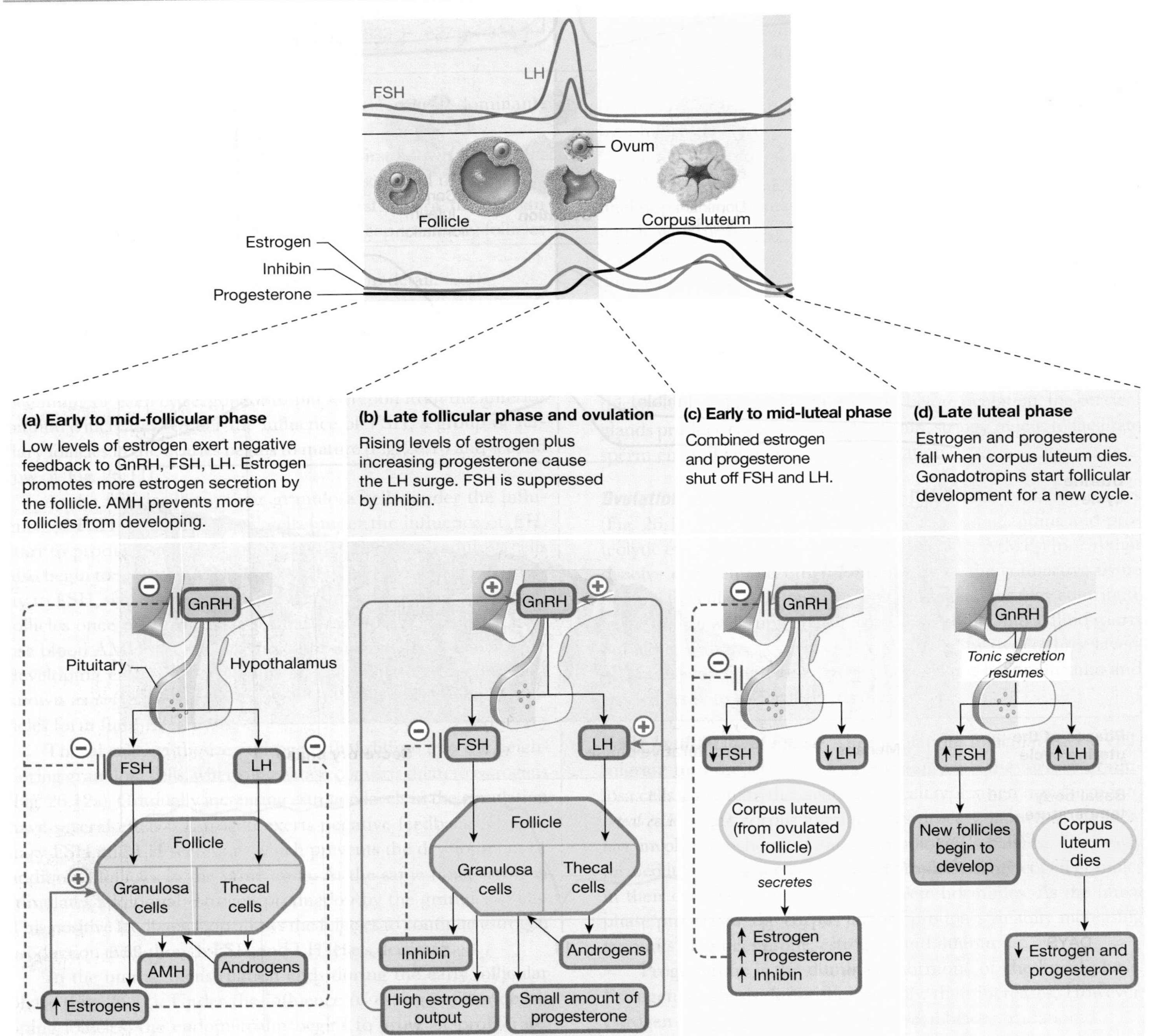
LH
FSH
Ovum
Follicle
Corpus luteum
Estrogen
Inhibin
Progesterone
(a) Early to mid-follicular phase
Low levels of estrogen exert negative feedback to GnRH, FSH, LH. Estrogen promotes more estrogen secretion by the follicle. AMH prevents more follicles from developing.
GnRH
Pituitary
Hypothalamus
FSH
LH
Follicle
Granulosa cells
Thecal cells
AMH
Androgens
↑ Estrogens
(b) Late follicular phase and ovulation
Rising levels of estrogen plus increasing progesterone cause the LH surge. FSH is suppressed by inhibin.
GnRH
FSH
LH
Follicle
Granulosa cells
Thecal cells
Inhibin
Androgens
High estrogen output
Small amount of progesterone
(c) Early to mid-luteal phase
Combined estrogen and progesterone shut off FSH and LH.
GnRH
↓FSH
↓LH
Corpus luteum (from ovulated follicle)
secretes
↑ Estrogen
↑ Progesterone
↑ Inhibin
(d) Late luteal phase
Estrogen and progesterone fall when corpus luteum dies. Gonadotropins start follicular development for a new cycle.
GnRH
Tonic secretion resumes
↑FSH
↑LH
New follicles begin to develop
Corpus luteum dies
↓ Estrogen and progesterone

menstruation. Because this change in the temperature setpoint occurs after ovulation, it cannot be used effectively to predict ovulation. However, it is a simple way to assess whether a woman is having ovulatory or *anovulatory* (non-ovulating) cycles.

Late Luteal Phase and Menstruation The corpus luteum has an intrinsic life span of approximately 12 days. If pregnancy does not occur, the corpus luteum spontaneously undergoes apoptosis. As the luteal cells degenerate, progesterone and estrogen production decrease (Fig. 26.12d). This fall removes the negative feedback signal to the pituitary and hypothalamus, so secretion of FSH and LH increases. The remnants of the corpus luteum become an inactive structure called a **corpus albicans** {*albus*, white}

Maintenance of a secretory endometrium depends on the presence of progesterone. When the corpus luteum degenerates and hormone production decreases, blood vessels in the surface layer of the endometrium contract. Without oxygen and nutrients, the surface cells die. About two days after the corpus luteum ceases to function, or 14 days after ovulation, the endometrium begins to slough its surface layer, and menstruation begins.

Menstrual discharge from the uterus totals about 40 mL of blood and 35 mL of serous fluid and cellular debris. There are usually few clots of blood in the menstrual flow because of the presence of *plasmin* [p. 525], which breaks up clots. Menstruation continues for 3–7 days, well into the follicular phase of the next ovulatory cycle.

Hormones Influence Female Secondary Sex Characteristics

Estrogens control the development of primary sex characteristics in females, just as androgens control them in males. Estrogens also control the most prominent female secondary sex traits: breast development and the female pattern of fat distribution (hips and upper thighs). Other female secondary sex characteristics, however, are governed by androgens produced in the adrenal cortex. Pubic and axillary (armpit) hair growth and libido (sex drive) are under the control of adrenal androgens.

RUNNING PROBLEM

Results of the temperature tracking for a few months reveal that Kate is not ovulating. Dr. Baker suspects Kate might have a condition known as *primary ovarian insufficiency* (POI, also called premature ovarian failure). POI has long been untreatable, preventing infertile couples from having their own baby. Dr. Baker orders more tests to confirm her diagnosis of POI. Kate and Jon are devastated by the news and begin to consider their remaining options: adoption or using an egg from an egg donor for assisted reproductive technologies (ART), more commonly called *in vitro* fertilization.

Q5: *The tests Dr. Baker orders include blood levels of FSH and estrogen. Because POI is failure of the ovaries to produce mature follicles, what do you predict the levels of FSH and estrogen (compared to normal) would be in a woman with POI?*

801 – 811 – 820 – **823** – 827 – 834

Concept Check

16. Name the phases of the ovarian cycle and the corresponding phases of the uterine cycle.
17. What side effects would you predict in female athletes who take anabolic steroids to build muscles?
18. Aromatase converts testosterone to estrogen. What would happen to the ovarian cycle of a woman given an aromatase inhibitor?
19. On what day of the menstrual cycle will a woman with the following cycle lengths ovulate?
    (a) 28 days
    (b) 23 days
    (c) 31 days

26.5 Procreation

Reproduction throughout the animal kingdom is marked by species-specific behaviors designed to ensure that egg and sperm meet. For aquatic animals that release gametes into the water, coordinated timing is everything. Interaction between males and females of these species may be limited to chemical communication by pheromones.

In terrestrial vertebrates, internal fertilization requires interactive behaviors and specialized adaptations of the genitalia. For example, the female must have an internal receptacle for sperm (the vagina in humans), and the male must possess an organ (the penis in humans) that can place sperm in the receptacle. The human penis is *flaccid* (soft and limp) in its resting state, not capable of penetrating the narrow opening of the vagina. In the male sex act, the penis first stiffens and enlarges during **erection**, and then releases sperm from the ducts of the reproductive tract during **ejaculation**. Without these events, fertilization and *procreation*, the production of offspring, cannot take place.

The Human Sexual Response Has Four Phases

The human sex act—also known as sexual intercourse, copulation, or **coitus** {*coitio*, a coming together}—is highly variable in some ways and highly stereotypical in other ways.

Human sexual response in both sexes is divided into four phases: (1) excitement, (2) plateau, (3) orgasm, and (4) resolution. In the excitement phase, various erotic stimuli prepare the genitalia for the act of copulation. For the male, excitement involves erection of the penis. For the female, it includes erection of the clitoris and vaginal lubrication. In both sexes, erection is a state of vascular congestion in which arterial blood flow into spongy erectile tissue exceeds venous outflow.

Erotic stimuli include sexually arousing tactile stimuli as well as psychological stimuli. Because the latter vary widely among individuals and among cultures, what is erotic to one person or in with a spermicidal cream, then inserted into the top of the vagina so they cover the cervix. One advantage to the diaphragm is that it is nonhormonal. When used properly and regularly, diaphragms are highly effective (97–99%). However, they are not always used because they must be inserted close to the time of intercourse, and consequently about 20% of women who depend on diaphragms for contraception are pregnant within the first year. Another female barrier contraceptive is the **contraceptive sponge**, which contains a spermicidal chemical.

The male barrier contraceptive is the **condom**, a closed sheath that fits closely over the penis to catch ejaculated semen. Males have used condoms made from animal bladders and intestines for centuries. Condoms lost popularity when oral contraceptives came into widespread use in the 1960s and 1970s, but in recent years, they have regained favor because they combine pregnancy protection with protection from many sexually transmitted diseases. However, latex condoms may cause allergic reactions, and there is evidence that HIV can pass through pores in some condoms currently produced. A female version of the condom is also commercially available. It covers the cervix and completely lines the vagina, providing more protection from sexually transmitted diseases.

Implantation Prevention Some contraceptive methods do not prevent fertilization but do keep a fertilized egg from establishing itself in the endometrium. They include **intrauterine devices (IUDs)** as well as chemicals that change the properties of the endometrium. IUDs are copper-wrapped plastic devices that are inserted into the uterine cavity, where they kill sperm and create a mild inflammatory reaction that prevents implantation. They have low failure rates (0.5% per year) but side effects that range from pain and bleeding to infertility caused by pelvic inflammatory disease and blockage of the Fallopian tubes. Some IUDs contain progesterone-like hormones.

Hormonal Treatments Techniques for decreasing gamete production depend on altering the hormonal milieu of the body. In centuries past, women would eat or drink various plant concoctions for contraception. Some of these substances actually worked because the plants contained estrogen-like compounds. Modern pharmacology has improved on this method, and now women can choose between oral contraceptive pills, injections lasting months, or a vaginal contraceptive ring.

The **oral contraceptives**, also known as *birth control pills*, first became available in 1960. They rely on various combinations of estrogen and progesterone that inhibit gonadotropin secretion from the pituitary. Without adequate FSH and LH, ovulation is suppressed. In addition, progesterones in the contraceptive pills thicken the cervical mucus and help prevent sperm penetration. These hormonal methods of contraception are highly effective when used correctly but also carry some risks, including an increased incidence of blood clots and strokes, especially in women who smoke.

Development of a male hormonal contraceptive has been slow because of undesirable side effects. Contraceptives that block testosterone secretion or action are also likely to decrease the male libido or even cause impotence. Both side effects are unacceptable to men who would be most interested in using the contraceptive. Some early male oral contraceptives irreversibly suppressed sperm production, which was also unacceptable. A number of clinical trials are currently looking at both hormonal and nonhormonal methods for decreasing male fertility.

Contraceptive vaccines are based on antibodies against various components of the male and female reproductive systems, such as antisperm or antiovum antibodies. However, clinical trials of human vaccines have been disappointing and vaccines may not be a practical contraceptive for humans.

Infertility Is the Inability to Conceive

While some couples are trying to prevent pregnancy, others are spending thousands of dollars trying to get pregnant. *Infertility* is the inability of a couple to conceive a child after a year of unprotected intercourse. For years, infertile couples had no choice but adoption if they wanted to have a child, but incredible strides have been made in this field since the 1970s. As a result, many infertile couples today are able to have children.

Infertility can arise from a problem in the male, the female, or both. Male infertility usually results from a low sperm count or an abnormally high number of defective sperm. Female infertility can be mechanical (blocked Fallopian tubes or other structural problems) or hormonal, leading to decreased or absent ovulation. One problem involving both partners is that the woman may produce antibodies to her partner's sperm. In addition, not all pregnancies go to a successful conclusion. By some estimates, as many as a third of all pregnancies spontaneously terminate—many within the first weeks, before the woman is even aware that she was pregnant.

Some of the most dramatic advances have been made in the field of **assisted reproductive technology (ART)**, strategies in which both sperm and eggs are manipulated. For *in vitro* fertilization, a woman's ovaries are manipulated with hormones to ovulate multiple eggs at one time. The eggs are collected surgically and fertilized outside the body. The developing embryos are then placed in the woman's uterus, which has been primed for pregnancy by hormonal therapy. Because of the expense and complicated nature of the procedure, multiple embryos are usually placed in the uterus at one time, which may result in multiple births. *In vitro* fertilization has allowed some infertile couples to have children, with a 2009 live-birth delivery rate in the United States averaging 37%. ART success varies significantly with age, ranging from 41% for women younger than 35 to 12% for women older than 40.

26.6 Pregnancy and Parturition

Now, we will return to a recently ovulated egg and some sperm deposited in the vagina and follow them through fertilization, pregnancy, and **parturition**, the birth process.

Fertilization Requires Capacitation

Once an egg is released from the ruptured follicle, it is swept into the Fallopian tube by beating cilia. Meanwhile, sperm deposited in the female reproductive tract must go through their final maturation step, **capacitation**, which enables the sperm to swim rapidly and fertilize an egg. The process involves changes in lipids and proteins of the sperm head membrane.

Normally, capacitation takes place in the female reproductive tract, which presents a problem for *in vitro* fertilization. Those sperm must be artificially capacitated by placing them in physiological saline supplemented with human serum. Much of what we know about human fertilization has come from infertility research aimed at improving the success rate of *in vitro* fertilization.

Fertilization of an egg by a sperm is the result of a chance encounter, possibly aided by chemical attractants produced by the egg. An egg can be fertilized for only about 12–24 hours after ovulation. Sperm in the female reproductive tract remain viable for 5–6 days. Apparently they bind to the epithelium of the Fallopian tube while awaiting a chemical signal from a newly ovulated egg.

Fertilization normally takes place in the distal part of the Fallopian tube. Of the millions of sperm in a single ejaculation, only about 100 reach this point. To fertilize the egg, a sperm must penetrate both an outer layer of loosely connected granulosa cells (the *corona radiata*) and a protective glycoprotein coat called the **zona pellucida** (**FIG. 26.14b**). To get past these barriers, capacitated sperm release powerful enzymes from the acrosome in the sperm head, a process known as the **acrosomal reaction**. The enzymes dissolve cell junctions and the zona pellucida, allowing the sperm to wiggle their way toward the egg.

The first sperm to reach the egg quickly finds sperm-binding receptors on the oocyte membrane and binds to the egg (Fig. 26.14c). The fusion of sperm and oocyte membranes triggers a chemical reaction called the **cortical reaction** that excludes other sperm. In the cortical reaction, membrane-bound **cortical granules** in the peripheral cytoplasm of the egg release their contents into the space just outside the egg membrane. These chemicals rapidly alter the membrane and surrounding zona pellucida to prevent **polyspermy**, in which more than one sperm fertilizes an egg.

To complete fertilization, the fused section of sperm and egg membrane opens, and the sperm nucleus sinks into the egg's cytoplasm. This signals the egg to resume meiosis and complete its second division. The final meiotic division creates a second polar body, which is ejected. At this point, the 23 chromosomes of the sperm join the 23 chromosomes of the egg, creating a zygote nucleus with a full set of genetic material.

Once an egg is fertilized and becomes a zygote, it begins mitosis as it slowly makes its way along the Fallopian tube to the uterus, where it will settle for the remainder of the **gestation** period {*gestare*, to carry in the womb}.

The Developing Embryo Implants in the Endometrium

The dividing embryo takes four or five days to move through the Fallopian tube into the uterine cavity (Fig. 26.14d). Under the influence of progesterone, smooth muscle of the tube relaxes, and transport proceeds slowly. By the time the developing embryo reaches the uterus, it consists of a hollow ball of about 100 cells called a **blastocyst**.

Some of the outer layer of blastocyst cells will become the **chorion**, an *extraembryonic membrane* that will enclose the embryo and form the placenta (**FIG. 26.15a**). The inner cell mass of the blastocyst will develop into the embryo and into three other extraembryonic membranes. These membranes include the **amnion**, which secretes *amniotic fluid* in which the developing embryo floats; the **allantois**, which becomes part of the umbilical cord that links the embryo to the mother; and the **yolk sac**, which degenerates early in human development.

Implantation of the blastocyst into the uterine wall normally takes place about seven days after fertilization. The blastocyst secretes enzymes that allow it to invade the endometrium, like a parasite burrowing into its host. As it does so, endometrial cells grow out around the blastocyst until it is completely engulfed.

As the blastocyst continues dividing and becomes an embryo, cells that will become the placenta form fingerlike **chorionic villi** that penetrate into the vascularized endometrium. Enzymes from the villi break down the walls of maternal blood vessels until the villi are surrounded by pools of maternal blood (Fig. 26.15b). The blood of the embryo and that of the mother do not mix, but nutrients, gases, and wastes exchange across the membranes of the villi. Many of these substances move by simple diffusion, but some, such as maternal antibodies, must be transported across the membrane.

The placenta continues to grow during pregnancy until, by delivery, it is about 20 cm in diameter (the size of a small dinner plate). The placenta receives as much as 10% of the total maternal cardiac output. The tremendous blood flow to the placenta is one reason sudden, abnormal separation of the placenta from the uterine wall is a medical emergency.

The Placenta Secretes Hormones During Pregnancy

As the blastocyst implants in the uterine wall and the placenta begins to form, the corpus luteum nears the end of its

CHAPTER 26

RUNNING PROBLEM

A few months after receiving the devastating news, Kate and Jon get a call from their doctor. "I have some potentially good news for you! A treatment has finally become available for women with POI and I am hopeful that it might work for you," explains Dr. Baker. In a recently developed procedure known as *in vitro* activation (IVA), the mother's ovaries are removed, cut into strips, and then inserted back into the abdomen after being treated with egg-stimulating drugs. After approximately six weeks, mature eggs are harvested surgically and fertilized *in vitro*. The zygotes are allowed to develop into early embryos before being frozen for later placement into the mother's uterus.

Q6: *Because of the time delay with IVA, the mother must be treated with hormones to prepare the uterine endometrium for implantation of the embryo. What hormones must Kate be given so that her uterus is able to receive an embryo?*

801 – 811 – 820 – 823 – **827** – 834

mellitus (GDM), with elevated blood glucose levels caused by insulin resistance, similar to type 2 diabetes. The cause is not clear. After delivery, glucose metabolism in most women with GDM returns to normal, but these mothers and their babies are at higher risk of developing type 2 diabetes later in life.

Estrogen and Progesterone Estrogen and progesterone are produced continuously during pregnancy, first by the corpus luteum under the influence of hCG and then by the placenta. With high circulating levels of these steroid hormones, feedback suppression of the pituitary continues throughout pregnancy, preventing another set of follicles from beginning development.

During pregnancy, estrogen contributes to the development of the milk-secreting ducts of the breasts. Progesterone is essential for maintaining the endometrium and also helps suppress uterine contractions. The placenta makes a variety of other hormones, including inhibin and prorenin, but the function of most of them remains unclear.

Pregnancy Ends with Labor and Delivery

Parturition normally occurs between the 38th and 40th week of gestation. What triggers this process? For many years, researchers developed animal models of the signals that initiate parturition, only to discover recently that there are no good nonprimate models that apply to humans. Parturition begins with **labor**, the rhythmic contractions of the uterus that push the fetus out into the world (**FIG. 26.16a**). Signals that initiate these contractions could begin with either the mother or the fetus, or they could be a combination of signals from both.

In many nonhuman mammals, a decrease in estrogen and progesterone levels marks the beginning of parturition. A decrease in progesterone levels is logical, as progesterone inhibits uterine contractions. In humans, however, levels of these hormones do not decrease until labor is well under way.

Another possible labor trigger is oxytocin, the peptide hormone that causes uterine muscle contraction. As a pregnancy nears full term, the number of uterine oxytocin receptors increases. However, studies have shown that oxytocin secretion does not increase until after labor begins. Synthetic oxytocin is often used to induce labor in pregnant women, but it is not always effective. Apparently, the start of labor requires something more than adequate amounts of oxytocin.

Another possibility is that the fetus somehow signals that it has completed development. One theory supported by clinical evidence is that corticotropin-releasing hormone (CRH) secreted by the placenta is the signal to begin labor. (CRH is also a hypothalamic releasing factor that controls release of ACTH from the anterior pituitary.) In the weeks prior to delivery, maternal blood CRH levels increase rapidly. In addition, women with elevated CRH levels as early as 15 weeks of gestation are more likely to go into premature labor.

Although we do not know for certain what initiates parturition, we do understand the sequence of events. In the days prior to the onset of active labor, the cervix softens ("ripens"), and ligaments holding the pelvic bones together loosen as enzymes destabilize collagen in the connective tissue. The control of these processes is not clear and may be due to estrogen or the peptide hormone **relaxin**, which is secreted by ovaries and the placenta.

Once the contractions of labor begin, a positive feedback loop consisting of mechanical and hormonal factors is set into motion. The fetus is normally oriented head down (Fig. 26.16a). At the beginning of labor, it repositions itself lower in the abdomen ("the baby has dropped") and the head pushes on the softened cervix (Fig. 26.16b).

Cervical stretch triggers uterine contractions that move in a wave from the top of the uterus down, pushing the fetus farther into the pelvis. The lower portion of the uterus stays relaxed, and the cervix stretches and dilates. Cervical stretch starts a positive feedback cycle of escalating contractions (Fig. 26.16d). The contractions are reinforced by secretion of oxytocin from the posterior pituitary [p. 207], with continued cervical stretch reinforcing oxytocin secretion.

Prostaglandins are produced in the uterus in response to oxytocin and CRH secretion. Prostaglandins are very effective at causing uterine muscle contractions at any time. They are the primary cause of menstrual cramps and have been used to induce abortion in early pregnancy. During labor and delivery, prostaglandins reinforce the uterine contractions induced by oxytocin (Fig. 26.16d).

As the contractions of labor intensify, the fetus moves down though the vagina and out into the world (Fig. 26.16c), still attached to the placenta. The placenta then detaches from the uterine wall and is expelled a short time later. Uterine contractions clamp the maternal blood vessels and help prevent excessive bleeding, although typically the mother loses about 240 mL of blood.

The Mammary Glands Secrete Milk During Lactation

A newborn has lost its source of maternal nourishment through the placenta and must rely on an external source of food instead. Primates, who normally have only one or two offspring at a time, have two functional mammary glands. A mammary gland is composed of 15–20 milk-secreting lobes (**FIG. 26.17a**). Each lobe branches into lobules, and the lobules terminate in clusters of cells called *alveoli* or *acini*. Each alveolus is composed of secretory epithelium that secretes into a duct, similar to the exocrine secretions of the pancreas [Fig. 21.14, p. 675]. Contractile *myoepithelium* surrounds the alveoli.

During puberty, the breasts begin to develop under the influence of estrogen. The milk ducts grow and branch, and fat deposits behind the glandular tissue. During pregnancy, the glands develop further under the direction of estrogen, growth hormone, and cortisol. The final development step also requires progesterone, which converts the duct epithelium into a secretory structure. This process is similar to progesterone's effect on the uterus, in which progesterone makes the endometrium into a secretory tissue during the luteal phase.

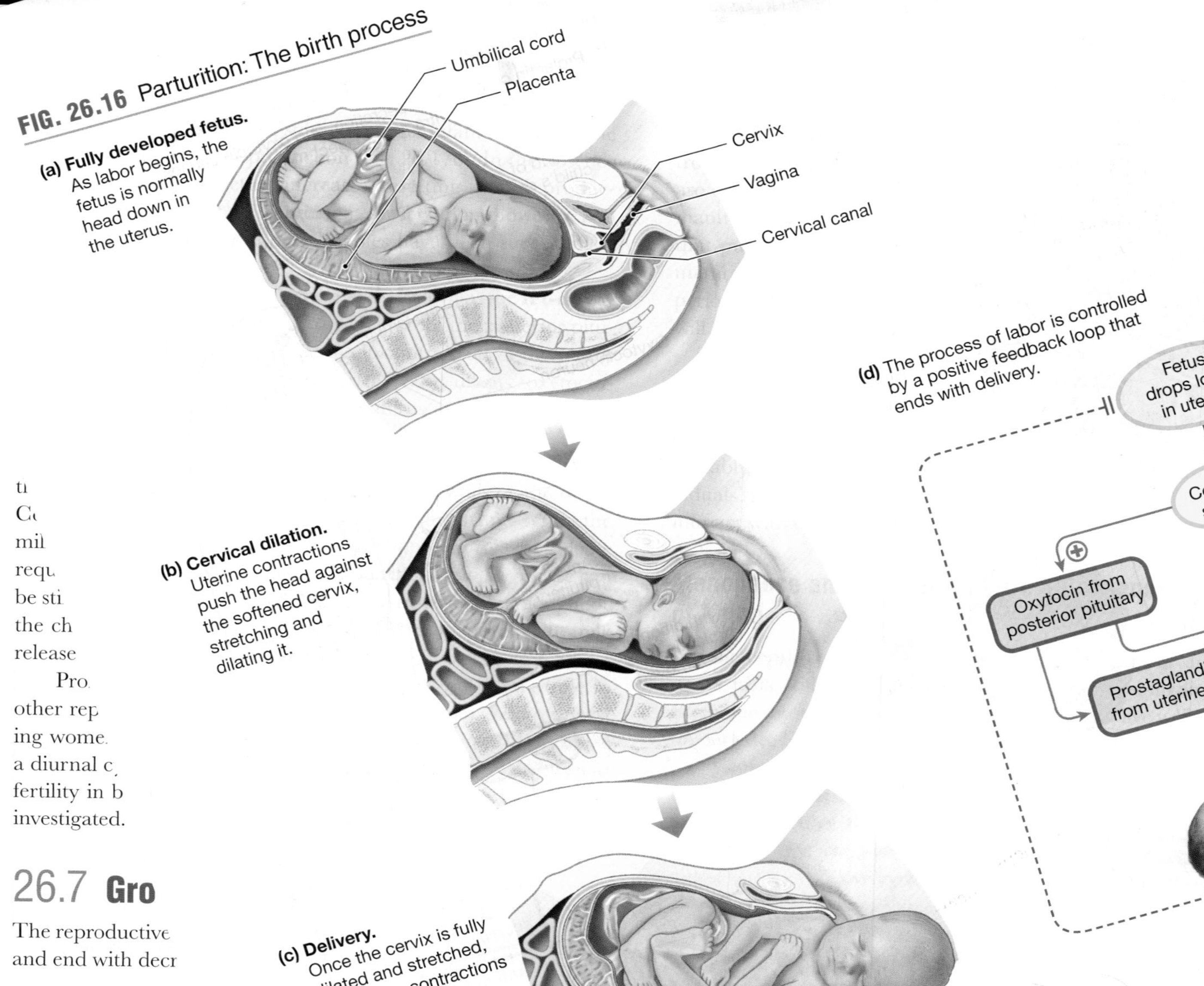
Umbilical cord
Placenta
Cervix
Vagina
Cervical canal
(d) The process of labor is controlled by a positive feedback loop that ends with delivery.
Fetus drops lower in uterus.
Cervical stretch
Oxytocin from posterior pituitary
Prostaglandins from uterine wall
Uteri contra

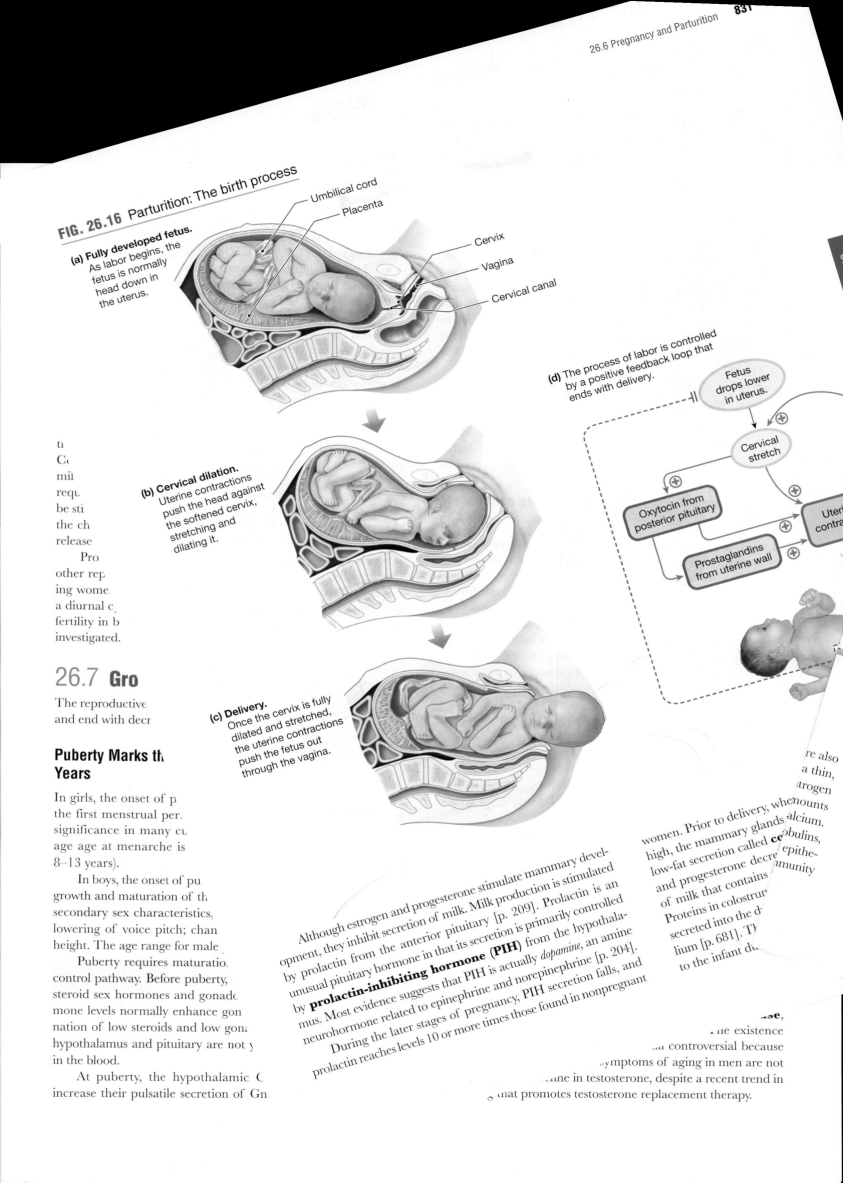

FIG. 26.16 Parturition: The birth process

(a) Fully developed fetus. As labor begins, the fetus is normally head down in the uterus.

(b) Cervical dilation. Uterine contractions push the head against the softened cervix, stretching and dilating it.

(c) Delivery. Once the cervix is fully dilated and stretched, the uterine contractions push the fetus out through the vagina.

ti
C(
mil
requ
be sti
the ch
release

Pro
other rep
ing wome
a diurnal c
fertility in b
investigated.

26.7 Gro

The reproductive
and end with decr

Puberty Marks th Years

In girls, the onset of p
the first menstrual per
significance in many cu
age age at menarche is
8–13 years).

In boys, the onset of pu
growth and maturation of th
secondary sex characteristics,
lowering of voice pitch; chan
height. The age range for male

Puberty requires maturatio
control pathway. Before puberty,
steroid sex hormones and gonado
mone levels normally enhance gon
nation of low steroids and low gon
hypothalamus and pituitary are not
in the blood.

At puberty, the hypothalamic C
increase their pulsatile secretion of Gn

Although estrogen and progesterone stimulate mammary development, they inhibit secretion of milk. Milk production is stimulated by prolactin from the anterior pituitary [p. 209]. Prolactin is an unusual pituitary hormone in that its secretion is primarily controlled by **prolactin-inhibiting hormone (PIH)** from the hypothalamus. Most evidence suggests that PIH is actually *dopamine*, an amine neurohormone related to epinephrine and norepinephrine [p. 204].

During the later stages of pregnancy, PIH secretion falls, and prolactin reaches levels 10 or more times those found in nonpregnant women. Prior to delivery, when
high, the mammary glands
low-fat secretion called **co**
and progesterone decre
of milk that contains
Proteins in colostrum
secreted into the d
lium [p. 681]. Th
to the infant d

re also
a thin,
strogen
nounts
alcium.
bulins,
epithe-
mmunity

se,
the existence
controversial because
symptoms of aging in men are not
line in testosterone, despite a recent trend in
that promotes testosterone replacement therapy.

832 CHAPTER 26 Reproduction and Development

FIG. 26.17 Lactation

(a) Mammary glands

Epithelial cells of the mammary glands secrete milk into the ducts of the gland. Contraction of the myoepithelium ... fluid out of the ducts through openings in th...

Review Questions

LEVEL ONE Reviewing Facts and Terms

1. Pressure gradients, solubility in water, alveolar capillary perfusion, blood pH, temperature.
2. *98%*. Remainder is dissolved in plasma.
3. P_{O_2}, temperature, pH, and the amount of hemoglobin available for binding (most important).
4. Four globular protein chains, each wrapped around a central heme group. Requires iron.
5. *medulla* and *pons*. Dorsal—neurons for inspiration; ventral—neurons for inspiration and active expiration. Central pattern generator—group of neurons that interact spontaneously to control rhythmic contraction of certain muscle groups.
6. Medullary chemoreceptors increase ventilation when P_{CO_2} increases. Carotid body chemoreceptors respond to P_{CO_2}, pH, and P_{O_2} < 60 mm Hg. P_{CO_2} is most important.
7. They include irritant-mediated bronchoconstriction and the cough reflex.
8. Partial pressure gradients
9. Decreased atmospheric P_{O_2}, decreased alveolar perfusion, loss of hemoglobin, increased thickness of respiratory membrane, decreased respiratory surface area, increased diffusion distance.

LEVEL TWO Reviewing Concepts

10. Start with Fig. 18.10.
11. Most oxygen is bound to hemoglobin, not dissolved in the plasma.
12. (a) Most O_2 is transported bound to hemoglobin, but most CO_2 is converted to HCO_3^-. (b) Concentration is amount of gas per volume of solution, measured in units such as moles/L. Partial pressure and concentration are proportional, but concentration is affected by the gas solubility and therefore is not the same as partial pressure.
13. decrease
14. Hypoxia—low oxygen inside cells. COPD—chronic obstructive pulmonary disease (includes chronic bronchitis and emphysema). Hypercapnia—elevated CO_2.
15. Oxygen is not very soluble in water, and the metabolic requirement for oxygen in most multicellular animals would not be met without an oxygen-transport molecule.
16. (a) *x*-axis—ventilation in L/min; *y*-axis—arterial p_{O_2}, in mm Hg. See Fig. 18.9. (b) *x*-axis—arterial p_{CO_2} in mm Hg; *y*-axis—ventilation in L/min. As arterial p_{CO_2} increases, ventilation increases. There is a maximum ventilation rate, and the slope of the curve decreases as it approaches this maximum.
17. (a) increases (b) increases
18. Normal, because p_{O_2} depends on the p_{O_2} of the alveoli, not on how much Hb is available for oxygen transport.
19. (a) See Fig. 18.17. (b) See Fig. 18.13.

LEVEL THREE Problem Solving

20. Increased dead space decreases alveolar ventilation. (a) increases, (b) decreases, (c) increases, (d) decreases
21. Person (a) has slightly reduced dissolved O_2 but at p_{O_2} = 80 Hb saturation is still about 95%. Most oxygen is transported on Hb but the increased p_{O_2} of 100 mm Hg cannot compensate for the decreased hemoglobin content.
22. (a) decrease, (b) decrease, (c) decrease
23. (a) Respiratory movements originate above the level of the cut, which could include any area of the brain. (b) Ventilation depends upon signals from the medulla and/or pons. (c) Respiratory rhythm is controlled by the medulla alone, but other important aspects of respiration depend upon signals originating in the pons or higher.
24. With chronic elevated P_{CO_2}, the chemoreceptor response adapts, and CO_2 is no longer a chemical drive for ventilation. The primary chemical signal for ventilation becomes low oxygen (below 60 mm Hg). Thus, when the patient is given O_2, there is no chemical drive for ventilation, and the patient stops breathing.
25. (a) Alveoli—96%; exercising cell—23% (b) At rest, Bzork only uses about 20% of the oxygen that his hemoglobin can carry. With exercise, his hemoglobin releases more than 3/4 of the oxygen it can carry.
26. All three lines show that as P_{CO_2} increases, ventilation increases. Line A shows that a decrease in P_{CO_2} enhances this increase in ventilation (when compared to line B). Line C shows that ingestion of alcohol lessens the effect of increasing P_{CO_2} on ventilation. Because alcohol is a CNS-depressant, we can hypothesize that the pathway that links increased P_{CO_2} and increased ventilation is integrated in the CNS.
27. Apical—faces airspace; basolateral—faces interstitial fluid. Apical side has ENaC and aquaporins; basolateral side has aquaporins and Na^+-K^+-ATPase. Na^+ enters the cell through ENaC, then is pumped out the basolateral side. (Cl^- follows to maintain electrical neutrality.) Translocation of NaCl allows water to follow by osmosis.

LEVEL FOUR Quantitative Problems

28. 1.65 mL O_2/gm Hb
29. 247.5 mL O_2/min
30. Nothing. The percent saturation of Hb is unchanged at any given P_{O_2}. However, with less Hb available, less oxygen will be transported.

CHAPTER 19

Concept Check Questions

1. hyperpolarizes (becomes more negative)
2. Force of contraction decreases.
3. *greater than*
4. *less than*
5. Alike: both represent movement from ECF into the lumen. Filtration is only into Bowman's capsule; secretion takes place along the rest of the tubule.
6. Glomerulus → Bowman's capsule → proximal tubule → loop of Henle → distal tubule → collecting duct → renal pelvis → ureter → urinary bladder → urethra
7. The body would run out of plasma in under an hour.
8. Osmotic pressure is higher in efferent arterioles due to same amount of protein in a smaller volume.
9. Mean arterial pressure is 119 mm Hg and GFR is 180 L/day.
10. Renal blood flow and GFR decrease.
11. With fewer plasma proteins, the plasma has lower-than-normal colloid osmotic pressure opposing GFR, so GFR increases.
12. Creatinine clearance = (1.5 mg creatinine/mL urine × 1.1L L urine/day)/1.8 mg creatinine/100 mL plasma = 92 L/day. GFR = 92 L/day.

Figure Questions

Fig. 19.2: 1. (a) Bowman's capsule, (b) proximal tubule, loop of Henle, distal tubule, collecting duct, (c) proximal tubule, distal tubule, collecting duct, (d) collecting duct. 2. (a) $18/180 = 10\%$, $1.5/180 = 0.8\%$

Fig. 19.3: $E = F - R + S$. 79 mmol/day $= 720 - R + 43$. $R =$ 684 mmol reabsorbed per day.

Fig. 19.4: 120 mL/min $\times$ 1440 min/day $=$ 172,800 mL/day filtered $=$ 172.8 L. 172.8 L $=$ 20% of plasma flow. Plasma flow $=$ 864 L/day.

Fig. 19.6: Capillary blood pressure, GFR, and renal blood flow all increase.

Fig. 19.9: The transport rate at 3 mg/mL is 3 mg/min; at 5 and 8 mg/mL, it is 4 mg/min. The transport rate is 2 mg/min at a plasma concentration of 2 mg/mL.

Fig. 19.11: Pressure is lower because blood flowing out of the glomerulus loses pressure as it moves along the peritubular capillaries.

Try It! Following insulin injections, glucose excretion decreased. If blood glucose concentrations are in the normal range, there should be no glucose in the urine because all filtered glucose is reabsorbed. If blood glucose concentrations exceed the renal threshold, carriers saturate and glucose remains in the tubule. The significance of reabsorbing all glucose is that it is an important nutrient and the only fuel source for the brain.

Review Questions

LEVEL ONE Reviewing Facts and Terms

1. color (concentration), odor (infection or excreted substances), clarity (presence of cells), taste (presence of glucose), and froth (presence of proteins)
2. regulation of extracellular fluid volume (to maintain adequate blood pressure), regulation of osmolarity, maintenance of ion balance (neuron function), regulation of pH (proteins denature if pH not maintained), excretion of wastes and foreign substances (to prevent toxic effects), and production of hormones (that regulate RBC synthesis, Ca^{2+} and Na^+ balance)
3. 20–25%
4. nephrons through ureters to urinary bladder (storage), leaving through the urethra
5. (a), (e), (b), (g), (f), (d), (c), (h)
6. Glomerular capillary endothelium, basal lamina, and epithelium of Bowman's capsule. Blood cells and most plasma proteins are excluded.
7. Capillary hydrostatic pressure promotes filtration. Fluid pressure in Bowman's capsule and colloid osmotic (oncotic) pressure of plasma oppose it. Net driving force is the sum of these pressures.
8. GFR—glomerular filtration rate. 125 mL/min or 180 L/day.
9. (a) Found where distal tubule passes between afferent and efferent arterioles. Composed of macula densa cells in the distal tubule and granular cells in arteriole wall. (b) Macula densa paracrine signals control autoregulation of GFR and renin secretion. (c) Alter the size of filtration slits. (d) Specialized epithelial cells that surround glomerular capillaries. Changes in slit size alter GFR. (e) An internal smooth muscle sphincter that is passively contracted and an external skeletal muscle sphincter that is tonically (actively) contracted. (f) Outer layer of the kidney that contains renal corpuscles, proximal and distal tubules, and parts of the loop of Henle and collecting ducts.
10. 70% occurs in the proximal tubule. Reabsorbed molecules go into the peritubular capillaries and the systemic venous circulation. If filtered and not reabsorbed, a molecule is excreted in the urine.
11. (a) 2, 3, 5; (b) 3, 4; (c) 4, 7; (d) 6; (e) 5, 7
12. penicillin, K^+, and H^+
13. creatinine
14. urination

LEVEL TWO Reviewing Concepts

15. Use Figs. 19.5 to 19.7.
16. (a) Filtration and secretion both move material from blood to tubule lumen, but filtration is a bulk flow process while secretion is a selective process. Excretion is also bulk flow but involves movement from the kidney lumen to the outside world. (b) Saturation—all transporter binding sites are occupied by ligand. Transport maximum—the maximum rate at which carriers are saturated by substrate. Renal threshold—plasma concentration at which saturation occurs. (c) Creatinine and inulin—compounds used to determine GFR. Penicillin and probenecid—xenobiotics that are secreted. (d) Clearance—rate at which plasma is cleared of a substance (mL plasma cleared of substance X/min). GFR—filtration rate of plasma (mL plasma filtered/min). Excretion—removal of urine, mL urine/min.
17. Allows rapid removal of foreign substances that are filtered but not reabsorbed.
18. Afferent arteriole constricts, GFR decreases. Efferent arteriole constricts, GFR increases.
19. See Fig. 19.15. Toilet training allows higher brain centers to inhibit the reflex until an appropriate time. Higher brain centers can also initiate the reflex.
20. Bladder smooth muscle contracts under parasympathetic control, so blocking muscarinic receptors decreases bladder contraction.

LEVEL THREE Problem Solving

21. See Fig. 19.8. Place transporters as described. Cl^- moves between the cells.
22. (a) Inulin is filtered, secreted, and excreted. No evidence for reabsorption is presented. (b) The line indicating net secretion will be close to the filtration line until the slope changes, after which the secretion line is horizontal (no further increase in rate due to saturation).
23. Dialysis fluid should resemble plasma without waste substances, such as urea. This will allow diffusion of solutes and water from the blood into the dialysis fluid, but diffusion will stop at the desired concentration. To remove excess water from the blood, you can make the dialysis fluid more concentrated than plasma.
24. Filtration line: Use several plasma concentrations of Z (0–140 mg Z/mL plasma) $\times$ GFR. The line will be a straight line beginning at the origin and extending upward to the right. Secretion reaches its maximum rate of 40 mg/min at 80 mg Z/mL plasma. Plot that point. Draw the secretion line from the origin to that point. Above the renal threshold, secretion rate does not change, so the line becomes horizontal. Excretion line: Add the filtration rate and secretion rate at a number of plasma concentrations of Z.

LEVEL FOUR Quantitative Problems

25. 1 mg X/mL plasma $\times$ 125 mL plasma/min $=$ 125 mg X filtered/*min*. Same values for inulin. Inulin excretion $=$ filtration $=$ 125 mg inulin excreted/min. Cannot say what the excretion rate of X is because there is insufficient information.
26. 1 L/min
27. First specimen clearance $=$ 1000 L plasma/day. Normally, creatinine clearance $=$ GFR. However, this value is not at all realistic for

12. Bile salts do not digest triglycerides. They emulsify them into small particles so that lipase can digest them.
13. The ASBT faces the lumen and brings Na^+ and bile acid into the enterocyte together. The OAT transports bile acid by itself out of the enterocyte and into the ECF.
14. Enzymes function best in a restricted range of pH. Stomach enzymes must be active at acidic pH; salivary and intestinal enzymes work best in alkaline pH.
15. They are activated by trypsin.
16. Damage to epithelial cells means that brush border sucrase may be less effective or absent. In these cases, sucrose digestion would be impaired, so it would be better to use a solution containing glucose because this sugar need not be digested before being absorbed.

Figure Questions

Fig. 21.1: Glands are the salivary glands. Organs are liver, gallbladder, and pancreas.

Fig. 21.5: 1. The myenteric plexus controls smooth muscle contraction; the submucosal plexus controls smooth muscle contraction and endocrine and exocrine secretions by secretory cells. 2. Stretch receptors respond to stretch, osmoreceptors to osmolarity, and chemoreceptors to products of digestion.

Fig. 21.10: 1. The vagal input is parasympathetic, which stimulates digestion. 2. Neurotransmitter is ACh, and receptor is muscarinic.

Review Questions

LEVEL ONE Reviewing Facts and Terms

1. (a) 2; (b) 3; (c) 4; (d) 7, 10; (e) 8; (f) 2, 3, 7; (g) 9
2. *absorption* and *digestion; secretion* and *motility*. By not regulating absorption and digestion, the body ensures that it will always absorb the maximum available nutrients.
3. Digestion—chemical or mechanical breakdown of nutrients (proteins). Absorption—transport from lumen to ECF (water). Secretion—transport from ECF to lumen (enzymes). Motility—movement of material through the digestive tract.
4. Layers (lumen outward): mucosa (epithelium, connective tissue, and smooth muscle), submucosa (connective tissue), musculature (smooth muscle), serosa (connective tissue).
5. Secretory epithelium (endocrine and exocrine) lines the stomach; absorptive epithelium with a few secretory cells lines the intestines.
6. Peyer's patches—nodes of lymphoid tissue. M cells—epithelial cells that transfer information from gut lumen to Peyer's patches.
7. Moves food through the GI tract and helps mix food with secretions. Results from contraction of longitudinal and circular muscle layers to create propulsive peristaltic movements or mixing segmental movements.
8. An inactive digestive proenzyme. Must have a segment of protein chain removed to activate. Examples: pepsinogen-pepsin, trypsinogen-trypsin.
9. (a) 8, 9; (b) 3; (c) 1, 3, 7; (d) 1, 7; (e) 6; (f) 2; (g) 4; (h) 5
10. (a) Increases surface area for enzymes to work; stomach and small intestine. (b) Motility and secretion along the length of the digestive tract. (c) Acidic pH in stomach helps break down food and digest microorganisms. Must be neutralized in the small intestine. (d) Size determines the surface area upon which enzymes can act.
11. *capillaries, hepatic portal system, liver, lymphatic, basement membrane (basal lamina)*
12. ENS: network of neurons within the GI tract that can sense a stimulus, integrate information, and create an appropriate response without integration or input from the CNS. Also interacts with the CNS through sensory and autonomic neurons.
13. Short reflexes—mediated entirely within the ENS; regulate secretion and motility. Long reflexes—GI reflexes integrated in the CNS.
14. Paracrines help mediate secretion and motility. Examples: serotonin (5-HT) and histamine.

LEVEL TWO Reviewing Concepts

15. Map 1: Use Figs. 21.6 to 21.7 and 21.16 to 21.18, then add details. Map 2: Use Fig. 21.19a.
16. (a) Mastication—chewing; deglutition—swallowing. (b) Villi—folds of intestine; microvilli—folds of cell membrane. Both increase surface area. (c) All patterns of GI muscle contraction. Migrating motor complex—move material from stomach to large intestine between meals. Peristalsis—progressive waves of contraction. Segmental contraction—contraction and relaxation of short intestinal segments. Mass movements—push material into rectum, triggering defecation. (d) Chyme—semidigested food and secretions, produced in the stomach. Feces—solid waste material that remains after digestion and absorption are complete; produced in the large intestine. (e) Short reflexes—integrated within the ENS. Long reflexes—integrated within the CNS. (f) Submucosal plexus—ENS in the submucosal layer. Myenteric plexus—ENS that lies between muscle layers of the GI tract wall. Vagus nerve—carries sensory and efferent signals between the brain and ENS. (g) Cephalic phase—digestive reflexes triggered by stimuli received in the brain. Gastric phase—short reflexes that begin with food entering the stomach. Intestinal phase—begins when chyme enters the small intestine.
17. (a) See Fig. 21.19c. (b) See Figs. 21.9c and 21.14c.
18. Both use similar neurotransmitters and neuromodulators (serotonin, VIP, NO). Enteric support cells are similar to CNS astroglia. GI capillaries are not very permeable, like blood-brain barrier. Both act as integrating centers.
19. See Tbl. 21.1 for specific hormones.
20. See Figs. 21.9c and 21.10.

LEVEL THREE Problem Solving

21. Hepcidin causes enterocytes to destroy ferroportin transporters. If hepcidin is absent or not functional, intestinal iron uptake cannot be down-regulated when iron levels become too high, so these patients have elevated plasma levels of iron.
22. severe diarrhea $\rightarrow$ loss of small intestine HCO_3^- $\rightarrow$ metabolic acidosis
23. (a) Ingestion of a fatty meal triggers contraction of the gallbladder to release bile salts, but the blocked bile duct prevented bile secretion, causing pain. (b) Micelle formation—decreased due to lack of bile salts. Carbohydrate digestion—decreased because pancreatic secretions with amylase not able to pass blockage. Protein absorption—decreased slightly because of low pancreatic secretion; however, brush border enzymes also digest protein, so digestion does not stop completely when the bile duct is blocked. Therefore, some digested proteins will be absorbed.
24. Apical membrane has ENaC (Na^+ leak channel) and K^+ leak channels. Basolateral membrane has the Na^+-K^+-ATPase. At high flow, saliva has more Na^+ and less K^+.

LEVEL FOUR Quantitative Problems

25. (a) MIT started out with equal concentrations in both solutions, but by the end of the experiment MIT was more concentrated on the serosal side. Therefore, MIT must be moving by active transport. (b) MIT moves apical to basolateral, which is absorption. (c) Transport across the apical membrane goes from bath into tissue. Tissue MIT is more concentrated than bath. Therefore, this must be active transport. (d) Transport across the basolateral membrane goes from tissue into the sac. Tissue MIT is more concentrated than sac fluid, so this must be passive transport.

CHAPTER 22

Concept Check Questions

1. Feeding center causes an animal to eat, and satiety center causes an animal to stop eating. Both centers are in the hypothalamus.
2. Might be abnormal tissue responsiveness—a target cell with no leptin receptors or defective receptors. There might also be a problem with leptin's signal transduction/second messenger pathway.
3. age, sex, lean muscle mass, activity, diet, hormones, and genetics
4. One g of fat contains more than twice the energy of 1 g of glycogen.
5. $C_6H_{12}O_6 + 6\ O_2 \rightarrow 6\ CO_2 + 6\ H_2O$
6. $RQ = CO_2/O_2 = 6/6 = 1$
7. GLUT transporters are passive facilitated diffusion transporters.
8. dL is the abbreviation for deciliter, or 1/10th of a liter (100 mL).
9. Bile acid sequestrants and ezetimibe leave bile salts and cholesterol in the intestinal lumen to be excreted, so possible side effects are loose, fatty feces and inadequate absorption of fat-soluble vitamins.
10. Glycogenesis is glycogen synthesis; gluconeogenesis is synthesis of glucose from amino acids or glycerol.
11. Amino acids used for energy become pyruvate or enter the citric acid cycle.
12. Plasma cholesterol is bound to a carrier protein and can't diffuse across the cell membrane.
13. The primary target tissues for insulin are liver, skeletal muscle, and adipose tissue.
14. If glucose uptake depended on insulin, the intestine, kidney tubule, and neurons could not absorb glucose in the fasted state. Neurons use glucose exclusively for metabolism and must always be able to take it up.
15. During fight-or-flight, skeletal muscles need glucose for energy. Inhibiting insulin secretion causes the liver to release glucose into the blood and prevents adipose cells from taking it up, making more glucose available for exercising muscle, which does not require insulin for glucose uptake.
16. No, you would not get the same result because you would not be ingesting the same amount of glucose. Table sugar is sucrose: half glucose and half fructose. Most soft drinks are sweetened with high-fructose corn syrup.
17. Insulin is a protein and is digested if administered orally.
18. Although dehydrated patients may have elevated plasma K^+ concentrations, their total *amount* of K^+ is below normal. If fluid volume is restored to normal with no K^+ added, a below-normal K^+ concentration results.
19. Sitagliptin inhibits DPP4, the enzyme that breaks down GLP-1 and GIP. Prolonging the action of these two gut hormones enhances insulin release and slows digestion, which gives cells time to take up and use absorbed glucose.
20. Possible side effects include hypoglycemia from excessive loss of glucose and dehydration from osmotic diuresis.
21. Norepinephrine binds to α-receptors to elicit vasoconstriction.
22. Researchers probably classified the neurons as sympathetic because of where they leave the spinal cord.
23. Water cooler than body temperature draws heat away from the body through conductive heat transfer. If this loss exceeds the body's heat production, the person feels cold.
24. A person exercising in a humid environment loses the benefit of evaporative cooling and is likely to overheat faster.

Figure Questions

Fig. 22.5: 1. (a) next to left arrow from G-6-P to glycogen, (b) next to arrow going from acetyl CoA to fatty acids, (c) next to right arrow from glycogen to G-6-P, (d) with electron transport system. 2. No, amino acids entering the citric acid cycle cannot be used to make glucose because the step from pyruvate to acetyl CoA is not reversible.

Fig. 22.6: Because they are made from 2-carbon acyl units.

Fig. 22.7: The decrease from 190 to 160 mg/dL has the greatest effect.

Fig. 22.10: hydrolysis

Fig. 22.15: acetylcholine on muscarinic receptor

Review Questions

LEVEL ONE Reviewing Facts and Terms

1. Metabolic—all pathways for synthesis or energy production, use, or storage. Anabolic—primarily synthetic; catabolic—break down large molecules into smaller ones.
2. Transport (moving molecules across membranes), mechanical work (movement of muscles), chemical work (protein synthesis).
3. The amount of heat required to raise the temperature of 1 L water by 1°C. In direct calorimetry, food is burned to see how much energy it contains.
4. The ratio of CO_2 produced to O_2 used in cellular metabolism. Typical RQ is 0.82.
5. BMR—an individual's lowest metabolic rate, measured at rest after sleep and a 12-hour fast. Higher in adult males because females have more adipose tissue with a lower respiration rate. Factors that affect BMR: age, physical activity, lean muscle mass, diet, hormones, and genetics.
6. Broken down for energy, used for synthesis, or stored.
7. Absorptive state—anabolic reactions and nutrient storage. Postabsorptive—mobilizes stored nutrients for energy and synthesis.
8. A group of nutrients (glucose, free fatty acids, and amino acids), mostly in the blood, available for cell use.
9. To maintain adequate glucose supply for the brain.
10. glycogen and adipose tissue fat
11. Proteins: protein synthesis, energy, and conversion to fat for storage. Fats: lipid synthesis, energy, and storage as fats.
12. Insulin decreases blood glucose and glucagon increases it.
13. amino acids and glycerol, gluconeogenesis.
14. Excessive breakdown of fatty acids, as occurs in starvation. Can be burned as fuel by neurons and other tissues. Many ketone bodies are strong acids and can cause metabolic acidosis.

15. Increased plasma glucose or amino acids and parasympathetic input stimulate, sympathetic input inhibits.
16. Type 1: absolute lack of insulin. Type 2: cells do not respond normally to insulin. Both: elevated fasting blood glucose levels. Type 1: body uses fats and proteins for fuel. Type 2: not as severe because the cells can use some glucose.
17. Low plasma glucose or increased plasma amino acids stimulate. Primary target—liver, which increases glycogenolysis and gluconeogenesis.
18. (a) Capillary endothelium enzyme that converts triglycerides into free fatty acids and monoglycerides. (b) Co-secreted with insulin; slows gastric emptying and gastric acid secretion. (c) "Hunger hormone" secreted by the stomach. (d) Hypothalamic peptide that increases food intake. (e) Protein components of lipoproteins. Apoprotein B on LDL-C facilitates transport into most cells. (f) "Satiety hormone" produced by adipocytes. (g) Loss of water in the urine due to high amounts of urine solutes. Hyperglycemia causes dehydration through osmotic diuresis. (h) Target cells fail to respond normally to insulin.
19. (a) stimulates, (b) inhibits, (c) stimulates, (d) stimulates, (e) stimulates

LEVEL TWO Reviewing Concepts

20. Use Figs. 22.5 and 22.8. Try different colors for each organ or hormone.
21. Glucagon and insulin cycle according to food intake, but both hormones are always present in some amount. So it appears that the ratio rather than an absolute amount of hormone determines the direction of metabolism.
22. (a) Glucose—monosaccharide. Glycogenolysis—glycogen breakdown. Glycogenesis—glycogen production from glucose. Gluconeogenesis—glucose synthesis from amino acids and fats. Glucagon—hormone that increases plasma glucose. Glycolysis—first pathway in glucose metabolism for ATP production. (b) Thermogenesis—heat production by cells. Shivering thermogenesis—muscle twitches produce heat as a by-product. Nonshivering thermogenesis occurs in all cells. Diet-induced thermogenesis—heat generated by digestive and anabolic reactions during the absorptive state. (c) Lipoproteins—transport molecules. Chylomicrons—lipoprotein complexes assembled in intestinal epithelium and absorbed into lymphatic system. Cholesterol—steroid component of cell membranes and precursor to steroid hormones. HDL-C—takes cholesterol into liver cells, where it is metabolized or excreted. LDL-C—elevated concentrations are associated with atherosclerosis. Apoprotein—protein component of lipoproteins. (d) Calorimetry—measurement of energy content and a means of determining metabolic rate. Direct calorimetry—measuring heat production when food is burned. Indirect calorimetry—measures oxygen consumption or CO_2 production. (e) Conductive heat loss—loss of body heat to a cooler object. Radiant heat loss—loss from production of infrared electromagnetic waves. Convective heat loss—upward movement of warm air and its replacement by cooler air. Evaporative heat loss—heat lost when water evaporates. (f) Absorptive state—following a meal, when anabolism exceeds catabolism. Postabsorptive state—catabolism exceeds anabolism.
23. (a) Hyperglycemia results from the lack of insulin production and failure of cells to take up and use glucose. (b) Glucosuria results when filtered glucose exceeds the kidney's capacity to reabsorb it. (c) Polyuria results from osmotic diuresis caused by glucosuria. (d) Ketosis results from increased fatty acid metabolism. (e) Dehydration is a consequence of polyuria due to osmotic diuresis. (f) Severe thirst is a consequence of dehydration.
24. If a person ingests a pure protein meal and only insulin is released, blood glucose concentrations might fall too low. Glucagon co-secretion ensures that blood glucose remains within normal levels.
25. See Fig. 22.1. The satiety center inhibits the feeding center.
26. See Figs. 22.22 and 22.23.

LEVEL THREE Problem Solving

27. Amino acids in excess of what's needed for protein synthesis are stored as glycogen or fat.
28. As insulin secretion (x-axis) increases, plasma glucose (y-axis) decreases.
29. Some other neurotransmitter besides acetylcholine (which binds to muscarinic receptors) is involved in the vasodilation reflex.
30. (a) [See Fig. 18.9 on p. 574.] Acidosis shifts the curve to the right. Low BPG would shift curve to the left [Fig. 18.9f]. The net effect would be close to normal oxygen binding. (b) As pH normalizes, curve shifts back to the left. With BPG still low, curve would be between the left shift for low BPG and normal. Oxygen release after treatment would therefore be less than normal.

LEVEL FOUR Quantitative Problems

31. (a) Height = 1.555 m and weight = 45.909 kg. BMI = 19, which is normal. (b) Answers vary.
32. Fat: 6 g × 9 kcal/g = 54 kcal. Carbohydrate: 30 g × 4 kcal/g = 120 kcal. Protein: 8 g × 4 kcal/g = 32 kcal. Total = 206 kcal. 54/206 = 26% of calories from fat.

CHAPTER 23

Concept Check Questions

1. The medulla secretes catecholamines (epinephrine, norepinephrine), and cortex secretes aldosterone, glucocorticoids, and sex hormones.
2. Androstenedione is a prohormone for testosterone. Testosterone is anabolic for skeletal muscle, which might give an athlete a strength advantage.
3. HPA = hypothalamic-pituitary-adrenal. CBG = corticosteroid-binding. globulin or transcortin
4. This immediate stress response is too rapid to be mediated by cortisol and must be a fight-or-flight response mediated by the sympathetic nervous system and catecholamines.
5. No, because cortisol is catabolic on muscle proteins.
6. Primary and iatrogenic hypercortisolism: ACTH is lower than normal because of negative feedback. Secondary hypercortisolism: ACTH is higher because of the ACTH-secreting tumor.
7. Addison's disease: high ACTH due to reduced corticosteroid production and lack of negative feedback.
8. ACTH is secreted during stress. If the stress is a physical one caused by an injury, the endogenous opioid β-endorphin can decrease pain and help the person continue functioning.
9. In peripheral tissues, T_4 is converted to T_3, which is the more active form of the hormone.
10. When mitochondria are uncoupled, energy normally captured in ATP is released as heat. This raises the person's body temperature and causes heat intolerance.
11. prolactin
12. Normal growth and development require growth hormone, thyroid hormone, insulin, and insulin-like growth factors.
13. Their epiphyseal plates have closed.

14. Hypercalcemia hyperpolarizes the membrane potential, which makes it harder for the neuron to fire an action potential.
15. The Na^+-Ca^{2+} exchanger is a secondary active transporter, and the Ca^{2+}-ATPase is an active transporter.
16. ATP and phosphocreatine store energy in high-energy phosphate bonds.
17. A kinase transfers a phosphate group from one substrate to another. A phosphatase removes a phosphate group, and a phosphorylase adds one.

Figure Questions

Fig. 23.1: 1. A baby born with deficient 21-hydroxylase would have low aldosterone and cortisol levels and an excess of sex steroids, particularly androgens. Low cortisol would decrease the child's ability to respond to stress. Excess androgen would cause masculinization in female infants. 2. Women, who synthesize more estrogens, would have more aromatase.

Fig. 23.2: ACTH = adrenocorticotropic hormone or corticotropin. CRH = corticotropin-releasing hormone. MSH = melanocyte-stimulating hormone.

Fig. 23.4: 1. The apical membrane faces the colloid, and the basolateral membrane faces the ECF. 2. I^- comes into the cell by secondary active transport (co-transport with Na^+). 3. and 4. Thyroglobulin moves between colloid and cytoplasm by exocytosis and endocytosis. 5. Thyroid hormones leave the cell by an unknown membrane transporter.

Fig. 23.7: A pituitary tumor hypersecreting TSH would cause hyperthyroidism and an enlarged thyroid gland. The pathway would show decreased TRH resulting from short-loop negative feedback from TSH to the hypothalamus, increased TSH caused by the tumor, elevated thyroid hormones, but no negative feedback from the thyroid hormones to the anterior pituitary because the tumor does not respond to feedback signals.

Fig. 23.15: Other cells that use CA and the anion exchanger include red blood cells, parietal cells, cells of the exocrine pancreas and intestine, and intercalated cells of the nephron.

Review Questions

LEVEL ONE Reviewing Facts and Terms

1. Zona glomerulosa (aldosterone), zona fasciculata (glucocorticoids), zona reticularis (sex steroids, primarily androgens).
2. (a) corticotropin releasing hormone (hypothalamus) → adrenocorticotropic hormone (anterior pituitary) → cortisol (adrenal cortex) feeds back to inhibit secretion of both CRH and ACTH. (b) growth hormone releasing hormone and somatostatin (hypothalamus) → growth hormone (anterior pituitary). Negative feedback by IGFs. (c) decreased blood Ca^{2+} → parathyroid hormone (parathyroid glands) → increases blood Ca^{2+} by increasing reabsorption of bone, among other effects → negative feedback inhibits secretion of PTH. (d) Thyrotropin releasing hormone (hypothalamus) → thyroid stimulating hormone (thyrotropin) (anterior pituitary) → triiodothyronine (T_3) and thyroxine (T_4) (thyroid gland) → negative feedback to hypothalamus and anterior pituitary
3. Conditions: adequate diet, absence of chronic stress, and adequate amounts of thyroid and growth hormones. Other important hormones: insulin, IGFs (somatomedins), and sex hormones at puberty.
4. Triiodothyronine (T_3) and tetraiodothyronine (T_4 or thyroxine). T_3 is the more active; most of it is made from T_4 in peripheral tissues.
5. (a) Include ACTH (cortisol secretion) and MSH (not significant in humans). (b) Bone loss that occurs when bone reabsorption exceeds bone deposition. (c) The inorganic portion of bone matrix, mostly calcium salts. (d) Steroid hormones that regulate minerals, i.e., aldosterone. (e) Spongy bone, with an open latticework. (f) Pro-opiomelanocortin, inactive precursor to ACTH and other molecules. (g) Growth zones in long bones, comprised of cartilage.
6. Functions: blood clotting, cardiac muscle excitability and contraction, skeletal and smooth muscle contraction, second messenger systems, exocytosis, tight junctions, strength of bones and teeth.
7. In this table, A indicates anabolism, C indicates catabolism, and CHO = carbohydrate.

HORMONE	PROTEIN	CHO	FAT
Cortisol	C (skeletal muscle)	C	C
Thyroid	A (children), C (adults)	C	C
GH	A	C	C
Insulin	A	A	A
Glucagon	C	C	C

LEVEL TWO Reviewing Concepts

8. (a) CRH and ACTH low, cortisol high; (b) CRH low, ACTH and cortisol high; (c) TRH and TSH low, thyroid high; (d) TRH high, TSH and thyroid hormones low
9. (a) Hypothalamic CRH stimulates anterior pituitary secretion of ACTH, which stimulates adrenal cortex (zona fasciculata) secretion of glucocorticoids such as cortisol. (b) Thyroid gland follicle cells secrete colloid from which thyroid hormones are produced; C cells secrete calcitonin. (c) Thyroid hormone synthesis is controlled by TSH, whose release is controlled by TRH. In the thyroid gland, tyrosine and iodine combine on thyroglobulin to make thyroid hormones. Thyroid-binding globulin (TBG) carries thyroid hormones in the blood. Target cell deiodinase removes iodine from T_4 to make T_3. (d) Growth hormone releasing hormone (GHRH) stimulates anterior pituitary secretion of growth hormone (GH or somatotropin). Somatostatin (GHIH) inhibits production of GH. Growth hormone binding protein binds about half the GH in the blood. Insulin-like growth factors (IGFs) from the liver act with GH to promote growth. (e) Dwarfism results from severe GH deficiency in childhood. Giantism results from hypersecretion of GH during childhood. Acromegaly is lengthening of jaw and growth in hands and feet, caused by hypersecretion of GH in adults. (f) Hyperplasia—increased cell number. Hypertrophy—increased cell size. (g) Osteoblasts—bone cells that secrete organic bone matrix. Osteocytes—inactive form of osteoblasts. Chondrocytes—cartilage cells. Osteoclasts—bone-destroying cells. (h) PTH increases blood Ca^{2+} by stimulating bone and renal reabsorption, and intestinal absorption of Ca^{2+}. Calcitriol (1,25-dihydroxycholecalciferol) is a vitamin D derivative that mediates PTH effect on intestinal absorption of Ca^{2+}. Calcitonin decreases bone reabsorption of Ca^{2+}. Estrogen promotes bone deposition.
10. Thyroid hormones have intracellular receptors, so you expect 60–90 minute onset of action. However, effects on metabolic rate are apparent within a few minutes and are thought to be related to changes in ion transport across cell and mitochondrial membranes.
11. Equivalents = ion's molarity × the number of charges/ion. 2.5 mmoles $Ca^{2+} \times 2 = 5$ mEq Ca^{2+}.
12. See Fig. 23.15. The cell uses carbonic anhydrase to make H^+ from $CO_2 + H_2O$. Apical membrane: H^+-ATPase secretes H^+. Basolateral membrane secretes HCO_3^- with Cl^--HCO_3^- antiporter.

LEVEL THREE Problem Solving

13. Physiological stress stimulates secretion of cortisol, which increases blood glucose. Increased insulin opposes this effect.
14. Normal response: dexamethasone → ACTH suppression → decrease in cortisol. Patient A: no response to dexamethasone suggests adrenal hypersecretion that is insensitive to ACTH. Patient B: Dexamethasone decreases cortisol, suggesting that the problem is in the pituitary.
15. Mr. A—elevated TSH. Ms. B—low TSH. Ms. C—elevated TSH. (a) Not possible to determine if the lab slip has the results of Mr. A or Ms. C without knowing the thyroid hormone levels. (b) Ms. B can be ruled out, because her TSH would be low if the tentative diagnosis is correct.
16. (a) People in all age groups showed vitamin D insufficiency at the end of winter. Deficiency was most pronounced in the 18–29 age group and least pronounced in the 50+ age group. At the end of summer, fewer subjects were deficient in vitamin D. Variables: season when blood collected, age group, and percent of people with vitamin D insufficiency. (b) Energy from the sun is required for precursors in the skin to be converted to vitamin D. Days are shorter in the winter and the sun is at a more oblique angle and its rays are weaker. Also, at northern latitudes like Boston, people spend less time outside during the winter. This explains the seasonal difference. Fewer than half the people tested were deficient, however, suggesting that most people consumed enough vitamin D. The biggest seasonal difference was in the 18–29 age group, who probably spent more time outside in the summer than members of other groups. (c) Taking multivitamin supplements containing vitamin D should reduce vitamin D insufficiency.
17. Osteopetrosis is characterized by excessive bone formation due to loss of osteoclast function. This can close up the normal holes (foramena) in bone and compress nerves running through the openings, causing changes in vision and hearing. Bone growth can also fill in the central marrow space of bones, leading to decreased production of red and white blood cells and platelets.

LEVEL FOUR Quantitative Problems

18. (a) 5 mgCa^{2+}/L plasma × 125 mL plasma filtered/min × 1440 min/day = 900 mg Ca^{2+}filtered/day. (b) To remain in Ca^{2+} balance, he must excrete 170 mg/day. (c) 900 mg filtered − 170 mg excreted = 730 mg reabsorbed. 730/900 = 81%.
19. Graph A: x-axis = plasma parathyroid hormone concentration, hormone concentration, y-axis = plasma Ca^{2+}concentration. Graph line goes up to the right. Graph B: x-axis = plasma Ca^{2+}concentration, y-axis = plasma parathyroid hormone concentration. parathyroid hormone concentration. Graph starts high at low x-axis values and goes down to the right.

CHAPTER 24

Concept Check Questions

1. Antibodies can be moved across cells by transcytosis or released from cells by exocytosis.
2. The child developed antibodies to bee venom on first exposure, so if the child will have a severe allergic reaction to bee venom, it will occur on the second exposure. Allergic reactions are mediated by IgE. After the first bee sting, IgE antibodies secreted in response to the venom are bound to the surface of mast cells. At the second exposure, bee venom binding to the IgE causes the mast cells to degranulate, resulting in an allergic reaction that may be severe enough to cause anaphylaxis.
3. When capillary permeability increases, proteins move from plasma to interstitial fluid. This decreases the colloid osmotic force opposing capillary filtration, and additional fluid accumulates in the interstitial space (swelling or edema).
4. The AB recipient has no A or B antibodies and will not react to RBCs of any blood type.
5. The type O recipient's anti-A antibodies will cause agglutination of the type A blood cells.

Figure Questions

Fig. 24.4: Either the lymphocytes live for a long time or they can reproduce themselves outside the thymus gland.

Fig. 24.11: The antigen will activate clone 1.

Fig. 24.17: Steps 3 and 5 are cell-mediated immunity. Steps 1 and 4 are humoral immunity.

Fig. 24.19:

		Mother		
		A	**B**	**O**
Father	**A**	**A**	AB	AO
		A	AB	A
	B	A	BB	BO
		A	B	B
	O	A	BO	OO
		A	B	O

Review Questions

LEVEL ONE Reviewing Facts and Terms

1. The body's ability to defend itself against disease-causing pathogens. Memory—immune cells remember prior exposure to an antigen and create a stronger immune response. Specificity—antibodies that target specific antigens.
2. Thymus gland, bone marrow, spleen, lymph nodes, and diffuse lymphoid tissues
3. Protect the body against foreign pathogens; remove dead or damaged tissues and cells; recognize and remove abnormal "self" cells.
4. Detect the pathogen, recognize it as foreign, organize a response, and recruit assistance from other cells. If a pathogen cannot be destroyed, it may be suppressed.
5. (a) Severe IgE-mediated allergic reaction with widespread vasodilation, circulatory collapse, and bronchoconstriction. (b) To clump together. When blood cells are exposed to an antibody, an antibody-antigen reaction may cause the blood cells to agglutinate. (c) Outside the blood vessels. Many immune reactions are extravascular. (d) Release of cytoplasmic granule chemicals into the ECF. (e) Opsonins that coat pathogens; released in the early stages of injury or infection. (f) One cell of a clone divides to make many identical cells. (g) Ability of the immune system to find and destroy abnormal cells (especially cancerous).
6. They are all names given to specialized tissue macrophages before scientists recognized that they were the same cell type.
7. It consists of monocytes and macrophages, which ingest and destroy invaders and abnormal cells
8. (a) 5, (b) 1, (c) 3, (d) 6, (e) 2, (f) 4

9. Physical: skin, mucous membranes and mucus. Mechanical: respiratory mucociliary escalator, coughing and sneezing, GI motility. Chemical: lysozymes, opsonins, enzymes, and antibodies.
10. B lymphocytes secrete antibodies; T lymphocytes and NK cells kill infected cells. T lymphocytes bind to antigen presented by MHC complexes; NK cells can also bind to antibodies coating foreign cells.
11. The ability of the body's immune system to ignore the body's own cells (self-antigen). Occurs because T lymphocytes that react with "self" cells are eliminated by clonal deletion. If self-tolerance fails, the body makes antibodies against itself (autoimmune disease).
12. the study of brain-immune interactions
13. Stress—nonspecific stimulus that disturbs homeostasis. Stressor—stimulus that causes stress. General adaptation syndrome—stress response that includes activation of the adrenal glands (fight-or-flight response by adrenal medulla and cortisol secretion by the cortex).

LEVEL TWO Reviewing Concepts

14. Use figures and tables of the chapter to create the map.
15. When lymph nodes trap bacteria, activated immune cells create a localized inflammatory response with swelling and cytokine activation of nociceptors that create the pain sensation.
16. (a) Pathogen—any organism that causes disease. Microbes—microscopic organisms, pathogens or not. Pyrogens—fever-causing chemicals. Antigens—substances that trigger an immune response and react with products of the response. Antibodies—disease-fighting chemicals produced by the body. Antibiotics—drugs that destroy bacteria and fungi. (b) Infection—illness caused by pathogens, especially viruses or bacteria. Inflammation—nonspecific response to cell damage or invaders, including nonpathogens such as a splinter. Allergy—inflammatory response to a nonpathogenic invader, such as plant pollen. Autoimmune disease—body creates antibodies to its own cells. (c) Allergens—nonpathogenic substances that create allergic reactions. Bacteria—cellular microorganisms. Viruses—acellular parasites that must invade the host's cells to reproduce. (d) Chemotaxins—chemicals that attract immune cells. Cytokines—peptides made on demand and secreted for action on other cells. Opsonins—proteins that coat and tag foreign material so that it can be recognized by the immune system. Interferons—lymphocyte cytokines that aid in the immune. (e) Innate immunity—same as nonspecific, present from birth; adaptive—directed at specific invaders. Adaptive can be divided into cell-mediated and humoral (antibodies). (f) Immediate hypersensitivity response—mediated by antibodies; occurs within minutes of exposure to allergen. Delayed—may take several days to develop; mediated by helper T cells and macrophages. (g) All chemicals of the immune response. Membrane attack complex and perforin are membrane pore proteins. Perforins allow granzymes (cytotoxic enzymes) to enter the cell.
17. Active immunity comes from the body's own reaction to pathogens. Passive immunity is the acquisition of antibodies made by another organism (i.e., being given a shot of gamma globulin).
18. See Fig. 24.7. Fc region—determines antibody class; Fab region—antigen-binding sites that confer the antibody's specificity.
20. See Fig. 24.16.
21. See Fig. 24.17.
22. See Fig. 24.18.

LEVEL THREE Problem Solving

22. Type O—universal donor because these RBCs lack A or B surface antigens and do not trigger an immune response. Type AB—universal recipient because these RBCs have both A and B antigens and no A or B antibodies.
23. Maxie and baby are both OO. Snidley could be either BB or BO. Baby received an O gene from Maxie, and could have received the other O gene from Snidely. Thus, it is possible that Snidley is the father of Maxie's baby.
24. Emotional stress → increases cortisol secretion → immune system suppression. Also likely that students are spending more time inside and having closer contact with fellow students.
25. Barbara's immune cells recognize her connective tissue as an antigen, and attack it. Autoimmune diseases often begin in association with an infection and are thought to represent cross-reactivity of antibodies that developed because of the infection. Stress reduces the immune system's ability to suppress the autoantibodies and the inflammation they cause.
26. Increase in neutrophils—bacterial infection because neutrophils eat bacteria. Increase in eosinophils—parasitic infection because eosinophils kill parasites.

CHAPTER 25

Concept Check Questions

1. If venous P_{O_2} decreases, P_{O_2} in the cells is also decreasing.
2. The mean blood pressure line lies closer to the diastolic pressure line because the heart spends more time in diastole than systole.
3. The neurons are classified as sympathetic because of where they originate along the spinal cord.

Figure Questions

Fig. 25.6: 1. Arterial P_{O_2} remains constant because pulmonary ventilation is matched to blood flow through the lungs. 2. Although arterial P_{O_2} is constant, oxygen delivery to cells increases due to increased cardiac output (not shown). 3. Venous P_{O_2} drops as exercise increases because cells remove more oxygen from hemoglobin as oxygen consumption increases. 4. Arterial P_{CO_2} does not increase because increased production is matched by increased ventilation. 5. As the person begins to hyperventilate, arterial (and alveolar) P_{CO_2} decreases.

Fig. 25.7: Blood flow to an organ is calculated by multiplying cardiac output (L/min) times the percentage of flow to that organ. When rest and exercise values are compared, actual blood flow decreases only in the kidneys, GI tract, and "other tissues."

Fig. 25.8: Mean arterial pressure is cardiac output × resistance. If resistance is falling but MAP is increasing, then cardiac output must be increasing.

Review Questions

LEVEL ONE Reviewing Facts and Terms

1. ATP and phosphocreatine
2. *aerobic, both glucose and fatty acids*
3. Aerobic metabolism: requires O_2; glucose goes through glycolysis and citric acid cycle; produces 30–32 ATP/glucose through oxidative phosphorylation. Anaerobic: no O_2 used; glucose undergoes glycolysis to lactate; produces only 2 ATP/glucose.

FIG.C.1 Phases of cell division

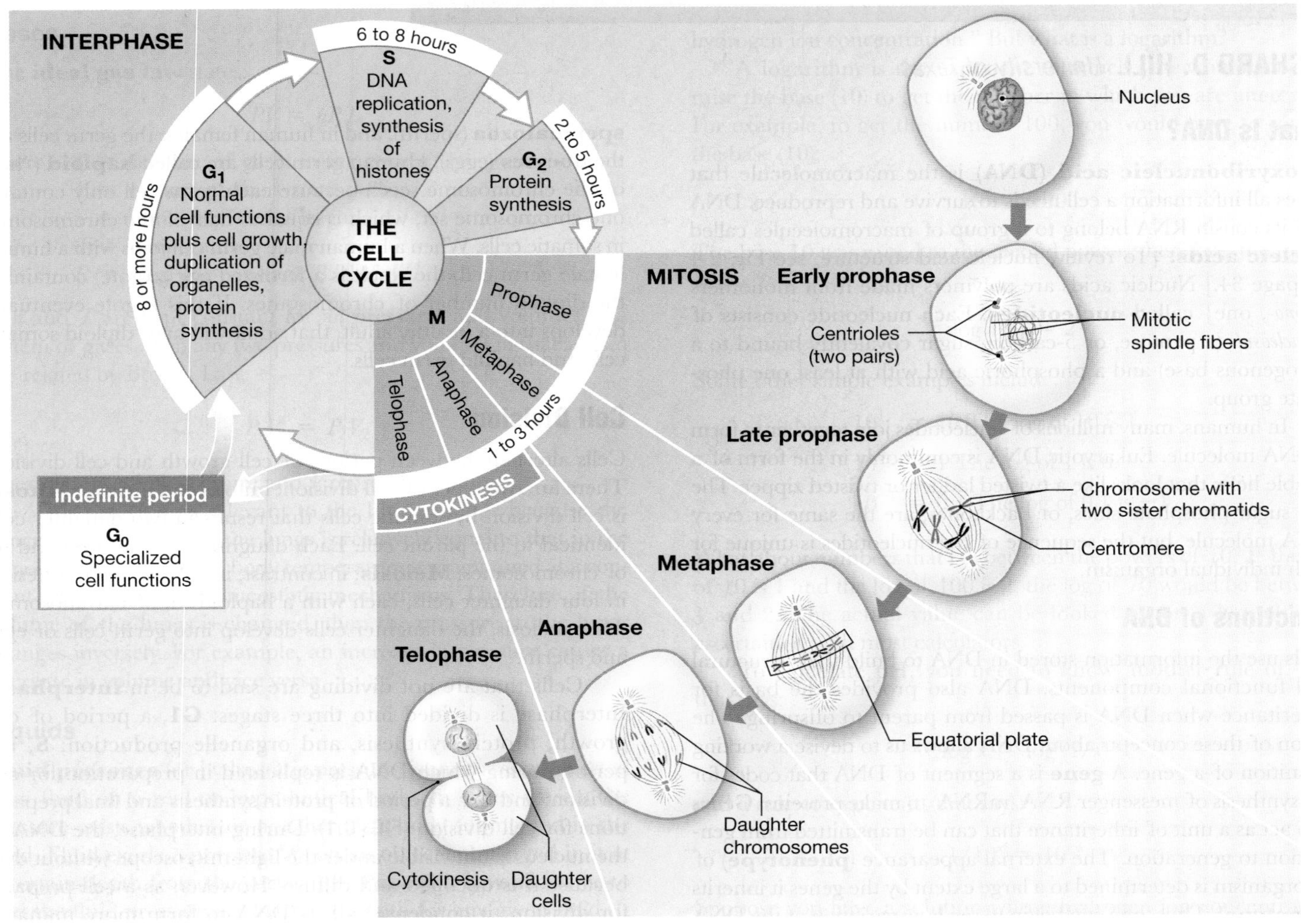
INTERPHASE
6 to 8 hours
S
DNA replication, synthesis of histones
2 to 5 hours
G2
Protein synthesis
G1
Normal cell functions plus cell growth, duplication of organelles, protein synthesis
8 or more hours
THE CELL CYCLE
M
Prophase
Metaphase
Anaphase
Telophase
1 to 3 hours
CYTOKINESIS
Indefinite period
G0
Specialized cell functions
Nucleus
MITOSIS
Early prophase
Centrioles (two pairs)
Mitotic spindle fibers
Late prophase
Chromosome with two sister chromatids
Centromere
Metaphase
Equatorial plate
Anaphase
Telophase
Daughter chromosomes
Cytokinesis
Daughter cells

The four main steps of mitosis are **prophase**, **metaphase**, **anaphase**, and **telophase** (Fig. C.1). The entire somatic cell cycle can be remembered by the acronym, IPMAT, in which the "I" stands for interphase and the other letters stand for the steps of mitosis that follow.

Prophase

During prophase, chromatin becomes condensed and microscopically visible as duplicate chromosomes. The duplicated chromosomes form **sister chromatids**, which are joined to each other at the **centromere**. The cell's centriole pair duplicates and the two centriole pairs move to opposing ends of the cell. The **mitotic spindle**, composed of microtubules, assembles between the centriole pairs. The nuclear membrane begins to break down and disappears by the end of prophase.

Metaphase

In metaphase, mitotic spindle fibers extending from the centrioles attach to the centromere of each chromosome. The 46 chromosomes, each consisting of a pair of sister chromatids, line up at the "equator" of the cell.

Anaphase

During anaphase, the spindle fibers pull the sister chromatids apart, so that an identical copy of each chromosome moves toward each pole of the cell. By the end of anaphase, an identical set of 46 chromosomes is present at each pole. At this point, the cell has a total of 92 chromosomes, double the diploid number.

Telophase

The actual division of the parent cell into two daughter cells takes place during telophase. In **cytokinesis**, the cytoplasm divides when an actin contractile ring tightens at the midline of the cell. The result is two separate daughter cells, each with a full diploid set of chromosomes. The spindle fibers disintegrate, nuclear envelopes form around the chromosomes in each cell, and the chromatin returns to its loosely coiled state.

DNA Replication

The information stored in DNA is encoded in the nucleotide sequence of the molecule. When nucleotides link together, the phosphate group of one nucleotide bonds covalently to the sugar

FIG.C.2 DNA: Levels of organization

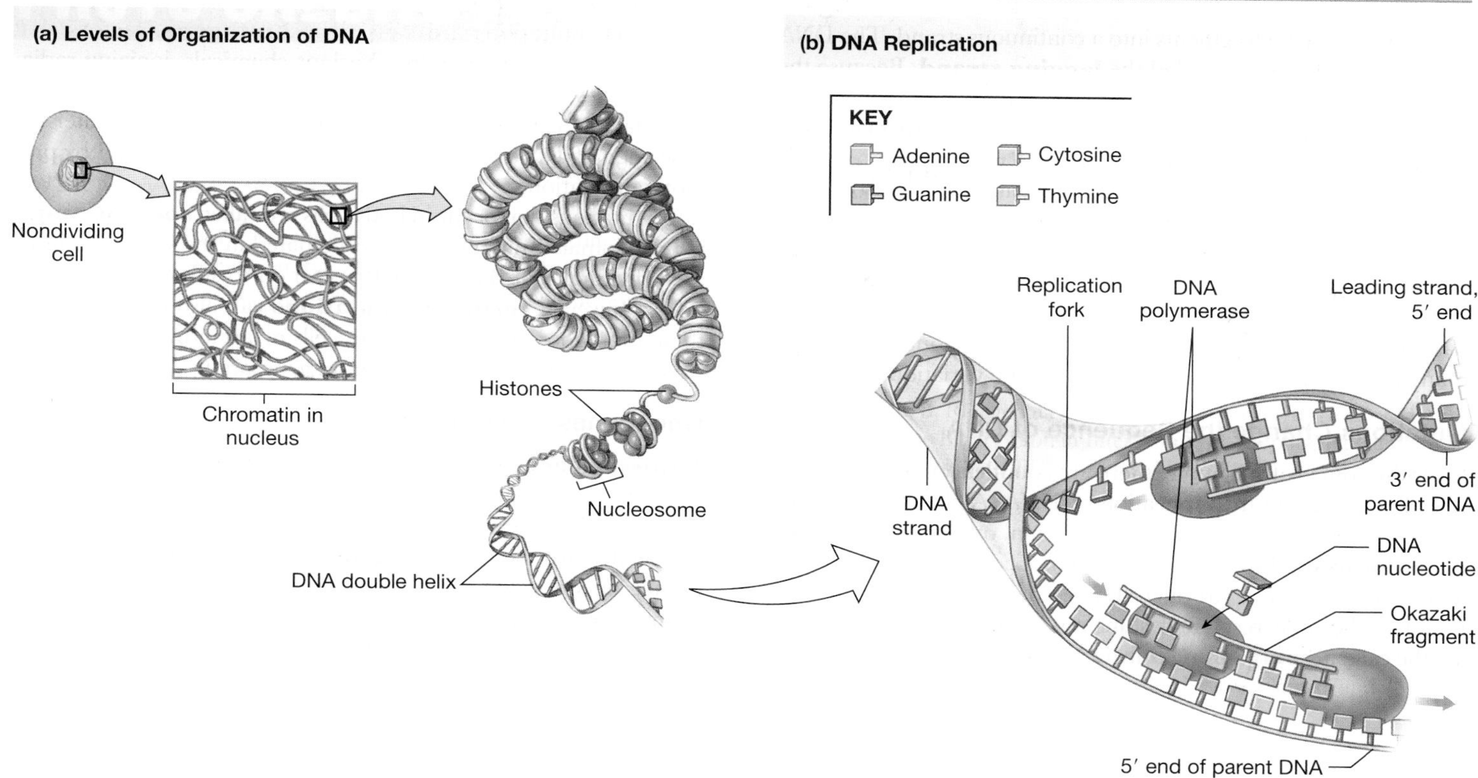
(a) Levels of Organization of DNA
Nondividing cell
Chromatin in nucleus
Histones
Nucleosome
DNA double helix
(b) DNA Replication
KEY
Adenine
Cytosine
Guanine
Thymine
Replication fork
DNA polymerase
Leading strand, 5′ end
DNA strand
3′ end of parent DNA
DNA nucleotide
Okazaki fragment
5′ end of parent DNA

group of the adjacent nucleotide. The end of the polymer that has an unbound sugar is called the 3′ ("three prime") end. The end of the polymer with the unbound phosphate is called the 5′ end. A DNA molecule has four types of nucleotides, distinguished by their nitrogenous bases.

The nitrogenous bases in nucleic acids are classified as either **purines** or **pyrimidines**. The purine bases are **guanine (G)** and **adenine (A)**. The pyrimidine bases are **cytosine (C)**, and either **thymine (T)**, found in DNA only, or **uracil (U)**, found in RNA only. To remember which DNA bases are pyrimidines, look at the first syllable. The word "pyrimidine" and names of the DNA pyrimidine bases all have a "y" in the first syllable.

The "rungs" of the DNA double helix are created when the nitrogenous bases on one DNA strand form hydrogen bonds with nitrogenous bases on the adjoining DNA strand. This phenomenon is called **base-pairing**. The base-pairing rules are as follows:

1. Purines pair only with pyrimidines.
2. Guanine (G) bonds with cytosine (C) in both DNA and RNA.
3. Adenine (A) bonds with thymine (T) in DNA or with uracil (U) in RNA.

The two strands of DNA are bound in **antiparallel** orientation, so that the 3′ end of one strand is bound to the 5′ end of the second strand. This organization has important implications for DNA replication.

DNA Replication Is Semi-Conservative

To be transmitted from one generation to the next, DNA must be replicated. Furthermore, the process of replication must be accurate and fast enough for a living system. The base-pairing rules for nitrogenous bases provide a means for making an appropriate replication system.

In DNA replication, special proteins unzip the DNA double helix and build new DNA by pairing new nucleotide molecules to the two existing DNA strands. The result of this replication is two double-stranded DNA molecules, such that each DNA molecule contains one DNA strand from the template and one newly synthesized DNA strand. This form of replication is called **semi-conservative replication**.

Replication of DNA is bidirectional. A portion of DNA that is "unzipped" and has enzymes performing replication is called a **replication fork** (Fig. C.2b). Replication begins at many points (**origins of replication**), and it continues along both parent strands simultaneously until all the replication forks join.

Nucleotides bond together to form new strands of DNA with the help of an enzyme called **DNA polymerase**. DNA polymerase can add nucleotides only to the 3′ end of a growing strand of DNA. For this reason, DNA is said to replicate in a 5′ to 3′ direction.

The antiparallel orientation of the DNA strands and the directionality of DNA polymerase force replication into two different modes: **leading strand replication** and **lagging strand replication**. The DNA polymerase can replicate continuously along only one parent strand of DNA: the parent strand in the 3′ to 5′ orientation. The DNA replicated continuously is called the **leading strand**.

The DNA replication along the other parent strand is discontinuous because of the strand's 5′ to 3′ orientation. DNA replication on this strand occurs in short fragments called

Stock Photo; **23.10a**: MedImage/Science Source; **23.12:** Pearson Education, Inc.; **23.15:** Prof. P.M. Motta/Univ. "La Sapienza", Rome/Science Source; **23.15:** Professor Pietro M. Motta/Science Source.

24 **CO:** CDC/SPL/Science Photo Library/Getty Images; **24.4**: Pearson Education, Inc; **24.8:** Pearson Education, Inc.

25 **CO:** Titz B, Rajagopala SV, Goll J, Ha ¨user R, McKevitt MT, et al. (2008) The Binary Protein Interactome of Treponema pallidum – The Syphilis Spirochete. PLoS ONE 3(5): e2292. doi:10.1371/journal.pone.0002292; **25.7:** Izf/Shutterstock; **25.7**: Izf/Shutterstock.

26 **CO:** AJPhoto/Science Source; **26.1a:** CNRI/Science Source; **26.9b**: Pearson Education, Inc.; **26.14a**: AJPhoto/Science Source; **26.16:** William C. Ober,Human Physiology: An Integrated Approach, 8e,©2019,Pearson Education, Inc., New York, NY.

Cover: Kent Wood/Science Source

GLOSSARY/INDEX

NOTE: Page numbers followed by a t refer to tables. Page numbers followed by an f refer to figures.

A

C

J

K

L

M

N

O

P

T

U

uterine tube. *See* Fallopian tube
uterus, 198f, 209f, 360f, 361t, 804f, 806, 815, 816f–817f, 828f, 831f
utricle One of the otolith organs of the vestibular apparatus (Ch 10), 335, 336f

V

V2 receptor for vasopressin, 623, 624f
vaccine, 771
vagina, 805f, 806, 816f–817f, 831f
vagotomy Operation that severs the vagus nerve (Ch 11), 359
vagus nerve Cranial nerve that carries sensory information and efferent signals to many internal organs including the heart and GI tract (Ch 9, 11, 14, 21), 186, 283, 286t, 359, 360f, 452, 579, 670f, 672f
Valsalva maneuver Abdominal contraction and forced expiratory movement against a closed glottis (Ch 21), 685
valve, heart, 480f. *See also* specific valve
valvular stenosis, 461
van der Waals force Weak attractive force that occurs between two polar molecules or a polar molecule and an ion (Ch 2), 38f, 39
vanilloid receptor, 320
variable resistance, 478, 485f
varicosity Swollen regions along autonomic axons that store and release neurotransmitter (Ch 8, 11), 228, 361, 362f, 367f, 368t, 401f
vasa recta Peritubular capillaries in the kidney that dip into the medulla and then go back up to the cortex, forming hairpin loops (Ch 19, 20), 591f, 592, 593f, 622f, 626f
vascular elements of the kidney, 592
vascular endothelial growth factor (VEGF) Growth factors that regulate angiogenesis (Ch 15), 480
vascular smooth muscle The smooth muscle of blood vessels (Ch, 12, 15), 400f, 478, 488, 489
vasculature The blood vessels (Ch 14), 434
vas deferens Tube that carries sperm from the epididymis to the urethra. Synonym: ductus deferens (Ch 26), 803f, 811, 812f–813f
vasectomy, 825
vasoactive intestinal peptide, 487t
vasoconstriction Contraction of circular vascular smooth muscle that narrows the lumen of a blood vessel (Ch 14, 15, 16, 19, 20, 22), 439, 478, 487t
vasodilation Relaxation of circular vascular smooth muscle that widens the lumen of a blood vessel (Ch 14, 15, 16, 18, 20, 22, 24), 439, 478, 486f, 487t
vasopressin (ADH, antidiuretic hormone), 198f, 207, 209f, 253, 286t, 487t, 631f, 632f, 639
  inhibition, 637t
  receptor, 181t, 622f
  release, 638f
  secretion, 624f, 637t
  and water reabsorption, 623
vasopressin, arginine (AVP), 621–623. *See also* vasopressin
vasovagal syncope Fainting due to a sudden decrease in blood pressure as a result of an emotional stimulus (Ch 15), 477, 495
vegetative nervous system, 356. *See also* autonomic nervous system
vein Blood vessels that return blood to the heart (Ch 11, 14, 15, 17, 21), 361f, 441f, 442f, 477f–480f, 485f, 493f, 497f, 657f
velocity of flow The distance a fixed volume will travel in a given period of time (Ch 14, 15), 439–440, 440f, 481t
venae cavae (singular: vena cava), 443, 477f, 497f
venipuncture, 480
venous blood, 565f, 576f
venous circulation, 498f–500f
venous constriction, 471f
venous P_{O2}, 791f
venous return The amount of blood that enters the heart from the venous circulation (Ch 14, 15), 467, 471f, 482
ventilation The movement of air between the atmosphere and the lungs (Ch 17, 18, 20, 25), 532, 642–643, 791f
  air flows, 544
  alveolar, 552f, 553–554, 565
  and alveolar blood flow, 554
  brain centers, 582–583
  CO_2, oxygen, and pH, 580–581
  and exercise, 790–791, 790f
  intrapleural pressure during, 547–548
  local control mechanisms, 554
  lung volumes, 542–544
  maximum voluntary, 553
  neural networks, brain stem control, 579f
  pH disturbances, compensates for, 642–643
  pressure gradients, 544
  reflex control of, 578f
  reflexes, 642–643
  regulation of, 578–583
  total pulmonary, 551–552, 552f
  types and patterns, 553t
ventral horn Region of the spinal cord that contains efferent nuclei (Ch 9), 281f
ventral respiratory group (VRG) Medullary neurons for active expiration and greater-than-normal inspiration (Ch 18), 579
ventral root Section of a spinal nerve that carries information from the central nervous system to the muscles and glands (Ch 9), 281f
ventricle
  of the brain, 274, 278f
  of the heart, 434
ventricular
  contraction, 461
  diastole, 461, 463f, 481f
  ejection, 461, 463f
  end-diastolic volume (mL), 464
  filling, 461
  systole, 461, 463f
vertebrae, 276f, 277, 284f

W

X

Y

Z

Anatomical Positions of the Body

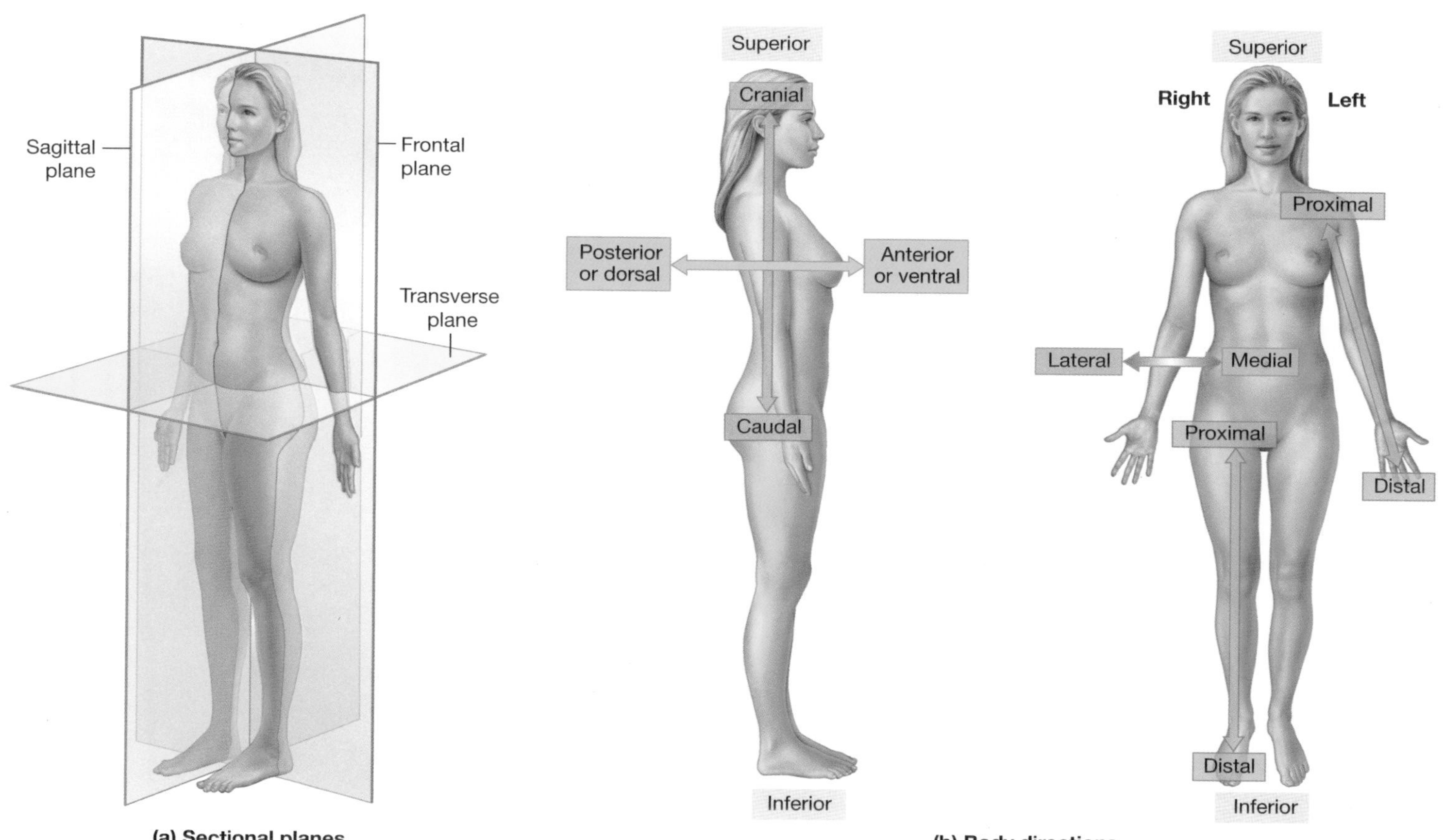
Sagittal plane
Frontal plane
Transverse plane
Superior
Cranial
Posterior or dorsal
Anterior or ventral
Caudal
Inferior
Superior
Right
Left
Proximal
Lateral
Medial
Proximal
Distal
Distal
Inferior

(a) Sectional planes

(b) Body directions

Anterior	(situated in front of): in humans, toward the front of the body (see VENTRAL).
Posterior	(situated behind): in humans, toward the back of the body (see DORSAL).
Medial	(middle, as in *median strip*): located nearer to the midline of the body (the line that divides the body into mirror-image halves)
Lateral	(side, as in a *football lateral*): located toward the sides of the body
Distal	(distant): farther away from the point of reference or from the center of the body
Proximal	(closer, as in *proximity*): closer to the center of the body
Superior	(higher): located toward the head or the upper part of the body
Inferior	(lower): located away from the head or from the upper part of the body
Prone:	lying on the stomach, face downward
Supine:	lying on the back, face up
Dorsal:	refers to the back of the body
Ventral:	refers to the front of the body
Ipsilateral:	on the same side as
Contralateral:	on the opposite side from

Measurements and Conversions

PREFIXES

deci-	(d)	1/10	0.1	1×10^{-1}
centi-	(c)	1/100	0.01	1×10^{-2}
milli-	(m)	1/1000	0.001	1×10^{-3}
micro-	(μ)	1/1,000,000	0.000001	1×10^{-6}
nano-	(n)	1/1,000,000,000	0.000000001	1×10^{-9}
pico-	(p)	1/1,000,000,000,000	0.000000000001	1×10^{-12}
kilo-	(k)		1000.	1×10^{3}

METRIC SYSTEM

1 meter (m)	=	100 centimeters (cm)	=	1000 millimeters (mm)
1 centimeter (cm)	=	10 millimeters (mm)	=	0.01 meter (m)
1 millimeter (mm)	=	1000 micrometers (μm; also called micron, μ)		
1 angstrom (Å)	=	1/10,000 micrometer	=	1×10^{-7} millimeters
1 liter (L)	=	1000 milliliters (mL)		
1 deciliter (dL)	=	100 milliliters (mL)	=	0.1 liter (L)
1 cubic centimeter (cc)	=	1 milliliter (mL)		
1 milliliter (mL)	=	1000 microliters (μL)		
1 kilogram (kg)	=	1000 grams (g)		
1 gram (g)	=	1000 milligrams (mg)		
1 milligram (mg)	=	1000 micrograms (μg)		

CONVERSIONS

1 yard (yd)	= 0.92 meter
1 inch (in)	= 2.54 centimeters
1 meter	= 1.09 yards
1 centimeter	= 0.39 inch
1 liquid quart (qt)	= 946 milliliters
1 fluid ounce (oz)	= 8 fluid drams = 29.57 milliliters (mL)
1 liter	= 1.05 liquid quarts
1 pound (lb)	= 453.6 grams
1 kilogram	= 2.2 pounds

TEMPERATURE

FREEZING

0 degrees Celsius (°C)	=	32 degrees Fahrenheit (°F)	=	273 Kelvin (K)

To convert degrees Celsius (°C) to degrees Fahrenheit (°F): (°C × 9/5) + 32

To convert degrees Fahrenheit (°F) to degrees Celsius (°C): (°F − 32) × 5/9

NORMAL VALUES OF BLOOD COMPONENTS

SUBSTANCE OR PARAMETER	NORMAL RANGE	MEASURED
Calcium (Ca^{2+})	4.3–5.3 meq/L	Serum
Chloride (Cl^{-})	100–108 meq/L	Serum
Potassium (K^{+})	3.5–5.0 meq/L	Serum
Sodium (Na^{+})	135–145 meq/L	Serum
pH	7.35–7.45	Whole blood
P_{O_2}	75–100 mm Hg	Arterial blood
P_{CO_2}	34–45 mm Hg	Arterial blood
Osmolality	280–296 mosmol/kg water	Serum
Glucose, fasting	70–110 mg/dL	Plasma
Creatinine	0.6–1.5 mg/dL	Serum
Protein, total	6.0–8.0 g/dL	Serum

Modified from W. R Ganong, *Review of Medical Physiology* (Norwalk: Appleton & Lange). 1995.